Unity of Forces in the Universe

Volume II

A. Zee

Professor of Physics
University of Washington

World Scientific

World Scientific Pub. Co. Pte. Ltd.
P.O. Box 128
Farrer Road
Singapore 9128

The author and publisher are indebted to the original authors and publishers of the various journals and books for their assistance and permission to reproduce the selected papers found in this volume.

ISBN 9971-950-14-6
9971-950-15-4 pbk

Printed by Richard Clay (S.E.Asia) Pte.Ltd.

To G. G.

CONTENTS

VOLUME I

Foreword v
A Note to the Reader vii
I. Introduction 1
II. Review of the Standard 123 Theory 4
III. Grand Unification 10
IV. *SO*(10) 16
V. Exceptional Unification 18
VI. Reality and Complexity of the World 20
VII. Proton Decay 22
VIII. Family Problem and Orthogonal Unification 26
IX. Fermion Mass Hierarchy 29
Appendix A: Instant Review of Group Theory 31
Appendix B: Two-component Formalism 40

REPRINTED PAPERS

III. *Grand Unification*

1) Unity of all elementary-particle forces
by H. Georgi and S. L. Glashow, *Phys. Rev. Lett.* **32,** 438 (1974) 46
2) Lepton number as the fourth "color"
by J. Pati and A. Salam, *Phys. Rev.* D**10,** 275 (1974) 50
3) Hierarchy of interactions in unified gauge theories
by H. Georgi, H. R. Quinn and S. Weinberg, *Phys. Rev. Lett.* **33,** 451 (1974) 66
4) Aspects of the grand unification of strong, weak and electromagnetic interactions
by A. J. Buras, J. Ellis, M. K. Gaillard, and D. V. Nanopoulos, *Nucl. Phys.* B**135,** 66 (1978) 70

IV. *SO*(10)

1) The state of the art – gauge theories
by H. Georgi, in *Proceedings of the American Institute of Physics,* edited by C. E. Carlson, Meetings at William and Mary College, 1974 98

2) Unified interactions of leptons and hadrons
by H. Fritzsch and P. Minkowski, *Ann. Phys.* **93,** 193 (1975) 106
3) Color embeddings, charge assignments, and proton stability in unified gauge theories
by M. Gell-Mann, P. Ramond and R. Slansky, *Rev. Mod. Phys.* **50,** 721 (1978) 180

V *Exceptional Unification*
1) E_6 gauge field theory model revisited
by F. Gürsey and M. Serdaroğlu, *Nuovo Cimento* **65**A, 337 (1981) 206
2) Quark-lepton symmetry and mass scales in an E_6 unified gauge model
by Y. Achiman and B. Stech, *Phys. Lett.* **77**B, 389 (1978) 223

VI. *Reality and Complexity of the World*
1) Towards a grand unified theory of flavor
by H. Georgi, *Nucl. Phys.* B**156,** 126 (1979) 230

VII. *Proton Decay*
1) Non-conservation of baryon number
by A. Zee, in *Proceedings of the 1980 Guangzhou Conference on Theoretical Particle Physics,* Vol. 1, p. 287 (Science Press, Beijing, 1980) 240
2) Conservation or violation of $B — L$ in proton decay
by F. Wilczek and A. Zee, *Phys. Lett.* **88**B, 311 (1979) 270
3) Baryon- and lepton-nonconserving processes
by S. Weinberg, *Phys. Rev. Lett.* **43,** 1566 (1979) 274
4) Operator analysis of nucleon decay
by F. Wilczek and A. Zee, *Phys. Rev. Lett.* **43,** 1571 (1979) 279
5) Local $B — L$ symmetry of electroweak interactions, majorana neutrinos, and neutron oscillations
by R. N. Mohapatra and R. E. Marshak, *Phys. Rev. Lett.* **44,** 1316 (1980) 282

VIII. *Family Problem and Orthogonal Unification*
1) Complex spinors and unified theories, in *Supergravity*
by M. Gell-Mann, P. Ramond and R. Slansky, edited by P. Van Nieuwenhuizen and D. Z. Freedman (North-Holland Pub. Co., 1979) 288
2) Families from spinors
by F. Wilczek and A. Zee, *Phys. Rev.* D**25,** 553 (1982) 295

IX. *Fermion Mass Hierarchy*
1) Electromagnetic and weak masses by S. Weinberg, *Phys. Rev. Lett.* **29,** 388 (1972) 310
2) Attempts to calculate the electron mass by H. Georgi and S. Glashow, *Phys. Rev.* D**7,** 2457 (1973) 315
3) *SO*(10) model of fermion masses by S. Barr, *Phys. Rev.* D**24,** 1895 (1981) 322
4) Calculable masses in grand unified theories by M. J. Bowick and P. Ramond, *Phys. Lett.* **103**B, 338 (1981) 327
5) Hierarchical fermion masses in *SU*(5) by R. Barbieri, D. V. Nanopoulos and D. Wyler, *Phys. Lett.* **103**B, 433 (1981) 332

Appendix A: Instant Review of Group Theory
1) Group theory for unified model building by R. Slansky, *Phys. Reports* **79,** 1 (1981) 338

VOLUME II

X. A Short Course in Cosmology 465
XI. Genesis of Matter 478
XII. Introduction to the Theory of Galaxy Formation 483
XIII. Neutrinos and Galaxies 493
XIV. Monopoles and Inflation 501
XV. Hierarchy, Technicolor, Supersymmetry, and Variations 506
XVI. Invisible Axions 509
XVII. Composite Quarks and Leptons 511
XVIII. Gravity and Grand Unification 515
Postscript 521

REPRINTED PAPERS

X. *A Short Course in Cosmology*
1) Expanding universe and the origin of elements by G. Gamow, *Phys. Rev.* **70,** 572 (1946) 526
2) The origin of chemical elements by R. A. Alpher, H. Bethe and G. Gamow, *Phys. Rev.* **73,** 803 (1948) 528
3) The origin of elements and the separation of galaxies by G. Gamow, *Phys. Rev.* **74,** 505 (1948) 530
4) The evolution of the universe by G. Gamow, *Nature* **162,** 680 (1948) 532

5) Proton-neutron concentration ratio in the expanding universe at the stages preceding the formation of the elements
by C. Hayashi, *Prog. Theor. Phys.* **5,** 224 (1950) 535

6) Physical conditions in the initial stages of the expanding universe
by R. A. Alpher, J. W. Follin, Jr., and R. C. Herman, *Phys. Rev.* **92,** 1347 (1953) 547

7) The mystery of the cosmic helium abundance
by F. Hoyle and R. Tayler, *Nature* **203,** 1108 (1964) 562

8) Density of relict particles with zero rest mass in the universe
by V. F. Shvartsman, *JETP Lett.* **9,** 184 (1969) 565

9) Cosmological limits to the number of massive leptons
by G. Steigman, D. N. Schramm and J. E. Gunn, *Phys. Lett.* **66**B, 202 (1977) 568

10) Big-bang nucleosynthesis as a probe of cosmology and particle physics
by K. A. Olive, D. N. Schramm, G. Steigman, M. S. Turner, and J. Yang, *Astrophys. J.* **246,** 557 (1981) 571

XI. *Genesis of Matter*

1) Violation of *CP* invariance, C asymmetry, and baryon asymmetry of the universe
by A. D. Sakharov, *Zh. Ek. Teor. Fiz.* **5,** 32 (1967) [English translation: *JETP Lett.* **5,** 24 (1967)] 586

2) Unified gauge theories and the baryon number of the universe
by M. Yoshimura, *Phys. Rev. Lett.* **41,** 281 (1978); *erratum,* **42,** 740 (1979) 589

3) Baryon number of the universe
by S. Dimopoulos and L. Susskind, *Phys. Rev.* D**18,** 4500 (1978) 593

4) Matter-antimatter accounting, thermodynamics, and black-hole radiation
by D. Toussaint, S. Treiman, F. Wilczek, and A. Zee, *Phys. Rev.* D**19,** 1036 (1979) 603

5) Cosmological production of baryons
by S. Weinberg, *Phys. Rev. Lett.* **42,** 850 (1979) 613

6) Mechanisms for cosmological baryon production
by D. V. Nanopoulos and S. Weinberg, *Phys. Rev.* D**20,** 2484 (1979) 617

7) Magnitude of the cosmological baryon asymmetry
by S. M. Barr, G. Segrè, and H. A. Weldon, *Phys. Rev.* D**20,** 2494 (1979) 627

8) The development of baryon asymmetry in the early universe
by E. W. Kolb and S. Wolfram, *Phys. Lett.* **91**B, 217 (1980) 632

9) Hierarchy of cosmological baryon generation
by J. N. Fry, K. A. Olive and M. S. Turner, *Phys. Rev. Lett.* **45,** 2074 (1980) 637

XII. *Introduction to the Theory of Galaxy Formation*

1) Gauge-invariant cosmological perturbations
by J. M. Bardeen, *Phys. Rev.* D**22,** 1882 (1980) 642

2) Tenacious myths about cosmological perturbations larger than the horizon size
by W. H. Press and E. T. Vishniac, *Astrophys. J.* **239,** 1 (1980) 666

3) The theory of the large scale structure of the universe
by Ya. B. Zel'dovich, *International Astronomical Union* No. **79,** p. 409 (1977) 677

4) The black-body radiation content of the universe and the formation of galaxies
by P. J. E. Peebles, *Astrophys. J.* **142,** 1317 (1965) 690

5) Fluctuations in the primordial fireball
by J. Silk, *Nature* **215,** 1155 (1967) 700

6) Primeval adiabatic perturbation in an expanding universe
by P. J. E. Peebles and J. T. Yu, *Astrophys. J.* **162,** 815 (1970) 702

7) Core condensation in heavy halos: A two-stage theory for galaxy formation and clustering
by S. D. M. White and M. J. Rees, *Mon. Not. R. Astron. Soc.* **183,** 341 (1978) 724

8) Galaxy formation in an intergalactic medium dominated by explosions
by J. P. Ostriker and L. L. Cowie, *Astrophys. J.* **243,** L127 (1981) 742

XIII. *Neutrinos and Galaxies*

1) Rest mass of muonic neutrino and cosmology
by S. S. Gershtein and Ya. B. Zel'dovich, *JETP Lett.* **4,** 120 (1966) 748

2) An upper limit on the neutrino rest mass
by R. Cowsik and J. McClelland, *Phys. Rev. Lett.* **29,** 669 (1972) 751

3) Dynamical role of light neutral leptons in cosmology
by S.Tremaine and J. E. Gunn., *Phys. Rev. Lett.* **42,** 407 (1979) 753

4) Massive neutrinos and the large-scale structure of the universe
by J. R. Bond, G. Efstathiou and J. Silk, *Phys. Rev. Lett.* **45,** 1980 (1980) 757

5) Massive neutrinos and galaxy formation
by F. R. Klinkhamer and C. A. Norman,
Astrophys. J. Lett. **243,** L1 (1981) 762
6) Formation of galaxies and clusters of galaxies
in the neutrino dominated universe
by H. Sato and F. Takahara, *Prog. Theor. Phys.*
66, 508 (1981) 766
7) On the linear theory of density perturbations in a
neutrino + baryon universe
by I. Wasserman, *Astrophys. J.* **248,** 1 (1981) 784
8) The formation of galaxies from massive neutrinos
by M. Davis, M. Lecar, C. Pryor, and E. Witten,
Astrophys. J. **250,** 423 (1981) 796
9) Cosmological impact of the neutrino rest mass
by A. G. Doroshkevich, M. Yu. Khlopov,
R. A. Sunyaev, A. S. Szalay, and Ya. B. Zel'dovich,
in the 10th Texas Symposium on *Relativistic
Astrophysics,* edited by R. Ranaty & Frank C. Jones,
Annals of the New York Academy of Sciences
375, 32 (1981) 805
10) Formation of structure in a neutrino-dominated
universe
by J. R. Bond and A. S. Szalay, to appear in
Proceedings of Neutrino 1981, Maui, Hawaii. 816
11) Some astrophysical consequences of the existence
of a heavy stable neutral lepton
by J. E. Gunn, B. W. Lee, I. Lerche,
D. N. Schramm, and G. Steigman,
Astrophys. J. **223,** 1015 (1978) 831

XIV. *Monopoles and Inflation*
1) Inflationary universe: A possible solution to
the horizon and flatness problems
by A. H. Guth, *Phys. Rev.* D**23,** 347, (1981) 850
2) Phase transitions in gauge theories and cosmology
by A. D. Linde, *Rep. Prog. Phys.* **42,** 389 (1979) 861
3) A new inflationary universe scenario: A possible
solution of the horizon, flatness, homogeneity,
isotropy and primordial monopole problems
by A. D. Linde, *Phys. Lett.* **108**B, 389 (1982) 910
4) Cosmology for grand unified theories with
radiatively induced symmetry breaking
by A. Albrecht and P. J. Steinhardt, *Phys.
Rev. Lett.* **48,** 1220 (1982) 915

XV. *Hierarchy, Technicolor, Supersymmetry, and Variations*

1) Dynamics of spontaneous symmetry breaking in the Weinberg-Salam theory by L. Susskind, *Phys. Rev.* D**20,** 2619 (1979) — 920
2) Implications of dynamical symmetry breaking: An addendum by S. Weinberg, *Phys. Rev.* D**19,** 1277 (1979) — 927
3) Composite/fundamental Higgs mesons I: Dynamical speculations by H. Georgi and I. N. McArthur, *Nucl. Phys.* B**202,** 382 (1982) — 931
4) Dynamical breaking of supersymmetry by E. Witten, *Nucl. Phys.* B**188,** 513 (1981) — 946

XVI. *Invisible Axions*

1) Weak-interaction singlet and strong *CP* invariance by J. E. Kim, *Phys. Rev. Lett.* **43,** 103 (1979) — 990
2) A simple solution to the strong *CP* problem with a harmless axion by M. Dine, W. Fischler and M. Srednicki, *Phys. Lett.* **104**B, 199 (1981) — 995
3) *SU*(5) and the invisible axion by M. B. Wise, H. Georgi and S. L. Glashow, *Phys. Rev. Lett.* **47,** 402 (1981) — 999

XVII. *Composite Quarks and Leptons*

1) Naturalness, chiral symmetry, and spontaneous chiral symmetry breaking by G. 't Hooft, in *Recent Developments in Gauge Theories,* p. 135 (Plenum Press, 1980) — 1004
2) Family structure with composite quarks and leptons by I. Bars, *Phys. Lett.* **106**B, 105 (1981) — 1027

XVIII. *Gravity and Grand Unification*

1) The $N=8$ supergravity theory. I. The Lagrangian by E. Cremmer and B. Julia, *Phys. Lett.* **80**B, 48 (1978) — 1034
2) A grand unified theory obtained from broken supergravity by J. Ellis, M. K. Gaillard and B. Zumino, *Phys. Lett.* **94**B, 343 (1980) — 1039
3) Search for a realistic Kaluza-Klein Theory by E. Witten, *Nucl. Phys.* B**186,** 412 (1981) — 1045
4) Gravity as a dynamical consequence of the strong, weak, and electromagnetic interactions by A. Zee, in the *Proceedings of the Erice Conference* (1981) — 1062

CHAPTER X

A SHORT COURSE IN COSMOLOGY

The interface between particle physics and cosmology has emerged over the last few years as a very active area of research. In particular, grand unification offers us the exciting prospect of actually calculating the matter content of the Universe from first principles. (A discussion of this topic will be postponed to the next Chapter.)

Several years ago, most particle physicists were unfamiliar with cosmology. But the situation has changed dramatically with the realization that grand unified theories can resolve some outstanding problems of cosmology. Most particle theorists are now familiar with at least the rudiments of cosmology. For the benefit of those who are not, in Chapters X through XIII we give what amounts to a brief course in cosmology.

Needless to say, we cannot possibly hope to give a detailed treatment. Fortunately, a number of texts and reviews are available[1–7]. (Again, the list given is by no means complete.) The spirit of our discussion is to exhibit the essence of physics while avoiding detailed calculations as much as possible. Very few exact equations will be given. The symbol "$\sim$" will be used liberally. Relations given by this symbol could easily be off by a couple of orders of magnitude.

We assume the reader knows that a homogeneous, isotropic Universe is described by the Robertson-Walker metric

$$\mathrm{d}\tau^2 = \mathrm{d}t^2 - R^2(t)\left\{\frac{\mathrm{d}r^2}{1-kr^2} + r^2\mathrm{d}\Omega^2\right\}.$$

A scaling of the constant k could be absorbed into the $R(t)$. Thus, k could be chosen to either $+1$, 0, or -1. Sometimes it is convenient not to carry out this scaling. We will often refer to R loosely as the radius or the "size" of the Universe. Einstein's theory fixes the time evolution of the Universe to be

$$\frac{\dot{R}^2}{R^2} = \frac{8\pi}{3}G\rho - \frac{k}{R^2}. \tag{1}$$

Energy conservation relates the energy density ρ to the pressure p.

$$\frac{\mathrm{d}}{\mathrm{d}R}(\rho R^3) = -3pR^2. \tag{2}$$

Einstein's cosmological equation could actually be derived from essentially Newtonian considerations[8]. For a handy mnemonic and heuristic derivation, consider a sphere of radius R and density ρ. Then its kinetic energy $\sim(\rho R^3)\dot{R}^2$

and its potential energy $\sim G(\rho R^3)^2/R$. Equating kinetic and potential energies we find

$$\frac{\dot{R}^2}{R^2} \sim G\rho. \tag{2'}$$

The curvature term k/R^2 in Eq. (1) becomes important only late in the Universe's evolution as we will see, so Eq. (1) is usually adequate, especially in discussing the early Universe. A more precise Newtonian derivation (to be given in an Appendix to this Chapter) reproduces Eq. (1), whose three terms may be identified as kinetic energy, potential energy, and total energy respectively.

We will exploit dimensional analysis as a way of avoiding long derivations. In doing this, we find it convenient to use natural units in which $\hbar = 1 = c$ and to measure masses in GeV units. Thus, length and time are given in units of GeV^{-1}. Occasionally we need to convert to more familiar units. The conversion factors are

$$\begin{aligned} 1\ \text{GeV}^{-1} &\sim 6 \times 10^{-25}\ \text{sec} \\ &\sim 2 \times 10^{-14}\ \text{cm} \\ 1\ \text{GeV} &\sim 2 \times 10^{-24}\ \text{g}. \end{aligned}$$

For historical reasons, temperature is often measured in degrees. We have the conversion

$$1\ \text{eV} \sim 10^4\ {}^\circ\text{K}.$$

It is useful to express Newton's gravitational constant G as $G = M_{pl}^{-2}$ where the Planck mass $M_{pl} \sim 10^{19}$ GeV. We note

$$\begin{aligned} M_{pl}^{-1} &\sim 6 \times 10^{-44}\ \text{sec} \\ &\sim 2 \times 10^{-33}\ \text{cm} \end{aligned}$$

known as the Planck time and the Planck length respectively.

As an application of dimensional analysis we "derive" some important results from statistical mechanics. In a gas of relativistic particles such as photons maintained at temperature T the number density of particles n and the energy density are given by

$$n \sim T^3 \tag{3}$$

$$\rho \sim T^4. \tag{4}$$

These relations are easily obtained since T is the only dimensional variable available and n and ρ have dimensions M^3 and M^4 respectively. By dimensional analysis, the entropy density s is essentially n.

Loosely speaking[9], R^3 is the "scale volume" of the Universe; so the total entropy or (what is essentially the same thing, the total number of relativistic particles) in a Universe dominated by relativistic particles is

$$N \sim T^3 R^3. \tag{5}$$

Since entropy is conserved $T \propto 1/R$. The Universe cools as it expands.

Next, let us have the facts, please. We need some observational input. Fortunately, for our purposes there are only a few relevant cosmological facts.

(1) The Universe is at present expanding at a certain rate of $H_o \sim 10^2$ km $\sec^{-1}$ Mpc^{-1}. (A parsec is about 3 light years $\sim 3 \times 10^{18}$ cm.) The present age of the Universe τ is roughly the same order as $H_o^{-1} \sim 10^{10}$ years. Given Einstein's theory the relation between τ and the Hubble parameter H_o can be made more precise. Actually, after half a century of hard work, H_o is still not known precisely. In the astronomical literature, the Hubble parameter is written as $H_o = 10^2\, h_o$ km sec^{-1} Mpc^{-1} where the "ignorance factor" $h_o \lesssim 1$ and $\gtrsim 1/2$. (The misleading term "Hubble constant" is sometimes used but the rate of expansion in fact depends on time.)

(2) The discovery and measurement of the cosmic microwave background has established that the Universe is pervaded with photons with a black body distribution corresponding to a temperature T_0 of 2.7 °K − 2.9 °K. There are thus, according to Eq. (3), about 400 photons per cm^3. We can certainly see no farther than the distance light has travelled since the Universe was born, namely about 10^{28} cm. This may be taken as the effective size or "horizon" of the Universe. Thus, there are roughly 10^{87} photons within the horizon.

(3) The number density of nucleons in the Universe is much less precisely known. It is a difficult problem in astronomy. Typically, gravitational mass is inferred by analyzing rotations of galaxies or clusters of galaxies and by applying the virial theorem. A mass-to-light ratio is then constructed. The mass density of far away regions is then estimated by multiplying the mass-to-light ratio with the observed luminosity density of that region. Unfortunately, particle physicists and astronomers also use different notations. Astronomers define

$$\Omega_N \equiv \frac{\rho_N}{\rho_c} \equiv \frac{8\pi G \rho_N}{3H_o^{\,2}}. \tag{6}$$

ρ_N is the nucleon mass density. It will become clear later that ρ_c is the critical density which will close the Universe. Since H_o is itself uncertain the data is usually given in the combination $\Omega_N h_o^{\,2}$. Particle physicists generally use

$$\eta = n_B/n_\gamma, \tag{7}$$

the ratio of the baryon number density to the photon number density. The translation is given by

$$\eta \sim 3 \times 10^{-8}\, \Omega_N h_o^{\,2}. \tag{8}$$

The present data may be summarized as (see Chapter XIII)

$$\eta \sim (0.3 - 10) \times 10^{-10}. \tag{9}$$

For temperatures far below 10^{15} GeV baryon number is conserved to a high accuracy so η is essentially a constant since n_B and n_γ both fall like R^{-3} as the Universe expands. (The number of photons produced by stars in the history of the Universe is completely negligible.) Since m_N/T_γ at present is about $\sim 10^{13}$, even with such a small η, matter energy density ρ_N dominates radiation energy density ρ_γ. However, radiation dominated when the temperature

$$T^{\gamma}_{\text{dom}} \gtrsim \eta m_N \sim 10^{-1}\ \text{eV} \sim 10^3\ {}^\circ\text{K}. \tag{10}$$

(4) The microwave background is remarkably isotropic. Observation suggests $\Delta T_o/T_o \lesssim 10^{-4}$. (A dipole anistropy of 10^{-3} is observed but is believed to be due to the motion of the earth through the microwave background.) The Universe is homogeneous over large scales. More precisely, the Universe was homogeneous way back when photons decoupled from matter. (See later.) On the other hand, the Universe is inhomogeneous. Firstly, matter is concentrated in stars which cluster into galaxies. There is a hierarchy of clustering: Galaxies in their turn tend to cluster and so on. Secondly, recent deep sky surveys have found big (~ 10 Mpc) "holes" in space—regions devoid of galaxies[10]. Finally, some quadrupole deviations from isotropicity in the microwave background have been observed.

(5) The relative abundance of elements in the Universe has been measured by geologists and astronomers for well over a century. By mass, the Universe is about 70% H, 30% He^4, with all other elements amounting to about 1 or 2%.

Cosmology seeks to explain these five basic facts.

In the early Universe, the temperature is high and many species of particles are relativistic. We must multiply the formulas in Eqs. (3) and (4) by a factor g where g counts the total number of degrees of freedom of all the relativistic species. More precisely,

$$g = \sum g_{\text{bosons}} + 7/8 \sum g_{\text{fermions}}. \tag{11}$$

Of course, g depends on the temperature regime. For $T \sim$ a few MeV only the electron (and positron), the various neutrinos, and the photon contribute to g. For rough estimates in this era we will often ignore the factor of g. For $T \gtrsim 10^{15}$ GeV, g could be of order 10^2 if we can count on the particle content of the grand unified theories. For some effects, g is important, as we will see below and in Chapter XIII.

In a "radiation" dominated era the evolution of the Universe is thus governed by

$$\frac{\dot{R}^2}{R^2} \sim G\left(\frac{N^4}{g}\right)^{1/3} \frac{1}{R^4} - \frac{k}{R^2} \tag{12}$$

or equivalently by

$$\frac{\dot{T}^2}{T^2} \sim gGT^4 - k\left(\frac{g}{N}\right)^{2/3} T^2. \tag{12'}$$

After the Universe cools below T^{γ}_{dom}, the evolution of the Universe is governed by

$$\frac{\dot{R}^2}{R^2} \sim (G\eta m_N N)\frac{1}{R^3} - \frac{k}{R^2} \tag{13}$$

since the energy density $\rho = mn_B = \eta n_\gamma m$.

We can interpret Eqs. (12) and (13) as describing a point particle of unit mass moving in a one-dimensional potential of the form $-(const)/R^2$ or $-(const)/R$ with a total energy $-k$. Assuming that $\dot{R} > 0$ at a given instant, we see that if $k < 0$, R will keep growing but if $k > 0$, R will reach a maximum value and decrease thereafter. Thus, the Universe is open or closed according to whether k is negative or positive. We now understand the rationale for defining the quantities Ω and ρ_c (Eq. (6)): If $\Omega > 1$ the Universe will be closed. (H_o is of course just $\dot{R}/R$ at present.)

Since N is enormous, the curvature term in Eqs. (12), (12′), and (13) is completely negligible for high T and small R. Neglecting the curvature term we integrate Eq. (12′) to obtain the "temperature clock"

$$t \sim g^{-1/2}\frac{M_{pl}}{T^2} \sim (2.4 \times 10^{-6}\ \text{sec})g^{-1/2}\left(\frac{1\ \text{GeV}}{T}\right)^2. \tag{14}$$

It is usually more convenient to specify the temperature at which various events occurred. However, in the astrophysical literature, one often speaks of events in the history of the Universe as having occurred at certain redshifts. The redshift parameter z is related to $R(t)$ in the Robertson-Walker cosmography by

$$1 + z = \frac{R(t_o)}{R(t)}.$$

$R(t_o)$ denotes the present scale size. The parameter z is observational, in contrast to the theoretical concept $R(t)$.

Let us follow the evolution of the Universe at a temperature of say a few tens of MeV, since "traditional" cosmology[11] covers events from this point on. At this temperature, the Universe contains protons, neutrons, electrons, positrons, photons, and various neutrinos. The baryons are of course nonrelativistic while all the other particles are relativistic. Since $\eta m_N/T \sim 10^{-8}$ the nucleons contribute a negligible fraction to ρ. These particles are kept in thermal equilibrium with each other by various electromagnetic and weak processes involving them such as $\nu\bar{\nu}\leftrightarrow e^+e^-$, $\nu e^-\leftrightarrow \nu e^-$, $\nu n\leftrightarrow pe^+$, $\gamma\gamma\leftrightarrow e^+e^-$, $\gamma p\rightarrow\gamma p$, etc. But meanwhile the Universe is expanding. The condition for thermal equilibrium is just that the relevant reactions must be faster than the expansion rate. Inspecting Eqs. (12), (12′) or (14) we see that the characteristic expansion rate

$$\frac{1}{t} \sim \frac{T^2}{M_{pl}} \tag{15}$$

drops with T. But as we will see presently the weak interaction rate drops even faster so at some point the neutrinos will drop out of thermal equilibrium. The reaction rate per neutrino is given by $n\langle v\sigma\rangle$. Here $n \sim T^3$ is the number density of electrons and neutrinos. The thermally averaged interaction cross-section $\langle v\sigma\rangle$ can be easily estimated by dimensional analysis. The amplitude is proportional to Fermi's constant $G_F \sim \alpha/M_W^2 \sim 10^{-5}\, m_N^{-2}$. The cross-section has a dimension of $1/M^2$ and so $\langle v\sigma\rangle \sim G_F^2\, T^2$ since T is the only relevant quantity of dimension M

available. Putting it together, the reaction rate involving neutrinos is

$$n\langle v\sigma\rangle \sim G_F^2 T^5. \tag{16}$$

Thus, at high T, weak interaction processes are fast enough. But as T drops, at some characteristic decoupling temperature T_{dec} neutrinos "decouple"—they lose thermal contact with the electrons. Comparing Eqs. (15) and (16) we see that

$$T_{dec}^{\nu} \sim \left(\frac{1}{M_{pl}G_F^2}\right)^{1/3} \sim \left(10^{10}\,\frac{m_N}{M_{pl}}\right)^{1/3} m_N \sim 1\text{ MeV}. \tag{17}$$

(The precise value is about 3.5 MeV for ν_e, and about 6 MeV for ν_μ and ν_τ. We are not far off.) According to Eq. (14) this happened at around 1 sec after the big bang. After this time, the temperature of the neutrino gas T_ν simply drops like $1/R$ as the Universe expands.

The much stronger electromagnetic and strong interactions continue to keep the protons, neutrons, positrons, and photons in equilibrium. Again by dimensional analysis we can estimate the typical electromagnetic cross-section involving relativistic particles to be $\langle v\sigma\rangle \sim \alpha^2/T^2$. Thus, the reaction rate $\alpha^2 T$ is much larger than the expansion rate T^2/M_{pl} for $T \ll \alpha^2 M_{pl}$. The reaction rate per nucleon is $\sim T^3(\alpha^2/M_N^2)$ which is larger than the expansion rate as long as $T > m_N^2/(\alpha^2 M_{pl}) \sim$ a very low temperature. (For the cross section we find the non-relativistic form α^2/m_N^2 again by dimensional analysis.) The nucleons are thus maintained in kinetic equilibrium. The average kinetic energy per nucleon is 3/2 T. One must be careful to distinguish between kinetic equilibrium and chemical equilibrium. Reactions like $\gamma\gamma \to p\bar{p}$ having long been suppressed by a Boltzmann factor since only very energetic photons from the tail end of the distribution can produce an anti-nucleon. There are essentially no anti-nucleons around.

For $T > m_e \sim 0.5$ MeV $\sim 5 \times 10^9$ °K, the number of electrons, positrons, and photons are comparable, $n_{e-} \sim n_{e+} \sim n_\gamma$. The exact ratios are of course easily supplied by inserting the appropriate "g-factors". Since the Universe is electrically neutral, $n_{e-} - n_{e+} = n_{\text{protons}}$ and so there is a slight 10^{-10} excess of electrons over positrons.

As T drops below m_e the process $\gamma\gamma \to e^+e^-$ is severely suppressed by the Boltzmann factor $e^{-m_e/T}$ since only very energetic photons in the "tail-end" of the Bose distribution can participate. Thus positrons and electrons annihilate rapidly via $e^+e^- \to \gamma\gamma$ and are not replenished, leaving a small number of electrons $n_{e-} \sim n_p \sim 10^{-10}\, n_\gamma$. The annihilation of e^-e^+ heats up the photons relative to the neutrinos. The amount of heating may be calculated exactly by requiring entropy conservation. Thus, after annihilation, we find, referring to Eq. (11),

$$\frac{T_\gamma^{\text{after}}}{T_\gamma^{\text{before}}} = \left(\frac{T_\gamma}{T_\nu}\right)_{\text{after}} = \left(\frac{g_{\text{before}}}{g_{\text{after}}}\right)^{1/3} = \left(\frac{11}{4}\right)^{1/3} \sim 1.4. \tag{18}$$

This ratio has remained constant ever since. Unfortunately, there is no known way to measure the temperature of the ν background of the Universe.

We next come to the crowning achievement of big bang cosmology — nucleosynthesis[12–15]. Gamow and his collaborators[11] realized in the late 1940's that in the hot big bang nuclear reactions could build nuclei out of the "primordial" (see the next Chapter) protons and neutrons. The calculated abundance of He^4 agrees remarkably well with observations. We outline how the calculation goes.

At temperatures when weak processes are still fast compared to the expansion of the Universe, neutrons and protons are converted back and forth rapidly (by $n + e^+ \leftrightarrow p + \bar{\nu}$, $p + e^- \leftrightarrow n + \nu$, etc.) and so the neutron to proton number ratio follows the Boltzmann factor $e^{-(m_n - m_p)/T}$. As the temperature falls, there are fewer neutrons since the neutron is heavier. The weak interaction rate per nucleon is of order $G_F^2 T^5$ and so these reactions are significant till a temperature of a couple of MeV. (This is of course the same order as T^{ν}_{dec} calculated earlier.) As the temperature drops further, the decay $n \to pe^- \ \bar{\nu}$ begins to play a role.

The significant fact from nuclear physics is that the deuteron has a low binding energy Δ_d of order 2.2 MeV. Thus, for T greater than a certain critical temperature, any deuteron formed is knocked apart almost immediately by photo-dissociation:

$$\gamma + d \to p + n .$$

Since there are so many photons per nucleon, the temperature has to drop substantially below Δ_d before deuterons can be formed. Roughly, the number density of energetic photons in the tail-end of the Bose distribution capable of dissociating the deuteron has to drop below the nucleon density:

$$\frac{n_\gamma^{\text{dissociating}}}{n_B} \sim \frac{1}{\eta} e^{-\Delta_d/T} \lesssim 1 .$$

Thus, deuteron formation, and thus nucleosynthesis, starts when the temperature drops below $T_d \sim \Delta_d/\log \eta^{-1} \sim \Delta_d/23 \sim 10^9$ °K. (We have taken η to be $\sim 10^{-10}$.) There are too few nucleons around so that many body processes such as $p + p + n + n \to He^4$ are not significant. Nuclei must be built up step-wise via two body strong and electromagnetic processes: $pn \to d\gamma$, $nd \to H^3\gamma$, $dd \to He^3 n$, $He^3 n \to He^4 \gamma$, $H^3 d \to He^4 n$, etc. The "deuteron bottleneck" thus delays nucleosynthesis till $T \lesssim 0.1$ MeV. But once the bottleneck is passed, nucleosynthesis proceeds rapidly and essentially all the neutrons around at that time are incorporated into He^4.

Incidentally, it is believed that nuclei heavier than He^4 are synthesized later in stars. The absence of stable nuclei of mass number 5 and 8 leads to two other bottlenecks[16]. The nucleon density in stars is high enough so that the gaps at 5 and 8 could be bridged. For instance, two He^4 nuclei could form the unstable Be^8 followed quickly by the formation of C^{12} by a Be^8-He^4 reaction. The nucleon density in the early Universe is too low. (Using Eqs. (3) and (9) we estimate the nucleon density at $T \sim 0.1$ MeV to be $\lesssim 10^{18}/\mathrm{cm}^3$. A typical star like our sun contains 10^{57} nucleons and has a radius $\sim 10^{11}$ cm.)

The actual calculation[12–15] of nucleosynthesis is of course quite detailed. But we see from our discussion that the abundance of He^4, denoted by Y, depends on η, the half-life of neutron τ, and the number of neutrino species N_ν, given that the various weak and nuclear reaction rates have been well measured in the laboratory. Clearly, Y increases with increasing η and τ.

That big bang nucleosynthesis can tell us about N_ν was realized by Shvartzman[17]. The point is that each neutrino species contributes T^4 to the energy density, more neutrino species speed up the Universe and so the weak processes converting nucleons $n \leftrightarrow p$ drop out earlier, leading to a higher n/p ratio[18]. (To see that more species speed up the expansion, consider time reversal invariance.) Hence, an observed upper limit on Y will put an upper limit on N_ν. This is the classic example on how cosmology could constrain particle physics.

However, the situation is actually more complicated. If N_ν is large enough, the Universe expands so fast that even reactions producing He^4 (such as $H^2 + H^3 \rightarrow He^4 + n$) cannot keep up. Y has a maximum as a function of N_ν. Obviously, this effect is most significant for small η. (The maximum Y occurs at $N_\nu \sim 80$ for $\eta \sim 6 \times 10^{-11}$ and at $N_\nu \sim 15$ for $\eta \sim 1.5 \times 10^{-11}$.) The maximum value of Y decreases with decreasing η.

Not only is the precise value of η uncertain, but perhaps surprisingly the neutron life-time τ is also uncertain. The current experimental value is about 10.6 ± 0.2 min. The observational determination of Y is no cinch, either, since one has to subtract out the He^4 formed later in stars. Since stars produce both helium and heavier elements (known collectively as "metals" in astrophysical circles), one usually tries to determine the helium abundance of metal-poor regions of interstellar clouds. Or one can rely on stellar nucleosynthesis calculations. Anyhow, an upper limit for Y is usually cited to be about 0.25. For $\eta \lesssim 2.9 \times 10^{-11}$ this value implies no bound on N_ν at all.

It is a wide-spread belief in the particle physics community that helium abundance places stringent limits on the number of families (Chapter VIII). However, because of all the uncertainties we have just discussed, the book is not yet closed on the subject. Relevant plots of Y versus η, τ, and N_ν are nicely displayed in Ref. 19. We should also emphasize that one should include in N_ν only neutrinos with mass $\ll 10$ MeV.

Of course, detailed nucleosynthesis calculations[14] also yield expected abundances for deuterium d and He^3. From our discussion, it is clear that the deuterium abundance depends sensitively on η. For large η, most of the deuterium would have been cooked at nucleosynthesis, eventually into He^4. From measurements of absorption lines due to the intersteller medium, a deuterium abundance by mass of $\sim 2 \times 10^{-5}$ was deduced, corresponding to a rather low value for η. But arguments based on d and He^3 abundances suffer from uncertainties concerning whether the observed deuterium comes from primordial nucleosynthesis. The primordial deuterium could have been burned up. Or deuterium could have been produced "recently", after big bang nucleosynthesis. These are clearly issues which can only be settled by detailed calculations. Eventually, all these observable numbers will be pinned down and one would be able to determine the "unobservable" number N_ν. (Indeed, even as these words are written, a new analysis[20] suggests that N_ν can be at most four.)

This discussion implicitly assumes that the lepton number density n_L is small compared to n_γ. The picture of baryongenesis to be discussed in the next Chapter suggests that $n_L \sim n_B$. If for some reason n_L is comparable to n_γ, then the large $n_\nu - n_{\bar{\nu}} \sim n_L$ neutrino degeneracy would affect the neutron-proton number ratio before nucleosynthesis.[21]

After nucleosynthesis the temperature continues to drop until it is cool enough for hydrogen atoms to form. We would expect this to occur when the temperature T drops below $\Delta \sim 10$ eV $\sim 10^5$ °K, the binding energy of hydrogen atom. When $T \lesssim \Delta$, those photons capable of ionizing the hydrogen atoms which have formed comprise a fraction of order one (i.e., $\sim \frac{1}{2}, \frac{1}{4}$) of all the photons. However, as explained before in connection with deuteron photodissociation, since there are so many photons per baryon, T has to drop substantially below Δ. The "recombination era" starts when the temperature drops below

$$T_{\rm rec} = \frac{\Delta}{\log \eta^{-1}} \sim \frac{\Delta}{23} \sim 4000\ ^\circ\text{K} \quad \text{for} \quad \eta \sim 10^{-10}.$$

(The traditional terminology "recombination" is possibly misleading in this play-by-play description of the Universe from the big bang. The electrons and the protons were never combined in the first place.)

Recombination has the profound effect that photon coupling to matter drops dramatically. Matter now exists in electrically neutral units, and is essentially "transparent" to photon propagation. The pressure of the Universe suddenly decreases and this will have profound implications on the formation of galaxies (see Chapter XII). Photons decouple from matter and have been red-shifting ever since. Thus the Universe has expanded by a factor of (4000/3) $\sim 10^3$ since recombination. We will use the terms "decoupling" and "recombination" interchangeably.

According to eq. (10) matter dominates the energy density of the Universe when the temperature drops below $T_{\rm dom} = \eta m_N \sim 10^3$ °K. Again, we take $\eta \sim 10^{-10}$. Note that $T_{\rm rec}$ is about the same order as $T_{\rm dom}$. Whether or not the Universe is matter dominated or radiation dominated at decoupling depends on the value of $\eta = n_B/n_\gamma$. For η as low as 10^{-10} the Universe is radiation dominated at decoupling, as we just saw. But for $\eta > 10^{-9}$ the Universe is matter dominated at decoupling.

Below $T_{\rm dom}$ we enter into the matter dominated era in which the Universe expands {eq. (13)} according to

$$R(t) \propto t^{2/3}.$$

An important concept in cosmology is the distance over which a photon has travelled since the big bang. This distance, called the horizon size, determines at any given time the size of causally related regions. The horizon size is essentially given by {eq. (14)}

$$d_{\rm H} \sim t \sim g^{-1/2} Mpl/T^2.$$

In contrast, the scale size $R(t)$ of the early Universe grows like $t^{1/2}$ {eq. (12)}. Thus, at early times, the size of causally-connected domains is much smaller than the effective size of the Universe[22]. It is puzzling how the observed homogeneity and isotropy of the Universe came about. This is known as the horizon problem—the Universe was expanding too fast after the big bang. (Incidentally, this also explains why general relativistic effects are unimportant in the early Universe. The causality domains are small compared to the radius of curvature.)

Different regions of the sky are observed to have the same microwave temperature to a high degree of accuracy. However, these different regions were not causally related when photons went out of contact with matter. Similarly, the abundance of He^4 does not appear to vary over the observational regions. But nucleosynthesis was going on in causally disconnected regions. The horizon problem represents an outstanding puzzle in cosmology[23].

The horizon problem is one of several "problems" or "puzzles" for the standard big bang cosmology. These problems are all related in one way or another. We now state some of these; we will come back to them in Chapter XIV.

We start with the "Longevity Problem". The Universe is very old. The Universe is supposedly described by a theory in which the only characteristic time scale is the Planck time. Why hasn't the Universe, if it is closed, turned over and recontracted (or, if it is open, reached the curvature dominated expansion era) in a time of the order of the Planck time? Instead, the Universe has already lived 10^{60} units of Planck time. Equivalently, we have the "Frigidity Problem". The Universe is very cold. Why is the present temperature not of the order of the Planck temperature? Or, if one prefers, we also have the "Vastness Problem". The Universe is very big.

Referring to the evolution equation {eqs. (12), (12′), and (13)}, we see that a characteristic time in the evolution of the Universe, whether open or closed, can be defined as the time when the two terms on the right-hand side are comparable. After this time, in an open Universe the curvature term dominates. In a closed Universe, the curvature term becomes important but soon relinquishes control again. From eqs. (12), (12′), and (13) we see clearly that all the problems stated above can be understood in terms of the "Population Problem". The Universe contains a lot of photons. The number of photons within the present horizon N is order 10^{87}. Why is this dimensionless number, which is essentially the entropy of the visible Universe, not of order unity, which is in some sense a more "reasonable" number? The Universe does not become curvature dominated until the temperature has dropped to $\sim T_{pl}/N^{1/3}$.

Alternatively, one can also speak of the "Flatness Problem"[24]. The Universe is very flat. This is basically again the statement that the curvature term is negligible compared to the other two terms, fantastically so, at the Planck time. In Newtonian language, one can ask "Who adjusted, at the Planck time, the kinetic energy and the potential energy to be so nearly equal and opposite?" As an analogy, think of throwing a ball upward. If the initial kinetic energy is low, the ball will quickly fall back to earth. If the kinetic energy is high, the ball will quickly escape. The initial kinetic energy must be finely adjusted to very nearly cancel the potential energy for the ball to travel for several light-years, say, before falling back.

Different people prefer different statements of this problem. This author finds it most physical to think in terms of why the Universe contains so many photons. It is, however, illuminating to describe the situation in many equivalent ways.

Actually, one can also argue that the entropy, or the population, of the Universe is very small. The entropy[25] of a black hole of mass M is given by GM^2. This is believed to be in fact the maximum allowable entropy for a system of given mass and size[26]. If the mass of the Universe within the horizon is squeezed into one giant black hole, the resulting entropy will be of order $GM^2_{\text{Universe}} \sim G(10^{88} m_N 10^{-10})^2 \sim 10^{118}$. Why is the observed entropy of the Universe so much less?

No list of cosmological problems is complete without including that most perplexing puzzle in physics: the cosmological constant puzzle. Contemporary particle theory involves a cascade of spontaneous symmetry breaking, from grand unification breaking to chiral breaking. Each one of these symmetry breakings generates an enormous cosmological constant in contradiction with observation. The problem is all the more fascinating since it lies at the interface between gravity and the other three interactions, and its resolution will surely shed light on how gravity is connected with the other interaction. (For related discussions, see Chapters XIV and XVIII.)

We close this short review of cosmology by quoting Lemaître: "The evolution of the world can be compared to a display of fireworks that has just ended: some few red wisps, ashes, and smoke. Standing on a cooled cinder, we see the slow fading of the suns, and we try to recall the vanished brilliance of the origin of the worlds".

APPENDIX TO CHAPTER X

Consider infinite Newtonian space filled with matter of uniform density ρ. Clearly, Newton's equation for the gravitational potential

$$\nabla^2\phi = 4\pi G\rho \tag{1}$$

has no time-independent solution for ρ uniform.

In a time-dependent solution, the matter moves and one must write down Euler's fluid equation

$$\frac{\partial \boldsymbol{v}}{\partial t} + (\boldsymbol{v}\cdot\nabla)\boldsymbol{v} = -\nabla\phi \tag{2}$$

which of course represents just Newton's other equation $F = ma$ in disguise. Since we are talking about non-relativistic matter, the pressure is negligible. Also, matter is conserved so that

$$\frac{\partial \rho}{\partial t} + \nabla(\rho\boldsymbol{v}) = 0. \tag{3}$$

A spatially uniform solution takes the form

$$\rho = C/R^3(t). \tag{4}$$

C denotes a constant. This form is completely general; it just says that ρ is a function of t only. The equation of continuity {eq. (3)} now determines

$$\boldsymbol{v} = \boldsymbol{x}\left(\frac{\dot{R}(t)}{R(t)}\right). \tag{5}$$

Newton's equation determines the gravitational potential ϕ to be

$$\phi = \frac{1}{2}x^2(4\pi G\rho/3). \tag{6}$$

Finally, putting all this into Euler's equation we determine the unknown function $R(t)$ to be given by

$$\frac{\ddot{R}}{R} = -4\pi G\rho/3. \tag{7}$$

The first integral of this equation gives Einstein's equation for a matter-dominated Universe. The solution {cf. Eq. (5)} describes a Hubble expansion of the Universe.

When one is faced with this problem of a uniform distribution of gravitating matter initially at rest, one's first reaction might be that there is no preferred point for the matter to contract to. The Hubble picture resolves this apparent puzzle nicely; every point is a preferred point. To understand the Hubble expansion (or contraction), imagine hollowing out a sphere around an arbitrary point O. Then, by a well-known theorem due to none other than Newton, the matter inside the sphere does not feel the gravitational presence of all the other matter outside the sphere. By spherical symmetry, the matter inside the sphere will contract towards point O. But since the point O was arbitrarily chosen, one arrives at the Hubble picture. More precisely, a mass point on the surface of this hypothetical sphere feels a gravitational attraction given by $(4\pi R^3/3)\rho/R^2$ and so accelerates at the rate $\ddot{R}$. By Newton's law, one gets eq. (7).

Thus, Newton could have discovered the Hubble expansion of the Universe by applying his own theorem and equations! Why he did not is a fascinating question. Presumably, the notion of an expanding (or contracting) Universe would have sounded so far-fetched in the seventeenth century (if anybody broached the possibility at all) that the psychological barrier to following the equations to their logical conclusions must have been overwhelming[27]. In fact, remarkably enough, some two hundred years later, another great mind of physics saw[28] the Hubble expansion implied by his equation and rejected it. As is well-known, Einstein modified his equation so as to stop the expansion of the Universe, a move he regarded later as the biggest blunder of his life. Whether or not Newton saw that his equations imply the expansion of the Universe will presumably never be known.

There is a moral to the story somewhere.

REFERENCES

1. S. Weinberg, *Gravitation and Cosmology*, J. Wiley & Sons, (New York, N.Y., 1972).
2. P. J. E. Peebles, *The Large Scale Structure of the Universe*, (Princeton University Press, 1980); *Physical Cosmology* (1971).
3. A. Dolgov and Ya. B. Zel 'dovich, *Rev. Mod. Phys.* **53,** 1 (1981).
4. M. S. Turner, *Proc. Second Workshop on Grand Unification* (Ann Arbor, 1981); *Proc. of Neutrino* '81 (1981); *Weak Interaction as Probes of Unification*, edited by G. Collins *et al.* (1980).
5. G. Steigman, *Proc. First Workshop on Grand Unification* (1980); *Proc. of Rome Conference* (1980); *Unification of Fundamental Interactions*, Erice (1981).
6. D. N. Schramm and R. V. Wagoner, *Ann. Rev. Nucl. Sci.* **27,** 37 (1977).
7. For an excellent popular overview, see S. Weinberg, *The First Three Minutes*, (Basic Books, 1977). For a somewhat more advanced treatment, see J. Silk, *The Big Bang*, W. H. Freeman & Co., San Francisco.
8. This was realized by McCrea and Milne in the thirties. An excellent discussion may be found in Sec. 15.1 of Weinberg (Ref. 1).
9. For a more precise discussion on N, see G. Steigman's *Erice talk* (Ref. 5).
10. R. P. Kirschner, A. G. Oemler, and P. L. Schechter, *Astron. J.* **84,** 951 (1979).
11. G. Gamow *et al.*, papers X. 1, 2, 3, 4; C. Hayashi, paper X. 5. (Gamow had assumed that the Universe started out as a neutron gas. Hayashi was the first to recognize that at high temperatures the weak interaction determines the neutron to proton ratio.); R. A. Alpher, J. W. Follins, Jr., R. C. Herman, Paper X. 6. See also R. A. Alpher and R. Herman, *Nature* **162,** 774 (1948) and *Phys. Rev.* **75,** 1089 (1949).
12. F. Hoyle and R. Tayler, Paper X. 7.
13. P. J. E. Peebles, *Ap. J.* **146,** 542 (1966).
14. R. V. Wagoner, W. A. Fowles, and F. Hoyle, *Ap. J.* **148,** 3 (1967).
15. Ya. B. Zel 'dovich, *Adv. in Astron. Astrophys.* **3,** 241 (1965).
16. E. Fermi and A. Turkevich, unpublished.
17. V. Shvartzman, Paper X. 8.
18. G. Steigman, D. N. Schramm, and J. E. Gunn, Paper X. 9.
19. K. Olive et al., Paper X. 10.
20. J. Yang, M. Turner, G. Steigman, D. Schramm, and K. Olive, to be published.
21. A. D. Linde, *Phys. Lett.* **83B,** 311 (1979); D. N. Schramm and G. Steigman, *ibid.* **87B,** 141 (1979).
22. W. Rindler, *Mon. Not. R. Astro. Soc.* **116,** 663 (1956).
23. Solutions of the horizon problem based on modifying the theory of gravity have been proposed by A. Zee, *Phys. Rev. Lett.* **44,** 703 (1980), F. W. Stecker, *Ap. J.* **235,** 21 (1980).
24. This was emphasized by R. H. Dicke and P. J. E. Peebles, *General Relativity*, ed. S. W. Hawking and W. Israel, (Cambridge University Press, 1979).
25. S. W. Hawking, *Comm. Math. Phys.* **43,** 199 (1975); J. D. Beckenstein, *Phys. Rev.* **D7,** 2333 (1973).
26. J. D. Beckenstein and G. W. Gibbons, preprint NSF-1TP-80-38.
27. Another possibility is that Newton was preoccupied with how to generate the observed inhomogeneities. See Ref. 1 in Chapter XII.
28. Actually, in his 1917 paper Einstein did not mention the expanding Universe. Rather, he plugged in the metric of a static Universe and noted that his equations are not satisfied.

CHAPTER XI

GENESIS OF MATTER

There are two striking facts about the Universe we live in:

(1) The Universe is not empty;

(2) The Universe is almost empty.

It behooves physicists to understand these two facts.

When we say the Universe is not empty, we as physicists mean that the values of some conserved quantum numbers are not equal to zero in the Universe. (This, of course, represents the only physically sensible and intrinsic definition of the word "empty".) For the moment, we will take baryon number B and lepton number L to be conserved. When Dirac's theoretical prediction of anti-matter was verified it was extremely natural and tempting to speculate[1–3] that the Universe is symmetric between matter and anti-matter, so that the total baryonic and leptonic charges of the Universe vanish exactly. After all, the total electric charge Q of the Universe is known to be zero to an extremely high accuracy. We just happen to live in a region dominated locally by matter. This rather attractive picture, however, is not supported[4] by evidences accumulated over the years[5]. The observable Universe appears to be non-empty.

The popular image of the Universe is one vast emptiness dotted here and there by a few galaxies. With the discovery of the microwave background, this emptiness is quantified by the ratio $\eta \equiv n_B/n_\gamma \sim 10^{-10}$. (See Chapters X and XIII.) The Universe is almost empty. The Universe is only slightly contaminated by matter.

(To proponents of symmetric cosmology, however, the observed value of η appears rather large. One of the severe difficulties of symmetric cosmology is that a plausible mechanism for separating matter and anti-matter has never been found. The following simple argument[4] shows that statistical fluctuation is insufficient. Some causal domains, by statistical fluctuations, would have slightly more baryons, while others have more antibaryons. We expect the excess to be given roughly by $N_B \sim \pm\sqrt{N_\gamma}$ where N_γ is the number of photons in the domain. Thus, when baryons and anti-baryons are annihilated as the Universe cooled, each domain will end up with only $\sim 10^{10}$ baryons (or anti-baryons), given that $N_B/N_\gamma \sim 1/N_B \sim 10^{-10}$. Since a typical star like the sun contains 10^{57} nucleons, a more efficient mechanism is clearly needed.)

The discovery of CP violation in 1964 showed that the laws of physics are not exactly symmetric between matter and anti-matter, and opened the door to a possible understanding of the matter-anti-matter asymmetry in the Universe[6]. However, if baryon number is absolutely conserved as was generally believed[7], then the observed baryon contamination of the Universe has to be put in at the "beginning". The ratio η represents an initial condition and its smallness cannot be understood from first principles.

With the advent of grand unification, baryon non-conservation was no longer regarded as an unfounded speculation. (See Chapter VII.) Several groups[8–16] independently recognized that grand unification allows us to calculate η in terms of other fundamental parameters, such as the measure of CP violation. We thus can have an "esthetically appealing" cosmology in which the Universe started out empty and gradually developed a preponderance of matter over anti-matter.

It is worth emphasizing that, not only does grand unification predict baryon non-conservation, but it also predicts that the baryon non-conservation amplitude becomes substantial at temperatures higher than the unification scale. This is essential since observations have already established the proton lifetime to be more than 10^{20} times the present age of the Universe. Thus, it is not enough to simply add a term like $(qqql)$ to a non-grand-unified theory.

For the Universe to develop a matter-anti-matter asymmetry, three conditions must be satisfied. (1) The laws of physics must be asymmetric between matter and anti-matter; (2) the relevant physical processes were out of equilibrium so that there was an arrow of time; (3) baryon numbers must be violated. That these three conditions are necessary was clearly stated in Refs. (8) and (13), for instance, but was not fully appreciated in some of the early papers on the subject.

A pedagogical introduction is perhaps provided by the following toy example given in Ref. (13). Consider an empty box at temperature $\gtrsim m_K$. The box would contain an appreciable population of neutrinos, photons, electrons and positrons; π and K mesons, and so forth. The K mesons are known to decay preferentially into states containing e^+ over states containing e^- by roughly one part in 10^3. CP and electron number are both violated. One would be naive, however, to conclude that the box would contain more e^+ than e^-. After all, Boltzmann assured us that the equilibrium density of a particle is determined completely by its mass and Schwinger, Lüders, and Pauli all assured us that CPT invariance implies that $m_{e^+} = m_{e^-}$. The point is of course that processes such as $\nu e^+ \to \pi^+ K^\circ$ remove any excess positrons from K decays. Now, if one suddenly opens the box, the particles would stream out and the eventual K meson decays would indeed lead to more positrons than electrons[17]. Note that lepton number is conserved in this example.

The expanding Universe provides us with precisely such a setting in which physical processes go out of equilibrium (see Chapter X). A very important point, particularly emphasized[18] in Refs. (13) and (14), is that no baryon asymmetry would develop in processes involving only massless particles such as $u + d \to \bar{u} + e^+$. As the Universe expands, the momentum of a massless particle and hence its energy simply red-shifts like $1/R$. Thus, if the particles were in equilibrium at one point, they would simply maintain an equilibrium density[19], corresponding to a temperature T decreasing like $1/R$. Indeed, we already used this fact in our discussion in Chapter X.

An important conceptual point is that, as long as unitarity holds, Boltzmann's H-theorem is valid even if time reversal invariance is violated.[20]

We thus conclude that processes involving a massive particle ($M > T$) are needed to generate a baryon asymmetry. One possible scenario[13,14] involves the drifting decay of a massive X particle, which could be either a gauge boson or a

Higgs boson in a grand unified theory. At high temperatures $T \gg M$, X is supposed to be as abundant as photons. At some temperature T_d the X particles decay. This critical temperature is determined by

$$1/\tau \sim (GgT_d^4)^{1/2} \tag{1}$$

where τ denotes the X lifetime. The rest of the notation is as in Chapter X. The right-hand side is the inverse of the time elapsed since the big bang. For the scenario to work, the condition $T_d < M$ must be satisfied, since otherwise inverse decays[21] will replenish the X population. This yields a lower bound on the X mass[14]. If, on the average, each X decay produces ε more baryons than anti-baryons, then eventually

$$n_B/n_\gamma \sim \varepsilon/g(T_d). \tag{2}$$

The factor $g(T_d)$ counting the number of degrees of freedom present at the decay accounts for the subsequent heating of the photon gas (see Chapter X).

It is exciting that one day we may be able to actually calculate how much matter the Universe should contain. When Gamow first proposed primordial nucleosynthesis, he exulted that he is letting his "imagination fly beyond all limits", and remarked that nuclear abundances represent the "most ancient archaelogical document" known[22]. Physics is a remarkable subject; only three decades later people are extending Gamow's cosmology to temperatures some nineteen orders of magnitude higher than what he had in mind. Matter is the ultimate fossil.

Unlike the situation in nucleosynthesis, where nuclear reaction rates were more-or-less measurable in the laboratories, the relevant reaction rates in baryosynthesis have to be estimated theoretically. Calculations[23] of the parameter ε in the simplest version of $SU(5)$ typically yield too small a value. Our present understanding also does not relate directly and precisely CP violations as measured in the K-meson system to CP violations in baryon number non-conserving process. It would be an amusing triumph for physics if one could actually "predict" that, given laboratory measurements of the K-meson decays, the Universe contains matter and not anti-matter. Unfortunately, at the moment, a determination of the sign and magnitude of n_B/n_γ in terms of known K-meson parameters is still lacking.

Numerical studies of the Boltzmann evolution of the baryon number density have been performed[24,25].

We have outlined here the basic scenario for the genesis of matter. The picture has been extended and elaborated in a number of ways. We have not mentioned any of the complications so far. For instance, a careful analysis by Yoshimura[26] suggests that the gauge bosons in $SU(5)$ fail to satisfy the mass bound mentioned above in the drifting decay scenario. Instead, Higgs boson decays may provide the correct mechanism. The decays of heavy fermions have also been considered[27–29] in the context of $SU(10)$[28] and left-right symmetric grand unified theories[30]. Another potentially important effect involves the washing-out or dilution of any existing baryon asymmetry. For instance, if the $SU(2) \times U(1) \to U_{\text{em}}(1)$ transition is first order (see Chapter XIV), the baryon to photon ratio would be vastly diluted[31]. The literature on the subject is vast and we cannot possibly hope to give a complete review here[32].

The grand unification picture on the origin of matter has far-reaching implications (for a summary, see paper VII.1). The nature of the "primordial" density perturbations leading to galaxy formation (Chapter XII) might be determined. It also opens up the question of understanding the observed homogeneity and isotropy of the Universe. Long before grand unification, Misner[33], and Barrow and Matzner[33] showed that if the Universe started out in a chaotic state and subsequently smoothed itself out, the entropy per baryon would be much larger than the observed value $n_\gamma/n_B \sim 10^{10}$. However, if n_γ/n_B is, in fact, determined by microphysics, then one can no longer deduce from the observed value of n_γ/n_B that the Universe cannot have started in a chaotic state[34].

We close by noting that Steigman began his influential review[4] with a paraphrased remark of T. H. Huxley that "the terrible tragedies of science are the horrible murders of beautiful theories by ugly facts". As is often the case in physics, the beautiful theory which Steigman was referring to, symmetric cosmology, has been replaced by a vastly more beautiful theory which may even fit the facts. Huxley was wrong; there is no tragedy ever in physics.

REFERENCES

1. M. Goldhaber, *Science* **124,** 218 (1956); R. A. Alpher and R. C. Herman, *ibid.* **128,** 904 (1958).
2. H. Alfvén and O. Klein, *Ark. Fys.* **23,** 187 (1962).
3. R. Omnès, *Phys. Reps.* **C3,** 1 (1970).
4. A definitive review was given by G. Steigman, *Ann. Rev. Astron. Astrophysics* **14,** 339 (1976).
5. The present author was introduced to this subject by chance when he rented an apartment belonging to one of Omnès collaborators during a visit to Paris in the early seventies.
6. J. Cronin, *private communication.*
7. For a historical review, see M. Goldhaber, *Proc. Am. Philosophical Soc.* **119,** 24 (1975); M. Goldhaber, P. Langacker, R. Slansky, *Science* (1980).
8. In a remarkably prescient, but somehow forgotten, paper, A. D. Sakharov (paper XI.1) outlined an explanation of the baryon asymmetry of the Universe. See also *Zh. Eksp. Teor. Fiz.* **76,** 1172 (1979). (English translation: *JETP Lett.* **49,** 594 (1979).)
9. For an early remark, see S. Weinberg, *Brandeis Summer Lectures*, ed. by S. Deser and K. Ford (Prentice-Hall, N.J., 1964), p. 482.
10. V. A. Kuzmin, *JETP Letters* **12,** 228 (1970); A. Y. Ignatiev. N. V. Krasnikov, V. A. Kuzmin, and A. N. Tavkhelidge, *Phys. Lett.* **76B,** 436 (1979). This work is based on ref. (8) but is outside the grand unification framework.
11. M. Yoshimura, paper XI.2.
12. S. Dimopoulos and L. Susskind, paper XI.3.
13. D. Toussaint, S. Treiman, F. Wilczek, and A. Zee, paper XI.4.
14. S. Weinberg, paper XI.5.
15. J. Ellis. M. K. Gaillard, and D. V. Nanopoulos, *Phys. Lett.* **80B,** 360 (1979); *erratum* **82B,** 464 (1979).

16. The subsequent literature on the subject is vast. Consult the bibliography and the references cited in the reviews in ref. (32).
17. This example is imperfect in that we must assume the pions are stable.
18. See also S. Barr, *Phys. Rev.* **D19,** 3803 (1979).
19. For a related discussion, see E. L. Schücking and E. A. Spiegel, in *Comments on Astrophysics & Space Physics* **11,** 121 (1970); and P. T. Landsberg, W. C. Saslaw and A. J. Haggett, *Mon. Not. R. Astr. Soc.* **154,** 7 (1971).
20. C. N. Yang and C. P. Yang, unpublished preprint ITP-SB-78-51; A. Aharony, in *Modern Developments in Thermodynamics* (Wiley, 1973); see in particular footnote 3 of paper XI.5.
21. In particle physics, one is not used to thinking about inverse decays (such as $p + e^- + \bar{\nu} \rightarrow n$). But in equilibrium, according to Boltzmann, these processes have to be important.
22. See G. Gamow, paper X.4.
23. There have been numerous calculations. Two early calculations are D. V. Nanopoulos and S. Weinberg, paper XI.6; S. M. Barr, G. Segre, and H. A. Weldon, paper XI.7.
24. E. W. Kolb and S. Wolfram, paper XI.8 and *Nucl. Phys.* **B172,** 224 (1980).
25. J. N. Fry, K. A. Olive, M. S. Turner, *Phys. Rev.* **D22,** 2953, 2977 (1980), and paper XI.9.
26. M. Yoshimura, *Phys. Lett.* **88B,** 294 (1979).
27. T. Yanagida and M. Yoshimura, *Phys. Rev. Lett.* **45,** 71 (1980).
28. J. A. Harvey, E. W. Kolb, D. B. Reiss and S. Wolfram, *Nucl. Phys.* **B177,** 456 (1981); and CALT 68-850, to appear in *Nucl. Phys. B.*
29. R. Barbieri, D. V. Nanopoulos, and A. Masiero, *Phys. Lett.* **98B,** 456 (1981).
30. A. Masiero and R. N. Mohapatra, *Phys. Lett.* **103B,** 343 (1981), A. Masiero and T. Yanagida, preprint MPI-PAE/PTh 53/81 (1981); A. Masiero, in *Proceedings of Unification Conference*, Erice, (1981).
31. E. Witten, *Nucl. Phys.* **B177,** 477 (1981).
32. Fortunately, a number of excellent reviews are available; for instance, M. Yoshimura, in *Grand Unified Theories and Related Topics*, World Scientific, 1981; P. Langacker, *Phys. Rep.* **72,** 185 (1981); M. Turner, in *Proceedings of the 1981 Les Houches Summer School.*
33. C. W. Misner, *Ap. J.* **151,** 431 (1968); J. D. Barrow, *Nature* **272,** 211 (1978), and references therein.
34. M. Turner, *Nature* **281,** 549 (1979).

CHAPTER XII

INTRODUCTION TO THE THEORY OF GALAXY FORMATION

The existence of massive neutrinos may have a dramatic impact on cosmology. Particularly interesting is the effect on galaxy formation.

The theory of galaxy formation started when Newton remarked to Bentley in a letter that a uniform distribution of matter is inherently unstable under gravity[1]. The theory is developed further by Jeans. We cannot hope to give a detailed discourse on the subject here, so we will content ourselves with a brief account of the essential features of galaxy formation theory[2].

According to the classic Jeans theory, in a uniform distribution a local density fluctuation will clump if the relevant gravitational attraction overwhelms the pressure. In symbols, the criterion reads

$$\frac{G(\rho R^3)^2}{R} \gtrsim pR^3 \tag{1}$$

or equivalently

$$R \gtrsim \frac{M_{pl} p^{1/2}}{\rho} \equiv R_J. \tag{2}$$

Here ρ denotes the mass density of matter, p the pressure, R the length scale of the fluctuation. Eq. (2) informs us that to clump, R must be greater than R_J, called the Jeans length. The mass contained in the Jeans length is called the Jeans mass

$$M_J \sim \rho R_J^3 \sim p^{3/2} M_{pl}^3/\rho^2. \tag{3}$$

Before photon decoupling, in the radiation era, the mass density and the pressure of the Universe are both dominated by photons:

$$p \sim \rho \sim T^4. \tag{4}$$

The Jeans mass is then

$$M_J \sim (M_{pl}/T)^2 M_{pl} \sim \left(\frac{1\ \text{GeV}}{T}\right)^2 \mathrm{M}_\odot. \tag{5}$$

(We recall that a solar mass $\mathrm{M}_\odot \sim 10^{57}\, m_N$.) At $T \sim 1$ eV $\gtrsim$ the recombination temperature which, as we recall from Chapter X, is about 1/10 the binding energy of the hydrogen atom, we find, as an order-of-magnitude estimate

$$M_J^{\text{before}} \sim 10^{18}\ \mathrm{M}_\odot. \tag{6}$$

After recombination, matter exists in the form of atoms and becomes transparent to the free propagation of photons. The pressure, now due to atoms, drops precipitiously to

$$p = \frac{NT}{V} = n_B T. \tag{7}$$

Recalling the observed photon to baryon ratio $n_B \sim 10^{-10}\, n_\gamma \sim 10^{-10}\, T^3$ (see Chapter X), we see that the pressure drops by roughly ten orders-of-magnitude going through recombination. We also have

$$\rho \sim m_N n_B.$$

Thus we expect the Jeans mass to drop; indeed

$$M_J^{\text{after}} \sim (T^3/n_B)^{1/2}(M_{pl}/m_N)^2 M_{pl} \sim 10^5\ \mathrm{M}_\odot. \tag{8}$$

We have implicitly assumed that recombination and the transition from the radiation era to the matter era occur at around the same time. If $\eta \equiv n_B/n_\gamma$ is smaller, then the Universe would still be dominated by radiation in ρ when matter "recombined" and photons decoupled. The important point is that, going through recombination, the pressure P drops by a factor η^{-1}, while the energy density ρ essentially does not change.

The mass $10^5\ \mathrm{M}_\odot$ corresponds to typical globular clusters[3]. The classic Jeans theory says that density fluctuations $\delta\rho/\rho$ on mass scales $\gtrsim 10^5$ GeV are unstable and grow. Galaxies are observed to have masses typically around $10^{11}\ \mathrm{M}_\odot$. One of the goals of galaxy formation theory is to understand how this particular mass value emerges. One must go beyond Jeans theory to discuss how a given density fluctuation $\delta\rho/\rho$ grows. Lifschitz and others showed that $\delta\rho/\rho$ grows like a power of time $\sim t^{2/3}$ rather than exponentially, essentially because the growth of the instability is competing with an expanding background.

A very heuristic derivation of this result goes as follows: A density fluctuation may be thought of as curvature fluctuation; the denser regions are more curved. In the fundamental equation

$$\frac{\dot{R}^2}{R^2} = \left(\frac{8\pi}{3}\right) G\rho - \frac{k}{R^2}, \tag{9}$$

a "curvature fluctuation" is like

$$\delta k/R^2 \sim \left(\frac{8\pi}{3}\right) G\delta\rho.$$

Thus while $\rho \propto R^{-3}$ (in the matter dominated era) $\delta\rho/\rho$ goes like $R^{-2}/R^{-3} \sim R \sim t^{2/3}$. Needless to say, this argument is at the "hand-waving" level but serves handily at least as a mnemonic. A more precise derivation will be given in an Appendix to this Chapter. Actually, because of the huge freedom represented by local coordinate invariance in general relativity, the definition of cosmological perturbations is far from clear, since by coordinate transformations a given perturbation will be transformed. Density is not a scalar. Even a perturbation in a scalar quantity will not be invariant, since a coordinate transformation also changes the point in the background spacetime corresponding to a given point in physical spacetime to which that perturbation is referred. This difficulty has bedeviled workers in this field for decades. The confusion was finally cleared up recently by J. Bardeen[4] who formulated cosmological perturbation in a gauge invariant way, and by Press and Vishniac[5].

In the literature, pictures on what the fluctuations in the era before photon decoupling look like are discussed[6]. In the adiabatic picture, fluctuations in matter density are correlated with fluctuations in radiation density so that n_B/n_γ is constant. In the isothermal picture, there is no fluctuation in the photon temperature at all[7] (papers XII.1, 2).

This is where grand unification offers valuable input. Clearly, the baryon-genesis scenario discussed in the last chapter favors the adiabatic picture. The ratio n_B/n_γ is determined by microphysics. However, this conclusion might be altered if the Universe started out with substantial large scale shear[8].

The subsequent evolution is quite complicated. We will discuss here how things roughly go in these two pictures.

In the adiabatic picture, due to photon diffusion, fluctuations on mass scales less than a characteristic mass M_S, the Silk mass, are damped out[9]. Basically, if the photon mean free path is larger than the length scale of the fluctuation, photons diffuse out of the fluctuation and the perturbation dissipates. Consider a perturbation with a mass scale M. From eq. (5) we see that the Jeans mass M_J rises rapidly before decoupling. When $M_J > M$, there is an acoustic regime in which the perturbation is stable against collapse and simply oscillates. For M of the order of a galactic mass $\sim 10^{11}\ M_\odot$ this happens, according to eq. (5), between the time when the temperature drops below $\sim \frac{1}{3} \times 10^4$ eV $\sim \frac{1}{3} \times 10^8$ °K and recombination $\sim$4000°K.

We now give a very rough and heuristic calculation of this damping effect due to photon diffusion. At the temperature regime in question, photons are scattered by non-relativistic electron with the Thomson cross-section

$$\sigma \sim \alpha^2/m_e^2. \tag{10}$$

The mean free time of the photon between scattering is thus

$$\tau \sim (n\sigma)^{-1}, \tag{11}$$

where the number density of electrons $n \sim n_B = \eta\ n_\gamma \sim \eta T^3$. Putting in some (precise) numbers we find

$$\tau = 5 \times 10^7\ \text{sec}\left(\frac{10^5\ ^\circ\text{K}}{T}\right)^3 \tag{12}$$

of the order of a year when $T \sim 10^5$ °K. The electrons transfer energy by Coulomb scattering off other electrons (or protons). Since the kinetic energy of an electron is $\sim T$, the distance of closest approach r in a Coulomb scattering is given by $T \sim e^2/r$, and so the relevant cross-section $\sigma \sim e^4/T^2$. So the mean free time of the electron

$$\tau_e \sim (nv\sigma)^{-1} \sim \left[n\left(\frac{T}{m_e}\right)^{1/2}\frac{e^4}{T^2}\right]^{-1}$$

is shorter than the photon mean free time by a factor $(T/m_e)^{3/2}$.

Photons have difficulty keeping up with the acoustic oscillation. In Appendix B to this Chapter, we show that if the pressure lags behind the density fluctuations by a characteristic time τ, the damping rate of a sound wave is given by

$$\Gamma \sim \tau(c_s k)^2 \tag{13}$$

with c_s the speed of sound and k the inverse of the wavelength. Consider a fluctuation on the mass scale M. Then $M \sim nm\, k^{-3}$, since the baryon mass density is nm. We will assume that this era just before recombination is dominated by radiation, as is the case for a small $\eta \sim 10^{-10}$. Then c_s is of order one and temperature depends on time by $t^2 \sim (GT^4)^{-1}$. So, putting all this together we find the damping rate

$$\Gamma \sim \left(\frac{m}{M}\right)^{2/3}\left(\frac{1}{\sigma\eta^{1/3}}\right)\frac{1}{T}. \tag{14}$$

Thus, by recombination the proto-galactic fluctuation will be damped by an exponential factor

$$\exp\left[-\int^{t_{\rm rec}} \Gamma\, dt\right] \equiv \exp[-(M_s/M)^{2/3}]. \tag{15}$$

This defines the critical mass M_s. The integral is easily done since we know how temperature depends on time. We find

$$M_s \sim m\sigma^{-3/2}G^{-3/4}\eta^{-1/2}T_{\rm rec}^{-9/2}. \tag{16}$$

Putting in $\sigma \sim \alpha^2/m_e^2$ we rewrite this in a more reasonable form

$$M_s \sim m(\alpha^3\eta^{1/2})^{-1}\left(\frac{m_e}{T_{\rm rec}}\right)^3\left(\frac{M_{pl}}{T_{\rm rec}}\right)^{3/2} \tag{17}$$

which comes out to be about $\sim 10^{13}$–$10^{14}\ M_\odot$.

An alternative but equivalent way of deriving M_s involves calculating the photon diffusion time. A fluctuation on the mass scale M has a length scale of the order $k^{-1} \sim (M/nm)^{1/3}$. The time it takes for photons to diffuse out of this fluctuation is then

$$t_{\rm diffuse} \sim \tau\left(\frac{k^{-1}}{c\tau}\right)^2 \sim \frac{1}{\tau}\left(\frac{M}{nm}\right)^{2/3}. \tag{18}$$

We explicitly indicate the speed of light c in the mean free path $l = c\tau$ and use the fact that in N scatterings a diffusing photon travels a distance $\sim N^{1/2}l$. The fluctuation does not survive if photons diffuse out in a time less than the characteristic time scale of the Universe $t \sim (G^{1/2}T^2)^{-1}$, i.e.

$$M \lesssim nm\left(\frac{\tau}{G^{1/2}T^2}\right)^{3/2}. \tag{19}$$

Simplifying, we obtain eq. (17).

Our discussion is clearly heuristic and serves to introduce the reader to the original literature[9]. A detailed calculation is quite involved, and includes effects such as heat conduction which we ignored. It turns out that eq. (16) is essentially correct if the Universe is radiation dominated at recombination. In the astrophysical literature, M_s is given numerically as

$$M_s \simeq 10^{12}\,M_\odot(\Omega h^2)^{-2}[1 + 0.04\,(\Omega h^2)^{-1}]^{-3}. \tag{20}$$

Thus, in the adiabatic picture, only proto-galactic fluctuations on mass scales larger than $\sim 10^{14}\,M_\odot$ survive photon diffusion damping until decoupling. After decoupling, the Jeans mechanism causes these proto-galactic lumps to collapse. Since pressure is ineffective, the collapse is supersonic, turbulent, and asymmetrical. In general, the collapse will be faster along one direction than along the other two, so one expects these lumps to collapse to objects shaped like pancakes or disks. Further instability and growth of density fluctuation within the "pancakes" lead to the formation of galaxies. Incidentally, the adiabatic picture is thus also known as the "pancake" or fragmentation scenario in the literature. In fact, galactic super-clusters have typically $10^{14}-10^{15}\,M_\odot$. This picture also explains nicely the big "holes in space" recently observed (see Chapter X). The spaces between the pancakes should be devoid of galaxies.

In the isothermal picture, galaxy formation evolves differently. Before decoupling, the baryon density fluctuation δn_B finds itself superposed on a uniform radiation background. Since there are so many photons per baryon, any tendency for the mass density δn_B to move encounters tremendous radiation pressure. Thus, in the isothermal picture, essentially nothing happens to the fluctuations until decoupling—they neither grow nor damp. At decoupling, the Jeans mass drops dramatically to $\sim 10^5\,M_\odot$. All fluctuations on mass scales greater than $10^5\,M_\odot$ grow. One generally assumes that the initial fluctuation amplitude decreases for larger mass scale (perhaps like a featureless power law $\delta\rho/\rho \propto M^{-n}$). Thus, the fluctuation at the lowest mass scale will go nonlinear first and so one expects objects of $\sim 10^5\,M_\odot$ to form first. Structure is then built up by clumping objects on larger and larger scales. Thus, the isothermal picture is also known as the aggregation or hierarchical model.

Fluctuations can grow only after matter dominates the energy density of the Universe. We have shown that $\delta\rho/\rho$ grows like $t^{2/3}$ (i.e., like the inverse of the background photon temperature) in the matter-dominated era. Therefore, knowing when galaxies formed and assuming that galaxies formed when $\delta\rho/\rho$ became order 1, one can deduce how large an initial $\delta\rho/\rho$ is needed. If we assume that at recombination the Universe is matter dominated, $\delta\rho/\rho$ at recombination is $\sim 10^{-3}$ since T_γ has dropped by a factor of $\sim 10^3$. However, if η is very low so that the Universe does not become matter-dominated until long after recombination, then $\delta\rho/\rho$ has to be larger. Roughly, there is not enough time for the fluctuations to grow. This problem is known as the "low η" or "low Ω squeeze".

If indeed $\delta\rho/\rho \sim 10^{-2}$ or 10^{-3} just after recombination, then in the adiabatic picture, temperature fluctuation $\delta T/T$ was also of order 10^{-2} and 10^{-3}. This appears to be in conflict with the observed anisotropy in the microwave background of not more than 10^{-4}. It would appear that the adiabatic picture, favored by grand unification, is in some difficulty in this regard. In the next Chapter we will see how massive neutrinos might resolve this difficulty. In contrast, in the isothermal picture, one expects much smaller fluctuations in the observed microwave background, induced[10] by fluctuations in the gravitational potential generated by $\delta\rho/\rho$.

This survey is meant to introduce the non-expert as rapidly as possible to the theory of galaxy formation, but can hardly be expected to reflect the full glory of the subject. We have not described the important processes of gravitational clustering, condensation, cooling, and fragmentation. (The reader is referred to Paper XII.7 for a sampling of the subject) Explosions in the intergalactic medium may also be important[11]. Galaxy formation begets more galaxy formation by "explosive amplification" (Paper XII.8).

In order to read the astrophysical literature, some readers may need to know that 1 parsec is about 3×10^{18} cm $\sim$ 3 light years.

APPENDIX A TO CHAPTER XII

We sketch here the Newtonian theory of the growth of small fluctuations[12]. It represents a straightforward application of linear perturbation theory to the solutions of Newton's equations given in the Appendix to Chapter X. (The equations in that Appendix are referred to by a prime after the equation number.) Let us denote the zeroth order solutions given in Eqs. (4′), (5′), and (6′) as ρ_0, $\boldsymbol{v}_0$, ϕ_0, and $P_0 = 0$. Denote the small deviations from the zeroth order quantities by ρ_1, $\boldsymbol{v}_1$, ϕ_1 and $P_1 = c_s^2\,\rho_1$. (Hence c_s denotes the speed of sound.) We simply linearize Eqs. (1′), (2′), and (3′). The resulting linear equations are homogeneous in space and so one can decompose the first-order small quantities in plane-wave modes:

$$\rho_1(\boldsymbol{x}, t) = \rho_1(t)e^{i\boldsymbol{q}.\boldsymbol{x}/F(t)} \tag{1}$$

and similarly for ϕ_1 and $\boldsymbol{v}_1$. Since the background is expanding, the wave vector will depend on time, and so we have included the function $F(t)$. Physically, we expect that $F(t) = R(t)$, since the wave vector counts the number of nodes and is simply red-shifted by the expansion. Indeed, when we insert the plane-wave forms into the linear equations the resulting equations for $\rho_1(t)$, $\phi_1(t)$, and $\boldsymbol{v}_1(t)$ depend explicitly on $\boldsymbol{x}$ unless $F(t) = R(t)$ as expected. One also decomposes $\boldsymbol{v}_1$ into a transverse and longitudinal mode:

$$\boldsymbol{v}_1(t) = \boldsymbol{v}_T(t) + i\boldsymbol{q}\varepsilon(t).$$

Here $\boldsymbol{q}\cdot\boldsymbol{v}_T = 0$.

One finds, after some straightforward arithmetic, that

$$\dot{v}_T + (\dot{R}/R)v_T = 0 \tag{3}$$

$$\dot{\varepsilon} + (\dot{R}/R)\varepsilon = \left(-\frac{c_s^2}{\rho R} + \frac{4\pi G R}{q^2}\right)\rho_1 \tag{4}$$

$$\dot{\rho}_1 + (3\dot{R}/R)\rho_1 = \left(\frac{\rho q^2}{R}\right)\varepsilon. \tag{5}$$

The transverse mode simply decays like $1/R$, according to eq. (3). We are interested in the compressional longitudinal mode. Before one combines the two coupled equations (4) and (5), it is good tactics to factor out of $\rho_1(t)$ the $1/R^3$ "red-shift" dependence.
Thus, one defines

$$\delta(t) \equiv \frac{\rho_1(t)}{\rho(t)} \tag{6}$$

i.e., the density fluctuation amplitude $\delta\rho/\rho$ used in the text.

Eliminating $\varepsilon(t)$ one finally obtains

$$\ddot{\delta} + (2\dot{R}/R)\dot{\delta} + \left[\frac{c_s^2 q^2}{R^2} - 4\pi G\rho\right]\delta = 0. \tag{7}$$

Since the size of the fluctuation L is $\sim R/q$, the condition determining the sign of the quantity in the square bracket in eq. (7) is precisely the Jeans' criterion [eq. (2) of the text, this Chapter].

In the matter-dominated era but before the curvature term in Einstein's equation becomes important we have

$$\frac{\dot{R}^2}{R^2} \simeq \left(\frac{8\pi}{3}\right) G\rho$$

so that $R \propto t^{2/3}$ and $4\pi G\rho = 2/3t^2$. Parameterize the adiabatic equation of state by $P \propto \rho^\gamma$; then $c_s^2 = (\partial P/\partial\rho) \propto \rho^{\gamma-1} \propto t^{2-2\gamma}$. So the product of $(c_s^2 q^2/R^2)$ and $t^{2/3-2\gamma}$ is time independent and will be denoted by Λ^2. Then Eq. (7) becomes

$$\ddot{\delta} + \left(\frac{4}{3t}\right)\dot{\delta} + \left(\frac{\Lambda^2}{t^{2\gamma-2/3}} - \frac{2}{3t^2}\right)\delta = 0. \tag{8}$$

After recombination (we are always in the matter-dominated era in this discussion) $\gamma = 5/3$. For large time t, the Λ^2 term is negligible. One easily solves eq. (8) to get a growing mode and a dying mode

$$\delta \propto t^{2/3} \quad \text{and} \quad \delta \propto t^{-1}. \tag{9}$$

This is exactly what we had in the text.

If n_B/n_γ is large, matter domination occurs before recombination. There is an era in which matter dominates ρ but radiation dominates P so that $\gamma = 4/3$. Solving eq. (8) one finds two modes. For $\Lambda^2 > (5/6)^2$, both modes oscillate. For $(5/6)^2 > \Lambda^2 > 2/3$, both modes are damped. For $\Lambda^2 < 2/3$, one mode grows and the other is damped. We note that $\Lambda^2 < 2/3$ is just Jeans' criterion.

APPENDIX B TO CHAPTER XII

CRUDE THEORY OF ACOUSTIC ATTENUATION

A small amplitude longitudinal compressional wave (traveling in the x-direction) is described by linearizing the equation of continuity

$$\frac{\partial \rho_1}{\partial t} + \rho_0 \frac{\partial v}{\partial x} = 0 \tag{1}$$

and Euler's equation

$$\frac{\partial v}{\partial t} = -\frac{1}{\rho_0}\frac{\partial P_1}{\partial x}. \tag{2}$$

(See Appendix to Chapter X.) We consider linear perturbation around an equilibrium situation with density ρ_0, pressure P_0 and $v = 0$ by writing $\rho = \rho_0 + \rho_1$, $P = P_0 + P_1$ with ρ_1, P_1 the same order as v. Attenuation occurs when the pressure response lags behind the density fluctuation. Let us write phenomenologically

$$P = P_0 + c^2\left[(\rho - \rho_0) + \tau\frac{\partial \rho}{\partial t}\right]. \tag{3}$$

The lag time is denoted by τ. Inserting this expression, which clearly violates time reversal invariance and so leads to damping, into eqs. (1) and (2) we find

$$\frac{\partial^2 \rho}{\partial t^2} = c^2\left(\frac{\partial^2 \rho}{\partial x^2} + \tau\frac{\partial^3 \rho}{\partial x^2 \partial t}\right). \tag{4}$$

So the dispersion relation is

$$\omega^2 = c^2k^2(1 + i\tau\omega). \tag{5}$$

In the absence of response lag, $c = (\partial P/\partial \rho)^{1/2}$ gives the speed of sound. For small τ, the characteristic damping rate is the imaginary part of ω

$$\Gamma \simeq \tfrac{1}{2}(\tau\omega)\omega \simeq \tfrac{1}{2}\tau(ck)^2. \tag{6}$$

REFERENCES

1. Newton wrote: "It seems to me, that if the matter of our sun and planets, and all the matter of the universe, were evenly scattered throughout all the heavens, and every particle had an innate gravity towards all the rest, and the whole space throughout which this matter was scattered, was finite, the matter on the outside of this space would by its gravity tend towards all the matter on the inside, and by consequence fall down into the middle of the whole space, and there compose one great spherical mass. But if the matter were evenly disposed throughout an infinite space, it could never convene into one mass; but some of it would convene into one mass and some into

another, so as to make an infinite number of great masses, scattered great distances from one to another throughout all that infinite space. And thus might the sun and fixed stars be formed".
2. See the review and texts cited in refs. 1–3 in Chapter X.
3. P. J. E. Peebles and R. H. Dicke, *Ap. J.* **154,** 891 (1968).
4. J. M. Bardeen, Paper XII.1.
5. W. H. Press and E. T. Vishniac, Paper XII.2.
6. The terminology "adiabatic" and "isothermal" perturbations were first introduced by Ya. B. Zel'dovich, *Usp. Fiz. Nauk.* **89,** 647 (1966) (English translation: *Sov. Phys. Usp.* **9,** 602 (1967)). References to earlier Soviet literature may be found in this nice review paper.
7. Ya. B. Zel'dovich, paper XII.3; References to earlier work of the Zel'dovich school may be found in this paper. See also Ya. B. Zel'dovich, *Astron. Astrophys.* **5,** 84 (1970); P. J. E. Peebles, Paper XII.4.
8. The literature includes M. Turner and D. N. Schramm, *Nature* **279,** 303 (1979); J. Barrow and M. Turner, *Nature* **291,** 469 (1981); J. R. Bond, E. W. Kolb, J. Silk, Berkeley preprint 1981; W. H. Press, *Phys. Scripta* **21** (1980); Ya. B. Zel'dovich, *Mon. Not. R. A. Soc.* **192,** 663 (1980); C. J. Hogan, *Nature* **286,** 360 (1980).
9. J. Silk, Paper XII.5. P. J. E. Peebles and J. T. Yu, Paper XII.6; *Ap. J.*, **151,** 459 (1968): A. G. Doroshkovich, Ya. B. Zel'dovich, and I. D. Novikov, *Soviet Astron. – AJ*, **11,** 233 (1967); G. B. Field, *Ap. J.*, **165,** 29 (1971): K. Tomita, H. Nariai, H. Sato, T. Matsuda, and H. Takeda, *Prog. Theor. Phys.* **43,** 1511 (1970); H. Sato, *Prog. Theor. Phys.* **45,** 370 (1971); S. Weinberg, *Ap. J.*, **168,** 175 (1971).
10. P. J. E. Peebles, *Ap. J., Lett.* **243,** L119 (1981).
11. J. P. Ostriker and L. L. Cowie, Paper XII. 8.
12. See e.g., Ref. (1) of Chapter X.

CHAPTER XIII
NEUTRINOS AND GALAXIES

Massive neutrinos may have profound effects on cosmology.

Although non-vanishing neutrino masses may be postulated independently of grand unification, it is in grand unified theories such as $SO(10)$ (see Chapters II, IV) or $SU(5)$ with broken[1] $B - L$ that neutrino masses arise "naturally". In the simplest version of $SO(10)$, a small neutrino mass is necessarily induced when one gives the unobserved right-handed neutrino field a large Majorana mass[2] (Chapters II, IV). In this sense, $SO(10)$ requires the left-handed neutrino to be massive. In contrast, neutrino masses are typically optional in other models and one can put in neutrino masses as one pleases.

Clearly, since grand unified theories may be reduced to the $SU(3) \times SU(2) \times U(1)$ theory, any mechanism for generating neutrino masses can also be considered at the $SU(3) \times SU(2) \times U(1)$ level. For instance, the $SO(10)$ "see-saw" mechanism could have been implemented[3] years ago by introducing a right-handed $SU(3) \times SU(2) \times U(1)$ singlet neutrino field ν_R into the standard Glashow-Salam-Weinberg theory, adding a Majorana mass term $\nu_R C \nu_R$ to the Lagrangian, and coupling ν_R to ν_L by the doublet Higgs field. This was already explained in Chapter II. The important point is of course that at the $SU(3) \times SU(2) \times U(1)$ level nothing compels us to introduce right-handed neutrino fields.

We should emphasize that a given mechanism for generating neutrino masses could have rather different physical consequences depending on whether it is implemented at the grand unified level or at the "123" level. The relevant quantum numbers B, L, $B - L$ have different status at different levels. In $SO(10)$, $B - L$ is a gauge generator. In $SU(5)$, $B - L$ is a linear combination of a gauge generator and a generator of a $U(1)$ global symmetry. (See, e.g., Paper VII.2.) An attractive feature of $SU(3) \times SU(2) \times U(1)$ is that conservation of B and L follows automatically from the gauge symmetry. However, neither the current corresponding to B nor the current corresponding to L is anomaly-free[4]. The theory singles out the combination $B - L$, which does have an associated conserved current. Finally, in the $SU(3) \times SU_L(2) \times SU_R(2) \times U(1)$ theory, the $U(1)$ gauge symmetry generates $B - L$. The breaking of $B - L$ is related to gauge symmetry breaking.[5]

We may classify mechanisms for generating neutrino masses according to whether right-handed neutrino fields are introduced or not. In the absence of right-handed neutrino fields, left-handed neutrino masses are necessarily $\Delta L = 2$ and Majorana[6]. $B - L$ must be broken. With right-handed neutrino fields, one can in principle generate a Dirac mass term $\nu_L \nu_R$ by the Higgs mechanism. This is however very unattractive since there is no natural reason for this Dirac mass to be small. (See Chapter II.) The "see-saw mechanism"[2] solves this problem nicely and leads to neutrino masses which are predominantly Majorana.

We saw in Chapter X that in the present Universe the number density of neutrinos n_ν is comparable to the photon number density n_γ. This conclusion, derived for massless neutrinos, holds for massive neutrinos also, provided that $m_\nu < 1$ MeV so that when the neutrinos decouple they are relativistic and effectively massless. After decoupling, the ratio n_ν/n_γ remains essentially constant. Since the present baryon energy density is about $\eta m_N n_\gamma$, neutrinos would dominate the mass density of the Universe if $m_\nu \gtrsim \eta m_N$. A more precise evaluation yields the contribution of relic neutrinos to the mass density as

$$\Omega_\nu h_0^2 = \left(\frac{T_0}{2.7\,^\circ\mathrm{K}}\right)^3 \left(\frac{\sum \frac{1}{2} g_\nu m_\nu}{100\ \mathrm{eV}}\right). \tag{1}$$

Here $g_\nu = 2$ for a two-component Majorana neutrino. The rest of the notation is as in Chapter X. Thus, as Gershstein and Zel'dovich[7], (Paper XIII.1), Cowsik and McClelland[8], (Paper XIII.2) and Marx and Szalay[9] first pointed out, an upper bound on $\Omega_\nu h_0^2$ may be translated into an upper bound on $\Sigma g_\nu m_\nu$.

Let us briefly review the observational data[10] on Ω. Estimates of the luminous inner parts of galaxies yield

$$\Omega_{\mathrm{lum}} \gtrsim 0.012.$$

The observed rotational curve of galaxies suggests

$$\Omega_{\mathrm{rot}} \sim 0.04.$$

Studies of the rotational dynamics of binary galaxies and small groups of galaxies yield

$$\Omega_{\mathrm{BSG}} \approx 0.04 \text{ to } 0.07.$$

Studies of clusters of galaxies and of X-ray emissions from gas within such clusters suggest

$$\Omega_{\mathrm{cluster}} \approx 0.4.$$

On the other hand, the theory of big bang nucleosynthesis coupled with observed abundances of He^4, Li^7, and H^2 (and standard assumptions on the number of neutrino species, see Chapter X) requires $\eta \lesssim 7 \times 10^{-10}$ which translates into a rather low upper bound for Ω:

$$\Omega_{\mathrm{BBN}}\, h_0^2 \lesssim 0.02 \left(\frac{T_0}{2.7\,^\circ\mathrm{K}}\right)^3.$$

The inequalities

$$\Omega_{\mathrm{lum}} < \Omega_{\mathrm{BBN}} \lesssim \Omega_{\mathrm{BSG}} < \Omega_{\mathrm{cluster}} \tag{2}$$

suggest that baryons may not dominate the masses contained in galactic clusters. While Ω_{lum} almost certainly is dominated by nucleons, Ω_{BSG} and $\Omega_{\mathrm{cluster}}$ may not be. Whatever contributes to the mass represented in Ω_{BSG} and $\Omega_{\mathrm{cluster}}$ is not luminous. This is known as the "missing light problem" (and more often, perhaps somewhat misleadingly, as the "missing mass problem").

Massive neutrinos (and/or other massive feebly interacting particles) may resolve this discrepancy[7–9]. If $\Sigma\frac{1}{2}g_\nu m_\nu \gtrsim 1$ eV, $\Omega_\nu \gtrsim \Omega_{\rm lum}$ and neutrinos dominate the mass density of the Universe. If $\Omega_\nu \geq 1$ massive neutrinos close the Universe. Finally, a rather conservative upper bound, $\Omega \lesssim 2$, comes from a study of the deceleration parameter $q_0 = -R\ddot{R}/\dot{R}^2$ of the Universe. This bound translates into the conservative bound $\Sigma\frac{1}{2}g_\nu m_\nu \lesssim 200$ eV.

More recently, comparison of the observed deviations of the velocities of field galaxies from the general Hubble velocity and numerical simulations suggest[11] that $\Omega \lesssim 0.1$. In astronomical jargon, field galaxies are those galaxies which do not belong to clusters. Taken literally, this limit on Ω would impose a stringent upper bound on m_ν.

A simple argument due to Tremaine and Gunn[12] (paper XIII.3) places a lower limit on neutrino masses if neutrinos were the dominant mass sources in galaxies and clusters of galaxies. The point is that neutrinos are fermions and so there can be no more than one neutrino per unit volume of phase space. The mass due to neutrinos in a gravitating system of size R is thus less than $M_\nu \sim m_\nu R^3 \int d^3p \sim m_\nu R^3 (m_\nu v)^3$. Here v denotes the typical velocity of the neutrinos and is given by $v^2 \sim GM/R$ according to the virial theorem. Thus, for M_ν to be $\gtrsim M =$ the mass of the system in question, we must have

$$m_\nu \gtrsim (G^3 M R^3)^{-1/8}. \tag{3}$$

The critical neutrino mass decreases with the size and mass of the astrophysical systems in question and works out to be about ~ 20 eV for galactic halo, $\sim 14 h_0^{1/2}$ eV for binary galaxies and small groups of galaxies, $\sim 5 h_0^{1/2}$ eV for rich clusters of galaxies. Thus, depending on m_ν, neutrinos may cluster in groups of galaxies but may be not particularly concentrated in galactic halos. An equivalent way of stating the Tremaine-Gunn bound is as follows. If the neutrinos are too light, the ones at the top of the Fermi sea would be fast enough to escape the gravitating pull of the system.

Massive neutrinos may affect[13–19] (papers XIII.4–10) drastically the picture of galaxy formation outlined in Chapter XII. The point is that massive neutrinos can start clumping long before photons decoupled, since neutrinos, interacting so weakly, can freely stream through distribution of particles. Charged particles are prevented by photons from clumping, but not neutrinos.

To estimate the mass contained in a typical clump of neutrinos, we introduce the concept of a neutrino diffusion mass $M_{H\nu}$, which is essentially the extension of the concept of Jeans mass to neutrinos. First, we remark that a fluctuation in the neutrino density on a scale that is small compared to the horizon will simply diffuse away, since neutrinos stream freely. Let $M_{H\nu}$ denote the amount of neutrino masses contained within the horizon at a critical time when the neutrinos become non-relativistic. This critical time corresponds to a temperature $T \sim m_\nu$; after this time neutrinos are slow moving and no longer stream freely. Thus

$$M_{H\nu} \sim n_\nu m_\nu \,(\text{horizon size})^3. \tag{4}$$

The horizon is, of course, just that distance scale over which photons could have travelled since the big bang. We recall from Chapter X that the horizon size $\sim M_{pl}/T^2$ in the radiation era. Thus, inserting this fact, $n_\nu \sim T^3$, and $T \sim m_\nu$ into eq. (4), we obtain

$$M_{H\nu} \sim M_{pl}^3/m_\nu^2 \sim (1\ \text{eV}/m_\nu)^2\ 10^{18}\ M_\odot. \tag{5}$$

Fluctuations on scales less than $M_{H\nu}$ would have diffused away. The rough estimate given here agrees with more detailed calculations.[13–19]

The general idea is that massive neutrinos clump into objects on mass scales $\gtrsim M_{H\nu}$. These then provide the potential wells which matter can then fall into. The point is that neutrinos clumping does not have to wait till decoupling and matter dominance. Thus, even if Ω_{nucleons} turn out to be small, the "low Ω squeeze" problem mentioned at the end of Chapter XII would be avoided. We also mentioned there that the adiabatic picture has some difficulty in accounting for the observed isotropy of the microwave background. If clumping started early, then the initial $\delta\rho/\rho$ does not have to be as large as $\sim 10^{-3}$. The adiabatic picture, favored by baryongenesis in grand unification, is thus "improved" by massive neutrinos.

Incidentally, it is usually assumed that the "primordial" fluctuation spectrum is some rapidly falling function of mass, typically a "scale-invariant" power law $\delta\rho/\rho \propto M^{-n}$ for some n. Thus, massive neutrinos generate a fluctuation spectrum rather sharply peaked about $M_{H\nu}$.

In order for $M_{H\nu}$ to be of the order of a typical galactic mass $\sim 10^{11}\ M_\odot$, one needs $m_\nu \sim$ few keV. However, we have also the (rather conservative) upper bound $m_\nu \lesssim 200$ eV mentioned above. So neutrinos cannot clump directly into galaxies, but into clumps of mass $> 10^{14}\ M_\odot$ or $10^{15}\ M_\odot$ which corresponds roughly to the observed scale of super-clusters. In the adiabatic picture, the baryon matter in these clumps subsequently fragments into galaxies and clusters, much as in the Zel'dovich pancake scenario. Neutrinos form an extended halo around superclusters.

In the isothermal picture, $\delta T = 0$, so the only density fluctuations on scales $< 10^{14}\ M_\odot$ are in the baryons. But baryon density fluctuation $\delta\rho_B$ is small compared to the dominant neutrino background:

$$\delta\rho/\rho \sim \delta\rho_B/\rho_\nu = (\rho_B/\rho_\nu)(\delta\rho_B/\rho_B).$$

Furthermore, in a neutrino dominated Universe the growth of $\delta\rho/\rho$ is reduced from $t^{2/3}$ to t^p where

$$p = \frac{1}{6}\left[\left(1 + \frac{24\rho_B}{\rho_\nu}\right)^{1/2} - 1\right] \sim \rho_B/\rho_\nu.$$

(see e.g., paper XIII.4.) Thus, a very large $\delta\rho_B/\rho_B$ is needed in this scenario. The subsequent growth is by aggregation as in the standard isothermal picture (see Chapter XII), starting from small clumps. The neutrinos are "dragged" along and thus may cluster on scales from galactic halos to superclusters.

Recently, Bond, Szalay, and Turner[20] suggested that particles which are even more weakly coupled than neutrinos may trigger galaxy formation. The story unfolds along essentially the same line as for neutrinos, but with some important differences. To tell the story, let us denote by T_{dec} the temperature at which these X particles decouple, by T_{NR} the temperature at which these X particles become non-relativistic, and by T_{dom} the temperature at which X dominates the mass density of the Universe. The weaker the X interacts with the rest of the world, the higher T_{dec} will be. By the time the temperature falls to $T_{NR} \sim m_x/3$, the number density ratio n_x/n_γ could be substantially less than unity, since a number of species could have annihilated into photons. In that case, T_{dom} may be substantially lower than T_{NR}. For neutrinos, $T_{dom} \sim T_{NR}$. The important point is that density fluctuations of X's cannot grow until T_{dom}.

To render this story quantitative, we note that after decoupling

$$\frac{n_x}{n_\gamma} \sim g_x \left\{\frac{T_x}{T_\gamma}\right\}^3 . \tag{6}$$

Here g_x counts the degrees of freedom in X. The temperature of the X gas T_x is equal to T_γ just after decoupling, but by the time X becomes non-relativistic because of heating of the photon gas,

$$T_x \sim g_*^{-1/3} T_\gamma .$$

We denote by g_* the effective degrees of freedom of the particles which have annihilated between $T_x = T_{dec}$ and $T_x = T_{NR}$. Thus, n_x/n_γ falls to

$$\frac{n_x}{n_\gamma} \sim \frac{g_x}{g_*} . \tag{7}$$

This ratio remains constant thereafter and is very small if the X's are very weakly interacting so that T_{dec} is very high. (In Ref. (12) m_x is taken to be in the keV range (see below) so that the photon gas was reheated for the last time by $T_x = T_{NR}$.) We have

$$\rho_x/\rho_\gamma \sim (m_x g_x)/(T_\gamma g_*) \tag{8}$$

so that

$$T^\gamma_{dom} \sim m_x g_x/g_* . \tag{9}$$

The amount of X mass contained within the horizon {from eq. (8) and Chapter X} is given by

$$M_{HX} \sim \left(\frac{M_{pl}}{T_\gamma^2}\right)^3 m_x \left(\frac{g_x}{g_*}\right) T_\gamma^3 . \tag{10}$$

As the Universe cools M_{HX} grows like T_γ^{-3}.

The Jeans mass, which controls the clumping of the X's, is given by {eq. (3) of Chapter XII}

$$M_{HX} \sim P^{3/2} M_{pl}^3/\rho^2 \sim (M_{pl}^3/m_x^2)(T_x^3/n_x)^{1/2} . \tag{11}$$

After the X's become non-relativistic, $\rho \sim n_x m_x$ and $P = n_x T_x$. We must be careful to note that after $T_x = T_{NR}$, T_x is red-shifting like $1/R^2$ while T_γ, as usual, falls like $1/R$. Thus, after $T_x = T_{NR}$

$$T_x/T_\gamma \sim T_\gamma/(g_*^{2/3} m_x). \tag{12}$$

The Jeans mass decreases like $T_\gamma^{3/2}$.

For a massive "collision-less" particle (such as massive neutrinos and very weakly interacting exotic particles), the concept of free-streaming mass M_{FSX} is relevant. After $T_x = T_{NR}$, the X's are non-relativistic; so the distance d_X over which a free-streaming X particle can traverse is less than the horizon distance. M_{FSX} denotes the amount of X mass contained within the X horizon (< the photon horizon). We give here a very rough calculation (in the era before the X's dominate the mass density). The distance travelled after $T_x \sim T_{NR}$ is

$$\mathrm{d}_x \sim R(t) \int_{NR} \mathrm{d}t'\, v(t')/R(t'). \tag{13}$$

After $T_x = T_{NR}$, the typical velocity $v \sim p/m_x$ of the X's are red-shifted down like $1/R \propto T_\gamma$. Thus,

$$v \sim \frac{T_\gamma}{T_\gamma^{NR}} \sim \frac{T_\gamma}{g_*^{1/3} m_x}. \tag{14}$$

Inserting this relation into eq. (14) we find

$$\begin{aligned} \mathrm{d}_x &\sim \frac{1}{T_\gamma} \int_{T_\gamma^{NR}}^{T_\gamma} T_\gamma \,(T_\gamma/g_*^{1/3} m_x)\,(M_{pl}/T_\gamma^3)\,\mathrm{d}T_\gamma \\ &\sim (M_{pl}/g_*^{1/3} m_x T_\gamma) \log\,(g_*^{1/3} m_x/T_\gamma). \end{aligned} \tag{15}$$

Here we have used the temperature-time relation given in eq. (14) of Chapter X. The distance a typical X particle has traversed from the big bang until $T_x \sim T_{NR}$ is of order $M_{pl}/(g_*^{2/3} m_x^2)$ and is negligible compared to d_x for T_γ much smaller than m_x. In that case,

$$M_{FSX} \sim (M_{pl}^3/m_x^2)\,(g_x/g_*^2) \log g_*^{1/3} m_x/T_\gamma)^3 \tag{16}$$

grows slowly as the Universe cools.

The point is that a density fluctuation of X's of a given mass M, even after M_{JX} has fallen below M, cannot grow until the X's dominate the mass density of the Universe. This occurs when $T_\gamma^{dom} \sim m_x g_x/g_*$, at which we have the Jeans mass

$$M_{JX} \sim (M_{pl}^3/m_x^2)\,(g_x/g_*^2), \tag{17}$$

the free-streaming mass

$$M_{FSX} \sim M_{JX}\,(\log g_*^{4/3}/g_x)^3 \tag{18}$$

and the horizon mass

$$M_{HX} \sim M_{JX}\,(g_*^4/g_x^3). \tag{19}$$

Bond *et al.*[20] considered an X particle with $m_x \sim 1$ keV, $g_x \sim 3/2$, and so weakly interacting that it decouples at $T_{dec} > 10^2$ GeV. The gravitino, the partner of the graviton in supergravity (see Chapter XVIII), may have precisely these properties. Hence $\hat{g}_*$ may be quite large $\sim 10^2$ and so $M_{HX} > M_{FSX} > M_{JX}$. Initial density fluctuations of mass scale M between M_{JX} and M_{HX} suffer "stunted growth" until the X's dominate. The resulting fluctuation spectrum should be relatively flat, extending from roughly $M_{JX} \sim 10^{10}\ M_\odot$ up to $M_{HX} \sim 10^{16}\ M_\odot$.

If gravitinos exist, the same sort of considerations which lead to eq. (1) imposes an upper bound[21] of about 1 keV on the gravitino mass m_g. (Very massive gravitinos are also allowed because of Boltzmann suppression[22].) By imposing the vanishing of the net cosmological constant in a class of supersymmetric theories, one can relate m_g to the scale of spontaneously broken supersymmetry. The cosmological upper bound on m_g translates into an upper bound of about 10^3 TeV on the scale of supersymmetry breaking[23].

We should mention that parts of this discussion must be modified for a model of neutrino masses due to Gelmini and Roncadelli[24]. A Higgs triplet with $Y/2 = +1$ is added to standard $SU(3) \times SU(2) \times U(1)$. The Higgs triplet breaks the global $B - L$ symmetry spontaneously, thus producing a Nambu-Goldstone boson, called the Majoron M. (See also Chapter XVI.) Since M is massless, neutrinos could decay or annihilate in pairs: $\nu' \to \nu + M$, $\nu'\nu' \to \nu\nu$, $\nu\nu \to MM$. These processes cause neutrinos to disappear from the Universe. The astrophysical and cosmological consequences of this model are thus amusingly different.[25,26]

Finally, we mention that Lee and Weinberg[27] and Dolgov et al.[28] pointed out that the reasoning leading to eq. (1) can be extended to give a lower bound of a few GeV on a heavy "neutrino". (These are heavy neutral leptons introduced in some variations of the standard "123" theory.) They do not appear in the standard $SU(5)$ grand unified theory. The point is that when these heavy leptons ("heptons") decouple they are already non-relativistic, so that the Boltzmann factor severely limits their equilibrium number density. An upper bound on Ω is thus translated into a lower bound on the mass of these heavy neutrinos. Subsequently, bounds on the mass and lifetime of weakly interacting particles have been extensively studied in the literature.[29–32]

REFERENCES

1. S. L. Glashow, HUTP-79/A079 *Cargese Lectures*; F. Wilczek and A. Zee, *Phys. Lett.* **88B,** 311 (1979); A. Zee, *Phys. Lett.* **93B,** 389 (1980); L. N. Chang and N. P. Chang, *Phys. Lett.* **92B,** 103 (1980); *Erratum,* **94B,** 551 (1980).
2. M. Gell-Mann, P. Ramond, R. Slansky, *Aspen talk,* 1979; T. Yanagida, *KEK Lecture notes* (1979).
3. Y. Chikashige, R. N. Mohapatra, R. Peccei, *Phys. Lett.* **98B,** 265 (1981).
4. Weak interaction instantons can thus induce violations of *B* and *L*. G. 't Hooft, *Phys. Rev. Lett.* **37,** 8 (1976).
5. R. N. Mohapatra and G. Senjanovic, *Phys. Rev. Lett.* **44,** 912 (1980).
6. For example, A. Zee, ref. (1).
7. S. S. Gershtein and Ya. B. Zel'dovich, *JETP Lett.* **4,** 174 (1966). (paper XIII.1.)
8. R. Cowsik and J. McClelland, (Paper XIII.2), also *Astrophy. J.* **180,** 7 (1973).

9. G. Marx and A. S. Szalay, *Proceedings of the "Neutrino 72" Conference; Acta Phys. Hung.* **35,** 113 (1974).
10. For a review, see D. N. Schramm and G. Steigman, *J.* of *Gen. Rel. and Gravity,* **13,** 101 (1981).
11. W. C. Saslaw and S. J. Aarseth, *Ap. J.* **253,** 470 (1982).
12. S. Tremaine and J. E. Gunn, paper XIII.3.
13. J. R. Bond, G. Efstathiou, and J. Silk, paper XIII.4.
14. F. Klinkhamer and C. A. Norman, paper XIII.5.
15. H. Sato and F. Takahara, paper XIII.6.
16. I. Wasserman, paper XIII.7.
17. M. Davis, et al., paper XIII.8.
18. A. G. Doroshkevich et al., paper XIII.9.
19. J. R. Bond and A. S. Szalay, paper XIII.10.
20. J. R. Bond, A. Szalay, and M. Turner, *Phys. Rev. Lett.*, to appear.
21. H. Pagels and J. Primack, *Phys. Rev. Lett.* **48,** 223 (1982).
22. S. Weinberg, *private communications*; see Refs. (24) and (25) below.
23. A. Das and D. Z. Freedman, *Nucl. Phys.* **B120,** 221 (1977); S. Deser and B. Zumino, *Phys. Rev. Lett.* **38,** 1433 (1977); H. Pagels, in *Orbis Scientia,* (1981); Ref. 21; R. Barbieri, S. Ferrara, and D. V. Nanopoulos, TH. 3159-CERN, September 1981.
24. G. Gelmini and M. Roncadelli, *Phys. Lett.* **99B,** 411 (1981).
25. H. Georgi, S. L. Glashow, S. Nussinov, *Nucl. Phys. B* (to be published).
26. G. Gelmini, S. Nussinov, and M. Roncadelli, Max-Planck preprint MPI-PAE/PTh 59/81.
27. B. W. Lee and S. Weinberg, *Phys. Rev. Lett.* **39,** 165 (1977).
28. A. D. Dolgov, M. I. Vysotsky, and Ya. B. Zel'dovich, *JETP Lett.* **26,** 200 (1977).
29. D. A. Dicus, E. W. Kolb, V. L. Teplitz, *Phys. Rev. Lett.* **39,** 168 (1977).
30. J. E. Gunn et al., paper XIII.11.
31. G. Steigman, C. L. Sarazin, H. Quintana, J. Faulkner, *Astron. J.* **83,** 1050 (1978).
32. J. E. Gunn, B. W. Lee, I. Lerche, D. N. Schramm, and G. Steigman, *Astrophys. J.* **223,** 1015 (1978); P. Hut, *Phys. Lett.* **69B,** 85 (1977); D. A. Dicus, E. W. Kolb, and V. L. Teplitz, *Phys. Rev. Lett.* **39,** 168 (1977); D. A. Dicus, E. W. Kolb, and V. L. Teplitz, *Astrophy. J.* **221,** 327 (1978); D. A. Dicus, E. W. Kolb, V. L. Teplitz, and R. V. Wagoner, *Phys. Rev.* **D17,** 1529 (1978); F. W. Stecker, *Phys. Rev. Lett.* **45,** 1460 (1980); R. Kimble, S. Bowyer, and P. Jakobsen, *Phys. Rev. Lett.* **46,** 80 (1981); D. Lindley, *Mon. Not. R. Astr. Soc.* **188,** 15p (1979); Y. Raphaeli and A. S. Szalay, UCSB-ITP preprint 81–52, to be published in *Phys. Lett. B* (1981); A. L. Melott and D. W. Sciama, *Phys. Rev. Lett.* **46,** 1369 (1981); A. D. Dolgov and Ya. B. Zel'dovich, *Rev. Mod. Phys.* **53,** 1 (1981) and refs. therein; K. Sato and M. Kobayashi, *Prog. Theor. Phys.* **58,** 1775 (1977); P. Hut and K. A. Olive, *Phys. Lett.* **87B,** 144 (1979); S. W. Falk and D. N. Schramm, *Phys. Lett.* **79B,** 511 (1978); and R. Cowsik, *Phys. Rev. Lett.* **39,** 784 (1977).

CHAPTER XIV

MONOPOLES AND INFLATION

Polyakov and 't Hooft discovered[1] that the topology of gauge theories is such that, when a simple gauge group G is broken to H at a mass scale M, magnetic monopoles exist if H contains a $U(1)$ factor. The monopole has mass of the order M/α with α the relevant gauge coupling squared. In $SU(5)$ grand unification, there are thus monopoles with mass $\sim 10^{16}$ GeV. (Note that stable topological monopoles do not exist in the $SU(3) \times SU(2) \times U(1)$ theory with its non-simple gauge group.) As the early Universe cools past the grand unification scale, the statistical random alignment of the Higgs field which breaks $SU(5)$ to $SU(3) \times SU(2) \times U(1)$ will lead to "twists" and "knots" in the gauge fields. These knots are the monopoles. One expects perhaps one monopole per horizon volume at temperature $T \sim 10^{14}$ or 10^{15} GeV. According to eq. (15) of Chapter X there are only about $\sim 10^{13}$ photons within the horizon at that time.

Kibble, Zel'dovich and Khlopov, and Preskill[2] have estimated the present density of monopoles to be given by $n_M/n_\gamma \sim 10^{-10}$. In contrast, the observational bound on n_M/n_γ from the mass density of the Universe gives $n_M/n_\gamma \lesssim 10^{-24}$. This discrepancy by 14 orders of magnitude imposes a stringent bound on grand unification and cosmology.

A number of suggestions[3] to avoid a monopole dominated Universe have been suggested. A particularly amusing possibility[4] involves constructing a theory with the breaking chain

$$SU(5) \to H \to SU_c(3) \times U_{\text{e.m.}}(1)$$

where H does not contain a $U(1)$ factor. The first step of the breaking occurs at about $\sim 10^{15}$ GeV and the second step at a scale $T_c \sim 1$ TeV. The photon is thus a "recent" arrival on the cosmological scene, after the Universe has cooled below T_c. The density of monopoles will be suppressed by an enormous Boltzmann factor $\sim \exp(-M_m/T_c)$. Unfortunately, the grand unification theory required appears rather complicated and contrived.

An exciting solution to the monopole problem is based on the suggestion[5] that the Universe has gone through an inflationary phase. While this scenario was originally motivated by the monopole abundance problem, it also resolves the cosmological problems discussed at the end of Chapter X. (Paper XIV. 1.) Thus, the inflationary Universe has generated considerable excitement.

The paper by Guth (paper XIV.1) reprinted here gives a lucid introduction to the subject. Thus, we need to do no more than sketch the basic idea here.

Phase transitions are classified as first order, second order, and so on according to how the effective potential[6] or free energy $V(\phi)$ depends on the order parameter ϕ as an external parameter like temperature T changes. A second order phase transition is typified by the familiar Landau-Ginsburg function $V(\phi, T) = a(T - T_c)\phi^2 + b\phi^4$ with a, b positive. As T decreases through T_c, the point $\phi = 0$ changes smoothly from being a minimum of V to a maximum. The new minimum occurs at $|\phi| \propto (T_c - T)^{1/2}$. In a first order transition, $V(\phi, T)$ starts out at high T with a global minimum at $\phi = 0$, say. For simplicity, let us assume $V(\phi, T)$ has only one minimum at high T. As T decreases below a temperature T_{c2}, another minimum for $\phi \sim \phi_o$ may develop. As T decreases further below T_c, the minimum for $\phi \sim \phi_o$ becomes the true, i.e. global minimum, while the minimum at $\phi = 0$ turns into merely a local, i.e., a "false" minimum. As T decreases below T_c, the minimum at $\phi = 0$ suddenly becomes a local maximum. (See Figure 10 in paper XIV.2, which should be contrasted with Figure 8, illustrating a second order phase transition.) Kirzhnits and Linde first emphasized that phase transitions in gauge theories may be strongly first order (see the review by Linde, paper XIV.2). Let the notation be such that the gauge symmetry is unbroken at $\phi = 0$ but broken at $\phi = \phi_o$ so that the states represented by $\phi = 0$ and $\phi = \phi_o$ will be referred to as the symmetric and asymmetric vacuum, respectively.

Suppose the breaking of grand unification symmetry corresponds to a first order transition. At large T, the Universe is in the unbroken symmetric phase $\phi = 0$. (The terminology "vacuum" may be potentially confusing. The Universe is, of course, filled with massless particles with energy density $\sim g(T)T^4$. One means that the field theory is constructed on the symmetric or the asymmetric vacuum, as the case may be.) As T drops below T_c, the false symmetric vacuum at $\phi = 0$ is separated from the true asymmetric vacuum at $\phi = \phi_o$ by a potential barrier and so is metastable. Thus, the Universe may be trapped in the metastable false vacuum for a long time. This is known as supercooling. The false symmetric vacuum decays into the true asymmetric vacuum by quantum tunnelling[7] and/or by thermal fluctuation. The transition thus occurs by the nucleation of bubbles[8] in which $\phi \sim \phi_o$. These bubbles expand rapidly into the surrounding $\phi = 0$ vacuum. Latent heat is released and the Universe finds itself finally in the asymmetric phase. How long the Universe finds itself trapped in the false metastable vacuum depends on the bubble nucleation rate and hence on the precise shape and height of the potential barrier. This chain of events corresponding to a first order phase transition, of course occurs in such familiar processes such as the cooling and heating of water.

In order to solve the monopole abundance problem, Kibble[2] and Guth and Tye[9] first proposed that the breaking of $SU(5)$ may involve a first order phase transition. Monopoles are formed when the bubbles collide and coalesce, since the orientations of the Higgs field in group space are uncorrelated between different bubbles. One expects the order of one monopole per bubble. From the fact that the bubble size cannot exceed the horizon size one may easily estimate the ratio of monopoles to photons at the temperature T_{coal} when bubbles coalesce to be

$$\frac{n_M}{n_\gamma} \gtrsim g(T_{\text{coal}})^{3/2}\left(\frac{T_{\text{coal}}}{M_{\text{pe}}}\right)^3. \tag{1}$$

(See paper XIV.2.) Thus, the concentration of monopoles may possibly be diluted by arranging for substantial supercooling. However, the discussion of Guth and Tye misses an important effect, as we shall see presently.

But what about the cosmological constant associated with the phase transition? As discussed in Chapter X, the cosmological constant problem represents one of the great unsolved puzzles in physics. Without any understanding whatsoever of why the observed cosmological constant is so fantastically small, one might adopt a purely phenomenological attitude and add a suitable constant to the Lagrangian so that the effective potential in the asymmetric vacuum is zero. (In quantum field theory this is conceptually no different from setting the electron mass to whatever its observed value is.)

Guth[5], Sato[10], and earlier Kazanas[11] pointed out that this procedure leads to a dramatic revision of the evolutionary history of the Universe, if the phase transition is first order. When the Universe was trapped in the metastable false vacuum, a huge cosmological constant corresponding to a constant energy density ρ_o (of order ϕ_o^4) rapidly overwhelms the radiation energy density $\sim g(T)T^4$. Referring to eq. (1′) in Chapter X, we see that the Universe expands exponentially like

$$R(t) \propto e^{t/\tau} \tag{2}$$

as was noted by de Sitter in 1917. The characteristic time scale τ is given by

$$\tau \sim (G\rho_o)^{-\frac{1}{2}} \sim t_{pl}\left(\frac{M_{pl}}{10^{14}\ \text{GeV}}\right)^2 . \tag{3}$$

If the Universe is trapped in the metastable state for a long time, it could have expanded enormously. This is known as the inflationary Universe. Guth[5] observed that the horizon problem would be solved since two points which after inflation appear to be causally disconnected were, in fact, connected before inflation. The entropy or population problem is also solved. When the metastable Universe finally undergoes the transition into the asymmetric vacuum $\phi \sim \phi_o \sim 10^{14}$ GeV at a temperature T_s, the vast amount of latent heat released heats the Universe back up to a temperature of the order of the critical temperature $T_c \sim \phi_o$. The entropy density increases by a factor of order $(T_c/T_s)^3$. If T_c/T_s is of order 10^{28} or larger, the observed entropy of the Universe may be accounted for naturally. Since $T_c \sim 10^{14}$ GeV, T_s must be $\lesssim 0.1$°K, a fantastically small temperature.

The inflationary scenario, while attractive, unfortunately is beset by a number of difficulties, as pointed out by Guth[5] and others[10,11,12]. These difficulties may be referred to as the "graceful exit" problem. Estimates of the nucleation rate turn out to be so small that the gradual bubble formation may never trigger an end to inflation. The nucleation rate has to be small in order for the Universe to stay in the inflationary phase for a long time. To exit at 0.1 °K the potential barrier must have some peculiar temperature dependence. Even though the bubbles may expand at the speed of light, they may never fill the exponentially inflating Universe. In the Robertson-Walker metric the coordinate radius of the bubble $r(t)$ expands according to $dr/dt = R^{-1}(t)$. With $R(t) \sim e^{t/2\tau}$, a bubble formed at coordinate time t_o will have a limiting finite coordinate size

$$r(t = \infty) = r(t_o) + \text{const } \tau e^{-t_o/2\tau}. \tag{4}$$

Some people describe this situation picturesquely by saying that the Universe would end up resembling "Swiss cheese". Even if the bubbles managed to collide, the collision of the walls might be expected to produce gross inhomogeneities and anisotropies. (If the inhomogeneities are small enough, they may be desirable for galaxy formation.)

While this book was being prepared, Linde[13] (paper XIV.3) and Albrecht and Steinhardt[14] (Paper XIV.4) proposed a new improved inflationary Universe. We give a brief outline of this scenario, following Linde. Linde takes the scale invariant Coleman-Weinberg potential. In this case, the temperature $T_{c1} = 0$. For any $T \neq 0$, $\phi = 0$ remains a local minimum. Near $\phi = 0$, the potential is roughly of the form

$$V(\phi,T) \simeq \rho_o + g^2T^2\phi^2,$$

where g^2 is some gauge coupling squared. For $T \ll \phi_o$, this gives a small barrier at $\phi \sim \phi_1 \ll \phi_o$ which is quite low compared to ρ_o. Thermal fluctuations could heave the system over the barrier. This happens at $T_1 \sim 10^8$ GeV according to one estimate[14]. The point is that it takes so long for the Universe to "slide" down from $\phi_1 \sim 10^8$ GeV to $\phi_0 \sim 10^{14}$ GeV that the Universe inflates by a really fantastic factor (10^{10^4}, according to Linde[13]). The observable Universe is, in fact, inside one single bubble. One avoids having to arrange for the bubbles to collide.

This whole subject has been fast developing. Additional work[15,16,17] has already been stimulated. No doubt, by the time these words see print, further developments will have taken place. At this writing, no completely consistent inflationary scenario has yet been worked out. Bardeen[16] has emphasized that the standard treatment of vacuum decay assumes local thermodynamic equilibrium, which however does not hold. Also, in the Linde scenario, the generation of sufficient inhomogeneities for galaxy formation may pose a problem.[16]

In concluding this chapter, we must emphasize that the inflationary scenario is predicted on the rather unsatisfactory phenomenological procedure of adjusting the present cosmological constant to zero. Presumably some profound new physics, lying at the interface between gravity and the other three interactions, is at work in driving the cosmological constant to zero. This new physics, whatever it may be, may affect drastically the inflationary scenario. Also, with regards to the new improved scenario, one may feel particularly uneasy about the addition of a compensating cosmological constant, since the Coleman-Weinberg potential is motivated by scale invariance. Finally, precise details of the scenario may be quite different if the gauge symmetry is broken dynamically rather than by explicit Higgs field.

Incidentally, in the inflationary Universe scenario, any baryon concentration generated (or put in) before inflation started would have been diluted by the enormous entropy production. It is crucial that after inflation terminated, the Universe was reheated back to a temperature high enough for baryongenesis to proceed again.

Note added: As this typescript is going to press, some preliminary evidence for the detection of a magnetic monopole was reported (B. Cabrera, *Phys. Rev. Lett.* **48,** 1378 (1982)).

REFERENCES

1. G. t' Hooft, *Nucl. Phys.* **B79,** 276 (1974); A. M. Polyakov, *JETP Lett.* **20,** 194 (1974). For the properties of $SU(5)$ monopoles, see C. P. Dokos and T. N. Tomaras, *Phys. Rev.* **D21,** 2940 (1980).
2. T. W. B. Kibble, *J. Phys. A.* **9,** 1387 (1976).
3. M. Daniel, G. Lazarides, and G. Shafi, *Phys. Lett.* **91B,** 72 (1980); *Nucl. Phys.* **B170,** 156 (1980). J. N. Fry and D. N. Schramm, *Phys. Rev. Lett.* **44,** 1361 (1980); G. Lazarides and Q. Shafi, *Phys. Lett* **94B,** 149 (1980); G. Lazarides, M. Magg, and Q. Shafi, TH. 2856-CERN; A. D. Linde, Lebedev preprint 125.
4. P. Langacker and S. Y. Pi, *Phys. Rev. Lett.* **45,** 1 (1980).
5. A. H. Guth, paper XIV.1; A. H. Guth and E. Weinberg, *Phys. Rev.* **D23,** 876 (1981).
6. S. Coleman and E. Weinberg, *Phys. Rev.* **D7,** 1888 (1973).
7. M. Voloshin, I. Yu Kobzarev, L. B. Okun, *Yad. Fiz.* **20,** 1229 (1974); S. Coleman, *Phys. Rev.* **D15,** 2929 (1977).
8. A. Linde, *Phys. Lett.* **70B,** 306 (1977).
9. A. Guth and H. Tye, *Phys. Rev. Lett.* **44,** 631, 963 (1980).
10. K. Sato, *Mon. Not. Roy. Astron. Soc.* **195,** 467 (1981); *Phys. Lett.* **99B,** 66 (1981).
11. D. Kazanas, *Astrophys. J.* **241,** L59 (1980).
12. J. Bardeen (unpublished).
13. A. Linde (Paper XIV.3).
14. A. Albrecht and P. Steinhardt, (Paper XIV.4).
15. A. Albrecht, P. Steinhardt, M. Turner and F. Wilczek, to be published.
16. J. Bardeen, to be published.
17. A. Linde, Lebedev preprint, 1982.

CHAPTER XV
HIERARCHY, TECHNICOLOR, SUPERSYMMETRY, AND VARIATIONS

The triumph of the $SU(5)$ theory in satisfying the existing bounds on proton life-time also brings with it the so-called hierarchy problem[1]. The fundamental scale of the theory, at 10^{15} GeV, is so much larger than the other two symmetry scales of the theory: A scale of $\sim 10^2$ GeV at which weak interaction breaks itself and a scale of ~ 1 GeV at which strong interaction breaks chiral symmetry. Since we only know how to break the grand unified symmetry and the weak interaction symmetry by hand using the Higgs mechanism, we have to adjust the appropriate parameters in the Higgs sector of the Lagrangian to one part in $(10^{13})^2 = 10^{26}$! This strikingly "unnatural" adjustment[2] is known as the hierarchy problem.

The hierarchy problem naturally brings us to one of the outstanding problems in particle physics today, namely that of dynamical symmetry breaking. The most promising scheme, dubbed technicolor, is based on the notion of infrared-slavery[3], in some sense the other side of the coin of asymptotic freedom. The asymptotic freedom equation

$$\frac{1}{g^2(M)} = \frac{1}{g^2(\mu)} + b \log \frac{M^2}{\mu^2} \tag{1}$$

also informs us that as the mass scale M decreases $g^2(M)$ increases. Taking the equation literally, one finds that $g^2(M)$ becomes strong-coupling at a mass scale of order[4]

$$M \sim \mu e^{-1/(2bg^2(\mu))}. \tag{2}$$

Of course, the renormalization group equation (1) is derived for small g^2; nevertheless, the estimate in Eq. (2) may be taken as an order-of-magnitude indication of the strong-coupling scale. It is plausible to assume that when an interaction becomes strong it breaks the symmetry of that interaction. We can now see how a hierarchy in scales could arise. Given two interactions with the same coupling at a scale μ the ratio of the scales at which these two symmetries are broken is given roughly by

$$\frac{M'}{M} \sim e^{-1/(2(b'-b)g^2(\mu))}. \tag{3}$$

A modest difference in the group-theoretic factors b' and b could be exponentially amplified to give a large ratio M'/M. The astute reader will notice that this is essentially just the same calculation given in Chapter IV. The scheme is very appealing since it represents essentially the only known mechanism for generating an exponentially large ratio.

The original proposal of technicolor[5,6] (Papers XV.1 and 2) was meant to explain the ratio of the weak interaction scale and the chiral scale and the notion is logically independent of grand unification. Nevertheless, it is an attractive possibility that the breaking at 10^{15} GeV may also be due to some coupling becoming strong. In this connection, it may be worth remarking that the relevant coupling constant squared[7] associated with a "large" gauge symmetry (say, $SU(N)$ or $SO(N)$, N large) is not g^2 but g^2N.

Amusing toy-models have been constructed[8] exhibiting a hierarchical succession of symmetry breaking with the help of an additional dynamical assumption. However, no realistic model has yet been found.

Attempts to incorporate technicolor into the grand unification framework have not been successful (see, e.g., Paper VIII.2.)

In trying to implement technicolor, quite aside from grand unification, people have encountered a number of difficulties[9]. This is largely due to the fact that technicolor, in its original form, preserves the chiral symmetry of quarks and leptons. Very recently, an interesting dynamical proposal[10] to avoid some of these difficulties have been proposed (see Paper XV.3). While this approach leads to rather complicated and unesthetic models and generates new difficulties of its own, it nevertheless may be a promising line of investigation. (We reprint Paper XV.3 for this reason, but must warn the reader that this paper, more so than the other papers reprinted, is very speculative.)

Quite recently, various authors[11] have tried to incorporate sypersymmetry[12] into grand unification. As another way of formulating the hierarchy problem, one may say that the Higgs field responsible for breaking the weak interaction is almost massless compared to the Higgs field which breaks the grand unified symmetry. Now, outside of supersymmetry, we do not know of a "natural" scheme for keeping scalar fields massless. In contrast, fermions may be kept massless by chiral symmetries. Supersymmetry links fermions and bosons and therefore offers a natural mechanism for keeping a Higgs field light. The general philosophy and physics is set forth[13] nicely in Paper XV.4. To actually implement this idea, one must make sure that the predictions for $\sin^2\theta$ and the proton life-time are not rendered inconsistent with experiment[14]. This is a topic of current research interests, and as these words are being written a sizable and growing literature already exists[15]. It is fair to say that models constructed so far are all extremely complicated and unattractive. We should also mention in this connection an ingenious scheme[16] of generating an "inverse hierarchy".

The existence of a large hierarchy in grand unification disturbs some people. If we believe in the simplest $SU(5)$ theory, we have a remarkable situation in which we live at low energies amidst the debris of a symmetry broken smashingly at 10^{15} low. No "new" physics appears between 10^2 GeV and 10^{15} GeV, the so-called Glashow's desert. It is not clear whether one should be disturbed. After all, while life may be linear, physics is definitely logarithmic. A number of people have been motivated nevertheless to construct alternatives to grand unification. One either has to dismiss the agreement of the predicted value of $\sin^2\theta$ with the experimental value as a random coincidence or constructs a theory predicting $\sin^2\theta = 1/4$ at a relatively low mass scale (say 1 TeV). Perhaps more seriously, one also has to give up the snug fit of the fermion spectrum in $SU(5)$. To lower the grand unification scale, one can try to arrange to have an absolutely stable proton. Of the many schemes proposed, we mention only the so-called "petite unification" model.[17]

It is fair to say that none of the variant schemes, some of them incorporating technicolor and supersymmetry, is as compact and tight as the original $SU(5)$ and/or the $SO(10)$ scheme.

REFERENCES

1. E. Gildener, *Phys. Rev.* **D14,** 1667 (1976).
2. Y. Kazama and Y. P. Yao, Fermilab-PUB-81/18 THY.
3. D. Gross and F. Wilczek, *Phys. Rev.* **D8,** 3633 (1973); S. Weinberg, *Phys. Rev. Lett.* **31,** 494 (1973).
4. D. Gross and A. Neveu, *Phys. Rev.* **D10,** 3235 (1974).
5. L. Susskind, Paper XV.1.
6. S. Weinberg, Paper XV.2.
7. G. 't Hooft, *Nucl. Phys.* **B72,** 461 (1974).
8. S. Dimopoulos, S. Raby, and L. Susskind, *Nucl. Phys.* **B173,** 208 (1980).
9. For example, S. Dimopoulos and L. Susskind, *Nucl. Phys.* **B155,** 237 (1979); E. Eichten and K. Lane, *Phys. Lett.* **90B,** 125 (1980); S. Dimopoulos and J. Ellis, *Nucl. Phys.* **B182,** 505 (1981); see also the review by E. Farhi and L. Susskind, *Phys. Rev.* **74,** 277 (1981).
10. H. Georgi and S. L. Glashow, *Phys. Rev. Lett.* **47,** 1511 (1981); H. Georgi and I. McArthur, Paper XV.3; H. Georgi, HUTP-81/A057.
11. M. Veltman, *Acta. Phys. Pol.* **B12,** 437 (1981); S. Dimopoulos and S. Raby, *Nucl. Phys.* **B192,** 353 (1981); S. Dimopoulos, S. Raby, and F. Wilzcek, *Phys. Rev.* **D24,** 1681 (1981); N. Sakai, preprint TU/81/225, to appear in *Z. Phys. C* (1981); S. Dimopoulos and H. Georgi, *Nucl. Phys.* **B193,** 150 (1981); H. P. Nilles and S. Raby, SLAC-Pub-2743 (1981).
12. For a review, see P. Fayet and S. Ferrara, *Phys. Rev.* **32C,** 250 (1977); J. Wess, *Princeton Report*, to be published.
13. E. Witten, Paper XV.4.
14. S. Weinberg, HUTP-81/A047; N. Sakai and T. Yanagida, MPI-PAE/PTH 55/81; S. Dimopoulos *et al.*, Ref. 11; a review has been given by J. Ellis, D. V. Nanopoulos, and S. Rudaz, CERN preprint (1982).
15. P. Fayet, XVI Rencontre de Moriond, First Session; N. Sakai, Tohoku preprint TU/81/225; R. N. Cahn, I. Hinchliffe and L. J. Hall, LBL-13726 (1981); L. Alvarez Gaumé, M. Claudson and M. Wise, HUTP-81/A063; M. Dine and W. Fischler, Princeton preprint (1981); M. Dine, W. Fischler and M. Srednicki, *Nucl. Phys.* **B189,** 575 (1981); L. Hall and I. Hinchliffe, LBL-14020 (Feb. 1982), and Refs. (11) and (14).
16. E. Witten, *Phys. Lett.* **105B,** 267 (1981).
17. P. Q. Hung, A. J. Buras and J. D. Bjorken, Fermilab-PUB-81/22-THY.

CHAPTER XVI

INVISIBLE AXIONS

Back in Chapter II we mentioned that one of the outstanding puzzles of contemporary physics is the absence of CP violation in the strong interaction[1]. As explained in Chapter II, an attractive solution involves introducing a global $U(1)_{PQ}$ symmetry which when broken leads to the prediction that a light boson exists. Furthermore, the mass and couplings of this so-called axion are fairly well-determined theoretically[2,3]. However, the axion as originally described by Weinberg[2] and Wilczek[3] has not been observed experimentally.

It turns out that grand unification may offer a more-or-less natural solution of this strong CP problem. The story began with an observation originally due to Kim[4] and elaborated later by Dine, Fischler, and Srednicki[5]. The point is the following: were the axion a true Nambu-Goldstone boson it would be massless and it would interact with matter through a derivative coupling. More precisely, the axion field A could only couple to matter in the combination $(1/f)\,\partial_\mu A$, where f is the mass scale at which the Peccei-Quinn $U(1)_{PQ}$ symmetry is broken. These are just the same well-known considerations which were applied to pions twenty years ago. However, the fact is that the axion is not a true Nambu-Goldstone boson. $U(1)_{PQ}$ is not a true global symmetry of the Lagrangian. So the statements above must be modified slightly[6]. The axion is massive. Its mass could be estimated by comparing it to the pion mass:[2,3,7] $m \sim f_\pi m_\pi/f$. However, its coupling is still inversely proportional to f. Note that since a true Nambu-Goldstone boson cannot couple to the scalar flavor-conserving density, the induced axion coupling to the scalar flavour-conserving density will be of order $\bar{\theta} f_\pi/f$. Thus, with f very large, one has a very weakly coupled light axion. Kim[4] originally suggested making f of the order 10^5 GeV.

The relevant phenomenological constraint on a weakly coupled light axion comes from axion emission from the helium core of a red supergiant star[8,9]. It turns out that f must be larger than $\sim 10^9$ GeV. Dine *et al.*[5] realised that with f so large the axion would have escaped detection and suggested that the mechanism responsible for breaking grand unification symmetry also breaks $U(1)_{PQ}$. Wise, Georgi and Glashow[10] constructed a specific $SU(5)$ model based on this idea. With f at the grand unification scale $\sim 10^{15}$ GeV, one can estimate the axion mass $\sim 10^{-8}$ eV, its pseudo-scalar couplings $\sim 10^{-16}$, its scalar couplings $\sim 10^{-31}$. Numbers like these justify calling this particle the "invisible axion".[11]

While there is no reason whatsoever that experimental capabilities should always be matched to theory, it is somewhat unsatisfying and disquieting that the strong CP puzzle is solved because of the existence of a particle which no conceivable experiment could detect. The superheavy gauge bosons of grand unification might reveal their existence indirectly through proton instability. In this sense, one can argue that the invisible axion has already indirectly revealed its existence.

Intertwined with this development is the realization that spontaneously broken global symmetry may not necessarily be phenomenologically unacceptable. This point was particularly emphasized by Chikashige, Mohapatra, and Peccei[12] who broke lepton number spontaneously in a model. The assiciated Goldstone boson, called the Majoron, is so weakly coupled to matter that it is essentially "invisible". (See Chapter XIII.)

REFERENCES

1. R. Peccei and H. Quinn, Ref. 23, Chapter II.
2. S. Weinberg, Ref. 24, Chapter II.
3. F. Wilczek, Ref. 24, Chapter II.
4. J. E. Kim, Paper XVI.1.
5. M. Dine, W. Fischler and M. Srednicki, Paper XVI.2.
6. A nice discussion was given by H. Georgi, in *Grand Unified Theories and Related Topics*, Kyoto Summer School 1981, (World Scientific Press).
7. W. A. Bardeen and S-H. H. Tye, *Phys. Letters* **74B,** 229 (1978); J. Kandaswamy et al., *Phys. Rev.* **D17,** 1430 (1978).
8. K. Sato and H. Sato, *Prog. Theo. Phys.* **54,** 912 (1975).
9. D. A. Dicus, R. Kolb, V. Teplitz and R. V. Wagoner, *Phys. Rev.* **D18,** 1892 (1978).
10. M. B. Wise, H. Georgi, S. Glashow, paper XVI.3.
11. Further literature includes H. Georgi, L. J. Hall and M. B. Wise, *Nucl. Phys.* **B192,** 409 (1981); D. B. Reiss, *Phys. Lett.* **B109,** 365 (1982).
12. Y. Chikashige, R. N. Mohapatra, and R. D. Peccei, *Phys. Lett.* **98B,** 265 (1981).

CHAPTER XVII

COMPOSITE QUARKS AND LEPTONS

The notion that quarks and leptons may be composites naturally suggests itself. The notion is of course logically independent of grand unification, but it may be hoped that composite theories might solve some of the problems, such as the family problem, on which grand unification has so far failed to shed any light.

The literature[1] on composite models has grown to be quite large. Unfortunately, there is as yet no composite model which has either advanced our understanding or made clear-cut predictions. In particular, we know of no model which fixes unequivocally the number of families or at least explains why there must be more than one family. We feel that a composite model should, at the least, shed some light on the family problem.

There is quite a diversity of models. In some, elementary scalars bind with elementary fermions to form quarks and leptons. In others, some of the gauge bosons are themselves composite[2]. Some models do not respect the anomaly matching conditions to be discussed below. Many models are orthogonal to grand unification. In the following discussion, we choose to focus on a class of theories which may allow the incorporation of grand unification and which involves elementary gauge bosons. Opinions differ on this last question. Some feel that gauge fields with their geometrical underpinning should somehow be elementary. Certainly, the photon has remained elementary as physicists "peeled off successive layers of the onion". Attempts to construct composite photon and graviton models have not been particularly successful in the past. At the moment, there is no convincing argument for either side of this question. In this regard, the following observation[3] may be relevant: if the underlying dynamics were to produce an effective theory which is structurally independent of physics at high mass scales, that is to say, an effectively renormalizable theory, then this theory must resemble quite closely our present day theories with its non-Abelian gauge interactions, to the extent that there is only one class of renormalizable theories known to date.

Within the set of particle physicists who believe in composite quarks and leptons there is a subset who, following 't Hooft[4], feels that the striking fact that the Compton wavelength of the quarks and leptons is much larger than the upper limit to their sizes is crucial and must be explained in terms of a symmetry, presumably a chiral symmetry. Equivalently put, quarks and leptons are essentially massless compared to the mass scale Λ of the physics which forms quarks and leptons. A lower limit to this scale Λ may be estimated[5] by examining processes like $\mu \to e + \gamma$ and the agreement between theory and experiment on the magnetic moments of leptons. Estimates of Λ range from $\gtrsim 1$ TeV to

$\gtrsim 10^2$ TeV, depending on the processes examined. The dynamical puzzle is how quarks and leptons, which are essentially massless compared to Λ, can be formed out of constituents confined over a distance scale of order Λ^{-1}. Thus, if quarks and leptons turn out to be indeed composites, Nature is not simply repeating herself. The physics of making massless composites looks quite different from the physics which makes hadrons out of quarks (which in turn differs from the physics making atoms and molecules).

The above discussion represents an elaborate way of saying that we are trying to make quarks and leptons composites even though they are known experimentally to be point-like down to very small distances.

Let us call the massless fermions which make up quarks and leptons "rizoms"[6]. The rizoms listen to a gauge interaction based on a group G_{HC} which we refer to as "hypercolor." Rizoms belong to some anomaly free representation R of G_{HC}. We presume that G_{HC} is confining and binds rizoms into hypercolor singlets. The fermionic bound states are identified as quarks and leptons. We assume that a chiral symmetry keeps the composites massless. The observed quark and lepton masses (including that of the top quark!) would have to be accounted for by a slight explicit breaking of this chiral symmetry due to some other effects. At this stage of the theory one may perhaps ignore this potentially troublesome problem. Eventually, of course, one must understand the fermion mass hierarchy (Chapter IX).

We must emphasize that by assuming the existence of a chiral symmetry one has not solved in any way the dynamical puzzle of producing massless composites. However, an unbroken chiral symmetry is the only known way of insuring the presence of massless fermion composites.

In general, the rizom theory will have an "automatic" global symmetry group G_{HF} which we refer to as "hyperflavor". Thus, if the representation R is reducible and contains the irreducible representations $R_1, R_2, \ldots R_k$ repeated $N_1, N_2, \ldots N_k$ times, $G_{\mathrm{HF}} = SU(N_1) \times SU(N_2) \cdots \times SU(N_k) \times (U(1))^{k-1}$. (Note that, of the k $U(1)$'s expected naively, one is broken down by non-perturbative effects in the hypercolor interaction to a discrete group.)

In order to have a non-trivial G_{HF}, one has to abandon the hope, at least at this stage, of eventually putting all fundamental fermions into a single irreducible representation (see Chapter VIII).

Without a full understanding of hypercolor dynamics we cannot predict what the spectrum of massless bound fermions will consist of. In general, hypercolor may break G_{HF} dynamically down to some subgroup. Suppose for simplicity that this chiral group G_{HF} does not break and insures the presence of massless composites. 't Hooft pointed out that the anomalies in the global symmetry G_{HF} when evaluated at the rizom level and the composite level must match[4,7,8] (paper XVII.1). This anomaly matching condition tells us the possible spectrum of the massless bound fermions. Unfortunately, without a detailed understanding of dynamics, we cannot decide which of the possible spectra the theory actually produces.

In his paper, 't Hooft also imposes additional conditions. These are based on questionable dynamical assumptions and are clearly on a different footing than the anomaly matching condition[9]. Many workers on composite models feel that these additional constraints should be dropped or replaced by other conditions[10,11,12].

In particular, 't Hooft imposed the decoupling condition that if the chiral symmetry is explicitly broken by giving one of the rizoms a mass m, the remaining chiral symmetry must permit all the composites which contain the massive rizom to become massive. As $m \to \infty$, all the composites which contain the massive rizom indeed must all become infinitely massive according to a rigorous theorem[13] in field theory. However, a phase transition may occur so that for m below some critical mass some composites containing the massive rizom may be massless. Indeed, explicit models of this phenomenon are known[10,11].

We are invited to gauge some subgroup G of G_{HF} and to identify G as the grand unification group times possibly a technicolor group and/or a horizontal family group. Or, if we prefer, G could contain only $SU(3) \times SU(2) \times U(1)$. G represents the gauge symmetries "already known" to us. This framework, outlined by 't Hooft, thus repeats the familiar pre-grand-unified situation of the strong, weak, and electromagnetic interactions in the hadronic world. The reader would have noted that a "non-hyper" version of the preceding discussion, with the substitutions $G_{HC} \to SU_C(3)$, $G_{HF} \to SU(n) \times SU(n) \times U(1)$ with n quark flavors, and $G \to SU(2) \times U(1)$, describes the hadronic world. The difference is of course that in this case the chiral group G_{HF} is broken, explicitly and dynamically.

Of course G has to be anomaly-free on the rizoms. Otherwise, one would be obliged to introduce "spectator" fermions: singlet fields under G_{HC} which transform non-trivially under G. There is in principle nothing wrong with a picture in which some of the observed quarks and leptons are composites and the rest are elementary "spectators," but most people would feel that it is rather contrived to introduce spectators. One should keep in mind, however, that at the quark to hadron level Nature in fact does introduce spectators, namely the familiar leptons, to cancel anomalies.

In this framework, it may be possible to have composite quarks and leptons without giving up the successes of grand unification, including the prediction of $\sin^2\theta$ and the seamless fit of the fermion spectrum. In a sense, the framework we outlined represents the least imaginative avenue in that it merely follows the familiar story of forming hadrons out of quarks. The correct theory may well lie outside this framework.

A number of instructive toy models[14] have been constructed within this framework. Of all the published models within this framework which claim to be more or less realistic we prefer one constructed by Bars[15] (paper XVII.2). Nevertheless, one must admit that this model looks rather contrived. Bars chooses G_{HC} to be $SU(4)$ and takes R to consist of sixteen **4**'s, eight $\bar{\mathbf{4}}$'s, and one $\overline{\mathbf{10}}$. This rather haphazard-looking collection of representations is anomaly-free. One can have eight $SO(10)$ families. Basically, of the three sets of rizoms, one set "carries" grand unification quantum numbers, the second set carries family number, and the third is needed to form the desired hypercolor singlets.

We prefer to have a vector-like theory[16,17] in which $G_{HC} = SU(N)$, N odd, and $R = n(N + \bar{N})$'s so that the hypercolor theory is automatically anomaly free.

While quarks and leptons are essentially massless compared to the composite scale, their masses, from the electron to the top quark, span an enormous range. A successful composite model must not only fix the number of families but it should also account for the observed masses. Thus the chiral symmetry has to be broken slightly, perhaps by explicit scalar fields. Weinberg[17] has pointed out that one advantage of the composite models is precisely that quarks and leptons can acquire mass without the aid of explicit scalar fields.

REFERENCES

1. A fairly extensive list of references may be found in the reviews by M. Peskin, *Proceedings of the 1981 International Symposium on Lepton and Photon Interactions*; by J. Preskill, Harvard preprint HUTP-81/A051; and by H. Terazawa, *Proceedings of the 1981 Tokyo Symposium.*
2. For example, see H. Harari, *Phys. Lett.* **86B,** 83 (1979); M. A. Shupe, *Phys. Lett.* **94B,** 54 (1980); L. F. Abbott and E. Farhi, *Phys. Lett.* **101B,** 69 (1981); Y. P. Kuang and H. Tye, Cornell preprint (1982); H. Fritzsch and G. Mandelbaum, *Phys. Lett.* **102B,** 309 (1981).
3. This statement, in its various versions, is often referred to as Veltman's theorem.
4. G. 't Hooft (paper XVII.1).
5. For instance, R. Barbieri, L. Maiani, and R. Petronzio, *Phys. Lett.* **96B,** 63 (1980); S. J. Brodsky and S. D. Drell, *Phys. Rev.* **D22,** 2236 (1980).
6. Unfortunately, there is not a standard terminology in the literature. We arrived at this term by interviewing several Greek physicists, in particular G. Lazarides. The best term of Greek origin in this connection is, of course, "proton".
7. A discussion of what amounts essentially to the anomaly matching condition was also given by A. Zee, *Phys. Lett.* **95B,** 290 (1980). The importance of an unbroken chiral symmetry was however not fully appreciated in this paper.
8. For further discussion of the anomaly conditions see S. Dimopoulos, S. Raby and L. Susskind, *Nucl. Phys.* **B173,** 208 (1980); R. Barbieri *et al.*, ref. (4); T. Banks, S. Yankielowicz and A. Schwimmer, *Phys. Lett.* **96B,** 67 (1980); Y. Frishman, A. Schwimmer, T. Banks and S. Yankielowicz, *Nucl. Phys.* **B177,** 157 (1981); T. Banks and A. Schwimmer, Trieste preprint ICTP 80/81-6; I. Bars and S. Yankielowicz, *Phys. Lett.* **101B,** 159 (1981); R. Casalbuoni and R. Gatto, *Phys. Lett.* **93B,** 47 (1980); S. Coleman and E. Witten, *Phys. Rev. Lett.* **45,** 100 (1980); G. Farrar, *Phys. Lett.* **96B,** 273 (1980); R. Chanda and P. Roy, *ibid.*, **99B,** 453 (1981); and E. Cohen and Y. Frishman, *Phys. Lett.* **109B,** 35 (1982). Further references may be found in the reviews cited in ref. (1).
9. The anomaly condition may be proven on rather rigorous grounds. S. Coleman and B. Grossman, Harvard preprint 1982.
10. I. Bars, *Phys. Lett.* **109B,** 73 (1982).
11. J. Preskill and S. Weinberg, *Phys. Rev.* **D24,** 1059 (1981).
12. R. K. Kaul and R. Rajaraman (*Phys. Lett.* **110B,** 385 (1982)) showed that the three conditions imposed by 't Hooft are not independent.
13. T. Appelquist and J. Carazzone, *Phys. Rev.* **D11,** 2856 (1975).
14. See, in particular, S. Dimopoulos *et al.*, ref. (8).
15. I. Bars, paper XVII.2.
16. For instance, one might consider the simplest case in which $N=3$. For n large enough, one would be able to fit $SO(10)$ into $SU(n) \times SU(n) \times U(1)$. However, models of this type suffer from severe difficulties (X. Y. Li and A. Zee, unpublished, 1981).
17. S. Weinberg, *Phys. Lett.* **102B,** 401 (1981); see also H. Harari and N. Selberg, *Phys. Lett.* **98B,** 269 (1981); **100B,** 41 (1981).

CHAPTER XVIII
GRAVITY AND GRAND UNIFICATION

Finally, we come to the question which has baffled generations of physicists: How does gravity tie in with the rest of physics? Three of the four fundamental interactions are now unified, but gravity remains outside the fold.

There are a number of tantalizing hints and problems. Gravity has been known for almost seventy years to be based on local coordinate invariance. The developments of the last decade or so have established fairly convincingly that the other three interactions are also based on local invariances. The Kaluza-Klein approach (see below) ties together these two kinds of gauge symmetries. However, gravity does look different from the other three interactions. It is mediated by a spin two gauge boson. Supersymmetry offers a way to relate particles with different spins. The Einstein-Hilbert Lagrangian looks different from the Yang-Mills Lagrangian; it is not the square of a field strength. The analog of $F_{\mu\nu}$ is $R_{\mu\nu\lambda\sigma}$ in some sense.

A major puzzle in contemporary physics involves the cosmological constant[1]. The observed upper bound on the cosmological constant is $\sim 10^{-46}$ (GeV)^4. But contemporary theories of the strong, weak, and electromagnetic interactions involve a hierarchy of symmetry breaking. Each time a symmetry is broken at mass scale μ a cosmological constant of order μ^4 is induced. The solution of the cosmological constant puzzle will most likely shed light on the question of how gravity is linked to the other three interactions.

It is intriguing that the simplest $SU(5)$ theory sets its scale at 10^{14}–10^{15} GeV, some four orders of magnitude below the Planck scale gravity. $SU(5)$ is presumably not a complete theory. As we have mentioned, a number of major problems are not unsolved in $SU(5)$. Does this mean that $SU(5)$ represents merely the broken remnant of a more complete theory in the same way that $SU(3) \times SU(2) \times U(1)$ follows from $SU(5)$? Is this theory set at the Planck scale? If not, why is there a gap between the ultimate grand unified theory and gravity?

The preceding remarks bear upon the philosophy to be adopted in building grand unification models. Should we apply the usual esthetic criteria of physics and choose among grand unification models the most elegant one? An equally valid view might be that since the physics at 10^{15} GeV represents merely the broken manifestations of a more fundamental theory one should feel free to invent arbitrarily complex schemes[2] which can resolve some of the problems of the simple $SU(5)$ theory.

There have been numerous attempts to unify gravity with the other interactions since the days of Einstein. We cannot possibly discuss all of these here. We limit ourselves to three approaches which have been discussed in the recent literature: (1) Supergravity, (2) Kaluza-Klein theories, and (3) Dynamically induced gravity. These approaches are not necessarily mutually exclusive. In particular, approaches (1) and (2) overlap significantly.

Supergravity[3,4] is the gauge theory of global supersymmetry and represents a natural extension of supersymmetry. Since the supersymmetry algebra involves the Poincaré group generators P^{μ} and $M^{\mu\nu}$ it is perhaps not surprising that there is a link with gravity. Some would even argue that it is unnatural to consider theories which are merely globally supersymmetric. In general, the supersymmetry algebra may contain N Majorana spinor charges. The corresponding gauge theory is known as $SO(N)$ supergravity and possesses a manifest $SO(N)$ symmetry.

So far, supergravity has failed to make convincing contact with phenomenology. After all, there is no experimental evidence even for supersymmetry. The difficulty is that these theories are very restrictive in their internal symmetry and particle content.

The internal symmetry $SO(N)$ does not appear to be large enough[5] since $N \leqslant 8$. (This theorem is easy to see since the Majorana generators change helicity by half-a-unit and there are eight half-unit steps between -2 and $+2$. For $N > 8$, particles with spin $\geqslant 5/2$ will appear.) In particular, there is no room in the spectrum for $W^{\pm}$, the muon and the τ. The obvious suggestion is that these are composites of the particles found in the theory. This situation was dramatically changed with the brilliant discovery of Cremmer and Julia[6–8] that $N = 8$ supergravity contains a "hidden" local $SU(8)$ symmetry and a global non-compact group $E(7, 7)$. The phenomenon of "hidden" symmetries is a fascinating one in itself. Our interest here is in the possibility of embedding grand unification[9–11] in $SO(8)$ supergravity opened up by the work of Cremmer and Julia. We must caution the reader, however, that while this approach appears interesting, the theories proposed are far from complete and are not universally accepted.

The few remarks above hardly do justice to supergravity, but detailed discussion of supergravity would take us far afield and is in any case beyond the competence of this writer. The practitioners are known to assert that supergravity theory is so mathematically beautiful that it must be correct. Perhaps one reason that the community has not embraced the theory enthusiastically is that the mathematical manipulations involved tend to be extremely complex.

We next turn to the Kaluza-Klein approach. Two major concepts lie at the foundation of modern physics: local coordinate invariance and local internal symmetry. The former leads to the theory of gravity while the latter leads to the gauge theory of strong, weak and electromagnetic interactions. The remarkable discovery of Kaluza is that if we suppose that space-time is embedded in a space with dimension higher than four these two concepts may not be independent—the latter may be derived from the former. The physics is so astonishing and the mathematics is so elegant that it is easy to believe that Nature does in fact use the Kaluza mechanism at some level. Whether or not this level is already accessible to present-day physics of course remains to be seen.

The Kaluza mechanism, while profound, is easy to explain.[12] Imagine, as in the original version, that the world is actually five-dimensional but the fifth dimension is curled up in a tiny circle. In other words, each space-time point is actually a tiny circle, with a radius so much smaller than any length scale explored experimentally that we have all been fooled and have mistaken the circle for a point. Denote the coordinates by

$$x^{\hat{\mu}} = \{x^\mu, x^5\}, \hat{\mu} = 1 \cdots 5, \mu = 1 \cdots 4.$$

Here $x^5 \equiv \theta$ runs from 0 to 2π. Consider an infinitesimal coordinate transformation

$$\begin{aligned} x^\mu &= x'^\mu \\ x^5 &= x'^5 + \lambda(x'). \end{aligned} \tag{1}$$

In other words, we rotate the circles by an infinitesimal space-time dependent amount. The metric $g_{\hat{\mu}\hat{\nu}}$ contains a piece $g_{\mu 5}$ which transforms as a four-vector under the usual coordinate transformation. Under the transformation (1) we find that

$$g_{\mu 5} \to g_{\hat{\sigma}\hat{\rho}} \frac{\partial x^{\hat{\sigma}}}{\partial x'^\mu} \frac{\partial x^{\hat{\rho}}}{\partial x'^5} = g_{\sigma 5} \delta^\sigma_\mu + g_{55} \partial_\mu \lambda. \tag{2}$$

Setting $g_{55} = 1$ and calling $g_{\mu 5}$, A_μ, we see that A_μ transforms like a gauge potential:

$$A_\mu \to A_\mu + \partial_\mu \lambda. \tag{3}$$

Gauge transformation may be cast as a special case of coordinate transformation!

Furthermore, the gravitational action in five-dimensions $\int d^4x dx^5 \sqrt{\hat{g}} \hat{R}$ reduces to the action $\int d^4x \sqrt{g}(1/l^2 R + F_{\mu\nu}F^{\mu\nu})$ in four dimensions. The appearance of Maxwell's action is not so surprising if we recall that $\hat{R}$ involves two derivatives and is invariant under all five-dimensional coordinate transformations. The fact that Maxwell's action appears with the correct sign is however not so evident a priori. The relative scale between Einstein's action and Maxwell's action is set by the length scale of the Kaluza circles, which clearly must have radii the order of Planck's length. It is perfectly understandable then that experimentalists all believe the world is four-dimensional. It is also worth remarking on charge conjugation in the Kaluza framework. Of the three fundamental discrete symmetries of physics, charge conjugation stands apart from parity and time reversal, which are in some sense more intuitively accessible. Some people find this a bit mysterious. A beautiful feature of the Kaluza theory is that charge conjugation may be interpreted as a space-time transformation as well.

Over the decades, the theory has been developed by a number of people, including Klein, Jordan, Thiery, Trautman, Cho, Freund, Cremmer, Julia, Scherk, Schwarz, Luciani, Palla, Thirring, Tanaka, Koerner, and no doubt many others. In particular, there apparently exists a body of work in Kaluza's homeland which is perhaps not as well known in the West as it should be[13]. Despite its elegance, the Kaluza concept has been more or less regarded as an oddity and has languished outside the mainstream of physics. Historically, this is partly due to the fact that the symmetry of the thirties, isospin, and extensions of it later, clearly does not appear geometrical in origin. Interest in Kaluza theory has been ignited once again in recent years thanks largely to the discovery that the underlying symmetry is indeed geometrical, local, and gauged and that flavor symmetries such as isospin are approximate and rather accidental. Also, supergravity has naturally led people to consider higher space-time dimensions.

The crucial question is then whether the original Kaluza concept may be generalized from Abelian Maxwell to non-Abelian Yang-Mills. Perhaps not surprisingly, the answer to this question is a resounding yes. All one has to do is to replace the circle, which Kaluza attached to each space-time point, by some general manifold M. The circle is intimately connected to $U(1)$, the Abelian group of electromagnetism. It turns out if the manifold M permits the action of the group G on it one obtains, instead of electromagnetism, a guage theory based on G.

The Kaluza program is very ambitious. Once the group and the manifold are chosen, gravity and gauge fields both follow from the geometry and the fermion spectrum is completely determined by the structure of the manifold. One might hope in this way for a "deep" solution of the family problem.

Even more ambitious is the super-Kaluza program in which one starts with supergravity. In supergravity, one is obliged to put in spin 3/2 fermions in a definite way. These spin 3/2 fermions when reduced to 4-dimensions will contain spin 1/2 fermions. In this context Witten[14] has noted an interesting numerology. For $G = SU(3) \times SU(2) \times U(1)$ the maximal subgroup is $U(2) \times U(1)$. The minimal manifold having an $SU(3) \times SU(2) \times U(1)$ gauge theory in the Kaluza framework is $SU(3) \times SU(2) \times U(1)/U(2) \times U(1) = CP^2 + S^2 + S^1$ which is seven dimensional. Together with Minkowski space-time this suggests that the basic theory should be $4 + 7 = 11$ dimensional. But workers on supergravity have long known that the maximal dimension for supergravity is precisely eleven! Whether or not this is a pure coincidence remains to be determined. Note that there is no room for $SU(5)$ in this discussion. The manifold $SU(5)/U(4) = \mathrm{CP}^4$ is 8-dimensional. However, one may not have to start with supergravity.

Unfortunately, the theory must overcome a large number of difficulties before it could be considered phenomenologically relevant. To mention but one of these difficulties[15], we note that we tend to end up with equal numbers of $V + A$ and $V - A$ fermions, since the geometry does not distinguish between left and right. This is the same problem which plagues us in Chapter VIII.

As we saw above, the Kaluza-Klein approach overlaps considerably with supergravity. In fact, the discussion of Cremmer and Julia[6,7] starts by reducing the supergravity theory in 11 dimensions to 4 dimensions. Of course, one can logically discuss the Kaluza-Klein theory quite independently of supergravity.

We now come to the idea of dynamically generated gravity, which is motivated by a rather different set of considerations compared to the two preceding approaches. For a detailed discussion, we refer the reader to various review articles[15–18] which also contain extensive lists of references. See Paper XVII.4 for a brief introduction to the subject. We confine ourselves here to some motivating remarks.

We begin with the simple but perhaps profound observation that in Yang-Mills theory the non-Abelian structure of the gauge field $F_{\mu\nu} = \partial_\mu A_\nu - \partial_\nu A_\mu - i[A_\mu, A_\nu]$ forces A_μ to have the dimension of ∂_μ. Thus, the Yang-Mills Lagrangian $F^2_{\mu\nu}$ has dimension four regardless of the dimension of space-time. In contrast, the Einstein-Hilbert Lagrangian R involves the scalar curvature and has dimension two in space-time of any dimension. Thus, in this sense, Yang-Mills theory is perfectly matched to the four dimensional space-time in which we find ourselves. It is tempting to speculate that this fact may be intimately connected to the observation that the world is, indeed, four dimensional.

Why is our theory of gravity matched to two, rather than four-dimensional space-time? We would like to attach some significance to this apparent mismatch. More precisely, scale invariance forbids Einstein's Lagrangian for gravity. As was first fully appreciated by Coleman and E. Weinberg[19], it is extremely attractive to impose scale invariance on the fundamental Lagrangian of the world. In a scale invariant theory with n dimensionless coupling constants, all dimensionless ratios of physical quantities are calculable in terms of $(n - 1)$ dimensionless couplings. This is because the renormalization procedure introduces a mass scale μ whose choice is arbitrary and for which one dimensionless coupling may be traded through "dimensional transmutation". If the fundamental Lagrangian is indeed scale invariant then the ratio of Newton's constant to some other dimensional physical quantity such as the proton mass should be calculable.

In fact, a formal expression for G can be easily derived[20,21] as follows:

$$\frac{1}{16\pi G} = \frac{i}{96}\int \mathrm{d}^4x \cdot x^2 \, \langle 0 | \, T^* T(x) T(0) | 0 \rangle .$$

Here T denotes the trace of the stress-energy tensor.

In this approach, gravity may be regarded, in some sense, as the dynamical consequence of the other three interactions. "Let there be light, and the apple fell". We find this philosophy rather appealing in that it obviates the need for a marriage of gravity with the other three interactions. Gravity, mediated by a spin-two field, does look quite different from the other three interactions, which have now been revealed to be all mediated by spin-one fields.

The philosophy behind supergravity is in some sense the exact opposite of the one advocated here in that it seeks to determine the other three interactions starting with Einstein's theory of gravity. Regrettably, the philosophy of induced gravity also appears to be incompatible with the Kaluza-Klein idea, which is in turn not unrelated to supergravity.

Unfortunately, none of the three approaches discussed here resolves the cosmological constant puzzle. Supergravity offers the most promise. In exactly supersymmetric theory, the contributions of fermions and bosons to the vacuum energy tend to cancel[22]. However, given that supersymmetry is manifestly broken, it is difficult to see how this cancellation could be maintained to the fantastic degree of accuracy required phenomenologically. In the Kaluza-Klein approach a cosmological constant is typically needed to close the unobserved dimensions into a compact manifold[23]. Finally, in the induced gravity approach, scale invariance forbids both the cosmological constant and the Einstein-Hilbert term. With the breaking of scale invariance, both of these terms may be induced.

We believe that a true understanding of gravity and its connection to the other three interactions would not be possible without a solution of the cosmological constant puzzle.

REFERENCES

1. A. D. Linde, *JETP Lett.* **30,** 447 (1979); Y. B. Zel'dovich, *Usp. Fiz. Nauk.* **95,** 209 (1968) [*Sov. Phys. Usp.* **11,** 381 (1968)]; J. Dreitlein, *Phys. Rev. Lett.* **33,** 1243 (1974); M. Veltman, *ibid.*, **34,** 777 (1975).
2. We thank H. Georgi for a spirited discussion on this point.
3. D. Z. Freedman, P. van Nieuwenhuizen and S. Ferrara, *Phys. Rev.* **D13,** 3214 (1976).
4. S. Deser and B. Zumino, *Phys. Letters* **62B,** 335 (1976).
5. M. Gell-Mann, unpublished.
6. E. Cremmer and B. Julia, paper XVIII.1.
7. E. Cremmer and B. Julia, *Nucl. Phys.* **B159,** 141 (1979).
8. E. Cremmer, preprint LPTENS 81/18, lectures delivered at the Spring School on Supergravity, Trieste, 1981 and the Summer Institute in Theoretical Physics, Seattle, 1981. This paper and ref. 7 contain a more leisurely discussion of the material in ref. 6.
9. J. Ellis, M. K. Gaillard and B. Zumino, paper XVIII.2.
10. J. Ellis, M. K. Gaillard, L. Maiani and B. Zumino, in *Unification of the Fundamental Particle Interactions*, ed. S. Ferrara *et al.*, 1980.
11. For earlier work in this direction, see P. H. Frampton, *Phys. Rev. Lett.* **46,** 881 (1961); J. E. Kim and H. S. Song, Seoul National University preprint (1981); J. P. Derendinger, S. Ferrara and C. A. Savoy, *Nucl. Phys.* **B188,** 77 (1981).
12. A rather pedagogical approach may be found in A. Zee, *Proceedings of the 1981 Kyoto Summer School*, edited by M. Konuma *et al.*.
13. See, for instance, M. W. Kalinowski, Institute of Theoretical Physics, Warsaw, Poland, preprints IFT/9/81, 10/81, 2/80 I and II, and 8/81.
14. E. Witten, Paper XVIII.3.
15. A. Zee, Paper XVIII.4.
16. A. Zee, *Grand Unified Theories and Related Topics*, edited by M. Konuma and T. Maskawa, p. 143.
17. S. L. Adler, in *The High Energy Limit*, edited by A. Zichichi.
18. S. L. Adler, *Rev. of Mod. Phys.* (to be published). Of the four reviews mentioned this is the most encyclopedic and exhaustive, and contains an excellent discussion of a number of technical details.
19. S. Coleman and E. Weinberg, *Phys. Rev.* **D7,** 1888 (1973).
20. S. Adler, *Phys. Rev. Lett.* **44,** 1567 (1980); *Phys. Lett.* **95B,** 241 (1980).
21. A. Zee, *Phys. Rev.* **D23,** 858 (1981); and *Phys. Rev. Lett.* **48,** 295 (1982).
22. B. Zumino, *Nucl. Phys.* **B89,** 535 (1975).
23. For instance, J. F. Luciani, *Nucl. Phys.* **B125,** 111 (1978).

POSTSCRIPT

We are including this postscript in an attempt to bring the book up-to-date as the typescript goes to press. This postscript was written in the early Fall of 1982, some six months after the bulk of the manuscript was written. In the intervening months the literature on grand unification and related topics has grown steadily, as one might expect. However, there has been no dramatic theoretical development. The deep problems in the field continue to resist solution. The origin of family structure, for instance, remains as obscure as ever.

The literature on marrying grand unification and global supersymmetry has grown quite large by now.[1] A number of people have begun investigating the effects of including supergravity (local supersymmetry) in model building.[2]

The inflationary universe scenario is being vigorously developed.[3] The spectrum of density fluctuations leading possibly to galaxy formation is being studied in this scenario. Interesting constraints on grand unification may emerge[3] from work along this line.

Another interesting development stems from the realization[4] that, in certain grand unified theories, strings and domain walls may appear and influence cosmology. In particular, theories with a Peccei-Quinn symmetry tend to involve domain wall production. A number of proposals[5] have been made to avoid these walls. It was hoped that the condition that domain walls are absent would provide a welcomed constraint on the theory. The variety of suggestions, however, indicates that there is still considerable freedom in building models. Cosmological considerations on the energy density contained in invisible axions have led to an interesting upper limit[6] on the scale at which Peccei-Quinn symmetry is broken, rendering the viability of the axion idea more suspect.

These are but some of the highlights. The literature also includes many papers on composite models, for instance. However, at our present level of understanding dynamics, much of these works amount to little more than counting quantum numbers. There is also an ongoing discussion on the roles of massive neutrinos in cosmology.[7]

Finally, it should be mentioned that the candidate events for proton decay observed at the Kolar Gold Field have now been reported in the literature.[8]

REFERENCES

(Note: The following list is clearly not intended to be complete by any means.)

1. See recent issues of the leading journals.
2. For instance, E. Cremmer, S. Ferrara, L. Girardello, and A. van Proeyen, Ref. TH. 3312-CERN (1982); R. Arnowitt, P. Nath and A. H. Chamseddine, NUB 2565 and 2570 (1982); S. Weinberg *Phys. Rev. Lett.* **48,** 1176 (1982); B. Ovrut and J. Wess, IAS preprint (1982); J. Bagger and E. Witten, Princeton preprint (1982); L. Ibáñez, Ref. TH. 3374-CERN (1982).

3. V. A. Rubakov, M. V. Sazhin and A. V. Veryaskin, *Phys. Lett.* **115B,** 189 (1982); S. W. Hawking, *ibid.,* **115B,** 295 (1982); A. Guth and S.-Y. Pi, CTP 1014 (1982); J. M. Bardeen, P. J. Steinhardt and M. S. Turner, preprint in preparation.
4. T. W. B. Kibble, *Phys. Rep.* **67,** 183 (1980); T. W. B. Kibble, G. Lazarides and Q. Shafi, ICTP/81/82-18 (1982); P. Sikivie, *Phys. Rev. Lett.* **48,** 1156 (1982); A. Vilenkin and A. E. Everett, *Phys. Rev. Lett.* **48,** 1867 (1982).
5. G. Lazarides and Q. Shafi, *Phys. Lett.* B (to appear); H. Georgi and M. B. Wise, HUTP-82/A037; S. M. Barr, D. B. Reiss and A. Zee, *Phys. Lett.* B (to appear); Y. Fujimoto, K. Shigemoto and Z.-Y. Zhao, Trieste preprint 1982; S. M. Barr, X. C. Gao and D. Reiss, UW 40048-22-P2 (1982); S. Dimopoulos, P. H. Frampton, H. Georgi and M. B. Wise, HUTP-82/A040; B. Grossman, RU82/B/39 (1982).
6. L. F. Abbott and P. Sikivie, preprint 1982; J. Preskill, M. B. Wise and F. Wilczek, HUTP-82/A048 (1982); M. Dine and W. Fischler, UPR-0201T (1982).
7. See the contributions of Y. Raphaëli and A. Quadir to the *Proceedings of the Varenna School on Gamow Cosmology,* August 1982.
8. M. R. Krishnaswamy *et al., Phys. Lett.* **115B,** 349 (1982).

REPRINTED PAPERS

X. *A Short Course in Cosmology*

1) Expanding universe and the origin of elements
by G. Gamow, *Phys. Rev.* **70,** 572 (1946) 526
2) The origin of chemical elements
by R. A. Alpher, H. Bethe and G. Gamow,
Phys. Rev. **73,** 803 (1948) 528
3) The origin of elements and the separation of galaxies
by G. Gamow, *Phys. Rev.* **74,** 505 (1948) 530
4) The evolution of the universe
by G. Gamow, *Nature* **162,** 680 (1948) 532
5) Proton-neutron concentration ratio in the expanding universe at the stages preceding the formation of the elements
by C. Hayashi, *Prog. Theor. Phys.* **5,** 224 (1950) 535
6) Physical conditions in the initial stages of the expanding universe
by R. A. Alpher, J. W. Follin, Jr., and R. C. Herman, *Phys. Rev.* **92,** 1347 (1953) 547
7) The mystery of the cosmic helium abundance
by F. Hoyle and R. Tayler, *Nature* **203,** 1108 (1964) 562
8) Density of relict particles with zero rest mass in the universe
by V. F. Shvartsman, *JETP Lett.* **9,** 184 (1969) 565
9) Cosmological limits to the number of massive leptons
by G. Steigman, D. N. Schramm and J. E. Gunn,
Phys. Lett. **66**B, 202 (1977) 568
10) Big-bang nucleosynthesis as a probe of cosmology and particle physics
by K. A. Olive, D. N. Schramm, G. Steigman, M. S. Turner, and J. Yang, *Astrophys. J.* **246,** 557 (1981) 571

Expanding Universe and the Origin of Elements

G. Gamow
The George Washington University, Washington, D. C.
September 13, 1946

It is generally agreed at present that the relative abundances of various chemical elements were determined by physical conditions existing in the universe during the early stages of its expansion, when the temperature and density were sufficiently high to secure appreciable reaction-rates for the light as well as for the heavy nuclei.

In all the so-far published attempts in this direction the observed abundance-curve is supposed to represent some equilibrium state determined by nuclear binding energies at some very high temperature and density.[1-3] This point of view encounters, however, serious difficulties in the comparison with empirical facts. Indeed, since binding energy is, in a first approximation, a linear function of atomic weight, any such equilibrium theory would necessarily lead to a rapid exponential decrease of abundance through the entire natural sequence of elements. It is known, however, that whereas such a rapid decrease actually takes place for the first half of chemical elements, the abundance of heavier nuclei remains nearly constant. Attempts have been made[2] to explain this discrepancy by the assumption that heavy elements were formed at higher temperatures, and that their abundances were already "frozen" when the adjustment of lighter elements was taking place. Such an explanation, however, can be easily ruled out if one remembers that at the temperatures and densities in question (about 10^{10}°K, and 10^6 g/cm^3) nuclear transformations are mostly caused by the processes of absorption and re-evaporation of free neutrons so that their rates are essentially the same for the light and for the heavy elements. Thus it appears that the only way of explaining the observed abundance-curve lies in the assumption of some kind of unequilibrium process taking place during a limited interval of time.

The above conclusion finds a strong support in the study of the expansion process itself. According to the general theory of expanding universe,[5] the time dependence of any linear dimension l in it is given by the formula

$$\frac{dl}{dt}=\left(\frac{8\pi G}{3}\rho l^2-\frac{C^2}{R^2}\right)^{\frac{1}{2}}, \tag{1}$$

where G is the Newton constant, ρ the mean density, and R (real or imaginary) a constant describing the curvature of space. It may be noticed that the above expression represents a relativistic analog of the familiar classical formula

$$v=\left(2\cdot\frac{4\pi l^3}{3}\rho\cdot\frac{G}{l}-2E\right)^{\frac{1}{2}} \tag{2}$$

for the inertial expansion-velocity of a gravitating dust sphere with the total energy E per unit mass. The imaginary and real values of R correspond to an unlimited expansion (in case of superescape velocity), and to the expansion which will be ultimately turned into a contraction by the forces of gravity (subescape velocity). To use some definite numbers, let us consider in the present state of the universe (considered as quite uniform) a cube containing, say, 1 g of matter. Since the present mean density of the universe is $\rho_{\text{present}}\cong 10^{-30}$ g/cm^3, the side of our cube will be: $l_{\text{present}}\cong 10^{10}$ cm. According to Hubble,[6] the present expansion-rate of the universe is 1.8×10^{-17} cm/sec. per cm, so that $(dl/dt)_{\text{present}}\cong 1.8\times 10^{-7}$ cm/sec. Substituting the numerical values in (1) we obtain

$$1.8\times 10^{-7}=(5.7\times 10^{-17}-C^2/R^2)^{\frac{1}{2}}, \tag{3}$$

showing that at the present stage of expansion the first term under the radical (corresponding to the potential energy of gravity) is negligibly small as compared with the second one. For the numerical value of the (constant) radius of curvature we get from (3): $R=1.7\times 10^{17}\sqrt{-1}$ cm or about 0.2 imaginary light year.

In the past history of the universe, when l was considerably smaller, and ρ correspondingly larger, the first term in (1) was playing an important role corresponding physically to the slowing-down effect of gravity on the original expansion. The transition from the slowed down to the free expansion took place at the epoch when the two terms were comparable, i.e., when l was about one thousandth of its present value. At this epoch the gravitational clustering of matter into stars, stellar clusters, and galaxies, probably must have taken place.[7]

Applying our formula (2) with $C^2/R^2=-3.3\times 10^{-14}$ to the earlier epoch when the average density of masses in the universe was of the order of 10^6 g/cm^3 (as required by the conditions for the formation of elements), we find that at that time $l\cong 10^{-2}$ cm, and $dl/dt\cong 0.01$ cm/sec. This means that *at the epoch when the mean density of the universe was of the order of 10^6 g/cm^3, the expansion must have been proceeding at such a high rate, that this high density was reduced by an order of magnitude in only about one second.* It goes without saying that one must be very careful in extrapolating the expansion formula to such an early epoch, but, on the other hand, this formula represents nothing more than the statement of the law of conservation of energy in the inertial expansion against the forces of gravity.

Returning to our problem of the formation of elements, we see that *the conditions necessary for rapid nuclear reactions were existing only for a very short time,* so that it may be quite dangerous to speak about an equilibrium-state which must have been established during this period. It is also interesting to notice that the calculated time-period during which rapid nuclear transformations could have taken place is considerably shorter than the β-decay period of free neutrons which is presumably of the order of magnitude of one hour. Thus if free neutrons were present in large quantities in the beginning of the expansion, the mean density and temperature of expanding matter must have dropped to comparatively low values *before* these neutrons had time to turn into protons. We can anticipate that neutrons forming this comparatively cold cloud were gradually coagulating into larger and larger neutral complexes which later turned into various atomic species by subsequent processes of β-emission. From this point of view the decrease of relative abundance along the natural sequence of elements must be understood as being caused by the longer time which was required for the formation of heavy neutronic complexes by the successive processes of radiative capture. The present high abundance of hydrogen must have resulted from the competition between the β-decay of original neutrons which was turning them into inactive protons, and the coagulation-process through which these neutrons were being incorporated into heavier nuclear units.

It is hoped that the further more detailed development of the ideas presented above will permit us to understand the observed abundance-curve of chemical elements giving at the same time valuable information concerning the early stages of the expanding universe.

[1] v. Weizsäcker, Physik. Zeits. 39, 633 (1938).
[2] Chandrasekhar and Henrich, Astrophys. J. 95, 288 (1942).
[3] G. Wataghin, Phys. Rev. 66, 149 (1944).
[4] Goldschmidt, *Verteilung der Elemente* (Oslo, 1938).
[5] R. Tolman, *Relativity, Thermodynamics and Cosmology* (Oxford Press, New York, 1934).
[6] Hubble, *The Realm of the Nebulae* (Yale University Press, New Haven, 1936).
[7] G. Gamow and E. Teller, Phys. Rev. 55, 654 (1939).

The Origin of Chemical Elements

R. A. ALPHER*
Applied Physics Laboratory, The Johns Hopkins University, Silver Spring, Maryland

AND

H. BETHE
Cornell University, Ithaca, New York

AND

G. GAMOW
The George Washington University, Washington, D. C.

February 18, 1948

As pointed out by one of us,[1] various nuclear species must have originated not as the result of an equilibrium corresponding to a certain temperature and density, but rather as a consequence of a continuous building-up process arrested by a rapid expansion and cooling of the primordial matter. According to this picture, we must imagine the early stage of matter as a highly compressed neutron gas (overheated neutral nuclear fluid) which started decaying into protons and electrons when the gas pressure fell down as the result of universal expansion. The radiative capture of the still remaining neutrons by the newly formed protons must have led first to the formation of deuterium nuclei, and the subsequent neutron captures resulted in the building up of heavier and heavier nuclei. It must be remembered that, due to the comparatively short time allowed for this process,[1] the building up of heavier nuclei must have proceeded just above the upper fringe of the stable elements (short-lived Fermi elements), and the present frequency distribution of various atomic species was attained only somewhat later as the result of adjustment of their electric charges by β-decay.

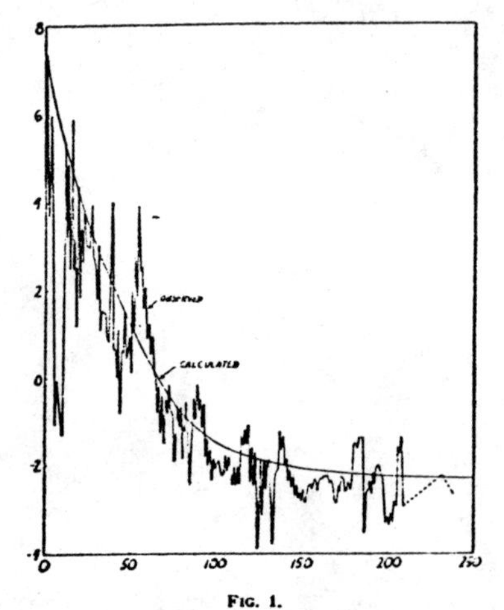

FIG. 1.
Log of relative abundance
Atomic weight

Thus the observed slope of the abundance curve must not be related to the temperature of the original neutron gas, but rather to the time period permitted by the expansion process. Also, the individual abundances of various nuclear species must depend not so much on their intrinsic stabilities (mass defects) as on the values of their neutron capture cross sections. The equations governing such a building-up process apparently can be written in the form:

$$\frac{dn_i}{dt}=f(t)(\sigma_{i-1}n_{i-1}-\sigma_i n_i) \quad i=1,2,\cdots 238, \qquad (1)$$

where n_i and σ_i are the relative numbers and capture cross sections for the nuclei of atomic weight i, and where $f(t)$ is a factor characterizing the decrease of the density with time.

We may remark at first that the building-up process was apparently completed when the temperature of the neutron gas was still rather high, since otherwise the observed abundances would have been strongly affected by the resonances in the region of the slow neutrons. According to Hughes,[2] the neutron capture cross sections of various elements (for neutron energies of about 1 Mev) increase exponentially with atomic number halfway up the periodic system, remaining approximately constant for heavier elements.

Using these cross sections, one finds by integrating Eqs. (1) as shown in Fig. 1 that the relative abundances of various nuclear species decrease rapidly for the lighter elements and remain approximately constant for the elements heavier than silver. In order to fit the calculated curve with the observed abundances[3] it is necessary to assume the integral of $\rho_n dt$ during the building-up period is equal to 5×10^4 g sec./cm^3.

On the other hand, according to the relativistic theory of the expanding universe[4] the density dependence on time is given by $\rho\cong10^6/t^2$. Since the integral of this expression diverges at $t=0$, it is necessary to assume that the building-up process began at a certain time t_0, satisfying the relation:

$$\int_{t_0}^{\infty} (10^6/t^2)dt \cong 5\times 10^4, \tag{2}$$

which gives us $t_0 \cong 20$ sec. and $\rho_0 \cong 2.5\times 10^5$ g sec./cm^3. This result may have two meanings: (a) for the higher densities existing prior to that time the temperature of the neutron gas was so high that no aggregation was taking place, (b) the density of the universe never exceeded the value 2.5×10^5 g sec./cm^3 which can possibly be understood if we use the new type of cosmological solutions involving the angular momentum of the expanding universe (spinning universe).[5]

More detailed studies of Eqs. (1) leading to the observed abundance curve and discussion of further consequences will be published by one of us (R. A. Alpher) in due course.

* A portion of the work described in this paper has been supported by the Bureau of Ordnance U. S. Navy, under Contract NOrd-7386.

[1] G. Gamow, Phys. Rev. 70, 572 (1946).
[2] D. J. Hughes, Phys. Rev. 70, 106(A) (1946).
[3] V. M. Goldschmidt, *Geochemische Verteilungsgesetz der Elemente und der Atom-Arten*. IX. (Oslo, Norway, 1938).
[4] See, for example: R. C. Tolman, *Relativity, Thermodynamics and Cosmology* (Clarendon Press, Oxford, England, 1934).
[5] G. Gamow, Nature, October 19 (1946).

The Origin of Elements and the Separation of Galaxies

G. Gamow
George Washington University, Washington, D. C.
June 21, 1948

THE successful explanation of the main features of the abundance curve of chemical elements by the hypothesis of the "unfinished building-up process,"[1,2] permits us to get certain information concerning the densities and temperatures which must have existed in the universe during the early stages of its expansion. We want to discuss here some interesting cosmogonical conclusions which can be based on these informations.

Since the building-up process must have started with the formation of deuterons from the primordial neutrons and the protons into which some of these neutrons have decayed, we conclude that the temperature at that time must have been of the order $T_0 \cong 10^9$ °K (which corresponds to the dissociation energy of deuterium nuclei), so that the density of radiation $\sigma T^4/c^2$ was of the order of magnitude of water density. If, as we shall show later, this radiation density exceeded the density of matter, the relativistic expression for the expansion of the universe must be written in the form:

$$\frac{d}{dt}\lg l = \left(\frac{8\pi G}{3}\frac{\sigma T^4}{c^2}\right)^{\frac{1}{2}} \tag{1}$$

where l is an arbitrary distance in the expanding space, and the term containing the curvature is neglected because of the high density value. Since for the adiabatic expansion T is inversely proportional to l, we can rewrite (1) in the form:

$$-\frac{d}{dt}\lg T = \frac{T^2}{c}\left(\frac{8\pi G\sigma}{3}\right)^{\frac{1}{2}} \tag{2}$$

or, integrating:

$$T^2 = \left(\frac{3}{32\pi G\sigma}\right)^{\frac{1}{2}}\cdot\frac{c}{t}. \tag{3}$$

For the radiation density we have:

$$\rho_{\text{rad.}} = \frac{3}{32\pi G}\cdot\frac{1}{t^2}. \tag{4}$$

These formulas show that the time t_0, when the temperature dropped low enough to permit the formation of deuterium, was several minutes. Let us assume that at that time the density of matter (protons plus neutrons) was $\rho_{\text{mat.}}{}^\circ$. Since, in contrast to radiation, the matter is conserved in the process of expansion, $\rho_{\text{mat.}}$ was decreasing as $l^{-3} \sim T^3 \sim t^{-\frac{3}{2}}$. The value of $\rho_{\text{mat.}}{}^\circ$ can be estimated from the fact that during the time period Δt of about 10^2 sec. (which is set by the rate of expansion), about one-half of original particles were combined into deuterons and heavier nuclei. Thus we write:

$$v\Delta t n\sigma \cong 1 \tag{5}$$

where $v = 5\cdot 10^8$ cm/sec. is the thermal velocity of neutrons at 10^9 °K, n is the particle density, and $\sigma \cong 10^{-29}$ cm^2 the capture cross section of fast neutrons in hydrogen. This gives us $n \cong 10^{18}$ cm^{-3} and $\rho_{\text{mat.}}{}^\circ \cong 10^{-6}$ g/cm^3 substantiating our previous assumption that matter density was negligibly small compared with the radiation density. (Thus we have $\rho_{\text{mat.}}{}^\circ\cdot\Delta t \cong 10^{-4}$ g$\cdot$cm$^{-3}\cdot$sec. and not 10^{+4} g$\cdot$cm^{-3} sec. as was given incorrectly in the previous paper[2] because of a numerical error in the calculations.)

Since $\rho_{\text{rad.}} \sim t^{-2}$ whereas $\rho_{\text{mat.}} \sim t^{-\frac{3}{2}}$ the difference by a factor of 10^6 which existed at the time 10^9 sec. must have vanished when the age of the universe was $10^2\cdot(10^6)^2 = 10^{14}$ sec. $\cong 10^7$ years. At that time the density of matter and the density of radiation were both equal to $[(10^6)^2]^{-2} = 10^{-24}$ g/cm^3. The temperature at that epoch must have been of the order $10^9/10^6 \cong 10^3$ °K.

The epoch when the radiation density fell below the density of matter has an important cosmogonical significance since it is only at that time that the Jeans principle of "gravitational instability"[3] could begin to work. In fact, we would expect that as soon as the matter took over the principal role, the previously homogeneous gaseous substance began to show the tendency of breaking up into separate clouds which were later pulled apart by the progressive expansion of the space. The density of these individual gas clouds must have been approximately the same as the density of the universe at the moment of separation, i.e., 10^{-24} g/cm^3. The size of the clouds was determined by the condition that the gravitational potential on their surface was equal to the kinetic energy of the gas particles. Thus we have:

$$\frac{3}{2}kT = \frac{4}{3}\pi R^3\rho\frac{Gm_H}{R} = \frac{4\pi Gm_H\rho}{3}R^2. \tag{6}$$

With $T \cong 10^3$ and $\rho \cong 10^{-24}$ this gives $R \cong 10^{21}$cm $\cong 10^3$ light years.

The fact that the above-calculated density and radii correspond closely to the observed values for the stellar galaxies strongly suggests that we have here a correct picture of galactic formation. According to this picture the galaxies were formed when the universe was 10^7 years old, and were originally entirely gaseous. This may explain their regular shapes, resembling those of the rotating gaseous bodies, which must have been retained even after all their diffused material was used up in the process of star formation (as, for example, in the elliptic galaxies which consist entirely of stars belonging to the population II).

It may also be remarked that the calculated temperature corresponding to the formation of individual galaxies from the previously uniform mixture of matter and radiation, is close to the condensation points of many chemical elements. Thus we must conclude that some time before or soon after the formation of gaseous galaxies their material separated into the gaseous and the condensed (dusty) phase. The dust particles, being originally uniformly distributed through the entire cloud, were later collected into smaller condensations by the radiation pressure in the sense of the Spitzer-Whipple theory of star formation.[5] In fact, although there were no stars yet, there was still plenty of high intensity radiation which remained from the original stage of expanding universe when the radiation, and not the matter, ruled the things.

In conclusion I must express my gratitude to my astronomical friends, Dr. W. Baade, Dr. E. Hubble, Dr. R. Minkowski, and Dr. M. Schwartzschield for the stimulating discussion of the above topics.

[1] G. Gamow, Phys. Rev. **70**, 572 (1946).
[2] R. Alpher, H. Bethe, and G. Gamow, Phys. Rev. **73**, 803 (1948).
[3] J. H. Jeans, *Astronomy and Cosmogony* (Cambridge University Press, Teddington 1928).
[4] W. Baade, Astrophys. J. **100**, 137 (1944).
[5] L. Spitzer, Jr., Astrophys. J. **95**, 329 (1942); F. L. Whipple, Astrophys. J. **104**, 1 (1946).

680 NATURE October 30, 1948 Vol. 162

THE EVOLUTION OF THE UNIVERSE

By Dr. G. GAMOW
George Washington University, Washington, D.C.

THE discovery of the red shift in the spectra of distant stellar galaxies revealed the important fact that our universe is in the state of uniform expansion, and raised an interesting question as to whether the present features of the universe could be understood as the result of its evolutionary development, which must have started a few thousand million years ago from a homogeneous state of extremely high density and temperature. We conclude first of all that the relative abundances of various atomic species (which were found to be essentially the same all over the observed region of the universe) must represent the most ancient archæological document pertaining to the history of the universe. These abundances must have been established during the earliest stages of expansion when the temperature of the primordial matter was still sufficiently high to permit nuclear transformations to run through the entire range of chemical elements. It is also interesting to notice that the observed relative amounts of natural radioactive elements suggest that their nuclei must have been formed (presumably along with all other stable nuclei) rather soon after the beginning of the universal expansion. In fact, we notice that natural radioactive isotopes with the decay periods of many thousand million years (such as uranium-238, thorium-232 and samarium-148) are comparatively abundant, whereas those with decay periods measuring only several hundred million years are extremely rare (as uranium-235 and potassium-40). If, using the known decay periods and natural abundances of these isotopes, we try to calculate the date when they have been about as abundant as the corresponding isotopes of longer life, we find that it must have been a few thousand million years ago, in general agreement with the astronomically determined age of the universe.

The early attempts to explain the observed relative abundances of the elements[1,2] were based on the assumption that the present distribution represents a 'frozen equilibrium state' corresponding to some very high temperature and density in an early stage of universal expansion. Such equilibrium theories lead, however, to the result that the logarithm of the relative abundance must be a linear function of the nuclear binding energy, which in its turn is known to be a linear function of atomic weight. Thus, according to that picture, we would expect a rapid exponential decrease of relative abundances all the way from hydrogen to uranium, in direct contradiction to the observed distribution (circles in Fig. 1), which shows a rapid decrease for the first half of the natural sequence of elements, but levels up almost to a constant value in the second half.

As the result of this difficulty, I suggested[3] that the observed abundances do not correspond to any equilibrium state at all, but, quite on the contrary, represent a dynamical building-up process which was arrested in a certain stage of its development by a rapid expansion of the universe.' According to this point of view, one should imagine the original state of matter as a very dense over-heated neutron gas which could have originated (if one lets one's imagina-

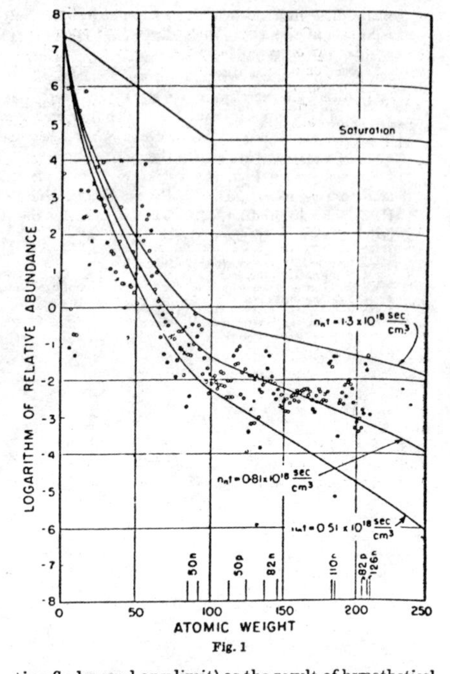

Fig. 1

tion fly beyond any limit) as the result of hypothetical universal collapse preceding the present expansion. In fact, the extremely high pressures obtaining near the point of complete collapse (singular point at $t = 0$) would have squeezed the free electrons into the protons, turning the matter into the state of over-heated neutron fluid. When the expansion began, and the density of neutron gas dropped, the neutrons would be expected to begin decaying again into protons, and more and more complex nuclear aggregates could be built up as the result of the union between the newly formed protons and the neutrons still remaining. Such a building-up process must have started when the temperature of the neutron–proton mixture dropped below a few times 10^{10} °K., which corresponds to the mutual binding energies of these nuclear particles. The equations governing such a gradual building-up process can evidently be written in the form:

$$\frac{dn_i}{dt} = \lambda_{i-1}\, n_{i-1} - \lambda_i\, n_i \; (i = 1, 2, 3, \ldots), \quad (1)$$

where n_i is the number of atomic nuclei of atomic weight i, and λ's are the coefficients depending on the collision frequency, and the capture cross-sections for fast neutrons in the nuclei of various atomic weight.

The numerical study of these equations was carried out by R. Alpher[4,5], who used the recent experimental data on the capture cross-sections of fast neutrons. These cross-sections are known to increase very rapidly by a factor of several hundred for the first half of the atomic weights, and to remain more or less constant for the second half; a fact which is

No. 4122 October 30, 1948 NATURE 681

of paramount importance for understanding the general shape of the abundance curve. Suppose that the building-up process goes at a constant density, ρ, for a limited period of time, Δt; the resulting abundances will depend apparently only on the product $\rho\Delta t$, which determines the total number of collisions between the developing nucleus and the free neutrons.

The results of the computations carried out by R. Alpher are shown by continuous curves in Fig. 1, with the values of $n_n \Delta t = \frac{\rho}{m}\Delta t$ marked on each curve. It is seen that for very large values of $n_n \Delta t$, the process reaches saturation, and the number of the various nuclei becomes inversely proportional to the corresponding capture cross-sections. For smaller values of $n_n \Delta t$, heavier nuclei do not have time to build up to the saturation point, and we see from the figure that for $\rho\,\Delta t \cong 1{\cdot}3 \times 10^{-6}$ gm. cm.$^{-3}$ sec. ($n_n\Delta t \cong 0{\cdot}8 \times 10^{18}$ sec. cm.$^{-3}$) the calculated curve stands in very good agreement with the observational data.

The agreement obtained is amplified by the fact that the observed abundances show the abnormally high values for the isotopes containing the completed shells of neutrons or protons (as indicated along the atomic weight axis in Fig. 1); in fact, it is known that such nuclei possess abnormally small capture cross-sections, which would cause the accumulation of the material at these particular atomic weights. Since the building-up process must have been accomplished within a time period comparable with the decay period of neutrons, we have $\Delta t \cong 30$ min. $\cong 2 \times 10^{3}$ sec., from which it follows that during that period the density of matter must have been of the order of magnitude 10^{-6} gm. cm.$^{-3}$. On the other hand, since the temperature must have been of the order of 10^{9} °K., the mass-density of radiation, aT^4/c^2, was comparable with the density of water. Thus we come to the important conclusion that, at that time, the expansion of the universe was governed entirely by radiation and not by matter.

In this case the relativistic formula for the expansion can be written in the form[6],

$$dl/dt = \sqrt{\frac{8\pi G}{3}\,\frac{aT^4}{c^2}\,l^2}, \tag{2}$$

where l is an arbitrary distance in the expanding space, and the constant term containing the radius of curvature is neglected because of the high value of the density. Remembering that for the adiabatic expansion of the radiation $T \sim 1/l$, we can integrate (2) into the form:

$$T = \sqrt[4]{\frac{3c^2}{32\pi Ga}}\cdot\frac{1}{t^{1/2}} = \frac{2{\cdot}14 \times 10^{10}}{t^{1/2}}\ \text{°K.} \tag{3}$$

For the mass-density of radiation we have:

$$\rho_{\text{rad.}} = \frac{3}{32\pi G}\cdot\frac{1}{t^2} = \frac{4{\cdot}5 \times 10^{5}}{t^2}\ \text{gm.cm.}^{-3}. \tag{4}$$

For the density of matter we must evidently write:

$$\rho_{\text{mat.}} = \frac{\rho_0}{t^{3/2}}\ \text{gm.cm.}^{-3}, \tag{5}$$

where ρ_0 is to be determined from the conditions of the nuclear building-up process. It can be done in the simplest way by considering the building up of deuterons by proton–neutron collisions. Writing X for the concentration of neutrons (with $X(0) = 1$), and Y for the concentration of protons (with $Y(0)=0$), we obtain the equations:

$$\frac{dX}{dt} = -\lambda X - \frac{XY}{m}\rho v\sigma,$$
$$\frac{dY}{dt} = +\lambda X - \frac{XY}{m}\rho v\sigma, \tag{6}$$

where v is the thermal velocity, and σ the capture cross-section which (for the energies in question) can be sufficiently accurately represented by the formula[7]:

$$\sigma = \frac{2^{3/2}\pi e^2\hbar\varepsilon^{5/4}}{m^3c^5}\cdot\frac{1}{E^{3/2}}, \tag{7}$$

$\varepsilon = 2{\cdot}19$ MeV. being the binding energy of the deuteron. Expressing v and E through the temperature, and using (3), we can rewrite (6) as:

$$\frac{dX}{d\tau} = -X - \frac{\alpha XY}{\tau},$$
$$\frac{dY}{d\tau} = +X - \frac{\alpha XY}{\tau}, \tag{8}$$

where $\tau = \lambda t$ and:

$$\alpha = \frac{2^{13/2}\pi^{5/4}e^2\hbar\varepsilon^{5/2}G^{1/4}a^{1/4}}{3^{5/2}m^{9/2}c^{11/2}k}\cdot\rho_0. \tag{9}$$

In order that the equations (8) should yield $Y \cong 0{\cdot}5$ for $\tau \to \infty$ (since hydrogen is known to form about 50 per cent of all matter), the coefficient α must be set equal to $0{\cdot}5$. The change of X and Y with time in this case is shown in Fig. 2, which also indicates the corresponding variation of temperature. Assuming $\alpha = 0{\cdot}5$, we find from equation (9) that $\rho_0 = 0{\cdot}72 \times 10^{-3}$, which fixes the dependence of material density on the age of the universe.

Once we have $\rho_{\text{rad.}}$ and $\rho_{\text{mat.}}$ as functions of time, we can follow the physical processes taking place during the further expansion of the universe, and in particular calculate the masses and sizes of the con-

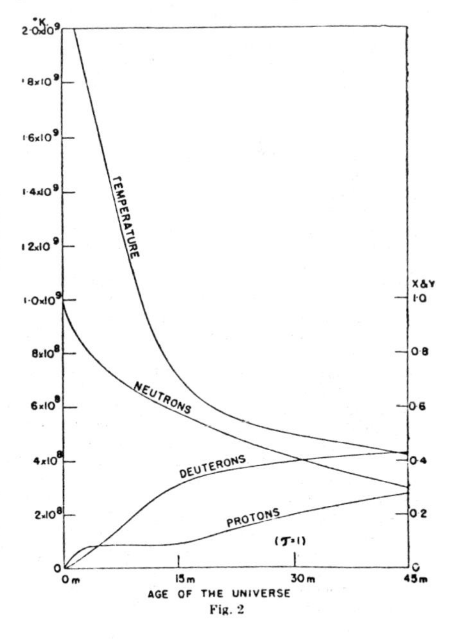

Fig. 2

densations of that primordial gas which must have originated sooner or later according to Jeans' principle of gravitational instability[8]. Jeans' classical formula gives the diameter D of the condensations which will be formed in a gas of temperature T and density ρ in the form:

$$D^2 = \frac{10\pi}{9mG\rho} \cdot \frac{3}{2} kT. \qquad (10)$$

Using the expressions (3) and (5), we get:

$$M = \rho D^3 = \frac{2^{31/8} 5^{7/4} \pi^{5/4} e \hbar^{5/4} \varepsilon^{5/4}}{3^{17/8} m^{15/4} c^{5/4} G^{7/4}} \qquad (11)$$

(where a has been expressed through other fundamental constants).

It is interesting to notice that the time-factor cancels out in the calculation of M, so that the mass of the condensations comes out the same, independent of the epoch when they were formed. It seems, however, reasonable to assume that the effect of gravitational instability became important only when the mass-density of radiation became comparable with the density of matter, since it is hard to imagine a 'gravitational condensation of pure radiation'. Using (4) and (5), we find that $\rho_{\text{rad.}} = \rho_{\text{mat.}} = 3 \times 10^{-26}$ gm.cm.$^{-3}$ at $t = 3{\cdot}9 \times 10^{15}$ sec. $= 1{\cdot}3 \times 10^{8}$ years, at which point $T = 340°$ K. For this value of t we obtain:

$$D = \frac{2^{45/8} 5^{1/4} \pi^{7/4} e^3 \hbar^{3/4} \varepsilon^{15/4}}{3^{27/8} m^{29/4} c^{35/4} G^{5/4}}. \qquad (12)$$

Substituting numerical values, we have:

$$\begin{aligned} M &= 5{\cdot}5 \times 10^{40} \text{ gm.} = 2{\cdot}7 \times 10^{7} \text{ sun-masses} \\ D &= 1{\cdot}3 \times 10^{22} \text{ cm.} = 13{,}000 \text{ light-years,} \end{aligned} \qquad (13)$$

which must represent the masses and the diameters of the original galaxies.

The above estimate of galactic masses falls short by a factor of about one hundred from the mass-values of galaxies obtained from astronomical data. But it must be remembered that the simple Jeans' formula used in these calculations does not take into account the effect of radiation pressure, and also is applicable only to the gravitational condensations in non-expanding space. The effect of additional radiation pressure (which is quite important according to the previous considerations) and the tearing force of expansion will lead to considerably larger condensation masses. The detailed study of this question will require, however, the extension of Jeans' classical arguments for the case of a mixture of gas and radiation in the expanding space. At the present stage one should be satisfied with the fact that, by such comparatively simple and rather natural considerations, *masses and sizes comparable to those of stellar galaxies can be expressed in terms of fundamental constants, and the basic quantities of nuclear physics.*

We may add that, according to the above picture, the galaxies have been originally formed in the purely gaseous form (including a certain amount of solid dust particles), which must account for the regular shapes of rotating bodies. The formation of individual stars within the galactic bodies must have taken place at a somewhat later stage, probably along the lines of the Spitzer–Whipple theories[9,10]. When stars were formed by the condensation process within the rotating gaseous mass, their tangential velocities, being equal to the original velocities of the gas-masses, were clearly not high enough to maintain them on circular Kepler orbits, so that *the newly formed stars must have been moving along elongated elliptical orbits with the points of maximum elongation in the places of their origin.* This situation must have remained essentially unchanged even when all the material of originally gaseous galaxies was used up in the formation of stars.

These considerations give a simple explanation of the otherwise mysterious fact that the elliptical galaxies and the central bodies of spirals rotate 'as solid bodies' with the linear velocities proportional to the distance from the axis. In fact, according to our picture, the maximum Doppler displacements observed at various distances from the axis correspond to the velocities of stars passing through 'aphelion' at these particular distances, and, according to the previous argument, are equal to the velocities which the gas-masses must have had in these regions prior to their condensation into the stars.

[1] v. Weizsäcker, C., *Phys. Z.*, 39, 633 (1938).

[2] Chandrasekhar, S., and Henrich, L. R., *Astrophys. J.*, 95, 288 (1942).

[3] Gamow, G., *Phys. Rev.*, 70, 572 (1946).

[4] Alpher, R. A., Bethe, H. A., and Gamow, G., *Phys. Rev.*, 73, 803 (1948).

[5] Alpher, R. A., *Phys. Rev.* (in the press).

[6] Tolman, R. C., "Relativity, Thermodynamics and Cosmology" (Clarendon Press, Oxford, 1934).

[7] Bethe, H. A., "Elementary Nuclear Physics" (John Wiley and Sons, 1947).

[8] Jeans, J., "Astronomy and Cosmogony" (Cambridge University Press, 1928).

[9] Spitzer, jun., L., *Astrophys. J.*, 95, 329 (1942).

[10] Whipple, F., *Astrophys. J.*, 104, 1 (1946).

Progress of Theoretical Physics, Vol. 5, No. 2, March~April, 1950.

Proton-Neutron Concentration Ratio in the Expanding Universe at the Stages preceding the Formation of the Elements.

Chushiro HAYASHI.

Department of Physics, Naniwa University.

(Received January 12, 1950)

§ 1. Introduction.

In the theory of the origin of the elements by Gamow, Alpher, and colaborators[1], primordial matter (ylem) of the universe, which afterwards has been cooled down owing to the expansion of the universe and has formed the elements through nuclear reactions such as radiative capture and beta-decays, is assumed to consist solely of neutrons. At early stages, however, of high temperatures ($kT \gtrsim mc^2$, m being the electron mass) in the expanding universe before the formation of the elements, induced beta-processes caused by energetic electrons, positrons, neutrinos and antineutrinos, in addition to the natural decay of neutrons, such as

$$n+e^+ \rightleftarrows p+a\nu, \tag{1a}$$

$$n+\nu \rightleftarrows p+e^-, \tag{1b}$$

$$n \rightleftarrows p+e^-+a\nu, \tag{1c}$$

must have proceeded, their rates being faster at higher temperatures, and had a effect on the proton-neutron concentration ratio. At still higher temperatures $kT \gtrsim \mu c^2$ (μ is the mesons' mass), where large number of mesons are expected to be in existence, n-p conversion process induced by mesons would have been much more rapid owing to their stronger interactions with nucleons than the processes induced by light particles. Consequently, the n-p ratio must have been determined by the rates of such processes and those of changes in temperature and density in the universe resulting from its expansion.

We shall be based on the relativistic theory of the expanding universe, which are shown by Gamow as having a possibility to explain the origins both of the elements and the galactic nebulae.[1),5)] Then, the expansion and contraction rates of the universe at the stages of high compression are given by[2)]

$$\frac{1}{l}\frac{dl}{dt} = \pm\left(\frac{8\pi}{3}G\rho\right)^{\frac{1}{2}}, \tag{2}$$

where l is an arbitrary proper length of a volume containing a given amount of matter and ρ the total mass density. We restrict ourselves in the case where radiational mass density is greater than material one,

$$\rho_r = aT^4/c^2 > \rho_m, \tag{3}$$

which is satisfied if $\rho_m < 1\,\text{g/cm}^3$ at $T = 10^9\,°\text{K}$, including the case of Gamow and Alpher where the temperature and material density at the beginning of the elements formation are given by $T \simeq 10^9\,°\text{K}$ and $\rho_m = 10^{-5} \sim 10^{-8}\,\text{g/cm}^3$. In this case, conservations of energy and matter give for the variations of temperature and material density, for $kT < Mc^2$ (M being the nucleon mass),

$$T \sim 1/l, \quad \text{and} \quad \rho_m \sim 1/l^3. \tag{4}$$

We shall study the changes in the concentrations of proton, neutron, electron, positron, photon, neutrino, and antineutrino, which are all assumed to be free, in conditions of varying temperature and density which are determined macroscopically by Eqs. (2) and (4). It must be noticed that, when we consider a volume containing a given mass of matter in the universe which is isotropic and homogeneous, numbers of neutrinos and antineutrinos which are missed through the boundaries are just gained from the surroundings, and then we must take into account their effects on the beta-processes and on the statistics of all the particles in the case of equilibrium.

In § 2, rates of reactions between those particles at constant temperature and density are studied and it is shown that, except for the beta-processes, they are very fast compared with the rates of the temperature change of the universe. Accordingly, these particles, for instance electron pairs, can be treated as in thermodynamical equilibrium with radiations. In § 3, rates of beta-processes are calculated, and in § 4 we obtain equations which determine the concentrations of neutron, proton, neutrino, and antineutrino, taking into account the contraction or expansion of the universe. These equations are reduced to a simple form due to a circumstance that concentrations of neutrinos are much larger than that of nucleons in the condition (3), in § 5. At temperatures higher than about $2 \times 10^{10}\,°\text{K}$ thermodynamical equilibrium is being continually established, but at lower temperatures induced beta-processes are largely reduced owing to lack of energetic light particles and there occurs a time lag in the *n-p* ratio departing from its equilibrium value, or, in other words, *n-p* ratio becomes "frozen" when the expansion of the universe becomes faster than the beta-processes. The numerical results show that, for the universe satisfying the condition (3), *n-p* ratio at the beginning of the elements formation is nearly 1 : 4 almost irrespective of its initial values as long as initial temperatures are so high ($T \gtrsim 2 \times 10^{10}\,°\text{K}$) that equilibrium has once been attained. Further remarks are discussed in § 6.

§ 2. Rates of Processes at High Temperatures.

Reactions between nucleons, electrons, photons, and neutrinos proceed in the direction to approach the statistical equilibrium. For the sake of simplicity these particles are all assumed to be free. We compare rates of processes (Coulomb-, Compton-, and nuclear scatterings, pair creations, etc.) at constant temperature and density with that of the temperature change in the universe which is given by Eq. (2). The latter is conveniently characterized by the time during which the temperature changes by its order of magnitude. Putting in Eq. (2) $\rho=aT^4/c^2$, we have

$$\tau_T=T/\left|\frac{dT}{dt}\right|=\left(\frac{8\pi}{3}\frac{Ga}{c^2}\right)^{-\frac{1}{2}}T^{-2}. \tag{5}$$

In the following we consider the case where the temperature is not so high as to allow of the existence of mesons but is higher than that at the time of the elements formation, i. e., $10^{12}>T>5\times10^{8}$°K. An average life of any one particle for a single collision or reaction with other particles having concentration n_i is given by $1/v\sigma n_i$, where v is the relative velocity corresponding to the temperature and σ is the cross section. Such lives for nuclear scattering between nucleons, for Coulomb scattering between charged particles, and for electromagnetic processes such as Compton scattering and pair creation between photons and electron pairs are found to be much shorter than τ_T in the above temperature range and in the universe in which condition (3) is satisfied as long as ρ_m is greater than 10^{-10}g/cm^3 at $T=10^{9}$°K. Consequently, we can consider that these particles and photons obey the Maxwell's and Planck's distribution laws at temperature T, respectively, and are in thermodynamical equilibrium with each other, apart from the concentrations of neutron and proton and the energy distributions and concentrations of neutrino and antineutrino, which are controlled by beta-processes having much smaller cross sections.

As an important example the rate of reactions between photons and electron pairs is shown below. At such high temperatures electron pairs are found to be created exclusively by two quanta collisions

$$h\nu+h\nu' \rightarrow e^{+}+e^{-}.$$

The number of pairs created per cm^3 per sec is given by

$$\frac{dn_{pair}}{dt}=\frac{1}{2c}\iint d\Omega d\Omega'\iint d\nu d\nu'\frac{B(\nu)}{h\nu}\frac{B(\nu')}{h\nu}\sin^2\frac{\phi}{2}\,\sigma\left(h\nu^{\frac{1}{2}}\nu'^{\frac{1}{2}}\sin\frac{\phi}{2}\right),$$

$$B(\nu)=2h\nu^3c^{-2}(e^{h\nu/kT}-1)^{-1},\quad \sigma(h\nu)=2\pi\left(\frac{e^2}{mc^2}\right)^2(2\theta C^{-2}+2\theta C^{-4}$$

$$-\theta C^{-6}-SC^{-3}-SC^{-5}),\quad C=\frac{h\nu}{mc^2}=\cosh\theta,\quad S=\sinh\theta.$$

where Ω is a solid angle, ϕ is the angle between the directions of propagation of the two quanta, and σ is the cross section obtained by Breit and Wheeler.[3] After elementary calculations we have

$$\frac{dn_{pair}}{dt} \cong \pi c\left(\frac{e^2}{mc^2}\right)^2\left(\frac{kT}{hc}\right)^6 \times \begin{cases} 8\pi^3\left(\frac{mc^2}{kT}\right)^3 e^{-2mc^2/kT}, & (kT \ll mc^2), \\ \frac{8}{9}\pi^6\left(\frac{mc^2}{kT}\right)^2\left(\log\frac{2kT}{mc^2}-1\right), & (kT \gg mc^2). \end{cases} \tag{6}$$

The order of magnitude of the time required to attain equilibrium is given by $n_{pair,eq} / \frac{dn_{pair}}{dt}$, where $n_{pair,eq}$ is the number of pairs in equilibrium which is given by Eqs. (7), (8) and (9) below, and this is found to be much smaller than τ_T of the universe. Thus, electron pairs are continually in equilibrium with radiations, following the change in temperature without any time lag.

Numbers of electrons and positrons per cm^3 are given by the theory of statistics as

$$n_{e\mp} = \frac{8\pi}{h^3}\int_0^\infty e^{\pm\lambda - E/kT} p^2 dp = 8\pi\left(\frac{kT}{hc}\right)^3 x^2 K_2(x) e^{\pm\lambda}, \tag{7}$$

where

$$K_n(x) = \frac{\pi i}{2} e^{\frac{1}{2}n\pi i} H_n^{(1)}(ix) = \begin{cases} \left(\frac{\pi}{2x}\right)^{\frac{1}{2}} e^{-x}\left\{1+\frac{4n^2-1}{8x}+\cdots\right\}, & (x \gg 1), \\ \frac{1}{2}\left\{\frac{(n-1)!}{(x/2)^n} - \frac{(n-2)!}{(x/2)^{n-2}} + \cdots\right\}, & (x \ll 1), \end{cases}$$

$$x = mc^2/kT, \tag{8}$$

and λ is determined by the condition that the total charges are zero, i.e., $n_{e^-} - n_{e^+} = n_p$, n_p being the number of protons per cm^3,

$$\lambda = \sinh^{-1}\left\{n_p / 16\pi\left(\frac{kT}{hc}\right)^3 x^2 K_2(x)\right\}. \tag{9}$$

In our case λ is found to be very small compared with unity, confirming that degeneracies of electrons and positrons can be neglected.

§ 3. Rates of Beta-processes.

According to the Fermi's theory of beta-decay, processes, which change neutrons to protons and *vice versa* and are energetically possible, are given by (1a), (1b), and (1c). In the following, energy distributions of neutrinos and antineutrinos are assumed to obey the Maxwell's law, apart from their concentrations, for the following reasons: at temperatures higher than 10^{12}°K they will be in thermodynamical equilibrium through rapid processes involving the mesons; in view of their motions in the expanding universe given by the theory of relativity,[2] they obey the Maxwell's law continually if they do initially and their numbers do not change in course of time; and in our case numbers of neutrinos and antineutrinos can be considered as much larger than that of nucleons, as shown later. Then, their energy distributions are expressed as ,

$$n^{\nu}_{a\nu}(E)dE=\frac{n^{\nu}_{a\nu}}{2(kT)^3}e^{-E/kT}E^2dE\,,\qquad(10)$$

where n_ν and $n_{a\nu}$ are their numbers per cm³, and their rest mass is assumed to be zero.

Numbers of reactions per cm³ per sec in the leftward and rightward processes in (1a); (1b); and (1c) can be written as $An_pn_{a\nu}$, $A'n_nn_{e^+}$; $Bn_pn_{e^-}$, $B'n_nn_\nu$; and $Cn_pn_{e^-}n_{a\nu}$, $C'n_n$, respectively. In particular C' is the decay constant of the neutron. $An_pn_{a\nu}$ can be calculated in the following way.[4] The probability that an antineutrino of energy E is captured by protons per sec. is given by

$$w_{a\nu}(E)=(g^2n_p/2\pi\hbar^4c^3)cp_+E_+,\quad E_+=(m^2c^4+c^2p_+^2)^{\frac{1}{2}}=E-Q,\qquad(11a)$$

$$Q=M_nc^2-M_pc^2\,,\qquad(12)$$

where g is the Fermi's interaction constant and E_+ is the energy of the emitted positron. Then, we have

$$An_pn_{a\nu}=\int_{Q+mc^2}^{\infty}w_{a\nu}(E)n_{a\nu}(E)dE\,,\qquad(13a)$$

where $n_{a\nu}(E)$ is given by Eq. (10). For $A'n_nn_{e^+}$ we have in the same way,

$$w_{e^+}(E)=(g^2n_n/2\pi\hbar^4c^3)E_{a\nu}^2,\quad E_{a\nu}=E+Q\,,\qquad(11a')$$

$$A'n_nn_{e^+}=\int_{mc^2}^{\infty}w_{e^+}(E)n_{e^+}(E)dE,\qquad(13a')$$

where $n_{e^+}(E)$ is given by Eq. (7). In the same way we obtain B and B',

$$w_{e^-}(E)=(g^2n_p/2\pi\hbar^4c^3)E_\nu^2,\quad E_\nu=E-Q\,,\qquad(11b)$$

$$Bn_pn_{e^-}=\int_{Q}^{\infty}w_{e^-}(E)n_{e^-}(E)dE\,;\qquad(13b)$$

$$w_\nu(E)=(g^2n_n/2\pi\hbar^4c^3)cp_-E_-,\quad E_-=E+Q\,,\qquad(11b')$$

$$B'n_nn_\nu=\int_{0}^{\infty}w_\nu(E)n_\nu(E)dE\,.\qquad(13b')$$

To calculate C and C' we consider antineutrinos as holes in the negative energy states of neutrino,

$$n+\nu^*\rightleftarrows p+e^-,$$

where asterisk denotes negative energy states. The probability that an electron of energy E is captured by protons per sec is given by

$$w_{e^-}(E)=(g^2n_p/2\pi\hbar^4c^3)E_{\nu*}^2\times\frac{n_{a\nu}}{2(kT)^3}e^{-E_{a\nu}/kT}E_{a\nu}^2\Big/\frac{8\pi}{h^3c^3}E_{a\nu}^2\,,\qquad(11c)$$

$$E_{\nu*}=-E_{a\nu}=E-Q<0\,,$$

where the last factor in $w_{e^-}(E)$ means the probability that the state with negative energy $E_{\nu *}$ is vacant, i. e., anti-neutrino of positive energy $E_{a\nu}$ is present. Then, we have

$$Cn_p n_e - n_{a\nu} = \int_{mc^2}^{Q} w_{e^-}(E) n_{e^-}(E) dE . \tag{13c}$$

C' for the reverse process can be obtained from C by the principle of detailed balancing, but we have directly

$$w_{\nu *}(E^*) = (g^2 n_n / 2\pi \hbar^4 c^3) c p_- E_-,$$
$$E_- = Q + E^* = Q - E_{a\nu}, \tag{11c'}$$

$$C' n_n = \int_{mc^2 - Q}^{0} w_{\nu *}(E^*) (8\pi / h^3 c^3) \times E^{*2} dE^*, \tag{13c'}$$

where $(8\pi/h^3c^3)E^{*2}$ is the density of the negative energy states E^*.

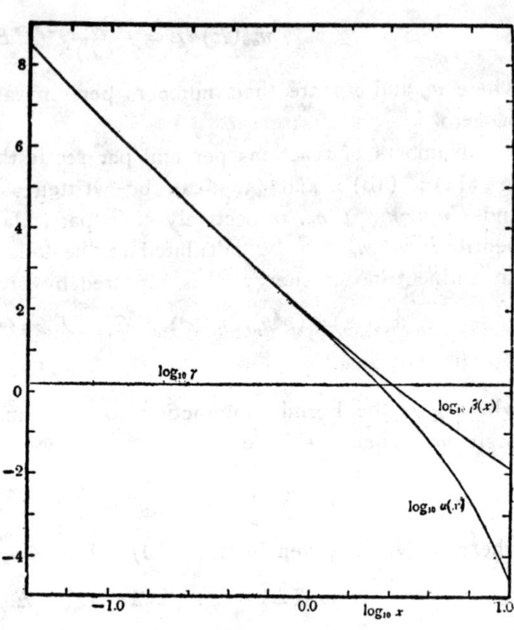

Fig. 1.

Accordingly, we obtain A, A', etc. as functions of temperature only,

$$A = \frac{1}{\tau_0} \frac{1}{8\pi} \left(\frac{hc}{kT}\right)^3 \frac{1}{2} e^{-qx} \alpha(x), \qquad A' = \frac{1}{\tau_0} \frac{1}{8\pi} \left(\frac{hc}{kT}\right)^3 \frac{1}{x^2 K_2(x)} \alpha(x) ; \tag{14a}$$

$$B = \frac{1}{\tau_0} \frac{1}{8\pi} \left(\frac{hc}{kT}\right)^3 \frac{1}{x^2 K_2(x)} e^{-qx} \beta(x), \quad B' = \frac{1}{\tau_0} \frac{1}{8\pi} \left(\frac{hc}{kT}\right)^3 \frac{1}{2} \beta(x) ; \tag{14b}$$

$$C = \frac{1}{\tau_0} \frac{1}{8\pi} \left(\frac{hc}{kT}\right)^3 \frac{1}{x^2 K_2(x)} \frac{1}{8\pi} \left(\frac{hc}{kT}\right)^3 \frac{1}{2} e^{-qx} \gamma, \quad C' = \frac{1}{\tau_0} \gamma ; \tag{14c}$$

where

$$q = Q/mc^2, \ \tau_0 = 2\pi^3 \hbar^7 / g^2 m^5 c^4, \tag{15}$$

$$\alpha(x) = \int_1^\infty e^{-xy} (y+q)^2 (y^2-1)^{\frac{1}{2}} y dy = \frac{3K_3(x)}{x^2} + \frac{K_2(x)}{x} + 2q \left\{ \frac{3K_2(x)}{x^2} + \frac{K_1(x)}{x} \right\} + q^2 \frac{K_2(x)}{x}$$

$$= \begin{cases} 24x^{-5}(1 + \frac{1}{2}qx + \frac{1}{12}q^2x^2) - x^{-3}(1 + qx + \frac{1}{2}q^2x^2) + \cdots, & (x \ll 1) \\ (\pi/2)^{\frac{1}{2}} (q+1)^2 x^{-\frac{3}{2}} e^{-x} + \cdots, & (x \gg 1) ; \end{cases} \tag{16a}$$

$$\beta(x) = \int_0^\infty e^{-xy} \{ (y+q)^2 - 1 \}^{\frac{1}{2}} (y+q) y^2 dy = \begin{cases} 24x^{-5}(1 + \frac{1}{2}qx + \frac{1}{12}q^2x^2) - x^{-3} + \cdots, & (x \ll 1) \\ 2q(q^2-1)^{\frac{1}{2}} x^{-3} + \cdots, & (x \gg 1) \cdot \end{cases} \tag{16b}$$

$$\gamma=\int_1^q (q-y)^2(y^2-1)^{\frac{1}{2}} y dy. \tag{16c}$$

When we put $q=2.5$, γ becomes 1.51. Values of $\alpha(x)$ and $\beta(x)$ which are calculated for $q=2.5$ are shown in Fig. 1.

§ 4. Reactions in the Expanding Universe.

Changes in the concentrations of neutron, proton, neutrino, and antineutrino in the contracting or expanding universe are given by the following equations. Here, we consider the two cases:

(i) neutrinos and antineutrinos are physically distinguishable, say by their magnetic moments,

(ii) they are identical, say the form of the Hamiltonian for the beta-decay is symmetrical in neutrino and antineutrino.

In the case (i) these equations are

$$\frac{d}{dt}\left(\frac{n_p}{\rho_m}\right)=\frac{1}{\rho_m}\left(-An_pn_{a\nu}+A'n_nn_{e^+}-Bn_pn_{e^-}+B'n_nn_\nu-Cn_pn_{e^-}n_{a\nu}+C'n_n\right), \tag{17.1}$$

$$\frac{d}{dt}\left(\frac{n_\nu}{\rho_m}\right)=\frac{1}{\rho_m}\left(Bn_pn_{e^-}-B'n_nn_\nu\right),$$

with two conditions

$$(n_n+n_p)/\rho_m=1/M, \text{ and } (n_p+n_\nu-n_{a\nu})/\rho_m=\text{constant}, \tag{18.1}$$

where the relation between t and T is given by Eqs. (2) and (4).

In the case (ii) they are

$$\frac{d}{dt}\left(\frac{1}{\rho_m}\begin{Bmatrix}n_p\\n_\nu\end{Bmatrix}\right)=\frac{1}{\rho_m}\left(-An_pn_\nu+A'n_nn_{e^+}\mp Bn_pn_{e^-}\pm B'n_nn_\nu-Cn_pn_{e^-}n_{a\nu}+C'n_n\right), \tag{17.2}$$

where upper and lower signs correspond to equations for n_p and n_ν, respectively, with a condition that

$$(n_n+n_p)/\rho_m=1/M. \tag{18.2}$$

Introducing non-dimensional variables

$$y=\frac{n_p-n_n}{n_p+n_n}, \quad \text{and} \quad z_{a\nu}^{\nu}=n_{a\nu}^{\nu}/16\pi\left(\frac{kT}{hc}\right)^3, \tag{19}$$

and using Eq. (7), above equations become, in the case (i)

$$\tau_0\frac{dy}{dt}=\alpha(x)\{(1-y)e^{-\lambda}-(1+y)z_{a\nu}e^{-qx}\}+\beta(x)\{(1-y)z_\nu-(1+y)e^{\lambda-qx}\}$$
$$+\gamma\{(1-y)-(1+y)z_{a\nu}e^{\lambda-qx}\}, \tag{20.1}$$

$$R\tau_0 \frac{dz_\nu}{dt} = -\beta(x)\{(1-y)z_\nu - (1+y)e^{\lambda - qx}\},$$

$$z_{a\nu} - z_\nu = (y+1)/R + z_c\,, \tag{21.1}$$

where z_c is a constant which is determined by initial conditions, and in the case (ii)

$$\tau_0 \frac{dy}{dt} = \alpha(x)\{(1-y)e^{-\lambda} - (1+y)ze^{-qx}\} + \beta(x)\{(1-y)z - (1+y)e^{\lambda - qx}\}$$

$$+\gamma\{(1-y) - (1+y)ze^{\lambda - qx}\}, \tag{20.2}$$

$$R\tau_0 \frac{dz}{dt} = \alpha(x)\{\cdots\ \} - \beta(x)\{\cdots\ \} + \gamma\{\cdots\ \},$$

with

$$R = \frac{2M}{\rho_m} 16\pi \left(\frac{kT}{hc}\right)^3 = \frac{15}{4\pi^4} \frac{\rho_r}{\rho_m} \frac{Mc^2}{kT}\,, \tag{22}$$

which measures the ratio of concentrations of photon (or, as shown later, neutrino and antineutrino) and nucleon, and is a very large constant for the universe satisfying the condition (3). It is shown in the next section that these somewhat complicated equations can be simplified substantially to a linear form.

§ 5. Solutions for the N-P Ratio.

First the equilibrium states are considered. If the change of temperature in the universe is much slower than the beta-processes considered above, energy distributions and concentrations of all the particles vary in accordance with their equilibrium values corresponding to a temperature and density at each stage. In this case detailed balancings are always established in all the processes, and in particular curly brackets on the righthand sides of Eqs. (20.1) or (20.2) must all vanish simultaneously. Then, we have in the case (i)

$$n_n/n_p \equiv (1-y)/(1+y) = e^{\lambda - Q/kT}/z_\nu,$$

$$z_\nu \cdot z_{a\nu} = 1, \quad \text{and} \quad z_{a\nu} - z_\nu = (1+y)/R + z_c\,. \tag{23.1}$$

In view of the facts that in our case $0 \leq (1+y)/R \leq 2/R \ll 1$ and that, in the hole theory of the neutrino, $n_{a\nu}$ is the number of holes in the negative energy states, it will be natural to assume that $z_c \approx 0$ and to take $z_{a\nu} \approx z_\nu$ since $z_{a\nu}$ and z_ν can not be both smaller than unity. In such a case Eqs. (23.1) reduce to the following ones which are obtained in the case (ii):

$$n_n/n_p \equiv (1-y)/(1+y) = e^{\lambda - Q/kT}, \quad \text{and} \quad z = 1. \tag{23.2}$$

In solving Eqs. (20.1) or (20.2) which are general as yet, we put $\lambda = 0$ since λ is smaller than 2×10^{-4} at $T = 10^{9}$°K even if ρ_m is as large as 1 g/cm^3 and much smaller at higher temperatures, as shown by Eq. (9), and further we take $z_\nu = z_{a\nu}(=z)$ in the case (i) since they do not change appreciably after the equilibrium state has once been attained, as shown later. From these equations and Fig. 1 we see that times during which y and z change by their orders of magnitude at given high temperatures ($x<1$) are given by

$$\tau_y = \tau_0/u(x), \quad \text{and} \quad \tau_z = R\tau_0/u(x), \tag{24}$$

respectively. At temperatures where they are both smaller than τ_T given by Eq. (5), equilibrium states, which are considered above, will be attained immediately even if there are initially appreciable deviations from them. Such temperatures are given by

$$1/x \equiv kT/mc^2 \geq 2 \times R^{\frac{1}{3}}. \tag{25}$$

Further, at temperatures higher than 10^{12}°K, z will attain its equilibrium value much faster through processes involving mesons which are in existence in much larger numbers (almost the same numbers as of photons) than nucleons.

It is shown, as follows, that we can put $z=1$ (equilibrium value) throughout in solving Eqs. (20.1) or (20.2). Expanding y and z in powers of $1/R$,

$$y = y_0 + y_1/R + y_2/R^2 + \cdots, \quad \text{and} \quad z = z_0 + z_1/R + z_2/R^2 + \cdots, \tag{26}$$

and equating the terms with same powers of $1/R$ in Eqs. (20.1) or (20.2), we obtain first

$$dz_0/dt = 0. \tag{27}$$

Putting $z_0 = 1$ further, we have

$$\tau_0 dy_0/dt = -\{y_0 - \tanh(qx/2)\}(1+e^{-qx})(\alpha+\beta+\gamma), \tag{28}$$

$$\tau_0 dz_1/dt = \{y_0 - \tanh(qx/2)\}(1+e^{-qx}) \times \begin{cases} \beta, & \text{in the case (i)}, \quad (29.1) \\ (\beta-\alpha-\gamma), & \text{in the case (ii)}, \quad (29.2) \end{cases}$$

$$\tau_0 dy_1/dt = -y_1(\alpha+\beta+\gamma)(1+e^{-qx}) + z_1\{\beta(1-y_0) - (\alpha+\gamma)(1+y_0)e^{-qx}\}, \text{ etc.} \tag{30}$$

From these equations one can see that above expansions are good approximations in our case where R is a very large constant. This corresponds to a circumstance that numbers of neutrinos and antineutrinos, which are equal to that of photons in orders of magnitude in the case $z=1$, are far greater than that of nucleons in the universe satisfying the condition (3). Accordingly, the n-p ratio is almost completely determined by Eq. (28) and there are no essential differences between the cases (i) and (ii).

In order to obtain an accurate relation between t and x, we must include in the total mass density those contributions from neutrinos and electron pairs in addition to that from photons. They are given by, from Eqs. (10) and (7), as

$$\rho_{a\nu}^{\nu} = \frac{1}{c^2}\int \frac{n_{a\nu}^{\nu}}{2(kT)^3} e^{-E/kT} E^3 dE = \frac{48\pi}{c^2} \frac{(kT)^4}{(hc)^3} z_{a\nu}^{\nu}, \tag{31}$$

$$\rho_e^{\mp}=\frac{8\pi}{(hc)^3}\int e^{\pm\lambda-E/kT}Ep^2dp=\frac{48\pi^4}{c^2}\frac{(kT)^4}{(hc)^3}e^{\pm\lambda}J(x), \tag{32}$$

where

$$J(x)=x^3\left(\frac{K_3(x)}{8}+\frac{K_1(x)}{24}\right)=\begin{cases}1-\frac{1}{12}x^2+\cdots & ,\ (x\ll 1),\\ \left(\frac{\pi}{2}\right)^{\frac{1}{2}}x^{\frac{5}{2}}e^{-x}\left(\frac{1}{6}+\frac{9}{16}\frac{1}{x}+\cdots\right), & (x\gg 1).\end{cases} \tag{33}$$

Since n-p ratio is mainly determined in the region $x\ll 1$, we put in further calculations $J(x)=1$, and $\lambda=0$ as before. Then, the total mass density is expressed as

$$\rho=(1+r)aT^4/c^2, \tag{34}$$

$$r=\begin{cases}4\times 90/\pi^4 & \text{in the case (i)}, \quad (35.1)\\ 3\times 90/\pi^4 & \text{in the case (ii)}. \quad (35.2)\end{cases}$$

From Eqs. (2) and (4) we obtain

$$x^2=\frac{|t|}{t_m},\ t_m=\frac{1}{2}\left(\frac{k}{mc^2}\right)^2\left\{\frac{8\pi}{3}\frac{Ga}{c^2}(1+r)\right\}^{-\frac{1}{2}}=6.6(1+\gamma)^{-\frac{1}{2}}\text{sec.} \tag{36}$$

Then, Eq. (28) becomes, omitting a suffix in y,

$$\frac{dy}{dx}=\pm\frac{2t_m}{\tau_0}x(1+e^{-qx})(\alpha(x)+\beta(x)+\gamma)(\tanh\frac{qx}{2}-y),\ \text{for } t\lesseqgtr 0. \tag{37}$$

Using a function with an arbitrary constant x_0,

$$f(x)=(2t_m/\tau_0)\int_{x_0}^{x}(1+e^{-qx})(\alpha(x)+\beta(x)+\gamma)x\,dx, \tag{38}$$

general solutions of Eq. (37) are written as follows:

in the case $t<0$, i. e., the universe being contracting,

$$y(x)=y_-(x)+Ae^{f(x)},$$

$$y_-(x)=e^{f(x)}\int_x^{\infty}e^{-f(x)}\frac{df(x)}{dx}\tanh\frac{qx}{2}dx, \tag{39}$$

in the case $t>0$, i. e., the universe being expanding,

$$y(x)=y_+(x)+Be^{-f(x)},$$

$$y_-(x)=e^{-f(x)}\int_0^{x}e^{f(x)}\frac{df(x)}{dx}\tanh\frac{qx}{2}dx, \tag{40}$$

where A and B are arbitrary constants, and y_- and y_+ are the particular solutions which are unity at infinity, tend to $\tanh(qx/2)$ as x approaches to zero, and are finite everywhere.

Numerical calculations are carried out by taking $q=2.5$ and $\tau_0/\gamma=30$ min., and using the value of r given by Eq. (35.2), there being no essential difference between using Eqs. (35.1) and (3.52) in view of the uncertainty in the value of τ_0/γ (decay life of neutron). The results are shown in Fig. 2 together with the equilibrium value $y_{eq}=\tanh(qx/2)$. In the region $x \lesssim 0.1$, general solutions tend to y_- and y_+ rapidly, with a consequence that n-p ratio at lower temperatures in the case $t>0$ are almost entirely independent of its initial values. With decreasing temperature the rate of induced beta-processes, i. e., $\alpha(x)+\beta(x)$, is largely reduced owing to decreases in numbers of energetic electrons and neutrinos, and there occur departures in the n-p ratio from its equilibrium value ("freezing" sets in). The value of x, where these deviations become appreciable, is roughly estimated by putting $\tau_y=\tau_T$, where τ_y and τ_T are given by Eqs. (24) and (5), and is found to be nearly 0.5. In the region $x>3$, γ becomes larger than $\alpha(x)+\beta(x)$ and natural decay of neutrons is the main process.

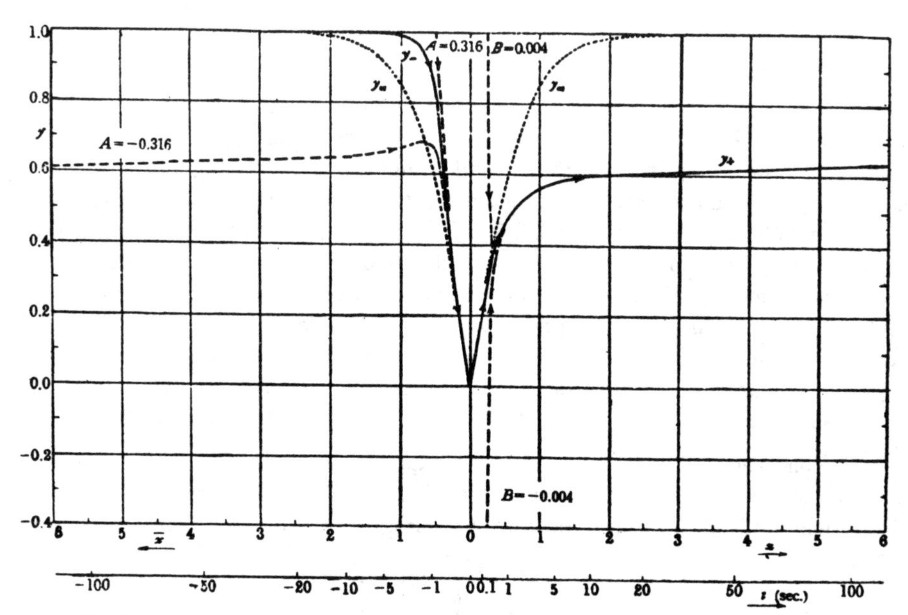

Fig. 2. n-p ratios as functions of $x \equiv mc^2/kT$ and time t of the universe. Dotted line is equilibrium value: $y_{eq}=\tanh(qx/2)$, and broken curves show $y_-+Ae^{f(x)}$ and $y_++Be^{-f(x)}$ where x_0 is taken as unity in Eq. (38).

§ 6. Concluding Remarks.

In the universe in which $\rho_r>\rho_m$ and at high temperatures where elements are not allowed to be in existence, the n-p ratio follows the curves shown in Fig. 2. It can be seen in particular that, if the existing laws of physics, microscopic and macroscopic, are valid at least up to temperatures $\sim 2\times 10^{10}$°K, the n-p

ratio at the beginning of the elements formation, i. e., at $x \gtrsim 1$, is nearly 1 : 4, whatever the physical conditions at higher temperatures, especially at the epoch $t=0$ when the universe is singular according to the current theory, may be.

It is known that at present hydrogen and helium together form about 97 percent of all matter. If we assume that formation of nuclei heavier than He^4 can be neglected, and that reactions involving beta-processes such as $n \to p+e^-$, $p+p \to H^2 + e^+$, and $H^3 \to He^3 + e^-$, which are much slower than other nuclear transmutations such as gamma-ray or particle emission unless material density is extremely low, are not effective during the formation process, He^4 is built up from original neutrons and protons, after all, as $2n+2p \to He^4$, whatever the routes of formation may be, for instance $n+p \to H^2$, $H^2+H^2 \to H^3+p$, $H^3+H^2 \to He^4+n$, or $n+p \to H^2$, $H^2+n \to H^3$, $H^3+p \to He^4$. Consequently, the hydrogen-helium abundance ratio (in number) resulting from the initial n-p ratio 1 : 4 becomes 6 : 1, whereas recent observed values in stellar atmospheres and meteorites range from 5 : 1 to 10 : 1.

Under the original assumption of Gamow that "ylem" consists solely of neutrons, it is difficult to explain the fact that the building-up processes of the elements jump over the "crevasses" of unstable mass numbers 5 and 8, as shown by Fermi and Turkevitch.[5] However, the existence of an appreciable amount of original protons may relieve this situation owing to increasing probabilities of capture by light nuclei of proton, deuteron, triton, and helium nuclei, since the formation process is expected to begin at higher temperatures. Such a situation, together with the above conclusion on the hydrogen and helium abundances, seems to encourage further calculations on the formation of light elements, at least up to C^{12}. In this connection, to treat the building-up processes at initial stages, especially formation of deuterons, more accurately, it will be necessary to take into account the effects of their reverse processes, for instance $A_z + n \leftarrow (A+1)_z + h\nu$, caused by radiations having high concentrations at high temperatures.

In conclusion, the author wishes to express his cordial thanks to Prof. H. Yukawa for his continual interests and advices, to Prof. G. Gamow for his kind suggestions and advices, and to Prof. Z. Shirogane for his continual encouragements.

References.

1) See the references in the paper: R. A. Alpher and R. C. Herman, Phys. Rev. **75** (1949), 1089.
2) R. C. Tolman, "*Relativity, Thermodynamics and Cosmology*" (1934).
3) G. Breit and J. A. Wheeler, Phys. Rev. **46** (1934), 1087.
4) G. Gamow and M. Schönberg, Phys. Rev. **59** (1941), 537.
5) G. Gamow, Rev. Mod. Phys. **21** (1949), 267.

Reprinted from THE PHYSICAL REVIEW, Vol. 92, No. 6, 1347–1361, December 15, 1953
Printed in U. S. A.

Physical Conditions in the Initial Stages of the Expanding Universe*,†

RALPH A. ALPHER, JAMES W. FOLLIN, JR., AND ROBERT C. HERMAN
Applied Physics Laboratory, The Johns Hopkins University, Silver Spring, Maryland

(Received September 10, 1953)

The detailed nature of the general nonstatic homogeneous isotropic cosmological model as derived from general relativity is discussed for early epochs in the case of a medium consisting of elementary particles and radiation which can undergo interconversion. The question of the validity of the description afforded by this model for the very early super-hot state is discussed. The present model with matter-radiation interconversion exhibits behavior different from non-interconverting models, principally because of the successive freezing-in or annihilation of various constituent particles as the temperature in the expanding universe decreased with time. The numerical results are unique in that they involve no disposable parameters which would affect the time dependence of pressure, temperature, and density.

The study of the elementary particle reactions leads to the time dependence of the proton-neutron concentration ratio, a quantity required in problems of nucleogenesis. This ratio is found to lie in the range $\sim$4.5:1–$\sim$6.0:1 at the onset of nucleogenesis. These results differ from those of Hayashi mainly as a consequence of the use of a cosmological model with matter-radiation interconversion and of relativistic quantum statistics, as well as a different value of the neutron half-life.

I. INTRODUCTION

THE nonstatic homogeneous isotropic cosmological model which satisfies the equations of general relativity has received a great deal of attention. However, the detailed nature of the model does not appear to have been examined at the extremely high temperatures and densities characteristic of the very early stages of the expanding universe. This question has been examined in the present paper and the dependence of the temperature and density on time has been determined for the case where the radiation density (taken to include photons, neutrinos, electrons, positrons, and mesons) is much greater than the density of matter (nucleons). For initial conditions compatible with present astrophysical observations, one can demonstrate that the radiation density exceeded the density of matter for about the first hundred million years in the expansion.

We have carried our study of this problem back to a temperature of $\sim$100 Mev ($\sim 1.2\times10^{12}$°K), corresponding to an epoch of $\sim 10^{-4}$ sec. For temperatures below this value one can treat reactions among elementary particles with some confidence. Furthermore, below $\sim$100 Mev the energy stored in the gravitational field is a negligible part of the total energy so that the question of using a correct unified field theory, including the quantization of the field equations, can be avoided. Finally, at $\sim$100 Mev one has a state of thermodynamic equilibrium among all the known constituent particles and radiation so that a knowledge of the previous history of the universe is not required. As part of the detailed study of the cosmological model we have examined the reactions among the elementary particles present and followed their course in the universal expansion. As will be seen, all reaction rates, except those involving the neutrino, are sufficiently high to maintain thermodynamic equilibrium. An examination of the kinetics of the reactions between nucleons and neutrinos has yielded the relative concentrations of protons and neutrons as a function of time. The only parameters involved in the cosmological model are the nucleon density and radius of curvature. At the early times prior to element formation, neither of these parameters affects the course of events because of the very high total density and because the nucleon density is neglible compared with the density of radiation. The nucleon density becomes of importance at later times in considering element formation, while the radius of curvature becomes of interest only at times of the order of a hundred million years.

The foregoing detailed considerations of the early stages of the universal expansion bear significantly on the neutron-capture theory of element formation. This theory has been concerned principally with understanding the general trend in the distribution of the cosmic abundances of the chemical elements with atomic weight.[1,2]

* This work was supported by the U. S. Bureau of Ordnance, Department of the Navy, under Contract NOrd-7386.

† Preliminary accounts of this work were presented at the Symposium on the Abundance of the Elements held at Yerkes Observatory, Williams Bay, Wisconsin, November 6–8, 1952, under the joint sponsorship of the National Science Foundation and the University of Chicago, and at the 1953 Washington Meeting of the American Physical Society, Phys. Rev. 91, 479 (1953).

[1] R. A. Alpher and R. C. Herman, Revs. Modern Phys. **22**, 153 (1950); Phys. Rev. **84**, 60 (1951).

[2] R. A. Alpher and R. C. Herman, *Annual Review of Nuclear Science* (Annual Review of Nuclear Science, Inc., Stanford, 1953), vol. 2, p. 1. The simple-neutron capture theory has satisfactorily reproduced for all but the lightest elements the observed approximately exponential decrease in abundance with increasing atomic weight up to $A\sim100$, as well as the approximate constancy of abundance for the heavier elements. Briefly, in the neutron-capture theory, as thus far developed, the various nuclear species were supposed to have been formed from nucleons by the successive radiative capture of fast neutrons with adjustment of nuclear charge by intervening β decay during the early stages of the expansion of the universe. The primordial material or ylem was taken to be a mixture of neutrons and radiation. As the universal expansion proceeded the neutrons underwent free decay, so that by the time the universal temperature had decreased to a value

For the lightest elements the processes of neutron capture and β decay, while adequate to explain the formation of the heavier elements, must be supplemented by thermonuclear reactions involving protons, deuterons, and other light nuclei. The very light element reactions were examined in some detail by Fermi and Turkevich,[1] using the cosmological model previously employed for the neutron-capture theory, with a finite starting time and a primordial mixture of neutrons and radiation. This improved light-element calculation did not satisfactorily resolve what remains the principle difficulty of the theory, namely, the deduction of the specific nuclear reactions and physical conditions which might carry the formation chain of reactions through and beyond atomic weight 5. To resolve this and other difficulties in the theory, it will apparently be necessary to remove many of the simplifying restrictions. In particular, the assumption of a starting time must be replaced by detailed consideration of element-building reactions increasing in importance from very early times in the universal expansion as the rates of various dissociative processes diminish with decreasing temperature. Moreover, one should include all possible reactions among the elementary particles, since these reactions, which are important at very high temperatures, may influence the physical conditions that control the element-building processes.

The elementary-particle reactions determine the ratio of the relative concentrations of protons and neutrons, a quantity which plays a vital role in predicting the general trend of abundances according to the neutron-capture theory. As has already been mentioned, in previous calculations the proton-neutron abundance ratio has been taken to be that resulting from free decay of the primordial neutrons during the period from the start of the expansion up to the starting time selected for element-building reactions. A more detailed calculation was made by Hayashi,[3] who determined the value of the proton-neutron ratio resulting from spontaneous and induced β processes among protons and neutrons in the presence of electron pairs and neutrinos in the early stages of the expansion. Whereas on the basis of the crude assumption of neutron decay only, one obtains a proton-neutron ratio of ~1:7, Hayashi's calculation gave ~4:1 by the starting time for element-building reactions. With this latter value of the ratio, it has not yet proven possible to represent the cosmic abundance distribution in atomic weight on the basis of the simple neutron-capture theory,[2] a theory which contains only one arbitrary parameter, *viz.*, the density of matter at the start of the element-building epoch, and which involves only neutron-capture reactions. In part because of this difficulty and because it seemed worth while to investigate the effect of certain modifications on Hayashi's calculation of the final value of the proton-neutron ratio, the work described in the remainder of this paper was carried out. Among the changes involved in the present study are the use of relativistic quantum statistics instead of Boltzmann statistics, a modified cosmological model for early epochs as required by the interconversion of matter and radiation, which as we have already indicated is of considerable interest for its own sake, and the use of the value of the neutron half-life recently reported by Robson[4] which differs materially from the older value employed by Hayashi.

It seems most likely that element synthesis is intimately connected with questions of cosmology. In the present work we consider the sequence of events up to the time when the rate of element formation became significant. As we shall see later in detail, all the constituents remained in thermodynamic equilibrium as the universe expanded and cooled to a temperature of ~10 Mev. At ~10 Mev the neutrinos were essentially frozen out of the equilibrium. By ~0.3 Mev the proton-neutron ratio was almost entirely determined by the free decay of the neutron. It remains for future study to re-examine the formation of the elements by thermonuclear reactions as a subsequent part of the picture developed here. A detailed chronology is given in a later portion of this paper [see Sec. V].

II. THE COSMOLOGICAL MODEL

The theory of element formation by non-equilibrium thermonuclear reactions has been developed as an integral part of the very early stages of the expanding universe. Detailed calculations of the necessary rate processes require a knowledge of the temporal behavior of temperature, density, and rate of expansion during these early epochs. The cosmological model that has been used previously for this purpose is the most general nonstatic model satisfying the requirements of general relativity, exhibiting homogeneity and isotropy, and which is composed of a perfect fluid with no interconversion of matter and radiation.[1,2] The rate of expansion and, implicitly, the rate of change of temperature in the expansion for this model, with no restrictions on the composition of the perfect working fluid, are given in relativistic units by the following differ-

where nuclei would be thermally stable, an appreciable number of protons had been generated. Then the capture of neutrons by protons provided the first step in the formation of the successively heavier elements. More specifically, the temperature for the beginning of building-up reactions was taken to be ~0.1 Mev (corresponding to a specific starting time for element building in the cosmological model used, in which $T=1.52\times10^{10}t^{-\frac{1}{2}}$°K). This choice was dictated by the magnitude of the binding energy of the deuteron on the one hand and by the lack of evidence in the abundance data for any resonance neutron capture on the other hand. At the starting time, neutron decay had led to a proton-neutron ratio of ~1:7.

One of the approximations involved thus far in calculations with the neutron-capture theory has been the smoothing of available data on fast neutron radiative capture cross sections as a function of atomic weight. Moreover, reactions other than radiative neutron capture among the very lightest elements have been ignored.

[3] C. Hayashi, Progr. Theoret. Phys. (Japan) **5**, 224 (1950).

[4] J. M. Robson, Phys. Rev. **83**, 349 (1951).

ential equations:[5]

$$-\frac{e^{-g}}{R_0^2}-\frac{d^2g}{dt^2}-\frac{3}{4}\left(\frac{dg}{dt}\right)^2+\Lambda=8\pi p_0, \quad (1a)$$

$$\frac{3e^{-g}}{R_0^2}+\frac{3}{4}\left(\frac{dg}{dt}\right)^2-\Lambda=8\pi\rho_{00}, \quad (1b)$$

and

$$e^{\frac{1}{2}g(t)}=l/l_0=R/R_0, \quad (1c)$$

where l and R are proper distance and radius of curvature, respectively, given in units of l_0 and R_0, Λ is the cosmological constant, and p_0 and ρ_{00} are proper pressure and density. The quantities p_0 and ρ_{00} are functions of temperature and of l, and hence implicitly of time. Equation (1b) may also be rewritten, by using Eq. (1c), in the following form:

$$\frac{dl}{dt}=+\left(\frac{8\pi}{3}\rho_{00}l^2-\frac{l_0^2}{R_0^2}+\frac{\Lambda l^2}{3}\right)^{\frac{1}{2}}, \quad (2)$$

with the plus sign taken to indicate expansion. We have taken $\Lambda=0$ in keeping with current practice.[6] As can be easily shown, the constant term l_0^2/R_0^2 in Eq. (2) may be neglected in the application of this model to early epochs. This is equivalent to neglecting l^2e^{-g}/R_0^2 in Eq. (1). If $\rho_{00}\propto l^{-n}$ where $n>2$, then for sufficiently early times l will be so small that one has $8\pi\rho_{00}l^2/3 \propto 8\pi l^{2-n}/3\gg l_0^2/R_0^2$. Hence, for early epochs in the expansion one may replace Eq. (2) by

$$\frac{dl}{dt}=\frac{l}{2}\frac{dg}{dt}=+\left(\frac{8\pi}{3}\rho_{00}l^2\right)^{\frac{1}{2}}. \quad (3)$$

As has already been mentioned, the cosmological model, which is discussed in this paper, taken together with the presently observed smoothed-out matter density in the universe as well as the estimated age, are consistent with the supposition that during the early epochs of interest the matter density was much smaller than the radiation density (i.e., $\sim 1:10^6$). The neutron-capture theory of element formation[1] requires that the radiation density greatly exceeded the density of matter during the early epochs of the universal expansion. In this previous work it was not necessary to consider the interconversion of matter and radiation since for the epochs considered the temperature was already below that required to maintain a significant electron-pair density. Hence, the working fluid for the cosmological model was taken as black-body radiation, containing a trace of matter, and expanding adiabatically according to $T\propto 1/l$. It has been shown[1] that for early epochs Eq. (3) leads to the following expressions for the radiation density, ρ_γ, the matter density, ρ_m, the total density, ρ_{total}, the temperature, T, and proper distance, l, with $\rho_\gamma\gg\rho_m$:

$$\rho_{\text{total}}\cong\rho_\gamma\cong[3/(32\pi G)]t^{-2}=4.48\times10^5t^{-2}\ \text{g/cm}^3, \quad (4)$$

$$\rho_m=\rho_0t^{-\frac{3}{2}}\ \text{g/cm}^3, \quad (5)$$

$$T=(c^2\rho_\gamma/a_\gamma)^{\frac{1}{4}}=1.52\times10^{10}t^{-\frac{1}{2}}\,^\circ\text{K}, \quad (6)$$

and

$$l=(32\pi G\rho_{\gamma''}l_0^4/3)^{\frac{1}{4}}t^{\frac{1}{2}}, \quad (7)$$

where G is the gravitational constant, c is the velocity of light, a_γ is the Stefan-Boltzmann constant, t is the time in seconds from the "start" of the expansion, $\rho_{\gamma''}$ is the density of radiation when $l=l_0$, and ρ_0 is a constant. As will be seen, the above equations are still valid in the case discussed in the present paper providing that $t\gtrsim 100$ sec, ρ_γ is eliminated from Eq. (4), the constant 1.52 in Eq. (6) becomes 1.45 for Majorana neutrinos and 1.38 for Dirac neutrinos, and $\rho_{\gamma''}$ in Eq. (7) is replaced with the value of ρ_{total} when $l=l_0$. The quantity ρ_0 is the one arbitrary parameter in the simple neutron-capture theory. It has been adjusted in previous calculations[2] so that the density of matter at the start of the element-forming processes would lead to the observed cosmic abundance distribution. The cosmological model at early epochs described by Eqs. (4) and (6) was adopted by Hayashi as a basis for his calculation of the proton-neutron ratio.

While we have assumed a homogeneous and isotropic model of the universe in agreement with present astronomical evidence, it should be pointed out that this restriction is not necessary in the present considerations. Homogeneity is required only over a region of radius equal to ct since nothing further away can affect the cosmology or the elementary particle reactions to be discussed. At the universal age of ~ 600 seconds corresponding to the end of the period of this study, the nuclear mass enclosed in the sphere of influence is $\sim 10^{34}$ g, that is, ~ 5 solar masses, and is much less at earlier epochs. Another way of looking at this result is that lengths greater than ct, in particular R_0 and any gradient of R_0, must be negligible because of the finite velocity of propagation of disturbances.

[5] R. C. Tolman, *Relativity, Thermodynamics and Cosmology* (Clarendon Press, Oxford, 1934).

[6] A very small value of Λ may be used to adjust the present age of this model, although it is of no consequence during the early epochs of interest here [see G. Gamow, Revs. Modern Phys. 21, 367 (1949)]. In this connection it may be of interest to note that while Eqs. (1) and (2) contain a density singularity at zero time, they also implicitly contain the conclusion that the duration or age of the expansion from this singularity is finite. Taking $\Lambda=0$ and neglecting terms containing $1/R_0^2$ for early epochs, one can show that this age is given in cgs units by the following integral:

$$\mathfrak{a}=\int_0^{\mathfrak{a}}dt=\left(\frac{3}{8\pi G}\right)^{\frac{1}{2}}\int_{\rho(\mathfrak{a})}^{\infty}\frac{c^2d\rho}{\rho^{\frac{1}{2}}(p+\rho c^2)},$$

where p and ρ are total pressure and density. Since the pressure is positive,

$$\mathfrak{a}\leqq\left(\frac{3}{8\pi G}\right)^{\frac{1}{2}}\int_{\rho(\mathfrak{a})}^{\infty}\frac{d\rho}{\rho^{\frac{3}{2}}}=\left(\frac{3}{2\pi G\rho}\right)^{\frac{1}{2}}.$$

A lower bound on the duration may be obtained by noting that for a relativistic fluid $0\leqq p\leqq\rho c^2/3$. Hence

$$\mathfrak{a}\geqq\left(\frac{3}{8\pi G}\right)^{\frac{1}{2}}\int_{\rho(\mathfrak{a})}^{\infty}\frac{3d\rho}{4\rho^{\frac{3}{2}}}=\left(\frac{8}{3\pi G\rho}\right)^{\frac{1}{2}},$$

so that $\mathfrak{a}$ is bounded.

As already mentioned, the cosmological model outlined above was a sufficient approximation in previous calculations of the neutron-capture theory[1] because the temperature taken for the start of element formation was well below the electron rest mass equivalent and, therefore, reactions among elementary particles and photons could be ignored. One has only to consider that the nucleons and nuclei formed remained in thermal equilibrium with the expanding radiation field. These previous calculations, which were based on the time scale described by Eq. (6), continue to be valid provided ρ_0 is adjusted as required to fit the time scale to be described in this paper. The adjustment required is insignificant.

The study of the induced and inverse β processes involving neutrons and protons prior to any appreciable element formation concerns much earlier epochs and therefore much higher temperatures. In this case one must consider positrons, electrons, neutrinos, antineutrinos (if distinguishable from neutrinos), and radiation. The equation of state for radiation only, implicit in Eqs. (4)–(6), may no longer be an adequate approximation. We shall suppose that this mixture of elementary particles and photons is at a sufficiently high temperature for equilibrium to be maintained, but we shall not require temperatures so high as to require nucleon pairs. Furthermore, the nucleons present are assumed to have a negligible effect on pressure, density, and temperature, since even for temperatures as low as ~0.1 Mev the nucleon density is many orders of magnitude less than the radiation density.

The density and pressure of the constituents of the medium may be obtained from the Fermi-Dirac and Bose-Einstein distribution laws for the number of particles in the energy range dE at E, *viz.*,

$$N(E)dE=\frac{4\pi}{h^3}\Sigma'|\mathbf{p}|E[\exp(E/kT)\pm1]^{-1}dE, \quad (8)$$

where $|\mathbf{p}|$ is the momentum, and the summation, $\sum'$, is over charge and spin states. In the present calculation the number of particles and photons is not conserved so that a degeneracy parameter is not required. The density and pressure, according to Eq. (8), are given by

$$\rho(T)=\frac{4\pi}{c^2h^3}\Sigma'\int_{mc^2}^{\infty}|\mathbf{p}|E^2[\exp(E/kT)\pm1]^{-1}dE, \quad (9a)$$

and

$$p(T)=\frac{4\pi}{3h^3}\Sigma'\int_{mc^2}^{\infty}|\mathbf{p}|^3[\exp(E/kT)\pm1]^{-1}dE, \quad (9b)$$

where

$$|\mathbf{p}|=(1/c)(E^2-m^2c^4)^{\frac{1}{2}}. \quad (9c)$$

In particular, for electrons and positrons one obtains,

for T → ∞ x → 0

with the transformation $E=m_ec^2\cosh\theta$, the following:

$$\rho_{e^-}=\rho_{e^+}=\frac{a_e}{c^2}\int_0^\infty\frac{\sinh^2\theta\cosh^2\theta d\theta}{1+\exp(x\cosh\theta)}\ \text{g/cm}^3, \quad (10a)$$

$$p_{e^-}=p_{e^+}=\frac{a_e}{3}\int_0^\infty\frac{\sinh^4\theta d\theta}{1+\exp(x\cosh\theta)}\ \text{dynes/cm}^2. \quad (10b)$$

In these equations

$$a_e=8\pi m_e^4c^5/h^3; \quad (11a)$$

m_e and h are the electron rest mass and Planck's constant, respectively;

$$x=m_ec^2/(kT) \quad (11b)$$

defines temperature in units of the electron rest mass, and k is Boltzmann's constant. Spin states have been counted in the above expressions, and the total electron energy includes rest mass.

In order to carry out numerical calculations, it should be noted that the definite integrals in the expressions for $p_e(x)$ and $\rho_e(x)$ can be expanded in series of modified Bessel functions $K_i(nx)$. One can write

$$f_0=\int_0^\infty[1+\exp(x\cosh\theta)]^{-1}d\theta=\sum_{n=1}^{\infty}(-1)^{n+1}K_0(nx), \quad (12)$$

$$f_1=\int_0^\infty\sinh^2\theta[1+\exp(x\cosh\theta)]^{-1}d\theta$$

$$=x^{-1}\sum_{n=1}^{\infty}(-1)^{n+1}n^{-1}K_1(nx), \quad (13)$$

and

$$f_2=\int_0^\infty\sinh^4\theta[1+\exp(x\cosh\theta)]^{-1}d\theta$$

$$=3x^{-2}\sum_{n=1}^{\infty}(-1)^{n+1}n^{-2}K_2(nx), \quad (14)$$

so that

$$\rho_e(x)=\rho_{e^-}+\rho_{e^+}=(2a_e/c^2)(f_1+f_2), \quad (15)$$

and

$$p_e(x)=p_{e^-}+p_{e^+}=(2a_e/3)f_2. \quad (16)$$

In the high temperature limit, $kT\gg mc^2$, which is equivalent to setting $m=0$ in Eqs. (9), the density and pressure for all Bose-Einstein particles approach those for photons except for factors which depend on spin and charge states. Similarly, for Fermi-Dirac particles the density and pressure approach those for neutrinos, again except for a factor which accounts for differing charge and spin states.

For radiation, taking into account the two states of polarization, one obtains the following from the Bose-Einstein integral:

$$\rho_\gamma=\frac{a_\gamma}{c^2}T^4=\left(\frac{\pi^4a_e}{15c^2}\right)x^{-4}\ \text{g/cm}^3, \quad (17)$$

and

$$p_\gamma = \tfrac{1}{3}\rho_\gamma c^2 \text{ dynes/cm}^2, \tag{18}$$

where

$$a_\gamma = 8\pi^5 k^4/15c^3h^3. \tag{18a}$$

For neutrinos we consider two cases,[7] namely, neutrinos and antineutrinos indistinguishable ($\nu \equiv \nu^*$) and distinguishable ($\nu \not\equiv \nu^*$). For the temperature range in which the neutrinos are in thermal equilibrium with the other constitutents of the medium, the Fermi-Dirac integral gives

for $\nu \equiv \nu^*$:

$$\rho_\nu = \tfrac{7}{8}\rho_\gamma, \tag{19}$$

and

$$p_\nu = \tfrac{7}{8}p_\gamma; \tag{20}$$

for $\nu \not\equiv \nu^*$:

$$\rho_\nu = \rho_{\nu^*} = \tfrac{7}{8}\rho_\gamma, \tag{21}$$

and

$$p_\nu = p_{\nu^*} = \tfrac{7}{8}p_\gamma, \tag{22}$$

so that the neutrino pressure and density in the latter case are twice those in the former. It should be noted that the results stated in Eqs. (19)–(22) are predicated on the assumption that no type of particle is degenerate in the present problem. The simple expressions for the neutrino density and pressure given in Eqs. (19)–(22) hold for all Fermi-Dirac particles in the limit of sufficiently high temperature, i.e., there is a contribution to the density of $(7/16)\rho_\gamma$ for each degree of freedom. Similarly for Bose-Einstein particles there is a contribution to the density of $\tfrac{1}{2}\rho_\gamma$ for each degree of freedom.

It can be shown from Eq. (9) that for a Fermi-Dirac particle of mass, m_i,

$$\rho_i(x) \propto \left(\frac{m_i}{m_e}\right)^4 \rho_e[(m_i/m_e)x],$$

with the proportionality factor depending on the previously mentioned spin and charge states. Thus all Fermi-Dirac particles exhibit the same behavior provided that an appropriate shift is made in the temperature scale. A similar result can be obtained for Bose-Einstein particles. The qualitative behavior of ρ_i *versus* T after suitable normalization of the temperature scales is essentially the same for fermions and bosons.

The neutrino contribution given by Eqs. (19)–(22) to the total pressure and density requires modification for the temperature range of interest in calculating the proton-neutron ratio as a function of the time. At very high temperatures the neutrino component maintains itself in equilibrium with the other constituents of of the medium through interaction with mesons. When the medium has expanded and cooled somewhat below a temperature equivalent to the rest mass of the lightest meson, the neutrinos freeze in and continue to expand and cool adiabatically as would a pure radiation gas. After this freeze-in the neutrino temperature will differ from that of the other components of the medium. It will be seen that the freeze-in must have occurred at a temperature higher than is required for neutrons and protons to be very nearly in thermodynamic equilibrium. For the temperature region of interest, then, we must deal with nucleons, electrons, positrons, and radiation at one temperature, and neutrinos at another temperature. The calculation of the neutron-proton ratio does not require that a specific freeze-in temperature be given, but only that neutrinos be frozen in before an appreciable fraction of the electron pairs start to annihilate.

It is of some interest to examine in more detail the freezing in of neutrinos during the period from ~15 to ~5 Mev. Non-equilibrium reactions involving neutrinos become important only below ~5 Mev. When the temperature was well above the rest mass equivalent of mesons, the neutrinos maintained equilibrium through interaction with mesons. At such temperatures the contribution of mesons to the density was $3.25\rho_\gamma$, while the total contribution due to photons, electrons, positrons, and neutrinos was $3.625\rho_\gamma$ or $4.50\rho_\gamma$, for $\nu \equiv \nu^*$ and $\nu \not\equiv \nu^*$, respectively.[8] Since the meson rest energy is distributed uniformly among the lighter particles when the mesons annihilate, it is clear that the number of neutrinos will about double when meson annihilation occurs. Now the bulk of mesons will annihilate when the temperature in the universal expansion has dropped significantly below that equivalent to $m_\mu c^2$ (~108 Mev) or $m_\pi c^2$ (~138 Mev), down to 10 Mev. At 10 Mev the Boltzmann factors for μ and π mesons are $\sim 2\times 10^{-5}$ and $\sim 10^{-6}$, respectively. This temperature decrease, as will be seen later when the time scale for the cosmological model is calculated, requires a duration of $\sim 10^{-2}$ sec in the universal expansion.

The meson reactions $\pi^\pm \rightleftharpoons \mu^\pm + \nu$ and $\mu^\pm \rightleftharpoons e^\pm + 2\nu$ are very fast, even if one neglects induced decay, having lifetimes of $\sim 2\times 10^{-8}$ sec and $\sim 2\times 10^{-6}$ sec, respectively. Since the concentrations of neutrinos and mesons are comparable, the reaction rate $1/(2\times 10^{-8})$ per second per neutrino is $\sim 10^6$ times the equilibrium rate (due to annihilation) of $\sim 1/10^{-2}$ per second per neutrino. Hence between ~100- and ~10-Mev thermal equilibrium holds. By 5 Mev, however, the Boltzmann factor $\exp(-m_\pi c^2/kT) \cong \exp(-138/5)$ has reduced the reaction rate to insignificance even though there is a

[7] Recently, theoretical arguments in favor of distinguishability, i.e., against the Majorana theory of neutral particles, have been given by E. R. Caianiello, Phys. Rev. **86**, 564 (1952). However, we consider both cases throughout this paper because it does not appear to be a settled question at this time. [See also C. S. Wu, Physica **18**, 989 (1952).]

[8] As has been shown, in the high temperature limit the Fermi-Dirac μ^+ and μ^- mesons each contribute $(7/8)\rho_\gamma$, while the Bose-Einstein π^+, π^-, and π^0 mesons each contribute $(1/2)\rho_\gamma$ for a total of $3.25\rho_\gamma$. Electrons and positrons each contribute $(7/8)\rho_\gamma$, neutrinos contribute $(7/8)\rho_\gamma$ or $2(7/8)\rho_\gamma$ according as $\nu \equiv \nu^*$ or $\nu \not\equiv \nu^*$, and photons contribute ρ_γ for a total of $3.625\rho_\gamma$ or $4.50\rho_\gamma$. The numerical factors obtained here depend on the discussion following Eq. (17).

good deal more time available for reactions to take place due to the reduced rate of cooling in the universal expansion. Hence the residual mesons cannot transfer a significant amount of rest mass energy to the neutrino gas, although almost all the meson rest mass energy is uniformly distributed.

Having described the nature of the medium, we can now proceed to determine the universal expansion rate for the period of interest in this problem. The rate of expansion for early times, Eq. (3), can be written in cgs units as

$$\frac{1}{2}\frac{dg}{dt}=\frac{1}{l}\frac{dl}{dt}=\frac{d\ln l}{dt}=\left(\frac{8\pi G}{3}\right)^{\frac{1}{2}}\rho^{\frac{1}{2}}, \quad (23)$$

where ρ, the total density, may now be written

$$\rho(x,l)=\rho(x)+\rho(l), \quad (24)$$

for the following reason. The quantity

$$\rho(x)=\rho_{e^+}+\rho_{e^-}+\rho_\gamma \quad (24a)$$

depends only on the temperature, while

$$\rho(l)=\rho_\nu \text{ (or } \rho_\nu+\rho_{\nu^*}) \quad (24b)$$

depends only on the proper distance l, since the neutrinos are expanding adiabatically as a radiation gas after freeze-in. We can, in fact, write $\rho_\nu \propto l^{-4}$, so that Eq. (23) can be rewritten as follows, after differentiation with respect to time:

$$\frac{d^2g}{dt^2}=\left(\frac{8\pi G}{3\rho}\right)^{\frac{1}{2}}\left(\frac{\partial\rho}{\partial\ln x}\frac{d\ln x}{dt}+\frac{\partial\rho}{\partial\ln l}\frac{d\ln l}{dt}\right). \quad (25)$$

Substituting for $\rho^{\frac{1}{2}}$ from Eq. (23) and using Eq. (24) yields

$$\frac{d^2g}{dt^2}=\frac{8\pi G}{3}\left[\frac{d\rho(x)}{d\ln x}\frac{d\ln x}{d\ln l}-4\rho(l)\right]. \quad (26)$$

If now we add Eqs. (1a) and (1b), neglect terms containing $1/R_0^2$, and convert to cgs units, we obtain

$$\frac{d^2g}{dt^2}=-\left(\frac{8\pi G}{c^2}\right)(p+\rho c^2)$$
$$=-\left(\frac{8\pi G}{c^2}\right)[p(x)+p(l)+\rho(x)c^2+\rho(l)c^2]. \quad (27)$$

If we equate Eqs. (26) and (27) and note that $(4/3)\rho(l)c^2=p(l)+\rho(l)c^2$, then we obtain

$$\frac{d\ln l}{d\ln x}=\frac{-c^2}{3[p(x)+\rho(x)c^2]}\frac{d\rho(x)}{d\ln x}, \quad (28)$$

where

$$\frac{d\rho(x)}{d\ln x}=\frac{d\rho_e(x)}{d\ln x}-\frac{3}{c^2}[p_\gamma(x)+\rho_\gamma(x)c^2]. \quad (28a)$$

This result is independent of the presence of the frozen-in neutrinos. Equation (28) can now be integrated in the following manner. From Eqs. (10) one obtains

$$\frac{dp_e(x)}{d\ln x}=-(p_e+\rho_e c^2), \quad (29)$$

or, more conveniently,

$$c^2x^4\frac{d\rho_e}{d\ln x}=-3x^4(p_e+\rho_e c^2)+\frac{d}{d\ln x}[x^4(p_e+\rho_e c^2)]. \quad (29a)$$

Employing Eq. (29a) in Eqs. (28) yields the desired result, namely,

$$\frac{d\ln l}{d\ln x}=1-\frac{\frac{d}{d\ln x}\{x^4[p(x)+\rho(x)c^2]\}}{3x^4[p(x)+\rho(x)c^2]}. \quad (30)$$

Since the adiabatic expansion of a radiation universe leads[1] to $d\ln l/d\ln x=1$, the second term in Eq. (30) represents a correction to the description of the cosmological model previously used with the neutron-capture theory, a correction which accounts for the interconversion of matter and radiation. Equation (30) may be integrated to yield

$$\ln l=\ln x-\tfrac{1}{3}\ln\{x^4[p(x)+\rho(x)c^2]\}+\text{constant}. \quad (31)$$

Finally, using Eqs. (10)–(14), one can write Eq. (25) as

$$\frac{d\ln l}{d\ln x}=1+\frac{2a_e(f_0+2f_1)}{3[p(x)+\rho(x)c^2]}. \quad (32)$$

As will become evident, Eq. (32) is required in order to obtain the explicit time dependence of the temperature.

The neutrino temperature T_ν (or $x_\nu=m_ec^2/kT_\nu$) may be determined from Eq. (30) by recalling that during the period of interest the neutrinos expand and cool adiabatically, so that $x_\nu=f(l)$ only, and in fact, $x_\nu\propto l$. Then it follows that Eq. (31) can be written as

$$\ln x_\nu=\ln x-\tfrac{1}{3}\ln\{x^4[p(x)+\rho(x)c^2]\}+\text{constant}. \quad (33)$$

The constant of integration can be evaluated by noting that for small x (high temperatures) the neutrino temperature approaches the temperature of the rest of the medium. In fact, as $x\to 0$, $x_\nu\to x$ so that the integration constant becomes

$$\text{constant}=\tfrac{1}{3}\ln\{x^4[p(x)+\rho(x)c^2]\}\,|_{x=0}. \quad (34)$$

From the definition of p_γ and ρ_γ it is evident [see discussion following Eq. (22)] that, for any x, $\rho_\gamma c^2x^4=\text{constant}=\pi^4a_e/15$, and $p_\gamma x^4=\frac{1}{3}\rho_\gamma c^2x^4$. Since

$$\lim_{x\to 0}p_ex^4=2(7/8)p_\gamma x^4,\quad \lim_{x\to 0}\rho_e c^2x^4=2(7/8)\rho_\gamma c^2x^4,$$

TABLE I. Time scales, rate coefficients, and quantum statistical integrals.[a]

T (°K)	x	t(sec) $\nu\equiv\nu^*$	t(sec) $\nu\not\equiv\nu^*$	t(sec) rad. model[b]	$\frac{15x^4}{\pi^4}f_0$	$\frac{15x^4}{\pi^4}f_1$	$\frac{15x^4}{\pi^4}f_2$	K_n	K_p
∞	0	0	0	0	0	0	0.87500	∞	∞
5.930×10^{10}	0.1	0.04	0.04	0.06	2.2105×10^{-5}	1.2535×10^{-3}	0.87311	5.07×10^{6}	3.94×10^{6}
2.965×10^{10}	0.2	0.14	0.12	0.26	2.6866×10^{-4}	4.9011×10^{-3}	0.86754	1.91×10^{5}	1.12×10^{5}
1.482×10^{10}	0.4	0.55	0.50	1.05	2.9583×10^{-3}	1.8300×10^{-2}	0.84626	7.58×10^{3}	2.51×10^{3}
9.884×10^{9}	0.6	1.27	1.14	2.36	1.1131×10^{-2}	3.7625×10^{-2}	0.81339	1.17×10^{3}	2.34×10^{2}
7.413×10^{9}	0.8	2.26	2.02	4.20	2.6998×10^{-2}	6.0185×10^{-2}	0.77190	3.09×10^{2}	37.8
5.930×10^{9}	1.0	3.58	3.21	6.57	5.1303×10^{-2}	8.3590×10^{-2}	0.72406	1.18×10^{2}	9.37
2.965×10^{9}	2.0	15.65	14.12	26.28	0.25584	0.16512	0.46121	6.88	4.00×10^{-2}
1.977×10^{9}	3.0	39.72	36.05	59.11	0.41847	0.16423	0.25402	2.38	9.49×10^{-4}
1.482×10^{9}	4.0	78.99	72.04	105.20	0.43426	0.12227	0.12828	1.91	2.69×10^{-5}
1.186×10^{9}	5.0	134.15	122.64	164.25	0.35354	7.7674×10^{-2}	6.1253×10^{-2}	1.74	...
9.884×10^{8}	6.0	203.84	186.56	236.49	0.24783	4.4664×10^{-2}	2.8128×10^{-2}	...	...
8.472×10^{8}	7.0	285.68	261.76	321.90	0.15696	2.3982×10^{-2}	1.2552×10^{-2}	...	...
7.413×10^{8}	8.0	378.40	347.03	420.44	9.2363×10^{-2}	1.2248×10^{-2}	5.4786×10^{-3}	...	...
6.589×10^{8}	9.0	481.59	442.06	532.17	5.1402×10^{-2}	6.0209×10^{-3}	2.3499×10^{-3}	...	...
5.930×10^{8}	10.0	595.56	546.96	657.02	2.7379×10^{-2}	2.8716×10^{-3}	9.9368×10^{-4}	...	...
0	∞	∞	∞	∞	0	0	0	1.63	0

[a] The universal constants employed in these calculations are those given by J. A. Bearden and H. M. Watts, Phys. Rev. 81, 73 (1951). Note that the limiting values at high temperatures do not include any contributions from mesons.
[b] This column gives the time scale for the pure radiation model described by Eqs. (4)–(7).

it follows that Eq. (33) can be written as

$$\left(\frac{x_\nu}{x}\right)^3=\left(\frac{T}{T_\nu}\right)^3=\frac{2.75(p_\gamma+\rho_\gamma c^2)}{p(x)+\rho(x)c^2},\qquad(35)$$

from which the neutrino temperature can be determined for any value of x. For the sake of completeness it should be noted that

$$\lim_{x\to\infty}p_e x^4=0,\quad \lim_{x\to\infty}\rho_e c^2x^4=0,$$

while the quantities $p_\gamma x^4$ and $\rho_\gamma c^2x^4$ are constants for all x, as just described.

One other relationship which we shall require is that between temperature and time. This is obtained by multiplying Eq. (23) by $d\ln x/d\ln l$, as evaluated from Eq. (32), with the following result:

$$\frac{d\ln x}{dt}=\frac{d\ln x}{d\ln l}\frac{d\ln l}{dt}=\left(\frac{8\pi G}{3}\rho\right)^{\frac{1}{2}}\frac{d\ln x}{d\ln l}.\qquad(36)$$

The integration of Eq. (36) (performed to an accuracy better than 0.1 percent on a Maddida, a digital differential analyzer built by Northrup Aircraft, Inc.) for the two cases $\nu\equiv\nu^*$ and $\nu\not\equiv\nu^*$ gives the time in the universal expansion as a function of x, and these quantities are given in Table I. For comparison, Table I also contains the time as a function of x for the expanding cosmological model containing radiation only [see Eq. (6)]. Other quantities given in Table I are the series of modified Bessel functions f_0, f_1, and f_2, as defined in Eqs. (12)–(14), which are used in computing pressure and density. In Table II are given the total density ρ

TABLE II. Neutrino temperature; total, radiation, and electron-pair densities; and universal expansion rates.[a]

T (°K)	x	x_ν/x	ρ_γ (g/cm³)	ρ/ρ_γ ($\nu\equiv\nu^*$)	ρ/ρ_γ ($\nu\not\equiv\nu^*$)	ρ_e/ρ_γ	$\frac{d\ln l}{d\ln x}$	$\frac{d\ln l}{dt}$(sec⁻¹) ($\nu\equiv\nu^*$)	$\frac{d\ln l}{dt}$(sec⁻¹) ($\nu\not\equiv\nu^*$)
∞	0	1.0000	∞	3.625	4.500	1.750	1.0000	∞	∞
5.930×10^{10}	0.1	1.0002	7.226×10^{7}	3.623	4.497	1.749	1.0005	14.51	16.17
2.965×10^{10}	0.2	1.0009	4.516×10^{6}	3.617	4.488	1.745	1.0018	3.625	4.039
1.482×10^{10}	0.4	1.0037	2.822×10^{5}	3.591	4.454	1.729	1.0073	0.9031	1.006
9.884×10^{9}	0.6	1.0082	5.575×10^{4}	3.549	4.396	1.702	1.0161	0.3990	0.4441
7.413×10^{9}	0.8	1.0145	1.764×10^{4}	3.490	4.317	1.664	1.0280	0.2226	0.2475
5.930×10^{9}	1.0	1.0224	7.226×10^{3}	3.416	4.217	1.615	1.0424	0.1409	0.1566
2.965×10^{9}	2.0	1.0821	4.516×10^{2}	2.891	3.529	1.253	1.1350	3.241×10^{-2}	3.581×10^{-2}
1.977×10^{9}	3.0	1.1616	89.20	2.317	2.798	0.836	1.2129	1.290×10^{-2}	1.417×10^{-2}
1.482×10^{9}	4.0	1.2407	28.22	1.870	2.240	0.501	1.2357	6.517×10^{-3}	7.132×10^{-3}
1.186×10^{9}	5.0	1.3044	11.56	1.580	1.882	0.278	1.2054	3.834×10^{-3}	4.184×10^{-3}
9.884×10^{8}	6.0	1.3478	5.575	1.411	1.676	0.146	1.1501	2.516×10^{-3}	2.742×10^{-3}
8.472×10^{8}	7.0	1.3736	3.009	1.319	1.564	0.073	1.0966	1.787×10^{-3}	1.946×10^{-3}
7.413×10^{8}	8.0	1.3876	1.764	1.271	1.508	0.035	1.0568	1.343×10^{-3}	1.463×10^{-3}
6.589×10^{8}	9.0	1.3947	1.101	1.248	1.479	0.017	1.0313	1.052×10^{-3}	1.145×10^{-3}
5.930×10^{8}	10.0	1.3981	0.7226	1.237	1.466	0.008	1.0165	8.480×10^{-4}	9.231×10^{-4}
0	∞	1.4010	0	1.227	1.454	0	1.0000	0	0

[a] The universal constants employed in these calculations are those given by J. A. Bearden and H. M. Watts, Phys. Rev. 81, 73 (1951). Note that the limiting values at high temperatures do not include any contributions from mesons.

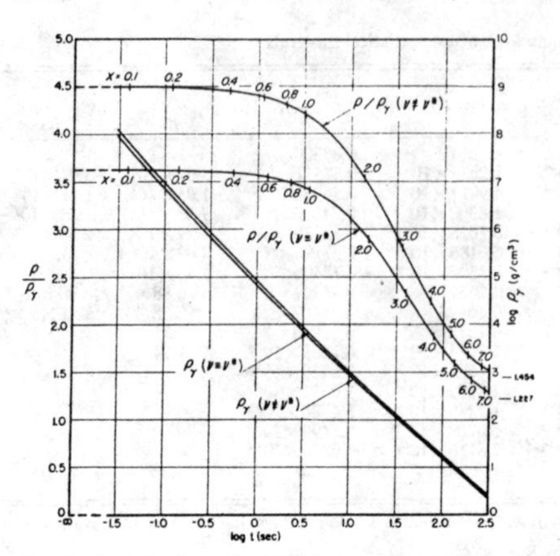

FIG. 1. Total density ρ, in units of photon density ρ_γ, and ρ_γ *versus* time during the very early epochs of the expanding universe for Majorana and Dirac neutrinos. The corresponding temperatures are given in terms of $x=m_ec^2/kT$. The ρ/ρ_γ curves are extrapolated to $t=0$ without regard to the presence of mesons and other elementary particles.

(i.e., the density of electrons, positrons, neutrinos, and radiation) and the density of electrons plus positrons, ρ_e, in units of the radiation density, ρ_γ, the neutrino temperature expressed as $x_\nu/x=T/T_\nu$, the differential quotient $d\ln l/d\ln x$, and finally the expansion rate $d\ln l/dt$.

Several interesting features of this cosmological model are evident upon examination of Tables I and II. First, the temperature drops much more rapidly in the nonstatic model with interconversion of matter and radiation than it does in the model of adiabatically expanding radiation only.[9] However the cases of distinguishable and indistinguishable neutrinos differ very little in this respect. The total density ρ does not drop off very greatly until the universe has cooled to about the electron rest mass equivalent. At this point the large density contribution of electron pairs begins to decrease sharply [see Table II] as the pairs disappear by annihilation into the radiation field which has fewer degrees of freedom. This behavior is demonstrated in Fig. 1. Next, if one recalls that the expanding model of radiation only is represented by $d\ln l/d\ln x=1$, then one can see in Table II that the maximum deviation from this model due to the interconversion of matter and radiation is $\sim$24 percent, and this occurs at 0.125 Mev ($\sim m_ec^2/4$). This deviation represents a more rapid expansion rate than in the pure radiation model and, in fact, the expansion rate $d\ln l/dt$ is higher in the model with interconversion by just the factor $(\rho/\rho_\gamma)^{\frac{1}{2}}$. Finally, it should be noted that x_ν/x (a quantity which does not depend on whether $\nu\equiv\nu^*$ or $\nu\not\equiv\nu^*$) differs from unity by less than one percent until x has increased to about 0.7 or $kT\cong 0.73$ Mev. At $x=0.1$, where $kT=5$ Mev, the deviation is 0.02 percent. It is then quite clear that selecting say 5–10 Mev as the freeze-in temperature for neutrinos is not only reasonable but quite an adequate approximation. It should be noted that the mathematical limits approached as $x\rightarrow\infty$ for all the quantities given in Tables I and II are included for the sake of completeness. However, the behavior of the cosmological model discussed changes at longer times, $x\sim 10^7$, when the density of matter exceeds the density of radiation.

In the next section we shall calculate the relative concentrations of neutrons and protons as a function of the time in the universal expansion. This ratio, it will be recalled, plays a most important role in determining the relative abundances of the nuclear species as calculated according to the simple neutron-capture theory of element formation including the effects of thermonuclear reactions. As has been stated, this theory quite clearly requires that during the early epochs in the universal expansion the density of nucleons, and of the nuclear species formed, should be negligibly small compared with the radiation density. Consequently, the physical conditions in the expanding model as described in Tables I and II, in which nucleon density is taken to be negligible, are used as a basis for examining the various non-equilibrium reactions between neutrons and protons.

III. THE NEUTRON-PROTON RATIO

In this section we shall examine the reactions which may occur among neutrinos, electrons, positrons, and nucleons in the very early stages of the cosmological model described in the previous section. In particular we shall calculate the ratio of the concentrations of protons and neutrons as a function of time, a ratio upon which the results of the neutron-capture theory of element formation strongly depend.

The nuclear reactions which must be considered in determining the proton-neutron ratio are the following:[10]

$$n+e^+\rightleftharpoons p+\nu, \tag{37a}$$

$$n+\nu\rightleftharpoons p+e^-, \tag{37b}$$

$$n\rightleftharpoons p+e^-+\nu. \tag{37c}$$

[9] There are perhaps slight shifts in the expansion time scale, much too small to appear with the number of significant figures given in Tables I and II. These are caused by the presence of and annilation of mesons, nucleons, gravitons, etc., between $x=0$ and $x=0.1$, should such particles exist during this epoch, and should the relativistic cosmology apply at the extreme conditions existing during this very brief early period. There is, however, serious doubt that the cosmology applies and, since we are interested only in the epochs of temperature lower than $x=0.1$, we can for the present perhaps ignore this problem and accept the insignificant additive constant in the time scale. See reference 1 as well as A. Einstein, *The Meaning of Relativity* (Princeton University Press, Princeton, 1945).

[10] The question of the existence of the antineutrino is of no concern in determining the individual reaction rates (see reference 7). In the absence of charge, mass, and magnetic moment, the absorption of a neutrino from a negative energy state is in all respects equivalent to the emission of an antineutrino.

The probability per second w for these reactions may be obtained from the Fermi theory of β decay. For the reaction $n+e^+\rightarrow p+\nu$, one has, per electron

$$w_{A'}=\frac{2\pi}{\hbar}(|\psi_p(0)|\cdot|\psi_\nu(0)|\cdot|\mathfrak{M}|g)^2\frac{dn}{dE}\ \mathrm{sec}^{-1}, \quad (38)$$

where the expectation value at the origin for the product particles, proton, and neutrino, depends on $|\psi_p(0)|^2|\psi_\nu(0)|^2$, the ψ are taken as plane wave states, $\mathfrak{M}$, the matrix element, is taken as unity for lack of a better estimate, g is the Fermi coupling constant and the quantity dn/dE is the energy density of available final states. We can, therefore, write

$$w_{A'}=\frac{2\pi g^2}{\hbar\Omega^2}\frac{dn}{dE}\ \mathrm{sec}^{-1}, \quad (39)$$

where Ω is any finite normalization volume. For the neutrino the number of available states per unit energy in the volume Ω is the difference between the total and occupied number of states, *viz.*,

$$\frac{4\pi(2i_\nu+1)|\mathbf{p}_\nu|}{2\pi\hbar^3}\{1-[1+\exp(E_\nu/kT_\nu)]^{-1}\}\frac{d|\mathbf{p}_\nu|}{dE_\nu}, \quad (40a)$$

while for each neutron the number of available final states in this particular reaction is $(2i_p+1)$. Since all neutrons are equivalent, one may write for the total number of final states in Ω:

$$\Omega n_n(2i_p+1)=2\Omega n_n. \quad (40b)$$

In Eqs. (40) the neutrino and proton spins are denoted by i_ν and i_p, $|\mathbf{p}_\nu|$ is the neutrino momentum and n_n the number of neutrons per unit volume. Since dn/dE n Eq. (39) is the product of terms given by Eq. (40a) and Eq. (40b) one can write

$$w_{A'}=\frac{4g^2n_n|\mathbf{p}_\nu|\exp(E_\nu/kT_\nu)}{\pi\hbar^4[1+\exp(E_\nu/kT_\nu)]}\frac{d|\mathbf{p}_\nu|}{dE_\nu}\ \mathrm{sec}^{-1}. \quad (41)$$

The number of such reactions, $n+e^+\rightarrow p+\nu$, per second per unit volume, is given by

$$A'n_nn_{e^+}=\int_{m_ec^2}^{\infty}w_{A'}n_{e^+}(E_{e^+})dE_{e^+}, \quad (42)$$

where

$$E_\nu=E_{e^+}+Q; \quad (42a)$$

$$Q=(m_n-m_p)c^2 \quad (42b)$$

is the neutron-proton energy difference and n_{e^+} is the concentration of positrons per unit energy at E_{e^+}. The lower limit of integration is the threshold energy, which in this case is the electron rest energy. Using the relation $E_\nu=|\mathbf{p}_\nu|c$ and replacing n_{e^+} by means of the Fermi-Dirac integral, *viz.*,

$$n_{e^+}(|\mathbf{p}_{e^+}|)d|\mathbf{p}_{e^+}|=\frac{8\pi|\mathbf{p}_{e^+}|^2d|\mathbf{p}_{e^+}|}{(2\pi\hbar^3)[1+\exp(E_{e^+}/kT)]}, \quad (43)$$

where $|\mathbf{p}_{e^+}|$ is given by Eq. (9c), one can write Eq. (42) in the form

$$A'n_nn_{e^+}=a_0n_n\int_{m_ec^2}^{\infty}\frac{E_{e^+}(E_{e^+}+Q)^2(E_{e^+}^2-m_e^2c^4)^{\frac{1}{2}}\exp[(E_{e^+}+Q)/kT_\nu]}{\{1+\exp[(E_{e^+}+Q)/kT_\nu]\}\{1+\exp[E_{e^+}/kT]\}}dE_{e^+}, \quad (44)$$

where Eq. (42a) has been used to eliminate E_ν, and

$$a_0=(4g^2)/(\pi^3c^6\hbar^7). \quad (44a)$$

It should be recalled that the neutrino temperature $T_\nu\neq T$, where the latter is the temperature of the remainder of the medium. Equation (44) may be rewritten by taking

$$x=m_ec^2/(kT),\quad x_\nu=m_ec^2/(kT_\nu),$$
$$q=Q/(m_ec^2),\quad \epsilon=E_e/(m_ec^2), \quad (45)$$

with the result that for the reaction $n+e^+\rightarrow p+\nu$,

$$A'n_nn_{e^+}=m_e^5c^{10}a_0n_nI_{A'}\ \mathrm{sec}^{-1}\ \mathrm{cm}^{-3}, \quad (46)$$

where

$$I_{A'}=\int_1^\infty\frac{\epsilon(\epsilon+q)^2(\epsilon^2-1)^{\frac{1}{2}}\exp[(\epsilon+q)x_\nu]d\epsilon}{\{1+\exp[\epsilon x]\}\{1+\exp[(\epsilon+q)x_\nu]\}}. \quad (46a)$$

Rates for the other reactions in Eq. (37a) and Eq. (37b) may be obtained from similar considerations, since they are all of second degree. One obtains the following:

for $p+\nu\rightarrow n+e^+$,

$$An_pn_\nu=m_e^5c^{10}a_0n_pI_A\ \mathrm{sec}^{-1}\ \mathrm{cm}^{-3}, \quad (47)$$

where

$$I_A=\int_1^\infty\frac{\epsilon(\epsilon+q)^2(\epsilon^2-1)^{\frac{1}{2}}\exp(\epsilon x)d\epsilon}{\{1+\exp[\epsilon x]\}\{1+\exp[(\epsilon+q)x_\nu]\}}, \quad (47a)$$

and where

$$E_{e^+}=E_\nu-Q; \quad (47b)$$

for $n+\nu\rightarrow p+e^+$,

$$B'n_nn_\nu=m_e^5c^{10}a_0n_nI_{B'}\ \mathrm{sec}^{-1}\ \mathrm{cm}^{-3}, \quad (48)$$

where

$$I_{B'}=\int_q^\infty\frac{\epsilon(\epsilon-q)^2(\epsilon^2-1)^{\frac{1}{2}}\exp(\epsilon x)d\epsilon}{\{1+\exp[\epsilon x]\}\{1+\exp[(\epsilon-q)x_\nu]\}}, \quad (48a)$$

and where

$$E_\nu=E_{e^-}-Q; \quad (48b)$$

for $p+e^-\rightarrow n+\nu$,

$$Bn_pn_{e^-}=m_e^5c^{10}a_0n_pI_B\ \mathrm{sec}^{-1}\ \mathrm{cm}^{-3}, \quad (49)$$

where

$$I_B=\int_q^\infty \frac{\epsilon(\epsilon-q)^2(\epsilon^2-1)^{\frac{1}{2}}\exp[(\epsilon-q)x_\nu]d\epsilon}{\{1+\exp[\epsilon x]\}\{1+\exp[(\epsilon-q)x_\nu]\}}, \quad (49a)$$

and where

$$E_\nu=E_{e^-}-Q. \quad (49b)$$

The reaction rates for free neutron decay and the inverse process, Eq. (37c), require a slightly different calculation. Thus for the reaction $p+e^-+\nu\rightarrow n$, we note that the quantity dn/dE in Eq. (39) is given by just the number of protons present in the volume Ω, so that

$$w_C=(2\pi g^2/\hbar)n_p(2i_n+1),$$

and

$$Cn_pn_{e^-}n_\nu=\int_{m_ec^2}^Q w_C(E_{e^-})n_{e^-}(E_{e^-})n_\nu(Q-E_{e^-})dE_{e^-}, \quad (50)$$

where

$$E_{e^-}=Q-E_\nu. \quad (50a)$$

In Eq. (50) $n_{e^-}(E_{e^-})$ is the concentration of electrons per unit energy at E_{e^-} and $n_\nu(Q-E_{e^-})$ is the concentration of neutrinos per unit energy at $(Q-E_{e^-})$, the argument of n_ν being that required for energy balance in this reaction. Finally, one can rewrite Eq. (50) replacing n_{e^-} and n_ν by means of the Fermi-Dirac integral [see Eq. (43)] and using Eqs. (45), as

$$Cn_pn_{e^-}n_\nu=m_e^5c^{10}a_0n_pI_C \text{ sec}^{-1}\text{ cm}^{-3}, \quad (51)$$

where

$$I_C=\int_1^q \frac{\epsilon(\epsilon-q)^2(\epsilon^2-1)^{\frac{1}{2}}d\epsilon}{\{1+\exp[\epsilon x]\}\{1+\exp[(q-\epsilon)x_\nu]\}}. \quad (51a)$$

For the reaction $n\rightarrow p+e^-+\nu$, we note that dn/dE in Eq. (39) is the product of three quantities, *viz.*, the available states per unit energy in the volume Ω for protons, electrons, and neutrinos. For protons the number of available states is the number of neutrons in the volume Ω, *viz.*, $\Omega n_n(2i_n+1)$, while for electrons and neutrinos one can use the form of Eq. (40a) which gives this quantity for Fermi-Dirac particles. Since formally the reaction rate for free neutron decay is

$$C'n_n=\int_{m_ec^2}^Q w_{C'}dE, \quad (52)$$

it follows from Eq. (39) after some manipulation that

$$C'n_n=m_e^5c^{10}a_0n_nI_{C'} \text{ sec}^{-1}\text{ cm}^{-3}, \quad (53)$$

where

$$I_{C'}=\int_1^q \frac{\epsilon(\epsilon-q)^2(\epsilon^2-1)^{\frac{1}{2}}\exp[(q-\epsilon)x_\nu]\exp(\epsilon x)d\epsilon}{\{1+\exp[\epsilon x]\}\{1+\exp[(q-\epsilon)x_\nu]\}}. \quad (53a)$$

The foregoing reaction rates have been used in the equations developed below which describe the time dependence of neutron and proton concentrations. Let N_j be the number of nucleons of species j in the arbitrary finite volume V. The time derivative of N_j can be expressed formally for two- and three-body processes as

$$\frac{dN_j}{dt}=\sum_{\alpha,\beta}K_{\alpha\beta}n_\alpha N_\beta+\sum_{\alpha,\beta,\gamma}K_{\alpha\beta\gamma}n_\alpha n_\beta N_\gamma, \quad (54)$$

where

$$n_\alpha=N_\alpha/V. \quad (54a)$$

We note that

$$\frac{dn_j}{dt}=\frac{1}{V}\frac{dN_j}{dt}-\frac{n_j}{V}\frac{dV}{dt}, \quad (55)$$

where, since $V\propto l^3$,

$$\frac{1}{V}\frac{dV}{dt}=\frac{3}{l}\frac{dl}{dt}. \quad (55a)$$

For the cosmological model described in Sec. II,

$$\frac{1}{l}\frac{dl}{dt}=\left(\frac{8\pi G}{3}\rho\right)^{\frac{1}{2}},$$

where ρ is the total density, so that

$$\frac{dn_j}{dt}=\sum_{\alpha,\beta}K_{\alpha\beta}n_\alpha n_\beta+\sum_{\alpha,\beta,\gamma}K_{\alpha\beta\gamma}n_\alpha n_\beta n_\gamma-3n_j(8\pi G\rho/3)^{\frac{1}{2}}. \quad (56)$$

Consequently, we can write for neutrons the following rate equation:

$$dn_n/dt=An_pn_\nu-A'n_nn_{e^+}+Bn_pn_{e^-}-B'n_nn_\nu+Cn_pn_{e^-}n_\nu-C'n_n-3n_n(8\pi G\rho/3)^{\frac{1}{2}}. \quad (57)$$

Equation (57) can be rewritten using some of Eqs. (46)–(53) as

$$dn_n/dt=m_e^5c^{10}a_0[n_pK_p-n_nK_n]-3n_n(8\pi G\rho/3)^{\frac{1}{2}}, \quad (58)$$

where

$$K_p=I_A+I_B+I_C, \quad (58a)$$

and

$$K_n=I_{A'}+I_{B'}+I_{C'}. \quad (58b)$$

The limits of integration in the six integrals involved in K_p and K_n make it possible to combine certain pairs, with the result that

$$K_p=\int_1^\infty \frac{\epsilon(\epsilon^2-1)^{\frac{1}{2}}}{1+\exp[\epsilon x]}\left\{\frac{(\epsilon+q)^2\exp[\epsilon x]}{1+\exp[(\epsilon+q)x_\nu]}+\frac{(\epsilon-q)^2\exp[(\epsilon-q)x_\nu]}{1+\exp[(\epsilon-q)x_\nu]}\right\}d\epsilon, \quad (59)$$

and

$$K_n=\int_1^\infty \frac{\epsilon(\epsilon^2-1)^{\frac{1}{2}}}{1+\exp[\epsilon x]}\left\{\frac{(\epsilon+q)^2\exp[(\epsilon+q)x_\nu]}{1+\exp[(\epsilon+q)x_\nu]}+\frac{(\epsilon-q)^2\exp[\epsilon x]}{1+\exp[(\epsilon-q)x_\nu]}\right\}d\epsilon. \quad (60)$$

The equation describing the time rate of change of proton concentration can be written in a manner analogous to Eq. (58) as follows:

$$dn_p/dt = m_e{}^5c^{10}a_0[n_nK_n - n_pK_p] - 3n_p(8\pi G\rho/3)^{\frac{1}{2}}. \quad (61)$$

As shall be seen below, Eqs. (58) and (61) can be combined to give a single equation for the proton-neutron ratio, with the effect of the universal expansion not appearing explicitly.

The rate coefficients K_p and K_n have been evaluated numerically using Eqs. (59) and (60) for the range of values of x of interest. The values of x_ν corresponding to x have been taken from Table II, with[11]

$$q = 1 + (m_n - m_p)/m_e = 2.53.$$

It can be shown from Eqs. (59) and (60) that for small x, i.e., $x<1$ where $x_\nu \to x$,

$$\lim_{x_\nu \to x} K_p = e^{-qx}K_n. \quad (62a)$$

Furthermore, for large values of x, one finds that

$$\lim_{x\to\infty} K_p = 0, \quad (62b)$$

and

$$\lim_{x\to\infty} K_n = \int_1^\infty \epsilon(\epsilon - q)^2(\epsilon^2-1)^{\frac{1}{2}}d\epsilon = 1.6318. \quad (62c)$$

The limit approached by K_n in Eq. (62c) for $x\to\infty$ is just the term $C'/(m_e{}^5c^{10}a_0)$ where $C'n_n = dn_n/dt$ describes free neutron decay and $C'(=\lambda)$ is the neutron-decay constant. One can, therefore, select the Fermi coupling constant g, which is the only undetermined constant in a_0 [see Eq. (44a)], so that the value of C' is the observed neutron decay constant.[12] The neutron half-life measured recently by Robson[13] is $\tau_{\frac{1}{2}} = 12.8 \pm 2.5$ minutes. The values of $m_e{}^5c^{10}a_0$ and of g corresponding to the measured half-life limits are given in Table III.

In Table I are given values of the dimensionless quantities K_n and K_p, evaluated numerically from

TABLE III. Values of Fermi constant for various neutron half-lives.

Neutron half-life (min)	10.3	12.8	15.3
$m_e{}^5c^{10}a_0$(sec^{-1})	4.627×10^{-4}	5.531×10^{-4}	6.876×10^{-4}
g(erg cm^3)	1.01×10^{-49}	1.11×10^{-49}	1.23×10^{-49}

[11] For $(m_n-m_p)c^2$ we have used 0.782 Mev, as given by D. M. Van Patter, Massachusetts Institute of Technology, Technical Report No. 57, January 1952 (unpublished), while m_ec^2 has been taken as 0.5110 Mev, as given by J. W. M. DuMond and E. R. Cohen, Phys. Rev. 82, 555 (1951).

[12] Although the numerical constants in a_0 which depend on spin, etc., have been carefully evaluated in building up the rate coefficients, it may be noted that the equality $C'n_n/(m_e{}^5c^{10}a_0) = 1.6318$, with $C'=\lambda$ known from experiment, automatically yields a value for a_0, so that g, spin factors, etc., need not be separately specified.

[13] See reference 4. Recently L. M. Langer and R. J. D. Moffat, Phys. Rev. 88, 689 (1952), obtained the value $\tau_{\frac{1}{2}} = 10.4 \pm 0.6$ min indirectly from studying tritium decay. This value and Robson's value agree within the probable errors.

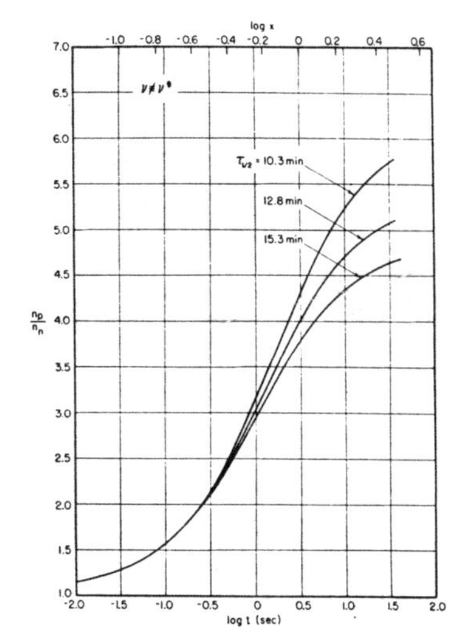

FIG. 2. The proton-neutron concentration ratio *versus* time and temperature $(x = m_ec^2/kT)$ in the case of the Dirac neutrino (distinguishable neutrino and antineutrino) for the Robson neutron half-life value of 12.8 min, plus-and-minus the probable error.

Eqs. (59) and (60) in the range required for the present calculation. It may be noted that for x slightly greater than 3, K_n is close to the free decay value of 1.63 while K_p is negligibly small. Also for $x<1$ the relationship $K_p \cong K_n \exp(-qx)$ holds quite closely.

Equations (58) and (61) describing neutron and proton concentrations can be combined by defining

$$\phi(t) = n_n/(n_n + n_p). \quad (63)$$

Taking the time derivative of ϕ and employing Eqs. (58) and (61) as required, one can write

$$d\phi/dt = m_e{}^5c^{10}a_0[K_p(1-\phi) - K_n\phi], \quad (64)$$

where a_0 is given by Eq. (44a). The actual integrations were done with $\ln x$ as independent variable where one writes

$$\frac{d\phi}{d\ln x} = \frac{d\phi}{dt}\frac{d\ln l}{d\ln x}\frac{dt}{d\ln l}. \quad (64a)$$

The quantity $d\ln l/d\ln x$ is given by Eq. (30) and calculated values are given in Table I, while values of $d\ln l/dt$, determined from Eq. (23), are also given in Table I

Equation (64) has been integrated numerically for the six cases of interest, *viz*, $\nu\equiv\nu^*$ and $\nu\not\equiv\nu^*$ taking three values of the neutron half-life, namely, the mean value and the mean value plus-and-minus the probable error, as given by Robson.[4] The integration procedure was such as to give a final accuracy in ϕ of the order

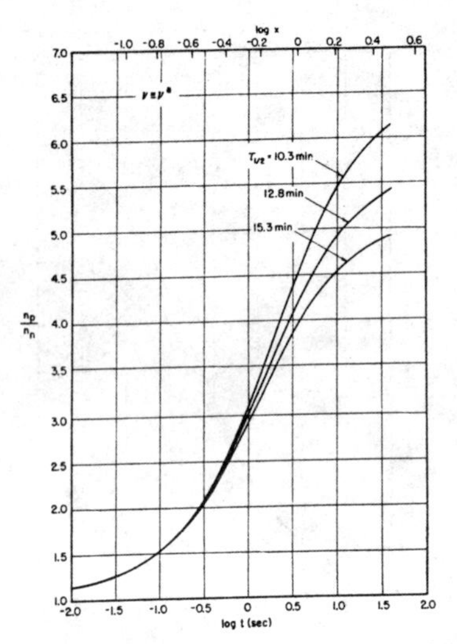

FIG. 3. The proton-neutron concentration ratio *versus* time and temperature ($x=m_ec^2/kT$) in the case of the Majorana neutrino (indistinguishable neutrino and antineutrino) for the Robson neutron half-life value of 12.8 min, plus-and-minus the probable error.

of the accuracy of the coefficients, i.e., $\sim$1 percent. The solutions we have obtained are presented in Figs. 2 and 3, where the proton-neutron ratio is plotted *versus* x and *versus* the time scale appropriate to the type of neutrinos involved. The integrations were carried from $x=0.1$ toward larger x, the initial value of $\phi=n_n/(n_n+n_p)$ being taken as the equilibrium value, i.e., $\sim[1+\exp(qx)]^{-1}$, since at $x=0.1$ the rate coefficients K_n and K_p are large and show negligible deviation from their respective equilibrium values. The integration interval was 0.1 in $\ln x$, with the first approximation at each step being the equilibrium value which Eq. (64) would predict at the given value of x. It may be noted in Figs. 2 and 3 that by $x=4$ further change in the proton-neutron ratio is almost entirely caused by free neutron decay.

A comparison of these results with those of Hayashi indicates that the major difference between our calculation and his may arise from the difference in neutron half-life used, *viz.*, Hayashi used 20.8 min, while the remainder of the differences, amounting to perhaps 20 percent in the proton-neutron ratio, arise from the use of relativistic quantum statistics, a more detailed cosmological model, and different temperatures for neutrinos and the rest of the medium in the present calculation. The proton-neutron ratio obtained by Hayashi by the time free neutron decay predominated was $\sim$4:1, whereas in the present calculation values from $\sim$4.5:1 to $\sim$6.0:1 are obtained depending on the half-life taken for the neutron and the type of neutrino considered.

It is interesting to note that if all the neutrons available at the start of element synthesis were used in making helium nuclei only, then the ratio of hydrogen to helium abundances corresponding to the range of proton-neutron ratios computed above would be from $\sim$7:1 to $\sim$10:1. Since some of the neutrons decay and some are involved in making the heavier elements, the above ratios would be minimum values of the initial universal H/He ratios. These values are consistent with the range of values obtained from astronomical data, *viz.*, $\sim$5:1 to $\sim$30:1 as found in planetary nebulae, stellar atmospheres, and theoretical stellar models.[14]

IV. DISCUSSION

In the preceding sections we have discussed quantitatively the physical conditions in the initial stages of the universal expansion. It now seems pertinent to mention some of the small physical effects whose influence on the present calculations has been neglected and also to comment on some of the limitations of the cosmological model when extrapolated to very early epochs.

The first question to be considered is whether or not the various processes, such as pair production, Compton and Coulomb scattering, etc., occur at sufficiently rapid rates to maintain equilibrium. A qualitative criterion as first described by Hayashi[3] is to compare the time required for the concentration of a constituent to change by about its own value with the time required for the universal temperature to change by about its own value. Characterizing these as relaxation times, τ, one finds from Eq. (6) and Table I that the relaxation time for temperature is given with sufficient accuracy for the present purpose by the following expression:

$$\tau_T\sim-\frac{dt}{d\ln T}=-\left(\frac{3c^2}{8\pi Ga_\gamma}\right)^{\frac{1}{2}}T^{-2}=2t. \tag{65}$$

To take a specific example one can calculate from the equilibrium concentration of electron pairs and their rate of production by photon-photon collisions the relaxation times, τ_{pair}, for pair production-annihilation. The result for $x\ll1$ is

$$\tau_{\text{pair}}\sim-\frac{dt}{d\ln n_{\text{pair}}}\sim\frac{144x\hbar^3}{\pi^3m_ee^4}, \tag{66}$$

where e is the charge on the electron. Hence

$$\frac{\tau_{\text{pair}}}{\tau_T}\cong10x^3\left(\frac{Gm^2}{\hbar c}\right)^{\frac{1}{2}}\left(\frac{e^2}{\hbar c}\right)^{-2}\sim10^{-20}x^3\ll1. \tag{67}$$

[14] A. Underhill, Symposium on the Abundance of Elements held at Yerkes Observatory, Williams Bay, Wisconsin, November 6–8, 1952 (unpublished), under the joint sponsorship of the National Science Foundation and the University of Chicago.

A similar result is obtained for other reactions not involving neutrinos, for which the change in the coupling coefficient $e^2/\hbar c$ does not greatly change the order of magnitude of the ratio. This is not the case for neutrino interactions. Hence all processes not involving neutrinos proceed at sufficiently rapid rates to maintain equilibrium.

The question of electron degeneracy is most easily examined by considering the requirement of electrical neutrality.[3] If one integrates Eq. (43) with a degeneracy parameter, ζ, included, then for high temperatures the electron or positron concentration can be written as

$$n_{e\mp}\sim(2/\pi^2)(kT/\hbar c)^3 e^{\pm\zeta}, \tag{68}$$

and

$$n_{e^-}+n_{e^+}\sim(4/\pi^2)(kT/\hbar c)^3 \cosh\zeta. \tag{68a}$$

If the condition of electrical neutrality is imposed then $n_{e^-}-n_{e^+}=n_p$ and

$$\sinh\zeta=\frac{\pi^2 n_p x^3}{4(mc/\hbar)^3}\sim\frac{n_p}{n_{e^-}+n_{e^+}}. \tag{69}$$

As has been shown[1,2] the nucleon concentration during the early stages of the expanding universe is very small compared with the density of radiation ($1:10^6$) and, therefore, also small compared with the electron-positron pair concentration. It follows then from Eq. (69) that the parameter ζ is very small and, therefore, the degeneracy of electrons or positrons properly has been neglected.

The charge on the electrons and positrons gives rise to a Coulomb interaction energy which contributes to the total energy of the medium. The reasonableness of neglecting this interaction energy can be seen from the following. The average distance between, say, electrons is found from Eq. (68), taking $\zeta\ll 1$ as

$$(1/n_{e^-})^{\frac{1}{3}}\sim\hbar/|\mathbf{p}|, \tag{70}$$

i.e., the de Broglie wavelength. It follows that the Coulomb energy, E_c, for two electrons is

$$E_c\sim-\frac{e^2}{(\hbar/|\mathbf{p}|)}=-\left(\frac{e^2}{\hbar c}\right)|\mathbf{p}|c\sim-\frac{1}{137}E_T, \tag{71}$$

where E_T is the mean thermal energy per electron. Because of this Coulomb interaction energy there will be a slight tendency for a given charge to have more nearest neighbors with charge of opposite sign. The fractional charge excess per nearest neighbor at the distance $\hbar/|\mathbf{p}|$ may be expected to be of the order $\exp[-E_c/kT]-1\sim 1/137$. Therefore, the contribution of Coulomb energy due to nearest neighbors to the total energy of the medium is $\sim E_c/137\cong E_T/(137)^2$ times the mean number of nearest neighbors. Assuming this number to be of the order of 10, the contribution of the Coulomb energy is $<10^{-3}E_T$, and can, therefore, be neglected.

The contribution of specifically nuclear forces is negligible because the nucleon density is very small compared with nuclear density. Furthermore, the energy evolution of nuclear reactions also can be neglected because it is itself small compared with the already small contribution of the low density of nucleons.

The foregoing small effects bear mainly on the cosmological model which has been discussed in Sec. II. There are also several questions of this kind which concern the calculation of the rates of the nuclear reactions in Eq. (37) which were determined in Sec. III. An examination of these reactions shows that of the six rates only $B'n_n n_r$ and $C'n_n$, Eqs. (48) and (53), involve two charged product particles. For these, one should more correctly include a factor in the reaction probability, w, to take into account the effect of Coulomb forces. In general this factor is given by[15]

$$2\pi\eta[1-\exp(-2\pi\eta)]^{-1}, \tag{72}$$

where

$$\eta=Ze^2E_e(\hbar c^2|\mathbf{p}_e|)^{-1},$$

Z is the nuclear charge, and E_e and $|\mathbf{p}_e|$ are electron energy and momentum, respectively. The effect of this correction on the two integrals in $C'n_n$ and $B'n_n n_r$ has been estimated and found to be less than one percent. Thus the effect of the Coulomb forces can be completely neglected in these cases.

As has been mentioned the matrix elements for the nuclear reactions stated in Eqs. (37) have been taken equal to unity for lack of a better estimate.[16] There seems to be little doubt that free neutron decay is a super-allowed transition since the decay rate is consistent with those of other light element β emitters. Furthermore, it would seem likely that the matrix elements for all the reactions considered here would remain about equal in the event that one included effects such as nucleon recoil.

It should also be pointed out that in calculating reaction rates we have considered that the nucleons, i.e., the heavy particles, are at rest. This approximation, which is customarily made, leads to a negligible error.

In addition to the above questions there are a number of more general points which may bear on the validity of the theory presented in this paper. One such question concerns the extrapolation of physical theories back to extremely high temperatures and densities. For example, some quantum field theories introduce a cutoff in, say, the electric field at the value it would have on the surface of the classical electron in order to avoid high-energy difficulties. This cutoff is introduced by appropriate modification of the field equations and, therefore, of the distribution of states in momentum space. When the mean electric field is equal to the

[15] See for example, G. Gamow and C. L. Critchfield, *Theory of Atomic Nucleus and Nuclear Energy Sources* (Clarendon Press, Oxford, 1949).

[16] E. Fermi, *Nuclear Physics* (University of Chicago Press, Chicago, 1950).

TABLE IV. Timetable of events in the early epochs of the expanding universe.

Temperature (Mev)	Remarks Neutrino ≡ antineutrino	Neutrino ≢ antineutrino
>100	Region of doubtful validity of the field equations where ρ_γ exceeds nuclear density.	
~ 100	Thermodynamic equilibrium prevails.	
	$\rho_\gamma \cong 1.2\times 10^{13}$ g/cm^3 $\rho_\mu = (7/4)\rho_\gamma$, $\rho_\pi = (3/2)\rho_\gamma$ $\rho_\nu = (7/8)\rho_\gamma$, $\rho_e = (7/4)\rho_\gamma$ $t \cong 6.3\times 10^{-5}$ sec	Same as for $\nu \equiv \nu^*$ except $\rho_\nu = (7/4)\rho_\gamma$ $t \cong 5.9\times 10^{-5}$ sec
$\sim 100 - \sim 10$	Mesons annihilate converting energy into photons, electrons, and neutrinos.	
~ 10	Neutrinos are freezing-in during this period.	
	$\rho_\gamma \cong 1.2\times 10^{9}$ g/cm^3 $\rho_\mu \sim 10^{-6}\rho_\gamma$, $\rho_\pi \sim 10^{-6}\rho_\gamma$ $\rho_\nu = (7/8)\rho_\gamma$, $\rho_e = (7/4)\rho_\gamma$ $t \cong 8.7\times 10^{-3}$ sec	Same as for $\nu \equiv \nu^*$ except $\rho_\nu = (7/4)\rho_\gamma$ $t \cong 7.8\times 10^{-3}$ sec
$\sim 10 - \sim 2$	Continued adiabatic expansion of universe with $T_\nu \cong T$ despite negligible interaction of neutrinos with medium.	
~ 2	Start of electron-positron annihilation.	
	$\rho_\gamma \cong 1.9\times 10^{6}$ g/cm^3 $\rho_\mu = \rho_\pi \sim 0$ $\rho_\nu \cong (7/8)\rho_\gamma$, $\rho_e = (7/4)\rho_\gamma$ $t \cong 0.22$ sec	Same as $\nu \equiv \nu^*$ except $\rho_\nu \cong (7/4)\rho_\gamma$ $t \cong 0.20$ sec
$\sim 2 - \sim 0.05$	Electron-positron annihilation, converting energy into photons. Neutrinos cool adiabatically relative to remaining particles, the latter maintaining thermodynamic equilibrium. [See Tables I and II for more details during this epoch.] The neutron-proton abundance ratio reaches the free decay value, 4.5:1–6.0:1, at $T\sim 0.2$ Mev. Nucleogenesis begins at $T \sim 0.2$ Mev.	
~ 0.05	Nucleogenesis is well under way.	
	$\rho_\gamma \cong 0.72$ g/cm^3 $\rho_\nu \cong 0.24\rho_\gamma$, $\rho_e \sim 0$ $t \cong 600$ sec	$\rho_\gamma \cong 0.72$ g/cm^3 $\rho_\nu \cong 0.47\rho_\gamma$, $\rho_e \sim 0$ $t \cong 550$ sec
~ 0.03	Nucleogenesis essentially complete except for charge adjustment by β decay. $t \sim 30$ min	
~ 0.03 Mev $-\sim 1$ kev	Thermonuclear reactions among some of the light elements, *viz.*, Li, Be, B, D with H, continue during this period.	
~ 0.015 ev	At $t \sim 10^8$ yr, $T \sim 170°$K and $\rho \sim 10^{-26}$ g/cm^3, galaxies probably form.	

foregoing cutoff, one has

$$\sim [e/(e^2/m_e c^2)^2]^2 \cong \rho_\gamma c^2,$$

which leads to a temperature of ~ 15 Mev. However, recent advances in quantum field theory obviate such a high-energy cutoff. In fact, if such a cutoff exists it is probably an order of magnitude higher. This is evidenced by the quantitative agreement between theory and the observed Lamb shift, for example. Cutoffs in momentum space must be larger than present day experimental energies, i.e., $kT > 100$ Mev, since observed bremsstrahlung and pair production, etc., agree with theory quite well.

Another pertinent question is the possible contribution of equilibrium concentrations of "gravitational quanta" to the total density. Although at equilibrium the "graviton" density would be expected to be equal to the photon density, one must consider at what temperatures such equilibria can be maintained. We may apply Eq. (67) to the present situation and replace the coupling coefficient $(e^2/\hbar c)^2$ by the product of $(Gm^2/\hbar c)$ with an electronic or mesonic coupling coefficient whose value will be in the range $\sim 1 - \sim 10^{-2}$. Then, since $(Gm^2/\hbar c) \cong 10^{-45}$ with $m = m_e$, one finds $\tau_{grav}/\tau_T \sim 10^{22}$ at ~ 1 Mev. In order for $\tau_{grav}/\tau_T \sim 1$, i.e., for the "gravitons" to maintain equilibrium, the temperature must be $\sim 10^4$ Mev.[17] It is difficult to see how the introduction of many-body processes would reduce this temperature drastically. One does not know how many different kinds of particles exist in the range $\sim 10^2 - \sim 10^4$ Mev but on the basis of the presently known types of particles one can determine an upper limit to the graviton contribution. We can compute the ratio of graviton density to that of neutrinos down to the temperature at which neutrinos freeze-in, since beyond this temperature the ratio remains constant. From the analog to Eq. (35) one has, if $\mathfrak{F}_i$ represents degrees of freedom for each constituent present, i.e., $\mathfrak{F}_i \rightarrow \rho_i/\rho_\gamma$ as $T \rightarrow \infty$, the following relationship:

$$\frac{\rho_{grav}}{\rho_\gamma} = \left(\frac{\sum_i \mathfrak{F}_i \text{ at } T_\nu'}{\sum_i \mathfrak{F}_i \text{ at } T_{grav}'} \right)^{4/3}, \tag{73}$$

where T_ν' and T_{grav}' are the freeze-in temperatures of neutrinos and gravitons, respectively. From the presently known elementary particles which would exist at these temperatures one can estimate from Eq. (73) that $\rho_{grav}/\rho_\gamma < 0.1$ at T_ν'. During the subsequent expansion down to $T \sim 0.1$ Mev, ρ_{grav}/ρ_γ diminishes by a factor of ~ 4, just as ρ_ν/ρ_γ diminishes [see Table II]. At no time does the upper bound of the graviton contribution to the density exceed 2 or 3 percent, and the total contribution is probably much smaller.

Finally, it seems pertinent to comment on the question as to whether the density of nucleons relative to the density of radiation can be calculated at some very early time on the basis of theoretical considerations with complete symmetry between nucleons and antinucleons or whether it is a free initial condition. In particular can the nucleon density be the result of a statistical fluctuation in the competition between different processes of nucleon annihilation such as

$$p^+ + p^- \rightleftharpoons 2h\nu$$
$$\rightleftharpoons \pi^+ + \pi^-, \text{ etc.},$$

where p^+ and p^- are proton and antiproton, respec-

[17] It should be noted that in the coupling coefficient the quantity m must be taken to be the relativistic mass of the interacting particles, i.e., $(Gm^2/\hbar c) = (Gm_e^2/\hbar c)x^{-2}$. Also note that the numerical results given here for extreme physical conditions are at best rather crude approximations.

tively, and as yet unknown high-energy processes such as

$$p^+ \rightleftharpoons \text{mesons} + e^\pm, \text{ etc.?}$$

An examination of this question on statistical grounds yields a probable residual density of nucleons approximately equal to $\rho_\gamma/N^{\frac{1}{2}}$, where N is the number of nucleons initially present in any given finite volume under consideration in co-moving coordinates.[18] If we take an initial volume corresponding to the presently observable universe, the residual number of nucleons is found to be less than would be required to form the earth. It appears that the situation described above is untenable and that the initial nucleon concentration must be specified arbitrarily. This result is in agreement with present thinking in elementary particle physics which does not allow for single nucleon annihilation processes. In addition, it should be pointed out that no catalytic type of reaction (e.g., $2p^+ + p^- \rightarrow 2p^+ + \mu^-$) can vitiate the above statistical arguments because of the finite propagation velocity of disturbances noted in Sec. II.

V. CONCLUSION

The problem discussed in this paper has been concerned with the detailed nature of the general nonstatic homogeneous isotropic expanding cosmological model derived from general relativity as well as the elementary particle reactions which occur during early epochs. The study of the elementary particle reactions leads to a knowledge of the time dependence of the proton-neutron concentration ratio which is required in the problem of nucleogenesis. While the problem of element origin stimulated the present study, the results concerning the cosmological model are of interest in themselves. On the basis of the new physical conditions which have been discussed here, it would appear necessary to reexamine the specific reactions among the lighter nuclei, particularly as regards the missing species at $A = 5$.

In order to summarize, we have presented the above calculations in abbreviated form as a timetable of events in the very early stages of the expanding universe, through the period of residual thermonuclear reactions[19] and galaxy formation.[20] In Table IV are given for various temperatures the corresponding epochs according to the expanding cosmological model involving the interconversion of matter and radiation, the densities of the various constituents according to the appropriate relativistic quantum statistics, as well as remarks concerning some of the principal physical phenomena that occur during these various early stages. This tabulation, it will be noted, covers both distinguishable and indistinguishable neutrinos.

Finally, we should like to point out that all of the results presented in this paper follow uniquely from general relativity, relativistic quantum statistics, and β-decay theory without the introduction of any free parameters, so long as the density of matter is very small compared with the density of radiation.

VI. ACKNOWLEDGMENTS

We wish to thank Mrs. Betty Grisamore, Mrs. Kathryn Stevenson, and Mr. Charles V. Bitterli for their assistance in some of the numerical work, Miss Shirley Thomas for typing this manuscript, and Miss Doris Rubenfeld for assistance with the illustrations.

[18] This can be seen from the following arguments. Let the numbers of protons, antiprotons, neutrons, and antineutrons in any finite co-moving volume V be equal and equal to N. Let α be the probability per particle for any of these particles to transmute to mesons at high temperature. We shall suppose that such transmutations occur first in the expansion, and that annihilation occurs later. This situation yields the largest residual density. Then on the average $4\alpha N$ particles transmute to mesons. The standard deviation σ in the number transmuting is then

$$\sigma = [4\alpha N(1-\alpha)]^{\frac{1}{2}},$$

which is a maximum of $N^{\frac{1}{2}}$ for $\alpha = \frac{1}{2}$. One may expect that in any volume V the excess of nucleons over antinucleons, or conversely, will be of the order of σ, i.e., of the order of $N^{\frac{1}{2}}$. The concentration of these residual nucleons at a later time when the initial volume V_0 has expanded to V_1 is given by $n_{\text{nuc}} = (N^{\frac{1}{2}}/V_0)(V_0/V_1)$. To a rough approximation the number of photons originally in V_0, a number approximately equal to N initially, has remained constant down to V_1, so that $V_0/V_1 \cong n_{\gamma_1}/n_{\gamma_0}$, where n_γ is photon concentration. Hence one can write

$$n_{\text{nuc}} \cong N^{\frac{1}{2}} n_{\gamma_1}/(V_0 n_{\gamma_0}) \cong n_{\gamma_1}/N^{\frac{1}{2}}, \quad \text{or} \quad \rho_{\text{nuc}} \sim \rho_{\gamma_1}/N^{\frac{1}{2}}.$$

[19] Alpher, Herman, and Gamow, Phys. Rev. 74, 1198 (1948).

[20] G. Gamow, Kgl. Danske Videnskab. Selskab, Mat.-fys. Medd. 27, No. 10 (1953).

1108 NATURE September 12, 1964 VOL. 203

THE MYSTERY OF THE COSMIC HELIUM ABUNDANCE

By PROF. F. HOYLE, F.R.S., and DR. R. J. TAYLER
University of Cambridge

It is usually supposed that the original material of the Galaxy was pristine material. Even solar material is usually regarded as 'uncooked', apart from the small concentrations of heavy elements amounting to about 2 per cent by mass which are believed on good grounds to have been produced by nuclear reactions in stars. However, the presence of helium, in a ratio by mass to hydrogen of about 1 : 2, shows that this is not strictly the case. Granted this, it is still often assumed in astrophysics that the 'cooking' has been of a mild degree, involving temperatures of less than 10^8 K, such as occurs inside main-sequence stars. However, if present observations of a uniformly high helium content in our Galaxy and its neighbours are correct, it is difficult to suppose that all the helium has been produced in ordinary stars.

It is the purpose of this article to suggest that mild 'cooking' is not enough and that most, if not all, of the material of our everyday world, of the Sun, of the stars in our Galaxy and probably of the whole local group of galaxies, if not the whole Universe, has been 'cooked' to a temperature in excess of 10^{10} K. The conclusion is reached that: (i) the Universe had a singular origin or is oscillatory, or (ii) the occurrence of massive objects has been more frequent than has hitherto been supposed.

The values of the helium to hydrogen number ratio determined for various objects are shown in Table 1.

Table 1

	He/H
Orion nebula (ref. 1)	0·091
NGC 604 in *M* 33 (ref. 2)	0·102
Small magellanic cloud (ref. 3)	0·11
B stars (ref. 4)	0·16
Planetary nebulæ (ref. 5)	0·09–0·19
Solar cosmic rays (refs. 6 and 7)	0·091
Solar evolution (ref. 8)	0·09

The first five determinations were made by spectroscopic methods. It has been found that the ratios of carbon, oxygen, magnesium and silicon in solar cosmic rays are in good agreement with well-determined spectroscopic values, and it is therefore argued that the composition of the cosmic rays reflects the true solar composition, and hence that the He/O value in the cosmic rays reflects the true solar helium/oxygen ratio. Given reliable spectroscopic values for all elements other than helium, and given accurate opacities within the Sun, it is possible to determine He/H from the requirement that solar evolution must be such that the present-day luminosity is arrived at after $4·5–5 \times 10^9$ years, the known age of the solar system. It is of immediate interest that two such different methods as these should arrive at closely the same result for the solar He/H value, particularly as this result—if accepted—is sufficient to establish that the Universe did not have a singular origin, nor can the Universe be oscillatory.

A high helium abundance in some particular star or nebula need not be taken as proof of a high primeval value, since helium may have been produced locally by nuclear reactions. For this reason low values of He/H are of more interest in relation to the original composition of the Galaxy than high values. However, O'Dell's high value of $0·18 \pm 0·03$ for the planetary nebula *M* 15 is of special interest because O'Dell also finds a low value for the ratio O/H, suggesting that the material of this nebula may not have been adulterated by the products of nuclear reactions. Adulteration by hydrogen-burning could have occurred; but not adulteration by helium-burning, so the situation remains rather uncertain.

We begin our argument by noticing that helium production in ordinary stars is inadequate to explain the values in Table 1, if they are general throughout the Galaxy, by a factor of about 10. Multiplying the present-day optical emission of the Galaxy, $\sim 4 \times 10^{43}$ ergs sec^{-1}, by the age of the Galaxy, $\sim 3 \times 10^{17}$ sec, and then dividing by the energy production per gram, $\sim 6 \times 10^{18}$ ergs g^{-1}, for the process H $\rightarrow$ He, gives $\sim 2 \times 10^{42}$ g ($10^9 M_\odot$). This is the mass of hydrogen that must be converted to helium in order to supply the present-day optical output of the Galaxy for the whole of its lifetime. Allowance for emission in the ultra-violet and in the infra-red increases the required hydrogen-burning, but probably not by a factor more than ~ 3. Since the total mass of the Galaxy is $\sim 10^{11} M_\odot$ the value of the He/H to be expected from H $\rightarrow$ He inside stars is only $\sim 0·01$. While it is true that the Galaxy may have been much more luminous in the past than it is now, there is no evidence that this was the case.

Next, we shift the discussion to a 'radiation origin' of the Universe, in which the rest mass energy density is less than the energy density of radiation. The relation between the temperature T_{10}, measured in units of 10^{10} K, and the time t in seconds can be worked out from the equations of relativistic cosmology and is:

$$T_{10} = 1·52\, t^{-1/2} \tag{1}$$

In the theory of Alpher, Bethe and Gamow[9] the density was given by:

$$\rho \approx 10^{-4} T_{10}^3 \text{ g cm}^{-3} \tag{2}$$

a relation obtained from the following considerations. The material is taken at $t = 0$ to be entirely neutrons. At $t \simeq 10^3$ sec, $T_{10} \simeq 0·05$, approximately half the neutrons have decayed. If the density is too low the resulting protons do not combine with the remaining neutrons and very little helium is formed. On the other hand, if the density is too high there is a complete combination of neutrons and protons, and with the further combination of the resulting deuterium into helium very little hydrogen remains as t increases. Thus only by a rather precise adjustment of the density, that is, by (2), can the situation be arranged so that hydrogen and helium emerge in approximately equal amounts.

It was pointed out by Hayashi[10] and Alpher, Follin and Herman[11] that the assumption of material initially composed wholly of neutrons is not correct. The radiation field generates electron-positron pairs by:

$$\gamma + \gamma \rightleftarrows e^- + e^+ \tag{3}$$

and the pairs promote the following reactions:

$$n + e^+ \rightleftarrows p + \bar{\nu} \tag{4}$$

$$p + e^- \rightleftarrows n + \nu \tag{5}$$

The situation evidently depends on the rates of these reactions. It turns out that for sufficiently small t the balance of the reactions is thermodynamic. This means that not only are protons generated by (4) and (5), but also that the energy densities of the pairs and of the neutrinos must be included in the cosmological equations. At $T_{10} \simeq 10^2$, even μ-neutrinos are produced and these too should be included. The effect of these new contributions to the energy density is to modify (1) to:

$$T_{10} \simeq 1·04\, t^{-1/2} \tag{6}$$

The values of σv for (4), (5), read from left to right, are:

$$\pi^2\left(\frac{\hbar}{mc}\right)^3\left(\frac{W \pm W_0}{mc^2}\right)^2\left(\frac{\ln 2}{(f\tau)_{\rm lab}}\right) \quad (7)$$

in which the well-known Coulomb factor has been taken as unity, and where the symbols have the following significance: m, electronic mass; W, the energy, including rest mass, of the positron or electron; W_0, energy difference between the neutron and proton ($2\cdot54\ mc^2$); $\pm$ apply to (4) and (5) respectively; $(f\tau)_{\rm lab}$, the $f\tau$-value for free neutron decay (1,175 sec).

To obtain the rates of reactions (4) and (5), read from left to right, multiply (7), using the appropriate sign, by the number of positrons/electrons with energies between W and $W + dW$ and then integrate the product with respect to W from zero to infinity. The corresponding values for the rates of (4) and (5) read from right to left can then easily be obtained by noticing that in thermodynamic equilibrium:

$$\frac{n}{p} = \exp - \frac{W_0}{kT} \quad (8)$$

where n, p represent the densities of neutrons and protons.

If, however, one is content with accuracy to within a few per cent it is sufficient simply to write $W \doteq mc^2 + qkT$ in (7) and to multiply by the total number density of pairs. The value then chosen for q is that which makes $mc^2 + qkT$ equal to the average electron energy at temperature T; this is a slowly varying function of T and is given by Chandrasekhar[12]. The pair density is:

$$\frac{1}{\pi^2}\left(\frac{kT}{\hbar c}\right)^3\left(\frac{mc^2}{kT}\right)^2 K_2\left(\frac{mc^2}{kT}\right) \quad (9)$$

where K_2 is the modified Bessel function of the second kind and second order. For $T_{10} \sim 1$ this expression can be closely approximated by the simple form $30\ a\ T^3/\pi^4 k$. The number density of neutrino-antineutrino pairs (of both kinds) is $15\ a\ T_3/\pi^4 k$.

Adopting this simplified procedure, the reaction rate for $n + e^+ \rightarrow p + \bar{\nu}$ is easily seen to be:

$$0\cdot071\ T_{10}^3\ (1 + 0\cdot476\ qT_{10})^2 \text{ per neutron per sec} \quad (10)$$

The effect of $n + \nu \rightarrow p + e^-$ is approximately to double the rate at which neutrons are converted to protons. Using equation (8), the rates of the inverse reactions are obtained by multiplying equation (10) by $\exp - W_0/kT = \exp - 1\cdot506/T_{10}$. Hence the following differential equation determines the variation of $n/(n + p)$ with time:

$$\frac{d}{dt}\left(\frac{n}{n+p}\right) = 0\cdot142 T_{10}^3\ (1 + 0\cdot476 qT_{10})^2 \times \left[\frac{n}{n+p}\left(1 + \exp\frac{-1\cdot506}{T_{10}}\right) - \exp\frac{-1\cdot506}{T_{10}}\right] \quad (11)$$

To express this in a form convenient for numerical integration use T_{10} as the independent variable. With the aid of equation (6):

$$\frac{d}{dT_{10}}\left(\frac{n}{n+p}\right) = 0\cdot308(1 + 0\cdot476 qT_{10})^2 \times \left[\frac{n}{n+p}\left(1 + \exp\frac{-1\cdot506}{T_{10}}\right) - \exp\frac{-1\cdot506}{T_{10}}\right] \quad (12)$$

Equation (12) can be integrated from a sufficiently high temperature, at which the neutrons and protons are almost in thermodynamic balance, down to the temperature at which the pairs disappear and deuterons are formed. The results are insensitive to starting temperature if it is chosen above $T_{10} = 2\cdot5$. When the protons and neutrons are in thermodynamic balance the right-hand side of (12) is zero and the initial value of $n/(n + p)$ is chosen to make this right-hand side zero.

An important question evidently arises as to the precise value of T_{10} down to which equation (12) should be integrated. Rather surprisingly, it appears the deuterium combines into helium, through $D(D,n)He^3(n,p)T(p,\gamma)He^4$, at a temperature as high as $T_{10} = 0\cdot3$, in spite of the small binding energy of deuterium. (The concentration of deuterium used in establishing this conclusion was just that which exists for statistical equilibrium in $n + p \rightleftarrows D + \gamma$.) Hence, equation (12) must not be integrated to T_{10} below 0·3. We estimate that integration down to $T_{10} = 0\cdot5$ probably gives the most reliable result, because (12) overestimates the rate of conversion of neutrons to protons below $T_{10} = 0\cdot5$.

Mr. J. Faulkner has solved the equation for several starting temperatures. Provided $T_{10} > 2\cdot5$ initially, he finds $n/(n + p) = 0\cdot18$ at $T_{10} = 0\cdot5$, giving:

$$\frac{\rm He}{\rm H} = \frac{n}{2(p - n)} \simeq 0\cdot14 \quad (13)$$

a result in good agreement with the calculations of Alpher, Follin and Herman[11]. Allowing for the approximation in our integration procedure we estimate that this value is not more uncertain than $0\cdot14 \pm 0\cdot02$. It should be particularly noted that, unlike the result of Alpher, Bethe and Gamow, this value depends only slightly on the assumed material density; essentially this result is obtained provided the density is high enough for deuterons to be formed in a time short compared to the neutron half-life and low enough for the rest mass energy density of the nucleons to be neglected in comparison with the energy density of the radiation field.

Before comparing this result with observation we note that variations of the cosmological conditions which led to equation (6) all seem as if they would have the effect of increasing He/H. If the rest mass energy density were not less than the sum of the energy densities of radiation, pairs and neutrinos, the Universe would have to expand faster at a given temperature in order to overcome the increased gravity, the time-scale would be shorter and the coefficient on the right-hand side of equation (12) would be reduced. Similarly, if there were more than two kinds of neutrino the expansion would have to be faster in order to overcome the gravitational attraction of the extra neutrinos, and the time-scale would again be shorter; and the smaller the coefficient on the right-hand side of equation (12) the larger the ratio He/H turns out to be.

We can now say that if the Universe originated in a singular way the He/H ratio cannot be less than about 0·14. This value is of the same order of magnitude as the observed ratios although it is somewhat larger than most of them. However, if it can be established empirically that the ratio is appreciably less than this in any astronomical object in which diffusive separation is out of the question, we can assert that the Universe did not have a singular origin. The importance of the value 0·09 for the Sun is clear; should this value be confirmed by further investigations the cosmological implication will be profound. (A similar situation arises in the case of an oscillating universe. The maximum temperature, achieved at moments of maximum density, must be high enough for all nuclei to be disrupted, that is, $T_{10} > 1$. Otherwise, after a few oscillations all hydrogen would be converted into heavier nuclei, and this is manifestly not the case.)

It is reasonable, however, to argue in an opposite way. The fact that observed He/H values never differ from 0·14 by more than a factor 2, combined with the fact that the observed values are of necessity subject to some uncertainty, could be interpreted as evidence that the Universe did have a singular origin (or that it is oscillatory). The difficulty of explaining the observed values in terms of hydrogen-burning in ordinary stars supports this point of view. So far as we are aware, there is only one strong counter to this argument, namely, that there is nothing really special to cosmology in the foregoing discussion. A similar result for the He/H ratio will always be obtained if matter is heated above $T_{10} = 1$, and if the time-scale of the process is similar to that given by equation (6). In this connexion it may well be important that the physical conditions inside massive objects or superstars simulate a radiation Universe. Hoyle, Fowler, Burbidge and Burbidge[13] were led, for reasons independent of those of the present article, to consider temperatures exactly in the region $T_{10} \simeq 1$. These authors give the following differential equation between the time t and the density ρ, in such a superstar:

$$dt = \left(\frac{1}{24\pi G\rho}\right)^{1/2} \frac{d\rho}{\rho} \tag{14}$$

and also the following relation for an object of mass M:

$$\rho = 2{\cdot}8 \times 10^{8}\left(\frac{M_{\odot}}{M}\right)^{1/2} T_{10}^{3}\ \text{g cm}^{-3} \tag{15}$$

Eliminating ρ and $d\rho$ we have:

$$dt = \left(\frac{M}{2{\cdot}44 \times 10^{4} M_{\odot}}\right)^{1/4} \frac{dT_{10}}{T_{10}^{2.5}} \tag{16}$$

whereas the differential form of equation (6) is:

$$dt = -\,2{\cdot}08\,\frac{dT_{10}}{T_{10}^{3}} \tag{17}$$

The difference of sign arises because equation (16) was given for a contracting object. For re-expansion of an object the sign must be reversed, so that the time-scales are identical if $M \simeq 5 \times 10^{5} M_{\odot}$. It may be significant that this is about the largest mass in which the temperature $T_{10} \simeq 1$ can be reached without the object being required to collapse inside the Schwarzschild critical radius. If collapse inside this radius followed by re-emergence be permitted, larger masses can be considered. The time-scale is then increased above equation (17) and this has the effect of giving a smaller He/H ratio than that calculated above. If an object is inside the Schwarzschild radius and neutrinos do not escape from it, the conditions are closely similar to the cosmological case. On the other hand, the calculation must be slightly changed for objects that do not enter the Schwarzschild radius, since neutrinos are certainly not contained within them. Thus if the same time-scale were used, that is, $M \simeq 10^{5} - 10^{6} M_{\odot}$, absence of neutrinos would reduce the right-hand side of equation (12) by a factor of 2. A corresponding calculation leads to $n/(n + p) = 0{\cdot}22$, also in reasonable agreement with observation, especially as all material need not have passed through massive objects. However, a more detailed discussion of massive objects will be required to decide whether the required amount of helium cannot only be produced but also ejected from them.

This brings us back to our opening remarks. There has always been difficulty in explaining the high helium content of cosmic material in terms of ordinary stellar processes. The mean luminosities of galaxies come out appreciably too high on such a hypothesis. The arguments presented here make it clear, we believe, that the helium was produced in a far more dramatic way. Either the Universe has had at least one high-temperature, high-density phase, or massive objects must play (or have played) a larger part in astrophysical evolution than has hitherto been supposed. Clearly the approximate calculations of this present article must be repeated more accurately, but we would stress two general points: (1) the weak interaction cross-sections turn out to be just of the right order of magnitude for interesting effects to occur in the time-scale available; (2) for a wide range of physical conditions (for example, nucleon density) roughly the observed amount of helium is produced.

[1] Mendez, M. E., thesis, California Institute of Technology (1963).
[2] Mathis, J. S., *Astrophys. J.*, **136**, 374 (1962).
[3] Aller, L. H., and Faulkner, D. J., *Publ. Astro. Soc. Par.*, **74**, 219 (1962).
[4] Aller, L. H., *The Abundance of the Elements* (New York: Interscience Publ., 1961).
[5] O'Dell, C. R., *Astrophys. J.*, **138**, 1018 (1963).
[6] Gaustad, J. E., *Astrophys. J.*, **139**, 406 (1964).
[7] Biswas, S., and Fichtel, C. E., *Astrophys. J.*, **139**, 941 (1964).
[8] Sears, R. L., *Helium Content and Neutrino Fluxes in Solar Models* (preprint, Cal. Inst. Tech., 1964).
[9] Alpher, R. A., Bethe, H. A., and Gamow, G., *Phys. Rev.*, **73**, 803 (1948).
[10] Hayashi, C., *Prog. Theo. Phys.* (Japan), **5**, 224 (1950).
[11] Alpher, R. A., Follin, J. W., and Herman, R. C., *Phys. Rev.*, **92**, 1347 (1953).
[12] Chandrasekhar, S., *An Introduction to the Study of Stellar Structure* (Chicago, 1939).
[13] Hoyle, F., Fowler, W. A., Burbidge, G. R., and Burbidge, E. M., *Astrophys. J.*, **139**, 909 (1964).

DENSITY OF RELICT PARTICLES WITH ZERO REST MASS IN THE UNIVERSE

V. F. Shvartsman
Institute of Applied Mathematics, Moscow State University
Submitted 20 January 1969
ZhETF Pis. Red. 9, No. 5, 315 - 317 (5 March 1969)

So far, one cannot exclude the possible existence in the universe of a large number of difficult-to-observe particles with zero rest mass (DZP), left over from the superdense phase (neutrinos, gravitons, etc). Estimates of their density, based on the gravitational influence during the later expansion stages, leads to a value [1]

$$\rho_{m=0} < 3\rho_c \approx 5\times10^{-29}\ \mathrm{g/cm^3}$$

(ρ_c = critical density). On the other hand, the gravitational action of similar particles in the early expansion changes, by changing the rate of the expansion, changes the course of the nuclear reactions in the primordial matter [4, 5]. In the present paper we develop this idea further and take into account the fact that the helium content in a number of stars is known to be less than 40%; we then obtain for the case of the Friedmann model with small or zero "specific" lepton charge[1)] a stronger limit on the DZP density:

$$\rho_{m=0} < 5\rho_\gamma \approx 2\times10^{-33}\ \mathrm{g/cm^3}, \tag{1}$$

where ρ_γ is the density of the background radio emission with temperature T = 2.7°K.

Let us turn to the formula relating the temperature T of the primordial matter with the time t elapsed from the start of the expansion:

$$\rho_\Sigma = \kappa\sigma T^4/c^2 = 3/32\pi G t^2. \tag{2}$$

Here ρ_Σ - density of all of the matter, σ - Stefan-Boltzmann constant, G - gravitational constant, and κ - dimensionless quantity characterizing the ratio of the density of all the particles to the density of the gamma quanta (equilibrium γ, e^+, e^-, ν_e, $\bar\nu_e$, ν_μ, and $\bar\nu_\mu$ correspond to $\kappa = 9/2$). We note that the change of κ involves a change in the connection between T and t.

At high temperatures, the reactions

$$\begin{aligned} e^+ + n &\rightleftarrows p + \bar\nu \\ \nu + n &\rightleftarrows p + e^- \end{aligned} \tag{A}$$

ensure an equilibrium ratio of the neutron concentration n to the proton concentration p:

$$n/p = \exp(-\Delta mc^2/kT).$$

Here Δm is the mass difference between the neutron and the proton.

---

1) We refer to the lepton charge of electronic neutrinos and antineutrinos; its smallness is connected with the condition $|\rho_{\nu e} - \rho_{\bar\nu e}| \ll \rho_{\nu e} + \rho_{\bar\nu e}$ (ρ = density). We recall that here $\rho_{\nu e} + \rho_{\bar\nu e} = (7/8)(8/11)^{4/3}\rho_\gamma$. A similar fraction is obtained for the muonic neutrinos and antineutrinos under suitable limitations on their lepton charge.

The time of establishment of equilibrium is

$$\tau = 1/\sigma n c \simeq \text{const}/T^5, \tag{3}$$

where σ is the cross section and n_ν is the density of the neutrino; the electron rest mass is neglected. During the course of the expansion, an instant $t \simeq \tau$ is reached at which the equilibrium no longer has time to be established: "quenching" of the neutrons takes place [2]. The corresponding temperature can be readily estimated from Eqs. (2) and (3):

$$T \simeq \text{const}\cdot\kappa^{1/6}.$$

Thus, larger κ correspond to higher quenching temperatures: the number of remaining neutrons is larger. An examination of the subsequent processes shows that almost all the neutrons have time to combine with the protons and to form helium nuclei; the corresponding estimates can be found in the table. ρ_n stands for the present-day nucleon density, which determines the specific entropy (see [2]), and ρ is the density of the difficultly-observed matter with zero rest mass, while the index "q" indicates the instant of quenching. We have already

$\rho_{m=0}/\rho_\gamma$	$(\kappa/4{,}5)_q$	$(n/p)_q$	He^4		
			$\rho_H = 3\cdot10^{-29}$	$\rho_H = 3\cdot10^{-30}$	$\rho_H = 3\cdot10^{-31}$
0.45	1	0.16	0,29	0.27	0.25
1.45	1.9	0.20	0.36	0.34	0,31
5.45	5.3	0.26	0.48	0.45	0,43
10.45	9,6	0.29	0.54	0.50	0.48

mentioned that the hydrogen content in the protostars certainly exceeded 60%, and consequently $\rho_{m=0} < 5\rho_\gamma$.

To which particles does this limit pertain?

1. To muonic neutrinos and antineutrinos. The deviation of their density from the equilibrium value is connected with the possibility of a larger leptonic (muonic) charge per unit of the co-moving volume of the universe.

2. To gravitons. The latter do not have any time at all to enter in equilibrium with the primordial matter [3], and their amount is determined by the initial conditions.

3. To still-unknown ultra-weakly-interacting particles left from the superdense phase[1)].

The question of the electronic neutrinos and antineutrinos calls for a separate analysis. The point is that the presence of a chemical potential in ν_e ($\bar{\nu}_e$) leads not only to an increase of their density, but also to a direct change of the dynamics of the reactions (A), and this can cause cancellation of the indicated mechanism and lead to an arbitrarily low helium content [4]; the latter does not contradict the observations. In other words, if the specific leptonic (electronic) charge of the universe is large, the flux of DZP may exceed 2 x 10^{-33} g/cm^3.

In conclusion, we call attention of the adherents of the variable gravitational constant to the fact that G enters in formula (2) in the same manner as κ, and that variability of G would lead to an entirely different behavior of the nuclear processes in the primordial matter.

[1)] We note that the limitation on the number of species of the undiscovered DZP is weaker than the limitation on their density; the earlier the particles leave the equilibrium state, the smaller the fraction of the total energy that they acquire [3].

The author is sincerely grateful to Ya. B. Zel'dovich for interest in the work and significant remarks, and to R. A. Syunyaev for useful discussions.

[1] Ya. B. Zel'dovich and Ya. A. Smorodinskii, Zh. Eksp. Teor. Fiz. 41, 907 (1961) [Sov. Phys.-JETP 14, 647 (1962)].
[2] Ya. B. Zel'dovich and I. D. Novikov, Relyativistskaya astrofizika (Relativistic Astrophysics), Nauka, 1967.
[3] Ya. B. Zel'dovich, Usp. Fiz. Nauk 89, 647 (1966) [Sov. Phys.-Usp. 9, 602 (1966)].
[4] R. V. Wagoner and W. A. Fowler, Astroph. J. 148, 3 (1967).
[5] G. Dautcour and G. Wallis, Fortschr. Physik 16, 545 (1968).

Volume 66B, number 2 PHYSICS LETTERS 17 January 1977

COSMOLOGICAL LIMITS TO THE NUMBER OF MASSIVE LEPTONS

Gary STEIGMAN
National Radio Astronomy Observatory[1] *and Yale University*[2], *USA*

David N. SCHRAMM
University of Chicago, Enrico Fermi Institute (LASR), 933 E 56th, Chicago, Ill. 60637, USA

James E. GUNN
University of Chicago and California Institute of Technology[2], *USA*

Received 29 November 1976

If massive leptons exist, their associated neutrinos would have been copiously produced in the early stages of the hot, big bang cosmology. These neutrinos would have contributed to the total energy density and would have had the effect of speeding up the expansion of the universe. The effect of the speed-up on primordial nucleosynthesis is to produce a higher abundance of ^{4}He. It is shown that observational limits to the primordial abundance of ^{4}He lead to the constraint that the total number of types of heavy lepton must be less than or equal to 5.

Possible interpretations of recent observations (for example, the anomalous eμ events produced in e^+e^- annihilation [1] suggest the existence of leptons more massive than muons. It would be expected that each such lepton would have associated with it a corresponding neutrino-antineutrino pair. It will be shown in this note that the existence of these neutrinos has observable implications in the standard big bang cosmology and that present observations place limits on the number of heavy lepton types.

It is interesting that in V-A theories of the weak interaction there is a direct correspondence between the number of lepton types and the number of quarks. (This correspondence does not hold in all theories such as some of the recent vector-like models of the weak interaction.) Therefore, a limit to the number of lepton types may also be a limit to the number of quarks. Since asymptotic freedom [2] does not hold if there are more than 16 quark color triplets and since in V-A theories each lepton and its neutrino are related to a pair of quark color triplets, then it would be quite interesting if it can be shown that the total number of lepton types is less than 8. In fact, we will show that present observations imply less than 5 heavy leptons (less than 7 lepton types including e and μ).

It has been known for some time [3] that the number of particle types can produce observable effects in the standard hot big bang model of the universe. For the present context, note that in the early, hot stages of a big bang cosmology, any massive leptons would have been copiously produced and, in particular, their associated neutrinos would have been as abundant as the electron neutrinos and muon neutrinos. As a result, these new neutrinos would have contributed significantly to the total energy density [‡1]. During the early stages, the universe is in "free expansion" (the time reversal of free fall) so that the expansion time scale (age) and the energy density are related by $\rho \propto t^{-2}$. The effect of "extra" zero-mass particles is to increase the density ($\rho \to \rho' \equiv \xi^2\rho$) and, thus, to decrease the time scale ($t \to t' = \xi^{-1}t$).

A change in the expansion rate can significantly change the abundances of the elements produced by primordial nucleosynthesis [4]. The dominant effect

[1] The National Radio Astronomy Observatory is operated by Associated Universitites, Inc. under contract with the National Science Foundation.
[2] Permanent address.

[‡1] For massless particles (or, for $Mc^2 \ll kT$) the energy density varies as the fourth power of the temperature whereas for massive particles ($Mc^2 \gg kT$) the energy density varies as the cube of the temperature. Thus, in the early, hot epochs, the energy density is dominated by the contribution from the massless particles.

of the speed-up is to alter the neutron-to-proton ratio. The neutron-to-proton ratio has its equilibrium value ($n/p = \exp(-\Delta Mc^2/kT)$) as long as the rates of the various weak interactions (e.g., $n + e^+ \rightleftarrows p + \bar{\nu}_e$, $p + e^- \rightleftarrows n + \nu_e$, $n \rightleftarrows p + e^- + \bar{\nu}_e$) are fast compared with the expansion rate. An increased expansion rate ($\xi > 1$) forces the weak interactions out of equilibrium at a higher temperature and, thus, leads to a higher neutron-to-proton ratio [‡2]. The primary effect of an increased neutron-to-proton ratio is to increase the abundances of deuterium, helium-3 and helium-4. Since ^{4}He is produced but not easily destroyed in the course of galactic evolution, the observed ^{4}He abundance provides an upper limit to the primordial abundance and, hence, provides a limit to the speed-up of the expansion. From the chain: "new" leptons → increased density → speed-up → increased ^{4}He abundance, we may obtain an upper limit to the number of unknown leptons.

In this letter we demonstrate that the observed ^{4}He abundance provides a significant constraint to the number of massive leptons [‡3]. For the "standard" big bang model, the primordial abundances only depend on the density of nucleons at the temperatures at which nucleosynthesis occurs. A convenient parameter is the ratio of the present mass density in nucleons (g cm^{-3}) to the cube of the photon (i.e., black body) temperature (in units of 10^9 K).

$$h \equiv \rho_N / T^3_{\gamma 9} \,. \tag{1}$$

In terms of the usual density parameter Ω_0 ($\Omega_0 \equiv \rho/\rho_c$ where ρ_c is the critical density) and the present Hubble parameter h_0 (in units of 100 km/s/$M\rho_c$),

$$h \approx 10^{-3} \Omega_0 h_0^2 \,, \tag{2}$$

(Note that the entropy parameter originally introduced by Wagoner et al. [4] is larger than h by a factor 11/4 ($= T^3_\gamma / T^3_\nu$).) For the normal range of h (10^{-5}–10^{-3}), the ^{4}He abundance (by mass) resulting from the standard model for big bang nucleosynthesis lies within the range: $0.23 \lesssim Y \lesssim 0.27$. To account for the effect of speed-up we may adapt a formula given by Wagoner [4]

$$Y = 0.333 + 0.0195 \log h + 0.380 \log \xi \,. \tag{3}$$

Recall now that the square at the speed-up factor ξ is determined by the ratio of energy densities with and without the neutrinos associated with the massive leptons. Accounting for Fermi statistics and for the antineutrinos, we may relate ξ^2 to the number (ΔN_L) of "new" massive leptons.

$$\xi^2 = (\rho'/\rho) = 1 + \tfrac{7}{36} \Delta N_L \,. \tag{4}$$

(Actually ξ varies slightly depending on whether one is discussing the expansion rate before or after the annihilation of particle-antiparticle pairs; however, such changes are small and eq. (4) is sufficient for present purposes.) Before presenting more accurate results it is valuable to note by comparing eqs. (3) and (4) that, for a fixed value of h, the addition of ΔN_L leptons produces a change in the ^{4}He abundance which is roughly

$$\Delta Y \approx 0.016 \, \Delta N_L \,, \tag{5}$$

(provided that $7\Delta N_L \ll 36$).

We have taken as a good upper limit to the primordial abundance of ^{4}He [5] : $Y \lesssim 0.29$. However, it should be remembered that the exact value of Y is quite uncertain. In the first column of table 1 we list various values of h and in the second column the corresponding helium abundance with only electrons and muons but no massive leptons ($\Delta N_L = 0$, $\xi = 1$). In the third column we list that value of the speed-up (ξ_{max}) which results in a helium abundance of $Y = 0.29$ and, in the fourth column appears the number of additional leptons (ΔN_L) which would give (see eq. (4)) the speed-up ξ_{max}.

Table 1

$\log h$	Y ($\xi = 1$)	$\xi_{max}[Y(h, \xi_{max}) \equiv 0.29]$	ΔN_L
−3.0	0.274	1.10	1.1
−3.5	0.265	1.17	1.9
−4.0	0.255	1.24	2.7
−4.5	0.245	1.31	3.7
−5.0	0.235	1.39	4.8

‡2 Note that the neutrinos associated with the heavy leptons do not affect the reactions which determine the neutron-to-proton ratio.

‡3 Note that the primordial abundance of ^{7}Li is determined by a delicate balance of several production and destruction reactions and is thus very sensitive to changes in the expansion rate (see Wagoner, ref. [4]). We are in the process of studying the effect of a speed-up on the ^{7}Li abundance using detailed reaction network calculations and we will present our results in a subsequent publication.

From table 1 we see for any reasonable density ($h \gtrsim 10^{-5}$) that there must be fewer than ~5 new leptons or else too much ^{4}He would have been produced primordially. If the universe has a high density ($h \gtrsim 10^{-3.5}$), the constraint on the number of massive leptons is quite restrictive.

However, even for a low density universe, the maximum ΔN_L is still small enough that asymptotic freedom would not in any way be threatened. Note also that this dependence of ΔN_L on density might eventually provide an additional constraint on the density of the universe if several heavy lepton types were detected.

While mentioning constraints on the density, it is worthwhile noting that the existence of heavy leptons does not seriously effect the density limits derived from the primordial abundance of deuterium [6]. It can be shown that if the big bang produces significant amounts of deuterium, then the amount of deuterium left after the big bang is controlled by the rate of the D + D reaction. A faster expansion rate would allow less time for destruction and more D would be left. In fact, in the region of interest a particular deuterium abundance corresponds to a constant value for the ratio ξ/h. For a fixed deuterium abundance, an increased ξ corresponds to a higher density but, since the maximum value for ξ is ~1.4, then the maximum effect on the density is also $\lesssim 1.4$ whereas the density required to close the universe is more than an order of magnitude [6] greater than that implied by standard big bang deuterium production.

In summary, it has been shown that the standard big bang model for the universe implies a correspondence between the number of lepton types and the abundance of helium. Since the abundance of helium in the universe seems to be less than ~29% by mass, this implies that there are no more than 7 lepton types. Of course, it should be remembered that the standard big bang model is by no means rigorously verified but is merely the simplest model which seems consistent with observations.

We would like to acknowledge some extremely useful conversations with R. Epstein on the deuterium effects, M. Kislinger on some of the implications of this work for particle theory and M. Rees on historical perspective. This work was begun while one of us (JEG) held the George Ellery Hale Visiting Professorship at the University of Chicago and JEG thanks the university for that privilege. Two of us (DNS and GS) would like to acknowledge the hospitality of the Aspen Center for Physics where some of this work was done while participating in the Astrophysics Workshop which is supported by NASA grant NGR 06-018-001. This research was supported in part by NSF grant AST 74-21216 at the University of Chicago.

References

[1] M. Perl, Proc. 1976 Neutrino Conference at Aachen.
[2] H.D. Politzer, Phys. Rev. Lett. 30 (1973) 1346;
D.J. Gross and F. Wilczek, Phys. Rev. Lett. 30 (1973) 1343.
[3] V.F. Shvartsman, JEPT Lett. 9 (1969) 184.
[4] R.V. Wagoner, W.A. Fowler and F. Hoyle, Astrophys. J. 148 (1967) 3;
P.J.E. Peebles, Astrophys. J. 146 (1966) 542;
R.V. Wagoner, Astrophys. J. 179 (1973) 343.
[5] M. Peimbert, Ann. Rev. Astron. Astrophys. 13 (1976) 113.
[6] D.N. Schramm and R.V. Wagoner, Physics Today 27 (1974) 40;
J.R. Gott et al., Astrophys. J. 194 (1974) 543;
H. Reeves et al., Astrophys. J. 179 (1973) 909.

THE ASTROPHYSICAL JOURNAL, 246:557–568, 1981 June 15

BIG-BANG NUCLEOSYNTHESIS AS A PROBE OF COSMOLOGY AND PARTICLE PHYSICS

KEITH A. OLIVE[1], DAVID N. SCHRAMM[1], GARY STEIGMAN[2], MICHAEL S. TURNER[1], AND JONGMANN YANG[1]
Received 1980 November 3; accepted 1980 December 19

ABSTRACT

The mass fraction of ^{4}He synthesized in the big bang, Y_P, depends upon the neutron half-life $\tau_{1/2}$, the ratio of baryons to photons η, and the number of two-component neutrino species N_ν. New observational and experimental data have led us to reexamine the constraints on cosmology and particle physics which follow from primordial nucleosynthesis. We find that η must lie in the range $10^{-9.9\pm1}$, implying that baryons alone cannot close the universe; the related ratio of the baryon number to the specific entropy must lie in the range $10^{-10.8\pm1}$. If baryons provide most of the mass which binds binary and small groups of galaxies, then N_ν must be ≤ 4. However, if massive neutrinos (or other nonbaryonic matter) provide this mass, then at present no firm limit can be placed on N_ν. If the universe is dominated by nonbaryonic matter, than there is no contradiction between the predictions of primordial nucleosynthesis and the observations of ^{4}He, provided that $Y_P \gtrsim 0.15$.

Subject headings: abundances — cosmology — nucleosynthesis

I. INTRODUCTION

A careful study of the physics of the early universe can provide information of value in determining the present and future evolution of the universe and may lead to significant constraints on models of elementary particle physics (Steigman 1979; Turner and Schramm 1979). Through the abundances of the light elements ^{2}H, ^{3}He, ^{4}He, and ^{7}Li, primordial nucleosynthesis offers a unique probe of the early evolution of the universe (for recent reviews, see Schramm and Wagoner 1977; Wagoner 1980). For several reasons, the primordial abundance of ^{4}He is of outstanding value as a probe of the physics of the early universe. The presently observed abundance of ^{4}He is large, so that the expected contamination from ^{4}He produced in the course of stellar and galactic evolution is small. Furthermore, in the standard (Robertson-Walker-Friedmann) cosmology, the primordial abundance of ^{4}He produced is large, so that it could not be destroyed easily without producing excessively large abundances of heavier elements. As a result, a good case can be made that the present abundance of ^{4}He is dominated by (and places a good upper limit on) the primordial abundance. Finally, it is important that the predicted abundance of ^{4}He is least sensitive to the most poorly constrained cosmological parameter: η, the ratio of baryons to photons. (Note that we use baryon and nucleon interchangeably, so that by baryon we only mean protons or neutrons). In contrast, the observed abundances of the other light elements are very small, and since they are more easily produced (with the exception of ^{2}H) and destroyed during stellar and galactic evolution, it is quite difficult to derive reliable primordial abundances for ^{2}H, ^{3}He, and ^{7}Li. For these reasons, they are of less value than ^{4}He in testing the standard model or in constraining particle physics. However, the other light elements may be used (with caution) to provide significant constraints on the present nucleon density (Wagoner 1973; Reeves *et al.* 1973; Steigman 1975; Austin and King 1977; Yang *et al.* 1981).

Recent work has illustrated the value of this approach to particle physics and cosmology. For example, Yang *et al.* (1979, hereafter YS^{2}R; also, Steigman, Schramm, and Gunn 1977; Shvartsman 1969) have shown that, at most, one new "flavor" of two-component neutrinos (in addition to the e, μ, and τ neutrinos) is permitted if the primordial abundance (by mass) of ^{4}He, Y_P, ≤ 0.25 (and if baryons dominate the mass of small groups of galaxies). In addition, Olive, Schramm, and Steigman (1981, hereafter OSS; also, Steigman, Olive, and Schramm 1979, hereafter SOS) have noted that right-handed counterparts (if they exist) to the familiar left-handed neutrinos must interact much more weakly in order for them to have decoupled much earlier in the evolution of the universe (and hence to have not affected nucleosynthesis). Of more direct interest to cosmology is the limit $\eta < 4\times10^{-10}$ on the nucleon-to-photon ratio derived by YS^{2}R for $Y_P \leq 0.25$. Provided that the present value of the Hubble parameter $H_0 \geq 50$ km s^{-1}Mpc^{-1}, this corresponds to a nucleon density of less than 6% of the critical density ($\Omega_N \leq 0.06$).

[1]Astronomy and Astrophysics Center, The University of Chicago.
[2]Bartol Research Foundation, University of Delaware.

The strength and importance of these constraints to particle physics and cosmology demand a careful scrutiny of the relevant assumptions and the underlying physics. In fact, there have been recent developments in particle physics, nuclear physics, and cosmology which, when coupled with new data on the helium abundance, make this an appropriate time for just such a reexamination.

To provide a framework for the subsequent discussion, let us briefly review the physics of helium production in the standard model (for further details, discussion of underlying assumptions and of the evidence which supports the validity of the model, see Schramm and Wagoner 1977; Wagoner 1980). The buildup of the heavier elements in primordial nucleosynthesis is suppressed by the gaps at mass 5 and 8 (no stable nuclei) and by the coulomb barriers which become insurmountable as the universe expands and cools. Since ^{4}He is the most tightly bound of the light nuclei, virtually all the neutrons available are eventually bound into ^{4}He. The primordial abundance of ^{4}He is determined, therefore, by the neutron abundance at the time nucleosynthesis begins in earnest. There are several factors which determine the neutron abundance at this time. As the universe expands and cools, the ratio of neutrons to photons decreases, attempting to "track" the equilibrium value $[\exp(-\Delta mc^2/kT)]$. The reactions which are responsible for this are the usual charged-current weak interactions,

$$n+e^+ \leftrightarrow p+\bar{\nu}_e, \quad n+\nu_e \leftrightarrow p+e^-, \tag{1a}$$

and the occasional β-decay,

$$n \rightarrow p+e^- + \bar{\nu}_e. \tag{1b}$$

At a given temperature, how well the neutron abundance tracks the equilibrium value is determined by the competition between the weak interaction rate Γ_{wk} and the expansion rate of the universe $\sim t^{-1}$. As the temperature decreases, eventually Γ_{wk} becomes equal to and subsequently falls below t^{-1}, and reactions (1a) and (1b) effectively cease to occur. When this happens ($T_f \approx 10^{10}$ K), the neutron-to-proton ratio "freezes out" at a value of $\sim 1/6$; then it very slowly decreases because of occasional neutron decays until nucleosynthesis, when $n/p \sim 1/7$. Nucleosynthesis does not begin until the temperature decreases to about 10^9 K because of the "deuterium bottleneck." The relatively weak binding of the deuteron (2.22 MeV) and the large photon-to-baryon ratio ($\approx 10^{10}$) conspire to keep the deuterium abundance small until $T \approx 10^9$ K. At this point, the bottleneck breaks and essentially all the neutrons end up in ^{4}He.

In the standard model of primordial nucleosynthesis (see Wagoner 1980 for a detailed discussion) which we have briefly described above, the mass fraction of helium produced depends upon three parameters: (1) $\tau_{1/2}$, the neutron half-life; (2) η, the baryon-to-photon ratio; and (3) N_ν, the number of two-component neutrino species ($\nu_e, \nu_\mu, \nu_\tau, \ldots$). The neutron half-life determines the rates for reactions (1a) and (1b); if $\tau_{1/2}$ increases, then Γ_{wk} decreases and falls below the expansion rate t^{-1} earlier. The neutron-to-proton ratio freezes out at a higher temperature and, hence, value; thus, more helium is produced ultimately.

The baryon-to-photon ratio determines when the deuterium bottleneck breaks; a larger value of η (fewer photons per baryon) results in the bottleneck breaking earlier, at a higher value of the neutron-to-proton ratio, once again resulting in more helium production. The dependence of Y_P upon η is rather weak. The number of baryons has remained constant since nucleosynthesis (baryon number is conserved at these low energies, $\lesssim 10$ MeV$\ll 10^{14}$ GeV), while the number of photons has increased (in a calculable way) because of $e^{\pm}$ annihilations at a temperature of $\approx 3\times 10^9$ K, so that the baryon-to-photon ratio then is simply related to η. The number densities of photons and baryons today are

$$n_{\gamma 0} = 400.(T_0/2.7\text{ K})^3\text{ cm}^{-3}, \tag{2a}$$

$$n_b = 1.13\times 10^{-5}\,\Omega_N h_0^2\text{ cm}^{-3}, \tag{2b}$$

where T_0 is the present temperature of the microwave background, $\Omega_N = \rho_N/\rho_c$, ρ_N is the baryon (nucleon) mass density, $\rho_c \equiv 3H_0^2/8\pi G$ is the present critical density, and $H_0 = 100\, h_0$ km s^{-1}Mpc^{-1} is the present value of the Hubble parameter. The quantities $\Omega_N h_0^2$ and ρ_N are related to η by

$$\Omega_N h_0^2 = 3.53\times 10^7 \eta (T_0/2.7\text{ K})^3, \tag{3a}$$

$$\rho_N = (6.64\times 10^{-22}\text{ g cm}^{-3})\eta (T_0/2.7\text{ K})^3. \tag{3b}$$

Again, we emphasize that, although nucleosynthesis has often has been discussed in terms of its dependence upon Ω_N, h_0, T_0, and ρ_N, its *only* dependence upon these quantities is through its dependence upon η.

Finally, N_ν affects the energy density of the universe and, therefore, the expansion rate ($t^{-1} \sim \rho^{1/2}$). The energy density ρ depends on the number of relativistic particle species present and, as N_ν increases, ρ and the expansion rate become larger (for a given temperature). Hence Γ_{wk} becomes equal to t^{-1} at a higher temperature, so that the neutron-to-proton ratio freezes out at a higher value, and more helium is synthesized.

In principle, Y_P also depends upon the many other nuclear reaction rates which go into the big-bang nucleosynthesis code. We updated the reaction rates in Wagoner's (1973) version of the code using the rates of Fowler, Caughlan, and Zimmerman (1975) and of

Caughlan and Fowler (1980) and found no significant changes in the primordial abundances of ^{2}H, ^{3}He, and ^{4}He produced ($|\Delta X|/X<0.5\%$ for ^{2}H and ^{3}He, and $|\Delta Y_P|/Y_P \ll 0.1\%$). However, the production of ^{7}Li increased by a factor of ~ 3; this is actually a welcome change since previous attempts to explain the present abundance of ^{7}Li by primordial nucleosynthesis were not successful because the primordial production was a factor of ~ 10 too small. Taking into account both the numerical errors associated with the integration scheme in the code and the uncertainties in the nuclear reaction rates, the uncertainty in Y_P is <0.004 (Wagoner 1969), so that in practice $Y_P \equiv Y_P(\tau_{1/2}, \eta, N_\nu)$.

In summary, Y_P increases with increasing values of $\tau_{1/2}$, η, and N_ν. Knowledge of any three of these quantities (Y_P, $\tau_{1/2}$, η, and N_ν) constrains the fourth, so that primordial nucleosynthesis can be used to determine (or constrain) important cosmological and particle physics quantities. Recent changes in the "accepted" values (or, ranges of values) for $\tau_{1/2}$, η, and Y_P have motivated the present reinvestigation. In § II we review, in some detail, the current best estimates for all of these parameters. In § III we reexamine the constraints on η and N_ν and briefly discuss the concordance of the standard model of primordial nucleosynthesis. Our results are summarized in § IV, where we also address the issue of how these limits can be tested and improved.

II. ESTIMATES OF THE VARIOUS PARAMETERS

a) The Lifetime of the Neutron

The rate, Γ_{wk}, of the reactions (1a) and (1b) depends on the measured value of the neutron half-life. Until recently, the accepted value was $\tau_{1/2}=10.61\pm0.16$ minutes (Christensen *et al.* 1972). Our work here was stimulated, in part, by the report of a new determination by Bondarenko *et al.* (1978) which yielded a significantly different value: $\tau_{1/2}=10.13\pm0.09$ minutes. There are, however, other data which are not in support of this new, shorter half-life. Kugler, Paul, and Trinks (1978, 1979) found a value of $\tau_{1/2}=10.62$ minutes and a lower limit of $\tau_{1/2}>10.5\pm0.8$ minutes. Very recently, Byrne *et al.* (1980) reported a preliminary value of $\tau_{1/2}=10.82\pm0.20$ minutes. As techniques for storing neutrons are refined, an accurate (to better than ±0.1 minutes) determination of $\tau_{1/2}$ should be forthcoming. In our subsequent analysis we consider half-lives in the range

$$10.13 \le \tau_{1/2} \le 10.82 \text{ minutes}, \qquad (4)$$

and we present results for $\tau_{1/2}=10.13$, 10.61, 10.82 minutes.

We mention in passing that Wilkinson (1980) has recently reviewed the experimental data on $\tau_{1/2}$. From theoretical and experimental considerations, he infers a value of $\tau_{1/2}=10.43\pm0.12$ minutes; from experimental data only, he infers a value of $\tau_{1/2}=10.69\pm0.13$ minutes. His recommended value for $\tau_{1/2}$ is 10.56 ± 0.09 minutes. These values are all within the range of neutror half-lives that we consider.

b) The Primordial Abundance of ^{4}He

Since ^{4}He is produced during stellar and galactic evolution, the abundance derived from objects at the present epoch is an upper limit to the primordial abundance. In their recent review of the observational and theoretical situation, YS^{2}R suggested that $Y_P \le 0.25$ provided a good upper limit to the primordial abundance of ^{4}He. The discussion here is an extension and updating of their analysis.

There is a general agreement among a large number of observers that, for the "average", normal metal abundance ($Z \approx 0.02$), galactic H II region, $Y=0.30\pm0.02$. In this context it should be noted that the Sun is "underabundant"; the models of Bahcall *et al.* (1980) suggest that $Y_\odot=0.24$, and the analysis of Mazzitelli (1979) implies $Y_\odot=0.23\pm0.01$. For H II regions formed from material which has undergone less stellar processing ($Z<0.02$), the observed ^{4}He abundance is lower, $Y=0.23\pm0.02$; the most recent results are summarized in Table 1. Other data also support this result, which is good evidence for $Y_P \le 0.25$. For example, Thum, Mezger, and Pankonin (1980) find $Y \approx 0.23-0.25$ by observing radio recombination lines from H II regions more than 10 kpc from the galactic center. From a comparison of the numbers of red giants and horizontal branch stars, Renzini (1977) derives for globular clusters $Y=0.22\pm0.04$. From a consideration of the number of blue versus red Cepheids, Castellani (1980) derives for globular clusters $Y \approx 0.23$.

The observational data accumulated recently (see Table 1), if confirmed by further study, suggest that $Y_P \le 0.23$ (i.e., there are increasing numbers of H II regions with $Y \approx 0.23$). It is possible, but with more uncertainty, to try to extrapolate from this data to a truly primordial value. Faulkner (1967) originally suggested that the increase in helium abundance should correlate with the increase in the abundance of the heavier elements. The recent data, in our opinion, do not present a convincing case for a well-defined ΔY versus ΔZ correlation. Nonetheless, we summarize in Table 2 recent estimates of Y_P and $\Delta Y/\Delta Z$ as derived by various authors. Taken at face value, the data seem to suggest that $0.20 \le Y_P \le 0.24$. We also mention that Schmid-Burgk (1981) has used the radio combination line observations of Thum, Mezger, and Pankonin (1980) and $\Delta Y = 3\Delta Z$ to extrapolate to $Y_P \approx 0.21-0.234$. Some models of galactic chemical evolution (Chiosi 1979) do suggest $\Delta Y \approx 3\Delta Z$ in agreement with Peimbert's (1975) estimate from the observational data. However, it is possible that, before the heavy elements were synthesized, an earlier genera-

TABLE 1

HELIUM ABUNDANCE DETERMINATIONS

Object	Y	References
Galactic H II regions	0.32 ± 0.01	(1)
Orion	0.28 ± 0.01	(1), (2), (3), (4)
LMC, SMC	0.25 ± 0.01	(1), (3), (4), (5)
Young galaxies	0.23 ± 0.02	(6)
Low luminosity galaxies	0.24 ± 0.01	(1)
Emission-line dwarf galaxies ...	$0.25^{+0.01}_{-0.02}$	(2)

REFERENCES.—(1) French 1979; (2) Kinman and Davidson 1980; (3) Lequeux *et al.* 1979; (4) Dufour, Shields, and Talbot 1980; (5) Peimbert and Torres-Peimbert 1976; (6) Talent 1980.

TABLE 2

ESTIMATES OF THE PRIMORDIAL ABUNDANCE OF HELIUM

Y_P	$\Delta Y/\Delta Z$	References
0.216 ± 0.020	2.7	(1)
0.228 ± 0.014	2.8 ± 0.6	(2)
0.216 ± 0.015	3.2 ± 0.7	(3)
0.216	2.8 ± 2.7	(4)
0.225 ± 0.015	...	(5)

REFERENCES.—(1) Peimbert and Torres-Peimbert 1976; (2) Lequeux *et al.* 1979; (3) French 1979; (4) Talent 1980; (5) Peimbert 1980.

tion of stars produced some amount of ^{4}He (so that $\Delta Y \gg 3\Delta Z$); even in these models, the upper limit $Y_P \leq 0.23$ would still remain intact.

In surveying the data, we have found no convincing evidence for $Y_P \lesssim 0.20$. For example, although French (1979) finds for the very low-metal abundance object I Zw 18, $Y = 0.17 \pm 0.04$, Peimbert and Torres-Peimbert (1976) found $Y = 0.24$, Lequeux *et al.* (1979) derived $Y = 0.23$, and Kinman and Davidson (1980) derived $Y = 0.22 \pm 0.03$. Clearly, this object needs to be remeasured.

Our assessment of the data is that, while $Y_P \leq 0.25$ provides a good upper limit to the primordial abundance of ^{4}He, evidence is accumulating which suggests $Y_P \leq 0.23$ might provide a better limit. In our subsequent discussion, we also consider the possibility that Y_P is as large as 0.27, although present observations (see Table 1) seem to indicate that this is unlikely. In addition, we comment on the consequences of $Y_P \leq 0.22$.

c) The Temperature of the Background Radiation

The ^{4}He abundance depends on the ratio of baryons to photons, η, which, in the standard model, has remained constant since just after nucleosynthesis ($T \lesssim 0.1$ MeV). If the present density of relic photons is known, then estimates of η from nucleosynthesis may be used to infer the present density in nucleons (cf. eqs. [2] and [3]). Although most data are consistent with a blackbody spectrum for the background radiation, there is the suggestion that the spectrum may deviate slightly from a true blackbody beyond the peak (Woody *et al.* 1975; Woody and Richards 1979). These latter observations, which, if correct, would be of extreme importance to cosmology, await independent confirmation. The remaining data are consistent with a blackbody spectrum corresponding to a present photon temperature in the range,

$$2.7 \leq T_0 \leq 3.0 \text{ K}, \tag{5}$$

(Thaddeus 1972; Hegyi, Traub, and Carleton 1974; Woody *et al.* 1975; Danese and DeZotti 1978). From equation (3a) and $T_0 \leq 3$ K, an upper limit on the nucleon density Ω_N is obtained in terms of η and h_0,

$$\Omega_N \leq 4.84 \times 10^7 \eta h_0^{-2}. \tag{6}$$

d) The Present Value of the Hubble Parameter

In recent years there have been several independent determinations of H_0 by various groups using different techniques (Sandage and Tammann 1976; de Vaucouleurs and Bollinger 1979; Branch 1979; Kirshner and Kwan 1974; Aaronson *et al.* 1980). Although the internal errors in each determination are small, the vast discrepancy among these results suggests residual systematic errors. The range of probable values at present is

$$50 \leq H_0 \leq 100 \text{ km s}^{-1}\text{Mpc}^{-1}; \tfrac{1}{2} \leq h_0 \leq 1. \tag{7}$$

For the standard cosmology, without a cosmological constant, the largest value of H_0 corresponds to an embarrassingly young universe

$$t_0 < H_0^{-1} \approx 10\, h_0^{-1} \times 10^9 \text{ yr}. \tag{8}$$

The ages of globular cluster stars ($\gtrsim 13 \times 10^9$ yr for $Y_P \lesssim 0.25$) suggest that $h_0 \lesssim 3/4$. It is more difficult to constrain the lower end of the range in equation (7). Here, though, it should be mentioned that recent work on deriving H_0 from supernovae suggest that $H_0 \approx 60 \pm 10$ km s^{-1}Mpc^{-1} (Branch 1980). For our estimates in this paper, we use the range of values given in equation (7). From equation (6) and $h_0 \geq 1/2$, an upper limit on the nucleon density, Ω_N, is obtained in terms of η alone,

$$\Omega_N \leq 1.94 \times 10^8 \eta. \quad (9)$$

e) Estimates of Ω_N

The standard "dynamical" approach to determining the average mass density (cf. Faber and Gallagher 1979, hereafter FG) does not distinguish between nucleons and other forms of gravitating material (e.g., massive neutrinos); we will return to this crucial distinction. The average mass density associated with galaxies is derived from the luminosity density through the intermediary of mass-to-light ratios

$$\langle \rho \rangle = \left\langle \frac{M}{L} \right\rangle \mathcal{L}. \quad (10)$$

From $\mathcal{L}$ and the critical density ($\rho_c = 2.78 \times 10^{11}\, h_0^2\, M_\odot$ Mpc^{-3}), a critical mass-to-light ratio may be defined:

$$\left(\frac{M}{L}\right)_c = \frac{\rho_c}{\mathcal{L}}. \quad (11)$$

It is convenient to compare the observed mass-to-light ratios on various scales with the critical value. It must be cautioned, though, that not all such comparisons are meaningful since it is not clear to what extent a given M/L ratio is characteristic of the bulk of the universe.

From a recent deep survey, Kirshner, Oemler, and Schechter (1979, hereafter KOS) derived

$$\mathcal{L}\,(\mathrm{KOS}) = (1.90 \pm 0.28) \times 10^8\, h_0 L_\odot \mathrm{Mpc}^{-3}. \quad (12)$$

This compares surprisingly well with the value obtained by Felten (1977) from older data,

$$\mathcal{L}\,(\mathrm{Felten}) = 1.72 \times 10^8\, h_0 L_\odot\ \mathrm{Mpc}^{-3}, \quad (13)$$

but is in some conflict with the recent result of Davis, Geller, and Huchra (1978, hereafter DGH),

$$\mathcal{L}\,(\mathrm{DGH}) = 1.20 \times 10^8\, h_0 L_\odot\ \mathrm{Mpc}^{-3}. \quad (14)$$

Since KOS and DGH find large-scale inhomogeneities in the distribution of galaxies, the present surveys may not be deep enough to yield the "true" average luminosity density. Hereafter, we use $\mathcal{L} = 2 \times 10^8\, h_0 L_\odot$ Mpc^{-3}, which corresponds to

$$\left(\frac{M}{L}\right)_c = 14 \times 10^2\, h_0 \left(\frac{M_\odot}{L_\odot}\right). \quad (15)$$

Keep in mind though that the uncertainty in this estimate may be as large as a factor of 2 or 3 because of the possible uncertainty in $\mathcal{L}$. In Table 3 we record the observed mass-to-light ratios from systems of different scales (FG); also given is $\Omega \equiv (M/L)/(M/L)_c$.

Since we find most galaxies in binaries or small groups, the estimate $0.04 \lesssim \Omega \lesssim 0.13$ is probably significant. Most galaxies are not found in rich clusters, so that the value $M/L = 700\, h_0$ found from Coma, for example, need not be representative of mass on such scales. The cluster entry in Table 3, however, is taken from applications of the "cosmic virial theorem" and may provide a more reliable estimate.

Given the various uncertainties, we are of the opinion that $\Omega \geq 0.04$ (derived from binaries and small groups) is a good estimate of the lower limit to the density; this corresponds to (cf. equation [9]) $\eta \gtrsim 2 \times 10^{-10}$. It is important to ask whether or not the mass density of the universe is dominated by nucleons. As emphasized most recently by Schramm and Steigman (1981*a*, *b*), massive neutrinos may dominate the universe. Indeed, from constraints derived from nucleosynthesis (YS^{2}R), Schramm and Steigman (1981*a*, *b*) showed that nucleons could

TABLE 3

MASS-TO-LIGHT RATIOS FOR ASTROPHYSICAL SYSTEMS

Scale	M/L (Solar Units)	Ω[a]
Solar neighborhood[b]	2 ± 1	$(0.0014 \pm 0.0007)/h_0$
Galaxies[c]	$(8-20)\, h_0$	0.006–0.014
Binaries and small groups ...	$(60-180)\, h_0$	0.04–0.13
Clusters of galaxies[d]	$(300-1000)\, h_0$	0.2–0.7
Hot gas in clusters[e]	...	$\gtrsim 0.007\, h_0^{-3/2}$

[a] $\Omega \equiv (M/L)/(M/L)_c$.

[b] With young stars removed, these numbers would double.

[c] The inner, luminous parts of spirals and elliptials.

[d] Peebles 1979; DGH.

[e] From X-rays, $M_{\mathrm{GAS}} \approx 0.1\, M_{\mathrm{TOT}}$.

not dominate the mass in clusters; perhaps neutrinos do (Cowsik and McClelland 1972; Szalay and Marx 1974).

If neutrinos (or other, nonbaryonic matter) dominate, it becomes very difficult to estimate the contribution of nucleons. As a firm but rather uninteresting lower limit, we could take the solar neighborhood material (see Table 3), which is surely dominated by nucleons:

$$\eta \gtrsim \eta_{sn} \gtrsim 0.14 \times 10^{-10}. \quad (16)$$

Going further, it is not unreasonable to assume that the luminous inner parts of galaxies are also dominated by nucleons. The mass-to-light ratios measured for the central regions of galaxies range from 8 h_0 to 20 h_0 (Table 3); this range does not reflect observational uncertainties but, instead, real variations in the mass-to-light ratios for different types of galaxies, produced by the different stellar populations present. Spiral galaxies tend to have M/L ratios closer to 8 h_0, while elliptical galaxies tend to have M/L ratios closer to 20 h_0. Using the extreme lower value of 8 h_0, we find that

$$\eta \gtrsim \eta_G \gtrsim 0.29 \times 10^{-10}. \quad (17)$$

It is possible that further analysis might permit one to use an appropriately weighted mass-to-light ratio rather than the lower limit. For example, if most of the luminosity were known to come from galaxies with $M/L \approx 15\ h_0$, then we could conclude that $\eta \gtrsim 0.55 \times 10^{-10}$.

Finally, we note that the X-ray–emitting, hot gas in rich clusters is surely nucleons. Unfortunately, the estimates of the amount of this gas are very uncertain. If $\Omega_{Gas} \gtrsim 0.007\ h_0^{-3/2}$, then $\eta \gtrsim 1 \times 10^{-10}$. We note here that, for $\eta < 1 \times 10^{-10}$, a large abundance of deuterium is produced primordially: $X_D \gtrsim 3 \times 10^{-4}(YS^2R)$. Unless the primordial abundance has been reduced by more than an order of magnitude (requiring more than 90% of all nucleons to have been cycled through stars), this also suggests (in a less compelling way) $\eta \gtrsim 10^{-10}$.

III. WHAT CAN WE LEARN FROM BIG-BANG NUCLEOSYNTHESIS?

a) Bounds on η, Ω_N, and kn_B/s

First, consider the constraint on η obtained from Y_P, $\tau_{1/2}$, and N_ν. Since Y_P increases with increasing η, $\tau_{1/2}$, and N_ν, an upper limit on Y_P and lower limits on N_ν and $\tau_{1/2}$ allow us to place an upper limit on η. The upper bounds on η obtained by using $N_\nu \gtrsim 3$, $\tau_{1/2} \gtrsim 10.82$, 10.61, 10.13 minutes, and $Y_P \leq 0.23$, 0.25, 0.27 are summarized in Table 4A. For the interesting ranges $\tau_{1/2} \geq 10.13$ minutes and $Y_P \leq 0.25$, η must be less than 10^{-9}. (We note that this limit applies to the value of η during nucleosynthesis; if substantial entropy production has occurred since nucleosynthesis, η must be even smaller, and this limit is still valid.) Recall that in § II an absolute lower bound on η of $10^{-10.8}$ was found by using the mass-to-light ratio obtained from the solar neighborhood. Combining these two constraints, we find that η should lie in the range $10^{-9.9 \pm 0.9}$. This is nearly an order of magnitude smaller than the usual estimate $\eta = 10^{-9 \pm 1}$.

Although primordial nucleosynthesis permits one to deduce η directly, in order to infer ρ_N (or Ω_N) one must also specify T_0 (and h_0), cf. equations (3a), (3b), (6), and (9). Using equation (9), the upper bounds on η in Table 4A are translated into the upper bounds on Ω_N which are summarized in Table 4B. With $\tau_{1/2} \geq 10.13$ minutes and $Y \leq 0.25$, Ω_N must be less than 0.19, which is a factor of ~5 from closing the universe by nucleons alone. The only entry in Table 4B which allows the universe to be closed by nucleons is for $\tau_{1/2} = 10.13$ minutes and $Y_P = 0.27$, but a primordial helium mass fraction of 0.27 is highly unlikely considering the present observations.

TABLE 4A

UPPER BOUND ON η FROM PRIMORDIAL NUCLEOSYNTHESIS

η_{max}	$\tau_{1/2} \geq 10.13$ min.	≥ 10.61 min.	≥ 10.82 min.
$Y_P \leq 0.27$ ...	$10^{-8.0}$	$10^{-8.4}$	$10^{-8.6}$
$Y_P \leq 0.25$ ...	$10^{-9.0}$	$10^{-9.4}$	$10^{-9.5}$
$Y_P \leq 0.23$ ...	$10^{-9.8}$	10^{-10}	10^{-10}

TABLE 4B

UPPER BOUND ON Ω_N FROM PRIMORDIAL NUCLEOSYNTHESIS

$\Omega_{N\max}$	$\tau_{1/2} \geq 10.13$ min.	≥ 10.61 min.	≥ 10.82 min.
$Y_P \leq 0.27$ ...	2.0	0.75	0.51
$Y_P \leq 0.25$ ...	0.19	0.076	0.055
$Y_P \leq 0.23$ ...	0.032	0.021	0.018

Since the number of photons per comoving volume in the universe has not remained constant (e.g., it increased because of e^+e^- annihilations when $kT \sim 0.5$ MeV), η is not the most useful quantity to use in discussing cosmological baryon generation. Assuming the expansion to be adiabatic, the specific entropy per comoving volume is conserved ($s/k \sim$ number of relativistic particles). When the baryon number is effectively conserved (i.e., when the rate of baryon nonconserving reactions is much less than the expansion rate, $kT \lesssim 10^{14}$ GeV), the ratio of the baryon number to the specific entropy, kn_B/s, remains constant. Today the bulk of the entropy resides in the cosmic photon and neutrino backgrounds (and possibly other as of yet unknown particle backgrounds). Assuming three two-component neutrino species (ν_e, ν_μ, ν_τ) with temperature $T_\nu = (4/11)^{1/3}\ T_0$ (e.g., see Weinberg 1972, chap. 15), η and kn_B/s are related by (Fry, Olive, Turner 1980)

$$kn_B/s = 0.14\,\eta. \qquad (18)$$

(We note that, by considering only the entropy contributed by the photons, we obtain $kn_B/s_\gamma = 0.28\ \eta$.) The requirement that η be in the range $10^{-9.9\pm0.9}$ restricts kn_B/s to be in the range $10^{-10.8\pm0.9}$. We should mention that an additional two-component neutrino species with $T_\nu = (4/11)^{1/3}\ T_0$ only increases s/k (decreases kn_B/s) by $\approx 15\%$. A relativistic relic species which interacts more weakly (e.g., right-handed neutrinos) would contribute even less since it would have decoupled earlier and would have a lower temperature than the usual left-handed neutrinos.

b) Limits on the Number of Neutrino Types

At present most theories of elementary particles are quark-lepton symmetrical, i.e., for each quark pair [e.g., $(u\text{-}d)$] there is a corresponding lepton pair (ν_e-e^-), so that counting the number of neutrino types allows one to count the number of quark and lepton pairs (or generations). In order not to spoil asymptotic freedom in standard QCD (e.g., without technicolor), there can be no more than eight generations (corresponding to $N_\nu \leq 8$). The limits on N_ν derived from big-bang nucleosynthesis are potentially much more restrictive (YS^{2}R). In the following discussion, we review the present situation especially in light of the uncertainties in $\tau_{1/2}$ and the neutrino rest mass.

Because Y_P increases with increasing η, $\tau_{1/2}$, and N_ν, lower bounds on $\tau_{1/2}$ and η, and an upper bound on Y_P, serve to constrain the number of neutrino types. Of the three quantities which we need know, the most elusive is η. As discussed in § II, studies of binary galaxies and small groups of galaxies (BSG) strongly suggest that $\Omega \gtrsim 0.04$. However, only if baryons dominate the mass of BSG can we conclude that this bound implies $\Omega_N \gtrsim 0.04$. In this case with $h_0 \geq 1/2$, η must be $\geq 2.0 \times 10^{-10}$, and the limits on N_ν for $\tau_{1/2} = 10.70$ minutes have been discussed by YS^{2}R.

If the usual neutrinos (e, μ, τ) have a small rest mass, of the order of 10 eV, they may cluster and dominate the mass of BSG and of clusters of galaxies (Schramm and Steigman 1981a, b). In this situation, we must turn to other, less restrictive, lower bounds on η. However, if their rest mass is less than a few electron volts, they cannot cluster sufficiently to account for a substantial fraction of the mass in BSG (Tremaine and Gunn 1979). In this case, the lower bound on η from BSG is valid, unless other unknown particles (heavy neutral leptons, monopoles, etc.) provide the observed mass in BSG. The other limits on η come from: (1) hot gas in clusters and deuterium production ($\Omega_N h_0^2 \gtrsim 0.005$, $\eta > 1.0 \times 10^{-10}$); (2) mass-to-light ratios for the central regions of galaxies ($\Omega_N h_0^2 \gtrsim 0.0014$, $\eta \gtrsim 0.29 \times 10^{-10}$); and (3) mass-to-light ratios for the solar neighborhood ($\Omega_N h_0^2 \gtrsim 0.0007$, $\eta \gtrsim 0.14 \times 10^{-10}$). We will first review the limits on N_ν, assuming that baryons provide the mass inferred in BSG, and then discuss the situation in which massive neutrinos ($m \sim 10$ eV) or other particles dominate the mass of BSG (where we must use the less restrictive lower bounds for η).

In the first case ($\Omega_N \gtrsim 0.04$, $\eta \gtrsim 2.0 \times 10^{-10}$), the limits on N_ν obtained here differ from those obtained by YS^{2}R only in their dependence upon $\tau_{1/2}$ (Peebles 1971; Tayler 1979). In Figures 1a, 1b, and 1c, the upper limit on the number of neutrino types is shown as a function of η and $\tau_{1/2}$ for $Y_P \leq 0.23$, 0.25, 0.27. The results for $\eta \geq 2.0 \times 10^{-10}$ and $\tau_{1/2} = 10.13$, 10.61, 10.82 minutes are summarized in Table 5. The results differ from YS^{2}R in that lower neutron half-lives allow the possibility of more neutrino types (other parameters being fixed). With $Y_P \leq 0.25$ and $\tau_{1/2} \geq 10.13$ minutes, only four neutrino types are allowed; if the upper limit on primordial helium is relaxed to 0.27 and $\tau_{1/2} \geq 10.13$ minutes, there can be as many as six neutrino types (two neutrino types, ν_e and ν_μ, are known with certainty, and there is good evidence for a third neutrino type, ν_τ).

If we suppose that the mass which is inferred from observations of BSG is not predominantly baryons, then the lower bounds on η are much worse, as are the corresponding limits on N_ν. In fact, the lower bounds on η presently obtained from solar neighborhood and galactic mass-to-light ratios ($\eta = 0.14 \times 10^{-10}$ and 0.29×10^{-10}) result in *no limits* on N_ν (for $Y_P \gtrsim 0.21$). Let us explain briefly. Increasing the number of neutrino species speeds up the expansion rate, causing the n/p ratio to "freeze out" earlier and at a higher value, and ultimately more ^{4}He is synthesized. Hence Y_P increases with N_ν. However, if the expansion rate is increased enough, the amount of helium produced begins to drop simply because there is not enough time for the helium-producing reactions to run to completion (this was first pointed out by Peebles 1966). For this reason,

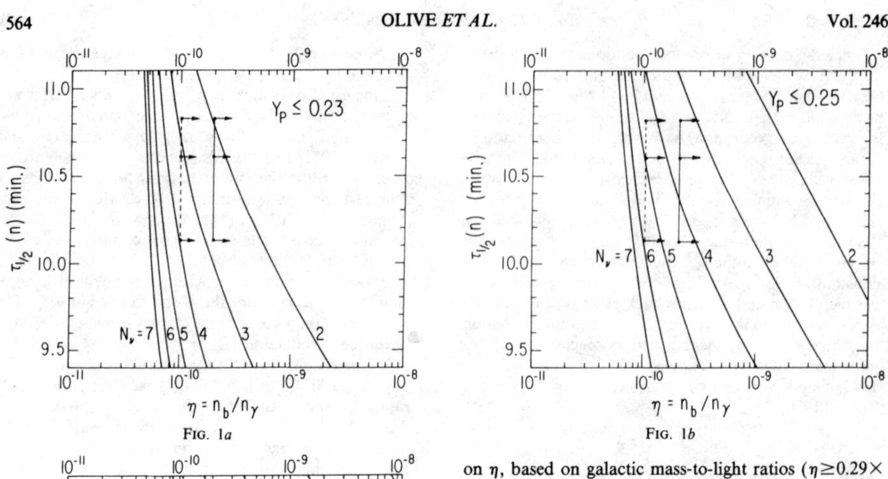

FIG. 1*a*

FIG. 1*b*

$N_\nu = 7$ 6 5 4 3 2

$Y_P \leq 0.27$

$\tau_{1/2}$ (n) (min.)

$\eta = n_b/n_\gamma$

FIG. 1*c*

FIG. 1*a*.—Contours of the allowed number of two-component neutrino species in the $(\eta, \tau_{1/2})$-plane for $Y_P \leq 0.23$. The points and arrows correspond to $\tau_{1/2} = 10.13$, 10.61, 10.82 minutes and $\eta \geq 1.0 \times 10^{-10}$ (lower bound from hot gas), 2.0×10^{-10} (lower bound from BSG). To determine the number of allowed species, pick a point in $(\eta, \tau_{1/2})$-space, and read the value of N_ν. 1*b*.—Same as 1*a* for $Y_P \leq 0.25$. 1*c*.—Same as 1*a* for $Y_P \leq 0.27$.

the primordial helium fraction as a function of N_ν (or expansion rate) achieves a maximum value (see Figs. 2*a* and 2*b*). For $\eta = 0.14 \times 10^{-10}$ (solar neighborhood constraint), the maximum helium production is 12%, 12.5%, or 12.7% ($\tau_{1/2} = 10.13$, 10.61, 10.82 minutes), and for $\eta = 0.29 \times 10^{-10}$ (galactic constraint), the maximum is 20.3%, 21.0%, or 21.3% ($\tau_{1/2} = 10.13$, 10.61, 10.82 minutes). Unless Y_P can be constrained to be less than ~21% or ~12%, these lower bounds on η result in no limits on N_ν. We remind the reader that the lower bound on η, based on galactic mass-to-light ratios ($\eta \geq 0.29 \times 10^{-10}$), was obtained by assuming $M/L = 8\ h_0$, the extreme lower value (see Table 3). If most of the light in the universe were shown to be from galaxies with $M/L \approx 15\ h_0$, then we could use the lower bound $\eta \gtrsim 0.55 \times 10^{-10}$. In this case for $Y_P \leq 0.25$ and $\tau_{1/2} = 10.61$ minutes, at most nine neutrino types are allowed (see Fig. 2*a*).

The lower limits on η from hot gas in clusters and from primordial deuterium production are both less certain and give $\eta \gtrsim 1.0 \times 10^{-10}$. Once again there are limits on N_ν, and they are summarized in Table 5. For $Y_P \leq 0.25$ and $\tau_{1/2} \geq 10.13$ minutes, N_ν must be ≤ 6. With Y_P as high as 0.27 and $\tau_{1/2} \geq 10.13$ minutes, there is room for eight two-component neutrinos. Finally, we note that if some of the assumptions of the standard model are not valid (e.g., absence of large anisotropies during nucleosynthesis or negligible production of en-

TABLE 5

LIMITS ON N_ν DERIVED FROM PRIMORDIAL NUCLEOSYNTHESIS

	$Y_P \leq 0.23$	$Y_P \leq 0.25$	$Y_P \leq 0.27$
$\eta \gtrsim 2.0 \times 10^{-10}$ (BSG)			
$\tau_{1/2} \geq 10.82$ min. ...	2	3	5
$\tau_{1/2} \geq 10.61$ min. ...	2	3	5
$\tau_{1/2} \geq 10.13$ min. ...	2	4	6
$\eta \gtrsim 1.0 \times 10^{-10}$ (Hot Gas)			
$\tau_{1/2} \geq 10.82$ min. ...	2	4	7
$\tau_{1/2} \geq 10.61$ min. ...	3	5	7
$\tau_{1/2} \geq 10.13$ min. ...	4	6	8
$\eta \gtrsim 0.29 \times 10^{-10}$ (Galactic M/L's)			
No limit on N_ν until $\eta \geq 0.36 \times 10^{-10}$ (for $Y_P \leq 0.23$).			

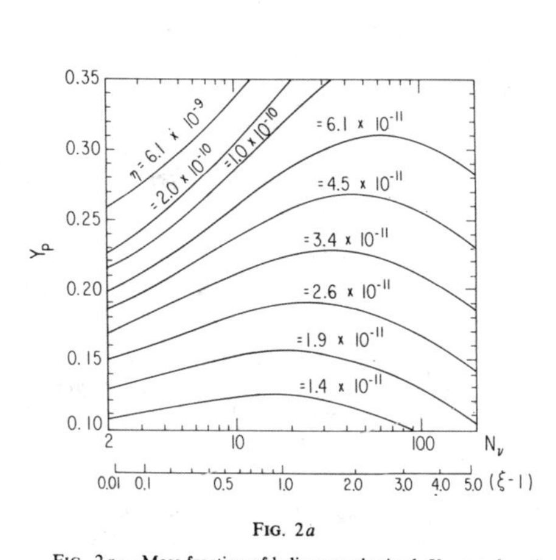

FIG. 2a

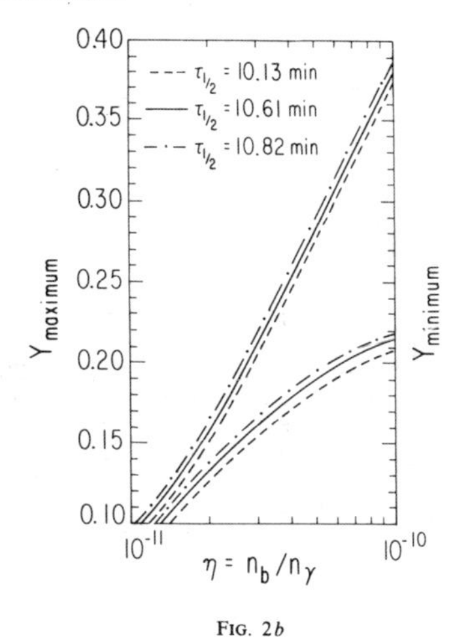

FIG. 2b

FIG. 2a.—Mass fraction of helium synthesized, Y_P, as a function of N_ν (or ξ) for $\tau_{1/2} = 10.61$ min. and various values of η. The speed-up factor, ξ, $\equiv$(expansion rate)/(expansion rate for $N_\nu = 2) = (\rho'/\rho)^{1/2}$. As N_ν (or ξ) increases, Y_P increases since n/p freezes out at a higher value (see § I). Eventually, the expansion is so rapid that the reactions which synthesize ^{4}He do not have time to run to completion, and Y_P decreases with increasing N_ν; hence, as a function of N_ν (or ξ) Y_P reaches a maximum value. Note that, for $\eta = 0.14, 0.29 \times 10^{-10}$, the maximum value of Y_P is 0.125, 0.21, and unless Y_P can be constrained to be less than $\sim$0.12, 0.21, these lower bounds on η result in no limits on N_ν.

FIG. 2b.—The maximum value of Y_P produced (as a function of N_ν, see Fig. 2a) as a function of η for $\tau_{1/2} = 10.13$, 10.61, 10.82 min. and the minimum value of Y_P produced ($N_\nu = 2$) also as a function of η for $\tau_{1/2} = 10.13$, 10.61, 10.82 min. The actual mass fraction of ^{4}He synthesized (in the standard model) must lie between the two sets of curves.

tropy after nucleosynthesis), then these limits may still be valid and may be even stronger. Anisotropy tends to speed up the expansion, hence causing more ^{4}He to be synthesized. If significant entropy is produced after the epoch of nucleosynthesis, then the value of η during nucleosynthesis is even larger than our lower bounds indicate, again strengthening our limits on N_ν. For further discussion of these issues, we refer the reader to Schramm and Wagoner (1977).

c) *Massive Neutrinos*

Of importance to our rediscussion of primordial nucleosynthesis is the possibility that neutrinos might have a small but finite rest mass. As long as a neutrino species is relativistic during the epoch of nucleosynthesis ($m_\nu < 1$ MeV), it will contribute to the energy density and affect nucleosynthesis just as a massless species would, so that our results are still valid (with possible slight modification to be discussed below). The e-neutrino and μ-neutrino both satisfy this condition ($m_\nu < 1$ MeV) and have been shown to be distinct species. However, the properties of the τ-neutrino are not well known, and it is possible that its mass is as large as 250 MeV or that it is not a separate species and, thus, it might not contribute to the energy density of the universe during nucleosynthesis.

The limits we have discussed apply to two-component neutrinos. There are two varieties of massive neutrinos: (1) two-component Majorana neutrinos ($\nu \equiv \bar{\nu}$), in which case our limits are unaffected, or (2) four-component Dirac neutrinos. In this case, if the right-handed components have interactions strong enough so that they remain in thermal contact with the rest of the universe at least until $T \sim 10^{12}$ K (e.g., couple with the same strength as their left-handed counterparts), then there are effectively twice the number of two-component neutrino species, and the limits in Figures 1a, 1b, and 1c, and Table 5 must be halved. However, if they interact more weakly than left-handed neutrinos, then they will decouple earlier, will have a lower temperature during nucleosynthesis, and will contribute less to the energy density than a left-handed neutrino. Shapiro, Teukolsky, and Wasserman (1980) and Bond, Efstathiou, and Silk (1980) have shown that right-left interactions are not sufficiently strong to keep the neutrinos in thermal contact until $T \approx 10^{12}$ K. Unless there exist purely right-handed interactions of sufficient strength, the right-handed components will have decoupled so early that their contribution to the energy density of the universe during nucleosynthesis is negligible. Therefore, if neutrinos are massive, be they Dirac or Majorana, the limits in Figures 1*a*, 1*b*, and 1*c*, and Table 5 are more than likely applicable. For further discussion of these issues, we refer the reader to OSS, SOS, or Olive and Turner (1981*b*).

d) *Concordance of the Standard Model*

A primordial ^{4}He abundance of $Y_P \leq 0.25$ is consistent with the observations of low-metal abundance H II regions, although recent data suggest that $Y_P = 0.23$ might provide a better upper limit. Some extrapolations from the data suggest that $Y_P \leq 0.22$ (see Table 2), and such low values lead one to question the consistency of the standard model (Stecker 1980; see also Olive and Turner 1981*a* and Stecker 1981). Are very low values of Y_P in conflict with the standard model? In order to address this question, we show in Figure 3 the low η and Y_P portion of the Y_P versus η curve for $\tau_{1/2} = 10.13$ ($N_\nu = 3$), 10.82 ($N_\nu = 3$), and 10.61 ($N_\nu = 2, 3, 4$) minutes. In Table 6 we tabulate Y_P for $\tau_{1/2} = 10.13$, 10.61, 10.82 minutes, $N_\nu = 2$, 3, and $\eta = 0.14$, 0.29, 1.0, 2.0×10^{-10} (as discussed in § III*c* $N_\nu = 2$ is still a possibility).

From Table 6 we see that a primordial helium abundance as low as $Y_P = 0.22$ may still be consistent with a nucleon-dominated universe and the standard model. For a nonbaryon-dominated universe (e.g., neutrino dominated), the constraints on η are less stringent. Using the mass in galaxies inferred from mass-to-light ratios ($\eta \geq 0.29 \times 10^{-10}$), the standard model can produce Y_P as low as 0.15; with $\eta \geq 0.14 \times 10^{-10}$ (M/L ratios for the solar neighborhood), Y_P can be as small as 0.10. In summary, given our present uncertainty with regard to η, the standard model does not contradict the observations unless Y_P were found to be ≤ 0.15 or, perhaps, even as low as 0.10. Unless a lower bound on η can be derived which is significantly higher, or alternately an upper bond on Y_P which is significantly lower, these results show that the standard model (perhaps neutrino dominated) is in no serious trouble at the present time.

IV. CONCLUSIONS—SUMMARY

Big-bang nucleosynthesis offers at present the most powerful probe of the early universe. In particular, the helium produced depends upon three parameters: the neutron half-life, $\tau_{1/2}$; the number of two-component neutrino species, N_ν; and the baryon-to-photon ratio, η. Nucleosynthesis and mass-to-light ratios constrain η to the range $10^{-9.9 \pm 1}$, which is nearly an order of magnitude smaller than the usual estimates. The upper bound on η constrains Ω_N to less than ≈ 0.2 (for $h_0 \gtrsim 1/2$), implying that nucleons alone cannot close the universe. With the assumption that most of the entropy in the universe today is in the photon and neutrino backgrounds (with $N_\nu = 3$), the baryon-to-photon ratio translates into a ratio of the baryon number to the specific entropy of $kn_B/s = 10^{-10.8 \pm 1}$. Since this ratio remains constant when the baryon number is effectively conserved, it is the input parameter for models of cosmological baryon generation.

If one assumes that the mass which binds BSG is primarily in the form of baryons, then only three

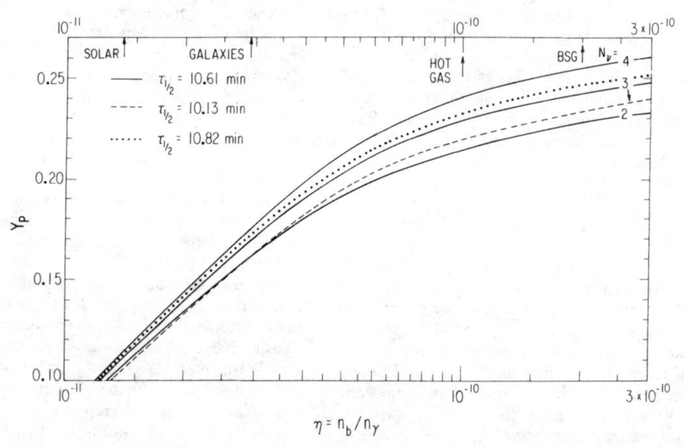

FIG. 3.—The mass fraction of ^{4}He synthesized, Y_P, as a function of η for $\tau_{1/2} = 10.13$ ($N_\nu = 3$), 10.82 ($N_\nu = 3$), and 10.61 ($N_\nu = 2,3,4$) min. Only for reasons of clarity are the curves for $\tau_{1/2} = 10.13$ ($N_\nu = 2,4$) min. and 10.82 ($N_\nu = 2,4$) min. not shown.

TABLE 6
MASS FRACTION OF ^{4}He SYNTHESIZED

Minimum η	N_ν	$\tau_{1/2}=10.13$ min.	$\tau_{1/2}=10.61$ min.	$\tau_{1/2}=10.82$ min.
0.14×10^{-10} (Solar)	2	0.104	0.109	0.111
	3	0.108	0.113	0.115
0.29×10^{-10} (Galaxies) ...	2	0.154	0.161	0.164
	3	0.162	0.169	0.172
1.0×10^{-10} (Hot gas)......	2	0.208	0.216	0.220
	3	0.220	0.230	0.234
2.0×10^{-10} (BSG)	2	0.219	0.228	0.231
	3	0.234	0.242	0.245

($\tau_{1/2}\gtrsim10.61$ minutes) or at most four ($\tau_{1/2}\gtrsim10.13$ minutes) neutrino species are permitted. Allowing for the possibility that this mass is not baryonic (e.g., neutrinos of mass $\gtrsim10$ eV may cluster and dominate the mass of BSG), and adopting a lower limit on η inferred from the hot gas in clusters or from the abundance of deuterium, then as many as eight neutrino types are allowed. However, if only the most conservative bounds on η are used (mass-to-light ratios for the solar neighborhood or for central regions of galaxies), then at present there are no limits on the number of neutrino types. This situation could change, for example, if Y_P were shown to be less than $\approx21\%$ or if further analysis allowed one to deduce, from the mass-to-light ratios for central regions of galaxies (see § II*e*), a lower bound on η which was higher.

The issue of which lower bound on η to use will be at least partially resolved by the experimental determination of neutrino masses. If there exists a neutrino species with mass $\gtrsim10$ eV, it may cluster and dominate the mass of BSG, making it impossible to use the BSG lower bound on η. If no neutrino species has a mass greater than a few electron volts, then neutrinos will not cluster and will not dominate the mass of BSG. In this case, we would have no reason to believe that anything but baryons contribute significantly to the mass of BSG, and the limit, $N_\nu\lesssim4$ (obtained by YS^{2}R), should still be valid.

Implicit in our discussion has been the assumption of the validity of the standard model for nucleosynthesis. Recent observations which indicate low helium abundances have raised some questions concerning this assumption. As we have discussed in § III*d*, unless the primordial helium fraction is determined to be less than $\sim15\%$ (or perhaps as low as 10%), the standard model is not necessarily in serious trouble—perhaps just the assumption that nucleons dominate the mass of the universe. For example, if $Y_P\approx0.22-0.23$, $N_\nu=3$, and $\tau_{1/2}=10.61$ minutes, then the nucleon abundance $\eta\approx0.8-1\times10^{-10}$, indicating that baryons cannot dominate the mass of BSG; perhaps massive neutrinos do (Schramm and Steigman 1981*a*, *b*).

The primordial abundance of ^{4}He provides a beacon illuminating the strong interconnections between particle physics and cosmology. Recall that Y_P depends upon and therefore provides a probe of $\tau_{1/2}$, η, and N_ν. Laboratory experiments with confined neutrons should soon provide an accurate value for $\tau_{1/2}$. The increasing body of high quality, observational data on less-evolved extragalactic systems (lower than galactic abundances for the heavy elements) should lead to a better estimate for Y_P. These advances will help us to better constrain η and N_ν. It is already known that baryons alone cannot close the universe ($\Omega_N\lesssim0.2$), and if nucleons dominate the mass on scales of BSG ($\Omega_N\gtrsim0.04$), N_ν is small ($\lesssim3$ or 4). Although N_ν could be quite large in a nonbaryon-dominated universe, continued studies of hot gas in clusters and the use of the other light elements (^{2}H, ^{3}He, and ^{7}Li) produced by primordial nucleosynthesis to constrain η (Yang *et al.* 1981) should lead to a better lower limit to η and, therefore, to a reliable upper limit to N_ν. In contrast, a laboratory measurement of the decay width of the Z^0 (the neutral gauge boson of the Weinberg-Salam-Glashow theory of the electroweak interactions) will provide a direct measurement of N_ν which may then be used to constrain η. Better knowledge of all the quantities (Y_P, $\tau_{1/2}$, η, and N_ν) will provide yet another test of our best probe of the early universe, big-bang nucleosynthesis.

Much of this work was done during the cosmology and particle physics workshop at the Aspen Center for Physics (Summer 1980). We thank Robert V. Wagoner for generously providing us with his nucleosynthesis code and for several valuable discussions. This work was supported in part by M.S.T.'s Enrico Fermi Fellowship, K.A.O.'s Hertz Fellowship, NSF AST 78-20402, DOE AC02-80ER10773 (all at The University of Chicago), and DOE ER78-8-02-5007 (at Bartol Research Foundation).

REFERENCES

Aaronson, M., Mould, J., Huchra, J., Sullivan, W. T., Schommer, R. A., and Bothun, G. D. 1980, *Ap. J.*, **239**, 12.
Austin, S. M., and King, C. H. 1977, *Nature*, **269**, 782.
Bahcall, J. N., *et al.* 1980, *Phys. Rev. Letters*, **45**, 945.
Bond, J. R., Efstathiou, G., and Silk, J. 1980, preprint.
Bondarenko, L. N., Kurguzov, V. V., Prokof'ev, Yu. A., Rogov, E. V., and Spivak, P. E. 1978, *JETP Letters*, **28**, 303.
Branch, D. 1979, *M.N.R.A.S.*, **186**, 609.
_____. 1980, private communication.
Byrne, J., Morse, J., Smith, K. F., Shaikh, F., Green, K., and Greene, G. L. 1980, *Phys. Lett. B*, **92**, 274.
Castellani, V. 1980, Paper presented at the 1980 Santa Cruz Workshop on Astronomy and Astrophysics.
Caughlan, G. R. and Fowler, W. A. 1980, private communication.
Chiosi, C. 1979, *Astr. Ap.*, **80**, 252.
Christensen, C. J., Nielsen, A., Bahnsen, A., Brown, W. K., and Rustad, B. M. 1972, *Phys. Rev. D*, **5**, 1628.
Cowsik, R., and McClelland, J. 1972, *Phys. Rev. Letters*, **29**, 669.
Danese, L., and DeZotti, G. 1978, *Astr. Ap.*, **68**, 157.
Davis, M., Geller, M. J., and Huchra, J. 1978, *Ap. J.*, **221**, 1 (DGH).
de Vaucouleurs, G., and Bollinger, G. 1979, *Ap. J.*, **233**, 433.
Dufour, R., Shields, G., and Talbot, R. 1980, research reported at the 1980 Santa Cruz Workshop on Astronomy and Astrophysics.
Faber, S. M. and Gallagher, J. S. 1979, *Ann. Rev. Astr. Ap.*, **17**, 135 (FG).
Faulkner, J. 1967, *Ap. J.*, **147**, 617.
Felten, J. E. 1977, *A. J.*, **82**, 861.
Fowler, W. A., Caughlan, G. R., and Zimmerman, B. A. 1975, *Ann. Rev. Astr. Ap.*, **13**, 69.
French, H. B. 1979, "Galaxies with the Spectra of Giant H II Regions", Lick Obs. Bull. No. 863.
Fry, J. N., Olive, K. A., and Turner, M. S. 1980, *Phys. Rev. D*, **22**, 2953.
Hegyi, D. J., Traub, W. A., and Carleton, N. P. 1974, *Ap. J.*, **190**, 543.
Kinman, T. D., and Davidson, K. 1980, "Spectroscopic Observations of Ten Emission-Line Dwarf Galaxies", preprint.
Kirshner, R. P., and Kwan, J. 1974, *Ap. J.*, **193**, 27.
Kirshner, R. P., Oemler, A., Jr., and Schechter, P. L. 1979, *Ap. J.*, **84**, 951 (KOS).
Kugler, K. J., Paul, W., and Trinks, U. 1978, *Phys. Lett. B*, **72**, 422.
_____. 1979, *IEEE Trans.*, **26**, 3152.
Lequeux, J., Peimbert, M., Rayo, J. F., Serrano, A., and Torres-Peimbert, S. 1979, *Astr. Ap.*, **80**, 155.
Mazzitelli, I. 1979, *Astr. Ap.*, **79**, 251.
Olive, K. A., Schramm, D. N., and Steigman, G. 1981, *Nucl. Phys. B.*, in press (OSS).
Olive, K. A., and Turner, M. S. 1981*a*, *Phys. Rev. Letters*, **46**, 516.
_____. 1981*b*, *Phys. Lett. B*, submitted.
Peebles, P. J. E. 1966, *Ap. J.*, **146**, 542.
_____. 1971, *Physical Cosmology* (Princeton: Princeton University Press).
_____. 1979, *A. J.*, **84**, 730.
Peimbert, M. 1975, *Ann. Rev. Astr. Ap.*, **13**, 113.
_____. 1980, Paper presented at the 1980 Santa Cruz Workshop on Astronomy and Astrophysics.
Peimbert, M., and Torres-Peimbert, S. 1976, *Ap. J.*, **203**, 581.
Reeves, H., Audouze, J., Fowler, W. A., and Schramm, D. N. 1973, *Ap. J.*, **179**, 909.
Renzini, A. 1977, in *Advanced Stages in Stellar Evolution*, ed. P. Bovier and A. Maeder (Sauverny: Geneva Observatory), p. 149.
Sandage, A., and Tammann, G. A. 1976, *Ap. J.*, **210**, 7.
Schmid-Burgk, J. 1981, Talk given at Erice School on Nuclear Astrophysics, to appear in *Prog. Nucl. Part. Phys.*, in press.
Schramm, D. N., and Steigman, G. 1981*a*, "A Neutrino Dominated Universe", First Prize Essay, Gravity Research Foundation, to appear in *Gen. Rel. Grav.*, in press.
_____. 1981*b*, *Ap. J.*, **243**, 1.
Schramm, D. N., and Wagoner, R. V. 1977, *Ann. Rev. Nucl. Sci.*, **27**, 37.
Shapiro, S. L., Teukolsky, S. A., and Wasserman, I. 1980, *Phys. Rev. Letters*, **45**, 669.
Shvartsman, V. G. 1969, *JETP Letters*, **9**, 184.
Stecker, F. W. 1980, *Phys. Rev. Letters*, **44**, 1237.
_____. 1981, *Phys. Rev. Letters*, **46**, 517.
Steigman, G. 1975, Talk given at the Harvard Neighborhood Meeting on Cosmology.
_____. 1979, *Ann. Rev. Nucl. Sci.*, **29**, 313.
Steigman, G., Olive, K. A., and Schramm, D. N. 1979, *Phys. Rev. Letters*, **43**, 239 (SOS).
Steigman, G., Schramm, D. N., and Gunn, J. E. 1977, *Phys. Lett. B*, **66**, 202.
Szalay, A. S., and Marx, G. 1974, *Acta Physica Hungaricae*, **35**, 113.
Talent, D. L. 1980, Ph.D. thesis, Rice University.
Tayler, R. J. 1979, *Nature*, **282**, 559.
Thaddeus, P. 1972, *Ann. Rev. Astr. Ap.*, **10**, 305.
Thum, C., Mezger, P. G., and Pankonin, V. 1980, *Astr. Ap.*, **87**, 269.
Tremaine, S., and Gunn, J. E. 1979, *Phys. Rev. Letters*, **42**, 407.
Turner, M. S., and Schramm, D. N. 1979, *Phys. Today*, **32**, 42.
Wagoner, R. V. 1969, *Ap. J. Suppl.*, **18**, 247.
_____. 1973, *Ap. J.*, **179**, 343.
_____. 1980, "The Early Universe," in *Physical Cosmology*, Les Houches, *École D'Été de Physique Théorique*, Session XXXII, 1979 (New York: North-Holland).
Weinberg, S. 1972, *Gravitation and Cosmology* (New York: Wiley).
Wilkinson, D. H. 1980, in *Proceedings of the Erice School on Nuclear Astrophysics*.
Woody, D. P., Mather, J. C., Nishioka, N., and Richards, P. L. 1975, *Phys. Rev. Letters*, **34**, 1036.
Woody, D. P., and Richards, P. L. 1979, *Phys. Rev. Letters*, **42**, 925.
Yang, J., Schramm, D. N., Steigman, G., and Rood, R. T. 1979, *Ap. J.*, **227**, 697 (YS^2R).
Yang, J., Turner, M. S., Steigman, G., Schramm, D. N., and Olive, K. A. 1981, in preparation.

K. A. Olive, D. N. Schramm, M. S. Turner, and J. Yang: The University of Chicago, Enrico Fermi Institute, Astronomy and Astrophysics Center, 5640 South Ellis Avenue, Chicago, IL 60637

G. Steigman: The Bartol Research Foundation of the Franklin Institute, University of Delaware, Newark, DE 19711

Addendum

In the article reprinted above we have emphasized that the bound on the number of light neutrino species $N_\nu \le 4$ relies on the lower bound to the baryon-to-photon ratio, $\eta \geq 2 \times 10^{-10}$. As we also pointed out this lower bound to η is based upon the self-gravitating mass in binary galaxies and small groups of galaxies (BSG), which does not necessarily have to be baryonic (e.g., it could be massive neutrinos, gravitinos, etc.). If the self-gravitating mass in BSG is not predominantly baryonic, then the bound $\eta \geq 2 \times 10^{-10}$ is not valid, and the limit $N_\nu \le 4$ no longer holds. However, in our most recent analysis of primordial nucleosynthesis[1] we have used the primodial production of D and ^3He together with the present abundances of D and ^3He to obtain essentially the same bound on η ($\eta \geq 2 \times 10^{-10}$) - a bound which is independent of the dynamics of BSG. There are two obvious implications: (1) The bound $\eta_\nu \le 4$ is secure; (2) A reasonable fraction of the self-gravitating mass of BSG must be baryonic.

[1]J. Yang, M.S. Turner, G. Steigman, D.N. Schramm, and K.A. Olive, Enrico Fermi Institute preprint, submitted to Astrophys. J. (1982).

XI. *Genesis of Matter*

1) Violation of *CP* invariance, C asymmetry, and baryon asymmetry of the universe
by A. D. Sakharov, *Zh. Ek. Teor. Fiz.* **5,** 32 (1967) [English translation: *JETP Lett.* **5,** 24 (1967)] 586

2) Unified gauge theories and the baryon number of the universe
by M. Yoshimura, *Phys. Rev. Lett.* **41,** 281 (1978); *erratum,* **42,** 740 (1979) 589

3) Baryon number of the universe
by S. Dimopoulos and L. Susskind, *Phys. Rev.* D**18,** 4500 (1978) 593

4) Matter-antimatter accounting, thermodynamics, and black-hole radiation
by D. Toussaint, S. Treiman, F. Wilczek, and A. Zee, *Phys. Rev.* D**19,** 1036 (1979) 603

5) Cosmological production of baryons
by S. Weinberg, *Phys. Rev. Lett.* **42,** 850 (1979) 613

6) Mechanisms for cosmological baryon production
by D. V. Nanopoulos and S. Weinberg, *Phys. Rev.* D**20,** 2484 (1979) 617

7) Magnitude of the cosmological baryon asymmetry
by S. M. Barr, G. Segrè, and H. A. Weldon, *Phys. Rev.* D**20,** 2494 (1979) 627

8) The development of baryon asymmetry in the early universe
by E. W. Kolb and S. Wolfram, *Phys. Lett.* **91**B, 217 (1980) 632

9) Hierarchy of cosmological baryon generation
by J. N. Fry, K. A. Olive and M. S. Turner, *Phys. Rev. Lett.* **45,** 2074 (1980) 637

VIOLATION OF CP INVARIANCE, C ASYMMETRY, AND BARYON ASYMMETRY OF THE UNIVERSE

A. D. Sakharov
Submitted 23 September 1966
ZhETF Pis'ma 5, No. 1, 32-35, 1 January 1967

The theory of the expanding Universe, which presupposes a superdense initial state of matter, apparently excludes the possibility of macroscopic separation of matter from antimatter; it must therefore be assumed that there are no antimatter bodies in nature, i.e., the Universe is asymmetrical with respect to the number of particles and antiparticles (C asymmetry). In particular, the absence of antibaryons and the proposed absence of baryonic neutrinos implies a non-zero baryon charge (baryonic asymmetry). We wish to point out a possible explanation of C asymmetry in the hot model of the expanding Universe (see [1]) by making use of effects of CP invariance violation (see [2]). To explain baryon asymmetry, we propose in addition an approximate character for the baryon conservation law.

We assume that the baryon and muon conservation laws are not absolute and should be unified into a "combined" baryon-muon charge $n_c = 3n_B - n_\mu$. We put:

$n_\mu = -1$, $n_K = +1$ for antimuons μ_+ and $\nu_\mu = \mu_0$,

$n_\mu = +1$, $n_K = -1$ for muons μ_- and $\nu_\mu = \mu_0$,

$n_B = +1$, $n_K = +3$ for baryons P and N,

$n_B = -1$, $n_K = -3$ for antibaryons P and N

This form of notation is connected with the quark concept; we ascribe to the p, n, and λ quarks $n_c = +1$, and to antiquarks $n_c = -1$. The theory proposes that under laboratory conditions processes involving violation of n_B and n_μ play a negligible role, but they were very important during the earlier stage of the expansion of the Universe.

We assume that the Universe is neutral with respect to the conserved charges (lepton, electric, and combined), but C-asymmetrical during the given instant of its development (the positive lepton charge is concentrated in the electrons and the negative lepton charge in the excess of antineutrinos over the neutrinos; the positive electric charge is concentrated in the protons and the negative in the electrons; the positive combined charge is concentrated in the baryons, and the negative in the excess of μ-neutrinos over μ-antineutrinos).

According to our hypothesis, the occurrence of C asymmetry is the consequence of violation of CP invariance in the nonstationary expansion of the hot universe during the superdense stage, as manifest in the difference between the partial probabilities of the charge-conjugate reactions. This effect has not yet been observed experimentally, but its existence is theoretically undisputed (the first concrete example, Σ_+ and Σ_- decay, was pointed out by S. Okubo as early as in 1958) and should, in our opinion, have an important cosmological significance.

We assume that the asymmetry has occurred in an earlier stage of the expansion, in which the particle, energy, and entropy densities, the Hubble constant, and the temperatures were of the order of unity in gravitational units (in conventional units the particle and energy densities were $n \sim 10^{98}$ cm^{-3} and $\epsilon \sim 10^{114}$ erg/cm^3).

M. A. Markov (see [5]) proposed that during the early stages there existed particles with maximum mass on the order of one gravitational unit ($M_0 = 2 \times 10^{-5}$ g in ordinary units), and called them maximons. The presence of such particles leads unavoidably to strong violation of thermodynamic equilibrium. We can visualize that neutral spinless maximons (or photons) are produced at $t < 0$ from contracting matter having an excess of antiquarks, that they pass "one through the other" at the instant $t = 0$ when the density is infinite, and decay with an excess of quarks when $t > 0$, realizing total CPT symmetry of the Universe. All the phenomena at $t < 0$ are assumed in this hypothesis to be CPT reflections of the phenomena at $t > 0$. We note that in the cold model CPT reflection is impossible and only T and TP reflections are kinematically possible. TP reflection was considered by Milne, and T reflection by the author; according to modern notions, such a reflection is dynamically impossible because of violation of TP and T invariance.

We regard maximons as particles whose energy per particle ϵ/n depends implicitly on the average particle density n. If we assume that $\epsilon/n \sim n^{-1/3}$, then ϵ/n is proportional to the interaction energy of two "neighboring" maximons $(\epsilon/n)^2 n^{1/3}$ (cf. the arguments in [6]). Then $\epsilon \sim n^{2/3}$ and $R_0^0 \sim (\epsilon + 3p) = 0$, i.e., the average distance between maximons is $n^{-1/3} \sim t$. Such dynamics are in good agreement with the concept of CPT reflection at the point $t = 0$.

We are unable at present to estimate theoretically the magnitude of the C asymmetry, which apparently (for the neutrino) amounts to about $[(\bar{\nu} - \nu)/(\bar{\nu} + \nu)] \sim 10^{-8} - 10^{-10}$.

The strong violation of the baryon charge during the superdense state and the fact that the baryons are stable in practice do not contradict each other. Let us consider a concrete model. We introduce interactions of two types.

1. An interaction between the quark-muon transformation current and the vector boson field $a_{i\alpha}$, to which we ascribe a fractional electric charge $\alpha = \pm 1/3, \pm 2/3, \pm 4/3$ and a mass $m_a \sim (10 - 10^3) m_p$. This interaction produces reactions $q \to a + \bar{\mu}$, $q + \mu \to a$, etc. The interaction of the first type conserves the fractional part of the electric charge and therefore the actual number of quarks minus the number of antiquarks ($= 3n_B$) is conserved in processes that include the a-boson only virtually.

We estimate the constant of this interaction at $g_a = 137^{-3/2}$, from the following considerations: The vector interaction of the a-boson with the μ-neutrino leads to the presence of a certain rest mass in the latter. The upper bound of the mass μ_0 is estimated in [7] on the basis of cosmological considerations. If we assume a flat cosmological model of the Universe and assume that the greater part of its density $\rho \sim 1.2 \times 10^{-29}$ g/cm^3 should be ascribed to μ_0, then the rest mass of μ_0 is close to 30 eV. The given value of g_a follows then from the hypothetical formula

$$\frac{m_{\mu_0}}{m_e} = \frac{g_a^2}{e^2} \sim (137)^{-2}.$$

We note that the presence in the Universe of a large number of μ_0 with finite rest mass should lead to a number of very important cosmological consequences.

2. The baryon charge is violated if the interaction described in Item 1 is supplemented with a three-boson interaction leading to virtual processes of the type $a_{\alpha_1} + a_{\alpha_2} + a_{\alpha_3} \to 0$. At the advice of B. L. Ioffe, I. Yu. Kobzarev, and L. B. Okun', the Lagrangian of this interaction is assumed to be dependent on the derivatives of the a-field, for example,

$$L_2 = g_2(\sum_\alpha f_k^i f_j^k f_i^j + \text{h.c.}), \quad f_{ik} = R_0 t a_i.$$

Inasmuch as L_2 vanishes when two tensors coincide, in this concrete form of the theory we should assume the presence of several types of a-fields. Assuming $g_2 = 1/M_0^2$ and $M_0 = 2 \times 10^{-5}$ g, we have strong interaction at $n \sim 10^{98}$ cm^{-3} and very weak interaction under laboratory conditions. The figure shows a proton-decay diagram including three vertices of the first type,

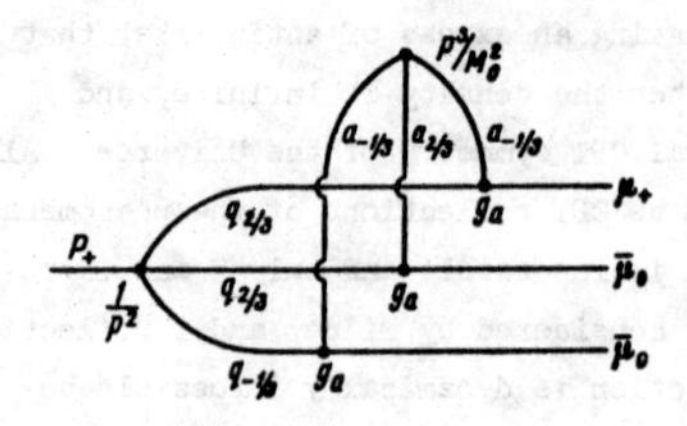

one vertex of the second, and the vertex of proton decay into quarks, which we assume to contain the factor $1/p_q^2$ (due, for example, to the propagator of the "diquark" boson binding the quarks in the baryon). Cutting off the logarithmic divergence at $p_q = M_0$, we find the decay probability

$$\omega \sim \frac{m_p^5 g_a^6 [\ln(M_0/m_a)]^2}{M_0^4}.$$

The lifetime of the proton turns out to be very large (more than 10^{50} years), albeit finite.

The author is grateful to Ya. B. Zel'dovich, B. Ya. Zel'dovich, B. L. Ioffe, I. Yu. Kobzarev, L. B. Okun', and I. E. Tamm for discussions and advice.

[1] Ya. B. Zel'dovich, UFN 89, 647 (1966) (Review), Soviet Phys. Uspekhi 9 (1967), in press.

[2] L. B. Okun', UFN 89, 603 (1966) (Review), Soviet Phys. Uspekhi 9 (1967), in press.

[3] M. A. Markov, JETP 51, 878 (1966), Soviet Phys. JETP 24 (1967), in press.

[4] A. D. Sakharov, JETP Letters 3, 439 (1966), transl. p. 288.

[5] Ya. B. Zel'dovich and S. S. Gershtein, JETP Letters 4, 174 (1966), transl. p. 120.

Unified Gauge Theories and the Baryon Number of the Universe

Motohiko Yoshimura
Department of Physics, Tohoku University, Sendai 980, Japan
(Received 27 April 1978)

I suggest that the dominance of matter over antimatter in the present universe is a consequence of baryon-number–nonconserving reactions in the very early fireball. Unified guage theories of weak, electromagnetic, and strong interactions provide a basis for such a conjecture and a computation in specific SU(5) models gives a small ratio of baryon- to photon-number density in rough agreement with observation.

It is known that the present universe is predominantly made of matter, at least in the local region around our galaxy, and there has been no indication observed[1] that antimatter may exist even in the entire universe. I assume here that in our universe matter indeed dominates over antimatter, and I ask within the framework of the standard big-bang cosmology[2] how this evolved from an initially symmetric configuration, namely an equal mixture of baryons and antibaryons. Since the baryon number is not associated with any fundamental principle of physics,[3] such an initial value seems highly desirable. I find in this paper that generation of the required baryon number is provided by grand unified gauge theories[4] of weak, electromagnetic, and strong interactions, which predict simultaneous violation of baryon-number conservation and *CP* invariance. More interestingly, my mechanism can explain why the ratio of the baryon- to the photon-number density in the present universe is so small, roughly of the order[2] of 10^{-8}–10^{-10}.

The essential point of my observation is that in the very early, hot universe the reaction rate of baryon-number–nonconserving processes, if they exist, may be enhanced by extremely high temperature and high density. In gauge models discussed below, the relevant scale of temperature is given by the grand unification mass around 10^{16} GeV where fundamental constituents, leptons and quarks, begin to become indistinguishable. This mass is high enough to make futile virtually all attempts to observe proton decay in the present universe: proton lifetime $\gg 10^{30}$ year.[5] Instead, if my mechanism works, we may say that a fossil of early grand unification has remained in the form of the present composition of the universe.

The laws obeyed by the hot universe at temperatures much above a typical hadron mass (~ 1 GeV) might, at first sight, appear hopelessly complicated because of many unknown aspects of hadron dynamics. Recent developments of high-energy physics, however, tell that perhaps the opposite is the case. At such high temperatures and densities hadrons largely overlap and an appropriate description of the system is given in terms of pointlike objects—quarks, gluons, leptons, and any other fundamentals. The asymptotic freedom[6] of the strong interaction and weakness of the other interactions further assure[7] that this hot universe is essentially in a thermal equilibrium state made of almost freely moving objects. I shall assume that this simple picture of the universe is correct up to a temperature close to the Planck mass, $G_N^{-1/2} \sim 10^{19}$ GeV, except possibly around the two transitional regions where spontaneously broken weak-electromagnetic and grand-unified gauge symmetries become restored.[8]

In such a hot universe the time development of the baryon-number density $N_B(t)$ is given by

$$\frac{dN_B}{dt} = -3\frac{\dot{R}}{R} N_B + \sum_{a,b} (\Delta n_B)\langle \sigma v \rangle N_a N_b. \tag{1}$$

Here R is the cosmic scale factor; $\langle \sigma v \rangle$ is the thermal average of the reaction cross section for $a+b \to$ anything times the relative velocity of a and b; Δn_B is the difference of baryon number between the final and initial states of this elementary process; N_a is the number density of a in thermal equilibrium. At the temperatures I am considering here, the energy density is dominated by highly relativistic particles and the following relation holds[2]:

$$\dot{R}/R = -\dot{T}/T = (8\pi\rho G_N/3)^{1/2}, \tag{2}$$

where ρ is the energy density

$$\rho = d_F \pi^2 T^4/15. \tag{3}$$

I use units such that the Boltzmann constant $k=1$. The effective number of degrees of freedom d_F is counted as usual, $\frac{1}{2}$ and $\frac{7}{16}$, respectively, for each boson or fermion species and spin state. To solve Eq. (1) with (2) and (3) given, it is convenient to rewrite (1) in the form

$$\frac{dF_B}{dT} = -(8\pi^3 G_N d_F/45)^{-1/2}(\tfrac{3}{8})^2 F_\gamma^{\,2}\delta, \tag{4}$$

where $\delta = \sum (\Delta n_B)\langle \sigma v \rangle$; $F_B = N_B/T^3$; $N_a = 3N_\gamma/8$; $F_\gamma = N_\gamma/T^3$ with N_γ the photon-number density. I used the fact, valid in the following example, that only massless fermions participate in baryon-number–nonconserving reactions.

To obtain a nonvanishing baryon number one must break the microscopic detailed balance (more precisely, reciprocity), because otherwise the inverse reaction would cancel the baryon number gained. This necessity of simultaneous violation of baryon-number conservation and CP or T invariance has a further consequence that the amount of generated baryon number may be severely limited. This is because the detailed balance is known[9] *not* to be broken by Born terms, and one must deal with higher-order diagrams.

As an illustration of the ideas presented above, I shall work in grand unified models based on the group SU(5), which are direct generalizations of the original Georgi-Glashow model[4] to allow more than six flavors of quarks and heavy leptons sequentially in accord with recent experimental observations.[10] Fundamental fermions are thus classified as follows:

$$\underline{5}, \quad (\varphi)_{iR} = \begin{pmatrix} l_1 & \\ & q_3^{\,a} \\ l_2 & \end{pmatrix}_{iR} \tag{5a}$$

$$\underline{10}, \quad (\psi)_{iL} = \begin{pmatrix} & q_1^{\,a} & \\ l_3 & & C\bar{q}_4^{\,a} \\ & q_2^{\,a} & \end{pmatrix}_{iL}, \tag{5b}$$

with several sequences ($i = 1 - n_S$). Here $(l_1 l_2)$ and $(q_1 q_2)$ form $SU(2)_W$ doublets and the others singlets; $a = 1, 2, 3$ are three colors. The baryon number is assigned as $\frac{1}{3}$ to any $SU(3)_C$ triplet and $-\frac{1}{3}$ to any antitriplet.

There are two characteristic mass scales, $m(W)$ and $m(\tilde{W})$, in this class of models with W a representative of ordinary weak bosons and with $\tilde{W}$ that of colored weak bosons. Existence of the two extremely different mass scales reflects two Higgs systems, $\tilde{H}(\underline{24})$ and (presumably several of) $\tilde{H}(\underline{5})$, being responsible for the breaking of SU(5) and $[SU(2)\otimes U(1)]_W$,[11] respectively. At the temperatures that most concern us, $m(W) \ll T \lesssim m(\tilde{W})$, the universe is effectively $[SU(2)\otimes U(1)]_W \otimes SU(3)_C$ symmetric and all fermions remain massless, which makes subsequent computations easier.

The baryon nonconservation is caused by exchange of $\tilde{W}$ coupled to fermions,

$$(g/2\sqrt{2})\sum_i [(\bar{l}_1\bar{l}_2)\gamma_\alpha (q_3^{\,c})_R + \bar{l}_3\gamma_\alpha (q_2^{\,c}, -q_1^{\,c})_L + \epsilon_{abc}(\bar{q}_1^{\,a}\bar{q}_2^{\,a})\gamma_\alpha (C\bar{q}_4^{\,b})_L]_i \begin{pmatrix} \tilde{W}_1^{\,c} \\ \tilde{W}_2^{\,c} \end{pmatrix}_\alpha + (\text{H.c.}), \tag{6}$$

and by exchange of colored Higgs $H_i(\underline{5})$ contained in the full Yukawa coupling,

$$L_Y = \tfrac{1}{2} f_{ij}\bar{\psi}_i^{\,\alpha\beta}[H_1^{\,\alpha}(\varphi_j^{\,\beta})_R - H_1^{\,\beta}(\varphi_j^{\,\alpha})_R] + \tfrac{1}{4} h_{ij}\epsilon_{\alpha\beta\gamma\delta\eta}\psi_i^{\,\alpha\beta} C(\psi_j^{\,\gamma\delta})_L H_2^{\,\eta} + (\text{H.c.}), \tag{7}$$

where the two Higgs bosons of $\underline{5}$, H_1 and H_2, may or may not coincide. The Yukawa coupling constants in (7) are related to fermion masses when $SU(2)\otimes U(1)$ is broken,

$$(f_{ij}) = (2\sqrt{2}G_F)^{1/2}\cos\zeta M_1, \tag{8a}$$

$$(h_{ij}) = (2\sqrt{2}G_F)^{1/2}\sin\zeta U^T M_2 U, \tag{8b}$$

where G_F is the Fermi constant, $G_F m_N^2 \simeq 10^{-5}$; M_1 and M_2 are diagonalized mass matrices for quarks of charge $\frac{2}{3}$ and $-\frac{1}{3}$, respectively [masses of charged leptons are equal to those of $q(-\frac{1}{3})$ except for renormalization effects[12]]; U is a unitary matrix that generalizes the Cabibbo rotation for more than three flavors.[13] To allow CP nonconservation also in the Higgs sector as in the Weinberg model[14] of CP nonconservation, I introduce the complex dimensionless parameters, α and β, in propagators by

$$i\langle T(H_1{}^{\dagger a} H_2{}^a)\rangle_{q=0} = \begin{cases} \alpha/m^2(W) \text{ for } a = 1, 2 \\ \beta/m^2(W) \text{ for } a = 3-5, \end{cases} \quad (9)$$

where q is the momentum involved in propagation. This is possible only if more than three Higgs bosons $H_i(\underline{5})$ exist with complex, trilinear and quartic couplings to $\tilde{H}(\underline{24})$.

I now calculate the quantity δ in Eq. (4) by keeping only two-body reactions. Here it is reasonable to suppose that masses of Higgs bosons are of the order of $m(W)$ for uncolored and of $m(\tilde{W})$ for colored ones and fermion masses are $\ll m(W)$, and hence $\alpha, \beta = O(1)$. Remarkably, I found after summing over all sequences that δ vanishes if α and β are real. I can actually prove that this is true to any order of perturbation. The result of this computation is particularly simple when $m(W) \ll T \ll m(\tilde{W})$. The dominant contribution to δ of (4) comes from interference of the diagrams of Figs. 1(a) and 1(b) and, after the thermal average is taken, leads to

$$\frac{\delta}{T^2} \simeq \frac{3g^2}{8\pi^2 m^4(\tilde{W})} \operatorname{Im}\beta\alpha^* [\operatorname{tr}hh^\dagger \operatorname{tr}f^2 - \operatorname{tr}hh^\dagger f^2]. \quad (10)$$

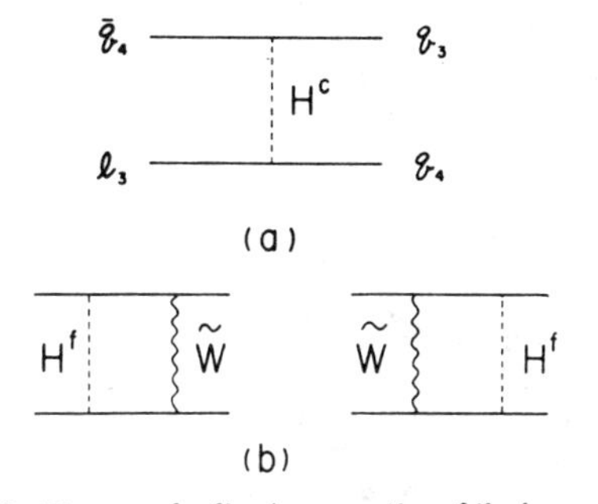

FIG. 1. Diagrams leading to generation of the baryon number. H^c (H^f) means a colored (uncolored) Higgs boson.

In this computation I ignored the Pauli exclusion effect in the final state caused by occupied thermal fermion states. Correct inclusion of this effect would not affect the result (10) drastically. Under the reasonable assumption $F_B \ll F_\gamma$ or $N_B \ll N_\gamma$, F_γ on the right-hand side of (4) may be approximated by the value at an initial temperature T_i where $N_B(T_i) = 0$, and the rate equation (4) is integrated to

$$\frac{N_B(T)}{N_\gamma(T)} \simeq -(8\pi^3 G_N d_F/5)^{-1/2}(\tfrac{3}{8})^2 \frac{\delta}{T^2} N_\gamma(T_i). \quad (11)$$

During the evolution of universe from this high temperature T down to the recombination temperature (~4000°K) all fundamental constituents in the initial universe annihilate each other or go out of thermal contact, and the ratio (11) is roughly conserved to give finally the present value $(N_B/N_\gamma)_0$.

To obtain a rough quantitative idea of this ratio, I shall make a drastic extrapolation of formula (11) up to $T_i = m(\tilde{W})$. In the case of six flavors ($n_S = 3$) the best guess[12] for the parameters of the model is $g^2/4\pi = 0.022$, $m(\tilde{W}) = 2 \times 10^{16}$ GeV. I also use the large difference of quark mass scales, $m(b) \gg m(s)$, $m(d)$ and $m(t) \gg m(c)$, $m(u)$, and the small mixing parameters[15] ($\lesssim 0.1$) in U of (8b). Combining (8), (10), and (11) and putting in some numerical factors, I find

$$\left(\frac{N_B}{N_\gamma}\right)_0 \approx 0.12 \frac{(d_F G_N)^{-1/2}}{m(\tilde{W})} \epsilon, \quad (12a)$$

$$\epsilon \simeq \frac{g^2}{\pi^2} G_F{}^2 m^2(t) m^2(b)(\sin\theta_2 + \sin\theta_3)^2 A, \quad (12b)$$

where $A = \operatorname{Im}\beta\alpha^* \sin^2\zeta \cos^2\zeta \approx O(1)$ in general. For definiteness, mixing angles are set by $\sin\theta_2 = \sin\theta_3 = 0.1$, which is consistent with present data.[15] Furthermore, $d_F = 63$ (69) with three H_i of $\underline{5}$ and real (complex) $\tilde{H}$ of $\underline{24}$, and hence

$$\left(\frac{N_B}{N_\gamma}\right)_0 \approx 2\times 10^{-9} \left[\frac{m(t)}{10 \text{ GeV}}\right]^2 \left[\frac{m(b)}{5 \text{ GeV}}\right]^2 A. \quad (13)$$

An estimate of this quantity[2] deduced from data ranges from 10^{-8} to 10^{-10}, which agrees with (13). The numerical value of (13) should not be taken too seriously because of the very crude approximations assumed, but it is hard to imagine that neglected corrections would alter (13) by more than 100.

Although my result (13) depends on the specific SU(5) model taken, the formula (12a) is presumably more general than this example provided that $\epsilon \approx$ (CP-invariance violation parameter)/137.

To this extent the small ratio of the number densities appears to be a general consequence of grand unification in the earliest history of our universe. The possibility that this fundamental parameter of cosmology is related to those of elementary-particle physics seems intriguing and deserves much investigation.

---

[1]See e.g., T. Weeks, *High Energy Astrophysics* (Chapman and Hall, London, 1969).

[2]For a review, see S. Weinberg, *Gravitation and Cosmology* (Wiley, New York, 1972), Chap. 15.

[3]However, for a possibility of elevating the baryon number to an absolute conservation law in grand unified models, see M. Yoshimura, Progr. Theor. Phys. 58, 972 (1977); M. Abud, F. Buccella, H. Ruegg, and C. A. Savoy, Phys. Lett. 67B, 313 (1977); P. Langacker, G. Segrè, and A. Weldon, Phys. Lett. 73B, 87 (1978). Also M. Gell-Mann, P. Ramond, and R. Slansky, to be published.

[4]H. Georgi and S. L. Glashow, Phys. Rev. Lett. 32, 438 (1974); H. Fritzsch and P. Minkowski, Ann. Phys. (N.Y.) 93, 193 (1975); F. Gürsey and P. Sikivie, Phys. Rev. Lett. 36, 775 (1976); K. Inoue, A. Kakuto, and Y. Nakano, Progr. Theor. Phys. 58, 630 (1977).

[5]F. Reines and M. F. Crouch, Phys. Rev. Lett. 32, 493 (1974).

[6]H. D. Politzer, Phys. Rev. Lett. 30, 1346 (1973); D. J. Gross and F. Wilczek, Phys. Rev. Lett. 30, 1343 (1973).

[7]J. C. Collins and M. J. Perry, Phys. Rev. Lett. 34, 1353 (1975); M. B. Kislinger and P. D. Morley, Phys. Lett. 67B, 371 (1977).

[8]D. A. Kirzhnits and A. D. Linde, Phys. Lett. 42B, 471 (1972); S. Weinberg, Phys. Rev. D 9, 3357 (1974); C. Bernard, Phys. Rev. D 9, 3312 (1974); L. Dolan and R. Jackiw, Phys. Rev. D 9, 3320 (1974).

[9]See, e.g., J. M. Blatt and V. F. Weisskopf, *Theoretical Nuclear Physics* (Wiley, New York, 1952), p. 530.

[10]S. Herb *et al.*, Phys. Rev. Lett. 39, 252 (1977); W. R. Innes *et al.*, Phys. Rev. Lett. 39, 1240, 1640(E) (1977); M. L. Perl, in *Proceedings of the International Symposium on Lepton and Photon Interactions at High Energies, Hamburg, 1977*, edited by F. Gutbrod (DESY, Hamburg, Germany, 1977).

[11]S. Weinberg, Phys. Rev. Lett. 19, 1264 (1967); A. Salam, in *Elementary Particle Physics*, edited by N. Svartholm (Almquist and Wiksels, Stockholm, 1968), p. 367.

[12]H. Georgi, H. R. Quinn, and S. Weinberg, Phys. Rev. Lett. 33, 451 (1974); A. J. Buras, J. Ellis, M. K. Gaillard, and D. V. Nanopoulos, to be published.

[13]M. Kobayashi and T. Maskawa, Progr. Theor. Phys. 49, 652 (1973).

[14]S. Weinberg, Phys. Rev. Lett. 37, 657 (1976).

[15]J. Ellis, M. K. Gaillard, and D. V. Nanopoulos, Nucl. Phys. B109, 213 (1976).

ERRATA

UNIFIED GAUGE THEORIES AND THE BARYON NUMBER OF THE UNIVERSE. Motohiko Yoshimura [Phys. Rev. Lett. 41, 281 (1978)].

In the computation of δ of Eq. (10), Feynman diagrams containing triangle loops were omitted. A mistake was made by incorrectly ignoring a finite discontinuity that remains after the vertex renormalization. This correction leads to a conclusion that contributions of massless fermions to the baryon asymmetry vanish to the order of $g^2hh^\dagger f^2$ in the approximation of this paper. This agrees with a recent result of D. Toussaint *et al.* (to be published) based on two-body unitarity. More satisfactory computation, including effects near the unification temperature, will be dealt with in a separate paper. The author should like to thank Dr. J. Arafune, Dr. S. M. Barr, and Dr. S. Weinberg for critical comments.

Baryon number of the universe

Savas Dimopoulos*
Enrico Fermi Institute, University of Chicago, Chicago, Illinois 60637

Leonard Susskind
Stanford Linear Accelerator Center, Stanford University, Stanford, California 94305

(Received 9 June 1978)

We consider the possibility that the observed particle-antiparticle imbalance in the universe is due to baryon-number, C, and CP nonconservation. We make general observations and describe a framework for making quantitative estimates.

I. INTRODUCTION

Evidence exists[1] that the universe contains many more particles than antiparticles. A quantitative measure of this particle excess is given by the number of baryons within a unit thermal cell of size $R = T^{-1}$. Such a cell contains a single black-body photon.[2] In current cosmological theories it is a box, expanding according to[2]

$$R^{-1}\frac{d}{dt}R(t) \approx \left(\frac{8\pi}{3}G\rho\right)^{1/2}. \tag{1.1}$$

In the very early universe there was approximately 1 of every species of particle within a unit cell. However, the unit cell today contains only 10^{-9} baryons and essentially no antibaryons. If baryon number is conserved then the unit cell has always contained a baryon number of order 10^{-9}.

One cannot rule out the possibility that the universe was created with net baryon number and no explanation is needed. However, to quote Einstein: "If that's the way God made the world then I don't want to have anything to do with Him." In fact modern theories of particle interactions suggest that baryon number is not strictly conserved.[3,4] If this is true then today's baryon number is as much dependent on dynamical processes as on initial conditions. Indeed Yoshimura[5] has made the exciting suggestion that baryon-number violation can combine with CP noninvariance to produce a calculable net baryon number even though the universe was initially baryon neutral. Yoshimura has also made estimates[5] which indicate that this may be quantitatively plausible.

There are three interesting reasons to believe that baryon number is not exactly conserved:

(1) Black holes can swallow baryons.[6]

(2) Quantum-mechanical baryon-number violations have been discovered by '*t* Hooft in the *standard* Weinberg-Salam theory.[4]

(3) Superunified theories of strong, electromagnetic, and weak interactions naturally violate baryon number at superhigh energy.[3]

Although baryon-number violations are minute at ordinary energy, in cases (2) and (3) they may become significant at sufficiently high temperature.

Baryon-number violations is not enough to create an excess of baryons. The process itself must be particle-antiparticle asymmetric.[5] Otherwise the sign of the effect will be random and cancel in different cells. In this case the total baryon excess would be of the order of the square root of the total number of photons. However, the total number of photons in the observed universe is $\sim 10^{88}$ and the baryon number is $\sim 10^{79}$.

The required particle-antiparticle asymmetry is known to exist. Indeed charge conjugation is maximally violated in ordinary weak interactions. Were this the only asymmetry, CP invariance would destroy any possible effect because total baryon number changes sign under CP as well as C. Luckily CP violations are known to exist.[7]

CPT invariance also imposes a very interesting constraint on the expansion rate of the universe. As we shall see, CPT invariance ensures vanishing baryon density in thermal equilibrium. Therefore the expansion rate must remain rapid enough to prevent the baryon-number-violating forces from coming to equilibrium.

In this paper we will discuss how baryon-number, C, and CP nonconservation can conspire with the early Hubble expansion to produce an observable baryon excess.

As we shall see, the baryon excess may originate at or close to the very earliest times, $\sim 10^{-43}$ sec. At that time the temperature, energy density, and local space-time curvature are assumed to be of order unity in units of the Planck mass. The metric in Planck units is of the Robertson-Walker type[2]

$$(ds)^2 = dt^2 - R(t)^2 dx_i dx_i ,$$

where $R(t) \sim 1$ at the Planck time $t = 1$.

Let us follow the evolution of a single unit coordinate cell of dimensions $\Delta x_i = 1$. At the earliest of times it is a cube of unit volume (10^{-100} cm^3) in Planck units. We will assume that quantum fluctuations and gravitational interactions between gravitons and matter rapidly bring the universe to equilibrium at a temperature of unity. It follows that our unit cell initially contains about one elementary particle of each species. In current unified theories this means ~100 particles (photons, leptons, gravitons, intermediate bosons, quarks, vector gluons, Higgs bosons, superheavy bosons,...).

As the unit cell evolves it expands and cools. The process is not too different from the slow expansion of a box containing radiation. As in this case, the entropy within the cell is not significantly changed during the expansion. Roughly speaking this implies that the number of particles within that cell is the same today as it was at creation. Of course, by now, the only particles left are photons, neutrinos, and any excess protons and electrons. The others all annihilated or decayed when the temperature decreased below their mass.

The excess, expressed as a baryon number in the unit coordinate cell, is a number of order

$$n_B = \frac{N_B}{N_\gamma} n_\gamma ,$$

where $N_B/N_\gamma \approx 10^{-9}$ and n_γ is the number of photons in the unit cell today. Assuming it is of the order of the number of elementary particle types, we must account for 10^{-7} baryons per box.

The estimates made in later sections for the baryon excess are too uncertain to be taken seriously. In addition to particle physics uncertainties, the properties of the initial conditions at creation are unknown and can influence the result. Our estimates are made for the most pessimistic case which we call "chaotic initial conditions." Such an initial condition is described by a density matrix ρ which is diagonal in baryon number and symmetric under the interchange of baryons and antibaryons. It is the sort of initial condition which would describe equilibrium if the earliest interactions respected baryon-number, C and CP invariance.

II. *CPT* AND EQUILIBRIUM

It is self-evident that if C or CP are symmetries of the equations of motion then no global baryon excess can result from baryon-number-violating processes. To illustrate the constraints imposed by CPT in an expanding universe we discuss some examples.

Consider a complex scalar field $\phi(x)$ in an expanding universe described by the metric

$$(ds)^2 = (dt)^2 - R(t)^2 (d\vec{x})^2 . \tag{2.1}$$

The action for this model is taken to be

$$s = \int d^4x \sqrt{-g}\, [g^{\mu\nu} \partial_\mu \phi \partial_\nu \phi^* - V(\phi)] , \tag{2.2}$$

where

$$V(\phi) = \lambda(\phi\phi^*)^n (\phi + \phi^*)(\alpha\phi^3 + \alpha^*\phi^{*3}) \tag{2.3}$$

and α is a complex phase. The baryon current density is

$$B_\mu = \sqrt{-g}\, i\phi \overleftrightarrow{\partial}_\mu \phi^* . \tag{2.4}$$

Note that $V(\phi)$ violates baryon-number conservation, C invariance ($\phi \to \phi^*$), and CP invariance.

The Hamiltonian for this model is

$$H(t) = \int d^3x \left[\frac{\pi\pi^*}{R^3(t)} + R(t) |\nabla\phi|^2 + R^3(t) V(\phi) \right] . \tag{2.5}$$

This Hamiltonian is invariant under the following CPT transformation[8]:

$$\phi(x) \to \phi(-x) , \tag{2.6}$$

$$\pi(x) \to -\pi(-x) . \tag{2.7}$$

The baryon number

$$B_\mu(x) = \begin{cases} i(\phi\pi - \pi^*\phi^*), & \mu = 0 \\ \sqrt{-g}\, i\phi \overleftrightarrow{\nabla}\phi^*, & \mu = i \end{cases} \tag{2.8}$$

changes sign under (2.6) and (2.7).

The CPT transformation is a symmetry of the spectrum of the instantaneous Hamiltonian but not of the equation of motion because of the explicit time dependence of H.

Now consider the case where the universe expands so slowly that at every instant it is in thermal equilibrium with respect to the instantaneous Hamiltonian $H(t)$. The density matrix at time t is

$$\rho(t) = \exp[-\beta(t) H(t)] . \tag{2.9}$$

Since CPT conjugate states carry equal energy but opposite baryon charge B the expectation value of B vanishes,

$$\langle B \rangle = \mathrm{Tr}(e^{-\beta(t) H(t)} \hat{B}) = 0 . \tag{2.10}$$

Therefore the only hope of generating baryon excess is for the baryon-number-violating interactions to remain out of thermal equilibrium. This implies that the rate of expansion of the universe has to be faster than the baryon-number-violating reaction rates.

Now we will discuss a second model to illustrate the possibility of baryon-number generation if we are out of equilibrium.

Consider a time-independent Hamiltonian $H = H_0 + V$. H_0 is baryon-number, C, and CP conserving and V is a small perturbation which violates these quantum numbers. Suppose that at time $t = 0$ the system is in thermal equilibrium with respect to the Hamiltonian H_0,

$$\rho(0) = e^{-\beta H_0}. \tag{2.11}$$

Under the action of the full Hamiltonian the density matrix at time t has evolved to

$$\rho(t) = e^{-iHt}e^{-\beta H_0}e^{+iHt}.$$

The mean baryon number is

$$\langle B(t)\rangle = \mathrm{Tr}[e^{-\beta H_0}\hat{B}(t)]/\mathrm{Tr}\rho\,, \tag{2.12}$$

where

$$\hat{B}(t) = e^{iHt}\hat{B}e^{-iHt}. \tag{2.13}$$

The CPT invariance of H and H_0 implies that $\langle B(t)\rangle$ is an odd function of time

$$\begin{aligned}\langle B(t)\rangle &= \mathrm{Tr}e^{-\beta H_0}\hat{B}(t) &(2.14)\\ &= \mathrm{Tr}[\theta e^{-\beta H_0}\theta^{-1}\theta\hat{B}(t)\theta^{-1}]\\ &= \mathrm{Tr}\{e^{-\beta H_0}[-\hat{B}(-t)]\}\\ &= -\mathrm{Tr}[e^{-\beta H_0}\hat{B}(-t)]\\ &= -\langle B(-t)\rangle\,, &(2.15)\end{aligned}$$

where $\theta = CPT$. This antisymmetry of $\langle B\rangle$ with time is the *only* constraint implied by CPT.

Interesting information can also be extracted by looking at the rate of change of B,

$$\langle\dot{B}(t)\rangle = i\,\mathrm{Tr}\{e^{-iHt}[e^{-\beta H_0}, V]e^{iHt}\hat{B}\}. \tag{2.16}$$

If we approximate e^{-iHt} by e^{-iH_0t} then $\langle\dot{B}\rangle$ must vanish since $[B, H_0] = 0$. This implies $\langle\dot{B}\rangle$ is at least second order in V and first order in time, $\langle\dot{B}\rangle \sim t$. But since $\langle B\rangle$ is an odd function of t, $\langle\dot{B}\rangle$ must be even and cannot be of order t. It follows that $\langle\dot{B}\rangle$ is at least second order in t and $\langle B\rangle$ is third order:

$$\langle B(t)\rangle \sim t^3. \tag{2.17}$$

That baryon-number excess vanishes to first order in V is to be expected. The nontrivial part of the time translation operator U is anti-Hermitian to first order. Therefore amplitudes changing B by opposite amounts have equal magnitude and cancel. The relation $\langle B\rangle \sim t^3$ shows that baryon excess builds up slowly in the beginning.

In this example, a period of time will elapse during which $\langle B\rangle$ is not zero. Eventually the interactions in V will restore the system to true thermal equilibrium with vanishing $\langle B\rangle$. If, however, the baryon-number-violating force is switched off after a finite time the system will retain a finite net baryon excess.

The process of early expansion can disturb thermal equilibrium and lead to a temporary excess. If the universe expands and cools sufficiently rapidly the baryon-number-violating forces may not have time to come back to equilibrium. This is especially true if the reaction rates for these processes are rapidly falling with decreasing temperature. In order to estimate if this is so we consider the quantity $\dot{R}/R$ which measures the rate of expansion of the universe. The condition for equilibrium is

$$\frac{\dot{R}}{R} < \text{reaction rate}\,. \tag{2.18}$$

The expansion rate in the radiation-dominated epoch is given by

$$\frac{\dot{R}}{R} \approx T^2\,, \tag{2.19}$$

where the temperature T and time are in units $c = \hbar = G = 1$.

The dependence of the reaction rate on temperature can be obtained from dimensional considerations. For example, in a renormalizable theory with all mass scales much lower than T the reaction rate must be proportional to T. This is because coupling constants in renormalizable theories are dimensionless. Accordingly the condition for equilibrium is

$$T^2 < T \tag{2.20}$$

or

$$T < 1\,. \tag{2.21}$$

Therefore the condition for thermal equilibrium in renormalizable theories is increasingly satisfied as the universe cools. This continues as long as explicit masses can be ignored. From these arguments it is easy to see that ordinary strong electromagnetic and weak interactions are in thermal equilibrium from superhigh temperatures ($\sim 10^{15}$ GeV) down to ordinary temperatures (~ 1 GeV).

In superunified theories baryon-number-violating processes are effectively nonrenormalizable Fermi interactions below energies $\sim 10^{18}$ GeV. This energy corresponds to the mass $\tilde{M}$ of the superheavy bosons which mediate the process. The effective Fermi coupling constant is

$$\tilde{G} \approx \frac{\alpha}{M^2} \sim 10^{-38}\ \mathrm{GeV}^{-2}. \tag{2.22}$$

The reaction rate is proportional to $\tilde{G}^2$ and by dimensional arguments is

$$(\text{reaction rate}) \approx \tilde{G}^2T^5\,.$$

The condition for equilibrium becomes

$$T^2 < G^2T^5 \text{ (in Planck units)}$$

or

$$T > \left(\frac{\tilde{M}^4}{\alpha^2}\right). \tag{2.23}$$

For $\tilde{M} \sim M_{\text{Planck}}$ it is unlikely that the baryon-number-violating forces were ever in equilibrium.

Note that the baryon-number violations are of order α at temperatures $\sim 10^{18}$ GeV. Effectively we are in a situation where these interactions are switched on for a brief time interval and are then switched off. These considerations indicate that the possibility of generating baryon excess is viable.

III. MODELS WITH BARYON-NUMBER VIOLATION

By a unified theory[3] we mean a theory in which the strong, weak, and electromagnetic gauge invariance are embedded in a simple unifying group. Such theories involve a single coupling constant of the order of the electric charge. Both leptons and quarks appear in the same multiplets. Therefore quarks can turn into leptons by the emission of vector bosons called $\tilde{W}$. For example, in the SU_5 theory of Georgi and Glashow the process shown in Fig. 1 is possible. This process implies that a proton can decay into a positron and photons.

In order to suppress the decay of the proton, the mass of the $\tilde{W}$ must be made large. Consistency with the empirical bounds on the lifetime of the proton requires

$$\tilde{M} > 10^{15}\ \text{GeV}. \tag{3.1}$$

We will assume $\tilde{M}$ is approximately the Planck mass and set it equal to unity. This assumption simplifies our discussion.

At energies below $\tilde{M}$ the baryon-number-violating processes are effectively described by four-Fermi interactions. The coupling constant is approximately

$$G = \frac{\alpha}{\tilde{M}^2} = \alpha \tag{3.2}$$

in Planck units. The baryon-number-changing interactions obviously are unimportant for temperatures very much smaller than $\tilde{M}$.

The other ingredient needed for baryon excess is CP violation.[7] In principle the observed CP violation could arise spontaneously[9] or from explicit asymmetry of the Lagrangian.[10] If it arises spontaneously then it disappears at temperatures well above 1 TeV. In this case the CP and baryon processes cannot combine to yield an excess.

We will assume that a CP violation, perhaps unrelated to observed CP violation, exists at the superheavy scale. We might suppose that this breaking is also spontaneous. However, in this case it could not be effective in producing an excess. The reason is because the radius of an event horizon is very small at the time when the baryon excess is produced. This means that uncorrelated domains of different CP directions must occur with small spatial extent. Within these domains the baryon excess will have opposite sign and therefore cancel. Thus we must have an explicit CP violation in the part of the Lagrangian which is relevant at superheavy scales. This does not exclude the idea[9] that the observed CP violation is spontaneous.

For definiteness we will assume explicit four-Fermi vertices which break both CP and baryon-number conservation.

A second source of baryon-number violation has been discovered in the standard Weinberg-Salam theory. In this model the baryon-number violation is of purely quantum-mechanical origin.[4] There exists a discrete infinity of classical degenerate vacuums[11] labeled by the "winding number" n. Quantum-mechanical transitions between these classical vacuums can occur by tunneling through an energy barrier. These events are called instantons. The physics is analogous to tunneling between the minima of a periodic potential. As 't Hooft first noted,[4] each instanton event is accompanied by a change in baryon number. A change in lepton number also occurs in order to compensate the electric charge. The tunneling amplitued at zero temperature is proportional to[4]

$$e^{-8\pi^2/g^2},$$

which is of the order of 10^{-93}. At very high temperatures $T > 250$ GeV two qualitatively new things happen. First, the Higgs vacuum expectation value goes away.[12] Second, there exists a lot of thermal energy available. This can be used to overcome the potential barrier.

To estimate the importance of this effect we must compare the barrier height with the available thermal energy. Consider an instanton of space-time radius ρ. For temperatures $\gg 250$ GeV the expectation value of the Higgs potential vanishes and the action of an instanton is roughly what it would be for pure Yang-Mills theory:

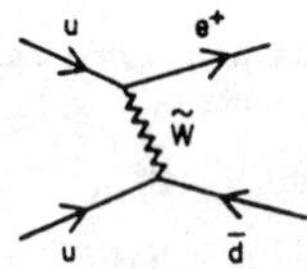

FIG. 1. Baryon-number-violating process occurring in the SU_5's unified theory.

$$\text{action} = 8\pi^2/g^2 . \tag{3.3}$$

The tunneling barrier is estimated by dividing this action by the duration of the event ρ,

$$V = 8\pi^2/g^2\rho . \tag{3.4}$$

[We remind the reader that Eq. (3.4) only applies above the transition temperature for the Higgs field to disappear.]

Equation (3.4) suggests that we can always lower the barrier as small as we like by considering arbitrarily large instantons. This is not so. The reason is that a tunneling event is a coherent process in which the instanton density $F_{\mu\nu}\tilde{F}_{\mu\nu}$ is of a definite sign over the size of the tunneling region. Thus ρ cannot exceed the coherence length which is given by the Debye screening length in the gauge-field plasma.[13] This is given by the plasmon Compton wavelength which for pure Yang-Mills theory is

$$\lambda_{\text{plasma}} \approx \left(\frac{gT}{\sqrt{6}}\right)^{-1} = \rho_{\max} . \tag{3.5}$$

The thermal energy within such a volume is $\sim(\pi^2/2)\rho_{\max}^3 T^4 \sim (\pi^2/2) T^4 \lambda_p^3$. The condition that this thermal energy overcomes the barrier V is

$$\frac{\pi^2}{2} T^4 \lambda_p^3 > \frac{8\pi^2}{g^2} \frac{1}{\lambda_p}$$

or (3.6)

$$(18 - 8g^2)\frac{\pi^2}{g^4} \geq 0.$$

This appears to be satisfied for the coupling constants characteristic of weak-electromagnetic theories.

These crude estimates only suggest the possibility that baryon-number-violating interactions are not suppressed at $T > 250$ GeV. Quantitative calculations are needed to decide the importance of this effect. In particular, the effects of fermions will probably suppress the tunneling. For the remainder of this paper we will ignore this quantum-mechanical source of baryon-number violation, although it is possible for it to seriously alter the results of this paper.

IV. BARYON GENERATION MECHANISM IN FIELD THEORY

In this section we will describe field-theoretic methods for computing the baryon-number excess in an expanding universe. For definiteness we will consider a model in which both baryon and *CP* violation are mediated by superheavy bosons of mass $\sim M_{\text{Planck}}$. In practice this means that these interactions are described as four-Fermi couplings.

We are going to consider a field theory in an expanding universe described by the metric

$$ds^2 = (dt)^2 - R(t)^2 (d\vec{x})^2 \tag{4.1}$$

$$= (dt)^2 - t\,(d\vec{x})^2 . \tag{4.2}$$

The choice $R = \sqrt{t}$ is appropriate to a radiation-dominated epoch. We will illustrate such a system by considering a scalar field with action

$$S = \int d^4x\sqrt{-g}\left[g^{\mu\nu}\frac{\partial\phi}{\partial x^\mu}\frac{\partial\phi^*}{\partial x^\nu} + V(\phi)\right]. \tag{4.3}$$

Now the metric in Eq. (4.2) is of the conformally flat type meaning that by a change of variables it can be brought to the form

$$ds^2 = \rho^2(x)[(dx_0)^2 - (d\vec{x})^2] . \tag{4.4}$$

In particular, if we change variables from t to $\tau = (2t)^{1/2}$ then

$$ds^2 = \tau^2(d\tau^2 - d\vec{x}^2) . \tag{4.5}$$

Now the reader can verify that if the field ϕ is replaced by

$$s = \rho^{-1}\phi , \tag{4.6}$$

then the free part of the Lagrangian becomes

$$S = \int d^3x\,d\tau\left[\left(\frac{ds}{d\tau}\right)^2 - (\nabla s)^2\right]$$

$$+ \text{pure divergence}. \tag{4.7}$$

Furthermore, if a renormalizable ϕ^4 interaction is present in V then it is replaced by s^4. If on the other hand nonrenormalizable terms such as ϕ^{4+2n} are present they are replaced by

$$V(s) = \frac{s^{4+2n}}{\tau^{2n}} . \tag{4.8}$$

Thus, in the new time coordinate, the free and renormalizable terms in the action take their flat-space form and appear to be τ independent. The nonrenormalizable terms appear time dependent with rapidly falling coefficients.

Similar results hold for more general theories. If we consider the usual type of theory containing scalar spinor and vector fields ϕ, ψ, A_μ and define conformal fields by

$$\begin{aligned} &\phi \to \rho^{-1}\phi ,\\ &\psi \to \rho^{-3/2}\psi ,\\ &A_\mu \to A_\mu , \end{aligned} \tag{4.9}$$

then the free and renormalizable terms take their flat-space form. The nonrenormalizable Fermi couplings are replaced by their flat-space counterparts times the factor $1/\tau^2$. Thus the form that the action for our model takes is

$$S=\int d^3x\,d\tau\left(L_0+\frac{1}{\tau^2}L_I\right), \quad (4.10)$$

where L_0 is a renormalizable τ-independent Lagrangian containing all the usual interactions and L_I is a four-Fermi coupling containing the superheavy mediated effects.

We will make two cautionary remarks before proceeding to study baryon-number excess generation. The first is that the flat-space form for renormalizable theories ignores mass effects. Since we only use it for very high temperatures this is no problem. The second remark concerns ultraviolet divergences. The above analysis was purely classical and fails when renormalization is accounted for. However, because the unified coupling is small at the Planck length, the failure only involves very weakly varying logarithms. In fact, these effects would show up as logarithms of τ multiplying the renormalizable interactions. They are completely unimportant for our problem.

Let us now return to the baryon excess problem. We write the Hamiltonian resulting from Eq. (4.10) as

$$H=H_0+V(\tau), \quad (4.11)$$

where H_0 is baryon-number and CP conserving. $V(\tau)$ contains the violating terms and scales like τ^{-2}.

Suppose the initial density matrix at the Planck time $\tau=1$ is given by $\rho(1)$. The expectation value of the baryon number at this time is

$$\langle B(1)\rangle=\operatorname{Tr}\rho(1)\hat{B}. \quad (4.12)$$

At a later time τ the value of $\langle B\rangle$ is

$$\begin{aligned}\langle B(\tau)\rangle&=\operatorname{Tr}\rho(1)U^\dagger(\tau)\hat{B}U(\tau)\\&=\operatorname{Tr}U(\tau)\rho(1)U^\dagger(\tau)\hat{B}, \end{aligned} \quad (4.13)$$

where $U(\tau)$ is the time translation operator from $\tau=1$ to τ.

For the case that $V(\tau)$ is τ independent (renormalizable interactions) we may immediately conclude that as $\tau\to\infty$ $\langle B\rangle\to 0$. This is because a field theory with time-independent Hamiltonian will eventually come to thermal equilibrium and we have seen that CPT ensures $B=0$ in this case.

On the other hand, if $V(\tau)\to 0$ fast enough we can use ordinary perturbation theory in V to compute the baryon-number excess as $\tau\to\infty$. To do this we use the standard interaction-picture formalism to obtain

$$U(\tau)=U_0(\tau)U_V(\tau)\,,\quad U_0(\tau)=e^{-iH_0(\tau-1)},$$

$$U_V(\tau)=T\exp\left[-i\int_1^\tau V_I(\tau')d\tau'\right], \quad (4.14)$$

$$V_I(\tau)=U_0^\dagger(\tau)V(\tau)U_0(\tau).$$

Thus using $[B,U_0]=0$,

$$\langle B(\tau)\rangle=\operatorname{Tr}\rho(1)U_V^\dagger(\tau)\hat{B}U_V(\tau). \quad (4.15)$$

Graphical rules are derived in Appendix A for the evaluation of (4.15). The following features emerge from analysis of these rules:

(1) For the case $V(\tau)\sim 1/\tau^2$ each order has a finite limit as $\tau\to\infty$. These limits give an order-by-order expansion of the final baryon-number excess.

(2) The first order in which a nonvanishing excess occurs depends on certain features of $\rho(1)$. In particular, if $[\rho(1),B]=0$ then the first order vanishes.

(3) If in addition to $\rho(1)$ being diagonal in baryon number it is CP symmetric then the second order also vanishes. Thus in the case of initially chaotic conditions, baryon-number excess is a third-order effect. Thus, since we suppose [see Eq. (3.2)] that $V\sim\alpha$, baryon-number excess will be $\sim\alpha^3$ for an initially chaotic ρ.

We are currently constructing Feynman rules for the evaluation of Eq. (4.15). These rules will be applied to some unified models in a future paper.

V. SCALAR TOY MODEL

Consider the model introduced in Sec. II [Eq. (2.2)]. In conformal coordinates the action becomes

$$S=\int d^3x\,d\tau\left[\left|\frac{\partial\phi}{\partial\tau}\right|^2-|\nabla\phi|^2-\frac{\lambda}{\tau^{2n}}(\phi\phi^*)^n(\phi+\phi^*)\right.$$

$$\left.\times(\alpha\phi^3+\alpha^*\phi^{*3})-g(\phi\phi^*)^2\right], \quad (5.1)$$

where we have added the renormalizable term $g\phi^4$ to represent all the renormalizable interactions. In this section we will make some very crude approximations which reduce the system to a single degree of freedom.

First we shall assume that the initial density matrix is in thermal equilibrium at a temperature ~ 1. If we ignore the small ($\sim\alpha$) nonrenormalizable couplings then the system will remain in equilibrium at this temperature for all τ. (Note that in transforming to the original coordinates the temperature becomes $1/\tau$ since it scales like energy.) Thus the average value of $|\phi|$ will remain constant of order unity. Indeed the first simplification will be to replace $|\phi|$ by unity.

The other drastic simplification will be to focus on a single unit coordinate cell over which ϕ will be assumed spatially constant. Setting $\phi=e^{i\theta}$ we obtain a system described by the Lagrangian

$$\mathcal{L}=\left(\frac{d\theta}{d\tau}\right)^2-\tau^{-2n}V(\theta)\,. \tag{5.2}$$

The baryon number of a unit cell is given by Eq. (2.4):

$$B=R^3(t)i\phi\overleftrightarrow{\partial}_t\phi^*$$
$$=2\frac{d\theta}{d\tau}\,. \tag{5.3}$$

Equation (5.2) describes a pendulum in a time-dependent unsymmetric potential and Eq. (5.3) says that the baryon number of a single cell is given by its angular velocity. The CPT invariance of the original instantaneous Hamiltonian corresponds to the time-reversal invariance of the pendulum.

The approximation of ignoring the interaction of neighboring cells is surely too severe to correctly describe the high-temperature nonequilibrium properties of the subsystem. In particular, it is impossible for the single pendulum to relax to thermal equilibrium if it is disturbed. For example, if the pendulum is given a hard "clockwise" swing it will forever continue to rotate so that $\dot\theta \not\to 0$. But in thermal equilibrium $\langle\dot\theta\rangle=0$ by the same arguments which we used to prove $\langle B\rangle=0$.

By ignoring the surrounding heat bath we have eliminated the possibility of dissipation. A simple method for incorporating it is to introduce a dissipative damping term into the equation of motion. Thus we write the equation of motion

$$\frac{d^2\theta}{d\tau^2}+\tau^{-2n}\frac{\partial V}{\partial\theta}+f(\tau)\frac{d\theta}{d\tau}=0\,. \tag{5.4}$$

The computation of friction coefficients in nonequilibrium statistical mechanics typically involves the computation of the absorptive (imaginary) part of some thermal Green's function.[14] That is to say, we calculate the width of some excitation which propagates in the medium.

In the case of electrical resistance we calculate the absorptive part of the plasmon propagator.[14] In our case, a nonzero baryon charge must dissipate as equilibrium is restored. Accordingly we must compute the width of the charge-carrying excitation described by the field ϕ due to baryon-number-violating processes. In the model field theory with interaction $V(\phi)=\lambda(\phi^*\phi)^n(\phi+\phi^*)(\alpha\phi^3+\alpha^*\phi^{*3})$ the relevant width is described by graphs shown in Fig. 2. Dimensional arguments require the temperature-dependent width to be

$$\gamma(T)\approx\lambda^2T^{4n+1}\,. \tag{5.5}$$

Thus if the number of baryons in the unit cell is B, the number lost by dissipation is

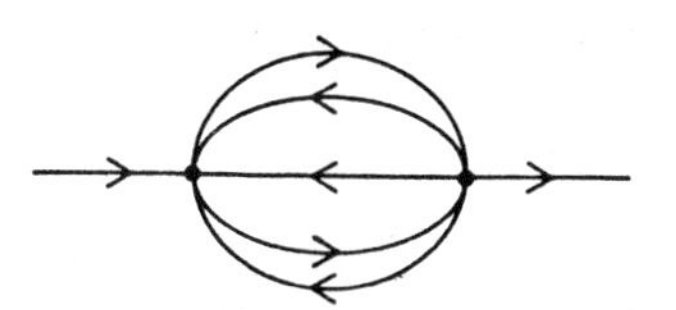

FIG. 2. Graph contributing to baryon dissipation.

$$\left(\frac{d}{dt}B\right)_{\rm dis}=-B\gamma=-B\lambda^2T^{4n+1} \tag{5.6}$$

or

$$\frac{dB}{dt}_{\rm dis}=-\frac{\lambda^2B}{\tau^{4n}}\,. \tag{5.7}$$

Recalling that B is identified with $d\theta/d\tau$ we interpret Eq. (5.7) to mean that the coefficient f in Eq. (5.4) is λ^2/τ^{4n}:

$$\frac{d^2\theta}{d\tau^2}+\tau^{-2n}\frac{\partial V}{\partial\theta}+\frac{\lambda^2}{\tau^{4n}}\frac{d\theta}{d\tau}=0\,. \tag{5.8}$$

Equation (5.8) defines the toy model.

To see how the toy model can lead to an asymmetric distribution of baryons and antibaryons consider a $V(\theta)$ which looks like Fig. 3, i.e., it has no point of reflection symmetry. Now suppose the initial probability density in θ and $\dot\theta$ is uniform in θ and symmetric under $\dot\theta\to-\dot\theta$. We observe that a particle has a large probability to get a small kick to the left and a small probability for a large kick to the right. Thus the probability distribution becomes asymmetric. However, to first order in time no average change in $\dot\theta$ occurs. This is because the average force $\partial V/\partial\theta$ vanishes for a uniform distribution in θ. In fact $\langle\dot\theta\rangle$ only becomes nonzero in order τ^5. Furthermore, the first nonvanishing order in V is third order.

If the universe were a nonexpanding box at fixed temperature then no net baryon excess could be maintained at long times. Indeed the toy model is consistent with this. In a nonexpanding universe the form of the toy model is

$$\frac{d^2\theta}{dt^2}+\lambda T^{2n+2}\frac{\partial V}{\partial\theta}+\lambda^2T^{4n+1}\frac{d\theta}{dt}=0\,. \tag{5.9}$$

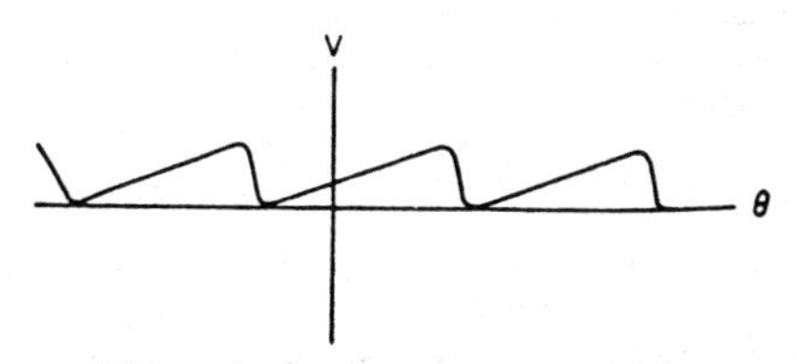

FIG. 3. A potential which violated $V(\theta)$.

Let us suppose after a long time that the baryon number $T^2 d\theta/dt$ is constant. Then

$$\lambda T^{2n+2}\frac{\partial V}{\partial\theta}+\lambda^2 T^{2+n+1}\dot{\theta}=0. \quad (5.10)$$

Integrating this over a period and using the periodicity of $V(\theta)$ we see that the baryon number has to vanish. Note that both the periodicity of the potential and the existence of the friction term are important in reaching this conclusion.

Now we find the conditions that will allow a nonvanishing baryon number at large times. Multiplying Eq. (5.8) by τ^{2n} and integrating over a period we obtain

$$\oint \tau^{2n}\frac{dK}{d\tau}d\tau=-2\lambda^2\oint\frac{K}{\tau^{2n}}\,d\tau, \quad (5.11)$$

where $K=\frac{1}{2}(d\theta/d\tau)^2$. After a long time, this equation effectively becomes

$$\frac{dK}{d\tau}=-2\lambda^2\frac{K}{\tau^{4n}}\,. \quad (5.12)$$

For the renormalizable case $n=0$ we see that baryon number is exponentially damped. This agrees with our previous expectations. For $n>\frac{1}{4}$ this equation has solutions for which the baryon number tends to a constant. Thus we see that for nonrenormalizable theories the friction term can be neglected, and baryon-number excess occurs as $\tau\to\infty$.

VI. CONCLUDING REMARKS

In this paper we have argued that a baryon-number excess may be produced in an expanding universe even though the initial conditions are symmetric. For the case of unified theories the excess is developed at times of order 10^{-40} sec while the temperature is comparable to the Planck mass. An admittedly oversimplified model yields a small number of baryons per unit cell of the order α^3.

The conclusion that the effect is $\sim\alpha^3$ does not appear to be general. It is a consequence of replacing the superheavy interactions by four-Fermi interactions. While this helps us visualize the process, it is not entirely consistent. This is because the main action occurs at energies of order $\tilde{M}$ and not much lower energies. Therefore it is important to open up the "black box" hiding the superheavy-boson exchange. As far as we can tell there are then order-α^2 effects. This is somewhat too large empirically but we must keep in mind that there are effects which we ignored which decrease N_B/N_γ. We have treated the universe expansion as if it were a reversible process with respect to the ordinary interactions. In fact there are possible sources of irreversibility which can heat up the system.[15] Eventually this heat must appear as photons.

Unfortunately this optimistic picture which emerges in unified theories may be drastically changed if the baryon-number-violating tunneling events are really important. The point is that the rates for these processes are of the renormalizable type for $T>250$ GeV. Thus they can allow the system to return to equilibrium and may wash out any excess which developed at superhigh temperature.

Of course as the temperature goes below 250 GeV the tunneling processes also go out of equilibrium. In principle the observed baryon-number excess could be attributed to this final stage of baryon-number violation. In this case the number of baryons in the universe is independent of the initial conditions and the details of the particular unified model.

ACKNOWLEDGMENTS

We would like to thank R. Wagoner for many critical insights without which we could not have written this paper. We are also indebted to E. Liang, D. Sciama, and G. Steigman for interesting comments. One of us (S. D.) would like to thank Sid Drell and Dirk Walecka for hospitality at SLAC and Stanford University. This work was supported in part by the National Science Foundation under Contract No. 76-16992.

APPENDIX A

Graphical rules for computing $\langle B(\tau)\rangle$. Consider a theory of fermions interacting with baryon-number-, C- and CP-violating four-Fermi forces. The Hamiltonian of this theory in the expanding universe in terms of the conformal coordinates is

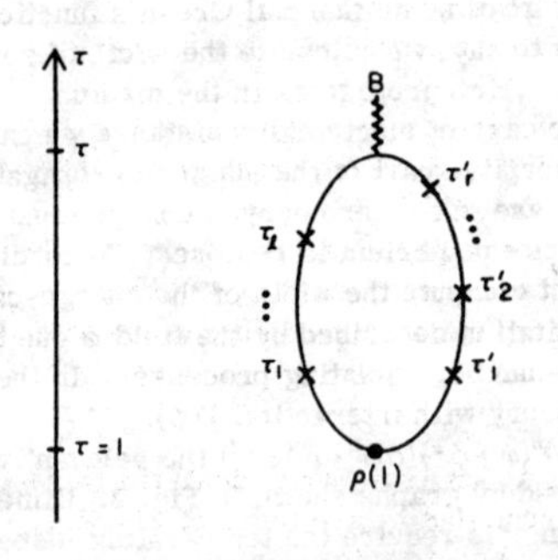

FIG. 4. Graphical notations. Solid lines represent propagating state vectors. Crosses represent the action of V. The balck dot represents the initial density matrix and the wavy line represents the measurement of baryon number.

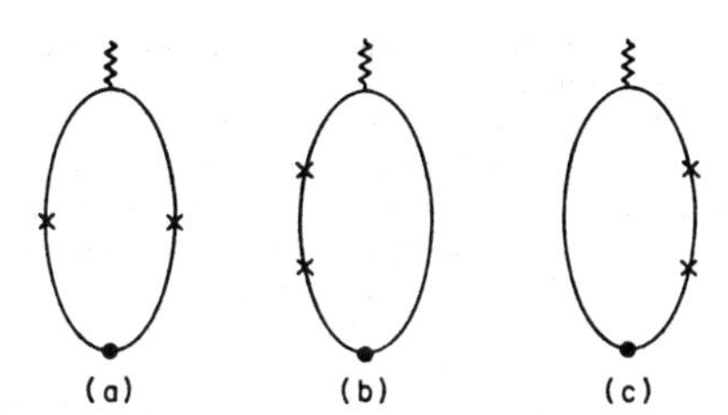

FIG. 5. The second-order contributions to $\langle B(\tau)\rangle$.

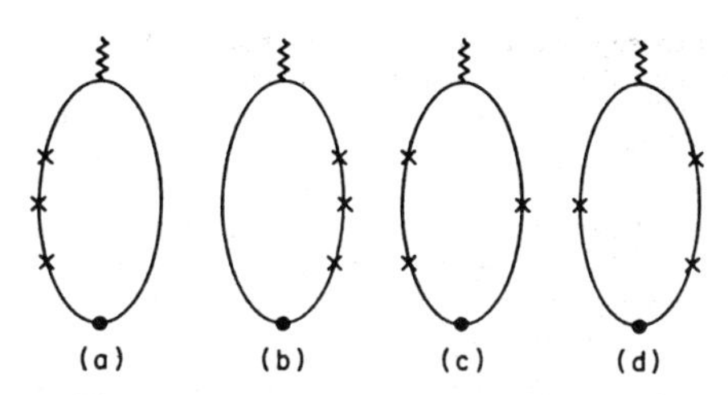

FIG. 6. The third-order contributions to $\langle B(\tau)\rangle$.

of the form

$$H = H_0 + V(\tau).$$

The baryon-number-violating piece $V(\tau)$ is of the form

$$V(\tau) = \frac{\alpha}{\tau^2}\int d^3x(\bar{\psi}\Gamma\psi)^2 \equiv \frac{v}{\tau^2}.$$

The graphical rules for the evaluation of $\langle B(\tau)\rangle$ can be deduced from the expression

$$\langle B(\tau)\rangle = \mathrm{Tr}\rho(1)U^\dagger_{V_I}(\tau)U^\dagger_{H_0}(\tau)\hat{B}U_{H_0}(\tau)U_{V_I}(\tau), \quad \text{(A1)}$$

where

$$U_{H_0}(\tau) = T\exp\left[-i\int_1^\tau H_0(\tau')d\tau'\right] = e^{-iH_0(\tau-1)},$$

$$U_{V_I}(\tau) = T\exp\left[-i\int_1^\tau V_I(\tau')d\tau'\right], \quad \text{(A2)}$$

$$V_I(\tau) = U^\dagger_{H_0}(\tau)V(\tau)U_{H_0}(\tau).$$

Since H_0 conserves baryon number, expression (A1) simplifies to

$$\langle B(\tau)\rangle = \mathrm{Tr}\rho(1)U^\dagger_{V_I}(\tau)\hat{B}U_{V_I}(\tau). \quad \text{(A3)}$$

The graphical rules for the evaluation of this quantity are the following:

(1) Draw the closed loop shown in Fig. 4 in order $l+r$.

(2) For each cross on the right write $ie^{iH_0(\tau'-1)} \times ve^{-iH_0(\tau'-1)}$. For each cross on the left write $-ie^{iH_0(\tau-1)}\,ve^{-iH_0(\tau-1)}$.

(3) Write down the terms indicated in Fig. 4 in anticlockwise order and take the trace.

(4) Carry out the time integrations with weight $1/\tau^2$. Respect time ordering.

Do the same for the $l+r+1$ graphs appearing in order $l+r$. Note that the lines in Fig. 4 are not particle lines. They represent propagation of states.

APPENDIX B

Here we will show explicitly that for the model discussed in Appendix A the second-order contributions to $\langle B(\tau)\rangle$ vanish. We shall label each state solely by its baryon number $|n\rangle$. The CPT conjugate state will be denoted by $|-n\rangle$, and by CPT invariance, $\rho_n(1) = \rho_{-n}(1)$. Since $[B,\rho(1)] = 0$, $\rho_{nm}(1) \equiv \rho_{nm} = \rho_n\delta_{nm}$. Since B is CPT odd, $B_{-n} = -B_n$. The second-order contributions to $\langle B(\tau)\rangle$ arise from the graphs of Fig. 5. The contribution of graph (a) is

$$i(-i)\int_1^\tau \frac{d\tau_1}{\tau_1^2}\int_1^\tau \frac{d\tau_2}{\tau_2^2}\rho_n e^{i\epsilon_n\tau_1}v_{nm}e^{-i\epsilon_m\tau_1}B_m e^{i\epsilon_m\tau_2} \times v_{mn}e^{-i\epsilon_n\tau_2} = 0.$$

In deriving this we used the CPT invariance of the Hamiltonian H,

$$\epsilon_n = +\epsilon_{-n} \text{ and } |v_{nm}|^2 = |v_{-m,-n}|^2.$$

The contribution of graph (b) is

$$(-i)^2\int_1^\tau \frac{d\tau_2}{\tau_2^2}\int_1^{\tau_2}\frac{d\tau_1}{\tau_1^2}\rho_n B_n e^{i\epsilon_n\tau_2}v_{nm}e^{-i\epsilon_m\tau_2}e^{i\epsilon_m\tau_1} \times v_{mn}e^{-i\epsilon_n\tau_1} = 0.$$

This vanishes for the same reason with graph (a). The vanishing of the second-order contribution to $\langle B(\tau)\rangle$ is not a general feature of all models. It only happens because the explicit time dependence of $V(\tau)$ can be factored out.

APPENDIX C

In this appendix we write down the third-order contributions to $\langle B(\tau)\rangle$ for the model of Appendix A. The graphs contributing are those of Fig. 6.

Graph (a) contributes

$$(-i)^3\int_1^\tau \frac{d\tau_3}{\tau_3^2}\int_1^{\tau_3}\frac{d\tau_2}{\tau_2^2}\int_1^{\tau_2}\frac{d\tau_1}{\tau_1^2}\rho_n B_n e^{i\epsilon_n\tau_3}v_{nm}e^{-i\epsilon_m(\tau_3-\tau_2)}v_{me}e^{-i\epsilon_e(\tau_2-\tau_1)}v_{en}e^{-i\epsilon_n\tau_1}$$

$$= (-i)^3\int_1^\tau \frac{d\tau_3}{\tau_3^2}e^{i\tau_3(\epsilon_n-\epsilon_m)}\int_1^{\tau_3}\frac{d\tau_2}{\tau_2^2}e^{i\tau_2(\epsilon_m-\epsilon_e)}\int_1^{\tau_2}\frac{d\tau_1}{\tau_1^2}e^{i\tau_1(\epsilon_e-\epsilon_n)}\rho_n B_n v_{nm}v_{me}v_{en}.$$

Graph (b) contributes

$$i^3\int_1^{T}\frac{d\tau_3}{\tau_2^{\,2}}\int_1^{\tau_3}\frac{d\tau_2}{\tau_2^{\,2}}\int_1^{\tau_2}\frac{d\tau_1}{\tau_1^{\,2}}\rho_n B_n e^{i\epsilon_n\tau_1}v_{nm}e^{-i\epsilon_m(\tau_1-\tau_2)}v_{me}e^{-\epsilon_e(\tau_2-\tau_3)}v_{en}e^{-i\epsilon_n\tau_3}$$

$$=i\int_1^{T}\frac{d\tau_3}{\tau_3^{\,2}}e^{i\tau_3(\epsilon_e-\epsilon_n)}\int_1^{\tau_3}\frac{d\tau_2}{\tau_2^{\,3}}e^{i\tau_2(\epsilon_m-\epsilon_e)}\int_1^{\tau_2}\frac{d\tau_1}{\tau_1^{\,2}}e^{i\tau_1(\epsilon_n-\epsilon_m)}\rho_n B_n v_{nm}v_{me}v_{en}.$$

Graph (b), of course, is just the complex conjugate of graph (a).

Graph (c) yields

$$(-i)^2 i\int_1^{T}\frac{d\tau_1'}{\tau_1'^2}\,e^{i(\epsilon_n-\epsilon_m)\tau_1'}\int_1^{T}\frac{d\tau_2}{\tau_2^{\,2}}e^{i(\epsilon_m-\epsilon_e)\tau_2}\int_1^{\tau_2}\frac{d\tau_1}{\tau_1^{\,2}}e^{i(\epsilon_e-\epsilon_n)\tau}\,\rho_n v_{nm}\beta_m v_{me}v_{en}.$$

Graph (d) yields the complex conjugate of (c),

$$i^2(-i)\int_1^{T}\frac{d\tau_2'}{\tau_2'^2}e^{i\tau_2'(\epsilon_m-\epsilon_e)}\int_1^{\tau_2'}\frac{d\tau_1'}{\tau_1'^2}e^{i\tau_1'(\epsilon_n-\epsilon_m)}\int_1^{T}\frac{d\tau_1}{\tau_1^{\,2}}e^{i\tau_1(\epsilon_e-\epsilon_n)}\rho_n v_{nm}v_{me}B_e v_{en}.$$

These expressions do not vanish in general. They, of course, vanish if we assume C- or CP-invariant matrix elements for v.

---

*Present address: Physics Department, Columbia University, New York, N. Y. 10027.

[1] G. Steigman, Ann. Rev. Astron. Astrophys. 14, 339 (1976).

[2] S. Weinberg, *Gravitation and Cosmology* (Wiley, New York, 1972).

[3] H. Georgi and S. L. Glashow, Phys. Rev. Lett. 32, 438 (1974); J. C. Pati and A. Salam, Phys. Rev. D 8, 1240 (1973); 10, 275 (1974); F. Gürsey and P. Sikivie, Phys. Rev. Lett. 36, 775 (1976).

[4] G. 't Hooft, Phys. Rev. Lett. 37, 8 (1976); Phys. Rev. D 14, 3432 (1976); A. A. Belavin *et al.*, Phys. Lett. 59B, 85 (1975).

[5] M. Yoshimura, Phys. Rev. Lett. 41, 281 (1978). 179, 1978 (unpublished).

[6] S. Hawking, Commun. Math. Phys. 43, 199 (1975); R. M. Wald, *ibid.* 45, 9 (1975), L. Parker, in *Strong Gravitational Fields*, edited by F. P. Esposito and L. Witten (Plenum, New York, 1977), pp. 107–226.

[7] T. D. Lee, R. Oehme, and C. N. Yang, Phys. Rev. 106, 340 (1957); T. D. Lee and C. S. Wu, Annu. Rev. Nucl. Sci. 16, 511 (1966).

[8] We use Schwinger's definition of CPT transformation which differs from the standard (Wigner) definition by a Hermitian conjugation.

[9] T. D. Lee, Phys. Rev. D 8, 1226 (1973); Phys. Rep. 3C, 143 (1974).

[10] S. Weinberg, Phys. Rev. Lett. 37, 657 (1976).

[11] R. Jackiw and C. Rebbi, Phys. Rev. Lett. 37, 172 (1976); C. Callan, R. Dashen, and D. Gross, Phys. Lett. 63B, 334 (1976).

[12] D. A. Kirzhnits and A. D. Linde, Phys. Lett. 42B, 471 (1972); L. Dolan and R. Jackiw, Phys. Rev. D 9, 3320 (1974); S. Weinberg, *ibid.* 9, 3357 (1974).

[13] M. B. Kislinger and P. D. Morley, Phys. Rev. D 13, 2765 (1976); 13, 2771 (1976).

[14] A. L. Fetter and J. D. Walecka, *Quantum Theory of Many Particle Systems* (McGraw-Hill, N.Y., 1971).

[15] See, for example, E. P. T. Liang, Phys. Rev. D 16, 3369 (1977).

Matter-antimatter accounting, thermodynamics, and black-hole radiation

D. Toussaint, S. B. Treiman, and Frank Wilczek
Princeton University, Princeton, New Jersey 08540

A. Zee
University of Pennsylvania, Philadelphia, Pennsylvania 19174
(Received 16 August 1978)

We discuss several issues bearing on the observed asymmetry between matter and antimatter in the content of the universe, in particular, the possible role in this of Hawking radiation from black holes, with allowance for weak C- and T-violating interactions. We show that the radiation, species by species, can be asymmetric between baryons and antibaryons. However, if baryon number is microscopically conserved there cannot be a net flux of baryon number in the radiation. Black-hole absorption from a medium with net baryon number zero can drive the medium to an asymmetric state. On the other hand, if baryon conservation is violated, a net asymmetry can develop. This can arise through asymmetric gravitational interactions of the radiated particles, and conceivably, by radiation of long-lived particles which decay asymmetrically. In the absence of microscopic baryon conservation, asymmetries can also arise from collision processes generally, say in the early stages of the universe as a whole. However, no asymmetries can develop (indeed any "initial" ones are erased) insofar as the baryon-violating interactions are in thermal equilibrium, as they might well be in the dense, high-temperature stages of the very early universe. Thus particle collisions can generate asymmetries only when nonequilibrium effects driven by cosmological expansion come into play. A scenario for baryon-number generation suggested by superunified theories is discussed in some detail. Black-hole radiation is another highly nonequilibrium process which is very efficient in producing asymmetry, given microscopic C, T, and baryon-number violation.

I. INTRODUCTION AND SUMMARY

An annoying feature of big-bang cosmology, as currently formulated, is the seeming necessity to specify nonzero values for baryon and perhaps lepton numbers. (The net electric charge, on the other hand, must be very nearly equal to zero.) It would be more attractive to suppose that the initial state is symmetric between matter and antimatter. The problem becomes more acute if baryon number is not microscopically conserved, and if its violation becomes large at the high temperatures characteristic of the early universe. For then when thermal equilibrium is established at early times baryons and antibaryons are equally numerous. In either case, if we start at early times with a situation symmetric between matter and anitmatter, we must understand how the matter and anitmatter later separated, on a scale which is sufficiently large to accord with the observation of local asymmetry,[1] or else we must understand how a universe which is symmetric early on can evolve in time into one which is asymmetric. We shall examine the latter alternative here, with special attention to the role of Hawking radiation from black holes.

Even if baryon number is microscopically conserved, it is not completely obvious that the net flux of baryon number in black-hole radiation[2,3] is zero, given the observed fact of CP violation in weak interactions. Neither baryon-number conservation for the CPT theorem is directly relevant—it is well known that the baryon number of a black hole is ill defined,[4] and the fact that a black hole is radiating gives us an arrow of time (so that the CPT theorem is not applicable). We shall show nevertheless that the net flux of baryon number is zero. Both a quantum-mechanical and a thermodynamic argument are offered for this. This shows that if baryon number is microscopically conserved then black holes can only tend to give matter-antimatter symmetry—no matter what goes in, equal numbers of baryons and antibaryons come out. It is worth remarking, however, that species by species the radiation need not be symmetric, e.g., one may have more Λ than Λ particles. We shall supply an explicit example of this. Only the net flux connected with absolutely conserved quantum numbers is forced to be zero.

To summarize, if baryon number is microscopically conserved then the "transcendence" of baryon number in a black hole can only make an initially asymmetric situation more symmetric; it cannot give a net baryon-antibaryon asymmetry starting from a symmetric situation.

On the other hand, *absorption* from a symmetric medium can lead to an asymmetric condition. For example, if the hole which preferentially absorbs Λ rather than $\bar{\Lambda}$ is surrounded by a medium containing equal amounts of Λ and $\bar{\Lambda}$, it will lead to a medium containing mostly $\bar{\Lambda}$.

The situation becomes more interesting if baryon number is not microscopically conserved. As we have mentioned, in this case a dynamical understanding of how matter-antimatter asymmetry can arise from a symmetric situation is not only esthetically desirable but also physically necessary if the baryon-violating interactions ever establish thermal equilibrium in the very early stages of the universe. Our earlier result that the flux of particles and antiparticles need not balance species by species, but only in the net flux of a conserved quantity, means that if baryon number is microscopically violated then there can be net baryon-number flux in black-hole radiation.

Our analysis leads to the conclusion that a net imbalance of baryons and antibaryons will arise in particle production by cosmological expansion if baryon number, C and T are violated microscopically. Cosmological particle production, but without symmetry violation, has been discussed by Parker[5] and others.

The above-mentioned effects result from interactions between matter and gravitational fields, more precisely from corrections to the minimal coupling of gravity to matter induced by C-, T-, and baryon-number-violating interactions. Even in the presence of such symmetry violations in thermal equilibrium the number of baryons and antibaryons must be equal—the number of particles is governed by the Boltzmann factor, and baryon and antibaryon have equal mass by the CPT theorem. One can, however, generate asymmetries if there are nonequilibrium processes. Nonequilibrium situations may arise because of cosmological expansion. Another, very powerful, method of generating nonequilibrium situations is through black-hole radiation, an explosive process.[2,3] We will discuss these possibilities in detail below.

The contents of the paper are as follows. In Sec. II, we shall discuss how asymmetry may develop kinetically in nonequilibrium processes. A simple thought experiment involving K mesons is used to illustrate this. The kinetic mechanisms involving cosmological expansion and black-hole radiation are compared and contrasted. A scenario suggested by superunified gauge theories is described. Some speculations regarding a possible black-hole-dominated phase of the universe are presented. In Sec. III, we prove our theorem that if baryon number is microscopically conserved then the net flux of baryon number in black-hole radiation is zero. We show by example that if, on the contrary, baryon number is not microscopically conserved then a net flux can arise. The generalization of this result to cosmological particle production is mentioned.

From the above it should be clear that we cannot claim to have a theory of the baryon-antibaryon asymmetry which gives the magic number $n_B/n_\gamma \approx 10^{-8}$. Our knowledge of microscopic baryon nonconservation (if any) and T violation is much too uncertain for that. We hope, however, to have established some rules of the game so that such questions may be rationally discussed, and to have shown that the baryon-antibaryon asymmetry need not be a "given" of cosmology but could arise dynamically from regular physical processes.

II. KINETIC PROCESSES

A. A thought experiment with K mesons

Consider a blackbody with temperature $T \geq m_K$ comparable to the K-meson mass radiating into empty space. It will radiate K^0 and $\overline{K}^0$ mesons equally. In free space the mesons will decay. The decay of K mesons is a classic story.[6] For our purposes it is sufficient to recall that the time development of K^0 and $\overline{K}^0$ are described by

$$|K^0(t)\rangle = \frac{(1+|\epsilon|^2)^{1/2}}{\sqrt{2}\,(1+\epsilon)}\left(e^{-\gamma_S t/2}|K_S^0\rangle + e^{-\gamma_L t/2}|K_L^0\rangle\right), \tag{2.1}$$

$$|\overline{K}^0(t)\rangle = \frac{(1+|\epsilon|^2)^{1/2}}{\sqrt{2}\,(1-\epsilon)}\left(e^{-\gamma_S t/2}|K_S^0\rangle - e^{-\gamma_L t/2}|K_L^0\rangle\right), \tag{2.2}$$

where ϵ is a parameter which measures T violation ($\epsilon = 0$ if time-reversal symmetry is good). For our illustrative purposes we ignore the $K_L - K_S$ mass difference. Using CPT and the $\Delta S = \Delta Q$ rule we find that the amplitudes for semileptonic K_S^0, K_L^0 decay are given by

$$\langle \pi^- e^+ \nu | K_S^0\rangle = f(1+\epsilon), \tag{2.3a}$$

$$\langle \pi^+ e^- \overline{\nu} | K_S^0\rangle = f^*(1-\epsilon), \tag{2.3b}$$

$$\langle \pi^- e^+ \nu | K_L^0\rangle = f(1+\epsilon), \tag{2.3c}$$

$$\langle \pi^+ e^- \overline{\nu} | K_L^0\rangle = -f^*(1-\epsilon), \tag{2.3d}$$

where f is a form factor which for our purposes we take to be constant. Combining these formulas we find that the total yields from the various possible semileptonic decays, integrated over time, are proportional to

$K^0 \to \pi^- e^+ \nu$:

$$\left(\frac{1}{\gamma_S} + \frac{1}{\gamma_L} + \frac{4}{\gamma_S + \gamma_L}\right), \tag{2.4a}$$

$K^0 \to \pi^+ e^- \bar{\nu}$:

$$\left|\frac{1-\epsilon}{1+\epsilon}\right|^2 \left(\frac{1}{\gamma_S} + \frac{1}{\gamma_L} - \frac{4}{\gamma_S + \gamma_L}\right), \tag{2.4b}$$

$\bar{K}^0 \to \pi^- e^+ \nu$:

$$\left|\frac{1+\epsilon}{1-\epsilon}\right|^2 \left(\frac{1}{\gamma_S} + \frac{1}{\gamma_L} - \frac{4}{\gamma_S + \gamma_L}\right), \tag{2.4c}$$

$\bar{K}^0 \to \pi^+ e^- \bar{\nu}$:

$$\left(\frac{1}{\gamma_S} + \frac{1}{\gamma_L} + \frac{4}{\gamma_S + \gamma_L}\right). \tag{2.4d}$$

Since experimentally $\epsilon = 1.5 \times 10^{-3}$ and $\gamma_S^{-1} \ll \gamma_L^{-1}$ we find that in the radiation there is a preponderance of positrons over electrons by roughly a part in 10^3.

Needless to say, lepton number, which is microscopically conserved, remains zero. With the preponderance of e^+ over e^- comes a preponderance of ν over $\bar{\nu}$. In equilibrium, the back-reactions $e^+\pi^-\nu \to K$, etc., would restore the Boltzmann distribution, with equal numbers of e^+ and e^-.

(This example is imperfect because the $\pi^{\pm}$ mesons will eventually decay into $e^{\pm}$, restoring the balance. So our asymmetry is short lived. This problem, however, is practical rather than conceptual. One could, for instance, imagine a world with T violation and a stable π meson.)

This thought experiment illustrates how an asymmetry between matter and antimatter may arise from the interplay of C and T violation and a nonequilibrium process. If our blackbody radiator is replaced by a radiating black hole and the K mesons by some heavy mesons whose decays violate C, T, and baryon number, we see how a baryon-antibaryon asymmetry might be induced. More explicitly, suppose there is a meson M and antimeson $\bar{M}$ with the decay channels $M \to p + \bar{p}$, $p + e$, $\bar{M} \to p + \bar{p}$, $\bar{p} + \bar{e}$. M and $\bar{M}$ mix through the $p + \bar{p}$ channel, and a proper analysis of their decays would proceed much like the case of K mesons. If C and T are violated, a net baryon number would arise from the decay in free space of an equal mixture of M and $\bar{M}$.

The radiation of a blackbody into free space is of course closely related to the behavior of blackbody radiation subject to rapid expansion. Therefore the same process, production of an imbalance from the interplay of baryon-number-, C-, and T-violating interactions with a nonequilibrium process, could be driven by rapid cosmological expansion in the early stages of the universe.

B. General discussion of kinetics of expansion

As we have just seen, in the presence of baryon-number-, C-, and T-violating interactions, nonequilibrium processes can generate a net baryon number starting from zero baryon number. We have in mind two mechanisms which may lead to cosmologically significant disequilibrium: black-hole radiation and cosmological expansion. Black-hole radiation is to a first approximation radiation into empty space, so its kinetics is very simple. Cosmological expansion is much more complicated —one must investigate the behavior of matter at a high and rapidly changing temperature. We shall now discuss this more complicated case. The two mechanisms are compared and contrasted in Sec. II D.

We shall show in Sec. II C that insofar as particle masses are negligible (i.e., the temperature is much higher than the rest mass of the particles present) no asymmetry can arise. Furthermore, as we have mentioned, no asymmetry can arise from equilibrium processes. We therefore are concerned with estimating when a massive particle goes out of equilibrium. Its subsequent decays (or reactions) can then generate asymmetries.

For definiteness we consider a meson of mass M which couples to two-fermion (quark or lepton) channels with electromagnetic strength. In our estimates, we keep only Born terms. This is in the spirit of asymptotic freedom, which is presumably very good at the relevant high temperatures. This discussion, of course, could be generalized. Let the temperature be T. The characteristic expansion time is then $(G^{1/2}T^2)^{-1}$, where G is the gravitational constant. As T decreases below M, the reactions which create the heavy mesons cease. If the decrease of temperature is slow enough annihilation reactions leading to decrease in the number of heavy mesons will proceed, decreasing the density of mesons in line with the Boltzmann factor $e^{-M/T}$. However, if the annihilation reactions are slow compared to the characteristic time, then the distribution of mesons is no longer in equilibrium. Let us estimate the temperature at which this occurs.

(i) Pair annihilation: Two heavy mesons may annihilate into two quarks or leptons. The rate of this per meson in equilibrium is $\sim(\alpha/\pi)^2(T^{3/2}/M^{1/2})e^{-M/T}$ for $M > T$ and $\sim(\alpha/\pi)^2 T$ for $M \leq T$, and becomes comparable to the characteristic inverse expansion time when

$$\left(\frac{\alpha}{\pi}\right)^2 \frac{1}{M^{1/2}} T_d^{3/2} e^{-M/T} \sim G^{1/2} T_d^2 \quad (M \geq T), \tag{2.5a}$$

$$\left(\frac{\alpha}{\pi}\right)^2 T_d \sim G^{1/2} T_d^2 \quad (M \leq T). \tag{2.5b}$$

Numerically, this gives $T_d \approx M/8$ for $M = 10^6$ GeV, $T_d \approx M/4$ for $M = 10^{10}$ GeV. In either case the density $\sim e^{-M/T}$ of mesons is small at the decoupling time. When $M \approx 10^{12}$ GeV, T_d becomes comparable to M, and when $M \geq 10^{13}$ GeV the reaction is *never* in equilibrium.

(ii) Annihilation against light particles: A heavy meson and a quark or lepton may annihilate into a quark or lepton plus (say) a photon. This decouples roughly when

$$\left(\frac{\alpha}{\pi}\right)^2 \frac{1}{M^2} T_d^{\,3} \sim G^{1/2} T_d^{\,2} \quad (M \geq T), \tag{2.6a}$$

$$\left(\frac{\alpha}{\pi}\right)^2 T_d \sim G^{1/2} T_d^{\,2} \quad (M \leq T). \tag{2.6b}$$

which gives $T_d \sim 10^{-3}$ GeV for $M = 10^5$ GeV, $T_d \sim 10^7$ GeV for $M = 10^{10}$ GeV. In either case, this process dominates the pair annihilation and shows that very few heavy mesons are present at the decoupling time. When $M \geq 10^{13}$ GeV, Eq. (2.6) cannot be satisfied for any value of T.

(iii) Decay: The decay of the heavy mesons will occur with a rate $(\alpha/\pi)M$. If $M \geq T_d$, the rate dominates the previous two until decoupling. In any case, decays will dominate once $T \leq (\pi/\alpha)M$.

(iv) Inverse decay: Production of heavy mesons by inverse decays occurs at a rate $\sim(\alpha/\pi)M e^{-M/T}$ $(M \gtrsim T)$ or $\sim(\alpha/\pi)T$ $(M \lesssim T)$. At early enough times particles of mass up to $(\alpha/\pi)G^{-1/2} \approx 10^{16}$ GeV will be brought into equilibrium by this process..

If $M \geq 10^{16}$ GeV we have a very peculiar situation. The distribution of heavy mesons is never brought into equilibrium. Furthermore, such mesons decay before they interact once $T \leq (\pi/\alpha)M$. For these reasons it is problematical to estimate how many such mesons would be created in the history of the universe. If we assume, however, that in the earliest stages the density of mesons is not as singular as T^3, we can estimate the contribution from their decays to the net baryon number. The mesons are produced [from reaction (iv) above] at roughly the rates

$$r \sim \left(\frac{\alpha}{\pi}\right) \frac{T^6}{M^2 + T^2} e^{-M/T} \tag{2.7}$$

per unit time per unit volume at temperature T. Suppose the meson decays into baryons more than antibaryons by a fraction ϵ. A baryon produced at temperature T contributes proportional to $(T_0/T)^3$ to the present density, where T_0 is the present temperature. Putting it all together, the contribution of heavy-meson decay to the present baryon density is approximately $(dt = dT/G^{1/2}T^3)$

$$n_B \approx \int_M^{(\pi/\alpha)M} \epsilon \left(\frac{\alpha}{\pi}\right) \frac{T^6}{M^2 + T^2} \left(\frac{T_0}{T}\right)^3 \frac{dT}{G^{1/2} T^3}$$

$$\approx \frac{\epsilon(\alpha/\pi) T_0^{\,3}}{G^{1/2} M}. \tag{2.8}$$

The limits on the integral are determined by the exponential cutoff in (2.7), and the temperature at which annihilation dominates decay. The contribution to n_B/n_γ is then about $\epsilon(\alpha/\pi)/G^{1/2}M$. Such a number could be close to the desired 10^{-9} for $\epsilon = 10^{-6}$, $M = 10^{18}$ GeV as might be suggested by ideas of superunification.[7,8] Notice that in this picture the baryons are produced at such an early time that the standard big-bang scenario for the later stages is unaffected.

In this discussion (for $M \gtrsim 10^{16}$ GeV) we have assumed that initially there were no heavy mesons. An alternative scenario [developed by one of the authors (F. W.) in conversations with S. Weinberg[9]] assumes that by some (quantum gravitational?) mechanism the heavy mesons do follow a Boltzmann distribution at the highest temperatures $T \gg M$. As we have seen, the fastest processes involving the heavy mesons are decays; they become important at times $t_D \approx [(\alpha/\pi)M]^{-1}$ or temperatures $T_D = G^{-1/4} t_D^{\,-1/2} \approx [G^{-1/2}(\alpha/\pi)M]^{1/2}$. Until this time the heavy mesons initially present simply red-shift freely and therefore are overabundant compared to the Boltzmann distribution for a massive particle. In fact the number of heavy mesons is just t. e same as the number of photons. The heavy mesons then decay away (if $M \gtrsim T_D$ back reactions creating the heavy mesons are unimportant); the net asymmetry per photon produced is then remarkably simple:

$$\frac{n_B}{n_\gamma} \approx \epsilon \frac{n_{\text{heavy}}}{n_\gamma} \approx \epsilon\,. \tag{2.9}$$

Finally, we will briefly discuss the regime $M \lesssim 10^{16}$ GeV, which may be the most interesting case.[8] A very crude, preliminary analysis of this case is as follows. Because massless particles generate no asymmetry, we concentrate again on the heavy mesons. Now these are *forced* into equilibrium at early times. Asymmetry can be generated when the temperatures reach $T \approx M$, where there are significant numbers of heavy mesons (roughly equal to the number of photons) and cosmological expansion drives their distribution out of equilibrium (see part C below). The asymmetry produced will be the asymmetry per decay multiplied by a parameter characterizing the "nonequilibrium" character of expansion, i.e.,

$$\frac{n_B}{n_\gamma} \approx \epsilon \frac{t_c(T=M)}{t(T=M)} \approx \epsilon \frac{\alpha}{\pi} G^{1/2} M\,, \tag{2.10}$$

where $t_c \approx [(\alpha/\pi)T]^{-1}$ is a characteristic interaction time and $t = \dot{R}/R \approx (G^{1/2}T^2)^{-1}$ is a characteristic expansion time. Any baryon number produced will be partially thermalized (driven to zero) by later interactions.[10] We believe that all these effects may be accurately taken into account using the fluctuation-dissipation theorems of statistical mechanics. Calculations of this kind, together with a calculation of ϵ, will be presented in a subsequent paper.

C. Absence of asymmetry for massless particles

One can prove a little theorem that reactions among massless particles do not give particle-antiparticle asymmetries from cosmological expansion. This means that such asymmetries will be characterized by parameters $\sim m/T$ to a power, in addition to other parameters of smallness, when the relevant masses m are much less than the temperature.

It is important to distinguish between kinetic and chemical equilibrium. Kinetic equilibrium, distribution of energies and momenta according to the Boltzmann factor, is enforced by the presence of any collisions at all. Chemical equilibrium may be established only by much slower interactions and in fact never reached in an expanding universe. In the case at hand, baryon-number-violating processes may be rare and although in true equilibrium the baryon number would be zero such equilibrium may never be established. On the other hand, kinetic equilibrium should be a good approximation in the early universe.

Taking into account collisions and cosmological expansion, we have for the densities

$$\frac{dn_i(p_i)}{dt} = \sum_{\substack{jkl \\ p_j p_k p_l}} [-\phi_{p_i p_j p_k p_l} |\langle kp_k lp_l | T | ip_i jp_j\rangle|^2 n_i(p_i) n_j(p_j) + \phi_{p_k p_l p_i p_j} |\langle ip_i jp_j | T | kp_k lp_l\rangle|^2 n_k(p_k) n_l(p_l)] - Kp_i \frac{\partial n_i(p_i)}{\partial p_i} . \tag{2.11}$$

Here the Latin indices denote particle type, and $K = \dot{R}/R$ is the expansion rate. The last term indicates the effect of the cosmological expansion, which for massless particles is a simple red-shift. ϕ is the phase space (and statistics) factor, and T is the usual scattering matrix. We claim (2.11) is solved by

$$n_i(p_i) = e^{-p_i/T(t)}, \qquad \frac{dT}{dt} = -KT. \tag{2.12}$$

Indeed, with the particle number independent of species the first two terms on the right-hand side of (2.11) cancel by the completeness relation $T^\dagger T = TT^\dagger$, and the equality is a trivial calculation. This shows that in the approximation of massless particles cosmological expansion does not induce particle-antiparticle asymmetry. We disagree in this with the calculation of Yoshimura.[11] For massive particles the proof does not work because the expansion term (red-shift) is not simply $-Kp\partial n/\partial p$.

From a deeper point of view the essential ingredients of the preceding argument are the existence of thermal equilibrium and the fact that free expansion of a gas of massless particles is adiabactic. Therefore, free expansion of a gas of massless particles will always reproduce the equilibrium distribution.

We note parenthetically that the use of the completeness relation $TT^\dagger = T^\dagger T$ as a substitute for detailed balance in the proof of the existence of thermal equilibrium is completely general, not tied to massless particles. This follows immediately from the analysis of Ref. (12), although it is not noted there.

D. Comparison of black-hole radiation and cosmological expansion kinetics

Let us compare the two kinetic mechanisms, black-hole radiation and cosmological expansion.

(1) Black-hole radiation of particles is governed simply by the mass of the particles. Thus even very weakly interacting particles (such as we might need to give small baryon-number violation) can be copiously produced if they are light. Then their decay could give large asymmetries, as in our K-meson example. We never run into the problem of particles whose production cannot be estimated, as in Sec. II B. On the other hand since the characteristic temperature $T = 1/8\pi GM$ of a black hole of mass M is inversely proportional to the mass, very heavy particles do not get produced until the black hole has lost most of its mass. For instance, $T = 10^5$ GeV requires a black-hole mass $M = 10^7$ g. Only with large asymmetries in the decay and very many small black holes could one generate a sufficient baryon number from such particles.

(2) Particles which annihilated rapidly in pairs but whose decays violated baryon number would generate an asymmetry by the black-hole mechanism but would not survive long enough to produce asymmetry in cosmological expansion.

(3) There are deep questions of cosmology connected with each mechanism: How hot does it get in the early universe? Is there a limiting temperature for hadronic matter? What was the spectrum of black holes in the early universe? An interesting speculation concerning the second question is that the present stage of the universe results from a cosmological "bounce" from a pre-

vious collapsing stage. It is then an interesting question which could be investigated mathematically whether the collapse results in a large population of black holes whose subsequent explosion triggers the big bang. This idea, if it can be realized, would have some other advantages:

(a) The violation of the weak energy condition associated with black holes allows one to circumvent the singularity theorem and could conceivably lead to a "bounce."

(b) In this picture the universe might never get very hot (except locally, at the surface of black holes). Thus problems associated with restoration of symmetries at high temperatures and subsequent formation of domain walls need not arise.

(c) Most attractive conceptually, this type of cosmology completely eliminates the necessity of specifying initial conditions in addition to ordinary physical laws.

Finally, an important remark of a general nature: It has often been objected against cyclic cosmologies that each cycle increases the entropy, so that the present finite value of the entropy per baryon would militate against an infinite number of cycles. In theories of the type discussed in this paper, this objection is baseless. The entropy per baryon is determined dynamically by physical processes, at extreme temperatures, which are the same for each cycle. The value of the entropy per baryon is therefore independent of the number of cycles.

III. INTERACTIONS WITH FIELDS

A. Theorem on matter-antimatter balance

We now show that in a world with microscopic baryon-number conservation the net flux of baryon number in Hawking radiation is zero. We shall set up slightly more machinery than is strictly necessary to prove this, so that later we can easily analyze the effect of dropping the assumption of microscopic baryon conservation.

Only interactions odd under C and T can possibly generate asymmetric radiation. (Terms odd under parity are of no interest. They might generate a preponderance of particles over antiparticles in one hemisphere around a rotating black hole, but there would be a reversed imbalance in the other hemisphere and no net asymmetry.) To allow for such effects we take into account matter interactions, in particular the weak C- and T-violating interactions: For example, loop diagrams which describe quark-graviton scattering, corrected by exchange of virtual W bosons and involving C- and T-violating quark–W-boson vertices.

When the gravitational field is weak we can summarize the matter field interactions by means of an effective local Lagrangian. The corrections to the "free field" Lagrangian will of course be very complicated, including strong-interaction effects as well as the weak C- and T-violating effects of interest here. These corrections will alter the details of the Hawking radiation, but cannot introduce any asymmetry between matter and antimatter. We focus therefore only on the C- and T-violating interactions between matter and gravity. These introduce an asymmetry between particles and antiparticles in their propagation through the background gravitational field in the vicinity of the black hole. For illustrative purposes only, let us consider a set of (three or more) complex scalar fields Φ_i, $1 \leq i \leq N$. Restricting ourselves for simplicity to terms bilinear in these fields, we may take the following Lagrangian as representative of the effects under discussion:

$$\mathcal{L} = \sqrt{g}\,(g_{\mu\nu}\partial_\mu \Phi_i^* \partial_\nu \Phi_i - m_i{}^2 \Phi_i^* \Phi_i - \Phi_i^* V_{ij} \Phi_j R_{\alpha\beta\gamma\delta} R_{\alpha\beta\gamma\delta}). \tag{3.1}$$

Here $R_{\alpha\beta\gamma\delta}$ is the curvature tensor, and V_{ij} is a matrix whose details depend on the details of the weak interactions [recall that (3.1) is not to be taken as a fundamental Lagrangian, but as an effective Lagrangian summarizing the weak corrections to gravity]. The matrix V_{ij} is of course Hermitian but not necessarily real and can lead to C and T violation. The Lagrangian (3.1) may serve to illustrate both the cases where there is a conserved "baryon number" and where there is not: The total number of Φ quanta of all types is conserved, since $\mathcal{L}$ is invariant under the multiplication of all the Φ_i by a common phase, but the number of quanta of species $1, 2, \ldots, N$ are not separately conserved. For the remainder of this section we will analyze the case of a microscopically conserved quantum number and call the quanta of Φ_i baryons, the quanta of Φ_i^* antibaryons.

The rate of Hawking radiation is a product of two factors. One is a universal factor, the same for particles and antiparticles and irrespective of type i; roughly speaking, it corresponds to blackbody emission at the event horizon. The other factor describes transmission of the emitted objects through the gravitational field outside the event horizon. The net effect corresponds to graybody emission, with graybody coefficients that depend on the transmission. In our example the transmission phenomena for baryons involves transformations among the various types, and similarly for antibaryons. The former effects are governed by the field equations for Φ_i, the latter by the equations for Φ_i^*.

To evaluate the transmission coefficients we

should now solve for the probability, given the Lagrangian (3.1), of particles and antiparticles emitted at the horizon to emerge at infinity. We shall abstract from this a model problem which contains the essence of the phenomena (it amounts to a generalization of the radial equation one would derive by separating the field equations in appropriate coordinates). The model problem is that of one-dimensional transmission through a position-dependent matrix potential, governed by the Lagrangian

$$\mathcal{L}=\dot{\Phi}_i^*\dot{\Phi}_i-\frac{\partial}{\partial x}\Phi_i^*\frac{\partial}{\partial x}\Phi_i-\Phi_i^*(x)V_{ij}(x)\Phi_j(x). \tag{3.2}$$

All of our remarks about C, T, and baryon number apply equally as well to the Lagrangian of Eq. (3.2) as to that of Eq. (3.1). We assume (for simplicity) that $V_{ij}(x)$ vanishes as $|x|\to\infty$.

Let us introduce a $2N$-component complex vector

$$v\equiv\begin{bmatrix}a_1\\ \cdot\\ \cdot\\ \cdot\\ a_N\\ b_1\\ \cdot\\ \cdot\\ \cdot\\ b_N\end{bmatrix}\equiv\begin{pmatrix}a\\ b\end{pmatrix}, \tag{3.3}$$

which describes the situation where we have an outgoing wave $a_1e^{ik(x-t)}$ and an incoming wave $b_1e^{ik(-x-t)}$ as $x\to-\infty$ associated with a particle of species 1, outgoing wave $a_2e^{ik(x-t)}$ and incoming wave $b_2e^{ik(-x-t)}$ as $x\to-\infty$ for species 2, etc. Similarly we introduce

$$w\equiv\begin{bmatrix}c_1\\ \cdot\\ \cdot\\ \cdot\\ c_N\\ d_1\\ \cdot\\ \cdot\\ \cdot\\ d_N\end{bmatrix}\equiv\begin{pmatrix}c\\ d\end{pmatrix}, \tag{3.4}$$

which describes an outgoing wave $c_1e^{ik(x-t)}$ and an incoming wave $d_1e^{ik(-x-t)}$ for species 1 as $x\to+\infty$, and so forth. There is a linear relationship between the fields at $\pm\infty$ if the fields are governed by the Lagrangian Eq. (3.2); we write this as

$$\begin{pmatrix}c\\ d\end{pmatrix}=\begin{pmatrix}A & B\\ C & D\end{pmatrix}\begin{pmatrix}a\\ b\end{pmatrix}\equiv\mathfrak{M}\begin{pmatrix}a\\ b\end{pmatrix}. \tag{3.5}$$

The baryon number flux $|a|^2-|b|^2$ at $-\infty$ must be equal to the baryon number flux $|c|^2-|d|^2$ at $+\infty$, for any choice of the vectors a, b. This implies

$$\begin{pmatrix}1 & 0\\ 0 & -1\end{pmatrix}=\begin{pmatrix}A^\dagger & C^\dagger\\ B^\dagger & D^\dagger\end{pmatrix}\begin{pmatrix}1 & 0\\ 0 & -1\end{pmatrix}\begin{pmatrix}A & B\\ C & D\end{pmatrix}, \tag{3.6}$$

or, after a little algebra, the three matrix equations

$$A^\dagger A-C^\dagger C=1, \tag{3.7a}$$

$$D^\dagger D-B^\dagger B=1, \tag{3.7b}$$

$$A^\dagger B=C^\dagger D. \tag{3.7c}$$

The transmission probabilities are readily expressed in terms of A. Indeed, the total baryon transmission for a particle of species j is governed by requiring that the outgoing waves at $-\infty$ be only a unit flux of species j and that there be no incoming wave at $+\infty$, i.e.,

$$\begin{pmatrix}t^{(j)}\\ 0\end{pmatrix}=\begin{pmatrix}A & B\\ C & D\end{pmatrix}\begin{pmatrix}\vec{e}_j\\ r^{(j)}\end{pmatrix}, \tag{3.8}$$

where $\vec{e}_j$ is the vector with unit entry in the jth slot and zeros elsewhere. Multiplying both sides by

$$\mathfrak{M}^{-1}=\begin{pmatrix}A^\dagger & -C^\dagger\\ -B^\dagger & D^\dagger\end{pmatrix} \tag{3.9}$$

leads to

$$A^\dagger t^{(j)}=\vec{e}_j,\quad t^{(j)}=(A^\dagger)^{-1}\vec{e}_j\,. \tag{3.10}$$

[It follows from (3.7a) that A is invertible.] Thus the transmission probability for a unit incoming wave of particle species j is

$$|t^{(j)}|^2=\vec{e}_jA^{-1}(A^\dagger)^{-1}\vec{e}_j\,. \tag{3.11}$$

And the net flux for all species j is

$$\sum_j|t^{(j)}|^2=\mathrm{tr}A^{-1}(A^\dagger)^{-1}\,. \tag{3.12}$$

We must compare this to the total flux of antiparticles. If the particles are governed by the potential $V_{ij}(x)$, the antiparticles are governed by $V_{ji}(x)=V_{ij}^*(x)$, as one can see by comparing the equations of motion for Φ_i and Φ_i^* following from the Lagrangian Eq. (3.2). Let us call the corresponding transmission matrix for antiparticles

$$\overline{\mathfrak{M}}=\begin{pmatrix}\overline{A} & \overline{B}\\ \overline{C} & \overline{D}\end{pmatrix}. \tag{3.13}$$

Then one can show by manipulating the relevant Schrödinger equations and boundary conditions that

$$A=\overline{D}^*,\qquad D=\overline{A}^*. \tag{3.14}$$

Before deriving this from the Schrödinger equation, we should remark that the same results follow from the deeper principle of CPT invariance. Namely, the amplitude for an outgoing particle of type i at $-\infty$ to propagate to a particle of type j

at $+\infty$ is the complex conjugate of the amplitude for an incoming antiparticle of type j at $-\infty$ to propagate to an antiparticle of type i at $+\infty$. Writing this verbal statement in equations gives simply Eq. (3.14).

To see Eq. (3.14) directly from the Schrödinger equation, notice that the matrix A is defined as the solution of the problem

$$\frac{\partial^2\psi_i^{(j)}}{\partial t^2}-\frac{\partial^2\psi_i^{(j)}}{\partial x^2}+V_{ik}\psi_k^{(j)}=0\ , \tag{3.15a}$$

subject to

$$\psi_i^{(j)}(x,t)\to e^{ik(x-t)}\delta_{ij}+\text{incoming waves},\quad x\to-\infty \tag{3.15b}$$

$$\psi_i^{(j)}(x,t)\to e^{ik(x-t)}A_{ij},\quad x\to+\infty. \tag{3.15c}$$

Similarly $\overline{D}$ is the solution of the problem

$$\frac{\partial^2\psi_i^{(j)}}{\partial t^2}-\frac{\partial^2\psi_i^{(j)}}{\partial x^2}+V_{ki}\psi_k^{(j)}=0, \tag{3.16a}$$

subject to

$$\psi_i^{(j)}(x,t)\to e^{ik(-x-t)}\delta_{ij}+\text{outgoing waves},\quad x\to-\infty \tag{3.16b}$$

$$\psi_i^{(j)}(x,t)\to e^{ik(-x-t)}\overline{D}_{ij},\quad x\to+\infty. \tag{3.16c}$$

[Notice the transposition of V in Eq. (3.16a), which arises because the antiparticles are governed by the equation of motion for Φ^* derived from the Lagrangian (3.2).] Now complex conjugating all the equations [(3.16a)–(3.16c)] and changing the sign of time (which leaves the equation of motion invariant) maps a solution of [(3.16a)–(3.16c)] onto a solution of [(3.15a)–(3.15c)], with $A_{ij}=\overline{D}^*_{ij}$. This proves (3.14).

Now the total flux of particles minus the total number of antiparticles, starting from unit amplitude for all species at $-\infty$, is, from (3.12) and (3.14),

$$\operatorname{tr}A^{-1}(A^\dagger)^{-1}-\operatorname{tr}\overline{A}^{-1}(\overline{A}^\dagger)^{-1}=\operatorname{tr}A^{-1}(A^\dagger)^{-1}-\operatorname{tr}(D^{*\dagger})^{-1}D^{*-1}. \tag{3.17}$$

Use of the unitarity equations (3.7) and cyclic invariance of the trace gives

$$\operatorname{tr}A^{-1}(A^\dagger)^{-1}=\operatorname{tr}(A^\dagger)^{-1}A^{-1}=\operatorname{tr}[1-(A^\dagger)^{-1}C^\dagger CA^{-1}],$$

$$\operatorname{tr}(D^{*\dagger})^{-1}(D^*)^{-1}=\operatorname{tr}(D^\dagger)^{-1}D^{-1}=\operatorname{tr}[1-(D^\dagger)^{-1}B^\dagger BD^{-1}]. \tag{3.18}$$

But according to Eq. (3.7c) $(A^+)^{-1}C^+=BD^{-1}$, so we see that the final traces in both equations are equal. Thus the total flux of particles is equal to the total flux of antiparticles.

This proof[13] requires several comments:

(1) We have assumed that the potential turns off at both $\pm\infty$. In the realistic black-hole case the potential does turn off at $-\infty$ (the event horizon) but not at $+\infty$ (i.e., the particles may have nonvanishing mass). The proof actually works for this situation too, with only notational modifications.

(2) Although we couched our derivation in the language of one-particle quantum mechanics, clearly the ingredients of the proof are very general: basically CPT invariance and unitarity.

(3) The proof involves use of manipulations inside the trace which are not general matrix identities— which is a particularly obscure way of saying that it does not imply equality of particle and antiparticle flux species by species.

B. Thermodynamic argument

We now show that the theorem on matter-antimatter balance just proved follows also from a thermodynamic argument. It is reassuring to see that thermodynamic principles retain their vitality even in a world with T violation.

Consider the box depicted in Fig. 1, consisting of two compartments at temperature T and chemical potential zero separated by a semipermeable membrane characterized by the potential $V_{ij}(x)$. Suppose that baryon number is conserved, and that baryons (summed over species) penetrate $V_{ij}(x)$ from left to right more readily than antibaryons (summed over species). It follows from the PCT theorem that from right to left this membrane will allow anitbaryons to penetrate more readily than baryons. We see that such a membrane would act as a Maxwell demon for baryon number, separating baryons from antibaryons. In time, the two sides of the box would become distinguishable, contrary to the zeroth law of thermodynamics (uniqueness of thermal equilibrium). Alternatively one could set up a pipe connecting the two compartments, and extract work from the diffusion gradient, violating the second

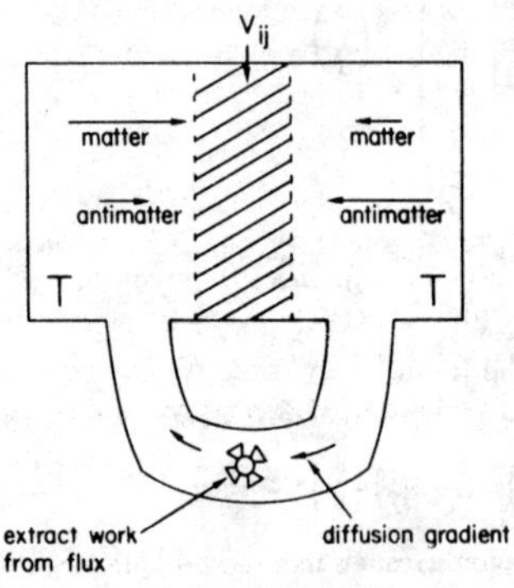

FIG. 1. Box illustrating the thermodynamic argument for zero net baryon number. See text.

law. It follows that thermodynamics does not permit a membrane with these properties, in agreement with the previous quantum-mechanical argument.

Notice that a net flux of any nonconserved quantity across the membrane is permitted. Such fluxes do not lead to different equilibrium situations in the two compartments, because information concerning the nonconserved quantum numbers is completely lost in thermal equilibrium. We now establish that such fluxes do indeed occur.

C. Example of imbalance

For our example we will use notations as before with two species 1, 2 of scalar fields and the potential

$$V(x) = M_1\delta\left(x+\frac{\alpha}{2}\right) + M_2\delta\left(x-\frac{\alpha}{2}\right), \qquad (3.19)$$

where M_1 and M_2 are two-by-two Hermitian matrices. We assume species 1 is distinguished from species 2 by interactions [other than the corrections to gravity summarized by Eq. (3.19)] whose details need not concern us—e.g., species 1 might be strongly interacting while species 2 is not. We shall call particles of species 1 "baryons" and particles of species 2 "leptons"—so if M_1 or M_2 has off-diagonal components baryon and lepton number are not separately conserved.

By standard quantum-mechanical methods it is not difficult to compute the transmission matrix for this problem; in our previous notations the matrix A is (up to an overall phase)

$$A = 1 - i(M_1 + M_2) + (\alpha - 1)M_2M_1, \qquad (3.20)$$

where $\alpha = e^{-2ika}$ is the phase change for a round trip between the two δ functions, and we have absorbed a factor $1/k$ into the definition of M_1 and M_2. In the present case the transmission probability of baryons is given not by the trace of $A^{-1}(A^\dagger)^{-1}$, but by its 11 component,

$$T_B \equiv \sum_j |t_1^{(j)}|^2 = [A^{-1}(A^\dagger)^{-1}]_{11} = (A^\dagger A)^{-1}{}_{11}, \qquad (3.21)$$

as one readily sees from Eq. (3.10). One could of course solve for this directly given Eq. (3.20), but much simpler expressions result if one expands T_B in a power series in M_1, M_2:

$$(A^\dagger A)^{-1} = 1 - \alpha M_2 M_1 - \alpha^* M_1 M_2 - M_1^2 - M_2^2 - \cdots. \qquad (3.22)$$

For antiparticles we use the complex-conjugate potential to Eq. (3.20), which is equivalent to replacing M_1 and M_2 by their transposes. Then

$$(\bar{A}^\dagger \bar{A})^{-1} = (1 - \alpha M_1 M_2 - \alpha^* M_2 M_1 - M_1^2 - M_2^2 - \cdots)^T. \qquad (3.23)$$

Letting

$$M_1 = \begin{pmatrix} a_1 & c_1 \\ c_1^* & b_1 \end{pmatrix}, \qquad (3.24)$$

$$M_2 = \begin{pmatrix} a_2 & c_2 \\ c_2^* & b_2 \end{pmatrix}, \qquad (3.25)$$

we find, assembling the formulas,

$$T_B - T_{\bar{B}} = (A^\dagger A)^{-1}{}_{11} - (\bar{A}^\dagger \bar{A})^{-1}{}_{11} = (\alpha - \alpha^*)(c_1^* c_2 - c_1 c_2^*). \qquad (3.26)$$

This expression for the net flux of baryon number does not vanish except for special values of α, c_1, c_2.

Notice that the asymmetry vanishes if $\alpha = 1$. In this case one has effectively only one δ function in the potential and one associated matrix, namely $M_1 + M_2$. This matrix may be made real by a redefinition of the phase of Φ_2, so that there is actually no time-reversal noninvariance in this case. Similar remarks imply that the asymmetry must vanish if C_1 or C_2 vanishes, or if they have the same phase, so that the form of the asymmetry in Eq. (3.26) is almost dictated *a priori*.

We have now demonstrated that a net baryon number flux will arise in Hawking radiation if (and only if) there is microscopic baryon nonconservation, and C and T violation.

The species-by-species imbalance found here also implies that a symmetric medium (consisting of, say, an equal number of p's and $\bar{p}$'s) surrounding a black hole will evolve asymmetrically.

D. Cosmological particle production

Particles may also be produced by the time-changing gravitational fields associated with the expansion of the universe.[5] The analysis we have performed for black-hole radiation carries over essentially unchanged to this case. It is only necessary to replace the space-dependent potential $V_{ij}(x)$ by a time-dependent potential $V_{ij}(t)$. We conclude that in this case also a net baryon asymmetry arises if and only if there is a microscopic C-, T-, and baryon-number-violating interaction.[14]

ACKNOWLEDGMENT

We thank S. Bludman, S. Deser, J. Hartle, S. Hawking, D. Sciama, B. Simon, S. Weinberg, and C. N. Yang for helpful discussions.

This work was supported in part by the U. S. Department of Energy under Contract Nos. AT(E11-1)3071 and EY-76-C02-3072, and by the National Science Foundation under Contract No. PHY-78-01221. The work of F. W. and A. Z. was also supported in part by the Alfred P. Sloan Foundation.

[1]For a survey of the evidence, see G. Steigman, Annu. Rev. Astron. Astrophys. 14, 339 (1976).

[2]S. Hawking, Nature 248, 30 (1974).

[3]S. Hawking, Commun. Math. Phys. 43, 199 (1975).

[4]J. Bekenstein, Phys. Rev. D 5, 1239 (1972).

[5]L. Parker, Phys. Rev. D 3, 346 (1971).

[6]See, e.g., E. Commins, *Weak Interactions* (McGraw-Hill, New York, 1973), Chap. 11.

[7]J. Pati and A. Salam, Phys. Rev. D 8, 1240 (1973); 10, 275 (1974).

[8]H. Georgi and S. Glashow, Phys. Rev. Lett. 32, 438 (1974); H. Georgi, H. Quinn, and S. Weinberg, *ibid.* 33, 451 (1974).

[9]This scenario is presented in more detail in the paper by S. Weinberg, report (unpublished). We are grateful to S. Weinberg for numerous discussions of these ideas, and especially for pointing out the simple formula (2.9).

[10]S. Dimopoulos and L. Susskind, Phys. Rev. D 18, 4500 (1978).

[11]M. Yoshimura, Phys. Rev. Lett. 41, 381 (1978). This paper came to our attention during the preparation of the present manuscript. Although we disagree with some of its technical content, it did stimulate our thinking, especially in Secs. II B and II D.

[12]R. C. Tolman, *Principles of Statistical Mechanics* Clarendon Press, Oxford, 1938). After preparation of this manuscript we learned of a much more detailed discussion by A. Aharony, in *Modern Developments in Thermodynamics*, edited by Gal-Or (Wiley, New York, 1973).

[13]Essentially the same theorem has been proved in a completely different context by P. Davies and B. Simon [Princeton report, 1978 (unpublished)] using slightly different methods. We are very grateful to B. Simon for bringing this result to our attention.

[14]This answers a question posed by L. Parker, in *Asymptotic Structure of Space-Time*, edited by F. L. Esposito and L. Witten (Plenum, New York, 1977).

Cosmological Production of Baryons

Steven Weinberg
Lyman Laboratory of Physics, Harvard University, and Harvard-Smithsonian Center for Astrophysics, Cambridge, Massachusetts 02138
(Received 27 October 1978)

Departures from thermal equilibrium which are likely to occur in an expanding universe allow the production of an appreciable net baryon density by processes which violate baryon-number conservation. It is shown that the resulting baryon to entropy ratio can be calculated in terms of purely microscopic quantities.

It is an old idea[1] that the observed excess of matter over antimatter in our universe may have arisen from physical processes which violate the conservation of baryon number. Of course, the rates of baryon-nonconserving processes like proton decay are very small at ordinary energies, but if the slowness of these processes is due to the large mass of intermediate vector of scalar "X bosons" which mediate baryon nonconservation, then at very high temperatures with $kT \simeq m_X$, the baryon-nonconserving processes would have rates comparable with those of other processes. However, even if there are reactions which do not conserve C, CP, T, and baryon number, and even if these reactions proceed faster than the expansion of the universe, there can be no cosmological baryon production once the cosmic distribution functions take their equilibrium form, until the expansion of the universe has had a chance to pull these distribution functions out of equilibrium. This can easily be seen from the generalized Uehling-Uhlenbeck equation[2] for a homogeneous isotropic gas,

$$dn(p_1)/dt = \sum_{kl} \int dp_2 \cdots dp_k dp_1' \cdots dp_l' \times \{\Gamma(p_1' \cdots p_l' \to p_1 \cdots p_k) n(p_1') \cdots n(p_l')[1 \mp n(p_1)] \cdots [1 \pm n(p_k)] - \Gamma(p_1 \cdots p_k \to p_1' \cdots p_l') n(p_1) \cdots n(p_k)[1 \pm n(p_1')] \cdots [1 \pm n(p_l')]\}, \tag{1}$$

where n is the single-particle density in phase space; p labels the three-momentum and any other particle quantum numbers, including baryon number; and Γ is a rate constant, equal, for $k = l = 2$, to the cross section times the initial relative velocity. The factors $1 \pm n(p)$ represent the effect of stimulated emission or Pauli suppression for bosons or fermions, respectively. If at any instant, $n(p)$ takes its equilibrium form, then $n(p)/[1 \pm n(p)]$ is an exponential of a linear combination of the energy and any other conserved quantities; so for any allowed reaction with $\Gamma \neq 0$, we have

$$n(p_1') \cdots n(p_l')[1 \pm n(p_1)] \cdots [1 \pm n(p_k)] = n(p_1) \cdots n(p_k)[1 \pm n(p_1')] \cdots [1 \pm n(p_l')]. \tag{2}$$

Under T invariance, Γ would be symmetric, and the two terms in the integrand of Eq. (1) would cancel. But even without T invariance, unitarity always gives

$$0 = \sum_l \int dp_1' \cdots dp_l' [1 \pm n(p_1')] \cdots [1 \pm n(p_l')][\Gamma(p_1 \cdots p_k \to p_1' \cdots p_l') - \Gamma(p_1' \cdots p_l' \to p_1 \cdots p_k)], \tag{3}$$

so that the p' integrals in (1) still cancel.[3] For an expanding gas there are also terms in Eq. (1) which represent the effects of dilution and red shift, and these terms can produce departures from equilibrium, but of course they have no direct effect on the baryon number per co-moving volume.

This note will describe a mechanism for production of a cosmic baryon excess, based on the departures from thermal equilibrium which are likely to have occurred in the early universe. It is assumed here that all particles have masses below (though not necessarily far below) the Planck mass $m_P \equiv G^{-1/2} = 1.22\times 10^{19}$ GeV. For simplicity, it will be assumed that the only superheavy particles with masses above 1 TeV or so are the X bosons which mediate baryon nonconservation; however, it would not be difficult to incorporate superheavy fermions with masses $m \simeq m_X$ in these considerations. Aside from gravitation itself, all interactions are supposed to have dimensionless coupling constants. For the interaction of X bosons with fermions, this coupling is denoted g_X. Finally, it will also be assumed that $\alpha_X{}^2 N \ll 1$, where $\alpha_X = g_X{}^2/4\pi$, and N is the number of helicity states of all particle species. Under these assumptions, we can trace the following chain of events[4]:

(1) At very early times, when $kT \approx m_P$, the interactions of gravitons were so strong that thermal equilibrium distributions would have been established at least approximately for all particle species; for instance, by graviton-graviton collisions.[5] (Of course, we do not know how to calculate detailed reaction rates at these times, but we can be confident that gravitational interactions were strong, because this is indicated by lowest-order calculations, and it is only the strength of the interactions that invalidates such calculations.) If gravitational interactions conserved baryon number at $kT \gtrsim m_P$, then the universe could have begun with a nonvanishing value for the baryonic chemical potential; I assume here that this is not the case.

(2) As kT fell below m_P, gravitational interactions became ineffective. The rates for X-boson decay, baryon-nonconserving collisions (or, for $kT \gtrsim m_X$, all collisions) and cosmic expansion may be estimated as[6]

$$\Gamma_X \simeq \alpha_X m_X{}^2 N/[(kT)^2 + m_X{}^2]^{1/2}, \tag{4}$$

$$\Gamma_C \simeq \alpha_X{}^2 (kT)^5 N/[(kT)^2 + m_X{}^2]^2, \tag{5}$$

$$\dot{R}/R \equiv H = 1.66 (kT)^2 N^{1/2}/m_P. \tag{6}$$

With $\alpha_X{}^2 N \ll 1$ and $m_X < m_P$, both Γ_X and Γ_C were much less than H at $kT \simeq m_P$. However, as long as kT remained above all particle masses, the expansion preserved the equilibrium form of all particle distributions, with red-shifted temperature $T \propto 1/R$.

(3) The X bosons began to decay when $\Gamma_X \simeq H$. If at this time $kT > m_X$, the collisions of the decay products with each other or with ambient particles would have rapidly recreated the X bosons through the inverse of the decay process, thus reestablishing equilibrium distributions. In order to produce any appreciable baryon excess, it is necessary that $kT \lesssim m_X$ when $\Gamma_X \simeq H$, so that the Boltzmann factor $\exp(-m_X/kT)$ could block inverse decay. Equation (4) then gives $\Gamma_X \simeq H$ at a temperature

$$kT_D \simeq (N^{1/2} \alpha_X m_X m_P)^{1/2}, \tag{7}$$

so that the condition $m_X \gtrsim kT_D$ yields a lower bound on m_X

$$m_X \gtrsim N^{1/2} \alpha_X m_P. \tag{8}$$

(For gauge bosons we expect $\alpha_X \simeq \alpha$, so (8) requires $m_X \gtrsim 10^{17} N^{1/2}$ GeV, while for Higgs bosons α_X is presumably in the range of 10^{-4} to 10^{-6}, and the lower bound on m_X would be of order 10^{13} to $10^{15} N^{1/2}$ GeV.) Note also that (5), (6), and (8) give $\Gamma_C \ll H$ for all temperatures. This justifies the neglect of X-boson production or annihilation in reactions other than X decay and its inverse, and insures that any baryon excess produced when the X bosons decayed would have survived to the present time.

Before the X bosons decayed, at temperatures just above T_D, their number density was $n_{XD} = \zeta(3)(kT_D)^3 N_X/\pi^2$, where N_X is the total number of X (and $\bar{X}$) spin states. Also, the total entropy density of all other particles was $s_D = 4\pi^2 k (kT_D)^3 \times N/45$, with N now understood to include factors of 7/8 for fermion spin states. If the mean net baryon number produced in X or $\bar{X}$ decay is ΔB per decay, and if one can ignore the entropy released in X-boson decay, then the ratio of baryon number to entropy after the X bosons decayed was

$$kn_B/s = kn_{XD}\Delta B/s_D = 45\zeta(3)(N_X/N)\Delta B/4\pi^4. \tag{9}$$

If one assumes the subsequent expansion to be adiabatic, both n_B and s would have scaled as R^{-3}, so that Eq. (9) would give the ratio of baryon number to entropy of the present universe.

Strictly speaking, one should take into account the entropy contributed by the X-boson decay products when they finally thermalize. This increases the energy density by a factor

$$\lambda = 1 + \frac{m_X n_{XD}}{\pi^2 (kT_D)^4 N/30} = 1 + \frac{30\zeta(3) N_X m_X}{\pi^4 N k T_D}$$

$$\simeq 1 + (N_X/N)(m_X/N^{1/2}\alpha_X m_P)^{1/2}, \tag{10}$$

and so decreases the ratio of baryon number to entropy by a factor $\lambda^{-3/4}$. However, this effect can be ignored if $N \gg N_X$.

The crucial quantity ΔB in Eq. (9) can be determined from the branching ratios for X-boson decay. For instance, suppose that an X boson decays into two channels with baryon numbers B_1 and B_2 and branching ratios r and $1-r$. The antiparticle will then decay into channels with baryon numbers $-B_1$ and $-B_2$, with the same total rate, but with different branching ratios $\bar{r}$ and $1-\bar{r}$. The mean net baryon number produced when X or $\bar{X}$ decays is then

$$\Delta B = \tfrac{1}{2}[rB_1 + (1-r)B_2 - \bar{r}B_1 - (1-\bar{r})B_2]$$

$$= \tfrac{1}{2}(r-\bar{r})(B_1 - B_2). \qquad (11)$$

CPT invariance gives $r=\bar{r}$ in the Born approximation. If the leading contribution to $r-\bar{r}$ arises from an interference of graphs with a total of l loops, then one expects $r-\bar{r}$ to be of order $\epsilon(\alpha_X/2\pi)^l$, where ϵ is whatever small angle characterizes CP violation. Of course, to be more definite, a detailed model of baryon nonconservation is needed. However, in any given model, Eqs. (9) and (11) give a precise prediction for the ratio of baryon number to entropy kn_B/s, which may be compared with the observed value[7] 10^{-8} to 10^{-10}.

The above discussion has assumed a homogeneous isotropic expansion, in which the entropy stays fixed except for the small effects of bulk viscosity.[7] However, it is also possible to deal with gross departures from thermal equilibrium that might be produced by cosmic inhomogeneities. As any part of the universe relaxes toward equilibrium, the rate at which its entropy increases will be proportional to the difference between the entropy and its maximum value S_{max}. Baryon production vanishes in the equilibrium configuration with $S=S_{max}$, so the rate of increase of baryon number will also be proportional to $S-S_{max}$. Thus, the ratio of the baryon-number production to the entropy production will be given by the ratio of the coefficients of $S-S_{max}$ in dB/dt and dS/dt, and independent of the amount of the initial departure from thermal equilibrium. If most of the entropy and baryon number of the universe were created in this way, then it is this ratio that would have to be compared with the experimental value of 10^{-8} to 10^{-10}.

Note added.—(1) Any X bosons which can mediate baryon-nonconserving reactions are necessarily much heavier than the Z^0 or $W^\pm$; so their interactions can be analyzed using the weak and electromagnetic gauge group SU(2)⊗ U(1) as well as the strong gauge group SU(3) as if they were all unbroken symmetries. In this way one finds in general there are just three kinds of bosons which can couple to channels consisting of a pair of ordinary fermions, with these channels not all having equal baryon numbers: They are an SU(3) triplet SU(2) singlet X_S of scalar bosons with charge $-\frac{1}{3}$; an SU(3) triplet SU(2) doublet X_V of vector bosons with charges $-\frac{1}{3}$, $-\frac{4}{3}$; and an SU(3) triplet SU(2) doublet of X_V' of vector bosons with charges $\frac{2}{3}$, $-\frac{1}{3}$; plus their corresponding SU(3)-$\bar{3}$ antibosons. For all these bosons, the decay channels are $X\to gl, \overline{qq}$ and $\bar{X}\to \bar{g}\bar{l}, qq$, with q and l denoting general quarks and leptons. Hence $B_1 = +\frac{1}{3}$ and $B_2 = -\frac{2}{3}$ in Eq. (11). This analysis incidentally shows that lowest-order baryon-number–nonconserving interactions always conserve baryon number *minus* lepton number, so nucleons may decay in lowest order into antileptons, but not leptons.

(2) Detailed calculations have been carried out with Nanopoulos[8] to estimate the difference in the branching ratios $r,\bar{r}$ for $X\to gl$ and $X\to \bar{g}\bar{l}$ that arises from the interference of tree graphs with one-loop graphs. In general, a difference between r and $\bar{r}$ could arise from one-loop graphs in which a scalar or vector boson is exchanged between the final fermions, even when all fermion masses are negligible compared with the temperature, provided that CP invariance is violated in the Lagrangian, or is already spontaneously broken at these high temperatures. In various grand unified theories there are relations among the various couplings of Higgs or gauge bosons to fermions, which eliminate most of these contributions to $r-\bar{r}$. However, there will still be a contribution to $r-\bar{r}$ in X_S decay from the exchange of X_S bosons of different species. Since Higgs-boson exchange is naturally weaker than $W^\pm$ or Z^0 exchange at ordinary energies, it is possible that the CP-invariance violation is maximal in the coupling of fermions to Higgs bosons, including X_S bosons. In this case, $r-\bar{r}$ is of order $\alpha_H/2\pi \approx 10^{-6}$. With $B_1 - B_2 = 1$ and $N_X/N \approx 10^{-2}$, Eqs. (11) and (9) then give a ratio of baryon number to entropy of order 10^{-9}.

(3) The masses of superheavy gauge bosons were estimated in grand unified gauge theories to be of order 10^{16} GeV, by Georgi, Quinn, and Weinberg.[9] (As shown there, this estimate applies for arbitrary simple grand unified gauge groups, under reasonable general assumptions on the spectrum of fermions. The same assumptions yielded a Z^0-γ mixing angle with $\sin^2\theta \simeq 0.2$.) Presumably the Higgs-boson masses are of the same order. Decay and inverse-decay processes arising from the gauge coupling of vec-

tor bosons to each other and to Higgs bosons and fermions will bring all these particles into thermal equilibrium at a temperature given by Eq. (7) [with $N \approx 100$, $\alpha_X \approx 10^{-2}$, $m_X \approx 10^{16}$ GeV] as of order 10^{17} GeV. Hence there is no need to invoke gravitational processes at the Planck temperature to establish initial equilibrium distributions, and any preexisting baryon imbalance would have been wiped out at $kT \simeq 10^{17}$ GeV. As the temperature dropped below 10^{16} GeV all superheavy gauge bosons and some of the superheavy Higgs bosons would have disappeared. However, if the lightest superheavy bosons happen to be X_S bosons, then these bosons would have survived as the temperature fell below their mass, because the only decay channels open then would have been two-fermion states, and $\alpha_X \ll \alpha$ for Higgs-fermion couplings. The decay of these scalar bosons when the temperature finally dropped to $kT_D \approx 10^{14}$ GeV $\ll m(X_S)$ would then produce the baryon excess estimated in Note (2).

I am grateful for valuable conversations with J. Ellis, D. Nanopoulos, A. Salam, L. Susskind, F. Wilczek, C. N. Yang, and M. Yoshimura.

---

[1]For one example, see S. Weinberg, in *Lectures on Particles and Fields*, edited by S. Deser and K. Ford (Prentice-Hall, Englewood Cliffs, N. J., 1964), p. 482. The subject has been considered in recent papers by M. Yoshimura, Phys. Rev. Lett. 41, 381 (1978); S. Dimopoulos and L. Susskind, Phys. Rev. D 18, 4500 (1978); B. Toussaint, S. B. Treiman, F. Wilczek, and A. Zee, Phys. Rev. D 19, 1036 (1979). After this work was completed I also became aware of a discussion by A. Yu. Ignatiev, N. V. Krosnikov, V. A. Kuzmin, and A. N. Tavkhelidze, Phys. Lett. 76B, 436 (1978); and new reports have appeared by S. Dimopoulos and L. Susskind, Stanford University Report No. ITP-616, (to be published), and J. Ellis, M. K. Gaillard, and D. V. Nanopoulos, CERN Report No. Ref. TH-2596 (to be published). The approach followed and the conclusions reached here differ from those of Yoshimura, for reasons indicated below; from Ignatiev *et al.* because they adopt a different picture of baryon nonconservation (without superheavy X bosons); and from Ellis, Gaillard, and Nanopoulos, for reasons indicated in their erratum (to be published). The assumptions and general approach followed here is similar in many respects to that of Dimopoulos and Susskind and Section 2 of Toussaint *et al.* A major difference is that by following the baryon production scenario in detail, a formula is obtained here, Eq. (9), which gives the ratio of baryon number to entropy in terms of purely microscopic quantities.

[2]E. A. Uehling and G. E. Uhlenbeck, Phys. Rev. 43, 552 (1933).

[3]A very general version of this argument in the context of the "master" equation was given about a decade ago in an unpublished work of C. N. Yang and C. P. Yang. Also see A. Aharony, in *Modern Developments in Thermodynamics* (Wiley, New York, 1973), pp. 95–114, and references cited therein. I first learned of this argument for the special case of massless distinguishable particles from the original version of the paper of Toussaint *et al.*, Ref. 1. For indistinguishable particles, the factors $1 \pm n(p')$ in Eq. (3) arise from the effects of the ambient bosons or fermions on identical virtual particles in these reactions; in old-fashioned perturbation theory, the ambient particles generate a product of $1 \pm n$ factors for each intermediate state. These factors were omitted in the unitarity relation as given by Aharony, so that it was not possible in his paper to see how the Uehling-Uhlenbeck equation yields a vanishing rate of change for equilibrium distributions. Equation (3) shows that the physical processes considered by Yoshimura (Ref. 1) cannot produce an appreciable net baryon density if all relevant channels are taken into account, as pointed out by Toussaint *et al.*, Ref. 1.

[4]This scenario was developed in the course of conversations with F. Wilczek, and is also discussed in Section 2 of Toussaint *et al.*, Ref. 1. I am very grateful to F. Wilczek for numerous discussions of these ideas.

[5]Horizon effects may prevent complete establishment of thermal equilibrium at $kT \simeq m_P$; G. Steigman, private communication.

[6]It we keep track of all factors of 2π from Fourier integrals and 4π from solid-angle integrals, but set all other numerical constants equal to unity, then factors 4 and $8/\pi$ would appear in the right-hand sides of Eqs. (4) and (5), respectively. The powers of $m_X^2/[(kT)^2 + m_X^2]$ in Eqs. (4) and (5) are inserted to take account of time dilation and the virtual X-boson propagator, respectively.

[7]See, e.g., S. Weinberg, *Gravitation and Cosmology-Principles and Applications of the General Theory of Relativity* (Wiley, New York, 1972), Chap. 15.

[8]S. Weinberg and D. V. Nanopoulos, to be published.

[9]H. Georgi, H. Quinn, and S. Weinberg, Phys. Rev. Lett. 33, 451 (1974).

PHYSICAL REVIEW D VOLUME 20, NUMBER 10 15 NOVEMBER 1979

Mechanisms for cosmological baryon production

D. V. Nanopoulos and S. Weinberg
Lyman Laboratory of Physics, Harvard University, Cambridge, Massachusetts 02138
(Received 29 June 1979)

General formulas are given for the mean net baryon number produced in the decay of superheavy scalar or vector bosons. These results are used to make rough numerical estimates of the cosmological baryon abundance that would result from such decay processes in the very early universe.

I. INTRODUCTION

The universe appears to have a baryon-number density that is nonzero but small. Quantitatively, assuming that all galaxies are composed of matter rather than antimatter,[1] the ratio of the baryon-number density n_B to the dimensionless entropy density s/k of the 3 °K microwave background is of order[2] 10^{-8} to 10^{-10}. If baryon number is conserved and the expansion of the universe is essentially adiabatic, then the quantity $n_B k/s$ is a constant, which governs the whole course of cosmic evolution. Thus it is an important matter to learn why this ratio is not zero, and why, though not zero, it is so small.

Recently a number of authors[3-10] have considered the possibility that the cosmic baryon-number excess was produced by physical baryon-number-nonconserving processes, which are cosmologically insignificant at present, but may have occurred at significant rates in the very early universe. It has become clear that in order to produce an appreciable baryon excess, it is necessary not only that some reactions violate baryon-number and *CP* conservation, but also that these reactions occurred at a time when the expansion of the universe had already pulled the cosmic particle distributions out of the equilibrium form.

The simplest way that this can happen[11] is for an equilibrium distribution to be established[12] for some heavy "X boson" at $kT \gg m_X$, with equal numbers of X and its antiparticle $\bar{X}$, and for equilibrium then to be lost when kT drops below m_X, because the decay rates of X and $\bar{X}$ are less than the rate of expansion of the universe at that time. When the X and $\bar{X}$ finally decay, at temperatures $kT \ll m_X$ which are low enough to prevent inverse decay, the baryon-entropy ratio produced will be[8]

$$kn_B/s = 45\zeta(3)(N_X/N)\Delta B/2\pi^4\,,$$

where ΔB is the mean baryon number produced in the decay of a single X or $\bar{X}$ boson, and N_X and N are the (suitably weighted) numbers of species of X bosons and of all particles with masses $m \lesssim m_X$, respectively.

In order to calculate the crucial quantity ΔB, we need a specific theory of baryon nonconservation. A class of such theories has been provided over the last few years by the grand unified gauge models, which unite the strong with the weak and electromagnetic interactions.[13] There is as yet no one grand unified model that clearly is realized in nature, so we choose here to work in a more general theoretical framework. Our main assumption is that there is some simple grand unified gauge group, whose spontaneous breakdown at the grand unification scale leaves unbroken only the gauge groups SU(3) and SU(2)× U(1) of the observed strong and weak and electromagnetic interactions.

As recognized some time ago,[14] this general framework provides a natural explanation for the fact that baryon-nonconserving processes are so slow at ordinary energies. The masses of those gauge bosons of the grand unified group which are not associated with SU(3) or SU(2) × U(1), and in particular of the bosons which mediate baryon nonconservation, are roughly of the order of the critical energy M where the strong and weak and electromagnetic couplings merge into the single coupling of the grand gauge group. But the decrease of the strong-interaction coupling is so slow that M must be enormous, and the proton lifetime, which is proportional to M^4, must be correspondingly long. Specifically, if we fix the ratios of the SU(3) and SU(2)× U(1) couplings at M by the assumption that there is a representation of some grand unified gauge group consisting solely (or chiefly) of quark-lepton families like those already observed, and take the observed values of e and the quantum-chromodynamic scale parameter Λ as an input, then M is found[14,15] to be of order 10^{16} GeV. (The same analysis[14,15] yields a Z^0-γ mixing parameter $\sin^2\theta$ between 0.19 and 0.21, only a little lower than the present experimental value $\sin^2\theta = 0.23 \pm 0.02$.)

These considerations lead us to assume that the superheavy vector and scalar bosons that mediate baryon-nonconserving reactions have masses in the range of 10^{14} to 10^{16} GeV.[16] For vector bosons, this is probably too low to allow the production of an appreciable baryon excess. As remarked in Refs. 6 and 8, a gauge boson with mass $m_X > kT$ will have a decay rate of order $\alpha m_X N$, so that these bosons decay when $\alpha m_X N$ becomes equal to the cosmic expansion rate $H = 1.66\ (kT)^2 N^{1/2}/m_P$, where $m_P = 1.22 \times 10^{19}$ GeV. This occurs at a temperature $kT \simeq (N^{1/2}\alpha m_X m_P)^{1/2}$, which is smaller than m_X only if m_X is above a value $N^{1/2}\alpha m_P \approx 10^{17} N^{1/2}$ GeV. On the other hand, for Higgs bosons we must replace α with $G_F \bar{m}^2/4\pi$, where $\bar{m}$ is an rms quark or lepton mass; for $\bar{m} \simeq 2$ GeV, the Higgs bosons will decay at temperatures kT which are below their mass m_X provided that m_X is greater than $N^{1/2} G_F \bar{m}^2 m_P / 4\pi \approx 3 \times 10^{13} N^{1/2}$ GeV. We will consider the decays of both superheavy gauge and Higgs bosons here, but it is the Higgs-boson decays that seem most relevant for cosmological baryon production.

At energies of the order of the superheavy gauge and Higgs bosons, it is a very good approximation to neglect the spontaneous breakdown of SU(2) $\times$ U(1) to U(1)$_{em}$, so that particle states and interactions can be analyzed using SU(3)$\times$SU(2)$\times$U(1) as if it were unbroken. In this way, it has been possible to classify the vector and scalar bosons that can mediate baryon nonconservation in general theories.[17] This classification is reviewed in Sec. II, and SU(3)$\times$SU(2)$\times$U(1) is used to give explicit forms for the most general baryon-violating boson-fermion interactions that can arise in renormalizable theories.

In Sec. III, we use the results of Sec. II to give general results for the mean baryon excess ΔB produced per X or $\bar{X}$ boson decay. This calculation is aided by a general theorem proved in an appendix, which indicates that graphs of first order in baryon-violating interactions but of arbitrary order in baryon-conserving interactions make no contribution to ΔB. We find that in general ΔB will receive its leading contributions from the interference of tree graphs with one-loop graphs in which a boson with baryon-violating interactions is exchanged between the fermions in the final state.[18]

Finally, in Sec. IV we apply this analysis in simple cases, and obtain rough numerical estimates for ΔB and kn_B/s. Our conclusions are stated in Sec. V.

II. PARTICLE SPECIES AND INTERACTIONS

The processes of interest to us in this paper occur at enormous temperatures, very much higher than the masses of the $W^\pm$ and Z^0. At such temperatures, it is an excellent approximation to neglect the spontaneous breakdown of SU(2) $\times$U(1) to electromagnetic gauge invariance and treat SU(2) $\times$U(1) as well as SU(3) color as an unbroken symmetry. In this section we will describe the SU(3) $\times$SU(2)$\times$U(1) classification[17] of the particle species that will be of relevance to us, and we will give general expressions for their mutual interactions.

First, there are the "ordinary" leptons and quarks. These apparently form sequences, with left-handed fermion fields

$$l_{aL} \equiv \begin{pmatrix} \nu_a \\ e_a \end{pmatrix}_L (1, 2, \tfrac{1}{2}),\quad \bar{e}_{aL}(1, 1, -1),$$
$$q_{aL} \equiv \begin{pmatrix} u_a \\ d_a \end{pmatrix}_L (3, 2, -\tfrac{1}{6}),\quad \bar{u}_{aL}(\bar{3}, 1, +\tfrac{2}{3}),\quad \bar{d}_{aL}(\bar{3}, 1, -\tfrac{1}{3}). \tag{1}$$

$$e_1 = e,\quad e_2 = \mu,\quad e_3 = \tau, \ldots$$
$$u_1 = u,\quad u_2 = c,\quad u_3 = t, \ldots$$
$$d_1 = d,\quad d_2 = s,\quad d_3 = b, \ldots\ .$$

In the usual notation, subscripts L and R indicate multiplication with $\frac{1}{2}(1+\gamma_5)$ and $\frac{1}{2}(1-\gamma_5)$, and the numbers in parentheses give the SU(3) multiplicity, the SU(2) multiplicity, and the value of the U(1) quantum number $Y \equiv T_3 - Q$.

The only renormalizable interactions of a vector field V^μ with a pair of fermion fields (here including antifermion fields) are of the form $V^{\mu\dagger}\psi_{1R}^T\gamma_\mu\psi_{2L}$. Therefore, we can make a complete list of all vector bosons that can couple to a pair of ordinary fermions by multiplying together all left-handed fields of leptons, quarks, antileptons, and antiquarks with all right-handed fields and adding up their SU(3)$\times$SU(2)$\times$U(1) quantum numbers. In a similar way, the only renormalizable interactions of a scalar field S with a pair of fermion fields are of the form $S^\dagger\psi_{1L}^T\psi_{2L}$ or $S^\dagger\psi_{1R}^T\psi_{2R}$, so we can catalog all scalar bosons that couple to ordinary fermions by multiplying all left-handed fields of leptons, quarks, antileptons, and antiquarks with each other, and the same for the right-handed fermion fields.

These lists of possible vector or scalar fields have an interesting feature[17] that greatly simplifies discussions of baryon nonconservation. Almost all of the scalar and vector fields that can couple to a pair of ordinary fermions couple only to channels with a single value of the baryon number and a single value of the lepton number. Such bosons can be assigned a baryon number and a lepton number in such a way that these quantities are conserved in the boson-fermion interactions. The only bosons which couple to two-fermion channels with varied baryon and/or lepton numbers are

$(3, 2, \frac{5}{6})$ vectors X_V: charges $-\frac{1}{3}, -\frac{4}{3}$,

$(3, 2, -\frac{1}{6})$ vectors X'_V: charges $\frac{2}{3}, -\frac{1}{3}$,

$(3, 1, \frac{1}{3})$ scalars X_S: charge $-\frac{1}{3}$,

plus the corresponding antibosons. Using SU(3) $\times$SU(2)$\times$U(1), we easily see that the coupling of these bosons to ordinary fermions must take the form

$$g_{\chi,ab}(\bar{l}_{Laj}\gamma_\mu d^c_{Rb\alpha})V^\mu_{\chi\alpha j}+\text{H.c.}, \tag{2}$$

$$h_{\chi,ab}\epsilon_{jk}\epsilon_{\alpha\beta\gamma}(\bar{u}^c_{Ra\alpha}\gamma_\mu q_{Lb\beta j})V^\mu_{\chi\gamma k}+\text{H.c.}, \tag{3}$$

$$j_{\chi,ab}(\bar{q}_{La\alpha j}\gamma_\mu e^c_{Rb})V^\mu_{\chi\alpha j}+\text{H.c.}, \tag{4}$$

$$g'_{\eta,ab}(\bar{l}_{Laj}\gamma_\mu u^c_{Rb\alpha})V'^\mu_{\eta\alpha j}+\text{H.c.}, \tag{5}$$

$$h'_{\eta,ab}\epsilon_{jk}\epsilon_{\alpha\beta\gamma}(\bar{d}^c_{Ra\alpha}\gamma_\mu q_{Lb\beta j})V'^\mu_{\eta\gamma k}+\text{H.c.}, \tag{6}$$

$$F^{(1)}_{\xi,ab}(\bar{q}_{La\alpha j}l^c_{Lbk})S_{\xi\alpha}\epsilon_{jk}+\text{H.c.}, \tag{7}$$

$$F^{(2)}_{\xi,ab}(\bar{u}^c_{Ra\alpha}d_{Rb\beta})S_\xi\epsilon_{\alpha\beta\gamma}+\text{H.c.}, \tag{8}$$

$$G^{(1)}_{\xi,ab}(\bar{u}_{Ra\alpha}e^c_{Rb})S_{\xi\alpha}+\text{H.c.}, \tag{9}$$

$$\tfrac{1}{2}G^{(2)}_{\xi,ab}(\bar{q}^c_{La\alpha j}q_{Lb\beta k})S_{\xi\gamma}\epsilon_{jk}\epsilon_{\alpha\beta\gamma}+\text{H.c.} \tag{10}$$

In the notation used here, χ, η, and ξ label various species of X_V, X'_V, and X_S bosons of each SU(3) $\times$ SU(2) $\times$ U(1) type, a and b label fermions in the sequences (1), α, β, and γ are SU(3) indices, j and k are SU(2) indices, $\epsilon_{\alpha\beta\gamma}$ and ϵ_{jk} are the totally antisymmetric SU(3) and SU(2) tensors, with $\epsilon_{123} \equiv \epsilon_{12} \equiv +1$, and c denotes the Lorentz-invariant complex conjugation of fermion fields. The anticommutativity of fermion fields yields

$$G^{(2)}_{\xi,ab}=G^{(2)}_{\xi,ba}\,. \tag{11}$$

So far, we have made no use of grand unified gauge theories. Such theories impose relations among the various vector and scalar couplings in Eqs. (2)–(10). As an example, let us explore the consequences of the assumption that the grand unified gauge group contains SU(5)[13] as a subgroup (not necessarily less strongly broken than the rest of the group) and that the left-handed fermions in (1) fall into the representations $\underline{\bar{5}}$ and $\underline{10}$ of SU(5).

X_V. The X_V bosons couple to fermion pairs forming the SU(5) representations $\underline{\bar{5}}\times\underline{5}$ in Eqs. (2) and $\underline{\overline{10}}\times\underline{10}$ in Eqs. (3) and (4). Thus these bosons must belong to the SU(5) representations $\underline{24}$ or $\underline{75}$. If they all belong to the $\underline{24}$ representations, then the $\underline{10}\times\underline{\overline{10}}$ couplings are related by

$$h_{\chi,ab}=j_{\chi,ab}. \tag{12}$$

Further, if there is only one species of X_V bosons, which forms part of the multiplet of SU(5) gauge bosons [as is the case in grand unified theories[13] based on SU(5) and SO(10)], then the couplings are further constrained by

$$g_{ab}=-\tfrac{1}{2}h_{ab}=-\tfrac{1}{2}j_{ab}=g_0\delta_{ab}. \tag{13}$$

X'_V. The X'_V bosons couple to fermion pairs forming the SU(5) representations $\underline{5}\times\underline{10}$, so they must belong to the SU(5) representations $\underline{\overline{10}}$ or $\underline{40}$. If they all belong to the $\underline{\overline{10}}$ representation, then their couplings are constrained by

$$g'_{\eta,ab}=-h'_{\eta,ab}. \tag{14}$$

Further, if the grand unified gauge group contains SO(10) (Ref. 13) as a subgroup, and if there is only one species of X'_V bosons, which forms part of the multiplet of SO(10) gauge bosons, then

$$g'_{ab}=-h'_{ab}=g_0\delta_{ab}. \tag{15}$$

Of course, in an SU(5) theory, there is no X'_V.

X_S. The X_S bosons couple to fermion pairs forming the SU(5) representations $\underline{5}\times\underline{10}$ in Eqs. (7) and (8), and $\underline{10}\times\underline{10}$ in Eqs. (9) and (10). Hence these bosons must belong to the SU(5) representations $\underline{\bar{5}}$, $\underline{45}$, $\underline{50}$. If they belong solely to the $\underline{5}$ representation, then their couplings are related by

$$F^{(1)}_{\xi,ab}=F^{(2)}_{\xi,ab}\equiv F_{\xi,ab}, \tag{16}$$

$$G^{(1)}_{\xi,ab}=-G^{(2)}_{\xi,ab}\equiv G_{\xi,ab}\,. \tag{17}$$

III. BARYON PRODUCTION IN BOSON DECAY

We want to calculate the mean baryon number produced in the decays of one of the X_V, X'_V, or X_S bosons and the corresponding antibosons. Each of the bosons X_V, X'_V, and X_S has decay modes of the type $X\to QL$ and $X\to\bar{Q}\bar{Q}$, where Q denotes an arbitrary quark and L denotes an arbitrary lepton; the antibosons have decay modes $\bar{X}\to\bar{Q}\bar{L}$ and $\bar{X}\to QQ$. The branching ratios for $X\to QL$, $X\to\bar{Q}\bar{Q}$, $\bar{X}\to\bar{Q}\bar{L}$, and $\bar{X}\to QQ$ will be denoted r, $1-r$, $\bar{r}$, and $1-\bar{r}$, respectively. The mean net baryon number produced in X and $\bar{X}$ decay is then

$$\Delta B=\tfrac{1}{2}[\tfrac{1}{3}r-\tfrac{2}{3}(1-r)-\tfrac{1}{3}\bar{r}+\tfrac{2}{3}(1-\bar{r})]=\tfrac{1}{2}(r-\bar{r}). \tag{18}$$

Hence our task is to calculate the difference in the branching ratios for boson and antiboson decay.

In carrying out this calculation, we are guided by the theorem proved in the Appendix, which shows that $r-\bar{r}$ can receive no contribution from graphs which are of first order in baryon-violating interactions, even if the graphs involve an arbitrary number of baryon-conserving interactions.[19] We therefore calculate the decay amplitudes for $X_V\to QL$, $X'_V\to QL$, and $X_S\to QL$, including both tree graphs and the one-loop graphs in which a X_V, X'_V, or X_S boson is exchanged between the final fermions. The relevant Feynman diagrams are shown in Figs. 1 and 2. A straightforward calculation gives the $X\to QL$ decay amplitudes (in the notation of Sec. II)

$$A(V_{\chi\alpha j}\to l_{La j}+d_{Rb\alpha})=(g_\chi)_{ab}-\sum_\eta (g_\eta' h_\chi h_\eta'^\dagger)_{ab} I_{VV}(m_\eta/m_\chi)-\sum_\xi (F_\xi^{(2)\dagger} h_\chi F_\xi^{(1)})_{ba} I_{VS}(m_\xi/m_\chi), \tag{19}$$

$$A(V_{\chi\alpha j}\to q_{La\alpha j}+e_{Rb})=(j_\chi)_{ab}+\sum_{\chi'} (h_{\chi'}^\dagger h_\chi j_{\chi'})_{ab} I_{VV}(m_{\chi'}/m_\chi)+\sum_\xi (G_\xi^{(2)\dagger} h_\chi^T G_\xi^{(1)})_{ab} I_{VS}(m_\xi/m_\chi), \tag{20}$$

$$A(V'_{\eta\alpha j}\to l_{La j}+u_{Rb\alpha})=(g_\eta')_{ab}-\sum_\chi (g_\chi h_\eta' h_\eta^\dagger)_{ab} I_{VV}(m_\chi/m_\eta)+\sum_\xi (F_\xi^{(2)*} h_\eta' F_\xi^{(1)})_{ba} I_{VS}(m_\xi/m_\eta), \tag{21}$$

$$\begin{aligned}A(S_{\xi\alpha}\to q_{La\alpha j}+l_{Lbk})=\epsilon_{jk}\Big[F_{\xi,ab}^{(1)}&-\sum_\chi (h_\chi^\dagger F_\xi^{(2)} g_\chi^T)_{ab} I_{SV}(m_\chi/m_\xi)+\sum_\eta (h_\eta'^\dagger F_\xi^{(2)T} g_\eta'^T)_{ab} I_{SV}(m_\eta/m_\xi)\\ &-\sum_{\xi'} (G_{\xi'}^{(2)\dagger} G_\xi^{(2)} F_{\xi'}^{(1)})_{ab} I_{SS}(m_{\xi'}/m_\xi)\Big],\end{aligned} \tag{22}$$

$$A(S_{\xi\alpha}\to u_{Ra\alpha}+e_{Rb})=G_{\xi,ab}^{(1)}+2\sum (h_\chi^* G_\xi^{(2)} j_\chi)_{ab} I_{SV}(m_\chi/m_\xi)-\sum_{\xi'} (F_{\xi'}^{(2)*} F_\xi^{(2)T} G_{\xi'}^{(1)})_{ab} I_{SS}(m_{\xi'}/m_\xi). \tag{23}$$

Here I_{VV}, I_{VS}, I_{SV}, and I_{SS} are the Feynman integrals for vector exchange in vector decay, scalar exchange in vector decay, vector exchange in scalar decay, and scalar exchange in scalar decay, respectively. For the corresponding antiparticle processes, we may simply replace all coupling constants with their complex conjugates. By taking the difference of the absolute-value squares of the amplitudes (19)–(23) for particles and antiparticles, summing over fermion labels a and b [and, for (22), j and k as well], and dividing by the corresponding sums that appear in the tree approximation for the total rate, one obtains the difference in branching ratios

$$\begin{aligned}r(V_\chi\to QL)-r(\overline{V}_\chi\to\overline{Q}\overline{L})=&\,4[\mathrm{Tr}(g_\chi^\dagger g_\chi)+2\,\mathrm{Tr}(h_\chi^\dagger h_\chi)+\mathrm{Tr}(j_\chi^\dagger j_\chi)]^{-1}\\ &\times\Big[\sum_\eta \mathrm{Im\,Tr}(g_\chi^\dagger g_\eta' h_\chi h_\eta'^\dagger)\,\mathrm{Im}I_{VV}(m_\eta/m_\chi)+\sum_\xi \mathrm{Im\,Tr}(g_\chi^* F_\xi^{(2)\dagger} h_\chi F_\xi^{(1)})\,\mathrm{Im}I_{VS}(m_\xi/m_\chi)\\ &-\sum_{\chi'} \mathrm{Im\,Tr}(j_\chi^\dagger h_{\chi'}^\dagger h_\chi j_{\chi'})\,\mathrm{Im}I_{VV}(m_{\chi'}/m_\chi)-\sum_\xi \mathrm{Im\,Tr}(j_\chi^\dagger G_\xi^{(2)\dagger} h_\chi^T G_\chi^{(1)})\,\mathrm{Im}I_{VS}(m_\xi/m_\chi)\Big],\end{aligned} \tag{24}$$

$$\begin{aligned}r(V_\eta'\to QL)-r(\overline{V}_\eta'\to\overline{Q}\overline{L})=&\,4[\mathrm{Tr}(g_\eta'^\dagger g_\eta')+2\,\mathrm{Tr}(h_\eta'^\dagger h_\eta')]^{-1}\\ &\times\Big[\sum_\chi \mathrm{Im\,Tr}(g_\eta'^\dagger g_\chi h_\eta' h_\chi^\dagger)\,\mathrm{Im}I_{VV}(m_\chi/m_\eta)-\sum_\xi \mathrm{Im\,Tr}(g_\eta'^* F_\xi^{(2)*} h_\eta' F_\xi^{(1)})\,\mathrm{Im}I_{VS}(m_\xi/m_\eta)\Big],\end{aligned} \tag{25}$$

$$\begin{aligned}r(S_\xi\to QL)-r(\overline{S}_\xi\to\overline{Q}\overline{L})=&\,4[2\,\mathrm{Tr}(F_\xi^{(1)\dagger}F_\xi^{(1)})+2\,\mathrm{Tr}(F_\xi^{(2)\dagger}F_\xi^{(2)})+\mathrm{Tr}(G_\xi^{(1)\dagger}G_\xi^{(1)})+2\,\mathrm{Tr}(G_\xi^{(2)\dagger}G_\xi^{(2)})]^{-1}\\ &\times\Big[2\sum_\chi \mathrm{Im\,Tr}(F_\xi^{(1)\dagger} h_\chi^\dagger F_\xi^{(2)} g_\chi^T)\,\mathrm{Im}I_{SV}(m_\chi/m_\xi)-2\sum_\eta \mathrm{Im\,Tr}(F_\xi^{(1)\dagger} h_\eta' F_\xi^{(2)T} g_\eta'^T)\,\mathrm{Im}I_{SV}(m_\eta/m_\xi)\\ &+2\sum_\xi \mathrm{Im\,Tr}(F_\xi^{(1)\dagger} G_{\xi'}^{(2)\dagger} G_\xi^{(2)} F_{\xi'}^{(1)})\,\mathrm{Im}I_{SS}(m_{\xi'}/m_\xi)\\ &-2\sum_\chi \mathrm{Im\,Tr}(G_\xi^{(1)\dagger} h_\chi^* G_\xi^{(2)} j_\chi)\,\mathrm{Im}I_{SV}(m_\chi/m_\xi)\\ &+\sum_{\xi'} \mathrm{Im\,Tr}(G'^{(1)\dagger} F_{\xi'}^{(2)*} F_\xi^{(2)T} G_{\xi'}^{(1)})\,\mathrm{Im}I_{SS}(m_{\xi'}/m_\xi)\Big].\end{aligned} \tag{26}$$

The imaginary parts of the integrals I_{VV}, I_{VS}, etc., are easily calculated; we give here only the results for scalar and vector exchange in scalar boson decay:

$$\mathrm{Im}I_{SS}(\rho)=-\frac{1}{16\pi}\,[1-\rho^2\ln(1+1/\rho^2)], \tag{27}$$

$$\mathrm{Im}I_{VS}(\rho)=-\frac{1}{8\pi}\ln(1+1/\rho^2), \tag{28}$$

where ρ is the ratio of the masses of the exchanged boson and the decaying boson.

We see that in general the branching-ratio difference $r-\overline{r}$ can receive nonzero contributions from

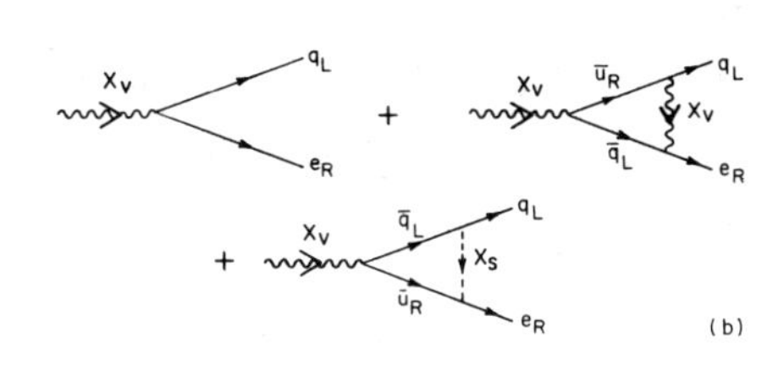

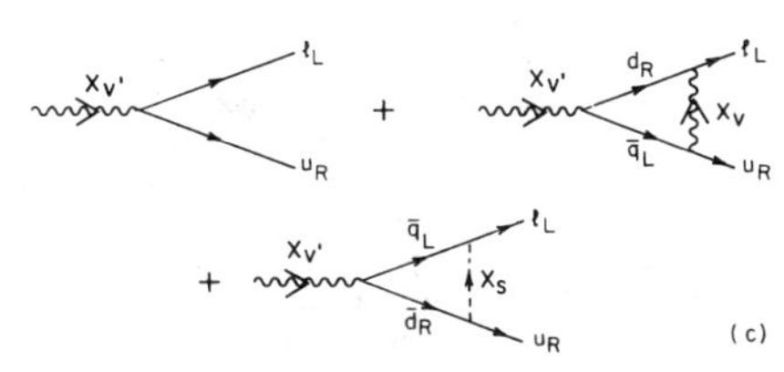

FIG. 1. Feynman diagrams for the decay of X_V and X_V' bosons into quark plus lepton. In the notation used here, l_L, e_R, q_L, d_R, u_R stand for generic quarks and leptons distinguished by their SU(3) ×SU(2) ×U(1) transformation properties, as described in Sec. II.

exchange of an X_S, X_V, or X_V' boson in the decay of any of the X_S, X_V, or X_V' bosons. However, there are a number of special cases in which the branching-ratio difference cancels for an individual boson or for some set of bosons. First, note that the value of $r-\bar{r}$ for any given species of boson receives no contribution from the exchange of the same species of boson. [The traces $\mathrm{Tr}(j_\chi^\dagger h_\chi^\dagger h_\chi j_{\chi'})$, $\mathrm{Tr}F_\xi^{(1)\dagger}G_\xi^{(2)\dagger}G_\xi^{(2)}F_{\xi'}^{(1)})$, and $\mathrm{Tr}(G_\xi^{(1)\dagger}F_{\xi'}^{(2)*}F_\xi^{(2)T}G_{\xi'}^{(1)})$ are real if $\chi=\chi'$ or $\xi=\xi'$, respectively.] Hence there is no baryon production unless there are at least two species of X bosons. More generally, if some set of bosons had equal masses, spins, and lifetimes, then in calculating the cosmological baryon production we would have to add up the branching-ratio differences $r-\bar{r}$ for each of these species; inspection of Eqs. (24)–(26) shows that this sum would vanish because the exchange of an X_1 boson in X_2-boson

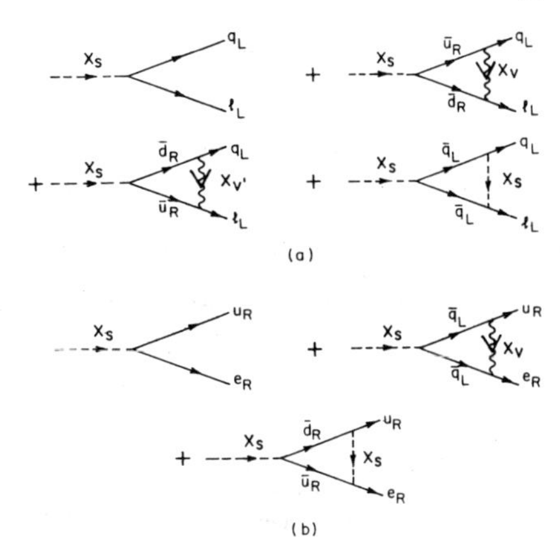

FIG. 2. Feynman diagrams for the decay of X_S bosons into quark plus lepton. Notation same as in Fig. 1.

decay would be canceled by the exchange of an X_2 boson in X_1-boson decay. Hence there is no baryon production unless some of the species of X bosons have different masses, spins, and/or lifetimes. (Of course, there is in any case no reason to expect equal masses and lifetimes for different X-boson species.) Finally, if we suppose that there is a grand unified gauge group which contains SU(5) at least as a subgroup, that the $(3,2,\frac{5}{6})$ X_V bosons belong to the gauge multiplet of SU(5), and that the $(3,2,-\frac{1}{6})$ X_V' bosons are either absent or part of the gauge multiplet of SO(10), then the couplings will be constrained by Eqs. (12)–(17), and almost all of the traces appearing in Eqs. (24)–(26) will be real. The only remaining complex traces in this case are the SS terms in Eq. (26), which give

$$r(S_\xi\to QL)-r(\bar{S}_\xi\to\overline{Q}\overline{L})$$
$$=+4[4\,\mathrm{Tr}(F_\xi^\dagger F_\xi)+3\,\mathrm{Tr}(G_\xi^\dagger G_\xi)]^{-1}\times\sum_{\xi'}\mathrm{Im}\,\mathrm{Tr}(F_\xi^\dagger G_{\xi'}^\dagger G_\xi F_{\xi'})\,\mathrm{Im}I_{SS}(m_{\xi'}/m_\xi). \tag{29}$$

In accordance with our previous remarks, we see that this would vanish if there were just a single species[25] of X_S boson, and would vanish when summed over ξ if there were any number of X_S bosons, all with equal masses and lifetimes. However, (29) indicates that baryon production is to be expected in X_S-boson decay in even the simplest grand unified gauge theories, provided there are at least two species[20] of X_S bosons with different masses or lifetimes. This is reassuring, for as discussed in Sec. I, it is chiefly the decay of the X_S bosons that is expected to yield an appreciable baryon excess.

IV. NUMERICAL ESTIMATES

We will now use the general results of the previous section to make a rough numerical estimate of the baryon abundance that is likely to be produced cosmologically in specific models.

As shown in Ref. 8, the delayed decay of superheavy bosons and antibosons, at temperatures sufficiently far below their mass, will produce a cosmic baryon-entropy ratio

$$kn_B/s = 0.28(N_X/N)\Delta B\,, \tag{30}$$

where N_X is the number of helicity states of all such bosons and antibosons, N is the number of helicity states of all lighter particles (including a factor of $\frac{7}{8}$ for fermions), and ΔB is the mean net baryon production per boson or antiboson decay. The numerical factor in (30) is a ratio of integrals over blackbody distributions, given analytically by $45\zeta(3)/4\pi^4$. The number N_X is unknown, but N is at least 100, so it seems reasonable to take the ratio in the range

$$N_X/N \approx 10^{-2} \text{ to } 10^{-1}. \tag{31}$$

The quantity ΔB is given by Eq. (18) as

$$\Delta B = \tfrac{1}{2}(r - \bar{r})_{\rm av}\,, \tag{32}$$

where r and $\bar{r}$ are the branching ratios for the quark-lepton and antiquark-antilepton modes of the bosons and antibosons, respectively.

In estimating this branching ratio difference, let us first consider the contribution of X_S-boson exchange in X_S-boson decay. From inspection of either Eq. (26) or (29) and Eq. (32), we may infer that the net baryon production per X_S- or $\bar{X}_S$-boson decay in this case is

$$(\Delta B)_{SS} \approx \Gamma^2 \epsilon (\mathrm{Im} I_{SS})_{\rm av}\,, \tag{33}$$

where Γ is a typical value of the Yukawa couplings $F_\xi^{(n)}$ and $G_\xi^{(n)}$ and ϵ is a phase angle characterizing the average strength of CP violation in the interaction of X_S bosons with fermions, or in the X_S-boson propagator.

To estimate Γ, we will assume that the X_S bosons interact about as strongly with any fermion as do the $(1, 2, -\frac{1}{2})$ doublets (ϕ^+, ϕ^0), whose vacuum expectation values give masses to the quarks, leptons, $W^\pm$, and Z^0. That is,

$$\Gamma \approx \bar{m} G_F^{1/2}, \tag{34}$$

where $\bar{m}$ is the rms value of quark and lepton masses and G_F is the Fermi coupling constant. [For instance, if the scalar bosons formed just a single SU(5) quintet, consisting of one $(3, 1, \frac{1}{3})$ X_S boson plus one $(1, 2, -\frac{1}{2})$ ϕ doublet, then we would have Yukawa couplings $F^{(1)} = F^{(2)} = m_E 2^{1/4} G_F^{1/2} = m_D 2^{1/4} G_F^{1/2}$ and $G^{(1)} = -G^{(2)} = m_U 2^{1/4} G_F^{1/2}$, where m_E, m_D, and m_U are the mass matrices of the leptons and quarks of e type, d type, and u type, respectively. Of course, in this particularly simple case, CP could not be violated in X_S-boson interactions.] In estimating $\bar{m}$, we must keep in mind that the values of quark masses at very high energies are likely to be less than their "observed" values at ordinary energies by a factor of order 3 to 4.[15] Taking the b and t quark masses (at ordinary energies) as $m_b = 4.75$ GeV and $m_t = 10$ to 20 GeV, we find

$$\bar{m} \approx 1.1 \text{ to } 2.5 \text{ GeV}. \tag{35}$$

Equation (35) then gives

$$\Gamma^2 \approx 10^{-5} \text{ to } 10^{-4}. \tag{36}$$

In estimating the average value of the integral $\mathrm{Im} I_{SS}$, we must take into account the exchange of each X_S boson in the decay of each other. Equation (26) shows that the exchange of X_{S1} in the decay of X_{S2} makes a contribution to the branching-ratio difference $r - \bar{r}$ which is of opposite sign to the contribution of X_{S2} exchange in the decay of X_{S1}. If one X_S boson is somewhat heavier than all the others, then the dominant contribution to $r - \bar{r}$ comes from the exchange of the lighter bosons in the decay of the heavier one, and $(\mathrm{Im} I_{SS})_{\rm av} \simeq \mathrm{Im} I_{SS}(0) = -\frac{1}{32}\pi$. If several of the heavier X_S bosons are of comparable mass, then some cancellation will occur, but there is no special reason to expect complete cancellation. As a reasonable lower bound on $\mathrm{Im} I_{SS}$ we will take $\frac{1}{4}$ of its value $\mathrm{Im} I_{SS}(1) = 0.19/16\pi$ for equal mass. This gives

$$|\mathrm{Im} I_{SS}| \simeq 10^{-3} \text{ to } 10^{-2}. \tag{37}$$

To estimate ϵ, we must rely on what we know of CP violation at ordinary energies. The violation of CP can be either intrinsic or spontaneous, and in either case, it can operate through gauge boson exchange,[22] Higgs boson exchange,[21] or both. In the case of gauge boson exchange, the CP violation can be traced to phases in the quark mass matrix, which appear in the quark-$W^\pm$ interaction after the quark fields are redefined to make the quark mass matrix real and diagonal. The phases in the quark mass matrix would not contribute to observed violations of CP if there were just four quarks,[22] so since we do not know the strength of the mixing of the b and t quarks with the four lighter quarks, all that we can deduce from the observed strength of the CP violation in K_L^0 decay is that the phases in the quark mass matrix would have to be in the range of 10^{-2} to 1 rad.[23] These phases would have to arise from an intrinsic CP violation in the coupling of scalar fields to quarks or from a CP violation in the scalar field vacuum expectation values, due either to an intrinsic CP violation in the scalar self-interaction or to a spontaneous breakdown of CP invariance. On the other hand, if the CP violation at ordinary energies is due to Higgs boson exchange, then these effects would be naturally suppressed relative to ordinary weak interactions by factors $(m_{\rm quark}/m_{\rm Higgs})^2$, so the phases in the Higgs boson exchange would have to be close to 1 rad.[21] These phases can arise from an intrinsic CP violation in the coupling of scalar fields to quarks or from a CP violation in the scalar propagator, due to either an intrinsic CP violation in the self-coupling of scalar fields or a spontaneous breakdown of CP invariance.

What does this tell us about CP violation in X_S-boson interactions? If the CP violation at ordinary energies is intrinsic, then we expect a similar CP violation in the couplings of X_S bosons to quarks and leptons and in the X_S-boson propagators, so that

$$|\epsilon|\approx 10^{-2} \text{ to } 1. \tag{38}$$

On the other hand, if the CP violation at ordinary energies arises spontaneously in the breakdown of SU(2)×U(1) to U(1), then we would expect this CP violation to disappear at temperatures above about 300 GeV.[24] However, whether the CP violation at ordinary energies is intrinsic or spontaneous, it is possible that there is an entirely different CP violation in X_S interactions, due to a spontaneous breaking of CP in the breakdown of the grand unified gauge group to SU(3)× SU(2)× U(1). We know nothing about the magnitude of such a CP violation, and in lieu of better information we will take (38) as our estimate of ϵ.

If we now use Eqs. (36)–(38) in Eq. (33), we find a mean net baryon number produced in X or $\bar{X}$ decay:

$$|\Delta B|\approx 10^{-10} \text{ to } 10^{-6}. \tag{39}$$

With Eqs. (30) and (31), this gives a baryon-entropy ratio

$$|kn_B/s|\approx 10^{-13} \text{ to } 10^{-7}. \tag{40}$$

Now let us consider the contribution to ΔB of X_V or X_V' exchange in X_S decay. We assume now that the grand unified gauge group is sufficiently complicated so that the 1st, 2nd, and 4th traces in the numerator of Eq. (26) are not all automatically real. From Eqs. (26) and (32), we have

$$\Delta B_{SV}\approx g^2\epsilon'(\mathrm{Im}\,I_{SV})_{\rm av}, \tag{41}$$

where g is a typical value of the vector-boson coupling constants g_X, h_X, j_X, g_η', or h_η' and ϵ' in a phase characterizing the CP violation in the coupling of vector or scalar bosons to quarks and leptons or in the vector-boson propagator. In any kind of grand unified theory, we expect $g^2/4\pi$ to be comparable with (though somewhat larger than[14,15]) the fine-structure constant α, so

$$g^2\approx 10^{-1}. \tag{42}$$

It is difficult to estimate ϵ' because the possible CP violation in the X_V or X_V' couplings or propagators has no direct analog at experimentally accessible energies. However, ϵ' can, like ϵ, receive contributions from CP violation in the coupling of X_S bosons to fermions, so we shall take for $|\epsilon'|$ the same estimate as for $|\epsilon|$

$$|\epsilon|\approx 10^{-2} \text{ to } 1. \tag{43}$$

For the average value of $\mathrm{Im}\,I_{SV}$, we take a rounded estimate

$$|(\mathrm{Im}\,I_{SV})_{\rm av}|\approx 10^{-3} \text{ to } 10^{-1} \tag{44}$$

corresponding to a ratio of vector- to scalar-boson masses in the range 0.3 to 6 in Eq. (28).

The mean baryon excess from X_V or X_V' exchange in X_S-boson decay is now given by (41)–(44) as

$$|(\Delta B)_{SV}|\approx 10^{-6} \text{ to } 10^{-2}. \tag{45}$$

Hence, in theories with a sufficiently complicated group structure, we expect a baryon-entropy ratio given by Eqs. (45) and (30) as

$$|kn_B/s|\approx 10^{-9} \text{ to } 10^{-3}. \tag{46}$$

The baryon production ΔB associated with exchange of a scalar or vector boson in X_V or X_V' decay may be estimated as roughly comparable to the value of ΔB for exchange of the same boson in X_S decay. We will not go into this in detail here, as X_S decay seems more promising than X_V or X_V' decay as a mechanism for cosmological baryon production.

V. CONCLUSIONS

We have seen that the delayed decay of a black-body distribution of X_S bosons at temperatures below their mass may be expected to produce a baryon-entropy ratio at least of order 10^{-13} to 10^{-7}, provided that there are enough species of X_S bosons. In sufficiently complicated theories, baryon number can also be produced in X_V or X_V' exchange processes, yielding a larger baryon-entropy ratio, of order 10^{-9} to 10^{-3}.

These ranges of possible baryon-entropy ratios overlap the values $kn_B/s\approx 10^{-10}$ to 10^{-8} that are allowed by astronomical observations.[2] However, the range of theoretical values is clearly far too broad for us to be able to conclude that X-boson decay really is the source of the observed cosmic abundance of baryons. In the absence of a specific grand unified theory, all that we can conclude now is that X-boson decay is a plausible mechanism for cosmological baryon production.

Let us mention one last point: We have made no attempt here to predict the *sign* of the baryon excess produced cosmologically. Of course, whatever kinds of particles survive the early universe would inevitably be called "matter," not "antimatter." The only real question is whether "matter," as defined by CP-violating cosmological baryon production processes, is the same as "matter," as defined by the observed CP violations in K_L decay.

It is not impossible that this question could some day be answered. For instance, phases in the interaction of scalar fields with quarks can contribute to the CP violation in both X-boson decay and K_L^0 decay. (Recall that these phases produce phases in the quark mass matrix, which produce phases in the interaction of $W^{\pm}$ bosons with quarks of definite mass,[22] which can contribute to CP violation in K_L^0 decay.) If such phases furnish the dominant contribution to CP violation in both K_L^0 and X decay, and if some grand gauge group relates the phases in the couplings of $(1,2,-\frac{1}{2})$ ϕ doublets and $(3,1,\frac{1}{3})$ X_S bosons to quarks, then it might be possible to relate the sign of the CP violation in K_L^0 decay and X_S-boson decay, provided we can learn how to calculate K_L^0-decay amplitudes despite the complication of strong interactions. But this must clearly wait until we have in hand a specific grand unified gauge theory.

Note added in proof. (1) After this paper was completed, we received a report by S. Barr, G. Segré, and H. A. Weldon, which deals in a similar way with the problem of calculating the cosmological baryon production. The topics dealt with in these papers are also discussed by P. Cox and A. Yildiz, Harvard Report No. HUTP-79/A019 (unpublished). (2) There are two additional kinds of boson which can have baryon-nonconserving interactions with pairs of ordinary fermions and/or antifermions. They are an SU(3)-triplet SU(2)-singlet scalar X_S' with charge $-\frac{4}{3}$, which can decay into the channels $d_R e_R$ and $\bar{u}_R \bar{u}_R'$, and an SU(3)-triplet SU(2)-triplet scalar X_S'', which can decay into the channels $q_L l_L$ and $\bar{q}_L \bar{q}_L'$. These cannot contribute to nucleon decay (because Fermi statistics require their two-quark decay channels to consist of quarks from different generations), and they were omitted in Ref. 8, note (1). The existence of these bosons would provide additional mechanisms for cosmological baryon production: interference between the Born approximation and X_S' or X_S'' exchange in X_S decay, and interference between the Born approximation and X_S exchange in X_S' or X_S'' decay. Our numerical estimates in Sec. IV apply also to these contributions.

ACKNOWLEDGMENT

We are grateful to G. Segré and A. Yildiz for helpful comments. This research was supported in part by the National Science Foundation under Grant No. PHY77-22864.

APPENDIX

This appendix will consider baryon-violating decays in the approximation that the decay amplitude is calculated to first order in the baryon-violating interaction H', but to all orders in other interactions. It will be shown that in this approximation, TCP invariance requires that the rate for decay of a particle X into all final states with a given value B of the baryon number equals the rate for the corresponding decay of the antiparticle $\bar{X}$ into all states with baryon number $-B$.[19] As discussed in the text, this theorem indicates that in calculations of cosmological baryon production, we must consider graphs which are at least of second order in the baryon-violating interactions.

To first order in the baryon-violating interaction H', the decay amplitude for a baryon-violating decay of a particle X to some final state f may be written

$$A(X \to f) = (\psi_f^{\text{out}}, H'\psi_X),$$

where ψ_f^{out} and ψ_X are eigenstates of the baryon-conserving part of the Hamiltonian, with outgoing-wave boundary conditions in ψ_f^{out}. (Since ψ_X is a one-particle state, there is no distinction between ψ_X^{out} and ψ_X^{in}.) According to TCP invariance, the amplitude for the corresponding antiparticle decay process is

$$A(\bar{X} \to \bar{f}) \equiv (\psi_{\bar{f}}^{\text{out}}, H'\psi_{\bar{X}}) = (\psi_X, H'\psi_f^{\text{in}})$$

with bars denoting the TCP conjugates of the various states. Inserting a complete set of "out" states gives then

$$A(\bar{X} \to \bar{f}) = \sum_g (\psi_X, H'\psi_g^{\text{out}})(\psi_g^{\text{out}}, \psi_f^{\text{in}})$$

$$= \sum_g A(X \to g)^* S_{gf}^0 ,$$

where S^0 is the S matrix in the absence of the baryon-violating interaction H'. The total rate for $\bar{X}$ decay into all states $\bar{f}$ with a given value $-B$ for the baryon number $B(\bar{f})$ is then

$$\bar{\Gamma}(-B)=\sum_{\bar{f}:B(f)=-B}\rho_{\bar{f}}|A(\bar{X}\to\bar{f})|^2$$

$$=\sum A(X\to g)^*A(X\to h)\sum_{f:B(f)=B}\rho_{\bar{f}}S^0_{gf}S^{0*}_{gh},$$

where $\rho_{\bar{f}}$ is a phase-space factor. Now TCP further tells us that all masses are equal in the corresponding processes $X\to f$ and $\bar{X}\to\bar{f}$, so the phase-space factors are equal:

$$\rho_{\bar{f}}=\rho_f\,.$$

Also, S^0 is unitary in the space of states with a given baryon number, so

$$\sum_{f:B(f)=B}\rho_f S^0_{gf}S^{0*}_{hf}=\begin{cases}\rho_g\delta_{gh}: & B(g)=B,\\ 0\;: & B(g)\neq B,\end{cases}$$

and therefore

$$\bar{\Gamma}(-B)=\sum_{g:B(g)=B}|A(X\to g)|^2\rho_g\,.$$

But this is the total rate $\Gamma(+B)$ for X decay with final states with baryon number $+B$, so $\bar{\Gamma}(-B)$ equals $\Gamma(+B)$ in this approximation, as was to be proved.

---

[1]For a discussion of the evidence regarding the possible cosmological abundance of antimatter, see G. Steigman, Annu. Rev. Astron. Astrophys. 14, 339 (1976).

[2]In this estimate, we take a range of values for the baryonic mass density of the universe which would give a deceleration parameter q_0 between 0.02 (corresponding to a low estimate of the mass density actually observed in galaxies) and 2 (corresponding to the upper bound on nonlinearity in the red-shift–distance relation), with a Hubble constant taken as 50 (km/sec)/Mpc.

[3]M. Yoshimura, Phys. Rev. Lett. 41, 281 (1978); 42, 746(E) (1979); Tohoku University Reports Nos. TU/79/192 and TU/79/193 (unpublished).

[4]S. Dimopoulos and L. Susskind, Phys. Rev. D 18, 4500 (1978); Phys. Lett. 81B, 416 (1979).

[5]A. Yu. Ignatiev, N. V. Krosnikov, V. A. Kuzmin, and A. N Tavkhelidze, Phys. Lett. 76B, 436 (1978).

[6]B. Toussaint, S. B. Treiman, F. Wilczek, and A. Zee, Phys. Rev. D 19, 1036 (1979).

[7]J. Ellis, M. K. Gaillard, and D. V. Nanopoulos, Phys. Lett. 80B, 360 (1979); 82B, 464(E) (1979).

[8]S. Weinberg, Phys. Rev. Lett. 42, 850 (1979).

[9]N. J. Papastamatiou and L. Parker, Phys. Rev. D 19, 2283 (1979).

[10]Implications of cosmological baryon production for the nature of cosmological inhomogeneities are discussed by M. S. Turner and D. N. Schramm, Nature 279, 303 (1979); M. S. Turner, *ibid.* (to be published); C. J. Hogan, *ibid.* (to be published); R. W. Brown and F. W. Stecker, NASA Report No. TM 80291 (unpublished); J. J. Aly, Institute of Astronomy, Cambridge, U. K. report (unpublished); W. H. Press and E. T. Vishniac, Harvard-Smithsonian Center for Astrophysics Reports Nos. 1147 and 1148 (unpublished); J. D. Barrow, Oxford Dept. of Astrophysics report (unpublished).

[11]This is the scenario adopted in Ref. 8. It is also treated in Sec. II of Ref. 6, and in the second article of Ref. 4. One of us (S. W.) owes much to conversations with F. Wilczek on this matter.

[12]A mechanism for establishing an initial thermal equilibrium distribution of scalar bosons at very early times was discussed in Ref. 8, Note (3). In this article we will simply assume that an equilibrium distribution of X bosons (with equal numbers of X and $\bar{X}$) was established before the temperature dropped below the X-boson mass, but we will not rely on any particular picture of how this came about.

[13]The first specific grand unified gauge model was that of J. C. Pati and A. Salam, Phys. Rev. D 8, 1240 (1973); 10, 275 (1974). They noted that baryon nonconservation occurs naturally in their model and similar models because leptons and quarks appear in the same gauge multiplet. In the present work, we adopt a somewhat different view of the strong interactions from that of Pati and Salam: We assume that the only colored gauge bosons with masses below the grand unification mass scale are the eight gluons of quantum chromodyanmics. The simplest grand unified gauge model of this type is the SU(5) model of H. Georgi and S. L. Glashow [Phys. Rev. Lett. 32, 483 (1974)]; other leading models of this type include the SO(10) model of H. Georgi [in *Particles and Fields—1974* proceedings of the 1974 meeting of the Division of Particles and Fields of the American Physical Society, Williamsburg, edited by Carl Carlson (AIP, N.Y., 1975)]; H. Fritzsch and P. Minkowski [Ann. Phys. (N.Y.) 93, 193 (1975)], H. Georgi and D. V. Nanopoulos [Phys. Lett. 82B, 392 (1979); Nucl. Phys. B155, 52 (1979); and the models based on exceptional groups by F. Gürsey, P. Ramond, and P. Sikivie [Phys. Lett. 60B, 177 (1975)], F. Gürsey and P. Sikivie [Phys. Rev. Lett. 36, 775 (1976)], P. Ramond [Nucl. Phys. B110, 214 (1976)], etc.

[14]H. Georgi, H. R. Quinn, and S. Weinberg, Phys. Rev. Lett. 33, 451 (1974).

[15]A. Buras, J. Ellis, M. K. Gaillard, and D. V. Nanopoulos, Nucl. Phys. B135, 66 (1978).

[16]In recent calculations that take into account mass-dependent terms and two-loop corrections in the renormalization-group equations, it was found that the superheavy gauge boson masses are 50–100 times less than the nominal grand unified mass scale calculated in Refs. 14 and 15; see T. J. Goldman and D. A. Ross, Phys. Lett. 84B, 208 (1979); D. Ross, Nucl. Phys. B140, 1 (1978); W. Marciano, Phys. Rev. D 20, 274 (1979).

[17]Reference 8, Note (1). Also see S. Weinberg in *Proceedings of the Einstein Centennial Symposium at Jerusalem*, 1979 (unpublished).

[18]There are other one-loop graphs, in which a virtual X boson is exchanged between the initial X boson and one of the final fermions. These graphs are not con-

sidered here because in the decay of the lighter X bosons they have no absorptive part, and hence cannot contribute to the branching ratio differences calculated in Sec. III.

[19]This is a special case of a theorem mentioned by S. Weinberg, Phys. Rev. 110, 782 (1958).

[20]Note that in an SO(10) theory, the scalar bosons that can couple to fermions would have to belong to the 10, 120, or 126 representations. With only one scalar multiplet, such a theory would yield the unacceptable result that the mass matrices of the charge $+\frac{2}{3}$ and charge $-\frac{1}{3}$ quarks would have to be equal (for the case of 10) or proportional (for the case of 120 or 126), so that in addition to having wrong quark mass relations, such a theory would have no Cabibbo mixing. Grand unified theories with a single scalar multiplet also lead to incorrect quark-lepton mass relations. These results are avoided if there are several 10's (see, e.g., Georgi and Nanopoulos, Ref. 13), in which case there are also several X_S bosons. In addition, if the observed CP violation at ordinary energies is due to exchange of Higgs bosons (as in Ref. 21), then there must be several $(1,2,-\frac{1}{2})$ doublets (ϕ^+,ϕ^0), in which case we also expect several species of X_S boson. We emphasize, however, that the converse need not hold: For instance, if there were just two ϕ doublets and two X_S bosons, then CP would not be violated at ordinary energies by Higgs-boson exchange, but it could be violated in X_S boson decay by X_S-boson exchange. Thus, even if it were found that CP violation at ordinary energies is due to the exchange of $W^{\pm}$ and not Higgs bosons, we could not conclude that the X_S-boson exchange terms in (26) or (29) must vanish.

[21]T. D. Lee, Phys. Rev. D 8, 1226 (1973); Phys. Rep. 9C, 143 (1974); S. Weinberg, Phys. Rev. Lett. 37, 657 (1976).

[22]M. Kobayashi and K. Maskawa, Prog. Theor. Phys. 49, 652 (1973); L. Maiani, Phys. Lett. 68B, 183 (1976); S. Pakvasa and and H. Sugawara, Phys. Rev. D 14, 305 (1976); J. Ellis, M. K. Gaillard and D. V. Nanopoulos, Nucl. Phys. B109, 213 (1976).

[23]J. Ellis, M. K. Gaillard, D. V. Nanopoulos, and S. Rudaz, Nucl. Phys. B131, 285 (1977).

[24]D. A. Kirzhnits and A. D. Linde, Phys. Lett. 42B, 471 (1972); S. Weinberg, Phys. Rev. D 9, 3357 (1974); L. Dolan and R. Jackiw, *ibid*. 9, 3320 (1974); C. Bernard, *ibid*. 9, 3312 (1974); J. Iliopoulos and N Papanicolaou, Nucl. Phys. B111, 209 (1976). It is not, howver, inevitable that spontaneously broken CP invariance must be restored at high temperature; see R. M. Mohapatra and G. Senjanović, Phys. Rev. Lett. 42, 1651 (1979), and references quoted therein.

[25]The calculations of Ref. 7 show that baryon production can occur with just a single species of X_S boson, but in a higher order of perturbation theory.

PHYSICAL REVIEW D VOLUME 20, NUMBER 10 15 NOVEMBER 1979

Magnitude of the cosmological baryon asymmetry

Stephen Barr, Gino Segrè, and H. Arthur Weldon
Department of Physics, University of Pennsylvania, Philadelphia, Pennsylvania 19104
(Received 6 June 1979)

We have examined the magnitude of the cosmological baryon asymmetry arising in several of the standard models of CP violation. Agreement with the experimental baryon to photon number ratio $n_B/n_\gamma \approx 10^{-8}$ is obtained in models where superheavy Higgs mesons decay with complex amplitude into other Higgs mesons. By contrast, in the Kobayashi-Maskawa model $n_B/n_\gamma \approx O(10^{-20})$.

I. INTRODUCTION

It was generally believed that the explanation for the matter-antimatter asymmetry of the universe lay in the simple fact that the universe originated with a nonzero net baryon number, which was conserved for all time. Recent examinations of this question[1-7] have led, however, to a possible new understanding of this asymmetry as having evolved from an originally symmetric state The chief new ingredient has been the introduction of grand unified models of strong, weak, and electromagnetic interactions. In these models quarks and leptons are placed on a similar footing and interactions are present which violate baryon (and lepton) number. As we shall discuss briefly there are at least two scenarios one can envision for the early universe ($T > M_{\text{Planck}}$): Either the net baryon number n_B is originally zero or baryon-number-violating interactions lead to an equilibrium state in which n_B becomes zero. In either case the universe must evolve from an $n_B \approx 0$ state into the present universe in which[8] $n_B \approx 10^{80}$. This is to be compared to the photon number, $n_\gamma \approx 10^{88}$, so that

$$n_B/n_\gamma \approx 10^{-8}. \tag{1.1}$$

There are three key ingredients[2] necessary for the evolution from the $n_B \approx 0$ state to the present asymmetric universe:

(a) microscopic violation of baryon number,
(b) CP (or equivalently T) violation,
(c) departure from thermal equilibrium.

The first occurs, as mentioned earlier, in grand unified models and is mediated by the interaction of superheavy, color-triplet bosons.[9] These, which we generically call X, may be either gauge bosons or Higgs bosons; they couple to both the ql and the $\overline{qq}$ fermion channels (q is a quark and l a lepton) and have characteristic masses of order $10^{15\pm 2}$ GeV.[10] CP violation is model dependent and may be introduced in a variety of ways, as we shall illustrate. The third key ingredient, departure from thermal equilibrium, was shown by Toussaint, Treiman, Wilczek, and Zee[2] to require something other than the scattering of ordinary (i.e., effectively massless) fermions since such processes have no mass threshold. Toussaint et al.[2] and Weinberg[6] have pointed out that the needed departure from equilibrium could be due to the decay of X particles as the temperature T falls below their mass M_X.

A possible scenario is the following[11]: (1) Starting at a temperature T of the order of the Planck mass M_P, X-mediated collisions have a rate Γ_c which is faster than H, the expansion rate of the universe. Since these processes include baryon-number-violating interactions, any baryon asymmetry originally present in the big bang will be effectively erased. During this period, characterized by $M_P > T > M_X$, the X decay rate and the X production rate by inverse decay are small compared to H. (2) As the temperature falls both H and Γ_c decrease while the X decay rate Γ_x increases. When $\Gamma_x \gtrsim H$ we reach a regime in which X decay is important. For M_X sufficiently heavy, e.g., $M_X \sim 10^{15}$ GeV, the inverse decay will always be smaller than the expansion rate and any baryon asymmetry due to X decays will persist throughout the later evolution.[12]

To relate the asymmetry to microscopic parameters we suppose, for example, that the X baryons decay into two channels with branching ratios r, $1-r$ and baryon number B_1, B_2; $\overline{X}$ bosons decay with different branching ratios $\overline{r}, 1-\overline{r}$ into channels with baryon number $-B_1$, $-B_2$. The average baryon number produced by decays of X and $\overline{X}$ is

$$\Delta B = \tfrac{1}{2}(r - \overline{r})(B_1 - B_2)\,. \tag{1.2}$$

The observed baryon asymmetry is determined by ΔB and the density of X's at the time of decay:

$$\frac{n_B}{n_\gamma} = \left(\frac{N_X}{N}\right)\Delta B\,, \tag{1.3}$$

where N is the total number of helicity states and N_X is the number of type X.

With the above setting it is the task of particle physics to explain the magnitude of n_B/n_γ. For definiteness we shall consider the standard SU(5) grand unified model of Georgi and Glashow.[13] Typically $N_X/N \approx 10^{-2}$ in this model so that we need $\Delta B \approx 10^{-6}$ in order to obtain a value for n_B/n_γ in agreement with (1.1). Any tree approximation to the decay amplitude will give $r = \bar{r}$ because of CPT. We therefore look for an interference between a tree amplitude g_0 and an amplitude with one (or more) quantum loops. If we denote the one-loop amplitude by the product of a Feynman integral $I(s+i\epsilon)$ and a coupling strength g_1 then

$$\Delta B \propto \int d\omega |g_0 + g_1 I(s+i\epsilon)|^2 - \int d\omega |g_0^* + g_1^* I(s+i\epsilon)|^2$$

$$= 4 \int d\omega \, \mathrm{Im}(g_1 g_0^*) \, \mathrm{Im}[I(s+i\epsilon)] \,. \qquad (1.4)$$

Thus we require both complex couplings and an s-channel discontinuity to obtain $\Delta B \neq 0$.

The necessary CP violation can be introduced into X decay in a number of ways. We will show that the values obtained for n_B/n_γ are usually too small by several orders of magnitude because the needed complex coupling constants are Yukawa couplings, whose magnitude is typically $O(G_F{}^{1/2}M_q) \lesssim 10^{-2}$. One model, originally proposed by Weinberg,[14] in which the CP violation is introduced via Higgs quartic self-couplings [which can be $O(1)$], has the potential to give a larger value of n_B/n_γ. The relevant processes are the decay of superheavy Higgs bosons into lighter Higgs bosons, which then decay to fermions. We show that one can obtain through this mechanism the experimental value of $n_B/n_\gamma \approx 10^{-8}$. In fact, a value as large as 10^{-4} is possible. The chain of decays we propose, in which the CP violation occurs directly in the Higgs self-couplings, may occur in a wide variety of models.

II. THE KOBAYASHI-MASKAWA (KM) MODEL

If CP is not imposed on the Lagrangian then the Yukawa couplings will be complex. Kobayashi and Maskawa[15] observed that in a model with six or more quarks this will generally lead to CP violation in the $K^0-\bar{K}^0$ mass matrix. In SU(5) we have

$$\mathcal{L}_{\mathrm{Yuk}} = (f_1^\dagger)_{mn} \bar{\psi}_{m,\alpha\beta} \chi_n^\alpha \phi^\beta + (f_2)_{mn} \epsilon_{\alpha\beta\mu\nu\lambda} \psi_m^{\alpha\beta} C \psi_n^{\mu\nu} \varphi^\lambda + \mathrm{H.c.} \qquad (2.1)$$

where m, n are summed over the generations of fermion and C is the charge-conjugation matrix. Each generation consists of a right-handed, five-dimensional representation χ^α and a left-handed, ten-dimensional representation $\psi^{\alpha\beta}$. We divide the SU(5) indices $\alpha, \beta = 1, 2, \ldots, 5$ into flavor indices $i, j, k = 1, 2$ and color indices $a, b, c = 3, 4, 5$.

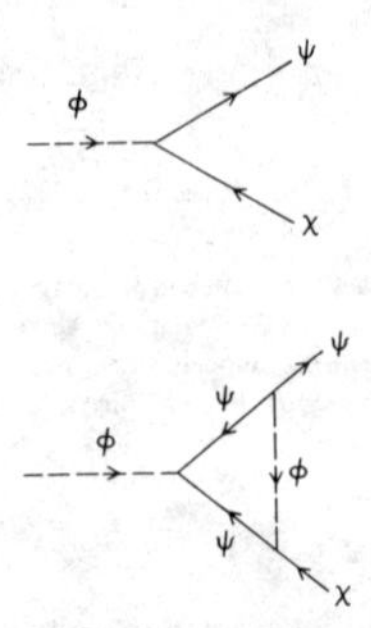

FIG. 1. A tree and one-loop diagram whose interference fails to give a ΔB in the KM model.

The color-triplet, Higgs boson φ^a is superheavy (~10^{15} GeV) and decays into both ql and $\overline{qq}$. If r is the branching ratio into ql and $1-r$ the branching ratio into $\overline{qq}$, one might expect that the tree and one-loop diagrams in Fig. 1 would interfere to give $r \neq \bar{r}$. However, the tree diagram is proportional to $f_1^\dagger$ and the one-loop diagram to $f_2^\dagger f_2 f_1^\dagger$ so that summing over fermions gives

$$\Delta B \propto \mathrm{Im}\,\mathrm{Tr}(f_2^\dagger f_2 f_1^\dagger f_1) = 0 \,.$$

To obtain a complex interference with the tree diagram one must actually go to a three-loop diagram[5] such as Fig. 2, which gives

$$\Delta B(\varphi \text{ decay}) \sim \frac{\mathrm{Im}\,\mathrm{Tr}(f_2^\dagger f_1 f_1^\dagger f_2 f_2^\dagger f_2 f_1^\dagger f_1)}{16\pi(8\pi^2)^2[\mathrm{Tr}(f_1^\dagger f_1) + \mathrm{Tr}(f_2^\dagger f_2)]} \,. \qquad (2.2)$$

Even if we optimistically put all the $f_i \approx 10^{-2}$ and assume the imaginary part is maximal, this gives a hopelessly small $\Delta B \sim 10^{-18}$.

The color-triplet vector bosons W_i^a are also superheavy and decay into ql and $\overline{qq}$. However, because the gauge couplings are real and diagonal in the original Lagrangian basis, it still requires eight Yukawa couplings to have an asymmetry

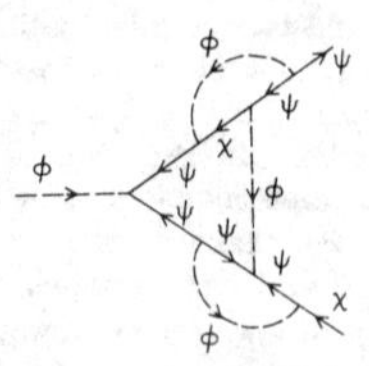

FIG. 2. A typical contribution to ΔB from ϕ decay in the KM model.

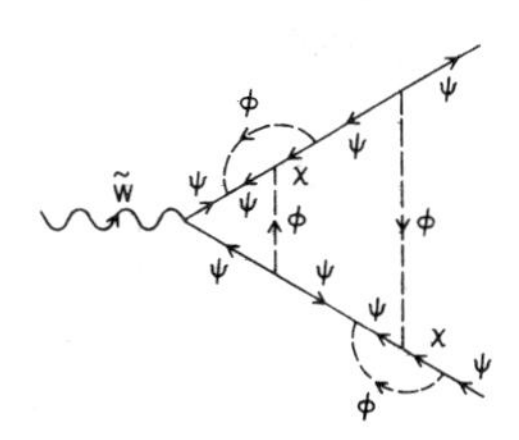

FIG. 3. A typical contribution to ΔB from $\tilde{W}$ decay in the KM model.

One such diagram is shown in Fig. 3. It involves precisely the same trace as (2.2) but has one more loop so that

$$\Delta B(\tilde{W}\text{ decay}) \sim \frac{g^2\,\mathrm{Im}\mathrm{Tr}(f_2^\dagger f_1 f_1^\dagger f_2 f_2^\dagger f_2 f_1^\dagger f_1)}{16\pi(8\pi^2)^3 g^2}\,.$$

This seems to be a general result, valid in all models of CP violation, that the vector-boson decay gives a smaller asymmetry ($f^2/8\pi^2$) than the scalar decay. In the later models we will therefore not discuss the $\tilde{W}$ interactions at all.

III. THE WEINBERG (THREE-HIGGS) MODEL

Weinberg pointed out in a four-quark model that the quartic self-couplings of Higgs mesons could violate CP invariance if there were three or more Higgs multiplets.[14] To ensure the natural conservation of quark flavors in neutral currents, only two of the Higgs can couple to fermions:

$$\mathcal{L}_{\text{Yuk}} = (f_1^\dagger)_{mn}\bar{\psi}_{m,\alpha\beta}\chi_n^\alpha\varphi_1^\beta + (f_2)_{mn}\epsilon_{\alpha\beta\mu\nu\lambda}\psi_m^{\alpha\beta}C\psi_n^{\mu\nu}\varphi_2^\lambda + \text{H.c.} \qquad (3.1)$$

The superheavy color triplets φ_1^a and φ_2^a have different fermionic decays. In particular, φ_1^a has four decay modes ($u^a e^-$, $d^a\nu$, $\bar{u}_b\bar{d}_c$, $\bar{u}_c\bar{d}_b$ with a, b, c cyclic) but φ_2^a has only three modes ($u_a e^-$, $\bar{u}_b\bar{d}_c$, $\bar{u}_c\bar{d}_b$) because ν appears only in the χ multiplet. The average baryon numbers produced by φ_1^a and φ_2^a, respectively, are

$$B_1 = \tfrac{1}{4}\left(\tfrac{1}{3}+\tfrac{1}{3}-\tfrac{2}{3}-\tfrac{2}{3}\right) = -\tfrac{1}{6}\,,$$
$$B_2 = \tfrac{1}{3}\left(\tfrac{1}{3}-\tfrac{2}{3}-\tfrac{2}{3}\right) = -\tfrac{1}{3}\,. \qquad (3.2)$$

As yet we have not specified the source of CP violation. If this were just an extended Kobayashi-Maskawa scheme then the baryon asymmetry would be much smaller than (2.2) because $\varphi_1 \neq \varphi_2$ prevents the existence of even three-loop diagrams such as Fig. 2. However, in the Weinberg scheme there is a third Higgs field φ_3, that does not couple to fermions, with interactions

$$V(\varphi) = M_r^2(\varphi_r^\dagger\varphi_r) + a_{rs}(\varphi_r^\dagger\varphi_r)(\varphi_s^\dagger\varphi_s) + b_{rs}(\varphi_r^\dagger\varphi_s)(\varphi_s^\dagger\varphi_r) + c_{rs}(\varphi_r^\dagger\varphi_s)(\varphi_r^\dagger\varphi_s)\,, \qquad (3.3)$$

where r, s are summed from 1 to 3. Hermiticity requires that a_{rs} and b_{rs} be real and symmetric but only that c_{rs} be Hermitian. One can always define away the phase in c_{12}, for example, but the product $c_{12}c_{23}c_{31}$ is generally complex after all redefinitions. The mass terms for the color triplets φ_1^a, φ_2^a, φ_3^a come from the superheavy vacuum expectation value ($\sim 10^{15}$ GeV) of the $\underline{24}$ and are automatically diagonal because of the discrete symmetries necessary to ensure that $\varphi_1, \varphi_2, \varphi_3$ have distinct Yukawa couplings. The vacuum expectation values of the φ are so small ($\sim$100 GeV) that we may safely ignore their contribution to masses of the color triplets.

A particularly simple scenario is obtained if we choose the superheavy masses to satisfy $M_3 > M_2 > M_1$.[16] Then φ_3^a has two decay channels: φ_1^a plus two massless Higgs mesons or φ_2^a plus two massless Higgs mesons; and these channels go to different baryon numbers B_1 and B_2. The decays of φ_3^a will violate CP invariance because of the complex c_{rs} in (3.3). The Born amplitude for $\varphi_3^a \to \varphi_1^a + \varphi_1^j + \bar{\varphi}_3^j$ is c_{13} and will interfere with the one-loop diagram in Fig. 4(a) to give

$$r_{3\to 1} - \bar{r}_{3\to 1} \propto 4\int d\omega\,\mathrm{Im}(c_{12}c_{23}c_{31})\,\mathrm{Im}[I_2(s+i\epsilon)]\,, \qquad (3.4)$$

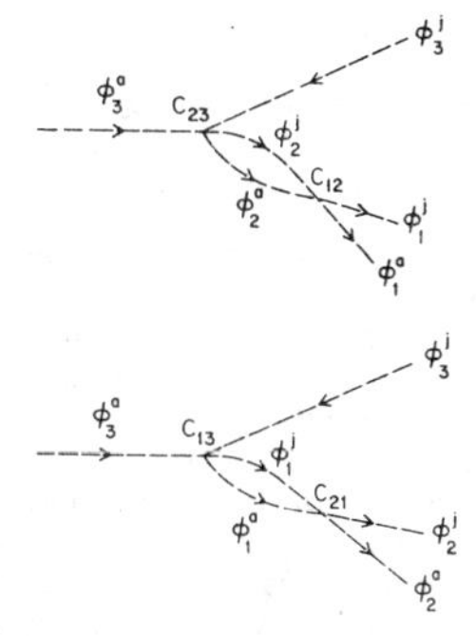

FIG. 4. Contributions to ΔB from ϕ_3 decay in the Weinberg (three-Higgs) model with $M_3 > M_2 > M_1$.

where I is the one-loop Feynman integral. Similarly, the Born amplitude for $\varphi_3^a \to \varphi_2^a + \varphi_2^j + \bar{\varphi}_3^j$ is c_{23} and will interfere with the one-loop diagram in Fig. 4(b) to give

$$r_{3\to 2} - \bar{r}_{3\to 2} \propto -4 \int d\omega \,\mathrm{Im}(c_{12}c_{23}c_{31})\,\mathrm{Im}[I_1(s+i\epsilon)]. \tag{3.5}$$

Explicit calculation verifies that

$$r_{3\to 1} + r_{3\to 2} = \bar{r}_{3\to 1} + \bar{r}_{3\to 2}$$

as guaranteed by the CPT theorem.

To calculate ΔB we evaluate the integrals over three-body phase space in (3.4) and obtain

$$\Gamma_{3\to 1} - \bar{\Gamma}_{3\to 1} = \frac{\mathrm{Im}(c_{12}c_{23}c_{31})}{64(2\pi)^4 (M_3)^3}\left[(M_3)^4 - (M_2)^4 + 4M_1^2M_3^2 - 4M_1^2M_2^2 - 2(M_1^2M_2^2 + M_1^2M_3^2 + M_2^2M_3^2)\ln\left(\frac{M_3}{M_2}\right)^2\right]. \tag{3.6}$$

For simplicity we shall neglect M_1 and M_2 with respect to M_3. In this approximation the total decay rate is

$$\Gamma_3 \approx \left[|c_{13}|^2 + |c_{23}|^2 + |b_{13}|^2 + |b_{23}|^2 + O\left(b_{ij}g^2\left(\frac{M_3}{\bar{M}}\right)^2\right) + O\left(g^4\left(\frac{M_3}{\bar{M}}\right)^4\right)\right]\frac{M_3}{32(2\pi)^3},$$

and using $\frac{1}{2}(B_1 - B_2) = \frac{1}{12}$ we obtain

$$\Delta B \approx \left(\frac{1}{48\pi}\right)\frac{\mathrm{Im}(c_{12}c_{23}c_{31})}{|c_{13}|^2 + |c_{23}|^2 + |b_{13}|^2 + |b_{23}|^2}.$$

This can give quite a large baryon asymmetry. The upper bound is attained when $b \ll c$ so that $\Delta B \leq c/96\pi$ or

$$\frac{n_B}{n_\gamma} < 10^{-4}c.$$

Quartic couplings such as c_{rs} are presumably less than 1 so as not to invalidate perturbation theory,[17] but this still gives a comfortable upper bound of 10^{-4} for the baryon asymmetry.

To summarize, the asymmetry arises from the fact that though the number of φ_3 and $\bar{\varphi}_3$'s are equal, their decay leads to an excess of φ_1 over $\bar{\varphi}_1$ and a matching excess of $\bar{\varphi}_2$ over φ_2 because of CP violation. Since both are out of equilibrium the imbalance persists when φ_1 and φ_2 decay to fermionic states with different baryon number.

IV. CONCLUSIONS

We have shown, within the framework of a given scenario for the evolution of the universe, that a satisfactory value for n_B/n_γ can be obtained and that this ratio is quite sensitive to various ingredients of gauge theories, namely how CP is introduced and how the Higgs mesons couple. In appendices A and B we discuss some other modes of CP violation.

It would be impressive if one could not only obtain the magnitude of n_B/n_γ, but predict the sign of the asymmetry to understand why the universe is made of matter instead of antimatter. We have tried, but failed to relate the sign of the asymmetry to the parameters of CP violation, as determined by K-meson decay. The problem is that the arbitrary sign of the Higgs-meson couplings prevents a direct comparison.

ACKNOWLEDGMENTS

We would like to thank Professor A. Zee for encouragement and helpful discussions. The work of S. M. B. and G. S. was supported in part by the U. S. Department of Energy under Contract No. EY-76-C-02-3071, that of H. A. W. by the National Science Foundation.

APPENDIX A: RELAXING THE CONSTRAINT OF FLAVOR CONSERVATION

In the previous examples we have allowed one Higgs meson to couple to $\bar{\psi}\chi$ and only one Higgs meson (either the same or different) to couple to $\psi C\psi$ in order that all neutral Higgs-meson exchanges would automatically conserve strangeness, charm, and other quark flavors.[18] If, however, we relax this condition it may still be possible that Yukawa couplings such as

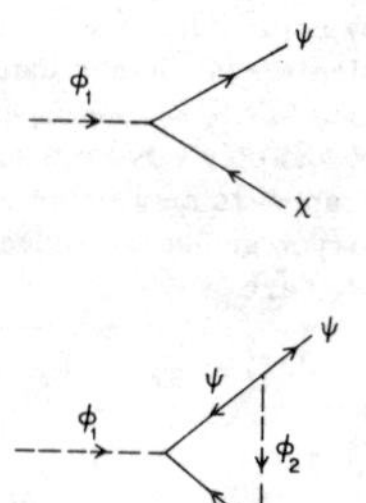

FIG. 5. A simple tree and one-loop diagram whose interference does give $\Delta B \neq 0$ when flavor conservation is abandoned.

$$\mathcal{L}_{\text{Yuk}}=\bar{\psi}_{m,\alpha\beta}\chi_n^{\alpha}[(f_1^{\dagger})_{mn}\varphi_1^{\beta}+(g_2^{\dagger})_{mn}\varphi_2^{\beta}]$$
$$+\epsilon_{\alpha\beta\mu\nu\lambda}\psi_m^{\alpha\beta}C\psi_n^{\mu\nu}[(g_1)_{mn}\varphi_1^{\lambda}+(f_2)_{mn}\varphi_2^{\lambda}]+\text{H.c.}$$

only violate flavor conservation in an acceptably small way. [Note that (3.1) corresponds to $g_1=g_2=0$.]

In such a model it may be possible to obtain a reasonable baryon asymmetry. For example, the tree diagram in Fig. 5 is proportional to $f_1^{\dagger}$ and the one-loop diagram to $f_2^{\dagger}g_1g_2^{\dagger}$ so that summing over fermions gives

$$\Delta B(\varphi_1 \text{ decay}) \approx \frac{\text{Im Tr}(f_2^{\dagger}g_1g_2^{\dagger}f_1)}{16\pi[\text{Tr}(f_1^{\dagger}f_1)+\text{Tr}(g_1^{\dagger}g_1)]} .$$

For large Yukawa couplings $f \approx g \approx 10^{-2}$ this might give an asymmetry as large as required by (1.1). (Of course, larger Yukawa couplings could result from hitherto unobserved heavy quarks.)

APPENDIX B: SPONTANEOUS *CP* VIOLATION AT LARGE *T*

Up until now all *CP* violation has come from explicit violation in the Lagrangian via complex couplings. It is also possible for *CP* to be an invariance of the Lagrangian that is broken spontaneously by the vacuum expectation values $\langle\varphi\rangle \approx 100$ GeV. Usually at $kT \gg 100$ GeV there is no spontaneous symmetry breaking (i.e., $\langle\varphi\rangle = 0$), however, it is possible to arrange the signs of certain quartic couplings so that the spontaneous symmetry breaking, and also the *CP* violation, persists at arbitrarily high temperature with $\langle\varphi\rangle \propto kT$.[19] This scheme requires a minimum of three Higgs fields: Two acquire vacuum expectation values with different phases and the third has some quartic couplings negative so that the vacuum expectation values of the radiatively corrected potential will grow with kT.

Finally the *CP* noninvariance arises because the physical (mass eigenstate) Higgs fields are complex linear combinations of the original fields. Consequently the physical Higgs particles have complex Yukawa couplings to the fermions and the situation effectively reduces to that of Appendix A. The baryon asymmetry is also similar,

$$\Delta B \approx O\left(\frac{f^2}{16\pi}\right)\left(R\frac{kT}{M_X}\right)^4,$$

except for the suppression factor due to the Higgs mixing. For such decays kT is only slightly smaller than M_X but R is a model-dependent ratio of Higgs couplings.

[1]M. Yoshimura, Phys. Rev. Lett. 41, 281 (1978); 42, 746(E) (1979).

[2]D. Toussaint, S. B. Treiman, F. Wilczek, and A. Zee, Phys. Rev. D 19, 1036 (1979); D. Toussaint and F. Wilczek, Phys. Lett. 81B, 238 (1979); S. M. Barr, Phys. Rev. D 19, 3803 (1979).

[3]S. Dimopoulos and L. Susskind, Phys. Rev. D 18, 4500 (1978); Phys. Lett. 81B, 416 (1979).

[4]A. Yu. Ignatiev, N. V. Krosnikov, V. A. Kuzmin, and A. N. Tavkhelidze, Phys. Lett. 76B, 436 (1978).

[5]J. Ellis, M. K. Gaillard, and D. V. Nanopoulos, Phys. Lett. 80B, 360 (1979).

[6]S. Weinberg, Phys. Rev. Lett. 42, 850 (1979).

[7]M. Yoshimura, Tohoku University Report No. TU/79/192 (unpublished) and Report No. TU/79/193 (unpublished).

[8]G. Steigman, Annu. Rev. Astron. Astrophys. 14, 339 (1976).

[9]Grand unified models do not have to violate baryon number. See M. Gell-Mann, P. Ramond, and R. Slansky, Rev. Mod. Phys. 50, 721 (1978); P. Langacker, G. Segrè, and H. A. Weldon, Phys. Rev. D 18, 552 (1978).

[10]A. J. Buras, J. Ellis, M. K. Gaillard, and D. V. Nanopoulous, Nucl. Phys. B135, 66 (1978).

[11]This general picture has been developed in Refs. 2 and 6. Yoshimura (Ref. 7) has refined both the high-temperature phase, by calculating the effect of fermion-fermion scattering, and the low-temperature phase, by strengthening the limits on M_X for an asymmetry to develop. Formula (1.2) is from Ref. 6.

[12]If M_X is too small both the direct and the inverse decay rates will exceed the expansion rate during some temperature interval. During this period X and $\bar{X}$ will remain in thermal equilibrium. When they finally depart from equilibrium there will be too few of them left to produce a significant asymmetry.

[13]H. Georgi and S. L. Glashow, Phys. Rev. Lett. 32, 438 (1974).

[14]S. Weinberg, Phys. Rev. Lett. 37, 657 (1976).

[15]M. Kobayashi and K. Maskawa, Prog. Theor. Phys. 49, 652 (1973).

[16]The mass splitting need not be large. If, for example, $M_3=2M_2=4M_1$, the particles are essentially degenerate as far as evolution is concerned so that the interval $M_3 > kT > M_1$ is negligible.

[17]C. E. Vayonakis, Lett. Nuovo Cimento 17, 383 (1976); M. Veltman, Acta. Phys. Pol. B8, 475 (1977); B. W. Lee, C. Quigg, and H. Thacker, Phys. Rev. Lett. 38, 883 (1977).

[18]S. L. Glashow and S. Weinberg, Phys. Rev. D 15, 1958 (1977). For a discussion of flavor conservation when more Higgs mesons are coupled see R. Gatto, G. Morchio, and G. Strocchi, Scuola Normale Superiora (Pisa) report (unpublished); G. Segrè and H. A. Weldon, Ann. Phys. (N.Y.) (to be published).

[19]S. Weinberg, Phys. Rev. D 9, 3357 (1974); R. N. Mohapatra and G. Senjanović, University of Maryland report (unpublished).

Volume 91B, number 2 PHYSICS LETTERS 7 April 1980

THE DEVELOPMENT OF BARYON ASYMMETRY IN THE EARLY UNIVERSE

Edward W. KOLB [1]
W.K. Kellogg Radiation Laboratory, California Institute of Technology, Pasadena, CA 91125, USA

and

Stephen WOLFRAM [2]
Theoretical Physics Laboratory, California Institute of Technology, Pasadena, CA 91125, USA

Received 19 November 1979

The development of an excess of baryons over antibaryons due to *CP* and baryon number violating reactions during the very early stages of the big bang is calculated in simple models using the Boltzmann equation. We show that it is necessary to solve the coupled Boltzmann equations in order to determine the final baryon number in any specific model.

There are observational and theoretical indications that the local preponderance of baryons over antibaryons extends throughout the universe (at least since the time when the temperature $T \approx 100$ MeV) with an average ratio of baryon to photon densities [1] $n_B/n_\gamma \equiv Y_B \approx 10^{-9}$. If baryon number ($B$) were absolutely conserved in all processes, this small baryon excess must have been present since the beginning of the universe. However, many grand unified gauge models [2] require superheavy particles (typically with masses $m_X \approx 10^{15}$ GeV $\equiv 1\ \Pi$ eV) which mediate baryon- and lepton-number (L) violating interactions. Any direct evidence for these must presumably come from an observation of proton decay. In the standard hot big bang model [1], the temperature T (of light particle species) in the early universe fell with time t according to (taking units such that $\hbar = c = k = 1$) $T \sim (m_P/2t)^{1/2}$, where

$$m_P = (45/8\pi^3)^{1/2} m_{\mathcal{P}}/\xi(T)^{1/2} \approx 5 \times 10^3/\xi^{1/2}\ \text{MeV},$$

and $m_{\mathcal{P}} = G^{-1/2} \approx 10^{19}$ GeV is the Planck mass, while ξ gives the effective number of particle species in equilibrium ($\xi = \frac{1}{2}$ ($\frac{7}{16}$) for each ultrarelativistic boson (nondegenerate fermion) spin state). At temperatures $T \gtrsim m_X$, B-violating interactions should have been important, and they should probably have destroyed or at least much diminished any initial baryon excess. (This occurs even when, for example, $B - L$ is absolutely conserved, since then an initial baryon excess would presumably be accompanied by a lepton excess, so as to maintain the accurate charge neutrality of the universe.) It is interesting (and in some models necessary) to postulate that B-violating interactions in the very early universe could give rise to a calculable baryon excess even from an initially symmetrical state. For this to be possible, the rates for reactions producing baryons and antibaryons must differ, and hence the interactions responsible must violate C and CP invariance. We describe here a simple but general method for calculating B generation in any specific model. We clarify and extend previous estimates [3]. A detailed account of our work is given in ref. [4].

Let $M(i \to j)$ be the amplitude for transitions from the state i to j, and let $\bar{i}$ be the CP conjugate of i. Then CPT invariance demands $M(i \to j) = M(\bar{j} \to \bar{i})$,

[1] Work supported in part by the National Science Foundation [PHY76-83685].
[2] Work supported in part by the Department of Energy [DE-AC-03-79ER0068] and a Feynman fellowship.

Volume 91B, number 2 PHYSICS LETTERS 7 April 1980

while C, CP invariance would require $M(i \to j) = M(\bar{i} \to \bar{j}) = M(j \to i)$. Unitarity (transitions to and from i must occur with total probability 1) demands [‡1] (e.g., ref. [5]) $\Sigma_j |M(i \to j)|^2 = \Sigma_j |M(j \to i)|^2$; combining this with the constraint of CPT invariance yields (the sum over j includes all states and their antistates)

$$\sum_j |M(i \to j)|^2 = \sum_j |M(\bar{i} \to j)|^2 = \sum_j |M(j \to \bar{i})|^2$$

$$= \sum_j |M(j \to i)|^2 . \qquad (1)$$

In thermal equilibrium (and in the absence of chemical potentials representing nonzero conserved quantum numbers) all states j of a system with a given energy are equally populated. Then the last equality in eq. (1) shows that transitions from these states (interactions) must produce i and $\bar{i}$ in equal numbers; thus no excess of particles over antiparticles may develop in a system in thermal equilibrium, even if CP is violated. In addition, the first equality in eq. (1) shows that the total cross sections for destroying particles and antiparticles must be equal. Since in thermal equilibrium no excess of i over $\bar{i}$ may develop, this implies that any initial excess must be destroyed.

The phase space distribution $f_i(p)$ (number per unit cell $d^3p\, d^3x$ [‡2] for a species i develops with time (on average) according to a Boltzmann transport equation. A closed system with no external influences obeys Boltzmann's H-theorem [which holds regardless of T (i.e., CP) invariance (e.g., ref. [4])], so that from any initial state the $f_i(p)$ evolve (on average) to their equilibrium forms for which $f_{\bar{i}}(p) = f_i(p)$, and no baryon excess may survive.

However, in an expanding universe, extra terms must be added to the Boltzmann equations, and if some participating particles are massive [‡3], a baryon excess may be generated; the relaxation time necessary to destroy the excess often increases faster than the age of the universe [‡4].

Eq. (1) requires that the total rates for processes with particle and antiparticle initial states be equal. CP violation allows the rates for specific conjugate reactions to differ; unitarity nevertheless requires $[T = i(1 - S), SS^\dagger = S^\dagger S = 1]$ [‡5]:

$$|M(i \to j)|^2 - |M(\bar{i} \to \bar{j})|^2 = |T_{ij}|^2 - |T_{ji}|^2$$

$$= 2\,\mathrm{Im}\left\{\left[\sum_n T_{in}(T_{jn})^\dagger\right] T_{ji}^*\right\} - \left|\sum_n T_{in}(T_{jn})^\dagger\right|^2 . \qquad (2)$$

Hence the fractional difference between conjugate rates must be at least $O(\alpha)$ where α is some coupling constant [‡6]. Moreover, the loop diagrams giving CP violation must allow physical intermediate states n. (These loop corrections must be usually also B violating to give a difference in rates when summed over all final states $\overset{(-)}{j}$ with a given $(-)B$ [4,6].)

Let $\overset{(-)}{b}$ be an "(anti)baryon" with $B = (\overset{+}{-})\frac{1}{2}$. For simplicity we assume here that all particles (including photons) obey Maxwell–Boltzmann statistics and have only one spin state. In our first (very simple) model, we consider C, CP, B violating $2 \leftrightarrow 2$ reactions involving $\overset{(-)}{b}$ and a heavy neutral particle ϕ; we take their rates to be (this parametrization ensures unitarity and CPT invariance)

$$|M(bb \to \bar{b}\bar{b})|^2 = \tfrac{1}{2}(1 + \zeta)|M_0|^2 ,$$

$$|M(bb \to \phi\phi)|^2 = |M(\phi\phi \to \bar{b}\bar{b})|^2 = \tfrac{1}{2}(1 - \zeta)|M_0|^2 ,$$

$$|M(\bar{b}\bar{b} \to bb)|^2 = \tfrac{1}{2}(1 + \bar{\zeta})|M_0|^2 ,$$

$$|M(\bar{b}\bar{b} \to \phi\phi)|^2 = |M(\phi\phi \to bb)|^2 = \tfrac{1}{2}(1 - \bar{\zeta})|M_0|^2 , \qquad (3)$$

‡1 Here we assume Maxwell–Boltzmann particles; the extra $(1 \overset{+}{-} f)$ factors accounting for stimulated emission (Pauli exclusion effects) in the creation of bosons (fermions) are compensated by corresponding terms in the Boltzmann equation [4].

‡2 We assume homogeneity and isotropy, so that $f(\boldsymbol{p}, \boldsymbol{x}) = f(\boldsymbol{p}) = f(p)$.

‡3 It is not necessary that these participate directly in B-violating reactions.

‡4 In the simple models discussed below, this phenomenon occurs if the universe is homogeneous and always cools faster than $T \sim m_P/(t m_P)^{1/5}$; in practice any quark excess will be contained in baryons where their probability for collisions remains constant rather than falling as in a homogeneous expanding universe.

‡5 This constraint applies only if no initial or final particles may mix with their antiparticles (as in the K^0 system). CP-violating mixing requires a difference $M(i \to j) - M(\bar{i} \to \bar{j}) \neq 0$ in amplitudes rather than rates.

‡6 Regardless of perturbation theory, CP violation is asymptotically suppressed by powers of $\log s$, where $\sqrt{s}$ is the invariant mass of the initial state.

Volume 91B, number 2 PHYSICS LETTERS 7 April 1980

where $\zeta - \bar{\zeta} = O(\alpha)$ measures the magnitude of *CP* (and *C*) violation. The number of a species i per unit volume $n_i \equiv \int d^3p\,(2\pi)^{-3} f_i(p)$ decreases with time even without collisions in an expanding universe according to (R is the Robertson–Walker scale factor; dots denote time derivatives)

$$dn_i/dt = N_i\, d(1/V)/dt$$

$$= -(3\dot{R}/R)n_i = (3\dot{T}/T)n_i = -(3T^2/M_P)n_i\,. \qquad (4)$$

The n_i are also changed by collisions; the (average) time development of the ϕ and baryon number ($n_B \equiv n_b - n_{\bar{b}}$) densities is given by the Boltzmann equations [$Y_i \equiv n_i/n_\gamma$, where γ is a massless particle; $|M_0|^2 = O(\alpha^2)$]

$$n_\gamma \dot{Y}_\phi = \dot{n}_\phi + (3\dot{R}/R)n_\phi$$

$$= 2\Lambda_{12}^{34}\{f_b(p_1)f_b(p_2)|M(bb \to \phi\phi)|^2$$

$$+ f_{\bar{b}}(p_1)f_{\bar{b}}(p_2)|M(\bar{b}\bar{b} \to \phi\phi)|^2$$

$$- f_\phi(p_1)f_\phi(p_2)[|M(\phi\phi \to bb)|^2 + |M(\phi\phi \to \bar{b}\bar{b})|^2]\}, \qquad (5a)$$

$$n_\gamma \dot{Y}_B = \Lambda_{12}^{34}\{-f_b(p_1)f_b(p_2)$$

$$\times [2|M(bb \to \bar{b}\bar{b})|^2 + |M(bb \to \phi\phi)|^2]$$

$$+ f_{\bar{b}}(p_1)f_{\bar{b}}(p_2)[2|M(\bar{b}\bar{b} \to bb)|^2 + |M(\bar{b}\bar{b} \to \phi\phi)|^2]$$

$$+ f_\phi(p_1)f_\phi(p_2)[|M(\phi\phi \to bb)|^2 - |M(\phi\phi \to \bar{b}\bar{b})|^2]\}, \qquad (5b)$$

where the operator Λ represents suitable integration over initial and final state momenta. We assume that the $\overset{(-)}{b}$ undergo baryon-conserving collisions with a frequency much higher than the $O(\alpha^3)$ rate on which n_B changes (as is presumably the case in realistic models). They are therefore always in kinetic equilibrium with the rest of the universe, and hence Maxwell–Boltzmann distributed in phase space:

$$f_{\overset{(-)}{b}}(p) \approx \exp[-(E \mp \mu)/T]\,,$$

$$Y_B \equiv (n_b - n_{\bar{b}})/n_\gamma \approx 2\sinh(\mu/T)\,. \qquad (6)$$

μ is a baryon number chemical potential, which is changed only by B-violating processes, and would vanish if chemical equilibrium prevailed. Assuming $Y_B \ll 1$, one may use energy conservation in eq. (5) to write

$$f_{\overset{(-)}{b}}(p_1)f_{\overset{(-)}{b}}(p_2) \approx \exp[-(E_3+E_4)/T](1 \pm Y_B)$$

$$\approx f_\phi^{eq}(p_3)f_\phi^{eq}(p_4)(1 \pm Y_B)\,,$$

where $f_\phi^{eq}(p) = \exp(-E/T)$ is the equilibrium distribution of ϕ at temperature T: The equilibrium ϕ number density

$$n_\phi^{eq} = T^3/(2\pi^2)(m_\phi/T)^2 K_2(m_\phi/T)\,,$$

where K_2 is a modifed Bessel function [7] [as $m_\phi \to 0$, $n_\phi^{eq} \to T^3/\pi^2$; as $T \to 0$, $n_\phi^{eq} \to (m_\phi T/2\pi)^{3/2}\exp(-m_\phi/T)$]. Then substituting the parametrization (3) and performing phase space integrations, eq. (5) becomes

$$\dot{Y}_\phi \approx n_\gamma\langle\sigma_0 v\rangle\{2[1 - \tfrac{1}{2}(\zeta + \bar{\zeta})][(Y_\phi^{eq})^2 - Y_\phi^2]$$

$$- (\zeta - \bar{\zeta})(Y_\phi^{eq})^2 Y_B\}\,, \qquad (7a)$$

$$\dot{Y}_B \approx n_\gamma\langle\sigma_0 v\rangle\{\tfrac{1}{2}(\zeta - \bar{\zeta})[Y_\phi^2 - (Y_\phi^{eq})^2]$$

$$- [3 + \tfrac{1}{2}(\zeta + \bar{\zeta})](Y_\phi^{eq})^2 Y_B\}\,, \qquad (7b)$$

where $\langle\sigma_0 v\rangle$ is the cross section corresponding to $|M_0|^2$ averaged over a flux incoming particles in equilibrium energy distributions. Eq. (7b) exhibits the necessity of deviation from equilibrium for B generation, and the destruction of Y_B in equilibrium. It also demonstrates that if $Y_B = 0$ and $\zeta = \bar{\zeta}$, $\dot{Y}_B$ will always be zero. This simple model demonstrates all the conditions necessary for baryon generation.

We now turn to a slightly more realistic but more complicated model in which massive particles $\overset{(-)}{X}$ decay to $\overset{(-)}{b}$ with rates [$\gamma_X = O(\alpha)$]

$$|M(X \to bb)|^2 = |M(\bar{b}\bar{b} \to \bar{X})|^2 = \tfrac{1}{2}(1+\eta)\gamma_X\,,$$

$$|M(X \to \bar{b}\bar{b})|^2 = |M(bb \to \bar{X})|^2 = \tfrac{1}{2}(1-\eta)\gamma_X\,,$$

$$|M(\bar{X} \to bb)|^2 = |M(\bar{b}\bar{b} \to X)|^2 = \tfrac{1}{2}(1-\bar{\eta})\gamma_X\,,$$

$$|M(\bar{X} \to \bar{b}\bar{b})|^2 = |M(bb \to X)|^2 = \tfrac{1}{2}(1+\bar{\eta})\gamma_X\,. \qquad (8)$$

Note that if $\overset{(-)}{X}$ decays preferentially produce b, then *CPT* invariance implies that $\bar{b}$ are preferentially destroyed in inverse processes; thus $\overset{(-)}{X}$ decays and inverse decays (DID) alone would generate a net B even if all particles were in thermal equilibrium, in contravention of the theorem (1) [‡7]. However, the *CP* violation parameter $\eta - \bar{\eta}$ is $O(\alpha)$, and hence changes in n_B from DID are of the same order as $2 \to 2$ scattering processes, such as $bb \to \bar{b}\bar{b}$. It will turn out that *s*-channel exchange of nearly on-shell $\overset{(-)}{X}$ in $\overset{(-)}{b}\overset{(-)}{b} \leftrightarrow b\,b$ cancels the DID contribution to $\dot{Y}_B$ so as to recover $\dot{Y}_B = 0$ in thermal equilibrium. In direct analogy with eqs. (5) and (7), and using the assumption (6), the equation for the evolution of the $\overset{(-)}{X}$ number density $n_{\overset{(-)}{X}} \equiv Y_{\overset{(-)}{X}}\, n_\gamma$ becomes

$$\dot{Y}_X = -\langle\Gamma_X\rangle[(Y_X - Y_X^{eq}) - \bar{\eta}\, Y_B Y_X^{eq}] \;; \qquad (9)$$

the corresponding equation for $\dot{Y}_{\bar{X}}$ is obtained by charge conjugation ($Y_X \leftrightarrow Y_{\bar{X}}, Y_B \to -Y_B, \eta \leftrightarrow \bar{\eta}$). The $\langle\Gamma_X\rangle$ in eq. (9) is the total $\overset{(-)}{X}$ decay width multiplied by the time dilation factor m_X/E_X and averaged over the equilibrium X energy distribution [‡8]. The baryon concentration evolves according to

$$\dot{Y}_B = \langle\Gamma_X\rangle[\eta Y_X - \bar{\eta} Y_{\bar{X}} + (\eta - \bar{\eta})\, Y_X^{eq} - 2 Y_B Y_X^{eq}]$$

$$- (2/n_\gamma)\Lambda_{12}^{34}\{f_X^{eq}(p_1 + p_2)$$

$$\times\, [|M'(bb \to \bar{b}\bar{b})|^2 - |M'(\bar{b}\bar{b} \to bb)|^2]\}$$

$$- 2 Y_B \Lambda_{12}^{34}\{f_X^{eq}(p_1 + p_2)$$

$$\times\, [|M'(bb \to \bar{b}\bar{b})|^2 + |M'(\bar{b}\bar{b} \to bb)|^2]\} \,, \qquad (10)$$

where the first term is from DID (and does not separately vanish when $Y_{\overset{(-)}{X}} = Y_X^{eq}$), while the second two terms arise from $2 \to 2$ scatterings. The DID term accounts for sequential inverse decay and decay processes involving real $\overset{(-)}{X}$: these are therefore subtracted from the true $2 \to 2$ scattering terms by writing $|M'(i \to j)|^2 = |M(i \to j)|^2 - |M_{RIX}(i \to j)|^2$, where $M_{RIX}(i \to j)$ is the amplitude for $i \to j$ due to on-shell *s*-channel X exchange. In the narrow X width approximation,

$$|M_{RIX}(i \to j)|^2 \sim |M(i \to \overset{(-)}{X})|^2 |M(\overset{(-)}{X} \to j)|^2/\Gamma_X \;;$$

the presence of the Γ_X denominator renders it $O(\alpha)$. According to the theorem (1), the *CP* violating difference of total rates $|M(bb \to \bar{b}\bar{b})|^2 - |M(\bar{b}\bar{b} \to bb)|^2 = O(\alpha^3)$. Hence $|M'(bb \to \bar{b}\bar{b})|^2 - |M'(\bar{b}\bar{b} \to bb)|^2 = |M_{RIX}(\bar{b}\bar{b} \to bb)|^2 - |M_{RIX}(bb \to \bar{b}\bar{b})|^2 + O(\alpha^3) = O(\alpha^2)$, and the second term in eq. (10) becomes $-2\langle\Gamma_X\rangle(\eta - \bar{\eta})\, Y_X^{eq}$, thereby elegantly cancelling the first term in thermal equilibrium. Finally, therefore

$$\dot{Y}_B = \langle\Gamma_X\rangle\{(\eta - \bar{\eta})[\tfrac{1}{2}(Y_X + Y_{\bar{X}}) - Y_X^{eq}]$$

$$+ \tfrac{1}{2}(\eta + \bar{\eta})(Y_X - Y_{\bar{X}})\} - 2 Y_B[\langle\Gamma_X\rangle Y_X^{eq}$$

$$+ n_\gamma\langle v\sigma'(bb \to \bar{b}\bar{b}) + v\sigma'(\bar{b}\bar{b} \to bb)\rangle] \,. \qquad (11)$$

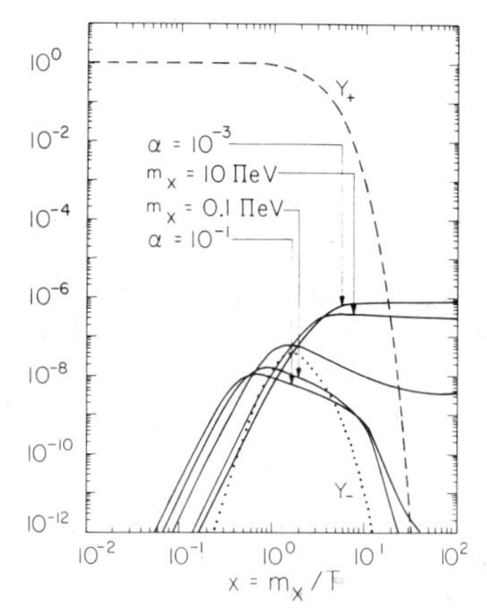

Fig. 1. The development of baryon number density (solid curves) as a function of inverse temperature in the model of eq. (11) for various choices of parameters (unless otherwise indicated, $\alpha = 1/40$ and $m_X = 1$ ΠeV $\equiv 10^{15}$ GeV). The dashed and dotted curves give $\frac{1}{2}(Y_X + Y_{\bar{X}})$ and $\frac{1}{2}(Y_X - Y_{\bar{X}})$, respectively. In all cases we have taken the *CP*-violation parameter $\eta - \bar{\eta} = 10^{-6}$ (even when α is changed). (Results depend only on m_X through the dimensionless combination m_X/m_P; here we take $\zeta = 100$ in the definition of m_P. Note that inhomogeneities in the early universe may be manifest in different expansion rates and hence different effective ζ for different regions. The final Y_B produced could vary considerably between the regions.)

‡7 This rather relevant point has also been noticed by Dolgov and Zeldovich [3], but was apparently neglected elsewhere.

‡8 Strictly, m_X/E_X should be averaged separately for the various terms of eq. (9); if X is in kinetic equilibrium, however, the averages are equal. Note that we have implicitly assumed all produced and decaying $\overset{(-)}{X}$ to be exactly on their mass shells. However, particularly at high T, the mean $\overset{(-)}{X}$ collision time $\ll 1/\Gamma_X$, so that the $\overset{(-)}{X}$ resonance is collision broadened, and produced or decaying $\overset{(-)}{X}$ may be far off shell. The m_X/E_X factor for inverse decays essentially arises from the fact that the incoming particles must subtend a sufficiently small angle to have invariant mass m_X; if produced $\overset{(-)}{X}$ are far off shell, the m_X/E_X in DID should disappear.

Volume 91B, number 2 PHYSICS LETTERS 7 April 1980

The differential eqs. (9) and (11) must now be solved with the initial condition $Y_X(t=0) = Y_X^{eq}(0)$, and possibly an initial baryon density Y_B. Fig. 1 shows the solutions with guesses for parameters based on the SU(5) model [2] [$m_X = 10^{15}$ GeV and 10^{14} GeV; $\alpha \approx 1/40$ (vector decays), or 10^{-3} (scalar decays)]. If all $\overset{(-)}{X}$ initially in thermal equilibrium decayed with no back reaction, the Y_B generated would be simply $\eta - \overline{\eta}$. For small α or large m_X/m_P this upper limit is approached. (At small $x \equiv m_X/T$, series solution of eqs. (10) and (11) gives $\frac{1}{2}(Y_X + Y_{\overline{X}}) \approx 1 - ax^5/20$, $\frac{1}{2}(Y_X - Y_{\overline{X}}) \approx (\eta - \overline{\eta})a^2x^8/160$, $Y_B \approx (\eta - \overline{\eta})ax^5/20$, where $a = m_P\Gamma_X/m_X^2$.) For $T \ll m_X$, baryon number is destroyed by $2 \to 2$ reactions with $\sigma \sim \alpha^2 T^2/m_X^4$ roughly like $Y_B(T) \sim \exp(\alpha^2 m_P T^3/m_X^4)$ ‡9, so that $Y_B \to$ constant as $T \to 0$, but if m_X is small, the final Y_B is much diminished from its value at higher T. The Y_B generated is always roughly linearly proportional to $\eta - \overline{\eta}$, but is a sensitive function of m_X/m_P and α; for realistic values of these parameters, a numerical solution is probably essential. Previous treatments of baryon number generation [3] have assumed that $Y_B = \eta - \overline{\eta}$, or $Y_B = 0$. Fig. 1 demonstrates that intermediate results are probable.

According to eq. (11), any baryon excess existing at the Planck time $t_P = 1/m_P$ should be diminished by inverse decays at $T \gg m_X$ so that $Y_B(t)/Y_B(t_P) \sim \exp(-\alpha m_X m_P/T^2)$; any initial Y_B should be reduced by a factor $\sim \exp(-m_P/m_X)$ before *CP* violating processes can generate Y_B at $T \lesssim m_X$. *B*-violating $2 \to 2$ scatterings at temperatures $m_P > T > m_X$ should reduce an initial Y_B by a factor $\sim \exp(-m_P \int_{m_P}^{m_X} \langle v\sigma \rangle \times \mathrm{d}T)$. One might expect that $\langle v\sigma \rangle \sim \alpha^2/m_X^2$ at high energies due to *t*-channel vector X exchange; however, the effective $\langle v\sigma \rangle$ presumably relevant for the Boltzmann equation is rather $\langle v\sigma_{eff} \rangle \sim \alpha^2 \lambda_D^2$ where the Debye screening length $\lambda_D \sim [(32\alpha)^{1/2}T]^{-1}$. In this approximation $2 \to 2$ and higher multiplicity collisions are probably no more effective at destroying an initial Y_B than are inverse decays.

We conclude therefore that *B*-violating reactions in the very early universe might well destroy any initial baryon number existing around the Planck time ($1/m_P$), requiring subsequent *B*- and *CP*-violating interactions to generate the observed baryon asymmetry. The methods described here allow a calculation of the resulting baryon excess in any specific model; the simple examples considered suggest that the observed Y_B should place stringent constraints on parameters of the model ‡10 [8].

‡9 If $T \sim m_P/(tm_P)^{1/5}$ as in footnote 8, then $Y_B \sim \exp(-1/T)$; the universe expands sufficiently slowly for Y_B to relax to zero.

We are grateful to many people for discussions, including A.D. Dolgov, S. Frautschi, William A. Fowler, G.C. Fox, S.E. Koonin, D.L. Tubbs and R.V. Wagoner. We thank T. Goldman for collaboration in the early stages of this work.

‡10 In grand unified models where there exist absolutely stable particles more massive than nucleons which appear as simple replications (cf. e, μ), the mechanism described above should generate roughly equal concentrations of these as of nucleons: observational constraints on the total energy density of the universe then suggest that no such particles exist.

[1] S. Weinberg, Gravitation and cosmology (Wiley, New York, 1972);
R.V. Wagoner, The early universe, Les Houches Lectures (1979), to be published.
[2] H. Georgi and S. Glashow, Phys. Rev. Lett. 32 (1974) 438;
A.J. Buras, J. Ellis, M.K. Gaillard and D.V. Nanopoulos, Nucl. Phys. B135 (1978) 66.
[3] A.D. Sakharov, Zh. Eksp. Teor. Fiz. Pis'ma 5 (1967) 32;
M. Yoshimura, Phys. Rev. Lett. 41 (1978) 281;
S. Dimopoulos and L. Susskind, Phys. Rev. D18 (1978) 4500; Phys. Lett. 81B (1979) 416;
D. Toussaint, S.B. Treiman, F. Wilcek and A. Zee, Phys. Rev. D19 (1979) 1036;
S. Weinberg, Phys. Rev. Lett. 42 (1979) 850;
J. Ellis, M.K. Gaillard and D.V. Nanopoulos, Phys. Lett. 80B (1979) 360;
A.D. Dolgov and Ya.B. Zeldovich, Cosmology and elementary particles, ITEP preprint (1979).
[4] E.W. Kolb and S. Wolfram, Baryon number generation in the early universe, Caltech preprint 0AP-547.
[5] V.B. Berestetskii, E.M. Lifshitz and L.P. Pitaevskii, Relativistic quantum theory (Pergamon, New York, 1971) p. 239.
[6] D.V. Nanopoulos and S. Weinberg, Mechanisms for cosmological baryon production, Harvard Univ. preprint 79/4023.
[7] I.S. Gradestyn and I.M. Ryzhik, Table of integrals, series and products (Academic Press, New York, 1965) p. 951.
[8] J. Harvey, E. Kolb, D. Reiss and S. Wolfram, Cosmological baryon number generation in gauge theories, Caltech preprint, in preparation;

VOLUME 45, NUMBER 25 PHYSICAL REVIEW LETTERS 22 DECEMBER 1980

Hierarchy of Cosmological Baryon Generation

J. N. Fry, Keith A. Olive, and Michael S. Turner
Astronomy and Astrophysics Center, Enrico Fermi Institute, The University of Chicago, Chicago, Illinois 60637
(Received 2 June 1980)

The hierarchy of baryon generation by multiple species of both gauge and Higgs bosons whose interactions do not conserve baryon number are considered. A procedure for computing the final baryon asymmetry in terms of the masses and coupling strengths of all the various species is presented. If the lightest gauge boson has a mass below 10^{15} GeV, then the final asymmetry depends only upon this boson and any lighter Higgs bosons.

PACS numbers: 98.80.Bp, 12.20.Hx, 11.10.Np, 95.30.Cq

Recent theoretical work on grand unified theories (GUT's) of particle interactions has provided a new solution to an old problem in cosmology: a fundamental explananation of the observed amount of matter in the universe, quantified as the baryon number-to-specific-entropy ratio kn_B/s, and the lack of any significant amount of antimatter in the observed portion of the universe up to the scale of clusters of galaxies.[1] The observed asymmetry has the value $kn_B/s = 10^{-10.8\pm1}$, the uncertainty primarily due to our poor knowledge of the baryon density.[2] It has been shown that, with (i) baryon-number-nonconserving processes mediated by superheavy bosons predicted by GUT's, (ii) C and CP nonconservations in the interactions of superheavy gauge or Higgs bosons, and (iii) a departure from thermal equilibrium (provided by the rapid expansion of the early universe), a baryon asymmetry can arise dynamically in an initially symmetrical universe.[3,4] Although two groups have considered in detail the effects of a single superheavy species,[5,6] an arbitrary GUT has more than one such species present. The purpose of this Letter is to outline a simple procedure which for many GUT's allows the straightforward calculation of the final asymmetry which evolves due to a whole hierarchy of superheavy species whose interactions do not conserve baryon number.

Towards that end, we first summarize the relevant results for the effects of a single species. In a general GUT which spontaneously breaks down to $[SU(3)]_c \otimes [SU(2)]_L \otimes U(1)$, there are five generic classes of superheavy bosons which couple to ordinary fermions and whose interactions do not conserve baryon number[7]: two isodoublet, color triplets of vector particles (XY with charges $\pm\frac{4}{3}$, $\pm\frac{1}{3}$; and $X'Y'$ with charges $\pm\frac{1}{3}$, $\mp\frac{2}{3}$); and three color triplets of scalar particles (H, an isosinglet with charge $\pm\frac{1}{3}$; S, an isosinglet with charge $\pm\frac{4}{3}$; and S', an isotriplet with charges $\pm\frac{4}{3}$, $\pm\frac{1}{3}$, $\mp\frac{2}{3}$). In a given GUT there may be several representatives of each class, each with its own mass and coupling strength. The XY and

$X'Y'$ bosons behave similarly and we refer to them generically as "gauge" bosons; the H, S, and S' Higgs bosons all behave similarly and we refer to them generically as "Higgs" bosons.

The effect of one given species on kn_B/s depends in general on its mass M, its coupling strength α, and on the size of the CP nonconservation in its decays. The dependence upon mass and coupling strength appears in the combination

$$K=(16\pi^3 g_*/45)^{-1/2}\alpha m_p/M$$
$$=(2.9\times10^{17}\text{ GeV})\alpha M^{-1}(160/g_*)^{1/2}, \qquad (1)$$

where g_* = total number of relativistic degrees of freedom $= g_{\text{bosons}} + \frac{7}{8} g_{\text{fermions}}$ [$g_* \simeq 160$ in the minimal SU(5)], $m_p = G^{-1/2} = 1.22\times10^{19}$ GeV, and typically, $\alpha_{\text{gauge}} = \frac{1}{45}$ and $\alpha_{\text{Higgs}} \simeq 10^{-4} - 10^{-6}$. K is approximately the ratio of the decay rate of the superheavy boson to the expansion rate when $T = M$. The important reactions for baryon generation are the decays (D) and inverse decays (ID) of superheavy bosons, and K determines whether or not these reactions are occuring on the expansion timescale. For $K \ll 1$, D and ID do not occur, and a departure from equilibrium results; for $K \gg 1$, D and ID do occur, and thermal equilibrium is almost maintained. We parametrize the CP nonconservation by ϵ, the mean net baryon number generated when a superheavy boson-antiboson pair decays.

Each species can generate a baryon asymmetry and/or damp a preexisting asymmetry, depending on these parameters. Damping occurs for $T \approx M$, primarily by ID, followed by D (e.g., $\bar{q}+\bar{q}\to X$, $X \to q+l$). Generation of an asymmetry is proportional to ϵ and occurs for $T \lesssim M$. Quantitatively, for gauge and Higgs bosons, respectively,

$$(kn_B/s)_i = (kn_B/s)_{i-1}\exp(-5.5K_i) + 1.5\times10^{-2}\epsilon_i/[1+(16K_i)^{1.3}], \qquad (2)$$

$$(kn_B/s)_i = (kn_B/s)_{i-1}\exp[-(0.26 \text{ to } 1.6)K_i] + 0.5\times10^{-2}\epsilon_i/[1+(3K_i)^{1.2}]. \qquad (3)$$

Here $(kn_B/s)_{i-1}$ is the preexisting asymmetry and $(kn_B/s)_i$ is the resultant asymmetry which develops because of species i, whose parameters are K_i, ϵ_i, and M_i. These are results of numerical integrations of the Boltzmann equations.[8]

The exponential damping term in Eqs. (2) and (3) represents the damping of a prior asymmetry by inverse decays. (In general, there are several modes of damping; this is the slowest of these.) However, because there are conserved quantities, certain types of asymmetries cannot be damped at all. These are modes accompanied by nonzero values of charge Q, weak isospin I_3, net baryon minus lepton number, $B-L$ (in some theories), and (for gauge particles only), "5-ness," which measures the net number of particles in the $\overline{5^*}$ representation of SU(5). We expect Q and I_3 to be identically zero from $t=0$ as these quantities are both gauged. Asymmetries with net 5-ness are not generated or damped by XY, $X'Y'$ bosons, but they can be by Higgs bosons. XY and $X'Y'$ bosons may also be able to damp asymmetries with nonzero 5-ness if they are aided by gauge or light Higgs bosons whose interactions do not nonconserve B but do not conserve 5-ness (e.g., W_R, the boson which mediates right-handed interactions). For simplicity we will assume that the net 5-ness of all asymmetries is zero, although it is possible that an asymmetry with net 5-ness could evolve and never be damped.[9] Any asymmetry with a nonzero value of a conserved quantity [e.g., in SU(5), $B-L$] must be either truly "initial" or generated by non-GUT processes, and cannot be damped. Such asymmetries would survive the GUT epoch and contribute to the asymmetry observed today.

The second term in Eqs. (2) and (3) represents the baryon asymmetry produced by species i. In the limit $K \ll 1$ (complete nonequilibrium), the saturation value which arises, $(kn_B/s) = 1.5\times10^{-2}\epsilon$ (gauge) or $(kn_B/s) = 0.5\times10^{-2}\epsilon$ (Higgs), is the familiar result of the out-of-equilibrium decay scenario.[4] In this limit, there is no significant damping and the saturation production simply adds to the initial asymmetry. In the limit $K \gg 1$ (equilibrium almost maintained), the baryon excess produced is $(kn_B/s) \approx 1.5\times10^{-2}\epsilon(16K)^{-1.3}$ (gauge) or $(kn_B/s) \approx 0.5\times10^{-2}\epsilon(3K)^{-1.2}$ (Higgs). The power laws are an approximation to the actual dependence $(Kz_f)^{-1}$, where z_f is the solution of $Kz_f^{7/2}\exp(-z_f) = 1$ (Ref. 5). In this limit any preexisting asymmetry which has no projection onto the eigenmodes with conserved quantities is completely erased, and the final asymmetry depends only on the parameters K_i and ϵ_i.

Although these results were obtained by integrating the Boltzmann equations for a single species from the Planck time to a final temperature $T \ll M_i$, we note that all effects occur $T \approx M_i$. Thus, we can apply (2) and (3) repeatedly to take into account the effects of all species ("impulse approximation"). We order all superheavy boson species according to decreasing mass, $M_i < M_{i-1}$. The universe is assumed to begin with an initial asymmetry $(kn_B/s)_0$, either as a true initial condition or produced by pre-GUT processes, which has zero projection on the conserved quantities Q, I_3, $B-L$, and 5-ness. As the universe expands and T decreases, interactions involving the most massive superheavy boson (species 1) cause the asymmetry to change at $T \approx M_1$ so that when its effect on kn_B/s is complete ($T < M_1$) the new baryon asymmetry is $(kn_B/s)_1$, determined by Eq. (2) or (3). This is then used as the input for species 2, etc., continuing until the effect of the lightest species has been taken into account. Thus, the general evolution of kn_B/s is reduced to a sequence of evolutions $(kn_B/s)_i$ with preexisting $(kn_B/s)_{i-1}$.[10]

Next, we mention two scenarios in which we can be more definite about the final outcome of the hierarchy. Note from (1) that as mass decreases (for fixed α), K increases and the damping of the previous asymmetry is more and more efficient. If the lightest gauge boson is lighter than 10^{15} GeV, then previously existing asymmetries are damped by a factor of at least 10^{12}, making it impossible for earlier asymmetries to contribute appreciably to the observed $kn_B/s \approx 10^{-10.8}$. The value of kn_B/s after this epoch, $T < M_g$, will be

$$(kn_B/s)_g = 1.5\times 10^{-2}\epsilon_g(16K_g)^{-1.3}. \quad (4)$$

Suppose now, as expected in SU(5), that there is one Higgs boson lighter than M_g but heavier than 3×10^{13} GeV. For $\alpha_H \lesssim 10^{-4}$, this gives $K_H < 1$, so that it will not reduce the gauge-produced asymmetry significantly, but will simply add on its own production. In this case,

$$(kn_B/s)_F = 1.5\times 10^{-2}\epsilon_g(16K_g)^{-1.3} + 0.5\times 10^{-2}\epsilon_H. \quad (5)$$

Figure 1 shows the results of actual numerical integration of the Boltzmann equations with two superheavy species for $M_g = 10^{15}$ GeV, $K_g = 10$, $\epsilon_g \approx 10^{-6}$, $M_H = 10^{14}$ GeV, $K_H = 0.1$, and $\epsilon_H \approx 10^{-8}$. These numerical results agree well with Eqs. (4) and (5), and illustrate the validity of the sequential treatment.

If on the other hand, this lighter Higgs boson is light enough or couples strongly enough so that $K_H \gtrsim 10$, then it can reduce the gauge-generated asymmetry to an extent such that the final asymmetry depends only on K_H and ϵ_H,

$$(kn_B/s)_f = 0.5\times 10^{-2}\epsilon_H(3K_H)^{-1.2}. \quad (6)$$

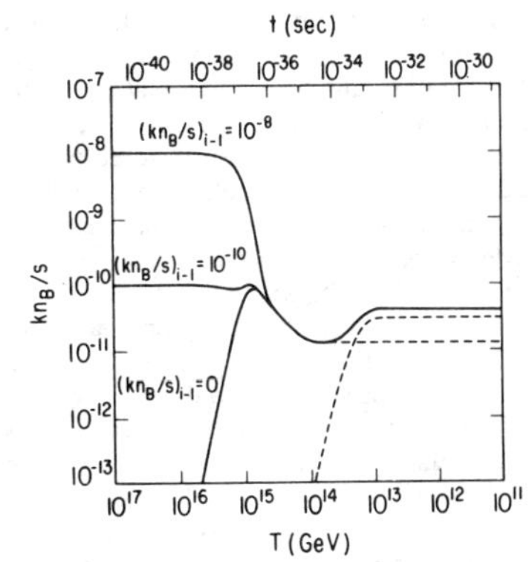

FIG. 1. The time development of kn_B/s under the effects of one gauge species ($M = 10^{15}$ GeV, $K=10$, $\epsilon_g \cong 10^{-6}$) and one Higgs species ($M = 10^{14}$ GeV, $K = 0.1$, $\epsilon_H \cong 10^{-8}$) for three different prior asymmetries. The gauge particle interactions totally erase the prior asymmetry and produce kn_B/s equal to the value which continues across as a broken line [cf., Eq. (4)]. The contribution of the Higgs particle simply adds on later, as shown by the continuation of the solid curve. The asymmetry generated by the Higgs particle alone is also shown as a broken line. Note that the solid curve for $T \lesssim 10^{13}$ GeV is the sum of the two broken curves [cf., Eq. (5)]. The curves were produced by numerically integrating the Boltzmann equations with two superheavy species.

Finally, we mention some possible complications which might invalidate our sequential treatment. First, one superheavy species (X) might decay primarily into a lighter superheavy species (X') and due to CP nonconservation in its decays create an X'-$\bar{X}'$ asymmetry.[11] If for the lighter species $K_{X'} \lesssim 1$, then even if $\epsilon_{X'} = 0$, decays of the X''s can create a baryon asymmetry (if $K_{X'} \gg 1$, our results above apply). Also, some theories [e.g., SU(10)] have superheavy Majorana neutrinos in addition to the usual fermions, and these particles may play an important role in baryon generation.[12]

In the last few years, the cosmological production of the baryon asymmetry has been the topic of much discussion and speculation. In fact, with the strong damping of prior asymmetries by inverse decays, the existence of a superheavy gauge boson lighter than 10^{15} GeV (which could be inferred from observations of proton decay) would make cosmological baryon generation almost a necessity, unless an initial asymmetry existed with a conserved quantity (e.g., $B-L$ or 5-ness). Such an observation would also require matter-antimatter separation by mechanisms invoked in baryon symmetrical universes without baryon generation to occur after $T=10^{15}$ GeV, since inverse decays can damp local as well as global asymmetries. This must happen on scales much larger than $\sim ct$, the distance over which separation could have occurred since damping ceased ($T\sim 10^{15}$ GeV, $t\sim 10^{-36}$ s). Although the metric might be horizonless because of quantum gravity effects,[13] the distance over which effects could propagate since $T\sim 10^{15}$ GeV does not differ significantly from $\sim ct$, the usual horizon size.

Although we do not know which (if any) GUT is correct, an even larger question is that of the C and CP nonconservations in the superheavy system. In Eq. (5), with $M_g \simeq 3\times 10^{14}$ GeV and $M_H \gtrsim 3\times 10^{13}$ GeV ($\alpha_H \lesssim 10^{-4}$), we need $\epsilon_g \approx 10^{-5.7}$ or $\epsilon_H \approx 10^{-8.5}$ to explain the observed asymmetry $kn_B/s \simeq 10^{-10.8}$. CPT and unitarity restrict $\epsilon \lesssim O(\alpha) \approx 10^{-2}$. In the minimal SU(5) (one $\underline{24}$ and one $\underline{5}$ of Higgs), it has been shown $\epsilon < 10^{-10}$ (Refs. 7, 11, and 14). However, the addition of just one more Higgs $\underline{5}$ again allows ϵ as large as $\sim 10^{-2}$ [the minimal SU(5) model also has difficulty explaining fermion mass ratios]. There is always the hope that ϵ can be related to the CP nonconservation in the K^0-$\bar{K}^0$ (or analogous) systems, tying together the two observed nonconservations of matter-antimatter symmetry.

This work was supported in part by the National Science Foundation under Grants No. AST 78-20402 and No. AST 78-20392. One of us (J.N.F.) is a recipient of a McCormick Fellowship. One of us (K.A.D.) is a recipient of a Hertz Fellowship, and one of us (M.S.T.) is a recipient of a Fermi Fellowship. We thank G. Segrè and E. Kolb for their helpful comments.

[1] G. Steigman, Annu. Rev. Astron. Astrophys. 14, 339 (1976).

[2] K. A. Olive, D. N. Schramm, G. Steigman, M. S. Turner, and J. Yang, Enrico Fermi Institute Report No. 80-42, 1980 (unpublished).

[3] M. Yoshimura, Phys. Rev. Lett. 41, 281 (1978); A. Ignatiev, N. Krasnikov, V. Kuzmin, and A. Tavkhelidze, Phys. Lett. 76B, 436 (1978); S. Dimopoulos and L. Susskind, Phys. Rev. D 18, 4500 (1979); J. Ellis, M. K. Gaillard, and D. V. Nanopoulos, Phys. Rev. Lett. 80B, 360 (1978).

[4] D. Toussaint, S. B. Treiman, F. Wilczek, and A. Zee, Phys. Rev. D 19, 1036 (1979); S. Weinberg, Phys. Rev. Lett. 42, 850 (1979).

[5] J. N. Fry, K. A. Olive, and M. S. Turner, Phys. Rev. D 22, 2953, 2977 (1980).

[6] E. W. Kolb and S. Wolfram, Phys. Lett. 91B, 217 (1980).

[7] D. Nanopoulos and S. Weinberg, Phys. Rev. D 20, 2484 (1979).

[8] These results are taken from Ref. 5, with ϵ defined differently by a factor of 2 and with a factor of $\frac{4}{3}$ applied to the Higgs case (allowing for the use of Maxwell-Boltzmann statistics). The quantitative differences between gauge and Higgs results are mainly due to different spin and isospin degeneracies.

[9] S. B. Treiman and F. Wilczek, Phys. Lett. 95B, 222 (1980).

[10] There is a slight complication if two species have nearly equal mass, $M_i \approx M_j$. If they are also of the same class (gauge or Higgs), then they can be replaced with one effective species with $K_{\rm eff} = K_i + K_j$ and $\epsilon_{\rm eff} = \epsilon_i + \epsilon_j$. If they are not of the same class (gauge or Higgs), and $K_i \ll K_j$ then the effects should be considered sequentially in the order species j followed by species i.

[11] S. Barr, G. Segrè, and H. A. Weldon, Phys. Rev. D 20, 2494 (1979).

[12] J. Harvey, E. Kolb, D. Reiss, and S. Wolfram, to be published.

[13] J. B. Hartle and B. L. Hu, Phys. Rev. D 21, 2756 (1980).

[14] G. Segrè and M. S. Turner have shown that one additional generation of heavy fermions (30–200 GeV) allows the observed asymmetry to be produced by Higgs bosons in minimal SU(5) [Enrico Fermi Institute Report No. 80-43, 1980 (unpublished)].

XII. *Introduction to the Theory of Galaxy Formation*

1) Gauge-invariant cosmological perturbations
by J. M. Bardeen, *Phys. Rev.* D**22,** 1882 (1980) 642

2) Tenacious myths about cosmological perturbations larger than the horizon size
by W. H. Press and E. T. Vishniac, *Astrophys. J.* **239,** 1 (1980) 666

3) The theory of the large scale structure of the universe
by Ya. B. Zel'dovich, *International Astronomical Union* No. **79,** p. 409 (1977) 677

4) The black-body radiation content of the universe and the formation of galaxies
by P. J. E. Peebles, *Astrophys. J.* **142,** 1317 (1965) 690

5) Fluctuations in the primordial fireball
by J. Silk, *Nature* **215,** 1155 (1967) 700

6) Primeval adiabatic perturbation in an expanding universe
by P. J. E. Peebles and J. T. Yu, *Astrophys. J.* **162,** 815 (1970) 702

7) Core condensation in heavy halos: A two-stage theory for galaxy formation and clustering
by S. D. M. White and M. J. Rees, *Mon. Not. R. Astron. Soc.* **183,** 341 (1978) 724

8) Galaxy formation in an intergalactic medium dominated by explosions
by J. P. Ostriker and L. L. Cowie, *Astrophys. J.* **243,** L127 (1981) 742

PHYSICAL REVIEW D VOLUME 22, NUMBER 8 15 OCTOBER 1980

Gauge-invariant cosmological perturbations

James M. Bardeen*
Institute for Advanced Study, Princeton, New Jersey 08540
(Received 7 April 1980)

The physical interpretation of perturbations of homogeneous, isotropic cosmological models in the early Universe, when the perturbation is larger than the particle horizon, is clarified by defining a complete set of gauge-invariant variables. The linearized perturbation equations written in these variables are simpler than the usual versions, and easily accommodate an arbitrary background equation of state, entropy perturbations, and anisotropic pressure perturbations. Particular attention is paid to how a scalar (density) perturbation might be generated by stress perturbations at very early times, when the non-gauge-invariant perturbation in the density itself is ill-defined. The amplitude of the fractional energy density perturbation at the particle horizon cannot be larger, in order of magnitude, than the maximum ratio of the stress perturbation to the background energy density at any earlier time, unless the perturbation is inherent in the initial singularity.

I. INTRODUCTION

The mathematical theory of perturbations in inhomogeneous, isotropic cosmological models has been worked over many times in the literature. References 1–7 are a selection of the more important comprehensive treatments. Nevertheless, troubling questions still remain about the physical interpretation of density perturbations at early times when the perturbation is larger than the particle horizon, which will here mean when the time for light to travel a characteristic wavelength of the perturbation is larger than the instantaneous expansion (Hubble) time. These questions are particularly relevant to attempts to explain the origin of perturbations which eventually give rise to galaxies through processes occurring at times when temperatures exceed a few hundred MeV and/or densities exceed nuclear densities, times when there is considerable latitude to speculate about the microscopic physics.[8,9]

The problem has to do with the freedom of making gauge transformations. In discussing perturbations one is dealing with two spacetimes—the physical, perturbed spacetime and a fictitious background spacetime, here described by a Robertson-Walker metric. Points in the background are labeled by coordinates x^k (Latin indices will range from 0 to 3, Greek indices from 1 to 3). A one-to-one correspondence between points in the background and points in the physical spacetime carries these coordinates over into the physical spacetime and defines a choice of gauge. A change in the *correspondence*, keeping the background coordinates fixed, is called a *gauge transformation*, to be distinguished from a coordinate transformation which changes the labeling of points in the background and physical spacetime together.

The perturbation in some quantity is the difference between the value it has at a point in the physical spacetime and the value at the *corresponding point* in the background spacetime. A gauge transformation induces a coordinate transformation in the physical spacetime, but it also changes the point in the background spacetime corresponding to a given point in the physical spacetime. Thus, even if a quantity is a scalar under coordinate transformations, the value of the *perturbation* in the quantity will *not* be invariant under gauge transformations if the quantity is nonzero and position dependent in the background.

The prime example is the perturbation in density (energy density or baryon density). Because the density is time dependent in the cosmological background, the value of the density perturbation is altered by any gauge transformation which changes the correspondence between hypersurfaces of simultaneity in the physical spacetime and the background spacetime. When the perturbation is well within the particle horizon, hypersurfaces of simultaneity are physically unambiguous (clocks can be synchronized by exchanging light signals) and the change in density perturbation between "reasonable" gauge choices is negligible. However, at early times there is no compelling physical reason to choose between gauges which give very different results for the time dependence of density perturbations, and several different values for the exponent in the power-law time dependence of what is physically the same mode of density perturbation can be found in the literature, each mathematically correct. Furthermore, if the gauge condition imposed to simplify the form of the metric leaves a residual gauge freedom, the perturbation equations will have spurious "gauge mode" solutions which can be completely annulled by a gauge transformation and have no physical reality.[2,10-12]

One interesting attempt to avoid these problems

was made by Hawking,[3] who formulated the perturbation equations in a completely covariant form, without any mention of metric tensor perturbations as such. This did not totally circumvent the problem of gauge ambiguity, since a choice of time slicing in the perturbed spacetime must still be made to define a density perturbation. The paper is flawed by a failure to recognize that hypersurfaces of constant proper time along the fluid-element world lines cannot be orthogonal to these world lines when pressure perturbations as well as density perturbations are present.

The Hawking line of analysis was completed successfully by Olson[13] for the special case of an isentropic perfect fluid and a background with zero spatial curvature. An equation was derived for a *gauge-invariant* variable, proportional to the intrinsic spatial curvature of hypersurfaces orthogonal to the fluid four-velocity everywhere. The density perturbation as such was defined relative to hypersurfaces of constant fluid-element proper time, and thus was subject to an ambiguity associated with the freedom of adjusting the origin of proper time differently for different fluid elements. The Hawking-Olson approach can perhaps be extended to allow for nonzero background curvature, entropy perturbations, and anisotropic pressure, but only with great difficulty.

This paper presents a complete gauge-invariant framework for studying the time development of physically general perturbations at early times, when the wavelength of the perturbation is larger than the particle horizon. The geometrical quantities are defined from the metric perturbations alone, without reference to the matter perturbations. A gauge-invariant formalism in some respects mathematically equivalent to this one has been developed by Gerlach and Sengupta,[14] but they did not consider any cosmological applications.

There are two independent gauge-invariant gravitational "potentials" for scalar (density) perturbations and one for vector (vorticity) perturbations. The appropriate combinations of Einstein equations give all these potentials directly from the matter perturbations through purely algebraic equations once the perturbations have been separated into spatial harmonics.

The gauge-invariant variables which give the mathematically simplest description of the matter dynamics are, for *scalar* perturbations, (1) the velocity amplitude which, when divided by the reduced wavelength of the perturbation, gives the time dependence of the rate of shear of the matter velocity field; (2) the perturbation in the "entropy" of the matter, specifically the excess of the actual fractional pressure perturbation over the adiabatic one; (3) the amplitude of the anisotropic stress associated with the perturbation, if any, and (4) the fractional energy density perturbation on the hypersurfaces orthogonal to the world lines of the matter. When written completely in terms of these gauge-invariant amplitudes, the equations of motion for the matter simplify to a form rather closely analogous to the corresponding *Newtonian* equations for perturbations in an expanding background, even when the wavelength of the perturbation is much larger than the particle horizon. The shear velocity and the comoving energy density perturbation formally respond to input from the entropy perturbation and the anisotropic stress, in the spirit of recent work by Press and Vishniac.[12] Analytic solution of these equations is straightforward when the background spatial curvature is negligible and the ratio of pressure to energy density in the background is independent of time.

The gauge invariance in itself does not resolve the physical ambiguity of what one means by an energy density perturbation or a spatial curvature perturbation when the perturbation wavelength is larger than the particle horizon. The physical perturbations relative to any well-defined set of hypersurfaces of simultaneity can be represented by appropriate combinations of the above gauge-invariant amplitudes. I will consider three such families of hypersurfaces. The *comoving hypersurfaces*, orthogonal to the world lines of the matter, have already been mentioned. The mathematically simplest gauge-invariant gravitational potentials represent directly the perturbations in the spatial curvature and the lapse function relative to *zero-shear hypersurfaces*, hypersurfaces for which the congruence of normal timelike world lines has zero shear; i.e., the traceless part of the extrinsic curvature tensor vanishes. A spatially uniform rate of expansion of the normal world lines characterizes *uniform-Hubble-constant hypersurfaces*. Only the last hypersurface condition carries over without modification to general perturbations, involving divergenceless vector fields and transverse traceless tensors. As long as the perturbations are linear the scalar perturbations can be treated separately.

Requiring the hypersurfaces of simultaneity to satisfy specific physical or geometric criteria is contrary to the usual approach to cosmological perturbations through a synchronous gauge.[2,7,15] In a synchronous gauge the physical properties of only one of the hypersurfaces, at one particular time, can be specified in advance; all the other hypersurfaces depend on the global solution to the perturbation equations. As a result the density perturbation in a synchronous gauge cannot be characterized by gauge-invariant amplitudes, and its value is in principle unrelated to the current physical state of the perturbations.

An important question is whether the perturbations that eventually give rise to galaxies need to be present in the singularity at $t=0$ or can somehow be produced at a later time in an initially homogeneous and isotropic universe. New physical mechanisms for generating inhomogeneities through gauge-theory symmetry breaking have been proposed in recent years.[8,9,16,17] The difficulty is that, with a conventional background equation of state, the comoving volume which produces a galaxy is far larger than the particle horizon at the times the symmetry breaking (or any other exotic process) is likely to occur. No causal process can then produce coherence on a galactic scale. There have been suggestions[18,19] that initially small *statistical* fluctuations on a galactic scale associated with large amplitude perturbations within the particle horizon at some early time would grow to a significant amplitude by the time the galactic scale comes inside the particle horizon. The conventional mythology is that the fractional energy density perturbation increases as the proper time t in a radiation-dominated background as long as the perturbation wavelength is larger than the particle horizon.

Local conservation of energy and momentum requires that any energy density perturbation arise as the result of a stress perturbation. Press and Vishniac[12] point out that for an isotropic stress (entropy) perturbation the fractional energy density perturbation *at the particle horizon* can be no larger than the maximum value of the ratio of stress perturbation to background energy density at earlier times. In this paper I generalize the result to allow for anisotropic stress perturbations and nonlinear excitation of density perturbations. The conclusion still holds that a given amplitude fractional energy density perturbation at the particle horizon can *only* arise from a stress perturbation with a comparable amplitude relative to the background energy density unless the perturbation is inherent in the initial singularity or unless the background equation state abolishes particle horizons at early times. While an anisotropic stress perturbation can produce an energy density perturbation of comparable amplitude (on a comoving hypersurface) at some early time, the fractional energy density perturbation belongs predominantly to the "decaying mode" and is no larger at the particle horizon.

The plan of the paper is as follows. The notation and description of the background and perturbed metric tensor and energy-momentum tensor in conventional terms, without any gauge restrictions, is established in Sec. II. In Sec. III I show how these quantities are affected by arbitrary gauge transformations and identify the gauge-invariant amplitudes describing the perturbation. The equations for the gauge-invariant amplitudes are derived in Sec. IV without any restrictions on the physical nature of the matter other than local conservation of energy and momentum, which is required in any case as an integrability condition on the Einstein equations. In Sec. V general analytic solutions in a background with constant ratio of pressure to energy density and effectively zero spatial curvature are used to discuss the generation of energy density perturbations from stress perturbations from the point of view of each of the hypersurface conditions mentioned above. Contact with the nonperturbative evolution of the geometry and the matter is made in Sec. VI, in order to understand the limits of validity of the linear perturbation analysis and nonlinear effects on the energy density perturbation. The results are summarized in Sec. VII.

The focus in this paper is on the physics. Mathematical questions regarding the existence of linear perturbations and the expansion of the perturbations in spatial harmonics have been largely answered by D'Eath.[20]

II. STANDARD FORMALISM

The background spacetime is described by some version of the Robertson-Walker metric

$$ds^2 = S^2(\tau)(-d\tau^2 + {}^3g_{\alpha\beta}dx^\alpha dx^\beta) . \tag{2.1}$$

The tensor ${}^3g_{\alpha\beta}$ is the metric tensor for a three-space of uniform spatial curvature K, with Riemann tensor

$${}^3R_{\alpha\beta\gamma\delta} = K({}^3g_{\alpha\gamma}{}^3g_{\beta\delta} - {}^3g_{\alpha\delta}{}^3g_{\beta\gamma}) , \tag{2.2}$$

where K is independent of time. The choice of coordinates in the background three-space is left arbitrary. Let a slash denote the covariant derivative of a three-tensor with respect to ${}^3g_{\alpha\beta}$ and a semicolon the covariant derivative in the physical spacetime. The scale factor $S(\tau)$ describes the expansion of the background as a function of the conformal time τ.

The unperturbed energy-momentum tensor must be formally that of a perfect fluid at rest relative to the above coordinates. The only nonzero components are

$$T^0_0 = -E_0, \quad T^\alpha_\beta = P_0\delta^\alpha_\beta , \tag{2.3}$$

where $E_0(\tau)$ is the background energy density and $P_0(\tau)$ is the background pressure. Let

$$w = P_0/E_0, \quad c_s^2 = dP_0/dE_0. \tag{2.4}$$

A nonzero cosmological constant can be considered part of this background energy-momentum tensor, contributing $+\Lambda$ to E_0 and $-\Lambda$ to P_0.

The time evolution of the background is governed by the equations[15]

$$(\dot{S}/S)^{\cdot} = -\tfrac{1}{6}(E_0 + 3P_0)S^2, \tag{2.5a}$$

$$(\dot{S}/S)^2 = \tfrac{1}{3}E_0 S^2 - K, \tag{2.5b}$$

$$\dot{E}_0/(E_0 + P_0) = -3\dot{S}/S, \tag{2.6}$$

where $\dot{S} \equiv dS/d\tau$, the derivative with respect to the *conformal* time, and units have been chosen so $c = 8\pi G = 1$.

Perturbations in various quantities can be classified, according to how they transform under *spatial* coordinate transformations in the *background* spacetime, as spatial scalars, vectors, and tensors. Furthermore, the homogeneity and isotropy of the background allows a separation of the time dependence and the spatial dependence, with the spatial dependence related to solutions of a generalized Helmholtz equation.[2] The representation in spatial harmonics is not always unique.[20]

Scalar harmonics are solutions of the scalar Helmholtz equation

$$Q^{(0)|\alpha}{}_{|\alpha} + k^2 Q^{(0)} = 0. \tag{2.7}$$

The wave number k sets the spatial scale of the perturbation relative to the comoving background coordinates. For zero background curvature the $Q^{(0)}$ can be taken as plane waves; solutions for nonzero spatial curvature are described by Harrison.[4] *Scalar perturbations* have a spatial dependence derived from one of the $Q^{(0)}$. A vector or tensor quantity, such as the three-velocity of the matter or the perturbation in the spatial metric tensor, which is associated with a scalar perturbation must be constructed from covariant derivatives of $Q^{(0)}$ and the metric tensor. The construction is unique, within a normalization, for any traceless, symmetric tensor. Define the vector

$$Q_\alpha^{(0)} = -(1/k)Q^{(0)}{}_{|\alpha}, \tag{2.8}$$

and the traceless, symmetric, second-rank tensor

$$Q_{\alpha\beta}^{(0)} = k^{-2}Q^{(0)}{}_{|\alpha\beta} + \tfrac{1}{3}\,{}^3g_{\alpha\beta}Q^{(0)}. \tag{2.9}$$

Higher-rank tensors can be useful in representing moments of the specific intensity of, say, the microwave background radiation field, but are not needed in this paper. All equations governing scalar perturbations are reducible to scalar equations by taking divergences, e.g.,

$$Q^{(0)\alpha}{}_{|\alpha} = kQ^{(0)}, \quad Q^{(0)\alpha\beta}{}_{|\alpha\beta} = \tfrac{2}{3}(k^2 - 3K)Q^{(0)}. \tag{2.10}$$

The divergenceless part of a vector field cannot be related to scalar harmonics, but instead must be proportional to a *vector harmonic* $Q^{(1)\alpha}$, a divergenceless vector field which is a solution of the vector Helmholtz equation

$$Q^{(1)\alpha|\beta}{}_{|\beta} + k^2 Q^{(1)\alpha} = 0. \tag{2.11}$$

The corresponding second-rank symmetric tensor, necessarily traceless but not divergenceless, is

$$Q^{(1)\alpha\beta} = -\tfrac{1}{2}k^{-1}(Q^{(1)\alpha|\beta} + Q^{(1)\beta|\alpha}). \tag{2.12}$$

A second-rank antisymmetric tensor can also be constructed from $Q_\alpha^{(1)}$, in contrast to $Q_\alpha^{(0)}$, since $Q_\alpha^{(1)}$ is not the gradient of a scalar.

Gravitational waves are described by a traceless, divergenceless tensor $Q_{\alpha\beta}^{(2)}$ which is a solution of

$$Q^{(2)\alpha\beta|\gamma}{}_{|\gamma} + k^2 Q^{(2)\alpha\beta} = 0, \tag{2.13}$$

and in linear perturbation theory are completely decoupled from the scalar and vector perturbations.

A completely general perturbation of the gravitational field can be written as a linear combination of perturbations associated with individual spatial harmonics as defined above, with no coupling between different harmonics. The gravitational wave tensor perturbations only couple to the anisotropic part of the stress tensor, the vector perturbations couple, in addition, to the divergenceless, vortical part of the velocity field of the matter, and scalar perturbations couple to perturbations in the density and isotropic pressure as well as perturbations in the irrotational part of the velocity field and the anisotropic stress.

From now on we assume the separation into individual harmonics has been made. Then a given quantity can be written as a linear combination of all the independent appropriate rank spatial tensors constructable from the fundamental harmonic, with the coefficients functions of time. The Einstein equations and matter evolution equations become ordinary differential equations in time for these coefficients.

Specific representations of the perturbations in the metric tensor and the energy-momentum tensor will now be defined separately for scalar, vector, and tensor perturbations.

A. Scalar perturbations

The conformal factor S^2 is removed from the metric tensor components before defining the perturbations. Let

$$g_{00} = -S^2(\tau)[1 + 2A(\tau)Q^{(0)}(x^\mu)], \tag{2.14a}$$

$$g_{0\alpha} = -S^2 B^{(0)}(\tau) Q^{(0)}_{\alpha}(x^\mu), \tag{2.14b}$$

$$g_{\alpha\beta} = S^2\{[1 + 2H_L(\tau)Q^{(0)}(x^\mu)]^3 g_{\alpha\beta}(x^\mu) + 2H_T^{(0)}(\tau)Q^{(0)}_{\alpha\beta}(x^\mu)\}. \tag{2.14c}$$

The representation of the energy-momentum tensor will also be completely general in the context of first-order perturbations. *Define* the rest frame for the matter to be the frame in which the energy flux vanishes. Let u^a be the four-velocity of this frame relative to the coordinate frame. The three-velocity associated with u^a is represented by

$$u^\alpha/u^0 = v^{(0)}(\tau)Q^{(0)\alpha}(x^\mu). \tag{2.15}$$

To first order the normalization $u_a u^a = -1$ gives

$$u^0 = S^{-1}[1 - AQ^{(0)}]. \tag{2.16}$$

In the rest frame of the matter the energy density is

$$E = -T^0_0 = E_0(\tau)[1 + \delta(\tau)Q^{(0)}(x^\mu)]. \tag{2.17}$$

In transforming back to the coordinate frame the mixed components T^a_b are unchanged to first order except for T^0_α and T^α_0. The stress tensor T^α_β is represented by an isotropic pressure

$$P \equiv \tfrac{1}{3}T^\alpha_\alpha = P_0(\tau) + P_0(\tau)\pi_L(\tau)Q^{(0)}(x^\mu) \tag{2.18}$$

and a traceless anisotropic stress, with

$$T^\alpha_\beta = P_0[1 + \pi_L Q^{(0)}]\delta^\alpha_\beta + P_0\pi_T^{(0)}(\tau)Q^{(0)\alpha}_\beta. \tag{2.19}$$

The remaining components are

$$T^0_\alpha = (E_0 + P_0)(v^{(0)} - B^{(0)})Q^{(0)}_\alpha, \qquad T^\alpha_0 = -(E_0 + P_0)v^{(0)}Q^{(0)\alpha}. \tag{2.20}$$

The perturbed isotropic pressure need not be related to the energy density in the same way as the background. The difference between the fractional pressure perturbation and that expected from the background pressure-energy density relation will be called the entropy perturbation,

$$\eta(\tau)Q^{(0)} = \left(\pi_L - \frac{E_0}{P_0}\frac{dP_0}{dE_0}\,\delta\right)Q^{(0)} = \frac{1}{w}\,(w\pi_L - c_s{}^2\delta)Q^{(0)}, \tag{2.21}$$

even though it may bear no relation to the true physical entropy, when this is a meaningful concept.

Perturbations in auxiliary quantities associated with the matter, such as a rest-mass density or specific intensity of radiation, can be defined in a similar fashion, and may be required to treat the internal dynamics of the matter, e.g., the interaction of matter in the narrow sense with electromagnetic or neutrino radiation and the propagation of this radiation.[21] However, in this paper we will only consider the overall dynamics of the matter as reflected in the equations $T^b_{a;b} = 0$.

B. Vector perturbations

Now all quantities which are scalars under spatial coordinate transformations in the background must be unperturbed, e.g., g_{00}, T^0_0, and u^0. The description of vector and tensor quantities is similar to the scalar case, but the spatial dependence is generated from a fundamental vector harmonic $Q^{(1)}_\alpha$. In the metric tensor,

$$g_{0\alpha} = -S^2(\tau)B^{(1)}(\tau)Q^{(1)}{}_\alpha(x^\mu), \tag{2.22a}$$

$$g_{\alpha\beta} = S^2[{}^3g_{\alpha\beta} + 2H_T^{(1)}(\tau)Q^{(1)}_{\alpha\beta}(x^\mu)]. \tag{2.22b}$$

In the energy-momentum tensor,

$$T^0_\alpha = (E_0 + P_0)(v^{(1)} - B^{(1)})Q^{(1)}_\alpha, \tag{2.23a}$$

$$T^\alpha_\beta = P_0\delta^\alpha_\beta + P_0\pi_T^{(1)}(\tau)Q^{(1)\alpha}_\beta, \tag{2.23b}$$

where

$$u^\alpha/u^0 = v^{(1)}(\tau)Q^{(1)\alpha}(x^\mu). \tag{2.24}$$

C. Tensor perturbations

The intrinsically tensor perturbations affect only the traceless part of the spatial metric and the traceless part of the stress tensor. For a particular tensor harmonic $Q^{(2)}_{\alpha\beta}$,

$$g_{\alpha\beta} = S^2[{}^3g_{\alpha\beta} + 2H_T^{(2)}(\tau)Q^{(2)}_{\alpha\beta}], \tag{2.25}$$

$$T^\alpha_\beta = P_0\delta^\alpha_\beta + P_0\pi_T^{(2)}(\tau)Q^{(2)\alpha}_\beta. \tag{2.26}$$

III. GAUGE TRANSFORMATIONS AND GAUGE-INVARIANT VARIABLES

As explained in Sec. I, a gauge transformation corresponds to a change of coordinates in the physical spacetime while the background coordinates are held fixed. Consistent with the perturbation analysis, only first-order effects of the coordinate transformation need be considered, and the spatial dependence of the transformation should correspond to the same harmonic that generates perturbations in the metric tensor and energy-momentum tensor.

A. Scalar perturbations

The most general possible gauge transformation associated with a scalar perturbation is the result of the coordinate transformation

$$\tilde{\tau} = \tau + T(\tau)Q^{(0)}(x^\mu), \tag{3.1a}$$

$$\tilde{x}^\alpha = x^\alpha + L^{(0)}(\tau)Q^{(0)\alpha}(x^\mu) \tag{3.1b}$$

with T and $L^{(0)}$ arbitrary functions of τ.

The changes in the metric tensor are computed from

$$g_{ab}(x^c) = \frac{\partial \bar{x}^k}{\partial x^a}\frac{\partial \bar{x}^l}{\partial x^b}\bar{g}_{kl}(\bar{x}^m). \quad (3.2)$$

In $\bar{g}_{kl}$ and g_{ab} the scale factors are related by

$$S(\bar{\tau}) \simeq S(\tau)[1 + (\dot{S}/S)TQ^{(0)}] \quad (3.3)$$

and

$${}^3g_{\alpha\beta}(\bar{x}^\mu) \simeq {}^3g_{\alpha\beta}(x^\mu) + L^{(0)}Q^{(0)\mu}\frac{\partial}{\partial x^\mu}{}^3g_{\alpha\beta}. \quad (3.4)$$

The metric derivatives in Eq. (3.4) combine with the coordinate derivative of $Q^{(0)\alpha}$ in $\partial\bar{x}^\alpha/\partial x^\beta$ to give covariant derivatives of $Q^{(0)\alpha}$. The final result for the changes in the amplitudes of the metric perturbations defined by Eqs. (2.14) is

$$\bar{A} = A - \dot{T} - (\dot{S}/S)T, \quad (3.5a)$$

$$\bar{B}^{(0)} = B^{(0)} + \dot{L}^{(0)} + kT, \quad (3.5b)$$

$$\bar{H}_L = H_L - (k/3)L^{(0)} - (\dot{S}/S)T, \quad (3.5c)$$

$$\bar{H}_T^{(0)} = H_T^{(0)} + kL^{(0)}. \quad (3.5d)$$

Now consider the matter perturbations. The new matter three-velocity is, by definition,

$$\bar{v}^{(0)}Q^{(0)\alpha} = d\bar{x}^\alpha/d\bar{\tau} \simeq dx^\alpha/d\tau + \dot{L}^{(0)}Q^{(0)\alpha},$$

so

$$\bar{v}^{(0)} = v^{(0)} + \dot{L}^{(0)}. \quad (3.6)$$

The energy density E is a coordinate scalar, but

$$E(\bar{\tau}) = E_0(\bar{\tau})[1 + \bar{\delta}Q^{(0)}]$$
$$\cong E_0(\tau)[1 + (\bar{\delta} + T\dot{E}_0/E_0)Q^{(0)}].$$

Thus the energy density perturbation does change by

$$\bar{\delta} = \delta + 3(1+w)(\dot{S}/S)T. \quad (3.7)$$

Equation (2.6) is used to eliminate $\dot{E}_0$. Similarly,

$$\bar{\pi}_L = \pi_L - T\dot{P}_0/P_0 = \pi_L + 3(1+w)\frac{c_s^2}{w}\frac{\dot{S}}{S}T. \quad (3.8)$$

The amplitude of the traceless part of the stress tensor $\pi_T^{(0)}$ is gauge invariant.

The usual way of dealing with this gauge freedom is to impose conditions on the form of the metric tensor and/or matter perturbations. Examples are the synchronous gauge $A = B^{(0)} = 0$,[2,7,15] the longitudinal gauge $H_T^{(0)} = B^{(0)} = 0$,[4] the comoving proper time gauge $A = v^{(0)} = 0$,[5] and the comoving time-orthogonal gauge $B^{(0)} = v^{(0)} = 0$.[22] Such a condition may or may not specify the gauge uniquely. For instance, the transformation from some other gauge to the synchronous gauge contains two free constants of integration in T and $L^{(0)}$. Any ambiguity in the gauge condition implies the existence of extra, unphysical gauge modes when the Einstein equations are solved.[2,11]

The general covariance of Einstein's theory of gravity guarantees complete freedom in the choice of gauge as long as one can demonstrate the existence of a gauge transformation to that gauge from arbitrary metric tensor components and/or matter variables. The corollary of this principle is that *only gauge-invariant quantities have any inherent physical meaning.* Gauge-dependent quantities, such as the energy density perturbation, have physical meaning only to the extent that, in a particular gauge, they can be identified with a gauge-independent quantity either exactly or approximately.

To have genuine physical significance, gauge-independent quantities should be constructed from the variables naturally present in the problem, here the perturbations in the metric tensor and energy-momentum tensor, without reliance on artificially introduced variables, such as the four-velocity of an *ad hoc* congruence of "observers."

First consider the amplitudes of the metric tensor perturbations. Only two independent gauge-independent quantities can be constructed from the metric tensor amplitudes alone, since there are two gauge functions and four metric tensor amplitudes. By inspection of Eqs. (3.5), these are conveniently taken as

$$\Phi_A \equiv A + \frac{1}{k}\dot{B}^{(0)} + \frac{1}{k}\frac{\dot{S}}{S}B^{(0)} - \frac{1}{k^2}\left(\ddot{H}_T^{(0)} + \frac{\dot{S}}{S}\dot{H}_T^{(0)}\right) \quad (3.9)$$

and

$$\Phi_H \equiv H_L + \tfrac{1}{3}H_T^{(0)} + \frac{1}{k}\frac{\dot{S}}{S}B^{(0)} - \frac{1}{k^2}\frac{\dot{S}}{S}\dot{H}_T^{(0)}. \quad (3.10)$$

The physical interpretation of these gauge-invariant potentials will be postponed until after we have considered the matter perturbations.

The simplest gauge-invariant matter "velocity" amplitude is obviously, from Eqs. (3.6) and (3.5d),

$$v_s^{(0)} \equiv v^{(0)} - \frac{1}{k}\dot{H}_T^{(0)}. \quad (3.11)$$

This has a direct physical interpretation in terms of the shear of the matter velocity field. The shear tensor is[15]

$$\sigma_{ab} = \tfrac{1}{2}P^k_a(u_{k;l} + u_{l;k})P^l_b - \tfrac{1}{3}P_{ab}u^k_{;k},$$

with $P_{ab} = g_{ab} + u_a u_b$. This vanishes in the background and to first order the only nonzero components are

$$\sigma_{\alpha\beta} = S(\dot{H}_T^{(0)} - kv^{(0)})Q^{(0)}_{\alpha\beta}.$$

The magnitude of the shear is then

$$\sigma \equiv (\tfrac{1}{2}\sigma_{kl}\sigma^{kl})^{1/2} = \left|\frac{k}{S}v_s^{(0)}\right|(\tfrac{1}{2}Q^{(0)\alpha\beta}Q^{(0)}_{\alpha\beta})^{1/2}. \quad (3.12)$$

The velocity amplitude $v_s^{(0)}$ is the velocity which, when divided by the proper reduced wavelength S/k, gives the time dependence of the rate of shear associated with the perturbation.

The energy density perturbation amplitude δ must be combined with other quantities to produce a gauge-invariant measure of the density perturbation, and one obvious criterion is that the gauge-invariant quantity reduce to δ as soon as the perturbation comes inside the particle horizon, $k^{-1}\dot{S}/S \ll 1$. There are two obvious possibilities. First, consider

$$\epsilon_m \equiv \delta + 3(1+w)\frac{1}{k}\frac{\dot{S}}{S}(v^{(0)} - B^{(0)}), \qquad (3.13)$$

which is gauge invariant by Eqs. (3.7), (3.6), and (3.5b). The amplitude ϵ_m is equal to δ in any gauge in which $v^{(0)} = B^{(0)}$, but this is just the condition that the matter world lines be orthogonal to the τ = constant spacelike hypersurface. Thus, *ϵ_m is the natural choice of gauge-invariant energy density perturbation amplitude from the point of view of the matter.* It is the density perturbation relative to the spacelike hypersurface which represents everywhere the matter local rest frame.

This is *not* the same as the density perturbation defined by Olson,[13] which compares energy densities at the same proper time calculated as an integral along each matter world line from the initial singularity. The Olson density perturbation definition has direct operational meaning for individual comoving observers only after the perturbation comes within the particle horizon. It has no gauge-invariant meaning which is local in time and can give a nonzero value for the density perturbation at times when the Universe is actually exactly homogeneous and isotropic, if a real perturbation was present temporarily at an earlier time.[23]

An alternative gauge-invariant density perturbation amplitude is

$$\epsilon_g \equiv \delta - 3(1+w)\frac{1}{k}\frac{\dot{S}}{S}\left(B^{(0)} - \frac{1}{k}\dot{H}_T^{(0)}\right). \qquad (3.14)$$

From the discussion of shear following Eq. (3.11) and the fact that $B^{(0)}$ is the three-velocity amplitude of world lines normal to the τ = constant hypersurface, one sees that ϵ_g measures the energy density perturbation relative to the hypersurface whose normal unit vectors have *zero shear*. This geometrically selected hypersurface is as close as possible to a "Newtonian" time slicing.

Of course, any linear combination of ϵ_m and ϵ_g is gauge invariant, and the physical significance of one such linear combination will be discussed in Sec. V [see Eq. (5.26)]. Here I focus on ϵ_m because first, it acts in the Einstein equations as the source for the gauge-invariant potential Φ_H [see Eq. (4.3)], and second, the equations governing the dynamics of the matter are more transparent physically when written in terms of ϵ_m. The difference

$$\epsilon_m - \epsilon_g = 3(1+w)\frac{1}{k}\frac{\dot{S}}{S}v_s^{(0)} \qquad (3.15)$$

is small once the perturbation is well inside the particle horizon, but is large at early times. Typically, $\epsilon_m \propto (k\tau)^2$ and ϵ_g is constant at $k\tau \ll 1$ for perturbations regular as $S \to 0$.

The zero-shear hypersurface can be invoked to give physical (geometrical) meaning to the gauge-invariant potentials Φ_A and Φ_H. In a gauge where each constant-τ hypersurface has normals with zero shear, i.e.,

$$B^{(0)} - \frac{1}{k}\dot{H}_T^{(0)} = 0,$$

Eqs. (3.9) and (3.10) greatly simplify to give

$$\Phi_A = A, \quad \Phi_H = H_L + \tfrac{1}{3}H_T^{(0)}.$$

But then Φ_A is the amplitude for the spatial dependence of the proper time intervals along the normals between two neighboring such zero-shear hypersurfaces (the lapse function), while the intrinsic scalar curvature of a zero-shear hypersurface is [see Eq. (A6)], to first order in the perturbation,

$$\mathcal{R}_{\text{zero shear}} = [6K + 4(k^2 - 3K)\Phi_H Q^{(0)}]/S^2. \qquad (3.16)$$

In this sense, Φ_H physically represents a "curvature perturbation." Sufficient conditions for the global perturbations of the spacetime geometry to be small are $\Phi_A Q^{(0)} \ll 1$, $\Phi_H Q^{(0)} \ll 1$, but these are not necessary conditions since other hypersurfaces may be less strongly warped by the perturbation (see Sec. V).

To complete the gauge-invariant description of the matter perturbations, note that the fractional isotropic pressure perturbation can be expressed in terms of the energy density perturbation and the entropy perturbation $\eta Q^{(0)}$ through Eq. (2.21). But from Eqs. (2.21), (3.7), and (3.8) η is gauge invariant.

Certain gauges greatly simplify the representation of the gauge-invariant variables. For instance, in the longitudinal gauge $H_T^{(0)} = B^{(0)} = 0$, Eqs. (3.9)–(3.11) and Eq. (3.14) become $\Phi_A = A$, $\Phi_H = H_L$, $v_s^{(0)} = v^{(0)}$, $\epsilon_g = \delta$. A gauge in which $\epsilon_m = \delta$ and the other amplitudes are relatively simple is $v^{(0)} - B^{(0)} = 0$, $H_T^{(0)} = 0$. Both these gauge conditions uniquely specify the gauge.

B. Vector perturbations

The gauge-transformation properties of vector perturbations are much simpler than for scalar perturbations, since now there is no gauge ambiguity about the time coordinate. The most general gauge transformation is

$$\tilde{x}^{\alpha}=x^{\alpha}+L^{(1)}(\tau)Q^{(1)\alpha}(x^{\mu}), \tag{3.17}$$

so

$$\tilde{B}^{(1)}=B^{(1)}+\dot{L}^{(1)}, \tag{3.18a}$$

$$\tilde{H}_T^{(1)}=H_T^{(1)}+kL^{(1)}, \tag{3.18b}$$

and the matter velocity transforms as

$$\tilde{v}^{(1)}=v^{(1)}+\dot{L}^{(1)}. \tag{3.19}$$

The only gauge-invariant combination of metric tensor amplitudes is

$$\Psi\equiv B^{(1)}-\frac{1}{k}\dot{H}_T^{(1)}. \tag{3.20}$$

This, times k/S, is the amplitude of the shear of the normals to the constant-τ hypersurface. What was naturally constrained to vanish in dealing with scalar perturbations is now gauge invariant and necessarily nonvanishing if vector perturbations are present.

As for the energy density perturbations in the scalar case, there are two alternative choices for gauge-invariant forms of the matter velocity perturbation. The one related to the *shear* by the vector perturbation analog of Eq. (3.12) is

$$v_s^{(1)}\equiv v^{(1)}-\frac{1}{k}\dot{H}_T^{(1)}. \tag{3.21}$$

The other is related to the vorticity tensor

$$\omega_{ab}=\tfrac{1}{2}P_a^k(u_{k;l}-u_{l;k})P_b^l,$$

the only nonzero components of which are

$$\omega_{\alpha\beta}=S(v^{(1)}-B^{(1)})kW_{\alpha\beta}$$
$$=S(v^{(1)}-B^{(1)})(Q_{\alpha|\beta}^{(1)}-Q_{\beta|\alpha}^{(1)}).$$

The magnitude, the intrinsic angular velocity of an individual fluid element

$$\omega=[\tfrac{1}{2}\omega_{\alpha\beta}\omega^{\alpha\beta}]^{1/2}$$
$$=\left|\frac{k}{S}(v^{(1)}-B^{(1)})\right|[\tfrac{1}{2}W_{\alpha\beta}W^{\alpha\beta}]^{1/2} \tag{3.22}$$

is, like the shear, directly measurable from the local behavior of the matter. Let

$$v_c\equiv v^{(1)}-B^{(1)}=v_s^{(1)}-\Psi. \tag{3.23}$$

It is v_c, the velocity relative to the normal to the constant-τ hypersurface, that is the source for Ψ in the Einstein equations, rather than $v_s^{(1)}$.

C. Tensor perturbations

Since no three-vector or scalar can be formed from a tensor harmonic $Q_{\alpha\beta}^{(2)}$, the amplitudes $H_T^{(2)}$ and $\pi_T^{(2)}$ as defined in Eqs. (2.25) and (2.26) are automatically gauge invariant.

IV. PERTURBATION EQUATIONS IN GAUGE-INVARIANT VARIABLES

The usual approach to the derivation of the equations governing linearized perturbations in cosmology has been to impose at the beginning a gauge condition to simplify the form of the metric and/or matter perturbations and then work directly with the metric tensor components and matter variables. However, a complete set of equations can be obtained directly in terms of the gauge-invariant variables defined in Sec. III. These equations are mathematically simpler and physically more transparent than the usual ones, particularly in comparison with the commonly used synchronous gauge. Spurious gauge modes are automatically excluded.

In the Appendix we give the perturbations in the Ricci tensor components in a general gauge. The perturbation in the Einstein tensor is

$$\delta G_b^a=\delta R_b^a-\tfrac{1}{2}\delta_b^a\delta R.$$

A. Scalar perturbations

For scalar perturbations, one gauge-invariant combination is

$$\delta G_0^0-\frac{3}{k^2}\frac{\dot{S}}{S}(\delta G_\alpha^0)^{|\alpha}=-2\frac{(k^2-3K)}{S^2}\Phi_H Q^{(0)}, \tag{4.1}$$

the other

$$\delta G_\beta^\alpha-\tfrac{1}{3}\delta_\beta^\alpha\delta G_\mu^\mu=-\frac{k^2}{S^2}(\Phi_A+\Phi_H)Q_\beta^{(0)\alpha}. \tag{4.2}$$

Upon equating δG_b^a with δT_b^a through the Einstein equations one finds from Eqs. (4.1), (2.17), (2.20), and (3.13)

$$2\frac{(k^2-3K)}{S^2}\Phi_H=E_0\epsilon_m \tag{4.3}$$

and from Eqs. (4.2) and (2.19)

$$-\frac{k^2}{S^2}(\Phi_A+\Phi_H)=P_0\pi_T^{(0)}. \tag{4.4}$$

Thus, both gauge-invariant amplitudes for the metric tensor perturbation are related algebraically in a very simple way to gauge-invariant amplitudes of the matter perturbations. Equation (4.3), in spite of appearances, really derives from the energy initial-value equation on the zero-shear hypersurface (see Sec. VI). For a perfect fluid, Eq. (4.4) gives $\Phi_A=-\Phi_H$.

All of the dynamics for scalar (and also vector) perturbations resides in the equations of motion of the matter, $T^k_{0;k}=0$ and $T^k_{\alpha;k}=0$. These are also written out in the Appendix for a general gauge. Straightforward manipulation of the momentum equation, with the help of Eqs. (3.9), (3.11), (3.13), and (2.21), gives the explicitly gauge-invariant form

$$\dot{v}_s^{(0)}+\frac{\dot{S}}{S}v_s^{(0)}=k\Phi_A+k(1+w)^{-1}(c_s^2\epsilon_m+w\eta)$$
$$-\tfrac{2}{3}k(1-3K/k^2)(1+w)^{-1}w\pi_T^{(0)}\,. \qquad (4.5)$$

The second term on the right-hand side of Eq. (4.5) is S times the acceleration in the rest frame of the matter due to the pressure gradient force, and the third term is S times the acceleration due to the divergence of the anisotropic part of the stress tensor. The inertial mass per unit volume is $E_0(1+w)$, and the *proper* wave number is k/S. However, the first term is *not* S times the gravitational acceleration in the matter rest frame. If it were, the left-hand side of Eq. (4.5) would be zero. Rather, it is the gravitational acceleration in the "shear-free frame" associated with the geometry perturbations, as discussed in Sec. III. Equation (4.5) has *exactly* the same form [except for the factor $(1-3K/k^2)$ in the third term on the right-hand side] as the corresponding *Newtonian* equation in an expanding background, with $v_s^{(0)}$ the analog of the Newtonian peculiar velocity and Φ_A the analog of the Newtonian gravitational potential.

The energy equation is less transparent because it is sensitive to first-order changes in the frame of reference in ways the momentum equation is not. Begin by considering Eq. (A4a) in a particular gauge, the comoving, time-orthogonal gauge $v^{(0)}=B^{(0)}=0$. In this gauge $\epsilon_m=\delta$, $v_s^{(0)}=-(1/k)\dot{H}_T^{(0)}$, and from Eq. (3.10)

$$\dot{H}_L=\dot{\Phi}_H+\tfrac{1}{3}kv_s^{(0)}-\frac{1}{k}\left(\frac{\dot{S}}{S}v_s^{(0)}\right)^{\cdot}\,. \qquad (4.6)$$

The obvious substitutions and elimination of $\dot{v}_s^{(0)}$ through Eq. (4.5) give a gauge-invariant equation for $\dot{\Phi}_H$, but a rather messy one. Now use Eqs. (4.3) and (4.4) to eliminate ϵ_m and Φ_A in favor of Φ_H and simplify with the help of the background Eqs. (2.5) and (2.6). After elimination of a common factor $2[k^2-3K+\frac{3}{2}(E_0+P_0)S^2]/E_0S^2$,

$$\dot{\Phi}_H+\frac{\dot{S}}{S}\Phi_H=-\frac{1}{2}\,\frac{(E_0+P_0)S^2}{k}v_s^{(0)}-\frac{P_0S^2}{k^2}\,\frac{\dot{S}}{S}\pi_T^{(0)}\,. \qquad (4.7)$$

Alternatively, write Eq. (4.7) in a more familiar form

$$[E_0S^3\epsilon_m]^{\cdot}=-(1-3K/k^2)(E_0+P_0)S^3kv_s^{(0)}$$
$$-2(1-3K/k^2)P_0S^2\dot{S}\pi_T^{(0)}\,. \qquad (4.8)$$

This has some resemblance to a *special* relativistic energy equation, in that $(k/S)(E_0+P_0)v_s^{(0)}$ is the divergence of the energy flux. However, ϵ_m is not the energy density perturbation in the appropriate frame, and the special relativistic equation would have $3\pi_L$ in place of $2\pi_T^{(0)}$.

Equations (4.3)–(4.5) and Eq. (4.8) combine into a single second-order equation for ϵ_m,

$$(E_0S^3\epsilon_m)^{\cdot\cdot}+(1+3c_s^2)\frac{\dot{S}}{S}(E_0S^3\epsilon_m)^{\cdot}+[(k^2-3K)c_s^2-\tfrac{1}{2}(E_0+P_0)S^2](E_0S^3\epsilon_m)$$
$$=(1-3K/k^2)\{-k^2(P_0S^3\eta)+\tfrac{2}{3}[k^2+3(1+3c_s^2)K](P_0S^3\pi_T^{(0)})+2(w-c_s^2)(E_0S^2)(P_0S^3\pi_T^{(0)})-2\dot{S}(P_0S^2\pi_T^{(0)})^{\cdot}\}\,, \qquad (4.9)$$

or a corresponding equation for Φ_H. A spatially homogeneous perturbation or the lowest inhomogeneous mode $k^2=3K$ in a closed universe require special treatment in that $Q_\alpha^{(0)}$ and/or $Q_{\alpha\beta}^{(0)}$ vanish identically, Φ_H, Φ_A, and $v_s^{(0)}$ are no longer gauge-invariant, and some of the above equations, including Eq. (4.9), are not applicable. See the discussion around Eq. (6.27). A homogeneous scalar perturbation is really no perturbation at all, but an inappropriate choice of background.

Formally, the entropy perturbation amplitude and the anisotropic stress amplitude are free functions of time which act as "sources" for the comoving energy density perturbation in Eq. (4.9). However, if the anisotropic stress is from a shear viscosity, $\pi_T^{(0)}$ should be proportional to $v_s^{(0)}$. In a simple kinetic-theory model the coefficient of viscosity is a density times a mean free path times a velocity. An upper limit to the density is E_0, a characteristic velocity is $w^{1/2}$, and a maximum effective mean free path is the *lesser* of the reduced wavelength of the perturbation S/k and the distance a particle can travel in a Hubble time, $w^{1/2}/(\dot{S}/S^2)$. At the upper limit the simple kinetic-theory picture breaks down and the relation between shear and anisotropic stress becomes nonlocal in time and space. Nevertheless, a reasonable estimate is

$$P_0\pi_T^{(0)} \sim \alpha E_0 \left\{\frac{S}{k}, w^{1/2}\frac{S^2}{\dot{S}}\right\}_{\min} w^{1/2}\left(\frac{k}{S}v_s^{(0)}\right), \quad (4.10)$$

where α is a dimensionless fudge factor unlikely to be larger than one. At early times, $k\tau<1$,

$$\pi_T^{(0)} \lesssim \frac{kS}{\dot{S}} v_s^{(0)}. \quad (4.11)$$

With this restriction, the shear-stress terms in Eqs. (4.8) and (4.9) can never dominate the early evolution.

An anisotropic stress as well as an entropy perturbation might also arise *ab initio* out of some speculative nonkinetic origin, perhaps inhomogeneous gauge-theory phase transitions of the vacuum.[8,9,16,17] This possibility will be discussed further in connection with specific solutions of Eq. (4.9) in Sec. V.

B. Vector perturbations

The equations for vector perturbations (see the Appendix) can be put in gauge-invariant form by inspection. The initial-value equation $\delta G^0_\alpha = \delta T^0_\alpha$ for the "frame-dragging potential" Ψ is [see Eq. (A2b)]

$$\frac{1}{2}\frac{(k^2-2K)}{S^2}\Psi Q_\alpha^{(1)} = (E_0+P_0)v_c Q_\alpha^{(1)}. \quad (4.12)$$

The equation of motion for the matter is

$$\dot{v}_c = \frac{\dot{S}}{S}(3c_s^2-1)v_c - k\frac{w}{1+w}\pi_T^{(1)}. \quad (4.13)$$

C. Tensor perturbations

There is just one equation for tensor perturbations, i.e.,

$$\frac{1}{S^2}\left(\ddot{H}_T^{(2)} + 2\frac{\dot{S}}{S}\dot{H}_T^{(2)} + (k^2+2K)H_T^{(2)}\right) = P_0\pi_T^{(2)}. \quad (4.14)$$

V. SOLUTIONS OF THE PERTURBATION EQUATIONS

A. Scalar perturbations

Explicit solutions of the gauge-dependent equations governing scalar perturbations have been obtained in the past[11] for the case of a perfect fluid ($\pi_T^{(0)}=0$), usually only for adiabatic perturbations ($\eta=0$), when the background equation of state satisfies $w=c_s^2=$const and the background spatial curvature $K=0$. The assumption $K=0$ is well justified for perturbations on the scale of clusters of galaxies or smaller, since then $k^2 \gg K$ and at least prior to recombination $(\dot{S}/S)^2 \gg K$ for any K in the range allowed by observation. The assumption $w=c_s^2=$const is made largely for mathematical convenience, though in the radiation-dominated phase of the universe following electron-position recombination and neutrino decoupling $w=\frac{1}{3}$ is a good approximation. Here the solutions will be extended to allow entropy and anisotropic pressure perturbations with arbitrary time dependence, and the physical significance of the solutions will be analyzed through their expression in gauge-invariant variables.

With $w=c_s^2=$const and $K=0$ the background equations (2.5) and (2.6) have the well-known solution

$$S \propto \tau^\beta, \quad \beta = 2/(3w+1) \quad (5.1)$$

and

$$E_0S^2 = 3(\dot{S}/S)^2 = 3\beta^2\tau^{-2}. \quad (5.2)$$

The parameter β ranges from 2 (for $w=0$) to 1 (for $w=\frac{1}{3}$) to $\frac{1}{2}$ (for $w=1$).

Define a new independent variable

$$x \equiv k\tau \quad (5.3)$$

and in Eq. (4.9) a new independent variable

$$f \equiv x^{\beta-2}\epsilon_m = \tfrac{2}{3}\beta^{-2}x^\beta\Phi_H. \quad (5.4)$$

Let a prime denote d/dx. Then Eq. (4.9) becomes

$$f'' + 2x^{-1}f' + [c_s^2 - \beta(\beta+1)x^{-2}]f = -x^{\beta-2}[w\eta - \tfrac{2}{3}w\pi_T^{(0)} + 2\beta x(x^{-2}w\pi_T^{(0)})']. \quad (5.5)$$

Given a solution for f, Eq. (4.8) determines $v_s^{(0)}$ as

$$v_s^{(0)} = -\frac{3}{2}\frac{\beta}{\beta+1}x^{2-\beta}f' - 3\frac{\beta^2}{\beta+1}x^{-1}w\pi_T^{(0)}. \quad (5.6)$$

First consider the homogeneous version of Eq. (5.5), with $w\eta = w\pi_T^{(0)}=0$. The solutions are obviously spherical Bessel functions, either j_β or the spherical Neumann function n_β, and

$$f = a j_\beta(c_s x) + b n_\beta(c_s x). \quad (5.7)$$

The simple form of the homogeneous solution, with nothing in it to mark the particle horizon at $x\sim 1$ as distant from the sound horizon at $c_s x \sim 1$, confirms the choice of ϵ_m, Φ_H, and $v_s^{(0)}$ as mathematically, and in some sense physically, natural gauge-invariant amplitudes to describe the perturbation.

Now take the limit $c_s x \ll 1$, but not necessarily $x \ll 1$. The oscillatory behavior at $c_s x > 1$ becomes power-law behavior. Renormalize the coefficients so

$$f \simeq c x^\beta + d x^{-\beta-1}, \quad (5.8)$$

with

$$c = \frac{\sqrt{\pi}\, c_s^\beta}{2^{\beta+1}\Gamma(\beta+3/2)}a, \quad d = -\frac{2^\beta\Gamma(\beta+1/2)}{\sqrt{\pi}\, c_s^{\beta+1}}b. \quad (5.9)$$

The standard gauge-invariant amplitudes at $c_s x \ll 1$ are

$$\epsilon_m \simeq cx^2 + dx^{-2\beta+1}, \quad \Phi_H \simeq \tfrac{3}{2}\beta^2(c + dx^{-2\beta-1}),$$
$$v_s^{(0)} \simeq \tfrac{3}{2}\beta\left(-c\,\frac{\beta}{\beta+1}x + dx^{-2\beta}\right). \tag{5.10}$$

The two independent modes may, at $c_s x \ll 1$, be characterized as a "growing mode" and a "decaying" mode by looking at the amplitude of the fractional energy density perturbation on comoving hypersurfaces. The amplitude Φ_H measures the distortion of the zero-shear hypersurfaces by the perturbation. The spatial curvature $\sim (k^2/S^2)\Phi_H$ is coherent over the reduced wavelength $\sim S/k$. For the growing mode this distortion is nonzero at $x=0$ and remains constant until the sound wave begins to oscillate. We shall see that this behavior is also characteristic of the distortion of the comoving hypersurfaces and uniform-Hubble-constant hypersurfaces, so geometrically the growing mode is a "constant" mode at $x \ll 1$. However, the apparent relative amplitude of the energy density perturbation is strongly hypersurface dependent at $x \ll 1$. Consider the amplitude ϵ_g, which measures the fractional energy density perturbation on zero-shear hypersurfaces. From Eq. (3.15),

$$\epsilon_g \simeq c(x^2 + 3\beta^2) + dx^{-2\beta-1}[x^2 - 3\beta(\beta+1)], \tag{5.11}$$

still assuming $(c_s x)^2 \ll 1$. For both modes ϵ_g is of order x^{-2} times ϵ_m at $x \ll 1$, though ϵ_g and ϵ_m coincide for all $x \gg 1$ even if $c_s \ll 1$.

About the only measure of the relative amplitude of the perturbation which is strictly hypersurface independent is the ratio of the rate of shear σ to the background expansion rate, because the shear vanishes in the background. The amplitudes of the shear and expansion are, respectively, $(k/S)\,v_s^{(0)}$ and $\dot{S}/S^2$. The ratio is

$$\xi \equiv (kS/\dot{S})v_s^{(0)} \simeq \frac{3}{2}\left(-c\,\frac{\beta}{\beta+1}x^2 + dx^{-2\beta+1}\right) \tag{5.12}$$

at $x \ll 1$, which has the same time dependence for each mode as ϵ_m. For $w < 1$, or $\beta > \frac{1}{2}$, the decaying homogeneous mode is unambiguously singular at $x=0$. The gauge-invariant amplitude Φ_H becomes singular as $x^{-(2\beta+1)}$ in the decaying mode, but we shall see that this overstates the physical strength of the singularity. On the comoving and the uniform-Hubble-constant hypersurfaces the distortion only becomes singular as $x^{-(2\beta-1)}$, similar to ϵ_m and the relative shear in Eq. (5.12).

The results so far are familiar ones,[4,11,22] though the explicit gauge-invariant form is new. However, relatively little attention has been paid to the generation of density perturbations at early times through stress perturbations. As mentioned in Sec. I and as is obvious from Eq. (4.9), an energy density perturbation can only arise from a nonadiabatic pressure perturbation or an anisotropic stress perturbation if the Universe began perfectly homogeneous. Local conservation of energy and momentum prevents any action directly on the energy and momentum densities.

Entropy perturbations have been considered before,[12] but it appears from Eq. (5.5) that anisotropic stress perturbations might be more interesting. At $x \ll 1$ the second anisotropic stress term is of order x^{-2} times the entropy perturbation term for comparable values of $w\eta$ and $w\pi_T^{(0)}$.

The solution of Eq. (5.5) for an arbitrary source is accomplished by variation of parameters or (equivalently) by constructing the Green's function. I will apply initial conditions that the perturbations vanish at $x=0$, but some of the homogeneous solution, Eq. (5.7), can always be added if desired. The derivative on the right-hand side of Eq. (5.5) can be integrated by parts. The identity

$$\frac{d}{dy}[y^{\beta+1}n_\beta(c_s y)] = c_s y^{\beta+1} n_{\beta-1}(c_s y), \tag{5.13}$$

and similarly for j_β, simplifies the result, which is

$$f(x) = c_s \int_0^x dy\, y^\beta \{(w\eta - \tfrac{2}{3}w\pi_T^{(0)})[j_\beta(c_s x)n_\beta(c_s y) - n_\beta(c_s x)j_\beta(c_s y)]$$
$$- 2\beta c_s y^{-1} w\pi_T^{(0)}[j_\beta(c_s x)n_{\beta-1}(c_s y) - n_\beta(c_s x)j_{\beta-1}(c_s y)]\}. \tag{5.14}$$

If $c_s x \ll 1$, the spherical Bessel functions may be replaced by

$$j_\beta(c_s x)n_\beta(c_s y) \simeq -(2\beta+1)^{-1}(c_s y)^{-1}(x/y)^\beta \text{ and } x \leftrightarrow y,$$
$$j_\beta(c_s x)n_{\beta-1}(c_s y) \simeq -(4\beta^2-1)^{-1}(x/y)^\beta, \quad \beta > \tfrac{1}{2},$$
$$n_\beta(c_s x)j_{\beta-1}(c_s y) \simeq -(c_s x)^{-2}(y/x)^{\beta-1}.$$

Keeping only the dominant terms, Eq. (5.14) simplifies to

$$f(x)\simeq(2\beta+1)^{-1}\left\{x^{\beta}\int_0^x dy\,y^{-1}\left[\frac{2}{3}\,\frac{\beta+1}{2\beta-1}w\pi_T^{(0)}-w\eta\right]+x^{-\beta-1}\int_0^x dy\,y^{2\beta}[-2\beta(2\beta+1)y^{-2}w\pi_T^{(0)}+w\eta]\right\}. \tag{5.15}$$

If $\beta=\frac{1}{2}$ the contribution of $w\pi_T^{(0)}$ to the growing mode is not infinite; rather, $n_{-1/2}$ contains a logarithm, and

$$\frac{2}{3}\,\frac{\beta+1}{2\beta-1}\to\tfrac{1}{2}\ln(1/y)\,. \tag{5.16}$$

In Eq. (5.15), the entropy perturbation contributes comparable amounts of growing mode and decaying mode to f and therefore to $\epsilon_m=x^{2-\beta}f$. The contribution to ϵ_m while the entropy perturbation is present is of order $x^2w\eta$, once the entropy perturbation has been on at roughly constant amplitude for at least one expansion time. Press and Vishniac[12] claim that an entropy perturbation couples *only* to the growing mode at $x\ll1$. The apparent contradiction arises because their definition of the "physical" energy density perturbation is based on uniform-Hubble-constant hypersurfaces; the relative strength of the growing and decaying modes is hypersurface dependent, as we shall soon see in detail. Still, if the entropy perturbation turns off at $x\ll1$, by the time the perturbation comes within the particle horizon at $x\sim1$ the decaying mode is insignificant compared with the growing mode. At $x\sim1$ the latter has an amplitude in ϵ_m which is roughly the maximum previous amplitude of $w\eta$, the ratio of the nonadiabatic pressure perturbation to the background energy density, averaged over one e-folding in the conformal time τ.

The anisotropic stress source term of order x^{-2} relative to the entropy perturbation source term contributes *only* to the decaying mode at $x\ll1$ in Eq. (5.15). Except for a modest enhancement if β is close to $\frac{1}{2}$, as indicated in Eq. (5.16), anisotropic stress and isotropic stress perturbations of the same amplitude generate comparable amounts of the growing mode in f, ϵ_m, and [see Eq. (5.4)] Φ_H. After an anisotropic stress perturbation has been on at roughly constant amplitude for one expansion time, Eq. (5.15) shows that the part of ϵ_m associated with the decaying mode is of order $w\pi_T^{(0)}$ and the part of ϵ_m associated with the growing mode is of order $x^2w\pi_T^{(0)}$. If the anisotropic stress then disappears at $x=x_1$, by the time $x=1$ the decaying-mode contribution to ϵ_m is of order $x_1^{+(2\beta-1)}w\pi_T^{(0)}$, small (unless $\beta\simeq\frac{1}{2}$) compared with the growing-mode contribution of order $w\pi_T^{(0)}$, if $x_1\ll1$.

The perturbation in ϵ_m is always small if $w\pi_T^{(0)}\ll1$. On the other hand, the amplitude Φ_H, which measures the global distortion of the zero-shear hypersurfaces, is of order $x^{-2}\epsilon_m$ and while the anisotropic stress is present can be larger than one even if the anisotropic stress is small compared with the background energy density, as long as the anisotropic stress is present at a very early time, with $x^2<w\pi_T^{(0)}$. Some gauge-theory symmetry breaking might take place near the Planck time.[8] At the Planck time the value of $x=k\tau$ corresponding to a mass M in the present universe is $x\sim10^{-26}(10^{12}M_\odot/M)^{1/3}$, so even though the comoving volume of a galactic mass is large compared with the particle horizon at this value of x, it is not inconceivable that a small statistical residual anisotropy on a galactic scale could make in a sense the perturbations nonlinear.

Does $\Phi_H>1$ really imply a physically significant nonlinearity, one that could perhaps couple the large amplitude decaying mode to the growing mode and give a value of ϵ_m at the particle horizon large compared with the maximum previous value of $w\pi_T^{(0)}$? To answer this question, at least in part, one should look carefully at the complete description of the perturbation on various types of hypersurfaces while the anisotropic stress is present or just after it turns off.

The description definitely is potentially nonlinear on the zero-shear hypersurfaces. The distortion amplitude $\phi_g\equiv\Phi_H$ is a measure of the amplitude of the spatial metric perturbations on the zero-shear hypersurfaces, and the amplitude $\alpha_g\equiv\Phi_A$ measures the fractional perturbation in the lapse function. Both these are of order $x^{-2}w\pi_T^{(0)}$, as is the fractional energy density perturbation amplitude ϵ_g from Eq. (5.11). The matter velocity amplitude on zero-shear hypersurfaces is $v_g\equiv v_s^{(0)}$, and from Eq. (5.6) is of order $x^{-1}w\pi_T^{(0)}$, though the shear to expansion rate ratio is of order $w\pi_T^{(0)}$ [see Eq. (5.12)].

Now consider the comoving hypersurfaces. We have already seen that the fractional energy density perturbation amplitude is only of order $w\pi_T^{(0)}$. The matter velocity relative to the comoving hypersurface is zero by definition. To get the geometrical amplitudes, consider a gauge transformation from the longitudinal gauge $H_T^{(0)}=B^{(0)}=0$, where $A=\Phi_A$, $H_L=\Phi_H$, $v^{(0)}=v_s^{(0)}$, to a comoving gauge with $\bar{v}^{(0)}-\bar{B}^{(0)}=0$. Equations (3.5b) and (3.6) give

$$T=k^{-1}(v^{(0)}-B^{(0)})=k^{-1}v_s^{(0)}\,. \tag{5.17}$$

Then from Eq. (3.5a) the fractional perturbation in the lapse function has an amplitude

$$\tilde{A}=\Phi_A-k^{-1}\left(\dot{v}_s^{(0)}+\frac{\dot{S}}{S}v_s^{(0)}\right)\equiv\alpha_m\,, \tag{5.18}$$

and the measure of the distortion of the intrinsic geometry is

$$\phi_m=\tilde{H}_L+\tfrac{1}{3}\tilde{H}_T=\Phi_H-\frac{\dot{S}}{S}k^{-1}v_s^{(0)}\,. \tag{5.19}$$

Equation (4.5) gives

$$\alpha_m=-(1+w)^{-1}(c_s^{\,2}\epsilon_m+w\eta)+\tfrac{2}{3}(1-3K/k^2)\frac{w}{1+w}\pi_T^{(0)}\,, \tag{5.20}$$

so α_m is of order $w\pi_T^{(0)}$ and $w\eta$. An equation for $\dot{\phi}_m$ can be derived from Eqs. (4.5), (4.7), and (2.5), with the result

$$\dot{\phi}_m=-Kk^{-1}v_s^{(0)}-(1+w)^{-1}\frac{\dot{S}}{S}[c_s^{\,2}\epsilon_m+w\eta-\tfrac{2}{3}(1-3K/k^2)w\pi_T^{(0)}]\,. \tag{5.21}$$

Each term on the right-hand side of Eq. (5.21) is at most of order $\tau^{-1}(w\eta, w\pi_T^{(0)})$ and therefore ϕ_m is also of order $w\eta$, $w\pi_T^{(0)}$, without any special assumption that w is constant.

On comoving hypersurfaces, then, all relative perturbation amplitudes are small up until the time the perturbation enters the particle horizon if the perturbation vanishes initially and subsequent stress perturbations are small ($w\eta\ll 1$, $w\pi_T^{(0)}\ll 1$), no matter how early the stress perturbations turn on, as long as the stress perturbations are on in full strength for only a reasonably finite number of e-foldings in τ. The larger amplitudes ϕ_ϵ, α_ϵ, ϵ_ϵ, and $v_s^{(0)}$ for zero-shear hypersurfaces then are *only* due to a large warping of the zero-shear hypersurfaces relative to comoving hypersurfaces.

Finally, consider the perturbations relative to uniform-Hubble-constant hypersurfaces. Again, the simplest route to the gauge-invariant amplitudes describing the matter and geometry from the point of view of this hypersurface is the gauge transformation from the longitudinal gauge to the gauge satisfying the hypersurface condition [see Eq. (A7)]

$$\tilde{\dot{H}}_L+\frac{k}{3}\tilde{B}^{(0)}-\frac{\dot{S}}{S}\tilde{A}=0\,. \tag{5.22}$$

The amplitude of change in the time coordinate is

$$\begin{aligned}T&=-3[k^2-3K+\tfrac{3}{2}(E_0+P_0)S^2]^{-1}\left(\dot{\Phi}_H-\frac{\dot{S}}{S}\Phi_A\right)\\&=\tfrac{3}{2}(E_0+P_0)S^2[k^2-3K+\tfrac{3}{2}(E_0+P_0)S^2]^{-1}k^{-1}v_s^{(0)}\,,\end{aligned} \tag{5.23}$$

after simplifying with the help of Eqs. (4.4) and (4.7). The intrinsic geometry of the uniform-Hubble-constant hypersurface is governed by the amplitude

$$\phi_h=\phi_m+[1+\tfrac{3}{2}(k^2-3K)^{-1}(E_0+P_0)S^2]^{-1}\frac{\dot{S}}{S}k^{-1}v_s^{(0)}\,. \tag{5.24}$$

Similarly, the lapse function perturbation amplitude is

$$\alpha_h=\alpha_m+k^{-1}S^{-1}\{[1+\tfrac{3}{2}(k^2-3K)^{-1}(E_0+P_0)S^2]^{-1}Sv_s^{(0)}\}^{\boldsymbol{\cdot}}\,. \tag{5.25}$$

The amplitude of the fractional energy density perturbation can be written as

$$\epsilon_h=[1+\tfrac{3}{2}(k^2-3K)^{-1}(E_0+P_0)S^2]^{-1}[\epsilon_m+3(1+w)\Phi_m]\,. \tag{5.26}$$

After simplification using the Einstein equations, the amplitude of the matter velocity relative to the hypersurface is

$$\begin{aligned}v_h&=\tilde{v}^{(0)}-\tilde{B}^{(0)}\\&=[1+\tfrac{3}{2}(k^2-3K)^{-1}(E_0+P_0)S^2]^{-1}v_s^{(0)}\,.\end{aligned} \tag{5.27}$$

In a constant-w background, and with $k^2\gg K$, the ubiquitous factor in Eqs. (5.24)–(5.27) becomes

$$[1+\tfrac{3}{2}(k^2-3K)(E_0+P_0)S^2]^{-1}=x^2/[3\beta(\beta+1)+x^2]. \tag{5.28}$$

The description of both the matter and the geometry changes character depending on whether the perturbation is larger or smaller than the particle horizon at $x\sim 1$. On the other hand, the description of the matter and geometry relative to the comoving hypersurface is in all respects oblivious to the particle horizon. I have already remarked on this for ϵ_m, but it also holds for ϕ_m and α_m as is obvious from Eqs. (5.20) and (5.21). Relative to the zero-shear hypersurfaces, the description of the *geometry* is unaffected by the particle horizon, since Φ_H acts as a quasi-Newtonian potential foe ϵ_m, but ϵ_ϵ does change character at the particle horizon [see Eq. (5.11)].

There are several measures of the relative amplitude of the perturbation whose physical interpretation is hypersurface dependent, but only one, the ratio ξ of matter shear rate to expansion rate [Eq. (5.12)], which has a hypersurface-independent *physical* significance. Since all are formally gauge invariant, the *mathematical* values of the amplitudes are independent of the gauge/hypersurface choice.

Table I gives the comparison of the hypersurface-dependent amplitudes with ξ for each of the above hypersurface conditions and for each of the two *homogeneous* modes of scalar perturbations,

TABLE I. Amplitudes of physical perturbations at $x \ll 1$ compared with amplitude ξ of the ratio of the shear rate of the matter to the background expansion rate.

Hypersurface	ϵ/ξ	v/ξ	ϕ/ξ	α/ξ
(a) "Growing" mode, $\xi \sim x^2$				
Uniform-Hubble-constant	$-\frac{2}{3}\frac{2\beta+1}{\beta}$	$\frac{1}{3}\frac{\beta}{\beta+1}\left(\frac{x}{\beta}\right)$	$-\frac{2\beta+1}{\beta}\left(\frac{x}{\beta}\right)^{-2}$	$\frac{2}{3}\frac{2\beta+1}{\beta(\beta+1)}$
Comoving	$-\frac{2}{3}\frac{\beta+1}{\beta}$	0	$-\frac{2\beta+1}{\beta}\left(\frac{x}{\beta}\right)^{-2}$	$\frac{1}{3}\frac{2-\beta}{\beta}$
Zero-shear	$-2\frac{\beta+1}{\beta}\left(\frac{x}{\beta}\right)^{-2}$	$\left(\frac{x}{\beta}\right)^{-1}$	$-\frac{\beta+1}{\beta}\left(\frac{x}{\beta}\right)^{-2}$	$\frac{\beta+1}{\beta}\left(\frac{x}{\beta}\right)^{-2}$
(b) "Decaying" mode, $\xi \sim x^{-(2\beta-1)}$				
Uniform-Hubble-constant	$\frac{2}{9}\frac{\beta}{2\beta-1}\left(\frac{x}{\beta}\right)^2$	$\frac{1}{3}\frac{\beta}{\beta+1}\left(\frac{x}{\beta}\right)$	$\frac{1}{3}\frac{\beta}{2\beta-1}$	$-\frac{2}{9}\frac{\beta}{(\beta+1)(2\beta-1)}\left(\frac{x}{\beta}\right)^2$
Comoving	$\frac{2}{3}$	0	$\frac{1}{3}\frac{\beta(2-\beta)}{(\beta+1)(2\beta-1)}$	$-\frac{1}{3}\frac{2-\beta}{\beta+1}$
Zero-shear	$-2\frac{\beta+1}{\beta}\left(\frac{x}{\beta}\right)^{-2}$	$\left(\frac{x}{\beta}\right)^{-1}$	$\left(\frac{x}{\beta}\right)^{-2}$	$-\left(\frac{x}{\beta}\right)^{-2}$

in the limit that $x \ll 1$. Of course, I also assume w is constant and $k^2 \gg K$. A couple of points are particularly noteworthy. First, for the growing mode the amplitude of the distortion in the intrinsic geometry (ϕ) is independent of time and roughly the same in all three hypersurfaces. The values of ϕ in all three hypersurfaces then predicts the value of the fractional energy density perturbation at the particle horizon, where ϵ_m, ϵ_g, and ϵ_h converge. Second, for the *decaying* mode, but not the growing mode, the amplitudes ϵ_h and α_h are down by a factor of x^2 relative to ϵ_m and α_m. The relatively small value of α_h means that the uniform-Hubble-constant hypersurfaces are close to being synchronous when $x^2 \ll 1$. The relatively small value of ϵ_h is the basis for the claim of Press and Vishniac[12] that an entropy perturbation couples only to the growing mode, but this claim is valid just for the one hypersurface condition. The value of ϕ_h is *not* suppressed relative to ϕ_m, so the perturbation as a whole has the same time dependence, $x^{-(2\beta-1)}$, from the point of view of the uniform-Hubble-constant hypersurfaces as from the point of view of the comoving hypersurfaces.

In conjunction with Table I, it may be helpful to give explicit results for how the entropy perturbation and the anisotropic stress contribute to the amplitude ξ. Assume that the stress perturbations are each on for n e-foldings in the scale factor S, i.e., $n\beta$ e-foldings in the conformal time τ, at a time when $kS/\dot{S} = x/\beta \ll 1$. During this time the relative stress amplitudes $w\eta$ and $w\pi_T^{(0)}$ are constant. The inhomogeneous stress perturbations disappear at $x = x_1$. The growing mode in ξ at $x > x_1$ is then

$$\xi = n\left[\frac{3}{2}\frac{\beta^2}{(\beta+1)(2\beta+1)}w\eta - \frac{\beta^2}{4\beta^2-1}w\pi_T^{(0)}\right]\left(\frac{x}{\beta}\right)^2 . \tag{5.29}$$

If $n \gg 1$, and $-\frac{1}{3} < w < 1$ ($\infty > \beta > \frac{1}{2}$), the ratio of the entropy perturbation part of the decaying mode in ξ to the corresponding part of the growing mode in ξ is $n^{-1}[(\beta+1)/(2\beta+1)](x_1/x)^{2\beta+1}$. The same ratio associated with the anisotropic stress is $3(n\beta)^{-1} \times (2\beta+1)(\beta/x_1)^2(x_1/x)^{2\beta+1}$.

While all relative amplitudes on the comoving and uniform-Hubble-constant hypersurfaces are small if excited by small stress perturbations with $w\eta \ll 1$ and $w\pi_T^{(0)} \ll 1$, it is perhaps conceivable that explicit nonlinear terms in the Einstein equations could generate enough growing mode at early times to give a large fractional energy density perturbation at the particle horizon. This possibility will be laid to rest in Sec. VI. Still, the strictly linear perturbation analysis will be shown to fail on uniform-Hubble-constant hypersurfaces at the same point it fails on zero-shear hypersurfaces, when $w\pi_T^{(0)} > x^2$, though only because the linear perturbation in fractional energy density is anomalously small, not because the hypersurface is strongly warped. The nonperturbative formulation of the

Einstein equations in Sec. VI also gives directly the perturbation equations in terms of ϵ_h, v_h, ϕ_h, and α_h.

So far, in discussing the solutions to the perturbation equations I have implicitly assumed that the background equation of state has some relation to that of a fluid, with $0 \leq w \leq 1$, so β is always of order unity. However, it is conceivable that the net pressure could have been negative in the early Universe. One possibility is that at early times the *vacuum* could have had a large energy density and pressure, i.e., a contribution to the energy-momentum tensor like that of a cosmological constant, with pressure equal to minus the energy density.[24] A phase change of the vacuum would be required at some point to reduce the effective cosmological constant to zero or to a small value consistent with the present Universe. Another possibility is that quantum fluctuations of nongravitational and/or gravitational fields could result in an effective negative pressure around the Planck time or before.[25] Zel'dovich *et al.*[16] propose that domain boundaries associated with gauge theories of weak interactions could produce an average $w = -\frac{2}{3}$.

In idealized models with w constant $\dot{S}/S$ is constant or decreasing as $S \to 0$ if $w \leq -\frac{1}{3}$, i.e., if the strong energy condition is violated.[26] Thus as $S \to 0$ the conformal time $\tau \to -\infty$, even though for $w > -1$ the proper time from the initial singularity is finite. If $w \leq -\frac{1}{3}$ at the beginning of the Universe, light signals can propagate arbitrarily far relative to the comoving background coordinates, and particle horizons in the strict sense do not exist. Nevertheless, it makes sense to talk of a perturbation being larger than the *effective* particle horizon while $k^{-1}\dot{S}/S > 1$. It is $k^{-1}\dot{S}/S$, equal to β/x if w is constant, that governs which terms in the perturbation equations are dominant. If $w < -\frac{1}{3}$ initially, $k^{-1}S/\dot{S}$ for a protogalaxy begins less than one, increases to a value much greater than one, and then, once $w > -\frac{1}{3}$, decreases and eventually becomes less then one again at relatively recent times.

A density perturbation established during an early epoch when communication over a wavelength is possible will persist through the period when the perturbation is larger than the effective particle horizon. Assuming a homogeneous solution, the equation

$$(E_0 S^2 \epsilon_m)^{\cdot\cdot} + 3(1+c_s{}^2)\frac{\dot{S}}{S}(E_0 S^2 \epsilon_m)^{\cdot} + [c_s{}^2 k^2 - 2(1+3c_s{}^2)K + (c_s{}^2 - w)E_0 S^2](E_0 S^2 \epsilon_m) = 0 \tag{5.30}$$

shows that $E_0 S^2 \epsilon_m$ is roughly constant for the growing mode while $k^{-1}\dot{S}/S \gg 1$. The change in the background equation of state from $w \leq -\frac{1}{3}$ to $w \sim \frac{1}{3}$ changes $E_0 S^2 \epsilon_m$ by only a factor of 2 or so. Since $E_0 S^2$ increases and decreases with $(\dot{S}/S)^2$, ϵ_m is very small when $k^{-1}\dot{S}/S \gg 1$ relative to its value when $k^{-1}\dot{S}/S \sim 1$. The "growth" of ϵ_m at early times in conventional backgrounds is a purely kinematic consequence of the time dependence of $E_0 S^2$, with no dynamic significance.

B. Vector perturbations

The solution of Eq. (4.13) for the vortical velocity amplitude v_c can be put in the form

$$S^4(E_0 + P_0)v_c = -\int_0^{k\tau} dy\, S^4 P_0 \pi_T^{(0)}, \tag{5.31}$$

assuming no perturbation initially. Since $S^4 P_0 \propto y^{2(\beta-1)}$, once the anisotropic stress has been on for more than a few expansion times,

$$v_c \sim -\frac{x}{2\beta - 1}\frac{w}{1+w}\pi_T^{(0)}. \tag{5.32}$$

After the anisotropic stress turns off,

$$v_c \propto [S^4(E_0 + P_0)]^{-1} \sim x^{-2(\beta-1)}. \tag{5.33}$$

At the particle horizon v_c is at best roughly comparable to the maximum previous value of $w\pi_T^{(1)}$, and then only if $\pi_T^{(1)}$ is at full strength at the particle horizon or if β is close to $\frac{1}{2}$.

The "frame-dragging potential" Ψ is

$$\Psi = 2(k^2 - 2K)^{-1} S^2 (E_0 + P_0) v_c \sim 4\beta(\beta+1)x^{-2} v_c\,. \tag{5.34}$$

Even if $w\pi_T^{(1)} \ll 1$ and $v_c \ll 1$ at all times, $\Psi \gtrsim 1$ is quite possible if the anisotropic stress is present when $kS/\dot{S} \lesssim w\pi_T^{(1)}$. While $\Psi > 1$ does mean that there is no gauge in which all the metric perturbations are small, there is a formulation of the Einstein equations (see Sec. VI) in which the equations remain approximately linear. The physical significance is that, as in the ergoregion around a rotating black hole, a timelike observer cannot be at rest relative to nonshearing spatial coordinates (with $\dot{H}_T^{(1)} = 0$).

C. Tensor perturbations

The homogeneous solution of Eq. (4.14) for tensor perturbations is well known (see Ref. 15), and the inhomogeneous solution raises no new physical questions. The gravitational wave amplitude $H_T^{(2)}$ can never exceed in order of magnitude the maximum previous value of $w\pi_T^{(2)}$, if $H_T^{(2)}$ vanishes at the initial singularity, as long as $E_0 S^2/k^2 \gg 1$.

VI. RELATIONSHIP TO NONLINEAR DYNAMICS

The standard Arnowitt-Deser-Misner (ADM) approach[27] to the nonlinear dynamics of the gravitational field, with or without the presence of matter, begins with the initial-value problem[28] on a spacelike hypersurface characterized by an *intrinsic metric tensor* $h_{\alpha\beta}$ and an *extrinsic curvature tensor* $\mathcal{K}_{\alpha\beta}$. The proper-time spacing between successive hypersurfaces is characterized by the *lapse function* N, a spatial scalar determined from the geometric condition specifying the spacelike hypersurfaces by, typically, an elliptic equation.[29] The dynamic equations evolve $\mathcal{K}_{\alpha\beta}$ and $h_{\alpha\beta}$ forward in time from one hypersurface to the next. It is instructive to consider the cosmological perturbation problem from this point of view, to understand better the physical meaning of the gauge-invariant variables, to derive the equations governing the amplitudes ϕ_h, α_h, ϵ_h, and v_h appropriate to uniform-Hubble-constant hypersurfaces, and to investigate the possible effect of nonlinearities on the development of the perturbations.

Let each hypersurface be labeled by a single value of the time coordinate τ, and let n_a be the unit future-directed four-vector normal to the hypersurface. Then

$$n_0=-N,\ n_\alpha=0,\ n^0=N^{-1},\ n^\alpha=N^{-1}N^\alpha . \tag{6.1}$$

The *shift vector* N^α, the coordinate three-velocity of the normal world line, describes how spatial coordinates are propagated from one hypersurface to the next. In this section a three-vector is a three-vector relative to the exact spatial metric $h_{\alpha\beta}$. The metric tensor of the spacetime g_{ab} is given by

$$g_{00}=-N^2+N^\alpha N_\alpha,\quad g_{0\alpha}=-N_\alpha,\quad g_{\alpha\beta}=h_{\alpha\beta}, \tag{6.2}$$

and the inverse is

$$g_{00}=-N^{-2},\quad g^{0\alpha}=-N^{-2}N^\alpha,\quad g^{\alpha\beta}=h^{\alpha\beta}-N^{-2}N^\alpha N^\beta. \tag{6.3}$$

Note $h^{\alpha\beta}$ is the inverse of $h_{\alpha\beta}$ and $N_\alpha\equiv h_{\alpha\beta}N^\beta$.

From the three-metric $h_{\alpha\beta}$ one calculates the spatial Ricci tensor $\mathcal{R}_{\alpha\beta}$ and the scalar intrinsic curvature $\mathcal{R}\equiv h^{\alpha\beta}\mathcal{R}_{\alpha\beta}$. The extrinsic curvature tensor describes the embedding of the hypersurface in the spacetime and is the natural focus of a geometrical condition picking out a particular family of hypersurfaces. Mathematically, it is given by[27]

$$\begin{aligned}\mathcal{K}_{\alpha\beta}&=-n_{\alpha;\beta}=-N\Gamma^0_{\alpha\beta}\\&=-\tfrac{1}{2}N^{-1}[h_{\alpha\beta,0}+N_{\alpha|\beta}+N_{\beta|\alpha}].\end{aligned} \tag{6.4}$$

The semicolon denotes a covariant derivative in the spacetime, the slash a covariant derivative (with respect to $h_{\alpha\beta}$) in the spacelike hypersurface. The comma indicates an ordinary partial derivative.

The correspondence with my representation of the cosmological perturbations is made more transparent by defining the *conformal metric tensor* $\tilde{h}_{\alpha\beta}$,

$$\tilde{h}_{\alpha\beta}\equiv h^{-1/3}h_{\alpha\beta}, \tag{6.5}$$

where h is the determinant of $h_{\alpha\beta}$. The background scale factor is not present in $\tilde{h}_{\alpha\beta}$, and the perturbation in $\tilde{h}_{\alpha\beta}$ is entirely due to the traceless part of the metric tensor perturbation. In a similar spirit, split $\mathcal{K}_{\alpha\beta}$ into the trace $\mathcal{K}$ and the traceless part

$$\bar{\mathcal{K}}_{\alpha\beta}\equiv\mathcal{K}_{\alpha\beta}-\tfrac{1}{3}h_{\alpha\beta}\mathcal{K}. \tag{6.6}$$

Equation (6.4) then becomes the two equations

$$h_{,0}/h=-2N\mathcal{K}-2N^\alpha{}_{|\alpha} \tag{6.7}$$

and

$$\begin{aligned}\tilde{h}_{\alpha\beta,0}+\tilde{h}_{\alpha\beta,\mu}N^\mu+\tilde{h}_{\alpha\beta}N^\mu{}_{,\beta}+\tilde{h}_{\mu\beta}N^\mu{}_{,\alpha}-\tfrac{2}{3}\tilde{h}_{\alpha\beta}N^\mu{}_{,\mu}\\=-2Nh^{-1/3}\bar{\mathcal{K}}_{\alpha\beta},\end{aligned} \tag{6.8}$$

which correspond to decomposition of $\mathcal{K}^\alpha_\beta$ in terms of perturbation amplitudes given in Eq. (A7).

The Einstein equations separate into the initial-value equations relating the extrinsic and intrinsic geometry of the constant-τ hypersurface to the matter energy density

$$\mathcal{E}\equiv n_aT^{ab}n_b=N^2T^{00} \tag{6.9}$$

and the momentum density

$$\mathcal{J}_\alpha\equiv-n_aT^a_\alpha=NT^0_\alpha, \tag{6.10}$$

and into the dynamic equations for the evolution of the extrinsic curvature in time.[29] The momentum initial-value equation

$$\bar{\mathcal{K}}^\beta_{\alpha|\beta}=\mathcal{J}_\alpha+\tfrac{2}{3}\mathcal{K}_{|\alpha} \tag{6.11}$$

is conventionally viewed[28] as an equation for the longitudinal part of $\bar{\mathcal{K}}^\beta_\alpha$ and the energy equation

$$\mathcal{R}=2\mathcal{E}+\bar{\mathcal{K}}_{\alpha\beta}\bar{\mathcal{K}}^{\alpha\beta}-\tfrac{2}{3}\mathcal{K}^2 \tag{6.12}$$

as an equation for the determinant of the spatial metric tensor h, given the conformal metric $\tilde{h}_{\alpha\beta}$, the transverse part of $h^{1/2}\bar{\mathcal{K}}^\beta_\alpha$ relative to $\tilde{h}_{\alpha\beta}$, and, as a hypersurface condition, $\mathcal{K}$.

The dynamic equations[29] involve the matter stress tensor

$$\mathcal{S}_{\alpha\beta}\equiv T_{\alpha\beta}. \tag{6.13}$$

Let the traceless part of $\mathcal{S}_{\alpha\beta}$ be $\bar{\mathcal{S}}_{\alpha\beta}$, and denote the trace by $\mathcal{S}$. Let $\bar{\mathcal{R}}_{\alpha\beta}$ be the traceless part of the spatial Ricci tensor. Then

$$\mathcal{K}_{,0}+N^\alpha\mathcal{K}_{,\alpha}=-\Delta N+N(\mathcal{R}+\mathcal{K}^2+\tfrac{1}{2}\mathcal{S}-\tfrac{3}{2}\mathcal{E}) \tag{6.14}$$

and

$$\bar{\mathcal{K}}^{\alpha}_{\beta,0}+N^{\mu}\bar{\mathcal{K}}^{\alpha}_{\beta,\mu}-N^{\alpha}_{,\mu}\bar{\mathcal{K}}^{\mu}_{\beta}+N^{\mu}_{,\beta}\mathcal{K}^{\alpha}_{\mu}$$
$$=-N^{|\alpha}{}_{|\beta}+\tfrac{1}{3}\delta^{\alpha}_{\beta}\Delta N+N(\bar{\mathcal{R}}^{\alpha}_{\beta}+\mathcal{K}\bar{\mathcal{K}}^{\alpha}_{\beta}-\bar{\mathcal{S}}^{\alpha}_{\beta}). \quad (6.15)$$

In both equations the left-hand side is the time derivative along the normal to the hypersurface, and Δ is the Laplacian operator in the hypersurface.

The physical dynamics of scalar and vector cosmological perturbations lies in the matter evolution equations, rather then Eqs. (6.14) and (6.15). The equation governing local energy conservation, $T^{0a}{}_{;a}=0$, gives the rate of change of the energy density $\mathcal{E}$ along the normal to the hypersurface,

$$\mathcal{E}_{,0}+N^{\alpha}\mathcal{E}_{,\alpha}=N\mathcal{K}(\mathcal{E}+\tfrac{1}{3}\mathcal{S})+N\bar{\mathcal{K}}^{\alpha\beta}\bar{\mathcal{S}}_{\alpha\beta}+N^{-1}(N^2\mathcal{J}^{\alpha})_{|\alpha}. \quad (6.16)$$

Similarly, the evolution of the momentum density $\mathcal{J}_{\alpha}$ is given by $T^{b}_{\alpha;b}=0$, or

$$\mathcal{J}_{\alpha,0}+N^{\beta}\mathcal{J}_{\alpha,\beta}+N^{\beta}_{,\alpha}\mathcal{J}_{\beta}=N\mathcal{K}\mathcal{J}_{\alpha}-(\mathcal{E}\delta^{\beta}_{\alpha}+\mathcal{S}^{\beta}_{\alpha})N_{|\beta}-N\mathcal{S}^{\beta}_{\alpha|\beta}. \quad (6.17)$$

Although redundant, it will prove useful to have an equation directly for the evolution of the spatial scalar curvature. Take a convective time derivative of Eq. (6.12) along the normal to the hypersurface and eliminate the time derivatives of $\mathcal{K}$, $\bar{\mathcal{K}}^{\alpha}_{\beta}$, and $\mathcal{E}$ with Eqs. (6.14)–(6.16). After simplification with the help of Eq. (6.12) again, the result is

$$\mathcal{R}_{,0}+N^{\alpha}\mathcal{R}_{,\alpha}=\tfrac{2}{3}N\mathcal{K}\mathcal{R}+2N^{-1}(N^2\mathcal{J}^{\alpha})_{|\alpha}+\tfrac{4}{3}\mathcal{K}\Delta N+2\bar{\mathcal{K}}^{\alpha}_{\beta}(N\bar{\mathcal{R}}^{\beta}_{\alpha}-N^{|\beta}{}_{|\alpha}). \quad (6.18)$$

There is considerable cancellation between the time derivatives of the separate terms on the right-hand side of Eq. (6.12).

The solution of the above equations requires a hypersurface condition to determine N and, of secondary importance in the present circumstances, some prescription for N^{α}. The most straightforward hypersurface condition in the nonlinear context is the uniform-Hubble-constant condition that $\mathcal{K}$ be spatially uniform on each hypersurface.[29] The choice of the time dependence of $\mathcal{K}$ is the choice of time coordinate τ; at least in spatially closed cosmologies each value of $\mathcal{K}$ picks out a more or less unique hypersurface in the spacetime. The value of $\mathcal{K}$ may also pick out a unique hypersurface in spatially open cosmologies.[30] On uniform-Hubble-constant hypersurfaces, then $\mathcal{K}_{,\alpha}=0$, $\mathcal{K}_{,0}$ can be specified, and Eq. (6.14) becomes an elliptic equation for N whose mathematical properties, including existence and uniqueness theorems, have been explored rather extensively (see Ref. 30).

The zero-shear hypersurface condition, in the present language $\bar{\mathcal{K}}_{\alpha\beta}=0$, is applicable only to scalar perturbations, since it is compatible with Eq. (6.11) only if $\mathcal{J}_{\alpha}$ is the gradient of a scalar. The obvious generalization, which I will call the *minimal shear hypersurface condition*, is

$$\bar{\mathcal{K}}^{\alpha\beta}{}_{|\alpha\beta}=0. \quad (6.19)$$

The longitudinal part of Eq. (6.11) becomes an equation for $\mathcal{K}$,

$$\Delta\mathcal{K}=-\tfrac{3}{2}\mathcal{J}^{\alpha}{}_{|\alpha}. \quad (6.20)$$

Then the right-hand side of Eq. (6.11) is the transverse part of the vector field $\mathcal{J}_{\alpha}$. An equation for N is more difficult to come by. The best procedure seems to be to combine a time derivative of Eq. (6.20) with a Laplacian of Eq. (6.14). Eliminate $\Delta\dot{\mathcal{K}}$ between the two equations and eliminate $\dot{\mathcal{J}}_{\alpha}$ using Eq. (6.17). The result is a complicated fourth-order equation for N. Acceptable solutions may not always exist in an open universe if highly nonlinear regions such as black holes are present, and may not be unique in closed universes. The linear scalar perturbation with $k^2=3K$ satisfies the minimal-shear condition for any amplitude of warping of the hypersurface since, as mentioned in Sec. IV, $Q^{(0)}_{\alpha\beta}$ vanishes identically.

The complexity of the minimal-shear hypersurface condition magically disappears for linear scalar perturbations. Then, with $\bar{\mathcal{K}}_{\alpha\beta}=0$, Eq. (6.15) reduces to Eq. (4.4) for the perturbation in N. Also, note that Eq. (6.12) reduces to Eq. (4.3) relating Φ_H to ϵ_m after the perturbation in $\mathcal{K}$ is eliminated in favor of $\mathcal{J}_{\alpha}$ through Eq. (6.20). The simplicity of Eq. (4.3) is somewhat deceptive.

The comoving hypersurface condition generalizes from $\mathcal{J}_{\alpha}=0$ for linear scalar perturbations to

$$\mathcal{J}^{\alpha}{}_{|\alpha}=0. \quad (6.21)$$

The equation for N, obtained by taking a divergence of Eq. (6.17), is only second order, simpler than in the minimal-shear case, but is still considerably more complicated than for uniform-Hubble-constant hypersurfaces.

The uniform-Hubble-constant hypersurface condition will be used in our analysis of nonlinear effects on the time development of perturbations. Somewhat different questions arise in the consideration of nonlinearities associated with scalar and vector perturbations, so these will be considered separately.

A. Scalar perturbations

The gauge-invariant scalar perturbation amplitudes appropriate to uniform-Hubble-constant hypersurfaces were defined in Eqs. (5.24)–(5.27). The perturbation equations directly in terms of these amplitudes are most easily derived from the equations of this section, rather than by reworking the equations of Sec. IV, and their derivation will clarify the later consideration of nonlinear effects.

On a uniform-Hubble-constant hypersurface $\phi_h = \tilde{H}_L + \frac{1}{3}\tilde{H}_T^{(0)}$ and $\alpha_h = \tilde{A}$. The hypersurface condition, Eq. (5.22), gives a gauge-invariant amplitude for the traceless part of the extrinsic curvature tensor

$$\bar{\mathcal{K}}^\alpha_\beta = -\frac{1}{S}(\dot{\tilde{H}}_T^{(0)} - k\tilde{B}^{(0)})Q_\beta^{(0)\alpha}$$
$$= -\frac{3}{S}\left(\dot{\phi}_h - \frac{\dot{S}}{S}\alpha_h\right)Q_\beta^{(0)\alpha}. \quad (6.22)$$

The two initial-value equations (6.11) and (6.12) become

$$-\frac{2}{S}k(1-3K/k^2)\left(\dot{\phi}_h - \frac{\dot{S}}{S}\alpha_h\right) = S(E_0+P_0)v_h \quad (6.23)$$

and

$$4S^{-2}(k^2-3K)\phi_h = 2E_0\epsilon_h. \quad (6.24)$$

Equations (6.14) and (6.24) combine to determine α_h in terms of ϕ_h and η,

$$[k^2 - 3K + \tfrac{3}{2}(E_0+P_0)S^2]\alpha_h$$
$$= -(1+3c_s^2)(k^2-3K)\phi_h - \tfrac{3}{2}P_0S^2\eta. \quad (6.25)$$

The geometric evolution equation (6.15) reduces to

$$\left(\dot{\phi}_h - \frac{\dot{S}}{S}\alpha_h\right)^{\cdot} + 2\frac{\dot{S}}{S}\left(\dot{\phi}_h - \frac{\dot{S}}{S}\alpha_h\right) - \tfrac{1}{3}k^2(\phi_h+\alpha_h) = \tfrac{1}{3}P_0S^2\pi_T^{(0)}. \quad (6.26)$$

Elimination of α_h with Eq. (6.25) gives a single equation in a single unknown ϕ_h, but the equation is much more complicated than Eq. (4.9) for ϵ_m. The solution is also not nearly as simple a function of time as the solution for ϵ_m/Φ_H given in Eq. (5.14); consider Eqs. (5.24), (5.19), and (5.6).

The uniform-Hubble-constant hypersurfaces do deal successfully with the spatial harmonics $k^2 = 3K$ in a closed universe. Since $Q_{\alpha\beta}^{(0)}$ vanishes identically,[2] Eq. (6.23) no longer applies. Equation (6.24) gives $\epsilon_h = 0$, and from Eqs. (6.25) and (6.17)

$$\alpha_h = -\frac{w}{1+w}\eta, \quad [S^4(E_0+P_0)v_h]^{\cdot} = 0. \quad (6.27)$$

The amplitude ϕ_h now depends on the way spatial coordinates are propagated from one hypersurface to the next through the hypersurface condition Eq. (5.22). The traceless part of the metric tensor perturbation and the spatial curvature perturbation vanish. The absence of any physical *adiabatic* mode when $k^2 = 3K$ was first recognized by Lifshitz and Khalatnikov.[2]

A relatively simple direct solution of Eqs. (6.23)–(6.26) is possible under the assumptions of Sec. V when the perturbation is large compared with the particle horizon, $x/\beta \ll 1$. To lowest order,

$$\epsilon_h \simeq \tfrac{2}{3}\beta^{-2}x^2\phi_h, \quad (6.28)$$

$$\alpha_h \simeq -\frac{w}{1+w}\eta, \quad (6.29)$$

and with

$$g \equiv \phi_h' - (\beta/x)\alpha_h \simeq \phi_h' + \frac{\beta w}{1+w}x^{-1}\eta, \quad (6.30)$$

Eq. (6.26) becomes

$$g' + 2(\beta/x)g \simeq (\beta/x)^2 w\pi_T^{(0)}. \quad (6.31)$$

With no perturbation at $x=0$, the solution is

$$g = \beta^2 x^{-2\beta}\int_0^x w\pi_T^{(0)} y^{2\beta-2}dy, \quad (6.32)$$

$$\phi_h = \int_0^x g(y)dy - \frac{\beta}{1+w}\int_0^x w\eta y^{-1}dy. \quad (6.33)$$

An entropy perturbation present only at $y \ll 1$ couples predominantly to the growing mode of energy density perturbation on uniform-Hubble-constant hypersurfaces, while anisotropic stress at $y \ll 1$ couples with comparable strength to both modes. The discussion in Sec. V showed rather different behavior for the energy density perturbation on comoving or zero-shear hypersurfaces. These and the hypersurface-independent ratio of shear rate to expansion rate displayed a coupling of the entropy perturbation to both growing and decaying modes and a coupling of the anisotropic stress predominantly to the decaying mode at early times. The situation is not as simple as presented by Press and Vishniac.[12]

The linear perturbations relative to the uniform-Hubble-constant hypersurfaces are small as long as $w\eta \ll 1$ and $w\pi_T^{(0)} \ll 1$. If the stress perturbations are present for at least one expansion time at $x \ll 1$, Eq. (6.33) gives $\phi_h \sim w\eta,\ w\pi_T^{(0)}$; Eq. (6.28) gives $\epsilon_h \sim x^2 w\eta,\ x^2 w\pi_T^{(0)}$; Eq. (6.25) gives $\alpha_h \sim w\eta,\ x^2 w\pi_T^{(0)}$; and from Eqs. (6.23) and (6.32) $v_h \sim x^2 g \sim x^3 w\eta,\ xw\pi_T^{(0)}$. Nevertheless, some second-order terms in the Einstein equations are larger than the linear perturbations whenever $w\pi_T^{(0)} > x^2$. Consider, for instance, the initial value Eq. (6.12). The linear perturbation in $\mathcal{R}$ is of order $(k^2/S^2)\phi_h \sim (k/S)^2[w\eta,\ w\pi_T^{(0)}]$, as is the linear perturbation in $\mathcal{E}$. The term $\bar{\mathcal{K}}_{\alpha\beta}\bar{\mathcal{K}}^{\alpha\beta}$, since $\bar{\mathcal{K}}^\alpha_\beta \sim (k/S)g \sim (k/S)[xw\eta,\ x^{-1}w\pi_T^{(0)}]$, is of order $(k/S)^2[x^2(w\eta)^2,\ x^{-2}(w\pi_T^{(0)})^2]$ times the first order perturbations. When $1 \gg w\pi_T^{(0)} > x^2$ the second-order term is not large in an absolute sense, since $\bar{\mathcal{K}}_{\alpha\beta}\bar{\mathcal{K}}^{\alpha\beta}/\mathcal{K}^2$ is of order $(w\pi_T^{(0)})^2$; the first-order perturbation in $\mathcal{E}$ is anomalously small. The key physical question is whether this breakdown of linearity on uniform-Hubble-constant hypersurfaces has an effect on the time development of the energy density perturbation, so that the fractional energy density perturbation as the perturbation enters the particle horizon is not predicted correctly by the linear theory.

A partial answer is that on *comoving* hypersurfaces the initial second-order perturbation is small compared with the corresponding first-order perturbation as long as $w\pi_T^{(0)} \ll 1$, independent of the value of x. The term $\bar{\mathcal{K}}_{\alpha\beta}\bar{\mathcal{K}}^{\alpha\beta}$ in Eq. (6.12), now the square of the shear of the matter, is still of order $(k/S)^2x^{-2}(w\pi_T^{(0)})^2$. However, the linear perturbations in $\mathcal{E}$ and $\mathcal{K}^2$ are now of order $(k/S)^2x^{-2}(w\pi_T^{(0)})^2$.

Can nonlinear effects couple the decaying mode of fractional energy density perturbation on comoving hypersurfaces to the growing mode so that at the particle horizon the fractional energy density is large compared with the original value of $w\pi_T^{(0)}$? At the particle horizon the fractional energy density perturbation is comparable to the distortion of the spacelike hypersurface, which is $(S/k)^2$ times the perturbation in scalar curvature $\mathcal{R}$. The time development of $\mathcal{R}$ is given by Eq. (6.18). From Eq. (6.14) the fractional perturbation in N is at most the order of the fractional perturbation in $\mathcal{E}$ or $\mathcal{K}$, which is the order of $w\pi_T^{(0)}$. The explicit second-order terms in Eq. (6.18) are of order $\tau^{-1}(k/S)^2(w\pi_T^{(0)})^2$, down by a factor of order $w\pi_T^{(0)}$ relative to the first-order terms. The nonlinearities are unable to alter appreciably the linear theory prediction for $\mathcal{R}$ as long as $w\pi_T^{(0)} \ll 1$.

The linear perturbation equations are really the same equations, but with variables regrouped, no matter what hypersurface condition is used to interpret the variables physically. The validity of the linear equations on the comoving hypersurfaces implies their validity on all sets of hypersurfaces as long as the physical interpretation of the linear perturbation amplitudes is qualified appropriately. For instance, if $w\pi_T^{(0)} > x^2$ in a linear perturbation calculation on uniform-Hubble-constant hypersurfaces (or zero-shear hypersurfaces), ϵ_h (or ϵ_g) is at first not the amplitude of the actual fractional energy density perturbation, but by the time the perturbation comes inside the particle horizon the discrepancy has disappeared. The dynamical origin of the nonlinear corrections to the fractional energy density perturbation is the work done by the anisotropic stress on the shearing volume element [see Eq. (6.16)].

B. Vector perturbations

The amplitude Ψ of the frame-dragging potential is a gauge-invariant measure of the amplitude of metric tensor perturbations associated with vortical motions of the matter. It is generated by anisotropic stress through Eqs. (4.12) and (4.13). In a standard background $\Psi \sim x^{-1}(w\pi_T^{(1)})$ from Eqs. (5.32) and (5.34). The possibility that $\Psi \gtrsim 1$ even if $w\pi_T^{(1)} \ll 1$ at all times was pointed out in Sec. V. The physical significance of this can best be understood in the context of the full Einstein equations as presented in this section.

One can verify that $w\pi_T^{(1)} \ll 1$ at all times does guarantee the validity of the linear equations in a gauge such that $\dot{H}_T^{(1)}$ vanishes. Then $\Psi = B^{(1)}$, and

$$N^\alpha = \Psi Q^{(1)\alpha} \quad . \tag{6.34}$$

The lapse function N, as a spatial scalar, is unperturbed to first order, and from Eq. (6.2)

$$g_{00} \simeq -S^2(1-\Psi^2 Q^{(1)\alpha}Q_\alpha^{(1)}) \, . \tag{6.35}$$

Indices on the spatial harmonics are always raised and lowered with the background metric.

If $\Psi > 1$ there are likely, depending on the precise normalization of $Q_\alpha^{(1)}$, to be regions where a physical observer or particle cannot be at rest relative to nonshearing spatial coordinates, analogous to an ergoregion in a stationary spacetime. However, g_{00} as such does not appear in the nonlinear equations (6.11)–(6.18), and these equations are not in any way singular when $g_{00} \geqslant 0$. That the second-order perturbation in N is small on uniform-Hubble-constant hypersurfaces can be verified from Eq. (6.14).

The dynamically significant nonlinearities arise in the same way as in the scalar case. From Eq. (A7)

$$\bar{\mathcal{K}}^\alpha_\beta = (k/S)\Psi Q_\beta^{(1)\alpha} \sim (k/S)x^{-1}w\pi_T^{(1)} \, , \tag{6.36}$$

which in Eq. (6.16) implies a nonlinear contribution to the fractional energy density perturbation of order $(w\pi_T^{(1)})^2$. Again, this belongs predominantly to the decaying mode at $x \ll 1$, and Eq. (6.18) shows that the nonlinear correction to $\mathcal{R}$ is small compared to the linear scalar perturbation for mixed scalar-vector perturbations. If the only linear perturbations are pure vector harmonics, the dominant nonlinearities in $\mathcal{R}$ come from the $\frac{4}{3}\mathcal{K}\Delta N$ term in Eq. (6.18), since $\bar{\mathcal{R}}^\alpha_\beta$ vanishes to first order for vector perturbations, and give $\phi_h \sim (S/k)^2\mathcal{R} \sim (w\pi_T^{(1)})^2$ for the induced scalar perturbation. This is roughly what one expects for the fractional energy density perturbation at $x \sim 1$.

Even though N^α may be large at $x \ll 1$, the convective terms involving N^α and a spatial gradient are of order $x\Psi \sim w\pi_T^{(1)}$ times the time-derivative term in Eqs. (6.14)–(6.18) as the perturbation is being generated.

C. Tensor perturbations

The nonlinear perturbation of the energy density follows the same pattern for tensor perturbations as for vector and scalar perturbations. If the completely transverse anisotropic stress is on for the order of one expansion time, Eq. (4.14) integrates to give $H_T^{(2)\prime} \sim x^{-1} w \pi_T^{(2)}$, so

$$\bar{\mathcal{K}}_\beta^\alpha \sim (k/S) x^{-1} w \pi_T^{(2)} . \qquad (6.37)$$

The potentially largest nonlinear contribution to the fractional energy density perturbation is the order of the square of the ratio of anisotropic stress to background energy density, both when the anisotropic stress is present (presumably at $x \ll 1$) and at the particle horizon, regardless of whether the anisotropic stress is associated with scalar, vector, or tensor harmonics or a mixture of all three.

VII. SUMMARY AND CONCLUSIONS

The primary gauge-invariant amplitudes for scalar (density) perturbations were chosen in Sec. III on the grounds of mathematical simplicity, based on a description of the geometry through the metric tensor and a description of the matter through the energy-momentum tensor. Three of these amplitudes have, to first order in the deviation from the homogeneous and isotropic background, a universal physical significance. The amplitude $v_s^{(0)}$, or $\xi \equiv (k/S) v_s^{(0)}$, is a measure of the shear of the curl-free part of matter velocity field. The amplitude η is a measure of the nonadiabatic, relative to the background pressure-energy-density relation, part of the perturbation in the isotropic pressure, while $\pi_T^{(0)}$ measures the completely longitudinal part of the anisotropic stress. In each case the physical quantity vanishes in the background.

On the other hand, the amplitude ϵ_m, while mathematically gauge invariant, has the physical significance of measuring the fractional energy density perturbation only on spacelike hypersurfaces orthogonal to the four-velocity of the frames in which the matter energy flux vanishes. The purely geometrical amplitudes Φ_H and Φ_A also have physical meaning with respect to a particular set of spacelike hypersurfaces, the hypersurfaces for which the normal unit vectors have zero shear. Specifically, Φ_H measures the amount of warping of the zero-shear hypersurfaces due to the presence of the perturbation, and Φ_A is the amplitude of the fractional perturbation in the lapse function.

The mathematical simplicity of the definition of these amplitudes carries over into the simple quasi-Newtonian structure of the equations which govern them. The nonadiabatic stress amplitudes η and $\pi_T^{(0)}$ are formally regarded as known functions of time which through the source terms in Eq. (4.9) for ϵ_m generate the perturbations in energy density shear, and spatial curvature. A more detailed treatment of the matter would require additional equations governing the evolution of individual components such as the microwave background radiation, neutrinos, and at very early times quarks and various gauge bosons, particularly if these components are thermally decoupled. It is straightforward to write these additional equations in gauge-invariant form, and presumably a complete theory of the matter would determine the time dependence of η and $\pi_T^{(0)}$, rather than leaving them as free functions. In the optically thick limit, $\eta Q^{(0)}$ is roughly the fractional perturbation in the photon-to-baryon ratio times the ratio of rest-mass energy density to total energy density in a standard radiation-plus-matter model.

The gauge-invariant variables and equations are closely related to the variables and perturbation equations in certain specific gauges. For instance, Φ_H and Φ_A are the metric perturbation amplitudes in the longitudinal gauge of Harrison, and ϵ_m is the density perturbation amplitude in the comoving gauge of Sakai.[11] Both Harrison[4] and Sakai[11] arrived at equations equivalent to the homogeneous version of Eq. (4.9). Unfortunately, most of the standard references, including the textbooks of Weinberg[15] and Peebles,[7] rely on the synchronous gauge. This gauge is unnecessarily complicated mathematically, since the presence of purely gauge modes means that the analog of Eq. (4.9) is a fourth-order equation, and the physical interpretation of the results is not straightforward. Of course, a calculation can be done in any gauge if done consistently and completely, and in practice difficulties of interpretation disappear once the perturbation is well within the particle horizon.

The physical interpretation of such hypersurface-dependent quantities as the fractional energy density perturbation is somewhat ambiguous even in the gauge-invariant formalism when the perturbation wavelength is larger than the particle horizon. In Secs. III and V I define in addition to ϵ_m gauge-invariant amplitudes ϵ_g and ϵ_h which measure the fractional energy density perturbation on zero-shear and uniform-Hubble-constant hypersurfaces, respectively. Which is considered the physically most appropriate is a matter of taste. A calculation done in a particular gauge or with a particular set of gauge-invariant variables can always be physically interpreted by computing the gauge-invariant amplitude appropriate to the preferred hypersurface.

The same is true of the geometrical amplitudes. While $\phi_\varepsilon \equiv \Phi_H$ measures the distortion of zero-shear hypersurfaces and $\alpha_\varepsilon \equiv \Phi_A$ the perturbation in the lapse function between zero-shear hypersurfaces, appropriate combinations of Φ_H and Φ_A as defined in Sec. V give the amplitudes of the distortion and the perturbation in the lapse function for comoving (ϕ_m and α_m) and uniform-Hubble-constant (ϕ_h and α_h) hypersurfaces.

The comparison between the values of the gauge-invariant amplitudes appropriate to these three hypersurfaces for each of the two independent homogeneous modes in a background with negligible spatial curvature and $w = P_0/E_0$ independent of time is given in Table I. The fractional energy density perturbation is particularly sensitive to the choice of hypersurface when $kS/\dot{S} = k\tau/\beta \ll 1$.

The hypersurface conditions discussed in this paper have the property that *each* hypersurface is picked out by a physical criterion based either on the extrinsic geometry of the hypersurface or on its relation to the matter. The geometric criteria are spatially global in character, since the lapse function relating one hypersurface to the next satisfies an elliptic equation. Even the comoving hypersurface condition is fundamentally global, since the separation of the longitudinal and vortical parts of the velocity field also requires the solution of an elliptic equation.

In contrast, the hypersurfaces might be chosen as surfaces of constant *proper* time along the world lines of a particular family of observers. If these observers are comoving with the matter one has the sort of hypersurface condition used by Sachs and Wolfe[5] and Olson[13] to define a density perturbation, even though Olson's dynamical calculation was performed with a variable (essentially equivalent to ϕ_m) defined on hypersurfaces orthogonal to the matter world lines. In a synchronous gauge the observers are freely falling and their world lines are orthogonal to the hypersurfaces of constant propertime.

The problem with these proper-time hypersurface conditions is that at any *one* time the choice of the hypersurface is completely arbitrary. There is a corresponding arbitrariness in the value of, say, the fractional energy density perturbation, and the description is intrinsically non-gauge-invariant. Press and Vishniac[12] avoid this problem, even though they calculate in a synchronous gauge, by continuously transforming to a uniform-Hubble-constant hypersurface to interpret their results. However, it would seem better to work directly with variables whose physical meaning is unambiguous.

A proper-time description may be useful in considering local physics in the presence of given inhomogeneities, e.g., in studying element formation in the early Universe. However, it is in appropriate when the dynamics of the inhomogeneities themselves are the focus of interest.

Is there any one best measure of the true amplitude of the perturbation? Once the perturbation is within the particle horizon the value of the fractional energy density perturbation on any of the standard hypersurfaces is a reasonable choice. At early times, when $kS/\dot{S} \ll 1$, the true amplitude should indicate the limits of validity of the linear perturbation analysis for both growing and decaying modes. All of the amplitudes compared in Table I are relative amplitudes in the sense some sort of breakdown of linearity is implied if one of them is larger than unity. A "good" hypersurface condition should not introduce any apparent nonlinearities which are *only* due to a large warping of the hypersurface. In this sense the zero-shear hypersurface condition is "bad," since for a decaying mode ϕ_ε and α_ε can exceed one even though all the relative amplitudes are small on comoving and uniform-Hubble-constant hypersurfaces.

On no hypersurface does the fractional energy density perturbation indicate the true amplitude for both homogeneous scalar modes at $kS/\dot{S} \ll 1$. The one hypersurface-invariant amplitude ξ is unsuitable because ξ is down by a factor $(kS/\dot{S})^2$ compared with the irreducible relative amplitude of the geometry perturbations for the growing mode. For the decaying mode ϕ_m is anomalously small compared with, say, ξ when $P_0/E_0 \ll 1$ (β close to 2). The one generally suitable choice is ϕ_h, the amplitude of warping of the uniform-Hubble-constant hypersurface. Both comoving and uniform-Hubble-constant hypersurfaces minimize (within a factor of order unity) the maximum relative amplitude, and ϕ_h is always comparable with, if not identical to, this maximum relative amplitude.

A further advantage of ϕ_h can be seen from Eq. (6.26). If no nonadiabatic stress perturbations are present ($w\eta = w\pi_T^{(0)} = 0$) at a time when $kS/\dot{S} \ll 1$, Eq. (6.25) shows that $\alpha_h = O[(kS/\dot{S})^2 \phi_h]$, and Eq. (6.26) reduces to

$$\ddot{\phi}_h + 2\frac{\dot{S}}{S}\dot{\phi}_h - \tfrac{1}{3}k^2\phi_h \simeq 0\ . \tag{7.1}$$

In a time $\Delta\tau \ll k^{-1}$, so light can travel only a small fraction of a wavelength, the growing mode in ϕ_h

is to a good approximation *constant* in amplitude *regardless of any change in the background equation of state*, even a change from $P_0/E_0 < -\frac{1}{3}$ to $P_0/E_0 > 0$ as discussed in connection with Eq. (5.30). Meanwhile, the fractional energy density perturbation as measured by ϵ_h or ϵ_m varies inversely with $E_0 S^2$ or $(\dot{S}/S)^2$ and can increase or decrease by many orders of magnitude. It is ϕ_h, rather than ϵ_m or ϵ_h, that gives the correct physical picture of the dynamics of the perturbation while the perturbation is large compared with the effective particle horizon. The perturbation amplitude can decay adiabatically but cannot grow except in *direct* response to a nonadiabatic stress perturbation. This conclusion may seem an obvious consequence of causality, but is obscured in many standard references which emphasize power-law growth of the fractional energy density perturbation at early times.

While the above considerations give preference to the uniform-Hubble-constant hypersurface for dealing with the nonlinear Einstein equations, we did point out in Sec. VI that the linear perturbation treatment is strictly valid to the fullest possible extent only on comoving hypersurfaces. On uniform-Hubble-constant hypersurfaces the second-order correction to the fractional energy density perturbation can exceed the linear term at $kS/\dot{S} \ll 1$ even though the overall perturbation amplitude is small. In this circumstance $\epsilon_h Q$ is not equal to the actual perturbation in the fractional energy density on uniform-Hubble-constant hypersurfaces, but the gauge-invariant linear perturbation equations still give the correct evolution and by the time $kS/\dot{S} \sim 1$, the fractional energy density perturbation is equal to $\epsilon_h Q$.

The exact inhomogeneous solutions to the perturbation equations obtained in Sec. V, assuming negligible spatial curvature and P_0/E_0 independent of time in the background, are new, though Press and Vishniac[12] did consider the effect of entropy perturbations at $kS/\dot{S} \ll 1$. The exact solution displayed in Eq. (5.14) reduces to Eq. (5.15) in this limit, assuming $-\frac{1}{3} < P_0/E_0 < 1$, so $\infty > \beta > \frac{1}{2}$. With $-\frac{1}{3} < w < 1$ the strong energy condition is satisfied and $kS/\dot{S}$ increases toward the future; the particle horizon expands rather than shrinks relative to the wavelength of the perturbation. While the adiabatic speed of sound is imaginary for $w < 0$, a single-component treatment of the matter is inappropriate when the net pressure is negative.

Over the whole range $-\frac{1}{3} < w < 1$ the analytic solution shows that the fractional energy density at the particle horizon cannot greatly exceed the maximum previous ratio of stress perturbation to background energy density, averaged over one e-folding in S, no matter how early the stress perturbation occurs. For w close to one ($\beta \sim \frac{1}{2}$), $(2\beta - 1)^{-1}$ should be replaced by $\ln(\dot{S}/kS)$ in the integral over the anisotropic stress. This logarithmic enhancement of the perturbation amplitude over the amplitude of the stress perturbation is at most a factor of order 10^2 even if the stress perturbation was excited as early as the Planck time.

As measured by ϕ_h an entropy or isotropic stress perturbation predominantly excites the growing mode (which really has constant amplitude until the perturbation comes within the particle horizon). An anisotropic stress perturbation excites comparable amounts of the growing mode and the decaying mode. Of course, at the particle horizon all that is left is the growing mode.

From the point of view of the fractional energy density perturbation it seems that small nonlinear coupling of the decaying mode excited by anisotropic stress to the growing mode could perhaps significantly enhance the amount of growing mode beyond what is expected from the linear theory. I showed rather conclusively in Sec. VI that as long as the amplitude of the stress perturbation relative to the background energy density and therefore ϕ_h is small, nothing of the sort can happen. The nonlinear effects of anisotropic stress associated with scalar (purely longitudinal), vector (semilongitudinal), and tensor (purely transverse) perturbations on the fractional energy density are similar even though the frame dragging associated with a vector perturbation can generate large metric perturbations at $kS/\dot{S} \ll 1$.

In summary, then, I conclude that there is no possibility of explaining the origin of galaxies through the dynamical evolution of perturbations which arise from genuinely small or "statistical" causes within the context of general relativity and the strong energy condition. Either rather large perturbations, with a relative amplitude of at least 10^{-3} or so, are present initially or correspondingly large stress perturbations appear at a later time. This conclusion is hardly new, but has been strengthened by consideration of completely general energy-momentum tensor perturbations and the nonlinear interaction of modes. Of course, it also requires consideration of the physical processes which act on the perturbation once it is inside the particle horizon, but for anything like a galaxy scale these are fairly well understood.[7]

The one real hope for a dynamical explanation of the origin of structure in the Universe is the abolition of particle horizons at early times, perhaps through quantum modifications to the energy-momentum tensor and/or the gravitational field equations which in effect violate the strong energy condition that $E_0 + 3P_0 > 0$.

ACKNOWLEDGMENTS

This research was supported in part by National Science Foundation Grant No. PHY 79-19884 at the Institute for Advanced Study and in part by the National Science Foundation Grant No. GP-15267 at the University of Washington. A draft discussing some aspects of the cosmological perturbation problem and which was a foundation for this paper was written while the author was a Senior Fellow at the Center for Theoretical Physics, University of Maryland. I would like to thank W. H. Press and P. J. E. Peebles for stimulating discussions.

APPENDIX

The general expressions are written down for the perturbed Ricci tensor components and the perturbed matter equations of motion in terms of the metric tensor perturbations and energy-momentum tensor perturbations defined in Eqs. (2.14)–(2.20). Also included are the intrinsic and extrinsic curvature tensors for the constant-τ spacelike hypersurface.

The Ricci tensor for scalar perturbations is

$$\delta R^0_0 = \frac{3}{S^2}\left[\ddot{H}_L + \frac{\dot{S}}{S}\dot{H}_L - \frac{\dot{S}}{S}\dot{A} + [k^2/3 - 2(\dot{S}/S)^{\bullet}]A + \frac{k}{3}\left(\dot{B}^{(0)} + \frac{\dot{S}}{S}B^{(0)}\right)\right]Q^{(0)}, \tag{A1a}$$

$$\delta R^0_\alpha = \frac{2}{S^2}\left[-k\dot{H}_L - \tfrac{1}{3}k(1-3K/k^2)\dot{H}^{(0)}_T + k\frac{\dot{S}}{S}A - KB^{(0)}\right]Q^{(0)}_\alpha, \tag{A1b}$$

$$\begin{aligned}\delta R^\alpha_\beta = \frac{1}{S^2}\Big[&\tfrac{4}{3}(k^2-3K)(H_L + \tfrac{1}{3}H^{(0)}_T) + \ddot{H}_L + 5\frac{\dot{S}}{S}\dot{H}_L - \frac{\dot{S}}{S}\dot{A} + (k^2/3 - 2S^{-2}(S\dot{S})^{\bullet})A \\ &+ \frac{k}{3}\left(\dot{B}^{(0)} + 5\frac{\dot{S}}{S}B^{(0)}\right)\Big]\delta^\alpha_\beta Q^{(0)} \\ &+ \frac{1}{S^2}\left[\ddot{H}^{(0)}_T + 2\frac{\dot{S}}{S}\dot{H}^{(0)}_T - k\left(\dot{B}^{(0)} + 2\frac{\dot{S}}{S}B^{(0)}\right) - k^2(H_L + \tfrac{1}{3}H^{(0)}_T + A)\right]Q^{(0)\alpha}_\beta .\end{aligned} \tag{A1c}$$

The Ricci tensor for vector perturbations:

$$\delta R^0_0 = 0, \tag{A2a}$$

$$\delta R^0_\alpha = \frac{k^2-2K}{2S^2}\left(B^{(1)} - \frac{1}{k}\dot{H}^{(1)}_T\right)Q^{(1)}_\alpha, \tag{A2b}$$

$$\delta R^\alpha_\beta = \frac{1}{S^2}\left[\ddot{H}^{(1)}_T + 2\frac{\dot{S}}{S}\dot{H}^{(1)}_T - \frac{k}{5}\left(\dot{B}^{(1)} + 2\frac{\dot{S}}{S}B^{(1)}\right)\right]Q^{(1)\alpha}_\beta . \tag{A2c}$$

The Ricci tensor for tensor perturbations:

$$\delta R^0_0 = 0, \tag{A3a}$$

$$\delta R^0_\alpha = 0, \tag{A3b}$$

$$\delta R^\alpha_\beta = \frac{1}{S^2}\left[\ddot{H}^{(2)}_T + 2\frac{\dot{S}}{S}\dot{H}^{(2)}_T + (k^2+2K)H^{(2)}_T\right]Q^{(2)\alpha}_\beta . \tag{A3c}$$

Equations of motion for scalar perturbations:

$$[E_0S^3\delta]^{\bullet} + (E_0+P_0)S^3(kv^{(0)} + 3\dot{H}_L) + 3P_0S^2\dot{S}\pi_L = 0, \tag{A4a}$$

$$[v^{(0)} - B^{(0)}]^{\bullet} + \frac{\dot{S}}{S}(1-3c_s^2)[v^{(0)} - B^{(0)}] - A - k\frac{w^{\bullet}}{1+w}\pi_L + \tfrac{2}{3}k(1-3K/k^2)\frac{w}{1+w}\pi^{(0)}_T = 0. \tag{A4b}$$

Equation of motion for vector perturbations:

$$[v^{(1)} - B^{(1)}]^{\bullet} + \frac{\dot{S}}{S}(1-3c_s^2)[v^{(1)} - B^{(1)}] + k\frac{w}{1+w}\pi^{(1)}_T = 0. \tag{A5}$$

Intrinsic curvature tensor on the constant-τ spacelike hypersurface for general perturbations:

$$\mathcal{R}^\alpha_\beta = \frac{1}{S^2}[2K + \tfrac{4}{3}(k^2-3K)(H_L + \tfrac{1}{3}H^{(0)}_T)Q^{(0)}]\delta^\alpha_\beta - \frac{k^2}{S^2}(H_L + \tfrac{1}{3}H^{(0)}_T)Q^{(0)\alpha}_\beta + \frac{(k^2+2K)}{S^2}H^{(2)}_T Q^{(2)\alpha}_\beta . \tag{A6}$$

Extrinsic curvature tensor for general perturbations:

$$\mathcal{K}^\alpha_\beta = -\frac{1}{S}\left[\frac{\dot{S}}{S} + \left(\dot{H}_L - \frac{\dot{S}}{S}A + \frac{k}{3}B^{(0)}\right)Q^{(0)}\right]\delta^\alpha_\beta - \frac{1}{S}[\dot{H}^{(0)}_T - kB^{(0)}]Q^{(0)\alpha}_\beta - \frac{1}{S}[\dot{H}^{(1)}_T - kB^{(1)}]Q^{(1)\alpha}_\beta - \frac{1}{S}\dot{H}^{(2)}_T Q^{(2)\alpha}_\beta . \tag{A7}$$

Neither the intrinsic curvature nor the extrinsic curvature are gauge invariant, but both are invariant under a purely spatial gauge transformation ($T=0$) and depend only on the instantaneous value of T for a time gauge transformation.

*Permanent address: Department of Physics, University of Washington, Seattle, Washington 98195.

[1]E. M. Lifschitz, J. Phys. (Moscow) 10, 116 (1946).
[2]E. M. Lifschitz and I. M. Khalatnikov, Adv. Phys. 12, 185 (1963).
[3]S. W. Hawking, Astrophys. J. 145, 544 (1966).
[4]E. R. Harrison, Rev. Mod. Phys. 39, 862 (1967).
[5]R. K. Sachs and A. M. Wolfe, Astrophys. J. 147, 73 (1967).
[6]G. B. Field, in *Galaxies and the Universe*, edited by A. Sandage, M. Sandage, and J. Kristian (Univeristy of Chicago Press, Chicago, 1975).
[7]P. J. E. Peebles, *Cosmology: The Physics of Large Scale Structure* (Princeton University Press, Princeton, 1980).
[8]A. H. Guth and S.-H. H. Tye, Phys. Rev. Lett. 44, 631 (1980).
[9]W. H. Press, Phys. Scr. 21, 702 (1980).
[10]G. B. Field and L. C. Shepley, Astrophys. Space Sci. 1, 309 (1968).
[11]K. Sakai, Prog. Theor. Phys. 41, 1461 (1969).
[12]W. H. Press and E. T. Vishniac, Astrophys. J. 239, 1 (1980).
[13]D. W. Olson, Phys. Rev. D 14, 327 (1976).
[14]U. H. Gerlach and U. K. Sengupta, Phys. Rev. D 18, 1789 (1978).
[15]S. Weinberg, *Gravitation and Cosmology* (Wiley, New York, 1972).
[16]Ya. B. Zel'dovich, L. B. Okun', and I. Yu. Kabzarev, Zh. Eksp. Teor. Fiz. 67, 3 (1974) [Sov. Phys.-JETP 40, 1 (1975)].
[17]T. W. B. Kibble, J. Phys. A 9, 1387 (1976).
[18]I. D. Novikov, Zh. Eksp. Teor. Fiz. 46, 686 (1964) [Sov. Phys.-JETP 19, 467 (1964)].
[19]P. J. E. Peebles, Nature 220, 237 (1968). Peebles does recognize the unphysical nature of these "statistical fluctuations."
[20]P. D. D'Eath, Ann. Phys. (N.Y.) 98, 237 (1976).
[21]P. J. E. Peebles and J. T. Yu, Astrophys. J. 162, 815 (1970).
[22]J. Silk, Astrophys. J. 151, 459 (1968).
[23]B. Carter, in 1979 Les Houches Summer School Proceedings (unpublished).
[24]E. W. Kolb and S. Wolfram (unpublished).
[25]M. V. Fischetti, J. B. Hartle, and B. L. Hu, Phys. Rev. D 20, 1757 (1979).
[26]S. W. Hawking and G. F. R. Ellis, *Large Scale Structure of Spacetime* (Cambridge University Press, Cambridge, 1973).
[27]R. Arnowitt, S. Deser, and C. W. Misner, in *Gravitation, An Introduction to Current Research*, edited by L. Witten (Wiley, New York, 1962). See also, C. W. Misner, K. S. Thorne, and J. A. Wheeler, *Gravitation* (Freeman, San Francisco, 1973).
[28]N. Ó Murchadha and J. W. York, Phys. Rev. D 10. 428 (1974).
[29]L. Smarr and J. W. York, Phys. Rev. D 17, 2529 (1978).
[30]D. M. Eardley and L. Smarr, Phys. Rev. D 19, 2239 (1979).

THE ASTROPHYSICAL JOURNAL, 239:1–11, 1980 July 1

TENACIOUS MYTHS ABOUT COSMOLOGICAL PERTURBATIONS LARGER THAN THE HORIZON SIZE[1]

WILLIAM H. PRESS AND ETHAN T. VISHNIAC
Harvard-Smithsonian Center for Astrophysics
Received 1979 September 18; accepted 1980 January 7

ABSTRACT

We review the linear pertubation theory of the Einstein–de Sitter ($k = 0$, Friedmann) big-bang cosmology in synchronous gauge, taking particular care to distinguish physical perturbations, which are locally measurable, from pure-gauge perturbations, which correspond to an unperturbed spacetime written in gauge-perturbed coordinates. Some new results are obtained about the growth of physical perturbations at early times, while they are still outside their horizon; and some commonly accepted rules for estimating the growth and decay of perturbations are shown to be false. Source terms corresponding to inhomogeneous perturbations in the equation of state (or, equivalently, to isothermal perturbations) are next included: we calculate the density perturbations that are induced, including both pressure terms and (higher-order) pressure-gradient terms. Here also, some uncritical beliefs are shown to be incorrect.

Subject headings: cosmology — relativity

I. INTRODUCTION

The theory of linear (i.e., small) perturbations of the expanding, isotropic, and homogeneous Friedmann cosmology springs into existence virtually full-grown with the work of Lifshitz (1946). Subsequent work by many others, including Lifshitz and Khalatnikov (1963), Hawking (1966), Sachs and Wolfe (1967), has elaborated on the theory; review papers with various emphases have been published (e.g., Harrison 1967; Field 1975); the theory is developed in detail in at least two major advanced texts, Weinberg (1972, hereafter W) and Peebles (1980, hereafter P).

Unfortunately, out of this literature there have developed a few wrong or misleading interpretations, even when there is general agreement on a body of results which are mathematically correct in the narrow sense. A certain number of wrong interpretations have insinuated themselves into the "lore" of the field; and these are, on occasion, used by practising astrophysical cosmologists. An area of particular confusion is that of perturbations whose spatial wavelengths are so large that they have not yet come within their particle horizon, i.e., such that

$$\lambda = 2\pi a(t)/q \gg ct ,$$

where t is the cosmological time, $a(t)$ the expansion factor, and q is the wavenumber in expanding coordinates. Since the purpose of this paper is more to lay to rest old fallacies than it is to present new results, we key our discussion as closely as possible to the relevant sections of the texts of P and W.

As a beginning, we invite the cosmologically minded reader to consider the following six propositions, and to form an opinion as to the truth or falsity of each. The propositions are to be taken in the context of the standard synchronous gauge (used in P and W), where metric perturbations appear only in the spatial components h_{jk} ($j, k = 1, 2, 3$).

Proposition 1*A*.—In a dust ($P = 0$) cosmology, density perturbations outside the horizon evolve like their Newtonian analogs inside the horizon, namely, as $t^{2/3}$ and t^{-1}.

Proposition 1*B*.—The density perturbation which evolves as t^{-1} outside its horizon describes a cosmology made inhomogeneous by spatial variations in the bang-time (nonsimultaneity of the initial, singular surface).

Proposition 2*A*.—In a radiation-dominated ($P = \frac{1}{3}\rho$) cosmology, the two modes of density perturbation evolve as t and t^{-1} while outside their horizon, and then match to linear combinations of the two independent phases of oscillating adiabatic disturbances as they come within their horizon.

Proposition 2*B*.—In a radiation-dominated cosmology, there are three density-perturbation modes outside the horizon, evolving as t, $t^{1/2}$, and t^{-1}; and as these come within the horizon, they match to two oscillatory, adiabatic modes (with one degree of degeneracy).

Proposition 3*A*.—In a radiation-dominated cosmology, an entropy perturbation with initially zero density perturbation will develop into a density perturbation which grows as $a \sim t^{1/2}$ (because photons redshift away,

[1] Supported in part by the National Science Foundation (PHY 78-09616).

leaving behind the inhomogeneous baryon excess) and which becomes of comparable magnitude ($\delta\rho/\rho \sim \delta s/s$) when baryon and radiation densities are about equal.

Proposition 3B.—An inhomogeneous pressure perturbation $\delta P/P$, in a radiation-dominated cosmology, will turn into an inhomogeneous density perturbation of about the same magnitude ($\delta\rho/\rho \sim \delta P/P$) in about one expansion time scale, and will subsequently grow $\sim t$ (growing mode in $P = \frac{1}{3}\rho$ cosmology).

In the remainder of this paper, we show that the above six propositions, in the context that they are raised, are *all false*, and we derive the corresponding correct results. In § II we establish the notation and set out the standard perturbation equations and some uncontroversial solutions; this section relies heavily on the standard texts. In § III we compute eigenvalues and eigenvectors for various modes. Gauge transformations of the perturbations are discussed in § IV. The six propositions are disproved in §§ V–VII. Some additional conclusions are in § VIII.

We do not wish to give the impression that we are the first ones to debunk these "tenacious myths." For example, Field and Shepley (1968), Sakai (1969), Olson (1976), and Carter (1980) all clearly state the necessity of removing gauge solutions of the perturbation equations before a physical interpretation is made. Bardeen (1969, 1980) takes the complementary approach of developing gauge-independent perturbation equations, completing the program begun by Hawking (1966) and corrected by Olson (1976). Bardeen's (1980) approach extends and generalizes the results of this paper in his different formalism. Our emphasis here is more specific than previous work, namely to clarify all six propositions above in one single unified treatment and within the framework of the standard (i.e., metric perturbations in synchronous gauge) formalism.

Nor should the reader gain the impression that our tenacious myths cast doubt on the great bulk of previous cosmological perturbation calculations (e.g., Peebles and Yu 1970, and subsequent work by a variety of investigators) which uses correct equations to evolve perturbations until they are well within their horizon (and therefore unambiguous in interpretation).

II. PERTURBATIONS OF AN EINSTEIN–DE SITTER SPACETIME

For simplicity, we will restrict ourselves to the marginally bound ($k = 0$; Einstein–de Sitter) cosmology, which is itself an accurate approximation of the general, $k = \pm 1$, Friedmann-Lemaître cosmologies at sufficiently early times. We take units with $c = 1$ (but not $G = 1$); greek indices range over 0 to 3; latin indices range over 1 to 3, and repeated latin indices are summed over even when both are lowered.

The metric consists of an unperturbed plus a perturbed piece,

$$g_{\mu\nu} = \text{diag}\,(1, -a^2, -a^2, -a^2) - a^2 h_{\mu\nu}\,, \tag{1}$$

where $a^2 = a^2(t)$ depends on time alone. The matter content of the universe is a perfect fluid,

$$T^{\mu\nu} = (\rho + P)u^\mu u^\nu - g^{\mu\nu}P\,, \tag{2}$$

with u^α (the fluid 4-velocity), P (the pressure), and ρ (the total mass-energy density) each written as a sum of background and perturbation parts:

$$u^\alpha = (1, 0, 0, 0) + u_1{}^\alpha(t, \boldsymbol{x})\,, \tag{3a}$$

$$P = P_0(t) + P_1(t, \boldsymbol{x})\,, \tag{3b}$$

$$\rho = \rho_0(t) + \rho_1(t, \boldsymbol{x})\,. \tag{3c}$$

It is convenient to use

$$w \equiv P_0/\rho_0\,, \tag{4}$$

$$c_s{}^2 \equiv dP_0/d\rho_0\,, \tag{5}$$

and to denote d/dt by a dot. (P § 85 uses v for our w.) Note that (from the first law of thermodynamics, e.g., W eq. [15.1.2])

$$\dot{\rho}_0/\rho_0 = -3(1 + w)\dot{a}/a \tag{6}$$

and also

$$\dot{w} = 3(1 + w)(w - c_s{}^2)\dot{a}/a\,. \tag{7}$$

The Friedmann equations for $a(t)$ give (W eq. [15.10.7])

$$(\dot{a}/a)^2 = \frac{8\pi}{3}G\rho_0\,, \tag{8}$$

which, in the case that w is constant (or constant to useful approximation), yields (by eq. [6] or P § 86)

$$a(t) \propto t^{2/(3+3w)} . \tag{9}$$

Turn now to the perturbation quantities u_1, P_1, ρ_1, and $h_{\mu\nu}$. One can always choose a perturbed coordinate system, i.e., a "synchronous" gauge, that makes

$$h_{00} = h_{0j} = 0 \tag{10}$$

(P, § 81; eq. [15.10.12]; see also § IV below). Likewise, we can take $u_1{}^0 = 0$, to the required linear order, since the normalization condition on 4-velocities $u^\alpha u_\alpha = 1$ requires corrections of only second order in $u_1{}^0$, if the quantities $u_1{}^j$ are specified to first order. For pertubations of any nonzero wavenumber (i.e., noninfinite scale length), $u_1{}^j$ can be decomposed uniquely into a piece that can be written as a gradient (and which has nonzero 3-divergence), and a piece can be written as a curl, and has zero 3-divergence,

$$u_1{}^j \equiv u_\parallel{}^j + u_\perp{}^j , \tag{11}$$

$$u_\perp{}^j{}_{,j} = 0 , \qquad u_\parallel{}^j = \Phi_{,j} \quad \text{for some } \Phi . \tag{12}$$

The perturbed metric h_{jk} satisfies the perturbed Einstein equations (including, if useful, the perturbed conservation law $T^{\mu\nu}{}_{;\nu} = 0$). The complete linear solution can be written as the sum of three independent terms:

$$h_{jk} = h^\times{}_{jk} + h^\perp{}_{jk} + h^\parallel{}_{jk} , \tag{13}$$

representing three orthogonal sets of modes.

The first, term, $h^\times{}_{jk}$ represents the purely gravitational degrees of freedom (gravitational waves), and does not couple to $u_1{}^j$, ρ_1, or P_1 as sources; these are zero for this set of modes. The modes can be required to satisfy (W eq. [15.10.38]; P § 83) transverse-traceless conditions and a homogeneous wave equation:

$$h_{jj} = 0 \quad \text{(traceless)} , \tag{14a}$$

$$h_{jk,k} = 0 \quad \text{(transverse)} , \tag{14b}$$

$$\frac{1}{a^2} h_{jk,ll} = \ddot{h}_{jk} + 3 \frac{\dot{a}}{a} \dot{h}_{jk} . \tag{15}$$

When w is constant (i.e., $c_s{}^2 = w$ for all time), the general solution to equation (15) is a superposition of wavenumber q (W eq. [15.10.44]):

$$h_{jk} \propto t^{(w-1)/(2+2w)} J_{\pm\nu}\left[\frac{(3+3w)|q|t}{(3w+1)a}\right] \exp(i\boldsymbol{q}\cdot\boldsymbol{x}) , \tag{16}$$

with

$$\nu = \frac{3}{2}\left(\frac{1-w}{3w+1}\right) . \tag{17}$$

The second set of modes, $h^\perp{}_{jk}$, represents rotational matter perturbations, so has nonzero $u_\perp{}^j$. The most general solution for this mode is (P § 90, W eq. [15.10.49])

$$u_\perp{}^j(t, \boldsymbol{x}) = \frac{F^j(\boldsymbol{x})}{a^5 \rho_0 (1+w)} , \tag{18a}$$

$$h^\perp{}_{jk} = -4G \int^t \frac{dt}{a^3} \int d^3x' \frac{F^j{}_{,k} + F^k{}_{,j}}{|\boldsymbol{x} - \boldsymbol{x}'|} , \tag{18b}$$

where $F^j(\boldsymbol{x})$ is an arbitrary 3-divergenceless function of space alone (which can be viewed as specifying $u_\perp{}^j$ at some initial time), and

$$P_1 = \rho_1 = 0 . \tag{19}$$

Note that $h^\perp{}_{jk}$ has no dynamic freedom, but only decreases its amplitude in prescribed fashion as the Universe expands.

The third mode, $h^{\|}{}_{jk}{}^{i}$, represents density perturbations and couples to $u_{\|}{}^{j}$, ρ_1, and P_1 as source terms. It is convenient to write

$$\theta \equiv a u_{\|}{}^{j}{}_{,j} \equiv \phi \dot{a}/a\,, \tag{20}$$

$$h \equiv h^{\|}\,, \tag{21}$$

$$\delta \equiv \rho_1/\rho_0\,, \tag{22}$$

$$\epsilon \equiv (P_1 - c_s{}^2\rho_1)/\rho_0\,. \tag{23}$$

Here $\epsilon(t, x^j)$ is the fractional pressure perturbation *excluding the piece that arises from density perturbations.* Nonzero ϵ corresponds to perturbations in the equation of state as a function of space and time. While P and W do not include ϵ explicitly, it will prove important in our analysis below. The most general solution of the density perturbation mode can then be shown to satisfy the equations

$$\ddot{h} + 2\frac{\dot{a}}{a}\dot{h} = 3\left(\frac{\dot{a}}{a}\right)^2(\delta + 3c_s{}^2\delta + 3w\epsilon)\,, \tag{24}$$

$$\dot{\delta} + (1 + w)(\theta - \tfrac{1}{2}\dot{h}) = 3\frac{\dot{a}}{a}[\delta(w - c_s{}^2) - \epsilon w]\,, \tag{25}$$

$$\dot{\theta} + \frac{\dot{a}}{a}\theta(2 - 3c_s{}^2) = -\frac{c_s{}^2}{a^2(1 + w)}\delta_{,jj} - \frac{w}{a^2(1 + w)}\epsilon_{,jj}\,. \tag{26}$$

(P § 85; W eqs. [15.10.50–53] with $\theta \equiv iq\cdot u_1$, $\delta \equiv \rho_1/\rho_0$, $0 = \chi = \eta$, and $h_{\text{here}} = -h_{\text{Weinberg}}/R^2$.)

The equations given thus far are general, in the sense that they apply equally well to scales smaller than or larger than the horizon size. At this stage, there appears to be the possibility of some dynamical freedom for $h^{\|}{}_{jk}$, indicated by the time derivatives in equation (24). In § IV below, we will see that this is *not* the case, that $h^{\|}{}_{jk}$ is determined by $u_{\|}{}^{j}$. The physical degrees of freedom (independent quantities that can be specified at some initial time and then evolved uniquely) on scales larger than the horizon will be seen to be precisely those that one expects on quasi-Newtonian grounds: an arbitrary initial density perturbation $\delta(x^j)$, an arbitrary initial 3-velocity field $u_1{}^j(x^k)$, and a superposed set of transverse-traceless gravitational waves, $h^{\times}{}_{jk}$ and $\dot{h}^{\times}{}_{jk}$ satisfying equations (14)–(15).

III. TIME-DEPENDENCE OF PERTURBATION MODES

If $c_s{}^2$ (eq. [5]) varies arbitrarily with time (due, for example, to the changing spectrum of fundamental particles with energy), then, by equation (7), w will also vary. In this case the universe has no particular scaling symmetry in time, and it does not make sense to talk of separating different modes of time development. In our universe there have, however, been epochs where $c_s{}^2$ has been very nearly constant over many expansion times, having either the value $\frac{1}{3}$ (radiation-dominated epoch) or 0 (matter-dominated epoch). In these epochs, by equation (7), $w \approx c_s{}^2 \approx$ constant, and equation (9) applies.

Equations (24)–(26) would now admit a homogeneous (in the sense of setting the source ϵ to zero) set of normal modes, but for the appearance of a term $\delta_{,jj}$ in equation (26). This pressure gradient term brings a special dimensional time t_h into the equations, namely the sound-crossing time

$$t_h \sim \lambda/c_s\,, \tag{27}$$

where $\delta_{,jj} \sim \delta/\lambda^2$ defines the scale λ.

If $\lambda \gg c_s t$ (which is always true in the matter-dominated case, $c_s = 0$, and true at sufficiently early times t for other cases), we can neglect the term in $\delta_{,jj}$ and derive a complete set of modes as follows:

Adopt a new time variable $\eta \equiv \ln a$, so that $d/dt = (\dot{a}/a)d/d\eta$. Denote $d/d\eta$ by a prime. Also define $\phi \equiv \theta a/\dot{a}$ (eq. [20]). Then equations (24)–(26). with $c_s{}^2 = w$, become

$$h'' + (\tfrac{1}{2} - \tfrac{3}{2}w)h' - 3(1 + 3w)\delta = 9w\epsilon\,, \tag{28}$$

$$\delta' + (1 + w)(\phi - \tfrac{1}{2}h') = -3w\epsilon\,, \tag{29}$$

$$\phi' - (\tfrac{9}{2}w - \tfrac{1}{2})\phi = 0\,. \tag{30}$$

Equations (28)–(30) are a fourth order set in η-time with constant coefficients. There are therefore four independent solutions, each varying as $\exp(\lambda\eta)$ for some eigenvalue λ, and four corresponding eigenvectors $(\delta, \phi, h, h')\exp(\lambda\eta)$. The time dependence in t-time of a mode with eigenvalue λ is (using eq. [9])

$$(\delta, \phi, h, h') \propto t^{2\lambda/(3+3w)}\,. \tag{31}$$

The four eigenmodes are computed in straightforward manner to be

$$\lambda = 0\,, \qquad [\delta, \phi, h, h'] \propto [0, 0, 1, 0]t^0\,, \tag{32a}$$

$$\lambda = -\tfrac{3}{2}(1+w)\,, [\delta, \phi, h, h'] \propto [\tfrac{1}{2}(1+w), 0, 1, -\tfrac{3}{2}(1+w)]t^{-1}\,, \tag{32b}$$

$$\lambda = 1 + 3w\,, \quad [\delta, \phi, h, h'] \propto [\tfrac{1}{2}(1+w), 0, 1, (1+3w)]t^{(2+6w)/(3+3w)}\,, \tag{32c}$$

$$\lambda = \tfrac{1}{2}(9w-1)\,, [\delta, \phi, h, h'] \propto [w(1+w)(9w-1), (3w+\tfrac{1}{2})(1-w)(9w-1), 2(1+3w)(1+w),$$
$$(1+3w)(1+w)(9w-1)]t^{(9w-1)/(3+3w)}\,. \tag{32d}$$

For later use, it will be convenient to list the above eigenmodes for the radiation-dominated ($w = \frac{1}{3}$) and matter-dominated ($w = 0$) cases:

$w = \frac{1}{3}$:

$$[\delta, \phi, h, h'] \propto [0, 0, 1, 0]\,, \tag{33a}$$

$$[\delta, \phi, h, h'] \propto [\tfrac{2}{3}, 0, 1, -2]t^{-1}\,, \tag{33b}$$

$$[\delta, \phi, h, h'] \propto [\tfrac{2}{3}, 0, 1, 2]t\,, \tag{33c}$$

$$[\delta, \phi, h, h'] \propto [\tfrac{8}{9}, 2, \tfrac{16}{3}, \tfrac{16}{3}]t^{1/2}\,; \tag{33d}$$

$w = 0$:

$$[\delta, \phi, h, h'] \propto [0, 0, 1, 0]\,, \tag{34a}$$

$$[\delta, \phi, h, h'] \propto [\tfrac{1}{2}, 0, 1, -\tfrac{3}{2}]t^{-1}\,, \tag{34b}$$

$$[\delta, \phi, h, h'] \propto [\tfrac{1}{2}, 0, 1, 1]t^{2/3}\,, \tag{34c}$$

$$[\delta, \phi, h, h'] \propto [0, -\tfrac{1}{2}, 2, -1]t^{-1/3}\,. \tag{34d}$$

Having written the homogeneous solutions, we can now consider the possibility of nonzero $\epsilon(t, \boldsymbol{x})$. Since equations (28)–(30) (or for that matter, the exact equations [24]–[26]) are linear in ϵ as well as in the dependent variables, the general solution is a superposition of Green's functions, each corresponding to a delta-function source at time t_0:

$$\epsilon(t) = \epsilon_0(x^k)\delta(t - t_0)\,. \tag{35}$$

From the exact equations (24)–(26) one can read off the discontinuities in δ, h, $\dot{h}$ or h', θ or ϕ at t_0:

$$[\delta, \phi, h, h']_{t_0\pm} = \left[-3w\epsilon_0\left(\frac{\dot{a}}{a}\right)_{t_0}, -\frac{w\epsilon_{0,jj}}{(a\dot{a})_{t_0}(1+w)}, 0, 9w\epsilon_0\left(\frac{\dot{a}}{a}\right)_{t_0}\right]. \tag{36}$$

The second, ϕ, component of equation (36) is smaller than the first and fourth component by $\sim(t/\lambda)^2$ (cf. eq. [27]), so in the approximation to which modes are defined at all, the discontinuity of equation (36) can be seen by inspection to consist purely of the modes equation (32a) and equation (32b), not at all of the modes equation (32c) and equation (32d). This will prove to be important in § VII below.

IV. GAUGE TRANSFORMATIONS OF THE PERTURBATION VARIABLES

We are now approaching the central issue, confusion about which results in most of the myths listed in § I. We want to be guided by one key principle: *If a spacetime is exactly a Friedmann cosmology, then it is by definition unperturbed.* In particular, if any of the modes that we have considered in § III are in fact the results of gauge transformations (infinitesimal coordinate transformations) of an exactly Friedmann spacetime, then we must reject these as entirely unphysical: no possible measurement or set of measurements could distinguish their presence or absence.

Consider an unperturbed Friedmann spacetime, described in the usual coordinates x^μ, and transform to the new coordinates $\hat{x}^\mu$ defined by

$$\hat{x}^\mu = x^\mu + \epsilon^\mu(x^\nu) \tag{37}$$

(the use of the symbol ϵ in this section is not to be confused with its use elsewhere in this paper). Then the metric $g_{\mu\nu}$ transforms as (see, e.g., Lightman *et al.* 1975, § 13.12)

$$a^2 h_{\mu\nu} \equiv \hat{g}_{\mu\nu} - g_{\mu\nu} = -g_{\mu\nu,\alpha}\epsilon^\alpha - g_{\mu\alpha}\epsilon^\alpha{}_{,\nu} - g_{\alpha\nu}\epsilon^\alpha{}_{,\mu}\,. \tag{38}$$

The density ρ, which is a 4-scalar, transforms as

$$\rho_1 \equiv \hat{\rho} - \rho = -\rho_{,\alpha}\epsilon^\alpha\,. \tag{39}$$

The 4-velocity of the fluid transforms as

$$u_1{}^{\mu} \equiv \hat{u}^{\mu} - u^{\mu} = -u^{\mu}{}_{,\alpha}\epsilon^{\alpha} + u^{\alpha}\epsilon^{\mu}{}_{,\alpha} . \tag{40}$$

The 4-divergence of the fluid u is a 4-scalar, so

$$\hat{\nabla}\cdot\boldsymbol{u} - \nabla\cdot\boldsymbol{u} = -(\nabla\cdot\boldsymbol{u})_{,\alpha}\epsilon^{\alpha} . \tag{41}$$

The perturbation quantity θ (eq. [20]) is a *coordinate 3-divergence* which differs from $\nabla\cdot\boldsymbol{u}$ by the volume expansion of the coordinates (as can be seen also in the term $\theta - \frac{1}{2}\dot{h}$ in eq.[25]). Therefore the gauge contribution to θ is

$$\theta = -(\nabla\cdot\boldsymbol{u})_{,\alpha}\epsilon^{\alpha} + \tfrac{1}{2}\dot{h}_{jj} , \tag{42}$$

where $a^2 h_{jj}$ is the 3-trace of the right-hand side of equation (38).

What is the most general allowed form for $\epsilon^{\mu}(x^{\nu})$? If we require $h_{\mu\nu}$ (eq. [38]) to satisfy equation (10), then (P § 81, W eqs. [15.10.9] ff.)

$$\epsilon^0 = \psi(x^k) , \tag{43a}$$

$$\epsilon^j = \int^t \frac{dt}{a^2}\psi_{,j} + \chi^j(x^k) , \tag{43b}$$

where ψ and χ are arbitrary functions of space alone. Equations (43) and (38)–(42) yield directly

$$h_{jk} = 2\frac{\dot{a}}{a}\psi\delta_{jk} + 2\psi_{,jk}\int^t \frac{dt}{a^2} + \chi^j{}_{,k} + \chi^k{}_{,j} , \tag{44}$$

$$h \equiv h_{jj} = 6\frac{\dot{a}}{a}\psi + 2\psi_{,jj}\int^t \frac{dt}{a^2} + 2\chi^j{}_{,j} , \tag{45}$$

$$\delta \equiv \frac{\rho_1}{\rho_0} = 3(1+w)\frac{\dot{a}}{a}\psi , \tag{46}$$

$$\theta = \psi_{,jj}/a^2 , \tag{47}$$

$$u_1{}^j = \psi_{,j}/a^2 . \tag{48}$$

Since $u_1{}^j$ (eq. [48]) is the spatial gradient of a scalar, its decomposition into $u_{\parallel}{}^j$ and $u_{\perp}{}^j$ (eq. [11]) is

$$u_{\perp}{}^j = 0 , \tag{49}$$

$$u_{\parallel}{}^j = \psi_{,j}/a^2 . \tag{50}$$

Similarly, one can verify that the right-hand sides of equations (44) and (45) contain no transverse-traceless part which would satisfy equations (14a) and (14b) (see, e.g., Misner, Thorne, and Wheeler 1973, Box 35.1), nor can it be cast into the functional form of equation (18b) (as is also evident from eqs. [49] and [18a]). Therefore, the gauge contribution of h_{jk} is entirely to the piece $h^{\parallel}{}_{jk}$ in equation (13).

One expects, now, that the pure gauge solutions (45), (46), and (47) should *exactly* satisfy equations (24)–(26). This is readily verified by direct substitution. From now on, we will use the term "pure gauge solution" to refer to the right-hand sides of equations (44)–(48) for any functions ψ and χ.

V. MATTER-DOMINATED LIMIT

If $P_0 = 0$ (i.e., $w = 0$), then the mode equations (28)–(30) are exactly equivalent to the exact equations (24)–(26), because only a pressure gradient term was dropped in going between the two sets. Therefore any solution of equations (24)–(26), including the pure gauge solution, must be *exactly* a linear combination of the modes, equations (34). By inspection, setting $\psi = 0$, $\chi^j \neq 0$, in equations (45)–(47) gives precisely the trivial mode (34a), which corresponds to a transformation of the spatial coordinates alone (eq. [43b]; P § 81; W eq. [15.10.55]). If $\chi^j = 0$, $\psi \propto \exp(iq_j x^j)$, then equations (45)–(47) can be readily shown to be the linear combination of modes (34b) and (34d):

$$(\delta, \phi, h, h') \propto (\tfrac{1}{2}, 0, 1, -\tfrac{3}{2})\left(\frac{t}{t_0}\right)^{-1} + (0, -\tfrac{1}{2}, 2, -1)\left(\frac{t}{t_0}\right)^{-1/3} , \tag{51}$$

where t_0 is the epoch such that

$$3\left(\frac{\dot{a}}{a}\right)^2 = -\frac{\nabla^2\psi}{a^2\psi} . \tag{52}$$

or (using $\nabla^2\psi = -q^2\psi$ and eq. [9])

$$t_0 = \left(\frac{4}{3}\right)^{1/2} \frac{a(t_0)}{q}, \tag{53}$$

which is evidently (to within a factor of order unity) the horizon time of the scale $2\pi a/q$. At times much before t_0, the pure gauge mode (51) is dominated by the t^{-1} part; at times much after t_0, it is dominated by the $t^{-1/3}$ piece.

We can now "mod out" the pure gauge mode from an arbitrary perturbation, and in so doing disprove Proposition 1A of § I: one natural way to remove gauge-freedom from the set of variables (δ, ϕ, h, h') is to require that physical measurements of δ and ϕ (or θ) at any epoch be reported in the coordinate system which *at that epoch* has

$$h = h' = 0\,. \tag{54}$$

Bardeen (1980) calls this choice the "uniform Hubble constant gauge." It is a natural coordinate system which would be constructed by an observer at some epoch who is ignorant of coordinate systems that may have been constructed by other observers in previous epochs, namely that coordinate system whose volume expansion is instantaneously uniform. To obtain δ and θ in this system, we add to the four modes (34a)–(34d) appropriate time-dependent amounts of the gauge modes combinations (34a) and (51), to produce conditions (57) holding for all time. There will obviously be two linearly independent such solutions, and any linear combination of them is also a solution. These solutions are *not* modes of equations (28)–(30), since they are the sum of modes with nonconstant coefficients. They exhibit, however, the two *physical* degrees of freedom that are inherent in the four modes (34a)–(34d). A short calculation gives the results

$$\delta = \frac{2s^{-1/3}}{2s^{2/3}+3}, \qquad \phi = \frac{3s^{1/3}}{2s^{2/3}+3} \quad \text{(dying solution)} \tag{55}$$

and

$$\delta = \frac{2s^{4/3}+s^{2/3}}{2s^{2/3}+3}, \qquad \phi = \frac{-2s^{4/3}}{2s^{2/3}+3} \quad \text{(growing solution)}\,, \tag{56}$$

where $s \equiv t/t_0$ and t_0 is given in equation (53). Note that the growing solution (56) evolves as $t^{2/3}$ for both $s \ll 1$ and $s \gg 1$. The dying solution (55) switches from $t^{-1/3}$ when it is outside its horizon, to t^{-1} when it comes within it. This disproves Proposition 1A.

VI. CASES OF NONZERO PRESSURE

That the pure gauge mode goes as t^{-1} at early times is more general than just the $P = 0$ case. In equations (45)–(47), the terms on the right-hand which contain $\dot{a}/a(\propto t^{-1})$ dominate the other terms (which contain $\psi_{,jj}/a^2$) at early times, i.e., far outside the horizon. In this limit, one can easily see that equations (45)–(47) go over to a constant multiple of equation (32b), the t^{-1} mode, in fact to $6\psi t\dot{a}/a$ times that mode. So, generically for all w, the t^{-1} mode is pure gauge at early times. This demonstrates the fallacy in Proposition 1B: the t^{-1} mode corresponds to a *mislabeling* of the constant time surfaces in an unperturbed cosmology, not to any actual (i.e., physically measurable) nonsimultaneity of the big bang. The small, neglected, differences between the t^{-1} mode (eq. [32b]) and the exact gauge solution (45)–(47) are only the same order as the pressure gradient terms, an order to which exact modes are not defined in any case. In the one case where the exact modes *are* defined, namely $w = 0$, we have seen that the small differences then become themselves a piece of another exact mode. That the Friedmann solution *does not permit* a physical perturbation corresponding to a nonsimultaneous big bang is a nontrivial statement about the nature of the cosmological singularity in general relativity, closely related to the Belinsky, Khalatnikov, Lifshitz (1970) results.

We can also now dispose of Propositions 2A and 2B. The two physical modes of density perturbation on scales far outside the horizon in a radiation-dominated cosmology are those of equations (33c) and (33d), which evolve as t and $t^{1/2}$ respectively. These modes match to linear combinations of the two phases of adiabatic oscillations once they come within their horizon. Mode (33a) is a trivial gauge mode both outside and inside the horizon. Mode (33b), evolving as t^{-1}, is the pure gauge mode at early times; far from matching to adiabatic oscillations inside the horizon, it goes over to a "mode" (actually pure gauge) with

$$h \propto 2 \ln t\,, \tag{57}$$

$$\theta \propto t^{-1}\,, \tag{58}$$

$$\delta \approx 0 \approx \theta - \tfrac{1}{2}\dot{h}\,, \tag{59}$$

which is the limiting case of the exact equations (45)–(47) for large t (using $a \propto t^{1/2}$).

It should, by now, be apparent that the most general gauge-free specification of an adiabatic cosmological perturbation at a given epoch consists of specifying, at that epoch (and in an instantaneous coordinate system with $h = \dot{h} = 0$), the perturbed density and velocity $\rho_1(x^k)$, $u_1{}^j(x^k)$, plus (if desired) an uncoupled, homogeneous, transverse-traceless gravitational wave $h^\times{}_{jk}$ and $\dot{h}^\times{}_{jk}$. Decomposing $u_1{}^j$ into $u_\parallel{}^j$ and $n_\perp{}^j$, one gets the single time dependence of equation (18a) from the single function $u_\perp{}^j$; and from the two functions ρ_1 and $u_\parallel{}^j$ $(=\Phi_{,j}$ for some $\Phi)$, one gets the two physical, linearly independent, modes of time dependence, e.g., equations (55)–(56) or equations (33c)–(33d), or two phases of oscillatory wave inside the horizon. Except for the uncoupled gravitational waves, there are no "extra" (i.e., non-Newtonian) degrees of freedom in the specification.

VII. PERTURBATIONS INDUCED BY PRESSURE INHOMOGENEITIES: ENTROPY PERTURBATIONS

In this section, we want to consider a cosmology which is exactly Friedmannian from $t = 0$ until some time t_i, when an inhomogeneous pressure perturbation is introduced. In the literature one finds several models approximating this ideal with, e.g., the pressure perturbation originating from $\sqrt{N}$ statistics of a population of mini-black holes (Mészáros 1975; Carr 1977, P §§ 94–95), or from the "freezing out" of massive X-bosons at some high temperature (Hogan 1979), or from some other sort of spontaneously broken gauge symmetry (Press 1980). Typically, the equation of state is perturbed in the direction of reducing the pressure in proportion to the amount of radiation mass-energy that is converted to (essentially) pressureless form, be it black holes, nonrelativistic particles, or whatever.

Since the thermodynamics of a mixture of nonrelativistic particles and photons is more or less identical to the thermodynamics of a mixture of nonrelativistic baryons and photons, it is convenient to take the photon-baryon gas as a canonical case. We are talking, then, about entropy perturbations in the photon-baryon mixture. The pressure is

$$P = \tfrac{1}{3}aT^4 = \tfrac{1}{3}An_\gamma{}^{4/3}\,, \tag{60}$$

where A is the appropriate constant and n_γ is the photon number density. The density is

$$\rho = An_\gamma{}^{4/3} + n_{\rm H}m_p \equiv \rho_r + \rho_m\,, \tag{61}$$

where $n_{\rm H}$ is the baryon number density, m_p is the proton mass, and ρ_r and ρ_m are the partial densities of radiation and matter. Defining the entropy per baryon

$$s \equiv n_\gamma/n_{\rm H}\,, \tag{62}$$

equations (60) and (61) imply (for small changes)

$$\delta P = \tfrac{4}{3}P(\delta s/s + \delta n_{\rm H}/n_{\rm H})\,, \tag{63}$$

$$\delta\rho = m_p\delta n_{\rm H} + 3\delta P\,. \tag{64}$$

Now from equations (23), (63), and (66), it follows that

$$\epsilon = \left(\frac{\delta P}{P}\right)_{\delta\rho=0} = \frac{\delta s}{s}\frac{\rho_m}{\rho_r}(1 + \tfrac{3}{4}\rho_m/\rho_r)^{-1}\,. \tag{65}$$

If t_e is the epoch when matter first becomes dominant over radiation, $\rho_m = \rho_r$, then equations (65) and (9) give approximately

$$\epsilon(t) \approx \frac{\delta s}{s}\left(\frac{t}{t_e}\right)^{1/2} \quad (t \ll t_e)\,, \qquad \epsilon(t) \approx \frac{4}{3}\frac{\delta s}{s} \quad (t \gg t_e)\,. \tag{66}$$

Let us now suppose that this pressure perturbation [multiplied, as usual, by $\exp(iq_jx^j)$] is turned on at time t_i. What is the perturbed state of the universe at time t, $t_i < t \ll t_e$? From equations (36), (33), and (66), but neglecting the ϕ component of equation (36) (see discussion following that equation), we have the Green's function solution

$$[\delta, \phi, h, h'] = \int_{t_i}^{t} dt_0\,\frac{\delta s}{s}\left(\frac{t_0}{t_e}\right)^{1/2}\frac{1}{2t_0}\left\{(1, 0, \tfrac{3}{2}, -3)\frac{t_0}{t} - (0, 0, \tfrac{3}{2}, 0)\right\} \tag{67}$$

(here $\dot{a}/a = [2t_0]^{-1}$ has been used). Performing the integration, we get, e.g.,

$$h = \frac{\delta s}{s}\left(\frac{t_i}{t_e}\right)^{1/2}\left[\frac{3}{2} - \frac{1}{2}\frac{t_i}{t} - \frac{t^{1/2}}{t_i{}^{1/2}}\right] \quad \text{(see below!)}\,. \tag{68}$$

$$\delta = \frac{1}{3}\frac{\delta s}{s}\left(\frac{t_i}{t_e}\right)^{1/2}\left[\frac{t^{1/2}}{t_i{}^{1/2}} - \frac{t_i}{t}\right] \quad \text{(see below!)}\,. \tag{69}$$

Equations (68) and (69) are exactly the equations (A17) and (A19) of Carr (1977), or equation (94.6) (and corresponding h) of P § 94. Equation (69) says that an initial fractional pressure perturbation $[(\delta s/s)(t_i/t_e)^{1/2}]$ becomes a density perturbation δ of about the same order after one expansion time scale ($t = 2t_i$, say), and then grows $\propto t^{1/2} \propto a$, as the radiation fluid redshifts away.

But wait a minute! We have just seemingly proved Propositions 3A and 3B, which were previously alleged to be false! The (here intentional) fallacy in equations (67)–(69) is that the Green's function modes integrated are precisely gauge modes, equations (33a) and (33b). If a radiation-dominated spacetime with h and δ given by equations (68) and (69) has its perturbing pressure turned off at some time $t = t_f$, then it is left in a state of being precisely Friedmannian, independent of t_f. The residual apparent δ, ϕ, h, h' are removable by a coordinate transformation. It would not be correct to say that the spacetime is exactly Friedmannian while the perturbing pressure is turned on, since the pressure itself appears in $T^{\mu\nu}$ and hence in the Riemann tensor. However, to the order considered thus far, the perturbing pressure induces *no* secular change in δ while it is turned on and its termination, however sudden, leaves behind *no* nonzero physical (i.e., nongauge) perturbation of the subsequent spacetime, not even a dying mode solution. The only difference between this spacetime and another one in which the pressure had never been perturbed is that the elapsed proper time along a matter world line from the singularity to a homogeneous constant density surface of time $t > t_f$, will vary from point to point, according to the different pressure history of each point. Carter (1980) has also pointed out this effect: that pressure inhomogeneities cause the surfaces of cosmological *time* (which exhibit the homogeneity and isotropy of the cosmology) to advance differently from the surfaces of constant *age* (proper time from the big bang).

Let us now go beyond the analysis of equations (67)–(69), and include the pressure gradient terms which are smaller by order t^2/λ^2. To do this, we need to subtract from equation (36) *not* a combination of gauge modes (33a)–(33b), which are not well defined to the necessary order in t^2/λ^2, but rather an *exact* gauge solution of the form equations (45)–(47). Choosing

$$\psi = \frac{-w\epsilon_0}{1+w} \tag{70}$$

and corresponding appropriate χ, one can write equation (36) as

$$(\delta, \phi, h, h')_\pm = \left[0, 0, 0, \frac{2w\epsilon_{0,jj}}{(1+w)(a\dot{a})t_0}\right] + [\text{pure gauge piece of form eqs. (45)–(47)}] \,. \tag{71}$$

That this is precisely the physical response that one would expect from an impulsive pressure gradient can be readily seen from a Newtonian (or special relativistic) side calculation: For a fluid with equations of motion,

$$\frac{\partial \rho}{\partial t} = -\nabla\cdot(\rho \boldsymbol{v}) \,, \tag{72}$$

$$\frac{\partial \boldsymbol{v}}{\partial t} = \frac{-\nabla P}{P+\rho} \,, \tag{73}$$

or, to perturbation order (using eqs. [4], [22], [23]),

$$\frac{\partial \delta}{\partial t} = -\nabla\cdot\boldsymbol{v} \,, \tag{74}$$

$$\frac{\partial(\nabla\cdot\boldsymbol{v})}{\partial t} = \frac{-w}{1+w}\nabla^2(\delta+\epsilon) \,, \tag{75}$$

the response to a delta-function ϵ_0 is then

$$(\delta, \nabla\cdot\boldsymbol{v})_\pm = \left(0, \frac{w}{1+w}\nabla^2\epsilon_0\right) . \tag{76}$$

Identifying $\nabla\cdot\boldsymbol{v}$ with $\phi - \frac{1}{2}\dot{h}$, ∇^2 with $\partial_j\partial_j/a^2$, and using $h' \equiv \dot{h}a/\dot{a}$, one sees that equations (76) and (71) are exactly equivalent.

Since the nongauge term in equation (71) is now entirely of order $\epsilon_{0,jj} \propto \lambda^{-2}$, it is accurate enough to write it as $\epsilon_{0,jj}$ times a sum of modes known only to lowest order, e.g., equations (33a)–(33d), with the result (now specializing to $w = \frac{1}{3}$, $\epsilon_0 \propto \exp[iq_jx^j]$)

$$\begin{aligned}(\delta, \phi, h, h')_\pm = \frac{-q^2\epsilon_0}{8(a\dot{a})_{t_0}}\left(\tfrac{2}{3}, 0, 1, 2\right)\frac{t}{t_0} &+ O(q^2\epsilon_0) \times (\text{gauge modes [33a]–[33b]}) \\ &+ O(\epsilon_0) \times (\text{gauge solutions [45]–[47]}) \,. \end{aligned} \tag{77}$$

Henceforth we can omit the various gauge terms as unphysical.

If the pressure perturbation, equation (66), is turned on at time t_i and turned off at time t_f, then the Green's function solution for the perturbation at time $t \geq t_f$ is (from eq. [77])

$$(\delta, \phi, h, h') = \int_{t_i}^{t_f} dt_0 \frac{\delta s}{s}\left(\frac{t_0}{t_e}\right)^{1/2} \frac{1}{8t_{hr}} (\tfrac{2}{3}, 0, 1, 2) \frac{t}{t_0} = \frac{1}{4}\frac{\delta s}{s}\left(\frac{t_f}{t_e}\right)^{1/2}\left(1 - \frac{t_i^{1/2}}{t_f^{1/2}}\right)\frac{t}{t_{hr}} (\tfrac{2}{3}, 0, 1, 2)\,, \tag{78}$$

where we have defined t_{hr} as the horizon time of a perturbation of wavenumber q in a radiation-dominated cosmology,

$$t_{hr} \equiv \left(\frac{a^2}{q^2}\right)\left(\frac{\dot{a}}{a}\right). \tag{79}$$

(This t_{hr} differs from t_0, eq. [53], by a factor of order unity; equation [79] has the advantage that the right-hand side can be evaluated at any epoch since it scales as $[t^{1/2}]^2 \times t^{-1} \propto t^0$.)

One can cast equation (78) in gauge independent form by computing δ in a gauge which has $h = h' = 0$ instantaneously at time t (adding appropriate amounts of eqs. [33a, b]).

This gives

$$\delta = \frac{1}{3}\frac{\delta s}{s}\left(\frac{t_f}{t_e}\right)^{1/2}\left(1 - \frac{t_i^{1/2}}{t_f^{1/2}}\right)\left(\frac{t}{t_{hr}}\right) \quad (t < t_{hr})\,, \tag{80}$$

which is valid up to time $t \approx t_e$ or $t \approx t_{hr}$, whichever comes first. If $t_{hr} \lesssim t_e$, so that the perturbation comes within its horizon before the universe becomes matter dominated, then equation (80) has a simple interpretation: *if it was once ever maintained for about an expansion time* ($t_i \lesssim t_f$), *then a fractional pressure perturbation* $\delta P/P\ [=(\delta s/s)(t_f/t_e)^{1/2}]$ *will produce a fractional density perturbation of about the same size, when it finally comes within its horizon.* This is the correct version of Proposition 3B, and it differs from the version stated in § I by a crucial factor t_f/t_{hr}, the entire alleged "growth factor" of Proposition 3B.

As for a correct version of Proposition 3A, we can set $t = t_f$ in equation (80) to discover that an entropy perturbation $\delta s/s$ will convert into a density perturbation as the *three-halves* power of t before it enters its horizon, and *one-half* power of t afterwards, but never achieving δ comparable to $\delta s/s$ until at least after the universe becomes matter dominated, $t \approx t_e$. (For times t, $t_{hr} < t < t_e$, the last factor in eq. [80] goes over to a constant, as the growing mode becomes oscillatory within its horizon.)

Therefore, to complete the analysis of Proposition 3A, we need to consider the case of perturbation scales so large that they come within their horizon only after $t = t_e$ when the universe goes matter-dominated. Here we must content ourselves with an approximate analysis, since the exact transition solution through $t \approx t_e$ is not expressible in terms of pure power laws. Roughly, at time t_e, w goes from $\frac{1}{3}$ to 0, so the Green's function, equation (71), goes to zero: there is no pressure gradient driving term. Therefore, t_f can be at most $\sim t_e$. Any (nongauge, growing mode) δ developed by time t_e subsequently grows as $(t/t_e)^{2/3}$ (mode 34c). Also, the horizon size of a wavenumber q is not t_{hr} given by equation (79) (which assumed $a \propto t^{1/2}$), but is rather t_{hm} which is easily shown to satisfy

$$\frac{t_{hm}}{t_{hr}} \approx \left(\frac{t_{hm}}{t_e}\right)^{1/3}. \tag{81}$$

Thus, in the case $t_{hm} > t_e$ (or $t_{hr} > t_e$, by eq.[81]), equation (80) becomes approximately

$$\delta \approx \frac{\delta s}{s}\left(\frac{t_f}{t_e}\right)^{1/2}\left(1 - \frac{t_i^{1/2}}{t_f^{1/2}}\right)\left(\frac{t}{t_{hm}}\right)^{2/3}. \tag{82}$$

Notice that t_e cancels out of the equation, except for normalizing the size of the pressure perturbation $\delta P/P \sim (\delta s/s)(t_f/t_e)^{1/2}$. The happy result is that the interpretation of equation (82) (when $t_{hr} > t_e$) is the same as that of equation (80) (when $t_{hr} < t_e$), namely, the italicized statement following equation (80) above. At time $t = t_e$ equation (85) differs from Proposition 3A by the crucial factor $(t_e/t_{hm})^{2/3}$. The equality $\delta s/s \sim \delta\rho/\rho$ is achieved not at the matter dominance time t_e, but only at the horizon time t_{hm}, which may be much later. And, it is worth repeating, this $\delta\rho/\rho$ is entirely due to pressure gradients transmitted at the speed of sound across the horizon size; it is *not* due to local processes of the sort envisaged in Propositions 3A and 3B.

VIII. DISCUSSION

Once the "tenacious myths" have been put aside, a clear conclusion emerges about the possibility of generating galaxy-forming or galaxy-clustering perturbations *spontaneously* by processes at very early times: it probably cannot be done. The necessary density perturbations must be $\gtrsim 1\%$ when they come within their horizon (so that presently observed structure can grow between the epoch of decoupling, $Z \sim 1000$, and the present). But, as we have seen, the pressure inhomogeneities *on the same scale* must have been of the same order at the early time. This then sets a lower bound on the scale where even 100% pressure fluctuations, statistically uncorrelated outside of each other's horizon, could do the job. That scale is evidently only 10^4 or so smaller in mass than the galaxy or cluster of galaxies mass that one is trying to explain, and thus came within its own horizon when the universe had, already, the modest temperature of 0.1–1 keV. It is hard to believe that there are unknown physical processes which could have given order-unity pressure *inhomogeneities* on these huge scales at these modest temperatures. Carr (1977) has suggested that the formation of massive black holes at $Z \sim 1000$, randomized in position by the gravitational-radiation recoil of their collapse event, is one possible way around the difficulty, at least conceptually. Another possible loophole, discussed elsewhere (Press 1980), is to consider a process *so* early, close to the Planck mass of 10^{19} GeV, that one might speculatively imagine a breakdown in the weak principle of equivalence, so that the $T^{\mu\nu}$ of gravitating matter is not conserved. In this case one can calculate the density perturbation generated by a spontaneously broken gauge symmetry, and one *does* then have benefit of the growth factor t_f/t_{hr} which is implicit in Proposition 3B. One is not today in a position to do more than model, speculative calculations along this line, unfortunately.

Both the t and $t^{1/2}$ modes grow toward the future; i.e., both are irregular as $t \rightarrow \infty$. Toward the past, as $t \rightarrow 0$, the t-mode is regular and remains bounded in all physically measurable quantities. The $t^{1/2}$-mode is, however, irregular in the past (P § 86): although the trace h and density δ go to zero as $t^{1/2}$ (eq. [33d]), the individual components h_{jk} become irregular as $t^{-1/2}$. This linear perturbation of Friedmann thus joins at early times onto some shearing, locally anisotropic but locally homogeneous, cosmological solution. It would be interesting to see the correspondence with the relevant velocity-dominated solutions of Eardley, Liang, and Sachs (1972); or, in cases where some gravitational wave mode is mixed in, with the Belinsky, Khalatnikov, Lifshitz (1970)–Misner (1969) solutions. We have not yet attempted this.

We thank Craig Hogan, Jim Peebles, Jim Bardeen, and Brandon Carter for helpful discussions and John Barrow and Michael Wilson for comments on the manuscript. We also acknowledge the pleasant hospitality of the Institute of Astronomy, Cambridge University, where this paper was written.

REFERENCES

Bardeen, J. M. 1969, unpublished paper.
———. 1980, *Ap. J.*, submitted.
Belinsky, V. A., Khalatnikov, I. M., and Lifshitz, E. M. 1970, *Usp. Fiz. Nauk*, **102**, 463–500 (English transl. in *Adv. Phys.*, **14**, 525).
Carr, B. J. 1977, *Astr. Ap.*, **56**, 377.
Carter, B. 1980, in *1979 Les Houches Summer School Proceedings*, in press.
Eardley, D., Liang, E. P. T., and Sachs, R. 1972, *J. Math. Phys.*, **13**, 99.
Field, G. B. 1975, in *Galaxies and the Universe* (Vol. **9** of Stars and Stellar Systems), ed. A. Sandage, M. Sandage, and J. Kristian (Chicago: University of Chicago Press).
Field, G. B., and Shepley, L. C. 1968, *Ap. Space Sci.*, **1**, 309.
Harrison, E. R. 1967, *Rev. Mod. Phys.*, **39**, 862.
Hawking, S. W. 1966, *Ap. J.*, **145**, 544.
Hogan, C. 1979, preprint.
Lifshitz, E. M. 1946, *Zh. Exp. Teoret. Fis.*, **16**, 587; also *J. Phys.* (*USSR*), **10**, 116.
Lifshitz, E. M., and Khalatnikov, I. M. 1963, *Advances Phys.*, **12**, 185.
Lightman, A. P., Press, W. H., Price, R. H., and Teukolsky, S. A. 1975, *Problem Book in Relativity and Gravitation* (Princeton: Princeton University Press).
Mészáros, P. 1975, *Astr. Ap.*, **38**, 5.
Misner, C. W. 1969, *Phys. Rev. Letters*, **22**, 1071.
Misner, C. W., Thorne, K. S., and Wheeler, J. A. 1973, *Gravitation* (San Francisco: Freeman).
Olson, D. W. 1976, *Phys. Rev. D*, **14**, 327.
Peebles, P. J. E. 1980, *Cosmology: The Physics of Large Scale Structure* (Princeton: Princeton University Press) (P).
Peebles, P. J. E., and Yu, J. T. 1970, *Ap. J.*, **162**, 815.
Press, W. H. 1980, *Phys. Scripta*, **21**, 702.
Sachs, R. K., and Wolfe, A. M. 1967, *Ap. J.*, **147**, 73.
Sakai, K. 1969, *Progr. Theoret. Phys.*, **41**, 1461.
Weinberg, S. 1972, *Gravitation and Cosmology* (New York: Wiley) (W).

WILLIAM H. PRESS and ETHAN T. VISHNIAC: Center for Astrophysics, 60 Garden St., Cambridge, MA 02138

THE THEORY OF THE LARGE SCALE STRUCTURE OF THE UNIVERSE

Ya. B. Zeldovich
Institute of Space Research
USSR Academy of Science
Moscow, USSR

INTRODUCTION

The God-father of psychoanalysis Professor Sigmund Freud taught us that the behaviour of adults depends on their early childhood experiences. In the same spirit, the problem of cosmological analysis is to derive the observed present day situation and structure of the Universe from certain plausible assumptions about its early behaviour. Perhaps the most important single statement about the large scale structure is that there is no structure at all on the largest scale - 1000 Mpc and more. On this scale the Universe is rather uniform, structureless and isotropically expanding - just according to the simplified pictures of Einstein-Friedmann........ Humason, Hubble.....Robertson, Walker. On the other hand there is a lot of structure on the scale of 100 or 50 Mpc and less. There are clusters and superclusters of galaxies.

Much work has been done on the classification of these bodies into "richness classes" and attempts have been made to deduce from observations a "mass function" giving the distribution of matter among clumps of various sizes and masses. There is a firmly established division between regions with enhanced, higher-than-average density of stars and radio-sources and regions with density lower than average. In recent years correlation functions have been used to characterize the relation between density enhancements and the linear scales of the distribution of galaxies in space.

A systematic effort of measuring the redshifts (optical and 21 cm radio) of thousands of galaxies has resulted in confirmation of Hubble's law. Surprisingly it is approximately valid for smaller distances than those characteristic of the density distribution. The redshift measurements have opened the way for disentangling the three-dimensional structure of the universe, as opposed to the two-dimensional projection of astronomical objects on the celestial sphere.

The present symposium has really opened up a new direction in the

M. S. Longair and J. Einasto (eds.), The Large Scale Structure of the Universe, 409–421.

search for the geometrical patterns governing the distribution of luminous matter in space. We heard about ribbons or filaments along which clusters of galaxies are aligned; the model of a honeycomb was presented with walls containing most of matter; the presence of large empty spaces (holes - not black holes of course) was emphasized. Cosmological theory must be aware of this information and try to use it in order to discriminate between various proposed schemes. Let us briefly characterize those schemes which seem to us most promising at the present time. There are two extreme assumptions which can be made about the initial density perturbations. The first concerns an ideal unperturbed metric connected by General Relativity with ideal homogeneity of the overall density in the early radiation dominated Universe. The perturbations consist of an inhomogeneous distribution of "matter" - of baryon excess - on a background of homogeneous radiation. Therefore the ratio γ/B (photons per baryon) varies from place to place. But the specific entropy of matter is proportional to this ratio and therefore those perturbations are called "entropy perturbations".

The second type of perturbation consists of common motion of photons and baryons. These perturbations conserve entropy and therefore they are called "adiabatic".

A departure from the main line of this report is permissible in the introduction. The actual value of the ratio γ/B which is of the order of 10^8 or 10^9, is most important for cosmology. The closed or open geometry of the Universe as a whole depends on this number.

Is it possible that in due time this number will be calculated by elementary particle theorists, taking into account the lack of exact symmetry between particles and antiparticles as indicated by laboratory experiments (so-called CP - violation, 1964) and also baryon non-conservation predicted by some theories? In this case it is conceivable, that the γ/B ratio is constant everywhere, just because the physical constants are everywhere the same. But this argument is not very strong. It is equally possible that γ/B depends on the interrelation of external physical constants and the local properties of the space-time metric; in this case the γ/B ratio must not be a constant.

Let us return to conventional cosmology. At the present moment we do not see any better policy than to make plausible assumptions about the size, amplitude and character of the initial perturbations, to develop logically and mathematically all the consequences of these assumptions and to compare them with observations.

In this report we shall investigate adiabatic perturbations - the second type, according to the classification given above. This investigation has been carried out during approximately the last ten years by our group, which includes Doroshkevich, Sunyaev, Novikov, Shandarin, Sigov, Kotok and others. We use important theoretical results obtained by Lifshitz, Bonnor, Silk, Peebles, Yu and others.

We consider several phases in the development of the perturbations:

1) acoustic oscillations of the radiation-dominated plasma and their attenuation before and during recombination;

2) the growth of small perturbations in the neutral gas;

3) the non-linear growth of perturbations leading to the formation of compressed gas layers - pancakes;

4) the further fate of pancakes, the interaction of pancakes, their decay into galaxies and protoclusters of galaxies.

The first two points are investigated using the "merry old" linear theory of perturbations. In 3) and 4) an approximate nonlinear theory is widely used and also numerical simulation. The statistical side of the problem is considered. Radio astronomical predictions are made. The main result is most encouraging: the adiabatic perturbation spectrum possesses a definite cut-off wavelength as already pointed out by Silk. We now see that this critical wavelength is also reflected in the cell structure of the Universe.

1. THE THEORY OF PERTURBATION GROWTH

A plausible featureless initial spectrum of density fluctuations in the radiation-dominated plasma is assumed. Due to photon viscosity and damping during recombination the final Fourier spectrum of growing perturbations in the neutral gas is given by

$$\overline{\left(\frac{\delta\rho}{\rho}\right)^2_K} = b_K^2 = a_K^2 K^n e^{-KR_c}$$

The critical length R_c depends on 1) the radiation density during recombination taking account of the specific effects of Ly-α reabsorption through the 2s $\rightarrow$ 1s+2γ metastable hydrogen decay, 2) the Compton cross-section for scattering of photons by electrons, 3) the matter density or γ/B ratio. The best calculations give the characteristic length (multiplied by $(1+z_{rec})$ in order to account for the expansion from recombination to the present epoch)

$$R_c = 8 \text{ Mpc for } \Omega = 1 \text{ and } R_c = 40 \text{ Mpc for } \Omega = 0.1.$$

The wavelength λ_c is determined by $\lambda_c = 2\pi/R_c$, $K_cR_c = 1$ so that $\lambda_c = 2\pi R_c$.

The index n and the average value of a_K^2 are adjusted to fit the observed picture. But independent of this adjustment, due to the exponential damping factor e^{-KR_c} we are sure that the surviving fluctuations are very smooth. It is immediately clear that in the adiabatic theory early formation of stars or globular clusters or even of galaxies is impossible. First large-scale density enhancements must grow, and only thereafter is their fragmentation in smaller units possible.

The second qualitative feature is gas motion under the influence of gravitation only. The pressure forces, which depend on gradients, are negligible on the large scales involved. In this case the growth of perturbations is especially simple: they grow in amplitude due to gravitational instability and increase in linear dimensions, conserving their form. The density perturbations and the peculiar velocity (the excess over the Hubble velocity) are given by

$$\frac{\delta\rho}{\rho} = f\left(\frac{r}{R}\right)\phi(t) \quad ; \quad \underline{u} = \underline{U}\left(\frac{r}{R}\right)\psi(t)$$

$$\phi(t) = t^{2/3} \propto (1+z)^{-1}, \quad \psi(t) = t^{1/3} \propto (1+z)^{-1/2} \quad \text{for } \Omega = 1$$

$$R = \text{constant} \times t^{2/3} \propto (1+z)^{-1}$$

Obviously the density field and velocity field are connected by the continuity equation

$$\frac{\partial}{\partial t}\left(\frac{\delta\rho}{\rho}\right) \propto f \propto \operatorname{div} \underline{u} \propto \operatorname{div} \underline{U}$$

and by the equation of motion in which the perturbation of the potential by the perturbed density is included.

It is important to realize that already in the linear theory the extra compression in places with positive and growing $\delta\rho/\rho$ is anisotropic: the three components of the divergence of the peculiar velocity are not equal

$$\frac{\partial u_x}{\partial x} \neq \frac{\partial u_y}{\partial y} \neq \frac{\partial u_z}{\partial z}$$

There is also shear, $\partial u_x/\partial x \neq 0$ etc. - but of course no vorticity $\partial u_x/\partial y - \partial u_y/\partial x = 0$ because the motion is due to potential (gravitational) forces. The anisotropy of the deformation due to peculiar velocities is easily understood by tidal action. The nearby density distribution distorts the motion at the point under consideration.

A natural way to build an approximate theory, exact in the linear region and also good enough in non-linear situations, is to use the Lagrangian formulation. The position of every particle in space (i.e. its Eulerian coordinates) $\vec{r}$ is given as a function of time t and the initial position (i.e. Lagrangian coordinate) of the particle $\vec{\xi}$.

The solution with growing perturbations only is written

$$\vec{r} = a(t)\left[\vec{\xi} + b(t)\vec{\psi}(\vec{\xi})\right]$$

The first term $a\vec{\xi}$ describes the Hubble expansion $\dot{a}/a = H$, the second term $ab\vec{\psi}(\vec{\xi})$ describes the displacement of every particle from its legitimate unperturbed position. b(t) is a growing function, $b(t) \propto t^{2/3}$. The perturbation due to gravitation $\vec{\psi}(\vec{\xi})$ is of potential type $\vec{\psi} = \text{grad}_\xi \phi$.

Analytical and numerical studies confirm that this is a good approximation - less than 20-30% errors occur in highly nonlinear situations; the proofs are in our original papers.

Given the formula for $\vec{r}(\vec{\xi},t)$, it is easy to write down the velocity of every particle

$$\vec{u} = \left.\frac{\partial \vec{r}}{\partial t}\right|_{\vec{\xi}} , \quad \vec{u}_{pec} = \vec{u} - H\vec{r} = a\dot{b}\,\vec{\psi}(\vec{\xi})$$

and also the density of matter

$$\rho = \rho(\vec{\xi},t) \propto \left(\frac{\partial^3 \vec{r}}{\partial \vec{\xi}^3}\right)^{-1}$$

Here $\partial^3\vec{r}/\partial^3\vec{\xi}$ is the Jacobian i.e. the determinant of the partial derivatives.

Using $\vec{\psi} = \mathrm{grad}_\xi \phi$ and choosing coordinate axes which diagonalize the deformations, using the notation

$$\frac{\partial^2\phi}{\partial \xi_1^2} = -\alpha \; ; \quad \frac{\partial^2\phi}{\partial \xi_2^2} = -\beta \; ; \quad \frac{\partial^2\phi}{\partial \xi_3^2} = -\gamma \; ; \quad \alpha > \beta > \gamma$$

we obtain

$$\rho = \bar{\rho}\,\frac{1}{(1-b\alpha)(1-b\beta)(1-b\gamma)}$$

With ϕ determined by the initial small perturbation field we find the particles where α has local maxima α_{m1}, α_{m2} .... The condition $1 - b(t_i)\alpha_{mi} = 0$ determines the moment when infinite density is obtained for the i-th particle.

From the density formula we see that this infinity is due to the intersection of trajectories of adjacent particles lying on the ξ_1 coordinate axis. At the moment t_i the perturbation along the other two axes ξ_2, ξ_3 is finite.

The approximate theory predicts the formation of thin dense gas clouds. They grow due to fresh gas falling onto their flat boundary and being compressed and heated by shock waves. They also spread sideways due to new intersections of trajectories of adjacent particles.

The picture outlined above was already known at the time of the Krakow IAU Symposium No.63, "Confrontation of cosmological theories with observational data". Qualitatively they are described in the report by Doroshkevich et al; formulae and detailed analyses were given in our original papers, and also in the book by Zeldovich and Novikov "Structure and Evolution of the Universe", published in Russian in 1975 and prepared for publication in English by Chicago University Press. These are mentioned in this report for the sake of completeness and to make it possible to read this report without using references.

Now we turn to the results obtained after the Krakow Symposium, partly published in Astronomical Journal (USSR) and partly in preprints of the Institute of Applied Mathematics. These results are most important in connection with optical and radio astronomical observations.

2. LATE PHASES OF THE DEVELOPMENT OF PERTURBATIONS AND CELL STRUCTURE

Numerical calculations were pushed to a late phase, when more than half the matter is brought together into the dense phase. During the lecture in Tallinn a movie was shown made by an electronic computer display. Here, in the written form, only a small number of selected pictures can be shown.

There is one movie (corresponding to Fig.1) calculated using the approximate nonlinear theory for two-dimensional perturbations. The initial spectrum has a sharp cut-off on the short wave - and also long wave end; it is flat (on the average) within the excited interval. The individual Fourier coefficients in this interval are taken at random according to a normal Gaussian distribution.

Comparing these calculations with others it must be stressed that the potential, the velocity and density contributions are calculated for a continuous medium, not for a finite number of discrete point masses. The calculations are not exact and the initial conditions somewhat artificial (two dimensions, flat spectrum). But these departures from the ideal calculation are not of the sort which arise when a finite number of discrete masses is considered.*

The pictures in the movie contain a finite number of points. But those points are test particles for visualising the motion and density distribution. The potential used in the calculation corresponds to a continuum or,in other words, to a calculation with an infinite number of particles with inertial and gravitational mass.

The calculation is continued to the moment after the first intersection of trajectories occurs. It is assumed that the particles are non-interacting and one layer can penetrate through another. The sticking together of particles, their physical, non-gravitational collisions and the formation of shockwaves are not included. Therefore the pancakes in this picture are somewhat thicker than would be found in a real gas-dynamical calculation. Still they are rather thin, distinctly different from the spherical or irregular clumps predicted in a simplified approach.

As time goes on,the pancakes spread laterally and they intersect. In Figure 1 a typical net structure is seen: matter is mostly concentrated in thin filaments, the inner regions are empty and divide up the network.

It seems plausible that a three-dimensional calculation will lead to a cell or honeycomb structure with matter concentrated in the walls

---

*It is $\sqrt{N}$ in two dimensions and $\sqrt[3]{N}$ in three dimensions which are important in incorporating shortwave perturbations involuntarily in N-body calculations just due to the discrete character of the mass distribution.

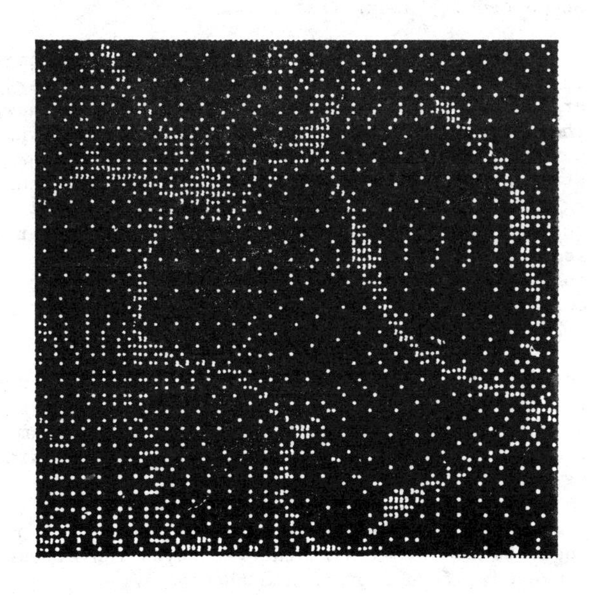

Approximate theory

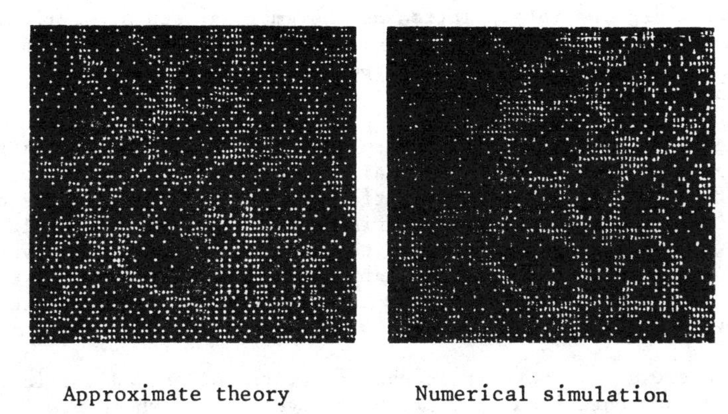

Approximate theory Numerical simulation

Figure 1

surrounding large disconnected empty spaces. The intersection of walls could give enhanced density along lines.

It is not yet clear if such types of structure, or at least its remnants are discernible in the observational data of Joeveer, Einasto, Tago and Seldner, Siebers, Groth and Peebles. It is well known that the human eye has the property of finding lines and other patterns in random assemblies of points. One example is the Sciapparelli channels on Mars, but even more striking are the constellations - figures of humans and beasts found by the ancients in the distribution of stars on the sky. Therefore one must be very cautious in interpreting the observations. One must find some mathematical algorithm to distinguish between superclusters as quasispherical clumps and the honeycomb structure. It is possible that correlation functions (two points, three points etc.) are not the best method for this particular task.

There are obvious difficulties: 1) the walls must fragment into separate galaxies and clusters of galaxies. The turbulence inside the pancakes and the gravitational interaction of the fragments must partly wash out the structure. 2) In investigations of the three-dimensional structure we use redshift as a measure of distance. Because of peculiar velocities, this procedure is not exact. These points need further investigation. But with all these uncertainties, one point must be stressed: the occurrence of cells in theoretical calculations is not an artefact due to the use of an approximate theory.

Calculations of another type were carried out and used to make the second part of the movie. The motion of $(128)^2 \approx 16000$ points in two dimensions was calculated numerically. The potential for every distribution of points was calculated using Poisson's equation $\Delta\phi = 4\pi G\rho$ with some smoothing and interpolation on the smallest scale. Periodicity on the largest scale was assumed: points intersecting from inside the wall of the square reappeared on the opposite wall. The periodicity condition was also used in the potential calculation.*

Again a flat spectrum of perturbations with cut-offs from both sides was used in formulating the initial conditions. The results of numerical simulations are practically indistinguishable from the results of the approximate theory. The characteristic pattern with thin walls and disconnected empty spaces depends on the cut-off of the short waves - this is our firm conclusion. It is confirmed by the fact that the average linear dimension of the empty spaces are approximately equal to the cut-off wavelength, $2\pi/K_{max}$.

Therefore this pattern is characteristic of the adiabatic theory with the exponential cut-off at short wavelengths, which results from the matter-radiation interaction.

---

*The force $\propto r^{-1}$ and potential $\propto \log r$ is characteristic of two dimensional gravitation; to be exact we are working with infinite bars, not points.

For entropy perturbations, there is no cut-off except the Jeans' mass for neutral hydrogen corresponding to masses of the order of globular clusters. Therefore probably no net or cell-structure will occur in this case. Corresponding numerical computations with sufficient accuracy are still lacking. Perhaps the two dimensional case will be easier to handle and be still meaningful.

We are optimistic about the prospects for discriminating between entropy and adiabatic perturbations by means of investigations of the large-scale structure of the Universe.

A three dimensional calculation was done for adiabatic perturbations with an exponentially cut-off spectrum, but it is the visualisation of results which is the bottleneck in this case. This report is written at the moment when this work is still in progress.

3. STATISTICAL PROPERTIES OF THE BIRTH OF PANCAKES AND COUNTS OF QSO

The birth of an individual pancake occurs at the moment of intersection of trajectories, i.e. mathematically speaking at the moment when the smallest denominator in the density expression $(1-b(t).\alpha)$ vanishes so that $\rho \to \infty$. Therefore we must find the local maxima of $\alpha = -\partial^2\phi/\partial\xi_1^2$. Due to the statistical character of the problem the answer is also given in statistical terms. The function $P(\alpha_m)$ gives the density of local maxima of given amplitude

$$dN = P(\alpha_n)\, d\alpha_n \qquad (cm^{-3})$$

In the case of a cut-off spectrum P is proportional to R_c^{-3}. The dependence on α_m is universal, given the normal Gaussian law of density perturbations. But $P(\alpha_m)$ is not a simple Gaussian function, because in calculations of α_m we are performing the nonlinear operation of diagonalisation of the $\partial^2\phi/\partial\xi_i\partial\xi_k$ matrix and we are choosing the maxima. Doroshkevich has obtained

$$P(\alpha_n) \propto \alpha^5 e^{-n\alpha^2} \qquad (\alpha > 1/\sqrt{n})$$

Using the connection between the amplitude of the maximum α_m and the moment t_m we can obtain the birth function F(t) or f(z) giving the number of pancakes born per unit comoving volume per unit time or unit of z.

The newborn pancakes have small vorticity and low temperature. The formation of compact objects and brightest galaxies is easiest just at the birth-place of the pancake and even before the growth of the ends of the pancake.

Therefore it is plausible to identify the birthrate of pancakes with the birthrate of brightest known compact objects - quasars. If the life time of quasars is short and independent of their absolute age, then at every epoch the concentration of quasars is proportional to their birthrate. The high power α^5 before the Gaussian factor leads the $(1+z)^7$ dependence of the density of quasars as an intermediate asymptote in the case of a flat Universe. At high z the power law is cut off by the Guassian exponent. By the choice of a single constant (corresponding to the amplitude of perturbations) it is possible to obtain a good fit of the pancake birth rate curve to Longair's results on radio source counts and Schmidt's data on quasar evolution.

Still, the similarity between the radio source and quasar evolution and the birthrate of pancakes should not be overestimated. The power laws involved refer to different regions of z. The birth of cold pancakes occurs from some high z (of the order of 10 or 20) up to $z \sim 4 \div 3$. It is well known that for $z < 4$ the gas is totally ionized; therefore even if pancakes are formed, their physical properties are totally different as compared with genuine pancakes formed from cold initial gas. On the other hand, the observed counts of radio sources and quasars refer to the range $0 < z < 4$; at $z > 4$ instead of evolution there is a cut-off or stagnation. This question needs further investigation.

Another statistical test concerns the two-point correlation curve. The adiabatic pancake theory does not contradict the most interesting part of Peebles' correlation function $\delta \sim \xi^{-1.7}$ in the region near $\delta \sim 1$. We refer to original papers for quantitative confirmation.

The general outlook seems to be that the adiabatic theory does not contradict the observations.

4. THE CRUCIAL TESTS AND FURTHER PROBLEMS

Still the absence of contradiction is not positive proof. In order to distinguish between the entropy and adiabatic theories one needs direct observation. The observation of very early globular clusters and galaxies at $z > 30$ to $z \sim 100$ or 200 would be strong evidence in favour of the entropy perturbation theory with further clumping of the initial small mass objects into clusters of galaxies. If hot gas clouds of primordial composition (H + He) are found, identifiable with pancakes, this would be a strong argument for the adiabatic theory. Fully ionized very hot gas could be detected by its X-ray emission and by distortions of the Planckian background radiation spectrum (cooling in the Rayleigh-Jeans region). The medium-hot hydrogen gives redshifted 21 cm radiation.

In any case, the controversy with the observed limits on $\Delta T/T$ of the relic radiation fluctuations must be solved - but this is needed for <u>all</u> variants. Entropy perturbations predict $\Delta T/T$ only 2 or 3 times less than adiabatic perturbations. The study of those perturbations which are directly connected with the structure of the Universe is the most rewarding part of the problem.

Extrapolating from Krakow through Tallinn to the next symposium somewhere in the early eighties one can be pretty sure that the question of the formation of galaxies and clusters will be solved in the next few years.

What remains is the wider question of the overall spectrum of perturbations including the smallest scale damped in the very early radiation dominated or hadronic era and of the longest perturbations, whose amplitude remains small even now. Is the power law spectrum without any characteristic length valid? New, indirect observational tests are needed. Still the major theoretical questions remain unsolved: what is the fundamental theory of the initial perturbations? And what is the ultimate reason for the homogeneous and isotropic expansion from the singularity which is the background for the perturbations?

REFERENCES

Doroshkevich, A.G., Sunyaev, R.A. and Zeldovich, Ya.B., 1974. "Confrontation of Cosmological Theories with Observational Data", M.S. Longair (ed.), Dordercht,Holland/Boston, USA.
Zeldovich, Ya.B. and Novikov, I.D., 1975. "Structure and Evolution of the Universe", Nauka, Moscow.
Zeldovich, Ya.B., 1970. Astron. & Astrophys., 5, 84.
Doroshkevich, A.G., 1970. Astrophysica, 6, 581
Sunyaev, R.A. and Zeldovich, Ya.B., 1972. Astron. & Astrophys., 20, 189.
Doroshkevich, A.G., Ryabenkyi, V.S., Shandarin, S.F., Astrophysica, 9, 257.
Doroshkevich, A.G., Shandarin, S.F., 1973. Astrophysica, 9, 549.
Doroshkevich, A.G., 1973. Astrophys. Lett., 14, 11.
Doroshkevich, A.G., Shandarin, S.F., 1974. Sov. Astron., 18, 24.
Shandarin, S.F., 1974. Astron. Zh. USSR, 51, 667.
Doroshkevich, A.G., Shandarin, S.F., 1975. Sov. Astron., 19, 4.
Sunyaev, R.A., Zeldovich, Ya.B., 1975. Mon. Not. Roy. Astron. Soc., 171, 375.
Doroshkevich, A.G., Shandarin, S.F., 1976. Mon. Not. Roy. Astron. Soc., 175, 15p.
Doroshkevich, A.G., Zeldovich, Ya.B., Sunyaev, R.A., 1976. In "Formation and Evolution of Galaxies and Stars", S.B. Pikelner (ed.), Nauka, Moscow.
Doroshkevich, A.G., Shandarin, S.F., 1976. Preprint IAM No.3.
Doroshkevich, A.G., Shandarin, S.F., 1977. Mon. Not. Roy. Astron. Soc., 179, 95p.
Doroshkevich, A.G., Shandarin, S.F., 1978. Mon. Not. Roy. Soc. Astron., 182 (in press).
Doroshkevich, A.G., Saar, E.M., Shandarin, S.F., 1977. Preprint IAM No.72.
Doroshkevich, A.G., Shandarin, S.F., 1977. Preprint IAM No.73.
Doroshkevich, A.G., Shandarin, S.F., 1977. Astron. Zh. USSR, 54, 734.
Doroshkevich, A.G., Shandarin, S.F., 1977. Preprint IAM No.84.

Doroshkevich, A.G., Saar, E.M., Shandarin, S.F., 1978. This volume.
Seldner, M., Siebers, B., Groth, E.J., Peebles, P.J.E., 1977. Astron. J., 82, 249.
Joeveer, M., Einasto, J., Tago, E., 1977. Preprint A-1, Struve Astrophysical Observatory Tartu.

DISCUSSION

Suchkov: There are quite distinct knots in your array of pancakes. Now, if the pancakes are destined to be superclusters or clusters of galaxies, what kind of future do you foresee for these knots?

Zeldovich: The numerical calculations need to be pushed further in order to obtain unambiguous answers. Possibly the filaments along which clusters of galaxies are aligned (if this effect is statistically verified) will be identified with intersection lines but it is not yet clear theoretically.

Chernin: What kind of relaxation could lead to the evolution of a flat pancake into a cluster like Coma with more or less spherical form?

Zeldovich: Pancake formation is due to compression on one axis, but this does not exclude less dramatic compression (without intersection) in one or two other directions. Therefore at least a part of pancakes can transform into rather dense clumps. Turbulence inside the pancake and also its curvature tend to make the clump thick. The last effect tending to make the cluster spherical is gravitational interaction.

On the other hand, there must also be pancakes which are expanding in the two directions tangential to the pancake surface and in this case one should observe Hubble's law in a region with strongly enhanced density. Of course, the Hubble constant for this region is different from the genuine long-range H; the local H is subject to quadrupole perturbations. One should ask Prof. de Vaucouleurs and Profs Sandage and Tammann if perhaps we are living in such a region.

Binney: One cannot but be impressed that Dr Zeldovich's beautiful film gives a better representation of the sky as published recently by Dr Peebles and collaborators than does that shown earlier by Dr Aarseth (Peebles et al. 1977). Further strong evidence in support of the picture, based on a spectrum biassed towards large masses, are the facts that both most rich clusters of galaxies and elliptical galaxies are as often as not nearly as aspherical as a slowly rotating body can be (Klingworth 1977, Rood and Chincarini 1974, Macgillivray 1976, Schipper and King 1977).

I should like to ask Dr Zeldovich, however, whether he believes large-scale shock formation is a necessary part of this picture. I ask this because I have difficulty in believing that the cold cosmic gas will fail to fragment soon after it starts to contract in one dimension. This will destroy the pressure-balance required across the centre of the pancake. My belief is that one may retain the cellular structure and the aspherical cluster formation even without large-scale shock formation. Certainly one cannot overemphasize the importance of anisotropic collapse on a large scale.

Zeldovich: Dr Binney is making a statement rather than a question. I should point out that the film was made by Doroshkevich, Shandarin, Sigov and Kotok; I would also add Einasto and Joeveer to the list of people observing large scale structure.

As to the origin of the structure: it is the cut-off of short wave perturbations which is most important. The cell structure remains (perhaps somewhat weaker, with thicker walls) in the collisionless case with trajectories continuing without break after intersection, i.e. in the absence of the shock. Concerning fragmentation, when the perturbations are small (linear regime) the exponent of the gravitational instability has no maximum; it is an increasing function of wavelength. The cut-off short wave perturbations do not outgrow those of long wavelength. The compression time before pancake formation is so short that it does not compensate the handicap due to short wave damping. We feel that the overwhelming part of fragmentation occurs after shock wave compression - if there are no primordial short wave entropy perturbations of course.

Reprinted from THE ASTROPHYSICAL JOURNAL
Vol. 142, No. 4, November 15, 1965

THE BLACK-BODY RADIATION CONTENT OF THE UNIVERSE AND THE FORMATION OF GALAXIES*

P. J. E. PEEBLES
Palmer Physical Laboratory, Princeton University, Princeton, N.J.
Received March 8, 1965; revised June 1, 1965

ABSTRACT

A critical factor in the formation of galaxies may be the presence of a black-body radiation content of the Universe. An important property of this radiation is that it would serve to prevent the formation of gravitationally bound systems, whether galaxies or stars, until the Universe has expanded to a critical epoch. There is good reason to expect the presence of black-body radiation in an evolutionary cosmology, and it may be possible to observe such radiation directly.

Assuming that the Universe is expanding and evolving, very likely most scientists would agree on the over-all picture for the evolution of the Universe. At a remote time in the past the Universe contained only dense gaseous material, with neither stars nor galaxies. As the Universe expanded from this state the material became organized into galaxies and clusters of galaxies, and the material within galaxies passed through the generations of stars. Now a central question is what were the physical processes, and what were the physical parameters and conditions that determined how galaxies formed, with the observed distributions of mass and size, and the observed tendency for galaxies to be distributed in clusters.

An approach to the problem may be based on the following important property of an evolutionary cosmology (Dicke, Peebles, Roll, and Wilkinson 1965).

1. Assuming that the Universe expanded from a sufficiently highly contracted phase, the early Universe would have been opaque to radiation. As a result the radiation field would have achieved thermal equilibrium with the matter—the Universe would have been filled with black-body radiation. This fireball radiation suffers the cosmological redshift, so that it is very much cooled by the expansion of the Universe, but it retains its thermal, black-body character.

It may be possible to observe this fireball radiation directly by means of a microwave radiometer. Recently, Penzias and Wilson (1965) have reported that at 7-cm wavelength there appears to be isotropic background radiation with intensity equivalent to $3.5° \pm 1°$ K. Further measurements at other wavelengths are necessary to establish that this radiation has a black-body spectrum, as expected for the cooled fireball from the big bang. The purpose of this article is to show that, if the Universe does contain black-body radiation of this general amount, this radiation must have had an important effect on the early evolution of matter leading to galaxy formation.

* This research was supported in part by the National Science Foundation and by the Office of Naval Research of the United States Navy.

It is important to distinguish this possible thermal radiation from the integrated background radiation due to the galaxies. Eddington (1926) estimated that the starlight in our own Galaxy amounts to 3° K. However, this is an effective temperature, such that σT^4 is the total starlight radiation energy flux. The radiation intensity spectrum is quite different from the black-body radiation considered here.

The role of the thermal, fireball radiation in the formation of galaxies is summarized in the following remarks.

2. The expansion of the Universe adiabatically decompresses and cools its contents. So long as the temperature exceeds about 4000° K, the material in the Universe is ionized and is sufficiently opaque to radiation that material and radiation would remain in thermal equilibrium. The temperature T varies with the mean density of matter ρ according to the formula

$$T \propto \rho^{1/3} . \tag{1}$$

3. In the expanding Universe any sufficiently large-scale perturbation to a homogeneous mass distribution grows more pronounced with time, eventually tending to form a gravitationally bound system. This is the familiar Jeans gravitational instability (Gamow 1948).

TABLE 1

FORMATION OF PROTOGALAXIES AND CLUSTERS OF GALAXIES

Present mean density of matter (gm/cm³)	2×10^{-29}	2×10^{-29}	7×10^{-31}	7×10^{-31}
Present black-body radiation temperature (° K)	3	0.3	3	0.3
Present age of the Universe (yr)	7×10^{9}	7×10^{9}	1×10^{10}	1×10^{10}
Formation of gravitationally bound gas clouds:				
Time of formation (yr)	1×10^{5}	3×10^{3}	7×10^{5}	2×10^{4}
Temperature (° K)	4000	5000	4000	4500
Radius (pc)	10	0.2	40	1
Density within cloud (protons/cm³)	3×10^{4}	5×10^{7}	1×10^{3}	1×10^{6}
Mass of cloud ($M_\odot$)	2×10^{6}	6×10^{4}	1×10^{7}	3×10^{5}
Formation of galaxies:				
Minimum mass of galaxy ($M_\odot$)	2×10^{6}	6×10^{4}	1×10^{7}	3×10^{5}
Protogalaxy mass ($M_\odot$):				
$n=0$	10^{9}	10^{9}	10^{9}	10^{9}
$n=1$	10^{10}	10^{10}	10^{10}	10^{10}
Maximum mass of a cluster of galaxies ($M_\odot$):				
$n=0$	10^{13}	10^{13}	10^{12}	10^{10}
$n=1$	10^{16}	10^{17}	10^{14}	10^{13}

4. With the Jeans instability alone we encounter a dilemma of initial conditions. The time at which a bound system forms depends critically on the details of the density perturbation evaluated at some chosen, initial time, and to form the observed galaxies it would be necessary to postulate extremely special initial conditions.

5. This unsatisfactory situation is avoided if it is assumed that the Universe contains black-body radiation. It will be shown that the radiation would prevent density perturbations from growing larger than the mean density itself until the Universe has expanded to a critical epoch.

6. After this epoch the Jeans instability leads to the formation of gravitationally bound gas clouds of well-defined size and mass. The properties of the gas clouds are listed in Table 1, where each column corresponds to definite assumptions about the present mean density and radiation temperature of the Universe. Subsequent evolution of the matter is not influenced by the black-body radiation.

7. The motion of the gas clouds is subject to a gravitational instability. This results in the formation of bound systems of gas clouds, which should collapse to form more massive protogalaxies.

This discussion is based on conventional general relativity and the homogeneous isotropic cosmological models. All quantities below will be expressed only in proper units, as measured with ordinary measuring rods and clocks and balances.

To obtain the critical condition mentioned in the fifth remark for forming a gravitationally bound system, suppose a spherical mass of gas has just achieved equilibrium, tending neither to expand nor collapse. As the Universe expands, the mean black-body electromagnetic-radiation energy density is decreasing, so that radiation is tending to flow out of the system. This is opposed by Thompson scattering of the radiation by the free electrons, if the temperature is above 4000° K. Thus, the radiation temperature within the bound system will exceed the mean radiation temperature by the amount

$$\frac{\Delta T}{T} \sim \frac{\sigma R^2 \rho H}{c m_p} . \tag{2}$$

In this equation it will be recalled that all quantities are measured in proper units, where R is the radius of the bound system, σ is the Thompson scattering cross-section for an electron, ρ is the density of matter within the system, H is Hubble's constant, and m_p is the mass of a proton.

Suppose first that the system is so large that $\Delta T/T \gg 1$. In this case radiation would be trapped inside the bound system, and for equilibrium the gravitational force would have to balance the radiation pressure of this trapped radiation. Ignoring the gas pressure, and assuming for the moment that the system is not too large, the condition for equilibrium is

$$\tfrac{4}{3}\pi G \left(\rho + \frac{2bT^4}{c^2}\right) R\rho \sim \frac{bT^4}{3R} , \tag{3}$$

where ρ and T are the density of matter and the radiation temperature within the system, and b is the radiation energy density constant ($b = 7.6 \times 10^{-15}$ erg/cm^3 °K^4). In situations of interest the mass density of matter will be substantially below that in radiation (Table 2) so that expression (3) reduces to

$$R \sim c/(G\rho)^{1/2} . \tag{4}$$

But Hubble's constant is

$$H = (8\pi G b T^4/3c^2)^{1/2} \tag{5}$$

and with the mass density ρ much less than the radiation density bT^4/c^2, we see that expression (4) implies

$$R \gg c/H . \tag{6}$$

This means that the system is larger than the visible Universe. Evidently the simple Newtonian approximation (3) would not apply if expression (6) were valid, but more important, we see that the system must be unstable against gravitational collapse, for the mass of the system satisfies $GM/Rc^2 > 1$. However, we do not believe that any appreciable part of the observable Universe has already collapsed. Thus, the system could not be large enough to contain the radiation pressure. Therefore, for equilibrium the system must be small enough to allow the radiation to escape ($\Delta T/T \ll 1$ in eq. [2])

$$R^2 \lesssim c m_p/\sigma \rho H . \tag{7}$$

If expression (7) is satisfied, the radiation will be very nearly uniformly distributed in space. Then assuming that the center of the gravitationally bound system is at rest in the comoving coordinate frame, and assuming that the (proper) size of the system is constant, the edge of the system will be moving with velocity HR relative to the comoving coordinate frame at that point. Since the radiation is moving with the comoving coordinate frame there will be a radiation drag force per electron at the edge of the system amounting to

$$F_r = \frac{\sigma b T^4 H R}{c}. \tag{8}$$

This is the radiation force for temperatures in the range $T \ll 10^{10}$ ° K (that is, non-relativistic electrons) to $T \gtrsim 4000°$ K (free electrons).

If the system satisfies Jeans's criterion for gravitational instability, pressure forces may be neglected, and equation (8) must be balanced by the gravitational force per proton,

$$F_g = \tfrac{4}{3}\pi G m_p R\left(\rho_b + \frac{2bT^4}{c^2}\right). \tag{9}$$

TABLE 2

CONDITION FOR A GRAVITATIONALLY BOUND SYSTEM*

Age of the Universe (yr)	7×10^5	5×10^3
Temperature (° K)	4×10^3	4×10^4
Mass density due to radiation (gm/cm³)	2×10^{-21}	2×10^{-17}
Mean mass density of matter:		
Gm/cm³	2×10^{-21}	2×10^{-18}
Protons/cm³	10^3	10^6
Matter density within a bound system (protons/cm³)	10^6	10^{12}
Maximum mass ($M_\odot$)	10^{14}	10^9

* Assumed present conditions $T = 3$ °K, $\rho = 7 \times 10^{-31}$ gm cm³.

Equating (8) and (9), the density ρ_b within the system divided by the mean density of matter in the Universe, ρ, is found to be

$$\rho_b/\rho = \frac{2bT^4}{\rho c^2}\left(\frac{3\sigma c H}{8\pi G m_p} - 1\right) = \frac{2bT^4}{\rho c^2}(0.067H/H_f - 1), \tag{10}$$

where in the final term we have assumed the present value of Hubble's constant is $H_f = 3.2 \times 10^{-18}$ sec^{-1} (100 km/sec Mpc).

The ratio (10) of the density ρ_b required for equilibrium to the mean matter density ρ is given in Table 2 for reasonable values of the mass density and radiation temperature in the present Universe. Also shown is the maximum mass of a bound system, as given by equation (7). It is evident that until the plasma recombines ($T \sim 4000°$ K) a bound system could have formed in this cosmology only if there were extremely large density fluctuations.

One could imagine a situation in which a system with density greater than the mean density is expanding just slightly less rapidly than the general expansion, so that the radiation drag and gravitational forces just balance. However, it is important to notice that this is an unstable balance. The radiation drag force is spherically symmetric, while within an elliptical mass distribution the gravitational force is larger, for the most part, along the minor axes. Thus the system tends to fragment, eventually into pieces so small that pressure forces can disperse them.

The significance of these remarks is that while $\rho_b/\rho \gg 1$ (eq. [10]) the Universe is stable against the development of large matter-density perturbations. On the other hand, given a smooth density distribution, small perturbations to the matter distribution tend to grow with time. It is important to recognize the distinction between these two cases. We shall show that small density perturbations ($\delta\rho/\rho \ll 1$) are not appreciably affected by the radiation drag. When the density excursions have grown comparable to the mean density radiation drag becomes an important factor, serving to prevent density perturbations from growing larger until ρ_b/ρ approaches unity.

To discuss the behavior of small density perturbations we write the matter density $\bar{\rho}(x,t)$ as

$$\bar{\rho}(x,t) = \rho(t)[1 + D(x,t)] , \tag{11}$$

where ρ is the mean matter density. This first-order perturbation problem was first discussed by Lifshitz (1946). Introducing the simplifying assumption that the linear dimensions of any perturbation are small compared with the radius of the visible Universe, it is shown in the Appendix that if the perturbation is resolved into Fourier components

$$D = d(t)e^{i\boldsymbol{k}\cdot\boldsymbol{x}} \tag{12}$$

such that the wavelength $\lambda = 2\pi/|\boldsymbol{k}|$ satisfies

$$\frac{1}{\lambda}\frac{d\lambda}{dt} = -H \tag{13}$$

(so that the wave is taking part in the general expansion) then the Fourier amplitude $d(t)$ satisfies

$$\frac{d^2\boldsymbol{d}}{dt^2} + \left(\frac{\sigma b T^4}{m_p c} + 2H\right)\frac{d\boldsymbol{d}}{dt} = \left(4\pi G\rho - \frac{8\pi^2 kT}{m_p \lambda^2}\right)\boldsymbol{d} . \tag{14}$$

This is valid if λ is small compared with the radius of the visible Universe.

From equation (14), if

$$\lambda > \lambda_c \equiv \left(\frac{2\pi kT}{G\rho m_p}\right)^{1/2} , \tag{15}$$

$\boldsymbol{d}(t)$ grows with time. This is Jeans's criterion. For smaller wavelengths $\boldsymbol{d}(t)$ oscillates, and the disturbance is dissipated by the damping terms (in $d\boldsymbol{d}/dt$).

To estimate the rate of growth of the perturbation $\boldsymbol{d}$ when expression (15) is valid, consider a cosmologically flat Universe. Under the assumption that the particle mass density is greater than the mass density in radiation, the age of the model since start of expansion from infinite density is $t = (6\pi G\rho)^{-1/2}$, and $H = 2/(3t)$. Then, neglecting the pressure and radiation drag terms in (14) (i.e., setting $T = 0$), we find $\boldsymbol{d} \propto t^{2/3}$. This is the result obtained by Lifshitz (1946). Similarly, if the mass density in radiation dominates, $\boldsymbol{d} \propto t^{0.61}$.

These results apply once the electrons have become non-relativistic. We see that the density perturbations grow slowly, as a power of time. We believe it is reasonable to assume that there has been adequate time for the growth of perturbations on the scale of galaxies and clusters of galaxies. This would not have been the case if it were assumed that at the epoch when the electrons first became non-relativistic the Universe was strictly homogeneous, with the density perturbations only random (thermal) fluctuations. However, the assumption of an exactly symmetrical Universe at this epoch appears quite overidealized. In the following discussion we shall assume that the early Universe was sufficiently irregular that appreciable density perturbations did form due to the gravitational instability, but not so irregular that parts of the Universe have already suffered gravitational collapse. The general question of the homogeneity of the early Universe will be discussed in detail elsewhere.

When the density excursions have grown to a value comparable with the mean density, perturbations can develop very rapidly. To understand this, consider a region, roughly uniform and spherical, in which the density is twice the mean value. At time t_0 let the material in the region be expanding at the general rate, so the speed v of material in the patch a distance r from the center is given by

$$(v/r)^2(t = t_0) = \tfrac{8}{3}\pi G[\rho(t_0) + \rho_r(t_0)] , \quad (16)$$

where we have written separately the densities of matter, ρ, and radiation, ρ_r. If we neglect for a moment the radiation drag, the subsequent motion is given approximately by

$$(v/r)^2 = \frac{8\pi G}{3}[2\rho(t) + \rho_r(t) - \rho(t_0)r_0^2/r^2]. \quad (17)$$

Thus the patch stops expanding when $2\rho(t) + \rho_r(t) = \rho(t_0)r_0^2/r^2$. If the mass density in radiation does not exceed that in matter (Table 2) this is an expansion of a factor of 2 in radius. The patch would then collapse by a factor $\sim$2 in radius to a stable, bound system. However, it is important that this situation is very much altered if $\rho_b/\rho \gg 1$ (eq. [10]). In this case, we have shown that the patch must fragment and be dispersed.

Now we can draw the following general picture. In the initial very contracted Universe, there was a more or less uniform distribution of ionized hydrogen. That part of any density perturbation with wavelength satisfying expression (15) would grow with time, while the rest of the perturbation decayed. The resulting pattern of density fluctuations takes part in the general expansion of the Universe (eq. [13]). Notice from expressions (1) and (15) that the critical Jeans wavelength expands with the pattern. The fully developed pattern of density perturbations is characterized by a power spectrum cut off at the Jeans wavelength. Assuming that the power spectrum (contribution to the variance of $\bar{\rho}$ per wavenumber increment) does not increase with wavelength faster than λ^3 for $\lambda > \lambda_c$, the characteristic dimension of density fluctuations is the Jeans length λ_c. That is, the density perturbation at any point is correlated with the density perturbation at points a distance less than λ_c away, and uncorrelated with points much more distant than λ_c. Density excursions are roughly comparable to the mean density. Peaks in the density pattern tend toward rapid growth, only to break, because of the radiation drag, and return to the general level.

When ρ_b/ρ (eq. [10]) approaches unity, at time t_c, patches of higher density, with dimensions of the order of λ_c, now are in a position to evolve toward gravitationally bound systems. If the mean density of matter is ρ_c at this time (t_c) the mass of one of the systems is of the order of

$$M_c \equiv \tfrac{4}{3}\pi\rho_c\lambda_c^3 . \quad (18)$$

Notice that

$$\frac{GM_c m_p}{2kT_c\lambda_c} = \frac{4\pi^2}{3}. \quad (19)$$

That is, the matter is formed at roughly an equilibrium configuration. This means that the bound systems have density and characteristic dimensions very roughly of the order of ρ_c and λ_c.

Assuming that the present thermal radiation temperature is 3° K (Penzias and Wilson 1965) and the present matter density is 7×10^{-31} gm cm^3 (Oort 1958) equation (18) implies a mass $M_c \sim 10^7 M_\odot$, roughly the mass of a dwarf galaxy (see Table 1).

With much of the matter in the Universe now concentrated in these discrete clouds, there is at any point a gravitational field, of the general order of GM_c/L, where L is the mean distance between clouds. A cloud thus tends to move toward nearby, higher-density regions, leading to clustering of the clouds.

The development of this gravitational instability depends on the power spectrum of the density fluctuations at the time t_c of formation of the gas clouds. With a flat power spectrum, the total power (i.e., contribution to the variance of $\bar{\rho}$) in wavelengths greater than λ goes as λ^{-3}. However, we can find no reason to expect a characteristic random (flat) spectrum, so for the purpose of a brief discussion of the formation of bound clusters of gas clouds, we shall characterize the power spectrum d_k^2 at time t_c as

$$d_k^2 \propto \lambda^n \,, \tag{20}$$

where the index n would vanish for a flat spectrum, and for boundedness $n < 3$. This is the power spectrum per increment of wavenumber k.

In the cosmological models in Table 1 the clusters of gas clouds are forming at a time when the mass density in radiation approximately may be neglected. With this assumption, we consider first a cosmologically flat Universe, such that the acceleration parameter, $q = 4\pi G\rho/(3H^2)$ is equal to $\frac{1}{2}$.

Let A be a spherical region which is expanding with the general expansion of the Universe and such that, within the volume of A, there would be a mass M of matter on the average. Then the actual mass within A at time t_c is uncertain, by the amount

$$(\delta M/M)_c \sim (M/M_c)^{-0.5+n/6} \,. \tag{21}$$

The functional form of expression (21) is obtained using expression (20) by integrating the density perturbation over A and averaging the square of this integral, with the assumption of random phases of the d_k. The normalization is obtained by noting that when $M \sim M_c$, $\delta M \sim M_c$.

As long as $\delta M/M \ll 1$, we have shown above that the uncertainty (21) grows as $t^{2/3}$. Now consider the epoch t such that

$$(t/t_c)^{2/3}(M/M_c)^{-0.5+n/6} = 1 \,. \tag{22}$$

At this time we can make the following assertions. Within spherical volumes equal to that of the region A the total mass varies by a factor $\sim$2. Within much larger volumes the mass is nearly constant. There exist systems, of mass

$$M \sim M_c(t/t_c)^{4/(3-n)} \,, \tag{23}$$

which are in the process of forming bound systems. In smaller subsystems this process occurred earlier, and the subsequent collapse, or adjustment to equilibrium is already well advanced. This process should be pictured as a roughly continuous progression along a hierarchy in the departure from the general expansion.

We see immediately from equation (23) that in this cosmology the maximum mass of a cluster of galaxies in the present Universe is

$$M \lesssim M_c(t_f/t_c)^{4/(3-n)} \,, \tag{24}$$

where t_f is the present age of the Universe. Adopting the value of Hubble's constant $H_f = 100$ km/sec Mpc, and assuming a cosmologically flat space, $q = \frac{1}{2}$, we have

$$\rho_f = 2 \times 10^{-29} \text{ gm cm}^3 \,. \tag{25}$$

This is the first density used in Table 1. The maximum mass of a cluster of galaxies as given by equation (24) is given in Table 1 for two different values of the index n in equation (20). It is seen that this limit can be consistent with Abell's (1961) conclusion that galaxies are ordered or clustered up to a scale $\sim 10^{16}\ M_\odot$.

Although it appears unlikely that the present value of q is very much larger than $\frac{1}{2}$,

q could be nearly equal to zero (Sandage 1961). Therefore, we have considered also the assumption

$$\rho_f = 7 \times 10^{-31} \text{ gm cm}^3 . \tag{26}$$

This is the estimated mean density of matter in galaxies (Oort 1958). It is interesting that in this limit of small mass density (small q) the expanding Universe becomes very nearly stable against small density perturbations. This follows directly from equation (14). Thus, the growth of the most massive organized clusters of matter is cut off at a time, t_3, when $2q$ departs from unity, and the maximum mass of a cluster of galaxies is given by equation (24) with t_f replaced by the effective time t_3. The resulting maximum masses for this case are shown in Table 1. It is seen that for any of the conditions assumed in the table a reasonable value of n may be chosen to obtain a reasonable upper mass limit.

In this theory a galaxy of normal size would be formed by accretion of many of the original gas clouds. In the manner just described, gravitational instability can lead to the formation of a bound system of clouds of gas, including the original gas clouds and more massive subsystems formed earlier. The clouds within the system may collide, radiate much of the energy of the collision, and fall toward the center of the system, thus leading to the formation of a massive central nucleus. A cloud which passes close to the nucleus may be captured due to gas drag, or it may be torn apart by tidal stresses, and the remnants captured by the nucleus. The situation is further complicated because the various subsystems, or gas clouds, would be in various stages of evolution, with masses ranging from M_c up to a mass approaching that of the total system.

Without attempting to provide a reasonably complete description of this process, we shall show only that the accretion necessary for the formation of galaxies can take place. Consider a gravitationally bound system, with mass $M = NM_c$. At time t_c the material within the system would have occupied a region with dimension

$$R_1 \sim N^{1/3}\lambda_c \tag{27}$$

and from equation (22) the system would have stopped expanding at time

$$t_2 \sim t_1 N^{(3-n)/4}$$

when the system had grown to a size

$$R_2 \sim R_1(t_2/t_1)^{2/3} \sim \lambda_c N^{(5-n)/6} . \tag{28}$$

As before, in obtaining equation (28) we have assumed an expansion parameter $q \sim \frac{1}{2}$.

Now we simplify the problem by supposing that the system contains original gas clouds only, and ask whether the system will have collapsed, due to collisions, by the present time t_f. The mean velocity of gas clouds in the system is

$$v \sim \left(\frac{GM_c N}{\lambda_c N^{(5-n)/6}}\right)^{1/2} \sim \left(\frac{GM_c}{\lambda_c}\right)^{1/2} \tag{29}$$

nearly independent of N. The cross-section for collision is λ_c^2. Using equations (28) and (29), the largest mass M for which the system could have collapsed by the present time satisfies

$$1 \sim \lambda_c^2 \left(\frac{GM_c}{\lambda_c}\right)^{1/2} \frac{N}{\lambda_c^3 N^{(5-n)/2}} t_f$$

or

$$M \sim M_c (G\rho_c t_f^2)^{1/(3-n)} . \tag{30}$$

This rough mass estimate, listed in Table 1, is of the same general order as observed galaxy masses, $\sim 10^{11} M_\odot$. The significance of this result is that there is time for gas clouds within a fairly massive system to have collided with each other to form a protogalaxy.

APPENDIX

THE DENSITY PERTURBATION EQUATION

We consider small perturbations away from a homogeneous, isotropic cosmological model. The calculation is simplified by confining attention to perturbations with dimensions small compared with the radius of the visible Universe. In this approximation Newtonian gravity theory applies (Dicke, Callan, and Peebles 1964). Furthermore, we shall take the temperature of matter and radiation to be a function of (world) time only. This is quite adequate in the linear perturbation case for the characteristic dimensions of density perturbations considered here. The calculation is similar to that of Bonnor (1957), but some new effects due to the radiation are taken into account.

In completely ionized hydrogen the electron density is $\bar{\rho}(x,t)/m_p$, where m_p is the mass of a proton and $\bar{\rho}(x,t)$ is the mass density. If the matter is moving with velocity $\boldsymbol{u}$ relative to the comoving coordinate frame, radiation, isotropic in the comoving frame, exerts a volume force $\sigma b T^4 \rho u/(m_p c)$. Therefore, the equations of motion for matter are

$$\bar{\rho}\left[\frac{\partial \boldsymbol{v}}{\partial t}+(\boldsymbol{v}\cdot\nabla)\boldsymbol{v}\right]=-\bar{\rho}\nabla\phi-\nabla p-\frac{\sigma b T^4}{m_p c}\bar{\rho}(\boldsymbol{v}-H\boldsymbol{r}). \tag{31}$$

This equation is expressed in coordinates, approximately Minkowski, at rest relative to the comoving coordinate frame at the origin, $\boldsymbol{r}=0$. The Newtonian gravitational potential satisfies

$$\nabla^2\phi = 4\pi G(\bar{\rho}+2bT^4/c^2). \tag{32}$$

The factor of 2 in the radiation energy density follows from the linearized form of Einstein's field equations (Tolman 1934). The equation of state is

$$p = 2\bar{\rho}(x,t)kT/m_p \tag{33}$$

and the continuity equation is

$$\frac{\partial\bar{\rho}}{\partial t}+\nabla\cdot(\bar{\rho}\boldsymbol{v})=0. \tag{34}$$

In a first-order perturbation calculation, we write

$$\boldsymbol{v} = \boldsymbol{r}H+\boldsymbol{u}(\boldsymbol{r},t), \quad \bar{\rho}=\rho(t)[1+D(\boldsymbol{r},t)], \quad \phi=\phi_0(t)+\psi(\boldsymbol{r},t). \tag{35}$$

The unperturbed variables satisfy

$$\rho \propto a^{-3}, \tag{36}$$

where $a(t)$ is the usual expansion parameter, and

$$\phi_0 = \tfrac{2}{3}\pi G(\rho_0+2bT^4/c^2)r^2, \qquad \ddot{a}/a = -\tfrac{4}{3}\pi G(\rho_0+2bT^4/c^2). \tag{37}$$

To first order in the perturbations (35), equations (31), (32), and (34) become

$$\frac{\partial \boldsymbol{u}}{\partial t}+H\left(\boldsymbol{u}+r\frac{\partial\boldsymbol{u}}{\partial r}\right)=-\nabla\psi-\frac{2kT}{m_p}\nabla D-\frac{\sigma b T^4}{m_p c}\boldsymbol{u}, \tag{38}$$

$$\nabla^2\psi = 4\pi G\rho D\,, \tag{39}$$

$$\frac{\partial D}{\partial t} + \nabla\cdot u + H r\,\frac{\partial D}{\partial r} = 0\,. \tag{40}$$

On taking the divergence of equation (38), and using expressions (39) and (40) and the formula

$$\nabla\cdot(r\partial u/\partial r) = r\,\frac{\partial}{\partial r}(\nabla\cdot u) + \nabla\cdot u\,, \tag{41}$$

we obtain

$$\left(\frac{\partial}{\partial t} + 2H + H r\frac{\partial}{\partial r}\right)\left(\frac{\partial D}{\partial t} + H r\,\frac{\partial D}{\partial r}\right) = 4\pi G\rho D + \frac{2kT}{m_p}\nabla^2 D - \frac{\sigma b T^4}{m_p c}\left(\frac{\partial D}{\partial t} + H r\,\frac{\partial D}{\partial r}\right). \tag{42}$$

This equation is simplified by transforming to new, comoving coordinates,

$$y = r/a(t)\,. \tag{43}$$

Then assuming a plane wave, $D = d(t)e^{ik_0 y}$, we obtain equation (14), where it should be noticed that the propagation vector k_0 is constant in comoving coordinates, so that the proper wavelength is expanding with the general expansion of the Universe (eq. [13]).

REFERENCES

Abell, G. O. 1961, *A.J.*, **66**, 607.
Bonnor, W. B. 1957, *M.N.*, **117**, 104.
Dicke, R. H., Callan, C., and Peebles, P. J. E. 1965, *Am. J. Phys.*, **33**, 105.
Dicke, R. H., Peebles, P. J. E., Roll, P. G., and Wilkinson, D. T. 1965, *Ap. J.*, **142**, 414.
Eddington, A. S. 1926, *The Internal Constitution of the Stars* (Cambridge: Cambridge University Press), p. 371.
Gamow, G. 1948, *Phys. Rev.*, **74**, 505.
Lifshitz, E. M. 1946, *J. Phys. USSR*, **10**, 116.
Oort, J. H. 1958, Solvay Conference on *Structure and Evolution of the Universe*, p. 163.
Penzias, A. A., and Wilson, R. W. 1965, *Ap. J.*, **142**, 419.
Sandage, A. 1961, *Ap. J.*, **133**, 355.
Tolman, R. C. 1934, *Relativity, Thermodynamics, and Cosmology* (London: Clarendon Press), p. 236.

NATURE. VOL. 215, SEPTEMBER 9, 1967 1155

LETTERS TO THE EDITOR

ASTRONOMY

Fluctuations in the Primordial Fireball

ONE of the overwhelming difficulties of realistic cosmological models is the inadequacy of Einstein's gravitational theory to explain the process of galaxy formation[1-6]. A means of evading this problem has been to postulate an initial spectrum of primordial fluctuations[7]. The interpretation of the recently discovered 3° K microwave background as being of cosmological origin[8,9] implies that fluctuations may not condense out of the expanding universe until an epoch when matter and radiation have decoupled[4], at a temperature T_D of the order of 4,000° K. The question may then be posed: would fluctuations in the primordial fireball survive to an epoch when galaxy formation is possible?

Misner[10] has recently pointed out that fluctuations ranging in size from a photon mean free path up to the event horizon will be damped out by neutrino viscosity during the 10^{10}–10^{11} °K epochs. At 10^{10} °K, however, the event horizon contains only $10^{-4}\, M_\odot$ in a small amplitude ($\delta\rho/\rho \ll 1$) fluctuation. The purpose of this communication is to demonstrate that a considerably more significant upper limit may be obtained by considering the effects of radiative diffusion on opaque fluctuations at subsequent epochs.

Over the temperature range 10^{10} °K $> T > T_D$, the maximum mass M_T contained in a transparent fluctuation (that is, of scale $\sim (\varkappa\rho)^{-1}$) is

$$\frac{4}{3}\pi \varkappa^{-3} \rho^{-2} \simeq 2\cdot3 \times 10^{-29}\, \rho_0^{-2}\, T^{-6}\, M_\odot$$

where $\varkappa$ is the opacity, assumed to be due to Thomson scattering by free electrons, and ρ_0 is the mean density of matter at the present epoch. In deriving this expression, we have assumed that the present value of the background radiation temperature is 3° K, and have used the relation $\rho T^{-3} = \text{constant}$. We consider initial fluctuations $> 10^{-4}\, M_\odot$, so that at some epoch subsequent to $T = 10^{10}$ °K these fluctuations are encompassed within the event horizon. Indeed, the number of baryons corresponding to a galaxy cluster of $\sim 10^{15}\, M_\odot$ is first contained within the event horizon at an epoch when $T \simeq 10^{15}\, \rho_0^{1/3}$ °K. At this temperature, $M_T \simeq 10\, M_\odot$, so that fluctuations of galactic dimensions are highly opaque.

We treat the damping of opaque fluctuations by radiative diffusion in the Newtonian approximation (an adequate approximation provided $aT^4 \ll \rho c^2$, where a is the radiation density constant, and equals $7\cdot6 \times 10^{-15}$ erg cm^{-3} deg^{-4}). The relevant equations are the equation of continuity

$$\frac{\partial\rho}{\partial t} + \nabla\cdot(\rho\mathbf{u}) = 0$$

the equation of motion

$$\frac{d\mathbf{u}}{dt} = \nabla\varphi - \frac{1}{\rho}\nabla p$$

Poisson's equation

$$\nabla^2\varphi = -4\pi G\rho$$

the energy equation

$$\rho\frac{d}{dt}\left(3\frac{kT}{m_p} + \frac{aT^4}{\rho}\right) - \frac{p}{\rho}\frac{d\rho}{dt} = \frac{c}{3}\nabla\cdot\left\{\frac{1}{\varkappa\rho}(\nabla aT^4)\right\}$$

and the equation of state

$$p = \frac{2\rho kT}{m_p} + \frac{1}{3}aT^4$$

This set of equations is linearized about the zeroth order solution (corresponding to an isotropic, homogeneous, expanding cosmological model of arbitrary curvature)

$$\mathbf{u} = \mathbf{r}\frac{\dot R}{R},\quad \rho R^3 = \text{constant},\quad \frac{\ddot R}{R} = -\frac{4\pi\rho G}{3},\quad TR = \text{constant}$$

where $R = R(t)$, $\rho = \rho(t)$. If we assume in addition that the perturbation may be expanded in plane waves of the order of $\exp(i\mathbf{n}\cdot\mathbf{r})$ and that the heat capacity of the radiation greatly exceeds that of the matter, we obtain

$$\frac{\partial^2 s}{\partial t^2} + 2\frac{\dot R}{R}\frac{\partial s}{\partial t} - \left(4\pi G\rho - \frac{kT}{m_p}\frac{4\pi^2}{\lambda^2}\right)s = -\frac{4aT^3}{3\rho}\frac{4\pi^2}{\lambda^2}T_1 \quad (1)$$

and

$$\left(\frac{\partial}{\partial t} + \frac{c}{3\varkappa\rho}\frac{4\pi^2}{\lambda^2}\right)\frac{T_1}{T} = 1/3\,\frac{\partial s}{\partial t} \quad (2)$$

where T_1 is the temperature perturbation, s is the relative density perturbation ($=\rho_1/\rho$) and $\lambda = 2\pi n/R$ is the co-moving wavelength. For the optically thin case ($T_1/T \ll 1$) the usual Jeans criterion is recovered[2,4]; that is, for instability we must have

$$\lambda > \left(\frac{2\pi kT}{m_p G\rho}\right)^{1/2}$$

In the optically thick limit, however, it can readily be shown by an asymptotic treatment of equations (1) and (2) that the dominant mode in s is damped on a time-scale

$$t_d \sim 5\frac{\varkappa\rho}{c}\frac{\lambda^2}{4\pi^2}$$

We shall apply this result to two models; one with $\rho_0 = 3 \times 10^{-29}$ g/cm^3, corresponding to an Einstein–de Sitter universe, and the other with $\rho_0 \simeq 10^{-30}$ g/cm^3, corresponding to an open universe of mean density consistent with observation[11].

For an Einstein–de Sitter universe, the expansion time is approximately given by $t \simeq 1\cdot6 \times 10^{18}\, T^{-3/2}$ sec, leading to

$$t_d/t = 2\cdot2 \times 10^{-9}\, M_\lambda^{2/3} T^{5/2} \rho_0^{1/3}$$

where $M_\lambda = 4/3\, \pi\rho\lambda^3$ is the mass contained in a small ($\rho/\rho \ll 1$) fluctuation. It follows that opaque fluctuations containing up to about $10^{12}\, M_\odot$ are damped (that is, $t_d/t \ll 1$). In other words, any primordial fluctuation of cosmogonical significance must be at least of proper diameter λ (where $M_\lambda \simeq 10^{12} M_\odot$) at the decoupling epoch, when galaxy formation may occur. Now the angular diameter subtended by a fluctuation of proper diameter λ is $H_0\lambda z_D/2c$, where z_D is the value of the red-shift at decoupling, and H_0 is the present value of Hubble's constant. We know that $1 + z_D \simeq T_D/3 \simeq 1\cdot33 \times 10^3$, and taking $H_0 = 100$ km/sec/Mparsec, we find that the angular diameter of a primordial fluctuation must be about half of a minute of arc.

In the case of an open universe, the expansion proceeds approximately linearly with time, and one has the relation $t \simeq 1\cdot2 \times 10^{18} T^{-1}$ sec. Therefore

$$t_d/t = 2\cdot9 \times 10^{-9} M_\lambda^{2/3} T^2 \rho_0^{1/3}$$

and it follows that the critical mass for a fluctuation to survive into an epoch when galaxy formation is possible is about $3 \times 10^{13}\, M_\odot$. One may then show* that this corresponds to an angular diameter of at least one minute of arc.

* For large z, the angular diameter of a source of proper diameter λ in an open cosmological model (with zero cosmological constant and $p \simeq 0$) is given approximately by the expression

$$\frac{z\,\dfrac{\lambda}{c}\sqrt{\dfrac{8\pi G}{3}\varrho_0}}{\sinh\left\{\dfrac{2}{H_0}\sqrt{\dfrac{8\pi G}{3}\varrho_0}\,\ln\sqrt{\dfrac{2}{\sigma_0}}\right\}},\ \text{where } \sigma_0 = \frac{4\pi G\varrho_0}{3H_0^2}$$

.1156

Measurement of the isotropy of the 3° K cosmic microwave background sets an upper limit on the scale and amplitude of fluctuations at the epoch when matter and radiation decouple. Indeed, angular variation in 3° K background is due essentially to intrinsic variations in the decoupling temperature, provided that no significant density fluctuations occur in the universe on a scale much greater than 1 Mparsec[6].

A recent observation[12] on the isotropy of the 3° K background radiation at a frequency of 10,690 Mc/s implies that the background radiation over a right ascension interval of about 4 h is structureless to less than 0·02 per cent on an angular scale of about 1°. Angular resolution of one minute of arc or better, however, is required in order to detect primordial inhomogeneities. When observations at this resolution are forthcoming, it should be possible to answer the question of whether or not galaxies may have formed from fluctuations present *ab initio*. Dr J. P. Wright has suggested to me that another critical factor in interpreting the anisotropy limit will be the mean separation distance between coherent fluctuations at decoupling. A linearized treatment does not allow such considerations; however, our calculation sets an upper limit to the scale of anisotropy expected from primordial fluctuations.

This research was supported in part by a contract from the US Air Force.

JOSEPH SILK

Harvard College Observatory,
Cambridge, Massachusetts.

Received July 20, 1967.

[1] Lifshitz, E., *J. Phys. (USSR)*, **10**, 116 (1946).
[2] Bonnor, W. B., *Mon. Not. Roy. Astron. Soc.*, **117**, 104 (1957).
[3] Layzer, D., *Ann. Rev. Astron. and Astrophys.*, **2**, 341 (1964).
[4] Peebles, P. J. E., *Astrophys. J.*, **142**, 1317 (1965).
[5] Hawking, S. W., *Astrophys. J.*, **145**, 544 (1966).
[6] Sachs, R. K., and Wolfe, A. M., *Astrophys. J.*, **147**, 73 (1967).
[7] Peebles, P. J. E., *Astrophys. J.*, **147**, 859 (1967).
[8] Gamow, G., in *Vistas in Astronomy* (edit. by Beer, A.), **2**, 1726 (1956).
[9] Dicke, R. H., Peebles, P. J. E., Roll, P. G., and Wilkinson, D. T., *Astrophys. J.*, **142**, 414 (1965).
[10] Misner, C. W., *Nature*, **214**, 40 (1967).
[11] Oort, J. H., *Solvay Conference on Structure and Evolution of the Universe* (edit. by Stoops, R.), 163 (Brussels, 1958).
[12] Conklin, E. K., and Bracewell, R. M., *Phys. Rev. Lett.*, **18**, 614 (1967).

THE ASTROPHYSICAL JOURNAL, 162:815–836, December 1970

PRIMEVAL ADIABATIC PERTURBATION IN AN EXPANDING UNIVERSE*

P. J. E. PEEBLES†
Joseph Henry Laboratories, Princeton University
AND
J. T. YU‡
Goddard Institute for Space Studies, NASA, New York
Received 1970 January 5; revised 1970 April 1

ABSTRACT

The general qualitative behavior of linear, first-order density perturbations in a Friedmann-Lemaître cosmological model with radiation and matter has been known for some time in the various limiting situations. An exact quantitative calculation which traces the entire history of the density fluctuations is lacking because the usual approximations of a very short photon mean free path before plasma recombination, and a very long mean free path after, are inadequate. We present here results of the direct integration of the collision equation of the photon distribution function, which enable us to treat in detail the complicated regime of plasma recombination. Starting from an assumed initial power spectrum well before recombination, we obtain a final spectrum of density perturbations after recombination. The calculations are carried out for several general-relativity models and one scalar-tensor model. One can identify two characteristic masses in the final power spectrum: one is the mass within the Hubble radius ct at recombination, and the other results from the linear dissipation of the perturbations prior to recombination. Conceivably the first of these numbers is associated with the great rich clusters of galaxies, the second with the large galaxies. We compute also the expected residual irregularity in the radiation from the primeval fireball. If we assume that (1) the rich clusters formed from an initially adiabatic perturbation and (2) the fireball radiation has not been seriously perturbed after the epoch of recombination of the primeval plasma, then with an angular resolution of 1 minute of arc the rms fluctuation in antenna temperature should be at least $\delta T/T = 0.00015$.

I. INTRODUCTION

a) Purpose

The possible discovery of radiation from the primeval fireball opens a promising lead toward a theory of the origin of galaxies. This primeval radiation would serve, first, to fix an epoch at which nonrelativistic bound systems like galaxies can start to develop (Peebles 1965*a*), and second, to impress on the power spectrum of initial density fluctuations characteristic lengths and masses (Gamow 1948; Peebles 1965*a*, 1967*a;* Michie 1967; Silk 1968). These characteristic features in the power spectrum hopefully result from all the complicated details of the evolution of the Universe *after* the initial power spectrum is arbitrarily set at some very early epoch. If one can make a reasonable argument for a coincidence of these features with observed phenomena, it will provide an important encouragement and guide to the further development of the theory. A more direct observational test of these processes might be provided by the residual small-scale fluctuations in the microwave background (Peebles 1965*b;* Sachs and Wolfe 1967; Silk 1968; Wolfe 1969; Longair and Sunyaev 1969), if we assume that this radiation has not been further scattered (Dautcourt 1969).

* Research supported in part at Princeton by the National Science Foundation and the Office of Naval Research of the U.S. Navy, and at the California Institute of Technology by the National Science Foundation [GP-15911 (formerly GP-9433) and GP-9114] and the Office of Naval Research [Nonr-220(47)].

† Alfred P. Sloan Fellow.

‡ NAS-NRC Postdoctoral Research Associate.

According to Zel'dovich (1967) there are two kinds of perturbations that are of interest: initial isothermal perturbations and initially adiabatic perturbations. It has been suggested that the globular clusters are the remnants of an isothermal perturbation in the early Universe (Peebles and Dicke 1968; Peebles 1969). Our purpose here is to discuss in some detail the evolution of adiabatic density fluctuations in the primeval-fireball picture.

An initially adiabatic perturbation evolves through four regimes: (*a*) When the age t of the Universe is much less than λ/c, where λ is the characteristic scale of the perturbation, a fractional perturbation $\delta\rho/\rho$ to the total mass density grows with time, but the entropy per nucleon is conserved (hence adiabatic). (*b*) When $\lambda \ll ct$, the perturbation oscillates like an acoustic wave. (*c*) As the Universe expands through the recombination phase, the photon mean free path becomes comparable to λ, and the oscillating wave is attenuated, leaving some residual perturbation in the matter distribution. (*d*) When $T \lesssim 2500°$ K, recombination is sufficiently complete that radiation drag on the matter may be neglected, and the residual perturbation may start to grow into bound systems like protogalaxies.

The above general scheme for initially adiabatic perturbations was already given by Lifshitz (1946). The very complicated regime (*c*) has been considered by a number of people in a variety of approximations, with the general conclusion that initially adiabatic perturbations on a characteristic mass scale $\lesssim 10^{11}$–10^{13} $\mathfrak{M}_\odot$ are strongly attenuated. This problem was first considered in approximations to first order in the photon mean free time t_c independently by Michie (1967), Peebles (1967*a*), and Silk (1968). It has since been considered by Bardeen (1968) in the first twenty moments of the radiation distribution function, and by Field (1970*a*), who solves the problem to all orders in t_c when the expansion of the Universe may be neglected. However, these approximation schemes run afoul of the enormous variation and rate of variation of the photon mean free path through the epoch of recombination. As a result, previous workers on this subject (Peebles 1967*a;* Michie 1967; Silk 1968; Field and Shepley 1968) could give only qualitative estimates of the different characteristic masses involved here. To obtain a more accurate description of the evolution through this complicated phase of recombination, we have resorted to direct numerical integration of the collision equation for the photon distribution function.

The more quantitative results of the present calculation are compared with the earlier estimates in § VII. We also discuss there the possible significance of these results. In § II we derive the differential equations to be integrated. It is impractical to integrate the collision equation numerically in the very early Universe because the photon mean free path t_c is so short, but here it becomes a good approximation to describe the radiation as a fluid with viscosity. This description of the radiation was used in all the previous work (Lifshitz 1946; Michie 1967; Silk 1968; Field and Shepley 1968), and is indeed a good approximation in this early epoch. The fluid description of radiation is equivalent to an expansion and integration of our collision equation to first order in t_c. In § III we give the resulting equations valid to first order in t_c, and we present solutions to these approximate equations under various limiting conditions. These results are used to start the numerical integration and to check numerical accuracy. In § IV we consider the residual perturbation to the microwave background. The numerical integrations are described in §§ V and VI.

b) Assumptions and Approximations

In the following calculations we use either conventional general-relativity theory, with cosmological constant Λ equal to zero, or the scalar-tensor theory (Brans and Dicke 1961). We start from a homogeneous, isotropic cosmological model, in which the present parameters are

$$H_0^{-1} = 1 \times 10^{10} \text{ years}, \quad T_0 = 2.7° \text{ K}. \tag{1}$$

We consider four cosmological models: (*a*) an open general-relativity model with present mass density $\rho_0 = 0.03\rho_c$, where ρ_c is the present density in the cosmologically flat general-relativity model; (*b*) the cosmologically flat general-relativity model; (*c*) a closed general-relativity model, $\rho_0 = 5\rho_c$; and (*d*) a cosmologically flat scalar-tensor model with coupling constant $\omega = 5$.

As pointed out by Sachs and Wolfe (1967), it is very convenient to assume that the metric for the unperturbed cosmological model satisfies the approximate formula

$$ds^2 = dt^2 - a(t)^2[(dx^1)^2 + (dx^2)^2 + (dx^3)^2] . \tag{2}$$

To justify this approximation we note that the present cosmological parameters satisfy

$$H_0^2 = \tfrac{8}{3}\pi G\rho_0 \pm \frac{c^2}{R^2 a_0^2} . \tag{3}$$

In the models we consider $\rho_0 \geq 0.03\rho_c$, so by equation (3)

$$\tfrac{8}{3}\pi G\rho_0 \gtrsim \frac{0.03c^2}{R^2 a_0^2} . \tag{4}$$

Therefore, at the epoch of recombination (redshift $Z \sim 1000$) we have

$$\tfrac{8}{3}\pi G\rho_r \gtrsim \frac{30c^2}{R^2 a_r^2} . \tag{5}$$

We are interested in perturbations with characteristic length $l = a_r r$ less than or comparable to the Hubble length at recombination,

$$a_r r \lesssim \frac{c}{(8\pi G\rho_r/3)^{1/2}} \lesssim a_r R/(30)^{1/2} , \tag{6}$$

where the second inequality follows from equation (5). Equation (6) shows that in cosmological models of interest it is a reasonable approximation to neglect $(r/R)^2$ compared with unity, as we have done in the line element (2). A similar argument applies to the closed model.

In the general-relativity cosmologies we use the connection between time and density given by Peebles (1968). For the scalar-tensor cosmology we express the scalar field $\phi_0(t)$ for the unperturbed model as

$$\phi_0 = \frac{4 + 2\omega}{3 + 2\omega}\frac{\lambda(t)}{G_0} , \tag{7}$$

where G_0 is Newton's constant, and $\lambda(t)$ has a present value of unity. The equations governing the rate of expansion of the model are

$$\left(\frac{1}{a}\frac{da}{dt} + \frac{1}{2\lambda}\frac{d\lambda}{dt}\right)^2 = \frac{8\pi G_0(3 + 2\omega)}{3\lambda(4 + 2\omega)}(\mathcal{E} + \rho) + \tfrac{1}{4}\left(1 + \frac{2\omega}{3}\right)\frac{1}{\lambda^2}\left(\frac{d\lambda}{dt}\right)^2 , \tag{8a}$$

$$\frac{d\lambda}{dt} = \frac{8\pi G_0 \rho t}{4 + 2\omega} , \tag{8b}$$

where ρ and $\mathcal{E}$ are the mass densities of matter and radiation. Equation (8b) is the result of a first integration of the field equation for ϕ, and we have set the constant of integration equal to zero. This corresponds to Dicke's solution of type 1 (Dicke 1968, Fig. 4). Equations (8) are integrated numerically to fit the boundary values $\lambda = 1$ and $(da/dt)/a = H_0$ now.

For simplicity we assume that the matter is pure hydrogen. We describe the matter as an ideal fluid with zero pressure and zero heat capacity. The heat capacity of the radiation is in fact some eight orders of magnitude larger than that of matter. The matter pressure defines a characteristic Jeans mass $\mathfrak{M} \sim 10^5\, \mathfrak{M}_\odot$. In the present paper we consider dimensions much larger than this, so the matter pressure is negligible. A measure of the relative importance of the off-diagonal terms in the stress tensors for the matter and radiation is given by the ratio of shear viscosities of matter and radiation,

$$\eta_p/\eta_\gamma \sim (l_p/l_\gamma)(\rho_p/\rho_\gamma)\bar{v}_p/c \sim 10^{-19}(T/T_0)^{3/2} . \tag{9}$$

Here the subscript p refers to the protons, the mean free path l_γ for the radiation is fixed by Thomson scattering by the free electrons, and the mean free path l_p for the protons is fixed by Coulomb scattering in the plasma. We conclude from equation (9) that the momentum transfer by the matter is negligible at all epochs of interest ($T \lesssim 10^{10}\,°$K). It might be mentioned finally that for the irrotational perturbations considered here the polarization field keeps the electrons and protons moving together to high accuracy.

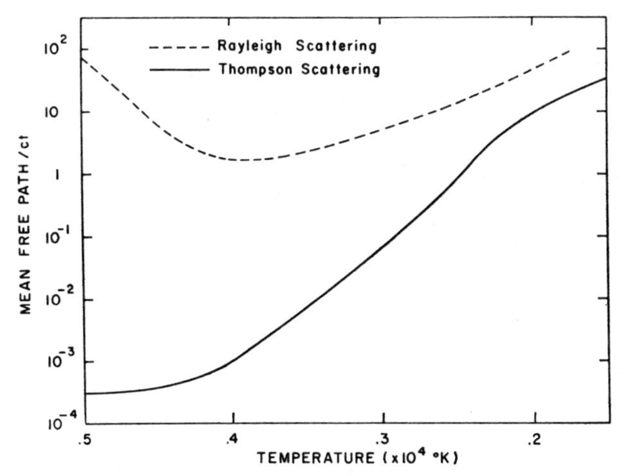

FIG. 1.—Mean free paths for Thomson scattering and Rayleigh scattering in the flat general-relativity model.

The radiation will be described by a distribution function f. In the numerical computation of f an important parameter is the mean free path for scattering the radiation. The most important process here is Thomson scattering by the free electrons; the next most important, Rayleigh scattering by atomic hydrogen. In Figure 1 we plot the mean free paths for these two processes. For Thomson scattering

$$l_\mathrm{T} = (\sigma n_e)^{-1} , \tag{10}$$

where n_e is the electron density. We obtain an average cross-section for Rayleigh scattering by integrating over the blackbody distribution,

$$\langle\sigma_\mathrm{R}\rangle = \int \sigma_\mathrm{R}(\nu) i_\nu d\nu \Big/ \int i_\nu d\nu ,$$

and we set

$$l_\mathrm{R} = (n_\mathrm{H}\langle\sigma_\mathrm{R}\rangle)^{-1} , \tag{11}$$

where n_H is the number density of hydrogen atoms. In the computation of the mean free path we use the approximate theory of plasma recombination in the primeval fireball as given by Peebles (1968), assuming zero primeval helium.

As shown in the figure, the ratio of mean free paths for Rayleigh and Thomson scattering reaches a minimum value of ~ 5 near the end of the epoch of recombination, when the mean free path for Thomson scattering is a factor of 10 larger than the Hubble radius. These results are based on the cosmologically flat model. If ρ_0 were smaller, the ratio of Rayleigh to Thomson mean free paths would be increased (because the residual ionization would be higher).

The next important process is free-free bremsstrahlung emission from the plasma. A direct calculation, using Kramers' approximate formula, gives the ratio of mean free path to Hubble radius just prior to recombination as

$$\frac{l_{ff}}{ct} \sim 10^5 .$$

The production of molecular hydrogen is found to be less than about one part in 10^5, and these molecules have negligible effect on the radiation.

We conclude that it is a reasonable approximation to take account of Thomson scattering only. The error in neglecting Rayleigh scattering only becomes appreciable (~ 20 percent of Thomson) once the Universe is already transparent.

To simplify the computation further, we introduce the additional approximation that Thomson scattering is isotropic (instead of varying as $\cos^2 \theta$) in the matter rest frame.

A basic assumption in all this discussion is that the very early Universe is only slightly perturbed away from the thermodynamic-equilibrium, homogeneous, and isotropic state. This means in particular that there is not supposed to be a primeval magnetic field. The picture breaks down once the earliest bound systems form. If this happens after the initial decoupling of matter and radiation is complete, it will not affect the computation of the density-perturbation transfer function. It is a separate and quite uncertain problem whether subsequent processes could scatter and smooth out the fireball radiation.

For initial values, it seems reasonable to assume that the perturbation is confined almost entirely to the most rapidly growing mode (Peebles 1967*b*). This is because it would be surprising to find at any epoch that the amplitude of the most rapidly growing mode was very much smaller than any of the others, for this would require a highly special perturbing influence. If we rule out this possibility at some very early epoch, it follows that at later epochs the most rapidly growing mode ought to be the dominant one. With this assumption the phase of the acoustic wave in the regime (*b*) mentioned above is fixed.

We consider separately two cases, (1) perturbation wavelength comparable to ct at recombination, and (2) wavelength much less than this characteristic value. In the first case, the transition between the regimes (*a*) and (*b*) mentioned above is of interest because it happens close to the recombination regime (*c*). In the second case, the transition from (*a*) to (*b*) has no interesting effect. Also, the perturbation here suffers a large number of oscillations prior to recombination, which makes it difficult to integrate the equations numerically and may make it unlikely that the phase predicted in the simple linear theory is preserved. Therefore in case (2) we simply start the integration in the regime (*b*), and we suppose here that the phase of each mode is chosen at random.

It might be argued on general grounds that our approach to the second case is the most reasonable one. As we attempt to trace the history of the Universe ever further back in time, our extrapolation surely is becoming more and more uncertain. Here we abandon all attempts to trace the expansion back earlier than $T \sim 10000°$ K, say, and we assume that at that epoch the adiabatic perturbation looks more or less like white noise.

The primeval neutrino density perturbation follows the matter and radiation through regime (*a*), and at the end of this regime the neutrino perturbation disperses because the mean free path is so long. As best we can see, this extra complication has no significant effect, so in our case (1) we ignore neutrinos altogether. In our case (2) the neutrinos simply increase slightly the rate of expansion of the unperturbed cosmological model, and we have included this effect.

Throughout this paper Greek indices $\alpha, \beta, \ldots$, have a range from 1 to 3, and Latin indices from 0 to 3. Units are chosen such that the velocity of light is unity. We choose time-orthogonal coordinates so that the components of the metric tensor are

$$g_{00} = 1\,, \quad g_{0\alpha} = 0\,, \quad g_{\alpha\beta} = -a(t)^2[\delta_{\alpha\beta} - h_{\alpha\beta}(x, t)]\,. \tag{12}$$

In the scalar-tensor cosmology we write the scalar field as

$$\phi = \phi_0(t)[1 + \psi(x, t)]\,, \tag{13}$$

where ϕ_0 is given by equation (7).

II. THEORY

a) Description of the Radiation

Relativistic transport theory has been discussed by a number of authors (e.g., Lindquist 1966). For our purposes the most convenient approach seems to be the following. We start with a photon-gas picture, which is the limit as $m \to 0$ of the motion of a gas of particles of mass m. It will be recalled that when $m \neq 0$ the equations of motion of a particle may be derived from the action principle

$$\delta \int \mathcal{L} dt = 0\,, \quad \mathcal{L} = m(g_{ij}v^i v^j)^{1/2}\,, \quad v^i = \frac{dx^i}{dt}\,, \tag{14}$$

where $dt = dx^0$ is coordinate time. It follows that the momenta canonically conjugate to the coordinates x^α are

$$p_\alpha = \frac{\partial \mathcal{L}}{\partial v^\alpha} = m u_\alpha\,. \tag{15}$$

With these momenta the geodesic equations of motion are

$$\frac{dp_i}{dt} = \tfrac{1}{2} g_{jk,i} p^j v^k\,, \quad v^i = p^i/p^0\,, \tag{16}$$

where v^i is the coordinate velocity. Equations (16) remain well defined in the limit $m \to 0$, with

$$g_{ij} p^i p^j = 0\,. \tag{17}$$

We use equations (16) and (17) to describe the motion of the radiation.

The radiation distribution function is defined to be the number of photons per unit volume in configuration and momentum space,

$$dN = f(x, p) dx^1 dx^2 dx^3 dp_1 dp_2 dp_3\,. \tag{18}$$

Since x^α and p_α are canonical coordinates, we know from Liouville's theorem that f is constant along the path of a particle. Since this is true whatever the choice of coordinates, f must be invariant against coordinate transformations.

b) Stress-Energy Tensor for the Radiation

For an observer at rest in our chosen time-orthogonal coordinate system, the photon energy is just p_0, and the observed bolometric radiation brightness per steradian is

$$\int f p_0^3 dp_0 \equiv \mathcal{E}(t)[1 + \delta(\theta, \phi)]/4\pi\,. \tag{19}$$

Here $\mathcal{E}(t)$ is the radiation density in the unperturbed cosmological model, and δ is the fractional perturbation to the brightness. We represent the spatial components of the photon four-momentum in our coordinate frame in terms of the usual direction cosines γ_α,

$$p_\alpha = -p_0 a(t) e \gamma_\alpha\,, \tag{20}$$

where e is chosen to satisfy equation (17). In terms of these variables the components of the energy-momentum tensor are, to first order,

$$(T_r)_{00} = \mathcal{E}(1 + \delta_r)\,, \quad \delta_r = \int \delta d\Omega/4\pi\,, \tag{21a}$$

$$(T_r)_{0\alpha} = -a\mathcal{E}f_\alpha\,, \quad f_\alpha = \int \delta\gamma_\alpha d\Omega/4\pi\,, \tag{21b}$$

$$(T_r)_{\alpha\beta} = a^2\mathcal{E}(\delta_{\alpha\beta}/3 - h_{\alpha\beta}/3 + \eta_{\alpha\beta})\,, \quad \eta_{\alpha\beta} = \int \delta\gamma_\alpha\gamma_\beta d\Omega/4\pi\,. \tag{21c}$$

One can obtain these expressions by direct computation from the covariant expression for the energy-momentum tensor, or one can simply recognize that equations (21) follow by coordinate transformation from the usual special-relativistic expressions in locally Minkowski coordinates.

c) Collision Equation for the Radiation

We describe the radiation by the coordinate position and time (x^α, t), the direction cosines γ_α, and the photon energy p_0 (all in the time-orthogonal coordinate system). Then the collision equation for the distribution function becomes

$$\frac{\partial f}{\partial t} + \frac{\partial f}{\partial x^\alpha}\frac{dx^\alpha}{dt} + \frac{\partial f}{\partial \gamma_\alpha}\frac{d\gamma_\alpha}{dt} + \frac{\partial f}{\partial p_0}\frac{dp_0}{dt} = \sigma n_e \frac{p'_0}{p_0}(f_+ - f)\,. \tag{22}$$

Here n_e is the number density of free electrons observed in the matter rest frame, f_+ is the distribution function for the scattered radiation, and p'_0/p_0 is the correction for coordinate time from proper time in the matter rest frame. To first order,

$$p'_0/p_0 = 1 - v^\alpha\gamma_\alpha\,, \tag{23}$$

where v^α is the proper matter velocity relative to our coordinate frame.

Because f and f_+ differ by terms of first order in the perturbation, we can set $p'_0/p_0 = 1$ in the right-hand side of equation (22). In the third term on the left side of equation (22) $\partial f/\partial\gamma_\alpha$ and $d\gamma_\alpha/dt$ both are of first order in the perturbation, so the terms may be dropped. In the fourth term we have from equation (12) and the equations of motion (16)

$$\frac{1}{p_0}\frac{dp_0}{dt} = -\frac{1}{a}\frac{da}{dt} + \tfrac{1}{2}\frac{\partial h_{\alpha\beta}}{\partial t}\gamma_\alpha\gamma_\beta\,. \tag{24}$$

On multiplying equation (22) by p_0^3, integrating over p_0, and using equations (16), (19), (20), (24), and the equation

$$\frac{d\mathcal{E}}{dt} = -\frac{4\mathcal{E}}{a}\frac{da}{dt}\,, \tag{25}$$

we obtain the first-order equation

$$\frac{\partial\delta}{\partial t} + \frac{\gamma_\alpha}{a}\frac{\partial\delta}{\partial x^\alpha} - 2\gamma_\alpha\gamma_\beta\frac{\partial h_{\alpha\beta}}{\partial t} = \sigma n_e\left(\frac{4\pi}{\mathcal{E}}\int f_+ p_0^3 dp_0 - \delta - 1\right). \tag{26}$$

We assume that the scattered radiation is isotropic in the matter rest frame. Since the distribution function is an invariant,

$$f_+(p_0, \gamma) = f'_+[p'_0(p_0, \gamma)] = \int f'(p'_0, \gamma')d\Omega'/4\pi\,, \tag{27}$$

where the primes refer to the locally Minkowski matter rest frame. With equations (23) and (27), the first term in the parentheses on the right-hand side of equation (26) becomes

$$\int f' d\Omega' p_0^3 dp_0/\mathcal{E} = (1 + 4\gamma_\alpha v^\alpha)\int f' p'_0 d^4p'/\mathcal{E}\,. \tag{28}$$

The integral on the right side of this equation is the radiation-energy density, and this is given by equation (21a) to terms of first order. With this result equation (26) finally becomes

$$\frac{\partial \delta}{\partial t} + \frac{\gamma_\alpha}{a}\frac{\partial \delta}{\partial x^\alpha} - 2\gamma_\alpha\gamma_\beta\frac{\partial h_{\alpha\beta}}{\partial t} = \sigma n_e(\delta_r + 4\gamma_\alpha v^\alpha - \delta) . \tag{29}$$

d) Description of the Matter

The matter is supposed to be an ideal fluid, with mass density

$$\rho_m(x, t) = \rho(t)[1 + \delta_m(x, t)] , \tag{30}$$

where $\rho(t)$ is the unperturbed value. The stress-energy tensor is

$$(T_m)_{ij} = \rho_m u_{mi} u_{mj} . \tag{31}$$

The motion of the matter is determined by the equations

$$(T_m)^{ij}{}_{;j} = g^i , \tag{32}$$

where in the locally Minkowski matter rest frame

$$g^0 = 0 , \quad g^\alpha = \sigma n_e (T_r)^{0\alpha} . \tag{33}$$

Here $(T_r)^{0\alpha}$ is the radiation-energy flux in this coordinate system. The covariant generalization of equations (33) is

$$g^i = \sigma n_e[(T_r)^{ij}u_{mj} - u_m{}^i(T_r)^{jk}u_{mj}u_{mk}] . \tag{34}$$

Letting $v^\alpha = a v_m{}^\alpha$ be the proper matter velocity measured by an observer at rest in the time-orthogonal coordinate system, and using equations (21), (32), and (34), we obtain, to first order,

$$\frac{dv^\alpha}{dt} + \frac{v^\alpha}{a}\frac{da}{dt} = \frac{\sigma n_e \mathcal{E}}{\rho}\left(f_\alpha - \tfrac{4}{3}v^\alpha\right) , \tag{35a}$$

$$\frac{\partial \delta_m}{\partial t} = \frac{1}{2}\frac{\partial h}{\partial t} - \frac{v^\alpha{}_{,\alpha}}{a} , \quad h = \Sigma h_{\alpha\alpha} . \tag{35b}$$

e) Plane-Wave Decomposition

Since the spatial coordinates x^α do not appear in the coefficients of the linear perturbation equations (29) and (35), we can decompose the perturbation into complex plane waves, for example,

$$\delta_m = \Sigma\delta(m; \boldsymbol{k}) \exp(i\boldsymbol{k}\cdot\boldsymbol{x}) . \tag{36}$$

In writing the differential equations for a single plane-wave component we generally omit the index $\boldsymbol{k}$, and we choose the coordinates such that the x^3 axis is along $\boldsymbol{k}$. Then by symmetry $h_{11} = h_{22}$, and $h_{12} = h_{23} = h_{13} = 0$. (We ignore gravitational radiation.) For a single plane-wave component, equation (29) is

$$\frac{d\delta}{dt} + \frac{ik\mu\delta}{a} + \Phi = \sigma n_e(\delta_r + 4\mu v - \delta) , \quad \mu \equiv \gamma_3 = \cos\theta ,$$

$$\Phi = (1 - 3\mu^2)\frac{dh_{33}}{dt} - (1 - \mu^2)\frac{dh}{dt} , \quad h = 2h_{11} + h_{33} . \tag{37}$$

Equations (21b) and (35) are

$$\frac{dv}{dt} + \frac{v}{a}\frac{da}{dt} = \frac{\sigma n_e \mathcal{E}}{\rho}\left(f - \tfrac{4}{3}v\right) , \tag{38a}$$

$$\frac{d\delta_m}{dt} = \frac{1}{2}\frac{dh}{dt} - \frac{ikv}{a}, \tag{38b}$$

and

$$f = \frac{1}{2}\int_{-1}^{+1} \delta\mu d\mu . \tag{38c}$$

In these equations we have dropped the index $\alpha = 3$.

In vector $\boldsymbol{k}$ in equation (36) is expressed in comoving coordinates. In the graphs displayed below the unit of k is chosen such that the comoving wavelength $2\pi/k$ is expressed in units of 10^{25} cm$/T$ at epoch T (° K). Associated with k is a characteristic time-independent mass defined to be the mass within a sphere with diameter equal to the wavelength,

$$\mathfrak{M}(k) = \frac{1}{6}\pi\rho_0\left(\frac{2\pi \times 10^{25}\text{ cm}}{kT_0}\right)^3 . \tag{39}$$

f) Gravitational Field Equations

In a plane-wave perturbation there are only two independent components to the perturbed metric. The field equations are obtained by substituting the metric into Einstein's equations or the scalar-tensor field equations, and using the matter and radiation stress-energy tensors (eqs. [21], [31]). For the general-relativity cosmologies we use the two field equations

$$\frac{d^2h}{dt^2} + \frac{2}{a}\frac{da}{dt}\frac{dh}{dt} = 8\pi G(\rho\delta_m + 2\mathcal{E}\delta_r) , \tag{40a}$$

$$\frac{dh_{33}}{dt} - \frac{dh}{dt} = -\frac{ia}{k}16\pi G(\mathcal{E}f + \rho v) . \tag{40b}$$

Equation (40a) is the 0–0 component of the field equations, and equation (40b) is the 0–3 component. The left-hand sides of these equations were first derived by Lifshitz (1946).

For the scalar-tensor cosmology equations (40) are generalized to

$$\begin{aligned}\frac{d^2h}{dt^2} + \frac{2}{a}\frac{da}{dt}\frac{dh}{dt} = \frac{8\pi G_0}{\lambda}&\left[\frac{3+2\omega}{2+\omega}\mathcal{E}(\delta_r - \psi) + \rho(\delta_m - \psi)\right] \\ &+ \frac{4(1+\omega)}{\lambda}\frac{d\psi}{dt}\frac{d\lambda}{dt} + 2\frac{d^2\psi}{dt^2} ,\end{aligned} \tag{41a}$$

$$\begin{aligned}\frac{dh_{33}}{dt} - \frac{dh}{dt} = -\frac{ia}{k}&\frac{16\pi G_0}{\lambda}\frac{(3+2\omega)}{(4+2\omega)}(\mathcal{E}f + \rho v) \\ &- 2(1+\omega)\frac{\psi}{\lambda}\frac{d\lambda}{dt} - \frac{2d\psi}{dt} + \frac{2\psi}{a}\frac{da}{dt} ,\end{aligned} \tag{41b}$$

and the equation for the perturbation to the scalar field is

$$\begin{aligned}\frac{d^2\psi}{dt^2} + \frac{d\psi}{dt}\left(\frac{2}{\lambda}\frac{d\lambda}{dt} + \frac{3}{a}\frac{da}{dt}\right) + \psi\left(\frac{1}{\lambda}\frac{d^2\lambda}{dt^2} + \frac{3}{\lambda a}\frac{da}{dt}\frac{d\lambda}{dt} + \frac{k^2}{a^2}\right) - \frac{1}{2\lambda}\frac{dh}{dt}\frac{d\lambda}{dt} \\ = \frac{8\pi G_0}{(4+2\omega)\lambda}\rho\delta_m .\end{aligned} \tag{41c}$$

The only difference between the general-relativity and scalar-tensor models is the gravitational field equations. Because we use the original Brans-Dicke (1961) formulation, the description of matter and radiation (eqs. ([37] and [38]) is the same in the two theories.

III. APPROXIMATE SOLUTIONS

a) Solution to First Order in t_c

Equations (37), (38), and (40) or (41) determine the time variation of linear perturbations within the framework of our general assumptions. In principle, these equations can be integrated numerically from a given initial time, but this becomes impractical at very early epochs, when the mean free time $t_c = (\sigma n_e)^{-1}$ is very short. We employ instead the standard iterative approximation to the collision equation in powers of t_c, to obtain, in the limit of small t_c, approximate equations which are used at the start of the integration.

To zeroth order in t_c equation (37) says

$$\delta = \delta_r + 4\mu v\,. \tag{42}$$

On substituting this equation into the left side of equation (37) we obtain the first-order equation for δ which, when substituted back into the left-hand side of equation (37), yields the second-order equation,

$$\begin{aligned} \delta &= \delta_r + 4\mu v - t_c\left[C - \frac{d}{dt}(Ct_c) - \frac{ik\mu}{a}t_c C\right], \\ C &= \frac{d\delta_r}{dt} + 4\mu\frac{dv}{dt} + \frac{ik\mu\delta_r}{a} + \frac{4ik\mu^2 v}{a} + \Phi\,. \end{aligned} \tag{43}$$

When this equation is integrated over μ, the left-hand side yields δ_r (eq. [21a]), so the quantity in brackets, integrated over μ, has to vanish. This result with equation (38b) yields the first of the desired equations,

$$\frac{d\delta_r}{dt} = \tfrac{4}{3}\frac{d\delta_m}{dt} - \tfrac{4}{3}Kt_c\,, \quad K = \frac{k^2\delta_r}{4a^2} - \frac{ik}{a}\frac{dv}{dt}\,. \tag{44}$$

Next we obtain an expression for f (eq. [21b]) by multiplying equation (43) by μ and integrating the result over μ. On using this expression in equation (38a), we find the equation of motion for the fluid,

$$\begin{aligned} (\rho + \tfrac{4}{3}\mathcal{E})\frac{dv}{dt} + \frac{\rho v}{a}\frac{da}{dt} &= -\tfrac{1}{3}\frac{ik\delta_r\mathcal{E}}{a} - \tfrac{4}{3}\frac{\mathcal{E}}{ik}\frac{d}{dt}Kat_c \\ &\quad + \frac{ikt_c\mathcal{E}}{a}\left[\tfrac{4}{5}\frac{ikv}{a} + \frac{d}{dt}\left(\tfrac{1}{3}\delta_r - \tfrac{4}{15}h_{33} - \tfrac{2}{15}h\right)\right]. \end{aligned} \tag{45}$$

Next we use equation (38b) to eliminate the velocity from this equation and obtain the second of the desired equations,

$$\begin{aligned} (1+R)\frac{d^2}{dt^2}(\delta_m - \tfrac{1}{2}h) &+ \frac{(1+2R)}{a}\frac{da}{dt}\frac{d}{dt}(\delta_m - \tfrac{1}{2}h) \\ &= -\frac{k^2\delta_r}{4a^2} + \frac{1}{a}\frac{d}{dt}(at_cK) - \frac{k^2t_c}{5a^2}\frac{d}{dt}(\delta_r + h_{33} - h)\,, \end{aligned} \tag{46}$$

where

$$R = 3\rho/4\mathcal{E}\,. \tag{47}$$

Finally, with equations (38b) and (45) we can reduce the second of equations (44) to the expression (valid to zeroth order in t_c)

$$K = \frac{R}{1+R}\left[\frac{k^2\delta_r}{4a^2} + \frac{1}{a}\frac{da}{dt}\left(\tfrac{1}{2}\frac{dh}{dt} - \frac{d\delta_m}{dt}\right)\right]. \tag{48}$$

b) Limiting Solutions

We present solutions to equations (44), (46), and (40) or (41) in two limiting cases of interest.

i) $R = 0$, $t_c = 0$, *and* $a/k \gg t$

This limit applies to the very early Universe. The solutions are

$$\delta_m = \tfrac{3}{4}\delta_r \propto t^n\,, \qquad n = -1, \tfrac{1}{2}, 1\,. \quad (49)$$

Different parts of these solutions have been previously obtained by Lifshitz (1946) and Hawking (1966). We use for initial values the most rapidly growing mode. For this mode, we have

$$\frac{d\delta_m}{dt} = \frac{\delta_m}{t}\,, \quad \delta_r = \tfrac{4}{3}\delta_m\,, \quad \frac{dh}{dt} = 2\,\frac{d\delta_m}{dt}\,. \quad (50)$$

Because in this solution the wavelength is greater than the horizon t, the reality of this density perturbation might be questioned. The point is that different parts of the Universe are evolving like independent Friedmann models, and different observers would find truly different variations of density with proper time. Equation (49) gives the fractional difference between the density histories seen by two observers separated by a distance greater than t.

In the scalar-tensor cosmology we have chosen initial values such that equation (8b) applies, which means that in the early radiation-dominated Universe λ approaches a constant. Also, in equation (41c) the source term for ψ is negligible in this limiting case, so we can choose the initial values

$$\psi = d\psi/dt = 0\,, \quad (51)$$

and equations (49) and (50) still apply.

ii) $R \lesssim 1$, $a/k \ll t$

This limit applies to the regime (*b*) mentioned in § 1. Since the perturbation wavelength is much less than ct, the gravitational fields in equation (46) may be neglected. Then the adiabatic approximation to equations (44) and (46) is

$$\delta_m \propto [\exp \textstyle\int (i\omega - \gamma)dt]/(1 + R)^{1/4}\,,$$

$$\gamma = \frac{k^2c^2t_c}{6a^2(1+R)^2}\,[R^2 + 0.8(R+1)]\,, \quad \omega = \frac{k}{a}\,[3(1+R)]^{-1/2}\,, \quad (52)$$

where we have assumed that the damping rate γ is much less than the frequency ω of the wave. The damping rate has been derived by Peebles (1967*a*) (in a circulated preprint, the value of γ is wrong because the viscosity of the radiation fluid was neglected) and by Field (1970*a*).

c) Joining Conditions

In the transition from numerical integration of the first-order equations (44) and (46) to the numerical integration of the collision equation we need starting values for the velocity and the radiation distribution function. These are fixed in terms of the variables in the first-order integration by equations (38b) and (43). In the second of these equations we use equation (45) to eliminate the rate of change of velocity. The resulting initial values are

$$v = \frac{ia}{k}\frac{d}{dt}\,(\delta_m - \tfrac{1}{2}h)\,,$$

$$\delta(\mu) = \delta_r + 4\mu v - t_c\Big[(1 - 3\mu^2)\,\frac{d}{dt}\,(\tfrac{4}{3}\delta_m + h_{33} - h) + \frac{4ia}{k}\,\mu K\Big]\,. \quad (53)$$

IV. RESIDUAL PERTURBATION TO THE MICROWAVE BACKGROUND

Following recombination of the primeval plasma, the initially adiabatic perturbation leaves a residual irregularity in the radiation field. This irregularity persists if the radiation is not subsequently scattered by intergalactic matter or by the matter in young galaxies, and it is our purpose here to give an expression for the expected effect in terms of observations of the microwave background. We consider here fluctuations on a small angular scale, of the order of 1° or less.

a) Time Dependence of the Radiation Perturbation

At some time t_f well after the period of recombination, we can assume that

$$\lambda_f = 2\pi a_f/k \ll t_f \ll t_c(t_f) , \qquad \rho \gg \mathcal{E} . \tag{54}$$

In this limit the gravitational field equations (40b) or (41b, c) reduce to

$$\frac{dh_{33}}{dt} = \frac{dh}{dt} , \tag{55}$$

and equation (37) becomes

$$\frac{d\delta}{dt} + \frac{ik\mu\delta}{a} = 2\mu^2 \frac{dh}{dt} . \tag{56}$$

To solve this equation we introduce the substitution

$$\delta = \bar{\delta} - \frac{2ia\mu}{k}\frac{dh}{dt} + \frac{2a}{k^2}\frac{d}{dt}\left(a\frac{dh}{dt}\right) . \tag{57}$$

With this change of functions equation (56) becomes

$$\frac{d\bar{\delta}}{dt} + \frac{ik\mu}{a}\bar{\delta} = 0 , \tag{58}$$

where we have dropped terms of order $(\lambda_f/t_f)^3$. We have kept the equation valid to this order because the residual perturbation to δ_m and $t dh/dt$ is much larger than the residual perturbation to δ.

By equations (57) and (58) the fractional perturbation to the radiation brightness at time t_1 is

$$\delta(t_1) = \bar{\delta}_f \exp(-ik\mu\tau) ,$$

$$\bar{\delta}_f = \delta_f + \frac{2ia_f\mu}{k}\frac{dh_f}{dt} - \frac{2a_f}{k^2}\frac{d}{dt}\left(a\frac{dh}{dt}\right)_f , \qquad \tau = \int_{t_f}^{t_1} \frac{dt}{a(t)} . \tag{59}$$

In principle we should correct $\delta(t_1)$ for the peculiar motion of our chosen time-orthogonal coordinate system. However, this generates only a slowly varying term (proportional to μ) in the brightness, so we ignore it.

b) Observed Temperature Fluctuation

Equation (59) is based on the assumption that curvature may be neglected. In the late stages of expansion, at time $t > t_1$, spatial curvature becomes important. To obtain the fractional perturbation to the radiation brightness at the present time t_0, we trace back along the path of a light ray to the epoch t_1. At this epoch we set up a Cartesian coordinate system with (x, y)-axes perpendicular to the light ray. If we observe along a direction fixed by the two orthogonal angles (ψ_1, ψ_2) relative to the chosen light ray, the observed brightness is

$$\delta(t_0, \psi_1, \psi_2) = \Sigma\delta(\boldsymbol{k}; t_1) \exp(i\boldsymbol{k}\cdot\boldsymbol{x}) , \tag{60}$$

where we are now summing over Fourier components (eq. [36]), $\boldsymbol{k}$, and we have distinguished the individual amplitudes by $\boldsymbol{k}$. We imagine that the observation scans only a small part of the sky, so that ψ_1, ψ_2 are small angles, and $\boldsymbol{x}$ has Cartesian components $(r\psi_1, r\psi_2, r)$, where r is the coordinate distance given by

$$r = R \sin \frac{1}{R} \int_{t_1}^{t_0} \frac{dt}{a(t)} \quad \text{(closed)} ,$$

$$r = \int_{t_1}^{t_0} \frac{dt}{a(t)} \quad \text{(flat)} , \tag{61}$$

$$r = R \sinh \frac{1}{R} \int_{t_1}^{t_0} \frac{dt}{a(t)} \quad \text{(open)} ,$$

for the different cosmological models (Landau and Lifshitz 1962). R is the coordinate radius of curvature (eq. [3]).

The brightness pattern (60) should be folded against the gain of the antenna. We assume that the gain is a Gaussian with resolving power $\Delta\psi$ independent of wavelength, so that the observed brightness is

$$\begin{aligned} \delta'_0(\psi'_1, \psi'_2) &= \int \delta(t_0, \psi_1, \psi_2) F(\psi_1 - \psi'_1, \psi_2 - \psi'_2) d\psi_1 d\psi_2 , \\ F(\psi_1, \psi_2) &= \frac{1}{2\pi(\Delta\psi)^2} \exp[-(\psi_1^2 + \psi_2^2)/2\Delta\psi^2] . \end{aligned} \tag{62}$$

Now we use equation (60) for δ in equation (62), and we find

$$\delta'_0(\psi_1, \psi_2) = \Sigma\delta(\boldsymbol{k}) \exp[-ir(k_x\psi_1 + k_y\psi_2 - k_z) - \tfrac{1}{2}(k_x^2 + k_y^2) r^2 \Delta\psi^2] . \tag{63}$$

If the individual Fourier components are assumed to have randomly chosen phases, the observed mean square fractional deviation of the brightness from the mean is

$$\langle |\delta'_0|^2 \rangle = \Sigma |\bar{\delta}_{kf}|^2 \exp[-(k_x^2 + k_y^2) r^2 \Delta\psi^2 . \tag{64}$$

As usual, we replace the sum by an integral,

$$\begin{aligned} \langle |\delta'_0|^2 \rangle &= \int_0^\infty G(k, \Delta\psi) dk , \\ G(k, \Delta\psi) &= \frac{V}{(2\pi)^2} \int_{-1}^{+1} d\mu k^2 |\bar{\delta}_{kf}|^2 \exp[-k^2(1 - \mu^2) r^2 \Delta\psi^2] , \end{aligned} \tag{65}$$

where the Fourier wave is fixed to be periodic in the large volume V. The normalization of equation (65) is fixed by the assumed amplitude of the residual perturbation to the matter distribution (§ VI).

In all this computation we have been able to concentrate on the radiation brightness integrated over frequency. This reduction of variables makes the numerical integration feasible, but it means that we lose detailed information of the spectrum of the residual irregularities in the background radiation. We can note, however, that as we follow the path of a photon back in time we end up at a spot in the primeval plasma near local thermal equilibrium. Because the photon mean free path is approximately independent of wavelength, the background looks like a mixture of blackbody spectra with weight independent of wavelength. Thus the rms fluctuation in antenna temperature is related to the rms fluctuation in brightness by the equation

$$\frac{\delta T}{T} \approx \tfrac{1}{4} \langle |\delta'_0|^2 \rangle^{1/2} . \tag{66}$$

V. RESULTS OF THE INTEGRATION: SHORT-WAVELENGTH LIMIT

In this section we consider that part of the perturbation with wavelength at recombination $\ll ct$. In this limit the perturbation to the gravitational field is unimportant, so we can drop h and h_{33} from the collision equation (37).

In accordance with the philosophy expressed in § I*a*, we start the integration in the regime (*b*), where the perturbation acts like an acoustic wave, and we assume that the phase of each acoustic mode is chosen at random. Then for each mode the quantity of interest is the amplitude

$$\delta_s = [\delta_m(1)^2 + \delta_m(2)^2]^{1/2} ; \tag{67}$$

where $\delta_m(1)$ and $\delta_m(2)$ differ in phase by 90°.

In the numerical integration the initial values are fixed by the adiabatic solution (52), and the initial amplitude is chosen so that δ_s is unity in the very early Universe. Because

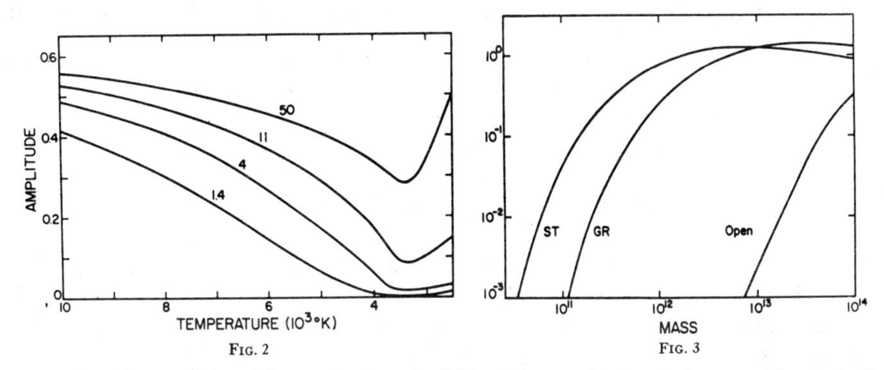

FIG. 2 FIG. 3

FIG. 2.—Time variation of the amplitude δ_s (eq. [67]) of the perturbation to the matter density in the cosmologically flat general-relativity model. The parameter is the characteristic mass (eq. [39]) in units of $10^{11}\,\mathfrak{M}_{\odot}$.

FIG. 3.—Transfer function for an open general-relativity model $\rho_0 = 0.03\rho_c$, the cosmologically flat general-relativity model (GR), and the flat scalar-tensor model (ST). The amplitude is normalized to unity when $T \lesssim 10^5$ ° K; the independent variable is the mass (eq. [39]) in units of solar masses.

this is a linear calculation, one is of course free to adjust the shape of the initial power spectrum and to adjust thereby the final result. The first-order equations (44) and (46) are integrated ahead in time until it becomes feasible to switch over to the direct integration of equations (37) and (38). We find by trial that the results are insensitive to the switching time. At recombination ($T \sim 2500°$ K) the radiation drag force rather abruptly becomes unimportant. Thereafter the perturbation to the mass density varies as

$$\delta_m = At^a + B/t , \tag{68}$$

where $a = \frac{2}{3}$ in the general-relativity models (Lifshitz 1946), and $a = (4 + 2\omega)/(4 + 3\omega)$ in the scalar-tensor model (Nariai 1969). This solution is used to carry the perturbation to the epoch $T = 2000°$ K.

For the cosmologically flat general-relativity model the amplitude δ_s is plotted as a function of time in Figure 2. Each curve belongs to a fixed comoving propagation vector k, and the parameter labeling the curves is the characteristic mass defined by equation (39). The sharp upturn of the curves at the right-hand side of the figure results from the residual matter velocity.

The amplitude δ_s at the final epoch $T = 2000°$ K is shown as a function of mass (eq. [39]) in Figure 3 for three cosmological models. These curves may be called the transfer functions for initially adiabatic perturbations with randomly distributed phase. The transfer function is larger in the scalar-tensor model because the model expands faster, so the decoupling of matter and radiation is more sharp. The transfer function for the open model is smaller because near recombination the photon mean free path is longer, so the dissipation is greater.

We discuss the possible significance of these results after we deal with the part of the perturbation with longer wavelength.

VI. RESULTS OF THE INTEGRATION: LONG WAVELENGTH

The numerical integration starts at the epoch $T = 10^8\,°$ K, where the Universe is radiation dominated, $R \leq 10^{-3}$ (eq. [47]). The initial values are fixed to the most rapidly growing mode (eqs. [50] and [51]). The first-order equations (44) and (46) are numerically integrated to the epoch $T = 10000°$ K, at which point we can, for the long wavelengths considered here, switch over to numerical integration of equations (37) and (38) along with the gravitational field equations. These equations are integrated to $T = 200°$ K.

We have several checks on the numerical accuracy of this scheme. The initial time variation is checked against the solution (50). When the integration reaches the regime (*b*), we can check against the solution (52). We verify that the equations to first order in t_c are a good approximation by shifting the transition to integration of the collision equation from the epoch 10000° to 7000° K. We verify that, in cases where the curvature may be neglected, the quantity $|\bar{\delta}_{kf}|^2$ (eqs. [59] and [65]) reaches a constant value by evaluating it at times earlier than 200° K.

a) Initial Conditions

Our assumptions fix the perturbation up to an initial amplitude for each wavelength. We choose for the initial power spectrum a sort of "cosmological white noise" spectrum indicated by the following argument.

The importance of a density perturbation in the early Universe is measured not only by the fractional density contrast $\delta\rho/\rho$ but also by the effect of the density contrast on the geometry—for any given $\delta\rho/\rho$ in the most rapidly growing mode one can always make the extent of the perturbation so large that it seriously affects the geometry, and even closes space back in on itself. A measure of the perturbation to the geometry due to a density perturbation with characteristic coordinate size r is the number

$$\epsilon = r/R_c\,, \tag{69}$$

where R_c is a coordinate measure of the local radius of curvature of space, and where in the radiation-dominated Universe the most rapidly growing mode for a density perturbation is (Peebles 1967*b*)

$$\delta_t = \delta\rho/\rho = 2t^2/a(t)^2 R_c^2\,. \tag{70}$$

We choose the initial power spectrum of density irregularities such that ϵ is independent of the scale of length on which we examine the geometry. That is, we write the variance of the density as

$$\langle\delta_t^2\rangle = \Sigma|\delta_{tk}|^2 \equiv \int \mathcal{P}(k)dk/k\,, \tag{71}$$

where

$$\mathcal{P}(k) \equiv Vk^3|\delta_{tk}|^2/2\pi^2 \tag{72}$$

is the contribution to the variance per logarithmic increment of the wavenumber k. This variance $\mathcal{P}(k)$ generates via equation (70) space curvature to the geometry observed on the coordinate length scale $r \sim k^{-1}$. Using equations (69) and (70), we write the initial power spectrum as

$$\mathcal{P}_i(k) = Vk^3|\delta_{tki}|^2/2\pi^2 \equiv 4(t_i\epsilon k/a_i)^4\,. \tag{73}$$

Now we choose the shape of the initial power spectrum $|\delta_{tki}|^2$ such that the characteristic number ϵ in equation (73) is independent of k. Also, we will be considering amplitudes such that $\epsilon \ll 1$, so that the perturbation to the geometry is in a sense small.

This choice of initial power spectrum, $|\delta_{tki}|^2 \propto k$, has two interesting and perhaps attractive features. First, the initial perturbation contains no built-in characteristic lengths. The perturbation to the geometry looks the same on each scale of size. This would not have been the case if we had started with conventional white noise, $|\delta_{tki}|^2$ constant, because this does violence to the geometry at long wavelengths unless a long-wavelength cutoff is introduced. If the power spectrum varied as k^2, say, then we would have had the same problem at short wavelengths.

The second feature is that, with the initial value (73) in regime (*a*), in regime (*b*) the irregularity in the matter distribution is independent of length. To see this, note that

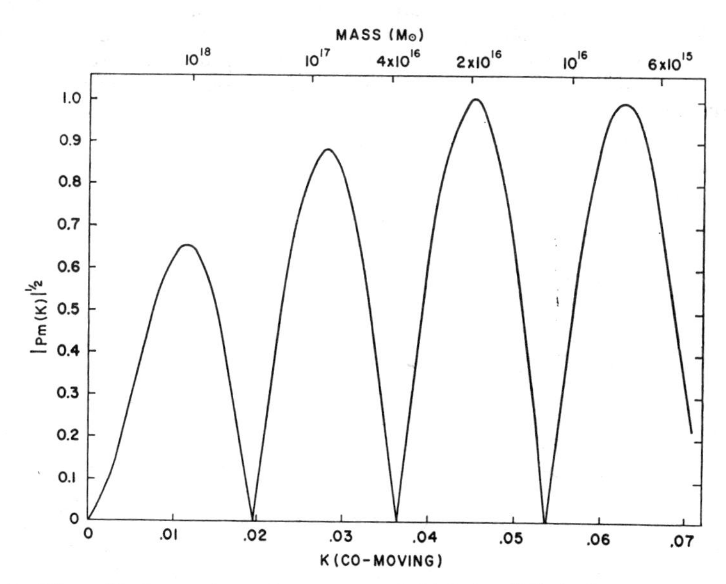

FIG. 4.—The residual mass-fluctuation spectrum $\mathcal{P}_m(k)^{1/2}$ (eq. [72]) in the open general-relativity model, $\rho_0 = 0.03\rho_c$ ($\rho_c = 1.8 \times 10^{-29}$ g cm^{-3}). The curve has been normalized to unity at maximum.

δ_{tk} grows in proportion to t until the proper wavelength $\lambda(t) = 2\pi a(t)/k$ becomes comparable to t. Because $a(t) \propto t^{1/2}$, we see that $|\delta_{tk}|^2$ grows by a total factor $\propto k^{-4}$, which cancels the factor k^4 in the right-hand side of equation (73), to make $\mathcal{P}(k)$ (eqs. [71] and [72]) independent of k.

The initial condition (73) determines the perturbation up to one normalizing factor (ϵ) which is fixed in the manner described below. We emphasize again that our numerical integration determines the transfer function in the framework of our assumptions, and it is quite a separate consideration that motivates the choice of starting values that seems reasonable to us. If different starting values seem more appropriate, they can of course be introduced by scaling the graphs presented below.

b) *Mass Density Fluctuation*

In Figures 4–7 we plot the mass-density-fluctuation spectrum well after recombination, as given by the function $\mathcal{P}_m(k)^{1/2}$, where

$$\mathcal{P}_m(k) = Vk^3|\delta_{m,k}|^2/2\pi^2 , \tag{74}$$

for the four cases of interest: the open, flat, and closed general-relativity models (cf. Figs. 4–7) and the flat scalar-tensor model. This is the contribution to the variance of the mass density per logarithmic increment of k. In the figures the normalization is arbitrarily fixed to peak value unity.

c) Residual Irregularity in the Microwave Background

In Figure 8 we plot $G(k, \Delta\psi)$ (eq. [65]) for the cosmologically flat general-relativity model. The area under the curve for fixed $\Delta\psi$ gives the variance of the brightness of the observed background when the resolving power is $\Delta\psi$. Notice that $G(k, \Delta\psi)$ is appreciable only near the first peak of $\mathcal{P}_m(k)$. As $\Delta\psi$ decreases, the curve moves to the right because one is sensitive to shorter wavelength (larger wavenumber). The shift is not large, however, because the residual radiation perturbation at shorter wavelength is so very small. We conclude from Figure 8 that the experimental search for small-scale irregularities in the microwave background provides a test for the first big peak in Figure 5 (if the radiation has not suffered further scattering). The same conclusion holds for the other three cosmological models.

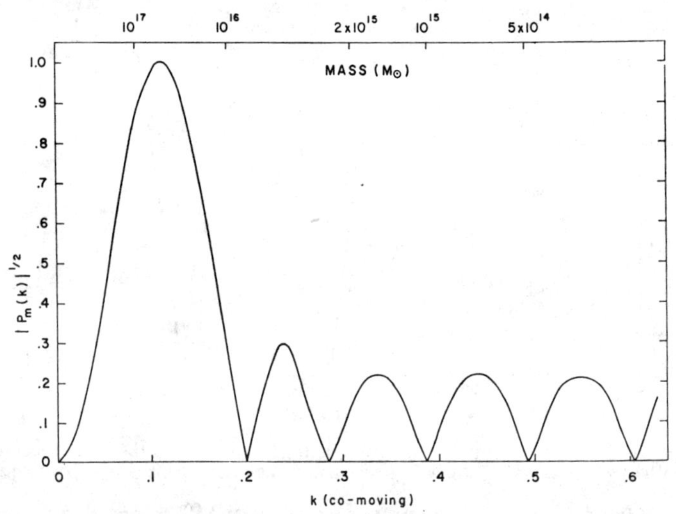

FIG. 5.—Same as Fig. 4 for the cosmologically flat general-relativity model, $\rho_0 = \rho_c$. The normalization is fixed to peak value unity.

In Figure 9 we plot the mean square fluctuation in the total brightness of the microwave background (eq. [65]; see also Table 1). In these curves the normalization has been fixed so that $\mathcal{P}_m(k)$ (Figs. 4–7) reaches the peak value unity at redshift $1 + Z_m = 10$. The time variation of the matter-density power spectrum is computed in the linear approximation, and our normalization means that at about redshift Z_m matter starts to fragment into separate and distinct bound systems with mass comparable to the mass function (eq. [39]) evaluated where $\mathcal{P}_m(k)$ is approaching unity. The observational limit shown on the figure is the upper limit estimated by Conklin and Bracewell after allowing for system noise. The results of the computation with this choice of Z_m are comparable to but smaller than this observational limit.

The above choice of Z_m may be too large. If Z_m were moved to a later epoch, it would reduce the required initial amplitude of the perturbation, hence reduce the mean square variation of the background. In Table 1 we list the factors by which the mean square variation of brightness must be multiplied when Z_m is reduced to smaller values (more recent epochs).

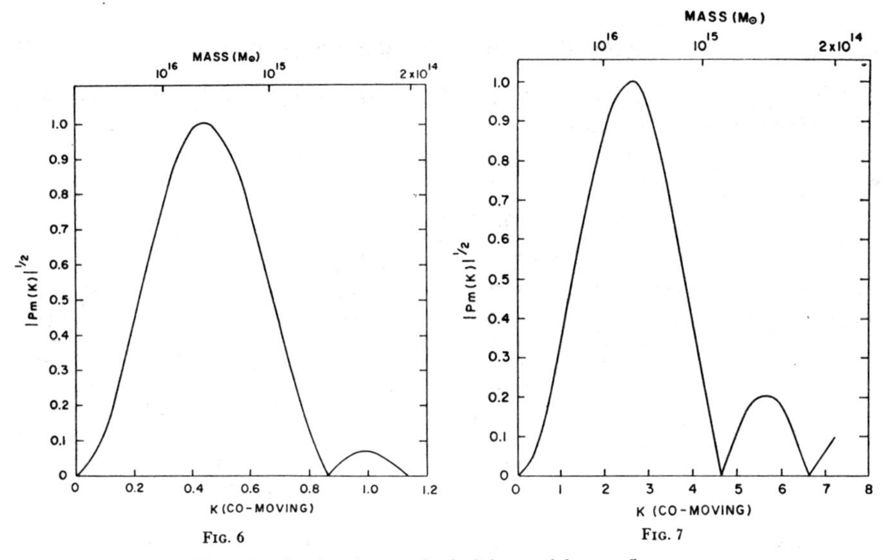

FIG. 6 FIG. 7

FIG. 6.—Same as Fig. 4 for the closed general-relativity model, $\rho_0 = 5\rho_c$.
FIG. 7.—Same as Fig. 4 for the flat scalar-tensor model ($\rho_0 = 1.24\rho_c$).

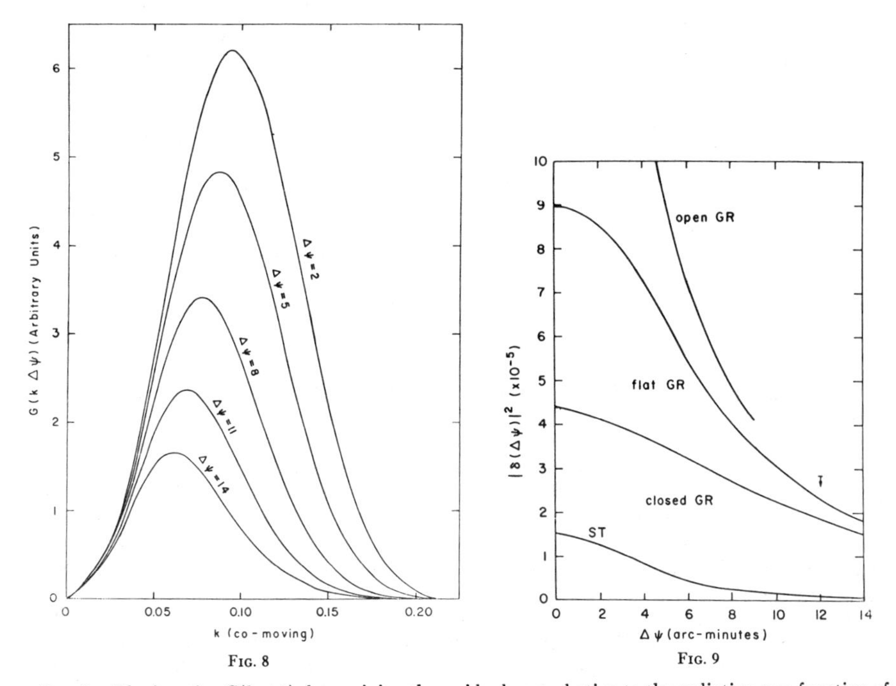

FIG. 8 FIG. 9

FIG. 8.—The function $G(k, \Delta\psi)$ determining the residual perturbation to the radiation as a function of angular resolution $\Delta\psi$ (eq. [65]). Unit of $\Delta\psi$ in the graph is minutes of arc. This curve applies to the cosmologically flat general-relativity model.

FIG. 9.—The mean square fluctuation of the observed bolometric brightness (eq. [65]) in the more dense cosmological models. The arrow is upper limit of Conklin and Bracewell (1967). The normalization is fixed so that $\mathcal{P}_m(k)$ reaches peak value unity at epoch $1 + Z_m = 10$.

We have shown no values for the expected fluctuations in the background in the open general-relativity model for angular resolution $\Delta\psi > 9'$. This is because the fluctuations are sensitive to very long wavelengths in this case, and our approximation of neglecting curvature no longer gives correct results.

VII. DISCUSSION

a) Comparison with Previous Results

It is of interest to compare the results of our numerical integration with the previous analytic estimates based on the assumption of a short photon mean free path for photons. Michie (1967) found that in a low-density model ($\rho_0 = 1 \times 10^{-30}$ g cm^{-3}), perturbations with characteristic mass $\lesssim 10^{11}\,\mathfrak{M}_\odot$ are strongly damped before recombination, while a moderate amount of growth is achieved for $\mathfrak{M} \gtrsim 10^{12}\,\mathfrak{M}_\odot$. Silk (1968) estimated that perturbations are damped up to a mass of about $5 \times 10^{11}\,\mathfrak{M}_\odot$ in a low-density model ($\rho_0 = 1 \times 10^{-30}$ g cm^{-3}) and about $7 \times 10^{10}\,\mathfrak{M}_\odot$ in the flat model. We can define a characteristic mass for damping at the point where the transfer function falls to one-third

TABLE 1

RESIDUAL PERTURBATION $10^5 \times \langle|\delta'_0|^2\rangle$* TO THE MICROWAVE BACKGROUND

$\Delta\psi$†	Open General-Relativity Model	Flat General-Relativity Model	Closed General-Relativity Model	Flat Scalar-Tensor Model
0	73.0	9.0	4.40	1.55
3	16.0	7.8	4.00	1.09
6	7.4	5.4	3.20	0.44
9	4.1	3.5	2.45	0.22
12	...	2.3	1.80	0.13
15	...	1.6	1.30	0.084
$1+Z_m=5$‡	0.61	0.25	0.23	0.20
$1+Z_m=2$	0.41	0.040	0.026	0.023

* Equation (65).
† Angular resolution in minutes of arc if $1+Z_m=10$, where the peak value of $\mathcal{P}_m$ at Z_m is unity.
‡ Correction factor to $\delta_0{}^2$ when Z_m is reduced to the indicated values.

its maximum value. This characteristic mass is $10^{14}\,\mathfrak{M}_\odot$ in the open model and $10^{12}\,\mathfrak{M}_\odot$ in the flat model (Fig. 3). Both are significantly larger than the corresponding analytic estimates. This is to be expected because the analytic approximation is inadequate at recombination.

Field and Shepley (1968) found that when the characteristic mass of a perturbation is greater than $9 \times 10^{15}\,\mathfrak{M}_\odot$ in a flat general-relativity model, the amplitude grows continuously. This critical mass corresponds approximately to the first peak in Figure 5. Our value for the mass at this peak is $\mathfrak{M} \sim 5 \times 10^{16}\,\mathfrak{M}_\odot$. These two characteristic masses are attributable to the same physical effect—the inability of pressure forces to stabilize the perturbation.

Residual perturbations to the microwave background have been computed by Longair and Sunyaev (1969). We find that the largest contribution to the residual perturbation comes from the first peak in the transfer function. For a mass $\mathfrak{M} \sim 5 \times 10^{16}\,\mathfrak{M}_\odot$ (corresponding to the first peak in the flat general-relativity model), Longair and Sunyaev find angular scale $\sim 20'$, and fractional perturbation $\delta T/T \sim 2 \times 10^{-3}$ to the microwave-background temperature. Our result (Fig. 9) yields characteristic angular scale (width at half-maximum) $\sim 7'$, and $\delta T/T \sim 1.7 \times 10^{-3}$ at this angular resolution, in agreement with Longair and Sunyaev.

b) Possible Significance

It is well to bear in mind that in this calculation the initial density fluctuations are invoked in an ad hoc manner because we do not have a believable theory of how they may have originated. Also, it is entirely possible that we have left out some relevant force, possibly that provided by a primeval magnetic field. Our calculation thus is at best exploratory; but we have remarked that one might consider the results of the exploration encouraging if, for example, the characteristic numbers one derives correspond to known phenomena.

In the more dense cosmological models we tentatively identify two characteristic masses associated with the evolution of an initially adiabatic perturbation. The larger characteristic mass is on the order of the mass within the Hubble radius ct at recombination. One can understand this as follows: When the wavelength is very large, pressure gradients are negligible and the perturbation grows at a rate independent of wavelength. Because we started with an approximation to ordinary white noise (eq. [73]), the density variance increases with increasing wavenumber. This is a property of white noise, not of the Universe. The function $\mathcal{P}_m(k)$ stops increasing with increasing k when the perturbation wavelength becomes comparable to ct at recombination because the radiation pressure can slow or reverse the curve $\delta_m(t)$, leading to the oscillating behavior shown in Figures 4–7. In the open cosmological model (Fig. 4) the first peak is not very prominent, and it is hard to see how one could attach any special meaning to it. On the other hand, in the more dense models the first peak is a prominent feature.

The first characteristic feature might be identified with the great rich clusters of galaxies. Indeed, in the closed general-relativity model and in the flat scalar-tensor model the mass at the peak amounts to $\sim 5 \times 10^{15}\,\mathfrak{M}_\odot$, moderately close to some estimates of the mass of the Coma cluster, $\sim 1 \times 10^{15}\,\mathfrak{M}_\odot$. Unhappily the mass at the peak and even the existence of the peak as a prominent feature depend on the cosmological model. For example, in the flat general-relativity model the mass has shifted to $\sim 5 \times 10^{16}\,\mathfrak{M}_\odot$. Thus until the cosmological parameters are better established we can only claim the possibility that the theory can produce a prominent feature with mass which can be comparable to the mass of a great cluster of galaxies.

This possible interpretation has some attractive features. The rich clusters may be a remarkably uniform class of objects, which may call for a special mode of formation (Abell 1962). On the other hand, the mass within the Hubble radius at recombination surely is an interesting parameter of the cosmological model, and one would like to hope that the value of this number has some special significance for the nature of the Universe. The interesting point is that this number may be close to the typical mass of a great cluster of galaxies (Alpher, Herman, and Gamow 1967; Field 1970*b*).

The second characteristic mass comes from the rather sharp onset of linear dissipation with decreasing wavelength (again in the more dense models). In the linear approximation the power $\mathcal{P}_m(k)$ is an oscillating function of k (depending on the number of waves up to the epoch of recombination). With the chosen initial values this curve is about constant from peak to peak at longer wavelengths (Fig. 5). If nonlinear effects do not disturb the phase of the perturbation (Peebles 1970), then each curve in Figure 3 represents the envelope of the rapidly oscillating function $\mathcal{P}_m(k)^{1/2}$ at shorter wavelength. It is apparent from Figure 3 that in the more dense models there is a sharp cutoff at large wavenumber (short wavelength) associated with linear dissipation. Again, the characteristic mass at the cutoff depends on the cosmological model, so again we can only claim the possibility of an interesting coincidence of numbers: *if* the cosmological model is sufficiently dense, the characteristic mass defined by linear dissipation agrees with the mass of a big galaxy, $\sim 10^{11}\,\mathfrak{M}_\odot$.

In view of the enormous range of known masses of galaxies one might question whether it is reasonable to attach a single characteristic mass to the phenomenon. Although one of us (P. J. E. P.) has in the past argued for the negative view, we are indebted to G. Abell for bringing the following suggestive point to our attention. In a rich cluster of galaxies the integrated galaxy luminosity function shows rather an abrupt change of slope at a characteristic absolute magnitude M^* (Abell 1962; Rood 1969). This means that the differential luminosity function has a local peak or plateau in the neighborhood of M^*. The brightest cluster member typically is a factor of 10 brighter than M^*. There are many more galaxies dimmer than M^* than there are galaxies brighter than M^*, but according to Abell's luminosity function the integrated luminosity of all the galaxies brighter than M^* is about equal to the total luminosity of all the dimmer members. If the mass-to-light ratio is about constant, the same remark applies to the integrated mass. The suggestion is then that when one counts on the basis of mass one may find that galaxies "typically" are large, with absolute magnitudes comparable to M^*.

The simple linear theory used here is not adequate to describe the fragmentation of the material into distinct bound systems. For this reason the mass estimates given here are only very rough approximations, and give no account of the expected dispersion of masses. Also, we can fix the time of fragmentation only in order of magnitude, as the epoch where the density contrast computed in the linear theory becomes comparable to unity.

If Figure 5 were taken literally, we would conclude that at about one single epoch, matter fragments into bound systems with masses ranging from $\sim 10^{11}\,\mathfrak{M}_\odot$ to $\sim 5 \times 10^{16}\,\mathfrak{M}_\odot$. That is, as the larger system starts to form out as a distinct cloud, it is itself fragmenting into smaller systems. This is not a capture hypothesis. Rather, it is supposed that galaxies and groups and clusters of galaxies form because there are initial irregularities on these scales. In this theory the order of formation is adjustable. It might be that protoclusters form first, and that galaxies fragment out during the initial collapse of the protocluster. With small adjustment of the initial power spectrum one can reverse this order without affecting the suggested interpretation of the two possibly characteristic masses.

For any choice of initial values it would be impossible to make small galaxies by the linear evolution of initially adiabatic perturbations (Peebles 1970). This is not necessarily a problem, however, for it is conceivable that a massive protogalaxy tends to fragment instead of collapsing to single system. There are also the initially isothermal perturbations. One interesting possibility is that the early Universe contains only adiabatic perturbations, but that the initial amplitude is large enough to cause nonlinear motions of matter and radiation prior to recombination. The result would be strong attenuation via shock waves. This attenuation must serve to smooth the radiation, but it could deposit the matter in an irregular fashion, producing an isothermal perturbation (Peebles 1970).

Finally, we discuss the expected irregularity of the microwave background on the assumption that this radiation has not been appreciably smoothed or perturbed by events after recombination. It will be recalled that this irregularity in the background is associated almost entirely with the first peak in the residual irregularity in matter density (Figs. 4–7). If this peak were identified with the great clusters of galaxies, we would have to fix the redshift at formation as $Z_m \gtrsim 1$. Then in the flat general-relativity cosmology the mean square fluctuation of antenna temperature at the beamwidth used by Conklin and Bracewell (1967) could be as small as 0.04 times the Conklin-Bracewell limit (Fig. 9 and Table 1). In the closed general-relativity model, which gives a somewhat more reasonable cluster mass, the mean square fluctuation could be 0.02 times this limit. In the flat scalar-tensor model the minimum mean square fluctuation would be reduced to 0.001 times this limit. If the angular resolution could be reduced to 1 minute of arc, the expected mean square fluctuation in antenna temperature would be increased by a factor of about 3 in the more dense general-relativity models and by a factor of about 10 in the scalar-tensor model.

If the observational accuracy could be improved by these factors, it would test the idea that the great clusters originated as initially adiabatic perturbations, although one must always bear in mind that the irregularities in the radiation background may have been smoothed or increased by subsequent processes (Dautcourt 1969; Longair and Sunyaev 1969).

In the open model the Conklin-Bracewell limit does seriously restrict the possible amplitude of adiabatic perturbations unless the radiation has been subsequently smoothed. On the other hand, in the computed spectrum of the density perturbations (Fig. 4) the first peak, which produces most of the irregularity, comes at a very large mass, $\sim 10^{18}\,\mathfrak{M}_{\odot}$. That is, it is already apparent that in this model our initial-value assumption (eq. [73]) is inadequate. This spectrum must be modified to reduce the power at very long wavelengths, for otherwise we would have produced bound systems on a mass scale much too big. If the first few peaks are thus de-emphasized, the residual perturbation to the microwave background is accordingly reduced. We conclude that in the open model we can find no ready interpretation of the observational limit on the irregularity in the microwave background because we cannot attach a possible observational interpretation to the first peak in the density-fluctuation curve (Fig. 5).

We have benefited from discussions with a number of people, including G. O. Abell, R. H. Dicke, E. Fackerell, G. B. Field, and K. S. Thorne. This work was done in part while one of us (P. J. E. P.) enjoyed the hospitality of W. A. Fowler at the California Institute of Technology. We would like to thank the referee for pointing out an error with reference to equation (64).

REFERENCES

Abell, G. O. 1962, *Problems of Extra-Galactic Research*, ed. G. C. McVittie (New York: Macmillan Co.), p. 213.
Alpher, R. A., Herman, R., and Gamow, G. 1967, *Proc. Nat. Acad. Sci.*, **58**, 2179.
Bardeen, J. 1968, *A.J.*, **73**, S164.
Brans, C., and Dicke, R. H. 1961, *Phys. Rev.*, **124**, 925.
Conklin, E. K., and Bracewell, R. N. 1967, *Nature*, **216**, 777.
Dautcourt, G. 1969, *M.N.R.A.S.*, **144**, 255.
Dicke, R. H. 1968, *Ap. J.*, **152**, 1.
Field, G. B. 1970*a*, in preparation.
———. 1970*b*, in *Star and Stellar Systems*, Vol. IX (Chicago: University of Chicago Press) (in press).
Field, G. B., and Shepley, L. C. 1968, *Ap. and Space Sci.*, **1**, 309.
Gamow, G. 1948, *Phys. Rev.*, **74**, 505.
Hawking, S. W. 1966, *Ap. J.*, **145**, 544.
Landau, L. D., and Lifshitz, E. M. 1962, *The Classical Theory of Fields* (New York: Pergamon Press).
Lifshitz, E. M. 1946, *J. Phys. USSR*, **10**, 116.
Lindquist, R. W. 1966, *Ann. Phys.* (N.Y.), **37**, 487.
Longair, M. S., and Sunyaev, R. A. 1969, *Nature*, **223**, 719.
Michie, R. W. 1967, Kitt Peak National Observatory preprint.
Nariai, H. 1969, *Progr. Theoret. Phys.*, **42**, 544.
Peebles, P. J. E. 1965*a*, *Ap. J.*, **142**, 1317.
———. 1965*b*, *Lectures in Applied Mathematics*, **8**, *Relativity Theory and Astrophysics*, p. 274.
———. 1967*a*, in *Proceedings of the Texas Conference on Relativistic Astrophysics* (in press).
———. 1967*b*, *Ap. J.*, **147**, 859.
———. 1968, *ibid.*, **153**, 1.
———. 1969, *ibid.*, **157**, 1075.
———. 1970, *Phys. Rev.*, *D*, **1**, 397.
Peebles, P. J. E., and Dicke, R. H. 1968, *Ap. J.*, **154**, 891.
Rood, H. J. 1969, *Ap. J.*, **158**, 657.
Sachs, R. K., and Wolfe, A. M. 1967, *Ap. J.*, **147**, 73.
Silk, J. 1968, *Ap. J.*, **151**, 459.
Wolfe, A. M. 1969, *Ap. J.*, **156**, 803.
Zel'dovich, Ya. B. 1967, *Soviet Phys.—Usp.*, **9**, 602.

Mon. Not. R. astr. Soc. (1978) 183, 341–358

Core condensation in heavy halos: a two-stage theory for galaxy formation and clustering

S. D. M. White and M. J. Rees *Institute of Astronomy, Madingley Road, Cambridge*

Received 1977 September 26

Summary. We suggest that most of the material in the Universe condensed at an early epoch into small 'dark' objects. Irrespective of their nature, these objects must subsequently have undergone hierarchical clustering, whose present scale we infer from the large-scale distribution of galaxies. As each stage of the hierarchy forms and collapses, relaxation effects wipe out its substructure, leading to a self-similar distribution of bound masses of the type discussed by Press & Schechter. The entire luminous content of galaxies, however, results from the cooling and fragmentation of residual gas within the transient potential wells provided by the dark matter. Every galaxy thus forms as a concentrated luminous core embedded in an extensive dark halo. The observed sizes of galaxies and their survival through later stages of the hierarchy seem inexplicable without invoking substantial dissipation; this dissipation allows the galaxies to become sufficiently concentrated to survive the disruption of their halos in groups and clusters of galaxies. We propose a specific model in which $\Omega \simeq 0.2$, the dark matter makes up 80 per cent of the total mass, and half the residual gas has been converted into luminous galaxies by the present time. This model is consistent with the inferred proportions of dark matter, luminous matter and gas in rich clusters, with the observed luminosity density of the Universe and with the observed radii of galaxies; further, it predicts the characteristic luminosities of bright galaxies and can give a luminosity function of the observed shape.

1 Introduction

A central issue in theories of galaxy formation is the relative importance of purely gravitational processes (*N*-body effects, clustering, etc.) and of gas-dynamical effects involving dissipation and radiative cooling. The large-scale distribution of galaxies is consistent with a smooth 'hierarchical clustering' picture lacking any preferred scale, and with the results of *N*-body simulations where no non-gravitational effects are included. On the other hand, dissipation must have played a role in the formation of disc galaxies (and perhaps of the luminous central parts of elliptical galaxies as well); uncondensed gas exists in clusters of galaxies and in individual galaxies; and the characteristic mass and size of galaxies finds no natural interpretation in a purely gravitational picture. We shall argue, in fact, that a purely dissipationless clustering process cannot be responsible for forming both individual galaxies and galaxy clusters.

We here develop a model which incorporates aspects of both these schemes: the distribution of the dominant mass component on all scales arises from purely gravitational clustering (Press & Schechter 1974); but the observed sizes and luminosity functions of galaxies are determined by gas-dynamical dissipative processes, for reasons which are a straightforward extension of arguments given by Binney (1977), Rees & Ostriker (1977) and Silk (1977).

A satisfactory theory of galaxy formation must account for the large amount of non-gaseous 'dark matter' which apparently provides $\gtrsim 80$ per cent of the virial mass in clusters like Coma and which may constitute massive halos around large galaxies. Indeed, the evidence is consistent with the view that all systems $\gtrsim 100$ kpc in size contain dark matter and luminous matter (i.e. visible stars and gas) in a universal ratio of (5–10):1. We shall not attempt here to assess the various arguments bearing on this question, but we do take the point of view that the discrepancy between the well-determined mass to light ratios found for rich galaxy clusters and for the main body of galaxies suggests that $\gtrsim 80$ per cent of the gravitating matter in the Universe may well be in some dark form that is still undetected except indirectly through its gravitational effects.

This segregation of high- and low-luminosity material seems incompatible with any theory which tries to build up galaxies and clusters from smaller units in an entirely dissipationless way, since one expects efficient mixing to occur during this process. Further, the luminous contents of galaxies appears too concentrated to be on a continuous hierarchy embracing clusters of galaxies (*cf.* Section 4). Finally the survival of galaxies as discrete subcondensations within clusters of short crossing time is inconsistent with dissipationless clustering, in which substructure is destroyed during the collapse of any bound unit. Numerical simulations of this process show neither stable clusters of clusters, nor structures with many subcondensations (*cf.* Section 2 and Appendix).

The need to invoke some dissipative process in galaxy formation is also suggested by the fact that the characteristic mass of a large galaxy is not the same as the 'turn-around' mass scale in a gravitational clustering model. Press & Schechter (1974) appreciated this problem, and postulated – *ad hoc* – that galaxies all formed at some well-defined past epoch by an unspecified process, the galactic luminosity function then being a scaled-down version of the cluster luminosity function. There is, of course, obvious evidence for gaseous dissipation in disc systems.

What does the dark mass consist of? Of the many possibilities, the most plausible candidates are low-mass stars, burnt-out remnants of high-mass stars, or the remnants of supermassive stars, any of which might have formed soon after the primordial plasma recombined (i.e. $z \simeq 1000$, $t \simeq 10^{13}\Omega^{-1/2}h^{-1}$ s the Hubble constant being taken as $100h$ km/(s Mpc)). Indeed, the conditions at this epoch (density $\sim 10^4\Omega h^2$ particles cm^{-3}, Jeans mass $\sim 10^6\Omega^{-1/2}h^{-1}M_\odot$) would seem much *more* propitious for gravitational instability and fragmentation than the present-day environments where star formation is usually assumed to occur. A plausible extrapolation to smaller mass scales of the fluctuation spectrum needed (in any theory) to explain galaxies and clusters suggests that the inhomogeneities on scales $\sim 10^6 M_\odot$ may already be of order unity, so that massive clouds would collapse and fragment immediately after decoupling from the background radiation field. Only in the special case of purely adiabatic fluctuations, which are attenuated by radiative viscosity on all scales $\lesssim 10^{13}M_\odot$, could this early fragmentation be suppressed. There are some processes of negative feedback which could limit the rate at which this fragmentation proceeded (e.g.

Compton drag following reheating); but these processes cannot readily have prevented most of the gas from condensing into bound objects by $z \simeq 100$. One cannot confidently say what mass range these objects would lie in (Rees 1977), but for the present discussion we need merely postulate that the dark mass behaves as an assemblage of point particles of individual mass $\lesssim 10^6 M_\odot$. In fact nothing in our discussion would change if the dark mass consisted of, for instance, massive neutrinos, or black holes which formed before recombination.

At redshifts $z \gtrsim 100$, the dark mass could not have been clustered on mass scales as large as galaxies. However, we shall take the point of view that it subsequently underwent hierarchical gravitational clustering (in the manner that can be simulated by N-body computations) and formed progressively larger systems; its present distribution can be inferred from the observed galaxy distribution and it constitutes the dark mass in clusters of galaxies and in galactic halos. We suggest, however, that the entire observed stellar content of galaxies condensed from residual gas whose distribution followed that of the dark material until cooling allowed it to settle within the gravitational potential wells of pre-existing 'halos'. A prerequisite for the occurrence of 'secondary' star formation is that the gas should have time to cool radiatively and fragment. This requirement sets a characteristic upper limit to galactic masses and sizes. When the clustering properties of the dark matter are matched to the observed galaxy covariance function, the inferred dimensions accord gratifyingly with the observations. We can, furthermore, make a preliminary attempt to derive a galactic luminosity function, on the assumption that low-mass galaxies condensed at early times when the hierarchical clustering of the dark matter did not extend to large masses and then retained their identity throughout the subsequent growth of the hierarchy. A consistent model can be obtained even on the restrictive assumption that all secondary star formation occurs with the same initial mass function (IMF). This contrasts with the situation in alternative theories: as noted above, dissipationless theories cannot explain why M/L is systematically lower in high-density regions; a purely gas-dynamical model would require an IMF dependency on density to account for this observation. The time-evolution of our model is illustrated in Fig. 1.

Within our picture, there is no process that can segregate the three components (dark material, luminous material and gas) on scales as large as a rich cluster. For consistency, we therefore require that the universal M/L be the same as that measured in clusters such as Coma ($M/L \simeq 400\,h$). This implies that $\Omega \simeq 0.2$. Assuming that the luminous material has $M/L \simeq 40\,h$, a typical value for the main body of elliptical galaxies, ~ 80 per cent of cosmic matter must have condensed at an early stage to form the dark mass; of the remainder, half is still uncondensed and the other half constitutes the luminous content of galaxies.

In Section 2 we first review the theory of 'self-similar' gravitational clustering in an expanding universe, which provides a model for the behaviour of the dark mass. The clustering builds up in a hierarchical fashion; the smaller scale virialized systems quickly merging into an amorphous whole when they are incorporated in a larger bound cluster. A detailed discussion of the disruption process is given in the Appendix. The fate of the residual gas within these transient potential wells is discussed in Section 3. Luminous galaxies up to a certain limiting size can form when this gas cools and becomes sufficiently concentrated in the centre of the potential well to be self-gravitating and liable to fragmentation. When a halo is disrupted in a larger system the luminous galaxy in its core can preserve its identity because dissipation has made it more concentrated than the surrounding dark material. Finally (Section 4) we present specific models for the evolution of the observed system of galaxies and clusters and for the galactic luminosity function, and we discuss the general implications of the proposed scenario.

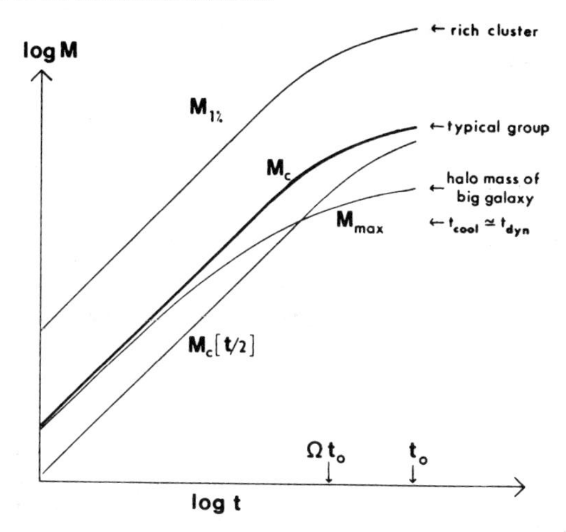

Figure 1. This diagram shows in a schematic way how 'dark' matter, assumed to have condensed into (stellar mass?) objects by, or soon after, the recombination time t_{rec}, becomes clustered on progressively larger scales. A universe with present age t_0 and $\Omega < 1$ is assumed. The heavy line shows the typical mass scale $M_c(t)$ for which $t = t_{dyn}$, where t_{dyn} is twice the turn-around time. If the inhomogeneities present at t_{rec} have amplitudes $\propto M^{-\alpha}$, then $M_c \propto t^{2/3\alpha} \propto (1+z)^{-1/\alpha}$ until $t \simeq \Omega t_0$, but the mass scale of clustering tends to grow more slowly thereafter (see (2.7)–(2.11) in text). Most bound units with $t_{dyn} \lesssim t/2$ will already have been incorporated into this scale of the hierarchy. There will be a spread in mass scales with a given overdensity (i.e. a given turn-around time); the line $M_{1\ \text{per cent}}(t)$, displaced *vertically* with respect to $M_c(t)$ by an amount that depends on α, indicates the mass such that 1 per cent of material at time t is bound in collapsed units $\geqslant M_{1\ \text{per cent}}$. The curves can be normalized so that $M_c(t_0)$ and $M_{1\ \text{per cent}}(t_0)$ fit the observed galaxy distribution. At any stage of the hierarchy, residual gas can condense in these halos to form luminous galaxies provided that it can cool before the halos are disrupted by merging into larger systems (see Section 3). The requirement $(t_{cool} + t_{dyn}) \leqslant t$ sets a characteristic upper limit $M_{max}(t)$ to the halo masses in which galaxies can have condensed by time t.

2 Dynamical aspects

2.1 GROWTH AND 'TURN-AROUND' OF DENSITY PERTURBATIONS: GRAVITATIONAL CLUSTERING

A viewpoint which is commonly adopted is that the present inhomogeneities in the Universe developed from initial density perturbations with a power-law spectrum such that, at some early time t_i (e.g. the recombination epoch t_{rec}, at a redshift $z \simeq 1000$) the characteristic amplitude on relevant mass scales M was of the form

$$\left\langle\left(\frac{\rho_i - \bar{\rho}_i}{\bar{\rho}_i}\right)^2\right\rangle^{1/2} = \left\langle\left(\frac{\delta\rho}{\rho}\right)^2\right\rangle_i^{1/2} \propto M^{-\alpha} \tag{2.1}$$

with $\alpha > 0$. If pressure effects are unimportant, then the density perturbation on each mass scale grows as $(\delta\rho/\rho) \propto (1+z)^{-1}$ until the amplitude becomes non-linear and the irregularity separates out into a non-expanding bound system (or until $(1+z) \lesssim \Omega^{-1}$, when linear growth terminates).

For a strictly spherical perturbation, there is a simple expression relating the turn-around time t_{turn} to the initial density perturbation (Sunyaev 1971; Gunn & Gott 1972). Provided that the initial redshift z_i satisfies $(1 + z_i) \gg \Omega^{-1}$, then the Hubble constant H_i at the initial epoch is related to H_0 by

$$H_i^2 = \Omega H_0^2 (1 + z_i)^3. \tag{2.2}$$

Defining ρ_{ci} to be $3H_i^2/8\pi G$, one finds

$$\frac{\rho_{ci} - \bar{\rho}_i}{\rho_{ci}} = \frac{1 - \Omega}{\Omega(1 + z_i)}. \tag{2.3}$$

Clearly, only perturbations with overdensities larger than this can ever become bound systems. For such perturbations, the turn-around time is related to the initial density ρ_i by

$$t_{\text{turn}} = \frac{\pi}{2H_i}\left(\frac{\rho_{ci}}{\rho_i - \rho_{ci}}\right)^{3/2}. \tag{2.4}$$

For any law of the form (2.1) there will be a mass scale m_i on which the initial fluctuations are of order unity; and so long as pressure effects are unimportant, the growth of density perturbations on scales $\gg m_i$ will proceed in an analogous fashion to the gravitational clustering of a set of point masses m_i which are initially expanding with the mean Hubble flow. The gravitational clustering of point masses was considered in an interesting paper by Press & Schechter (1974) and much of the present discussion is based on their work.

Laws of the form (2.1) relate the typical overdensity relative to the *mean* density to the mass scale; however, it is the overdensity relative to the *critical* density which determines the turn-around time and the final density. For $\Omega < 1$, this distinction is important for mass scales which turn around when $(1 + z) \lesssim \Omega^{-1}$. From (2.2)–(2.4) we find that, for scales just turning around now (i.e. $t_{\text{turn}} = t_0$),

$$\frac{\rho_i - \bar{\rho}}{\rho_i - \rho_{ci}} = 1 + (\Omega^{-1} - 1)\left(\frac{2H_0 t_0 \Omega^{1/2}}{\pi}\right)^{2/3}. \tag{2.5}$$

$H_0 t_0$ is a slowly-varying function of Ω (see, e.g. Gott *et al.* (1974)); for our chosen value $\Omega = 0.2$,

$$\left(\frac{\rho_i - \bar{\rho}}{\rho_i - \rho_{ci}}\right) = 2.55. \tag{2.6}$$

Assuming a self-similar clustering process and a law of the form (2.1) the physical characteristics of condensations depend on mass M according to the relations

$$t_{\text{turn}}/t_0 = (M/M_0)^{3\alpha/2}\, x^{-3/2} \tag{2.7}$$

$$\rho/\rho_0 = (M/M_0)^{-3\alpha}\, x^3 \tag{2.8}$$

$$r/r_0 = (M/M_0)^{1/3+\alpha}\, x^{-1} \tag{2.9}$$

$$\overline{v^2}/\overline{v_0^2} = T/T_0 = (M/M_0)^{2/3-\alpha}\, x \tag{2.10}$$

where, for a universe with $\Omega = 0.2$

$$x = [2.55 - 1.55(M/M_0)^\alpha]. \tag{2.11}$$

In these formulae, the suffix zero denotes the values appropriate to a mass M_0 for which $t_{\text{turn}} = t_0$. In (2.10) $\overline{v^2}$ is the virial velocity dispersion and T the corresponding temperature.

Statements such as (2.7)–(2.10) apply in an 'average' sense only. In fact there will be a spectrum of initial overdensities associated with regions of a given mass M; and therefore a spread in turn-around times for a given mass. A straightforward extension of the methods of Press & Schechter (1974) leads to the following expression for the mass spectrum of discrete bound systems at any given time (*cf.* Gott & Turner 1977)

$$N(M)\,dM \propto M^{\alpha-2} \exp\left[-(M/M_c)^{2\alpha}\right] dM \tag{2.12}$$

where M_c is a typical mass turning around at time t. At times corresponding to $(1+z) > \Omega^{-1}$, M_c increases as $t^{2/3\alpha}$, the basic point masses becoming incorporated in progressively larger units. In a low density ($\Omega < 1$) universe, the clustering process 'freezes' when $(1+z) \lesssim \Omega^{-1}$ and M_c levels off (compare equations (2.7) (2.11), and Fig. 1).

2.2 NORMALIZATION TO THE GALAXY COVARIANCE FUNCTION

If this process of gravitational clustering is an adequate model for what happened in the actual Universe, then the fiducial quantities in (2.7)–(2.10) can be determined from empirical estimates of the mass scale which is just turning around at the present time. The covariance function data provide the basis for such a determination, if we assume that the galaxies are distributed in the same way as gravitating matter in general. (This assumption is consistent with our general picture.) The empirical covariance function can be fitted by the power-law

$$\xi(r) = A(hr)^{-\gamma} \tag{2.13}$$

where the best (though somewhat uncertain) values of the parameters are $\gamma \simeq 1.77$, and $A = 4.5 \times 10^{44}$ cgs units (Peebles 1974).

The mean density within a sphere of radius r centred on a galaxy is given by

$$\langle\rho\rangle \simeq \frac{3}{3-\gamma}\,\bar{\rho}\xi(r) \quad (\xi(r) > 1). \tag{2.14}$$

Now the mean density within an object which is now at turn-around ($t_{\mathrm{turn}} = t_0 = 2 \times 10^{17} g^{-1/2} h^{-1}$ s) is $\rho_{\mathrm{turn}} = 5.5\,\rho_{\mathrm{co}}g$, where $\rho_{\mathrm{co}} = \Omega^{-1}\bar{\rho} = 3H_0^2/8\pi G$. The number g is 1 for $\Omega = 1$ and $\sim(1.5)^{-2}$ for $\Omega \ll 1$ (so that $\rho_{\mathrm{co}}g$ is the critical density in an Einstein–de Sitter universe with the same *age*, rather than the same H_0, as the actual Universe). The radius of a typical enhancement which is now at turn-around is therefore given by

$$A(hr_{\mathrm{turn}})^{-\gamma} = 5.5\left(\frac{3-\gamma}{3}\right)\Omega^{-1}g \tag{2.15}$$

i.e.

$$r_{\mathrm{turn}} = h^{-1}A^{1/\gamma}\left(5.5\,\frac{3-\gamma}{3}\,\Omega^{-1}g\right)^{-1/\gamma} \tag{2.16}$$

and the corresponding mass is

$$M_{\mathrm{turn}} = {}^4\!/\!_3\,\pi g \cdot 5.5\,\rho_{\mathrm{co}} r_{\mathrm{turn}}{}^3. \tag{2.17}$$

For $\Omega = 0.2$ and $h = 0.5$ numerical values are

$$M_{\mathrm{turn\,0}} = 5.5 \times 10^{13} M_\odot \tag{2.18}$$

$$r_{\mathrm{turn\,0}} = 4.4\ \mathrm{Mpc}. \tag{2.19}$$

If a sphere of mass $M_{\text{turn }0}$ virialized at a radius $\frac{1}{2}r_{\text{turn }0}$, its velocity dispersion would be

$$\langle v_0^2 \rangle^{1/2} = 385 \text{ km/s}. \tag{2.20}$$

These correspond to the dimensions of a typical sparse group of galaxies and we adopt them as fiducial values in what follows. Given these values, the spread in masses implied by (2.12) is sufficient to encompass rich clusters (*cf.* Section 4). The uncertainties in the procedure leading to (2.18)–(2.20) stem not only from imprecision in the data (2.13) but – more importantly – from the oversimplified model we have adopted for the actual non-linear clustering process. For the latter reason, there are also some grounds for doubting the precise applicability of the scaling relations (2.7)–(2.10).

2.3 NON-DISSIPATIVE HIERARCHICAL CLUSTERING

In so far as the clustering process can be modelled by a hierarchy with discrete steps (each separated by a factor 2 in t) we expect the following picture. Those units which turn around at a given time, t, will have a characteristic mass $M(t)$ (equation (2.7)) and a density of order $5.5\,\rho_c(t)$. These units will each be composed of three or four subunits which have just collapsed and so have turn-around times of order $t/2$ and densities $\sim 44\,\rho_c(t)$. At time $2t$, these units will in turn have collapsed and relaxed and will themselves be grouped in threes and fours, these new groups having binding energies per unit mass larger by a factor $2^{4/9\alpha - 2/3}$ than those of their predecessors. However, during (or soon after) their collapse the units will have destroyed their internal substructure; *a three or four member group which is more tightly bound than its constituent members will be transformed by relaxation effects into an amorphous system in* $\lesssim 1$ *crossing time.* This important result can be shown by rough analytic estimates (see Appendix) and is also found in N-body simulations (White 1976b; Aarseth, Gott & Turner 1978, in preparation).

We thus expect that a 'snapshot' of the Universe taken at time t would show the dark material to be clustered on a characteristic mass scale, which has been in existence for $\lesssim 1$ crossing time. A few subunits may be distinguishable within each subcluster, but *any* finer structure within the subunits would have been erased by collisions, mergers and tidal effects. A second snapshop taken at time $2t$ would show the same picture on the next level of the hierarchy: the masses $M(t)$ would now have collapsed and virialized, their substructure being erased in the process. Notice that in this picture one never expects to see clusters with more than a few members, nor to see clusters with distinct substructure and short crossing times. Because rich clusters do exist, and because many observed groups and clusters have $t_{\text{crossing}} \ll t_0$, a purely dissipationless picture cannot describe the formation of both galaxies and clusters (*cf.* White 1977).

At the present epoch, collapsed clusters have a characteristic scale $\sim 1.5 \times 10^{13} h^{-1} M_\odot$, with a tail towards higher masses (*cf.* Section 4). We wish to argue that the bulk of cosmic matter may indeed have behaved as described above, being distributed smoothly on scales $< 10^{13} h^{-1} M_\odot$. We suggest that this material constitutes the so-called 'missing mass' in clusters and the extensive halos of isolated galaxies; we further suggest that *all* the luminous matter seen in galaxies formed from residual gas that settled within the potential wells provided by the dark material at each stage of the clustering process and then collapsed to form stars. We must therefore next consider how the gravitational field of the clusters dark matter would influence any remaining gas. The main new features in the problem are dissipation and cooling.

3 The fate of gas

3.1 COOLING

Ionized gas can be supported by its own pressure in a potential well characterized by a mass M and radius r if its temperature T is given by

$$3\,kT = \frac{GMm_p}{r} \tag{3.1}$$

(we assume that the gas mass is $< M$, so we can ignore its self-gravitation). Gas within the potential well will be heated to $\sim T$ either by one strong shock or by a succession of weak ones during the violent relaxation that accompanies formation of the halo.

For the masses and radii relevant to galaxies, the temperatures (3.1) are well above 10^4 K, so any shocked or pressure-supported gas will indeed be ionized. It will then (even if it is pure H and He) cool radiatively: it cannot remain in equilibrium for more than a cooling timescale unless it can draw on a further supply of energy.

The cooling rate due to bremsstrahlung, recombination and collisionally-excited line emission, can be written as $\Lambda(T)\, n_e n_H$ erg/s cm^6; and the cooling time is

$$t_{cool} \simeq \frac{3\,kTm_p}{\rho_{gas}\Lambda\gamma(T)}. \tag{3.2}$$

Of obvious interest is the ratio of t_{cool} to the Hubble time. Also of interest is its relation to the dynamical or formation timescale t_{dyn}, which is equal to twice the turn-around time t_{turn} for the corresponding potential well. If $t_{cool} \lesssim t_{dyn}$, any gas within the potential well must cool and collapse to the centre, probably fragmenting into stars. If $t_{cool} > t_{dyn}$, the gas will be pressure-supported and will contract quasi-statically towards the centre, until eventually it becomes self-gravitating and able to fragment. Clearly this quasi-static contraction will not yet have proceeded far unless $t_{cool} \lesssim H_0^{-1}$. If the potential wells are due to agglomerations of dark material with typical mass $M(t)$ (*cf.* (2.7)) then they may themselves collide and merge into a mass on the next stage of the hierarchy after a further time $\sim t_{dyn}$. During these mergers the gas will be shock-heated and transformed into a single hotter and more rarified cloud, for which t_{cool}/t_{dyn} is even larger than it was before (and thus more unfavourable for condensation and fragmentation). Thus if $t_{cool}/t_{dyn} > 1$, condensation will be possible only in that small fraction of cases when a mass survives for an unusually long time before being incorporated into a larger system. A precise criterion would depend on the profile of the potential well and the detailed kinematics and dynamics, but we can make the approximate statement that gas can accumulate, cool (and possibly fragment) within clusters of dark mass provided that

$$t_{cool} \lesssim H^{-1}, \tag{3.3a}$$

and

$$t_{cool} \lesssim t_{dyn} \tag{3.3b}$$

but this can happen only in a small fraction of cases when (3.3b) is violated.

3.2 FRAGMENTATION INTO LUMINOUS GALAXIES

If the luminous content of galaxies (as opposed to their halos) forms by the collapse of gas within pre-existing potential wells, conditions (3.3) set an upper limit to the possible luminous mass. The argument is similar to that of Rees & Ostriker (1977) and Silk (1977), except that these authors considered *self*-gravitating gas clouds.

Condition (3.3) is *necessary* for fragmentation, but additional requirements must be satisfied if the gas is actually to be able to fragment into stars. Let us suppose that F is the fraction of cosmic material which is gaseous at any stage, and that a gas mass FM settles within the potential well due to a mass M of radius r. Even if the gas can cool, it will not fragment until it has accumulated in a region small enough for its local density (and hence self-gravity) to dominate that of the background halo. If it remained in a spherical cloud and the halo material had uniform density, this would require contraction to a radius such that

$$r_{gas} \lesssim F^{1/3} r. \tag{3.4}$$

This criterion is obviously a very rough one. If cooling instabilities allow the gas to become inhomogeneous, or if it collapses to a disc, then it may become liable to fragmentation even if r_{gas} exceeds the limit given by (3.4). On the other hand, if the halo material were centrally condensed, r_{gas} would need to be smaller to achieve a sufficient density enhancement (e.g. we would require $r_{gas} \lesssim Fr$ if the halo had an 'isothermal' r^{-2} density gradient with very small core radius).

A condition such as (3.4) in any case merely determines when fragmentation can *start*. As Larson and others have emphasized, the timescale for star formation is likely to be related to the free-fall timescale by some factor of order unity, but the amount of further gaseous dissipation and central concentration that develops *after* (3.4) is satisfied is exceedingly sensitive to this factor. We interpret (3.4) as defining the radius at which the gas-dynamical collapse calculations of Larson (1974a) first become applicable. The further contraction is likely to be greater when $t_{cool} \simeq t_{dyn}$ than when cooling and fragmentation can happen almost instantaneously. Once star formation has been triggered, there is an extra energy input (from young stars, supernovae, etc) which may be able to eject most of the gas before more than some fraction f has turned into stars (Larson 1974b).

3.3 GALAXY FORMATION DURING THE HIERARCHICAL CLUSTERING

We assume that, provided (3.3) holds, the potential wells which exist at any stage of the hierarchy will accumulate a core of luminous stars of total mass fFM. We further assume (*cf.* Larson 1974b) that f scales with the binding energy, so that

$$f \propto M/r. \tag{3.5}$$

The IMF of the luminous stars will be assumed to be the same for all values of M and all stages of the hierarchy. (It could have the form which – with suitable assumptions about the *rate* of gas–star conversion – accords with the observed colours and stellar populations of spirals and ellipticals.)

In this picture, low-mass systems of luminous stars will form early, before the characteristic mass M_{turn} of the hierarchical clustering has attained a high value. The low-mass halos that existed at these early times will by now have lost their identity and merged into larger amorphous systems, *but the luminous material that condensed in their centres may nevertheless have survived to the present day in identifiable stellar systems.* Provided that these luminous cores have become sufficiently concentrated during their cooling and fragmentation phases, they will be able to survive the violent relaxation which destroys their halos. The timescales for their subsequent dynamical evolution are then much larger than t_{dyn} (see Appendix).

When the halos of the first small galaxies are disrupted to form bigger units, the residual gas may again be able to cool and collapse to form a larger central galaxy. The model thus naturally predicts the existence of small satellites around big galaxies. If there is sufficient time before incorporation of the halo in a yet larger unit the central galaxy may swallow some of its larger satellites in the manner envisaged by Tremaine (1976), thus increasing its own luminosity relative to that of its satellite system.

The luminous cores thus survive as fossils of the earlier stages of the hierarchy and the luminosity function is *not* of the 'synchronic' form (2.12), but rather must be estimated by assuming that the number of galaxies which formed via condensation into halos of mass between M and $2M$ varies as M^{-1} (since a constant fraction $\sim \frac{1}{2}$ of all the mass in the Universe would at some stage have been incorporated in bound systems in any given mass range). A specific model for the luminosity function is constructed in Section 4.

The formation of the dark material may have resulted in the injection of pregalactic metals into the residual gas. Further, much of the gas is effectively recycled at each stage of the hierarchy, and so may be enriched. The X-ray detection of Fe in some rich clusters is thus no embarrassment to our scheme. Moreover the progressive enhancement provides a further reason why the more massive – and hence more recently formed – galaxies have higher metal abundance (*cf.* Larson 1974a, b). Note that differences in the star formation *rate* will affect the present stellar population and mass to light ratio of galaxies; a more efficient early conversion of gas to stars leads to a higher present M/L for a given IMF. Although we suggest that smaller galaxies condensed first, less efficient star formation could account for the persistence of gas (and for the presence of young stars) in some such systems.

4 A specific model

4.1 CHOICE OF PARAMETERS F AND f

In this section, we show that, if the amplitude of the clustering is normalized to agree with the covariance function, as discussed in Section 2 (equations (2.18)–(2.20)) then the processes described in Section 3 can lead to a system of galaxies whose luminosity function and characteristic parameters are consistent with observation.

The velocity v_0 (equation (2.20)) corresponds to a temperature

$$T_0 = 3.5 \times 10^6\,\mathrm{K}. \tag{4.1}$$

The temperature of typical bound condensations formed at earlier times will vary according to the scaling law (2.10). The corresponding scaling law for the cooling time is

$$\left(\frac{t_{\mathrm{cool}}}{t_{\mathrm{cool}\,0}}\right) = \frac{\Lambda(T_0)}{\Lambda(T)}\left(\frac{T}{T_0}\right)\left(\frac{\rho_0}{\rho}\right) \tag{4.2}$$

where the density and temperature are scaled according to (2.8) and (2.10) respectively. In what follows we consider two possible chemical compositions for the gas which formed the last generation of galaxies: a metal-free mixture of H and He, in the ratio 10:1 by number; and a mixture enriched to 10 per cent of the 'cosmic' metal abundance. Cooling curves for these two cases, assuming collisional ionization, are taken from Cox & Tucker (1969) and Raymond, Cox & Smith (1976).

The parameters F and f describing the amount of residual gas and its rate of conversion into stars can be determined from the observed properties of rich clusters of galaxies, since – on our hypothesis – the mass ratios of dark matter, luminous stars and gas in such systems should have their universal values. The fraction F_i of cosmic matter which remained gaseous after formation of the dark material is

$$F_i = \frac{\text{mass of gas + luminous mass in galaxies}}{\text{total mass}}. \tag{4.3}$$

The fractional depletion of this gas during subsequent star formation sets the value of f_{max} in the scaling law

$$f = f_{max}\left(\frac{M}{M_{max}}\right)^{2/3-\alpha} \tag{4.4}$$

which is derived from (3.5) under the assumption that all galaxies form sufficiently early for the x dependence in the scaling laws to be unimportant (we verify this later). M_{max} is then the maximum halo mass in which a luminous core can form and f_{max} is the corresponding efficiency for conversion of the contained gas into stars. Approximating the clustering by a discrete hierarchical process of N stages, we then have

$$\prod_{\substack{N \text{ stages of} \\ \text{hierarchy}}} (1 - \tfrac{1}{2}f_n) = \prod_{n=1,N}\left(1 - \frac{f_{max}}{2}(2^{2/3-4/9\alpha})^{n-1}\right)$$

$$= \frac{\text{gaseous mass}}{\text{gaseous mass + luminous mass}}. \tag{4.5}$$

The factor ½ in front of f_n corresponds to the assumption that half of the total mass participates in each stage of the hierarchy. (At early times, half the matter will be in overdense regions of mass M, and half in underdense regions.)

On the basis of observations, we take $F_i = 0.2$, and a value of 0.5 for the RHS of (4.5). In what follows, we also assume $h = 0.5$; and treat α, for the moment, as a free parameter.

4.2 THE MAXIMUM MASS AND LUMINOSITY FOR GALAXIES

In Fig. 2 we plot the dependence on α of various critical halo masses. The minimum halo mass, M_{min}, in which a luminous core can form is set when $\alpha \gtrsim 1/3$ by the condition that the formation time be later than recombination, and when $\alpha \lesssim 1/3$ by the condition that the virial temperature be $\gtrsim 10^4$ K. These masses define the lower limit of the hierarchy. They are quite strongly α dependent, but are $\lesssim 10^9 M_\odot$ for α in the range 0.05–0.6. (Under our assumptions, this limiting mass corresponds to the very low limiting luminosity $\sim 5 \times 10^6 f_i L_\odot$.) The maximum halo mass, M_{max}, in which a luminous core can form is set by the condition $t_{cool} = t_0$, and is of order $10^{13} M_\odot$ for all interesting values of α. The corresponding cores will have luminosities $L_{max} \sim 5 \times 10^{10} f_{max} L_\odot$. In fact, however, there will be a depletion of luminous cores when $t_{cool} > t_{dyn}$. We see from Fig. 2 that this condition gives typical masses (and luminosities) ~ 3 times lower than the condition $t_{cool} < t_0$.

We see from Fig. 2 that the inclusion of cooling due to metals does not affect these critical masses substantially. At $z \gtrsim 20$, Compton cooling on the microwave background would dominate radiative cooling. This has not been allowed for in the calculations leading to Fig. 2, but in fact even radiative cooling is always efficient enough to guarantee $t_{cool} < t_{dyn}$ throughout the domain where Compton cooling is important, so no conclusions are altered by its omission.

The line $t_{dyn} = t_0$ in Fig. 2 denotes the typical mass of collapsed clusters. This mass is

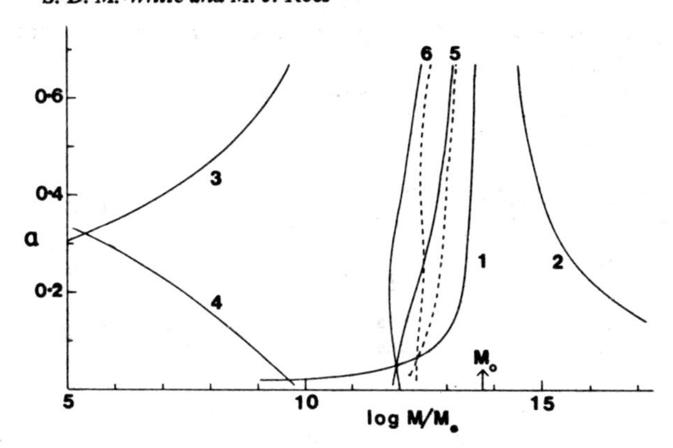

Figure 2. Various characteristic masses are here plotted, as a function of α (equation (2.1)), for the case when $\Omega = 0.2$, $h = 0.5$ and F_f (the fraction of cosmic matter still in gaseous form) is 0.1. The amplitude of (2.1) is normalized by choosing the scale mass M_0 to have the value (2.19). The curves are as follows: (1) The typical mass for which $t_{\rm dyn} = t_0$ (i.e. $t_{\rm turn} = t_0/2$). (2) The mass $M_{1\,\rm per\,cent}$ such that 1 per cent of the material is in units with $t_{\rm dyn} < t_0$. (3) The typical mass for which $t_{\rm dyn} = t_{\rm rec}$. (4) The typical mass for which $T_{\rm virial} = 10^4$ K. This sets the minimum halo mass within which gas can condense if H and He are the dominant cooling agents. (5) The typical mass for which $t_{\rm cool} = t_0$. The luminous cores of galaxies will have a characteristic maximum mass $0.1f$ times this value (see Table 1). (6) The typical mass for which $t_{\rm cool} = t_{\rm dyn}$. The continuous curves 5 and 6 correspond to cooling by H and He alone, and the dashed curves to the inclusion of heavy elements with 10 per cent of their 'cosmic' abundance.

related to M_c (equation (2.12)). If, as we assume in what follows, our normalization procedure picks out the mass scale such that half of all galaxies are at present in turning condensations of mass $> M_0$, one can show that (2.12) may be written more precisely as

$$N(M)\,dM \propto M^{\alpha-2} \exp\left[-0.23(M/M_c)^{2\alpha}\right] dM \tag{4.6}$$

where M_c is now the mass given by the line $t_{\rm dyn} = t_0$ in Fig. 2. In this case, 1 per cent of galaxies are in collapsed condensations of mass larger than that given by the line $t_{\rm dyn\,1\,per\,cent} = t_0$. (These masses are in fact related to M_0 by

$$M_c = 0.81^{1/\alpha} M_0 \quad \text{and} \quad M_{1\,\rm per\,cent} = 3.10^{1/\alpha} M_0;$$

the corresponding cluster luminosities are simply $L_c = {}^1/_{200} M_c$ and $L_{1\,\rm per\,cent} = {}^1/_{200} M_{1\,\rm per\,cent}$.) According to Gott & Turner (1977), the requirement that the distribution of cluster mass be broad enough to account for rich clusters as well as small groups leads to a value of α close to ⅓. Press & Schechter (1974) argued from numerical simulations that the effective α could not be larger than ½. Fig. 2 shows quite clearly that values of α much less than ⅓ are ruled out by the observed lack of collapsed systems significantly more massive than $10^{15} M_\odot$.

In Table 1 we present our calculated values for various characteristic parameters of the hierarchy for four different values of α; where two values are given for any quantity the upper value corresponds to the assumption of a metal-free gas mixture and the lower to the assumption of an enriched mixture. The number of stages in the equivalent discrete hierarchy, N, was calculated from the formula

$$N = \mathrm{Int}\left[\frac{3\alpha}{2}\frac{\log\,(M_{\rm max}/M_{\rm min})}{\log 2} + 1\right]. \tag{4.7}$$

Table 1.

α	N	$\log(M_{max}/M_\odot)$	$\log(L_{max}/L_\odot)$	$\log(r_{max}/1\,\text{kpc})$	$\log(v_{max}/1\,\text{km/s})$	$(t_{dyn\,max}/t_0)$	$\log(M_c/M_\odot)$	$\log(L_c/L_\odot)$
2/3	12	13.13	9.87	2.44	2.73	0.18	13.60	11.30
		13.21	9.95	2.54	2.72	0.23		
1/2	23	12.96	10.02	2.40	2.66	0.19	13.56	11.26
		13.07	10.13	2.52	2.66	0.25		
1/3	12	12.67	10.05	2.35	2.54	0.22	13.48	11.18
	13	12.92	10.30	2.55	2.57	0.33		
1/6	4	12.22	9.85	2.35	2.32	0.37	13.18	10.88
	5	12.71	10.33	2.64	2.42	0.59		

This is the number which we used in (4.5) to derive a value for f_{max}. We see from Table 1 that L_{max} is very insensitive to α or to the metal content of the gas; further, it is very close to the corresponding scale in the observed galaxy luminosity function. This agreement is one of the strong points of our scheme. The radius r_{max} in Table 1 is that of a homogeneous sphere with the same mass and energy as a typical halo of mass M_{max}. The large values found for r_{max} substantiate our earlier claim that the luminous parts of galaxies are too concentrated for galaxies and clusters to fit on a continuous hierarchy. They are, however, consistent with our contention that condition (3.4) specifies the initial radius for a gaseous collapse model of the type proposed by Larson (1974a). As expected the corresponding (three-dimensional) halo velocity dispersions, v_{max}, are similar to those observed in the luminous part of large elliptical galaxies, and we note that the formation timescales $t_{dyn\,max}$ corresponding to M_{max} are sufficiently small that our neglect of the x dependence in the scaling law for f (equation (4.4)) is indeed justified.

4.3 A MODEL FOR THE LUMINOSITY FUNCTION

Under our assumptions, the mass to light ratio is the same for all clusters, so the cluster mass function (4.6) can be converted directly to a *cluster* luminosity function

$$N(L)\,dL \propto L^{\alpha-2} \exp\left[-0.23(L/L_c)^{2\alpha}\right] dL. \tag{4.8}$$

To calculate the *galactic* luminosity function we need to consider the joint mass-density distribution of halos which turn around at any stage of the clustering process. This distribution is

$$N(m,\rho)\,dm\,d\rho \propto m^{-2}\,dm \exp\left(-0.23m^{2\alpha}\rho^{2/3}\right) m^{\alpha}\,d(\rho^{1/3}) \tag{4.9}$$

where m is the mass in units of M_{max} and ρ is the density in units of the corresponding typical scale density ρ_{max}.

The luminosity of a galaxy, L, is proportional to $fM \propto M^2/r \propto M^{5/3}\rho^{1/3}$, where we neglect the slow variation of F (from 0.2 to 0.1 over the whole range of the hierarchy). This gives

$$m = l^{3/5}\rho^{-1/5} \tag{4.10}$$

where $l = L/L_{max}$. Thus the joint luminosity-density distribution is

$$N(l,\rho)\,dl\,d\rho \propto l^{(3\alpha-8)/5}\rho^{-(3\alpha+7)/15} \exp\left(-0.23l^{6\alpha/5}\rho^{(10-6\alpha)/15}\right) dl\,d\rho. \tag{4.11}$$

A luminosity function can be calculated by integrating this function over those densities at each luminosity for which $t_{cool} < t_0$. We make the approximation $t_{cool} \propto T/\rho$, which is reasonable in the temperature range of interest, since $\Lambda(T)$ is fairly flat there. This gives

$$t_{cool}/t_0 = l^{2/5}\rho^{-4/5}, \tag{4.12}$$

the condition $t_{cool} < t_0$ then giving $\rho > l^2$. Thus

$$\phi(l)\,dl \propto l^{(3\alpha-8)/5}\,dl \int_{l^2}^{\infty} d\rho\rho^{-(3\alpha+7)/15} \exp\left(-0.23 l^{6\alpha/5}\rho^{(10-6\alpha)/15}\right). \tag{4.13}$$

Making the substitution $Y = \rho^{(8-3\alpha)/15} l^{(24\alpha-9\alpha^2)/(25-15\alpha)}$ in the integral reduces this to

$$\phi(l)\,dl \propto l^{-(8-3\alpha)/(5-3\alpha)}\,dl \int_{l^{(80-6\alpha-9\alpha^2)/(75-45\alpha)}}^{\infty} dY \exp\left(-0.23\, Y^{(10-6\alpha)/(8-3\alpha)}\right). \tag{4.14}$$

An adequate approximation to this expression over the whole range of l is obtained by setting the integral equal to the value of the integrand at the lower limit, yielding the final result

$$\phi(L)\,dL \propto L^{-(8-3\alpha)/(5-3\alpha)} \exp\left[-0.23(L/L_{max})^{(20+6\alpha)/15}\right]. \tag{4.15}$$

Because of the disruption of halos for which $t_{dyn} < t_{cool} < t_0$ before the contained gas has time to cool, the actual luminosity function will become somewhat steeper than (4.15) for a small range of luminosities below L_{max} (*cf.* Fig. 2). For $\alpha = 1/3$, (4.15) yields a power-law of slope -1.75 at the faint end. This is slightly steeper than Abell's (1975) value of -1.625, and substantially steeper than Schechter's (1976) value of -1.25. We note that our predicted slope would be flattened by any process which made low-mass galaxies relatively more vulnerable to disruption. (E.g. star formation may occur more rapidly when $t_{cool} \ll t_{dyn}$, making low-mass galaxies less centrally condensed than high-mass galaxies relative to the halos in which they form.)

With these provisos, the form and scale of our luminosity function compare quite well, at least at the bright end, with those of the fitting functions empirically derived by Schechter (1976). The agreement is as good as could be expected, given the schematic nature of the theory; note that some effects which may influence the upper end of the luminosity function have been neglected (Ostriker & Tremaine 1975; White 1976a). We stress that for a given α, the predicted *cluster* luminosity function (equation (4.8)) does not have the same shape or scale as $\phi(L)$.

5 Conclusions

There is much evidence that ~ 80 per cent of the matter in the Universe is not in gas or luminous stars, but is now in some dark form. On the (almost mandatory) assumption that this dark material condensed before or soon after recombination and clustered gravitationally on progressively larger scales, we have argued as follows:

(i) The dark material must now be in amorphous units whose mass spectrum spans the range from massive galactic halos to rich clusters of galaxies.

(ii) The luminous inner part of galaxies *cannot* have formed by purely dissipationless clustering. Rather, it most probably condensed from residual gas lying in the transient potential wells provided by the dark matter. By the present time, half this residual gas has been incorporated into luminous galaxies, the rest (perhaps enriched with heavy elements) remains uncondensed in intergalactic space. An upper limit to galactic luminosities is set by the requirement that the gas should have time to settle in a potential well, cool and fragment into luminous stars. This limit agrees adequately with the masses, luminosities and radii of large galaxies. The existence of giant galaxies surrounded by satellites, embedded in a common dark halo, is a natural consequence of our model. The model also explains why, in clusters such as Coma, the masses of the largest galaxies are so much less than that of the system as a whole.

(iii) On somewhat specific accumptions, a luminosity function can be derived which agrees reasonably well with observation.

Acknowledgments

We acknowledge helpful discussions with M. Fall, J. Gunn, B. Jones, J. Ostriker and J. Silk.

References

Abell, G. O., 1975. In *Galaxies and the Universe,* ed. Sandage, A. *et al.*, Chicago University Press.
Alladin, S. M., Potdar, A. & Sastry, K. S., 1974. In *Dynamics of stellar systems,* ed. Hayli, A., D. Reidel, Dordrecht, Holland.
Binney, J. J., 1977. *Astrophys. J.*, **215**, 483.
Cox, D. O. & Tucker, W. H., 1969. *Astrophys. J.*, **157**, 1157.
Gott, J. R., Gunn, J. E., Schramm, D. N. & Tinsley, B. M., 1974. *Astrophys. J.*, **194**, 543.
Gott, J. R. & Turner, E. L., 1977. *Astrophys. J.*, **216**, 357.
Gunn, J. E. & Gott, J. R., 1972. *Astrophys. J.*, **176**, 1.
Larson, R. B., 1974a. *Mon. Not. R. astr. Soc.*, **166**, 585.
Larson, R. B., 1974b. *Mon. Not. R. astr. Soc.*, **169**, 229.
Ostriker, J. P. & Tremaine, S. D., 1975. *Astrophys. J.*, **202**, L113.
Peebles, P. J. E., 1974. *Astrophys. J.*, **189**, L51.
Press, W. H. & Schechter, P., 1974. *Astrophys. J.*, **187**, 425.
Raymond, J. C., Cox, D. P. & Smith, B. W., 1976. *Astrophys. J.*, **204**, 290.
Rees, M. J., 1977. In *Evolution of galaxies and stellar populations,* p. 339, eds Larson, R. B. & Tinsley, B. M., Yale University Observatory Publications.
Rees, M. J. & Ostriker, J. P., 1977. *Mon. Not. R. astr. Soc.*, **179**, 451.
Schechter, P., 1976. *Astrophys. J.*, **203**, 297.
Silk, J. I., 1977. *Astrophys. J.*, **211**, 638.
Spitzer, L., 1958. *Astrophys. J.*, **127**, 17.
Spitzer, L. & Chevalier, R. A., 1973. *Astrophys. J.*, **183**, 565.
Sunyaev, R. A., 1971. *Astr. Astrophys.*, **12**, 190.
Tremaine, S. D., 1976. *Astrophys. J.*, **203**, 72.
Toomre, A., 1977. In *Evolution of galaxies and stellar populations,* p. 401, eds Larson, R. B. & Tinsley, B. M., Yale University Observatory Publications.
van Albada, T. S. & von Gorkom, J. H., 1977. *Astr. Astrophys.*, **54**, 121.
White, S. D. M., 1976a. *Mon. Not. R. astr. Soc.*, **174**, 19.
White, S. D. M., 1976b. *Mon. Not. R. astr. Soc.*, **177**, 717.
White, S. D. M., 1977. *Comm. Astrophys.*, **7**, 95.
White, S. D. M. & Sharp, N. A., 1977. *Nature*, **269**, 395.

Appendix: the disruption of substructure

When a bound unit first collapses in any dissipationless clustering process, it will be extremely inhomogeneous, being composed of a spectrum of smaller lumps which collapsed on shorter timescales than the system as a whole. A graphic example of this is given in White

(1976b). As the system collapses the subunits interact violently and lose their identity. This ironing out of substructure combines elements of at least three different dynamical processes. Encounters between sublumps give rise to strong tidal forces which increase the internal energy of individual lumps at the expense of their orbital motion through the system; this effect leads to the tidal evaporation and disruption of the lumps in the manner discussed by Spitzer (1958) for star clusters. The transfer of energy from orbital motion to internal motions during an encounter between two lumps can, however, result in their becoming bound to each other and merging into a single more diffuse object. This stickiness has been investigated in the context of galaxy–galaxy encounters by Alladin, Potdar & Sastry (1974), Toomre (1977) and van Albada & von Gorkom (1977). The third process which contributes to the destruction of substructure is dynamical friction; heavy subunits can rapidly give up their kinetic energy of motion through the cluster both to lighter subunits and to individual particles and as a result they settle to the centre of the system where they can disrupt and merge more easily. Elements of all these processes are discernible in the rapid destruction of subclustering in N-body simulations (*cf.* White 1976b and Aarseth, Gott & Turner 1978, in preparation). *In these simulations the substructure of any bound unit is rubbed out almost as soon as it collapses.* The rough analytic arguments given in this Appendix show that this important result should still be valid when the numbers of distinct particles making up the units and sub-units are far higher than can be simulated by N-body methods.

Spitzer (1958) shows that the change in internal energy of a lump of mass m_1 in an impulsive encounter at impact parameter D, pericentric distance p, velocity difference at infinity V_∞ and pericentric velocity difference V_p with another lump of mass m_2 is approximately

$$\Delta U_1 \simeq 4G^2 m_1 m_2^2 r_1^2/3p^4 V_p^2 \gtrsim 4G^2 m_1 m_2^2 r_1^2/3D^4 V_\infty^2 \tag{A1}$$

where, following Spitzer & Chevalier (1973), we take r_1 to be the half-mass radius of the lump. Assuming the system as a whole to have mass M, half-mass radius R and velocity dispersion V, we take $V_\infty = \sqrt{2}V$ in (A1) and integrate over all possible impacts to get:

$$\frac{dU_1}{dt} = 4.2\,\frac{G^2 m_1 r_1^2}{V}\int_0^\infty dm_2 n(m_2)\, m_2^2[r_1^2 + r_2^2]^{-1} \tag{A2}$$

where $n(m_2)\,dm_2$ is the number density of lumps in the range $(m_2, m_2 + dm_2)$ and where we have used a minor variation of the prescription of Spitzer & Chevalier (1973) to deal with the lower limit of the integration over impact parameters. Adopting $3M/8\pi R^3$ as a typical density for the system as a whole we find

$$\frac{dU_1}{dt} = 0.50\,\frac{G^2 m_1 r_1^2 QM}{VR^3}\left\langle \frac{m}{r^2 + r_1^2}\right\rangle \tag{A3}$$

where Q is the fraction of the mass of the system in lumps, and angular brackets denote a mass-weighted average over the lumps. If we assume $U_1 = Gm_1^2/4r_1$ (a good approximation for all likely density profiles for the lumps) and a formation time for the whole system $T_{\mathrm{dyn}} = 2\pi R/V$, we find that the disruption time is given by

$$t_{\mathrm{dis}\,1} = \frac{U_1}{dU_1/dt} = \frac{1}{25}\,\frac{1}{Q}\,\frac{m_1}{\langle m[1 + (r/r_1)^2]^{-1}\rangle}\,\frac{R}{r_1}\,T_{\mathrm{dyn}}. \tag{A4}$$

This clearly suggests that any object in which much of the mass is in fairly diffuse sublumps will destroy its substructure during, or shortly after, its collapse. In our hierarchy we expect $r/R \sim (m/4M)^{1/3}$ giving $t_{\rm dis} \simeq 0.13\, Q^{-1}(m/M)^{-1/3}\, T_{\rm dyn}$, and so all but the central regions of any sublumps will be disrupted when a unit collapses.

The approximations leading to (A4) neglect the orbital energy loss of the lumps. In fact two lumps are quite likely to capture each other during an encounter and to merge rapidly thereafter. We now estimate the timescale on which such mergers take place under the simplifying assumption that all lumps are similar. Defining v to be the internal velocity dispersion of a lump ($v^2 = Gm/2r$) we further assume a velocity-dependent capture cross-section

$$\sigma(V_\infty) = \begin{cases} 0 & V_\infty > v \\ 4\pi r^2[1 + 4(v^2/V_\infty^2)] & V_\infty < v \end{cases}. \qquad \text{(A5)}$$

The normalization and cut-off of this cross-section are suggested by N-body experiments (White 1978, in preparation); its velocity dependence merely accounts for gravitational focusing. Comparison of (A5) with the results of Alladin *et al.* (1974) suggest that it is a conservative criterion and underestimates the capture efficiency at low velocities. Assuming that the lumps have typical number density $3MQ/8\pi m R^3$ and a Gaussian distribution of orbital velocities with dispersion V, the expected time for a lump to capture another lump is given by

$$\frac{1}{t_{\rm cap}} = \frac{3MQ}{8\pi m R^3} \int_0^{v/V} dx \left(\frac{27}{4\pi}\right)^{1/2} x^2 \exp(-3x^2/4)\, 4\pi r^2 \left(1 + 4\frac{v^2}{V^2}x^{-2}\right) Vx. \qquad \text{(A6)}$$

This leads to

$$t_{\rm cap} = 0.032 \frac{1}{Q} \frac{mr^{-2}}{MR^{-2}} \left(\frac{v}{V}\right)^{-4} T_{\rm dyn}, \qquad \text{(A7)}$$

where we have assumed $v \lesssim V$ in approximating the integral over the velocity distribution. To apply this to substructure in our hierarchy, we take $r/R \simeq (m/4M)^{1/3}$, as before, and find $t_{\rm cap} = 0.032 Q^{-1}(m/M)^{-1}\, T_{\rm dyn}$, suggesting that merging of subunits will occur rapidly as any system collapses.

Any lump which escapes the initial violent relaxation of a system unscathed will subsequently experience dynamical friction as it moves through the cluster and loses its orbital energy to lighter objects. For a typical cluster model, the time for this friction to bring a lump into the cluster centre is (*cf.* White 1977)

$$t_{\rm fric} \simeq \frac{1}{8 \ln(R/r)} \frac{M}{m} T_{\rm dyn}. \qquad \text{(A8)}$$

Clearly all but the smallest lumps will quickly spiral to the centre where they will be incorporated in the general density profile of the system.

It is clear from equations (A4), (A7) and (A8) that the bound collapsed systems which form in a dissipationless clustering hierarchy cannot long retain any substructure. Thus in our model the halo of dark material around any luminous galaxy core must disrupt and merge with other halos as soon as it becomes part of a larger unit. Provided that the violent processes accompanying this initial merging do not bring the luminous cores into orbits too close to the centre of the final object, the dynamical timescales (equations (A4), (A7) and (A8)) are too long for the cores to undergo significant further evolution before the new system is incorporated in a yet larger object. When this happens the group of galaxy cores and its common halo are broken up together. Since the luminous galaxy cores form highly condensed subsystems within their halos, we expect that after the halos have merged the cores will be more concentrated to the cluster centre than the dark matter, but will still be well separated. In our model this concentration can explain why binary galaxies and small groups of galaxies appear to have lower mass to light ratios than rich clusters. We stress, however, that our understanding of the clustering and gas condensation processes is not good enough for us to be able to specify the exact conditions under which luminous cores can survive the disruption of their halos without merging at the centre of the resulting object. It is clearly necessary for the pregalactic gas to contract until it is self-gravitating and it seems probable that its final radius needs to be significantly less than the core radius of its halo. In any other situation it is difficult to escape the arguments of White & Sharp (1977) against the existence of isothermal halos in binary systems, and of White (1977) and Sections 1 and 2 against dissipationless galaxy formation in general.

THE ASTROPHYSICAL JOURNAL, 243:L127–L131, 1981 February 1

GALAXY FORMATION IN AN INTERGALACTIC MEDIUM DOMINATED BY EXPLOSIONS

JEREMIAH P. OSTRIKER AND LENNOX L. COWIE
Princeton University Observatory
Received 1980 July 21; accepted 1980 October 16

ABSTRACT

The explosive energy released at the death of massive stars in forming stellar systems will propagate into the intergalactic medium. There, under certain circumstances, a dense cooled shell will form with mass many times greater than the original "seed" system, whereupon gravitational instability and fragmentation of the shell can lead to the formation of new stellar systems. For $z \lesssim 5$ and for masses $10^8 \leq M/M_\odot \leq 10^{12}$, a very large amplification of original perturbations is possible, with galaxies naturally forming in small groups having velocity dispersions of order 200 km s^{-1}. Explosions before $z \approx 5$, which cool primarily because of the inverse Compton interaction with background radiation, will lead to production of very massive stars. The most interesting speculative predictions (not unique to this model) are that intergalactic space will be dusty, with significant absorption likely for objects seen from $z > 3$, that unvirialized groups should lie on two-dimensional surfaces, that black holes with masses 10^2–10^4 $M_\odot$ may be common, and that very large (100 Mpc radius) cavities may have been produced by early explosions.

Subject headings: cosmology — galaxies: formation — galaxies: intergalactic medium

I. INTRODUCTION

There exist characteristic values for the physical scales of galaxies, although no such scales are seen in larger astrophysical systems (see Peebles 1980). The evidence indicates that galaxies have typical masses of 10^{11}–10^{12} $M_\odot$, radii of order 10^4 pc, and form in small groups with velocity dispersion ~100 km s^{-1} at the epoch of $z < 5$. Gravity itself contains no preferred scales. It has been argued (see Carr and Rees 1979) that the scale sizes for luminous stars, planets, and other cosmic bodies can be derived from the basic physical constants. Star formation is studied by examining the ensemble of nonlinearly interacting physical processes occurring in the interstellar medium and calculating the rate of formation of gravitationally bound objects, assuming that all memory of initial spectrum of perturbations in the original collapsing galaxy has been lost.

The initiation of bound galactic systems condensing out of the intergalactic medium has been treated differently by considering an arbitrary spectrum of initial fluctuations, tracking the growth, decay, and interactions of various modes, and asking when the typical perturbation of galactic scale will have achieved a sufficient overdensity to collapse (see Gott 1977). In an alternative version (see Doroskevich, Sunyaev, and Zel'dovich 1970), larger-scale perturbations (10^{15} $M_\odot$) are followed which, at an epoch dependent on the initial amplitudes, form shocks and fragment to galactic scales.

Here we present a view of galaxy formation closer to the way star formation is viewed. We hope to show that there is a natural hydrodynamic instability by which galaxy formation begets more galaxy formation and that the resulting properties of galaxies are relatively independent of the number and size of the "seeds" which initiate the process. The basic idea, "explosive amplification," is as follows:

1. If a bound stellar system of mass M forms at an epoch $z < 100$, the massive stars will have a lifetime short compared to the current Hubble time.
2. The explosive energy E released at the death of these stars, measured in terms of an efficiency, $\epsilon \equiv E/Mc^2$ (see Bookbinder *et al.* 1980) is 10^{-5} to 10^{-4}.
3. If this energy propagates out into the intergalactic medium as an adiabatic blast wave which eventually cools when the shock velocity is v_c, then a mass will be swept up in the Oort "snowplow" phase of order $M \times \epsilon(c/v_c)^2$ or an amplification η of order 10^2–10^3.
4. If this happens in a time scale short compared to the current Hubble time, and if the dense, cooling shell is unstable to fragmentation and formation of new bound systems, there will be rapid amplification of the amount of bound material per unit volume.

All of the above conditions are satisfied at certain epochs. We enumerate the important physical processes, note the cosmological "windows" within which given mass scales will amplify significantly, describe the physical environment and fossil remains left by each epoch, and suggest observational checks of the overall picture.

A. Consider first $100 \gtrsim z \gtrsim 5$, the *Compton cooling epoch* when heated gas can be cooled to the blackbody radiation temperature by background radiation, but when the Compton drag force (Peebles 1971) is ineffective. Because T is low and ρ high in the cooling shells, the Jeans mass is very low and fragmentation will lead, not to normal galaxies, but to massive star formation. The remains of this epoch are condensed remnants and some chemical pollution.

B. At $z \approx 5$, the *radiative cooling epoch*, amplification of small-scale density fluctuations to galaxy dimensions proceeds rapidly. Masses greater than $\sim 10^{12}\ M_\odot$ cannot cool, and those less than $\sim 10^9\ M_\odot$ are not Jeans unstable. Objects of normal galactic scale are made. The expanding blast waves overlap when a fraction of order $1/\eta$ of the gas has been formed into galaxies.

C. For $5 \gtrsim z \gtrsim 4$, the IGM will resemble the present interstellar medium, with warm ($T = 10^4$ K) clouds embedded in a much hotter medium. Galaxy formation increases the pressure in the IGM, causing the most-massive clouds to collapse and the least-massive to evaporate. Most of the warm gas is processed into galaxies, but a comparable mass is left in a hot IGM.

D. For $4 \gtrsim z$, adiabatic expansion lowers the pressure more rapidly than explosions increase it. Galaxy formation decreases rapidly, and there results a spectrum of pressure-bounded low-mass clouds with properties close to those postulated by Sargent *et al.* (1980).

II. PHYSICAL PROCESSES

For definiteness, we adopt a present closure parameter $\Omega_0 \lesssim 0.2$ [utilizing an open Friedmann model, with density $\rho = 1.8 \times 10^{-29}\Omega_0 h^2(1+z)^3$ g cm^{-3} for $(1+z) < \Omega_0^{-1}$ with a closed model in prior epochs] and radiation temperature $T_{\rm BB} = 2.7(1+z)$ K, where h is the Hubble constant in units of 100 km s^{-1} Mpc^{-1}, and at any epoch, n_0 is the hydrogen particle density with helium equal to $0.1n_0$ by number. Let the energy input from a representative explosion be $10^{61}E_{61}$ ergs; then the shock radius in the adiabatic phase is $R = 3.3E_{61}^{1/5}[h^2\Omega_0(1+z)^3]^{-1/5}t_{10}^{2/5}$ Mpc, after $t_{10} \times 10^{10}$ yr. For 10^5 K $< T < 10^6$ K, the cooling time is for eras A and B, respectively,

$$\tau_{\rm cool\ 10} = \frac{240}{(1+z)^4}, \frac{0.03T_5^{2.2}}{h^2\Omega_0(1+z)^3}, \tag{1a, b}$$

owing to inverse Compton or gas-radiative processes (e.g., Cox and Tucker 1969). The conditions at the time of dense shell formation are determined by equating the cooling time to the age:

$$R_{\rm cool} = 30(1+z)^{-2.2}E_{61}^{+0.2}\Omega_0^{-0.2}h^{-0.4}\,,\ 3E_{61}^{+0.3}\phi^{-0.4}\ {\rm Mpc}\,; \tag{2a, b}$$

$$t_{\rm cool,10} = 240(1+z)^{-4}\,,\ 0.6E_{61}^{0.2}\phi^{-0.5}\ {\rm yr}\,; \tag{3a, b}$$

$$v_{\rm cool} = 5(1+z)^{+1.8}\Omega_0^{-0.2}h^{-0.4}E_{61}^{+0.2}\,,\ 200E_{61}^{0.05}\phi^{+0.11}\ {\rm km\ s^{-1}}\,; \tag{4a, b}$$

$$M_{\rm cool,11} = 3 \times 10^5(1+z)^{-3.6}E_{61}^{0.6}\Omega_0^{0.4}h^{0.8}\,,\ 200E_{61}^{0.88}\phi^{-0.21}\ M_\odot\,, \tag{5a, b}$$

where we define $\phi \equiv h^2\Omega_0(1+z)^3$. The postshock temperature at cooling is $14v_{\rm cool}^2$(km s^{-1}) K. The dimensionless amplification factor, $\eta \equiv (M_{\rm cool}/M_{\rm seed})$, and the Mach number, $\mathfrak{M}_{\rm cool}$ defined to be the velocity of the shock divided by the isothermal sound speed in a fully cooled shell, are

$$\eta = 4 \times 10^5(1+z)^{-3.6}\epsilon_{-4}^{0.6}M^{-0.4}{}_{\rm seed,11}\Omega_0^{+0.4}h^{0.8}\,,\ 300\epsilon_{-4}^{0.9}M^{-0.1}{}_{\rm seed,11}\phi^{-0.2}\,, \tag{6a, b}$$

$$\mathfrak{M}_{\rm cool} = 20(1+z)^{1.3}E_{61}^{+0.2}\Omega_0^{-0.2}(T_c/T_{\rm BB})^{-1/2}h^{-0.4}\,,\ 20E_{61}^{0.05}\phi^{0.1}T_{c4}^{-1/2}\,, \tag{7a, b}$$

where the temperature of the cool shell is $T_{c4} \times 10^4$ K in the radiative era.

The amplification factor is large and, in the radiative era, insensitive to epoch and seed mass. However, arbitrarily large explosions cannot cool in a Hubble time. The maximum energy explosion that can cool in era B will produce a radiative shell with mass

$$M_{\rm cool,max} = 2 \times 10^{14}h^{-0.4}\Omega_0^{1.7}(1+z)^{1.3}\ M_\odot\,. \tag{8}$$

This is larger than the mass of most galaxies and comparable to the mass of groups within which galaxies are formed. The cooling time for explosions producing shells smaller than $M_{\rm cool,max}$ is correspondingly shorter than the Hubble time [as $(M/M_{\rm cool,max})^{1/4}$] and the amplification large, so that in one Hubble time all seeds regardless of initial size will grow to the maximum possible size, finally producing shells of mass $\sim 10^{14}\ M_\odot$. For $M < M_{\rm cool,max}$ neglect of the Hubble expansion is justified.

In the inverse Compton cooled era, if $(1+z) > 9h^{0.4}\Omega_0^{0.2} \approx 6$, $t_{\rm cool}$ is shorter than the Hubble time. The earliest epoch at which amplification at a given mass can occur is specified by $\eta > 1$ or $M_{\rm max} = 1.2 \times 10^{25}(1+z)^{-9}\epsilon_{-4}^{+1.5}\Omega_0 h^2$, which for $z = 40$ is $\sim 10^{10}\ M_\odot$. The number of Compton cooling times available within a Hubble time increases so rapidly with z that, if a disturbance forms at $z > 10$ and remains isolated, it will reach a characteristic shell mass of $M_{\rm fin} \approx 10^{16.5}\epsilon_{-4}^{1.5}\Omega_0^{-0.8}h^{-1.6}\ M_\odot$ for any initial seed mass by $z \approx 6$. This corresponds to a present cavity size of $R_{\rm max} \approx 100$ Mpc.

Now consider the gravitational instability of a dense cool shell of radius R and velocity v_s. We assume, since seeds must still be forming at this epoch, that $\Omega(z_f) \approx 1$ and that $z_f \approx 5$. We base our arguments on rough energy considerations and conclude that there is a band of unstable wavelengths corresponding to a minimum and maximum galaxy mass. Define integrals through the dense shell $H \equiv \int ds$, $\Sigma_u \equiv \int P ds$, and $\Sigma_m \equiv \int \rho ds$. The total energy per unit mass of a pancake cut out of the shell, with radius $a(H < a < R)$ and mass $M(a)$, is $\epsilon = \frac{1}{4}(a/R)^2 v_s^2 - k\pi a G\Sigma_m + \frac{3}{2}(\Sigma_u/\Sigma_m)$, if the gravitational energy is $W \equiv -kGm(a)^2/a$. We rewrite this, noting that after cooling $v^2 = R^2/16t^2$, that $G\Sigma_m = R/18\pi T_E^2$ (T_E defined as the Einstein–de Sitter age of the universe) and $(\Sigma_u/\Sigma_m) = R^2/(16t^2\mathfrak{M}^2)$, so that

$$\epsilon = \frac{1}{64}\frac{a^2}{t^2} - \frac{kRa}{18T_E^2} + \frac{3}{32}\frac{R^2}{t^2\mathfrak{M}^2}, \tag{9}$$

which permits bound fragments ($\epsilon < 0$), if $\mathfrak{M} > 1.62(T_E/t)^2$, taking $k = 0.849$ for a flat disk. We define $(t/T_E)_{\rm crit} \equiv 1.274\mathfrak{M}^{-1/2}$, $x \equiv (t/T_E)/(t/T_E)_{\rm crit}$ and $\nu \equiv (-\epsilon t^2/a^2)^{1/2}$, a dimensionless growth rate. Equation (6) can be transformed to a relation between growth rate ν and pancake radius a for a blast wave of dimensionless age x and shell Mach number $\mathfrak{M}$:

$$\nu^2 = -\frac{1}{64} + \frac{R}{a}\frac{(3/2)^{1/2}}{16}\frac{x^2}{\mathfrak{M}} - \frac{3}{32}\left(\frac{R}{a}\right)^2\frac{1}{\mathfrak{M}^2}. \tag{10}$$

Bound systems exist if $x > 1$ with roots for the smallest and largest systems, $a_\pm$ and most rapidly growing mode a_m:

$$\frac{a_m}{R} = \frac{\sqrt{6}}{\mathfrak{M}x^2}, \quad \frac{a_\pm}{R} = \frac{a_m}{R}\frac{1}{1 \mp (1 - x^{-4})^{1/2}}. \tag{11}$$

For $x \gg 1$, these two modes are approximately $a_-/R = (a_m/2R)$, $a_+/R = 3.0(t/T_E)^2$. At cooling $x_c = 0.785(t/T_E)_c\mathfrak{M}_c^{1/2}$. In the inverse Compton era, explosions will tend to fragment on cooling, since $x_c \gtrsim 1$ for $20 \gtrsim (1+z)_c > 5$. For radiative cooling, the condition $x_c > 1$ will be satisfied if $(E/E_m) \gtrsim 0.01(1+z)^{-0.8}$, where E_m is the maximum mass that can cool in a Hubble time. The mass of the most rapidly growing fragment is $(M_m/M) = (a_m/R)^2/4 = 3/2(\mathfrak{M}^2x^4)^{-1}$ which, for $T_c/T_{\rm BB} = T_{c4} = 1$, is

$$M_m = 100(1+z)^{+1.2}h^{-1.6}E_{61}^{-0.2}\Omega_0^{-0.8}\,,\ 1 \times 10^{10}E_{61}^{-0.30}h^{-1.2}\Omega_0^{1.4}(1+z)^{0.2}\,M_\odot\,. \tag{12}$$

This gives fragments the size of dwarf ellipticals or Magellanic irregulars ($\sim 10^9\,M_\odot$) in the radiatively cooled era, but objects of order 10^3–$10^4\,M_\odot$ in the inverse Compton era.

Amplification will always proceed rapidly until $(t/T_E) \approx 1$, when the largest and slowest cooling explosions occur. Equations (2) and (12) show that in the radiative era this will always lead to bound groups with size of order 1 Mpc, velocity dispersion of order 200 km s^{-1}, and with the largest bound objects of order $10^{12}\,M_\odot$ made of fragments (some remaining as satellites) of order $10^9\,M_\odot$. Explosions occurring in the inverse Compton era produce very massive stars which may burn some fuel and then implode to black holes, contaminating their surroundings with debris characteristic of "little big bangs" (see Wheeler 1977; Wallace and Woosley 1980) and forming very large cavities.

III. OVERLAP AND THE GENERATION OF A HOT INTERGALACTIC MEDIUM

If there is a sufficient number of seeds, overlap between the hot bubbles generated in the final noncooling explosions will occur, resulting in the formation of a hot intergalactic medium with cool material remaining as embedded clouds, as in current models for the interstellar medium (Cox and Smith 1974; McKee and Ostriker 1977). At overlap, hot regions will be uniformly distributed between the ages at which the last and next outbursts occur, t_1 and t_2. Thus, on average the ratio of the mass in the hot bubble to that in the generating explosion is

$$M_{\rm hot}/M_{\rm gal} = \int_{t_1}^{t_2}[r(t)]^3 dt/R_1^3 t_2 = 5R_2^3/11R_1^3 = 5\eta/11\,,$$

where R_1 is the radius of the previous radiative shell, and R_2 the radius of the next. The fractional galaxy formation is then $11/5\eta \sim 0.2\%$. At overlap, most regions are within a factor of 2 of being radiative, and hence the temperature of the IGM then will be a few times 10^5 K. The medium is thermally unstable, with a time scale slightly greater than the current age of the universe. Thus, we may expect a spectrum of cold gas regions to form in the hot intergalactic gas and of order half the IGM to become dense clouds in the tenuous coronal medium. Geometric considerations suggest that the cold clouds will have a mass spectrum stretching to the mass of the overlapping bubbles dominated by the high-mass end of the spectrum. As the pressure in the IGM rises, these become gravitationally unstable, collapsing and releasing yet more energy into the IGM. This is a fast process on the cosmological time scale. The mass spectrum will be similar whether this or the prior amplification process is the dominant galaxy-producing machine; and the upper bound on the cloud masses should be less than, but of order of, the mass initially in the hot cavities. We adopt a value of $\Omega_{\rm gal} = 0.05F$, consistent with these arguments, giving the density of matter in galaxies at the epoch of formation $\rho_{\rm gal}(z_f) = 9 \times 10^{-31}(1+z_f)^3Fh^2$ g cm^{-3}. The pressure and temperature generated are

$$P_{\rm IGM}(z_f) = \tfrac{2}{3}\epsilon\rho_{\rm gal}(z_f)c^2 = 5 \times 10^{-14}(1+z_f)^3\epsilon_{-4}Fh^2 \text{ ergs cm}^{-3}\,, \tag{13}$$

$$T(z_f) = 1.3 \times 10^8\epsilon_{-4}F(0.15/\Omega_{\rm gas}) \tag{14}$$

with subsequent pressure and temperature scaling as $[(1+z)/(1+z_f)]^5$ and $[(1+z)/(1+z_f)]^2$. The spectrum of the clouds is then determined by the rise in pressure and temperature in the intergalactic medium. In order to survive, clouds must not gravitationally collapse and must not be destroyed by the hot intergalactic medium (Sargent *et al.* 1980). The constraints are most stringent at the epoch of formation at which the pressure and temperature of the IGM are highest when the cloud-mass spectrum is frozen. An upper bound to the cloud mass is given by the condition of stability to gravitational collapse (Mestel 1965): $M_{\rm crit} = 1.3C^4/(G^{3/2}P^{1/2})$, where C is the isothermal sound speed in the cloud and P is the external pressure. Thus,

$$M_{\rm cloud} \leq M_{\rm crit}(z) = 2 \times 10^8(1+z)^{-5/2}(1+z_f)^1\epsilon_{-4}^{-1/2}F^{-1/2}h^{-1}\ M_\odot\ , \tag{15}$$

or about $10^7\ M_\odot$ for $z_f = 5$. If clouds are too small, they may be evaporated thermally. We estimate an evaporation time (Cowie and McKee 1977):

$$t_{\rm evap} = 2 \times 10^9(\Omega/0.15)^{3/2}M_7^{1/3}(F\epsilon_{-4})^{1.8}(1+z_f)^{-1}\ , \tag{16}$$

where M_7 is the mass of the cloud in units of $10^7\ M_\odot$, limiting $M_{\rm cloud} \gtrsim 10^7\ M_\odot$.

The properties of the clouds are ($T_{\rm cloud} = 10^4$ K):

$$n_{\rm cloud} = 1.5 \times 10^{-2}F\epsilon_{-4}h^2(1+z_f)^{-2}(1+z)^5\ {\rm cm}^{-3}\ ;$$

$$R_{\rm cloud} = 2000M_7^{1/3}(1+z_f)^{+2/3}(1+z)^{-5/3}h^{-2/3}(F\epsilon_{-4})^{-1/3}\ {\rm pc}\ ;$$

$$N_{\rm cloud} = 10^{20}M_7^{1/3}(1+z_f)^{-4/3}(1+z)^{10/3}h^{4/3}(F\epsilon_{-4})^{5/3}\ {\rm cm}^{-2}\ , \tag{17}$$

close to those deduced by Sargent *et al.* (1980) for intergalactic clouds from statistics of Lα absorption lines in quasars, though the temperature of the intergalactic gas is a factor 30 higher.

IV. DISCUSSION

a) The Intergalactic Gas

Grain opacity.—In this model of galaxy formation, a considerable amount of metal-rich material must be injected into the IGM. Dust survives the expulsion at $z_f \lesssim 5$, and it may blanket observations at larger redshifts and render some of the following proposed observations impossible.

How significant is the opacity due to these grains? If we assume one L_* galaxy density per 0.01 Mpc^{-3} ejecting $2 \times 10^{11}\ M_\odot$ of gas at twice cosmic metallicity, the equivalent hydrogen density at cosmic value of the metals in the IGM is approximately 10^{-7} cm^{-3} now. Adopting $\tau(\lambda)/N_{\rm H} = 1.6 \times 10^{-22}\ \lambda_\mu^{-1}$ (Mathis, Rumpl, and Nordsieck 1977) for dust optical depth at wavelength λ_μ in microns for hydrogen column density $N_{\rm H}$,

$$\tau(z,\lambda) = 10^{-28.5}\left(\frac{5000\ \text{Å}}{\lambda_0}\right)\phi\int_0^z(1+z)^4\frac{cdt}{dz}\,dz \approx 0.1\left(\frac{5000\ \text{Å}}{\lambda_0}\right)\phi[(1+z)^3-1] \tag{18}$$

for the optical depth to an object at redshift z at the observed wavelength λ_0, where ϕ is the fraction of surviving grains. An equivalent normalization is obtained if one uses as calibration the metals seen in gas in the Coma cluster divided by the visual light of the cluster.

This optical depth is significant for high-redshift quasars. If $\phi = \frac{1}{3}$, the upper limits to τ found by McKee and Petrosian (1974) in the spectra of a sample of low-redshift quasars are generally consistent with the predicted opacity. From the Osmer (1979) sample of eight quasars having z between 3.0 and 3.5, we find that the observed spectral curvature gives $\tau_{\rm obs} = 2.75 \pm 1.51$ compared with $\tau_{\rm pred} = 2.82$. The computed τ's could be considered upper limits, since positive curvature may be intrinsic to the quasar spectra or owing to Lα absorption rather than extinction. However, the presence of this curvature at the theoretical level with the increase of τ with z suggests that the IGM may be dusty and distant objects significantly obscured.

If the absence of large-redshift quasars ($z > 3.5$) is due to dust obscuration, they should still be seen in X-rays, since the IGM is transparent for $E(z=1) > 1$ keV (Shapiro and Bahcall 1980). Visible-wavelength photons from the cooling shells are unimportant, but harder photons may be a powerful contributor to the diffuse background light at optical and UV wavelengths. If emitted at 10 eV, these are redshifted into bands where observational constraints are most severe. If a fraction y of the injected energy is radiated, the photon energy density at the present time is $\epsilon = 1.4 \times 10^{-14}yFh^2[6/(1+z_f)]$ ergs cm^{-3}; Dube, Wickes, and Wilkinson (1977) give an upper limit of 2.9×10^{-15} ergs cm^{-3} at 5.5×10^{14} Hz, giving $y < \frac{1}{6}$, just consistent with our estimate of the fraction of IGM formed into galaxies.

b) Numbers and Luminosities of Protogalaxies

The integrated luminosity of an L_* protogalaxy is around 5×10^{61} ergs (for $\epsilon \approx 10^{-4.5}$). Denoting the period of maximum luminosity as 10^7 yr $\times\ \tau_7$, we obtain $L_{\max} = 2 \times 10^{46}$(ergs s^{-1})$\tau_7^{-1}$. The total number of such galaxies on the sky is $[4\pi d_L^2/(1+z_f)]n_0c$, where d_L is the usual luminosity distance. At $z_f = 5$, $d_L = 5 \times 10^5$ Mpc, the number of sources per square degree is $4000\tau_7$ at a bolometric apparent magnitude of $22 + 2.5 \log \tau_7$; for $z_f = 2, d_L = 1.2 \times 10^4$ Mpc, the number is $500\tau_7$ per square degree at a magnitude of $19 + 2.5 \log \tau_7$. The images will be semistellar irrespective of z_f. In the absence of intergalactic dust, these are comparable to present observations (cf. Kron 1980; Davis and Wilkinson 1974). Extinction may substantially decrease the expected number, if $z_f = 5$, but not if $z_f = 2$.

c) Spatial Distribution of Galaxies

It is clear from the geometrical arguments presented in § II that unvirialized groups of galaxies (those with $t_{\rm crossing} > t_{\rm Hubble}$) should tend to lie on two-dimensional surfaces and have velocity dispersions of order 100–200 km s^{-1}.

d) The Dark Matter

A possible outcome of explosions in the Compton cooled epoch is production of a significant number of black holes in the mass range 10^3–10^4 $M_\odot$. These may be observable as gravitational lenses in front of distant objects (Press and Gunn 1973; Gott 1980) or as nearby accreting X-ray or XUV sources (Carr 1978).

A more definitive discussion of model predictions and test awaits the result of detailed numerical simulations now in progress.

We thank P. J. E. Peebles and R. Blandford for the benefit of their thoughts and C. McKee and R. McCray for significant contributions. J. P. O. was supported by National Science Foundation grant AST79-22074, and L. L. C. by National Aeronautics and Space Administration grant NGL-31-001-007.

REFERENCES

Bookbinder, J., Cowie, L. L., Krolik, J. H., Ostriker, J. P., and Rees, M. 1980, *Ap. J.*, **237**, 647.
Carr, B. J. 1978. *Comments Ap.*, **7**, 161.
Carr, B. J., and Rees, M. 1979, *Nature*, **278**, 605.
Cowie, L. L., and McKee, C. F. 1977, *Ap. J.*, **211**, 135.
Cox, D P., and Smith B. W. 1974, *Ap. J.* (*Letters*), **189**, L105.
Cox, D. P., and Tucker, W. H. 1969, *Ap. J.*, **157**, 1157.
Davis, M., and Wilkinson, D. 1974, *Ap. J.*, **192**, 251.
Doroskevich, A. G., Sunyaev, R. A., and Zel'dovich, Ya. B. 1979, *IAU Symposium 79, Large-Scale Structure of the Universe*, ed. M. S. Longair (Reidel: Holland), p. 213.
Dube, R. R., Wickes, W. C., and Wilkinson, D. T. 1977, *Ap. J.* (*Letters*), **215**, L51.
Gott, J. R. 1977, *Ann. Rev. Astr. Ap.*, **15**, 235.
———. 1980, preprint.
Kron, R. G. 1980, *Phys. Scripta*, **21**, 652.
Mathis, J. S., Rumpl, W., and Nordsieck, K. H. 1977, *Ap. J.*, **217**, 425.
McKee, C. F., and Ostriker, J. P. 1977, *Ap. J.*, **218**, 148.
McKee, C. F., and Petrosian, V. 1974, *Ap. J.*, **189**, 17.
Mestel, L. 1965, *Quart. J.R.A.S.*, **6**, 161.
Osmer, P. S. 1979, *Ap. J.*, **227**, 18.
Peebles, P. J. E. 1971, *Physical Cosmology* (Princeton: Princeton University Press).
———. 1980, *The Large-Scale Structure of the Universe* (Princeton: Princeton University Press).
Press, W. H., and Gunn, J. E. 1973, *Ap. J.*, **185**, 397.
Sargent, W. L. W., Young, P. J., Boksenberg, A., and Tytler, D. 1980, *Ap. J. Suppl.*, **42**, 41.
Shapiro, P., and Bahcall, J. 1980, preprint.
Wallace, R. K., and Woosley, S. E. 1980, *Lick Obs. Bul.*, No. 801.
Wheeler, J. C. 1977, *Ap. Space Sci.*, **50**, 125.

JEREMIAH P. OSTRIKER and LENNOX L. COWIE: Princeton University Observatory, Peyton Hall, Princeton, NJ 08544

XIII. *Neutrinos and Galaxies*

1) Rest mass of muonic neutrino and cosmology
by S. S. Gershtein and Ya. B. Zel'dovich,
JETP Lett. **4,** 120 (1966) 748

2) An upper limit on the neutrino rest mass
by R. Cowsik and J. McClelland, *Phys. Rev. Lett.*
29, 669 (1972) 751

3) Dynamical role of light neutral leptons in cosmology
by S.Tremaine and J. E. Gunn., *Phys. Rev. Lett.*
42, 407 (1979) 753

4) Massive neutrinos and the large-scale structure
of the universe
by J. R. Bond, G. Efstathiou and J. Silk,
Phys. Rev. Lett. **45,** 1980 (1980) 757

5) Massive neutrinos and galaxy formation
by F. R. Klinkhamer and C. A. Norman,
Astrophys. J. Lett. **243,** L1 (1981) 762

6) Formation of galaxies and clusters of galaxies
in the neutrino dominated universe
by H. Sato and F. Takahara, *Prog. Theor. Phys.*
66, 508 (1981) 766

7) On the linear theory of density perturbations in a
neutrino + baryon universe
by I. Wasserman, *Astrophys. J.* **248,** 1 (1981) 784

8) The formation of galaxies from massive neutrinos
by M. Davis, M. Lecar, C. Pryor, and E. Witten,
Astrophys. J. **250,** 423 (1981) 796

9) Cosmological impact of the neutrino rest mass
by A. G. Doroshkevich, M. Yu. Khlopov,
R. A. Sunyaev, A. S. Szalay, and Ya. B. Zel'dovich,
in the 10th Texas Symposium on *Relativistic
Astrophysics,* edited by R. Ranaty & Frank C. Jones,
Annals of the New York Academy of Sciences
375, 32 (1981) 805

10) Formation of structure in a neutrino-dominated
universe
by J. R. Bond and A. S. Szalay, to appear in
Proceedings of Neutrino 1981, Maui, Hawaii. 816

11) Some astrophysical consequences of the existence
of a heavy stable neutral lepton
by J. E. Gunn, B. W. Lee, I. Lerche,
D. N. Schramm, and G. Steigman,
Astrophys. J. **223,** 1015 (1978) 831

REST MASS OF MUONIC NEUTRINO AND COSMOLOGY

S. S. Gershtein and Ya. B. Zel'dovich
Submitted 4 June 1966
ZhETF Pis'ma 4, No. 5, 174-177, 1 September 1966

Low-accuracy experimental estimates of the rest mass of the neutrino [1] yield $m(\nu_e) < 200$ eV/c^2 for the electronic neutrino and $m(\nu_\mu) < 2.5 \times 10^6$ eV/c^2 for the muonic neutrino.

Cosmological considerations connected with the hot model of the Universe [2] make it possible to strengthen greatly the second inequality. Just as in the paper by Ya. B. Zel'-dovich and Ya. A. Smorodinskii [3], let us consider the gravitational effect of the neutrinos on the dynamics of the expanding Universe. The age of the known astronomical objects is not smaller than 5×10^9 years, and Hubble's constant H is not smaller than 75 km/sec-Mparsec = $(13 \times 10^9 \text{ years})^{-1}$. It follows therefore that the density of all types of matter in the Universe is at the present time [1)]

$$\rho < 2 \times 10^{-28} \text{ g/cm}^3.$$

The space surrounding us is filled presently with equilibrium radiation of temperature 3°K [4]. It is proposed that this is "relict" radiation and is proof of the high temperature possessed by the plasma during the pre-stellar high-density period.

At a temperature of the order of 3 MeV for ν_e and of the order of 15 MeV for ν_μ, complete thermodynamic equilibrium existed between ν, γ, e^+, and e^-. The number of other particles in this equilibrium is small, except perhaps gravitons, which, however, have no effect on the arguments that follow. In thermodynamic equilibrium, the ratio of the number of fermions and antifermions with spin 1/2 to the number of quanta is

$$[\nu_e] + [\bar{\nu}_e] = [\nu_\mu] + [\bar{\nu}_\mu] = [e^+] + [e^-] = 2\,\frac{\int (e^x + 1)^{-1} x^2 dx}{\int (e^x - 1)^{-1} x^2 dx}[\gamma] = 1.5[\gamma].$$

However, during the course of the cooling from $T > m_e c^2$ (for which these relations are written) to the present time, when $T \ll m_e c^2$, these relations change, since the annihilation of the e^+e^- increases the number of quanta without changing the number of neutrinos per unit of co-moving volume [5]. At the present time we can expect

$$[e^+] + [e^-] = 0, \quad [\nu_\mu] + [\bar{\nu}_\mu] = [\nu_e] + [\bar{\nu}_e] = 0.5[\gamma].$$

At 3°K we have $[\gamma] = 550$ g/cm^3, from which we obtain for the neutrino at the present time

$$[\nu_\mu] + [\bar{\nu}_\mu] = [\nu_e] + [\bar{\nu}_e] = 300 \text{ cm}^{-3}.$$

Comparing with the density limit given above, we obtain

$$m_0(\nu_\mu) < 7 \times 10^{-31} \text{ g} = 400 \text{ eV}/c^2$$

and the same for $m_0(\nu_e)$. Thus, we obtain no new information for the electronic neutrino; for the muonic neutrino, on the other hand, the cosmological considerations reduce the upper limit of the rest mass by three orders of magnitude.

In considering the question of the possible mass of the neutrino we have, naturally, used statistical formulas for the four-component ($m \neq 0$) particles. We know, however, that in accordance with the (V - A) theory, neutrinos having a definite polarization participate predominantly in weak interactions. Equilibrium for neutrinos for opposite polarization is established only at a higher temperature. This, incidentally, can change the limit of the mass by not more than a factor of 2.

A neutrino with non-zero rest mass can become annihilated in accordance with the diagram

$$\nu_\mu + \bar{\nu}_\mu \xrightarrow[\text{(weak)}]{} \mu^+ + \mu^- \xrightarrow[\text{(el.-mag.)}]{} \gamma + \gamma$$

if $m(\nu_\mu) > m(e^\pm)$, and also either into 3γ or, in the square of the weak interaction, into a $\nu_e + \bar{\nu}_e$ pair, if it is assumed the $m(\nu_\mu) > m(\nu_e)$. When $v < c$ the annihilation cross section behaves like $1/v$. Estimates show, however, that there is no time for noticeable annihilation to take place during the course of the cosmological expansion.

The momentum of the interaction particles changes during the course of expansion like $1/R$, where R is the linear scale, independently of the presence and magnitude of the particle rest mass. At the present time the neutrino momentum should be of the same order (somewhat smaller) as the momentum of the relict quanta, i.e., $\bar{p} \approx 5 \times 10^{-4}$ eV/c.

If the neutrinos have a rest mass, then their velocity and the speed of sound in the neutrino gas are of the order of p/m, i.e, say 30 km/sec at $m = 5$ eV/c^2 and 3 km/sec at $m = 50$ eV/c^2. Strong gravitational perturbations should be produced in such a gas by the galaxies. It is possible that a more detailed analysis of these processes will allow us to lower the foregoing estimate of the upper limit of the neutrino mass.

This note is the result of the stimulating circumstances of the Summer School in Balaton-vilagose, and we use this opportunity to express gratitutde to the organizers of the school.

[1] A. H. Rosenfeld, A. Barbaro Galtieri, W. H. Barkas, P. L. Bastien, I. Kírz, and M. Roos, Tables, Revs. Mod. Phys. 37, 633 (1965).

[2] G. Gamow, Phys. Rev. 70, 572 (1946); 74, 505 (1948); Revs. Mod. Phys. 21, 367 (1949); G. Gamow, Vistas in Astronomy 2, 1726 (1956); R. Dicke, P. J. E. Peebles, P. G. Roll, and D. T. Wilkinson, Astrophys. J. 142, 414 (1965); Ya. B. Zel'dovich, UFN 89, 647 (1966), Soviet Phys. Uspekhi 9, in press.

[3] Ya. B. Zel'dovich and Ya. A. Smorodinskii, JETP 41, 907 (1961), Soviet Phys. JETP 14, 647 (1962).

[4] A. A. Penzias and R. W. Wilson, Astrophys. J. 142, 419 (1965).

[5] P. J. E. Peebles, Phys. Rev. Lett. 16, 410 (1966).

1) We use the asymptotic formula

$$T = \pi/2H\sqrt{\rho/\rho_c}\ ;\quad \rho_c = 3H^2/8\pi\sigma\ ;\quad \rho = 3\pi/32\sigma T^2.$$

Other more complicated estimates based on an investigation of remote objects give a similar result:

$$q_0 = \rho/2\rho_c < 2.5;\quad H \leq 120\ \text{km/sec.Mparsec},$$

$$\rho_c \leq 2.5 \times 10^{-29}\ \text{g/cm}^3,\quad \rho < 1.25 \times 10^{-28}\ \text{g/cm}^3.$$

VOLUME 29, NUMBER 10 PHYSICAL REVIEW LETTERS 4 SEPTEMBER 1972

An Upper Limit on the Neutrino Rest Mass*

R. Cowsik† and J. McClelland
Department of Physics, University of California, Berkeley, California 94720
(Received 17 July 1972)

In order that the effect of graviation of the thermal background neutrinos on the expansion of the universe not be too severe, their mass should be less than 8 eV/c^2.

Recently there has been a resurgence of interest in the possibility that neutrinos may have a finite rest mass. These discussions have been in the context of weak-interaction theories,[1] possible decay of solar neutrinos,[2] and enumerating the possible decay modes of the K_L^0 meson.[3] Elsewhere, we have pointed out that the gravitational interactions of neutrinos of finite rest mass may become very important in the discussion of the dynamics of clusters of galaxies and of the universe.[4] Considerations involving massive neutrinos are not new[5,6]; an excellent review of the early developments in the field is given by Kuchowicz.[7] Here we wish to point out that the recent measurement[8] of the deceleration parameter, q_0, implies an upper limit of a few tens of electron volts on the sum of the masses of all the possible light, stable particles that interact only weakly.

In discussing this problem we take the customary point of view that the universe is expanding from an initially hot and condensed state as envisaged in the "big-bang" theories.[9] In the early phase of such a universe, when the temperature was greater than ~1 MeV, processes of neutrino production, which have also been considered in the context of high-temperature stellar cores,[10] would lead to the generation of the various kinds of neutrinos. In fact, similar processes would generate populations of other fermions and bosons as well, and conditions of thermal equilibrium allow us to estimate their number density[11]:

$$n_{Fi} = \frac{2s_i+1}{2\pi^2\hbar^3}\int_0^\infty \frac{p^2dp}{\exp[E/kT(z_{eq})]+1}, \tag{1a}$$

and

$$n_{Bi} = \frac{2s_i+1}{2\pi^2\hbar^3}\int_0^\infty \frac{p^2dp}{\exp[E/kT(z_{eq})]-1}. \tag{1b}$$

Here n_{Fi} is the number density of fermions of the ith kind, n_{Bi} is the number density of bosons of the ith kind, s_i is the spin of the particle (notice that in writing the multiplicity of states of the particles we have not discriminated against the neutrinos; since we are discussing neutrinos of nonzero rest mass, we have assumed that both the helicity states are allowed), $E = c(p^2+m^2c^2)^{1/2}$, k is Boltzmann's constant, and $T(z_{eq}) = T_r(z_{eq}) = T_F(z_{eq}) = T_B(z_{eq}) = T_m(z_{eq}) = \cdots$ is the common temperature of radiation, fermions, bosons, matter, etc. at the latest epoch, characterized by the red shift z_{eq}, when they may be considered to be in thermal equilibrium; $kT(z_{eq}) \approx 1$ MeV.

Since our discussion pertains to neutrinos and any hypothetical stable weak bosons,[2] we may assume that $kT(z_{eq}) \approx 1\text{ MeV} \gg mc^2$. In this limit Eqs. (1a) and (1b) reduce to

$$n_F(z_{eq}) \approx 0.0913(2s_i+1)[T(z_{eq})/\hbar c]^3, \tag{2a}$$

$$n_B(z_{eq}) \approx 0.122(2s_i+1)[T(z_{eq})/\hbar c]^3. \tag{2b}$$

As the universe expands and cools down, the neutrinos and such other weakly interacting particles survive without annihilation because of the extremely low cross sections[12] for these processes. Consequently, the number density decreases simply as $\sim V(z_{eq})/V(z) = (1+z)^3/(1+z_{eq})^3$. Noticing that $1+z = T_r(z)/T_r(0)$, the number densities of the various particles expected at the present epoch ($z=0$) are given by

$$n_{Fi}(0) = n_{Fi}(z_{eq})\left[\frac{1}{1+z_{eq}}\right]^3 \approx 0.0913(2s_i+1)\left[\frac{T_r(0)}{\hbar c}\right]^3 \tag{3a}$$

and

$$n_{Bi}(0) \approx 0.122(2s_i+1)\left[\frac{T_r(0)}{\hbar c}\right]^3. \tag{3b}$$

Taking $T_r(0) \approx 2.7$°K, we have

$$n_{Fi}(0) \approx 150(2s_i+1)\text{ cm}^{-3}, \tag{4a}$$

and

$$n_{Bi}(0) \approx 200(2s_i+1)\ \text{cm}^{-3}. \tag{4b}$$

These numbers are huge in comparison with the mean number density of hydrogen atoms in the universe; all the visible matter in the universe adds up to an average density of hydrogen atoms[9] of only $\sim 2\times 10^{-8}$ cm^{-3}. Notice that the expected density of the neutrinos and other weakly interacting particles is essentially independent of the temperature $T(z_{eq})$, of decoupling, and such other details; the measured temperature of the universal blackbody photons fixes the density of weak particles quite well.

Now, consider Sandage's[8] measurement of the Hubble constant H_0 and the deceleration parameter q_0 which together place a limit on ρ_{tot}, the density of all possible sources of gravitational potential in the universe.[9] His results, $H_0 = 50$ km sec^{-1} Mpc^{-1} $= 1.7\times 10^{-18}$ sec^{-1} and $q_0 = +0.94\pm 0.4$, imply

$$\rho_{tot} = 3H_0^2 q_0/4\pi G = (10\pm 4)\times 10^{-30}\ \text{g cm}^{-3} \approx (6\pm 2)\times 10^3\ (\text{eV}/c^2)\ \text{cm}^{-3} < 10^4\ (\text{eV}/c^2)\ \text{cm}^{-3}. \tag{5}$$

Here $G = 6.68\times 10^{-8}$ dyn cm^2 g^{-2} is the gravitational constant. If m_i were to represent the mass spectrum of the various neutrinos and other stable weakly interacting particles, we can combine Eqs. (4a), (4b), and (5) to obtain the limit

$$\rho_{weak} \approx \sum n_{Bi} m_i + n_{Fj} m_j \gtrsim 150(2s_i+1)m_i < \rho_{tot}$$

or (6)

$$\sum (2s_i+1)m_i \lesssim 66\ \text{eV}/c^2.$$

Here the summation is to be carried out over all the particle and antiparticle states of both fermions and bosons. Considering only the neutrinos and antineutrinos of the muon and electron kind each having a mass of m_ν, Eq. (6) leads to the result $m_\nu < 8$ eV/c^2.

This limit is obtained assuming big-bang cosmology to be correct; however, it depends only very weakly on the value of the deceleration parameter and other details of the cosmology. Thus, even when one allows for a large uncertainty in the cosmological parameters, the limits on the masses of neutrinos and other stable weakly interacting particles derived in this paper are still much lower than the direct experimental limits[13,14] of $m_{\nu\mu} < 1.5$ MeV/c^2 and $m_{\nu e} < 60$ eV/c^2.

Our thanks are due to Professor Eugene D. Commins, Professor J. N. Bahcall, Professor G. B. Field, and Professor P. Buford Price for many discussions.

*Work supported in part by the National Aeronautics and Space Administration under Grant No. NGR05-003-376.

†On leave from the Tata Institute of Fundamental Research, Bombay, India.

[1]K. Tennakone and S. Pakvasa, Phys. Rev. Lett. 27, 757 (1971), and 28, 1415 (1972).

[2]J. N. Bahcall, N. Cabibbo, and A. Yahil, Phys. Rev. Lett. 28, 316 (1972).

[3]S. Barshay, Phys. Rev. Lett. 28, 1008 (1972).

[4]R. Cowsik and J. McClelland, "Gravity of Neutrinos of Non-Zero Mass in Astrophysics" (to be published).

[5]M. A. Markov, *The Neutrino* (Nauka, Moscow, U. S. S. R., 1964).

[6]J. Bahcall and R. B. Curtis, Nuovo Cimento 21, 422 (1961).

[7]B. Kuchowics, *The Bibliography of the Neutrino* (Gordon and Breach, New York, 1967), and Fortschr. Phys. 17, 517 (1969).

[8]A. Sandage, Astrophys. J. 173, 485 (1972), and to be published.

[9]P. J. E. Peebles, *Physical Cosmology* (Princeton Univ. Press, Princeton, N. J., 1971).

[10]M. A. Ruderman, in *Topical Conference on Weak Interactions, CERN, Geneva, Switzerland, 1969* (CERN Scientific Information Service, Geneva, Switzerland, 1969), p. 111.

[11]L. D. Landau and E. M. Lifshitz, *Statistical Physics* (Addison-Wesley, Reading, Mass., 1969), 2nd ed., p. 324.

[12]T. de Graff and H. A. Tolhoek, Nucl. Phys. 81, 596 (1966).

[13]K. Bergkvist, Nucl. Phys. B39, 317 (1972).

[14]E. V. Shrum and K. O. H. Ziock, Phys. Lett. 37B, 115 (1971).

Dynamical Role of Light Neutral Leptons in Cosmology

Scott Tremaine
W. K. Kellogg Radiation Laboratory, California Institute of Technology, Pasadena, California 91125, and Institute of Astronomy, Cambridge, England

and

James E. Gunn
W. K. Kellogg Radiation Laboratory, Hale Observatories, California Institute of Technology, Pasadena, California 91125, and Carnegie Institution of Washington, Washington, D. C. 20005
(Received 31 May 1978)

Using the Vlasov equation, we show that massive galactic halos cannot be composed of stable neutral leptons of mass $\lesssim 1$ MeV. Since most of the mass in clusters of galaxies probably consists of stripped halos, we conclude that the "missing mass" in clusters does not consist of leptons of mass $\lesssim 1$ MeV (e.g., muon or electron neutrinos). Lee and Weinberg's hypothetical heavy leptons (mass ≈ 1 GeV) are not ruled out by this argument.

Observations of rich clusters of galaxies[1] show that most of their mass is in a form which emits very little light: the mass-to-blue-light ratio $M/L_B \approx 200(H/50)$ in solar units,[1] where H(km s^{-1} Mpc^{-1}) is the Hubble constant. This is the classical "missing-mass" problem, although really it is the light and not the mass which is missing. The problem persists in smaller systems. Groups of galaxies have $M/L_B \approx 140(H/50)$.[2] The Local Group has $M/L_B \approx 60(H/50)$ if $q_0 = 0$,[3] binary galaxies have $M/L_B \approx 65(H/50)$,[4] and even in our own Galaxy and $M31$ most of the mass probably lies at $r \gtrsim 10$ kpc, outside the visible regions.[5-7] We conclude that even for systems with radii as small as ~20 kpc, most of the mass lies in dark material unlike the stellar population with $M/L_B \approx 7(H/50)$,[8] which dominates the mass in the central regions of galaxies. It seems likely that the dark mass in binaries, groups, and clusters is material from the halos of galaxies that has been stripped off by tidal interactions.[9,10] More detailed and quantitative comparisons[11] based on expected values of M/L_B from stellar content yield the same conclusion; viz., that the ratio of "missing" to visible mass is similar for all systems large enough that separation would not occur by dissipative processes in the time available—and thus that the dark material is likely to be the same everywhere. These arguments and Occam's razor lead us to the assumption that the composition of the dark material is the same in galactic halos, binary galaxies, groups, and clusters.

In some gauge theories the electron and muon neutrinos have a small, nonzero rest mass,[12] and it has been proposed that the dark material consists of such particles.[13] Shortly after the big bang the neutrinos are relativistic and in thermal equilibrium, and so their momentum distribution is

$$n_{\nu_e}(\vec{p})\,d\vec{p} = n_{\bar{\nu}_e}(\vec{p})\,d\vec{p} = n_{\nu_\mu}(\vec{p})\,d\vec{p} = n_{\bar{\nu}_\mu}(\vec{p})\,d\vec{p} = \frac{g_\nu}{h^3}\,d\vec{p}\left\{\exp\left[\frac{p}{kT_\nu(z)}\right]+1\right\}^{-1}. \quad (1)$$

VOLUME 42, NUMBER 6 PHYSICAL REVIEW LETTERS 5 FEBRUARY 1979

In this equation $c=1$, $T_\nu(z)=T_\gamma(z)$ is the common temperature of the neutrinos and the radiation, and g_ν is the number of allowed helicity states. For neutrinos of nonzero rest mass two helicities are allowed, but the cross section for production of positive-helicity neutrinos is down by $\sim(m/kT_\nu)^2$ from the cross section for negative-helicity neutrinos, and as a result, the former decouple at a very high temperature, where the evolution of the universe is not well understood. Thus it is not clear whether both helicity states should be included in g_ν, and we only know that $1\le g_\nu \le 2$. Both helicity states are decoupled by $kT_\nu\approx 1$ MeV,[14] and, since the electron and muon neutrinos are known to be less massive than 1 MeV, they are still relativistic when they decouple. Thereafter their momenta red shift like $1+z$, so that Eq. (1) continues to hold with $T_\nu(z)\propto 1+z$. But $T_\gamma(z)$ is also $\propto 1+z$ except for a jump of $(\frac{11}{4})^{1/3}$ at e^+e^- recombination,[14] and so we have

$$T_\nu(0)=(\tfrac{4}{11})^{1/3}T_\gamma(0)=1.9\text{ K}. \tag{2}$$

From Eqs. (1) and (2) the present mass density in neutrinos is

$$\rho_\nu=[(m_{\nu_e}+m_{\bar\nu_e})g_{\nu_e}+(m_{\nu_\mu}+m_{\bar\nu_\mu})g_{\nu_\mu}]6\pi\zeta(3)\left(\frac{kT_\nu(0)}{h}\right)^3=(1.9\times10^{-31}\text{ g cm}^{-3})\left(\frac{m_{\nu_e}g_{\nu_e}+m_{\nu_\mu}g_{\nu_\mu}}{1\text{ eV}}\right). \tag{3}$$

In terms of the critical density $\rho_c=3H^2/8\pi G$,

$$\Omega_\nu=\frac{\rho_\nu}{\rho_c}=0.04\left(\frac{50}{H}\right)^2\frac{m_{\nu_e}g_{\nu_e}+m_{\nu_\mu}g_{\nu_\mu}}{1\text{ eV}}. \tag{4}$$

Neutrinos with a cosmologically interesting mass will have very small velocities: For $m\gtrsim 1$ eV, $v\approx kT_\nu(0)/m\lesssim 50$ km s^{-1}. Because this velocity is smaller than the typical velocity dispersion within galaxies or clusters, and because galaxies and clusters collapse in a dynamical time, the neutrinos will participate in the galaxy clustering. The ratio of the density distributed like the galaxies to ρ_c is called Ω^*. The best observational estimate of Ω^* is roughly 0.05.[15] Since $\Omega_\nu<\Omega^*$, we have

$$m_{\nu_e}g_{\nu_e}+m_{\nu_\mu}g_{\nu_\mu}\lesssim(1.2\text{ eV})(H/50)^2. \tag{5}$$

This mass is well below the current experimental upper limits on m_{ν_μ} and m_{ν_e} (cf. Ref. 13). This is the chain of argument which led Cowsik and McClelland[13] to the suggestion that massive neutrinos could supply the missing mass. If we look further, however, we can find a flaw. Because the neutrinos are noninteracting, the density of a fluid element in phase space is conserved (the Vlasov equation). Thus, the maximum fine-grained phase-space density is conserved. The maximum coarse-grained phase-space density is less well constrained: We know only that it must decrease with time, because the fluid can become "frothy" in phase space,[16] i.e., the regions of maximum fine-grained density may be mixed in with lower-density regions.

While neutrino background is homogeneous the maximum phase-space density is $2g_\nu h^{-3}$ from the four kinds of neutrinos, assuming $g_{\nu_e}=g_{\nu_\mu}\equiv g_\nu$ [cf. Eq. (1)]. We assume that the central regions of the bound systems formed by the neutrinos resemble isothermal gas spheres; then their velocity distribution is Maxwellian and the maximum phase-space density is $\rho_0 m_\nu^{-4}(2\pi\sigma^2)^{-3/2}$. Here ρ_0 is the central density and σ is the one-dimensional velocity dispersion, and we assume that $m_{\nu_e}=m_{\nu_\mu}\equiv m_\nu$. The core radius is defined by $r_c^2=9\sigma^2/4\pi G\rho_0$, and the requirement that the maximum phase-space density has decreased becomes

$$m_\nu^4>\frac{9h^3}{4(2\pi)^{5/2}g_\nu G\sigma r_c^2},\quad m_\nu>(101\text{ eV})\left(\frac{100\text{ km s}^{-1}}{\sigma}\right)^{1/4}\left(\frac{1\text{ kpc}}{r_c}\right)^{1/4}g_\nu^{-1/4}. \tag{6}$$

In comparison, the Pauli principle requires only that the present maximum phase-space density is $<4g_\nu h^{-3}$, giving a limit on m_ν which is similar to Eq. (6) but less severe by a factor $2^{1/4}$. Note, however, that the principle that the maximum phase-space density decreases is quite distinct from the Pauli principle. For example, our argument would also work for any hypothetical noninteracting Maxwell-Boltzmann particles. It does not work for bosons because their equilibrium phase-space density does not have a maximum.

For rich clusters of galaxies[1] $r_c\simeq 0.25(50/H)$ Mpc, $\sigma\simeq 10^3$ km s^{-1}. Thus, $m_\nu\gtrsim(3.6\text{ eV})(H/50)^{1/2}g_\nu^{-1/4}$, roughly consistent with the limit (5). Binary galaxies, however, have $r_c\lesssim 100$ $(50/H)$ kpc (the typical separation of the components) and $\sigma\approx 100$ km s^{-1} (Ref. 4), and inequalities (5) and (6) yield $m_\nu \lesssim (0.6\text{ eV})(H/50)^2 g_\nu^{-1}$ and $m_\nu \gtrsim (10\text{ eV})(H/50)^{1/2} g_\nu^{-1/4}$, which are inconsistent. The inconsistency is worse still for galactic halos, where $r_c \lesssim 20$ kpc, $\sigma \approx 150$ km s^{-1}, and $m_\nu \gtrsim (20\text{ eV}) g_\nu^{-1/4}$. We conclude that massive muon or electron neutrinos cannot supply the dark material in galactic halos or binary galaxies or, by extension, in clusters of galaxies.

The contradiction cannot be removed by assuming that the luminous parts of the galaxies condense first and suck the neutrinos into their potensial well, because we have not assumed that the neutrinos supply all the gravitational potential.

Notice that this argument is not based on the Jeans mass. In fact, the present Jeans mass for the neutrinos is

$$M_J \approx \rho[\pi v_s^2/G\rho]^{3/2} \approx 3\times 10^{13} M_\odot [m_\nu/1\text{ eV}]^{-7/2} g_\nu^{-1/2}, \qquad (7)$$

where we have replaced the "sound speed" v_s by $[kT_\nu(0)/m_\nu]^{1/2}$. Thus systems of galactic or larger mass could form if $m_\nu \lesssim 4$ eV; the phase-space argument shows, however, that they would be larger than the systems we observe.

Are there any loopholes? The most uncertain parameter is Ω^*. If we regard these arguments as setting a lower limit on Ω^*, then the most conservative limit is obtained when $g_\nu = 1$ and one species of neutrinos is much more massive than the other (say, $m_{\nu_\mu} \gg m_{\nu_e}$). Then the values $r_c \lesssim 20$ kpc, $\sigma \approx 150$ km s^{-1} appropriate for galactic halos require $\Omega^* \gtrsim 1.0(50/H)^2$, $m_\nu \gtrsim 24$ eV, if massive neutrinos are to make up the halos. This result is well outside the acceptable limits to Ω^* and H (cf. Ref. 15). A very simple argument is worth noting at this point. Particles as heavy as 24 eV behave completely classically at the phase-space densities in the great clusters and would cluster on these scales precisely as the galaxies do.[17] If Ω^* were as large as unity in such particles, the ratio of unseen to visible matter in the clusters would be larger than it is, corresponding to mass-to-light ratios in excess of 1500—a value never approached in observation.

In a cosmology with nonzero muon or electron lepton number the problem only gets worse. The reason is that the maximum phase-space density of neutrinos plus antineutrinos is independent of chemical potential μ_ν, while the spatial density of neutrinos plus antineutrinos increases if μ_ν is nonzero. Thus inequality (6) is unchanged but inequality (5) requires smaller neutrino masses. (If m_ν is very small or zero, the neutrinos do not cluster with the galaxies; in this case, if $\mu_\nu/kT_\nu \gg 1$, neutrinos could close the universe, but they could not supply the missing mass in clusters. Cosmological helium production is also impossible in such models.[14])

Finally, we consider the possibility that there are other stable neutral leptons besides the electron and muon neutrino. If the lepton mass is $\lesssim 1$ MeV all of our arguments are still valid [if there are *more* than two species with comparable masses, the inconsistency of inequalities (5) and (6) is still worse]. If their mass is $\gtrsim 1$ MeV, the assumption that the leptons are relativistic when they decouple is no longer valid; in this case Lee and Weinberg[18] have shown that leptons with mass $\gtrsim 1$ GeV can contribute $\Omega_\nu \approx \Omega^*$. Such heavy leptons, although reprehensible on etymological grounds, are not ruled out by our phase-phase-density arguments (see also Ref. 17).

In summary, we know that there is some dark material which dominates the dynamics of binary galaxies, and makes up most of the mass in the outer halos of isolated galaxies. We have shown that this material cannot be muon or electron neutrinos of nonzero rest mass, or any stable neutral lepton less massive than 1 MeV. Several arguments suggest that the dark material is universal, so that the same restrictions apply to the unknown component which provides most of the mass in groups and clusters of galaxies.

We feel that the most likely remaining possibilities are small black holes or very low-mass stars, or perhaps some much heavier stable neutral particles.

One of us (S.T.) thanks J. P. Ostriker and P. J. E. Peebles for discussions. This research was supported in part by the National Science Foundation Grant No. AST76-80801A01 and by the Alfred P. Sloan Foundation.

[1]The most recent review is N. A. Bahcall, Annu. Rev. Astron. Astrophys. 15, 505 (1977).

[2]J. R. Gott and E. L. Turner, Astrophys. J. 213, 309 (1977).

[3]J. E. Gunn, Comments Astrophys. Space Sci. 6, 7 (1975).

[4]E. L. Turner, Astrophys. J 208, 304 (1976).

[5]W. L. W. Sargent, in *The Evolution of Galaxies and Stellar Populations*, edited by B. M. Tinsley and R. B. Larson (Yale Univ. Observatory, New Haven, 1977), p. 427.

[6]D. N. C. Lin and D. Lynden-Bell, Mon. Not. Roy. Astron. Soc. 181, 59 (1977).

[7]M. S. Roberts and R. N. Whitehurst, Astrophys. J. 201, 327 (1975).
[8]S. M. Faber and R. E. Jackson, Astrophys. J. 204, 668 (1976).
[9]D. O. Richstone, Astrophys. J. 204, 642 (1976).
[10]J. E. Gunn, Astrophys. J. 218, 592 (1977).
[11]J. E. Gunn, J. Phys. (Paris), Colloq. 37, C6-183 (1976).
[12]H. Fritzsch, M. Gell-Mann, and P. Minkowski, Phys. Lett. 59B, 256 (1975).
[13]R. Cowsik and J. McClelland, Phys. Rev. Lett. 29, 669 (1972), and Astrophys. J. 180, 7 (1973).
[14]A general reference is S. Weinberg, *Gravitation and Cosmology* (Wiley, New York, 1972), Chap. 15, or P. J. E. Peebles, *Physical Cosmology* (Princeton Univ. Press, Princeton, N. J., 1971), Chap. 8.
[15]J. R. Gott, J. E. Gunn, D. N. Schramm, and B. M. Tinsley, Astrophys. J. 194, 543 (1974).
[16]D. Lynden-Bell, Mon. Not. Roy. Astron. Soc. 136, 101 (1967).
[17]J. E. Gunn, B. L. Lee, I. Lerche, D. N. Schramm, and G. Steigman, "Some Astrophysical Consequences of the Evolution of a Heavy Stable Neutral Lepton" (to be published).
[18]B. W. Lee and S. Weinberg, Phys. Rev. Lett. 39, 165 (1977).

VOLUME 45, NUMBER 24 PHYSICAL REVIEW LETTERS 15 DECEMBER 1980

Massive Neutrinos and the Large-Scale Structure of the Universe

J. R. Bond, G. Efstathiou, and J. Silk
Astronomy Department, University of California, Berkeley, California 94720
(Received 9 July 1980)

If neutrinos dominate the mass density of the Universe, they play a critical role in the gravitational instability theory of galaxy formation. For neutrinos of mass $30m_{30}$ eV, a maximum Jeans mass $M_{\nu m} \approx 4\times 10^{15} m_{30}^{-2} M_\odot$ is derived. On smaller scales, neutrino fluctuations are damped, and the growth of baryon fluctuations is greatly inhibited. Structures on scales $> M_{\nu m}$ may collapse, forming galaxies as in Zel'dovich's "pancake" theory.

PACS numbers: 98.50.Eb, 14.60.Gh, 95.30.Cq, 98.80.Ft

The idea that massive neutrinos may solve the missing-mass problem in galaxies and in clusters of galaxies, as well as dominating the mass content of the Universe, has been explored by many authors.[1,2] The recent experimental reports of electron-neutrino mass detection[3] and neutrino oscillations,[4] coupled with the natural appearance of a neutrino-flavor mass spectrum with splittings of order of those between the three leptoquark families in extensions of grand unified theories to SO(10),[5] suggest that we take seriously the astrophysical implications of a neutrino

VOLUME 45, NUMBER 24 PHYSICAL REVIEW LETTERS 15 DECEMBER 1980

mass. Here, we consider the role of light massive neutrinos in the early Universe and in galaxy formation.

The mean occupation number of the neutrino background in the standard big-bang model as a function of momentum p and time t is $n(p,t) = \{\exp[pc/T_\nu(t)]+1\}^{-1}$, which is valid even in the nonrelativistic (NR) regime, where it is neither Fermi-Dirac nor degenerate. The neutrino "temperature," $T_\nu(t)$, for left-handed neutrinos (ν_L) is now 1.9 K, red shifted from 1 MeV when they decoupled. If right-handed neutrinos (ν_R) couple only as in the Weinberg-Salam model extended to include a neutrino Dirac mass, rates for reactions such as $e^+e^- \to \nu_R\bar{\nu}_L$ [$\sim(m_\nu/T_\nu)^2$ times the $e^+e^- \to \nu_L\bar{\nu}_L$ rate] were always substantially less than the expansion rate.[6] The resulting number ratio of right-handed to left-handed neutrinos is $g-1 \lesssim 10^{-4}$, valid for all ν flavors. Subsequently, however, left and right helicities can mix through deflections in a gravitational field, though spin projection along a fixed axis is conserved.

We then obtain a neutrino plus antineutrino mass density in the NR epoch (in units of the cosmological closure density) of $\Omega_\nu = 0.93\bar{m}_{30}h^{-2}$, where h is Hubble's constant in units of 100 km s^{-1} Mpc^{-1}. (Mpc is 10^6 parsec.) If the average neutrino mass $\bar{m}_{30} = \frac{1}{3}(\sum gm_{30})$ (where the sum is over ν flavors) exceeds $\sim h^2$, the Universe is closed. Here m_{30} is the neutrino mass in 30-eV units. Observations of the deceleration parameter[7] suggest $\Omega_\nu \leqslant 2$, which slightly modifies the well-known constraint on light-neutrino masses,[1] $\bar{m}_{30} \lesssim 2h^2$.

Recent estimates[8] of the mean luminosity density in the Universe imply baryons in galaxies contribute a fraction $\Omega_B = (M/L)_G/2315h \sim 0.004$–0.007 to the mean density required for closure. The mass-to-blue-light ratio of a typical galaxy, $(M/L)_G$ (in solar units), due to "conventional" stars probably lies in the range $9h$–$16h$,[9] yielding the above spread. This gives the baryon density in luminous forms of matter, which is smaller than the mass density in neutrinos by the factor $\Omega_\nu/\Omega_B \sim (140\text{–}240)\bar{m}_{30}h^{-2}$. Neutrinos need only have $\bar{m} \geqslant 0.2h^2$ eV to dominate, unless most baryons reside in dark forms of matter, such as very-low-mass stars or black holes.[10]

In the primordial nucleosynthesis era, the neutrinos are extremely relativistic (ER), and, as discussed above, g is *not* 2, though there are two spin states, but ≈ 1. Thus, light element production remains essentially *unchanged* from the standard picture. This is not true if right-handed neutrinos couple, for example, through weak vector bosons[11] at a strength of $G_R \gtrsim 10^{-3}G_F$, where G_F is the usual Fermi coupling constant for left-handed neutrinos. The right decoupling temperature is then lower than the temperature at which copious quark-antiquark production occurs. Its red shift then results in the right- and left-handed neutrino temperatures maintaining approximate equality through the nuclear burning era. Thus, $g \approx 2$, which increases the expansion rate, and consequently raises the final helium and deuterium abundances above those of the standard model.[6] To get the observed deuterium abundance in the standard model with $g=1$ requires[12] $0.01 \leqslant \Omega_B h^2 \leqslant 0.02$; to explain low helium measurements may require $\Omega_B h^2$ to be as low as 0.003.[13]

When the red shift falls below $z_{NR} \approx 1.8\times10^5 m_{30}$, the neutrinos go NR. The neutrino distribution function then results in a velocity dispersion of the left-handed background $\langle v^2\rangle^{1/2} = (12\eta_5/\eta_3)^{1/2}T_\nu/m_\nu c \simeq 6m_{30}^{-1}(1+z)$ km s^{-1}, where η_k is the Riemann eta function. This is now smaller than the dispersion in rich clusters ($\sim$1500 km/s) and in galaxies ($\sim$250 km/s). Assume now that one neutrino flavor carries almost all of the mass[5]; our general conclusions do not depend upon this assumption. A slight modification of the Tremaine and Gunn[2] argument to one neutrino flavor shows that nondegenerate neutrinos cannot form the dark matter in galactic halos unless $m_\nu > 25$ eV. Such a high value is certainly not ruled out by cosmology; therefore it is possible that massive neutrinos are the dark matter in the halos of spiral galaxies and the "missing mass" in clusters. We thus consider the role which neutrinos play in large scale clustering.

The velocity dispersion and the mass density imply a Jeans mass of

$$M_{J\nu} = M_{\nu m}x^{-3/2}f(x), \quad x = (1+z_{NR})/(1+z),$$
$$f(x) = 26.6\left(\frac{x}{24\eta_5}\int_0^\infty \frac{y^4\,dy}{(e^y+1)(y^2+x^2)^{1/2}}\right)^{3/2}\left(\frac{1}{2x\eta_3}\int_0^\infty \frac{y^2\,dy\,(y^2+x^2)^{1/2}}{e^y+1}\right)^{-2}, \tag{1}$$

which peaks at the maximum Jeans mass $M_{\nu m} = 3.9\times10^{15}m_{30}^{-2}M_\odot$ attained at the red shift $z_m = 42\,900m_{30}$. The function $f(x)$ involves integrals of the distribution function which we evaluate numerically in the semirelativistic regime. In the ER regime, $f = 0.32x^{7/2}$, and so the Jeans mass rises as $(1+z)^{-2}$ as z

VOLUME 45, NUMBER 24 PHYSICAL REVIEW LETTERS 15 DECEMBER 1980

drops. In the NR regime, $f = 26.6$, and the Jeans mass falls as $(1+z)^{3/2}$ through the hierarchy of structural mass scales observed in the Universe until the present epoch, when it is $1.4\times 10^9 m_{30}^{-7/2} \times M_\odot$. (See Fig. 1.)

When neutrinos are ER, but decoupled from matter, perturbations decay on all length scales smaller than the horizon.[14] Further, in the NR regime, neutrinos Landau damp on all scales smaller than the Jeans length,[15] in the same manner as collisionless baryons.[16] We conclude that *all primordial neutrino perturbations will be erased on mass scales smaller than the peak value* $M_{\nu m}$ and on all scales smaller than the associated length $\lambda_{\nu m}(z) \approx m_{30}^{-2}(1+z_m)/(1+z)$ kpc. Neutrino damping does not drive the damping of baryon fluctuations.[14] However, adiabatic perturbations do damp prior to and during the recombination epoch by the viscous coupling of the baryons to the photons on all scales below the Silk mass in baryons,[17,18] $M_{BS} \approx 3\times 10^{13}\,\Omega_B^{-1/2} \times \Omega_\nu^{-3/4} h^{-5/2} M_\odot$, $\sim 10^{15} M_\odot$ for an Einstein-de Sitter universe with $\Omega_B = 0.01$ and $m_{30} = 1$. The maximum neutrino Jeans mass encloses a mass in baryons in our standard case $M_{Bm} \approx (\Omega_B/\Omega_\nu) M_{\nu m}$ $\approx 1.2\times 10^{14} m_{30}^{-3}(\Omega_B h^2/0.01) M_\odot$. Depending upon parameter choice, M_{BS} may be greater or less than M_{Bm}, but they are of same order.

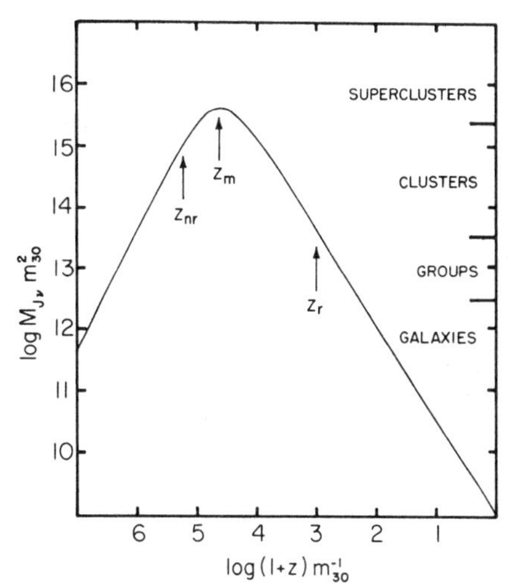

FIG. 1. The Jeans mass in solar mass units for one neutrino flavor vs red shift, along with approximate ranges for various large-scale structures. The neutrino mass in units of 30 eV is m_{30}. Perturbations on scales smaller than the maximum value of the Jeans mass are damped.

In the conventional picture of galaxy formation,[19] isothermal fluctuations (which do not suffer viscous damping) and all surviving adiabatic fluctuations grow at a rate $\sim(1+z)^{-1}$ in an Einstein-de Sitter universe after photon decoupling. In the presence of massive neutrinos, this result is drastically altered: *Baryon fluctuations smaller than the critical mass* M_{Bm} *grow extremely slowly in the linear regime until they first exceed the instantaneous Jeans mass.* The growth of perturbations of wavelength λ is described by the linearized Einstein field equations perturbed about a Friedmann background. After recombination, these reduce to an equation for the spatial Fourier transform of the fractional baryon and neutrino density fluctuations $\delta_B(\vec{k}, t)$ and $\delta_\nu(\vec{k}, t)$:

$$\frac{d^2\delta_B}{dt^2} + \frac{2\dot{a}}{a}\frac{d\delta_B}{dt} = 4\pi G\rho_T\left(\frac{\rho_B}{\rho_T}\delta_B + \frac{\rho_\nu}{\rho_T}\delta_\nu\right) \qquad (2)$$

which is valid in the linear regime, $\delta_B \ll 1$, $\delta_\nu \ll 1$, when matter pressure is neglected. Here, $a(t)$ is the Robertson-Walker scale factor and $\rho_T = \rho_\nu + \rho_B$. This is coupled to an integral equation for δ_ν obtained from the linearized Vlasov equation for the collisionless neutrino distribution function describing growth for $\lambda \geq \lambda_{J\nu}$ and damping for $\lambda \leq \lambda_{J\nu}$, where $\lambda_{J\nu}(z)$ is the instantaneous Jeans length.[15] Growth of neutrino density fluctuations occurs on scales $\lambda > \lambda_{\nu m}$ for $z < z_m$, but is suppressed in the radiation-dominated era.[20] Not until the epoch of photon decoupling can the baryons respond to the neutrino perturbations. Prior to recombination, no growth of baryon or photon fluctuations occurs on scales below the photon Jeans length, which is approximately the horizon size. After decoupling, δ_B rapidly responds to the neutrino perturbation (even if $\delta_B = 0$ initially), until $\delta_B \approx \delta_\nu$ on scales $\lambda > \lambda_{\nu m}$. Growth continues until $\delta_\nu \sim (1+z)^{-1}$ exceeds unity, when the enhanced self-gravity enables the first structures of mass $\sim M_{\nu m}$ to condense from the expanding background.

To consider scales $\lambda \leq \lambda_{\nu m}$, we take δ_ν to be zero below $\lambda_{\nu m}$ as our initial condition. For wavelengths $< \lambda_{J\nu}(z)$, we find the solutions in the Einstein-de Sitter case: $\delta_B \sim t^{p_\pm}$, where $p_\pm = \pm\frac{1}{6}[(1 + 24\rho_B/\rho_T)^{1/2} \mp 1]$. Only the p_- mode grows. If baryons dominate the mass density, $p_+ = \frac{2}{3}$, yielding the usual growing mode. If neutrinos dominate, p_+ is very small, $\sim\frac{1}{20}$ for $\Omega_B h^2 = 0.01$, and

$m_{30}=1$: Growth is negligible.

The growing part of the solution for $\lambda_{J\nu}(z) \leq \lambda \leq \lambda_{J\nu}(z_r)$ yields the fluctuation spectrum $\delta\rho_B/\rho_B(z) = \frac{3}{5}[\delta\rho_B/\rho_B(z_r)](\Omega_B/\Omega_\nu)[M/M_{J\nu}(z)]^{2/3}$. We write the fluctuation spectrum at z_r as $\delta\rho_B/\rho_B(z_r) = \delta\rho_B/\rho_B(z_m) = (M_c/M)^{n/6+1/2}$ which corresponds to a Fourier power spectrum $|\delta_B|^2 \propto k^n$. Note that the mass spectrum flattens by $M^{2/3}$ at $M < M_{J\nu}(z_r)$, which corresponds to a power spectrum $|\delta_B|^2 \propto k^{n-4}$. If $n<1$, fluctuations on scale $M_{\nu m}$ are the first to condense after the recombination epoch.

We now show that condensation of galactic or subgalactic masses requires that either galaxies were formed in the nonlinear neutrino regime, $\delta_\nu \gtrsim 1$, or the baryon fluctuations were substantially nonlinear on galactic scales, $\delta_B \gtrsim 1$, at recombination. The nonlinear evolution of neutrino structures of mass $M_{\nu m}$ is likely to resemble a variant of Zel'dovich's "pancake" theory of galaxy formation from adiabatic fluctuations.[21] According to this theory, the generic type of collapse is asymmetric, with preferential collapse along one axis occurring at red shift z_{nl}. The Zel'dovich theory faces a serious constraint due to the relatively large amplitudes required for the initial adiabatic fluctuations: $\delta_B \gtrsim 10^{-3}(1+z_{nl})$ at decoupling. Since neutrino fluctuations grow between z_m and z_r, the value of δ_B at decoupling is reduced to $\sim\delta_\nu(z_m) \lesssim 5\times10^{-5}(1+z_{nl})m_{30}^{-1}$. The associated temperature fluctuations in the cosmic background radiation are[18] $\lesssim \frac{1}{3}\delta_B(z_r) \approx \frac{1}{4}\delta_\nu(z_m)$ over scales $\sim\lambda_{\nu m}$ and are below recent observational upper limits ($\Delta T/T \lesssim 2\times10^{-4}$ over scales $\sim 4'$).[22]

A difficulty appears in the galaxy distribution where significant power would be expected over large scales, corresponding to M_{Bm}. The two-point galaxy correlation function[19] has the form $\xi(r) = (r_0/r)^{1.8}$ with $r_0 \approx 4h^{-1}$ Mpc, whence the mass scale below which clustering is nonlinear is $M_{Bnl} \approx 5\times10^{14}\Omega_B h^{-1}[r_0 h/(4\text{ Mpc})]^3 M_\odot$. There is, however, tentative evidence for larger-scale nonlinear structure.[23] If we require M_{Bnl} to exceed M_{Bm}, we require $m_{30} > 2.9[(4\text{ Mpc})/r_0]$. This requires $\Omega_\nu h^2 \sim 1$ and $M_{Bm} \approx 4.6\times10^{12}(\Omega_B h^2/0.01)M_\odot$. This mass scale corresponds to a group of ~ 80 galaxies.[19]

The mass limit deduced using the correlation function disagrees with the neutrino mass inferred from measured mass-to-light ratios if we assume that the luminous baryonic matter has not segregated from the dark neutrino matter. Since $M_{\nu m}/M_{Bm} \approx 800(M/L)^{-1}m_{30}h^{-1}$, and $M/L \approx 500h$ for rich clusters, we require $m_{30} \approx 0.6h^2$ and hence $M_{\nu m} \approx 10^{16}h^{-4}M_\odot$. The problem is still more severe in the case of groups of galaxies, where $M/L < 500h$. We conclude that substantial dissipation on cluster scales is required in order to segregate the luminous matter from the neutrinos. We cannot rule out this possibility. Since in such a picture, galaxy formation occurs after the collapse of the pancake, it is difficult to see how galaxy halos composed of neutrinos could form given the high velocity dispersion expected of this cluster material [$\sim 1300 m_{30}^{-1}(1+z_{nl})$ km/s].

The alternative possibility that $\delta_\nu \sim 0$ initially constrains the initial amplitudes of isothermal baryon density fluctuations. To achieve $\delta\rho_B/\rho_B \sim 1$ now on scales $\sim M_{Bnl}$, we infer that $\delta\rho_B/\rho_B \sim 0.02(0.01/\Omega_B)[(4\text{ Mpc})/r_0 h]^2 m_{30}^{-2}[M_{Bnl}/M_B]^{n/6+1/2}$ at z_r on scales corresponding to a mass in baryons of M_B. Hence, on galactic scales $\delta\rho_B/\rho_B \approx m_{30}^{-2}$ for $1 \lesssim n \lesssim 4$. A value of n in this range is required for isothermal fluctuations to account for the galaxy correlation function.[19] Such a theory seems unattractive (galaxies exist because galaxies have always existed).

In summary, the ratio of right- to left-handed neutrinos is likely to be $\ll 1$ and thus primordial nucleosynthesis is unaffected by the presence of a neutrino mass in the standard big-bang model. To avoid conflict with q_0 determinations implies that $m_\nu \lesssim 190$ eV, otherwise $\Omega_\nu > 2$. If the dominant mass density of the Universe is in massive neutrinos (e.g., $m_\nu > 1$ eV if $\Omega_B h^2 \sim 0.01$), the gravitational growth of density fluctuations is inhibited in the linear regime on mass scales $< M_{\nu m}$. Moreover, the large-scale structure displays a prominent feature at $M_{\nu m} \sim 4\times10^{15} m_{30}^{-2} M_\odot$, in which case comparison with the galaxy distribution requires $m_\nu > 85$ eV. Alternatively, nonlinear structure on galaxy scales must be present in the very early Universe. We conclude that confirmation of a neutrino mass will require a significant revision of the gravitational instability theory of galaxy formation as outlined here.

We thank our colleagues at Berkeley, especially G. Lake and M. Wilson, for numerous discussions. This work was supported by the National Science Foundation under Grants No. AST-79-15244 and No. AST-79-23243.

---

[1]R. Cowsik and J. McClelland, Phys. Rev. Lett. 29, 669 (1972); J. E. Gunn, B. W. Lee, I. Lerche, D. N. Schramm, and G. Steigman, Astrophys. J. 223, 1015 (1978), and references therein.

[2]S. Tremaine and J. E. Gunn, Phys. Rev. Lett. 42,

407 (1979).
[3]V. A. Lyubimov, E. G. Novikov, V. Z. Nozik, E. F. Tretyakov, and V. S. Kowsik, to be published.
[4]F. Reines, S. W. Sobel, and E. Pasierb, Phys. Rev. Lett. 45, 1307 (1980).
[5]P. Ramond, California Institute of Technology Report No. CALT-68-709, 1979 (to be published); E. Witten, Phys. Lett. 91B, 81 (1980).
[6]See also S. L. Shapiro, S. A. Teukolsky, and I. Wasserman, Phys. Rev. Lett. 45, 669 (1980).
[7]A. Sandage, Astrophys. J. 173, 485 (1972).
[8]M. Davis, J. J. Geller, and J. Huchra, Astrophys. J. 221, 1 (1978).
[9]S. M. Faber and J. S. Gallagher, Annu. Rev. Astron. Astrophys. 17, 135 (1979).
[10]See also D. N. Schramm and G. Steigman, to be published.
[11]M. A. B. Beǵ, R. V. Budny, R. Mohapatra, and A. Sirlin, Phys. Rev. Lett. 38, 1252 (1977).
[12]J. Yang, D. N. Schramm, G. Steigman, and R. T. Rood, Astrophys. J. 277, 697 (1979).
[13]F. W. Stecker, Phys. Rev. Lett. 44, 1237 (1980).
[14]P. J. E. Peebles, Astrophys. J. 180, 1 (1973).
[15]J. R. Bond, G. Efstathiou, and J. Silk, to be published.
[16]I. H. Gilbert, Astrophys. J. 144, 233 (1965).
[17]B. J. T. Jones, Rev. Mod. Phys. 48, 107 (1976).
[18]J. Silk and M. Wilson, Phys. Scr. 21, 708 (1980).
[19]S. M. Fall, Rev. Mod. Phys. 51, 21 (1979), and references therein.
[20]P. Mészáros, Astron. Astrophys. 37, 225 (1974).
[21]Ya. B. Zel'dovich, in *Large Scale Structure of the Universe*, I.A.U. Symposium No. 79, edited by M. Longair and J. Einasto (Reidel, Dordrecht, 1977), p. 409, and references therein.
[22]R. B. Partridge, Astrophys. J. 235, 681 (1980).
[23]R. P. Kirshner, A. G. Oemler, and P. L. Schechter, Astron. J. 84, 951 (1979).

THE ASTROPHYSICAL JOURNAL, 243:L1–L4, 1981 January 1

MASSIVE NEUTRINOS AND GALAXY FORMATION

FRANS R. KLINKHAMER AND COLIN A. NORMAN
Sterrewacht, Leiden, The Netherlands
Received 1980 June 10; accepted 1980 October 1

ABSTRACT

We investigate the cosmological implications of the finite neutrino rest mass ($m_\nu \gtrsim 1$ eV) indicated by recent experimental results. Neutrinos decouple while still relativistic at temperatures of ~1 MeV. If galaxies form from adiabatic fluctuations, the Zel'dovich amplitude for density perturbations is reduced to $\sim 10^{-4}$, consistent with present observational limits on the microwave background. The minimum neutrino-mass perturbation, M, corresponds to the particle horizon when the neutrinos become nonrelativistic, with $M \propto m_\nu^{-2}$. For one dominant neutrino mass of order 90 eV, the mass and virialized velocity dispersion of these minimum perturbations are characteristic of those suggested for dark halos associated with individual galaxies ($M \sim 10^{13} M_\odot$, $V \sim 500$ km s^{-1}). Such a model implies a closed or flat universe.

Subject headings: cosmology — galaxies: formation — neutrinos

I. INTRODUCTION

An outstanding problem in cosmology is to establish a primordial density fluctuation spectrum which provides a characteristic mass that can be directly related to that of galaxies. A completely new possibility arises if the neutrino has finite mass, evidence for which is given by Reines, Sobel, and Pasierb (1980) and Lyubimov *et al.* (1980). As discussed in detail below, a new mass scale can be associated with the horizon when finite mass neutrinos become nonrelativistic.

The detailed calculations of this mass scale are given in § V, but we first discuss relevant aspects of neutrino properties (§ II), the thermal history of the universe (§ III), and primordial density fluctuations (§ IV).

II. NEUTRINO PROPERTIES

There appears to be no "deep" physical reason why the spin-$\frac{1}{2}$ neutrino must have zero rest mass in contrast to the case for photons (cf. Taylor 1976). Experiments show that neutrinos exist in at least two types and have negative helicity. It appears that the basic quarks and leptons are grouped in three generations: (e^-, ν_e, u, d), (μ^-, ν_μ, c, s), and (τ^-, ν_τ, t, b) (e.g., Harari 1978).

Some years ago it was suggested, by analogy with Cabibbo mixing of quarks, that the operators ν_e and ν_μ in the lepton currents of the Weinberg-Salam theory of weak interactions are orthogonal superpositions of ν_1 and ν_2 neutrino fields with nonzero masses. This leads to oscillations $\nu_e \rightleftarrows \nu_\mu$ provided we exclude the trivial case $\Delta \equiv |m_{\nu_1}{}^2 - m_{\nu_2}{}^2| = 0$ and zero mixing angle (e.g., Bilenkii and Pontecorvo 1977).

Reines, Sobel, and Pasierb (1980) presented evidence for neutrino instability by comparing neutral-current reaction rates ($\bar{\nu} + d \rightarrow n + p + \bar{\nu}$), which allow all neutrino types, and charged current rates ($\bar{\nu}_e + d \rightarrow 2n + e^+$), which allow only the $\bar{\nu}_e$ type. The experimentally determined reaction rates differ significantly from the theoretical ones, indicating $\bar{\nu}_e$ instability over a distance of 11 m from the reactor to the experimental setup. Preliminary results give $\Delta \sim 1$ eV^2, but this depends strongly on the spectrum of $\bar{\nu}_e$ from the reactor. This shows that there is at least one neutrino with a mass $m_{\nu_1} \sim 1$ eV, and, if $m_{\nu_1} \gg 1$ eV, the mass of the other neutrino(s) is roughly the same. Lyubimov *et al.* (1980) have announced evidence for finite neutrino masses, inferred from β-decay of tritium, giving $14 \lesssim m_{\nu_e} \lesssim 46$ eV.

Earth-based experiments (Particle Data Group 1980) give upper limits for the (e, μ, τ) neutrino masses of (60 eV, 0.57 MeV, 250 MeV). The hot big-bang cosmology gives two constraints: (1) the helium abundance limits the number of degrees of freedom of thermal equilibrium particles at $T \sim 1$ MeV, implying there are at most ~3 left-handed neutrinos (Yang *et al.* 1979; see, however, Stecker 1980); and (2) the age of the universe or its mean density ($\lesssim 3 \times 10^{-29}$ g cm^{-3}) constrains the total neutrino mass

$$\Sigma_i \, m_{\nu_i} < 100 \text{ eV} ;$$

([2] cf. Cowsik and McClelland 1972). Hence, we will consider three types of left-handed neutrinos whose masses are necessarily of the Majorana type (Witten 1980*b*). For numerical estimates, we assume a mean neutrino mass

$$\tilde{M} \equiv \tfrac{1}{3}\Sigma_i \, m_{\nu_i} \sim 10\text{–}30 \text{ eV} , \quad \text{and} \quad m_{\nu_i} \sim m_{\nu_j} .$$

III. THERMAL HISTORY

We take as a basis for our estimates the standard hot big-bang model (Weinberg 1972) with nondegenerate neutrino content. An additional justification is that in several grand unified theories (GUTs), e.g., SU(5), and SO(10), the generated baryon and lepton numbers are equal.

The effects of finite neutrino mass occur at a complex

cosmological epoch, close to the epochs of recombination and equal energy content of radiation and baryons. We need a firm parametrical tightrope, which we choose to be the photon temperature, T. The cosmological neutrino history runs as follows: Neutrinos decouple from other leptons at a temperature of $\sim$1 MeV, when time scales of weak interactions are too long compared with the expansion time scale (Wagoner 1980). From that time on, they will expand freely. Because the neutrinos decouple while still relativistic ($kT \gg m_\nu c^2$), they will always remain roughly as abundant as photons. For three types of left-handed neutrinos, the total neutrino number density before decoupling is (Steigman 1979) $n_\nu = (3/8)g_\nu n_\gamma = (9/4)n_\gamma$, with $g = g_B + (7/8)g_F$, the effective number of degrees of freedom of relativistic bosons and fermions. After the e^+-e^- annihilation, the photons are heated by a factor $(11/4)^{1/3}$ as follows from entropy conservation of interacting particles, and the density of neutrinos of all types is

$$n_\nu = \frac{9}{11} n_\gamma . \qquad (1)$$

We now calculate the following energy densities (g cm^{-3}):

photons,

$$\rho_\gamma = 4.8 \times 10^{-33} n_\gamma T_{\text{eV}} ,$$

rest mass of neutrinos,

$$\rho_{\nu_m} = 1.5 \times 10^{-33} \tilde{M}_{\text{eV}} n_\gamma ,$$

relativistic neutrinos,

$$\rho_{\nu_r} = \frac{7}{16} g_\nu \left(\frac{4}{11}\right)^{4/3} \rho_\gamma = 0.68 \rho_\gamma , \qquad (2)$$

where subscript eV denotes quantities evaluated in electron volts. For the present densities with $\rho_{\gamma_0} = 6.85 \times 10^{-34}$ g cm^{-3}, we have $\rho_{\gamma_0}:\rho_{\nu_0}:\rho_{b_0}$ is $1:1.2 \times 10^4 (\tilde{M}/10 \text{ eV}):2.9 \times 10^2(\Omega_{b_0}/0.01)$, with Ω_{b_0} the ratio of baryon density to critical density $\rho_{cr} = (3H_0^2/8\pi G)$. Here, and in the following, we will take $h = 1$ in $H_0 = h$ 100 km s^{-1} Mpc^{-1}, and the present photon temperature $T_0 = 3$ K. Estimates for the visible matter give $\Omega_{b_0} \sim 0.01$. We find the epoch of equal neutrino rest mass and radiation energy density is at

$$T_{\gamma\nu} = 3.1 \left(\frac{\tilde{M}}{10 \text{ eV}}\right) \text{eV} , \qquad (3)$$

whereas equal energy densities in baryons and radiation occur at $T_{b\gamma} = 7.8\, \Omega_{b_0}$ eV. That the present energy densities of neutrinos and baryons are comparable appears to be accidental: $(n_b/n_\gamma) \sim 10^{-8}\Omega_{b_0}$ and $(m_b/m_\nu) \sim 10^8$. It has been proposed that GUTs may explain the first ratio (e.g., Fry, Olive, and Turner 1980), and one might speculate that GUTs may also explain the second ratio, if, for example, neutrino masses arise from radiative corrections with superheavy bosons (cf. Witten 1980*a*).

IV. GALAXY FORMATION: DENSITY FLUCTUATIONS

Prior to recombination, any density perturbation can be split into two compressional modes: (1) isothermal modes consisting of baryon fluctuations on a smooth background (hence varying n_b/n_γ), whose amplitudes are frozen by the radiation drag; and (2) adiabatic modes consisting of both baryon and radiation fluctuations (hence constant n_b/n_γ) oscillating as sound waves and strongly damped during recombination for perturbations with masses less than the Silk mass [$\sim 5 \times 10^{13}(\Omega h^2)^{-5/4} M_\odot$] from which a photon can random walk and escape in an expansion time scale (cf. Press and Vishniac 1980). Note that these comments on amplitudes only hold for perturbations smaller than the particle horizon.

When radiation is last scattered, it is somewhat Doppler shifted by matter perturbations which start to fall in, giving rise to small-angle temperature fluctuations in the cosmic background radiation (Davis and Boynton 1980). In addition, adiabatic perturbations have their own radiation fluctuations which give them larger temperature fluctuations than isothermal ones. Recent observational limits appear to just rule out adiabatic fluctuations ($\delta T/T \lesssim 10^{-4}$, Partridge 1980) unless reheating occurred. The major unsolved problem is to provide the large-amplitude (10^{-3} to $10^{-4} \gg$ thermal fluctuations) density perturbations when entering the horizon. Zel'dovich has advanced strong reasons to expect scale-independent metrical distortions which result in constant amplitudes of these perturbations (Zel'dovich 1972; Barrow 1980).

Recent developments set severe constraints on isothermal modes. It is most likely that baryon fluctuations, created at earlier times (e.g., by quantum gravity effects) are washed out at $T \sim M_x \sim 10^{15}$ GeV (Fry, Olive, and Turner 1980). Large-scale density fluctuations from before that period will be automatically of adiabatic character, since the ratio $kn_b/s \sim n_b/n_\gamma$, where s is the specific entropy per baryon, generated locally at $T \sim M_x$, is determined solely by microphysics. This argument does not hold if shear or spatial curvature fluctuations are present. GUT and graviton viscosities are ineffective for galaxy-sized perturbations independent of the behavior of the horizon (Ellis, Gaillard, and Nanopolous 1980).

V. ADIABATIC DENSITY FLUCTUATIONS: MASSIVE NEUTRINOS

Neutrinos will participate in any density fluctuation originating before $T \sim 1$ MeV. After decoupling, the neutrinos are a collisionless gas with only gravitational interactions. However, while the neutrinos are relativistic, all fluctuations entering the horizon will diffuse. In a relativistic fermion gas, the mean particle energy is $\sim 3kT$; hence the minimum scale of neutrino fluctuations is determined by the horizon at $T \sim (m/3)\nu$, and perturbations entering the horizon at $T \lesssim (m/3)\nu$ have their full Zel'dovich amplitude. Zakharov (1980) has considered density perturbations of relativistic collisionless particles and found small-scale fluctuations to damp rapidly, even faster than in the hydrodynamic case. These neutrino perturbations start growing at $T \sim (m/3)\nu$ instead of at recombination where $T \sim$ 0.3 eV. Neglecting curvature effects at these early times, we find that the Zel'dovich amplitude can be less than the required value at recombination by a factor of 10 $(m_\nu/10 \text{ eV})$. After recombination, the baryons are free to follow the enhanced neutrino fluctuations. The observational constraints on the microwave-background fluctuations now remain compatible with the adiabatic galaxy-formation model. Before proceeding to more quantitative results, we remark that Szalay and Marx (1976) used a minimum length of $\sim(\pi kT/G\rho_\nu m_\nu)^{1/2}$ at $kT \sim m_\nu c^2$. This is greater than our result for the size of an overdense perturbation $\lambda \sim \alpha ct \sim [(3\alpha^2 c^2)/(32\pi G\rho)]^{1/2}$. At time t the particle horizon is $2ct$ in proper distance. The relevant scale of the perturbation is not simply this maximum distance traveled by neutrinos, but is at most $\sim\alpha ct$, where $\alpha \sim 1$, because of the phase mixing process necessary for the perturbation damping (Bisnovaty-Kogan and Zel'dovich 1971).

The early history of the universe for photons and three types of decoupled left-handed neutrinos is described by Steigman (1979):

$$t(T) \sim 1.3 \times 10^{10} \left(\frac{T}{10 \text{ eV}}\right)^{-2} \text{s}. \quad (4)$$

To have condensations $(\delta \sim 1)$ at $z_i \sim 5$, we need an initial amplitude:

$$\delta_{\text{initial}} = \frac{1+z_i}{1+z^*} = 5.3 \times 10^{-5} N \left(\frac{\tilde{M}}{30 \text{ eV}}\right)^{-1} \left(\frac{1+z_i}{6}\right), \quad (5)$$

where $\tilde{M}$ is the mean mass of the three types of neutrinos, and the asterisk denotes the epoch when the neutrinos become nonrelativistic. We have introduced a parameter, N, which is 3 for $m_{\nu_i} \sim m_{\nu_j}$, and we will show below that for only one massive neutrino $N = 1$. The minimum perturbation mass, considering only neutrino rest masses, is

$$M_{\nu\,\min} \sim \frac{4\pi}{3}\left(\frac{N}{3}\right)^3 \tilde{M} n_\nu \left(\frac{\tilde{M}}{3}\right)\left[\alpha ct\left(\frac{\tilde{M}}{3}\right)\right]^3 \sim 7 \times 10^{12} N^3 \alpha^3 \left(\frac{\tilde{M}}{30 \text{ eV}}\right)^{-2} M_\odot. \quad (6)$$

We calculate the turnaround radius $(\delta \sim 1)$ at $z_i \sim 5$ from $5.5\rho_{\nu_m}(z_i)R^3_{\min} = M_{\nu\,\min}$, and the minimum halo radius will be approximately half of this value after virialization,

$$R_{h\,\min} = 130 N \left(\frac{\tilde{M}}{30 \text{ eV}}\right)^{-1} \alpha \left(\frac{1+z_i}{6}\right)^{-1} \text{kpc}, \quad (7)$$

and the corresponding velocity dispersion,

$$V_{h\,\min} = 480 N \left(\frac{\tilde{M}}{30 \text{ eV}}\right)^{-1/2} \alpha \left(\frac{1+z_i}{6}\right)^{1/2} \text{km s}^{-1}, \quad (8)$$

which gives the asymptotic velocity of a flat rotation curve $\sqrt{(2/3)}V_{h\,\min} \sim 400 \text{ km s}^{-1}$. Note that for only one massive neutrino (m_1) we have $M_{\nu\,\min} = \frac{4}{3}\Pi m_1 n_1 (m_1/3)[\alpha ct(m_1/3)]^3 = \frac{4}{3}\pi \tilde{M} n_\nu(\tilde{M})[\alpha ct(\tilde{M})]^3$. To make galaxy halos from neutrinos with

$$\Sigma_i\, m_{\nu_i} < 100 \text{ eV},$$

we require that one neutrino mass is $\sim$90 eV, much larger than the other two. This would not be compatible with Reines, Sobel, and Pasierb's (1980) preliminary results suggesting neutrino-mass spectra of types (1, 0, 0) eV up to $(\tilde{M}, \tilde{M}, \tilde{M})$. However, their result still allows the possibility of large Δ if the actual $\bar{\nu}_e$ spectrum from the reactor is not the one used for calculating the quoted value of Δ. Several GUT models for the generation of neutrino masses predict the m_ν to parallel the up quark masses (cf. Witten 1980*a;* some problems are discussed in Barbieri *et al.* 1980). This would automatically imply that there is one dominant neutrino mass. (Such a quark–neutrino-mass relation need not hold if the Higgs structure is enlarged to provide for substantial neutrino mixing [G. C. Branco, private communication].) We note in passing that massive neutrinos could reconcile "high" deuterium abundances, $\Omega_{b_0} \sim 0.01$, with a large mass-density of the universe in leptons, $\Omega_0 = 1.2(\tilde{M}/30 \text{ eV})h^{-2}$, but this may give problems with the age of the universe.

For these minimum neutrino perturbations (see [6]), the corresponding baryon fluctuations will be smoothed during recombination by radiative diffusion. The subsequent interaction between baryon fluctuations just after recombination and enhanced neutrino perturbations is rather uncertain. However, it seems very likely that the dissipative baryon material will be collected in the neutrino potential wells giving luminous galaxy cores in dark neutrino halos. We have tacitly assumed that the neutrino component is dominant.

Neutrino halos now separating out from the Hubble flow have entered the horizon at $1 + z = (1 + z^*)/(1 + z_i)$ (see eq. [5]) and hence will have masses a factor of $(1 + z_i)^p$ larger than perturbation (6), $(M_{H\nu_m} \propto z^{-p}$, $p = 1.5$ or 3 in a matter- or radiation-dominated universe; cf. [3]). Smaller perturbations which entered the horizon before T^* have their amplitudes damped as $\delta \propto t^{-3}$ (Zakharov 1980) and start to grow when the neutrinos become nonrelativistic. Minimal perturbations formed have masses a factor of $(1 + z_i)^{-1/2}$ times that of perturbation (6), and those not destroyed in the collapse of the larger systems might very well lead to smaller galaxies. We have thus found a range of masses typical of galaxies up to groups of galaxies. If $\Omega_0 < 1$, bound systems must have separated out at $z > \Omega_0^{-1} - 1$.

Hierarchical clustering leads to systems with typical masses larger by $\sim z_i^2$ ([6]; Press and Schechter 1974), but it may be difficult to account for superclustering in this way. One might expect that the other two neutrinos or superweakly interacting particles of low mass ($\lesssim 1$ eV) give supercluster masses, but in adiabatic perturbation schemes, it is difficult to see how these dynamically unimportant particles could determine these large-scale structures. If it is necessary to have a density contrast of ~ 1 for superclusters (10^{16} $M_\odot$), the initial amplitude upon entering the horizon should be larger than in equation (5) by a factor of $\sim 10^2$, and the Zel'dovich hypothesis would then be invalid.

Tremaine and Gunn (1979) have recently disclaimed the possibility of halos of massive neutrinos. Their argument is as follows: (1) from the present low temperature of the massive neutrinos they infer that these collapse in the same way as baryons; hence their contribution to the mean density of the universe must be less than that of luminous matter (an upper limit on the mean neutrino mass); (2) being collisionless, the maximum phase space density cannot increase, and comparing the isothermal distribution with the initial Fermi-Dirac distribution provides a lower limit on the neutrino mass; (3) for galaxy halos, the limits from (1) and (2) do not agree. Obviously our scenario disproves their assumption (1). If the neutrino halo is isothermal, their result (2) still holds for inferred core radius, R_c, and velocity dispersion, V,

$$m_\nu > 22\left(\frac{340 \text{ km s}^{-1}}{V}\right)^{1/4}\left(\frac{20 \text{ kpc}}{R_C}\right)^{1/2} \text{ eV}. \quad (9)$$

The velocity dispersion from our model (8) and the required neutrino mass are in agreement with the above values.

We have received preprints from two other groups who have made similar estimates but generally find larger characteristic masses. Bond, Efstathiou, and Silk (1980) adopt $\alpha = 2$ and take the maximum Jeans mass instead of that at T^*, giving a factor $\sim 10^2$ larger mass. Both choices are sensitive to the detailed physics of Landau damping. The difference of ~ 20 with Bisnovaty-Kogan, Lukash, and Novikov (1980) probably is due to (1) their use of $t^* \sim (G\rho_\nu)^{-1/2}$ which neglects the photon contribution and a numerical factor (see above); and (2) the ρ_ν^* used, whereas we consider only the contribution of the neutrino rest mass.

We emphasize that we have chosen to scale our equations with neutrino parameters giving characteristic masses compatible with those inferred for dark halos around individual galaxies. For smaller neutrino masses, halo-sized perturbations are damped, and one would need to invoke an alternative mechanism such as infall of cool neutrinos into pre-existing potential wells of baryons. In contrast, halos of massive neutrinos arise naturally in our scheme, in which dissipative baryon material aggregates in potential wells of massive neutrinos, whose size depends only on the neutrino mass.

The neutrino flux observable at the Earth from a massive halo of $\sim 10^{14}$ cm^{-2} s^{-1} is greater than the solar neutrino flux $\sim 10^{10}$–10^{11} cm^{-2} s^{-1}. The crucial difference is that, whereas the solar neutrinos have typical energies of $\sim$MeV, the halo neutrinos are nonrelativistic. Experimental detection of these halo neutrinos seems unfeasible, but the measurement of a dominant neutrino mass of ~ 90 eV would indirectly identify the major dynamical component in galaxy formation.

> Au milieu de la foule, errantes, confondues,
> Gardant le souvenir précieux d'autrefois,
> Elles cherchent l'écho de leurs voix éperdues.
>
> Charles Baudelaire, *Poèmes Divers*

We thank T. de Jong for giving us F. Reines's preprint, P. Hut and A. Masiero for discussions, and J. Silk and I. Novikov for exchanging preprints.

REFERENCES

Barbieri, R., Nanopolous, D. V., Morchio, G., and Strocchi, F. 1980, *Phys. Letters*, **90B**, 91.
Barrow, J. D. 1980, *Phil. Trans. Roy. Soc. London, A*, **296**, 273.
Bilenkii, S. M., and Pontecorvo, B. M. 1977, *Soviet Phys. Uspekhi*, **20**, 776.
Bisnovaty-Kogan, G. S., Lukash, V. N., and Novikov, I. D. 1980, preprint (SRI581).
Bisnovaty-Kogan, G. S., and Zel'dovich, Ya. B. 1971, *Soviet Astr.*, **14**, 758.
Bond, J. R., Efstathiou, G., and Silk, J. 1980, *Phys. Rev. Letters*, submitted.
Cowsik, R., and McClelland, J. 1972, *Phys. Rev. Letters*, **29**, 669.
Davis, M., and Boynton, P. 1980, *Ap. J.*, **237**, 365.
Ellis, J., Gaillard, M. K., and Nanopolous, D. V. 1980, preprint (TH. 2774-CERN).
Fry, J. H., Olive, K. A., and Turner, M. S. 1980, preprint (Fermi Inst. 80-07).
Harari, H. 1978, *Phys. Rep.*, **42C**, 235.
Lyubimov, V. A., Novikov, E. G., Nozik, V. Z., Tretyakov, E. F., and Kosik, V. S., 1980, preprint (ITEP 62).
Particle Data Group. 1980, *Rev. Mod. Phys.*, **52**, 51.
Partridge, R. B. 1980, *Phys. Scripta*, **21**, 624.
Press, W. H., and Schechter, P. 1974, *Ap. J.*, **187**, 425.
Press, W. H., and Vishniac, E. T. 1980, *Ap. J.*, **236**, 425.
Reines, F., Sobel, H. W., and Pasierb, E. 1980, *Phys. Rev. Letters*, submitted.
Stecker, F. W. 1980, *Phys. Rev. Letters*, **44**, 1237.
Steigman, G. 1979, *Ann. Rev. Nucl. Part. Sci.*, **29**, 313.
Szalay, A. S., and Marx, G. 1976, *Astr. Ap.*, **49**, 437.
Taylor, J. C. 1976, *Gauge Theories of Weak Interactions* (Cambridge: Cambridge University Press).
Tremaine, S., and Gunn, J. E. 1979, *Phys. Rev. Letters*, **42**, 407.
Wagoner, R. V. 1980, Les Houches Lectures.
Weinberg, S. 1972, *Gravitation and Cosmology* (New York: Wiley).
Witten, E. 1980*a*, *Phys. Letters*, **91B**, 81.
———. 1980*b*, preprint (HUTP 80/A031).
Yang, J., Schramm, D. N., Steigman, G., and Rood, R. T. 1979, *Ap. J.*, **227**, 697.
Zakharov, A. V. 1979, *Soviet Phys.—JETP*, **50**, 221.
Zel'dovich, Ya. B. 1972, *M.N.R.A.S.*, **160**, 1P.

FRANS R. KLINKHAMER and COLIN A. NORMAN: Sterrewacht, Postbus 9513, 2300 RA, Leiden, The Netherlands

Progress of Theoretical Physics, Vol. 66, No. 2, August 1981

Formation of Galaxies and Clusters of Galaxies in the Neutrino Dominated Universe

Humitaka SATO and Fumio TAKAHARA*)

Research Institute for Fundamental Physics
Kyoto University, Kyoto 606

(Received March 23, 1981)

We present a detailed picture on the formation of galaxies and clusters of galaxies in the neutrino dominated universe. The primordial density fluctuations in neutrino matter give rise to the formation of rich clusters of galaxies and not yet collapsed superclusters. For smaller scale, the isothermal density fluctuations in nucleon matter start to grow when the mass of neutrinos involved in the fluctuation exceeds the neutrino Jeans mass. These objects are identified with small groups of galaxies. In a cluster or a small group, whose motion deviates considerably from the cosmic expansion, the growth of density fluctuations is accelerated and leads to the formation of galaxies. The characteristic mass of galaxies is given as $10^{10}\sim10^{11}$ $M_\odot$ for nucleons and $10\sim40$ times larger for neutrinos.

§ 1. Introduction

Recent experimental suggestions on neutrino mass[1),2)] have stimulated the theoretical investigations of its effects on various astrophysical problems. One of the most significant effects of neutrino mass is the relevance to the missing mass problem in galaxies and clusters of galaxies and their formation process in the expanding universe. As has been shown by us[3)~5)] and other authors,[6)~9)] the instantaneous Jeans mass of the relic neutrinos takes the maximum value of $M_{J\nu,\max}\simeq10^{19}m_\nu^{-2}M_\odot$ at the epoch when neutrinos become non-relativistic, $z_{nr}\simeq2150\,m_\nu$, where m_ν denotes the mass of the heaviest stable neutrino in eV units. Since $M_{J\nu,\max}$ gives the typical mass scale of gravitational clustering of the relic neutrinos, the missing mass problem in a rich cluster of galaxies may be solved if $m_\nu\simeq20\sim30$ eV.

On the other hand, it is known that the density fluctuations in neutrino matter of the scale below the Jeans mass have been damped out by the phase randomization peculiar to collisionless systems. Then galaxies cannot been formed from the primordial density fluctuations in neutrino matter and the galaxy formation must be induced by other processes such as the growth of the density fluctuations in nucleon matter. However, Bond et al.[6)] have shown that the growth of density fluctuations in nucleon matter is severely suppressed by the presence of massive neutrinos, even if the mass of a nucleon density fluctuation exceeds the Jeans mass

*) Present address, Department of Physics, Kyoto University, Kyoto 606.

of nucleons.

At the first glance, this difficulty about galaxy formation seems to reject the neutrino dominated universe. In this paper, however, we show that this difficulty can be removed if we consider the growth of density fluctuations in a larger system like a rich cluster or a small group of galaxies, which has been decelerated faster than the background universe. This problem resembles the fragmentation of the collapsing cloud in the star formation. In fact we examine the fragmentation of a cluster or a small group of galaxies in the formation stage, extending Hunter's[10] analysis for the collapsing spherical cloud of uniform density; the inclusion of the expansion stage as well as the collapsing stage and the incorporation of two component fluid of neutrinos and nucleons. The mathematical details are described in the Appendices and the results are used in § 4.

Preceding § 4, we examine the growth of density fluctuations in neutrino matter and nucleon matter in the expanding universe, in §§ 2 and 3, respectively. In our scenario, rich clusters and superclusters are ascribed to the growth of neutrino density fluctuations, while small groups and galaxies are formed through the growth of isothermal density fluctuations in nucleon matter. Galaxies are concluded to be formed only in a clustered form. In § 5, we summarize the results and discuss the confrontation with observations.

§ 2. The growth of primordial neutrino density fluctuations in the expanding universe

In this section we briefly summarize the growth of the primordial density fluctuations in neutrino matter. The Jeans mass of the relic neutrinos is given in the non-relativistic regime as[4]

$$M_{J\nu}=8.0\times 10^{14} m_\nu^{-7/2}(1+z)^{3/2}\alpha^{-3}\mathrm{M}_\odot\,. \tag{1}$$

Here α is a numerical factor of the order of unity, which depends on a relation between the critical radius and the Jeans wave length. We will set $\alpha=2$ in the following numerical estimation. A density fluctuation begins to grow when the mass contained within it M_ν becomes larger than $M_{J\nu}$.

Another characteristic mass scale is the mass within the particle horizon $M_{\mathrm{Hor},\nu}$ defined as $4\pi\rho_\nu(a\chi_H)^3/3$, where a and ρ_ν are the cosmic scale factor and the neutrino rest mass density, respectively, and $\chi_H\equiv\int_0^t(cdt/a(t))$. Numerically $M_{\mathrm{Hor},\nu}$ is given as

$$M_{\mathrm{Hor},\nu}=3.8\times 10^{19} m_\nu^{-2}\left(\frac{1+z_{\mathrm{eq}}}{1+z}\right)^3\mathrm{M}_\odot \tag{2}$$

for $1+z\gtrsim 1+z_{\mathrm{eq}}\simeq 432 m_\nu$, where z_{eq} denotes the epoch of $\rho_\nu=\rho_r$, $\rho_r c^2$ being the radiation energy density. The behavior of these mass scales versus z is

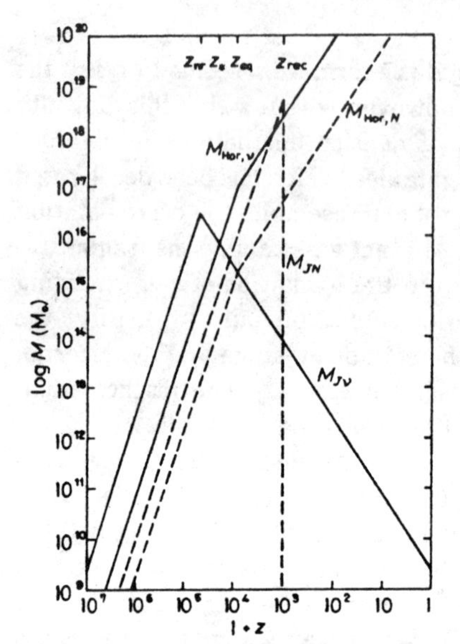

Fig. 1. Evolution of Jeans mass and horizon mass of neutrinos and nucleons versus $1+z$. We take $m_\nu = 20$ and $\Omega_{N0} h_{50}^2 = 0.02$.

shown in Fig. 1 as well as that of the Jeans mass of nucleon matter M_{JN} taking $\Omega_{N0} h_{50}^2 = 0.02$, Ω_{N0} and h_{50} being the nucleon density parameter and the Hubble constant H_0 in units of 50 km s^{-1} Mpc^{-1} at the present epoch. Later we introduce the neutrino density parameter at the present epoch $\Omega_{\nu 0}$ and the total density parameter is denoted by $\Omega_0 = \Omega_{\nu 0} + \Omega_{N0}$.

It is known that a neutrino density fluctuation is washed out if its scale is smaller than the horizon mass in the relativistic regime and below the Jeans mass in the non-relativistic regime. Before the Jeans mass becomes smaller than the horizon mass, the spatial scale of damping due to the phase randomization is determined by the distance which the neutrinos traverse after the decoupling until the epoch considered. In the regime when neutrinos become non-relativistic but $M_{J\nu} \gtrsim M_{Hor,\nu}$, this distance is smaller than $a\chi_H$, and the damping mass scale is smaller than $M_{Hor,\nu}$, However, the duration of this regime is very short as is seen in Fig. 1, if it ever exists. Then we can neglect a difference between the horizon mass and the damping mass effectively. As is seen in Fig. 1, $M_{J\nu}$ and $M_{Hor,\nu}$ intersect at $1+z_* \simeq 995 m_\nu$ or at $M_{J\nu} = M_{Hor,\nu} = M_*$, where M_* is given as

$$M_* = 7.8 \times 10^{15} (m_\nu / 20)^{-2} \, M_\odot . \tag{3}$$

The density fluctuation with mass M_ν grows when $M_\nu \geq M_{J\nu}$. For $M_\nu < M_*$, the fluctuations have been damped out when $M_\nu = M_{J\nu}$. For $M_* \leq M_\nu \leq M_{Hor,\nu}(z_{eq}) = 9.6 \times 10^{16} (m_\nu / 20)^{-2} \, M_\odot$, we may choose the starting epoch of growth as $z = z_{eq}$, effectively. We have no definite theory about the initial amplitude of fluctuation at the present, then we assume the power law type spectrum such as

$$\left(\frac{\delta\rho_\nu}{\rho_{cr}}\right)_i \simeq \left(\frac{\delta\rho_\nu}{\rho_\nu}\right)_i = K\left(\frac{M_\nu}{M_{Hor,\nu}(z_i)}\right)^{-p}, \tag{4}$$

where ρ_{cr} denotes the critical density at the initial epoch $z = z_i$.

Subsequent dynamical evolution of this density contrast is simply described if we assume a spherically symmetric density enhancement of uniform dust.[11),12)]

(See also Appendix A.) The density enhancement will first expand and turn to collapse finally. The time of total collapse is given as

$$t_{\rm coll}=\frac{\pi}{H_0\sqrt{\Omega_0}(1+z_i)^{3/2}}\left(\frac{\rho_{\rm cr}}{\delta\rho_\nu}\right)_i^{3/2}. \tag{5}$$

If $t_{\rm coll}$ is smaller than the present age of the universe t_0, it implies that a bound system has been formed until the present. Taking $z_i=z_{\rm eq}$, the condition $t_{\rm coll}\lesssim t_0$ becomes for $p=0$ in Eq. (4) as

$$K\gtrsim 3.25\times 10^{-4}(m_\nu/20)^{-1}(3H_0t_0\sqrt{\Omega_0}/2)^{-2/3}. \tag{6}$$

If Eq. (6) is satisfied for $p=0$ case, the mass range of the formed objects is rather wide, covering between $M_*=7.8\times 10^{15}(m_\nu/20)^{-2}M_\odot$ and at least $9.6\times 10^{16}(m_\nu/20)^{-2}M_\odot$. This range corresponds to from rich clusters up to superclusters of galaxies. On the other hand, if we choose $p=1/3$ in Eq. (4), the same index for the isothermal nucleon density fluctuation discussed in § 3, the condition $t_{\rm coll}\lesssim t_0$ is satisfies for $M_*\lesssim M_\nu\lesssim M^{\rm RC}_{{\rm max},\nu}$, where

$$M^{\rm RC}_{{\rm max},\nu}=2.8\times 10^{15}\left(\frac{K}{10^{-4}}\right)^3\left(\frac{3H_0t_0\sqrt{\Omega_0}}{2}\right)^2\left(\frac{m_\nu}{20}\right)M_\odot. \tag{7}$$

In this case, if K is chosen as $1.5\sim 2\times 10^{-4}$, the mass range of the formed objects becomes rather narrow and just corresponds to a rich cluster of galaxies. The larger scale density enhancement would be still in the forming stage of a supercluster. Observationally it seems that the relaxation has completed for a rich cluster but not for the larger clusterings. Therefore the spectrum with some positive value of p may be preferred for the initial density fluctuation.

One dimensional velocity dispersion of neutrinos in the relaxed system is calculated[4] for $p=1/3$ as

$$\sigma=9.2\times 10^7\left(\frac{M_\nu}{M_*}\right)^{1/6}\left(\frac{K}{10^{-4}}\right)^{1/2}\ {\rm cm/s}. \tag{8}$$

This σ agrees well with the observed velocity dispersion of galaxies in rich clusters. This may be a natural result of the violent relaxation, which results in the equipartition per unit mass rather than the equipartition per particle.[13]

§ 3. The growth of isothermal density fluctuations of nucleons in the expanding universe

In this section we examine the behavior of density fluctuations in nucleon matter of smaller scales than rich clusters. The density fluctuations in the adiabatic component have been damped out by photon viscosity below some critical mass which is nearly the same as the mass of rich clusters if $\Omega_{N0}h_{50}^2\simeq 0.01$

~ 0.1.[14),15)] From the observational upper limit of anisotropy in $3K$ radiation, $(\delta\rho_N/\rho_N)_i$ must be less than $\sim 10^{-3}$ and then $(\delta\rho_N/\rho_{cr})_i \ll (\delta\rho_\nu/\rho_{cr})_i$ for larger scales. On the other hand, the density fluctuation in the isothermal component may survive without dissipation until the recombination epoch with rather high amplitude if $\Omega_{N0} h_{50}^2 \simeq 0.01 \sim 0.1$.[15)] For the larger scales, it may also be ignored compared to the neutrino density fluctuation. But for the smaller scale like $M_N \simeq M_\nu(\Omega_{N0}/\Omega_{\nu 0}) \ll M_*(\Omega_{N0}/\Omega_{\nu 0})$, such fluctuations may work as a seed of a small group of galaxies, a binary galaxy or an individual galaxy.

As shown in the Appendices, a density fluctuation in nucleon matter can grow effectively only after the neutrino mass involved in this enhanced region exceeds the Jeans mass of neutrinos. Then we must take the starting epoch of growth as

$$1 + z_i = (1 + z_{rec})\left(\frac{M_\nu}{M_{J\nu}(z_{rec})}\right)^{2/3} \tag{9}$$

for $M_\nu \leq M_{J\nu}(z_{rec}) \simeq 8.8 \times 10^{13} (m_\nu/20)^{-7/2} M_\odot$ and

$$z_i = z_{rec} \tag{10}$$

for $M_* \geq M_\nu \geq M_{J\nu}(z_{rec})$, where $z_{rec} \approx 1000$ is the redshift factor at the recombination epoch.

Concerning the initial amplitude of fluctuations, we adopt, for definiteness, the picture of Gott and Rees,[16)] which is based on Zeldovich's hypothesis[17)] that the amplitude of density fluctuation has a constant value K when it enters the horizon. According to Gott and Rees,[16)] we assume the initial amplitude of the isothermal density fluctuation as

$$\left(\frac{\delta\rho_N}{\rho_N}\right)_i = \frac{K}{10^{-4}}\left(\frac{M_N}{3.5 \times 10^8 M_\odot}\right)^{-1/3}\left(\frac{\Omega_{N0} h_{50}^2}{0.02}\right)^{-2/3}. \tag{11}$$

For sufficiently small size, the isothermal fluctuation seems to be already in nonlinear regime at the recombination epoch. However, the velocity field of nucleons has not been disturbed so much and follows the general expansion, and moreover $\delta\rho_N/\rho_{cr}$ is still in linear regime. Then we can adopt Eq. (9) as the starting epoch of the growth of the fluctuation. The relatively high initial value of $(\delta\rho_N/\rho_N)_i$ may affect the ratio of ρ_ν/ρ_N in the formed objects, i.e., the ratio of the dynamical mass to the luminous mass. Generally, this ratio is smaller in small groups of galaxies and galactic halos than in rich clusters,[18)] which may be qualitatively consistent with the large value of $(\delta\rho_N/\rho_N)_i$.

Now we examine the subsequent growth of the density fluctuations given by Eq. (11). The condition that the collapse has completed until t_0 is written as

$$\frac{\pi}{H_0\sqrt{\Omega_0}(1+z_i)^{3/2}}\left(\frac{\rho_N}{\delta\rho_N}\right)_i^{3/2}\left(\frac{\Omega_0}{\Omega_{N0}}\right)^{3/2} \leq t_0\,, \tag{12}$$

where we approximate $(\delta\rho_N/\rho_{cr})_i \simeq (\delta\rho_N/\rho_N)_i(\Omega_{N0}/\Omega_0)$, the validity of which is seen in Appendix B. From Eqs. (9)~(12), we see the collapse condition is satisfied for $M^{SG}_{min,N} \lesssim M_N \lesssim M^{SG}_{max,N}$ and $M^{SG}_{min,\nu} \lesssim M_\nu \lesssim M^{SG}_{max,\nu}$, where

$$M^{SG}_{min,N} = 1.5\times 10^{13}\left(\frac{K}{10^{-4}}\right)^{-3}\left(\frac{m_\nu}{20}\right)^{-7}\left(\frac{\Omega_{N0}h_{50}}{0.02}\right)^{2}\left(\frac{3t_0H_0\sqrt{\Omega_0}}{2}\right)^{-2}\frac{\Omega_0}{30\Omega_{N0}}\,M_\odot\,, \quad (13)$$

$$M^{SG}_{min,\nu} \simeq \frac{\Omega_{\nu0}}{\Omega_{N0}}M_{min,N}$$

$$= 4.5\times 10^{14}\left(\frac{K}{10^{-4}}\right)^{-3}\left(\frac{m_\nu}{20}\right)^{-7}\left(\frac{\Omega_{N0}h_{50}}{0.02}\right)^{2}\left(\frac{3t_0H_0\sqrt{\Omega_0}}{2}\right)^{-2}\left(\frac{\Omega_0}{30\Omega_{N0}}\right)^{2}M_\odot\,, \quad (14)$$

$$M^{SG}_{max,N} = 5.8\times 10^{11}\left(\frac{K}{10^{-4}}\right)^{3}\left(\frac{\Omega_{N0}h_{50}}{0.02}\right)^{-2}\left(\frac{3t_0H_0\sqrt{\Omega_0}}{2}\right)^{2}\left(\frac{30\Omega_{N0}}{\Omega_0}\right)^{3}M_\odot \quad (15)$$

and

$$M^{SG}_{max,\nu} \simeq \frac{\Omega_{\nu0}}{\Omega_{N0}}M_{max,N}$$

$$= 1.7\times 10^{13}\left(\frac{K}{10^{-4}}\right)^{3}\left(\frac{\Omega_{N0}h_{50}}{0.02}\right)^{-2}\left(\frac{3t_0H_0\sqrt{\Omega_0}}{2}\right)^{2}\left(\frac{30\Omega_{N0}}{\Omega_0}\right)^{2}M_\odot\,. \quad (16)$$

In order that any collapsed object has been formed until t_0, $M^{SG}_{min,\nu}$ must be smaller than $M_{J\nu}(z_{rec}) \simeq 8.8\times 10^{13}(m_\nu/20)^{-7/2}\,M_\odot$ and $M_{max,\nu}$ must be larger than $M_{J\nu}(z_{rec})$. This is satisfied for $K \gtrsim 1.72\times 10^{-4}(m_\nu/20)^{-7/6}$. The mass range of the formed objects strongly depends on the values of m_ν and K as seen in Eqs. (13)~(16). The mass of the firstly formed object is $M_{J\nu}(z_{rec})$ and larger or smaller objects will be successively formed later. This feature is in clear contrast to the ordinary theory of galaxy formation in nucleon dominated universe, where smaller objects are formed earlier and larger one are formed later.

In this scenario, the mass of the firstly formed object is a characteristic mass of a small group of galaxies. Observationally most small groups of galaxies seem to be unvirialized yet and it is not certain whether they have completed the collapse. Then the value of K may not be so different from $1\sim 2\times 10^{-4}$. If this is true, individual galaxies have not been formed in the generally expanding universe. In our scenario, the galaxies are formed in rich clusters or in small groups, the motion of which considerably deviates from the general expansion of universe, as will be shown in the next section.

Finally we calculate the one-dimensional velocity dispersion of formed objects as

$$\sigma = 1.9\times 10^{7}\left(\frac{M_\nu}{8.8\times 10^{13}\,M_\odot}\right)^{1/2}\left(\frac{m_\nu}{20}\right)^{4/3}\left(\frac{\Omega_0}{30\Omega_{N0}}\right)^{-1/3}\left(\frac{K}{10^{-4}}\right)^{1/2}\left(\frac{\Omega_{N0}h_{50}^2}{0.02}\right)^{-1/3}\text{cm/s} \quad (17)$$

for $M_\nu < 8.8\times 10^{13}(m_\nu/20)^{-7/2}\ M_\odot$ and

$$\sigma = 1.9\times 10^7\left(\frac{M_\nu}{8.8\times 10^{13}\ M_\odot}\right)^{1/3}\left(\frac{m_\nu}{20}\right)^{1/6}\left(\frac{\Omega_0}{30\Omega_{N0}}\right)^{1/3}\left(\frac{K}{10^{-4}}\right)^{1/2}\left(\frac{\Omega_{N0}h_{50}^2}{0.02}\right)^{-1/3}\ \text{cm/s} \tag{18}$$

for $M_\nu \geq 8.8\times 10^{13}(m_\nu/20)^{-7/2}\ M_\odot$. This value of σ for small groups of galaxies is smaller than that of rich clusters by an order of magnitude and seems to be consistent with observations. But for galaxy mass scale Eq. (17) gives a too small value compared with observation.

§ 4. Growth of density fluctuations in a proto-cluster of galaxies

In this section we consider the behavior of the density fluctuations in a proto-cluster of galaxies or in a proto small group of galaxies, whose dynamics is given in Appendix A. Hereafter we call such a cluster or a group a parent cloud. We investigate the linear growth of density fluctuations in a parent cloud, extending Hunter's work[10] on star formation in a contracting cloud. The extension is done into two directions; one is the inclusion of the expanding stage as well as the contracting stage and the other is the treatment of the two component matters of neutrinos and nucleons.

The basic formulation is given in Appendix B. We expand the density ρ_i, the velocity $\boldsymbol{u}_i$ and the gravitational field Ψ around the given basic flow and denote the perturbed quantities by $\delta\rho_i$, $\delta\boldsymbol{u}_i$ and $\delta\Psi$, where the suffix i represents neutrinos $i=\nu$ and nucleons $i=N$. We assume that $\delta\rho_\nu=0$ and $\delta\rho_N\neq 0$ at the initial stage. Since we are considering smaller size fluctuations in a larger parent cloud which deviates from the general expansion of the universe, $\delta\rho_\nu$ has been washed out by the phase randomization but $\delta\rho_N$ in the isothermal component is assumed to exist as discussed in § 3.

The behavior of density fluctuations depends on the size of the perturbed region. If the neutrino mass in this region M_ν is smaller than the Jeans mass $M_{J\nu}$, the growth of fluctuation is severely suppressed even if M_N exceeds M_{JN} as discussed in Appendix C. Thus every fluctuation starts to grow when $M_\nu = M_{J\nu}$. In a parent cloud, $M_{J\nu}$ is given as

$$M_{J\nu} = M_{J\nu,i}(\sec^3\beta/\sec^3\beta_i), \tag{19}$$

where i denotes the value of the epoch when the parent cloud starts to grow and β represents a kind of time coordinate defined in Eq. (A·5) or (A·7). For a given mass M_ν, the condition $M_\nu \geq M_{J\nu}$ is satisfied in the range $\beta_1 \leq \beta \leq \beta_2$ $(=-\beta_1)$, β_1 and β_2 being determined by $\sec^3\beta_{1,2} = (M_\nu/M_{J\nu,i})\sec^3\beta_i$. It is to be noted that $M_{J\nu}$ taken the minimum value $M_{J\nu,\min} = M_{J\nu,i}\cos^3\beta_i$ at the maximum expansion stage

$\beta=0$ and $M_{J\nu,\min}$ turns out to be $\sim 10^{10}$ M$_\odot$, a characteristic mass of a galaxy.

As shown in Appendix B, the temporal behavior of $\delta\rho_\nu$ and $\delta\rho_N$ is given as

$$\frac{\delta\rho_N}{\rho_N}=\left(\frac{\delta\rho_N}{\rho_N}\right)_{\beta=\beta_1}\cdot\frac{\Omega_{N0}}{\Omega_0}\cdot\left[\frac{dH_2/d\beta}{(dH_2/d\beta)_{\beta=\beta_1}}+\frac{\Omega_{\nu 0}}{\Omega_{N0}}\right] \tag{20}$$

and

$$\frac{\delta\rho_\nu}{\rho_\nu}=\left(\frac{\delta\rho_N}{\rho_N}\right)_{\beta=\beta_1}\cdot\frac{\Omega_{N0}}{\Omega_0}\left[\frac{dH_2/d\beta}{(dH_2/d\beta)_{\beta=\beta_1}}-1\right], \tag{21}$$

in the range $\beta_1\leq\beta\leq\beta_2$, where the function $H_2(\beta)$ is defined in Eq. (B·16). The behavior of these density contrast is depicted in Fig. 2.

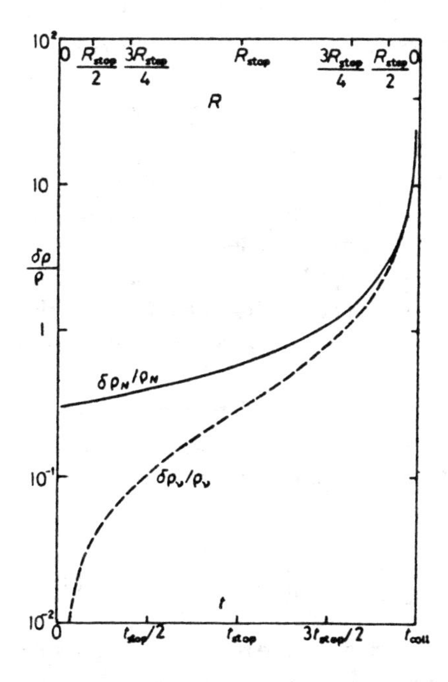

Fig. 2. An example of the evolution of the density contrast in a parent cloud. The abscissa is the time coordinates. The corresponding radius of the parent cloud is shown in the upper abscissa. For the parent cloud of 7.8×10^{15} M$_\odot$, the perturbation of $M_N=1.3\times 10^{10}$ M$_\odot$ and $M_\nu=3.9\times 10^{11}$ M$_\odot$ starts to grow at $\beta=-3\pi/8$ with initial amplitude of $\delta\rho_N/\rho_N=0.3$ ($K=10^{-4}$ in Eq. (11)) and $\delta\rho_\nu/\rho_\nu=0$.

(i) *Rich cluster of galaxies*

First, we discuss a fluctuation in a rich cluster with mass $M=M_*=7.8\times 10^{15}(m_\nu/20)^{-2}$ M$_\odot$ and with the initial density fluctuation of $(\delta\rho_\nu/\rho_{cr})_i\equiv\cos^2\beta_i=4.6\times 10^{-4}$ at $1+z_i=8640\times(m_\nu/20)$. For these parameters, $t_{coll}\simeq 0.6t_0$ and $M_{J\nu,\min}=2.2\times 10^{10}$ M$_\odot$. If we take Eq. (11) for $(\delta\rho_N/\rho_N)_{\beta=\beta_1}$, the density contrast at $\beta=0$ is given as

$$\left(\frac{\delta\rho_\nu}{\rho_\nu}\right)_{\beta=0}\simeq\left(\frac{\delta\rho_N}{\rho_N}\right)_{\beta=0}$$

$$\simeq\frac{K}{10^{-4}}\left(\frac{M_N}{3.5\times 10^8\ \mathrm{M_\odot}}\right)^{-1/3}\left(\frac{\Omega_{N0}h_{50}^2}{0.02}\right)^{-2/3}\frac{\Omega_{N0}}{\Omega_0}\cdot 5\sec^2\beta_1$$

$$\approx\frac{K}{10^{-4}}\left(\frac{M_N}{3.4\times10^{11}\,\mathrm{M}_\odot}\right)^{1/3}\left(\frac{\Omega_0}{30\Omega_{N0}}\right)^{-1/3}\left(\frac{m_\nu}{20}\right)^{4/3}\left(\frac{\Omega_{N0}h_{50}^2}{0.02}\right)^{-2/3}, \qquad (22)$$

assuming $\sec^2\beta_1\gtrsim\Omega_0/5\Omega_{N0}$, i.e., for $M_N\gtrsim1.1\times10^{10}\,\mathrm{M}_\odot$ and $M_\nu\gtrsim3.3\times10^{11}\,\mathrm{M}_\odot$. Then, if $K=10^{-4}$, this amplitude is already in nonlinear regime for $M_N\gtrsim3.4\times10^{11}\,\mathrm{M}_\odot$, and, if $K=2\times10^{-4}$, for $M_N\gtrsim4.2\times10^{10}\,\mathrm{M}_\odot$. This fact implies massive galaxies can be formed at the maximum expansion stage of a rich cluster. In the contracting stage of a rich cluster, the growth of small scale fluctuations is further accelarated and the fluctuation is expected to grow to form a bound system if the density contrast reaches the nonlinear regime until $\beta=\beta_2$, i.e., $(\delta\rho_N/\rho_N)_{\beta=\beta_2}\approx(\delta\rho_\nu/\rho_\nu)_{\beta=\beta_2}\gtrsim1$, which gives the minimum mass as $M^C_{\min,\nu}\approx6\times10^{10}\,\mathrm{M}_\odot$ and $M^C_{\min,N}\approx2\times10^9\,\mathrm{M}_\odot$.

The violent relaxation which leads to the virialization of neutrinos is considered to occur for $\beta\gtrsim\pi/4$ and $M_{J\nu}$ increases suddenly after it. In the course of relaxation, larger structure such as subclustering of galaxies may be destroyed. The structure of galaxies such as the separation of visible part and neutrino halo should be ascribed to the dissipative gas dynamics of nucleon matter. In the violent relaxation stage, galactic halo may be stripped from visible part. Then we can conclude that the typical mass scale formed by fragmentation in rich clusters will be $M_N\approx2\times10^9\sim10^{11}\,\mathrm{M}_\odot$, which is just a characteristic value for a typical galaxy.

The one dimensional velocity dispersion in the formed objects σ_g may be derived by a similar method in Ref. 4). Here we roughly estimate σ_g by assuming that the density of the parent cluster at the maximum expansion stage and that of the fragmented galaxies are the same; σ_g is given as

$$\sigma_g=\sqrt{\frac{2GM_g}{5r_{\max,g}}}=\sqrt{\frac{2GM_{\rm cl}}{5r_{\max,\rm cl}}\cdot\frac{M_g}{M_{\rm cl}}\cdot\frac{r_{\max,\rm cl}}{r_{\max,g}}}\approx\sigma_{\rm cl}\left(\frac{M_g}{M_{\rm cl}}\right)^{1/3}. \qquad (23)$$

Here M_g, $M_{\rm cl}$, $r_{\max,g}$, $r_{\max,\rm cl}$ denote the masses and radii at the maximum expansion stage of the fragmented galaxy and the parent cluster. Equation (23) gives typically $\sigma_g\sim10^7$ cm/s, which seems a little smaller than the observation. However, real velocity dispersion of stars in a galaxy is also affected by the subsequent dissipative processes.

(ii) *Small group of galaxies*

Next we discuss a possible formation process of small groups of galaxies resulting from an isothermal density fluctuation in nucleon matter. The object which is formed at first has the mass of $M_\nu=M_{J\nu}(z_{\rm rec})\simeq8.8\times10^{13}(m_\nu/20)^{-7/2}\,\mathrm{M}_\odot$ and $M_N\simeq2.9\times10^{12}(m_\nu/20)^{-7/2}(30\Omega_{N0}/\Omega_0)\mathrm{M}_\odot$, assuming the spectrum given in Eq. (11). If we take $m_\nu=20$, $\Omega_{N0}h_{50}^2=0.02$, $t_0H_0\sqrt{\Omega_0}=2/3$ and $\Omega_0/\Omega_{N0}=30$, such an object is in a collapsed stage, in the contracting stage or in the expanding stage corresponding to $K \gtrsim 1.72\times10^{-4}$, $1.72\times10^{-4} \gtrsim K \gtrsim 1.08\times10^{-4}$ and $K \lesssim 1.1\times10^{-4}$, respectively. We may set at the starting epoch

$$\cos^2\beta_i \equiv \left(\frac{\delta\rho_N}{\rho_{cr}}\right)_i = 1.6\times10^{-3}\left(\frac{K}{10^{-4}}\right) \tag{24}$$

for the parent small group of galaxies, and $M_{J\nu,\min}$ becomes $5.8\times10^9(K/10^{-4})^{3/2}\,M_\odot$.

By a similar analysis to Eq. (22), we obtain the density contrast of smaller scale fluctuations at the maximum expansion stage as

$$\left(\frac{\delta\rho_\nu}{\rho_\nu}\right)_{\beta=0} \simeq \left(\frac{\delta\rho_N}{\rho_N}\right)_{\beta=0} \approx \left(\frac{M_N}{2.4\times10^{10}\,M_\odot}\right)^{1/3} \tag{25}$$

for $M_N \gtrsim 2.9\times10^9(K/10^{-4})^{3/2}\,M_\odot$. If we assume K is large enough for the parent cluster to have completed the collapse, the minimum mass of formed galaxies is a few times $M_{J\nu,\min}$. Then we can conclude that a typical mass formed in a small group of galaxies is also $M_\nu \approx 10^{10} \sim 10^{12}\,M_\odot$ and $M_N \approx 10^9 \sim 10^{11}\,M_\odot$. In small groups, the violent relaxation may have not completed and moreover if it ever occurred the expected velocity dispersion is rather small as seen in Eqs. (17) and (18). Then subclustering such as binary pair has not been destroyed and galactic halos exist, which is consistent with observations.

§ 5. Summary and discussion

Our scenario presented in this paper should be confronted with the following observational features:

(i) A mass hierarchy of supercluster, rich cluster, small group and a galaxy.

(ii) A rich cluster and a galaxy are a relaxed system but a supercluster and a small group are a non-relaxed system.

(iii) The degree of missing mass is larger in a rich cluster than in a small group and a galactic halo.

According to our scenario, the first point is dependent on the mass of neutrino and the second point crucially depends on the magnitude of the initial fluctuation. About the third point a further study on the nonlinear stage is necessary. Thus all our explanations are still qualitative but general tendency seems to fit rather well to our scenario.

The primordial density fluctuation in the neutrino matter leads to the formation of rich clusters of galaxies with mass between $M_* = 7.8\times10^{15}(m_\nu/20)^{-2}\,M_\odot$ and $M^{RC}_{\max,\nu} \approx 2.3\times10^{16}(K/2\cdot10^{-4})^3(m_\nu/20)M_\odot$ with the one dimensional velocity dispersion $1.3\times10^8(M_\nu/M_*)^{1/6}\cdot(K/2\cdot10^{-4})^{1/2}$ cm/s, for the initial spectrum of $(\delta\rho_\nu/\rho_\nu)_{z_{eq}} = K\cdot(M_\nu/9.6\cdot10^{16}(m_\nu/20)^{-2}\,M_\odot)^{-1/3}$. We notice that these results strongly depends on the assumed initial fluctuation spectrum. The missing mass in a rich cluster is thus explained by massive neutrinos, giving the ratio of dynamical mass to visible mass as $\Omega_{\nu 0}/\Omega_{N0}$. For the fluctuations with $M_\nu > M^{RC}_{max,\nu}$, they are still in a contracting phase ($M_\nu < 4M^{RC}_{max,\nu}$) or in the expanding phase ($M_\nu > 4M^{RC}_{max,\nu}$) at the present, which we identify with superclusters of galaxies.

In order to form the smaller objects like $M_\nu < M_*$, we have to introduce another primordial density fluctuation, that is, isothermal density fluctuation in nucleon matter. This fluctuation can trigger the fluctuation in neutrino matter if M_ν contained in the perturbed region becomes larger than $M_{J\nu}$. By the growth of this fluctuation in the generally expanding universe, a small group or a loose cluster of galaxies is formed, whose mass scale is typically $M_{J\nu}(z_{rec}) \approx 8.8 \times 10^{13}(m_\nu/20)^{-7/2} M_\odot$.

Formation of individual galaxies is possible only in a cluster or a group, where the growth of fluctuation is much faster than that in the generally expanding universe. The mass scale of a galaxy corresponds to $M_{J\nu}$ at the maximum expansion of the cluster.

The ratio of M_ν/M_N may be smaller for smaller size objects for several reasons: First the value of $(\delta\rho_N/\rho_N)_i$ is larger for smaller objects and, for the scale of a galaxy, it takes on $0.1 \sim 1$. Second, in the collapsed stage nucleons can lose their kinetic energy by dissipative processes and condense further. Finally, the relaxed dynamical state of neutrinos is rather extended because of the limit of phase space density.[19] All these facts will tend to decrease the ratio M_ν/M_N, which is consistent with observations.

For our scenario, the existence of the primordial isothermal density fluctuations in nucleon matter is indispensable. Recently the origin of such fluctuations has been discussed on the GUT scheme of baryon number generation.[20]

Our scenario may give a critical impact on the study of galaxy distribution in terms of correlation functions. In usual picture, galaxies are formed first and subsequently large scale structure develops without any characteristic scale.[21] In such a picture, however, a smaller scale fluctuation completes collapse and relaxation earlier than a larger scale system. Therefore it seems difficult to reconcile with the observational feature that rich clusters appear to be relaxed but small groups do not. On the other hand, our scenario correctly reproduces this observational feature. We suggest that the correlation function analysis which discriminates a rich cluster and other structures is very important.

Our scenario is also different from Zeldovich's pancake theory.[22] In ours galaxies are formed before the virialization of neutrinos while they are formed after collapse in the pancake theory. In the pancake theory, the neutrinos will get very high velocity dispersion after collapse and it would be very difficult to form a neutrino halo around visible galaxies.

Acknowledgements

One of the authors (F.T.) thanks Soryushi Shogakukai for financial aid.

Appendix A

—— *The Dynamical Evolution of a Uniform Density Dust Sphere in the Expanding Universe* ——

The radius R of a uniform density dust sphere which starts to grow at $z=z_i$ with the initial density contrast $(\delta\rho/\rho_{cr})_i$ and the initial radius R_i is described as[11),12)]

$$\frac{R}{R_i}=\frac{1}{2}\left(\frac{\rho_{cr}}{\delta\rho}\right)_i\left[1-\cos\left(\sqrt{\left(\frac{\delta\rho}{\rho_{cr}}\right)_i}\theta\right)\right], \tag{A·1}$$

$$t=\frac{1}{2H_i}\left(\frac{\rho_{cr}}{\delta\rho}\right)_i^{3/2}\left[\sqrt{\left(\frac{\delta\rho}{\rho_{cr}}\right)_i}\theta-\sin\left(\sqrt{\left(\frac{\delta\rho}{\rho_{cr}}\right)_i}\theta\right)\right] \tag{A·2}$$

and

$$\frac{d\theta}{dt}=H_i\cdot\frac{R_i}{R}, \tag{A·3}$$

where H_i is the Hubble parameter at $z=z_i$, which is given for $z_i\gg1$ as

$$H_i=H_0\sqrt{\Omega_0}(1+z_i)^{3/2}. \tag{A·4}$$

To conform our problem to the expressions of Hunter's analysis,[10)] we use the variable β defined by

$$\beta=\frac{1}{2}\left(\sqrt{\left(\frac{\delta\rho}{\rho_{cr}}\right)_i}\theta-\pi\right). \tag{A·5}$$

Equations (A·1)~(A·3) are then transformed as

$$R=R_i\cdot\left(\frac{\rho_{cr}}{\delta\rho}\right)_i\cos^2\beta, \tag{A·6}$$

$$t=\frac{1}{H_i}\left(\frac{\rho_{cr}}{\delta\rho}\right)_i^{3/2}\left[\beta+\frac{\pi}{2}+\frac{1}{2}\sin 2\beta\right] \tag{A·7}$$

and

$$\frac{d\beta}{dt}=\frac{1}{2}\left(\frac{\delta\rho_\nu}{\rho_{cr}}\right)^{3/2}H_i\sec^2\beta. \tag{A·8}$$

The variable β runs from $\beta_i=-\cos^{-1}\sqrt{(\delta\rho/\rho_{cr})_i}$ to $\pi/2$; expanding stage for $\beta<0$ and collapsing stage for $\beta>0$, respectively. The density ρ is assumed to be uniform, then the radial velocity field u becomes proportional to the radial coordinates r, and these are given as

$$\rho = \rho_{\text{stop}} \sec^6 \beta \tag{A·9}$$

and

$$u = -\frac{\pi}{2} \frac{r}{t_{\text{stop}}} \sec^2 \beta \tan \beta , \tag{A·10}$$

respectively, where $\rho_{\text{stop}} = \rho_{cr,i}(\delta\rho/\rho_{cr})_i^3$ and $t_{\text{stop}} = (\pi/2H_i)(\rho_{cr}/\delta\rho)_i^{3/2} = \sqrt{3\pi/32G\rho_{\text{stop}}}$. Here the subscript stop means the quantities at the maximum expansion stage $\beta = 0$ and $R_{\text{stop}} = R_i(\rho_{cr}/\delta\rho)_i$. These expressions completely agree with those used by Hunter.[10)]

Appendix B

— Fragmentation of a Parent Cloud: The Case of $M_N \geq M_{JN}$ and $M_\nu \geq M_{J\nu}$ —

The linear perturbation to the basic flow given in Appendix A is investigated referring to Hunter's analysis[10)] for a contracting cloud. The basic equations are continuity equation and the equation of motion for nucleons and neutrinos and the Poisson equation:

$$\frac{d\boldsymbol{u}_i}{dt} = -\text{grad } \Psi , \tag{B·1}$$

$$\frac{d\rho_i}{dt} + \rho_i \text{ div } \boldsymbol{u}_i = 0 \tag{B·2}$$

and

$$\Delta \Psi = 4\pi G(\rho_N + \rho_\nu). \tag{B·3}$$

Expanding as

$$\boldsymbol{u}_i = \left(-\frac{\pi}{2} \frac{r}{t_{\text{stop}}} \sec^2 \beta \tan \beta, 0, 0\right) + \delta \boldsymbol{u}_i , \tag{B·4}$$

$$\rho_i = \rho_{\text{stop},i} \sec^6 \beta + \delta\rho_i \tag{B·5}$$

and

$$\Psi = -2\pi G\rho_{\text{stop}}\left(R^2 - \frac{r^2}{3}\right)\sec^6 \beta + \delta \Psi , \tag{B·6}$$

we get the equations for perturbed quantities as

$$\frac{\partial \phi_i}{\partial t} = \delta \Psi , \tag{B·7}$$

$$\frac{\partial \delta\rho_i}{\partial t} - 6\delta\rho_i \tan\beta \frac{d\beta}{dt} - \rho_{\mathrm{stop},i} \sec^{10}\beta D^2 \phi_i = 0 \tag{B·8}$$

and

$$D^2 \delta \Psi = 4\pi G \cos^4 \beta (\delta\rho_N + \delta\rho_\nu). \tag{B·9}$$

Here, differentiation with time is done at the fixed comoving coordinate $\eta = r \sec^2 \beta$ and D^2 is the Laplace operator in this comoving coordinates, and ϕ_i is the velocity potential defined as $\delta \boldsymbol{u}_i = -\mathrm{grad}\, \phi_i$, thus treating only the compressional mode.

Eliminating $\delta \Psi$ and $\delta\rho_i$, we obtain the equation for ϕ_i as

$$\frac{\partial^2 D^2 \phi_i}{\partial \beta^2} - \frac{6 \sec^2 \beta}{\rho_{\mathrm{stop}}} (\rho_{\mathrm{stop},N} D^2 \phi_N + \rho_{\mathrm{stop},\nu} D^2 \phi_\nu) = 0\,. \tag{B·10}$$

Assuming $D^2 \phi_i \neq 0$, we get from Eq. (B·10),

$$\left(\frac{\partial^2}{\partial \beta^2} - 6 \sec^2 \beta\right)(\rho_{\mathrm{stop},N} \phi_N + \rho_{\mathrm{stop},\nu} \phi_\nu) = 0 \tag{B·11}$$

and from Eq. (B·7)

$$\frac{\partial}{\partial \beta}(\phi_N - \phi_\nu) = 0\,. \tag{B·12}$$

From Eq. (B·12) we can put $\phi_N = \phi_\nu + f_3(\eta)$, where $f_3(\eta)$ is an arbitrary function of comoving coordinates. (For simplicity we denote $f_3(\eta)$, but in fact f_3 may be a function of angular coordinates as well as the radial coordinate η. This is also true for functions $f_i(\eta)$, $g_i(\eta)$, which will appear later.) Then we obtain the general solution of ϕ_i as

$$\phi_N = H_1(\beta) f_1(\eta) + H_2(\beta) f_2(\eta) + \frac{\rho_{\mathrm{stop},\nu}}{\rho_{\mathrm{stop}}} f_3(\eta) \tag{B·13}$$

and

$$\phi_\nu = H_1(\beta) f_1(\eta) + H_2(\beta) f_2(\eta) - \frac{\rho_{\mathrm{stop},N}}{\rho_{\mathrm{stop}}} f_3(\eta), \tag{B·14}$$

where

$$H_1(\beta) = 3 \sec^2 \beta - 2\,, \tag{B·15}$$

$$H_2(\beta) = (3 \sec^2 \beta - 2) \int_{-(\pi/2)}^{\beta} \frac{d\alpha}{\{3 \sec^2 \alpha - 2\}^2}$$

$$= \left(\frac{\beta}{4} + \frac{\pi}{8}\right)(3 \sec^2 \beta - 2) + \frac{3}{4} \tan \beta \tag{B·16}$$

and f_i are arbitrary functions of the comoving coordinates.

We can determine $\delta\Psi$ through Eq. (B·7), substitute it into Eq. (B·9) and obtain for $\delta\rho_N+\delta\rho_\nu$ as

$$\delta\rho_N+\delta\rho_\nu=\sec^4\beta\frac{d\beta}{dt}\left\{\frac{dH_1}{d\beta}g_1(\eta)+\frac{dH_2}{d\beta}g_2(\eta)\right\}, \tag{B·17}$$

where $D^2 f_i\equiv 4\pi G g_i$. On the other hand, substituting ϕ_i into Eq. (B·8) we can determine $\delta\rho_i$ as

$$\delta\rho_N=\sec^4\beta\frac{d\beta}{dt}\frac{\rho_{N,\text{stop}}}{\rho_{\text{stop}}}\left[\frac{dH_1}{d\beta}g_1(\eta)+\frac{dH_2}{d\beta}g_2(\eta)+\frac{6\rho_{\nu,\text{stop}}}{\rho_{\text{stop}}}\{\tan\beta\, g_3(\eta)+g_4(\eta)\}\right] \tag{B·18}$$

and

$$\delta\rho_\nu=\sec^4\beta\frac{d\beta}{dt}\frac{\rho_{\nu,\text{stop}}}{\rho_{\text{stop}}}\left[\frac{dH_1}{d\beta}g_1(\eta)+\frac{dH_2}{d\beta}g_2(\eta)-\frac{6\rho_{N,\text{stop}}}{\rho_{\text{stop}}}\{\tan\beta\, g_3(\eta)+g_4(\eta)\}\right], \tag{B·19}$$

where $D^2 f_3\equiv 4\pi G g_3$ and g_4 is an integration constant with respect to β. Dividing Eqs. (B·18) and (B·19) by the unperturbed densities, we get the formula for the density contrast as

$$\frac{\delta\rho_N}{\rho_N}=\frac{\pi}{4t_{\text{stop}}\rho_{\text{stop}}}\left[\frac{dH_1}{d\beta}g_1(\eta)+\frac{dH_2}{d\beta}g_2(\eta)+\frac{6\rho_{\nu,\text{stop}}}{\rho_{\text{stop}}}\{\tan\beta\, g_3(\eta)+g_4(\eta)\}\right] \tag{B·20}$$

and

$$\frac{\delta\rho_\nu}{\rho_\nu}=\frac{\pi}{4t_{\text{stop}}\rho_{\text{stop}}}\left[\frac{dH_1}{d\beta}g_1(\eta)+\frac{dH_2}{d\beta}g_2(\eta)-\frac{6\rho_{N,\text{stop}}}{\rho_{\text{stop}}}\{\tan\beta\, g_3(\eta)+g_4(\eta)\}\right]. \tag{B·21}$$

For the mode described by H_1, density contrast decreases in the expanding stage and increases in the collapsing stage. For the mode described by H_2, density contrast increases in both the stages. It is to be noted that the asymptotic behavior in $\beta\to-\pi/2$ is the same as that in the expanding universe; $\delta\rho/\rho\propto t^{2/3}$ for H_2 mode $\delta\rho/\rho\propto t^{-1}$ for H_1 mode. The growth behavior in the collapsing stage is similar for both modes. Then we consider only the mode described by H_2. The terms with g_3 and g_4 in Eqs. (B·18)~(B·21) appear in order to describe the different initial conditions for neutrinos and for nucleons. In the expanding stage, g_3 term describes the decrease of the density contrast for nucleons, while g_4 term keeps the density contrast. Then we set $g_3=0$ and $g_4=0$. If we impose such initial conditions as $\delta\rho_N/\rho_N=(\delta\rho_N/\rho_N)_1$ and $\delta\rho_\nu/\rho_\nu=0$ at $\beta=\beta_1$, the growth of density contrast is described by

$$\frac{\delta\rho_N}{\rho_N}=\left(\frac{\delta\rho_N}{\rho_N}\right)_1\cdot\frac{\rho_{N,\mathrm{stop}}}{\rho_{\mathrm{stop}}}\left[\frac{dH_2/d\beta}{(dH_2/d\beta)_1}+\frac{\rho_{\nu,\mathrm{stop}}}{\rho_{N,\mathrm{stop}}}\right] \tag{B·22}$$

and

$$\frac{\delta\rho_\nu}{\rho_\nu}=\left(\frac{\delta\rho_N}{\rho_N}\right)_1\cdot\frac{\rho_{N,\mathrm{stop}}}{\rho_{\mathrm{stop}}}\left[\frac{dH_2/d\beta}{(dH_2/d\beta)_1}-1\right]. \tag{B·23}$$

In the main text, we identify $\rho_{N,\mathrm{stop}}/\rho_{\mathrm{stop}}$ and $\rho_{\nu,\mathrm{stop}}/\rho_{N,\mathrm{stop}}$ with Ω_{N0}/Ω_0 and $\Omega_{\nu 0}/\Omega_{N0}$, respectively.

Appendix C

—— Linear Perturbation to a Parent Cloud:
The Case of $M_N \geq M_{JN}$ but $M_\nu < M_{J\nu}$ ——

If $M_\nu < M_{J\nu}$, we may approximate as $\delta\rho_\nu=\delta\boldsymbol{u}_\nu=0$ for simplicity, neglecting the response of neutrinos to the gravitational potential induced by nucleon density fluctuations. Then the basic equations become

$$\frac{\partial\phi_N}{\partial t}=\delta\Psi\,, \tag{C·1}$$

$$\frac{\partial\delta\rho_N}{\partial t}-6\delta\rho_N\tan\beta\frac{d\beta}{dt}+\rho_{N,\mathrm{stop}}\sec^{10}\beta D^2\phi_N=0 \tag{C·2}$$

and

$$D^2\delta\Psi=4\pi G\cos^4\beta\delta\rho_N\,. \tag{C·3}$$

Following similar procedures in Appendix B, we obtain the equation for ϕ_N as

$$\frac{\partial^2\phi_N}{\partial\beta^2}=6\frac{\rho_{\mathrm{stop},N}}{\rho_{\mathrm{stop}}}\sec^2\beta\phi_N\,. \tag{C·4}$$

If we transform the variable β to $s=\tan\beta$, Eq. (C·4) now becomes

$$(1+s^2)\frac{\partial^2\phi_N}{\partial s^2}+2s\frac{\partial\phi_N}{\partial s}-6\frac{\rho_{N,\mathrm{stop}}}{\rho_{\mathrm{stop}}}\phi_N=0\,. \tag{C·5}$$

The elementary solutions of Eq. (C·5) are the Legendre functions of the order of ν with imaginary argument, where $\nu(\nu+1)=6\rho_{N,\mathrm{stop}}/\rho_{\mathrm{stop}}$. In the expanding stage, real solutions are given by

$$\begin{aligned}F_1(s)&\equiv(-i)^{\nu+1}Q_\nu(is)\\&=\frac{\sqrt{\pi}\,\Gamma(\nu+1)}{\Gamma\left(\nu+\frac{3}{2}\right)}\frac{1}{(-2s)^{\nu+1}}F\left(\frac{\nu+1}{2},\frac{\nu}{2}+1,\nu+\frac{3}{2};-\frac{1}{s^2}\right)\end{aligned} \tag{C·6}$$

and

$$F_2(s)\equiv i^\nu\left\{P_\nu(is)-\frac{\tan\pi\nu}{\pi}Q_\nu(is)\right\}$$

$$=\frac{\Gamma\left(\nu+\frac{1}{2}\right)}{\sqrt{\pi}\Gamma(\nu+1)}\cdot(-2s)^\nu\cdot F\left(-\frac{\nu}{2},\frac{1-\nu}{2},\frac{1}{2}-\nu;-\frac{1}{s^2}\right), \qquad (C\cdot7)$$

where F is the hypergeometric function, and expression in terms of F is valid for $s^2>1$.

For the collapsing stage, two independent real solutions are $G_1(s)\equiv i^{\nu+1}Q_\nu(is)$ and $G_2(s)\equiv(-i)^\nu\{P_\nu(is)-((\tan\pi\nu)/\pi)Q_\nu(is)\}$. We note that the solution F_i for $s<0$ should be connected to a linear combinations of two solutions G_i for $s>0$ so as to satisfy the continuity of ϕ_N and $\partial\phi_N/\partial s$ at $s=0$.

Asymptotic behaviors of these solutions are as follows: In the expanding stage, F_1 describes the growing mode while F_2 describes the damping mode. For the F_1 mode $\delta\rho_N/\rho_N$ is given by

$$\frac{\delta\rho_N}{\rho_N}=\left(\frac{\delta\rho_N}{\rho_N}\right)_1\frac{dF_1/ds}{(dF_1/ds)_1}\frac{\sec^2\beta}{\sec^2\beta_1}. \qquad (C\cdot8)$$

For $-s\to\infty$, $\delta\rho_N/\rho_N\propto(-s)^\nu$, and the growth rate is very small if $\rho_{N,\mathrm{stop}}/\rho_{\mathrm{stop}}\ll1$. This asymptotic behavior proves to be the same one as found by Bond et al.[6] At the maximum expansion stage $\delta\rho_N/\rho_N$ has grown only to the value

$$\left(\frac{\delta\rho_N}{\rho_N}\right)_{\beta=0}=\left(\frac{\delta\rho_N}{\rho_N}\right)_1\times\frac{2\Gamma\left(\frac{\nu}{2}+1\right)\Gamma\left(\nu+\frac{3}{2}\right)\Gamma\left(\frac{1-\nu}{2}\right)\cos\frac{\nu\pi}{2}(-2s_1)^\nu}{\pi(\nu+1)\Gamma(\nu+1)}. \qquad (C\cdot9)$$

This growth factor becomes $(-2s_1)^\nu$ for $\nu\ll1$ and negligibly small.

In the collapsing stage, G_1 describes the damping mode while G_2 describes the growing mode. But the growth rate is small, and $\delta\rho_N/\rho_N$ is proportional to $s^{\nu+1}\sim s$ for large s, in clear contrast to the case of $M_\nu\geq M_{J\nu}$, where $\delta\rho_N/\rho_N$ behaves like as $\propto s^3$. Thus we may conclude that the density fluctuation with smaller scale than the neutrino Jeans mass is severely suppressed to grow.

References

1) F. Reines, H. W. Sobel and E. Pasieb, Phys. Rev. Letters **45** (1980), 1307.
2) V. A. Lubimov, E. G. Novikov, V. Z. Nozik, E. F. Tretyakov and V. S. Kosik, Phys. Letters **94B** (1980), 266.
3) H. Sato and F. Takahara, Prog. Theor. Phys. **64** (1980), 2029.
4) H. Sato and F. Takahara, Prog. Theor. Phys. **65** (1981), 374.
5) H. Sato, in *Proceedings of the Tenth Texas Symposium* (N. Y. Academy of Sciences, in press); Preprint RIFP 423 (1981).
6) J. R. Bond, G. Efstathiou and J. Silk, Phys. Rev. Letters **45** (1980), 1980.

7) A. G. Doroshkevich, M. Yu. Khlopov, R. A. Sunyaev, A. S. Szalay and Ya. B. Zeldovich, *Neutrino 80*, Erice, 1980.
8) A. L. Melott, Preprint (1980).
9) G. Gao and R. Ruffini, Phys. Letters **97B** (1980), 388.
10) C. Hunter, Astrophys. J. **136** (1962), 594.
11) K. Tomita, Prog. Theor. Phys. **42** (1969), 9.
12) J. E. Gunn and J. R. Gott III, Astrophys. J. **176** (1972), 1.
13) D. Lynden-Bell, Month. Notices Roy. Astron. Soc. **136** (1967), 101.
14) H. Sato, T. Matsuda and H. Takeda, Prog. Theor. Phys. Suppl. No. 49 (1971), 11.
15) J. Silk, *Confrontation of Cosmological Theories with Observational Data*, ed. M. S. Longair (Reidel, Dordrecht, 1974), p. 175.
16) J. R. Gott and M. Rees, Astron. Astrophys. **45** (1975), 365.
17) Ya. B. Zel'dovich, Month. Notices Roy. Astron. Soc. **160** (1972), 1.
18) S. M. Faber and J. S. Gallagher, Ann. Rev. Astron. Astrophys. **17** (1979), 135.
19) S. Tremaine and J. E. Gunn, Phys. Rev. Letters **42** (1979), 407.
20) J. D. Barrow and M. S. Turner, *Nature* (to be published); Preprint of the University of Chicago (1981).
21) P. J. E. Peebles, *The Formation and Dynamics of Galaxies*, ed. J. R. Shakeshaft (Reidel, Dordrecht, 1974), p. 55.
22) A. G. Doroshkevich, R. A. Sunyaev and Ya. B. Zel'dovich, *Confrontation of Cosmological Theories with Observational Data*, ed. M. S. Longair (Reidel, Dordrecht, 1974), p. 213.

THE ASTROPHYSICAL JOURNAL, 248:1-12, 1981 August 15

ON THE LINEAR THEORY OF DENSITY PERTURBATIONS IN A NEUTRINO+BARYON UNIVERSE

IRA WASSERMAN
Center for Radiophysics and Space Research, Cornell University*
Received 1980 July 25; accepted 1981 March 3

ABSTRACT

Various aspects of the linear theory of density perturbations in a universe containing a significant population of massive neutrinos are calculated. Because linear perturbations in the neutrino density are subject to nonviscous damping on length scales smaller than the effective neutrino Jeans length, the fluctuation spectrum of the neutrino density perturbations just after photon decoupling is expected to peak near the maximum neutrino Jeans mass. The gravitational effects of nonneutrino species are included in calculating the maximum neutrino Jeans mass, which is found to be $[M_J(t)]_{\max} \sim 10^{17} M_\odot/[m_\nu(\text{eV})]^2$, about an order of magnitude smaller than is obtained when nonneutrino species are ignored. An explicit expression for the nonviscous damping of neutrino density perturbations less massive than the maximum neutrino Jeans mass is derived. The linear evolution of density perturbations after photon decoupling is discussed. Of particular interest is the possibility that fluctuations in the neutrino density induce baryon density perturbations after photon decoupling and that the maximum neutrino Jeans mass determines the characteristic bound mass of galaxy clusters.

Subject headings: cosmology — neutrinos

I. INTRODUCTION

The possibility that neutrinos have finite rest masses (Pontecorvo 1968; Eliezer and Ross 1974; Gribov and Pontecorvo 1969; Bilenky and Pontecorvo 1976, 1978), which has received recent experimental support (Reines, Sobel, and Pasierb 1980), could have profound implications for cosmology. Although the inclusion of nonzero neutrino rest masses m_{ν_j} appears to have little effect on the standard model of the early universe (Dolgov and Zel'dovich 1980; Shapiro, Teukolsky, and Wasserman 1980), massive neutrinos, if sufficiently plentiful, could dominate the dynamics of the present day universe through their gravitational interactions with other forms of matter (Cowsik and McClelland 1972, 1973; Tremaine and Gunn 1979; Schramm and Steigman 1980*a*, *b*; Bond, Efstathiou, and Silk, 1980; Sato and Takahara 1980; Klinkhamer and Norman 1981; Melott 1980; Ching *et al.* 1980; Lu *et al.* 1980). In particular the average mass density of neutrinos is presently

$$\Omega_\nu \approx \frac{0.01}{h^2}\left(\frac{T_{\gamma,0}}{2.7\text{ K}}\right)^3 \sum_j m_{\nu_j}(\text{eV}) \tag{1}$$

in units of the closure density

$$\rho_c = \frac{3H_0^2}{8\pi G} \approx 2\times 10^{-29} h^2 \text{g cm}^{-2},$$

where h is Hubble's constant in units of 100 km s^{-1} Mpc^{-1}, $T_{\gamma,0}$ is the present temperature of the blackbody photon background, and the sum in equation (1) is over all neutrino flavors ($j = 1,2,3,\ldots ?$) less massive than ~ 1 MeV (Cowsik and McClelland 1972, 1973). The neutrino contribution to the total mass density of the universe may be comparable to and perhaps greater than Ω_B, the contribution due to baryons (Schramm and Steigman 1980*a*, *b*).

Since the residual thermal velocities of the relict neutrinos are of order (k_B is Boltzmann's constant)

$$\left(\frac{4}{11}\right)^{1/3} \frac{k_B T_{\gamma,0}}{m_{\nu_j}} \approx 50 \frac{(T_{\gamma,0}/2.7\text{ K})}{[m_{\nu_j}(\text{eV})]} \text{km s}^{-1},$$

which is less than the typical velocity dispersions in galaxies and clusters of galaxies, primordial massive neutrinos are expected to form gravitationally bound clumps along with baryons. Tremaine and Gunn (1979) have shown that violently relaxed, isothermal neutrino spheres with core radii and velocity dispersions typical of rich clusters of galaxies ($r_c \approx 125h^{-1}$ kpc, $\sigma_\upsilon \approx 1000$ km s^{-1}; cf. Bahcall 1977) cannot exist for neutrino flavors less massive than ~4 eV; the formation of isothermal halos around both individual and binary galaxies would require still more massive neutrinos, $m_{\nu_j} \gtrsim 10$–20 eV. Based on these considerations, Schramm and Steigman (1980, 1981) have argued that relatively light massive neutrinos (4 eV$\lesssim m_{\upsilon_j} \lesssim 10$–20 eV) could solve the missing mass problem for galaxy clusters without requiring unacceptably large mass-to-light ratios for individual galaxies and small groups and a cosmological baryon density larger than the tight constraints imposed by the primordial He^4 abundance (Yang, Schramm, Steigman, and Rood 1979; Steigman, Olive, and Schramm 1979; see also Melott 1980).

The formation and subsequent relaxation of gravitationally bound systems containing massive neutrinos as well as baryons would therefore appear to be a problem of primary astrophysical importance. In this paper we consider various aspects of the linear gravitational instability of a neutrino+baryon universe (previous work on this problem can be found in Schramm and Steigman 1981; Bond, Efstathiou, and Silk 1980; Sato and Takahara 1980; Klinkhamer and Norman 1981; see also Zakharov 1978).

Of particular interest is the possibility that neutrino fluctuations can induce perturbations in the baryons. Adiabatic fluctuations in the baryon density, which are likely to have arisen in the very early universe during the generation of the next baryon excess (cf. Weinberg 1979, and references therein; see also Turner and Schramm 1979), are damped out *exponentially* by radiative viscosity before photon decoupling if they contain a mass in baryons, M_B, less than

$$M_c \approx \frac{2.5\times 10^{12}}{\Omega_B^{1/2}\Omega^{3/4}h^{5/2}} M_\odot, \tag{2}$$

where $\Omega \equiv \Omega_B + \Omega_\nu$ (Silk 1967, 1968; Peebles and Yu 1970; Field 1971; Weinberg 1971). Substantial damping also occurs during recombination, resulting in a somewhat larger critical mass, estimated to be

$$M_c' \sim 10^{16} M_\odot . \tag{2'}$$

below which adiabatic perturbations are damped out (Press and Vishniac 1980). Neutrino perturbations are subject to an at most power-law nonviscous decay, which nonetheless leads to significant damping of density fluctuations smaller than the effective "Jeans length" for neutrinos (cf. § II*b* and, particularly, eq. [22]; see also Stewart 1972 and Peebles 1973 for discussions of this effect for massless neutrinos). The maximum neutrino "Jeans mass" (cf. eq. [11])

$$[M_J(t)]_{\max} \sim \frac{10^{17} M_\odot}{[m_\nu(\mathrm{eV})]^2}$$

therefore defines the mass scale of the neutrino perturbations that first condense out of the Friedmann background (assuming a primordial fluctuation spectrum that is a decreasing function of mass above $[M_J(t)]_{\max}$). For large enough values of the neutrino mass m_ν, $[M_J(t)]_{\max} \lesssim M_c'$ or, even, M_c. It is therefore possible that fluctuations in the baryon density on mass scales M_B smaller than the damping mass, M_c', could be generated *after* photon decoupling by the corresponding neutrino perturbations, which contain masses $M_\nu = (\Omega_\nu/\Omega_B)M_B$ in neutrinos. Fluctuations in the cosmological density of massive neutrinos may play an important role in the formation of galaxy clusters.

Because little is known about the (hypothetical) couplings of massive Dirac neutrinos we restrict our attention in this paper to massive left-handed neutrinos of the Majorana variety, which may arise naturally in some grand unified models (cf. Witten 1980). The thermal history of massive, left-handed Majorana neutrinos is identical with that of massless neutrinos (cf. Peebles 1971 and Weinberg 1972) until the neutrinos become nonrelativistic. Since little is known about the actual values of neutrino masses, we make no attempt to distinguish among the masses of the various neutrino flavors. Instead, we introduce a single typical or effective neutrino mass, m_ν, which could, in reality, be the mass of the heaviest type of neutrino if one flavor is much more massive than the others, or N_ν times the mass of each type of neutrino if all N_ν flavors have more or less the same mass, or some appropriately weighted sum of neutrino masses in intermediate cases.

The principal results of our paper are a value for the maximum neutrino "Jeans mass" $[M_J(t)]_{\max}$ (eq. [11]) and an explicit expression for the total nonviscous damping of neutrino density perturbations less massive than $[M_J(t)]_{\max}$ as a function of their mass, M_ν (eqs. [22], [23]). Our derivation of the effective neutrino Jeans mass, which is given in § II*a*, includes the gravitational effects of all particle species—photons and baryons as well as neutrinos. Previous calculations have either ignored nonneutrino species altogether, leading to an overestimate of $[M_J(t)]_{\max}$ (Bond, Efstathiou, and Silk 1980; Sato and Takahara 1980), or have set the dynamical time for the growth of neutrino perturbations equal to the Hubble time, $t_H \equiv [H(t)]^{-1}$ (where $H(t)$=Hubble's constant ["expansion rate"] at *any* time, t), throughout the history of the universe, leading to an underestimate of $[M_J(t)]_{\max}$ (Schramm and Steigman 1981; Klinkhamer and Norman 1981). Our calculation of the damping of neutrino density perturbations less *massive* than $[M_J(t)]_{\max}$ in § II*b* is the generalization to massive neutrinos of the earlier work of Stewart (1972) and Peebles (1973), who considered *massless* neutrinos only. In § III we discuss perturbations in the baryon density after photon decoupling on length scales smaller than the neutrino Jeans length but larger than the Jeans length for the baryons, while in § IV we treat perturbations larger than the Jeans lengths for both baryons and neutrinos. (similar considerations may be found in Bond *et al.* 1980). We discuss the relevance of our results to the formation of gravitationally bound systems in § V.

II. NEUTRINO PERTURBATIONS

The neutrino distribution at a given comoving position, $\boldsymbol{x}$, and a time t *after* neutrino decoupling is defined to be

$$f(\boldsymbol{x},\boldsymbol{p},t)=f_0(p,t)[1+\varepsilon(\boldsymbol{x},\boldsymbol{p},t)], \tag{3}$$

where $f_0(p,t)$ is the distribution function for the uniform neutrino background. The distribution function is normalized so that

$$\int \frac{d^3p}{(2\pi\hbar)^3} f(\boldsymbol{x},\boldsymbol{p},t)=n_\nu(\boldsymbol{x},t), \tag{4}$$

where $n_\nu(\boldsymbol{x},t)$ is the number density of neutrinos at $x^\mu=(\boldsymbol{x},t)$. For linear perturbations $\varepsilon(\boldsymbol{x},\boldsymbol{p},t)\ll 1$ by assumption.

If we set $\hbar=c=k_B=1$, the unperturbed neutrino distribution function is

$$f_0(p,t)=2\left\{\exp\left[\frac{pa(t)}{T_D a_D}\right]+1\right\}^{-1}, \tag{5}$$

where $p\equiv|\boldsymbol{p}|$, T_D and t_D are the temperature and age of the universe at neutrino decoupling, $a(t)$ is the cosmic scale factor, and $a_D\equiv a(t_D)$. The factor of 2 in equation (5) is added to include both + and − helicity neutrinos. Equation (4) holds for all neutrino flavors with masses $m_\nu \lesssim T_D$; since $T_D\sim 1$ MeV (cf. Peebles 1971 and Weinberg 1972) equation (5) is valid for the neutrinos of interest here, with masses $m_\nu \sim 1-100$ eV. (The abundances of any hypothetical neutrinos with masses $m_\nu > T_D$ would be suppressed by a Boltzmann factor, $\exp(-m_\nu/T_D)$, since they would have annihilated before t_D.)

a) The Effective "Jeans" Mass

Let

$$\varepsilon(\boldsymbol{x},\boldsymbol{p},t)=\int d^3k\, e^{i\boldsymbol{k}\cdot\boldsymbol{x}}\varepsilon(\boldsymbol{k},\boldsymbol{p},t);$$

$\varepsilon(\boldsymbol{k},\boldsymbol{p},t)$ is a measure of fluctuations on comoving scales of order $l_k=2\pi/k$, where $k\equiv|\boldsymbol{k}|$. For a given neutrino density perturbation there are four relevant time scales: (1) the average crossing time for neutrinos

$$t_{c,\nu}\sim\frac{l_k a(t)}{a_D}\langle v^{-1}\rangle,$$

where $\langle v^{-1}\rangle\equiv\langle E_\nu(p)/p\rangle$, with $E_\nu(p)\equiv(p^2+m_\nu^2)^{1/2}$, is the mean inverse thermal speed of the neutrinos; (2) the dynamical time

$$t_{g,\nu}\sim[8\pi G\rho_\nu(t)/3]^{-1/2}$$

for the growth of neutrino fluctuations under their own self-gravity alone; (3) the dynamical time

$$t_{g,\text{other}} \sim [8\pi G\rho_{\text{other}}(t)/3]^{-1/2}$$

for the growth of density perturbations due to the gravitational influence of all other particle species; and (4) the mean crossing or oscillation time

$$t_{c,\text{other}} \sim \frac{l_k a(t)}{a_D} \frac{1}{v_{\text{th}}}$$

for the other particle species, whose mean thermal speed is v_{th}.

A sufficient condition for the growth of neutrino perturbations via gravitational instability is $t_{c,\nu} > t_{g,\nu}$. However, gravitational instability may still occur even if $t_{c,\nu} < t_{g,\nu}$ as a result of the gravitational attraction of the "nonneutrino" particle species on the neutrinos.

To see how this comes about, consider the following extreme example.[1] At some arbitrary initial epoch, suppose that density fluctuations vanish for all particle species *except* neutrinos; however, $t_{c,\nu} < t_{g,\nu}$. If $t_{c,\text{other}}$ and $t_{c,\nu}$ are both $> t_{g,\text{other}}$ then, in $\sim$ one dynamical time, $t_{g,\text{other}}$, density perturbations comparable in magnitude to the initial density fluctuations in the neutrinos are induced in the other particle species. Roughly equal density perturbations are maintained in all particle species thereafter. These perturbations grow via gravitational instability with a dynamical time $\sim t_{g,\text{other}}$.[2]

The necessary condition for the growth of neutrino perturbations through gravitational instability is therefore

$$t_{c,\nu} > \min(t_{g,\nu}, t_{g,\text{other}}),$$

where $\min(x, y)$ is the smaller of x and y, provided $t_{c,\text{other}} > t_{g,\text{other}}$.

In the early universe, the neutrinos are relativistic for redshifts greater than

$$1+Z_\nu \equiv \frac{m_\nu}{(4/11)^{1/3} T_{\gamma,0}} = 6100 \frac{m_\nu(\text{eV})}{T_{\gamma,0}/2.7\text{ K}}. \tag{6}$$

During this epoch the energy density of each neutrino species is smaller than the photon energy density by a factor $(7/8)(4/11)^{4/3} \approx 0.23$, after e^+e^- annihilation at $T = m_e$ (cf. Peebles 1971; Weinberg 1972). For redshifts $Z \gtrsim 4.3\ 10^4 \Omega_B h^2 (T_{\gamma,0}/2.7\text{ K})^{-4}$, where $T_{\gamma,0}$ is the present cosmic background temperature, the photon energy density also exceeds the baryon density; the sound speed of the tightly coupled baryon-photon-electron fluid is $c_s \sim 1/\sqrt{3}$ at such early epochs. Therefore, in the photon-dominated era, as long as the neutrinos are relativistic, $t_{c,\nu} \sim t_{c,\text{other}}$ and $t_{g,\text{other}} < t_{g,\nu}$. The effective neutrino Jeans (rest) mass is then

$$M_J(t) \sim \frac{4\pi}{3} m_\nu \bar{n}_\nu(t) \left[\frac{3}{8\pi G\rho_\gamma(t)}\right]^{3/2},$$

where, if $\zeta(Z)$ is the Riemann zeta function (cf. Abramowitz and Stegun 1964),

$$\bar{n}_\nu(t) = \frac{3\zeta(3)}{2\pi^2}\left[\frac{a_D T_D}{a(t)}\right]^3$$

is the average number density of neutrinos at time t, and

$$\rho_\gamma(t) = \frac{\pi^2}{15}\left(\frac{11}{4}\right)^{4/3}\left[\frac{a_D T_D}{a(t)}\right]^4$$

[1] I thank E. E. Salpeter for suggesting this example to me.

[2] If $t_{c,\text{other}} > t_{g,\text{other}} > t_{c,\nu}$ the induced density fluctuations should be smaller than the initial neutrino density perturbations by a factor $\sim t_{c,\nu}/t_{g,\text{other}}$. The resulting density perturbations are therefore relatively small long after the initial epoch.

is the photon energy density. Numerically, using equation (6),

$$M_J(t) \sim \frac{10^{16} M_\odot}{[m_\nu(\mathrm{eV})]^2}\left(\frac{1+Z_\nu}{1+Z}\right)^3 \tag{7}$$

for relativistic neutrinos during the photon-dominated era.

At somewhat later epochs, the neutrinos are nonrelativistic, photon decoupling is complete ($Z \lesssim 1000$), and the energy density in photons is negligible ($Z \lesssim 4.3 \times 10^4 \Omega h^2$). During this epoch the sound speed of the baryons can be ignored, and, generally, $t_{c,\nu} \ll t_{c,B}$. If $\Omega_B > \Omega_\nu$, the effective neutrino Jeans mass is

$$M_J(t) \sim \frac{4\pi}{3}\left[\langle v^{-1}\rangle\right]^{-3} m_\nu \bar{n}_\nu(t)\left[\frac{3}{8\pi G \rho_B(t)}\right]^{3/2},$$

or, since

$$\langle v^{-1}\rangle \approx \frac{\zeta(2)}{3\zeta(3)}\frac{m_\nu a(t)}{T_D a_D} \tag{8}$$

for nonrelativistic neutrinos,

$$M_J(t) \sim \frac{6.4 \times 10^{18} M_\odot}{[m_\nu(\mathrm{eV})]^2}\left(\frac{\Omega_\nu}{\Omega_B}\right)^{3/2}\left(\frac{1+Z}{1+Z_\nu}\right)^{3/2}. \tag{9}$$

In the more interesting case of a neutrino-dominated universe, $\Omega_\nu > \Omega_B$, and we get

$$M_J(t) \sim \frac{4\pi}{3}\left[\langle v^{-1}\rangle\right]^{-3} m_\nu \bar{n}_\nu(t)\left[\frac{3}{8\pi G \rho_\nu(t)}\right]^{3/2}$$

$$\approx \frac{6.4 \times 10^{18} M_\odot}{[m_\nu(\mathrm{eV})]^2}\left(\frac{1+Z}{1+Z_\nu}\right)^{3/2} \tag{10}$$

(results similar to eq. [10] are to be found in Bond, Efstathiou, and Silk 1980; Sato and Takahara 1980; Klinkhamer and Norman 1981).

For intermediate epochs, when the neutrinos first become nonrelativistic and the universe switches over from photon to neutrino or baryon dominated, the analysis is somewhat more complicated. For definiteness, we consider the case $\Omega_\nu > \Omega_B$ only. The key point is that the neutrinos become nonrelativistic before the universe becomes neutrino dominated. At first, when the neutrinos are just becoming nonrelativistic but most of the mass of the universe is in photons, $M_J(t)$ levels off at a constant value

$$M_J(t) \sim \frac{4\pi}{3}\left(\langle v^{-1}\rangle\right)^{-3} m_\nu \bar{n}_\nu(t)\left[\frac{3}{8\pi G \rho_\gamma(t)}\right]^{3/2} \approx \frac{10^{17}}{[m_\nu(\mathrm{eV})]^2} M_\odot,$$

where we have used equation (8) for $\langle v^{-1}\rangle$. However, since $c_s \sim 1/\sqrt{3}$ for the photon-baryon-electron fluid, $t_{c,\mathrm{other}} < t_{g,\mathrm{other}}$ on all length scales $\lesssim$ the particle horizon; the gravitational effect of photons can be ignored on these scales. Once $(\langle v^{-1}\rangle)^{-1} t_{g,\nu}$ becomes smaller than the Jeans length for photons, which happens at a redshift $Z \approx 2100 m_\nu(\mathrm{eV})/(T_{\gamma,0}/2.7\ \mathrm{K})$, the effective neutrino Jeans length is determined by the condition $t_{c,\nu} = t_{g,\nu}$, and the neutrino Jeans mass decreases with time according to equation (10).[3] In a neutrino-dominated universe the maximum

[3] For $\Omega_B > \Omega_\nu$ the analysis is sensitively dependent on Ω_ν/Ω_B and m_ν (for example, if $\Omega_\nu \ll \Omega_B$ and m_ν is sufficiently small, neutrinos could remain relativistic even into the matter dominated era). Since, in any event, the most interesting cosmological implications of massive neutrinos arise when $\Omega_\nu > \Omega_B$, we omit a detailed discussion for $\Omega_B > \Omega_\nu$ here.

effective neutrino Jeans mass is therefore of order

$$[M_J(t)]_{\max} \sim \frac{10^{17}}{[m_\nu(\mathrm{eV})]^2} M_\odot, \tag{11}$$

which is of the same magnitude as galaxy cluster masses for $m_\nu \gtrsim 10$ eV. From equation (10) the present-day neutrino Jeans mass is

$$M_J(t_{\mathrm{now}}) \sim 1.4 \times 10^{13} M_\odot \frac{(T_{\gamma,0}/2.7\ \mathrm{K})^{3/2}}{[m_\nu(\mathrm{eV})]^{7/2}}, \tag{12}$$

which is on the order of galactic masses for $m_\nu \gtrsim 4$ eV.

b) Freely Propagating Neutrino Perturbations

Collisionless neutrinos move freely through density perturbations less massive than the effective neutrino Jeans mass without being appreciably deflected by the fluctuating gravitational field of the perturbation. As a result, small-scale neutrino density fluctuations are subject to a substantial nonviscous decay: neutrinos flow perferentially from overdense to underdense regions, thereby smoothing out neutrino density gradients (see Stewart 1972; Peebles 1973; Zakharov 1978).

As an illustrative example of this nonviscous decay, suppose, at some initial epoch, the density of massive neutrinos is $\bar{n}_\nu(1+\varepsilon_0)$ inside and $\bar{n}_\nu$ outside a sphere of radius R, with $\varepsilon_0 =$ a constant $\ll 1$. For simplicity, we ignore the (cosmological) expansion of the background and assume that the neutrinos are nonrelativistic with a mean thermal speed, v_0. After a time $\Delta t \gg R/v_0$ has elapsed, the perturbation has expanded to a radius $v_0 \Delta t$. Conservation of particle number requires that the density contrast has fallen to $\varepsilon_0(R/v_0\Delta t)^3$. However, the total rest mass inside the density perturbation has simultaneously grown to $M \approx (4\pi/3) m_\nu \bar{n}_\nu (v_0 \Delta t)^3$. Free expansion and decay of the neutrino density perturbation can therefore only continue for $\Delta t \sim$ one dynamical time, the smaller of $t_{g,\nu}$ and $t_{g,\mathrm{other}}$ (cf. § II*a*). After one dynamical time the total mass inside the perturbation exceeds the effective neutrino Jean mass, and the density contrast can grow through gravitational instability.

The above discussion can be generalized to the nonviscous decay of density perturbations of arbitrary shape on an expanding background of massive neutrinos with arbitrarily high mean thermal energies. Because $pa(t) =$ constant for freely propagating particles in a Robertson-Walker metric

$$\varepsilon(\boldsymbol{x}, \boldsymbol{p}, t) = \varepsilon[\boldsymbol{x}_0, \boldsymbol{p}a(t)/a(t_0), t_0], \tag{13}$$

where

$$\boldsymbol{x}_0 = \boldsymbol{x} - \int_{t_0}^{t} \frac{dt'}{a(t')} \frac{\boldsymbol{p}a(t)/a(t')}{E[pa(t)/a(t')]}.$$

Equation (13) is just the mathematical statement of the fact that the neutrinos are collisionless after t_D. The neutrinos in a six-dimensional (invariant) phase space element, $d\Gamma$, around position $\boldsymbol{x}$ and momentum $\boldsymbol{p}$ at time t are the same neutrinos that were in $d\Gamma$ around position $\boldsymbol{x}_0$ and momentum $\boldsymbol{p}a(t)/a(t_0)$ at time t_0. Taking the Fourier transform of equation (13), and defining $\Gamma_{\exp}(a') = (a')^{-1} da'/dt$, we get

$$\begin{aligned} \varepsilon(\boldsymbol{k}, \boldsymbol{p}, t) = {} & \varepsilon[\boldsymbol{k}, \boldsymbol{p}a(t)/a(t_0), t_0] \\ & \times \exp\left\{ -i\boldsymbol{k}\cdot\hat{\boldsymbol{p}} \int_{a(t_0)}^{a(t)} \frac{da'}{a'^2 \Gamma_{\exp}(a') [1 + m_\nu^2 a'^2/p^2 a^2(t)]^{1/2}} \right\}, \end{aligned} \tag{14}$$

where $\hat{\boldsymbol{p}} \equiv \boldsymbol{p}/p$.

The amplitude of the Fourier mode of the neutrino perturbations with wave vector $\boldsymbol{k}$ is

$$\delta_\nu(\boldsymbol{k},t)=\frac{\int d^3p\, f_0(p,t)\varepsilon(\boldsymbol{k},\boldsymbol{p},t)}{\int d^3p\, f_0(p,t)}. \tag{15}$$

For simplicity, let us assume that

$$\varepsilon[\boldsymbol{k},\boldsymbol{p}a(t)/a(t_0),t_0]=\varepsilon(\boldsymbol{k},t_0),$$

independent of the neutrino momenta, at time t_0. This form for the initial conditions suffices to illustrate the nonviscous damping of neutrino perturbations smaller than the Jeans length. For these initial conditions we find that, in general,

$$\delta_\nu(\boldsymbol{k},t)\approx\frac{\varepsilon(\boldsymbol{k},t_0)}{3\zeta(3)}\int_0^\infty\frac{dy\,y^2}{e^y+1}\,\frac{\sin 2\phi(y,t)}{\phi(y,t)}, \tag{16}$$

where the time-dependent phase function, $\phi(y,t)$, depends on whether or not the neutrinos are relativistic, and on whether the universe is radiation or neutrino dominated. In the radiation-dominated era, $\Gamma_{\rm exp}(a')\propto(a')^{-2}$, and

$$\phi(y,t)=\frac{ka_DT_Dt_0y}{m_\nu a^2(t_0)}\left\{\sinh^{-1}\left[\frac{m_\nu a(t)}{a_DT_Dy}\right]-\sinh^{-1}\left[\frac{m_\nu a(t_0)}{a_DT_Dy}\right]\right\}. \tag{17}$$

For extremely relativistic neutrinos in the radiation-dominated era equation (16) reduces to

$$\delta_\nu(\boldsymbol{k},t)\approx\varepsilon(\boldsymbol{k},t_0)\frac{\sin\{2kt_0[a(t)-a(t_0)]/a^2(t_0)\}}{2kt_0[a(t)-a(t_0)]/a^2(t_0)}, \tag{18}$$

a result that was previously derived by Stewart (1972) and Peebles (1973), who considered the nonviscous damping of density perturbations containing strictly massless neutrinos. For neutrinos that are relativistic at t_0 but not at t we get

$$\phi(y,t)\approx\frac{ka_DT_Dt_0y}{m_\nu a^2(t_0)}\ln\frac{2m_\nu a(t)}{a_DT_Dy} \tag{19}$$

in the radiation-dominated era. Assuming $\Omega_\nu>\Omega_B$, we find that in the neutrino-dominated era, $t>t_{\rm eq}$,

$$\phi(y,t)\approx\frac{3ka_DT_Dt_0y}{2m_\nu a^2(t_0)}\left[1-\left(\frac{1+\Omega Z}{1+\Omega Z_0}\right)^{1/2}\right], \tag{20}$$

where $\Omega=\Omega_\nu+\Omega_B$, and we have assumed that Z_0 is sufficiently large that $\Gamma_{\rm exp}[a(t_0)]\approx2/3t_0$. Note that for $t\gg t_0$, or equivalently $Z\ll Z_0$, $\phi(y,t)\to3ka_DT_Dy/2m_\nu a^2(t_0)$, which is independent of t, in the neutrino-dominated era.

From equations (18), (19), and (20) we see that the nonviscous decay of neutrino density perturbations less massive than $M_J(t)$ is most rapid for extremely relativistic neutrinos in the radiation-dominated era, in which case $\delta_\nu(\boldsymbol{k},t)\sim[a(t)]^{-1}$, apart from sinusoidal oscillations. The decay is least rapid in the matter-dominated era, during which $\delta_\nu(\boldsymbol{k},t)$ asymptotically approaches a constant value.

A typical perturbation, containing a mass $M_\nu<[M_J(t)]_{\max}$, becomes smaller than the neutrino Jeans length at a time t_0 when the neutrinos are relativistic and the universe is radiation dominated. If $\Omega_\nu>\Omega_B$ then after $t_{\rm eq}$ the neutrinos are nonrelativistic, and the universe is neutrino dominated. Assuming, as before, $\varepsilon(\boldsymbol{k},\boldsymbol{p},t_0)=\varepsilon(\boldsymbol{k},t_0)$, then at t_1, when the perturbation once again becomes more massive than $M_J(t)$, $\delta_\nu(\boldsymbol{k},t)$ is given by equation (16), with

$$\phi(y,t_1)\sim\frac{3}{2}\,\frac{ka_DT_Dt_0}{ma^2(t_0)}y\left[1-\left(\frac{1+Z_1}{1+Z_{\rm eq}}\right)^{1/2}+\frac{2}{3}\ln\frac{2m_\nu a(t_{\rm eq})}{a_DT_Dy}\right], \tag{21}$$

where we have used the relation $a^2(t) \approx (\text{constant}) \times t$, which is roughly valid between t_0 and t_{eq}. Since $kt_0/a(t_0) \approx 2\pi$ and $a_D T_D / m_\nu a^2(t_0) \sim 0.5\{[M_J(t)]_{max}/M_\nu\}^{1/3}$, so that $\phi(y, t_1)/y$ is large, we get

$$\frac{\delta_\nu(k, t_1)}{\delta_\nu(k, t_0)} \approx \left[48\zeta(3)\phi^4(1, t_1)\right]^{-1}, \tag{22}$$

with, from equation (21) with M_ν considerably smaller than $[M_J(t)]_{max}$,

$$\phi(1, t_1) \sim 15\left\{\frac{[M_J(t)]_{max}}{M_\nu}\right\}^{1/3}. \tag{23}$$

From equations (22) and (23) we see that the nonviscous decay of neutrino density perturbations less massive than $[M_J(t)]_{max}$ is considerable. Assuming a primordial fluctuation spectrum that decreases less rapidly than $M_\nu^{-4/3}$ below $[M_J(t)]_{max}$ and is a decreasing function of M_ν above $[M_J(t)]_{max}$, the predominant neutrino perturbations after photon decoupling are on mass scales $\sim [M_J(t)]_{max}$.

III. BARYON PERTURBATIONS

After photon decoupling, at a redshift $1 + Z_R \approx 10^3$, perturbations in the baryon density can grow via gravitational instability if they are larger than the Jeans mass

$$M_J^B \approx 10^6 M_\odot \left(\frac{1+Z}{1+Z_R}\right)^{3/2} \tag{24}$$

(cf. Peebles 1971; Weinberg 1972). The density contrast

$$\delta_B(k, t) \equiv \frac{\delta n_B(k, t)}{\bar{n}_B(t)},$$

where $\bar{n}_B(t)$ is the background baryon number density at time t, grows according to the usual equation (cf. Peebles 1971, Weinberg 1972; Bond, Efstathiou, and Silk 1980)

$$L_B(t)\delta_B(k, t) = \frac{\Omega_\nu}{\Omega}\delta_\nu(k, t), \tag{25}$$

with

$$L_\alpha(t) \equiv [4\pi G\rho(t)]^{-1}\left[\frac{d^2}{dt^2} + 2\Gamma_{exp}(a)\frac{d}{dt} - 4\pi G\rho(t)\frac{\Omega_\alpha}{\Omega}\right]. \tag{26}$$

Consider first the homogeneous equation $L_B(t)\delta_B(k, t) = 0$. At early times ($Z_R \gtrsim Z \gtrsim \Omega^{-1}$)

$$4\pi G\rho(t) \approx \frac{2}{3t^2},$$

and

$$2\Gamma_{exp}(a) \approx \frac{4}{3t},$$

so that

$$\frac{d^2\delta_B(k, t)}{dt^2} + \frac{4}{3t}\frac{d\delta_B(k, t)}{dt} - \frac{2\Omega_B}{3\Omega t^2}\delta_B(k, t) \approx 0,$$

which has the solutions

$$\delta_B^{\pm}(k,t)\propto t^{\alpha_\pm}, \tag{27a}$$

where

$$\alpha_\pm=-\frac{1}{6}\pm\frac{1}{6}\left(1+24\frac{\Omega_B}{\Omega}\right)^{1/2} \tag{27b}$$

(cf. Bond, Efstathiou, and Silk 1980). These solutions are exact if $\Omega=1$.

On length scales smaller than the neutrino Jeans length we can, according to the results of § II*b*, take $\delta_\nu(k,t)\approx\delta_\nu(k)$, a constant, in the matter-dominated era. With this approximation we get

$$\delta_B(k,t)\approx A_+(k)t^{\alpha_+}+A_-(k)t^{\alpha_-}-\frac{\Omega_\nu}{\Omega_B}\delta_\nu(k), \tag{28a}$$

where, if $\delta_B^R(k)$ and $\dot\delta_B^R(k)$ are the baryon density contrast and its time derivative at t_R, when photon decoupling is complete,

$$A_\pm(k)=\frac{\mp\alpha_\mp\left[\delta_B^R(k)+\Omega_\nu\delta_\nu(k)/\Omega_B\right]\pm\dot\delta_B^R(k)t_R}{t_R^{\alpha_\pm}(\alpha_+-\alpha_-)}. \tag{28b}$$

Since $\alpha_+<2/3$, its value for $\Omega_B=\Omega$ (i.e., $\Omega_\nu\to0$), the inclusion of massive neutrinos weakens the gravitational instability of the baryons at early times. However, if compressional perturbations in the baryon density on comoving length scales $\sim k^{-1}$ are absent at t_R because of viscous damping *before* photon decoupling, fluctuations in the neutrino density, $\delta_\nu(k)$, can give rise to growing perturbations in the baryon density, $\delta_B(k,t)$, *after* decoupling. In this case

$$\delta_B(k,t)\approx\frac{-\alpha_-\Omega_\nu\delta_\nu(k)/\Omega_B}{\alpha_+-\alpha_-}\left(\frac{t}{t_R}\right)^{\alpha_+}.$$

for $t\gg t_R$. The neutrino fluctuation, $\delta_\nu(k)$, required to generate a baryon density perturbation that reaches $\delta_B(k,t)\sim1$ at $t=t_c$ and then recollapses is therefore of order

$$\delta_\nu(k)\approx\frac{\Omega_B}{\Omega_\nu}\frac{(t_R/t_c)^{\alpha_+}}{1-\alpha_+/\alpha_-}.$$

As an example, suppose that galaxy formation is initiated by neutrino perturbations on mass scales $M_\nu\approx(\Omega_\nu/\Omega_B)M_B^G$, where M_B^G is the baryonic mass of a typical galaxy ($\sim10^{11}-10^{12}M_\odot$). If galaxies begin to recollapse at $t_c\sim10^8$ yr (cf. Field 1975; Gott 1977) and $\Omega_B/\Omega_\nu\sim0.1$, then, taking $t_R\approx2\times10^5(\Omega h^2)^{-1/2}$ yr (Peebles 1971), we find that the required neutrino density fluctuations are of order $\delta_\nu(k)\approx0.06(\Omega h^2)^{-0.07}$ (Bond *et al.* 1980 have considered the opposite case, $\delta_\nu^R(k)=0$ and $\delta_B^R(k)\neq0$, which may be appropriate if primordial *isothermal* perturbations are responsible for galaxy formation [see § V]).

IV. MUTUAL PERTURBATIONS

The most interesting behavior arises for perturbations that are larger than the Jeans lengths for *both* baryons and neutrinos. For these perturbations we have

$$L_B(t)\delta_B(k,t)=\frac{\Omega_\nu}{\Omega}\delta_\nu(k,t), \tag{29a}$$

and

$$L_\nu(t)\delta_\nu(k,t)=\frac{\Omega_B}{\Omega}\delta_B(k,t), \tag{29b}$$

where $L_\alpha(t)$ is given by equation (26). In addition to the usual growing and shrinking modes corresponding to the solutions $f_\pm(t)$ of the equation

$$\left[\frac{d^2}{dt^2}+2\Gamma_{\exp}(a)\frac{d}{dt}-4\pi G\rho(t)\right]f_\pm(t)=0$$

(cf. Peebles 1971; Weinberg 1972; see also Bond, Efstathiou, and Silk 1980), equations (29a) and (29b) have two new modes, proportional to t^0 = constant and

$$\psi(t)=\int\frac{dt}{a^2(t)}.$$

If we choose $f_+(t_i)=f_-(t_i)=\psi(t_i)=1$ at the epoch $t_i \gtrsim t_R$ when the perturbation first becomes more massive than the neutrino effective Jeans mass, then in the early universe ($Z_i \gtrsim Z \gtrsim \Omega^{-1}$),

$$f_+(t)\approx(t/t_i)^{2/3},$$

$$f_-(t)\approx(t/t_i)^{-1},$$

and

$$\psi(t)\approx(t/t_i)^{-1/3}.$$

Upon specifying $\delta^i_\alpha(k)$ and $\dot\delta^i_\alpha(k)$, the density contrast of component α and its time derivative at t_i, we get the general solution to equations (29a) and (29b)

$$\begin{bmatrix}\delta_B(k,t)\\ \delta_\nu(k,t)\end{bmatrix}=c_+(k)\begin{pmatrix}1\\1\end{pmatrix}f_+(t)+c_-(k)\begin{pmatrix}1\\1\end{pmatrix}f_-(t)$$

$$+c_0(k)\begin{pmatrix}-\Omega_\nu/\Omega_B\\1\end{pmatrix}+c_\psi(k)\begin{pmatrix}-\Omega_\nu/\Omega_B\\1\end{pmatrix}\psi(t), \tag{30a}$$

where

$$c_+(k)\approx\frac{3}{5}\left\{\frac{\Omega_B}{\Omega}\left[\delta^i_B(k)+t_i\dot\delta^i_B(k)\right]+\frac{\Omega_\nu}{\Omega}\left[\delta^i_\nu(k)+t_i\dot\delta^i_\nu(k)\right]\right\}, \tag{30b}$$

$$c_-(k)\approx\frac{2}{5}\left\{\frac{\Omega_B}{\Omega}\left[\delta^i_B(k)-\frac{3}{2}t_i\dot\delta^i_B(k)\right]+\frac{\Omega_\nu}{\Omega}\left[\delta^i_\nu(k)-\frac{3}{2}t_i\dot\delta^i_\nu(k)\right]\right\}, \tag{30c}$$

$$c_0(k)\approx\frac{\Omega_B}{\Omega}\left\{\left[\delta^i_\nu(k)+3t_i\dot\delta^i_\nu(k)\right]-\left[\delta^i_B(k)+3t_i\dot\delta^i_B(k)\right]\right\}, \tag{30d}$$

and

$$c_\psi(k)\approx 3t_i\frac{\Omega_B}{\Omega}\left[\dot\delta^i_B(k)-\dot\delta^i_\nu(k)\right]. \tag{30e}$$

For large t/t_i we get, from equation (30a),

$$\delta_B(k,t)\approx\delta_\nu(k,t)\approx c_+(k)f_+(t).$$

Once again we see that even if perturbations in the baryon density are absent at t_i, fluctuations in the density of massive neutrinos can initiate gravitational instability in the density of baryons. At large times, the behavior of initially absent baryon density fluctuations on length scales larger than the Jeans length for both baryons and massive neutrinos is indistinguishable from the behavior of baryon density perturbations of amplitude $\sim(\Omega_\nu/\Omega)\delta_\nu^i(k)(t_R/t_i)^{2/3}$ at photon decoupling in a universe containing no (or at least an insignificant density of) massive neutrinos. Nonzero baryon density perturbations can also induce growing fluctuations in a smooth distribution of massive neutrinos, which should be sufficient to initiate the formation of neutrino halos in bound systems less massive than $[M_J(t)]_{\max}$.

V. DISCUSSION

Initially adiabatic density perturbations can evolve according to two distinct scenarios after photon decoupling depending on the typical value of the neutrino rest mass, m_ν. For small values of m_ν, baryon density perturbations persist down to mass scales where neutrino density perturbations are absent: $[M_J(t)]_{\max} > M_c'$ for $m_\nu \lesssim 3$ eV and $[M_J(t)]_{\max} > M_c$ for $m_\nu \lesssim 10$ eV.[4] Just after photon decoupling, the density contrast grows slowly, $\propto(t/t_R)^{\alpha_+}$ with $\alpha_+ < 2/3$ (and possibly $\ll 2/3$ if $\Omega_B \ll \Omega_\nu$; cf. eqs (27a) and (27b) and Bond, Efstathiou, and Silk 1980); the growth rate only becomes $\propto t^{2/3}$ after t_i, when the neutrino effective Jeans mass falls below the total neutrino mass of the perturbations. According to the results of § IV, the baryon density contrast can then initiate gravitational instability in the neutrinos, acting as a seed for the eventual formation of a neutrino halo.

A possibly more interesting scenario arises if the neutrino mass is large, since $[M_J(t)]_{\max} < M_c'$ for $m_\nu \gtrsim 3$ eV and $[M_J(t)]_{\max} < M_c$ for $m_\nu \gtrsim 10$ eV. In this case neutrino density perturbations persist down to mass scales where fluctuations in the density of baryons are absent just after photon decoupling. Neutrino density perturbations on a neutrino mass scale $M_\nu \sim [M_J(t)]_{\max}$ can then induce growing baryon density fluctuations on a baryonic mass scale $M_B = (\Omega_B/\Omega_\nu)[M_J(t)]_{\max}$. If the primordial spectrum of the initially adiabatic perturbations is a decreasing function of the mass of the perturbations (which, however, decreases less rapidly than $M_\nu^{-4/3}$ below $[M_J(t)]_{\max}$; see § II*b*), then, in this picture the first systems to condense out of the Friedman background characteristically have total bound masses $\sim[M_J(t)]_{\max}$. Although for a neutrino mass $m_\nu \approx 10$ eV one finds $M_J(t) \sim 10^{15} M_\odot$, which is still several times larger than the core masses of galaxy clusters deduced from dynamical models, one should bear in mind that the total *bound* mass in a cluster of galaxies can exceed its core mass by factors ~ a few (Silk and Wilson 1979*a*, *b*; Hoffman, Olson, and Salpeter 1980). The maximum neutrino Jeans mass, $[M_J(t)]_{\max}$, could therefore determine the masses of clusters of galaxies if $m_\nu \approx 10$ eV. Note that we do not share the concern expressed by Bond, Efstathiou, and Silk (1980) that, for $m_\nu \lesssim 90$ eV, neutrino perturbations of mass $M_\nu \gtrsim [M_J(t)]_{\max}$ would lead to baryon condensations in the present-day universe on larger mass scales than those on which nonlinear clumping is observed to occur (cf. Davis and Peebles 1977). This is because the maximum neutrino Jeans mass derived here is more than an order of magnitude smaller than that advocated by Bond *et al.*, irrespective of the value of the neutrino mass.

As in the pancake theory (Zeldovich 1970), one must appeal to dissipative processes to account for the formation of galaxies from initially adiabatic density perturbations in a neutrino+baryon universe. Collisionless neutrinos cannot dissipate energy, as can baryons and electrons. Protogalaxies are therefore born with no neutrino content in this picture. Low velocity neutrinos, with (random speeds)$^2 \lesssim$ the binding energy per unit mass of a typical galaxy, can subsequently be accreted by the protogalaxies.

Although it has been argued that adiabatic density perturbations could arise naturally in the early universe (cf. Turner and Schramm 1979), primordial isothermal perturbations, involving fluctuations in the baryon density alone, remain a viable possibility (Zeldovich 1966; Peebles 1969). As is evident from the results of §§ III and IV, a significant cosmological density of massive neutrinos, with $\Omega_\nu \gtrsim \Omega_B$, slows the growth of isothermal perturbations on baryonic mass scales M_B after photon decoupling, until the neutrino Jeans mass falls below $(\Omega_\nu/\Omega_B)M_B$. The initial spectrum of primordial isothermal perturbations required to account for the observed clustering hierarchy of the present-day universe would therefore be quite different in a neutrino-dominated universe than in a baryon-dominated universe; in particular the required amplitude of primordial fluctuations on mass scales $M_B < (\Omega_B/\Omega_\nu)M_J(t_R)$ would be systematically larger. It is interesting to note that the present-day neutrino Jeans mass (eq. [11]) sets a not implausible lower bound on the masses of isothermal density perturbations that can grow appreciably between photon decoupling and the present epoch in a neutrino-dominated universe. From equations (1) and (12) we get a minimum mass in baryons

$$M_{\min} \sim \frac{1.4\times10^{15}\Omega_B h^2}{[m_\nu(\text{eV})]^{9/2}} M_\odot,$$

which is $\sim(10^7-10^9)M_\odot$ for $\Omega_B h^2 \sim 10^{-2}-10^{-1}$ and $m_\nu \sim (10-25)$ eV, reminiscent of the masses of dwarf galaxies.

[4]Here, we assume that

$$\frac{0.007-0.021}{h^{3/2}} \lesssim \Omega_B \lesssim \frac{0.014}{h^2}\left(\frac{T_{\gamma,0}}{2.7\text{ K}}\right)^3$$

where $T_{\gamma,0}$ is the present temperature of the blackbody background (Schramm and Steigman 1981).

Regardless of whether primordial density perturbations are isothermal or adiabatic, the potential cosmogonical significance of massive neutrinos would appear to be in determining the masses of the large-scale bound systems in the universe. At present little is known with certainty about either neutrino rest masses or the nature of the primordial density fluctuations responsible for the observable structure of the present-day universe. One can nonetheless reasonably speculate that the rest masses of the largest bound systems in nature are related to neutrino rest masses, the smallest nonzero rest masses in the microphysical realm.

It is a pleasure to thank O. Alvarez, M. Duncan, S. L. Shapiro, S. A. Teukolsky, and, especially, E. E. Salpeter for useful discussions on the implications of massive neutrinos for cosmology. I thank D. Stewart for aid in preparing the manuscript. This research was supported in part by NSF grant AST78-20708, and by a Chaim Weizmann Postdoctoral Fellowship.

REFERENCES

Abramowitz, M., and Stegun, I. A. 1964, *Handbook of Mathematical Functions* (Washington, D.C.: National Bureau of Standards).
Allen, C. W. 1973, *Astrophysical Quantities* (London: Althone).
Bahcall, N. A. 1977, *Ann. Rev. Astr. Ap.*, **15**, 505.
Bilenky, S. M., and Pontecorvo, B. 1976, *Phys. Letters*, **61B**, 248.
______. 1978, *Phys. Rev.*, **41C**, 225.
Bond, J. R., Efstathiou, G., and Silk, J. 1980, *Phys. Rev. Letters*, **45**, 1980.
Ching, C., Wu, Y., Ho, T., Chang, C., and Zou, Z. 1980, preprint ASITP-80-007.
Cowsik, R., and McClelland, J. 1972, *Phys. Rev. Letters*, **29**, 669.
______. 1973, *Ap. J.*, **180**, 7.
Davis, M., and Peebles, P. J. E. 1977, *Ap. J. Suppl.*, **34**, 425.
Dolgov, A. D., and Zel'dovich, Ya. B. 1980, preprint.
Eliezer, S., and Ross, D. A. 1974, *Phys. Rev.*, **D10**, 3088.
Field, G. B. 1971, *Ap. J.*, **165**, 29.
______. 1975, in *Stars and Stellar Systems*, Vol. **9**, *Galaxies and the Universe*, ed. Sandage, A., Sandage, M., and Kristian, J. (Chicago: University of Chicago Press).
Gott, J. R. 1977, *Ann. Rev. Astr. Ap.*, **15**, 235.
Gribov, V., and Pontecorvo, B. 1969, *Phys. Letters*, **28B**, 493.
Hoffman, G. L., Olson, D. W., and Salpeter, E. E. 1980, *Ap. J.*, **242**, 861.
Klinkhamer, F. R., and Norman, C. A. 1981, *Ap. J.* (*Letters*), **243**, 1.
Lu, T., Luo, L. F., Ostriker, J. P., and Yang, G. C. 1980, preprint.
Melott, A. L. 1980, preprint.
Peebles, P. J. E. 1969, *Ap. J.*, **157**, 1075.
______. 1971, *Physical Cosmology* (Princeton: Princeton University Press).
______. 1973, *Ap. J.*, **180**, 1.
Peebles, P. J. E., and Yu. J. T. 1970, *Ap. J.*, **162**, 815.
Pontecorvo, B. 1968, *Soviet Phys.—JETP*, **26**, 984.
Press, W. H., and Vishniac, E. T. 1980, *Ap. J.*, **236**, 323.
Reines, F., Sobel, H. W., and Pasierb, E. 1980, *Phys. Rev. Letters*, **45**, 1307.
Sato, H., and Takahara, F. 1980, Kyoto University preprint RIFP-400.
Schramm, D. N., and Steigman, G. 1980*a*, Gravity Research Foundation Gravity Prize essay.
______. 1981 *Ap. J.*, **243**, 1.
Shapiro, S. L., Teukolsky, S. A., and Wasserman, I. 1980, *Phys. Rev. Letters*, **45**, 669.
Silk, J. 1967, *Nature*, **215**, 1155.
______. 1968, *Ap. J.*, **151**, 459.
Silk, J., and Wilson, M. L. 1979*a*, *Ap. J.*, **228**, 641.
______. 1979*b*, *Ap. J.*, **233**, 769.
Steigman, G., Olive, K. A., and Schramm, D. N. 1979, *Phys. Rev. Letters*, **43**, 239.
Stewart, J. M. 1972, *Ap. J.*, **176**, 323.
Tremaine, S., and Gunn, J. E. 1979, *Phys. Rev. Letters*, **42**, 407.
Turner, M. S., and Schramm, D. N. 1979, *Nature*, **279**, 303.
Weinberg, S. 1971, *Ap. J.*, **168**, 175.
______. 1972, *Gravitation and Cosmology* (New York: Wiley).
______. 1979, *Phys. Rev. Letters*, **42**, 850.
Witten, E. 1980, *Phys. Letters*, **91B**, 81.
Yang, J., Schramm, D. N., Steigman, G., and Rood, R. T. 1979, *Ap. J.*, **227**, 697.
Zakharov, A. V. 1978, *Soviet Astr.—AJ*, **22**, 528.
Zel'dovich, Ya. B. 1966, *Soviet Phys.—USP*, **9**, 602.
______. 1970, *Astr. Ap.*, **5**, 84.

Note added in proof.—Some similar results to those reported in this paper have been obtained by A. G. Doroshkevich, Ya. B. Zel'dovich, R. A. Sunyaev, and M. Yu. Khlopov, *Soviet Astr. Letters*, **6**, 252 (1980).

IRA WASSERMAN: Center for Radiophysics and Space Research, Cornell University, Ithaca, NY 14853

THE ASTROPHYSICAL JOURNAL, 250:423–431, 1981 November 15

THE FORMATION OF GALAXIES FROM MASSIVE NEUTRINOS

MARC DAVIS, MYRON LECAR, AND CARLTON PRYOR
Harvard-Smithsonian Center for Astrophysics
AND
EDWARD WITTEN
Department of Physics, Princeton University
Received 1981 January 20; accepted 1981 May 19

ABSTRACT

Neutrinos with nonzero rest mass strongly influence galaxy formation in the early universe. If stable neutrinos have rest masses on the order of 100 eV, they close the universe, but they erase initial perturbations on mass scales less than $4\times10^{15}\ M_\odot$. However, if in addition there exist unstable neutrinos with rest masses on the order of 100 keV, they preserve and amplify initial perturbations on galactic mass scales ($10^{12}\ M_\odot$). These perturbations are picked up and further amplified by the lighter, stable neutrinos, as long as the heavy neutrinos decay somewhat after the lighter neutrinos go nonrelativistic. If the heavy neutrinos decay into light neutrinos, the decay products contribute about one-half of the present mass density in a hot unclustered background. The only alternative method of retaining initial perturbations until the light neutrinos become nonrelativistic is to introduce large amplitude initial fluctuations such as primordial black holes. If the light neutrinos close the universe, black hole seeds of size $10^9\ M_\odot$ would be required for galaxies of $10^{12}\ M_\odot$ to form. We point out that the neutrino damping mass is a steep function of the present neutrino temperature and that galaxy sized fluctuations would be preserved if $T_\nu<1.0$ K. However, the only model we can devise to effect this cooling is shown to be in serious violation of astrophysical constraints.

Subject headings: black holes — galaxies: formation — neutrinos

I. INTRODUCTION

Most grand unified theories of the strong, weak, and electromagnetic interactions predict that neutrinos have nonzero rest masses. It is plausible that the τ-neutrino is the most massive and the electron neutrino the least massive.

To set the mass scale of interest to cosmology, recall that the critical density (ρ_c) required to close the universe is:

$$\rho_c=\frac{3}{8\pi}\frac{H^2}{G}=1.88\times10^{-29}\,h^2\ \mathrm{g\ cm^{-3}}$$
$$=1.12\times10^{-5}\,h^2\,m_p\ \mathrm{cm^{-3}}$$
$$=10{,}540\,h^2\ \mathrm{eV\ cm^{-3}}, \quad (1)$$

where we set the Hubble constant $H=(100h)$ km s^{-1} Mpc^{-1}, and m_p is the mass of a nucleon. The number of blackbody photons is

$$n_\gamma=8\pi\left(\frac{kT}{hc}\right)^3 2\zeta(3)=399\ \mathrm{cm^{-3}} \quad (2)$$

with $T=2.7$ K.

The number of neutrinos (plus antineutrinos) of each type is

$$n_\nu=\tfrac{3}{4}\cdot\tfrac{4}{11}\cdot n_\gamma=109\ \mathrm{cm^{-3}}, \quad (3)$$

where the factor of 3/4 comes from Fermi statistics, and the factor of 4/11 is due to a heating of the γ's (but not the ν's) when the e^+-e^- annihilated.

Therefore, if the most massive, stable neutrino has a mass of $96.8h^2$ eV, it closes the universe.

Massive neutrinos offer appealing possibilities for resolving a number of astrophysical problems:

1. Neutrino oscillations could account for the anomalously low counting rates in Davis's solar neutrino experiment (Bahcall *et al.* 1980).

2. Big bang nucleosynthesis restricts the baryon density to less than about 3% of the closure density (Wagoner 1974; Yang *et al.* 1979; Schramm and Steigman 1981). More baryons would increase the helium and decrease the deuterium relative to what is observed.

On the other hand, measurements of distortions of the Hubble flow on scales of the Virgo supercluster indicate that the mean density of the universe is more than 40% of the closure density (Davis *et al.* 1980). Massive neutrinos could comprise the bulk of the gravitating mass without violating the nucleosynthesis constraints.

3. Individual galaxies are surrounded by massive dark halos, which have as much as 10 times the mass of the visible matter (Ostriker, Peebles, and Yahil 1974; Einasto, Kaasik, and Saar 1974). The density of the galactic halo is distributed like that of a self-gravitating isothermal sphere $[M(r) \propto r]$ in the outer regions. If that distribution is continued into the center, the central density (ρ_H) is

$$\rho_H = \frac{3}{4\pi}\frac{v^2}{Gr_c^2} = 3.38\times10^{-24}\frac{(v/300\ \mathrm{km\ s^{-1}})^2}{(r_c/10\ \mathrm{kpc})^2}\ \mathrm{g\ cm^{-3}}, \tag{4}$$

where r_c is the core radius.

As Tremaine and Gunn (1979) pointed out, neutrinos, being collisionless, preserve their maximum phase-space density. Thus,

$$\rho_\nu \le \frac{4\pi}{3}\left(\frac{mv}{h}\right)^3 m = 4.78\times10^{-22}\left(\frac{m}{100\ \mathrm{eV}}\right)^4\left(\frac{v}{300\ \mathrm{km\ s^{-1}}}\right)^3\ \mathrm{g\ cm^{-3}}. \tag{5}$$

The scaling factor is $4\pi/3$ rather than $8\pi/3$ because only half the states were occupied. Therefore, if the halos are neutrinos,

$$\left(\frac{m}{100\ \mathrm{eV}}\right) \gtrsim 0.3\left(\frac{v}{300\ \mathrm{km\ s^{-1}}}\right)^{-1/4}\left(\frac{r_c}{10\ \mathrm{kpc}}\right)^{-1/2}. \tag{6}$$

Note that for an isothermal sphere, the circular velocity is $\sqrt{2/3}\,v \approx 248\ \mathrm{km\ s^{-1}}$. We would find this coincidence (that the neutrino mass which just closes the universe is the same as needed to make galactic halos) compelling, if we could explain why $v \approx 300\ \mathrm{km\ s^{-1}}$.

Unlike baryons that are frozen in the radiation field prior to recombination, the neutrinos are free to cluster as soon as they become nonrelativistic. A 100 eV neutrino becomes nonrelativistic at $z \approx 2\times10^5$ and, if it did not cluster, would have a thermal velocity of $\sim 1\ \mathrm{km\ s^{-1}}$ today.

Although galaxy formation with neutrinos can begin at an earlier epoch than in the standard scenario, massive neutrinos impose a severe constraint on all theories. As shown in the next section, 100 eV neutrinos erase their primordial perturbations on scales less than 10^{15} $M_\odot$. Unless large baryon number fluctuations are postulated ab initio on galactic sizes, neutrino streaming results in very low density contrast on galactic scales and an extremely slow growth rate for the residual baryon fluctuations, thus making galaxy formation nearly impossible (Bond, Efstathiou, and Silk 1980).

We take note of the suggestion by Zel'dovich (1978), Doroshkevich, Saar, and Shandarin (1978), and Doroshkevich *et al.* (1980) that 10^{15} $M_\odot$ is the mass scale of rich clusters of galaxies, and that they would form directly by Jeans instabilities. However, on such scales, the neutrino velocities would be thousands of kilometers per second. Even if, later, baryon galaxies were formed by hydrodynamic shocks, we see no way to trap neutrinos around the baryon galaxies to form galactic halos.

To preserve and amplify perturbations on a galactic scale (10^{12} $M_\odot$) requires much more massive neutrinos. But if these more massive neutrinos existed today, they would overclose the universe by a large factor. Therefore, the more massive neutrino must be unstable.

To focus the discussion we will identify the massive unstable neutrino with ν_τ having a mass of keV to MeV energies. We identify the stable light neutrino with ν_μ, having a mass on the order of 100 eV. We assume ν_e is stable, and we neglect its mass.

In the following sections we discuss three separate means to form galaxies with the aid of the ν_τ and ν_μ, and the constraints imposed on these models by big bang nucleosynthesis, the requirement that the decay products not contradict present observations, and the requirement that the universe be matter dominated today.

The third scenario is presented as an obituary to discourage future investigations, because, while attractive, our particular model seriously violates several observational constraints.

II. THE GROWTH OF PERTURBATIONS

As is usual in studies of galaxy formation, we assume an initial perturbation spectrum on all scales and ask what scales grow and what scales decay. In a matter dominated universe, the Jeans mass is the dividing line; larger mass scales grow and smaller mass scales decay as damped sound waves. As the Jeans mass in a matter dominated universe decreases with time (as t^{-1}), one is assured that if ever a mass scale grows, it will continue to grow at later times. In a universe that alternates between radiation and matter domination, the Jeans mass is not a monotonic function of time. Furthermore, we are interested in the rest mass in galaxies today; but if one calculates the Jeans mass in a radiation dominated era, one includes a substantial contribution of energy density that has, by now, redshifted away.

A reliable estimate of the critical scale is the free-streaming neutrino horizon, or diffusion length, which is given by

$$d(t) = R_0\int_t^{t_0}\frac{v(t')\,dt'}{R(t')}, \tag{7}$$

where R is the scale factor of the universe, v is the neutrino velocity, t is the time when the neutrinos begin free-streaming, and t_0 is the present time. The mass contained in a diffusion length stops growing shortly after the neutrinos become nonrelativistic, and so d is insensitive to the upper limit. The mass scale, M, *above* which perturbations grow, is

$$M=\frac{4\pi}{3}\rho_0 d^3. \tag{8}$$

Since $(4\pi/3)\rho_0=1.16\times10^{12}\ \Omega h^2\ M_\odot\ \text{Mpc}^{-3}$ (Ω is the cosmological density parameter), we require $d\lesssim1$ Mpc for $M\lesssim10^{12}\ M_\odot$.

We evaluate d in detail in the Appendix, but we note here that if the bulk of the mass density is in the form of one type of massive neutrino, then $d\approx11\ (\Omega h^2)^{-1}$ Mpc.

If the mass of the τ-neutrino is increased by a factor of η, then its diffusion length decreases by a factor of η. But, as mentioned earlier, these more massive neutrinos must decay. In the next section, we follow the perturbations in a universe with a massive, unstable ν_τ.

III. A UNIVERSE WITH A SUPERMASSIVE, UNSTABLE ν_τ AND A MASSIVE STABLE ν_μ.

Suppose the mass of the τ-neutrino is η times the mass of the μ-neutrino. For $\eta\gtrsim15$, the primordial τ-neutrino perturbations will be preserved on galactic scales. However, the diffusion length for the μ-neutrino is larger, and so it will have erased its own perturbations on galactic scales. However, after they go nonrelativistic, the ν_μ will be Jeans unstable to clustering in the potential field of the ν_τ fluctuations on galactic mass scales. The Jeans length r_J of the ν_μ is given by

$$r_J=\frac{1}{2}\left(\frac{\pi}{3}\right)^{1/2} v_\mu/\sqrt{G\rho_\mu\eta}\,, \tag{9}$$

where the factor of $\eta^{-1/2}$ is due to the presence of the ν_τ. The mass of ν_μ within this Jeans length is

$$M_J=\frac{4\pi}{3}\rho_\mu r_J^3\approx5\times10^{14}\left(\frac{1+z}{10^5}\right)^{3/2}\eta^{-3/2}M_\odot, \tag{10}$$

so that shortly after the ν_μ go nonrelativistic ($z\approx10^5$) (and with $\eta\gtrsim100$), they will be Jeans unstable on sufficiently small mass scales and will quickly pick up the existing ν_τ perturbations.

For example, let δ_τ be the density contrast of the ν_τ and δ_μ be the density contrast of the ν_μ. Then

$$\ddot{\delta}_\mu+\frac{4}{3}\frac{1}{t}\dot{\delta}_\mu\approx4\pi G\rho_\tau\delta_\tau=\frac{2}{3}\frac{\delta_\tau}{t^2} \tag{11}$$

(since the ν_τ dominate the mass). Let the growing mode begin at $t=t_\mu$ with initial values $\delta(t_\mu)=\dot{\delta}(t_\mu)=0$. Then the solution of equation (11), with $\theta=t/t_\mu$, is

$$\delta_\mu(\theta)=\delta_\tau(t_\mu)(\theta^{2/3}-2\theta^{-1/3}-3)$$

$$=\delta_\tau(\theta)\left(1+\frac{2}{\theta}-\frac{3}{\theta^{2/3}}\right). \tag{12}$$

When $t/t_\mu=8$, $\delta_\mu=(1/2)\delta_\tau$ and is growing as $\theta^{2/3}$. Thus, the ν_μ readily cluster on scales smaller than their initial diffusion length, if the ν_τ preserve perturbations on smaller scales.

Suppose the ν_τ decay at epoch $z=z_d$. The decay products will be relativistic, and so, since the ν_τ dominated the mass density prior to decay, the universe will become radiation dominated again. Until the universe again becomes matter dominated at $z=z_M$, the perturbations in ν_μ stop growing. In fact, the perturbations damp slightly because the slower Hubble expansion rate of the radiation dominated epoch causes d_μ to slowly increase again even though the ν_μ are by now nonrelativistic.

To give the perturbations in ν_μ time to mimic the perturbations in ν_τ, we calculate d_μ starting at epoch $z=4z_D$ to the present. The result is approximately

$$d_\mu\approx\frac{c}{H_0}\frac{\rho_0}{\Omega_\nu^{3/2}}(1+z_M)^{1/2}\left(\ln4\eta+2\Omega_\nu^{1/2}\right)$$

$$\approx.016(1+z_M)^{1/2}\Omega_\nu^{-3/2}\left(\ln4\eta+2\Omega_\nu^{1/2}\right)\ \text{Mpc}, \tag{13}$$

where we have allowed for the possibility that, at present, the mass density of the clusterable ν_μ is $\Omega_\nu\rho_c$. The resulting M_D of clusterable matter versus z_D and η is plotted in Figure 1*a* for $\Omega_\nu=1$ and Figure 1*b* for $\Omega_\nu=1/2$. On scales smaller than M_D, the cutoff is not exponential but will scale approximately linearly with mass. We allow for the possibility of

$$\nu_\tau\to\nu_\mu+\nu_e+\bar{\nu}_e\quad(\Omega_\nu=\tfrac{1}{2}) \tag{14a}$$

or

$$\nu_\tau\to\nu_e+\nu_e+\bar{\nu}_e\quad(\Omega_\nu=1). \tag{14b}$$

The decay μ-neutrinos will be nonthermal as opposed to their primordial counterparts and would have average velocity $v=c/(1+z_M)$, far too hot to cluster in galaxies or clusters of galaxies. They could constitute an unclustered background field with perhaps half the total mass density of the universe.

The radiative decay

$$\nu_\tau\to\nu_\mu+\gamma \tag{14c}$$

is ruled out for ν_τ lifetimes greater than 10^{13} s and branching ratio $\gtrsim 10^{-2}$ by the absence of an optical or near-infrared background at the expected high flux levels (Kimble, Bowyer, and Jakobsen 1981; Hayakawa *et al.* 1978). Particle physics models in which triple neutrino decays predominate by this factor or more have been discussed by Dicus, Kolb, and Teplitz (1978) and by de Rújula and Glashow (1980).

After the decay products have redshifted by a factor η, the universe will again become matter dominated, at $z = z_M \approx z_D/\eta$. Once the universe becomes matter dominated for the last time, the neutrino streaming ceases, and perturbations again grow. However, if only a fraction f of the mass density is in clusterable form, then the growth rate of perturbations becomes $\delta\rho/\rho \sim t^{2m/3}$, $m = [(1+24f)^{1/2} - 1]/4$. If $f = \frac{1}{2}$, $m = 0.65$, so the growth rate can be slowed moderately. This scheme must obey several constraints:

1. If the ν_τ are not to change the rate of expansion during the time when the neutron-proton ratio is frozen-in ($T \sim 1$ MeV), the τ_ν must not dominate the mass at $z \sim 10^{+10}$. Since they will dominate the mass when

$$z \leq \frac{m_{\nu_\tau} c^2}{3.15\, kT_{\nu_0}} \approx \frac{\eta}{5.4 \times 10^{-6}}, \qquad (15)$$

we require $\eta < 5.4 \times 10^4$.

2. A lower limit to the decay times for the ν_τ is obtained from scaling the decay of the muon (cf. Dicus, Kolb, and Teplitz 1978):

$$t_d \geq 2.2 \times 10^{-6}\,\mathrm{s} \left(\frac{106\ \mathrm{MeV}}{m_\nu c^2}\right)^5 = \frac{3.5 \times 10^{24}}{\eta^5 \Omega_\nu^5}\ \mathrm{s}. \qquad (16)$$

The age of the universe (t_u) is

$$t_\mu \approx \tfrac{2}{3}(1+z_M)^{-3/2} H_0^{-1} \Omega_\nu^{-1/2} \eta^{-2}$$

$$= 2.1 \times 10^{17}\,\mathrm{s}\,(1+z_M)^{-3/2} \Omega_\nu^{-1/2} \eta^{-2}. \qquad (17)$$

Since $t_u \geq t_d$,

$$\eta > 256 (1+z_M)^{1/2} (\Omega_\nu)^{-3/2}. \qquad (18)$$

Some typical values of η and z_D are given in Table 1. As can be seen in Figure 1, the larger values of η lead to damped masses $> 10^{12}\ M_\odot$, so we are forced to small z_m. (The universe stays relativistic until $z \sim 10$.)

Note that in spite of this, the total time available for perturbations to grow is from $z \approx 2 \times 10^5\ \eta$ to z_d and from $z_M = z_d/\eta$ to the present. So perturbations grow by a factor of $\sim 2 \times 10^5$.

The lifetime of the ν_τ is obviously critical for this scenario, and it is not certain that equation (16) is applicable to the reactions that do not conserve tau lepton number. If the lepton has a longer lifetime than expected from equation (16), η is required to increase and/or z_m must decrease.

IV. GROWTH OF PERTURBATIONS AROUND A BLACK HOLE

The advantage of introducing initial perturbations via a black hole is to prevent them from being erased by neutrino diffusion. It has been suggested by a number of authors, independent of massive neutrinos (see, for example, Carr 1977 and Ryan 1972) that galaxies formed around "seed" black holes. As we show below, because the Jeans mass for 100 eV neutrinos does not drop to $10^{12}\ M_\odot$ until $z = 1570$, which is close to the decoupling z of 1000, the seed black hole scenario is not much modified by massive neutrinos. However, massive neutrinos, in this scenario, also do not inhibit galaxy formation. We take for the Jeans mass the expression

$$M_J = \frac{4\pi}{3} r_J^3 = \frac{\pi}{6}\rho\left(\frac{\pi}{3}\frac{v^2}{G\rho}\right)^{3/2}. \qquad (19)$$

We are interested only in nonrelativistic neutrinos in a matter dominated universe, so we let

$$\rho = \rho_0 (1+z)^3$$

and

$$V = c p_0 (1+z),$$

where $p_0 = 3.15 kT_{\nu_0}/m_\nu C^2$, whence

$$M_J = 2^{1/2}\left(\frac{\pi}{3}\right)^3 \frac{(cp_0)^3}{GH_0}(1+z)^{3/2}$$

$$= 1.6 \times 10^7\ M_\odot\ (1+z)^{3/2}.$$

Thus the Jeans mass is $10^{12}\ M_\odot$ when $z = 1570$.

TABLE 1

TYPICAL VALUES OF η AND z_D

z_m	$\eta(\Omega_\nu=1)$	$z_d(\Omega_\nu=1)$	$\eta(\Omega_\nu=1/2)$	$z_d(\Omega_\nu=1/2)$
1 ...	2.56×10^2	2.56×10^2	7.25×10^2	7.25×10^2
10 ...	8.10×10^2	8.10×10^3	2.29×10^3	2.29×10^4
25 ...	1.28×10^3	3.20×10^4	3.62×10^3	9.05×10^4
50 ...	1.81×10^3	9.05×10^4	3.13×10^3	2.57×10^5

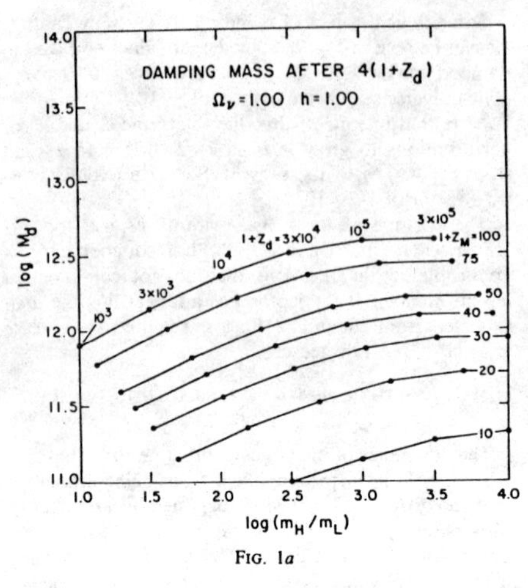

FIG. 1a

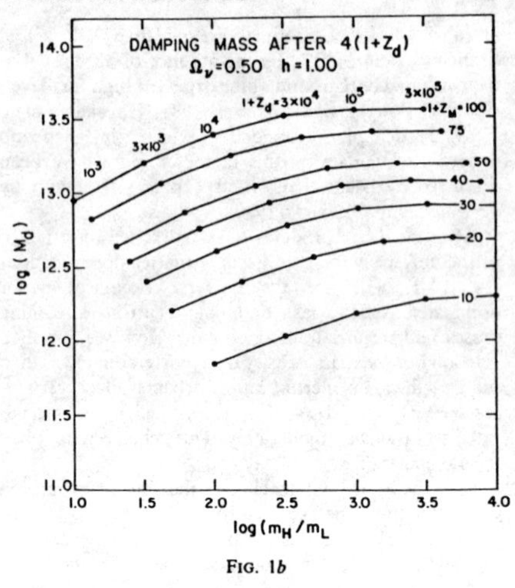

FIG. 1b

FIG. 1.—The damping mass is plotted versus $\eta = m_H/m_L$ for various z_m, the epoch when the universe became matter dominated for the second time. The curves are also labeled by the epoch of ν_τ decay, $z_D = z_m\eta$. Fig. 1a is for $\Omega h^2 = 1$, if the ν_τ decay results only in massless particles. Fig. 1b applies if $\Omega h^2 = 1$, but only half of the matter density is today cool enough to cluster, as would result if the ν_τ decays into $\nu_\mu + \nu_e + \bar{\nu}_e$.

We chose the mass of the seed black hole to make sure that a $10^{12}\ M_\odot$ patch of the universe reaches maximum expansion in the interval associated with $z = 1570$. Consider a patch of the universe of mass M and radius R_1, and with critical density. Insert, in the center of the patch, a black hole of mass m. Then, the energy per unit mass of a shell at R_1 is

$$\varepsilon = \frac{-Gm}{R_1}, \tag{20}$$

its maximum radius is $R_M = (M/m)R_1$, and the time to maximum expansion is

$$T_M = \frac{\pi}{2}\left(\frac{R_1^{\,3}}{2GM}\right)^{1/2}\left(\frac{M}{m}\right)^{3/2} = \frac{\pi}{2}\frac{1}{H(R_1)}\left(\frac{M}{m}\right)^{3/2}. \tag{21}$$

In the time interval T_M, the universe expands by a factor

$$\frac{R_2}{R_1} = 1 + z = \left(\frac{3\pi}{4}\right)^{2/3}\frac{M}{m}. \tag{22}$$

If we want to start when $M_J \lesssim 10^{12}\ M_\odot$, $z \lesssim 1570$ and $m \gtrsim 2\times10^9\ M_\odot$. This is about the same size black hole Carr required for a baryon universe.

V. INCREASING THE MASS OF THE μ NEUTRINO

In principle, a straightforward way to decrease the diffusion scale of the μ-neutrino is to increase its mass. If a very massive ν_τ were to go nonrelativistic and decay soon after the neutrinos had decoupled from the matter, and if this decay produced photons which thermalized, then the neutrino temperature today could be much lower than that of the microwave background. This would allow the ν_μ to be more massive for a given mass density, since their number density would be reduced. The effect is analogous to the heating of the photons when the electrons and positions annihilated, which in the standard model results in $T_\nu = (4/11)^{1/3}T_\gamma$.

If when the neutrinos go nonrelativistic the universe has the age t_{nr}, scale R_{nr}, and density ρ_{nr}, then their diffusion length is approximately given by

$$d_\mu \approx (R_0/R_{nr})(ct_{nr}) \approx (c/v_{\nu,0})\left(c/(G\rho_{nr})^{1/2}\right), \tag{23}$$

where $v_{\nu,0}$ is the neutrino velocity today. If the present mass density in neutrinos is held constant that velocity varies with $T_{\nu,0}$ as

$$v_{\nu,0} \propto p_{\nu,0}/m_\nu \propto T_{\nu,0}/T_{\nu,0}^{-3} \propto T_{\nu,0}^4. \tag{24}$$

Because the universe is radiation dominated at the time

$$\rho_{nr} \propto (R_0/R_{nr})^4 \propto T_{\nu,0}^{-16}. \quad (25)$$

If $T_{\nu,0}=r(4/11)^{1/3}T_{\gamma,0}$, then d_μ scales as r^4 and the diffusion mass scales as r^{12}. Since we have to reduce the diffusion mass by a factor of $\sim 10^3$, we require $r<1/2$. (A more careful calculation yields $r \lesssim 0.46$.)

A number of constraints combine to make this much heating of the photons relative to the neutrinos impossible. The neutrinos decouple right at the beginning of nucleosynthesis ($T \approx 2.6$ MeV according to Dicus *et al.* 1978), and thus the decay must occur after element production. For neutrinos with a mass of 5–10 MeV the Dicus *et al.* (1978) calculations show that it is just possible to get enough heating and preserve reasonable abundances. The effects of a photon-to-baryon ratio that is smaller by a factor of r^3 are offset by the more rapid expansion resulting from the presence of the nonrelativistic neutrino. These neutrinos also satisfy the requirement that the decay occur before $z \approx 10^7$ in order for the photons to thermalize.

However, Lindley (1979) has shown, with very conservative assumptions, that if the photons produced in the decay have an energy greater than 2.225 MeV, there are easily enough of them to destroy all the deuterium by photofission.

The scheme could survive with the boundary value of 5 MeV. But Falk and Schramm (1978) and Cowsik (1977) have pointed out that any neutrino less massive than ~ 10 MeV should be produced in large quantities in supernovae. If the neutrinos have a lifetime in the range of interest here (10^3–10^4 s), they escape from the star, and their decay would produce (from all supernovae taken together) more of a gamma ray background than is observed. It should be noted that this constraint can be tightened and made independent of the controversial supernova rate by the fact that supernovae are not observed as gamma ray burst sources.

One or the other of the above constraints can be evaded but not both at once. If a 10 MeV neutrino decays just before nucleosynthesis, it can produce only $r \approx 0.75$ and a diffusion mass of $\sim 2 \times 10^{14}$ $M_\odot$, a very minor reduction.

VI. CONCLUSIONS

If neutrinos have nonzero rest masses, and if the most massive neutrino (the τ-neutrino) has a mass on the order of 50–100 keV, it has a lifetime on the order of 100–3000 years and would have decayed by now according to our optimistic lifetime estimates.

Prior to its decay, it would have amplified initial perturbations on galactic mass scales, and 50–100 eV neutrinos (μ-neutrinos) would have had time to mimic these perturbations. Until the present, the μ-neutrinos could have amplified the initial perturbations on galactic mass scales by a factor of 1–2×10^5. This is to be compared with a baryon dominated universe where the growth factor is about 10^3. In addition, there is no need, in a neutrino dominated universe, to require isothermal (or entropy) perturbations on galactic mass scales. In the standard cosmological model, Silk damping during the recombination epoch erases adiabatic perturbations on galactic mass scales. Thus, it has been argued that primordial perturbations must have been isothermal, yet isothermal perturbations are quite contrary to the spirit of grand unified field theory in the early universe (Turner and Schramm 1979; Weinberg 1980).

The detailed nature of the decay of the heavy neutrino is the chief uncertainty in this model. If the lifetime is too long, the universe would be radiation dominated today, and if the decay occurs before the ν_μ become nonrelativistic, galactic size perturbations would be damped out. If the decay produces photons, the resulting background, which would not be thermalized if $z_D < 10^7$, would contain far more energy than the microwave background radiation and would have characteristic energies somewhere in the optical or near-infrared, where it would be observable. Present limits at 2 μm, for example, are some two orders of magnitude below the expected flux. If the decay of the heavy tau neutrino yields muon neutrinos, then these neutrinos, containing roughly half the mass density of the present universe, will today be too hot to cluster on any observed scale and will form a smooth background. They could help explain why estimates of mass-to-light ratios of galactic systems are monotonic functions of measurement scale, even for scales larger than clusters of galaxies.

If the initial perturbation was in the form of a massive (10^9 $M_\odot$) black hole, it would have clustered by infall galactic masses of 100 eV neutrinos by now. In this scenario, we do not require an unstable neutrino.

Finally, we have shown that the damping mass is a very strong function of the present neutrino temperature, a quantity not subject to direct observation. If some mechanism can be found for the cooling of the neutrino temperature to approximately 0.9 K, then neutrino streaming will not erase galactic size perturbations. However, a mechanism we once thought promising is in complete violation of a variety of astrophysical constraints. Perhaps there are alternative processes to further cool neutrinos, but for the present, this process is ruled out.

Our motivation has been to preserve fluctuations of 10^{12} $M_\odot$, which is the upper limit usually associated with one galactic mass. We note, however, that the space density of bright galaxies ($M<-18$) is $\sim 10^{-2}$ Mpc^{-3} and that the mass per bright galaxy is 2.7×10^{13} Ω $M_\odot$. Even with massive galactic halos, there is a large excess mass per galaxy if $\Omega > 0.1$. A mechanism for producing $1/r^2$ halos by tidally shearing off the outer regions of initial protogalaxies has been suggested by Dekel, Lecar, and Shaham (1978), and if this process is important, then perhaps only fluctuations of $10^{13}\ M_\odot$ and higher need be preserved. This would significantly ease the constraints on all the scenarios described here.

In summary, we have described scenarios that, by including an unstable τ-neutrino, facilitated galaxy formation. Although we chose the unstable particle to be the τ-neutrino, we note that another particle (perhaps not a neutrino at all) with similar mass, lifetime, and decoupling time would serve as well. However, without the massive, unstable particle, the lighter neutrinos by themselves seem to make galaxy formation on scales $\leq 10^{12}\ M_\odot$ almost impossible.

Future developments in high energy physics will hopefully provide solid information on the viability of these speculations.

We acknowledge the extremely useful comments of an anonymous referee. The research of M.D. is partially supported by NSF grant AST-8008996.

APPENDIX

THE DIFFUSION LENGTH d

The diffusion length is the horizon of free-streaming neutrinos. In the interval $t_1 \leq t \leq t_0$, a neutrino with a random velocity $v(t)$, moves a physical distance d, given by

$$d = R_0 \int_{t_1}^{t_0} \frac{v(t)\,dt}{R(t)} = R_0 \int_{R_1}^{R_0} \frac{v(R)\,dR}{R\dot{R}(R)}. \tag{A1}$$

As we will demonstrate shortly, $(R_1/R_0)d$ is interpreted as a Jeans length, i.e., the distance a neutrino with random velocity $v(t_1)$ travels in a "Hubble time." Thus, linear perturbations on scales smaller than $(R_1/R_0)d$ are damped by the excess neutrinos diffusing out of the region. It is convenient to set $R = xR_0$, where $x = 1/(1+z)$. Then d is written

$$d = \int_{x_1}^{1} \frac{v(x)\,dx}{x\dot{x}}. \tag{A2}$$

We set $v = \beta c$, and we approximate β by

$$\beta = (1 + p^{-2})^{-1/2}, \tag{A3}$$

with $p = \alpha k T_\nu / mc^2$ (α is a constant which we will specify below). Since $T_\nu = T_{\nu_0}/x$, we write

$$\beta = [1 + (x/p_0)^2]^{-1/2},$$

where

$$p_0 = \frac{\alpha k T_{\nu_0}}{mc^2}.$$

We also take $\dot{x}^2 = (8\pi/3)G\rho(x)x^2$. Finally, we let $\rho(x) = \rho_0 g(x)$ and write

$$d = \frac{c}{[(8\pi/3)G\rho_0]^{1/2}} \int_{x_1}^{1} \frac{dx}{[1 + (x/p_0)^2]^{1/2}[g(x)x^4]^{1/2}}. \tag{A4}$$

If the universe is matter dominated, $g \approx x^{-3}$; and if it is radiation dominated, $g \approx x^{-4}$. To approximate δ when the universe contains both matter and radiation, we set

$$g = \frac{1}{x^3} + \frac{\varepsilon}{x^4}. \tag{A5}$$

In a matter dominated universe ($\varepsilon = 0$), if the neutrinos are nonrelativistic ($x_1 \gg p_0$),

$$d = \frac{2cp_0}{[(8\pi/3)G\rho_0]^{1/2}} \frac{1}{x_1}. \tag{A6}$$

Since $v(x_1)=cp_0 x_1$ and $\rho(x_1)=\rho_0 x_1^{-3}$,

$$x_1 d=\frac{2v(x_1)}{\left[(8\pi/3)^{1/2} G\rho(x_1)\right]^{1/2}}\approx r_J(x_1), \tag{A7}$$

where r_J is the "Jeans length." The mass contained in d (evaluated at $x=x_0=1$) is the mass contained in r_J evaluated at $x=x_1$); i.e.,

$$M\sim\rho_0 d^3\sim\rho_0\frac{r_J^3(x_1)}{x_1^3}\sim\rho(x_1)r_J^3(x_1). \tag{A8}$$

In general, when the neutrinos are relativistic, or when the universe is radiation dominated, this equality does not hold, and we use d.

We now evaluate $\rho(x)/\rho_0=\delta$. To avoid integrating over the neutrino distribution function, we approximate the density in the following manner. For each species (ν_τ, ν_μ, ν_e, and photons), we set

$$\rho=nm_{\rm eff},$$

where $m_{\rm eff}$ is the "effective mass" for this species and

$$n=8\pi\left(\frac{kT_\nu}{hc}\right)^3\frac{3}{2}\zeta(3). \tag{A9}$$

For a neutrino with rest mass m, we set

$$m_{\rm eff}=m\sqrt{1+p^2}. \tag{A10}$$

When $p\gg 1$, $m_{\rm eff}=pm$, and since for a relativistic fermion,

$$\rho=8\pi\left(\frac{kT_\nu}{hc}\right)^3\frac{7}{8}\frac{\pi^4}{15}\frac{kT_\nu}{c^2}, \tag{A11}$$

we require

$$\frac{3}{2}\zeta(3)mp=\frac{3}{2}\zeta(3)m\frac{\alpha kT_\nu}{mc^2}=\frac{7}{8}\frac{\pi^4}{15}\frac{kT_\nu}{mc^2},$$

whence

$$\frac{3}{2}\zeta(3)\alpha=\frac{7}{8}\frac{\pi^4}{15}\quad(\alpha=3.15). \tag{A12}$$

We take m as the mass of the μ-neutrinos, ηm as the mass of the τ-neutrinos, and zero as the mass of the electron-neutrinos. Then we have the effective masses shown in Table 2.

The values of d used to crease Figure 1 were obtained by numerical integration of the appropriate density for the scenarios where ν_τ decays into either 3 massless ν_e's or half ν_e's and one half massive ν_μ's.

TABLE 2

EFFECTIVE MASSES

Particle	$m_{\rm eff}$
Tau neutrino	$\eta m\sqrt{1+(p^2/\eta^2)}$
Muon neutrino	$m\sqrt{1+p^2}$
Electron neutrino ...	$\alpha kT_\nu/c^2=mp$
Photon	$\frac{8}{7}[T_\gamma/T_\nu]\cdot\alpha kT_\nu/c^2$

REFERENCES

Bahcall, J. W., *et al.* 1980, *Phys. Rev. Letters*, **45**, 945.
Bond, J., Efstathiou, G., and Silk, J. 1980, *Phys. Rev. Letters*, **45**, 1980.
Carr, B. J., 1977, *Astr. Ap.*, **56**, 377.
Cowsik, R. 1977, *Phys. Rev. Letters*, **39**, 784.
Davis, M., Tonry, J., Huchra, H., and Latham, D. 1980, *Ap. J. (Letters)*, **238**, L113.
Dekel, A., Lecar, M., and Shaham, J. 1980, *Ap. J.*, **241**, 946.
de Rújula, A., and Glashow, S. 1980, *Phys. Rev. Letters*, **45**, 942.
Dicus, D., Kolb, E., and Teplitz, V. 1978, *Ap. J.*, **221**, 327.
Dicus, D., Kolb, E., Teplitz, V., and Wagoner, R. 1978, *Phys. Rev. D*, **17**, 1529.
Doroshkevich, A. G., Khlopov, M. Y., Synyaev, R. A., Szalay, A. S., and Zeldovich, Y. B. 1980, in Proc. 10th Texas Symp. Relativistic Astrophysics.
Doroshkevich, A. G., Saar, E. M., and Shandarin, S. F. 1978, in *IAU Symposium 79, The Large Scale Structure of the Universe*, ed. M. S. Longair and J. Einasto, (Dordrecht: Reidel), p. 423.
Einasto, J., Kaasik, A., and Saar, E. 1974, *Nature*, **250**, 309.
Falk, S. W., and Schramm, D. N. 1978, *Phys. Letters*, **79B**, 511.
Hayakawa, S., Ito, K., Matsumoto, T., Murakami, H., and Uyama, K. 1978, *Pub. Astr. Soc. Japan*, **30**, 369.
Kimble, R., Bowyer, S., and Jakobsen, P. 1981, *Phys. Rev. Letters*, **46**, 80.
Lindley, D. 1979, *M.N.R.A.S.*, **188**, 15P.
Ostriker, J. P., Peebles, P. J. E., and Yahil, A. 1974, *Ap. J. (Letters)*, **193**, L1.
Ryan, M. 1972, *Ap. J. (Letters)*, **177**, L79.
Schramm, D., and Steigman, G. 1981, *Ap. J.*, **243**, 1.
Turner, M. S., and Schramm, D. N. 1979, *Nature*, **279**, 363.
Tremaine, S., and Gunn, J. 1979, *Phys. Rev. Letters*, **42**, 407.
Wagoner, R. V. 1974, in *IAU Symposium 63, Confrontation of Cosmological Theories with Observational Data*, ed. M.S. Longair (Dordrecht: Reidel), p. 195.
Weinberg, S. 1980, *Phys. Scripta*, **21**, 773.
Yang, J., Schramm, D., Steigman, G., and Rood, R. 1979, *Ap. J.*, **227**, 697.
Zel'dovich, Y. B. 1978, in *IAU Symposium 79, The Large Scale Structure of the Universe*, ed. M. S. Longair and J. Einasto (Dordrecht: Reidel), p. 409.

MARC DAVIS, MYRON LECAR, and TAD PRYOR: Harvard-Smithsonian Center for Astrophysics, 60 Garden Street, Cambridge, MA 02138

EDWARD WITTEN: Joseph Henry Laboratory, P. O. Box 708, Princeton, NJ 08540

COSMOLOGICAL IMPACT OF THE NEUTRINO REST MASS

A. G. Doroshkevich,* M. Yu. Khlopov,* R. A. Sunyaev,†
A. S. Szalay,‡§ and Ya. B. Zeldovich*†

**Institute of Applied Mathematics*
Moscow, USSR

†*Space Research Institute*
Moscow, USSR

‡*Department of Atomic Physics*
Roland Eötvös University
1088 Budapest, Hungary

§*Astronomy Department*
University of California, Berkeley
Berkeley, California 94720

INTRODUCTION

In the last few years, several papers have been published on the cosmological effects of a finite neutrino rest mass. Gershtein and Zeldovich realized first that, in the Big Bang cosmology, even a small neutrino rest mass leads to a gravitationally dominating overall neutrino density.[1] From the age of the universe, the limit $m_\nu <$ 140 eV was estimated. More detailed calculations were made later by Marx and Szalay[2] and Cowsik and McClelland.[3] A new trend was initiated by Szalay and Marx,[4] who noticed that perturbation growth of neutrinos starts long before the decoupling of matter and radiation. Stable neutrino configurations, "neutrino superstars," were discussed by Markov,[5] Bludman,[6] and Marx and Kovessy-Domokos.[7]

Lee and Weinberg,[8] Dolgov *et al.*,[9] and Hut and Olive[10] applied the constraints of cosmology to get a lower bound on heavy (1 GeV) neutrino masses. Tremaine and Gunn realized the phase space limitations of neutrino clustering.[11] Gunn *et al.*[12] and Schramm and Steigman[13] revived the idea of the connection between m_ν and the missing mass.

The news of the experimental detection of the neutrino rest mass in the range 14 eV $< m_{\nu e} <$ 46 eV by Lyubimov *et al.*[14] prompted several authors to reconsider the cosmological consequences of the neutrino mass.[15–24]

The results will remain semiquantitative so long as the masses have not been accurately measured in the laboratory. Astronomy has to include the experimentally determined values of the neutrino masses, but it cannot be used to calculate or even prove their existence. On the other hand, cosmology with massive neutrinos is so attractive that, to paraphrase Voltaire, if it did not exist, it would have been necessary to invent it.

0077–8923/81/0375–0032 $1.75/1 © 1981, NYAS

COSMOLOGICAL NEUTRINOS

Most of the neutrinos present in the universe today were produced at the same time as most of the photons of the 3 K background radiation. The neutrinos and antineutrinos were in thermal equilibrium with the photons and the electron-positron gas at the temperature 10^{10} K. Assuming a lepton-symmetric universe, we find that their momentum distribution is described by the nondegenerate Fermi-Dirac distribution of extreme relativistic particles:

$$n(p) = \left[1 + \exp\left(\frac{\sqrt{m_\nu^2 + p^2}}{kT}\right)\right]^{-1} \simeq \left[1 + \exp\left(\frac{p}{kT}\right)\right]^{-1},$$

since $m_\nu \ll kT \simeq 1$ MeV.

At high temperatures, the neutrino rest mass does not affect the physical processes of the early universe (cosmological nucleosynthesis). Petcov and Khlopov have shown, however, that CP-violation effects in the oscillations of Majorana neutrinos with Dirac mass ($\delta m^2 \simeq 10^{-8}$–$10^{-10}$ eV^2) may lead to a considerable excess of ν_e over $\bar{\nu}_e$ at the moment when the n/p ratio freezes out.[25] This affects the ^{4}He abundance predictions, as a temporary nonzero chemical potential of ν_e,[26–28] which disappeared after a short time. The neutrino oscillations may influence the limits on the number of neutrino species in this way.[29–32]

When $kT \simeq 3$ MeV, the neutrinos came out of equilibrium, decoupling from the radiation. Their temperature decreased due to the adiabatic expansion of the universe. Soon after this moment, the electrons and positrons annihilated each other, giving all their energy and entropy to the photons. So the neutrino temperature today is

$$T_\nu(t) = (4/11)^{1/3}\, T_\gamma(t)\,[\text{today: } 2.14\ T_3\ \text{K}],$$

where $T_3 = (T/3\ \text{K})$.

The shape of the momentum distribution function was also conserved during the expansion,

$$n(p,t) = \left[1 + \exp\left(\frac{p}{kT_\nu(t)}\right)\right]^{-1} \simeq \left[1 + \exp\left(\frac{\sqrt{2m_\nu E_\nu}}{kT_\nu}\right)\right]^{-1}$$

in the nonrelativistic case $E_\nu \ll m_\nu$.

The number density for one neutrino degree of freedom is

$$n_i = 1.8\,\frac{4\pi}{h^3}\,(kT_\nu)^3 \qquad [\text{today: } 75\ T_3^3\ \text{cm}^{-3}].$$

The number density and temperature of the other two known neutrino species, ν_μ and ν_τ, are the same, since their decoupling occurred at about the same temperature.

The right-handed neutrinos—if any do, in fact, exist—decoupled much earlier. At that time, many particle types had been present; their annihilation resulted in a much lower temperature for the right-handed neutrinos; therefore, their present number density is at least four orders of magnitude lower than the density of the left-handed neutrinos.[22]

The total number density of neutrinos can be estimated by multiplying the following quantities:

$$n = \underset{(e,\mu,\tau)}{3} \times \underset{(\nu,\bar{\nu})}{2} \times \underset{(n_i)}{75}\, T_3^3\,\mathrm{cm}^{-3} = 450\, T_3^3\,\mathrm{cm}^{-3}.$$

The energy density of the neutrinos is given by the following integral:

$$\rho_\nu = \frac{4\pi}{h^3} g \int_0^\infty \sqrt{p^2 + m^2}\left[1 + \exp\left(\frac{p}{kT_\nu}\right)\right]^{-1} p^2\,dp,$$

where g is the number of the neutrino degrees of freedom.

If $p^2 \ll m^2$, the neutrinos are extremely relativistic (ER), $\rho_\nu \simeq T^4$; if $p^2 \ll m^2$, the neutrinos are nonrelativistic (NR), $\rho_\nu \simeq T^3$. The transition between the two cases occurs when $kT_\nu \simeq 1/3\, m_\nu$. It should be pointed out that NR neutrinos do not behave completely like a classical nonrelativistic gas, since their phase space density (conserved since the moment of decoupling) does not depend on their rest mass, so $[1 + \exp(p/kT)]^{-1}$ appears instead of the Maxwellian term in the momentum distribution.

The kinetic energy of the neutrinos rapidly decreased and their average velocity today is on the order of 6 km s^{-1},

$$\sqrt{v^2} = 3.6\, kT/m_\nu = 6.6\left(\frac{m_\nu}{30\ \mathrm{eV}}\right)^{-1} \mathrm{km\ s^{-1}}.$$

Today, the relic neutrinos are nonrelativistic, so their energy density is governed by their rest mass. The three neutrinos may have slightly different masses, but, since their cosmological number density is equal, an average mass can be used:

$$m_{30} = \frac{1}{3}\left(\frac{m_{\nu e} + m_{\nu\mu} + m_{\nu\tau}}{30\ \mathrm{eV}}\right) \quad \text{or} \quad \frac{1}{3}\left(\frac{m_1 + m_2 + m_3}{30\ \mathrm{eV}}\right),$$

allowing for the possibility of neutrino oscillations, where m_1, m_2, and m_3 are the masses of the mass eigenstates ν_1, ν_2, and ν_3.

The energy density of the neutrinos today is

$$\rho_\nu = 2.5 \times 10^{-29}\, m_{30} T_3^3\ \mathrm{g\ cm^{-3}},$$

$$\Omega_\nu = 5.1\, h_{50}^{-2}\, m_{30} T_3^3,$$

where $h_{50} = (H_0/50\ \mathrm{km\ s^{-1}\ Mpc^{-1}})$.

The age of the universe can be well approximated by the formula

$$t_0 = \frac{21 \times 10^9\ \mathrm{yr}}{h_{50} + \sqrt{2.8\, m_{30}\, T_3^3}}.$$

Relatively large values of m_ν (30–40 eV) are giving a t_0 far too small ($<8 \times 10^9$ yr). Measurements of the neutrino mass leading to such values may imply the existence of a nonzero cosmological constant. By introducing Λ, the cosmological constant, models with arbitrary t_0 can be constructed for any given value of m_ν. In order to obtain a reasonable age of the universe, the relation $\Lambda > \Lambda_1(m)$ has to be fulfilled. On the other hand, the very occurrence of the cosmological singularity requires $\Lambda_2(m) > \Lambda$. The $[\Lambda_2(m), \Lambda_1(m)]$ range is very narrow for $m_\nu = 40$ eV. At much larger m_ν and Λ, the age of the universe would exceed the observed lower limit only by a very artificial evolution curve, which is not compatible with the observed galaxy and quasar distribution, so $m_\nu < 150$ eV remains a definite upper limit for $\Lambda \neq 0$.[15]

The Linear Evolution of Neutrino Fluctuations

The observations definitely show the existence of large density perturbations at the scale of approximately 15–60 Mpc.[32,33] This is the average distance of superclusters; at larger scales, the density perturbations are smaller. There were perturbations of the same comoving scale at the moment of recombination, which grew in the linear regime until they reached $\delta\rho/\rho \simeq 1$. The exact form of the initial perturbation spectrum is not needed in the discussion given below.

In a universe gravitationally dominated by neutrinos, the growth of neutrino fluctuations is of particular importance. The growth of baryon perturbations is inhibited until recombination, due to their coupling to the photons. The neutrinos are collisionless, void of interaction from the time of their decoupling. They may be influenced by adiabatic perturbations only (i.e., simultaneous perturbations of the metrics, density, and velocity) so, in a neutrino-dominated universe, the adiabatic theory of galaxy formation is naturally preferred.

In the standard scenario (with no massive neutrinos involved) the growth of fluctuations as a function of their size or mass is best described using the Jeans length,[34] showing the competing effects of gravity and pressure:

$$\lambda_J = \left(\frac{\pi c_s^2}{\rho G}\right)^{1/2},$$

where c_s is the sound velocity. In the expanding universe, it is preferable to use the Jeans mass (the mass within a sphere of radius $\lambda_J/2$), invariant during expansion,

$$M_J = \frac{4\pi}{3}\rho\left(\frac{\lambda_J}{2}\right)^3.$$

All density fluctuations larger than the Jeans mass tend to grow, due to gravity, while perturbations below the Jeans mass oscillate with constant amplitude as ordinary sound waves.

Collisionless particles have no pressure, but they still have kinetic energy from their thermal motion. Using the velocity dispersion $\langle v^2\rangle/3$ instead of c_s^2, the calculation of the Jeans mass can be made in the same way, only the competition will not be between pressure and gravity but between the kinetic and gravitational energy of the particles.

In order to describe the growth of flucutations in the neutrinos, the time dependence of the neutrino Jeans mass $M_{J\nu}$ has to be determined. In the early period, when the neutrinos are still extremely relativistic, $M_{J\nu}$ is approximately equal to the mass within the horizon M_H. The Jeans mass thus increases proportionally to t, the age of the universe[22] (Figure 1).

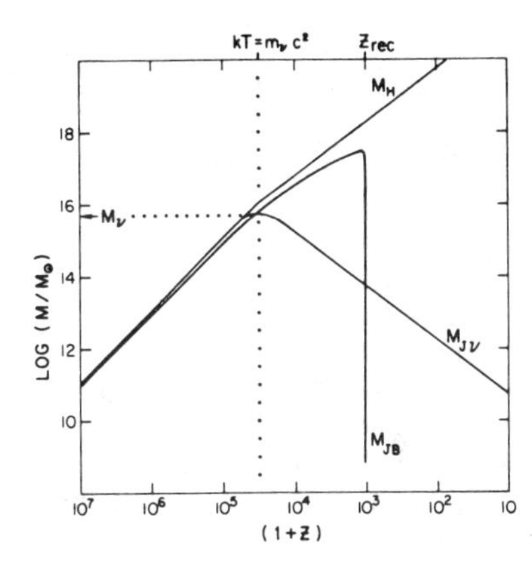

FIGURE 1. The dependence of the neutrino and baryon Jeans masses on the redshift z. The region below the neutrino Jeans mass corresponds to the damping of neutrino fluctuations. The smallest fluctuation with uninterrupted growth is M_ν, the maximum of $M_{J\nu}$. (From Bond *et al.*[22])

This picture will change only when the transition of neutrinos from ER to NR occurs at the temperature $kT_\nu = \frac{1}{3} m_\nu$. (Further on, the subscript ν will refer to this moment: the derelativisation of the neutrinos.) It should be emphasized that the total mass density of neutrinos is important for the evolution of the universe as a whole, so an average mass can be used, but the evolution of inhomogeneities is determined by the heaviest type of neutrino (m_ν will mean the mass of this particle henceforth).

After this transition period, the neutrinos become nonrelativistic and their velocity decreases quickly. The neutrino Jeans mass will also decrease from this moment, having a maximum M_ν when $kT_\nu = \frac{1}{3} m_\nu$.[4] This mass is given by the following relation:[19]

$$M_\nu = 4.0 \times 10^{15}\, m_{\nu 30}^{-2}\, M_0 \simeq M_{\mathrm{Pl}} \left(\frac{M_{\mathrm{Pl}}}{m_\nu}\right)^2,$$

where $M_{\mathrm{Pl}} = (\hbar c/G)^{1/2} = 2 \times 10^{-5}\,g$, the Planck mass. The value of the corresponding spatial scale is[24]

$$R_\nu = 12.5\,(1 + z)^{-1}\, m_{\nu 30}^{-1}\ \mathrm{Mpc}.$$

The curves on FIGURE 1 corresponding to $M_{J\nu}$ and M_H separate three regions corresponding to different behaviors of the perturbations. If the mass of a perturbation is larger than the horizon, the fluctuation is growing, as was shown by Lifshitz.[35] Fluctuations smaller than the horizon but larger than the Jeans mass are unstable gravitationally and growing. The collisionless neutrino fluctuations below the Jeans mass behave in a very different way from the baryons and are damped, as was first shown by Stewart.[36] This damping is due to the lack of interaction (very long mean free path). The displacement of neutrinos from more dense to less dense regions without interaction provides a mechanism for the smearing out of irregularities. The full analysis of this problem can be made numerically only, but the asymptotic case of short wavelengths ($kR_\nu \ll 1$) can be understood analytically.

The neutrino perturbations will be adiabatic—essentially, they are a small inhomogeneity of the temperature in the momentum distribution function. The free motion of the noninteracting neutrinos tends to smear out these irregularities on scales comparable to the average displacement. The net effect is a change in the spectrum of perturbation amplitudes: a power law decrease appears in the short wavelength part of the spectrum, proportional to $(kR_\nu)^{-8}$. An approximate interpolation formula describes the distortions of the spectrum:[16]

$$\left(\frac{\delta\rho}{\rho}\right)_{\rm dis} = \left(\frac{\delta\rho}{\rho}\right)_{\rm nondis} \times (1 + k^2R_\nu^2)^{-4}.$$

The result of this spectral distortion is that essentially all fluctuations smaller than M_ν are reduced in amplitude, so the perturbation spectrum has a maximum at M_ν corresponding to those fluctuations which reach the nonlinear scale first.

The assumption of adiabatic fluctuations has been used in the derivation of this formula. This assumption has another consequence: the fluctuations of all three constituents of the universe (neutrinos, baryons, and photons) have approximately the same initial amplitude when entering the horizon.

The fluctuations of neutrinos that have a mass larger than M_ν have an uninterrupted growth. Due to their coupling to photons, the fluctuations of baryons smaller than the horizon can start growing only after recombination. Starting from the same initial amplitude at the moment of recombination, the neutrinos have much larger fluctuation amplitudes than the baryons and photons of the background radiation. At the recombination, the baryon Jeans mass drops very quickly to the value $10^5\,M_0$. The growth of baryon fluctuations is accelerated by the large inhomogeneities formed in the neutrino density until the same amplitude is reached (FIGURE 2). This process thus provides large fluctuations in baryons and neutrinos and small $\delta T/T$ in the photon background.

In FIGURE 2, the growth of a fluctuation of mass M_ν has been plotted for three different cosmological scenarios. The initial amplitudes of the perturbations were normalized to give $\delta\rho/\rho = 1$ at $z = 0$. The solid lines correspond to ρ_B, the broken lines to ρ_ν.

1. $\Omega_\nu = 1$, $\Omega_B = 0.03$. The residual $\delta T/T$ is very small; $\delta T/T \simeq 3 \times 10^{-5}$.

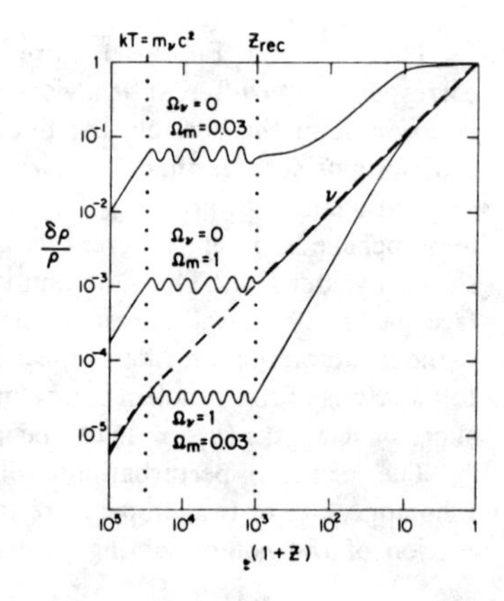

FIGURE 2. Growth of neutrino and baryon fluctuations of mass M_ν versus z. The dashed line corresponds to the neutrinos; the three solid lines show the growth of baryon fluctuations in three different cases. The initial amplitudes were normalized to give $\delta\rho_B/\rho_B = 1$ at $z = 0$. 1. $\Omega_\nu = 1$, $\Omega_B = 0.03$. The growth of $\delta\rho_B/\rho_B$ is accelerated by the neutrinos, $\delta T/T \simeq 3 \times 10^{-5}$. 2. $\Omega_\nu = 0$, $\Omega_B = 1$. The baryons have the standard $t^{2/3}$ growth after recombination. $\delta T/T \simeq 10^{-3}$. 3. $\Omega_\nu = 0$, $\Omega_B = 0.03$. In the low density universe, the growth of $\delta\rho_B/\rho_B$ is first inhibited by the larger radiation density, then by the Ω^{-1} effect. $\delta T/T \simeq 0.06$.

2. $\Omega_\nu \doteq 0$, $\Omega_B = 1$. The baryons have the standard $t^{2/3}$ growth after recombination; $\delta T/T \simeq 10^{-3}$.

3. $\Omega_\nu = 0$, $\Omega_B = 0.03$. The growth of $\delta\rho_B/\rho_B$ is first inhibited by the large radiation density,[37,38] then by the Ω^{-1} effect. The total fluctuation growth from $z = 1000$ to $z = 0$ is only 15.3;[39] $\delta T/T \simeq 0.06$ in this case.

This discussion enables us to present the strongest astronomical argument in favor of a nonzero neutrino rest mass. Observations of small-scale differential fluctuations of the background radiation temperature put an upper limit to these residual amplitudes,[40,41]

$$(\delta T/T)_{\rm obs} < 10^{-4}.$$

The cosmic abundance of deuterium is a sensitive measure of the baryon density. The observed value, $D/H = 2 \times 10^{-5}$,[42] implies that the baryon density of the universe has to be very low;[43]

$$\Omega_B < 0.06.$$

If the baryon density is low and $\Omega_\nu = 0$, the initial fluctuations have to be overwhelmingly large, which is not compatible with observations. If the baryon density is high, too little deuterium is produced. The neutrino rest mass provides a nice compromise: low baryon density, small $\delta T/T$, and a large-scale structure that is apparently on the correct scales.

Nonlinear Evolution of Neutrino Fluctuations

The nonlinear evolution of fluctuations in a neutrino-dominated universe is very important for the formation of galaxies and clusters of galaxies. Before going into the details of nonlinear fluctuation growth, let us briefly summarize the main features of neutrino perturbations in the late stages of linear evolution. The perturbations are adiabatic with a smooth spectrum at long wavelengths and a sharp cutoff at short wavelengths. The theory for the nonlinear growth of perturbations with a similar spectrum is well known.[44-46] The evolution of inhomogeneities in the nonlinear stage cannot be described in terms of Fourier amplitudes. More useful parameters are the characteristic perturbation scale, R_ν (the cutoff wavelength of the distorted specimen), and the overall amplitude of inhomogeneities,

$$\sigma^2 = \int \left(\frac{\delta\rho}{\rho}\right)^2_\kappa d^3\kappa.$$

The value of σ determines the moment when nonlinear structure is formed, and R_ν corresponds to the scales and masses of these objects. The growth rate and inner structure of these developing inhomogeneities is determined by the initial perturbation spectrum and the properties of the constituent matter. The value of σ depends on the particular form of the spectrum when $\delta\rho/\rho = 1$ is reached and is most sensitive to the shape around the cutoff point $k_\nu R_\nu = 1$.

In Zeldovich's classical pancake theory (without massive neutrinos), the rapid growth of density contrasts occur at the local maxima of the deformation tensor corresponding to the initial perturbations. The final result will be the formation of large flattened objects of high density: the pancakes. In the last stage of pancake formation, when the density and velocities have become very large, a shock wave is created. A particular feature of the theory is that all motion in the shock is essentially one-dimensional—normal to the plane of the pancake. Most of the kinetic energy of the particles (baryons in this case) will be dissipated in the shock. The characteristic scale for such objects is given by the cutoff in the initial spectrum, similar to the Silk mass,[47] but determined by dissipation in the recombination era in the neutrinoless case. Galaxies are formed at a later stage by the hydro-dynamical fragmentation of the pancakes.

The presence of massive neutrinos alters this picture quite significantly. The fluctuations of the neutrino density start growing first and the characteristic scale is determined by R_ν. The great difference is that neutrinos are collisionless, so no shock wave dissipation can occur. However, the absence of the high frequency components of the perturbation spectrum leads to pancake formation, too, so the similarity to pancake theory is preserved; here the problem is also one-dimensional, at least in the first approximation. Such a situation has been investigated numerically for a collisionless self-gravitating gas by Doroshkevich *et al.*[48] A one-dimensional violent relaxation of the Lynden-Bell type was observed.[49] Flattened gravitationally bound objects were formed, with boundaries given by the caustic surfaces of neutrino streams. The general analysis of formation and evolution of multistream flows in a collisionless medium was given by Arnold and Shandarin.[50] This phase is considered the first step in large-scale structure formation in the universe.

The real situation, where baryons are also present, is more complicated.[17,22,24] The baryon perturbations grow rapidly after the formation of neutral hydrogen atoms and are coupled to the neutrinos. They form the pancake together. The basic difference is that the baryons dissipate and lose their kinetic energy, while the neutrinos keep their velocities, which are on the order of 100–500 km s^{-1}, normal to the plane of the pancakes. This provides a natural segregation mechanism. The resulting structure will be a pancake sandwich: a thin baryon package in the inside and a neutrino pancake, approximately three times thicker, all around. The total mass that contracts into the pancake is dominated by the neutrinos; therefore, the amount of x and uv radiation from the dissipating baryons will be considerably stronger than that without neutrinos, and it may be responsible for the reheating of hydrogen in the space between clusters.[24] The baryon component of the pancake cools and becomes turbulent. This leads to the fragmentation of the pancake into clouds surrounded by hot gas. The clouds may evolve into gas-star complexes similar to dwarf galaxies. The formation of primary stars leads to enrichment of the gas (compressed into the pancake) by heavy elements. The dwarf galaxies cluster into galaxies, galaxies cluster into clusters of galaxies; all these processes take place simultaneously. In the high-velocity neutrino component (small tangential and high normal velocities), new types of instabilities appear—stream instabilities. Collisionless neutrinos can condense into gravitationally bound objects by violent relaxation. A large anisotropy of the initial conditions will result in a mild final anisotropy of the generated objects. This property is well preserved during the violent relaxation of the fragmented objects, as shown by the numerical *N*-body experiments of Aarseth and Binney;[51–52] this may account for the ellipticity of many nonrotating objects.[53]

The approximation of nondissipative collapse seems to be adequate for rich clusters. The role of dissipation in galaxies had been important, helping the condensation of the gas component in the central region, while neutrinos were left on the periphery of the galaxy, forming an invisible outer halo. If the missing mass is in the form of massive neutrinos,[54-56] it will have different features in clusters of galaxies, in groups of galaxies, and in individual galaxies, due to the difference in the potentials of these objects. The gravitational potential of a cluster is deep; the velocity dispersion is close to 1000 km s^{-1}. The strong tidal forces do not allow halos to form around individual galaxies, so a common massive neutrino background is formed. On the other hand, missing mass and extended neutrino halos may not arise in small galaxies (or, obviously, on smaller scales) where the virial velocities do not exceed 50 km s^{-1}. In galaxies and groups of galaxies with virial velocities of 100 km s^{-1}, there is no unambiguous answer to this problem.

The limitation from phase space restrictions on the masses of the bound objects was discussed by Tremaine and Gunn.[11] The idea is that the maximum of the phase space density (0.5 in our case) of collisionless particles cannot be increased during violent relaxation, which leads to a limit on the radius of a bound object with velocity dispersion σ formed of neutrinos with mass m_ν,

$$(r/1\ \mathrm{kpc}) > 4.63 \times m_{\nu 30}^{-2} \times (\sigma/100\ \mathrm{km\ s^{-1}})^{-1/2}.$$

The size of the halo may be several times larger than the visible size of our galaxy, so this inequality can easily be satisfied (for $m_{\nu 30} > 0.5$ and $\sigma = 300$ km s^{-1}, $r >$ 11 kpc).

The final conclusion of Tremaine and Gunn, that neutrinos cannot be responsible for the missing mass, is based on two assumptions. They assume that there is no time for dissipative separation of neutrinos and baryons, so that neutrinos have the same distribution as ordinary dissipating (and visible) matter. This is certainly not always valid in this picture, nor is their other assumption ($\rho_\nu < \rho_B$), since neutrinos are gravitationally dominant in the universe. Furthermore, deviations from the assumption that the neutrinos form isothermal spheres may also influence the formula by a numerical factor.

In the first stage of galaxy formation, when the density of the dissipating matter is small, it increases due to the effect of dissipation, without any sizable effect on the density of the neutrinos. Later, when the density of the dissipating matter becomes comparable with the neutrino density, the latter starts to increase according to the law derived by Clypin using adiabatic invariance,[57]

$$\frac{\rho_\nu(t)}{\rho_\nu(0)} = \left[\frac{\rho_B(t)}{\rho_B(0)}\right]^{3/4}.$$

This provides an estimate of the matter dominance in the central part of a galaxy and the neutrino dominance in the periphery. During violent relaxation, there will be only small regions where the baryons dominate the density, but, at these points, because of the slow relaxation, the density excess of baryons over neutrinos will increase further. On large scales, the neutrinos will always have a higher density. The larger the scales considered, the larger the part of the total mass that is in the form of neutrinos, so there is a tendency of missing mass ratios to grow towards the larger scales, as observed.

The distribution of the visible matter forms a cellular structure on large scales (superclusters) in this picture. Most of the matter is concentrated in pancakes and in the intersections of pancakes, forming filaments, with large empty space in between, in agreement with recent observations.

CONCLUSION

The nonzero neutrino mass provides a natural answer to several independent problems in astrophysics simultaneously.

1. The universe is gravitationally dominated by neutrinos, with all dynamical properties (H_0, t_0, q_0) determined by their mass density, $\Omega_\nu \simeq 1$, while the baryons can have a low density, $\Omega_B \simeq 0.03$, in agreement with the observations of luminous matter and the implications of the observed amount of deuterium.

2. The growth of neutrino perturbations starts very early, at $z_\nu = 1.8 \times 10^5\, m_{30}$, so the nonlinear stage can easily be reached. The growth of baryon fluctuations is accelerated by the neutrinos after recombination, so the fluctuations of the background radiation temperature can remain very small, $\delta T/T < 10^{-4}$, in agreement with observations.

3. The perturbations of neutrinos are adiabatic. The damping of these fluctuations in the linear regime determines a characteristic mass in a natural way. $M_\nu = 4 \times 10^{15}\, m_{30}^{-2}\, M_0$, a typical supercluster mass.

4. A cellular structure with filaments is predicted on the supercluster scale with characteristic separations on the order of 15–60 Mpc.

5. A specific distribution of missing mass is predicted, mostly associated with clusters and groups of galaxies and probably with individual galaxies. No considerable amount is expected on smaller scales.

It should be emphasized once more that the masses of neutrinos must be determined in the laboratory.

REFERENCES

1. GERHSTEIN, S. S. & YA. B. ZELDOVICH. 1966. JETP Lett. **4:** 174.
2. MARX, G. & A. S. SZALAY. 1972. Proc. Neutrino '72 **1:** 123. Technoinform. Budapest, Hungary.
3. COWSIK, R. & J. MCCLELLAND. 1972. Phys. Rev. Lett. **29:** 669.
4. SZALAY, A. S. & G. MARX. 1976. Astron. Astrophys. **49:** 437.
5. MARKOV, M. A. 1964. Phys. Lett. **10:** 122.
6. BLUDMAN, S. A. 1974. Proc. Neutrino '74 **1:** 284. AIP. New York.
7. MARX, G. & S. KOVESSY-DOMOKOS. 1964. Acta Phys. Hung. **17:** 171.
8. LEE, B. W. & S. WEINBERG. 1977. Phys. Rev. Lett. **39:** 165.
9. VYSOTSKY, M. I., A. D. DOLGOV & YA. B. ZELDOVICH. 1977. JETP Lett. **26:** 200.
10. HUT, P. & K. A. OLIVE. 1979. Phys. Lett. B **87:** 144.
11. TREMAINE, S. & J. E. GUNN. 1979. Phys. Rev. Lett. **42:** 407.
12. GUNN, J. E., B. W. LEE, I. LERCHE, D. N. SCHRAMM & G. STEIGMAN. 1978. Astrophys. J. **223:** 1015.
13. SCHRAMM, D. N. & G. STEIGMAN. 1981. Gen. Rel. Grav. **13:** 101.
14. LUBIMOV, V. A., E. G. NOVIKOV, V. Z. NOZIK, E. F. TRETYAKOV & V. S. KOZIK. 1980. Phys. Lett. B. **94:** 266.

15. ZELDOVICH, YA. B. & R. A. SUNYAEV. 1980. Pis'ma Astron. Zh. **6:** 451.
16. DOROSHKEVICH, A. G., M. YU. KHLOPOV, R. A. SUNYAEV & YA. B. ZELDOVICH. 1980. Pis'ma Astron. Zh. **6:** 457.
17. DOROSHKEVICH, A. G., M. YU. KHLOPOV, R. A. SUNYAEV & YA. B. ZELDOVICH. 1980. Pis'ma Astron. Zh. **6:** 465.
18. DOROSHKEVICH, A. G., M. YU. KHLOPOV, R. A. SUNYAEV, A. S. SZALAY & YA. B. ZELDOVICH. 1980. Proc. Neutrino '80. To be published.
19. BISNOVATY-KOGAN, G. S. & I. D. NOVIKOV. 1980. Astron. Zh. **57:** 899.
20. BISNOVATY-KOGAN, G. S., V. N. LUKASH & I. D. NOVIKOV. 1980. Proc. IAU. To be published.
21. SATO, H. & F. TAKAHARA. 1980. Preprint RIFP-400. Kyoto University, Kyoto.
22. BOND, J. R., G. EFSTATHIOU & J. SILK. 1980. Phys. Rev. Lett. **45:** 1980.
23. KLINKHAMER, F. R. & C. A. NORMAN. 1981. Astrophys. J. **243:** L1.
24. DOROSHKEVICH, A. G. & M. YU. KHLOPOV. 1981. Astron. Zh. To be published.
25. PETCOV, S. T. & M. YU. KHLOPOV. 1981. Phys. Lett. B **99:** 117.
26. BAUDET, G. & A. YAHIL. 1977. Astrophys. J. **218:** 253.
27. WAGONER, R., W. A. FOWLER & F. HOYLE. 1967. Astrophys. J. **148:** 3.
28. SZALAY, A. S. 1981. Phys. Lett. B **101:** 453.
29. STEIGMAN, G., D. N. SCHRAMM & J. E. GUNN. 1977. Phys. Lett. B. **66:** 202.
30. LINDE, A. D. 1979. Phys. Lett. B. **83:** 311.
31. SHVARTSMAN, V. F. 1969. JETP Lett. **9:** 184.
32. EINASTO, J., M. JOEVEER & E. SAAR. 1980. Nature (London) **283:** 47.
33. KIRSHNER, R. P., A. G. OEMLER & P. L. SCHECHTER. 1979. Astron. J. **84:** 951.
34. PEEBLES, P. J. E. 1971. Physical Cosmology. Princeton University Press. Princeton, N.J.
35. LIFSHITZ, E. M. 1946. JETP Lett. **16:** 587.
36. STEWART, J. M. 1972. Astrophys. J. **176:** 323.
37. GUYOT, M. & YA. B. ZELDOVICH. 1970. Astron. Astrophys. **9:** 227.
38. MESZAROS, P. 1974. Astron. Astrophys. **37:** 225.
39. SUNYAEV, R. A. & A. I. ROZGACHEVA. 1981. Pis'ma Astron. Zh. To be published.
40. PARTRIDGE, R. B. 1979. Phys. Scr. **21:** 624.
41. PARIJSKIJ, YU. N., Z. E. PETROV, L. N. CHERNOV. 1977. Pis'ma Astron. Zh. **3:** 483.
42. YORK, D. G. & J. B. ROGERSON. 1976. Astrophys. J. **203:** 378.
43. GOTT, J. R., J. E. GUNN, D. N. SCHRAMM & B. M. TINSLEY. 1974. Astrophys. J. **194:** 543.
44. ZELDOVICH, YA. B. 1970. Astron. Astrophys. **5:** 84.
45. SUNYAEV, R. A. & YA. B. ZELDOVICH. 1972. Astron. Astrophys. **20:** 189.
46. DOROSHKEVICH, A. G., S. F. SHANDARIN & E. SAAR. 1978. Mon. Not. R. Astron. Soc. **184:** 643.
47. SILK, J. 1968. Astrophys. J. **151:** 459.
48. DOROSHKEVICH, A. G., E. V. KOTOK, I. D. NOVIKOV, A. N. POLYUDOV, S. F. SHANDARIN & YU. S. SIGOV. 1980. Mon. Not. R. Astron. Soc. **192:** 321.
49. LYNDEN-BELL, D. 1967. Mon. Not. R. Astron. Soc. **136:** 101.
50. ARNOLD, V. I., S. F. SHANDARIN & YA. B. ZELDOVICH. 1981. Hydrodyn. Astrophys. To be published.
51. BINNEY, J. J. 1978. Mon. Not. R. Astron. Soc. **183:** 779.
52. AARSETH, S. J. & J. J. BINNEY. 1978. Mon. Not. R. Astron. Soc. **185:** 227.
53. ILLINGWORTH, G. 1977. Astrophys. J. **218:** L43.
54. OSTRIKER, J. P., P. J. E. PEEBLES & A. YAHIL. 1974. Astrophys. J. **193:** L1.
55. EINASTO, J. E., A. KAASIK & E. M. SAAR. 1974. Nature (London) **250:** 309.
56. GUNN, J. E. 1980. Philos. Trans. R. Soc. London Ser. A **296:** 313.
57. ZELDOVICH, YA. B., A. A. CLYPIN, M. YU. KHLOPOV & V. M. CHECHETKIN. 1980. Yad. Fiz. **31:** 1286.

FORMATION OF STRUCTURE IN A NEUTRINO-DOMINATED UNIVERSE *

J.R. Bond[1,3] and A.S. Szalay[1,2,4]

[1] Department of Astronomy, University of California, Berkeley
[2] Astronomy and Astrophysics Center, University of Chicago
[3] Present address: Department of Physics, Stanford University
[4] Permanent address: Dept. of Atomic Physics, Eotvos University, Hungary

ABSTRACT

We present a detailed calculation of the evolution of the neutrino perturbation spectrum due to damping on small scales and gravitational amplification on large scales in an expanding Universe. The severe damping we find naturally leads to the formation of Zeldovich pancakes. This theory predicts flattened superclusters with large holes in between, mass-to-light ratios which increase with size, and it can possibly account for the dark matter in galactic halos.

I. INTRODUCTION

Experimental and theoretical indications of a nonzero neutrino rest mass (Lubimov et al. 1980, Reines et al. 1980, Ramond 1979, Witten 1980) have led astrophysicists to reconsider the consequences of a neutrino mass for cosmology. Several astronomical puzzles can be solved by the introduction of this single free parameter; we summarize them here:

1. The total mass density of the Universe (normalized to critical density), Ω_T, can be large (even of order 1) (Gershtein and Zeldovich 1966, Marx and Szalay 1972, Cowsik and McClelland 1972, Schramm and Steigman 1981), yet the baryon density, Ω_B, could be low ($\lesssim 0.03$) as is inferred from primordial nucleosynthesis calculations (Gott et al. 1974, Olive at al. 1981).

2. Szalay and Marx (1976) recognized that the development of neutrino density fluctuations was characterized by a Jeans mass whose values correspond to large scale structures in the Universe if the neutrinos have a nonzero rest mass. (See also Doroshkevich et al. 1980, hereafter DKS^2Z, Bisnovaty-Kogan and Novikov 1980, Bond, Efstathiou and Silk 1980, hereafter BES, Klinkhamer and Norman 1981, Wasserman 1981). Theories of galaxy formation usually assume one of two types of initial perturbations: adiabatic or isothermal. In <u>both theories</u>, the neutrino-dominated Universe can account for a mass scale (determined by the collisionless damping of neutrinos) $M_{\nu m} \sim 4 \times 10^{15}\, m_{30}^{-2}\, M_\odot$ (BES) which is identified with superclusters if $m_{30} = m_\nu/30ev$ is of order unity.

3. The nonlinear evolution of neutrino perturbations results in the formation of Zeldovich pancakes which are highly flattened structures. Superclusters are observed to be quite nonspherical.

---

* To appear in Proceedings Neutrino-81.

4. With massive neutrinos, the large scale structure is predicted to be cellular, with large holes appearing in the distribution of galaxies on scales 40-100 Mpc, in agreement with observations.

5. The mass-to-light ratio of galaxies and clusters of galaxies is apparently larger for larger scale objects. This is a natural outcome of the presence of dissipationless dark matter, which massive neutrinos certainly would be.

6. The adiabatic theory of galaxy formation is preferred if the baryon asymmetry of the Universe is produced as a result of baryon nonconserving processes in the very early universe, although the isothermal theory cannot be ruled out (Bond, Kolb and Silk 1981, Barrow and Turner 1981). In the theory without massive neutrinos, in order to have nonlinear structure now, the induced fluctuations in the cosmic microwave background would have to be larger than is observed. A neutrino-dominated Universe beautifully sidesteps this difficulty (BES, DKS^2Z).

Here, we demonstrate how points 2-5 are satisfied. The plan of the paper is illustrated schematically by Figure 1. We assume an initial spectrum of adiabatic density perturbations arises in the early Universe. We follow their linear evolution in detail in Section II, and show that whereas amplification occurs on large scales, short wavelengths are filtered out. In Section III, the unfiltered scales are followed into the nonlinear amplification regime, and the essential features of pancake formation are presented. In Section IV, we compare the qualitative predictions of this theory with observations of large scale structure. In Section V, we sketch the modifications resulting if the initial perturbations are isothermal rather than adiabatic.

II. LINEAR DAMPING AND GROWTH

In the adiabatic picture of galaxy formation, primordial neutrino perturbations would exist. These may arise, for example, as a result of spatial fluctuations in the energy density at the Planck era; as these pass through the baryon synthesis epoch, the generated entropy-per-baryon is uniform over the perturbation, which leads to the relation $\delta_\nu=\delta_{\bar{\nu}}=\delta_\gamma=4\delta_B/3$. The apparent size of the horizon is exceedingly small at this era. It must be assumed that power in the fluctuations exists on comoving scales which correspond to the size of galaxies. We do not address how this can be, but take such a spectrum of large scale perturbations as our initial condition. The Universe can thus be considered as linearly perturbed about a smooth Friedmann-Robertson-Walker background. Lifshitz (1946) first demonstrated that in linear cosmological perturbation theory, each Fourier component of the perturbation is completely decoupled from the rest, and its behaviour depends only upon the background manifold.

We can thus follow the linear evolution of a density perturbation of wavelength λ, first when the wavelength is larger then the apparent size of the horizon, in which case it grows. As it enters the horizon the wave either damps if $\lambda < \lambda_{\nu m}$ or continues to grow if $\lambda > \lambda_{\nu m}$, where the critical length scale is $\lambda_{\nu m} \simeq 40\ m_{30}^{-1}$ Mpc at the present epoch, corresponding to the mass scale $M_{\nu m} = \pi m_\nu(n_\nu + n_{\bar{\nu}})\lambda_{\nu m}{}^3/6 \simeq 3.2 \times 10^{15}\ m_{30}{}^{-2}\ M_\odot$, where n_ν is the neutrino number density in the background (BES). The scale $\lambda_{\nu m *} = \lambda_{\nu m}/a$, which has the expansion of the Universe factored out, is the relevant quantity to describe damping; we also introduce the critical comoving wavenumber $k_{\nu m} = 2\pi/\lambda_{\nu m *}$.

The physical mechanisms responsible for damping and growth in a collisionless fluid, which the neutrinos certainly are once their temperature falls below ~ 1 MeV, can be simply understood pictorially (Figure 2). A collisional fluid has a pressure and a characteristic sound speed. The behaviour of sound waves in a self-gravitating medium may be described as a battle between the pressure, which would cause the waves to propagate at the sound speed, while maintaining constant amplitude, and gravity, which strives to condense the compression parts of the wave, causing the amplitude to grow. The wavelength boundary between the two types of behaviour is the Jeans length. In a collisionless fluid, the Jeans length is also meaningful, since gravity must overcome the random velocity dispersions of the neutrinos. However, unlike a collisional fluid, for which collisions keep the pulse profile intact (this includes the case of zero sound, for which the collective potential is short range), neutrinos move unimpeded from a compression region; fast neutrinos move a larger distance than slow ones, and the compression region damps at a rate dependant upon the neutrino velocity dispersion, and for this reason, and also due to the similarity to the method of analysis employed by Landau (1946) in his study of damping in plasmas, it is termed Landau damping.

While Landau damping is the dominant mechanism for the loss of structure on small scales, there is another mechanism which operates when the neutrinos are extremely relativistic and thus are all moving with speed c. As is illustrated in Figure 2c if there were only one dimension available for neutrino motion, then no decay of the wave amplitude would occur, since peaks would recur whenever the neutrinos moved through one wavelength; however, since neutrinos can move in 3-directions, neutrinos will be received from troughs as well as peaks, and though waves still recur, their amplitude decays due to this directional dispersion. This mechanism is especially important for short wavelengths, which enter the horizon long before the neutrinos become nonrelativistic, which occurs around a redshift $z_{nr} \simeq 57300\ m_{30}$.

We will demonstrate that damping is quite severe below the Jeans length. Thus, structure is wiped out if $k > k_{\nu m}$, where k is the comoving wavevector describing the perturbation, and $k_{\nu m}$ is the minimum value the comoving Jeans wavevector attains, which occurs at the redshift $1 + z_m \simeq 35000\ m_{30}$. The Universe then acts to filter high frequency neutrino waves, and amplifies low frequency waves ($k > k_{\nu m}$) as indicated in Figure 1. After this process is complete, the resulting spectrum (Figure 4) is then input for calculations of the nonlinear evolution of these density waves. We sketch here the method we use to calculate this spectrum in Bond and Szalay (1981) for the special case of a Universe consisting only of neutrinos.

The metric has the form

$$ds^2 = a^2(\tau)dx^\mu dx^\nu(\eta_{\mu\nu} + h_{\mu\nu})$$

Here a is the scale factor, $\tau = x^0 = \int dt/a$ is conformal time ($\sim t^{1/2}$ when relativistic particles dominate expansion, and $\sim t^{1/3}$ when nonrelativistic particles dominate), $\eta_{\mu\nu} = (1, -1, -1, -1)$ is the background metric for an Einstein-de Sitter Universe (k=0) which we assume for simplicity. The behaviour of the perturbed metric coefficients, $h_{\mu\nu}$, depend upon the choice of gauge. We adopt the synchronous gauge used by Lifshitz (1946), in which h_{oo} and h_{oi} are taken to be zero. We choose the direction of the comoving wavevector to lie along the z-direction.

Perturbations are then of three basic types, depending upon their azimuthal symmetry about the z-axis (Landau and Lifshitz 1971): tensor perturbations, which involve gravitational waves and anisotropic stresses, are expressed in terms of $(h_{11} - h_{22})/2$ and h_{12}; vector perturbations, which involve vorticity and anisotropic stresses, are expressed in terms of h_{13} and h_{23}; scalar perturbations, which involve fluctuations in the density, pressure, shear, and anisotropic stress, are expressed in terms of h_{33} and $(h_{11} + h_{22})/2$, or equivalently $h = h_{11} + h_{22} + h_{33}$. We are interested here solely in density perturbations, and hence in scalar perturbations.

In a Universe consisting only of neutrinos of the same mass, the relation between the metric and density perturbations is given by the Einstein equations, which reduce to

$$\frac{1}{2a^2}\left(h'' + \frac{a'}{a} h'\right) = 8\pi G\left[\left(\delta\rho_\nu + 3\delta P_\nu\right)/2\right] \tag{1}$$

$$\frac{ik}{2a^2}\left(h' - h_3'\right) = -8\pi G\delta H_\nu \tag{2}$$

where the prime denotes differentiation with respect to τ, and $h_3 \equiv h_{33}$. The fluctuations in the neutrino energy density, $\delta\rho_\nu$, energy current, δH_ν, and pressure, δp_ν, must be determined by the explicit solution of the Boltzmann transport equation for collisionless neutrinos, which relates them to h' and h_3'. The resulting closed system of ordinary integrodifferential equations is solved numerically, and solutions for representative wavenumbers are displayed in Figure 3. The scale factor is normalized to unity when the neutrinos become nonrelativistic at the redshift z_{nr}; actually, however, neutrinos are semirelativistic from $a \simeq 0.3$ up to ~ 3 where all the action occurs, and so we must explicitly take these effects into account.

Consider the evolution of the Fourier component of wavevector $k = 0.2\ k_{\nu m}$. Prior to the point at which one full wavelength enters the horizon (shown by the box), two types of behaviour are evident: in the extremely relativistic limit ($a < 1$) we can show $\delta_\nu \simeq 2h/3$, and thus growth follows $\delta_\nu \sim h \sim \tau^2 \sim a^2$ (since $a \sim \tau$, $\delta p_\nu \sim \delta\rho_\nu/3$, $8/3\ G\pi\rho_\nu a^2\tau^2 = 1$ in Eq. 1); in the nonrelativistic limit, we can show $\delta_\nu \simeq h/2$, and thus growth follows $\delta_\nu \sim h \sim \tau^2 \sim a$ (since $a \sim \tau\ ^2$, $\delta p_\nu \simeq 0$, and $8/3\pi G\rho_\nu a^2\tau^2 = 4$ in Eq. 1). This growth outside of the horizon depends upon the gauge we have chosen. Once the wave enters the horizon, the nonrelativistic growth law ($\delta_\nu \sim a$) continues to hold, since the wavenumber is much smaller than the minimum comoving Jeans wavevector $k_{\nu m}$. When even half a wavelength of the Fourier component with $k/k_{\nu m} = 1.3$ enters the horizon (denoted by a triangle), neutrinos can have propagated sufficiently far to begin damping, thus δ_ν dips. However, the velocity dispersion falls as a^{-1}, so eventually gravity wins, and growth continues, again following $\delta_\nu \sim a$ in an Einstein-de Sitter Universe. The wave with $k/k_{\nu m} = 3.0$ suffers extreme damping before its growth phase begins. Notice that zeros exist in this case, reflecting the fact that the propagating neutrinos can result in δ_ν oscillations superimposed upon the overall decay.

We now construct a spatial density spectrum associated with the values of these Fourier components. The usual method used is to assume random phases for the Fourier components, and to construct the root mean square overdensity average over a volume $V = 4/3(\pi\ell^3)$ (Peebles 1980). Then, the differential density spectrum is proportional to $k^{3/2}|\delta(k)|$, where we now interpret k as $2\pi/\ell$. Thus, $k^{3/2}|\delta(k)|$ gives the average relative contribution of scale $2\pi k^{-1}$ to a local density enhancement.

In Figure 4, this function can be seen to peak at $k/k_{\nu m} \simeq 0.85$, which corresponds to a mass scale for 3 massive neutrinos of $M = (0.85)^{-3} M_{\nu m}/\sqrt{3} \simeq 3 \times 10^{15}\ m_{30}^{-2}\ M_\odot$, where the $\sqrt{3}$ enters since 3 neutrino flavors rather than one is involved. Zeldovich argues that the natural initial spectrum for adiabatic perturbations should have $\delta \sim k^2$, and we have assumed this in constructing Fig.4. If instead we had assumed $\delta \sim k^\circ$ initially (white noise), the peak shifts to $k \simeq 0.5\ k_{\nu m}$. The final output spectrum clearly has in contrast to the initial spectrum almost no power at large k. The effects of damping beyond $k/k_{\nu m}$ are quite severe, with a falloff of ~ 200 by $k/k_{\nu m} = 3$.

Our calculation was made for an $\Omega = 1$ Universe, but since the damping process is complete by $z \simeq \Omega^{-1}$, the spectrum will be unmodified for smaller Ω. We (Bond and Szalay 1981) have also calculated the spectrum for the case when one flavor of neutrino has a larger ($\sim$ 30 eV) mass, the other two have small mass ($<$ 1 eV) in a realistic Universe by simultaneously following the evolution of density perturbations in the heavy neutrinos, the light ones and the photons. Baryons have not yet been included. This spectrum has the same shape as that in Fig.4, but peaks at $k/k_{\nu m} = 1.0$, in remarkable agreement with the BES prediction!

In an adiabatic theory, primordial fluctuations in the baryons are also damped on a scale which is approximately $M_{\nu m}$ (BES). The remaining large scale structures of mass $M > M_{\nu m}$ continue to expand as the Universe expands, but more slowly, and thus grow in density contrast. Expansion ceases and contraction begins when the wave becomes nonlinear ($k^{3/2}|\delta| > 1$).

III. NONLINEAR GROWTH: PANCAKE FORMATION

The theory of the nonlinear evolution of density perturbations with strong linear damping below some critical scale has been developed by Zeldovich (1970) and coworkers. They find that density enhancements on this scale are asymmetric, collapse proceeds preferentially in one direction, the asymmetry is therefore amplified, and highly flattened "pancakes" result.

Velocity perturbations are associated with the density perturbations through the continuity equation. About each point in the Universe, the velocity field on scales $> \lambda_{\nu m}$ in each direction is determined by a superposition of Fourier components, each with phases randomly determined. The velocity fields on these scales will then be on the average different in different directions. On smaller scales, the velocity field is approximately constant in each direction. Motion is thus asymmetric as a result of small scale damping. The case of equal velocities in all directions, which leads to spherically symmetric collapse of overdense regions, is a highly unlikely degenerate case. In the general case, motion in one direction will reach nonlinearity before the other two. The gravitational force along this axis becomes much higher than in the other two directions, making a one-dimensional model of nonlinear evolution a reasonable approximation.

The initial configuration on scale $k_{\nu m}$ in density evolves in the linear phase as $\delta(\zeta,t) \simeq b(t)\cos(k_{\nu m}\zeta)$, where ζ is the comoving coordinate, and $b \sim t^{2/3} \sim a \sim \tau^2$ describes the growth. If we neglect the neutrino velocity dispersion, which is small relative to the gravitationally induced velocity until late in the collapse, then the collapse is one of zero pressure, and the trajectory of a neutrino which is initially at the comoving point ζ is

$$z(t,\zeta) = a(t)(\zeta - k_{\nu m}{}^{-1}b(t)\sin k_{\nu m}\zeta) \qquad (3)$$

The first term is the Hubble expansion, the second is the deviation due to the gravitational influences of the overdense region. For this one dimensional pressure-free model, Eq. (3) is an exact solution to Poisson's equation. The evolution of density and velocity in this model is displayed in Figure 5. Conservation of mass ensures $dM = \rho_o a d\zeta = \rho(z,t)dz$, where ρ_o is the background density; thus, $\rho \sim (\partial z/\partial\zeta)^{-1} \sim (1 - b(t)\cos k_{\nu m}\zeta)^{-1}$ rises as $(1-b)^{-1}$ in the center, and infinite density is reached when $b = 1$. The formation of this caustic surface is due to the intersection of particle trajectories: faster neutrinos from higher up above the central plane overtake slower ones below them.

The baryons, which have fallen in with the neutrinos, build up pressure, and dissipate their kinetic energy in a shock wave as discussed by Sunyaev and Zeldovich (1972) and Doroshkevich, Shandarin, and Saar (1978). The collisionless neutrinos will pass through the caustics and separate from the baryons. The solution for the trajectories given by Eq. 3 is not applicable beyond this point, since in the center, three streams appear, two of large and opposite velocity, one of low velocity. Further, we should not continue to neglect the velocity dispersion of the neutrinos at this stage. Nonetheless, the gravitational effects on the neutrinos can be understood by continuing our pressure-free solution by making a one-dimensional N-body simulation which integrates the equations of motion (Bond, Szalay and White 1981, following Doroshkevich et al. 1980d). Figure 6 shows the phase space distribution of the neutrinos ($\dot{z} - \dot{a}z/a$ versus z/a) subsequent to caustic formation (T=1). At T=2, there are the three streams; by T=8, some of the neutrinos that earlier passed through the plane have turned around and fallen back. Thus, two more streams appear containing lower velocity neutrinos. The trajectories in phase space are spirals winding towards the center due to the orbital motion along linear paths through the plane. The inclusion of velocity dispersion will spread out the simple spirals seen in Figure 6 into a fuzzy distribution. The entire system is relaxing towards a state with neutrinos selfbound about the central plane leading to a sandwich-like structure: shocked baryons cooling in the central plane, surrounded by hot nondissipative neutrinos (DKS^2Z).

Galaxies are assumed to form in the breakup of the high-density baryon pancake (Doroshkevich 1980). A sufficiently large fraction of the neutrinos can have velocities low enough to be bound to these baryonic fragments (Bond, Szalay and White 1981). Large fragments have larger escape velocities than smaller ones, hence the mass fraction in bound neutrinos rises with size. This idea of the low velocity condensate predicts not only that neutrinos may form the dark matter in galactic halos, but that the mass-to-light ratio of astronomical objects rises with increasing scale.

IV. LARGE SCALE STRUCTURES

If the neutrino mass is of order 30 ev, the collapsed pancakes would be identified with superclusters. As redshift surveys of galaxies have demonstrated, superclusters do exist, are quite common in the Universe, are highly nonspherical, and there is some evidence that they are still expanding at least along one of the axes (Davis et al. 1981, Yahil et al. 1980, Tully 1981, de Vaucouleurs 1973, Gregory and Thompson 1978, Einasto et al. 1980, Kirshner, Oemler, and Schechter 1979, Ford et al. 1981). It is sometimes argued that the observed small deviations from the Hubble flow indicate that the collapse time of such systems is much larger than the present age of the universe, so the supercluster could not have gone through a collapse. We emphasize that this is not the case, since for pancakes there is no single collapse time. One must distinguish between the collapse time along the normal axis, which can be very short and the virial collapse time that can be much longer. For example, the pancake collapse may have occurred between $z \approx$ 5-10 and yet have not decoupled from the Hubble expansion in the other directions. From a galaxy near the pancake the velocities of other galaxies would appear to follow the unperturbed Hubble flow, in spite of the earlier nonlinear collapse (Dekel & Szalay 1981).

Our Galaxy is itself in the Local Supercluster plane. The value of Ω_T can be determined by comparing our infall velocity in the supercluster plane towards Virgo and the value of our peculiar acceleration as calculated from the density in this direction. All these calculations, which yield a low value of Ω, presume that the galaxy distribution in the Supercluster represents the overall mass distribution. It can be shown (Szalay and Silk 1981) that the nonsphericity and the dissipative separation of baryons in the pancake collapse can lead to a large overestimation of our acceleration, resulting in apparent Ω values between 0.2 and 0.3, even though the infall occurs in an $\Omega = 1$ universe.

Up to now we have been concerned with the local (mainly one dimensional) properties of pancakes. To determine the present structure of the universe arising from a given perturbation spectrum, all waves with their given amplitudes have to be superposed with random phases. At some places, local maxima of the perturbations develop and collapse into pancakes, leaving empty regions, holes void of matter on scales 40-100 Mpc. This feature is quite apparent in some of the recent redshift surveys of the galaxy distribution (Kirshner, Oemler, and Schechter 1979, Ellis et al. 1981, Davis et al. 1981, Gregory and Thompson 1978, Einasto et al. 1980). In the pancake picture, a large scale cellular structure is expected to exist in the universe, with possible strong density enhancements along the lines where neighboring pancakes intersect, which produces a filamentary object on the scale of superclusters (Clypin 1980). Arnold, Shandarin and Zeldovich (1981) have used catastrophe theory to generally prove that pancakes are the lowest order singularities expected in the adiabatic theory; they also discuss higher order singularities which should appear in addition to pancakes.

The very large scale structure of the Universe (> 100 Mpc) is determined by that part of the perturbation spectrum which has wavelengths much greater than the critical one, and it is still linear.

V. DISCUSSION

Primordial neutrino perturbations would be extremely small if entropy (isothermal) fluctuations formed the initial spectrum. Prior to recombination at $z \simeq 1000$, baryons are tightly coupled to photons, and the amplitude of baryon perturbations remains constant. However, as the energy density in radiation drops, the energy density in baryons becomes non-negligible, and curvature fluctuations arise due to the baryon self-gravity. Collisionless neutrinos respond to these: hence, δ_ν grows, driven by δ_B, provided $k < k_{\nu m}$ so that damping is not severe. Ultimately, δ_ν can exceed δ_B and continue to grow, driven now by the neutrino self-gravity. The initially isothermal spectrum has thus generated an adiabatic component on $> M_{\nu m}$ scales. After recombination, δ_B can then respond to the larger δ_ν, and grow until $\delta_B = \delta_\nu$.

On scales smaller than $M_{\nu m}$, neutrino perturbations initially do not exist in this isothermal picture, and the growth of baryon perturbations is either negligible (on the smallest scales) or highly suppressed in amplitude (BES). The first nonlinear scales would then be those that were already nonlinear at recombination ($\sim 10^6\ M_\odot$ which would likely fragment to form a pregalactic generation of stars), and later, at the high mass end, neutrino structures of mass $\sim M_{\nu m}$. The latter would evolve as in the pancake theory, except that now baryon structure exists on all smaller scales, and can fragment as collapse proceeds. Sato and Takahara (1980) have considered the fragmentation assuming the $M_{\nu m}$ structures are spherical in a hybrid adiabatic-isothermal theory. We have demonstrated that adiabatics arise naturally from isothermals on large scales, and explore the nature of fragmentation in highly aspherical collapses of $M_{\nu m}$ structures (Bond, Silk and Szalay 1981).

Landau damping on small scales and growth due to the Jeans instability on large scales in a neutrino-dominated Universe thus leads naturally in both adiabatic and isothermal theories to the appearance of superclusters, holes in space, cellular structure, and filaments which dominate the large scale structures of the Universe.

We would like to thank our many collaborators for stimulating discussions on massive neutrino cosmology. This research was supported in part by grants NSF AST-79-23243, NSF AST-79-15244, and NASA NGR 05-003-578 at Berkeley. One of us (A.S.) was also supported by DOE AC02-80ER10773 at Chicago and would like to acknowledge the hospitality of the Institute for Theoretical Physics, Santa Barbara, where some of the research was done.

REFERENCES

Arnold, V.I., S.F. Shandarin, Ya.B. Zeldovich. 1981. Hydrodyn. Astrophys. In press.
Barrow, J.D., M.S. Turner. 1981. Preprint.
Bisnovaty-Kogan, G.S., I.D. Novikov. 1980. Astr. Zh. 57, 899.
Bond, J.R., G. Efstathiou, J. Silk. 1980. Phys. Rev. Lett. 45, 1980.
Bond, J.R., E.W. Kolb, J. Silk. 1981. Preprint.
Bond, J.R., A.S. Szalay. 1981. In preparation.
Bond, J.R., J. Silk, A.S. Szalay. 1981. In preparation.

Bond, J.R., A.S. Szalay, S.D. White. 1981. In preparation.
Clypin, A.A. 1980. Private communication.
Cowsik, R., J. McClelland. 1972. Phys. Rev. Lett. 29, 669.
Davis, M., J. Huchra, D.W. Latham, J. Tonry. 1981. Harvard preprint.
Dekel, A., A.S. Szalay. 1981. In preparation.
de Vaucouleurs, G., de Vaucouleurs, A. 1973. Astron. Astrophys. 28, 109.
Doroshkevich, A.G., S.F. Shandarin, E. Saar. 1978. M.N.R.A.S. 184, 643.
Doroshkevich, A.G. 1980. Sov. Astron. 57, 259.
Doroshkevich, A.G., M.Yu. Khlopov, R.A. Sunyaev, A.S. Szalay, Ya.B. Zeldovich. 1980. Proc. Xth Texas Symposium on Relativistic Astrophysics. (See also Doroshkevich et al. 1980a,b,c).
Doroshkevich, A.G., M.Yu. Khlopov, R.A. Sunyaev, Ya.B. Zeldovich. 1980a. Pisma Astr. Zh. 6, 457.
Doroshkevich, A.G., M.Yu. Khlopov, R.A. Sunyaev, Ya.B. Zeldovich. 1980b. Pisma Astr. Zh. 6, 465.
Doroshkevich, A.G., M.Yu. Khlopov, R.A. Sunyaev, A.S. Szalay, Ya.B. Zeldovich. 1980c. Proc. Neutrino '80. Erice, Italy. In press.
Doroshkevich, A.G., E.V. Kotok, I.D. Novikov, A.N. Polyudov, S.F. Shandarin, U.S. Sigov. 1980d. M.N.R.A.S. 192, 321.
Einasto, J., M. Joeveer, E. Saar. 1980. Nature 283, 47.
Ellis, R. et al. 1981. Private communication.
Ford, H.C., R.J. Harms, R. Ciardullo, F. Bartko. 1981. Ap. J. 245, L53.
Gershtein, S.S., Ya.B. Zeldovich. 1966. JETP Lett. 4, 174.
Gott, J.R., J.E. Gunn, D.N. Schramm, B.M. Tinsley. 1974. Astrophys. J. 194, 543.
Gregory, S.A., L.A. Thompson. 1978. Ap. J. 222, 784.
Kirshner, R.P., A.G. Oemler, P.L. Schechter. 1979. Astron. J. 84 951.
Klinkhamer, F.R., C.A. Norman. 1981. Astrophys. J. Lett. To be published.
Landau, L.D. 1946. J. Phys. (USSR) 10, 25.
Landau, L.D., E.M. Lifshitz. 1971. The Classical Theory of Fields (New York: Pergamon Press).
Lifshitz, E.M. 1946. JETP Lett. 16, 587.
Lubimov, V.A., E.G. Novikov, V.Z. Nozik, E.F. Tretyakov, V.S. Kozik. 1980. Phys. Lett. B 94, 266.
Marx, G., A.S. Szalay. 1972. Proc. Neutrino '72 1, 123, Technoinform, Hungary.
Olive, K.A., D.N. Schramm, G. Steigman, M.S. Turner, J. Yang. 1981. Ap. J. 246, 557.
Peebles, P.J.E. 1980. The Large-Scale Structure of the Universe (Princeton: Princeton University Press).
Ramond, P. 1979. Caltech preprint CALT-68-709.
Reines, F., S.W. Sobel, E. Pasierb. 1980. Phys. Rev. Lett. 45, 1307.
Sato, H., F. Takahara. 1980. Preprint RIFP-400. Kyoto University, Kyoto.
Schramm, D.N., G. Steigman. 1981. Ap. J. 243, 1.
Sunyaev, R.A., Ya.B. Zeldovich. 1972. Astron. Astrophys. 20, 189.
Szalay, A.S., J. Silk. 1981. In preparation.
Szalay, A.S., G. Marx. 1976. Astron. Astrophys. 49, 437.
Tully, B. 1981. Proc. Neutrino '81
Wasserman, I. 1981. Ap. J., to appear.
Witten, E. 1980. Phys. Lett. 91B, 81.
Yahil, A., A. Sandage, G.A. Tammann. 1980. Ap. J. 242, 448.
Zeldovich, Ya.B. 1970. Astron. Astrophys. 5, 84.

FIGURE CAPTIONS

Fig.1. Schematics of fluctuation growth in a neutrino dominated Universe. Measurements of the denoted quantities allow us to gain information about the corresponding epoch of the Universe.

Fig.2. Illustration of the evolution of a neutrino density wave when the wavelength exceeds the Jeans length (a), the wavelength is less than the Jeans length and the neutrinos are nonrelativistic (b), $\lambda < \lambda_{J\nu}$, and the neutrinos moving with speed c are constrained to one dimension (c), or to two (d).

Fig.3. The evolution of various Fourier components of the density, $\delta(k,a)$ in terms of the scale factor $a = 57300\, m_{30}/(1+z)$, where z is the redshift. For convenience we have normalized all $\delta(k)$ to the same value when the waves are far outside the horizon. The triangles denote the point at which half a wavelength enters the horizon, the boxes denote, when a full wavelength enters. For $k/k_{\nu m} = 3.0$ the sign of δ changes as indicated.

Fig.4. The density spectrum on the mass scale $M = M_{\nu m}/\sqrt{3}\,(k/k_{\nu m})^{-3}$ for a cosmology consisting of 3 flavors of neutrinos all with the same mass. $\delta\rho/\rho(M) \sim (k/k_{\nu m})^{3/2}|\delta(k)|$ is plotted against scalesize. A Zeldovich initial spectrum ($\delta(k)\sim k^2$) has been assumed.

Fig. 5. One dimensional cross section of a pancake collapse. The upper graph shows the density, the lower the peculiar velocity vs the comoving coordinate at four different times, the last corresponding to the moment of infinite density. The perturbation shown is one damping wavelength, sinusoidal. Note, that infinite density is reached in a finite time.

Fig 6. Distribution of velocities of cold collisionless particles vs. the comoving coordinate after the pancake collapse at T=1.0. At the same position the velocity can have 3,5,... different values —multistream regions appear.

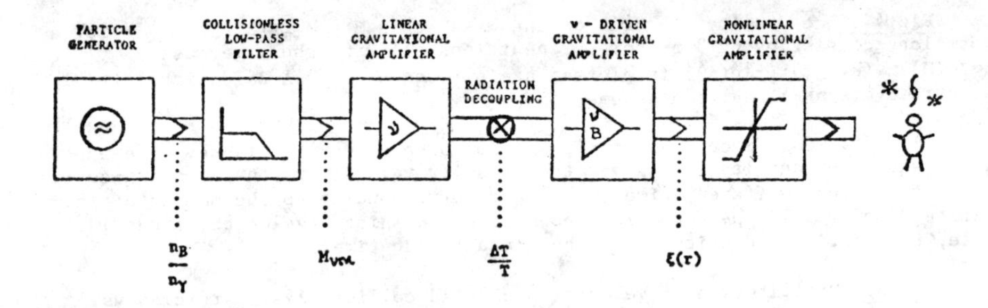
PARTICLE GENERATOR
COLLISIONLESS LOW-PASS FILTER
LINEAR GRAVITATIONAL AMPLIFIER
RADIATION DECOUPLING
ν - DRIVEN GRAVITATIONAL AMPLIFIER
NONLINEAR GRAVITATIONAL AMPLIFIER
$\frac{n_B}{n_\gamma}$
M_{VCM}
$\frac{\Delta T}{T}$
$\xi(r)$

FIG. 1

$\lambda > \lambda_{J\nu}$: GROWTH

$\lambda < \lambda_{J\nu}$: DAMP DUE TO VELOCITY DISPERSION (NON RELATIVISTIC NEUTRINOS)

EXTREMELY RELATIVISTIC NEUTRINOS

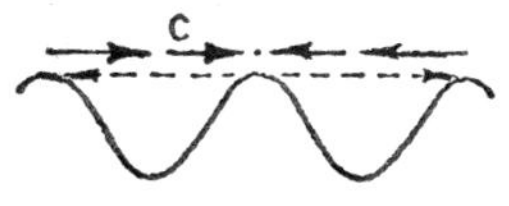

ONE DIMENSION :

NO DAMPING ; WAVES RECUR

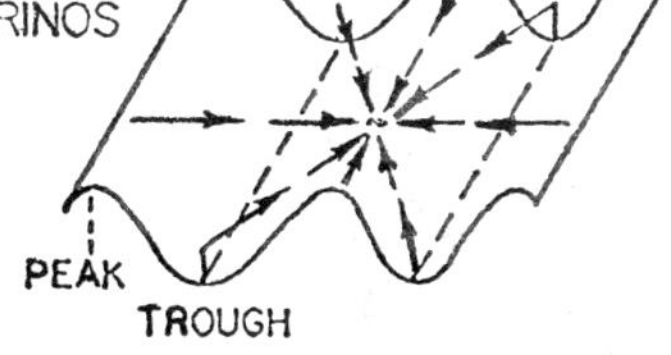

TWO DIMENSIONS :

WAVES RECUR BUT DECAY DUE TO DIRECTIONAL DISPERSION

FIG. 2

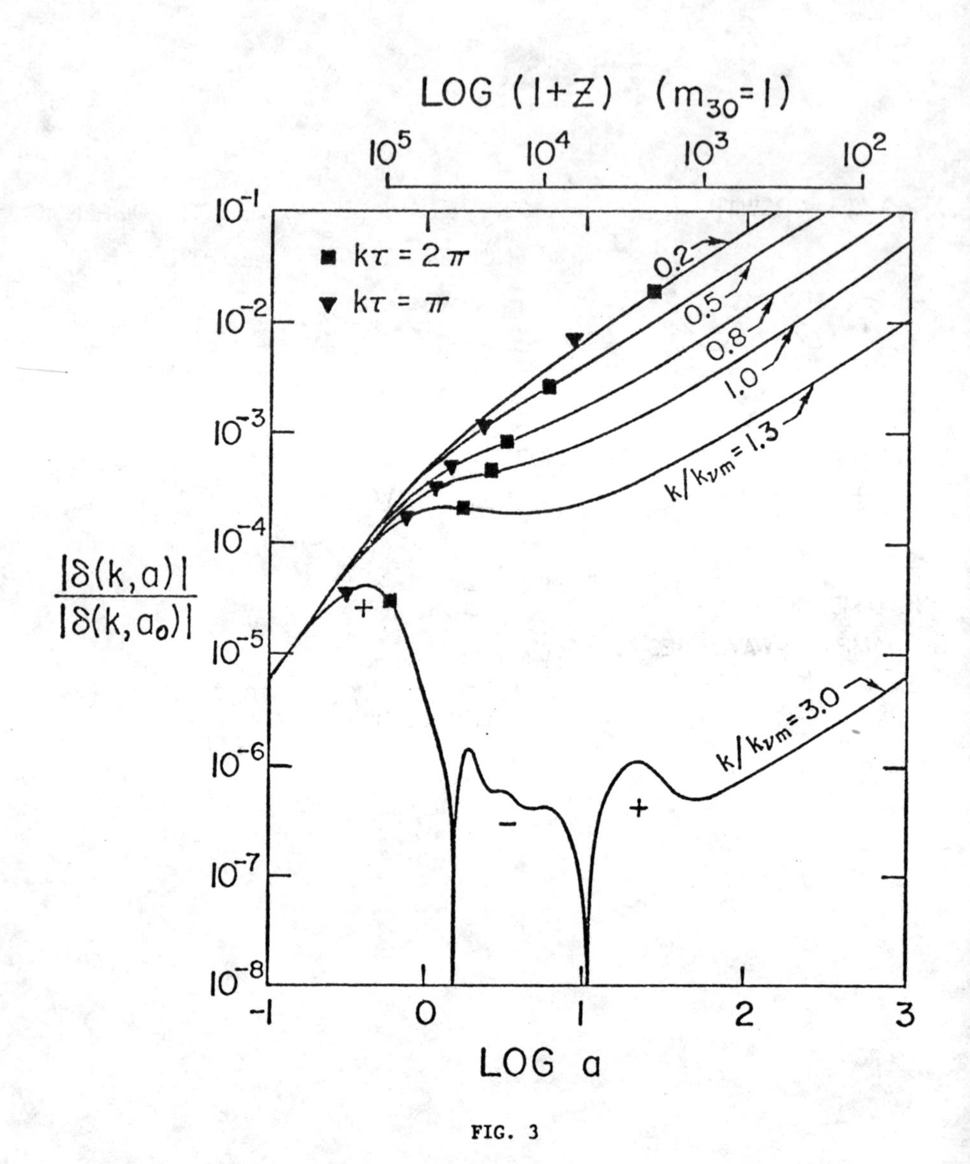

LOG (1+Z) (m_{30}=1)
10^5
10^4
10^3
10^2
■ $k\tau = 2\pi$
▼ $k\tau = \pi$
0.2
0.5
0.8
1.0
$k/k_{\nu m}$=1.3
$k/k_{\nu m}$=3.0
+
−
+
$\frac{|\delta(k,a)|}{|\delta(k,a_0)|}$
10^{-1}
10^{-2}
10^{-3}
10^{-4}
10^{-5}
10^{-6}
10^{-7}
10^{-8}
-1
0
1
2
3
LOG a

FIG. 3

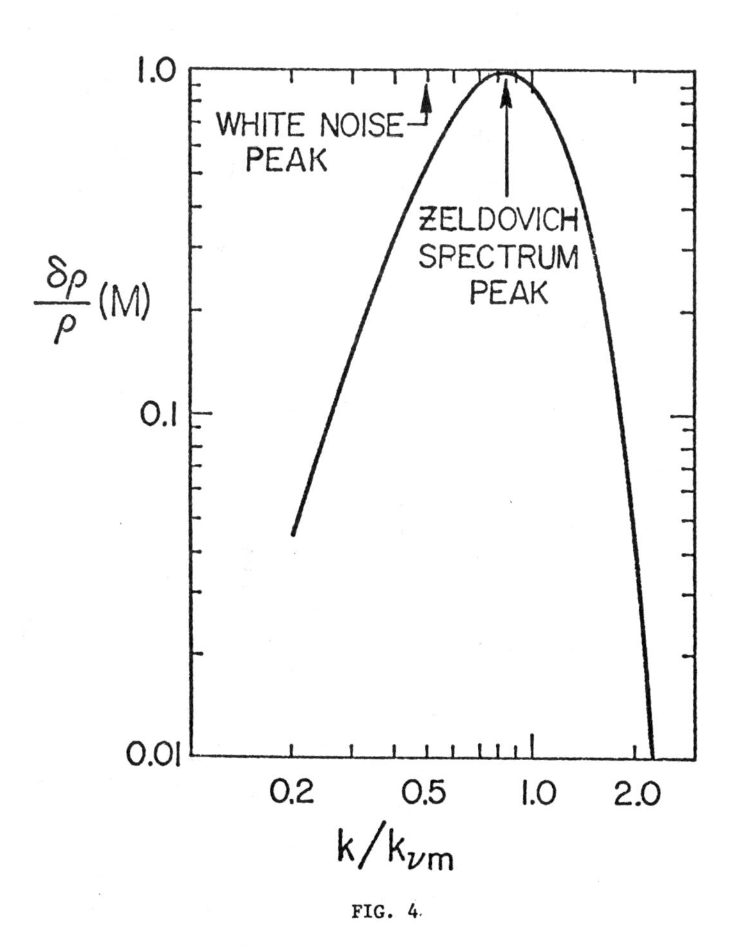
1.0
0.1
0.01
δρ/ρ (M)
WHITE NOISE PEAK
ZELDOVICH SPECTRUM PEAK
0.2
0.5
1.0
2.0
k/kνm

FIG. 4.

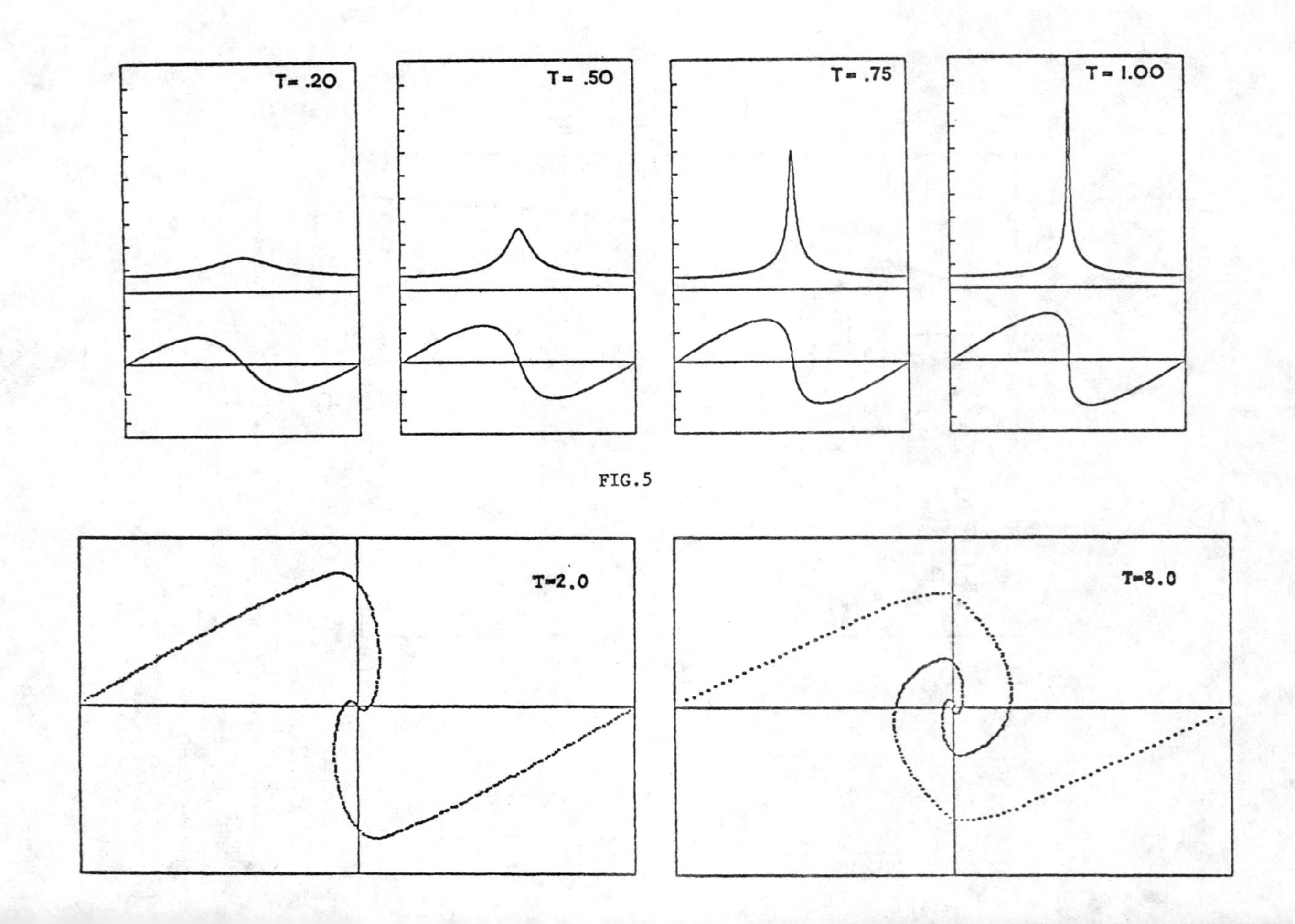

FIG.5

THE ASTROPHYSICAL JOURNAL, 223:1015–1031, 1978 August 1

SOME ASTROPHYSICAL CONSEQUENCES OF THE EXISTENCE OF A HEAVY STABLE NEUTRAL LEPTON

J. E. GUNN*
California Institute of Technology; and Institute of Astronomy, Cambridge, England
B. W. LEE†
Fermi National Accelerator Laboratory;‡ and Enrico Fermi Institute, University of Chicago
I. LERCHE
Enrico Fermi Institute and Department of Physics, University of Chicago
D. N. SCHRAMM
Enrico Fermi Institute and Departments of Astronomy and Astrophysics and Physics, University of Chicago
AND
G. STEIGMAN
Astronomy Department, Yale University
Received 1977 December 1; accepted 1978 February 14

ABSTRACT

Recently, high-energy particle theorists have constructed new extended gauge theories which may fit experiment somewhat better than previous already very successful theories. One of the predictions which is often discussed is the possible existence of a stable neutral lepton, probably with a mass of a few GeV/c^2. Following this motivation we here investigate some cosmological consequences of the existence of *any* stable, massive, neutral lepton, and show that it could well dominate the present mass density in the universe. The contribution to the mass density depends on the mass of the lepton, which should eventually be determined with high-energy accelerators. It is interesting that the more massive the lepton, the smaller its contribution to the present mass density. It is unlikely that these leptons affect big bang nucleosynthesis or condense into stellar size objects. However, such a lepton is an excellent candidate for the material in galactic halos and for the mass required to bind the great clusters of galaxies. Annihilation radiation from these structures should be detectable. At the end of the paper a brief mention is made of the astrophysical constraints on the mass-lifetime relationship if the neutral lepton is unstable.

Subject headings: cosmology — elementary particles

I. INTRODUCTION

The spectrum of leptons has always been, and continues to be, an intriguing puzzle in elementary particle physics. Until recently, the known species were the electron, the muon, their associated neutrinos ν_e and ν_μ, and their antiparticles. These neutrinos are assumed to be massless, since this naturally renders the chiral structure of the charged weak leptonic currents; experimentally there are reasonably stringent upper bounds on their masses: $m_{\nu_e} < 60$ eV and $m_{\nu_\mu} < 0.65$ MeV. Although in the past there was no convincing reason why the muon should exist in addition to the electron (and similarly for the muon neutrino), modern renormalizable unified gauge theories of the weak and electromagnetic interactions (Weinberg 1974; t'Hooft and Veltman 1973) have provided motivations for the existence of heavy leptons, both charged and neutral. These include quark-lepton symmetry, good high-energy behavior of scattering amplitudes, and the cancellation of certain anomalous divergences which would otherwise spoil the renormalizability of the theory. Typically, heavy leptons are unstable and decay into three lighter leptons via the exchange of a virtual gauge boson which gives rise to an effective current-current interaction. If the masses of the three decay leptons are negligible relative to that of the parent lepton, then the decay rate scales as the fifth power of the mass of this parent lepton. Thus, in general, as the mass of a lepton increases, its lifetime decreases quite rapidly.

In this paper we shall study the astrophysical consequences of the most extreme exception to this rule, namely the case of an absolutely stable heavy lepton. The initial motivation for this study was provided by a recently developed model of weak and electromagnetic interactions (Lee and Weinberg 1977*a*; Lee and Shrock 1977*a*, *b*)

* Alfred P. Sloan Foundation Fellow.

† The authors regret the tragic death of Dr. Lee in an auto accident 1977 June 16.

‡ Operated by the University Research Association, Inc., under contract with the Energy Research and Development Administration.

based upon the enlarged gauge group SU(3) × U(1). This model was initially constructed in order to account for the trimuon events observed by the Fermilab-Harvard-Pennsylvania-Rutgers-Wisconsin neutrino experiment (Benvenuti *et al.* 1977*a*, *b*). The most likely explanation of these unusual events seems to be the direct production and sequential decay of a charged heavy lepton M^-. In order to account for the observed rate it is most natural to invoke a full-strength weak coupling of ν_μ to M^-, via a new gauge boson, and thus necessarily to enlarge the gauge group beyond the standard Weinberg-Salam SU(2) × U(1) model (Weinberg 1967, 1971, 1972*a*, *b*; Salam 1967). The SU(3) × U(1) model successfully explains both the total rate and the kinematic characteristics of the trimuon events. Furthermore, it naturally ensures quark-lepton universality and $e = \nu$ universality, absence of right-handed currents in β and μ decay, and suppression of strangeness-changing neutral currents to order G_F^2 in the $K_L - K_S$ mass difference and $K_L \to \mu\bar{\mu}$ decay rate. Various discrete symmetries play a crucial role in maintaining these properties.

The heavy lepton production and cascade decay mechanism for trimuon production involves the decay of M^- into a neutral heavy lepton, M^0. An analysis of the trimuon data yields the approximate mass values $m_M = 7.0(+3.0, -1.0)$ GeV and $m_{M^0} = 3.5(+1.5, -0.4)$ GeV (private communication Shrock 1977). Thus, it is reasonable to propose that already two heavy leptons have manifested their presence in the trimuon data. It might be noted in passing, to bolster the reader's confidence in the seriousness and reality of leptons heavier than the muon, that experiments at the e^+-e^- colliding beam machine SPEAR have accumulated substantial evidence for the existence of a charged heavy lepton, called τ^-, with a mass of 1.9 GeV (Perl *et al.* 1975, 1976). This heavy lepton is also incorporated in the SU(3) × U(1) model. The M^-, M^0, and τ^- are all unstable.

For our purposes, one remarkable property of the model is of prime importance: as a result of an exact discrete symmetry in the theory (a kind of internal parity) the lightest of the massive leptons with odd internal parity is forbidden from decaying, i.e., it is absolutely stable. In the model this lepton is the electron-type partner to the M^0 and is accordingly labeled E^0.

Whatever the success or failure of this particular model, symmetries of this sort are likely to be needed in any attempt to expand the gauge group of weak and electromagnetic interactions since in order to preserve the universality of weak interaction strengths it is necessary to prevent the mixing of old and new leptons or quarks. The lightest particle which carries some new quantum number of this sort will of course be stable.

Although we were initially motivated by the SU(3) × U(1) model, *any stable neutral massive lepton would produce the astrophysical consequences we discuss in this paper*. Therefore we use the somewhat more general notation L^0 here to denote any absolutely stable neutral heavy lepton. Although the most recent theoretical ideas might identify L^0 with E^0, it may well be that future developments may give alternative motivation for a stable neutral heavy lepton. In any case the question should eventually be resolved with high-energy particle accelerators.

Lee and Weinberg (1977*b*, hereafter referred as LW2) showed that such neutral leptons must have a mass m_L greater than about ~ 2 GeV or less than ~ 50 eV (see also Cowsik and McClelland 1972);[1] otherwise they would be in conflict with the present observational constraints on the deceleration parameter q_0, and the lower limit to the age of the universe and the corresponding conservative upper limit on the density of the universe of $\sim 2 \times 10^{-29}$ g cm^{-3}. The argument is straightforward; first consider low-mass neutrinos ($M_L \ll 1$ MeV). At early epochs the universe was almost certainly hotter than $kT = 1$ MeV, so electron-positron pairs were as abundant as photons (see, for example, Weinberg 1972*c*). At such epochs ($kT > 1$ MeV), neutrino pairs would be copiously created and would be in thermodynamic equilibrium. At temperatures lower than $kT \approx 1$ MeV the weak interaction rate drops below the expansion rate and the neutrinos decouple from the ordinary matter. The present ratio of neutrinos of type i to photons is: $n_{\nu_i}/n_\gamma = \frac{3}{8} g_{\nu_i}(T_\nu/T_\gamma)^3$, where g_{ν_i} is the statistical weight of the type i neutrinos (e.g., for two helicity states and antineutrinos as well as neutrinos, $g_{\nu_i} = 4$). At present, $(T_\gamma/T_\nu)^3 \approx 11/4$ and $n_\gamma \approx 400$ cm^{-3} so that $n_{\nu_i} \approx 55\, g_{\nu_i}$. The present mass density contributed by such neutrinos would be $\rho_{\nu_i} \approx 9 \times 10^{-31}\, g_{\nu_i} m_i$ (eV) g cm^{-3}. For $\rho_0 \lesssim 2 \times 10^{-29}$ g cm^{-3} (equivalent to $\Omega = 1$, $H_0 = 100$ km s^{-1} Mpc^{-1}) we obtain: $\sum_i g_{\nu_i} m_i$ (eV) $\lesssim$ 220 (e.g., for $g_{\nu_i} = 4$, $m_i \lesssim 55$ eV).

The situation is somewhat different for high-mass neutrinos ($m_L \gg 1$ MeV). Such neutrinos become nonrelativistic when they are still in thermodynamic equilibrium. As a result their abundance relative to photons is low due to the presence of a Boltzmann factor: $n_\nu/n_\gamma \sim \exp(-m_L c^2/kT) \ll 1$. These neutrinos "freeze out" when their annihilation rate drops below the expansion rate. LW2 shows that the freezing temperature is of the order of 1/20 the neutrino mass for masses of a few to a few tens of GeV. The residual mass density of L^0 is shown by LW2 to be

$$\rho_L = 4.2 \times 10^{-28}\ \mathrm{g\ cm^{-3}}[m_L(\mathrm{GeV})]^{-1.85}\left(\frac{N_A}{N_F^{1/2}}\right)^{-0.95}, \qquad (1)$$

where N_A is the number of annihilation channels open and depends on the details of the $L^0\bar{L}^0$ interaction and N_F is the effective number of degrees of freedom allowed, counting 1/2 for each boson and 7/16 for each fermion species and spin state. Similarly the number density $n_L \propto [m_L(\mathrm{GeV})]^{-2.85}$. Clearly there is some uncertainty in N_A depending on the model, but m_L only goes as approximately $N_A^{1/2}$. Taking $N_A \approx 14$ and $N_F = 4.5$ gives, for $\rho < 2 \times 10^{-29}$, $m_L \gtrsim 2$ GeV. Figure 1 summarizes the relationship between m_L and ρ_L.

[1] Throughout this paper the subscript L will refer to the neutral lepton L^0 and not to any of its charged relatives.

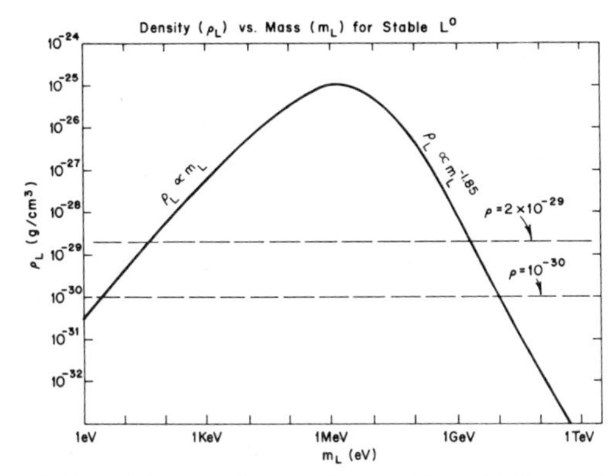

FIG. 1.—The cosmological relationship between density ρ_L and the mass of the stable L^0, m_L, making the assumptions given in the text.

The cosmological production of particles with masses this high and with concomitant freezing temperatures in the range around 1 GeV are plausible only if the multiplicity of hadron states is not exponential in the mass, as has been suggested (see, e.g., Hagedorn 1970; Huang and Weinberg 1970). If quarks are infinitely tightly bound, such a model may still be viable. In such a case the number of particles with masses larger than 2–3 GeV would be much smaller than the calculations of LW2 since the maximum temperature would then be the order of 170 MeV. Thus densities much smaller than 2×10^{-29} g cm^{-3} need not imply very high L^0 masses. Most of the subsequent discussion centers around L^0 masses of the order several GeV, but the reader must keep in mind that the uncertainties are large. It should be remembered that this mass question should eventually be resolvable with accelerator experiments. Notice that if massive neutrinos exist and if the universe reached equilibrium at temperatures $\gtrsim 10^{12}$ K, then such neutrinos *must* contribute to the mass density of the universe. Whether such a contribution is interesting depends on m_L. If m_L is very large (>10 GeV), then the contribution to the density would be small. However, for m_L between ~ 2 and ~ 10 GeV such neutrinos would be the dominant constituent of the present universe. It should also be noted that if the neutrinos cluster with the baryons in galaxies and clusters of galaxies so that the cosmological density limits from galaxies apply a more restrictive limit on ρ is possible (Gott *et al.* 1974, hereafter GGST). In such a case m_L would be $\gtrsim 10$ GeV and the neutrinos could provide the "missing mass" for galaxies and clusters but would not close the universe. If for some reason the neutrinos do not all cluster with the baryons and m_L is found to be the order of 2–5 GeV, then the neutrinos could close the universe. It will be shown that this latter possibility is unlikely (but, given the uncertainties about the physics of the early universe perhaps not impossible).

The only interactions that are important at the present epoch are large-scale gravitational ones and annihilation. For the latter we use the rate $\langle \sigma v \rangle_a = G_F{}^2 m_L{}^2 N_A / 2\pi$, with the Fermi coupling constant $G_L = 1.15 \times 10^{-5}$ GeV^{-2}, and N_A set equal to 14. Numerically, $\langle \sigma v \rangle_a \approx 1.4 \times 10^{-26} (m_L/2\ \text{GeV})^2$ cm^3 s^{-1}. In the simple picture presented in LW2, 9/14 of the decays go into hadron channels. We will use a picture in which half the decay energy emerges in photons, which for definiteness we will assume come out in an energy band E_0 centered at some fraction f of the mass m_L ($E_0 = f m_L c^2$) and of about that width. If roughly half the energy comes out in photons, this parametrization gives the average number of photons per annihilation, g_γ, as $\sim 1/2f$. Note also that a stable heavy neutral lepton (L^0) will interact with other matter through simple weak exchange of a neutral vector boson. Thus the $L^0 + p \rightarrow L^0 + p$, low-energy cross section is the order of a typical weak cross section. Then, for a "typical" galaxy of radius ~ 20 kpc and average density of roughly 1 H atom per cm^3, the collision probability is negligibly small—galaxies are transparent to such neutral leptons.

The bulk of the paper is arranged more or less chronologically in the cosmological setting. We consider the thermal history of the early universe, big bang nucleosynthesis, and the development of density perturbations and their growth. After the primeval hydrogen plasma combines, the density perturbations in the L^0 fluid and in the baryon fluid, which need not be correlated, develop into galaxies and clusters, and the extent to which the different species can segregate is investigated and shown to be probably very small. The effects of annihilation in the universe as presently seen are investigated, and we show that critical tests of the theory can be formulated if the "missing mass" is indeed heavy neutrinos with roughly the adopted annihilation rate and roughly the adopted decay scheme. For completeness we also show that other mechanisms, such as spin-flip radiation and Compton scattering, are far too weak to yield detection possibilities. A brief discussion is also given on the unlikely possibility of forming stellar size aggregates of heavy neutrinos. The paper concludes with a description of restrictions which might be placed on neutrino lifetimes if the symmetry were broken and the massive neutrinos were not absolutely stable.

II. THE EARLY UNIVERSE: THERMAL HISTORY

Since at temperatures much above the rest mass of L^0, the L^0's are simply one of a very large number of relativistic species present, the effects on the total mass density of the universe is very small. One can reasonably ask whether once the rest mass of the L^0 is important, the L^0 might dominate the mass density. It is clear from inspection of the expression for the mass density of such particles in equilibrium,

$$\rho = \text{const. } T^4 \int u^2 du \left[u^2 + \frac{m^2c^4}{K^2T^2}\right]^{1/2} \left[\exp\left(u^2 + \frac{m^2c^4}{K^2T^2}\right)^{1/2} + 1\right]^{-1} \tag{2}$$

that though ρ is in fact an increasing function of mc^2/kT, it never exceeds the value it has for $m = 0$ by very much and, of course, falls exponentially for large m. Since, again, there are many low-mass species present at $kT \sim m_L c^2$, the L^0, $\bar{L}^0$ mass density is always a small fraction of the total until, long after freezing, the radiation density drops once and for all below the cold rest-mass density of the residual L^0's. Stronger arguments can be made for the pressure: the pressure of the L^0's is *never* important. Since the number density is roughly the zero-mass thermal density of states, which is proportional to T^3 at $kT \gg m_L c^2$ and falls rapidly below that value for kT less than $m_L c^2$ (see the curves plotted in LW2), the particles never constitute a degenerate fluid (that is to say, neither the L^0's nor the $\bar{L}^0$'s constitute a degenerate fluid).

In fact, since $\langle \sigma v \rangle_a$ is independent of velocity, the distribution function would not change shape as a result of annihilation if there were not ongoing thermalization. We shall see, however, that the heavy neutrinos are in effective thermal contact with the rest of the universe until the temperature is somewhat lower than the freezing temperature, at which the L^0's are very nonrelativistic. The result is that a roughly Maxwellian distribution of velocities is set up in what for all practical purposes is a completely classical fluid of noninteracting particles.

We can estimate how long the L^0's are in thermal contact with the rest of the universe as follows. The momentum exchange in a collision with a relativistic thermal particle with $kT \ll m_L c^2$ is

$$\frac{\Delta p}{p} \approx kT(m_L kT)^{-1/2}. \tag{3}$$

The collision rate is

$$\lambda \approx n\sigma_t c\,,$$

where σ_t is the total collision cross section and n is the density of the light thermal particles (the only heavy particles besides the L^0's themselves at the relevant epoch are nucleons, whose number densities are down by about 10^8 compared to everything else [cf. Weinberg 1972*c*], as are the number densities of the L^0's). The thermalization time is the time required for the momentum to change of order itself by random encounters, or

$$\lambda \Delta t \left(\frac{\Delta p}{p}\right)^2 \sim 1\,, \tag{4}$$

which yields

$$t_{\text{th}} \approx \frac{m_L c}{NaT^4}\,, \tag{5}$$

where N is the number of species of relevant thermal particles. As mentioned above, the interaction depends on the simple exchange of vector boson. Thus the cross section for weakly interacting fermions with $mc^2 < kT < m_L c^2$ is

$$\sigma \sim G_F{}^2(kT)^2 \sim 3.8 \times 10^{-44}\left(\frac{T}{10^{10}\ \text{K}}\right)^2 \text{cm}^2\,. \tag{6}$$

The cross sections for interaction with photons is much smaller (see § VII*c*) since photons do not couple directly to the neutral L^0 but must couple through a charged virtual intermediary, either a lepton or a charged boson. Thus interactions with photons are not important. The relevant interactions at $T \lesssim 10^{12}$ K are with electrons and the various massless neutrinos and their antiparticles; N is about 3. Thus

$$t_{\text{th}} \sim 4 \times 10^3 \left(\frac{10^{10}\ \text{K}}{T}\right)^6 \left(\frac{m_L}{2\ \text{GeV}}\right) \text{seconds}\,. \tag{7}$$

The expansion rate is $\sim 1/2t$ for a radiation-dominated universe, where t is the age, so

$$t_{\text{exp}} \sim 2t \sim 2.2(10^{10}\ \text{K}/T)\ \text{seconds} \tag{8}$$

for $5 \times 10^9 \lesssim T \lesssim 10^{12}$ K. The L^0's are in thermal kinetic equilibrium as long as $t_{\rm th} \lesssim t_{\rm exp}$, or for

$$T \gtrsim 10^{11}(m_L/2 \text{ GeV})^{1/4} \text{ K}, \tag{9}$$

which is somewhat lower than the freezing temperature. At later times, the L^0's "cool" with the expansion (they do not interact), but the distribution function remains Maxwellian with a "temperature"

$$T_L \approx \left(\frac{T_\nu^2}{1.4 \times 10^{11} \text{ K}}\right) \text{K}, \tag{10}$$

where T_ν is the neutrino temperature, currently $(4/11)^{1/3}$ times the blackbody photon temperature, or about 2 K. The L^0's cool faster than radiation because they are nonrelativistic. We shall see that this temperature is meaningful only up until about the epoch of decoupling ($T_{\rm ph} \sim 4000$ K, $1 + z \sim 1500$), at which time it has fallen to about 5×10^{-5} K. Thus the L^0 fluid in the early universe becomes very cold.

III. BIG BANG NUCLEOSYNTHESIS

It has been shown that big bang nucleosynthesis places severe constraints on the baryon density of the universe (cf. review by Schramm and Wagoner 1977). In particular the amount of deuterium produced in big bang nucleosynthesis is very sensitive to the baryon density, with present indications being that the baryon density is more than an order of magnitude below the critical density necessary to close the universe. In addition, the ^{4}He yield, while weakly dependent on the baryon density, is very sensitive to the competition between the cosmological expansion rate and the weak interaction rate for the neutron decay processes $n + e^+ \leftrightarrow p + \bar{\nu}_e$ and $p + e^- \leftrightarrow n + \nu_e$. This latter fact was utilized by Steigman, Schramm, and Gunn (1977) in showing that the helium abundance limits the number of types of massless particles. ("Massless" here means $\ll 100$ keV.) This limit occurred because during and preceding big bang nucleosynthesis the density ρ of the universe is radiation dominated with $\rho \propto T^4$, the proportionality constant being dependent on the number of relativistic particle types. Therefore each new massless or low-mass ($\lesssim 1$ MeV) particle type would add to the density ρ, which would increase the expansion rate and thus increase the ^{4}He yield.

For massive neutrinos, the above scenario no longer occurs. At the temperatures relevant for big bang nucleosynthesis, 10^{11} to 10^9 K, the massive neutrinos do not significantly contribute to the density of the universe. Thus they have no effect on the expansion rate and cause *no significant effect* in big bang nucleosynthesis. Thus big bang nucleosynthesis proceeds in the same manner whether or not the massive neutrinos exist; and, if deuterium is cosmological in origin, its abundance still specifies the *baryon* density.

What *is* important is that the relation between the *total* matter density and the nucleosynthesis can be altered, since the L^0's can contribute to the mass density at present. The relations between the current baryon density and the nucleosynthesis are unaltered. Thus the present abundances of helium and deuterium seem to require a low baryon density now, of order 5×10^{-31} g cm^{-3} *or less* (see, for example, GGST for review and discussion). This requires a density parameter $\Omega \sim 0.08$ ($\Omega = \rho/\rho_c$, where ρ_c is the critical density of the universe), if all the mass is in baryons; but if a substantial density is in L^0's, Ω can be much larger, *argued from the deuterium and helium alone*; indeed, this argument cannot preclude closure ($\Omega > 1$), with most of the matter in the universe in L^0's. We will later fall back on dynamical arguments to show that, even with L^0's, Ω is probably less than 1, but it is important to note that, if they exist, a prime argument for a low-density universe no longer holds.

With regard to nucleosynthesis, it should be noted that if the L^0's provide the bulk of the dynamical mass density ($\Omega \sim 0.08$), then the baryon density is even less than 5×10^{-31} g cm^{-3}. This would imply that much more deuterium is synthesized in the big bang than is currently observed and consequently more astration is required to reach its observed value.

IV. DENSITY PERTURBATIONS IN THE EARLY UNIVERSE

Once the annihilations become rare, the L^0's only interact via gravity. It is very difficult for the L^0's to transfer their internal energy to the ordinary matter, and they will tend to get left behind in any collapse. For an interaction with a constant gravitational potential, it is clear that an L^0 which is initially not bound to gravitational system will remain unbound. It will pass a mass point in some hyperbolic orbit, thus creating a density enhancement $\rho^*/\rho \propto (r/r^*)^{3/2}$ near the mass, but each neutrino would have no effective way of dissipating energy to go from an unbound to a bound orbit. However, in a time-varying potential such as that produced in collapse, energy is not conserved and it is possible to capture initially unbound L^0's and, for initially bound L^0's, to partially follow the collapse.

In addition to the effect of the time-varying potential, it may be possible for a density perturbation of L^0's to be sufficiently large so as to be bound. It is known that in order to form galaxies in the universe, bound perturbations seem to be required. It is known that even for matter which only interacts gravitationally, there can be some collapse (the order of a factor 2) due to the time-changing effect of the collective gravitational potential. This type of collapse has come to be known as a violent relaxation (Lynden-Bell 1967). Regardless of whether one has a self-made condensate or a primordial density fluctuation, it can be seen that without some way to dissipate energy, the

L^0's will always be less compact than the collapsing normal matter and will tend to form an extended halo. Although the L^0's will not gravitationally collapse as much as ordinary matter, they can condense sufficiently that annihilation might again become important.

Since the existence of galaxies seems to imply the existence of primordial density perturbations, one question we need to address at the outset is whether small-scale but large-amplitude density perturbations can significantly alter the conclusions of LW2 about the annihilation and survival of the L^0 fluid. The L^0's are thermally produced, and the universe is thermal-radiation dominated in the era of their production. Therefore, temperature perturbations are not physical, since they merely represent a spatial variation of the zero of time (Rees, private communication). What may be important are curvature fluctuations. Imagine a small portion of the universe which locally has positive curvature, so that if it were big enough in the course of time it would collapse. In a matter-dominated universe, if the region has not begun its recollapse before the horizon grows to include neighboring regions of negative curvature (and hence lower density), the radiation can flow out, saving the perturbations from collapse but leaving behind any noninteracting nonrelativistic matter (the L^0's) with a large local density excess. Such perturbations do not grow so long as the universe is radiation-dominated, but they can have interesting effects if their magnitude is large.

First, it is clear that their relative amplitude cannot be much greater than unity, since a positive curvature region at turnaround has a density only a few times the mean (critical) density. It is indeed doubtful that they could be near unity, since if there are fluctuations in the amplitude of the curvature fluctuations, many would be large enough to result in an early collapse, with the perturbation being sufficiently large that the annihilation radiation could not escape. Such perturbations would result in the production of large numbers of black holes of mass

$$M \sim \rho(ct)^3 \sim \left(\frac{30aT^4}{c^2}\right)\left[\left(\frac{3}{32\pi G}\right)^{1/2}\left(\frac{30aT^4}{c^2}\right)\right]^3 \sim 6 \times 10^{30}\left(\frac{kT}{1\text{ GeV}}\right)^{-2} \text{ g} \tag{11}$$

for $T \gtrsim 0.5$ GeV$/k \sim 5 \times 10^{12}$ K, where the factor 30 in the density is the effective number of species present (see LW2; it is assumed again that there is in fact no maximum temperature). Thus a large fraction of the mass-energy of the universe would be dumped into black holes of masses large enough that they do not evaporate by the Hawking (1973) mechanism. They will accrete, but that is really irrelevant. What is important is that very soon the universe becomes matter-dominated, and the present ratio of matter density to radiation density is

$$\left(\frac{\rho_m}{\rho_{\text{rad}}}\right)_0 \gtrsim \frac{T_{\text{BH}}}{T_0} \approx 3 \times 10^{12}\left(\frac{kT_{\text{BH}}}{1\text{ GeV}}\right), \tag{12}$$

where T_{BH} is the temperature at black-hole formation. The ratio has an upper limit observationally of about 10^4.

Thus if the curvature fluctuations lead to collapse, they must do so in a vanishingly small fraction of the dense regions, and we can conclude that the rms amplitude is very small. This means that regions are not formed in which the annihilation proceeds much further than in other regions.

Let us now see what happens to perturbations in the L^0 density. In particular we need to examine the question of whether or not L^0 diffusion will damp out perturbations. The question comes down to how far the L^0's might travel in the time scale of perturbation growth. We will see that because the thermal velocities of the L^0's are low, they do not travel far and so damp out only small-scale fluctuations.

In the era of radiation dominance $R \propto t^{1/2}$, and one obtains

$$\Delta u = \frac{v_0 t_0}{R_0} \ln \frac{t}{t_0} \tag{13}$$

for the radial coordinate traversed at time t for an initial velocity v_0. As shown above, at thermal decoupling of the L^0's from the rest of the universe, the temperature was of order 10^{11} K$(m_L/2\text{ GeV})^{1/4}$. Thus

$$\left.\frac{v}{c}\right|_0 \approx \left(\frac{kT_0}{mc^2}\right)^{1/2} \sim 2.7 \times 10^{-2}\left(\frac{m_L}{2\text{ GeV}}\right)^{-1/4}, \tag{14}$$

$$t_0 \approx 1.1\left(\frac{T_0}{10^{10}\text{ K}}\right)^{-2} \sim 2.7 \times 10^{-3}\left(\frac{m_L}{2\text{ GeV}}\right)^{-1/2} \text{ seconds}. \tag{15}$$

At t_{eq}, the time at which the radiation density and mass densities are equal, we shall see that we get rapid growth of fluctuations. That epoch occurs at a redshift

$$(1 + z)_{\text{eq}} \approx \left.\frac{\rho_m}{\rho_{\text{rad}}}\right|_{\text{present}} = 10^4\Omega\left(\frac{H_0}{60\text{ km s}^{-1}\text{ Mpc}^{-1}}\right)^2, \tag{16}$$

or a temperature (for $T_0 = 2.9$ K, and a radiation field at present consisting of photons, ν_e, ν_μ, ν_τ, and their antiparticles)

$$T_{eq} = (1 + z)_{eq}T_0 3 \times 10^4\Omega(H_0/60)^2 \text{ K} \tag{17}$$

at an age

$$t_{eq} \approx \left(\frac{3}{32\pi G\rho_{eq}}\right)^{1/2} \approx 2 \times 10^{11}\Omega^{-2}\left(\frac{60}{H_0}\right)^4 \text{ seconds}. \tag{18}$$

Thus free diffusion occurs for a time t_{eq}, and

$$r_0 \approx R_0\Delta u \sim c\frac{v}{c}\bigg|_0 t_0 \ln\frac{t_{eq}}{t_0} \sim 7 \times 10^7 \text{ cm}\left(\frac{m_L}{2 \text{ GeV}}\right)^{-3/4}. \tag{19}$$

The corresponding mass (assuming the mass *now* is dominated by L^0's) is

$$m_{diff} \approx 4\rho_L|_{t_0} r_0^3 \approx 2.4 \times 10^{28}\Omega(H_0/60)^2 \text{ g}, \tag{20}$$

a small fraction of a solar mass. Therefore, perturbations of this size or larger will survive without diffusing away.

Let us assume there are primordial perturbations in the L^0 density of the type that seem to be necessary in the baryon density in order to form galaxies. We will now examine how these L^0 perturbations evolve and develop. Let us look at perturbations in which the radiation density is uniform, but the L^0 density is not: we shall see that this is the relevant picture for a large range of perturbations sizes (all but those corresponding to the very largest observed structures). Let $\rho_L = \rho_L(1 + \delta)$, with $\delta \ll 1$. The peculiar velocity of the L^0 fluid responds only to the gravitational potential induced by fluctuations in the L^0 fluid itself, since we assume the radiation to be uniform. The perturbation equations can be developed in the standard way (see, e.g., Weinberg 1972*c*, p. 578, for a treatment of this problem, and Mészáros 1974 for perturbations of this type in radiation-dominated universes) and yield

$$\ddot{\delta} + 2H\dot{\delta} = 4\pi G\bar{\rho}_L\delta, \tag{21}$$

where $H = \dot{R}(t)/R(t)$. Let $R_{eq} = R(t_{eq})$, and let

$$\tau_{eq}^2 = \left(\frac{3}{8\pi G\rho_{rad}}\right)_{eq} = \left(\frac{3}{8\pi G\rho_m}\right)_{eq}. \tag{22}$$

The quantity τ_{eq} is of the same order as t_{eq}. The field equations yield the equation for R,

$$\dot{R}^2 = \frac{8\pi GR^2}{3}(\rho_m + \rho_r) = \frac{R_{eq}^2}{\tau_{eq}^2}\xi^{-1} + \xi^{-2}, \tag{23}$$

where $\xi = R/R_{eq}$. After some manipulation we can cast equation (21) in the form

$$\frac{d^2\delta}{dt^2}(\xi + \xi^2) + \frac{d}{dt}(1 + 3\xi/2) = 3\delta/2. \tag{24}$$

One solution of this equation is $\delta = \frac{2}{3} + \xi$. The other solution (for a matter-dominated universe) is proportional to $\xi^{-3/2}$. This other solution is in fact monotonically decreasing in amplitude, as may be seen by the following simple argument. At large ξ, δ may be taken to be positive and has a negative first derivative and a positive second derivative. If for smaller ξ it is to have a positive first derivative, there must be a ξ such that $d\delta/d\xi = 0$. Let ξ_t be the largest such ξ. Clearly δ is positive there, since $d\delta/d\xi < 0$ for $\xi > \xi_t$. Clearly also $d^2\delta/d\xi^2$ must be *negative* there, since $d\delta/d\xi = 0$ for $\xi = \xi_t$. But the coefficients in equation (24) are all positive, so no such ξ_t can exist. For small ξ, the other solution goes like $\ln \xi^{-1}$, and must fit smoothly and monotonically on to const. $\times\ \xi^{-3/2}$. Thus the two modes are monotonically decaying and growing, respectively, but neither does very much for $\xi < 1$. Physically the reason is that the time scale for expansion is set by the radiation density; and since the matter responds only to matter fluctuations, there is simply not time for the perturbations to grow. All this development assumes the absence of random particle motions which affect only perturbations of order m_{diff} or smaller, so the approximation is good for all perturbations of interest. Thus, perturbations in the L^0 fluid cannot grow appreciably prior to t_{eq}.

The baryon perturbations are locked to the radiation by electron-scattering viscosity until the recombination epoch, or at $T \sim 4000$ K, which may be earlier or later than t_{eq} according as whether Ω is smaller or larger than $\Omega = 0.2(60/H_0)^2$, respectively. For $t > t_{eq}$, $\delta \propto \xi \propto t^{2/3}$ for the growing mode.

The baryon perturbations can in general be decomposed into an *adiabatic* component ($\delta\rho_B$ = const. $\delta\rho_{rad}$) and an *isothermal* one ($\delta\rho_{rad} = 0$). Since the sound speed in the medium before t_{eq} is $c/\sqrt{3}$, perturbations in the radiation density begin behaving like acoustic waves almost as soon as the horizon is bigger than the length scale of the perturbation, and the adiabatic perturbations are damped by photon viscosity, a well-known result (Silk 1967). The horizon at decoupling contains about $1.5 \times 10^{18}\Omega(H_0/60)^2 M_\odot$, so adiabatic perturbations on the scale of galaxies and even great clusters of galaxies had been propagating like waves for a long time prior to decoupling. Thus on average the cold L^0 fluid saw a uniform background, which motivated the development of a model for perturbations in the L^0 fluid alone. (Of course, the *isothermal* perturbations entail no radiation perturbations.) All but the largest of the adiabatic baryon perturbations are damped by photon viscosity (Silk 1967), but the L^0 perturbations do not couple to the photons and thus survive on all scales.

The effective damping of the adiabatic baryon perturbations is changed by the L^0's only if Ω is large, so that there is a period in which the universe is dominated by the L^0 fluid before recombination. The time scale is speeded thereby, so that smaller perturbations can survive for a given Ω_B, the fraction of the critical density presently in baryons. For smaller Ω_B the "classical" results hold (cf. Weinberg 1972*c*). In any case, it seems that only isothermal perturbations on the scale of galaxies and clusters survive. Remember, however, that L^0's only damp out their own perturbations on scales smaller than $M_{\rm diff} \sim 10^{-5} M_\odot$.

It is important to note that the perturbations are static in the *comoving* frame prior to $t_{\rm eq}$; their metric sizes grow with the expansion, and their density drops like $1/T^3$. Since one cannot by curvature perturbations produce density fluctuations in the L^0 fluid of large amplitude, one does not have to worry about perturbations in which the density of L^0's and $\bar{L}^0$'s is high enough to produce appreciable annihilation. It is instructive to note that the annihilation time is given by

$$\tau_a = \frac{1}{n_L\langle\sigma v\rangle_a} \approx \frac{7.3 \times 10^{24}\ \mathrm{s}}{n_L}\left(\frac{2\ \mathrm{GeV}}{m_L}\right)^2 \tag{25}$$

and is about the present Hubble time for a density of about $n_{\rm crit} = 10^8(2/m_L)^2\ \mathrm{cm}^{-3}$. At $t_{\rm eq}$, the number density of L^0's is

$$n_{\rm eq} = 6 \times 10^5\left[\Omega\left(\frac{H_0}{60}\right)^2\right]^4\left(\frac{2}{m_L}\right)\ \mathrm{cm}^{-3}, \tag{26}$$

much smaller than the value $n_{\rm crit}$ above for $\Omega \sim 0.1$, $H_0 \sim 60$, but note the very strong dependence on H_0.

V. THE DEVELOPMENT OF STRUCTURE AND THE VALUE OF Ω

In this section we will investigate the growth of perturbations to form the structure we see in the universe today. We will see that regardless of the initial correlation between L^0 and baryon perturbations, in the end they become correlated, and thus the L^0's will be associated with galaxies. To examine the development of structure, let us go to the era, $t_{\rm eq}$, when the radiation and matter densities are equal.

In the period between $t_{\rm eq}$ and the recombination time $t_{\rm rec}$ [recall that this period exists only for $\Omega \gtrsim 0.2(60/H)^2$], the perturbations in the L^0 density grow freely, and their amplitude increases by a factor $(1 + z)_{\rm eq}/(1 + z)_{\rm rec} \approx 5\Omega(H/60)^2$ (recall that the pressure effects are completely negligible, the velocity dispersion in the L^0's corresponding to a temperature of order 10^{-4} K at these epochs). Thus for large Ω, if there is any scale on which $\delta\rho_L/\rho_L$ is of order unity, bound structures will form with densities of order $\rho_{L,\rm eq}$. If these survive, they will be copious sources of annihilation radiation, but we will argue that they do not. There is some question, in fact, about whether they would exist—we have argued that it is difficult to make perturbations with large $\delta\rho_L/\rho_L$, and an attractive scheme has been forwarded in which all the density perturbations are in the *baryons* (Gott and Rees 1975) and arise associated with the same sort of curvature fluctuations we considered for the origin of the L^0 perturbations, with the amplitudes $\sim 10^{-4}$ when each new scale enters the horizon. In this picture the L^0 density perturbations on interesting scales would be very small, and nothing would happen until recombination.

The situation is sufficiently uncertain that one would like to find a general argument that indicates how the L^0 and baryon perturbations develop, given the distinct possibility that they are completely uncorrelated at recombination, and with (possibly also uncorrelated) velocity perturbations which are of the same order as the density perturbations.

We do know that the perturbation amplitudes are small for scales like galactic masses and larger; the density amplitude is related to the collapse time for a bound perturbation by (Gunn and Gott 1972)

$$t_c = \frac{\pi}{H_{\rm init}}(\delta_+)^{-3/2}, \qquad H_{\rm init} \sim H_0\Omega^{1/2}(1 + z_{\rm init})^{3/2}, \tag{27}$$

where $H_{\rm init}$ is the value of $\dot{R}/R$ at $t_{\rm init}$, the larger of $t_{\rm eq}$ and $t_{\rm rec}$, and δ_+ is the relative density excess then above the critical density $\rho_c = 3H_{\rm init}^2/8\pi G$. The average density at these epochs is nearly the critical density;

$$\delta_e = \frac{\rho - \rho_c}{\rho_c} = \frac{\Omega - 1}{\Omega(1 + z_{\rm init})}\,. \tag{28}$$

For a cluster like Coma [$t_c \sim 6 \times 10^9(60/H_0)$ yr], $\delta_+ \sim 5 \times 10^{-3}\Omega^{-1/2}$ while $\delta_e \sim 10^{-3}(\Omega - 1)/\Omega$. Thus even for $\Omega = 0.1$, the cluster excess above critical is twice as large as the excess of the critical density above the mean density. For a galaxy with a collapse time for 10^9 years, the perturbation is $\delta^+ \sim 1.8 \times 10^{-2}\Omega^{-1/2}$, about 6% for a low-density ($\Omega = 0.1$) model.

Thus initially the baryons and L^0's can each be considered to fair accuracy a uniform fluid, and the question is whether by virtue of their initial conditions they can separate.

For any reasonable value of $\Omega(>0.02)$ now, $\Omega_{\text{init}} = 1 + \delta_e$ is near unity, so $R \propto t^{2/3}$ during the initial development of the perturbation. Let us suppose that the density is dominated by L^0's; this assumption is really irrelevant but simplifies the discussion. Suppose we have a more or less spherical perturbation, which for the sake of definiteness has $\delta_+ > 0$, so that it is bound, and suppose the baryons are moving radially in the perturbation with a velocity different from the L^0's by an amount Δv_0 initially at an initial comoving radial coordinate u. We assume for definiteness that the perturbation has constant density inside some boundary r_m, so that it expands uniformly; clearly $r_m \propto t^{2/3}$ initially also, if the initial amplitude of the perturbation is small. Then the perturbation behaves like a small piece of a somewhat denser universe, and we can apply the same argument as we made to determine the diffusion size of the L^0 perturbations in the very early universe. Scale the coordinate u such that $u = r/r_m$ initially; thus $u = 1$ corresponds to the edge of the perturbation. Again,

$$r_m \dot{u} = \Delta v_0 (r_m)_0 / r_m \, . \tag{29}$$

This integrates to yield

$$u = u(t) - H_0 = 3\frac{v_0 t_0}{(r_m)_0}\left[1 - \left(\frac{t_0}{t}\right)^{1/2}\right] \tag{30}$$

But $v_0 = (\dot{r}_m)_0 u_0 = [2(r_m)_0/3t_0]u_0$, so

$$\frac{\Delta u}{u} = 2\frac{\Delta v_0}{v_0}\left[1 - \left(\frac{t_0}{t}\right)^{1.2}\right], \tag{31}$$

and a velocity perturbation of a few percent results in a *total* radial separation of only a few percent, and converges to a final value very rapidly. The above situation can be generalized without difficulty to the exact motion for a bound perturbation, with similar results. A crucial (well known) point is that gravitational perturbations never get much smaller than their maximum size. Coming to equilibrium merely involves generating random kinetic energy comparable to the gravitational energy, so the final gravitational energy is twice the total energy, whereas at maximum expansion it *is* the total energy. Thus, crudely, $r_{\text{final}} = \frac{1}{2}r_{\text{max}}$. Hence we conclude that the baryons and L^0's cannot separate by gravitational processes alone, and any structure which collapsed as a result of gravitational instability must contain matter which has a ratio of baryons to L^0's which is representative of the cosmic average ratio, provided it developed from small perturbations. The subsequent development of the perturbation is largely through the violent relaxation process described by Lynden-Bell (1967) which does not segregate by mass.

They *can* separate subsequently by virtue of the fact that if gas pressure is ever important and the densities are such that the gas (*baryon* gas, or course) can cool by radiative processes, the baryons can sink deeper into the potential well. Thus the L^0's can be left behind in more extended structures than *galaxies*, for which these conditions are satisfied, but cannot in *clusters* of galaxies. This implies that the L^0 to baryon ratio in clusters and perhaps in extended elliptical galaxies will have the cosmic ratio, but the disks of spirals will not and stars will have essentially no L^0's.

Galaxies in groups and clusters will experience dynamical friction against the L^0 background in these structures, but the effects are the same as they are for stars or gas, since the friction does not depend on the mass of the particles in the background fluid if these masses are much smaller than the galaxy mass. Some aspects of violent relaxation and dynamical friction have also been considered by Faulkner, Sarazin, and Steigman (1978) with regard to L^0's.

Gunn (1977) has noted that the mass-to-light ratio of structures large enough not to have suffered the dissipative separation discussed above have a mass-to-light ratio of about $120(H_0/60)$ in solar units (Zwicky photographic magnitude system) normalized to the stellar population of spiral galaxies. The luminosity density with the same normalization is $6.3 \times 10^7 (H_0/60) L_\odot$ Mpc^{-3} (Gott and Turner 1976), leading to a mass density of $7.5 \times 10^9 (H_0/60)^2 M_\odot$ Mpc^{-3}, $\rho = 5.1 \times 10^{-31}(H_0/60)^2$ g cm^{-3}, or an H_0-independent value of Ω of 0.08. The mass-to-light ratios of stellar material seem to only be in the range of 5–10, and this mass can be readily accounted for. Therefore, between 90 and 95% of the matter, even for this small value of Ω, is "missing." This mass, if it is assumed to be nondissipative and otherwise inert, was shown to form the roughly $1/r^2$ halos required for galaxies as the galaxy perturbation developed; these halos would be tidally stripped in the great clusters, and most of the halo mass would move with random velocities of the same order as the galaxy velocities in the cluster potential. Heavy noninteracting neutral particles present since the early universe, constrained by the arguments above to cluster with the matter, could not be better as stuff to constitute the dynamical missing mass.

Admittedly there are some uncertainties in the exact determination of limits on Ω from the dynamical mass associated with galaxies. Davis and Peebles (1977), using the BBGKY formalism, obtain a value for Ω closer to unity, although Aarseth, Gott, and Turner (1977) argue from n-body calculations that the observed clustering is also compatible with $\Omega \sim 0.1$. It is probably reasonable to say that Ω from the dynamical mass associated with galaxies is uncertain by about a factor of ~ 3, but seems to be below its critical value of unity. Therefore, if L^0's do provide the "missing" mass and cluster with galaxies, then the *L^0's would not close the universe.* The only way around this would be if, for some as yet unknown reason, the bulk of the L^0's were able to avoid the perturbations which produced the galaxies and clusters and remain unbound. It should be noted that if the L^0's do provide the dynamical missing mass and $\Omega \sim 0.1$, then $m_L \approx 10$ GeV rather than the 2 GeV lower limit mentioned before. Assuming our model for the annihilation cross section and that no Hagedorn (1970) limiting temperature occurs, then the critical test for L^0's being the missing mass is for m_L to be determined to be ~ 10 GeV. In fact, 10 GeV *may be a more realistic lower limit to m_L than* 2 GeV; however, there are sufficiently large uncertainties that for safety we will continue to write our expressions using the conservative 2 GeV normalization.

One other comment should be made before we leave the subject of perturbations. Baryon perturbations smaller than $10^6\ M_\odot$ or so, the Jeans mass at decoupling, cannot collapse because of ordinary gas pressure; the L^0 fluid suffers no such restriction, because of its much lower temperature. If there are large-amplitude L^0 perturbations on this scale (or smaller), they will collapse. It can be argued (Fall and Rees 1977)—and n-body calculations and observations of galaxies and clusters support the arguments—that very little such small-scale structure persists as larger aggregates form. Galaxies themselves are the interesting exception to this rule of large-scale structure destroying smaller-scale structure, and it seems almost certain that dissipative processes play a vital role in their formation. Thus we expect very few of these high-density, low-mass structures to survive; but if any do, they will be very interesting objects indeed, as we will discuss later.

VI. THE RESIDUAL ANNIHILATION RADIATION AND γ-RAY OBSERVATIONS

The observed γ-ray background has an isotropic component whose spectral flux may be approximated by a power law with (Fichtel, Simpson, and Thompson 1977)

$$\frac{d\mathcal{F}_\gamma}{dE} \approx 8.3 \times 10^{-2} E_{\rm mev}{}^{-2.85} \text{ photons (cm}^2 \text{ s sr MeV)}^{-1}. \tag{32a}$$

The corresponding integral flux is

$$\mathcal{F}_\gamma(E) \approx 0.045 E_{\rm mev}{}^{-1.85} \text{ photons (cm}^2 \text{ s sr)}^{-1}. \tag{32b}$$

We will investigate the extent to which residual L^0–$\bar{L}^0$ annihilation might account for that background; we will also consider whether discrete sources of annihilation radiation (e.g., galactic halos) can be seen against the background.

The following calculations are intended to provide estimates; in keeping with the spirit of these estimates, several simplifying assumptions will be made. For each L^0–$\bar{L}^0$ annihilation, we assume that g_γ γ-rays are produced at energy E_0; in analogy with p-$\bar{p}$ annihilation we assume that $2g_\gamma E_0 \approx m_L$. It is expected that the average photon would have energies like those of the bulk of a hadron shower ($E_0 \sim 200$ MeV), which implies for $M_L \sim Z$ GeV, $g_\gamma \approx (2f)^{-1} \approx 5$. In principle this point could eventually be checked experimentally. The annihilation radiation emissivity then is

$$\delta(E) = g_\gamma(\sigma v)_a n_L{}^2 \delta(E - E_0)[\text{photons (cm}^3 \text{ s MeV)}^{-1}]\,. \tag{33}$$

Consider the cosmological contribution to the background. The number density at any time in the past (corresponding to a redshift z) is

$$n_L = n_{\bar{L}} \approx 2 \times 10^{-5} M_L{}^{-3}(1 + z)^3 \text{ cm}^{-3}. \tag{34}$$

In equation (34) and subsequently, M_L is in GeV. For $(\sigma v)_a \approx 2.5 \times 10^{-28} N_A M_L{}^2$ (cm^3 s^{-1}), we obtain

$$j(E, z) = j_0(E)(1 + z)^6, \tag{35a}$$

where

$$j_0(E) \approx 10^{-37} g_\gamma N_A M_L{}^{-4}(E - E_0)[\text{photons (cm}^3 \text{ MeV)}^{-1}]\,. \tag{35b}$$

The present γ-ray flux at energy E is due to photons emitted in the past (at redshifts z) at energies $E(1 + z)$. To convert the appropriate time integral to an integral over redshift, we use (for a matter-dominated cosmology),

$$dt/dz = H_0{}^{-1}(1 + z)^{-2}(1 + \Omega_0 z)^{-1/2}. \tag{36}$$

It is convenient to replace the redshift integral by an integral over energies; this is accomplished by noting that $E' = E(1 + z)$. In this manner we obtain

$$\frac{d\mathcal{F}_\gamma}{dE} = \left(\frac{1}{4\pi}\right)\left(\frac{C}{H_0}\right)\left(\frac{1}{E^3}\right) \int_E^{E_m} j_0(E')E'^2\left[(1 - \Omega_0) + \frac{\Omega_0 E'}{E}\right]^{-1/2} dE'\,. \tag{37}$$

In (37), $E_M = E(1 + z_M)$; z_M is the redshift at which the universe is opaque to γ-rays. Substituting (35) in (37), we easily obtain for $E > E_0 > E_M$,

$$\frac{d\mathscr{F}_\gamma}{dE} \approx 1.2 \times 10^{-10}\left(\frac{60}{H_0}\right) g_\gamma \frac{N_A}{M_L^4}\frac{E_0^2}{E^3}\left[(1 - \Omega_0) + \Omega_0 \frac{E_0}{E}\right]^{-1/2} \text{ photons (cm}^2\text{ s sr MeV)}^{-1}. \quad (38)$$

For any production spectrum which is reasonably peaked around some energy E_0, equation (38) should provide a satisfactory order-of-magnitude estimate. For $E \gtrsim \Omega_0 E_0$, the integral flux we predict is

$$\mathscr{F}_\gamma(E) \approx 6 \times 10^{-11}\left(\frac{60}{H_0}\right) g_\gamma N_A M_L^{-4}(E_0/E)^2 \text{ photons (cm}^2\text{ s sr)}^{-1}. \quad (39)$$

Note that the spectral index of the predicted integral flux (~ 2) is in excellent agreement with that of the observed flux (1.85[+0.50, −0.35], Fichtel, Simpson, and Thompson 1977). (It is interesting to note that Stecker 1977 also showed that annihilation radiation could fit the γ-ray background; however, Stecker assumed baryon annihilation which has other problems [cf. Steigman 1976].) What of the magnitude of the predicted flux? For $H_0 = 60$, $N_A = 14$, E in MeV, and recalling that $2g_\gamma E_0 \approx M_L$ (with M_L in GeV), we obtain

$$\mathscr{F}_\gamma(E) \approx 2.2 \times 10^{-4}(g_\gamma M_L^2)^{-1} E_{\text{MeV}}^{-2} \text{ photons (cm}^2\text{ s sr)}^{-1}. \quad (40)$$

For $E_{\text{MeV}} \approx 100$, the observed flux is $F_\gamma^{\text{obs}} \approx 9 \times 10^{-6}$, whereas for $M_L \gtrsim 2$ the predicted flux is $F_\gamma^{\text{pred}} \approx 5.4 \times 10^{-9}\ g^{-1}$. Thus, γ-ray production in cosmological annihilation fails to account for the observed background by roughly three orders of magnitude.

We have seen that the average density of uniformly distributed L^0's is too low to contribute significantly to the observed, isotropic component of the γ-ray background. If L^0's form halos around galaxies, the density would be greatly enhanced. Let us now consider whether such discrete sources could be seen against the isotropic background. Consider the Galaxy itself. If the Galaxy is typical of other big spirals, it has a circular velocity at large radii of about 200 km s^{-1}. For a heavy halo of L^0's the corresponding density distribution is (Gunn 1977)

$$n_L = n_{\bar{L}} \approx v_c^2[8\pi G M_L r^2]^{-1}. \quad (41)$$

Since the central regions of galaxies are *not* dominated by high mass-to-light ratio stuff, this distribution does not extend to the center; the distribution must turn over at some core radius, a, which must be of the order of 10 kpc. We thus adopt for the L^0 distribution in galactic halos,

$$n_L = n_{\bar{L}} \approx n_0 a^2(a^2 + r^2)^{-1}, \quad (42a)$$

$$n^0 = \frac{v_c^2}{8\pi G M_L a^2} \approx 0.15 v_{200}^2 M_L^{-1} a_{10}^{-2} \text{ cm}^{-3}. \quad (42b)$$

The γ-ray emissivity is given in (33) where, now, we use (42) for n_L. The predicted flux will depend on the direction of observation and on the ratio of the core radius, a, to the distance, r_0, of the Sun from the galactic center. As a first estimate of the magnitude of the flux we compute F_0 defined by

$$\mathscr{F}_0(E) \equiv \frac{1}{4\pi}\int_E^\infty dE' \int_0^R dr j(E', r), \quad (43a)$$

$$\mathscr{F}_0(E) \approx \frac{g_\gamma(\sigma v)_a n_0^2 \mathscr{A}}{4\pi}\int_0^{R/a}(1 + x^2)^{-2}dx. \quad (43b)$$

In (43), R is the maximum extent of the halo; for $R/a \gg 1$, the integral in (43b) is $\approx \pi/4$ so that

$$\mathscr{F}_0 \approx \tfrac{1}{16} \times g_\gamma(\sigma v)_a n_0^2 \mathscr{A} \approx 1.1 \times 10^{-8} N_A v_{200}^4 a_{10}^{-3} g_\gamma \text{ photons cm}^{-2}\text{ s}^{-1}\text{ sr}^{-1}. \quad (44)$$

The flux expected in various directions may be related to $\mathscr{F}_0$; for $a_{10} \approx r_0 \approx 1$ (10 kpc), $v_{200} \approx 1$(200 km s^{-1}), and $N_A \approx 14$ we obtain

$$\mathscr{F}_\gamma(\text{center}) = \left(\frac{3}{2} + \frac{1}{\pi}\right)\mathscr{F}_0 \approx 2.8 \times 10^{-7} g_\gamma, \quad (45a)$$

$$\mathscr{F}_\gamma(\text{pole}) = \frac{1}{2\sqrt{2}}\mathscr{F}_0 \approx 5.4 \times 10^{-8} g_\gamma, \quad (45b)$$

$$\mathscr{F}_\gamma(\text{anticenter}) = \left(\frac{1}{2} - \frac{1}{\pi}\right)\mathscr{F}_0 \approx 2.8 \times 10^{-8} g_\gamma. \quad (45c)$$

Unless g_γ is very large ($\gtrsim 100$, in which case E_0 will be very small), the predicted flux is not in conflict with observations (for $E \approx 100$ MeV; see eq. [32b]). However, the unique distribution on the sky of halo annihilation γ-rays, as well as their spectral distribution (flat compared to the observed isotropic background) suggest the possibility that such a component may be discovered in future γ-ray observations.

The halos of other galaxies might be discrete sources of annihilation γ-rays; the sum of such discrete sources would contribute to the overall isotropic background. To estimate these effects, first compute the luminosity of a galactic halo in annihilation γ-rays:

$$L_\gamma = g_\gamma \int_{\text{halo}} d^3r\,[(\sigma v)_a n_L{}^2(r)]\,, \tag{46a}$$

$$L_\gamma = g_\gamma[\pi^2(\sigma v)_a n_0{}^2 a^3] \approx 2.3 \times 10^{40} g_\gamma \text{ photons s}^{-1}\,. \tag{46b}$$

In (46), the usual assumptions have been adopted: $a \approx 10$ kpc, $v \approx 200$ km s^{-1}, $N_A \approx 14$ (note that these results, as well as those in [44], are independent of M_L).

In our picture, all galaxies will be γ-ray sources; for example, M31 should be comparable to the Galaxy in luminosity and should be a source with a diameter somewhat larger than 1°.5. The total flux from M31 would be

$$\mathscr{F}_\gamma(\text{M31}) \approx 3.8 \times 10^{-10} g_\gamma \text{ photons (cm}^2\text{ s)}^{-1}\,. \tag{47}$$

This is well below the upper limit (Fichtel, Simpson, and Thompson 1977) of 10^{-6}. It is, thus, not surprising that the total contribution to the isotropic background from all galaxies is small:

$$\mathscr{F}_\gamma(\text{Total}) \approx \frac{L_\gamma}{4\pi} ng\left(\frac{c}{H_0}\right) \approx 4.4 \times 10^{-9} g_\gamma \left(\frac{H_0}{60}\right)^2 \text{ photons cm}^{-2}\text{ s}^{-1}\text{ sr}^{-1}\,. \tag{48}$$

Before passing on to other topics, it is worth investigating the expected fluxes from L^0's associated with other mass aggregations. Because of their low densities, the great clusters are inefficient radiators. Coma, for example, with $V_c \approx 1200$ km s^{-1}, $a \approx 100$ kpc, will have a total luminosity only ~ 100 times that of an individual galaxy. Even though the cluster galaxies almost certainly contain only a small fraction of the cluster mass, the galaxy contribution to the γ-rays probably exceeds that from the cluster as a whole.

It is interesting to note that if $L^0\bar{L}^0$ aggregates of globular cluster size were to survive subsequent dynamical evolution, then their γ-ray luminosity would be $\sim 10^{37}$ photons s^{-1}. It is not clear what their optical counterparts might be.

Although our estimates lead to somewhat disappointing conclusions, the situation is not entirely discouraging. Our crude calculations suggest that with a better model for γ-ray production in $L^0\bar{L}^0$ annihilation and more sensitive data (with respect to distribution on the sky as well as spectral distribution) a definitive test of the hypothesis of an L^0 halo around our Galaxy might be possible. In this regard it is worth noting that Stecker (1978) has carried out some model calculations which are essentially consistent with our general estimates.

VII. OTHER ASTROPHYSICAL ASPECTS OF L^0'S

In this section several interesting but probably undetectable aspects of the L^0's will be mentioned.

a) Magnetic Moment

Although at the present time we see no direct way to take advantage of the fact, it is nevertheless interesting to note that the L^0's have a magnetic moment. Because the L^0 is neutral, there is obviously no lowest order moment $\sim e/2m_L$ (where $c = \hbar = 1$) but from several single loop Feynman diagrams one can calculate following Fujikawa, Lee, and Sanda (1972) and Lee and Shrock (1977*b*) a moment

$$\mu_{L^0} = -\frac{3}{2}\frac{G_F m_L{}^2}{\pi^2\sqrt{2}}\left(\frac{e}{2m_L}\right) \equiv -a\left(\frac{e}{2m_L}\right), \tag{49}$$

where

$$a = 5.0 \times 10^{-6}\left(\frac{m_L}{2\text{ GeV}}\right)^2.$$

This moment is accurate to lowest order in $(m_L/m_w)^2$, where m_w is the mass of the intermediate vector boson [$m_w \sim 86$ GeV in the minimal SU(3) × U(1) with favored parameter values].

b) Spin-Flip Radiation

Placed in a magnetic field of strength which extends over a volume V, magnetic moments can spin-flip, producing radiation at about angular frequency $\omega = 2u_L B$. Let the spin-flip transition time be τ. Then the amount of radiation is

$$\tau \sim 2|\mu_L| B n_L V/\tau \text{ ergs s}^{-1} \tag{50}$$

at angular frequency ω. The transition time τ can be shown to be given by ($\alpha = e^2/4\pi \sim 1/137$):

$$\tau \sim \pi(2m_L)^2\alpha^{-1}(\mu_L B)^{-3} \sim 2.6 \times 10^{19}\left(\frac{2}{m_L}\right)\left(\frac{10^{13}\text{ gauss}}{B}\right) \text{ s}\,.$$

Thus

$$I \sim 3 \times 10^{-26}\left(\frac{m_L}{2}\right)\left(\frac{B}{10^{13}\text{ gauss}}\right)^4 n_L V \text{ ergs s}^{-1}\,.$$

Clearly, even for galactic densities $n_L \sim 1$ and for magnetic fields near neutron stars or magnetic white dwarfs, the intensity is still negligible.

c) Compton Scattering

Because the L^0 can look like a virtual electron ($\dot{e}$) or heavy "electron" (E^-) and an intermediate vector boson (W or U), it is possible for it to Compton scatter with the universal blackbody photons. In a typical gauge model such as the SU(3) × U(1) model of Lee and Weinberg the cross section for such a process can be estimated roughly to be

$$\sigma \sim \frac{\alpha^2}{\pi^3} G_F{}^2 m_E{}^{-2} \frac{m_L{}^4}{m_u{}^4}\,,$$

$$\sim 10^{-47}\left(\frac{m_E}{10\text{ GeV}}\right)^2\left(\frac{m_L}{2\text{ GeV}}\right)^2\left(\frac{80\text{ GeV}}{m_U}\right)^4 \text{ cm}^2\,,$$

where E^- is a charged unstable heavy lepton and U is a gauge boson, degenerate with the W which appear in this SU(3) × U(1) model. The mass of the U is determined in this model to be the same as $m_W \sim 86$ GeV, and $m_{E^-} \sim 10$ GeV. There would also be terms with $m_E{}^{-2}$ replaced by $\sim m_e{}^2$ or $\sim m_U{}^2$, but those terms are obviously small. The above cross section can be compared with that for normal Compton scattering off electrons $\sigma_{\text{comp}} \sim \pi\alpha^2/m_e 2 \sim$ $\sim 2.5 \times 10^{-25}$ cm^2. Notice that the present L^0–photon collisional lifetime is about

$$\tau \sim (\sigma c n_{\text{BB}})^{-1} \text{ s}\,;$$

with $n_{\text{BB}} \approx 4 \times 10^2$ photons cm^{-3} we have $\tau \approx 8 \times 10^{33}$ s, which is clearly negligible compared to the age of the universe ($\lesssim 10^{18}$ s). Even for the early universe it is clear that photon scattering was not occurring on a rapid time scale unless $T \gg 10^{10}$ K, and for those temperatures we have already seen that direct weak interactions dominate.

d) L^0 Stars

We have already seen that L^0's will probably not form condensed bound systems of stellar size since we know of no mechanism to dissipate the energy. We have also seen that, even if one could form such a bound system, the $L^0\bar{L}^0$ annihilation would prevent a significant density buildup and would prevent the L^0's from becoming degenerate. However, for the sake of speculation let us presume tht some condensation mechanism took place and some $L^0\bar{L}^0$ separation mechanism took place to prevent annihilation. (It is possible, for instance, that the net L^0 number is nonzero so there is an excess of L^0's over $\bar{L}^0$ or vice versa.) In such a situation the L^0's would then be the idealized fermion of the Oppenheimer-Volkoff calculation. This would mean that degeneracy would at most support a 0.7 $M_\odot$ L^0 star with GeV mass L^0's. It is also intriguing that because of the magnetic moment such an L^0 star might have an intrinsic magnetic field.

It is also interesting that if a regular (nondegenerate) star contains a nonsymmetric L^0 and/or $\bar{L}^0$ component, then a sufficient admixture of such L^0's can change both radiative transport and luminosity estimates of stars, and also influence the nucleosynthetic stellar evolution. The points here are that part of the radiation pressure necessary to hold a star up against its gravity could then arise in part from the neutrino annihilation photons, thereby relieving the burden on the core of the star to produce, by convential nucleosynthesis, the major part of the photon budget. This, in turn, means that the core can be cooler than under conventional scenarios of element production, which in turn means that reaction rates proceed more slowly. A star may then not undergo nuclear evolution as rapidly as normally presumed. Precise estimates clearly depend on the heavy neutrino admixture assumed *a priori*. This point is also discussed by Faulkner, Sarazin, and Steigman (1977). It should be remembered that at present we have no reason to believe in such asymmetries, nor do we see a plausible way to get bound L^0's into stellar size objects.

VIII. ASTROPHYSICAL LIMITS ON POSSIBLE HEAVY NEUTRINO DECAY TIME SCALES

While the SU(3) × U(1) formalism requires that the heavy neutrino (L^0) be stable, it is possible that this symmetry group is only approximate (it could, for instance, be an embedded part of a much larger group with a small, but nonzero, coupling to the whole group). It is then possible that the heavy neutrino is not absolutely stable. Supposing it can decay, what astrophysical consequences result? These depend on its decay products. As has been pointed out by Zel'dovich (1977) and Dicus, Kolb, and Teplitz (1977), it may be possible to use the astrophysical constraints to bracket the possible lifetime and decay products; in fact, much more stringent constraints than those discussed there can be placed. There are strong arguments against present radiation dominance in the universe (GGST, Mészáros 1974), but the γ-ray background limits are, as we shall see, much more restrictive for the current density of decay radiation.

The constraints we can impose fall into several different regimes depending on the decay time, let us consider them in turn.

a) Long Decay Times ($t \gtrsim 10^5$ years)

The redshift of the dominant decay period is less than a few thousand, and the universe is transparent to the decay photons. The intensity of the decay photons is roughly

$$6 \times 10^{-9}\frac{N}{4\pi}\frac{2}{m_L}[1 - \exp(-t/t_L)] = 16N[1 - \exp(-t/E_d)]\frac{2}{m_L} \text{ photons cm}^{-2}\text{ s}^{-1}\text{ sr}^{-1}. \tag{51}$$

If t/t_L is long, i.e., if the decay redshift is large, this intensity must be compared with the observed number above $E = E_0/(1 + z_d)$, where again E_0 is the average photon energy and z_d is the redshift corresponding to $t \sim t_L$. One obtains from the γ-ray background

$$16N\left(\frac{2}{m_L}\right) \lesssim 2 \times 10^{-5}\left[\frac{E_0}{100 \text{ MeV}(1 + z)}\right]^{-1.2}. \tag{52}$$

If $NE_0 \sim \phi m_L$, we can write

$$\phi \lesssim 5 \times 10^{-9}\left(\frac{E_0}{100 \text{ MeV}}\right)^{-0.2}(1 + z)^{1.2}, \tag{53}$$

which requires that the fraction of energy ϕ going into γ's is less than about 10^{-4}, which seems unlikely. For very low masses ($m \ll 1$ MeV) the decay will not be limited by the γ-ray background but can instead be restricted, for all but extremely small masses, by other wavelength backgrounds or by arguments that there cannot be a radiation-dominated universe during the epoch of galaxy formation. Thus we conclude that the range of decay times between about 10^5 years and the present age of the universe is ruled out for essentially any but extremely small masses for the L^0. For even larger decay times we can neglect the redshift of the decay photons, and the condition becomes

$$\frac{16Nt_0}{\tau_d(2/m_L)} \leq 2 \times 10^5\left(\frac{E_0}{100 \text{ MeV}}\right)^{-1.2}. \tag{54}$$

The particles are for all intents and purposes stable as far as dynamics is concerned, and the arguments for the mass of the L^0 remain unchanged—they must be more massive than a few GeV. We can write

$$\phi \leq 5 \times 10^{-9}\frac{\tau_d}{t_0}\left(\frac{E_0}{100 \text{ MeV}}\right)^{-0.2}. \tag{55}$$

If $\phi \sim 1$, τ must be longer than $10^8 t_0 \sim 10^{18}$ years! Thus if the decay time is longer than about 10^5 years, it must be very long indeed.

b) Intermediate Decay Times (2000 s $\lesssim t_L \lesssim 10^5$ years)

In this range we are in the curious situation that the universe is very opaque to the decay photons, and they come to thermal kinetic equilibrium with the charged particles present very quickly, but the photon production time is too long for the production of a blackbody photon distribution. In this temperature range ($10^4 \leq T \leq 10^8$) the dominant photon production is through bremsstrahlung, for which

$$\frac{d}{dt}\left(\frac{n_{ph}}{T^3}\right) \approx 1.6 \times 10^{-27}k^{-1}T^{-3.5}n_{e0}^{\;2}, \tag{56}$$

where n_{e0} is the present electron density. Since $t \sim 2 \times 10^{19}(1 + z)^{-2}$ s for the same temperature range, we have

$$\frac{d}{dt}\left(\frac{n_{\rm ph}}{T^3}\right) \approx 4 \times 10^{-5}\Omega^2(1 + z)^{-1/2} \tag{57}$$

or

$$(n_{\rm ph}/T^3)|_{\rm brems} \approx 8 \times 10^{-5}\Omega^2(1 + z)^{1/2}\,. \tag{58}$$

Now the value of this ratio for blackbody radiation is about 50 ($\sim a/k$), which says that bremsstrahlung cannot seriously change the blackbody photon number for $z \leq 4 \times 10^{11}\Omega^{-4}$. This is in fact not quite true, since for $1 + z \geq 10^8$, pair production begins to increase the electron density markedly, and other more efficient processes for making photons come into play. We know, however, that the present radiation spectrum is accurately blackbody with a temperature near 2.9 K. The distortion of the microwave background due to decay is going to be affected by the temperature at decay. The longer the lifetime, the lower the temperature at decay and thus the easier it is to create a distortion. The total mass-energy density, ρ_L, in the form of such particles prior to decay will follow the mass dependence in Figure 1. Therefore higher masses, which can contribute less energy density, will be restricted more for longer decay times. This is quantitatively shown in Figure 2 where it was assumed that the Zel'dovich distortion parameter for the microwave background was less than 0.02 (Jones and Steigman 1977).

For very low masses where $am_Lc^2 < kT_*$ (T_* is the temperature at the time of decay), it should be remembered that the number of density of L's prior to decay will be comparable to the number density of photons with the normal statistical factors ($n_L/n_\gamma = 6/11$). Thus, unless the branching ratio to radiative decay is extremely small (<0.03), the effect of any decay will be to change the number density of photons substantially and thus cause an unacceptable distortion. In particular there would be too many photons for a given kT.

c) *Short Decay Times* ($t_L \lesssim$ 2000 seconds)

For decay times shorter than about 200 seconds, whatever the mass range, there are no observable consequences that the L^0 existed except a probable distortion of the blackbody neutrino background; it can be confidently stated that that will remain a secret for quite some time.

For decay times between about 200 and 2000 seconds we need to look at the possible effect on big bang nucleosynthesis. It is useful to think of nucleosynthesis occurring in two distinct stages. Although this is clearly an oversimplification, short of doing a numerical computation, it is quite accurate.

Stage I: $T \gtrsim 10^{10}$ K.—During this stage, the weak interaction is in equilibrium ($T_\nu \approx T_\gamma$). The crucial n/p ratio is primarily determined during this stage. Of importance here is the competition between the expansion rate

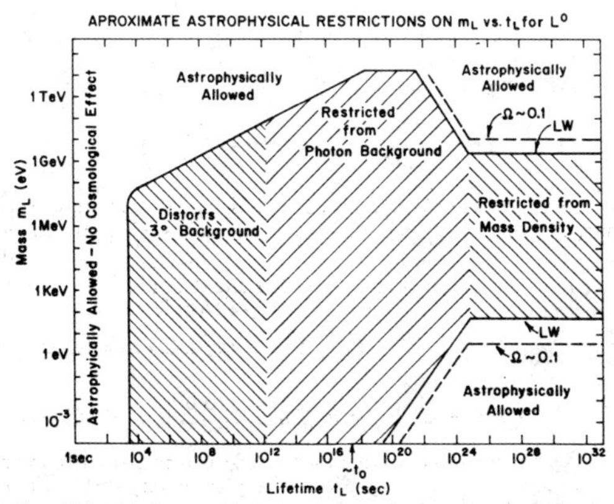

FIG. 2.—The restrictions on the relationship between the mass m_L and lifetime t_L for L^0's assuming the decay mode produces at least one photon. For long lifetimes the restrictions come from Fig. 1. For intermediate lifetimes (note $t_0 \approx$ age of the universe) the restrictions come from the various limits on the photon background (in particular the γ-ray background) if t_L is less than the radiation decoupling time ($\sim 10^5$ yr); but if t_L is greater than the time of big bang nucleosynthesis, then the restrictions come from the distortions in the present microwave background. No strong astrophysical constraints come for decays at the epoch of nucleosynthesis or earlier.

($\propto \rho^{-1/2}$) and the weak interaction rate (cf. the review by Schramm and Wagoner 1977). The entropy per baryon is not important during this stage.

Stage II: $T_\gamma \lesssim 2 \times 10^9$ K.—During this stage, $T_\gamma \approx (11/4)^{1/3} T_\nu$ and the light nuclei are built up via two-body reactions. Of importance here is the reaction rate (which depends on nucleon density) versus the expansion rate. A change in the expansion rate can be compensated, during this stage, by a change in the nucleon density (or, entropy per baryon).

We will now see that the heavy neutrinos can play no major role during Stage I. In particularly, for $T \gtrsim 10^{10}$, we will see that $\rho_L \ll \rho_{\rm rad}$; thus the expansion rate is virtually uncharged during this stage.

From Figure 1 it follows that the present density of L's, ρ_L, is $\lesssim 10^{-26} T_\gamma^3$; and we know that the present photon density $\rho_\gamma \sim 10^{-35} T_\gamma^4$. Thus for the time interval in question the density in L's is small ($\lesssim 0.036$) compared to the density of radiation and therefore the change in expansion rate will be small. In the limit of zero mass this will go to the results of Steigman, Schramm, and Gunn (1977), where it was shown that an additional zero-mass neutrino pair will increase the helium mass fraction by less than 0.02. When it is remembered that similar variations are possible by the uncertainty on the entropy per baryon alone, then it is clear that strict restrictions on L^0 lifetime cannot be placed in Stage I.

In Stage II ($T < 2 \times 10^9$) the expansion rate is competing with the nuclear reaction rates. However, any speedup in the expansion rate during Stage II has the same effect as lowering the baryon density (or raising the entropy per baryon). Since the baryon density is not well known, the deuterium and helium abundances cannot be used to isolate any such speedup.

The conclusion, therefore, is that if the heavy neutrino decays during nucleosynthesis, there is no unambiguous effect on primordial nucleosynthesis. If one does truly know the baryon density, then one can use the deuterium abundance to specify the entropy per baryon (cf. Schramm and Wagoner 1977) and thus restrict decays which affect this parameter. Recently Dicus *et al.* (1977) have used the above argument using what they believe to be the baryon density from galactic dynamics. However, we feel such restrictions may be premature. In any case, we are able to obtain comparable restrictions without precise assumptions on the baryon density.

Figure 2 summarizes what we believe to be the safest astrophysical limits on m_L and t_L.

IX. DISCUSSION AND CONCLUSION

We have investigated the consequences of the suggestion by LW2 that massive (few GeV) stable neutrinos L^0 and their antiparticles $\bar{L}^0$ may have been produced in the early universe and dominate the mass density in the universe at present. The L^0's remain in thermal kinetic equilibrium down to temperatures of about 10^{11} K, after which they decouple from the ambient radiation and cool as a nonrelativistic gas. Since annihilation has reduced their number density far below the thermal one, they constitute classical, albeit completely noninteracting Maxwellian gas.

Since they do not interact either with each other or with the baryonic matter present, and since at that time they make a negligible contribution to the mass density, they do not sensibly affect primordial nucleosynthesis, with the result that the relation between the helium and deuterium abundances and the baryon density at present is unchanged.

We showed that the only plausible mechanism for producing perturbations in the $\rho_{L^0}/\rho_{\rm rad}$ is through the presence of curvature perturbations, which must have small amplitude in order that the universe now not be dominated by black holes by a factor large compared to the limits on matter-to-radiation ratios. Hence annihilation rates calculated for a smooth distribution are almost certainly accurate.

Perturbations present in either ρ_L or the baryon density grow either after the universe becomes matter-dominated or after recombination, and we showed that even though the perturbations need not be correlated initially, both forms of matter respond to the total perturbation potential, and are effectively locked together gravitationally for all but very large-amplitude (and hence small-scale) perturbations. They can separate subsequently only if the baryonic matter can dissipate energy by radiation or other means. This can happen on galactic scales or smaller, but cannot happen on scales like the great clusters of galaxies or in any structure which is still expanding, like the Local Supercluster. Thus mass measurements in structures like these should measure the total mass of a mixture of baryons and L^0's in which their ratio is the primordial one—they are "fair samples" of the matter in the universe. Assuming such a case, standard dynamical arguments can be made which indicate that the density parameter Ω is less than 1, and the universe is not closed by the L^0's. The L^0's still serve well to account for the missing 90% of the mass required to make the extended halos of spiral galaxies and to bind the great clusters.

The production rate of γ-rays via annihilation in structures like galactic halos now is interestingly high, even though the annihilation times are very long. Our picture of the annihilation processes is very crude, however; and until a better model is forthcoming, we regard the predictions as tantalizing: the prospect of detecting the "missing" matter through its fundamental properties is a very exciting one.

We also pointed out some academically intriguing though probably unobservable aspects of L^0's such as their magnetic moment and spin flip radiation.

Finally, we investigated some of the consequences of the decay of the L^0 if it should not be absolutely stable. The basic conclusion is that, if the L^0 ever dominated the mass density in the universe, its decay time must be shorter than about 3000 seconds or longer than about 10^{18} years in order that it not produce a γ-ray background higher than observed, unacceptable distortions of the blackbody background, or do violence to helium production. We noted that for decay times less than about 100 seconds, the L^0 never can dominate the mass density, so it can be responsible for matter dominance and concomitant growth of perturbations in the early universe for only a very small interval, clearly of no importance dynamically.

If massive neutrinos are shown to exist in accelerator experiments, they will be present in the standard big bang model with no special ad hoc assumptions. Although other neutrino species can also be used to add to the density, special ad hoc assumptions (e.g., degeneracy) are needed to increase the density of such unobserved species much beyond that which occurs in the standard model, and this directly affects big bang nucleosynthesis. For the massive neutral stable leptons, equilibrium at $T \gtrsim 10^{12}$ K is all that is needed.

In other words: the question of the nature of the missing mass might well be decided by Earth-based particle accelerators rather than optical telescopes. Measurements of a stable heavy neutrino of about a few GeV mass will then have the astrophysical consequences pointed out above.

We wish to acknowledge most gratefully the aid of Steve Weinberg, who with B. L. initiated this paper and offered advice as a coauthor throughout most of the development, but commitments to axions and related things unfortunately took their toll on his time.

Special thanks also go to R. Shrock of Fermilab and Princeton who made significant contributions to the weak interaction physics in this paper, particularly with regard to the L^0 magnetic moment and Compton cross section. The γ-ray section of this paper profited greatly from discussions with Floyd Stecker who was extremely helpful in accurately interpreting the γ-ray observations for us. Floyd has gone on to do his own detailed modeling of γ's from $L^0\bar{L}^0$ annihilation which he is publishing as a separate paper elsewhere in this issue. One of us (D. N. S.) acknowledges a conversation with Ya. Zel'dovich who had independently thought about several of the points mentioned. Also I. L. is grateful to Jim Ipser for beneficial discussion. D. N. S. and G. S. also acknowledge conversations with R. Wagoner, V. Teplitz, and E. Kolb. J. E. G. and D. N. S. thank the Institute of Astronomy in Cambridge, UK, for hospitality while some of this work was carried out. The work reported here was supported by the National Science Foundation (J. E. G., D. N. S.), by the Energy Research and Development Agency (B. L.) and by the National Aeronautics and Space Administration (I. L., D. N. S.).

REFERENCES

Abers, A., and Lee, B. W. 1973, *Phys. Rept.*, **9**, 1.
Aarseth, S., Gott, J. R., and Turner, E. 1977, private communication.
Benvenuti, A., *et al.*, 1977*a*, *Phys. Rev. Letters*, **38**, 1110.
Benvenuti, A., *et al.*, 1977*b*, *Phys. Rev. Letters*, **38**, 1183.
Cowsik, R., and McClelland, J. 1972, *Phys. Rev. Letters*, **29**, 669.
Davis, M., and Peebles, J. 1977, *Ap. J. Suppl.*, **34**, 425.
Dicus, D. A., Kolb, E. N., and Teplitz, V. L. 1977, *Phys. Rev. Letters*, **39**, 168.
Dicus, D. A., Kolb, E. N., Teplitz, V. L., and Wagoner, R. V. 1977, preprint.
Fall, M., and Rees, M. 1977, preprint.
Faulkner, J., Sarazin, C., and Steigman, G. 1978, in preparation.
Fichtel, C. E., Simpson, G., and Thompson, R. 1977, NASA-Goddard preprint.
Fujikawa, K., Lee, B. W., and Sanda, A. I. 1972, *Phys. Rev.*, **D6**, 2923.
Gott, J. R., Gunn, J. E., Schramm, D. N., and Tinsley, B. M. 1974, *Ap. J.*, **194**, 543 (GGST).
Gott, J. R., and Rees, M. 1975, *Astr. Ap.*, **45**, 365.
Gott, J. R., and Turner, E. 1976, *Ap. J.*, **209**, 1.
Gunn, J. 1977, Caltech preprint.
Gunn, J., and Gott, G. R. 1972, *Ap. J.*, **176**, 1.
Hagedorn, R. 1970, *Astr. Ap.*, **5**, 184.
Hawking, S. 1973, *Nature*, **248**, 30.
Huang, K., and Weinberg, S. 1970, *Phys. Rev. Letters*, **25**, 895.
Jones, F., and Steigman, G. 1977, preprint.
Landau, L. D., and Lifshitz, E. M. 1958, *Statistical Physics* (New York: Pergamon Press).
Lee, B. W., and Shrock, R. E. 1977*a*, FERMILAB-PUB-77/21-THY (1977 February) to be published in *Phys. Rev. D.*
Lee, B. W., and Shrock, R. E. 1977*b*, FERMILAB-PUB-77/21-THY (1977 June).
Lee, B. W., and Weinberg, S. 1977*a*, *Phys. Rev. Letters*, **38**, 1237.
———. 1977*b*, *Phys. Rev. Letters*, **39**, 165 (LW2).
Lynden-Bell, D. 1967, *M.N.R.A.S.*, **136**, 101.
Mészáros, P. 1974, *Astr. Ap.*, **37**, 225.
Perl, M., *et al.* 1975, *Phys. Rev. Letters*, **35**, 1489.
Perl, M., *et al.* 1976, *Phys. Rev. Letters*, **63B**, 466.
Salam, A. 1967, in *Elementary Particle Physics*, ed. N. Svortholm (Stockholm: Almqvist and Wiksells: 1968), p. 367.
Sato, K. 1977, preprint.
Schramm, D. N., and Wagoner, R. V. 1977, *Ann. Rev. Nucl. Sci.*, **27**, 37.
Silk, J. 1967, *Nature*, **215**, 1155.
Stecker, F. W. 1977, *Ap. J.*, **212**, 60.
———. 1978, *Ap. J.*, **223**, 1032.
Steigman, G. 1976, *Ann. Rev. Astr. Ap.*, **14**, 339.
Steigman, G., Schramm, D. N., and Gunn, J. E. 1977, *Phys. Letters*, **66B**, 202.
t'Hooft, G., and Veltman, M. 1973, Diagrammar, CERN Rept. 73-9.
Weinberg, S. 1967, *Phys. Rev. Letters*, **19**, 1264.
———. 1971, *Phys. Rev. Letters*, **27**, 1688.
———. 1972*a*, *Phys. Rev. Letters*, **D5**, 1412.
———. 1972*b*, *Phys. Rev. Letters*, **D5**, 1962.
———. 1972*c*, *Gravitation and Cosmology: Principles and Applications of the General Theory of Relativity* (New York: Wiley).
———. 1974, *Rev. Mod. Phys.*, **46**, 255.
Zel'dovich, Ya. B. 1977, *Proc. Neutrino-77*, Caucausus, USSR.

JAMES E. GUNN: Department of Astronomy, California Institute of Technology, Pasadena, CA 91125

BEN W. LEE: deceased

IAN LERCHE: Division of Radiophysics, CSIRO, Epping, New South Wales, Australia

DAVID N. SCHRAMM: EFI–LASR, University of Chicago, 933 East 56th Street, Chicago, IL 60637

GARY STEIGMAN: Department of Astronomy, Yale University, New Haven, CT 06520

XIV. *Monopoles and Inflation*

1) Inflationary universe: A possible solution to the horizon and flatness problems by A. H. Guth, *Phys. Rev.* D**23,** 347, (1981) 850
2) Phase transitions in gauge theories and cosmology by A. D. Linde, *Rep. Prog. Phys.* **42,** 389 (1979) 861
3) A new inflationary universe scenario: A possible solution of the horizon, flatness, homogeneity, isotropy and primordial monopole problems by A. D. Linde, *Phys. Lett.* **108**B, 389 (1982) 910
4) Cosmology for grand unified theories with radiatively induced symmetry breaking by A. Albrecht and P. J. Steinhardt, *Phys. Rev. Lett.* **48,** 1220 (1982) 915

6 0 / 6

PHYSICAL REVIEW D VOLUME 23, NUMBER 2 15 JANUARY 1981

Inflationary universe: A possible solution to the horizon and flatness problems

Alan H. Guth*

Stanford Linear Accelerator Center, Stanford University, Stanford, California 94305

(Received 11 August 1980)

The standard model of hot big-bang cosmology requires initial conditions which are problematic in two ways: (1) The early universe is assumed to be highly homogeneous, in spite of the fact that separated regions were causally disconnected (horizon problem); and (2) the initial value of the Hubble constant must be fine tuned to extraordinary accuracy to produce a universe as flat (i.e., near critical mass density) as the one we see today (flatness problem). These problems would disappear if, in its early history, the universe supercooled to temperatures 28 or more orders of magnitude below the critical temperature for some phase transition. A huge expansion factor would then result from a period of exponential growth, and the entropy of the universe would be multiplied by a huge factor when the latent heat is released. Such a scenario is completely natural in the context of grand unified models of elementary-particle interactions. In such models, the supercooling is also relevant to the problem of monopole suppression. Unfortunately, the scenario seems to lead to some unacceptable consequences, so modifications must be sought.

I. INTRODUCTION: THE HORIZON AND FLATNESS PROBLEMS

The standard model of hot big-bang cosmology relies on the assumption of initial conditions which are very puzzling in two ways which I will explain below. The purpose of this paper is to suggest a modified scenario which avoids both of these puzzles.

By "standard model," I refer to an adiabatically expanding radiation-dominated universe described by a Robertson-Walker metric. Details will be given in Sec. II.

Before explaining the puzzles, I would first like to clarify my notion of "initial conditions." The standard model has a singularity which is conventionally taken to be at time $t=0$. As $t\to 0$, the temperature $T\to\infty$. Thus, no initial-value problem can be defined at $t=0$. However, when T is of the order of the Planck mass ($M_P\equiv 1/\sqrt{G}=1.22\times 10^{19}$ GeV)[1] or greater, the equations of the standard model are undoubtedly meaningless, since quantum gravitational effects are expected to become essential. Thus, within the scope of our knowledge, it is sensible to begin the hot big-bang scenario at some temperature T_0 which is comfortably below M_P; let us say $T_0=10^{17}$ GeV. At this time one can take the description of the universe as a set of initial conditions, and the equations of motion then describe the subsequent evolution. Of course, the equation of state for matter at these temperatures is not really known, but one can make various hypotheses and pursue the consequences.

In the standard model, the initial universe is taken to be homogeneous and isotropic, and filled with a gas of effectively massless particles in thermal equilibrium at temperature T_0. The initial value of the Hubble expansion "constant" H is taken to be H_0, and the model universe is then completely described.

Now I can explain the puzzles. The first is the well-known horizon problem.[2-4] The initial universe is assumed to be homogeneous, yet it consists of at least $\sim 10^{83}$ separate regions which are causally disconnected (i.e., these regions have not yet had time to communicate with each other via light signals).[5] (The precise assumptions which lead to these numbers will be spelled out in Sec. II.) Thus, one must assume that the forces which created these initial conditions were capable of violating causality.

The second puzzle is the flatness problem. This puzzle seems to be much less celebrated than the first, but it has been stressed by Dicke and Peebles.[6] I feel that it is of comparable importance to the first. It is known that the energy density ρ of the universe today is near the critical value ρ_{cr} (corresponding to the borderline between an open and closed universe). One can safely assume that[7]

$$0.01<\Omega_p<10\,, \tag{1.1}$$

where

$$\Omega\equiv\rho/\rho_{cr}=(8\pi/3)G\rho/H^2\,, \tag{1.2}$$

and the subscript p denotes the value at the present time. Although these bounds do not appear at first sight to be remarkably stringent, they, in fact, have powerful implications. The key point is that the condition $\Omega\approx 1$ is unstable. Furthermore, the only time scale which appears in the equations for a radiation-dominated universe is the Planck time, $1/M_P=5.4\times 10^{-44}$ sec. A typical closed universe will reach its maximum size on the order of this time scale, while a typical open universe will dwindle to a value of ρ much less than ρ_{cr}. A universe can survive $\sim 10^{10}$ years only by extreme fine tuning of the initial values of ρ and H, so that ρ is very near ρ_{cr}. For the initial conditions taken at

$T_0 = 10^{17}$ GeV, the value of H_0 must be fine tuned to an accuracy of one part in 10^{55}. In the standard model this incredibly precise initial relationship must be assumed without explanation. (For any reader who is not convinced that there is a real problem here, variations of this argument are given in the Appendix.)

The reader should not assume that these incredible numbers are due merely to the rather large value I have taken for T_0. If I had chosen a modest value such as $T_0 = 1$ MeV, I would still have concluded that the "initial" universe consisted of at least ~10^{22} causally disconnected regions, and that the initial value of H_0 was fine tuned to one part in 10^{15}. These numbers are much smaller than the previous set, but they are still very impressive.

Of course, any problem involving the initial conditions can always be put off until we understand the physics of $T \gtrsim M_P$. However, it is the purpose of this paper to show that these puzzles might be obviated by a scenario for the behavior of the universe at temperatures well below M_P.

The paper is organized as follows. The assumptions and basic equations of the standard model are summarized in Sec. II. In Sec. III, I describe the inflationary universe scenario, showing how it can eliminate the horizon and flatness problems. The scenario is discussed in the context of grand models in Sec. IV, and comments are made concerning magnetic monopole suppression. In Sec. V I discuss briefly the key undesirable feature of the scenario: the inhomogeneities produced by the random nucleation of bubbles. Some vague ideas which might alleviate these difficulties are mentioned in Sec. VI.

II. THE STANDARD MODEL OF THE VERY EARLY UNIVERSE

In this section I will summarize the basic equations of the standard model, and I will spell out the assumptions which lead to the statements made in the Introduction.

The universe is assumed to be homogeneous and isotropic, and is therefore described by the Robertson-Walker metric[8]:

$$d\tau^2 = dt^2 - R^2(t)\left[\frac{dr^2}{1-kr^2} + r^2(d\theta^2 + \sin^2\theta\, d\phi^2)\right], \tag{2.1}$$

where $k = +1$, -1, or 0 for a closed, open, or flat universe, respectively. It should be emphasized that any value of k is possible, but by convention r and $R(t)$ are rescaled so that k takes on one of the three discrete values. The evolution of $R(t)$ is governed by the Einstein equations

$$\ddot{R} = -\frac{4\pi}{3}G(\rho + 3p)R, \tag{2.2a}$$

$$H^2 + \frac{k}{R^2} = \frac{8\pi}{3}G\rho, \tag{2.2b}$$

where $H \equiv \dot{R}/R$ is the Hubble "constant" (the dot denotes the derivative with respect to t). Conservation of energy is expressed by

$$\frac{d}{dt}(\rho R^3) = -p\frac{d}{dt}(R^3), \tag{2.3}$$

where p denotes the pressure. In the standard model one also assumes that the expansion is adiabatic, in which case

$$\frac{d}{dt}(sR^3) = 0, \tag{2.4}$$

where s is the entropy density.

To determine the evolution of the universe, the above equations must be supplemented by an equation of state for matter. It is now standard to describe matter by means of a field theory, and at high temperatures this means that the equation of state is to a good approximation that of an ideal quantum gas of massless particles. Let $N_b(T)$ denote the number of bosonic spin degrees of freedom which are effectively massless at temperature T (e.g., the photon contributes two units to N_b); and let $N_f(T)$ denote the corresponding number for fermions (e.g., electrons and positrons together contribute four units). Provided that T is not near any mass thresholds, the thermodynamic functions are given by

$$\rho = 3p = \frac{\pi^2}{30}\mathfrak{N}(T)T^4, \tag{2.5}$$

$$s = \frac{2\pi^2}{45}\mathfrak{N}(T)T^3, \tag{2.6}$$

$$n = \frac{\zeta(3)}{\pi^2}\mathfrak{N}'(T)T^3, \tag{2.7}$$

where

$$\mathfrak{N}(T) = N_b(T) + \tfrac{7}{8}N_f(T), \tag{2.8}$$

$$\mathfrak{N}'(T) = N_b(T) + \tfrac{3}{4}N_f(T). \tag{2.9}$$

Here n denotes the particle number density, and $\zeta(3) = 1.20206\ldots$ is the Riemann zeta function.

The evolution of the universe is then found by rewriting (2.2b) solely in terms of the temperature. Againing assuming that T is not near any mass thresholds, one finds

$$\left(\frac{\dot{T}}{T}\right)^2 + \epsilon(T)T^2 = \frac{4\pi^3}{45}G\mathfrak{N}(T)T^4, \tag{2.10}$$

where

$$\epsilon(T) = \frac{k}{R^2T^2} = k\left[\frac{2\pi^2}{45}\frac{\mathfrak{N}(T)}{S}\right]^{2/3}, \tag{2.11}$$

where $S \equiv R^3 s$ denotes the total entropy in a volume specified by the radius of curvature R.

Since S is conserved, its value in the early universe can be determined (or at least bounded) by current observations. Taking $\rho < 10\rho_{cr}$ today, it follows that today

$$\left|\frac{k}{R^2}\right| < 9H^2 . \tag{2.12}$$

From now on I will take $k=\pm 1$; the special case $k=0$ is still included as the limit $R \to \infty$. Then today $R > \frac{1}{3}H^{-1} \sim 3\times 10^9$ years. Taking the present photon temperature T_γ as 2.7 °K, one then finds that the photon contribution to S is bounded by

$$S_\gamma > 3\times 10^{85} . \tag{2.13}$$

Assuming that there are three species of massless neutrinos (e, μ, and τ), all of which decouple at a time when the other effectively massless particles are the electrons and photons, then $S_\nu = 21/22 S_\gamma$. Thus,

$$S > 10^{86} \tag{2.14}$$

and

$$|\epsilon| < 10^{-58}\mathfrak{N}^{2/3} . \tag{2.15}$$

But then

$$\left|\frac{\rho-\rho_{cr}}{\rho}\right| = \frac{45}{4\pi^3}\frac{M_P^2}{\mathfrak{N}T^2}|\epsilon| < 3\times 10^{-59}\mathfrak{N}^{-1/3}(M_P/T)^2 . \tag{2.16}$$

Taking $T = 10^{17}$ GeV and $\mathfrak{N} \sim 10^2$ (typical of grand unified models), one finds $|\rho-\rho_{cr}|/\rho < 10^{-55}$. This is the flatness problem.

The ϵT^2 term can now be deleted from (2.10), which is then solved (for temperatures higher than all particle masses) to give

$$T^2 = \frac{M_P}{2\gamma t} , \tag{2.17}$$

where $\gamma^2 = (4\pi^3/45)\mathfrak{N}$. (For the minimal SU_5 grand unified model, $N_b = 82$, $N_f = 90$, and $\gamma = 21.05$.) Conservation of entropy implies $RT =$ constant, so $R \propto t^{1/2}$. A light pulse beginning at $t=0$ will have traveled by time t a physical distance

$$l(t) = R(t)\int_0^t dt' R^{-1}(t') = 2t , \tag{2.18}$$

and this gives the physical horizon distance. This horizon distance is to be compared with the radius $L(t)$ of the region at time t which will evolve into our currently observed region of the universe. Again using conservation of entropy,

$$L(t) = [s_p/s(t)]^{1/3} L_p , \tag{2.19}$$

where s_p is the present entropy density and $L_p \sim 10^{10}$ years is the radius of the currently observed region of the universe. One is interested in the ratio of volumes, so

$$\frac{l^3}{L^3} = \frac{11}{43}\left(\frac{45}{4\pi^3}\right)^{3/2}\mathfrak{N}^{-1/2}\left(\frac{M_P}{L_p T_\gamma T}\right)^3$$
$$= 4\times 10^{-89}\mathfrak{N}^{-1/2}(M_P/T)^3 . \tag{2.20}$$

Taking $\mathfrak{N} \sim 10^2$ and $T_0 = 10^{17}$ GeV, one finds $l_0^3/L_0^3 = 10^{-83}$. This is the horizon problem.

III. THE INFLATIONARY UNIVERSE

In this section I will describe a scenario which is capable of avoiding the horizon and flatness problems.

From Sec. II one can see that both problems could disappear if the assumption of adiabaticity were grossly incorrect. Suppose instead that

$$S_p = Z^3 S_0 , \tag{3.1}$$

where S_p and S_0 denote the present and initial values of $R^3 s$, and Z is some large factor.

Let us look first at the flatness problem. Given (3.1), the right-hand side (RHS) of (2.16) is multiplied by a factor of Z^2. The "initial" value (at $T_0 = 10^{17}$ GeV) of $|\rho-\rho_{cr}|/\rho$ could be of order unity, and the flatness problem would be obviated, if

$$Z > 3\times 10^{27} . \tag{3.2}$$

Now consider the horizon problem. The RHS of (2.19) is multiplied by Z^{-1}, which means that the length scale of the early universe, at any given temperature, was smaller by a factor of Z than had been previously thought. If Z is sufficiently large, then the initial region which evolved into our observed region of the universe would have been smaller than the horizon distance at that time. To see how large Z must be, note that the RHS of (2.20) is multiplied by Z^3. Thus, if

$$Z > 5\times 10^{27} , \tag{3.3}$$

then the horizon problem disappears. (It should be noted that the horizon will still exist; it will simply be moved out to distances which have not been observed.)

It is not surprising that the RHS's of (3.2) and (3.3) are approximately equal, since they both correspond roughly to S_0 of order unity.

I will now describe a scenario, which I call the inflationary universe, which is capable of such a large entropy production.

Suppose the equation of state for matter (with all chemical potentials set equal to zero) exhibits a first-order phase transition at some critical temperature T_c. Then as the universe cools through the temperature T_c, one would expect bubbles of the low-temperature phase to nucleate and grow. However, suppose the nucleation rate for this phase transition is rather low. The universe will continue to cool as it expands, and it will then supercool in the high-temperature phase. Suppose that this supercooling continues down to some temperature T_s, many orders of magnitude below T_c. When the phase transition finally takes place at temperature T_s, the latent heat is released. However, this latent heat is characteristic of the energy scale T_c, which is huge relative to T_s. The universe is then reheated to some temperature T_r which is comparable to T_c. The entropy density is then increased by a factor of roughly $(T_r/T_s)^3$ (assuming that the number $\mathfrak{N}$ of degrees of freedom for the two phases are comparable), while the value of R remains unchanged. Thus,

$$Z \approx T_r/T_s . \tag{3.4}$$

If the universe supercools by 28 or more orders of magnitude below some critical temperature, the horizon and flatness problems disappear.

In order for this scenario to work, it is necessary for the universe to be essentially devoid of any strictly conserved quantities. Let n denote the density of some strictly conserved quantity, and let $r \equiv n/s$ denote the ratio of this conserved quantity to entropy. Then $r_p = Z^{-3} r_0 < 10^{-84} r_0$. Thus, only an absurdly large value for the initial ratio would lead to a measurable value for the present ratio. Thus, if baryon number were exactly conserved, the inflationary model would be untenable. However, in the context of grand unified models, baryon number is not exactly conserved. The net baryon number of the universe is believed to be created by CP-violating interactions at a temperature of 10^{13}–10^{14} GeV.[9] Thus, provided that T_c lies in this range or higher, there is no problem. The baryon production would take place after the reheating. (However, strong constraints are imposed on the entropy which can be generated in any phase transition with $T_c \ll 10^{14}$ GeV, in particular, the Weinberg-Salam phase transition.[36])

Let us examine the properties of the supercooling universe in more detail. Note that the energy density $\rho(T)$, given in the standard model by (2.5), must now be modified. As $T \to 0$, the system is cooling not toward the true vacuum, but rather toward some metastable false vacuum with an energy density ρ_0 which is necessarily higher than that of the true vacuum. Thus, to a good approximation (ignoring mass thresholds)

$$\rho(T) = \frac{\pi^2}{30} \mathfrak{N}(T) T^4 + \rho_0 . \tag{3.5}$$

Perhaps a few words should be said concerning the zero point of energy. Classical general relativity couples to an energy-momentum tensor of matter, $T_{\mu\nu}$, which is necessarily (covariantly) conserved. When matter is described by a field theory, the form of $T_{\mu\nu}$ is determined by the conservation requirement up to the possible modification

$$T_{\mu\nu} \to T_{\mu\nu} + \lambda g_{\mu\nu} , \tag{3.6}$$

for any constant λ. (λ *cannot* depend on the values of the fields, nor can it depend on the temperature or the phase.) The freedom to introduce the modification (3.6) is identical to the freedom to introduce a cosmological constant into Einstein's equations. One can always choose to write Einstein's equations without an explicit cosmological term; the cosmological constant Λ is then defined by

$$\langle 0 | T_{\mu\nu} | 0 \rangle = \Lambda g_{\mu\nu} , \tag{3.7}$$

where $|0\rangle$ denotes the true vacuum. Λ is identified as the energy density of the vacuum, and, in principle, there is no reason for it to vanish. Empirically Λ is known to be very small ($|\Lambda| < 10^{-46}$ GeV^4)[10] so I will take its value to be zero.[11] The value of ρ_0 is then necessarily positive and is determined by the particle theory.[12] It is typically of $O(T_c^4)$.

Using (3.5), Eq. (2.10) becomes

$$\left(\frac{\dot{T}}{T}\right)^2 = \frac{4\pi^3}{45} G \mathfrak{N}(T) T^4 - \epsilon(T) T^2 + \frac{8\pi}{3} G \rho_0 . \tag{3.8}$$

This equation has two types of solutions, depending on the parameters. If $\epsilon > \epsilon_0$, where

$$\epsilon_0 = \frac{8\pi^2 \sqrt{30}}{45} G \sqrt{\mathfrak{N} \rho_0} , \tag{3.9}$$

then the expansion of the universe is halted at a temperature $T_{\min}$ given by

$$T_{\min}{}^4 = \frac{30 \rho_0}{\pi^2} \left\{ \frac{\epsilon}{\epsilon_0} + \left[\left(\frac{\epsilon}{\epsilon_0} \right)^2 - 1 \right]^{1/2} \right\}^2 , \tag{3.10}$$

and then the universe contracts again. Note that $T_{\min}$ is of $O(T_c)$, so this is not the desired scenario. The case of interest is $\epsilon < \epsilon_0$, in which case the expansion of the universe is unchecked. [Note that $\epsilon_0 \sim \sqrt{\mathfrak{N}} T_c{}^2/M_P{}^2$ is presumably a very small number. Thus $0 < \epsilon < \epsilon_0$ (a closed universe) seems unlikely, but $\epsilon < 0$ (an open universe) is quite plausible.] Once the temperature is low enough for the ρ_0 term to dominate over the other two terms on the RHS of (3.8), one has

$$T(t) \approx \text{const} \times e^{-\chi t} , \tag{3.11}$$

where

$$\chi^2 = \frac{8\pi}{3} G \rho_0 . \tag{3.12}$$

Since $RT = \text{const}$, one has[13]

$$R(t) = \text{const} \times e^{\chi t} . \tag{3.13}$$

The universe is expanding exponentially, in a false

vacuum state of energy density ρ_0. The Hubble constant is given by $H=\dot{R}/R=\chi$. (More precisely, H approaches χ monotonically from above. This behavior differs markedly from the standard model, in which H falls as t^{-1}.)

The false vacuum state is Lorentz invariant, so $T_{\mu\nu}=\rho_0 g_{\mu\nu}$. It follows that $p=-\rho_0$, the pressure is negative. This negative pressure allows for the conservation of energy, Eq. (2.3). From the second-order Einstein equation (2.2a), it can be seen that the negative pressure is also the driving force behind the exponential expansion.

The Lorentz invariance of the false vacuum has one other consequence: The metric described by (3.13) (with $k=0$) does not single out a comoving frame. The metric is invariant under an O(4,1) group of transformations, in contrast to the usual Robertson-Walker invariance of O(4).[14] It is known as the de Sitter metric, and it is discussed in the standard literature.[15]

Now consider the process of bubble formation in a Robertson-Walker universe. The bubbles form randomly, so there is a certain nucleation rate $\lambda(t)$, which is the probability per (physical) volume per time that a bubble will form in any region which is still in the high-temperature phase. I will idealize the situation slightly and assume that the bubbles start at a point and expand at the speed of light. Furthermore, I neglect k in the metric, so $d\tau^2=dt^2-R^2(t)d\vec{x}^2$.

I want to calculate $p(t)$, the probability that any given point remains in the high-temperature phase at time t. Note that the distribution of bubbles is totally uncorrelated except for the exclusion principle that bubbles do not form inside of bubbles. This exclusion principle causes no problem because one can imagine fictitious bubbles which form inside the real bubbles with the same nucleation rate $\lambda(t)$. With all bubbles expanding at the speed of light, the fictitious bubbles will be forever inside the real bubbles and will have no effect on $p(t)$. The distribution of all bubbles, real and fictitious, is then totally uncorrelated.

$p(t)$ is the probability that there are no bubbles which engulf a given point in space. But the number of bubbles which engulf a given point is a Poisson-distributed variable, so $p(t)=\exp[-\bar{N}(t)]$, where $\bar{N}(t)$ is the expectation value of the number of bubbles engulfing the point. Thus[16]

$$p(t)=\exp\left[-\int_0^t dt_1\lambda(t_1)R^3(t_1)V(t,t_1)\right], \qquad (3.14)$$

where

$$V(t,t_1)=\frac{4\pi}{3}\left[\int_{t_1}^{t}\frac{dt_2}{R(t_2)}\right]^3 \qquad (3.15)$$

is the coordinate volume at time t of a bubble which formed at time t_1.

I will now assume that the nucleation rate is sufficiently slow so that no significant nucleation takes place until $T\ll T_c$, when exponential growth has set in. I will further assume that by this time $\lambda(t)$ is given approximately by the zero-temperature nucleation rate λ_0. One then has

$$p(t)=\exp\left[-\frac{t}{\tau}+O(1)\right], \qquad (3.16)$$

where

$$\tau=\frac{3\chi^3}{4\pi\lambda_0}, \qquad (3.17)$$

and $O(1)$ refers to terms which approach a constant as $\chi t\to\infty$. During one of these time constants, the universe will expand by a factor

$$Z_\tau=\exp(\chi\tau)=\exp\left(\frac{3\chi^4}{4\pi\lambda_0}\right). \qquad (3.18)$$

If the phase transition is associated with the expectation value of a Higgs field, then λ_0 can be calculated using the method of Coleman and Callan.[17] The key point is that nucleation is a tunneling process, so that λ_0 is typically very small. The Coleman-Callan method gives an answer of the form

$$\lambda_0=A\rho_0\exp(-B), \qquad (3.19)$$

where B is a barrier penetration term and A is a dimensionless coefficient of order unity. Since Z_τ is then an exponential of an exponential, one can very easily[18,19,36] obtain values as large as $\log_{10}Z\approx 28$, or even $\log_{10}Z\approx 10^{10}$.

Thus, if the universe reaches a state of exponential growth, it is quite plausible for it to expand and supercool by a huge number of orders of magnitude before a significant fraction of the universe undergoes the phase transition.

So far I have assumed that the early universe can be described from the beginning by a Robertson-Walker metric. If this assumption were really necessary, then it would be senseless to talk about "solving" the horizon problem; perfect homogeneity was assumed at the outset. Thus, I must now argue that the assumption can probably be dropped.

Suppose instead that the initial metric, and the distribution of particles, was rather chaotic. One would then expect that statistical effects would tend to thermalize the particle distribution on a local scale.[20] It has also been shown (in idealized circumstances) that anisotropies in the metric are damped out on the time scale of ~10^3 Planck times.[21] The damping of inhomogeneities in the metric has also been studied,[22] and it is reasonable to expect such damping to occur. Thus, assuming that at least some region of the universe started at temperatures high compared to T_c, one would expect that, by the time the temperature in one of these regions falls to T_c, it will be *locally* homogeneous, isotropic, and in thermal equilibrium. By locally, I am talking about a length scale ξ which is of course less than the horizon distance. It will then be possible to describe this local region of the universe by a Robertson-Walker metric, which will be accurate at distance scales small compared to ξ. When the temperature of such a region falls below T_c, the inflationary scenario will take place. The end result will be a huge region of space which is homogeneous, isotropic, and of nearly critical mass density. If Z is sufficiently large, this region can be bigger than (or much bigger than) our observed region of the universe.

IV. GRAND UNIFIED MODELS AND MAGNETIC MONOPOLE PRODUCTION

In this section I will discuss the inflationary model in the context of grand unified models of elementary-particle interactions.[23,24]

A grand unified model begins with a simple gauge group G which is a valid symmetry at the highest energies. As the energy is lowered, the theory undergoes a hierarchy of spontaneous symmetry breaking into successive subgroups: $G \to H_n \to \cdots \to H_0$, where $H_1 = SU_3 \times SU_2 \times U_1$ [QCD (quantum chromodynamics) × Weinberg-Salam] and $H_0 = SU_3 \times U_1^{EM}$. In the Georgi-Glashow model,[23] which is the simplest model of this type, $G = SU_5$ and $n = 1$. The symmetry breaking of $SU_5 \to SU_3 \times SU_2 \times U_1$ occurs at an energy scale $M_X \sim 10^{14}$ GeV.

At high temperatures, it was suggested by Kirzhnits and Linde[25] that the Higgs fields of any spontaneously broken gauge theory would lose their expectation values, resulting in a high-temperature phase in which the full gauge symmetry is restored. A formalism for treating such problems was developed[26] by Weinberg and by Dolan and Jackiw. In the range of parameters for which the tree potential is valid, the phase structure of the SU_5 model was analyzed by Tye and me.[16,27] We found that the SU_5 symmetry is restored at $T > \sim 10^{14}$ GeV and that for most values of the parameters there is an intermediate-temperature phase with gauge symmetry $SU_4 \times U_1$, which disappears at $T \sim 10^{13}$ GeV. Thus, grand unified models tend to provide phase transitions which could lead to an inflationary scenario of the universe.

Grand unified models have another feature with important cosmological consequences: They contain very heavy magnetic monopoles in their particle spectrum. These monopoles are of the type discovered by 't Hooft and Polyakov,[28] and will be present in any model satisfying the above description.[29] These monopoles typically have masses of order $M_X/\alpha \sim 10^{16}$ GeV, where $\alpha = g^2/4\pi$ is the grand unified fine structure constant. Since the monopoles are really topologically stable knots in the Higgs field expectation value, they do not exist in the high-temperature phase of the theory. They therefore come into existence during the course of a phase transition, and the dynamics of the phase transition is then intimately related to the monopole production rate.

The problem of monopole production and the subsequent annihilation of monopoles, in the context of a second-order or weakly first-order phase transition, was analyzed by Zeldovich and Khlopov[30] and by Preskill.[31] In Preskill's analysis, which was more specifically geared toward grand unified models, it was found that relic monopoles would exceed present bounds by roughly 14 orders of magnitude. Since it seems difficult to modify the estimated annihilation rate, one must find a scenario which suppresses the production of these monopoles.

Kibble[32] has pointed out that monopoles are produced in the course of the phase transition by the process of bubble coalescence. The orientation of the Higgs field inside one bubble will have no correlation with that of another bubble not in contact. When the bubbles coalesce to fill the space, it will be impossible for the uncorrelated Higgs fields to align uniformly. One expects to find topological knots, and these knots are the monopoles. The number of monopoles so produced is then comparable to the number of bubbles, to within a few orders of magnitude.

Kibble's production mechanism can be used to set a "horizon bound" on monopole production which is valid if the phase transition does not significantly disturb the evolution of the universe.[33] At the time of bubble coalescence $t_{\rm coal}$ the size l of the bubbles cannot exceed the horizon distance at that time. So

$$l < 2t_{\rm coal} = \frac{M_P}{\gamma T_{\rm coal}^2} . \tag{4.1}$$

By Kibble's argument, the density n_M of monopoles then obeys

$$n_M \gtrsim l^{-3} > \frac{\gamma^3 T_{\rm coal}^6}{M_P^3} . \tag{4.2}$$

By considering the contribution to the mass density of the present universe which could come from 10^{16} GeV monopoles, Preskill[31] concludes that

$$n_M/n_\gamma < 10^{-24} , \tag{4.3}$$

where n_γ is the density of photons. This ratio changes very little from the time of the phase transition, so with (2.7) one concludes

$$T_{\rm coal} < \left[\frac{10^{-24}\pi^2}{2\zeta(3)}\right]^{1/3} \gamma^{-1} M_P \approx 10^{10}\ \text{GeV}\,. \quad (4.4)$$

If $T_c \sim 10^{14}$ GeV, this bound implies that the universe must supercool by at least about four orders of magnitude before the phase transition is completed.

The problem of monopole production in a strongly first-order phase transition with supercooling was treated in more detail by Tye and me.[16,34] We showed how to explicitly calculate the bubble density in terms of the nucleation rate, and we considered the effects of the latent heat released in the phase transition. Our conclusion was that (4.4) should be replaced by

$$T_{\rm coal} < 2\times 10^{11}\ \text{GeV}\,, \quad (4.5)$$

where $T_{\rm coal}$ refers to the temperature just before the release of the latent heat.

Tye and I omitted the crucial effects of the mass density ρ_0 of the false vacuum. However, our work has one clear implication: If the nucleation rate is sufficiently large to avoid exponential growth, then far too many monopoles would be produced. Thus, the monopole problem seems to also force one into the inflationary scenario.[35]

In the simplest SU_5 model, the nucleation rates have been calculated (approximately) by Weinberg and me.[19] The model contains unknown parameters, so no definitive answer is possible. We do find, however, that there is a sizable range of parameters which lead to the inflationary scenario.[36]

V. PROBLEMS OF THE INFLATIONARY SCENARIO[37]

As I mentioned earlier, the inflationary scenario seems to lead to some unacceptable consequences. It is hoped that some variation can be found which avoids these undesirable features but maintains the desirable ones. The problems of the model will be discussed in more detail elsewhere,[37] but for completeness I will give a brief description here.

The central problem is the difficulty in finding a smooth ending to the period of exponential expansion. Let us assume that $\lambda(t)$ approaches a constant as $t\to\infty$ and $T\to 0$. To achieve the desired expansion factor $Z > 10^{28}$, one needs $\lambda_0/\chi^4 < 10^{-2}$ [see (3.18)], which means that the nucleation rate is slow compared to the expansion rate of the universe. (Explicit calculations show that λ_0/χ^4 is typically much smaller than this value.[18,19,36]) The randomness of the bubble formation process then leads to gross inhomogeneities.

To understand the effects of this randomness, the reader should bear in mind the following facts.

(i) All of the latent heat released as a bubble expands is transferred initially to the walls of the bubble.[17] This energy can be thermalized only when the bubble walls undergo many collisions.

(ii) The de Sitter metric does not single out a comoving frame. The O(4,1) invariance of the de Sitter metric is maintained even after the formation of one bubble. The memory of the original Robertson-Walker comoving frame is maintained by the probability distribution of bubbles, but the local comoving frame can be reestablished only after enough bubbles have collided.

(iii) The size of the largest bubbles will exceed that of the smallest bubbles by roughly a factor of Z; the range of bubble sizes is immense. The surface energy density grows with the size of the bubble, so the energy in the walls of the largest bubbles can be thermalized only by colliding with other large bubbles.

(iv) As time goes on, an arbitrarily large fraction of the space will be in the new phase [see (3.16)]. However, one can ask a more subtle question about the region of space which is in the new phase: Is the region composed of finite separated clusters, or do these clusters join together to form an infinite region? The latter possibility is called "percolation." It can be shown[38] that the system percolates for large values of λ_0/χ^4, but that for sufficiently small values it does *not*. The critical value of λ_0/χ^4 has not been determined, but presumably an inflationary universe would have a value of λ_0/χ^4 below critical. Thus, no matter how long one waits, the region of space in the new phase will consist of finite clusters, each totally surrounded by a region in the old phase.

(v) Each cluster will contain only a few of the largest bubbles. Thus, the collisions discussed in (iii) cannot occur.

The above statements do not quite prove that the scenario is impossible, but these consequences are at best very unattractive. Thus, it seems that the scenario will become viable only if some modification can be found which avoids these inhomogeneities. Some vague possibilities will be mentioned in the next section.

Note that the above arguments seem to rule out the possibility that the universe was ever trapped in a false vacuum state, unless $\lambda_0/\chi^4 \gtrsim 1$. Such a large value of λ_0/χ^4 does not seem likely, but it is possible.[19]

VI. CONCLUSION

I have tried to convince the reader that the standard model of the very early universe requires the assumption of initial conditions which are very implausible for two reasons:

(i) *The horizon problem.* Causally disconnected regions are assumed to be nearly identical; in particular, they are simultaneously at the same temperature.

(ii) *The flatness problem.* For a fixed initial temperature, the initial value of the Hubble "constant" must be fine tuned to extraordinary accuracy to produce a universe which is as flat as the one we observe.

Both of these problems would disappear if the universe supercooled by 28 or more orders of magnitude below the critical temperature for some phase transition. (Under such circumstances, the universe would be growing exponentially in time.) However, the random formation of bubbles of the new phase seems to lead to a much too inhomogeneous universe.

The inhomogeneity problem would be solved if one could avoid the assumption that the nucleation rate $\lambda(t)$ approaches a small constant λ_0 as the temperature $T \to 0$. If, instead, the nucleation rate rose sharply at some T_1, then bubbles of an approximately uniform size would suddenly fill space as T fell to T_1. Of course, the full advantage of the inflationary scenario is achieved only if $T_1 \lesssim 10^{-28} T_c$.

Recently Witten[39] has suggested that the above chain of events may in fact occur if the parameters of the SU_5 Higgs field potential are chosen to obey the Coleman-Weinberg condition[40] (i.e., that $\partial^2 V/\partial\phi^2 = 0$ at $\phi = 0$). Witten[41] has studied this possibility in detail for the case of the Weinberg-Salam phase transition. Here he finds that thermal tunneling is totally ineffective, but instead the phase transition is driven when the temperature of the QCD chiral-symmetry-breaking phase transition is reached. For the SU_5 case, one can hope that a much larger amount of supercooling will be found; however, it is difficult to see how 28 orders of magnitude could arise.

Another physical effect which has so far been left out of the analysis is the production of particles due to the changing gravitational metric.[42] This effect may become important in an exponentially expanding universe at low temperatures.

In conclusion, the inflationary scenario seems like a natural and simple way to eliminate both the horizon and the flatness problems. I am publishing this paper in the hope that it will highlight the existence of these problems and encourage others to find some way to avoid the undesirable features of the inflationary scenario.

ACKNOWLEDGMENTS

I would like to express my thanks for the advice and encouragement I received from Sidney Coleman and Leonard Susskind, and for the invaluable help I received from my collaborators Henry Tye and Erick Weinberg. I would also like to acknowledge very useful conversations with Michael Aizenman, Beilok Hu, Harry Kesten, Paul Langacker, Gordon Lasher, So-Young Pi, John Preskill, and Edward Witten. This work was supported by the Department of Energy under Contract No. DE-AC03-76SF00515.

APPENDIX: REMARKS ON THE FLATNESS PROBLEM

This appendix is added in the hope that some skeptics can be convinced that the flatness problem is real. Some physicists would rebut the argument given in Sec. I by insisting that the equations might make sense all the way back to $t = 0$. Then if one fixes the value of H corresponding to some arbitrary temperature T_a, one always finds that when the equations are extrapolated *backward* in time, $\Omega \to 1$ as $t \to 0$. Thus, they would argue, it is natural for Ω to be very nearly equal to 1 at early times. For physicists who take this point of view, the flatness problem must be restated in other terms. Since H_0 and T_0 have no significance, the model universe must be specified by its conserved quantities. In fact, the model universe is completely specified by the dimensionless constant $\epsilon \equiv k/R^2T^2$, where k and R are parameters of the Robertson-Walker metric, Eq. (2.1). For our universe, one must take $|\epsilon| < 3\times 10^{-57}$. The problem then is the to explain why $|\epsilon|$ should have such a startlingly small value.

Some physicists also take the point of view that $\epsilon \equiv 0$ is plausible enough, so to them there is no problem. To these physicists I point out that the universe is certainly not described *exactly* by a Robertson-Walker metric. Thus it is difficult to imagine any physical principle which would require a parameter of that metric to be exactly equal to zero.

In the end, I must admit that questions of plausibility are not logically determinable and depend somewhat on intuition. Thus I am sure that some physicists will remain unconvinced that there really is a flatness problem. However, I am also sure that many physicists agree with me that the flatness of the universe is a peculiar situation which at some point will admit a physical explanation.

*Present address: Center for Theoretical Physics, Massachusetts Institute of Technology, Cambridge, Massachusetts 02139.

[1] I use units for which $\hbar = c = k$ (Boltzmann constant) = 1. Then 1 m = 5.068×10^{15} GeV^{-1}, 1 kg = 5.610×10^{26} GeV, 1 sec = 1.519×10^{24} GeV^{-1}, and 1 °K = 8.617×10^{-14} GeV.

[2] W. Rindler, Mon. Not. R. Astron. Soc. 116, 663 (1956). See also Ref. 3, pp. 489–490, 525–526; and Ref. 4, pp. 740 and 815.

[3] S. Weinberg, *Gravitation and Cosmology* (Wiley, New York, 1972).

[4] C. W. Misner, K. S. Thorne, and J. A. Wheeler, *Gravitation* (Freeman, San Francisco, 1973).

[5] In order to calculate the horizon distance, one must of course follow the light trajectories back to $t = 0$. This violates my contention that the equations are to be trusted only for $T \lesssim T_0$. Thus, the horizon problem could be obviated if the full quantum gravitational theory had a radically different behavior from the naive extrapolation. Indeed, solutions of this sort have been proposed by A. Zee, Phys. Rev. Lett. 44, 703 (1980) and by F. W. Stecker, Astrophys. J. 235, L1 (1980). However, it is the point of this paper to show that the horizon problem can also be obviated by mechanisms which are more within our grasp, occurring at temperatures below T_0.

[6] R. H. Dicke and P. J. E. Peebles, *General Relativity: An Einstein Centenary Survey,* edited by S. W. Hawking and W. Israel (Cambridge University Press, London, 1979).

[7] See Ref. 3, pp. 475–481; and Ref. 4, pp. 796–797.

[8] For example, see Ref. 3, Chap. 14.

[9] M. Yoshimura, Phys. Rev. Lett. 41, 281 (1978); 42, 746(E) (1979); Phys. Lett. 88B, 294 (1979); S. Dimopoulos and L. Susskind, Phys. Rev. D 18, 4500 (1978); Phys. Lett. 81B, 416 (1979); A. Yu Ignatiev, N. V. Krashikov, V. A. Kuzmin, and A. N. Tavkhelidze, *ibid.* 76B, 436 (1978); D. Toussaint, S. B. Treiman, F. Wilczek, and A. Zee, Phys. Rev. D 19, 1036 (1979); S. Weinberg, Phys. Rev. Lett. 42, 850 (1979); D. V. Nanopoulos and S. Weinberg, Phys. Rev. D 20, 2484 (1979); J. Ellis, M. K. Gaillard, and D. V. Nanopoulos, Phys. Lett. 80B, 360 (1979); 82B, 464 (1979); M. Honda and M. Yoshimura, Prog. Theor. Phys. 62, 1704 (1979); D. Toussaint and F. Wilczek, Phys. Lett. 81B, 238 (1979); S. Barr, G. Segre, and H. A. Weldon, Phys. Rev. D 20, 2494 (1979); A. D. Sakharov, Zh. Eksp. Teor. Fiz. 76, 1172 (1979) [Sov. Phys.—JETP 49, 594 (1979)]; A. Yu Ignatiev, N. V. Krashikov, V. A. Kuzmin, and M. E. Shaposhnikov, Phys. Lett. 87B, 114 (1979); E. W. Kolb and S. Wolfram, *ibid.* 91B, 217 (1980); Nucl. Phys. B172, 224 (1980); J. N. Fry, K. A. Olive, and M. S. Turner, Phys. Rev. D 22, 2953 (1980); 22, 2977 (1980); S. B. Treiman and F. Wilczek, Phys. Lett. 95B, 222 (1980); G. Senjanović and F. W. Stecker, Phys. Lett. B (to be published).

[11] The reason Λ is so small is of course one of the deep mysteries of physics. The value of Λ is not determined by the particle theory alone, but must be fixed by whatever theory couples particles to quantum gravity. This appears to be a separate problem from the ones discussed in this paper, and I merely use the empirical fact that $\Lambda \approx 0$.

[12] S. A. Bludman and M. A. Ruderman, Phys. Rev. Lett. 38, 255 (1977).

[13] The effects of a false vacuum energy density on the evolution of the early universe have also been considered by E. W. Kolb and S. Wolfram, CAL TECH Report No. 79-0984 (unpublished), and by S. A. Bludman, University of Pennsylvania Report No. UPR-0143T, 1979 (unpublished).

[14] More precisely, the usual invariance is O(4) if $k = 1$, O(3,1) if $k = -1$, and the group of rotations and translations in three dimensions if $k = 0$.

[15] See for example, Ref. 3, pp. 385–392.

[16] A. H. Guth and S.-H. Tye, Phys. Rev. Lett. 44, 631 (1980); 44, 963 (1980).

[17] S. Coleman, Phys. Rev. D 15, 2929 (1977); C. G. Callan and S. Coleman, *ibid.* 16, 1762 (1977); see also S. Coleman, in *The Whys of Subnuclear Physics*, proceedings of the International School of Subnuclear Physics, Ettore Majorana, Erice, 1977, edited by A. Zichichi (Plenum, New York, 1979).

[18] A. H. Guth and E. J. Weinberg, Phys. Rev. Lett. 45, 1131 (1980).

[19] E. J. Weinberg and I are preparing a manuscript on the possible cosmological implications of the phase transitions in the SU_5 grand unified model.

[20] J. Ellis and G. Steigman, Phys. Lett. 89B, 186 (1980); J. Ellis, M. K. Gaillard, and D. V. Nanopoulos, *ibid.* 90B, 253 (1980).

[21] B. L. Hu and L. Parker, Phys. Rev. D 17, 933 (1978).

[22] See Ref. 4, Chap. 30.

[23] The simplest grand unified model is the SU(5) model of H. Georgi and S. L. Glashow, Phys. Rev. Lett. 32, 438 (1974). See also H. Georgi, H. R. Quinn, and S. Weinberg, *ibid.* 33, 451 (1974); and A. J. Buras, J. Ellis, M. K. Gaillard, and D. V. Nanopoulos, Nucl. Phys. B135, 66 (1978).

[24] Other grand unified models include the SO(10) model: H. Georgi, in *Particles and Fields—1975*, proceedings of the 1975 meeting of the Division of Particles and Fields of the American Physical Society, edited by Carl Carlson (AIP, New York, 1975); H. Fritzsch and P. Minkowski, Ann. Phys. (N.Y.) 93, 193 (1975); H. Georgi and D. V. Nanopoulos, Phys. Lett. 82B, 392 (1979) and Nucl. Phys. B155, 52 (1979). The E(6) model: F. Gürsey, P. Ramond, and P. Sikivie, Phys. Lett. 60B, 177 (1975); F. Gürsey and M. Serdaroglu, Lett. Nuovo Cimento 21, 28 (1978). The E(7) model: F. Gürsey and P. Sikivie, Phys. Rev. Lett. 36, 775 (1976), and Phys. Rev. D 16, 816 (1977); P. Ramond, Nucl. Phys. B110, 214 (1976). For some general properties of grand unified models, see M. Gell-Mann, P. Ramond, and R. Slansky, Rev. Mod. Phys. 50, 721 (1978). For a review, see P. Langacker, Report No. SLAC-PUB-2544, 1980 (unpublished).

[25] D. A. Kirzhnits and A. D. Linde, Phys. Lett. 42B, 471 (1972).

[26] S. Weinberg, Phys. Rev. D 9, 3357 (1974); L. Dolan and R. Jackiw, *ibid.* 9, 3320 (1974); see also D. A. Kirzhnits and A. D. Linde, Ann. Phys. (N.Y.) 101, 195 (1976); A. D. Linde, Rep. Prog. Phys. 42, 389 (1979). ϵ-expansion techniques are employed by P. Ginsparg, Nucl. Phys. B (to be published).

[27] In the case that the Higgs quartic couplings are comparable to g^4 or smaller (g = gauge coupling), the phase structure has been studied by M. Daniel and C. E. Vayonakis, CERN Report No. TH.2860 1980

(unpublished); and by P. Suranyi, University of Cincinnati Report No. 80-0506 (unpublished).

[28]G. 't Hooft, Nucl. Phys. B79, 276 (1974); A. M. Polyakov, Pis'ma Zh. Eksp. Teor. Fiz. 20, 430 (1974) [JETP Lett. 20, 194 (1974)]. For a review, see P. Goddard and D. I. Olive, Rep. Prog. Phys. 41, 1357 (1978).

[29]If $\Pi_1(G)$ and $\Pi_2(G)$ are both trivial, then $\Pi_2(G/H_0) = \Pi_1(H_0)$. In our case $\Pi_1(H_0)$ is the group of integers. For a general review of topology written for physicists, see N. D. Mermin, Rev. Mod. Phys. 51, 591 (1979).

[30]Y. B. Zeldovich and M. Y. Khlopov, Phys. Lett. 79B, 239 (1978).

[31]J. P. Preskill, Phys. Rev. Lett. 43, 1365 (1979).

[32]T. W. B. Kibble, J. Phys. A 9, 1387 (1976).

[33]This argument was first shown to me by John Preskill. It is also described by Einhorn *et al.*, Ref. 34, except that they make no distinction between $T_{\rm cool}$ and T_c.

[34]The problem of monopole production was also examined by M. B. Einhorn, D. L. Stein, and D. Toussaint, Phys. Rev. D 21, 3295 (1980), who focused on second-order transitions. The structure of SU(5) monopoles has been studied by C. P. Dokos and T. N. Tomaras, Phys. Rev. D 21, 2940 (1980); and by M. Daniel, G. Lazarides, and Q. Shafi, Nucl. Phys. B170, 156 (1980). The problem of suppression of the cosmological production of monopoles is discussed by G. Lazarides and Q. Shafi, Phys. Lett. 94B, 149 (1980), and G. Lazarides, M. Magg, and Q. Shafi, CERN Report No. TH.2856, 1980 (unpublished); the suppression discussed here relies on a novel confinement mechanism, and also on the same kind of supercooling as in Ref. 16. See also J. N. Fry and D. N. Schramm, Phys. Rev. Lett. 44, 1361 (1980).

[35]An alternative solution to the monopole problem has been proposed by P. Langacker and S.-Y. Pi, Phys. Rev. Lett. 45, 1 (1980). By modifying the Higgs structure, they have constructed a model in which the high-temperature SU_5 phase undergoes a phase transition to an SU_3 phase at $T \sim 10^{14}$GeV. Another phase transition occurs at $T \sim 10^3$ GeV, and below this temperature the symmetry is $SU_3 \times U_1^{EM}$. Monopoles cannot exist until $T < 10^3$ GeV, but their production is negligible at these low temperatures. The suppression of monopoles due to the breaking of U_1^{EM} symmetry at high temperatures was also suggested by S.-H. Tye, talk given at the 1980 Guangzhou Conference on Theoretical Particle Physics, Canton, 1980 (unpublished).

[36]The Weinberg-Salam phase transition has also been investigated by a number of authors: E. Witten, Ref. 41; M. A. Sher, Phys. Rev. D 22, 2989 (1980); P. J. Steinhardt, Harvard report, 1980 (unpublished); and A. H. Guth and E. J. Weinberg, Ref. 18.

[37]This section represents the work of E. J. Weinberg, H. Kesten, and myself. Weinberg and I are preparing a manuscript on this subject.

[38]The proof of this statement was outlined by H. Kesten (Dept. of Mathematics, Cornell University), with details completed by me.

[39]E. Witten, private communication.

[40]S. Coleman and E. J. Weinberg, Phys. Rev. D 7, 1888 (1973); see also, J. Ellis, M. K. Gaillard, D. Nanopoulos, and C. Sachrajda, Phys. Lett. 83B, 339 (1979), and J. Ellis, M. K. Gaillard, A. Peterman, and C. Sachrajda, Nucl. Phys. B164, 253 (1980).

[41]E. Witten, Nucl. Phys. B (to be published).

[42]L. Parker, in *Asymptotic Structure of Spacetime*, edited by F. Esposito and L. Witten (Plenum, New York, 1977); V. N. Lukash, I. D. Novikov, A. A. Starobinsky, and Ya. B. Zeldovich, Nuovo Cimento 35B, 293 (1976).

Errata

Eq. (4.4) should read

$$T_{coal} < \left[10^{-24} \frac{2\zeta(3)}{\pi^2}\right]^{1/3} \gamma^{-1} M_p \approx 10^{10} \text{ GeV}$$

Addendum

Since the publication of this article, many of the papers which were cited as preprints have been published:

Ref. 9: G. Senjanovic and F. W. Stecker, Phys. Lett. 96B, 285 (1980).

Ref. 19: A. H. Guth and E. J. Weinberg, Phys. Rev. D23, 876 (1981).

Ref. 24: P. Langacker, Phys. Rep. 72C, 185 (1981).

Ref. 26: P. Ginsparg, Nucl. Phys. B170 [FS1], 388 (1980).

Ref. 27: M. Daniel and C. E. Vayonakis, Nucl. Phys. B180, 301 (1981).

Ref. 34: G. Lazarides, M. Magg, and Q. Shafi, Phys. Lett. 97B, 87 (1980).

Ref. 36: P. J. Steinhardt, Nucl. Phys. B179, 492 (1981).

Ref. 37: A. H. Guth and E. J. Weinberg, to be published in Nucl. Phys. B (1982).

Ref. 41: E. Witten, Nucl. Phys. B177, 477 (1981).

Rep. Prog. Phys., Vol. 42, 1979. Printed in Great Britain

Phase transitions in gauge theories and cosmology

A D LINDE

I E Tamm Department of Theoretical Physics, P N Lebedev Physical Institute, Academy of Sciences of the USSR, Moscow, USSR

Abstract

In this review we discuss phase transitions in super-dense matter, which consists of particles interacting in accordance with the unified gauge theories of weak, strong and electromagnetic interactions. It is shown that at a sufficiently large temperature a phase transition takes place after which almost all elementary particles in the hot super-dense matter become massless and weak interactions become long-range like electromagnetic interactions. Analogous phenomena may take place with an increase of fermion density in cold dense matter, and also in the presence of external fields and currents. Phase transitions in gauge theories lead to a time dependence of the masses of particles, of coupling constants and of the cosmological term in the expanding Universe, to the appearance of a domain structure of vacuum, to substance energy non-conservation, to a possibility of obtaining the 'hot' Universe starting with a 'cold' one, and to some other unusual effects important for cosmology and for elementary particle physics.

This review was received in April 1978.

Contents

	Page
1. Introduction	391
1.1. Gauge theories with spontaneous symmetry breaking	391
1.2. Phase transitions in SBGT	398
2. Effective potential and spontaneous symmetry breaking in quantum field theory	400
3. Symmetry restoration at high temperatures	404
3.1. Elementary theory of the phase transition	404
3.2. Effective potential at $T \neq 0$	406
3.3. Higher orders of perturbation theory	408
3.4. The phase transition in the Higgs model	409
3.5. The infrared problem in quantum statistics of gauge fields	413
4. Symmetry behaviour in external fields	417
4.1. Quasimagnetic massive vector fields	417
4.2. Magnetic and electric fields	418
5. Effects connected with the fermion density increase	420
5.1. Theories without neutral currents (σ model)	420
5.2. Theories with neutral currents (Weinberg–Salam model)	422
5.3. Symmetry behaviour at a simultaneous increase of fermion density and temperature	423
5.4. Condensation of the Yang–Mills fields in super-dense matter	425
6. SBGT and cosmology	426
6.1. Symmetry behaviour in the early Universe	426
6.2. Quarks in the Universe	428
6.3. Domain structure of vacuum	429
6.4. Substance energy non-conservation and the time-dependent cosmological term	430
6.5. Boiling of vacuum and an interplay between symmetry breaking in gauge theories and cosmology	432
6.6. Cold Universe?	433
7. Conclusions	434
Acknowledgments	435
References	435

1. Introduction

The discovery of the unified gauge theories of weak, strong and electromagnetic interactions at the beginning of the 1970s has opened up a new phase of development of the theory of elementary particles and of quantum field theory. For brevity, and in accordance with the accepted terminology, we shall call these theories SBGT (spontaneously broken gauge theories). A consistent description of weak and strong interactions was performed first by means of SBGT, and at present the quantum field theory treatment of weak and strong interactions is carried out almost exclusively in the framework of SBGT.

Therefore, it was natural to investigate which consequences SBGT lead to when describing matter at high temperature and density, when the effects connected with weak and strong interactions should be taken into account. The results of this investigation appear to be rather unexpected and interesting.

It appears that in super-dense matter, which consists of the particles interacting in accordance with SBGT, various phase transitions should occur, which lead to the restoration of the originally broken symmetry between weak, strong and electromagnetic interactions. As a result, almost all elementary particles become massless, weak and strong interactions become long-range like electromagnetic interactions, the energy of substance is not conserved due to the 'pumping' of energy from the non-observable classical scalar field, and a number of other striking and unusual effects take place. Being interesting in themselves, these effects are also very important for an understanding of the physical processes at early stages in the evolution of the Universe. In some cases the phase transitions in the early Universe may influence strongly the large-scale structure of the Universe and even the properties of elementary particles at the present time.

The existence of various phase transitions in super-dense matter may also appear to be important for the investigation of the properties of matter in the cores of neutron stars, for the theory of quark stars, and also for the investigation of the processes of multiparticle production in high-energy particle collisions.

In the present article we shall give a review of the results concerning phase transitions in gauge theories. Since many of the effects discussed in this review may seem rather unusual, we shall try to outline everywhere when possible the derivations of the main results, but for the details of calculations usually we shall refer the reader to the original literature. Before beginning the presentation of the main content of the review we shall remind readers of some general features of SBGT.

1.1. Gauge theories with spontaneous symmetry breaking

Before the discovery of SBGT the most popular theory of weak interactions was the theory involving the intermediate vector meson. Weak interactions in this theory were mediated by a heavy vector W_μ meson (figure 1(*a*)), whereas electromagnetic interactions were mediated by the massless vector (electromagnetic) field A_μ (figure 1(*b*)). This similarity stimulated many attempts to consider weak and electromagnetic interactions from a unique point of view by means of a unification of these interactions in some general symmetry group.

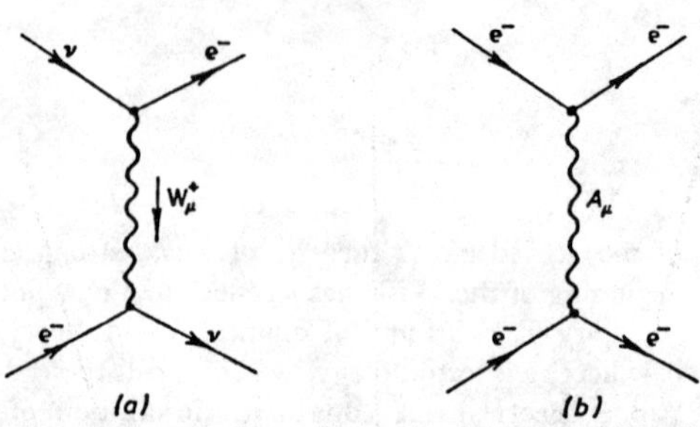

Figure 1. (*a*) Weak interaction between neutrino and electron mediated by the heavy W_μ meson. (*b*) Electromagnetic interaction between electrons mediated by the electromagnetic field A_μ.

However, this programme seemed to be unrealisable since the photons A_μ were massless while the W_μ mesons were massive. The same property also leads to the non-renormalisability of the theory of weak interactions. The reason for this non-renormalisability is that the Green function of the vector particle W_μ with the mass m and momentum k:

$$G_{\mu\nu}{}^{W}(k)=\left(\delta_{\mu\nu}+\frac{k_\mu k_\nu}{m^2}\right)\frac{1}{k^2+m^2} \tag{1.1}$$

(everywhere throughout this review we use Euclidean Green functions, $k^2=k_0{}^2+\mathbf{k}^2$) does not vanish at $k\to\infty$.

This leads to additional ultraviolet divergences, increasing with each new order of the perturbation theory. In quantum electrodynamics (QED) the corresponding difficulties are absent just due to the vanishing of $G_{\mu\nu}{}^{A}(k)$ at $k\to\infty$. For example, in the transverse gauge $\partial_\mu A_\mu=0$ in QED:

$$G_{\mu\nu}{}^{A}(k)=\left(\delta_{\mu\nu}-\frac{k_\mu k_\nu}{k^2}\right)\frac{1}{k^2}. \tag{1.2}$$

Nevertheless not only weak and electromagnetic, but also strong, interactions have been unified and the resulting models of weak, strong and electromagnetic interactions prove to be renormalisable. This success is connected with the use of theories with spontaneous symmetry breaking.

A simple example of a theory with spontaneous symmetry breaking is the theory of a scalar field ϕ with the Lagrangian:

$$L=\tfrac{1}{2}(\partial_\mu\phi)^2+\tfrac{1}{2}\mu^2\phi^2-\tfrac{1}{4}\lambda\phi^4. \tag{1.3}$$

For simplicity we suppose everywhere the coupling constant λ to be small (weak coupling): $0<\lambda\ll 1$. The 'wrong' sign of the mass term $\frac{1}{2}\mu^2\phi^2$ makes the Lagrangian (1.3) similar to that of the hypothetic superluminal particles—tachyons. This means that the ordinary solution of the Lagrange equation in the theory (1.3), which could be obtained from perturbation theory by expanding ϕ near $\phi=0$, is unstable. The simplest way to understand this is to consider the expression for the energy (for the effective potential, see §2) of a constant field $\phi=\sigma$, which at the classical level is given by:

$$V(\sigma)=-\tfrac{1}{2}\mu^2\sigma^2+\tfrac{1}{4}\lambda\sigma^4. \tag{1.4}$$

It is seen that due to the 'wrong' sign of the mass term $-\frac{1}{2}\mu^2\sigma^2$ the energy $V(\sigma)$

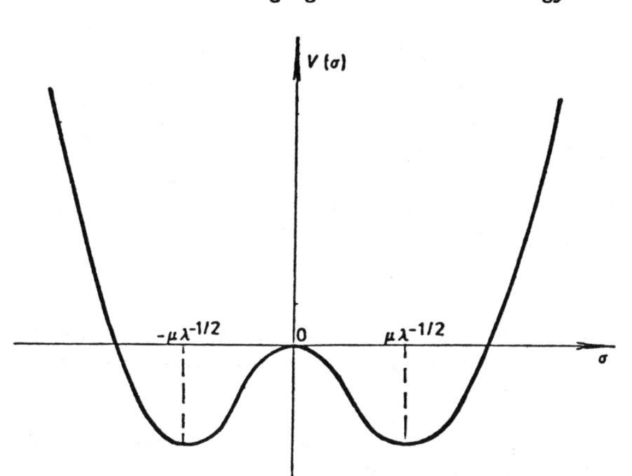

Figure 2. Effective potential $V(\sigma)$ in the model (1.3).

has a minimum not at $\sigma=0$ but at $\sigma=\pm\mu\lambda^{-1/2}$ (figure 2). Therefore, even if the system initially was in the symmetric state $\sigma=0$, very rapidly a transition to the state $\sigma=\pm\mu\lambda^{-1/2}$ should take place. This effect is just the so-called spontaneous symmetry breaking. In the language of the field operators this means that after the spontaneous symmetry breaking the vacuum expectation value $\langle 0|\phi|0\rangle$ does not vanish:

$$\langle 0|\phi|0\rangle\equiv\sigma=\pm\mu\lambda^{-1/2}\neq 0.$$

To return to the usual creation and annihilation operators with vanishing vacuum expectation values one should perform a shift:

$$\phi\to\phi+\sigma \tag{1.5}$$

where σ is the classical constant scalar field:

$$\sigma=\pm\mu\lambda^{-1/2} \tag{1.6}$$

and after the shift one has as usual:

$$\langle 0|\phi|0\rangle=0.$$

The Lagrangian (1.3) after the shift is given by:

$$L(\phi+\sigma)=\tfrac{1}{2}(\partial_\mu(\phi+\sigma))^2+\tfrac{1}{2}\mu^2(\phi+\sigma)^2-\tfrac{1}{4}\lambda(\phi+\sigma)^4$$
$$=\tfrac{1}{2}(\partial_\mu\phi)^2-\tfrac{1}{2}(3\lambda\sigma^2-\mu^2)\phi^2-\lambda\sigma\phi^3-\tfrac{1}{4}\lambda\phi^4+\tfrac{1}{2}\mu^2\sigma^2-\tfrac{1}{4}\lambda\sigma^4-\sigma(\lambda\sigma^2-\mu^2)\phi. \tag{1.7}$$

Taking equation (1.6) into account the last term in equation (1.7) vanishes and the mass term in equation (1.7) has a correct sign and corresponds to the ϕ particles with the mass squared:

$$m_\phi{}^2=2\mu^2. \tag{1.8}$$

One can obtain the same result by a somewhat more accurate method often used in what follows. Let us consider the vacuum average of the Lagrange equation for the theory (1.3):

$$\langle 0|\delta L/\delta\phi|0\rangle=\langle 0|(\Box\phi+\mu^2\phi-\lambda\phi^3)|0\rangle=0. \tag{1.9}$$

After the shift (1.5) this equation becomes:

$$\Box\sigma-(\lambda\sigma^2-\mu^2)\sigma-3\lambda\sigma\langle 0|\phi^2|0\rangle-\lambda\langle 0|\phi^3|0\rangle=0. \tag{1.10}$$

In the lowest order in λ the average $\langle 0|\phi^3|0\rangle$ vanishes, and the term $-3\lambda\sigma\langle 0|\phi^2|0\rangle$, which appears due to the vacuum fluctuations of the field ϕ, can be removed by the mass renormalisation (i.e. by adding the counter-term $\delta m^2\phi^2$ to the Lagrangian (1.3)). In this case equation (1.8) yields:

$$\Box\sigma+\sigma(\mu^2-\lambda\sigma^2)=0 \tag{1.11}$$

(for the quantum corrections to this equation see §2). Then, supposing $\sigma=$ constant (translationally invariant vacuum state), one obtains two possible solutions: $\sigma=0$ and $\sigma=\pm\mu\lambda^{-1/2}$. To obtain the excitation spectrum in the theory (1.3) one should perform in equation (1.10) an infinitesimal shift $\sigma\to\sigma+\delta\sigma$, where σ is one of the two possible constant solutions of equation (1.8). At $\sigma=0$ the equation for $\delta\sigma$ obtained in such a way is:

$$\Box\delta\sigma+\mu^2\delta\sigma=0 \tag{1.12}$$

which corresponds to the existence of tachyons with the mass squared $-\mu^2$. Therefore, the fluctuations $\delta\sigma(\boldsymbol{k})$ with the momentum $\boldsymbol{k}$ near $\sigma=0$ will grow with time as $\exp\,[(\mu^2-\boldsymbol{k}^2)^{1/2}t]$ (unstable vacuum state). Meanwhile, at $\sigma=\pm\mu\lambda^{-1/2}$ the excitation spectrum is given by the equation:

$$\Box\delta\sigma-2\mu^2\delta\sigma=0 \tag{1.13}$$

which corresponds to ordinary particles with the positive mass squared $2\mu^2$ (equation (1.8)) (stable vacuum).

Let us try to understand now how one can make vector mesons massive by means of spontaneous symmetry breaking without loss of renormalisability. For this purpose we shall consider here the Higgs model (Higgs 1964a, b, 1966, Kibble 1967, Guralnik *et al* 1964, Englert and Brout 1964), which describes the interaction of a massless vector field A_μ with a complex scalar field χ with a 'wrong' sign of the mass term $\mu^2\chi^*\chi$:

$$L=-\tfrac{1}{4}(\partial_\mu A_\nu-\partial_\nu A_\mu)^2+(\partial_\mu+\mathrm{i}eA_\mu)\chi^*(\partial_\mu-\mathrm{i}eA_\mu)\chi+\mu^2\chi^*\chi-\lambda(\chi^*\chi)^2. \tag{1.14}$$

This Lagrangian is invariant under the Abelian U(1) group of the gauge transformations:

$$A_\mu(x)\to A_\mu(x)+\frac{1}{e}\,\partial_\mu\zeta(x) \tag{1.15}$$

$$\chi(x)\to\chi(x)\exp\,(\mathrm{i}\zeta(x))$$

and, in particular, under the global rotation:

$$\chi\to\chi e^{\mathrm{i}\theta} \tag{1.16}$$

where $\theta=$ constant. Therefore, without loss of generality one may regard that after spontaneous symmetry breaking in the Higgs model (1.14) it is only the real component of the field χ that acquires some positive constant classical part σ:

$$\chi(x)=\frac{1}{\sqrt{2}}(\chi_1(x)+\mathrm{i}\chi_2(x))\to\frac{1}{\sqrt{2}}(\chi_1(x)+\sigma+\mathrm{i}\chi_2(x)) \tag{1.17}$$

and after the shift (1.17) $\langle 0|\chi_i|0\rangle=0$. According to the Goldstone theorem (Goldstone 1961, Goldstone *et al* 1962) in the gauges, which do not break the invariance of the Lagrangian (1.14) under the rotation (1.16), massless scalar particles should appear after spontaneous symmetry breaking (1.17). To verify it let us perform the shift (1.17) in the Lagrangian (1.14). After the shift the Lagrangian (1.14) looks as follows:

$$\begin{aligned}L=&-\tfrac{1}{4}(\partial_\mu A_\nu-\partial_\nu A_\mu)^2+\tfrac{1}{2}(\partial_\mu\chi_1)^2+\tfrac{1}{2}(\partial_\mu\chi_2)^2+\tfrac{1}{2}e^2((\chi_1+\sigma)^2+\chi_2^2)A_\mu^2\\&-e\sigma A_\mu\partial_\mu\chi_2+e(\chi_2\partial_\mu\chi_1-\chi_1\partial_\mu\chi_2)A_\mu-\tfrac{1}{2}(3\lambda\sigma^2-\mu^2)\chi_1^2-\tfrac{1}{2}(\lambda\sigma^2-\mu^2)\chi_2^2\\&-\lambda\sigma\chi_1(\chi_1^2+\chi_2^2)-\tfrac{1}{4}\lambda(\chi_1^2+\chi_2^2)^2-\sigma(\lambda\sigma^2-\mu^2)\chi_1+\tfrac{1}{2}\mu^2\sigma^2-\tfrac{1}{4}\lambda\sigma^4\\&+\frac{1}{2\alpha}(\partial_\mu A_\mu)^2\end{aligned}\tag{1.18}$$

where the last term is added to fix the gauge. It is seen that at the classical level the energy of the field σ is given by:

$$V(\sigma)=-\tfrac{1}{2}\mu^2\sigma^2+\tfrac{1}{4}\lambda\sigma^4\tag{1.19}$$

just like in the theory (1.3) (see (1.4)). Therefore the classical field σ, which appears after the spontaneous symmetry breaking, is equal to $\sigma=\mu\lambda^{-1/2}$ as in (1.3). The propagators of the fields A_μ and χ_i ,which correspond to the Lagrangian (1.18), are of a manifestly 'renormalisable' type. For example, in the transverse gauge $\partial_\mu A_\mu=0$ ($\alpha\to 0$ in (1.18)) these propagators are given respectively by:

$$G_{\mu\nu}{}^A=\left(\delta_{\mu\nu}-\frac{k_\mu k_\nu}{k^2}\right)\frac{1}{k^2+e^2\sigma^2}\tag{1.20}$$

$$G_{\chi_1}=\frac{1}{k^2+2\lambda\sigma^2}\tag{1.21}$$

$$G_{\chi_2}=\frac{1}{k^2}.\tag{1.22}$$

From (1.22) it is seen that after the symmetry breaking the field χ_2 becomes massless in accordance with the Goldstone theorem. The propagator of the field A_μ can be represented in the form:

$$G_{\mu\nu}{}^A=\left(\delta_{\mu\nu}+\frac{k_\mu k_\nu}{e^2\sigma^2}\right)\frac{1}{k^2+e^2\sigma^2}-\frac{k_\mu k_\nu}{k^2e^2\sigma^2}.\tag{1.23}$$

The first part of this Green function is the propagator of a vector particle with the mass $e\sigma=e\mu\lambda^{-1/2}$ (compare with (1.1)) while the second part of (1.23) corresponds to the propagation of a longitudinally polarised massless particle with indefinite metric. It can be shown that the contribution of this last particle to all physical processes is cancelled exactly by the corresponding contribution of the Goldstone particle χ_2. To illustrate this statement let us perform a change of variables in the Lagrangian (1.14):

$$\begin{aligned}&\chi(x)\to\frac{1}{\sqrt{2}}(\phi(x)+\sigma)\exp\left(i\zeta(x)/\sigma\right)\\&A_\mu(x)\to A_\mu(x)-\frac{1}{e\sigma}\partial_\mu\zeta(x)\end{aligned}\tag{1.24}$$

instead of the previously used shift (1.17). The Lagrangian (1.14) at $\sigma=\mu\lambda^{-1/2}$ is then transformed into:

$$L=-\tfrac{1}{4}(\partial_\mu A_\nu-\partial_\nu A_\mu)^2+\tfrac{1}{2}e^2\sigma^2A_\mu{}^2+\tfrac{1}{2}e^2\phi^2A_\mu{}^2+e^2\sigma\phi A_\mu+\tfrac{1}{2}(\partial_\mu\phi)^2-\lambda\sigma^2\phi^2$$
$$-\lambda\sigma\phi^3-\tfrac{1}{4}\lambda\phi^4+\mu^2/4\lambda \tag{1.25}$$

from which the auxiliary field $\zeta(x)$ has been completely transformed away.

The propagators of the fields A_μ and ϕ in this case are given by:

$$G_{\mu\nu}{}^A=\left(\delta_{\mu\nu}+\frac{k_\mu k_\nu}{e^2\sigma^2}\right)\frac{1}{k^2+e^2\sigma^2} \tag{1.26}$$

$$G_\phi=\frac{1}{k^2+2\lambda\sigma^2} \tag{1.27}$$

and no Goldstone mesons and the particles with indefinite metric are present.

It can be seen that the Lagrangian (1.25) is equal to the Lagrangian (1.18) (without the gauge term $(1/2\alpha)(\partial_\mu A_\mu)^2$) in the gauge $\chi_2=0$. This means that both the Goldstone field χ_2 and the longitudinal field with indefinite metric can be transformed away by choosing the gauge $\chi_2=0$. Therefore, after spontaneous symmetry breaking in the Higgs model instead of the massless vector field A_μ (two degrees of freedom) and the complex field χ with negative mass squared $-\mu^2$ (two degrees of freedom), one deals with the massive vector field A_μ (three degrees of freedom) with the mass $m_A=e\sigma=e\mu\lambda^{-1/2}$ and with the real scalar field ϕ with the mass $m_\phi=\sqrt{2\lambda}\sigma=\sqrt{2}\,\mu$ (one degree of freedom). The mechanism of the vector meson mass generation discussed above (the Higgs mechanism) serves as a basis for SBGT. Note that in the R gauges specified by the term $(1/2\alpha)(\partial_\mu A_\mu)^2$ in (1.18) one can show that the theory is renormalisable. However, in these gauges it is difficult to prove the unitarity of the theory in the subspace of the physical states (without the Goldstone field χ_2 and the field with indefinite metric). Meanwhile in the gauge $\chi_2=0$ (1.25) the unitarity is manifest, and therefore this gauge is called 'unitary gauge' (U gauge), but the renormalisability of the theory in this gauge is not obvious. However, the theories (1.18) and (1.25) are equivalent since the corresponding Lagrangians differ in the gauge conditions only and, consequently, the theory is both renormalisable and unitary. A more accurate formulation and the proof of this important statement were given by 't Hooft (1971), Slavnov (1972), Taylor (1971), Lee (1972), Lee and Zinn-Justin (1972), 't Hooft and Veltman (1972), Tyutin and Fradkin (1974), Kallosh and Tyutin (1973) and Ross and Taylor (1973).

Fermion masses in most of SBGT also appear due to spontaneous symmetry breaking. As an example we shall consider here a simplified version of the linear σ model, which is often used for the description of strong interactions (see, for example, Lee 1972). The Lagrangian of this model is given by:

$$L=\tfrac{1}{2}(\partial_\mu\phi)^2+\tfrac{1}{2}\mu^2\phi^2-\tfrac{1}{4}\lambda\phi^4+\bar\psi(\mathrm{i}\partial_\mu\gamma_\mu-g\phi)\psi=L_\phi+\bar\psi(\mathrm{i}\partial_\mu\gamma_\mu-g\phi)\psi \tag{1.28}$$

where ψ is the massless fermion field interacting with the scalar field ϕ and the Lagrangian L_ϕ is given by (1.3). After spontaneous symmetry breaking in the theory (1.28) the field ϕ, as in (1.3), acquires a non-vanishing constant classical part $\sigma=\pm\mu\lambda^{-1/2}$ (1.6). This leads to the appearance of the term $g\sigma\bar\psi\psi$ in the Lagrangian (1.28), which means that after the symmetry breaking the fermions acquire a mass

$$m_\psi=|g\sigma|=g\mu\lambda^{-1/2}.$$

The first realistic SBGT of weak and electromagnetic interactions has been suggested by Weinberg (1967) and Salam (1968). According to this model all vector particles and fermions before symmetry breaking are massless. After symmetry breaking only photons and neutrinos remain massless, whereas the intermediate vector mesons responsible for weak interactions acquire very large masses (~ 80 GeV). Even this first comparatively simple model proves to be fairly successful in the description of weak and electromagnetic interactions. However, the real interest in theories of this type appeared only at the beginning of the 1970s when it was shown that SBGT are renormalisable and therefore all quantum corrections in these theories can be computed unambiguously.

Further investigations have shown that some of the SBGT have serious advantages not only over the old theories of weak and strong interactions (which usually were unrenormalisable), but even over the ordinary quantum electrodynamics. More than twenty years ago it was argued that in most of the models in quantum field theory (and in particular in quantum electrodynamics) effective coupling constants become infinite at large momenta and the Green functions contain tachyon poles (Landau and Pomeranchuk 1955, Fradkin 1955). This leads to the vacuum instability with respect to spontaneous generation of infinitely strong classical fields (Linde 1977a, Kirzhnits and Linde 1978a,b), and therefore the models with such a pathological behaviour of coupling constants and Green functions are physically unacceptable. Fortunately a class of the non-Abelian asymptotically free gauge theories has been discovered (Gross and Wilczek 1973, Politzer 1973), in which effective coupling constants vanish at large momenta. In some of these theories tachyons are absent or disappear after symmetry breaking, and therefore such theories are free from the difficulties mentioned above (Voronov and Tyutin 1975). Another advantage of the asymptotically free SBGT is that the asymptotic freedom may serve as an explanation of approximate Bjorken scaling in strong interactions. All these (and many other) beautiful properties of gauge theories, together with the experimental discovery of neutral currents (of weak interactions mediated by neutral vector mesons) and of 'charmed' particles, which naturally appear in SBGT, have led most physicists into a belief that just the principle of spontaneous breaking of gauge invariance should be a basis for a theory of all fundamental interactions. For a detailed discussion of the properties of SBGT one can refer to excellent reviews by Abers and Lee (1973), Weinberg (1974b) and Fradkin and Tyutin (1974).

In recent years many attempts have been made to consider the theories in which some of the symmetries are broken not spontaneously, but dynamically (i.e. due to quantum effects), or even remain unbroken. The most interesting theory of this type is quantum chromodynamics (QCD), in which the Yang–Mills vector fields responsible for strong interactions are supposed to be massless, but the resulting 'strong' forces are believed to be short-range due to the effects connected with infrared instability and some special topological properties of the massless non-Abelian theories (see, for example, Wilson 1974, Polyakov 1977, 1978, Callan *et al* 1977, 't Hooft 1977). Unfortunately, in spite of a large number of different suggestions neither QCD nor realistic four-dimensional theories with dynamical symmetry breaking have been completely constructed so far. (As for the dynamical symmetry breaking in a very interesting approach of Coleman and Weinberg (1973), it was shown that this dynamical symmetry breaking is actually equivalent to a spontaneous one (Linde 1976a,d).) Therefore, in this review we shall give some comments concerning the possible phase transitions in QCD and in the models with dynamical symmetry breaking but the main content of the review will be concerned with gauge theories with spontaneous symmetry breaking.

1.2. Phase transitions in SBGT

The idea of spontaneous symmetry breaking which appears so useful in applications to quantum field theory actually was used long ago in solid-state physics and in quantum statistics applied to the theories of such phenomena as ferromagnetism, superfluidity, superconductivity, etc. In fact, any phase transition is a process with a change of symmetry which before (or after) the phase transition is spontaneously broken. The most instructive examples in this respect are the theories of Bose condensation and superconductivity.

Let us consider, for example, the occupation numbers in the Bose gas:

$$n_p = \langle a_p^+ a_p \rangle \tag{1.29}$$

which correspond to the number of particles with the momentum p in a system under consideration. Here a_p^+ and a_p are operators of creation and annihilation for a particle with the momentum p, and $\langle \ldots \rangle$ is the Gibbs average (see, for example, Landau and Lifshitz 1964, Abrikosov *et al* 1964, Fradkin 1965):

$$\langle \ldots \rangle = \frac{Sp[\exp(-H/T) \ldots]}{Sp[\exp(-H/T)]} \tag{1.30}$$

where H is the Hamiltonian of the system. At low temperatures almost all particles of the Bose gas are condensed in the state with $p=0$, i.e.:

$$\langle a_0^+ a_0 \rangle \sim N$$

where N is the overall number of particles in the system, $N \gg 1$. This means that:

$$[a_0 a_0^+] = 1 \ll \langle a_0^+ a_0 \rangle.$$

Therefore in the computation of thermodynamical (Gibbs) averages the non-commutativity of the operators a_0 and a_0^+ is inessential, and in the limit $N \to \infty$ one can deal with these operators as with ordinary C numbers. From this point of view one can treat the appearance of a classical (C-number) part σ of the field ϕ in SBGT as a Bose condensation of the particles of the field ϕ in a state with vanishing four-momentum $p=0$. We should note, however, that Bose condensation occurs in SBGT in spite of the non-conservation of the number of scalar particles. At the same time a condensation of an ideal Bose gas is caused by an 'overcrowding' of energy levels but it is impossible if the number of particles in the gas is not fixed. This is exactly the reason why Bose condensation of the photon gas does not take place. Bose condensation in the theories with spontaneous symmetry breaking occurs due to interaction between the scalar particles, and in this sense a more accurate analogue to SBGT is superconductivity theory. This analogy becomes particularly clear if one compares equation (1.14) with the expression for the energy of a superconductor in the phenomenological superconductivity theory of Ginzburg and Landau (1950) (see also the textbooks by De Gennes (1966) and Saint-James *et al* (1969)):

$$E = E_0 + \tfrac{1}{2}H^2 + \frac{|(\nabla - 2ieA)\psi|^2}{2m} - \alpha|\psi|^2 + \beta|\psi|^4. \tag{1.31}$$

Here E_0 is the energy of a normal metal without the magnetic field H, ψ is the classical Cooper-pair field, $2m$ is the mass of the Cooper pair, and α and β are some phenomenological parameters.

By the comparison of (1.14) and (1.31) it is seen that the Higgs model is nothing but a covariant generalisation of the phenomenological Ginzburg–Landau theory. As in the Higgs model, the 'wrong' sign of α leads to instability of the symmetric ('disordered' in the terminology of quantum statistics) state $\psi=0$, and to the appearance of the superfluid Bose condensate of the Cooper pairs (of the 'order parameter' $\psi\neq0$). After the symmetry breaking in expression (1.30) there arises the mass term of the field A, analogous to the term $\frac{1}{2}e^2\sigma^2A_\mu{}^2$ in (1.25). This fact is just responsible for the exponential decrease of the magnetic field inside superconductors (Meissner effect).

The analogy between superconductivity theory and SBGT appears to be extremely useful in the study of macroscopic consequences of SBGT. It is known, for example, that heating destroys superconductivity, since at sufficiently high temperatures the Bose condensate of the Cooper pairs 'evaporates'. Proceeding from the analogy between SBGT and superconductivity theory Kirzhnits (1972) (see also Kirzhnits and Linde 1972) has suggested that in the thermodynamic equilibrium systems of elementary particles, interacting in accordance with SBGT, at a sufficiently high temperature the Bose condensate of scalar particles disappears. This leads to the restoration of initial symmetry between weak, strong and electromagnetic interactions. As a result of this phase transition all particles, which acquire their masses due to spontaneous symmetry breaking, become massless again, and weak and strong interactions become long-range like electromagnetic interactions. These conclusions and the estimates of the critical temperature T_c of the phase transition have been further confirmed by the work of Weinberg (1974a), Dolan and Jackiw (1974) and Kirzhnits and Linde (1974a, b, 1976a).

As is known, superconductivity can be destroyed not only by heating but also by external currents and magnetic fields. Therefore, it was natural to expect that analogous effects should occur in SBGT as well (Kirzhnits 1972, Kirzhnits and Linde 1972). And indeed it proves that symmetry restoration in SBGT takes place in the presence of large external currents (Linde 1975a, Kirzhnits and Linde 1976a, Krive 1976) and in strong massive vector 'quasimagnetic' (see §4) fields (Kirzhnits and Linde 1974b, 1976a, Krive *et al* 1976b).

In the papers by Salam and Strathdee (1974, 1975) it was suggested that not only quasimagnetic fields but also an ordinary magnetic field can lead to symmetry restoration in SBGT. A further analysis of this problem has shown, however, that in most of the models with neutral currents symmetry restoration takes place not due to a magnetic field but due to the quasimagnetic fields created simultaneously by the magnetic-field sources (Kirzhnits and Linde 1976b, Linde 1976b).

Extreme conditions, which are necessary for all these effects to occur, can be achieved in cores of neutron stars, in processes of multiparticle production and also at early stages of the evolution of the Universe. In this review we shall be concerned mainly with the consequences of SBGT for cosmology; it will be seen that taking account of the phase transitions in SBGT leads to a substantial reconsideration of the standard viewpoint on the physical processes in the early Universe.

The contents of the present review is organised as follows. In §2 we consider in a more detailed way than in the introduction (by means of the effective potential method) the problem of spontaneous symmetry breaking in quantum field theory.

In §3 a theory of high-temperature symmetry restoration in SBGT is presented, and the infrared problem in quantum statistics of massless gauge fields is discussed. In §4 symmetry behaviour in external fields is investigated. In §5 we discuss the effects connected with the fermion charge density increase. Finally in §6 some consequences of the investigated phenomena for cosmology and for elementary particle physics are discussed.

2. Effective potential and spontaneous symmetry breaking in quantum field theory

As is mentioned in the introduction, spontaneous generation of the classical part (non-vanishing vacuum expectation) σ of the field ϕ takes place only if it is energetically advantageous. At the classical level the potential energy of the 'ordered' state $\sigma \neq 0$ in the theories (1.3), (1.14) and (1.28) is actually less than the energy of the disordered state $\sigma = 0$ (see, for example, (1.4)). It should be determined, however, whether quantum corrections may invalidate this conclusion.

To answer this question one should consider the so-called 'effective potential' $V(\sigma)$, the quantity which has the meaning of the potential energy of the field σ taking into account all the quantum corrections (Jona-Lasinio 1964, Coleman and Weinberg 1973, Jackiw 1974). For this purpose we shall consider first the simplest model (1.3) and represent $L(\phi+\sigma)$ (1.7) as a sum $L_0 + L_{\text{int}}$, where:

$$L_0 = \tfrac{1}{2}(\partial_\mu(\phi+\sigma))^2 - \tfrac{1}{2}(3\lambda\sigma^2 - \mu^2)\phi^2 + \tfrac{1}{2}\mu^2\sigma^2 - \tfrac{1}{4}\lambda\sigma^4.$$

To make the state with arbitrary σ equilibrium we shall add to L_{int} an external source term $I(x)(\phi(x)+\sigma(x))$ and introduce the generating functional of the connected Green functions $W(I)$ determined by the equation:

$$W(I) = W_0 + \ln Z(I).$$

Here $-W_0$ is the energy of the field σ in the classical approximation:

$$-W_0 = \tfrac{1}{2}(\dot{\sigma})^2 + \tfrac{1}{2}(\nabla\sigma)^2 - \tfrac{1}{2}\mu^2\sigma^2 + \tfrac{1}{4}\lambda\sigma^4 \tag{2.1}$$

and $Z(I)$ is the vacuum expectation value of the S matrix $S(I)$, where:

$$S(I) = T \exp\{\textstyle\int \mathrm{d}x[L_{\text{int}}(x) + I(x)(\phi(x)+\sigma(x))]\} \tag{2.2}$$

and the integration is performed over Euclidean four-dimensional space. It is well known that the quantity $-W(I)$ is equal to the total energy of the system in the presence of an external current I (see, for example, Coleman and Weinberg 1973, Abers and Lee 1973). To separate from $-W(I)$ the part $V(\sigma)$, which corresponds to the energy of the field σ, one should subtract from $-W(I)$ the source energy $-\int \mathrm{d}x\sigma(x)I(x)$:

$$V(\sigma) = -W(I) + \textstyle\int \mathrm{d}x I(x)\sigma(x). \tag{2.3}$$

For the time-independent $\sigma(x)$ ($\sigma(x) = \sigma(\mathbf{x})$) this quantity has the meaning of the potential energy of the field σ and is called the effective potential. Generally speaking, the most energetically favourable state may appear to be a state with a spatially inhomogeneous field $\sigma(\mathbf{x})$ (e.g. some periodic (crystalline) structure). Fortunately it can be shown that this is not the case for most of the models with weak coupling considered in the present review (see, however, §5.4).

Returning to the point-independent quantities I and σ one can obtain the following equations for the derivatives of the effective potential (Jona-Lasinio 1964, Coleman and Weinberg 1973, Jackiw 1974):

$$\mathrm{d}V/\mathrm{d}\sigma = I$$

$$\mathrm{d}^2V/\mathrm{d}\sigma^2 = G_\phi^{-1}(0)$$

where $G_\phi(k)$ is the Green function of the field ϕ with the momentum k. Therefore (as one could expect) in the absence of external current I the system may be only in the state which corresponds to an extremum of the potential energy of the field σ:

$$\mathrm{d}V/\mathrm{d}\sigma = 0. \tag{2.4}$$

This state will be stable only if:

$$\mathrm{d}^2V/\mathrm{d}\sigma^2 = G_\phi^{-1}(0) \geqslant 0. \tag{2.5}$$

This last inequality has a simple physical meaning. Up to higher-order corrections:

$$G_\phi^{-1}(0) = m_\phi^2 \tag{2.6}$$

where m_ϕ is the mass of the field ϕ. In this sense inequality (2.5) implies that only the states without tachyons can be stable. (In fact, this statement can be proved without any recourse to perturbation theory (Kirzhnits and Linde 1978b).)

Graphically the effective potential $V(\sigma)$ (2.3) is given by a set of all one-particle irreducible vacuum diagrams, corresponding to the Lagrangian $L(\phi+\sigma)$ (1.7) without the terms linear in ϕ (Jackiw 1974) (see figure 3). In this review we shall present the results of the calculation of $V(\sigma)$ in the one-loop approximation only (figure 3(a) and (b)) and analyse in which cases these results are reliable.

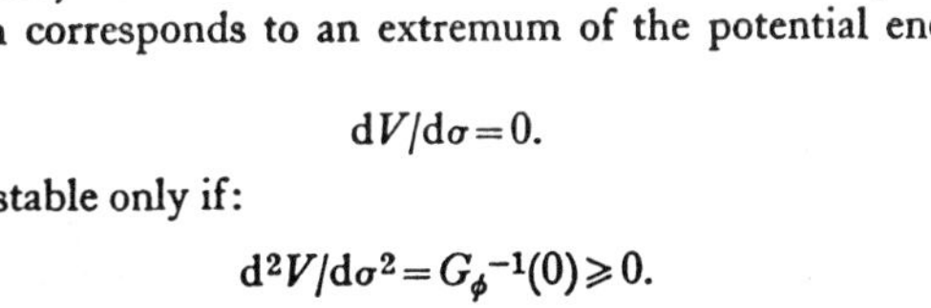

$$V(\sigma) = -\frac{\mu^2\sigma^2}{2} + \frac{\lambda\sigma^4}{4} + \ldots$$

(a) (b) (c)

Figure 3. Diagrams for $V(\sigma)$ in the model (1.3).

According to the above-mentioned rules in the one-loop approximation one obtains:

$$V(\sigma) = -\tfrac{1}{2}\mu^2\sigma^2 + \tfrac{1}{4}\lambda\sigma^4 + \frac{1}{2(2\pi)^4}\int \mathrm{d}^4k \ln\left(k^2 + m^2(\sigma)\right) \tag{2.7}$$

where $m^2(\sigma) = 3\lambda\sigma^2 - \mu^2$. The meaning of this expression for $V(\sigma)$ becomes particularly clear after the integration over k_0 in (2.7). The result (up to an infinite constant, which can be removed by the vacuum energy renormalisation) is given by:

$$V(\sigma) = -\tfrac{1}{2}\mu^2\sigma^2 + \tfrac{1}{4}\lambda\sigma^4 + \frac{1}{(2\pi)^3}\int \mathrm{d}^3k\left(k^2 + m^2(\sigma)\right)^{1/2}. \tag{2.8}$$

Thus in the one-loop approximation the effective potential in (1.3) can be considered as a sum of the 'classical' potential energy of the field (1.4) and of the σ-dependent shift in the vacuum energy density due to the zero-point oscillations of the field ϕ. (In this sense equation (2.8) for the effective potential $V(\sigma)$ in the theory (1.3) resembles the well-known Heisenberg–Euler effective Lagrangian of electromagnetic field (Heisenberg and Euler 1936).) The integral corresponding to the energy of the zero-point oscillations diverges. For the renormalisation of $V(\sigma)$ one can add to $L(\phi+\sigma)$ the counter-terms of the type $C_1(\partial_\mu(\phi+\sigma))^2$, $C_2(\phi+\sigma)^2$, $C_3(\phi+\sigma)^4$ and C_4. Actually, in our case, only the last three counter-terms are needed. The constants C_2 and C_3 will be determined by imposing the following normalisation conditions on $V(\sigma)$:

$$\left.\frac{\mathrm{d}V}{\mathrm{d}\sigma}\right|_{\sigma=\mu\lambda^{-1/2}}=0$$
$$\left.\frac{\mathrm{d}^2V}{\mathrm{d}\sigma^2}\right|_{\sigma=\mu\lambda^{-1/2}}=2\mu^2. \tag{2.9}$$

These normalisation conditions imply that the position of the minimum of $V(\sigma)$ at $\sigma\neq0$ and the curvature $\mathrm{d}^2V/\mathrm{d}\sigma^2$ in this minimum remain the same as in the classical theory. The constant C_4 fixes the vacuum energy at some given value of σ. Of course, one could use some other normalisation conditions (see, for example, Coleman and Weinberg 1973), but all the physical results should be equivalent (Linde 1976a).

The final expression for $V(\sigma)$ in (1.3), which can be obtained from (2.8) taking account of the conditions (2.9), is given by (Kirzhnits and Linde 1976a):

$$V(\sigma)=-\frac{\mu^2}{2}\sigma^2+\frac{\lambda}{4}\sigma^4+\frac{\lambda}{64\pi^2}(3\lambda\sigma^2-\mu^2)\ln\frac{3\lambda\sigma^2-\mu^2}{2\mu^2}+\frac{21\lambda\mu^2}{64\pi^2}\sigma^2-\frac{27\lambda^2}{128\pi^2}\sigma^4$$

It is seen that with $\lambda\ll1$ the quantum corrections become considerable at asymptotically large values of σ only (at $\lambda\ln\lambda\sigma^2/\mu^2\gtrsim1$), when an account of all higher orders of the perturbation theory in λ becomes necessary. Therefore one may think that in (1.3) at $\lambda\ll1$ everywhere up to the asymptotically large values of σ one can use the classical expression (1.4) for the effective potential. (Of course, in the case $\lambda\gtrsim1$ quantum corrections may become essential even at small σ (Chang 1975, 1976, Marguder 1976).)

A much more interesting situation appears in the theories with several different coupling constants. As an example we shall consider first the Higgs model (1.14). In this case the effective potential $V(\sigma)$ in the transverse gauge $\partial_\mu A_\mu=0$ is given by the diagrams shown in figure 4. At $e^2\ll\lambda$ the contribution of vector particles to

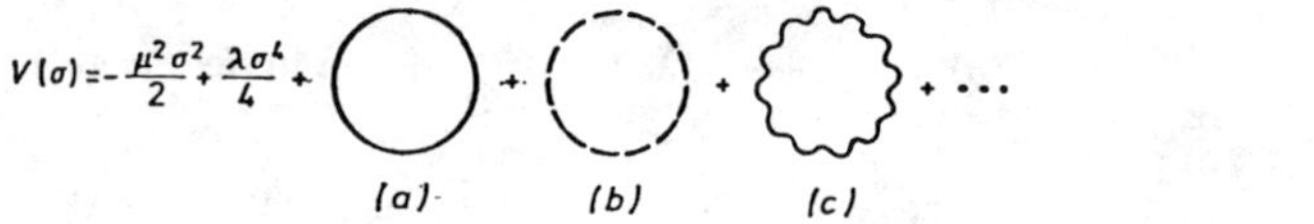

$V(\sigma)=-\frac{\mu^2\sigma^2}{2}+\frac{\lambda\sigma^4}{4}+$ (a) + (b) + (c) + ...

Figure 4. Diagrams for $V(\sigma)$ in the Higgs model. Full, broken and wavy lines correspond to the fields χ_1, χ_2 and A_μ, respectively.

$V(\sigma)$ (figure 4(*c*)) can be neglected, and the situation becomes similar to that considered above, but at $e^2\gg\lambda$, when the contribution of the scalar particles (figures 4(*a*) and (*b*)) can be neglected, the expression for $V(\sigma)$ takes the form (Linde 1976a):

$$V(\sigma)=-\frac{\mu^2\sigma^2}{2}\left(1-\frac{3e^4}{16\pi^2\lambda}\right)+\frac{\lambda\sigma^4}{4}\left(1-\frac{9e^4}{32\pi^2\lambda}\right)+\frac{3e^4\sigma^4}{64\pi^2}\ln\frac{\lambda\sigma^2}{\mu^2}. \tag{2.10}$$

From this equation it is seen that at $\lambda<3e^4/16\pi^2$ the effective potential acquires an additional ('dynamical') minimum at $\sigma=0$, and at $\lambda<3e^4/32\pi^2$ this minimum becomes even deeper than the ordinary ('classical') minimum at $\sigma=\mu\lambda^{-1/2}$ (see figure 5). Thus at $\lambda<3e^4/32\pi^2$ quantum corrections lead to the dynamical symmetry restoration in the Higgs model. Unlike in (1.3) these effects take place not due to the large logarithmic factor $\lambda \ln \lambda\sigma^2/\mu^2 \gtrsim 1$, but due to some particular relations between coupling constants λ and e^2 ($\lambda \sim e^4$), when 'classical' terms in expression (2.9) for the effective potential $V(\sigma)$ are of the same order as the lowest-order quantum corrections in e^2. The higher-order corrections are proportional to λ^2 and e^6 and can be neglected compared with the terms taken into account in equation (2.10). This means that the results discussed above are reliable. One can also

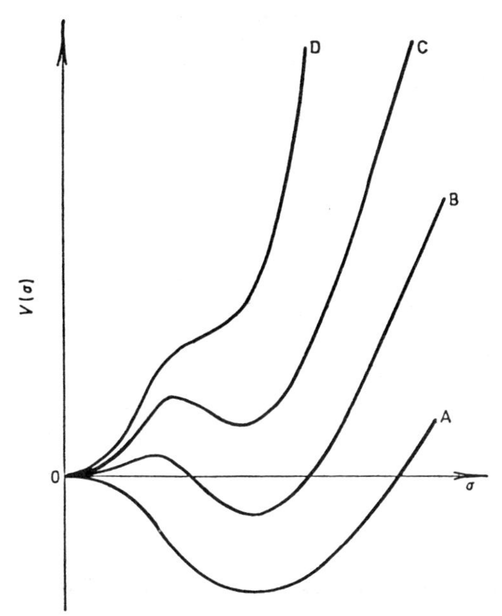

Figure 5. Effective potential $V(\sigma)$ in the Higgs model. A, $\lambda>3e^4/16\pi^2$; B, $3e^4/16\pi^2>\lambda>3e^4/32\pi^2$; C, $3e^4/32\pi^2>\lambda>0$; D, $\lambda=0$.

show that with the normalisation conditions (2.9) quantum corrections to the masses m_A and m_ϕ at $\sigma=\mu\lambda^{-1/2}$ are small:

$$m_A{}^2=e^2\sigma^2(1+O(e^2))$$

$$m_\phi{}^2=2\lambda\sigma^2(1+O(e^2)).$$

Therefore the results discussed above imply that symmetry breaking in the Higgs model is energetically favourable only if the Higgs meson mass m_ϕ is sufficiently large (Linde 1976a, Weinberg 1976):

$$m_\phi{}^2>\frac{3e^2}{16\pi^2}\,m_A{}^2. \tag{2.11}$$

We shall return to the discussion of this result in §6.

Whereas quantum corrections connected with the zero-point oscillations of the vector fields enforce dynamical symmetry restoration, the effects connected with fermions are quite the opposite. For example, in the simplified σ model (1.28) at large σ the effective potential in the one-loop approximation is given by (Krive and Linde 1976):

$$V(\sigma)=-\frac{\mu^2}{2}\sigma^2+\frac{\lambda}{4}\sigma^4+\frac{9\lambda^2-4g^4}{64\pi^2}\sigma^4\ln\frac{\lambda\sigma^2}{\mu^2}. \tag{2.12}$$

It is seen that at large σ the fermion contribution is negative, and at $3\lambda<2g^2$ the effective potential is unbounded from below (see figure 6).

Of course, for $\sigma\to\infty$ (at max $(\lambda, g^2)\ln\lambda\sigma^2/\mu^2\gtrsim 1$) the one-loop results become unreliable. Nevertheless, at $\lambda\ll g^2$ there exists some region of σ $[\sigma^2\sim(\mu^2/\lambda)\exp(\lambda/g^4)]$ for which $V(\sigma)<V(\mu\lambda^{-1/2})$ and the one-loop approximation is still reliable. Therefore, in the σ model (1.28) at $\lambda\ll g^2$, instead of the ordinary spontaneous symmetry breaking a much stronger dynamical symmetry breaking should take place (Krive and Linde 1976).

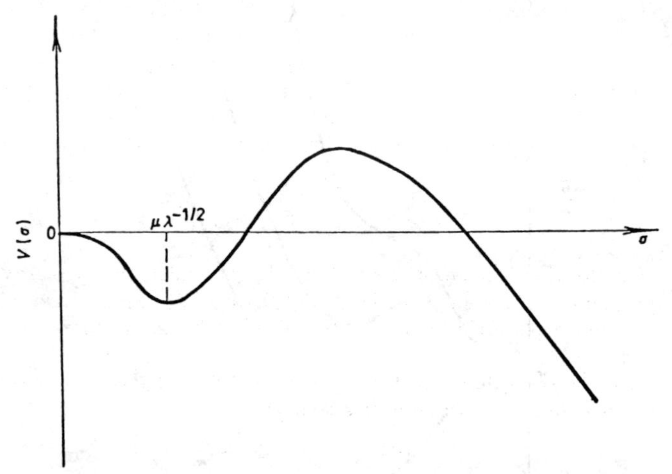

Figure 6. Effective potential $V(\sigma)$ in the σ model (1.28) at $\lambda\ll g^2$.

The above consideration shows that the quantum effects may be of crucial importance for the symmetry behaviour in gauge theories. (For a more detailed discussion of the dynamical effects in SBGT, see Linde (1976d).) However, these effects take place only at certain particular relations between coupling constants. Therefore, everywhere in this review (except for some cases which we shall mention explicitly) we shall consider the models with the relations between coupling constants for which the dynamical effects are small ($\lambda>e^4/16\pi^2$ in the Higgs model, $\lambda\gtrsim g^2$ in the σ-model, etc) and the effective potential is given by the 'classical' expressions of the type (1.4).

3. Symmetry restoration at high temperatures

3.1. Elementary theory of the phase transition

Now, after discussion of the main features of spontaneous symmetry breaking in quantum field theory, we shall begin an investigation of the symmetry behaviour in thermodynamic equilibrium systems of particles interacting in accordance with SBGT (Kirzhnits 1972, Kirzhnits and Linde 1972, 1974a, b, 1976a, Weinberg 1974a, Dolan and Jackiw 1974). We shall consider first a thermodynamic equilibrium system of the scalar particles ϕ with the Lagrangian (1.3):

$$L=\tfrac{1}{2}(\partial_\mu\phi)^2+\tfrac{1}{2}\mu^2\phi^2-\tfrac{1}{4}\lambda\phi^4.$$

These particles have no conserved charge and their number is not conserved as well. Therefore the only parameter characterising the thermodynamic equilibrium system of the ϕ particles in the temperature (T) of the system, which characterises the density of the particles in the momentum space:

$$n_p=\frac{1}{\exp(\omega_p/T)-1} \tag{3.1}$$

where $\omega_p=(p^2+m^2)^{1/2}$ is the energy of the particle with momentum p and mass m. According to (3.1) all the particles in the ground state at $T=0$ disappear and we return to the situation discussed in the previous section.

At a non-vanishing temperature all physically interesting quantities (thermodynamic potentials, Green functions, etc) in the system under consideration are given not by vacuum averages as in the field theory, but by the Gibbs averages defined by (1.20):

$$\langle\ldots\rangle=\frac{Sp[\exp(-H/T)\ldots]}{Sp[\exp(-H/T)]}$$

where H is the Hamiltonian of the system. In particular, the symmetry breaking parameter (density of the Bose condensate of the field ϕ) in the system is given not by the vacuum expectation value $\sigma=\langle 0|\phi|0\rangle$, but by the temperature-dependent quantity $\sigma(T)=\langle\phi\rangle$.

In order to investigate symmetry behaviour in (1.3) at $T\neq 0$, let us consider the Lagrange equation for the field ϕ in this theory:

$$(\Box+\mu^2-\lambda\phi^2)\phi=0 \tag{3.2}$$

and take the Gibbs average of (3.2) as has been done in the introduction for the case $T=0$. The resulting equation is (compare with (1.10)):

$$\Box\sigma(T)-(\lambda\sigma^2(T)-\mu^2)\sigma(T)-3\lambda\sigma(T)\langle\phi^2\rangle-\lambda\langle\phi^3\rangle=0. \tag{3.3}$$

In the lowest order in λ the quantity $\langle\phi^3\rangle$ vanishes as in field theory. To calculate $\langle\phi^2\rangle$ one should take into account that:

$$\langle\phi^2\rangle=\frac{1}{(2\pi)^3}\int\frac{d^3p}{2\omega_p}(2\langle a_p{}^+a_p\rangle+1).$$

Then using equations (1.29) and (3.1) and discarding the temperature-independent term $(2\pi)^{-3}\int d^3p/2\omega_p$, which can be eliminated by the mass renormalisation at $T=0$, one finds that:

$$\langle\phi^2\rangle=F(T,m_\phi)\equiv\frac{1}{2\pi^2}\int_0^\infty\frac{p^2\,dp}{(p^2+m_\phi{}^2)^{1/2}[\exp(p^2+m_\phi{}^2)^{1/2}/T-1]} \tag{3.4}$$

where m_ϕ is the mass of the field ϕ.

As will be seen, all interesting effects take place only at $T \gg m_\phi$, when one can neglect m_ϕ in (3.4). In this case:

$$\langle\phi^2\rangle = F(T, 0) = T^2/12 \tag{3.5}$$

and equation (3.3) looks like:

$$\Box\sigma(T) - (\lambda\sigma^2(T) - \mu^2 + \tfrac{1}{4}\lambda T^2)\sigma(T) = 0. \tag{3.6}$$

Supposing, as before, that $\sigma(T)$ = constant we get:

$$\sigma(T)(\lambda\sigma^2(T) - \mu^2 + \tfrac{1}{4}\lambda T^2) = 0. \tag{3.7}$$

This equation at a sufficiently low temperature T has two solutions:

$$\sigma(T) = 0 \qquad \sigma^2(T) = \frac{\mu^2}{\lambda} - \frac{T^2}{4}. \tag{3.8}$$

The second solution disappears at temperatures greater than the critical temperature:

$$T_c = 2\mu\lambda^{-1/2} = 2\sigma(0). \tag{3.9}$$

To obtain the excitation spectrum in (1.3) at $T \neq 0$ one should perform in equation (3.3) an infinitesimal shift $\sigma \to \sigma + \delta\sigma$, where σ is one of the two possible solutions of equation (3.7) (see the introduction). At $\sigma(T) = 0$ the corresponding equation for the fluctuations of the field σ is:

$$\Box\delta\sigma - (-\mu^2 + \tfrac{1}{4}\lambda T^2)\delta\sigma = 0$$

which implies that the mass of the scalar field excitations at $\sigma = 0$ is given by:

$$m_\phi^2 = -\mu^2 + \tfrac{1}{4}\lambda T^2. \tag{3.10}$$

The value of m_ϕ^2 (3.10) at $T < T_c$ is negative, and thus the disordered solution $\sigma(T) = 0$ is unstable at $T < T_c$.

Analogously it can be shown that for the solution $\sigma^2(T) = \mu^2/\lambda - T^2/4$:

$$m_\phi^2 = 3\lambda\sigma^2(T) - \mu^2 + \tfrac{1}{4}\lambda T^2 = 2\lambda\sigma^2(T). \tag{3.11}$$

For both solutions $\sigma(T) = 0$ and $\sigma(T) \neq 0$ the masses (3.10) and (3.11) vanish at $T = T_c$ and the mass squared of the field ϕ in the disordered state $\sigma(T) = 0$ becomes positive (the state $\sigma(T) = 0$ becomes stable) at the same temperature T_c, at which the ordered solution $\sigma(T) \neq 0$ disappears. This means that at the critical temperature $T = T_c$ (3.9) the phase transition with the symmetry restoration in (1.3) takes place.

Equations (3.8), (3.10) and (3.11) for $\sigma(T)$ and $m_\phi(T)$ are illustrated by figures 7(*a*) and (*b*). One should note that, with an increase of temperature, the symmetry breaking parameter $\sigma(T)$ in (1.3) decreases continuously, which corresponds to the second-order phase transition.

3.2. Effective potential at $T \neq 0$

The same results can also be obtained in another way by the generalisation of the effective potential method for the case $T \neq 0$. This generalisation proves to be very simple (Dolan and Jackiw 1974, Kirzhnits and Linde 1974a). In quantum statistics at $T \neq 0$ the effective potential $V(\sigma, T)$ is given by the same one-particle irreducible 'vacuum' diagrams as in field theory. The only difference is that at

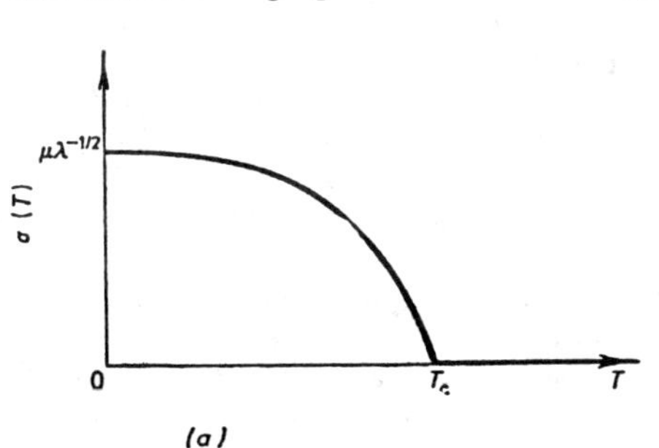

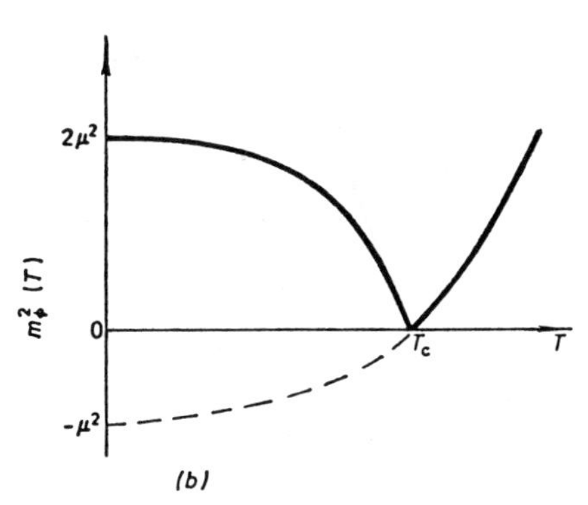

Figure 7. Temperature dependence of σ and m_ϕ in the model (1.3). Broken lines correspond to an unstable state $\sigma=0$ at $T<T_c$.

$T\neq 0$ the component k_0 of the momentum in all Euclidean integrals corresponding to these diagrams should be replaced by $2\pi nT$ for bosons and by $(2n+1)\pi T$ for fermions, and instead of the integration over k_0 one should perform a summation over all integer n: $\int dk_0 \to 2\pi T\Sigma_{n=-\infty}^{\infty}$. For example, the expression (2.6) for $V(\sigma)$ in (1.3) is transformed at $T\neq 0$ to the following:

$$V(\sigma, T)=-\frac{\mu^2}{2}\sigma+\frac{\lambda}{4}\sigma^4+\frac{T}{2(2\pi)^3}\sum_{n=-\infty}^{\infty}\int d^3k \ln\left[(2\pi nT)^2+k^2+m^2(\sigma)\right] \tag{3.12}$$

where $m^2(\sigma)=3\lambda\sigma^2-\mu^2$. For the renormalisation of this expression one can use the same counter-terms as in the field theory. Performing summation and integration in (3.12) one obtains the following expression for the renormalised effective potential $V(\sigma, T)$ at $T\gg m$, the terms of the higher orders in λ being neglected (Weinberg 1974a, Dolan and Jackiw 1974, Kirzhnits and Linde 1974a, b, 1976a):

$$V(\sigma, T)=-\frac{\mu^2}{2}\sigma^2+\frac{\lambda}{4}\sigma^4-\frac{\pi^2}{90}T^4+\frac{m^2(\sigma)}{24}T^2. \tag{3.13}$$

It can easily be seen that the equation $dV/d\sigma=0$ (2.4), which determines the equilibrium value of σ, in this case exactly coincides with (3.7). On the other hand, up to the higher orders in λ, $G_\phi^{-1}(0)=m_\phi^2$ as in the field theory. From this equation and equation (2.5) one can again obtain equations (3.10) and (3.11) for m_ϕ^2 and show that only the solutions with $m_\phi^2 \geqslant 0$ are stable according to the stability condition $d^2V/d\sigma^2 \geqslant 0$ (2.5).

By means of the 'energetical' approach used above one can give a very simple description of the high-temperature phase transition in the theory (1.3); namely, at $T=0$ the effective potential has a minimum at $\sigma=\mu\lambda^{-1/2}$. With an increase of temperature the energy difference between the minimum of $V(\sigma, T)$ at $\sigma\neq 0$ and the maximum at $\sigma=0$ decreases. At $T=T_c$ this energy difference vanishes

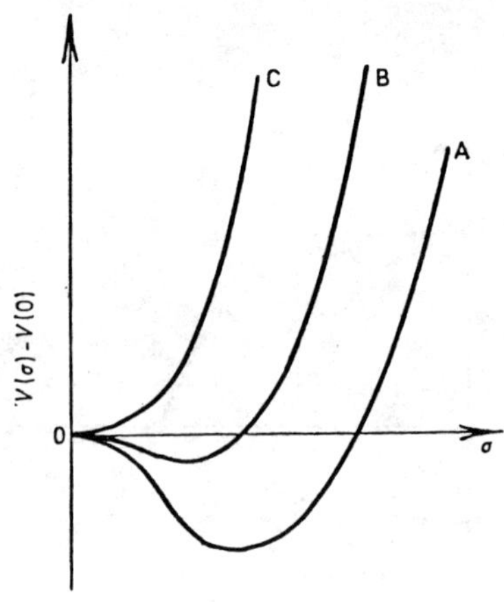

Figure 8. Effective potential $V(\sigma, T)$ in the model (1.3) at $\sigma>0$. A, $T=0$; B, $0<T<T_c$; C, $T>T_c$.

together with the minimum at $\sigma\neq 0$, and at $T>T_c$ the effective potential $V(\sigma, T)$ has its minimum at $\sigma=0$, i.e. the symmetry is restored (see figure 8).

3.3. Higher orders of perturbation theory

The investigation of the high-temperature symmetry restoration in SBGT performed above was based on the use of the lowest order of perturbation theory in λ. One may ask therefore whether the results obtained by this method are reliable.

This question is not quite trivial. For example, in the higher-order corrections to the effective potential at $T\neq 0$, besides small terms $\sim\lambda^n T^4$, $\lambda^n T^2 m^2(\sigma)$ the terms proportional to $m_\phi^{-n}(T)$ also appear. At small $m_\phi(T)$ such terms become large and should be taken into account.

To analyse this problem in a more detailed way let us consider the diagrams of the Nth order in λ for $V(\sigma, T)$ in (1.3) at $\sigma=0$ $(T>T_c)$. To obtain the correct behaviour of $V(\sigma, T)$ at small $m_\phi(T)$ one should use the self-consistent approximation in which the zero-temperature mass $m(\sigma)=(3\lambda\sigma^2-\mu^2)^{1/2}$ is replaced by the temperature-dependent mass $m_\phi(T)$ (Kirzhnits and Linde 1974a, 1976a). In this approximation:

$$V_{(0,\,T)}{}^N \sim (2\pi T)^{N+1}\lambda^N \int d^3p_1 \ldots d^3p_{N+1} \sum_{n_i=-\infty}^{\infty} \prod_{K=1}^{2N} [(2\pi r_K T)^2 + q_K{}^2 + m_\phi{}^2(T)]^{-1} \tag{3.14}$$

where q_K is a uniform linear combination of p_i, r_K is the corresponding combination of n_i, $i=1, \ldots, N+1$, $K=1, \ldots, 2N$. At $m_\phi \to 0$ the leading term in the sum over n_i is the term with all $n_i=0$ (and, consequently, $r_K=0$), since the factors, which contain $(2\pi r_K T)^2$, are not singular at $m_\phi \to 0$, $q_K \to 0$. The leading term is thus given by:

$$\tilde{V}_{(0,\,T)}{}^N \sim (2\pi T)^{N+1}\lambda^N \int d^3p_1 \ldots d^3p_{N+1} \prod_{K=1}^{2N} (q_K{}^2 + m_\phi{}^2(T))^{-1}$$
$$\sim \lambda^3 T^4 \left(\frac{\lambda T}{m_\phi(T)}\right)^{N-3} \tag{3.15}$$

It is seen that near T_c when $m_\phi \to 0$ the role of higher orders of perturbation theory can be significant even at small λ. Analogous terms singular at $m_\phi \to 0$ exist also at $\sigma \neq 0$ ($T < T_c$). Therefore, near T_c one should take into account all higher orders of perturbation theory, or use such well-known methods as ϵ expansion, $1/N$ expansion, etc (Wilson and Kogut 1974, see also Baym and Grinstein 1977). Fortunately, by the use of (3.10) and (3.11) it can be shown that such terms are important only in the small region near the critical temperature, in which:

$$|T - T_c| \lesssim \lambda T_c. \tag{3.16}$$

Everywhere except this small region near T_c higher-order corrections are small and our results concerning the phase transition in the model (1.3) are reliable.

3.4. *The phase transition in the Higgs model*

The methods developed above can be easily applied to more complicated theories with spontaneous symmetry breaking. For example, in the Higgs model (1.14) in the transverse gauge $\partial_\mu A_\mu = 0$, instead of equation (3.3) for the constant classical scalar field $\sigma(T)$ the following equation holds:

$$\left\langle \frac{\delta L}{\delta \phi} \right\rangle = \sigma(T)[\mu^2 - \lambda(\sigma^2(T) + 3\langle \chi_1{}^2 \rangle + \langle \chi_2{}^2 \rangle) + e^2 \langle A_\mu{}^2 \rangle] = 0. \tag{3.17}$$

Let us first suppose that, as in (1.3), the phase transition in the Higgs model takes place at $T \gg m_{\chi_i}$, m_A. Using the same methods which have been used in the investigation of the high-temperature symmetry behaviour in (1.3) one can show that at $T \gg m_{\chi_i}$, m_A:

$$\langle \chi_1{}^2 \rangle = \langle \chi_2{}^2 \rangle = -\tfrac{1}{3}\langle A_\mu{}^2 \rangle = T^2/12 \tag{3.18}$$

(compare with (3.4) and (3.5)). In this case equation (3.14) looks like:

$$\sigma\left(\lambda\sigma^2 - \mu^2 + \frac{4\lambda + 3e^2}{12} T^2\right) = 0 \tag{3.19}$$

from which it follows that the critical temperature in the Higgs model is given by (Weinberg 1974a, Dolan and Jackiw 1974, Kirzhnits and Linde 1974b, 1976a):

$$T_{c_1}{}^2 = \frac{12\mu^2}{4\lambda + 3e^2}. \tag{3.20}$$

According to (3.19) the value of $\sigma(T)$ depends on T continuously, i.e. we again deal with the second-order phase transition as in (1.3).

However, one can verify that our initial assumption that $T \gg m_{\chi_i}$, m_A in the region near the critical temperature is valid only at $\lambda \gg e^4$ (Kirzhnits and Linde 1974b, 1976a). It can be shown that at $\lambda \lesssim e^4$ $m_A(T_{c_1}) \approx e\mu\lambda^{-1/2} \gtrsim T_{c_1}$, and therefore at $T \sim T_{c_1}$ the vector particle contribution to (3.14), being proportional to $\langle A_\mu{}^2 \rangle = -3F(T, m_A)$ (compare with (3.4)), is very small. In this case the theory of the phase transition in the Higgs model becomes more complicated than in (1.3) (Kirzhnits and Linde 1974b, 1976a, Iliopoulos and Papanicolaou 1976); namely, in the temperature interval $T_{c_1} < T < T_{c_2}$ (where T_{c_2} is of the order of $m_A(T=0) = e\mu\lambda^{-1/2}$) there exist three different possible values of the field $\sigma(T)$: $\sigma_1(T) \neq 0$, $\sigma_2(T) \neq 0$ ($\sigma_1 > \sigma_2$) and $\sigma = 0$ (see figure 9). Correspondingly the effective potential

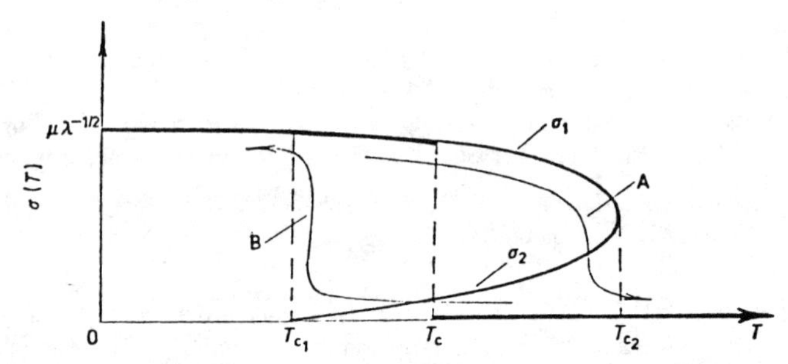

Figure 9. Temperature dependence of σ in the Higgs model at $3e^4/16\pi^2 < \lambda \lesssim e^4$. The thick line corresponds to a stable state of the system. Arrows show the behaviour of σ with an increase (A) and with a decrease (B) of temperature.

$V(\sigma, T)$ at $T_{c_1} < T < T_{c_2}$ has three extrema: two minima at $\sigma = \sigma_1(T)$ and at $\sigma = 0$ and a maximum at $\sigma = \sigma_2(T)$ (see figure 10). The state $\sigma = \sigma_2(T)$ is unstable at all temperatures. The state $\sigma = \sigma_1(T)$ is stable at low temperatures and becomes metastable at $T > T_c$, where the critical temperature T_c can be obtained from the equation:

$$V(\sigma_1(T_c), T_c) = V(0, T_c) \tag{3.21}$$

(see figure 10). Therefore at the temperature $T = T_c$ ($T_c > T_{c_1}$) the first-order (discontinuous) phase transition to the state $\sigma = 0$ takes place (figure 9).

The theory of this first-order phase transition is particularly simple at $3e^4/16\pi^2 \ll \lambda \ll e^4$. In this case it proves that $m_A(T_c) \gg T_c$ for the solution $\sigma = \sigma_1(T_c)$. As a result the vector particles give no contribution to $V(\sigma_1(T_c), T_c)$ and up to higher-order corrections in e^2, $\sigma_1(T_c) = \sigma_1(0) = \mu\lambda^{-1/2}$. Now let us take into account that at sufficiently high temperatures ($T \gg m$) every degree of freedom gives the contribution $-(\pi^2/90)T^4$ to $V(\sigma, T)$ (see (3.13)). At $\sigma = 0$ the massless vector field and the complex scalar field contribute (four degrees of freedom), and therefore $V(0, T_c) = -(4\pi^2/90)T^4$. On the other hand, at $\sigma = \sigma_1(T)$ the contribution $\sim T^4$

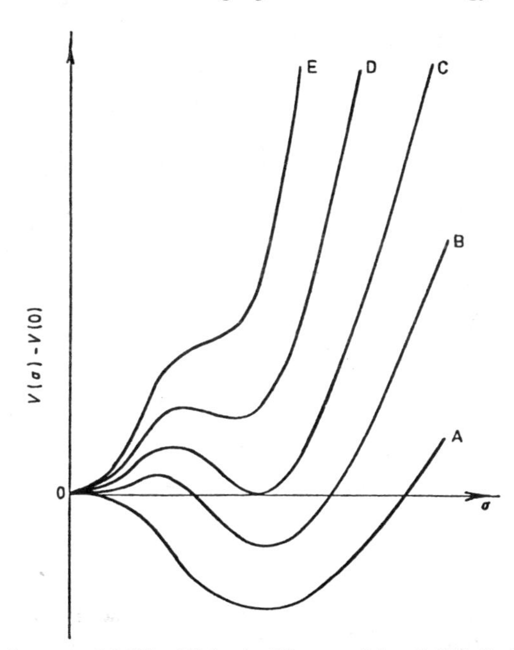

Figure 10. Effective potential $V(\sigma, T)$ in the Higgs model at $3e^4/16\pi^2 < \lambda \leqslant e^4$. A, $0 < T < T_{c1}$; B, $T_{c1} < T < T_c$; C, $T = T_c$: D, $T_c < T < T_{c2}$; E, $T > T_{c2}$.

is given only by the real scalar particles (one degree of freedom), i.e. $V(\sigma_1(T_c), T_c) = -\mu^4/4\lambda - \pi^2 T^4/90$. Then from equation (3.21) it follows that the critical temperature of the phase transition is given by (Kirzhnits and Linde 1974b, 1976a):

$$T_c = \left(\frac{15\lambda}{2\pi^2}\right)^{1/4} \mu. \tag{3.22}$$

At $\lambda \lesssim 3e^4/16\pi^2$ radiative corrections to $V(\sigma, T)$ become essential even at $T=0$ (see §2). As a result at $3e^4/32\pi^2 < \lambda < 3e^4/16\pi^2$ the curves for $V(\sigma, T)$ and $\sigma(T)$ acquire the form represented in figures 11 and 12.

Note that at $\lambda \to 3e^4/32\pi^2$ the critical temperature $T_c \to 0$ and at $\lambda < 3e^4/32\pi^2$ the symmetry in the Higgs model, as was already mentioned in §2, is restored even at $T=0$. Therefore, one could expect that we have a good chance of observing the effects connected with the symmetry restoration in SBGT in a laboratory if the critical temperature T_c is not too high. Unfortunately this is not quite true because of effects connected with the kinetics of the first-order phase transitions in SBGT.

These phase transitions proceed by the spontaneous formation of bubbles filled with matter in a new phase. The walls of the bubbles spread up with a velocity almost equal to that of light, and very soon all the Universe becomes filled with matter in the new phase. However, the formation of the bubbles is a barrier tunnelling process. Such a process may take place due to quantum fluctuations (Voloshin *et al* 1974, Coleman 1977) or due to thermodynamic ones (Linde 1977b), but in both cases it proves that the time necessary for the creation of at least one such bubble in all the Universe is usually much greater than the age of the Universe.

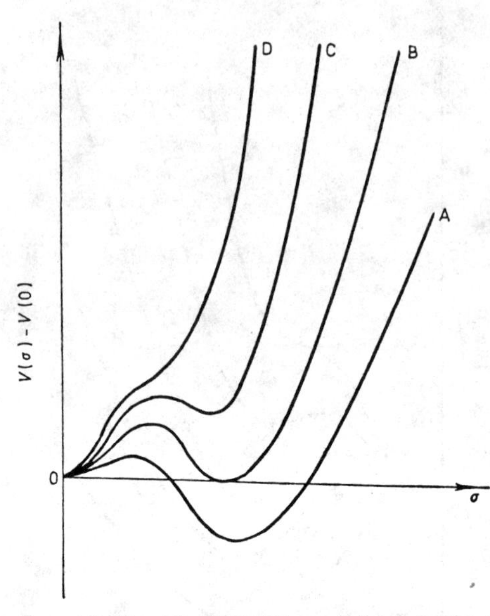

Figure 11. Effective potential $V(\sigma, T)$ in the Higgs model at $3e^4/32\pi^2 < \lambda < 3e^4/16\pi^2$. A, $0 < T < T_c$; B, $T = T_c$; C, $T_c < T < T_{c_2}$; D, $T > T_{c_2}$.

As a result, the first-order phase transitions in the gauge theories appear to be possible only in cases when the energetical barrier between two local minima of $V(\sigma, T)$ becomes extremely small. Therefore, the first-order phase transition in the Higgs model actually takes place at $T \approx T_{c_1}$ when the temperature decreases and at $T \approx T_{c_2}$ when the temperature increases (see figure 9). As a result, to realise the investigated phase transition 'in the laboratory' one should heat the system not up to the temperature T_c, which may be very small, but up to the relatively large temperature $T_{c_2} \sim m_A(T=0)$.

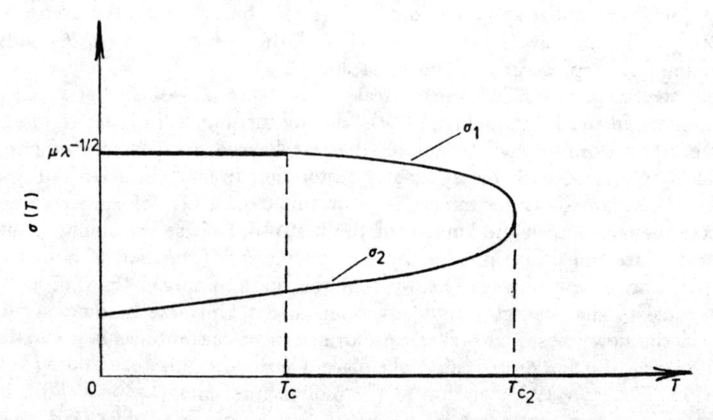

Figure 12. Temperature dependence of σ in the Higgs model at $3e^4/32\pi^2 < \lambda < 3e^4/16\pi^2$. The thick line corresponds to a stable state of the system.

The methods used above for the investigation of the symmetry behaviour in the Higgs model can be easily extended to more realistic theories with non-Abelian gauge fields and fermions. The estimates of the value of T_c in most models of weak and electromagnetic interactions give the critical temperature a value of the order of 10–1000 GeV (10^{14}–10^{16} K). Such temperatures have existed only in the early stages of the evolution of the Universe. In theories of strong interactions the critical temperature may be much less: $T_c \sim 100$ MeV–1 GeV (10^{12}–10^{13} K). Temperatures of this order can exist not only in the early Universe, but also inside the 'fireballs' which are created in the process of multiparticle production in the collisions of high-energy elementary particles†.

Strictly speaking, all these conclusions only concern the calculations in the lowest order of perturbation theory. As for the Higgs model, by the same methods as in §3.3, it can be shown that our results are actually reliable everywhere except in some small region near T_c. Massless vector particles, which appear after the symmetry restoration in this model, cause no difficulties of the type mentioned in §3.3 since these particles do not interact directly with each other. However, in the non-Abelian gauge theories of the Weinberg–Salam model type the reliability of the lowest-order results at $T \geqslant T_c$ is much less obvious. The reason is that after the symmetry restoration in non-Abelian theories the self-interacting Yang–Mills fields become massless. If the temperature corrections do not produce a mass for these fields (this question will be discussed in the next subsection), the corresponding infrared divergences may invalidate all perturbative results at $T \geqslant T_c$. In particular, one cannot be sure that the symmetry at $T \geqslant T_c$ is completely restored. In any case, however, it can be shown that at $T \geqslant T_c$ the value of $\sigma(T)$ is greatly diminished. For example, in the non-Abelian theories with $\lambda \sim g^2$ (where g is the constant of the self-interaction of the Yang–Mills fields) it can be shown that $\sigma(T_c) \lesssim g\sigma(0)$, $\sigma(T > T_c) \lesssim gT$ (Kirzhnits and Linde 1976b).

In conclusion, we would also like to note that the phase transitions of the type discussed above may take place not only in SBGT, but in the theories with dynamical symmetry breaking as well: see, for example, an investigation of the high-temperature phase transition in the two-dimensional Gross–Neveu model (Jacobs 1974, Hiro-O-Wada 1974, Harrington and Yildiz 1975, Dashen *et al* 1975).

3.5. The infrared problem in quantum statistics of gauge fields

Infrared divergences, which appear in the non-Abelian gauge theories with spontaneous symmetry breaking at $T > T_c$, exist also in quantum chromodynamics (QCD), in which the Yang–Mills fields responsible for strong interactions are supposed to be massless. One could hope, however, that due to the temperature corrections massless Yang–Mills fields in both the theories acquire some sufficiently large mass ($m \gg g^2T$, see below). This would solve the infrared problem in quantum statistics

† In the papers by Hagedorn (1965) and Frautschi (1971) it was argued that there may exist a limiting temperature $T_0 \sim 150$ MeV. These papers have been repeatedly criticised and the arguments in favour of the existence of a limiting temperature, in our opinion, are rather unconvincing, but in any case it can be shown that the same reasons which could lead to the existence of T_0 (an assumption concerning the exponential spectrum of elementary particles) lead also to the symmetry restoration in SBGT at some $T_c < T_0$ (Linde 1975b, Kirzhnits and Linde 1976a). Moreover, after the symmetry restoration the particle spectrum changes and there are no reasons to expect that in the theory with this new spectrum any limiting temperature exists (Cabibbo and Parisi 1975, Kirzhnits and Linde 1976b).

of gauge fields. To understand whether the massless Yang–Mills fields actually may become massive due to the temperature corrections we shall remind ourselves first of the corresponding results concerning the photon mass at $T \neq 0$.

As is well known, the photon Green function $G_{\mu\nu}(k)$ at $T \neq 0$ has a singular point at $k_0 = eT/3$, $\boldsymbol{k} = 0$ (Fradkin 1965). One could think (see, for example, Kislinger and Morley 1976a) that this means that photons at $T \neq 0$ have a mass $m = eT/3$, which serves as an infrared cutoff in the theory. Actually, however, analytic properties of the Green function $G_{\mu\nu}(k)$ at $T \neq 0$ are much more complicated than at $T = 0$ and the existence of a singularity at $k_0 \neq 0$, $\boldsymbol{k} = 0$ (as well as a singularity of $G_{\mu\nu}(k)$ at $k_0 = |\boldsymbol{k}| \neq 0$ (Shuryak 1978) is irrelevant to the problem of infrared divergences at $T \neq 0$.

Indeed, as is shown in §3.3, the leading infrared divergences in quantum statistics of the Bose particles are connected with the 'static' limit $k_0 = 0$, $\boldsymbol{k} \to 0$ of the corresponding Green functions (the term with all $r_i = 0$ in (3.14)). Properties of $G_{\mu\nu}(k)$ in quantum electrodynamics in this limit are well known. For example, in the Coulomb gauge $\partial_i A_i = 0$ the photon Green function $G_{\mu\nu}$ at $k_0 = 0$ has the following structure (Fradkin 1965):

$$
\begin{aligned}
G_{00}(\boldsymbol{k}) &= \frac{1}{\boldsymbol{k}^2 + \Pi_{00}(\boldsymbol{k})} \\
G_{i0}(\boldsymbol{k}) &= G_{0i}(\boldsymbol{k}) = 0 \\
G_{ij}(\boldsymbol{k}) &= \left(\delta_{ij} - \frac{k_i k_j}{\boldsymbol{k}^2}\right) \frac{1}{\boldsymbol{k}^2 + A(\boldsymbol{k})}
\end{aligned}
\tag{3.23}
$$

where $i, j = 1, 2, 3$. The quantity $\Pi_{00}(\boldsymbol{k})$ (which is equal to the corresponding component of the polarisation operator $\Pi_{\mu\nu}(k_0 = 0, \boldsymbol{k})$) does not vanish at $\boldsymbol{k} \to 0$. At a sufficiently low plasma density, $\Pi_{00}(0)$ is given by:

$$
\Pi_{00}(0) = \left(\frac{n_e}{4\pi e^2 T^2}\right)^2 \tag{3.24}
$$

where n_e is the density of pairs of charged particles at the temperature T (Landau and Lifshitz 1964). In the high-temperature limit (Fradkin 1965):

$$
\Pi_{00}(0) = e^2 T^2/3. \tag{3.25}
$$

Thus, according to (3.23)–(3.25), $G_{00}(k_0 = 0, \boldsymbol{k} \to 0) = \Pi_{00}^{-1}(0) \neq \infty$, and in this sense the component $G_{00}(k)$ at $k_0 = 0$, $\boldsymbol{k} \to 0$ behaves as the Green function of a massive field with the mass squared $m^2 = \Pi_{00}(0)$.

On the other hand, it can be shown that for all orders of perturbation theory $A(\boldsymbol{k}) \sim \boldsymbol{k}^2$ at $\boldsymbol{k} \to 0$ (Fradkin 1965). Therefore at $k_0 = 0$, $\boldsymbol{k} \to 0$ the Green function $G_{ij}(\boldsymbol{k})$ behaves like that of massless particles, $G_{ij} \sim 1/\boldsymbol{k}^2$.

The physical meaning of such a difference between the low-momentum behaviour of $G_{00}(k)$ and $G_{ij}(k)$ is very simple. One can easily verify that in the Coulomb gauge at $k_0 = 0$ the Green function of the electric field $\langle E_i E_j \rangle$ is proportional to G_{00}, and the Green function of the magnetic field $\langle H_i H_j \rangle \sim G_{ij}$. Electrostatic forces in a gas of charged particles become short-range due to the Debye screening ($G_{00}(0) = \Pi_{00}^{-1}(0) = \lambda_D^2$, where λ_D is the Debye length). If there are no magnetic charges in the theory, the magnetic forces cannot be screened and remain long-range, in accordance with the above results.

Note, however, that in some SBGT the monopoles actually exist ('t Hooft 1974, Polyakov 1974). In such theories magnetic forces also become screened. The theory of the magnetic screening in the low-density monopole–antimonopole plasma is completely analogous to that of electrostatic Debye screening; in particular, it can be shown that:

$$A(0)=\left(\frac{n_{\mathrm{M}}}{4\pi q^2 T^2}\right)^2 \tag{3.26}$$

where n_{M} is the density of the monopole–antimonopole pairs at $T\neq 0$ and q is the magnetic charge of the monopole (compare with (3.24)). In the dilute gas of monopoles:

$$n_{\mathrm{M}}=\frac{1}{4\pi^2}\int_0^\infty n_k k^2\,\mathrm{d}k=\frac{1}{4\pi^2}\int_0^\infty \frac{k^2\mathrm{d}k}{\exp\left[(k^2+m_{\mathrm{M}}^2)^{1/2}T^{-1}\right]}=\frac{c m_{\mathrm{M}}^2}{4\pi^2}\exp\left(-\frac{m_{\mathrm{M}}}{T}\right) \tag{3.27}$$

where n_k is the density of monopoles with momentum k, m_{M} is the mass of the monopole and c is some constant ($c\sim 1$). Finally one obtains the following expression for the 'mass' of the magnetic field at $T\neq 0$, $T\ll m_{\mathrm{M}}$ (Polyakov 1978, Linde 1978a):

$$m=\frac{c m_{\mathrm{M}}}{16\pi^3 q^2}\exp\left(-\frac{m_{\mathrm{M}}}{T}\right). \tag{3.28}$$

Therefore, in the presence of monopoles at $T\neq 0$ photons actually become massive. However, in most SBGT the mass of the monopole is extremely large, and therefore the photon mass becomes appreciable only at $T\sim T_{\mathrm{c}}$, when m_{M} vanishes. Moreover, at $T>T_{\mathrm{c}}$ the monopoles of the type discovered by 't Hooft and Polyakov disappear, and the photon Green function $G_{ij}(k)$ again behaves like that for massless particles in the limit $k_0=0$, $k\to 0$. Fortunately, the infrared problem in quantum electrodynamics (QED) can be easily solved since in QED there are no self-interacting massless Bose fields.

Now let us return to the quantum statistics of massless Yang–Mills fields. For this case one can also show that in the Coulomb gauge the Green function of the Yang–Mills field has the same structure as the Green function for photons (3.23) (except for the extra factor δ^{ab}, where a, b are isotopical indices). As in QED, it can be shown that in perturbation theory $\Pi_{00}(k\to 0)\neq 0$, $A(k\to 0)\sim k^2$ *if the diagrams for the polarisation operator* $\Pi_{\mu\nu}^{ab}(k)$ *are not singular at* $k_0=0$, $k\to 0$. In QED this requirement is always satisfied, but in the quantum statistics of massless Yang–Mills fields at $T\neq 0$ infrared divergences are so strong that the diagrams for $\Pi_{\mu\nu}^{ab}$ actually become singular at $k\to 0$. In the lowest order in g^2 the polarisation operator has only logarithmic singularity, and therefore $A(k\to 0)\sim k^2 \ln k^2$, i.e. as in QED the 'magnetic' part of the Yang–Mills field remains massless at $T\neq 0$. Dimensional estimates show that to the order g^4 the polarisation operator $\Pi_{\mu\nu}^{ab}$ may become singular enough to produce a 'mass' $m=A^{1/2}(0)$ of the Yang–Mills field, which would serve as an infrared cutoff in the theory. This mass, however, would be very small ($m\sim g^2T$), and order-by-order calculations in quantum statistics in this case would remain unreliable. Indeed, by the methods developed in §3.3 one can show that at $m\lesssim g^2T$ higher orders of perturbation theory become of the same order or greater than the lower ones. It is not excluded that the mass of the Yang–Mills fields at $T\neq 0$ can be generated by some non-perturbative effects of the type shown by the photon mass generation in the monopole–antimonopole plasma.

Unfortunately, at present it is not quite clear if the mass generated by the non-perturbative effects can be sufficiently large (greater than g^2T).

As was mentioned at the beginning of this subsection, the massless Yang–Mills fields exist not only in SBGT at $T > T_c$, but also in quantum chromodynamics (QCD). In recent years the main interest in the macroscopic consequences of QCD was connected with the possibility of quark liberation in super-dense matter and with the possible existence of super-dense but relatively cold quark stars, which were supposed to consist of free-quark matter (Collins and Perry 1975). Note that if one could obtain the thermodynamic potential and the effective gauge coupling constant g in QCD at some density, and if one could show that at this density the coupling constant g becomes small ($g \ll 1$), then it would be possible to use the renormalisation group equation and to prove that in super-dense matter $g \to 0$ and quarks behave as an ideal relativistic gas (Collins and Perry 1975, Kislinger and Morley 1976b). The only problem here is whether one can reliably obtain thermodynamic characteristics of a system containing massless Yang–Mills fields at any non-vanishing density and temperature.

At exactly zero temperature infrared divergences are absent from the diagrams for the thermodynamical potential of the quark matter in QCD. Therefore one could expect that many of the results concerning the cold quark matter (see, for example, Chapline and Nauenberg 1977, Freedman and McLerran 1977, Baluni 1978) are actually reliable. However, at any non-vanishing temperature the infrared divergences appear again. Moreover, the propagator of the massless Yang–Mills field at low momenta is drastically modified by the higher-order (and non-perturbative) effects. The true low-momentum behaviour of the improved propagator is now absolutely unknown even in quantum field theory (i.e. at vanishing density and temperature). As a result the calculation of the diagrams for the thermodynamic potential in QCD with the improved propagators of the Yang–Mills fields becomes ambiguous even at zero temperature.

A very interesting approach to the infrared problem in quantum statistics of gauge fields is contained in a recent paper by Polyakov (1978). In this paper it is concluded that at a sufficiently high temperature quarks in QCD become free and massless Yang–Mills fields acquire some relatively small mass (which was suggested to be $\sim g^2T$). This important statement has been obtained by means of a non-perturbative method which effectively takes into account all leading infrared divergences in quantum statistics of massless Yang–Mills fields. Unfortunately this method cannot be extended to the investigation of the relatively cold quark matter inside quark stars. Moreover, in his treatment of massless Yang–Mills fields, Polyakov has neglected non-leading infrared divergences. Such an approximation is sufficiently good for the study of massless infrared-stable theories of the type given by the theory $\lambda\phi^4$ (Wilson and Kogut 1974), but the applicability of this method to the infrared-unstable Yang–Mills theory does not seem quite clear.

To summarise our discussion we note that the existence of singularities of $G_{\mu\nu}{}^{ab}(k)$ at $k_0 \neq 0$ (Kislinger and Morley 1976a, Shuryak 1978) does not prove the existence of an infrared cutoff in quantum statistics of massless Yang–Mills fields, and *in this sense* massless Yang–Mills fields do not become massive due to lowest-order temperature corrections. Nevertheless the Yang–Mills fields may acquire some non-vanishing 'mass' $m = A^{1/2}(0)$ (which has the meaning of an infrared cutoff) due to higher-order corrections or due to non-perturbative effects. If, as one may expect, this mass is small ($m \lesssim g^2T$), then the higher-order corrections are large and all the lowest-order results concerning the quantum statistics of the gauge fields are unreliable. In this case some more elaborate methods should be used for the investigation of QCD at $T \neq 0$ and of SBGT at $T > T_c$ (see also the discussion at the end of §3.4).

Some results concerning quantum statistics of the massless non-Abelian gauge fields obtained in recent years by the usual methods (and in particular an equation of state $p = \frac{1}{3}\epsilon$ for the super-dense matter (Collins and Perry 1975)) seem very natural and may be true. However, one should recognise that these results will be finally proved (or disproved) only when the infrared problem in the quantum statistics of gauge fields will be solved. A solution of this problem may appear rather unexpected, for example the Yang–Mills fields in super-dense matter may prove not to be in a gaseous state, but in a crystalline one (Akhiezer *et al* 1978, Linde 1978b) (see also §5.4 of the present review). We hope that future investigations will shed some light on the questions touched upon in this section.

4. Symmetry behaviour in external fields

4.1. Quasimagnetic massive vector fields

As is mentioned in the introduction, most of the high-temperature effects in SBGT can be anticipated from the analogy between SBGT and superconductivity theory (Kirzhnits 1972, Kirzhnits and Linde 1972). This analogy also proves to be very useful for the investigation of symmetry behaviour in SBGT in the presence of external fields.

It is known that an external magnetic field destroys superconductivity. Symmetry restoration in an external quasimagnetic field $\mathscr{H} = \text{rot}\, \boldsymbol{A}$ in the Higgs model would be an exact analogue of this effect, but it is impossible to create a homogeneous external quasimagnetic field since the field A_μ in the Higgs model is massive. Fortunately the only fact which should be taken into account in the investigation of the behaviour of a superconductor in a magnetic field H is that some external currents or fields exist which should create the field H inside the superconductor *after* the destruction of superconductivity. Here we shall investigate an analogous situation in the Higgs model. We shall analyse the Higgs model in the presence of such external currents, which at $\sigma = 0$ should create the quasimagnetic field $\mathscr{H}$ in some spatial domain. The simplest way to realise this situation is to consider an ordinary magnetic coil with an electric current, which is also a neutral weak current according to the Weinberg–Salam model; at $\sigma = 0$ this coil created not only an ordinary magnetic field, but also a quasimagnetic field $\mathscr{H}$, corresponding to weak interactions: $\mathscr{H} = \text{rot}\, \boldsymbol{Z}$, where Z_μ is a heavy neutral intermediate vector meson (Weinberg 1967, Salam 1968).

For brevity and using the same terminology as in superconductivity theory we shall speak of the symmetry behaviour 'in an external quasimagnetic field $\mathscr{H}$'. One should recognise, however, that the quasimagnetic field actually appears inside the system only *after* the phase transition with symmetry restoration. The investigation of such a phase transition in SBGT is completely analogous to the behaviour of a superconductor in an external magnetic field, and here we shall present only the results of this investigation concerning the Higgs model (Kirzhnits and Linde 1974b, 1976a, see also Harrington and Shepard 1976).

In the Higgs model there exist three different critical fields, $\mathscr{H}_c$, $\mathscr{H}_{c_1}$, $\mathscr{H}_{c_2}$.

The thermodynamic critical field $\mathscr{H}_c$ is determined by the relation:

$$\mathscr{H}_c{}^2/2=\mu^4/4\lambda \tag{4.1}$$

and has the meaning of a maximum field for which a homogeneous ordered phase $\sigma=\mu\lambda^{-1/2}$ without a quasimagnetic field has lower energy than the disordered phase $\sigma=0$ with the quasimagnetic field inside the system. From (4.1) it follows that:

$$\mathscr{H}_c=\mu^2/\sqrt{2\lambda}. \tag{4.2}$$

The upper critical field $\mathscr{H}_{c_2}$ determines the lower boundary of the region in which the disordered state $\sigma=0$ can be stable or metastable. At $\mathscr{H}<\mathscr{H}_{c_2}$, $\sigma=0$ scalar particles have a tachyon spectrum. In the Higgs model (1.14):

$$\mathscr{H}_{c_2}=\mu^2/e. \tag{4.3}$$

From (4.2) and (4.3) it follows that:

$$\mathscr{H}_c/\mathscr{H}_{c_2}=e/\sqrt{2\lambda}=m_A/m_\phi. \tag{4.4}$$

Like superconductors, the Higgs model may be of the first type ($e^2>2\lambda$, $m_A>m_\phi$) or of the second type ($e^2<2\lambda$, $m_A<m_\phi$). In the first-type models, $\mathscr{H}_c>\mathscr{H}_{c_2}$ and the phase transition to the disordered state $\sigma=0$ takes place at $\mathscr{H}=\mathscr{H}_c$. In the second-type models ($\mathscr{H}_{c_2}>\mathscr{H}_c$) two different phase transitions take place with an increase of electric current in a magnetic coil. The first phase transition occurs at the current which at $\sigma=0$ would create inside the magnetic coil the quasimagnetic field:

$$\mathscr{H}_{c_1}=\frac{e^2}{2\lambda}\mu^2\left(\ln\frac{\sqrt{2\lambda}}{e}+0{\cdot}08\right). \tag{4.5}$$

However, actually at $\mathscr{H}=\mathscr{H}_{c_1}$ the phase transition takes place not to the state $\sigma=0$ but to some inhomogeneous state. Inside the magnetic coil there appear quasimagnetic flux tubes, each of which contains a quasimagnetic flux quantum $\Phi_0=2\pi/e$. Finally at $\mathscr{H}=\mathscr{H}_{c_2}$ the phase transition to the disordered state $\sigma=0$ takes place and the quasimagnetic field $\mathscr{H}$ completely penetrates into the system.

Analogous results have been obtained in the non-Abelian gauge theories as well (Krive *et al* 1976b). Unfortunately the critical strength of the quasimagnetic field $\mathscr{H}$ in most of the models is very large. Let us take for example $\sigma\sim100$ MeV, $m_\phi\sim1$ GeV (strong interactions), then $\mathscr{H}_c\sim2\times10^{18}$ G; for the weak interaction models $\mathscr{H}_c$ is even much greater.

4.2. Magnetic and electric fields

In our investigation of the symmetry behaviour inside a magnetic coil we have taken into account the quasimagnetic field, which is absent until the phase transition takes place, but we have neglected an ordinary magnetic field created by the same magnetic coil. The reason is that the quasimagnetic field $\mathscr{H}$ interacts with the condensate at the classical level (the quasimagnetic field acquires mass $\sim\sigma$), whereas the magnetic field affects the condensate only due to radiative corrections. Therefore, in most of the realistic models with neutral currents (which create the quasimagnetic fields) magnetic fields actually can be neglected in the analysis of the symmetry restoration (Linde 1975c, 1976b).

It is not excluded, of course, that there exist some realistic theories with neutral currents in which, for some reasons, the effects connected with the quasimagnetic fields are small. To get an idea of what we shall deal with in this case Salam and Strathdee have considered some models without neutral currents, and their first estimates of the critical magnetic field have been very encouraging: $H_c \sim 10^6$–10^{16} G (Salam and Strathdee 1974, 1975). These estimates have stimulated some attempts to discover the phase transitions in SBGT experimentally since the magnetic field inside heavy nuclei may be of the order of $H \sim 10^{15}$ G. Moreover, the parameter σ of the symmetry breaking depends only on the invariants $E^2 - H^2$ and $\boldsymbol{E}.\boldsymbol{H}$. Therefore, the electric field may also affect symmetry breaking, and in heavy nuclei this field may be even greater than the magnetic one: $E \sim 10^{16}$–10^{17} G.

Unfortunately a further analysis of this question has shown that the estimates $H_c - E_c \sim 10^6$–10^{16} G are too optimistic (Linde 1975c, 1976b). To verify it let us discuss again the first model considered by Salam and Strathdee (1974). The Lagrangian of this model is:

$$L = -\tfrac{1}{4}(G_{\mu\nu}{}^a)^2 + \tfrac{1}{2}(\nabla_\mu \phi^a)^2 + \tfrac{1}{2}\mu^2(\phi^a)^2 - \tfrac{1}{4}\lambda((\phi^a)^2)^2 \tag{4.6}$$

where the scalar and vector fields ϕ^a and $A_\mu{}^a$ are triplets with respect to O(3) symmetry ($a = 1, 2, 3$), and:

$$\begin{aligned} \nabla_\mu \phi^a &= \partial_\mu \phi^a + e\epsilon^{abc} A_\mu{}^b \phi^c \\ G_{\mu\nu}{}^a &= \partial_\mu A_\nu{}^a - \partial_\nu A_\mu{}^a + e\epsilon^{abc} A_\mu{}^b A_\nu{}^c. \end{aligned} \tag{4.7}$$

The symmetry being broken spontaneously to O(2), the component ϕ^3 of the scalar field acquires a non-zero vacuum expectation value:

$$\langle 0|\phi^3|0\rangle = \sigma = \mu\lambda^{-1/2}. \tag{4.8}$$

The corresponding component of the vector field $A_\mu{}^3$ does not acquire mass and can be identified with the electromagnetic potential.

To investigate symmetry behaviour in the model (4.6) in the presence of an external magnetic field $\boldsymbol{H} = \text{rot}\ \boldsymbol{A}^3$ one should introduce as before an effective potential $V(\sigma, H)$.

Let us suppose for simplicity that $e^2 \ll \lambda$. In this case one can neglect the contribution from the vector particles. To obtain the contribution from charged scalar particles with the mass m one should use the propagator:

$$G(k_0, k_H, n) = \frac{1}{k_0{}^2 - k_H{}^2 + (2n+1)eH + m^2} \tag{4.9}$$

instead of the ordinary propagator $G(k) = (k^2 + m^2)^{-1}$. Here k_H is the projection of the particle momentum $\boldsymbol{k}$ onto the direction of the field $\boldsymbol{H}$, $n = 0, 1, 2, \ldots$. The integral $\int d^4k$ in the one-loop expression for $V(\sigma)$ should be replaced by

$$2\pi eH \int_{-\infty}^{\infty} dk_0\, dk_H \sum_{n=0}^{\infty}.$$

The effect of all this is in the one-loop correction to $dV/d\sigma$, which at $H \gg m$ after renormalisation is given by:

$$\frac{dV^1(\sigma, H)}{d\sigma} = -\frac{\lambda e\sigma H}{8\pi^2} \int_0^\infty \frac{dx}{x^2}\left(1 - \frac{x}{\sinh x}\right) \equiv -\frac{\lambda e\sigma H}{8\pi^2} B \tag{4.10}$$

where $B \sim 1$. The equation for the equilibrium value of σ at $H \neq 0$ in this case has the following form (Linde 1976a):

$$\frac{dV}{d\sigma} = 0 = \sigma\left(\lambda\sigma^2 - \mu^2 - \frac{\lambda eH}{8\pi^2} B\right) \tag{4.11}$$

from which it is seen that an increase of H *increases* the symmetry breaking parameter σ.

One can easily verify that the H-dependent contribution from fermions, which could be added to the model (4.6), also increases symmetry breaking in this model. Only the vector particle contribution may lead to the symmetry restoration in (4.6) at a sufficiently large H.

Following from (4.10), a characteristic strength of the magnetic field which can substantially modify the symmetry breaking parameter σ in the model (4.6) is of the order of:

$$H_c \sim \frac{8\pi^2\mu^2}{\lambda e} = \frac{8\pi^2\sigma^2(H=0)}{e}. \tag{4.12}$$

If $\sigma(0) \sim 250$ GeV as in the Weinberg–Salam model, the characteristic field is of the order of $H_c \sim 10^{27}$ G. To make an estimate, which can serve as a lowest bound for H_c, let us take $\sigma(0) \sim 100$ MeV, as in some theories of strong interactions. In this case $H_c \sim 10^{19}$–10^{20} G (Linde 1975c, 1976b). This result is in agreement numerically with the result of some other model calculations by Salam and Strathdee (1976). The same estimate can also be obtained for the characteristic value of the electric field E. This means that the fields which can exist inside nuclei ($H \sim 10^{15}$ G, $E \sim 10^{16}$–10^{17} G) are still insufficient to give a considerable modification of symmetry breaking in most of the gauge theories.

Generally speaking, there exists some possibility of the phase transitions in SBGT at relatively small values of external fields. Such a phase transition could take place if for some special reasons the effective potential $V(\sigma)$ at $H=0$ has several different local minima, and the values of $V(\sigma)$ at these minima are almost equal. As was shown in §2, these conditions are satisfied, e.g. in the Higgs model at $\lambda \sim 3e^4/32\pi^2$. In this case the first-order phase transition between these local minima of $V(\sigma)$ may take place in the presence of very weak external fields. Unfortunately, as was mentioned in §3, the time which is necessary for a first-order phase transition in SBGT to take place is usually much greater than the age of the Universe. Therefore, one may expect that the most interesting effects connected with the phase transitions in external fields could take place only in the early Universe when such strong fields may actually have existed.

5. Effects connected with the fermion density increase

5.1. *Theories without neutral currents* (σ *model*)

In the previous section we have considered high-temperature symmetry behaviour in quantum field theory, the chemical potentials of all particles being equal to zero. In that case, at all temperatures the system under investigation contains equal amounts of particles and antiparticles, and at $T=0$ all particles disappear.

One may wonder, however, which physical effects take place in super-dense *cold* matter in SBGT. To examine this problem we shall consider below a dense gas of fermions with a chemical potential $\alpha \neq 0$. Following Lee and Wick (1974) we shall consider first the simplified σ model (1.28) with the Lagrangian:

$$L=\tfrac{1}{2}(\partial_\mu\phi)^2+\tfrac{1}{2}\mu^2\phi^2-\tfrac{1}{4}\lambda\phi^4+\bar{\psi}(\mathrm{i}\partial_\mu\gamma_\mu-g\phi)\psi.$$

At a low density of fermions ($\alpha\rightarrow 0$) the lowest energy state in the cold matter will be as before the state with $\langle\phi\rangle=\sigma\approx\mu\lambda^{-1/2}$. At large density it becomes important that with an increase of σ the fermion masses increase, which is energetically disadvantageous. Therefore, at a sufficiently large density the state with $\sigma=0$ becomes energetically favourable and symmetry restoration in (1.28) takes place.

To describe this effect quantitatively one should calculate the one-loop corrections to the effective potential due to the existence of the dense gas of fermions with the chemical potential α. The calculation of $V(\sigma)$ in this case goes like that in §2. The only difference is that one should add $\mathrm{i}\alpha$ to the component k_0 of the fermion momentum in the corresponding Euclidean integrals (see, for example, Fradkin 1965).

As will be seen, at $\lambda\sim g^2\ll 1$ symmetry restoration in (1.28) takes place at $\alpha\sim\sigma\gg m_\phi, m_\psi$. In this case the one-loop correction to $\mathrm{d}V/\mathrm{d}\sigma$ can be easily calculated (Lee and Wick 1974, Harrington and Yildiz 1974) and is given by:

$$\frac{\mathrm{d}V^1}{\mathrm{d}\sigma}=\tfrac{1}{2}g^2\left(\frac{9j^2}{\pi^2}\right)^{1/3}\sigma \tag{5.1}$$

where $j^2=j_0{}^2-\boldsymbol{j}^2$, $j_\mu=\langle\bar{\psi}\gamma_\mu\psi\rangle$ is the fermion current, and the fermion density j_0 in the rest frame of the medium ($\boldsymbol{j}=0$) is equal to (Landau and Lifshitz 1964):

$$j_0=\langle\bar{\psi}\gamma_0\psi\rangle=\alpha^3/3\pi^2.$$

The equation $\mathrm{d}V/\mathrm{d}\sigma=0$ in this case looks like:

$$\frac{\mathrm{d}V}{\mathrm{d}\sigma}=0=\sigma\left[\lambda\sigma^2-\mu^2+\tfrac{1}{2}g^2\left(\frac{9j^2}{\pi^2}\right)^{1/3}\right]. \tag{5.2}$$

From this equation it can be easily obtained that at $\boldsymbol{j}=0$, $j_0=j_0{}^\mathrm{c}$, where:

$$j_0{}^\mathrm{c}=\frac{2\sqrt{2}\,\pi}{3}\left(\frac{\mu}{g}\right)^3 \tag{5.3}$$

is the critical fermion density, a second-order phase transition with the symmetry restoration takes place. In the terminology of Lee and Wick this symmetry restoration is the phase transition from the 'normal' nuclear matter ($\sigma\neq 0$) to the 'abnormal' matter ($\sigma=0$).

At $\lambda\lesssim g^4$ this phase transition becomes the first-order one (like the high-temperature phase transition in the Higgs model at $\lambda\lesssim e^4$). Moreover, at $\lambda\lesssim g^4$ no external pressure is needed for this phase transition to occur. In this case it becomes energetically advantageous for fermions to collapse into 'super-nuclei' inside which $\sigma=0$ and the fermions are massless (Lee and Wick 1974, Lee and Margulies 1975).

The last result is particularly interesting, and it was even suggested that it is possible to produce such super-nuclei in heavy-ion collisions. A similar idea of local symmetry restoration inside a super-nucleus in application to quark matter was used in the SLAC bag model of quark confinement (Bardeen *et al* 1975). Unfortunately the problem of the Lee–Wick super-nuclei is not quite so clear. Indeed, as was mentioned in §2, at $\lambda\ll g^2$ the vacuum in (1.28) is unstable with respect to spontaneous generation of an extremely large classical field σ. At $g \ll 1$ from inequality $\lambda \lesssim g^4$ (which is needed for the super-nuclei formation) it follows that $\lambda \ll g^2$. This means that at $g \ll 1$, instead of the symmetry restoration obtained by Lee and Wick an extremely strong symmetry breaking should take place. Therefore the super-nuclei suggested by Lee and Wick may be stable only at $g \gtrsim 1$, when the lowest-order approximation used in their work is inapplicable. Moreover, one may argue (Krive and Chudnovsky 1978) that even if the super-nuclei could be formed, they would be unstable with respect to the pion condensation. From our point of view, however, the more important fact is that the symmetry restoration in the cold dense matter takes place only in the theories without neutral currents and neutral vector mesons of the type given by (1.28), or under some special circumstances for which the effects connected with the neutral currents are small (see the next section). The effects connected with the neutral vector mesons are actually small in 'normal' nuclear matter at a sufficiently low density, when the short-range interactions mediated by the relatively heavy vector mesons can be neglected. However, in the dense 'normal' matter as well as in the 'abnormal' matter, in which the interactions mediated by the vector mesons become long-range, the vector mesons should be taken into account. As will be shown in the next subsection, the effects connected with the neutral vector mesons are opposite to those considered above and lead to an increase of symmetry breaking in cold dense matter (Linde 1975a, 1976c).

5.2. *Theories with neutral currents* (*Weinberg–Salam model*)

High-density symmetry restoration in (1.28), discussed above, may seem rather unexpected from the point of view of the analogy between SBGT and superconductivity theory. Indeed, it is well known that an increase of an electric current $\boldsymbol{j}$ leads to the symmetry restoration in superconductivity theory (see, for example, De Gennes 1966). Therefore, one could expect that an increase of an external fermion current should lead to symmetry restoration in SBGT. In gauge theories the symmetry breaking parameter σ is a function of $j^2 = j_0^2 - \boldsymbol{j}^2$, and therefore an increase of a fermion charge density j_0 should lead to an *increase* of symmetry breaking.

The reason why we have obtained an opposite result in the study of the simplified σ model (1.28) is that this model is not a gauge theory and the current $j_\mu = \langle \bar{\psi}\gamma_\mu\psi \rangle$ in this model does not interact with a neutral vector field as distinct from the electric current in a superconductor. In the theories in which there exist some currents of the type $j_\mu = \langle \bar{\psi}\gamma_\mu\psi \rangle$ interacting with neutral vector mesons ('theories with neutral currents') an increase of the fermion charge density j_0 actually leads to an increase of symmetry breaking.

As an example we shall consider the Higgs model (1.14), extended by the inclusion of fermions:

$$L = -\tfrac{1}{4}(\partial_\mu A_\nu - \partial_\nu A_\mu)^2 + (\partial_\mu + \mathrm{i}eA_\mu)\chi^*(\partial_\mu - \mathrm{i}eA_\mu)\chi + \mu^2\chi^*\chi - \lambda(\chi^*\chi)^2 + \bar{\psi}(\mathrm{i}\partial_\mu\gamma_\mu - m)\psi - e\bar{\psi}\gamma_\mu\psi A_\mu. \quad (5.4)$$

Let us suppose that there exists a non-vanishing fermion current density $j_\mu = \langle \bar{\psi}\gamma_\mu\psi \rangle = \text{constant} \neq 0$. The current being constant, there is no reason to expect translational invariance breaking in this simple model (see however below (§5.4)). Therefore, we shall suppose the classical part of the physical fields χ_1 and A_μ to be constant in space and time. This means that in the transverse gauge $\partial_\mu A_\mu = 0$

we shall try to find a solution corresponding to (5.4) at $j_\mu \neq 0$ of the form:

$$\chi(x)=\frac{1}{\sqrt{2}}(\chi_1(x)+\sigma+i\chi_2(x))$$
$$A_\mu(x)=B_\mu(x)+C_\mu \tag{5.5}$$

where $\langle\chi_i\rangle=\langle B_\mu\rangle=0$ and the classical fields σ and C_μ are space–time-independent. In this case at the classical level the Lagrange equations for the fields σ and C_μ are:

$$\left\langle\frac{\delta L}{\delta\chi_1}\right\rangle=0=-\sigma(\lambda\sigma^2-\mu^2)+e^2C_\mu{}^2\sigma$$
$$\left\langle\frac{\delta L}{\delta A_\mu}\right\rangle=0=e^2C_\mu\sigma^2-ej_\mu. \tag{5.6}$$

The same equations could also be obtained by varying over σ and C_μ the *effective* Lagrangian L_{eff}, which is equal to the Lagrangian (5.4), the two last terms in (5.4) being replaced by $-ej_\mu A_\mu$. These equations are nothing but a covariant generalisation of the Ginzburg–Landau equations in the theory of superconductivity (De Gennes 1966). From (5.6) it follows that:

$$\sigma(\lambda\sigma^2-\mu^2)-j^2/\sigma^3=0 \tag{5.7}$$

where $j^2=j_0{}^2-\boldsymbol{j}^2$. Equation (5.7) implies that, like in the theory of superconductivity, an increase of $\boldsymbol{j}$ leads to the symmetry restoration in the Higgs model (5.4), whereas an increase of the fermion charge density j_0 increases the symmetry breaking (Linde 1975a, 1976c).

If a Lagrangian of the type (5.4) contained a term $\sim g\bar{\psi}\psi\phi$ as in (1.27), then on the left-hand side of equation (5.7) a term $\sim g^2\sigma\,(j^2)^{1/3}$ would appear, promoting symmetry restoration with an increase of j^2 (see (5.2)). However, at $g^2\ll 1$ this term can be neglected compared with the term j^2/σ^3 in (5.7). Moreover, even at $g^2\gtrsim 1$ the term j^2/σ^3, which appears in equation (5.7) due to the existence of neutral currents (the term $-\bar{\psi}\gamma_\mu\psi A_\mu$ in (5.4)), becomes the leading one at large density, i.e. the density increase at large j^2 always leads to an increase in symmetry breaking. This result, which was first obtained for the extended Higgs model (5.4) (Linde 1975a, 1976c, Kirzhnits and Linde 1976a) has also been confirmed by the investigation of some other gauge theories with neutral currents (Sato and Nakamura 1976, Krive and Chudnovsky 1976a, b, Källman 1977). In particular, it has been shown that in the Weinberg–Salam model an equation for σ coincides exactly with (5.7) if the quantity j_μ in (5.7) is understood as a neutrino current $j_\mu=\frac{1}{2}\langle\bar{\nu}\gamma_\mu(1+\gamma_5)\nu\rangle$ (Linde 1976c). Therefore, both in the Higgs model (5.4) and in the Weinberg–Salam model at sufficiently large j^2:

$$\lambda\sigma^6=j^2. \tag{5.8}$$

Let us now consider the case $\boldsymbol{j}=0$. Following from (5.7), the characteristic fermion density at which the parameter σ increases substantially is:

$$j_0\sim\mu^3/\lambda=\sqrt{\lambda}\,\sigma^3(0) \tag{5.9}$$

where $\sigma(0)\equiv\sigma(j_\mu=0)=\mu\lambda^{-1/2}$. To estimate the characteristic density j_0 we shall take $\sigma(0)\sim 250$ GeV as in the Weinberg–Salam model. In this case $j_0\sim\sqrt{\lambda}\times 10^{48}$ cm^{-3}. For $\sigma(0)\sim 100$ MeV, $\mu\sim 1$ GeV (strong interactions) and $j_0\sim 10^{39}$ cm^{-3}. The last value of fermion density is of the same order as the density in the cores of neutron stars (Zeldovich and Novikov 1971).

5.3. Symmetry behaviour at a simultaneous increase of fermion density and temperature

In the previous subsections it was shown that high-temperature effects lead to symmetry restoration in SBGT, whereas an increase of a fermion charge density usually leads to a further increase of the symmetry breaking. Therefore, to study symmetry behaviour in hot dense matter one should take into account the two opposed factors (temperature and fermion charge density) simultaneously. For this purpose we shall consider again the extended Higgs model (5.4) and take into account temperature corrections to the Lagrange equations (5.7) for σ and C_μ. In the lowest order in λ and e^2 the corresponding equations take the form:

$$\left\langle\frac{\delta L}{\delta\chi_1}\right\rangle = -\sigma[\lambda\sigma^2-\mu^2+\lambda(3\langle\chi_1^2\rangle+\langle\chi_2^2\rangle)-e^2C_\mu^2-e^2\langle B_\mu^2\rangle]+2e^2C_\mu\langle B_\mu\chi_1\rangle=0 \tag{5.10}$$

$$\left\langle\frac{\delta L}{\delta A_\mu}\right\rangle = e^2C_\mu\sigma^2+e^2C_\mu(\langle\chi_1^2\rangle+\langle\chi_2^2\rangle)+2e^2\sigma\langle B_\mu\chi_1\rangle+e\langle\chi_2\partial_\mu\chi_1-\chi_1\partial_\mu\chi_2\rangle-ej_\mu=0. \tag{5.11}$$

In a rough approximation one may neglect the non-diagonal terms in (5.10) (Linde 1976c). However, to get more detailed information concerning symmetry behaviour in the model (5.4) one should take into account that the non-diagonal terms in (5.10) and (5.11) at $j_\mu\neq 0$ do not vanish, since after the shift of variables (5.5) the Lagrangian (5.4) contains the non-diagonal terms $e^2\sigma C_\mu B_\mu\chi_1$ and $eC_\mu(\chi_2\partial_\mu\chi_1-\chi_1\partial_\mu\chi_2)$. For simplicity we shall suppose here that $\lambda\gg e^4$. In this case it can be shown that all the interesting effects take place at $T\gg m_{\chi_1}$, m_A, $(j^2)^{1/6}$, when:

$$\langle\chi_1^2\rangle=\langle\chi_2^2\rangle=-\tfrac{1}{3}\langle B_\mu^2\rangle=T^2/12 \tag{5.12}$$

(compare with (3.17)) and the non-diagonal terms in (5.10) can actually be neglected. As for the non-diagonal terms in (5.11), one can verify that they, together with the term $e^2C_\mu(\langle\chi_1^2\rangle+\langle\chi_2^2\rangle)$, are proportional to the polarisation operator of the field B_μ at zero momentum ($k_0=0$, $k\to 0$). In this limit the only non-vanishing component of the polarisation operator is $\Pi_{00}(0)=\tfrac{1}{3}e^2T^2$ (Fradkin 1965). As a result equations (5.10) and (5.11) at $j=0$ take the form:

$$\begin{aligned}\sigma[\lambda\sigma^2-\mu^2-e^2C_0^2+\tfrac{1}{12}T^2(3e^2+4\lambda)]&=0\\ j_0-eC_0(\sigma^2+\tfrac{1}{3}T^2)&=0\end{aligned} \tag{5.13}$$

from which it follows that:

$$\sigma\left(\lambda\sigma^2-\mu^2-\frac{j_0^2}{(\sigma^2+\frac{1}{3}T^2)^2}+\frac{3e^2+4\lambda}{12}T^2\right)=0. \tag{5.14}$$

At $T=0$ this equation coincides with equation (5.7), whereas at $j_0=0$ (5.14) coincides with equation (3.19). Let us consider the high-density limit, which will be most important for the cosmological applications of the above results. In this case the critical temperature T_c, at which symmetry restoration takes place in (5.4), is given by:

$$T_c{}^6 = \frac{108 j_0{}^2}{3e^2+4\lambda}. \tag{5.15}$$

In the Weinberg–Salam model the corresponding equation for σ is slightly more complicated than (5.14):

$$\sigma\left[\lambda\sigma^2-\mu^2-\frac{j_0{}^2}{\sigma^2+\frac{1}{3}T^2(1+8\cos^4\theta_W)}+\left(4\lambda+\frac{3e^2(1+2\cos^2\theta_W)}{\sin^2 2\theta_W}\right)\frac{T^2}{12}\right]=0. \tag{5.16}$$

Here j_0 is an excess of the neutrino density n_ν over the antineutrino density $n_{\bar\nu}$:

$$j_0 = n_\nu - n_{\bar\nu} = \tfrac{1}{2}\langle\bar\nu\gamma_0(1+\gamma_5)\nu\rangle \tag{5.17}$$

and θ_W is the Weinberg angle, which characterises the relative strength of weak and electromagnetic interactions (see, for example, Weinberg 1972); according to recent experimental data $\sin^2\theta_W \approx 0{\cdot}24$ (Holder *et al* 1977). In the high-density limit equation (5.16) yields the following expression for the critical temperature in the Weinberg model:

$$T_c{}^6 \approx \frac{j_0{}^2}{\lambda+3e^2}. \tag{5.18}$$

For further applications it is convenient to express temperature T through the photon density n_γ at this temperature (Landau and Lifshitz 1964):

$$n_\gamma = \frac{2\zeta(3)}{\pi^2}T^3 \approx 0{\cdot}244\,T^3. \tag{5.19}$$

Thus from (5.18) and (5.19) it follows that at large j_0 symmetry restoration in the Weinberg–Salam model takes place when the photon density n_γ becomes greater than the critical density $n_\gamma{}^c$, where:

$$n_\gamma{}^c \approx \frac{j_0}{4(\lambda+3e^2)^{1/2}} = \frac{n_\nu - n_{\bar\nu}}{4(\lambda+3e^2)^{1/2}}. \tag{5.20}$$

It is seen therefore that the phase transition in the Weinberg model takes place only at some definite relationship between the densities of photons and neutrinos. This fact will be important for our discussion of the cosmological consequences of the phase transitions in SBGT (see §6).

5.4. *Condensation of the Yang–Mills fields in super-dense matter*

In the previous subsections we have assumed that super-dense matter is spatially homogeneous, i.e. that super-dense matter should be in a gaseous or a liquid, but not in a crystalline state. The reason is that in those theories without Yang–Mills fields inhomogeneous states of super-dense matter consisting of ultra-relativistic particles are energetically unfavourable at the classical level. Therefore, crystallisation of super-dense matter usually becomes possible only in some special cases when radiative corrections are sufficiently large, see, for example, a theory of pion condensation (Migdal 1977) or a theory of vector field condensation in super-dense matter (Akhiezer *et al* 1978).

However, in the non-Abelian gauge theories crystallisation becomes possible even at the classical level (Linde 1978b). To outline the main features of this effect we shall consider here as an example the simplest (though non-realistic) non-Abelian O(3)-symmetric theory (4.6) without scalar fields but in the presence of fermions with a non-vanishing electric charge density j_0^3. The corresponding effective Lagrangian $L_{\rm eff}$ (see §5.2) is:

$$L_{\rm eff} = -\tfrac{1}{4}(G_{\mu\nu}^a)^2 - ej_0^3 A_0^3 \tag{5.21}$$

where $a = 1, 2, 3$, A_μ^3 is the electromagnetic potential (see §4.2).

One can easily verify that the classical Lagrange equations in (5.21) possess a standing wave solution:

$$\begin{aligned} A_1^1 &= A_1^2 = C \sin mZ \\ A_2^1 &= A_2^2 = C \cos mZ \\ A_3^a &= A_t^3 = A_0^{1,2} = 0 \end{aligned} \tag{5.22}$$

where

$$2eA_0^3 C^2 = j_0^3 \qquad m = eA_0^3. \tag{5.23}$$

Note that the covariant divergence of the current j_0^3 vanishes in the solution (5.22). The energy-momentum tensor $T_{\mu\nu}$ of the field (5.22) has time-independent gauge-invariant components $T_{12} = T_{21} = -2m^2C^2 \sin 2mZ$, which are periodic in one of the space coordinates. Therefore, the solution (5.22) and (5.23) corresponds to a one-dimensional Yang–Mills crystal. The electric charge density of this crystal is equal to $-2eA_0^3C^2$ and exactly compensates the fermion charge density j_0^3 (see equation (5.23)).

Analogous Yang–Mills crystals may also exist in those theories with spontaneous symmetry breaking. However, we have discussed above only one solution (5.22) of the non-linear Yang–Mills equations. To analyse whether the Yang–Mills fields in super-dense matter should actually be in a crystalline state or that some other type of the vector field condensation takes place, one should try to find other (periodic or non-periodic) solutions. An investigation of this problem is now in progress.

6. SBGT and cosmology

6.1. Symmetry behaviour in the early Universe

In this section we shall discuss briefly the most important consequences of the phase transitions in gauge theories.

The most evident (though the least investigated) possibility for using the results concerning phase transitions in SBGT is connected with superconductivity theory. This possibility is based on the often-used analogy between superconductivity theory and SBGT. Phenomenological properties of superconductors are well known, and one can use them to get many correct predictions of the macroscopic effects in SBGT. On the other hand, the theory of symmetry breaking and restoration in SBGT is in fact much simpler than that in superconductivity theory. For example, only about twenty years after the discovery of the microscopic theory of superconductivity (Bardeen *et al* 1957) it was shown that the transition from a superconductive to a normal state in certain cases is not a second-order phase transition, but a first-order one (Halperin *et al* 1974). In SBGT the same result has been independently obtained in one of the first papers on phase transitions in gauge theories (Kirzhnits and Linde 1974b).

In gauge theories, due to their relative simplicity, some effects have been obtained which are absent or have not been discovered so far in superconductivity theory. It was shown in particular that, under certain conditions, quantum fluctuations and external factors may substantially *increase* symmetry breaking in SBGT. If analogous possibilities should be discovered in superconductivity theory, this would lead to some progress in solving the problem of high-temperature superconductivity.

Another interesting possibility of using the results discussed above is connected with the theory of multiparticle production in collisions of high-energy elementary particles. In such collisions fireballs filled with elementary particles in a thermodynamically equilibrium state are created, and then these fireballs decay into many separate elementary particles. According to the statistical theory of multiparticle production (see, for example, a review of this theory by Feinberg (1972)) the temperature inside the fireballs may exceed several hundred MeV, and therefore this temperature may be sufficient for the phase transitions in the theory of strong interactions (Eliezer and Weiner 1976, Krive *et al* 1977).

However, the most interesting consequences of the phase transitions in gauge theories are connected with cosmology. According to the hot Universe theory, the Universe has been expanding and gradually cooling from a state with infinite temperature and density (Zeldovich and Novikov 1975, Weinberg 1972). Therefore, at a sufficiently small time t from the beginning of the expansion of the Universe the extreme conditions necessary for all phenomena considered in this review have been actually realised. As a result, at $t \to 0$ symmetry behaviour in SBGT was determined by the relation between the effects connected with an increase of temperature and the opposed effects connected with an increase of fermion charge density (Linde 1976a, Krive *et al* 1976a).

To illustrate this general statement let us consider symmetry behaviour in the Weinberg–Salam model at $t \to 0$. As follows from the discussion in §5.3 symmetry behaviour in the Weinberg–Salam model depends on the relative magnitude of the photon density n_γ and the fermion charge density j_0. In the course of the Universe evolution the ratio of the photon density n_γ to the charge density j_0 of leptons and baryons (specific entropy) remains practically constant (Weinberg 1972, Zeldovich and Novikov 1975). Therefore, in order to solve the problem whether the symmetry at $t \to 0$ was restored or not it is sufficient to compare the present fermion 'weak' charge density with the photon density. The baryon charge density $n_{\mathrm{B}} - n_{\bar{\mathrm{B}}} \approx n_{\mathrm{B}}$ at present is of the order of:

$$n_{\mathrm{B}} \sim 10^{-8}\, n_\gamma \tag{6.1}$$

and this is the reason why the effects connected with the baryon charge asymmetry of the Universe do not influence symmetry restoration in the Weinberg–Salam model at $t \to 0$. The lepton charge density of all known types of leptons except neutrinos is of the same order as n_{B} and therefore it also does not affect the symmetry restoration. Meanwhile, an excess of neutrinos over antineutrinos $j_0 = n_\nu - n_{\bar{\nu}}$ may be very large. The strongest constraint on the value of j_0 for the electron neutrinos follows from the theory of helium production in the early Universe: $j_0 \lesssim 10^3\ \mathrm{cm}^{-3}$ (Zeldovich and Novikov 1975). The value of j_0 for the muon neutrinos may be a few orders greater. On the other hand, the photon density in the Universe at present is of the order of $n_\gamma \sim 4\times 10^2$ cm^{-3}. If one neglects variation of n_γ/j_0 in the course of the evolution of the Universe (i.e. if specific entropy actually is a constant) then according to (5.20) one may conclude that at $t\to 0$ symmetry in the Weinberg–Salam model has not been restored only if at present:

$$j_0 \gtrsim 4(\lambda+3e^2)^{1/2}\, n_\gamma. \tag{6.2}$$

If one takes, for example, $\lambda\sim e^2\sim 10^{-1}$, the inequality (6.2) implies that:

$$j_0 \gtrsim 2{\cdot}5\, n_\gamma. \tag{6.3}$$

Such a possibility does not contradict present cosmological constraints on the value of j_0. Therefore, further on we shall study both possibilities $j_0 \gtrsim 2{\cdot}5\, n_\gamma$ and $j_0 \lesssim 2{\cdot}5\, n_\gamma$, which correspond respectively to symmetry restoration or symmetry breaking in the Weinberg–Salam model at $t\to 0$.

From (6.1) it follows that inequality (6.3) is equivalent to the following relation between the lepton (L) and baryon (B) charges of the Universe:

$$L \gtrsim 10^8\, B. \tag{6.4}$$

Thus, if the Universe is not so enormously charge-asymmetric, then the symmetry in the Weinberg–Salam model at $t\to 0$ was restored and all particles except the Higgs mesons were massless. The time from the beginning of the expansion of the Universe, at which the phase transition in the Weinberg–Salam model has taken place, is of the order of:

$$t\sim 10^{-7}\text{–}10^{-10}\ \text{s}.$$

If the inequalities (6.3) and (6.4) hold (strong charge asymmetry of the Universe), the symmetry was always broken, and at $t\to 0$ masses of all particles except photons and neutrinos were infinitely growing.

Now let us discuss some consequences of the phase transitions and of the unusual properties of super-dense matter in the early Universe.

6.2. *Quarks in the Universe*

Attempts to explain the reasons why free quarks up to now have not been discovered can be approximately divided into three large groups.

(i) Free quarks have infinite energy (permanent confinement of quarks): see, for example, papers by Wilson (1974), Polyakov (1977, 1978), Callan *et al* (1977), 't Hooft (1977) and Nambu (1974).

(ii) Free quarks have some extremely large but finite energy (partial confinement of quarks). The most known model of this type is the SLAC bag model (Bardeen *et al* 1975).

(iii) Free quarks decay immediately into leptons, but baryons built up from these quarks are practically stable (Pati and Salam 1973, 1974).

Permanent confinement of quarks may be achieved due to the infrared instability and some special topological properties of the massless Yang–Mills field theory ('electric confinement') (see Wilson 1974, Polyakov 1977, 1978, Callan *et al* 1977, 't Hooft 1977) or due to the existence of quasimagnetic flux tubes, which connect 'quasimonopoles' identified with quarks ('magnetic' confinement) (see Nambu 1974, Polyakov 1975, Mandelstam 1975, Linde 1976e). The 'magnetic' mechanism of quark confinement works only if symmetry in SBGT is broken, and therefore at a sufficiently high temperature $T > T_c$ the quarks become free. It was argued (Polyakov 1978) that 'electric' confinement also does not work at sufficiently large temperatures. This point, however, does not seem quite clear (see §3.5). In both cases, however, at the present time free quarks should be completely absent and therefore permanent confinement of quarks does not lead to any difficulties of the type discussed below.

In cases (ii) and (iii) at a sufficiently high temperature quarks should no doubt become free. As was shown by Zeldovich *et al* (1965) some of the stable quarks of type (ii), which were free in the early Universe, have not enough time to meet each other and form hadrons. These quarks would remain free at the present time and a lot of them would be discovered, for example, in the cosmic rays. On the other hand, free unstable quarks of type (iii) would almost completely decay into leptons, and as a result an extremely small amount of baryons would remain in the Universe (Okun' and Zeldovich 1976).

One cannot exclude, of course, that due to some special circumstances some of the theories of types (ii) and (iii) will not lead to these difficulties. However, at present we can see only one possibility of avoiding the problems peculiar to theories (ii) and (iii), namely, if the Universe is sufficiently strongly charge-asymmetric, then due to an increase of the fermion density in the early Universe the quark masses have been increasing as $t \to 0$ at the same rate as the temperature. As a result, if the ratio of the quark masses to the temperature was sufficiently large, the quarks never have been free and all the difficulties mentioned above disappear (Linde 1976c). Note, however, that the necessity to use such exotic assumptions as a strong charge asymmetry of the Universe to save theories (ii) and (iii) may serve as an argument against these theories in favour of the theories with permanent confinement of quarks.

6.3. Domain structure of vacuum

The kinetics of the process of symmetry breaking in the cooling Universe may be very complicated. Let us consider, for example following Zeldovich *et al* (1974), the process of symmetry breaking in the simplest model (1.3). In sufficiently far removed (causally unconnected) domains of the Universe the phase transition with the symmetry breaking may proceed from the disordered state $\sigma = 0$ *into two different states*: into the state with the field $\sigma = +\mu\lambda^{-1/2}$ or into the state with $\sigma = -\mu\lambda^{-1/2}$. The domains with the different signs of the field σ are separated from each other by thin walls ('kinks') inside which the field σ varies from $-\mu\lambda^{-1/2}$ to $+\mu\lambda^{-1/2}$. Such a domain structure is energetically unfavourable and the walls collapse or spread up to infinity with the velocity almost equal to the velocity of light c. Therefore the size of the domains filled with the constant homogeneous field σ of one of the two signs is of the order of ct (the so-called radius of the horizon), which is extremely large. However, the domain walls have such a large surface energy density that if at least one such a wall existed at present inside the horizon (and this seems to be unavoidable), the observable part of the Universe would be greatly anisotropic.

This result implies that most of the theories with spontaneous breaking of a discrete symmetry of the type (1.3) (which is symmetric with respect to the change $\phi \to -\phi$) contradict cosmological data (Zeldovich *et al* 1974). This conclusion is very important since it rules out many theories with spontaneous breaking of (discrete) CP invariance. One should note, however, that in some models with a discrete symmetry breaking it is possible to avoid the difficulties connected with the appearance of the domain walls. For this purpose one should assume, as in the preceding subsection, that the Universe is greatly charge-asymmetric (6.4). As was shown in §6.1, for some theories this may lead to the absence of the phase transition in the early Universe and, consequently, to the absence of the undesirable domain walls.

It is worth noting also that the same mechanism, which provides the domain wall formation after the phase transition with a discrete symmetry breaking, leads also to the vortex tube formation after the U(1) symmetry breaking in the Higgs model and to the creation of monopoles after the non-Abelian symmetry breaking (Kibble 1976). It can be shown that the formation of the vortex tubes does not lead to any considerable cosmological effects. However, some consequences of the creation of monopoles in the early Universe would be in contradiction with the cosmological data (see Zeldovich and Khlopov 1978).

6.4. Substance energy non-conservation and the time-dependent cosmological term

One of the most unexpected effects connected with the phase transitions in gauge theories is the substance energy non-conservation due to energy 'pumping' from the non-observable Bose condensate in the processes under consideration (Kirzhnits and Linde 1974a, 1976a).

The physical meaning of this effect may be easily understood if one takes into account that all matter in SBGT at a non-vanishing temperature (for simplicity we take here all chemical potentials and external fields equal to zero) can be uniquely divided into two parts: a set of interacting particles and the Bose condensate $\sigma(T)$. Note that the condensate is a Poincaré-invariant object, constant in space and time, which does not fix any preferred reference frame or preferred direction in space–time, and which does not influence a test particle since $\partial_\mu \sigma_\mu = 0$. Therefore the condensate $\sigma(T)$ at a given temperature can manifest itself only through its influence on the space–time curvature connected with a non-zero energy-momentum tensor of the condensate $g_{\mu\nu}\epsilon(\sigma)$, and in this sense the condensate does not differ from the ordinary vacuum of a quantum field theory. This is the reason why the division of all matter into a set of interacting particles and the condensate has the direct meaning of the division of matter into the observable part (i.e. substance) and non-observable part (vacuum).

In those theories without spontaneous symmetry breaking such a division is fixed once and forever. In our case the characteristics of the vacuum state are temperature-dependent. This fact is an inevitable consequence of spontaneous symmetry breaking and is connected with the appearance of the classical temperature-dependent field $\sigma(T)$ in the Lagrangian.

Let us consider, for example, the energy-momentum tensor, corresponding to the simplest theory (1.3):

$$\theta_{\mu\nu} = \partial_\mu \phi \partial_\nu \phi - \tfrac{1}{2} g_{\mu\nu} [(\partial_\alpha \phi)^2 - 2\sigma(\lambda\sigma^2 - \mu^2)\phi - (3\lambda\sigma^2 - \mu^2)\phi^2 - 2\lambda\sigma\phi^3 - \tfrac{1}{2}\lambda\phi^4] + g_{\mu\nu}[\tfrac{1}{4}\lambda\sigma^4 - \tfrac{1}{2}\mu^2\sigma^2 + \epsilon(0)] \quad (6.5)$$

where $\epsilon(0)$ is an arbitrary constant which can be subtracted from the Lagrangian (1.3) to fix the vacuum energy at $\sigma = 0$.

Averaging of $T_{\mu\nu}$ (where $T_{\mu\nu}$ is the operator part of $\theta_{\mu\nu}$) in the lowest order of perturbation theory gives a quantity with the obvious physical meaning of the energy-momentum tensor of interacting particles (some function of the occupation numbers n_p, vanishing at $n_p \to 0$). The C-number part of $\theta_{\mu\nu}$ gives the energy-momentum tensor of the non-observable Bose condensate $g_{\mu\nu}\epsilon(\sigma)$, and thus the overall energy-momentum tensor of matter $\langle\theta_{\mu\nu}\rangle$ can be divided into two parts:

$$\langle\theta_{\mu\nu}\rangle = \langle T_{\mu\nu}\rangle + g_{\mu\nu}\epsilon(\sigma) \tag{6.6}$$

where $\langle T_{\mu\nu}\rangle$ is the energy-momentum tensor of substance, and the condensate energy is given by:

$$\epsilon(\sigma) = \epsilon(0) + \tfrac{1}{4}\lambda\sigma^4(T) - \tfrac{1}{2}\mu^2\sigma^2(T). \tag{6.7}$$

Due to the dependence of $\epsilon(\sigma)$ on the condensate density $\sigma(T)$, the energy-momentum tensor of the condensate (of vacuum), $g_{\mu\nu}\epsilon(\sigma)$, varies in the processes in which the condensate density $\sigma(T)$ changes. As a result, in such processes only the overall energy-momentum tensor $\langle\theta_{\mu\nu}\rangle$ is conserved, but not the energy-momentum tensors of substance and vacuum separately. This means that in the processes where temperature changes with time, the first law of thermodynamics applied to substance is violated, i.e. the energy of the observable part of matter is non-conserved due to energy 'pumping' from the non-observable condensate (Kirzhnits and Linde 1974a, 1976a).

The physical meaning of what has been said becomes particularly clear if we consider a first-order phase transition in the Higgs model at $\lambda \ll e^4$. As is noted in §3.3 the first-order phase transition with symmetry breaking in the Higgs model takes place with a decrease of temperature near the point T_{c_1} (see figure 9). After the phase transition at $T \approx T_{c_1}$ the Bose condensate $\sigma(T) \approx \sigma(0) = \mu\lambda^{-1/2}$ appears. The substance energy density $\sim T^4$ at $T = T_{c_1}$ is much less than the energy density $\mu^4/4\lambda$ released due to the condensate formation (or, in other words, due to the vacuum reconstruction). Since before the phase transition the Bose condensate is absent and the substance energy is relatively small, an observer would see the creation of most of the substance at the point of the phase transition practically 'from nothing'. Note, however, that the effect of the non-conservation of the substance energy has nothing in common with stationary cosmology in which the creation of matter from nothing was postulated (Bondi and Gold 1948, Hoyle 1948). In our case the total energy-momentum tensor of matter $\langle\theta_{\mu\nu}\rangle$ is exactly conserved.

From the point of view of gravity theory the division of $\langle\theta_{\mu\nu}\rangle$ into $\langle T_{\mu\nu}\rangle$ and $g_{\mu\nu}\epsilon(\sigma)$ corresponds to the representation of the total energy-momentum tensor as a sum of the energy-momentum tensor of substance and the cosmological term (Linde 1974, Veltman 1974, 1975, Dreitlein 1974). The results discussed above imply in particular that in theories with spontaneous symmetry breaking the cosmological term $g_{\mu\nu}\epsilon(\sigma)$ is temperature-dependent (Linde 1974), and at $T > T_c$ the cosmological term becomes equal to $g_{\mu\nu}\epsilon(0)$. According to the present cosmological data $\epsilon(\sigma(T=0)) = \epsilon(\mu\lambda^{-1/2}) \lesssim 10^{-29}$ g cm^{-3}. On the other hand, in the Weinberg–Salam model $\sigma(T=0) \approx 250$ GeV, and if, for example, $\lambda \sim 10^{-2}$, then $\epsilon(0) = \tfrac{1}{4}\lambda\sigma^4(T=0) \sim 10^{25}$ g cm^{-3}. This means that in the early Universe the cosmological term was at least 10^{50} times greater than at present (Linde 1974, see also Bludman and Ruderman 1977). Of course, the energy of substance in the early Universe was also extremely large. Nevertheless, in certain cases the temperature-dependent cosmological term may become greater than $\langle T_{\mu\nu}\rangle$. In particular, near the point of the first-order phase transition with symmetry breaking in the Higgs model at

$\lambda \lesssim e^4$ the cosmological term is actually much greater than the substance energy-momentum tensor. In some cases the effects connected with the substance energy non-conservation and the time-dependence of the cosmological term in the expanding Universe may lead to important cosmological consequences (see §6.6).

6.5. Boiling of vacuum and an interplay between symmetry breaking in gauge theories and cosmology

In §2 it was pointed out that at certain relations between masses and coupling constants radiative corrections lead to symmetry restoration in the Higgs model (Linde 1976a). Analogous effects also take place in more realistic theories such as the Weinberg–Salam model. It proves (Weinberg 1976, Linde 1976b, d) that for the Weinberg angle $\sin^2 \theta_W \approx 0{\cdot}24$ the effective potential in the Weinberg–Salam model acquires an additional 'dynamical' minimum at $\sigma = 0$ if the Higgs meson mass m_ϕ is less than 9·31 GeV. At $m_\phi < 6{\cdot}55$ GeV this minimum becomes even deeper than the ordinary minimum at $\sigma = \mu\lambda^{-1/2}$ (see figure 5), i.e. the vacuum state with the broken symmetry becomes metastable. However, as was stressed in §3.3 the lifetime of such metastable vacuum states in SBGT usually exceeds the age of the Universe. Therefore a decay of a metastable vacuum state in SBGT (boiling of vacuum, see §3.3) may take place only at some special relations between masses and coupling constants at which the energy barrier between the stable and metastable vacua becomes very small (Voloshin *et al* 1974, Coleman 1977, Linde 1977b).

This important observation leads to some rather unexpected conclusions concerning symmetry breaking in the Weinberg–Salam model. Let us first suppose that the Universe is indeed greatly charge-asymmetric ($L \gtrsim 10^8 B$, see equation (6.4)). This implies that at $t \to 0$ the symmetry in the Weinberg–Salam model was broken. With a decrease of neutrino density in the course of the expansion of the Universe a 'dynamical' minimum of the effective potential has appeared at $\sigma = 0$. If the Higgs meson mass is sufficiently small ($m_\phi < 6{\cdot}55$ GeV), this minimum at the present time should become deeper than the minimum at $\sigma = \mu\lambda^{-1/2}$. However, an investigation of the phase transition from the state $\sigma = \mu\lambda^{-1/2}$ to the state $\sigma = 0$ shows that it is a tunnelling process which occurs in the course of the Universe evolution only if $m_\phi \lesssim 450$ MeV (Linde 1977b).

On the other hand, if the Universe is not so greatly charge-asymmetric, then initially the symmetry in the Weinberg–Salam model has been restored, and the new minimum of $V(\sigma)$ at $\sigma \neq 0$ appeared only after the cooling of the Universe. In this case the transition from the state $\sigma = 0$ to the state $\sigma = \mu\lambda^{-1/2}$ would proceed only if $m_\phi > 9{\cdot}3$ GeV (Linde 1977b).

At the present time the symmetry in the theory of weak and electromagnetic interactions is broken. This fact, together with the results discussed above, leads us to the following conclusions concerning the Weinberg–Salam model.

(i) The Higgs meson mass in the Weinberg model should exceed 450 MeV.

(ii) If, as one may expect, the Universe is not enormously charge-asymmetric, then $m_\phi > 9{\cdot}3$ GeV.

(iii) If the Higgs meson mass is in the range 450 MeV $< m_\phi < 9{\cdot}3$ GeV, then the symmetry in the Weinberg model may be broken at the present time only if the Universe is greatly charge-asymmetric. In other words, for certain relations between masses and coupling constants, *symmetry breaking in the Weinberg model is completely determined by the charge asymmetry of the Universe.* The discovery of the Higgs meson with a mass in the interval between 450 MeV and 9·3 GeV would imply the existence of a 'neutrino sea' in the Universe with an excess of neutrinos over antineutrinos $j_0 = n_\nu - n_{\bar{\nu}} \gtrsim 10^3\ \text{cm}^{-3}$. Such a correspondence between symmetry breaking in the Weinberg–Salam model and the charge asymmetry of the Universe seems to be a rather unexpected example of an interplay between micro- and macrophysics.

From the results obtained above one can draw a general conclusion concerning symmetry breaking in more complicated theories, namely that, if the effective potential corresponding to some theory has a number of deep enough local minima, then due to the large lifetime of the metastable vacuum states in quantum field theory the solution of the problem in which of the local minima the Universe should be at the present time is determined not by energetical considerations but by the physical processes at the early stages of the Universe evolution.

6.6. *Cold Universe?*

In order to demonstrate how much the existence of the phase transitions in SBGT may modify the present theory of the Universe evolution we shall consider below (Chibisov and Linde 1978) symmetry behaviour in cold dense matter, which consists of leptons and baryons with mutually compensated weak charge densities. In this case (charge-symmetric Universe) the effects connected with neutral currents (see §5.2) disappear and an increase of fermion density leads to symmetry restoration as in the σ model.

Let us suppose, contrary to the usual belief, that initially the Universe was cold. Nevertheless, due to an increase of fermion density at $t \to 0$ symmetry in the early Universe has been restored, and in the course of the Universe expansion a phase transition with symmetry breaking has taken place. As was mentioned in §5.1, at certain relations between coupling constants this phase transition is the first-order one. In this case, after the decay of the metastable vacuum $\sigma = 0$ all its energy $\epsilon(0)$ has been transformed into the thermal energy of substance, and therefore we again obtain an ordinary hot Universe, but only at $t > t_c$, where t_c is the time of the phase transition.

By an appropriate choice of the coupling constants in SBGT one may arrange things so that the phase transition with symmetry breaking will take place at a comparatively small fermion density. In this case, before the phase transition almost all the energy is concentrated in the metastable vacuum state $\sigma = 0$. Therefore, after the decay of the metastable vacuum almost all the substance energy appears 'from nothing' (substance energy non-conservation) and the ratio of the photon density n_γ to the baryon density in the hot matter (specific entropy) determined by the choice of the coupling constants in SBGT may be made arbitrarily large. In particular, at a certain choice of coupling constants one may get an experimental value of the specific entropy $s \sim 10^8$.

One should note that the first-order phase transition in SBGT proceeds by bubble formation (see §3.3) which, in principle, could lead to large inhomogeneities in the energy density in the early Universe and consequently to the creation of a large number of black holes. This may lead to a contradiction with the present cosmological data (Vainer and Naselsky 1977). However, in the theories with superstrong symmetry breaking (see, for example, Georgi and Glashow 1974) the phase transition with symmetry breaking takes place soon after the Planck time $t_{\text{Pl}} \sim 10^{-43}$ s. In this case all the black holes formed after the phase transition were extremely small and very soon they had evaporated due to the Hawking effect (Hawking 1975).

Since the phase transition takes place during the very early stages of the Universe evolution, all the observational consequences of our cold Universe model are the same as in the ordinary hot Universe theory.

In recent years there have been some other attempts to obtain all the observational consequences of the hot Universe theory starting from a cold Universe (see, for example, Zeldovich and Starobinsky 1976, Carr 1977). In these papers it is supposed that the cold Universe becomes hot due to some effects connected with primordial black holes, which are formed from initial density fluctuations in the early Universe. However, in this approach one needs to make several assumptions concerning the spectrum of the initial density fluctuations in the early Universe, which may seem rather unnatural, and some of the predictions of these models differ from those of the hot Universe theory (Carr 1977). In our case all the effects are determined by the properties of the elementary particle theory only, and no special assumptions concerning initial density fluctuations are needed.

Of course, all the effects discussed in this section take place only at certain conditions (weak charge symmetry of the Universe) and at some definite relations between coupling constants. In any case, however, the possibility of such a complete reconsideration of the theory of the Universe evolution seems to be a very interesting example of the new perspectives opened now in cosmology in connection with phase transitions in gauge theories.

7. Conclusions

The most interesting phenomena discovered in solid-state physics and quantum statistics, such as ferromagnetism, superfluidity and superconductivity, are connected with spontaneous symmetry breaking. It is not surprising, therefore, that the study of phase transitions in SBGT, being in the boundary region between elementary particle theory, quantum statistics and cosmology, leads to such non-trivial results as a time dependence of particle masses, coupling constants and of the cosmological term in the expanding Universe, to the appearance of a domain structure of vacuum, to substance energy non-conservation, to the correspondence between symmetry breaking in elementary particle theory and in cosmology, to the possibility of a description of all observed properties of the hot Universe supposing that initially the Universe was cold, etc. Note that all these effects do not appear as a consequence of some exotic hypotheses and assumptions. As is mentioned in the introduction, spontaneous symmetry breaking is a basic principle for practically all theories of weak, strong and electromagnetic interactions considered at the present time.

All the results discussed in the present review have been obtained only a few years ago and one may expect that the most interesting results concerning phase transitions in gauge theories are still to be obtained. In our review we have not discussed possible consequences of the phase transitions in SBGT for the theory of multiparticle production in collisions between high-energy elementary particles (Eliezer and Weiner 1976, Krive *et al* 1977). Everywhere throughout this review only the weak coupling case has been considered, whereas in the strong coupling regime a lot of new interesting effects may appear, such as the crystallisation of scalar and of Abelian vector fields, spontaneous generation of magnetic fields (ferromagnetism of super-dense matter), etc (Akhiezer *et al* 1978). Crystallisation of the non-Abelian gauge fields, which may take place even in the weak coupling case (Linde 1978b), also deserves further investigation. Some unsolved and very interesting problems arise in the study of the quantum statistics of massless Yang–Mills fields (see §3.5). Many questions are connected with the cosmological consequences of phase transitions in gauge theories. For example, one may wonder whether the large density fluctuations, which appear after the phase transitions in SBGT, could lead to galaxy formation. Another problem to be analysed is the possible connection between the phase transitions in SBGT and black hole formation in the early Universe. One may hope that further investigation of the macroscopic consequences of SBGT will be useful for a better understanding of the physical structure of gauge theories and will give us the possibility of looking from a new point of view at some other problems in theoretical physics.

Acknowledgments

A considerable part of this review is based on the results obtained in a collaboration with D A Kirzhnits. It is a pleasure to express my deep gratitude to D A Kirzhnits for his help and for many enlightening discussions. I am also thankful to G Chapline, E M Chudnovsky, E S Fradkin, R E Kallosh, I V Krive, A M Polyakov, A Salam, J Strathdee, E V Shuryak, I V Tyutin and Ya B Zeldovich for useful discussions of various problems touched upon in this review.

References

Abers G S and Lee B W 1973 *Phys. Rep.* **9C** 1–141
Abrikosov A A, Gorkov L P and Dzyaloshinski I E 1964 *Methods of Quantum Theory in Statistical Physics* (Englewood Cliffs, NJ: Prentice-Hall)
Akhiezer A I, Krive I V and Chudnovsky E M 1978 *Ann. Phys., NY* submitted
Baluni V 1978 *Phys. Rev.* D to be published
Bardeen W A, Chanowitz M S, Drell S D, Weinstein M and Yan T-M 1975 *Phys. Rev.* D **11** 1094–136
Bardeen J, Cooper L N and Schrieffer J R 1957 *Phys. Rev.* **108** 1175
Baym G and Grinstein G 1977 *Phys. Rev.* D **15** 2897–912
Bludman S A and Ruderman M A 1977 *Phys. Rev. Lett.* **38** 255–7
Bondi H and Gold T 1948 *Mon. Not. R. Astron. Soc.* **108** 252
Cabibbo N and Parisi G 1975 *Phys. Lett.* **59B** 67
Callan C, Dashen R and Gross D 1977 *Phys. Lett.* **66B** 375–81
Carr B J 1977 *Mon. Not. R. Astron. Soc.* **181** 293–309
Chang Sh-J 1975 *Phys. Rev.* D **12** 1071–88
—— 1976 *Phys. Rev.* D **13** 2778–88
Chapline G and Nauenberg M 1977 *Phys. Rev.* D **15** 2929–36
Chibisov G and Linde A D 1978 to be published
Coleman S 1977 *Phys. Rev.* D **15** 2929–36
Coleman S and Weinberg E 1973 *Phys. Rev.* D **7** 1888–910
Collins J C and Perry M J 1975 *Phys. Rev. Lett.* **34** 1353–6
Dashen R, Ma Sh-k and Rajaraman R 1975 *Phys. Rev.* D **11** 1499–508
De Gennes P G 1966 *Superconductivity of Metals and Alloys* (New York: Benjamin)
Dolan L and Jackiw R 1974 *Phys. Rev.* D **9** 3320–40
Dreitlein J 1974 *Phys. Rev. Lett.* **33** 1243–4
Eliezer S and Weiner R 1976 *Phys. Rev.* D **13** 87–94
Englert F and Brout R 1964 *Phys. Rev. Lett.* **13** 321

Feinberg E L 1972 *Phys. Rep.* **5C** 237–350
Fradkin E S 1955 *Zh. Eksp. Teor. Fiz.* **28** 750
—— 1965 *Proc. Lebedev Phys. Inst.* **29** 7–138 (Engl. trans. 1967 by Consultants Bureau, New York)
Fradkin E S and Tyutin I V 1974 *Riv. Nuovo Cim.* **4** 1–78
Frautschi S 1971 *Phys. Rev.* D **3** 2821–34
Freedman B A and McLerran L D 1977 *Phys. Rev.* D **16** 1169–85
Georgi H and Glashow S L 1974 *Phys. Rev. Lett.* **3** 2438
Ginzburg V L and Landau L D 1950 *Zh. Eksp. Teor. Fiz.* **20** 1064
Goldstone J 1961 *Nuovo Cim.* **19** 154
Goldstone J, Salam A and Weinberg S 1962 *Phys. Rev.* **127** 965
Gross D J and Wilczek F 1973 *Phys. Rev. Lett.* **30** 1343–6
Guralnik G S, Hagen C R and Kibble T W B 1964 *Phys. Rev. Lett.* **13** 585
Hagedorn R 1965 *Nuovo Cim. Suppl.* **43** 143
Halperin B I, Lubensky T C and Ma Sh-k 1974 *Phys. Rev. Lett.* **32** 292–5
Harrington B J and Shepard H K 1976 *Nucl. Phys.* B **105** 527–37
Harrington B J and Yildiz A 1974 *Phys. Rev. Lett.* **33** 324–7
—— 1975 *Phys. Rev.* D **11** 779–83
Hawking S W 1975 *Commun. Math. Phys.* **43** 199
Heisenberg W and Euler H 1936 *Z. Phys.* **98** 714
Higgs P W 1964a *Phys. Lett.* **12** 132
—— 1964b *Phys. Rev. Lett.* **13** 508
—— 1966 *Phys. Rev.* **145** 1156
Hiro-O-Wada 1974 *Lett. Nuovo Cim.* **11** 697
Holder M *et al* 1977 *Phys. Lett.* **72B** 254–60
't Hooft G 1971 *Nucl. Phys.* B **35** 167–88
—— 1974 *Nucl. Phys.* B **79** 279–84
—— 1977 *Preprint* Utrecht University
't Hooft G and Veltman M 1972 *Nucl. Phys.* B **50** 318
Hoyle F 1948 *Mon. Not. R. Astron. Soc.* **108** 372
Illiopoulos J and Papanicolaou N 1976 *Nucl. Phys.* B **111** 209–32
Jackiw R 1974 *Phys. Rev.* D **9** 1686–701
Jacobs L 1974 *Phys. Rev.* D **10** 3956–62
Jona-Lasinio G 1964 *Nuovo Cim.* **34** 1790
Källman C-G 1977 *Phys. Lett.* **67B** 195–7
Kallosh R E and Tyutin I V 1973 *Yad. Fiz.* **17** 190–209 (*Sov. J. Nucl. Phys.* **17** 98)
Kibble T W B 1967 *Phys. Rev.* **155** 1554
—— 1976 *J. Phys. A: Math., Nucl. Gen.* **9** 1387–98
Kirzhnits D A 1972 *Zh. Eksp. Teor. Fiz. Pis. Red.* **15** 471 (1972 *JETP Lett.* **15** 529)
Kirzhnits D A and Linde A D 1972 *Phys. Lett.* **42B** 471–4
—— 1974a *Zh. Eksp. Teor. Fiz.* **67** 1263–75 (1975 *Sov. Phys.–JETP* **40** 628–34)
—— 1974b *Preprint Lebedev Physical Institute No* 101
—— 1976a *Ann. Phys., NY* **101** 195–238
—— 1976b *Preprint Trieste* IC/76/28
—— 1978a *Phys. Lett.* **73B** 323–6
—— 1978b *Usp. Fiz. Nauk* submitted
Kislinger M B and Morley P D 1976a *Phys. Rev.* D **13** 2765–70
—— 1976b *Phys. Rev.* D **13** 2771–7
Krive I V 1976 *Yad. Fiz.* **24** 613–6
Krive I V and Chudnovsky E M 1976a *Zh. Eksp. Teor. Fiz. Pis. Red.* **23** 531–3
—— 1976b *Preprint Institute of Theoretical Physics, Kiev* ITP-76-131E
—— 1978 *Zh. Eksp. Teor. Fiz.* **74** 421–31
Krive I V, Fomin P I and Chudnovsky E M 1977 *Zh. Eksp. Teor. Fiz. Pis. Red.* **25** 215–8
Krive I V and Linde A D 1976 *Nucl. Phys.* B **117** 265–8
Krive I V, Linde A D and Chudnovsky E M 1976a *Zh. Eksp. Teor. Fiz.* **71** 826–39 (1976 *Sov. Phys.–JETP* **44** 435)
Krive I V, Pyzh V M and Chudnovsky E M 1976b *Yad. Fiz.* **23** 681–3
Landau L D and Lifshitz E M 1964 *Statistical Physics* (Moscow: Nauka)
Landau L D and Pomeranchuk I Ya 1955 *Dokl. Akad. Nauk* **102** 489
Lee B W 1972 *Phys. Rev.* D **5** 823–35

Lee B W and Zinn-Justin J 1972 *Phys. Rev.* D **5** 3121–60
Lee T D and Margulies M 1975 *Phys. Rev.* D **11** 1591–610
Lee T D and Wick G C 1974 *Phys. Rev.* D **9** 2291–316
Linde A D 1974 *Zh. Eksp. Teor. Fiz. Pis. Red.* **19** 320–2 (*JETP Lett.* **19** 183–4)
—— 1975a *Preprint Lebedev Physical Institute No* 25
—— 1975b *Preprint Lebedev Physical Institute No* 154
—— 1975c *Preprint Lebedev Physical Institute No* 166
—— 1976a *Zh. Eksp. Teor. Fiz. Pis. Red.* **23** 73–5 (*JETP Lett.* **23** 64–7)
—— 1976b *Phys. Lett.* **62B** 435–7
—— 1976c *Phys. Rev.* D **14** 3345–9
—— 1976d *Preprint Trieste* IC/76/26
—— 1976e *Preprint Trieste* IC/76/33
—— 1977a *Nucl. Phys.* B **125** 369–80
—— 1977b *Phys. Lett.* **70B** 306–8
—— 1978a *Preprint Lebedev Physical Institute No* 98
—— 1978b *Zh. Eksp. Teor. Fiz. Pis. Red.* **27** 470–2
Mandelstam S 1975 *Phys. Lett.* **53B** 476–8
Marguder S F 1976 *Phys. Rev.* D **14** 1602–6
Migdal A B 1977 *Usp. Fiz. Nauk* **123** 369–403
Nambu Y 1974 *Phys. Rev.* D **10** 4262–8
Okun' L B and Zeldovich Ya B 1976 *Comm. Nucl. Particle Phys.* **6** 69
Pati C and Salam A 1973 *Phys. Rev.* D **8** 1240–51
—— 1974 *Phys. Rev.* D **10** 275–89
Politzer H D 1973 *Phys. Rev. Lett.* **30** 1346–9
Polyakov A M 1974 *Zh. Eksp. Teor. Fiz. Pis. Red.* **20** 430
—— 1975 *Zh. Eksp. Teor. Fiz.* **68** 1975–90
—— 1977 *Nucl. Phys.* B **120** 429–57
—— 1978 *Phys. Lett.* **72B** 477–80
Ross D A and Taylor J C 1973 *Nucl. Phys.* B **51** 125
Saint-James D, Sarma G and Thomas E I 1969 *Type II Superconductivity* (Oxford: Pergamon)
Salam A 1968 *Elementary Particle Physics* (Stockholm: Almquist and Wiksells) p367
Salam A and Strathdee J 1974 *Nature* **252** 569
—— 1975 *Nucl. Phys.* B **90** 203–20
—— 1976 *Proc. Conf. on K-Meson Physics, Brookhaven*
Sato K and Nakamura T 1976 *Prog. Theor. Phys.* **55** 978–80
Shuryak E V 1978 *Zh. Eksp. Teor. Fiz.* **74** 408–20
Slavnov A A 1972 *Teor. Mat. Fiz.* **10** 153 (1972 *Theor. Math. Phys.* **10** 99)
Taylor J C 1971 *Nucl. Phys.* B **33** 436
Tyutin I V and Fradkin E S 1974 *Yad. Fiz.* **16** 835–53
Vainer B V and Naselsky P D 1977 *Pis. Astron. Zh.* **3** 147–51
Veltman M 1974 *Preprint* Rockefeller University, New York
—— 1975 *Phys. Rev. Lett.* **34** 777
Voronov B L and Tyutin I V 1975 *Yad. Fiz.* **23** 1316–23
Voloshin M B, Kobzarev I Yu and Okun' L B 1974 *Yad. Fiz.* **20** 1229–34
Weinberg S 1967 *Phys. Rev. Lett.* **19** 1264–6
—— 1972 *Gravitation and Cosmology* (New York: Wiley)
—— 1974a *Phys. Rev.* D **9** 3357–78
—— 1974b *Rev. Mod. Phys.* **46** 255–77
—— 1976 *Phys. Rev. Lett.* **36** 294–6
Wilson K 1974 *Phys. Rev.* D **10** 2445–59
Wilson K and Kogut J 1974 *Phys. Rep.* **12C** 75–199
Zeldovich Ya B and Khlopov M V 1978 *Phys. Lett.* submitted
Zeldovich Ya B and Novikov I D 1971 *Theory of Gravity and Evolution of Stars* (Moscow: Nauka)
—— 1975 *Structure and Evolution of the Universe* (Moscow: Nauka)
Zeldovich Ya B, Okun' L B and Kobzarev I Yu 1974 *Zh. Eksp. Teor. Fiz.* **67** 3–11
Zeldovich Ya B, Okun' L B and Pikelner S B 1965 *Usp. Fiz. Nauk.* **87** 115
Zeldovich Ya B and Starobinsky A A 1976 *Zh. Eksp. Teor. Fiz. Pis. Red.* **24** 616–8

Volume 108B, number 6 PHYSICS LETTERS 4 February 1982

A NEW INFLATIONARY UNIVERSE SCENARIO: A POSSIBLE SOLUTION OF THE HORIZON, FLATNESS, HOMOGENEITY, ISOTROPY AND PRIMORDIAL MONOPOLE PROBLEMS

A.D. LINDE
Lebedev Physical Institute, Moscow 117924, USSR

Received 29 October 1981

A new inflationary universe scenario is suggested, which is free of the shortcomings of the previous one and provides a possible solution of the horizon, flatness, homogeneity and isotropy problems in cosmology, and also a solution of the primordial monopole problem in grand unified theories.

There is now considerable interest in the cosmological consequences of symmetry breaking phase transitions, which occur in grand unified theories (GUTs) with the decrease of temperature at the very early stages of the evolution of the universe [1–3]. These phase transitions typically are strongly first order [4,5]. The lifetime of the supercooled symmetric phase $\varphi = 0$ (φ is the Higgs scalar field which breaks the symmetry) in some theories may be extremely large [2,3,6–8]. In that case the energy–momentum tensor of particles $\sim T^4$ in the phase $\varphi = 0$ almost vanishes in the course of the expansion of the universe, and the total energy–momentum tensor reduces to the vacuum stress tensor (cosmological term) $T_{\mu\nu}^{\mathrm{vac}} = g_{\mu\nu} V(0)$, where $V(\varphi)$ is the effective potential of the theory at vanishing temperature [2,3]. This leads to an exponentially fast expansion of the universe, $a \sim e^{Ht}$. Here a is the scale factor, and H is the Hubble constant at that time,

$$H = [(8\pi/3M_{\mathrm{P}}^2) V(0)]^{1/2},$$

where $M_{\mathrm{P}} \approx 10^{19}$ GeV is the Planck mass [9]. Then at some comparatively small temperature T_{c} the symmetry breaking phase transition takes place, all the vacuum energy $V_{(0)}$ transforms into thermal energy [2,3], the universe is reheated up to the temperature $T_1 \approx V_{(0)}^{1/4}$, and its further evolution proceeds in a standard way [10] [‡1].

A most detailed discussion of this scenario is contained in a very interesting paper of Guth [12], where it is shown that the existence of a sufficiently long period of exponential expansion (inflation) in the early universe would provide a natural solution of the horizon and flatness problems in cosmology and of the primordial monopole problem in grand unified theories [13].

Unfortunately, however, this scenario in the form suggested in ref. [12] leads to some unacceptable consequences, recognized by Guth himself and by other authors who have studied this problem later, see e.g. refs. [8,14–18].The phase transition from the symmetric vacuum state $\varphi = 0$ to the asymmetric state $\varphi = \varphi_0$ proceeds by creation and subsequent expansion of bubbles containing some nonvanishing fields φ. In ref. [12] it was implicitly assumed that inside these bubbles the scalar field φ rapidly grows to $\varphi = \varphi_0$, all energy of the bubbles becomes concentrated in their walls and thermalization occurs only after the collision of the walls. If this qualitative picture were correct, the exponential expansion would be finished at the temperature T_{c}, at which the phase transition occurs. For the flatness problem to be solved the universe (the scale factor a) should grow at least 10^{28}

‡1 For an alternative scenario of the exponential expansion at the very early stages of the evolution of the universe, which may occur due to quantum gravity effects, see ref. [11].

0 031-9163/82/0000–0000/$ 02.75

Volume 108B, number 6 PHYSICS LETTERS 4 February 1982

times during the exponential expansion period [12]. Since at this period the value of aT is constant, the critical temperature T_c should be 10^{28} times smaller than the temperature T_0, at which the exponential expansion starts. In the simplest SU(5) model [19] $T_0 \sim 10^{14}$ GeV, so that

$$T_c \lesssim 10^{-14}\ \text{GeV} \sim 0.1\ \text{K}. \tag{1}$$

No GUTs with such a fantastically small value of the critical temperature have been suggested.

There is also another problem with the above mentioned scenario. If the bubble wall collisions are necessary for the reheating of the universe, then after such a phase transition the universe becomes greatly inhomogeneous and anisotropic, which would contradict cosmological data [8,12,17,18].

In the present paper we would like to suggest an improved inflationary universe scenario, which is free of the above mentioned difficulties. With this purpose we shall consider the phase transitions in GUTs with the Coleman–Weinberg mechanism of symmetry breaking [20]. Phase transitions in such theories have been studied recently by many authors [14,15,21–23]. In our opinion, however, several important features of these phase transitions have escaped their attention. A detailed discussion of the phase transitions in GUTs with the Coleman–Weinberg mechanism of symmetry breaking will be contained in a subsequent publication. Here we shall only outline the main idea, which is essential for the understanding of the new inflationary universe scenario.

For definiteness let us consider the SU(5) grand unified theory [19], though most of what will be discussed here will not depend on the details of the model under consideration. The one-loop effective potential for the symmetry breaking SU(5) → SU(3) × SU(2) × U(1) in the Coleman–Weinberg version of this model at finite temperature T is [1–3, 14,15]

$$V(\varphi, T) = (18T^4/\pi^2) \times \int_0^\infty \mathrm{d}x\, x^2 \ln\{1 - \exp[-(x^2 + 25g^2\varphi^2/8T^2)^{1/2}]\} + (5625/512\pi^2)g^4(\varphi^4 \ln(\varphi/\varphi_0) - \varphi^4/4 + \varphi_0^4/4]\,, \tag{2}$$

where $\varphi_0 \sim 10^{14}$–10^{15} GeV, $g^2 \sim 1/3$ is the gauge coupling constant. At $T \gg \varphi_0$ the symmetry in this theory was restored, $\varphi = 0$ [1–3]. With a decrease of temperature the absolute minimum of $V(\varphi)$ appears at $\varphi \approx \varphi_0$. However at any $T \neq 0$ the point $\varphi = 0$ remains a local minimum of $V(\varphi, T)$, since near $\varphi = 0$

$$V(\varphi, T) = \tfrac{75}{16} g^2 T^2 \varphi^2 - (5625/512\pi^2)g^4\varphi^4 \ln(M_X/T) + (9/32\pi^2)M_X^4\,, \tag{3}$$

where $M_X^2 = \frac{25}{8} g^2 \varphi_0^2$. The phase transition with symmetry breaking proceeds from a strongly supercooled state $\varphi = 0$ at temperature T_c, which is many orders of magnitude smaller than φ_0 [14,15,21–23]. The shape of the potential $V(\varphi, T)$ for $T \ll \varphi_0$ is shown in fig. 1. The phase transition begins with the formation of bubbles of the field φ, which is a tunneling process [6]. One may argue that for $T_c \ll \varphi_0$ this process does not depend on the properties of $V(\varphi, T_c)$ at $\varphi \sim \varphi_0$, and the maximal value of the field φ inside the bubble immediately after its formation should be of the order of φ_1, where

$$V(\varphi_1, T_c) = V(0, T_c)\,, \tag{4}$$

$\varphi_1 \ll \varphi_0$, see fig. 1. Indeed, a detailed study of the bubble formation in this theory performed in refs. [24,25] by means of computer calculations shows that at the moment of the bubble formation the maximal value of the field φ inside the bubble equals approximately $3\varphi_1$. Therefore inside the bubble

$$\varphi \lesssim 3\varphi_1 = \frac{12\pi T_c}{5g}\left(\frac{2}{3\ln(M_X/T_c)}\right)^{1/2} \ll \varphi_0\,. \tag{5}$$

This means that the (negative) mass squared of the

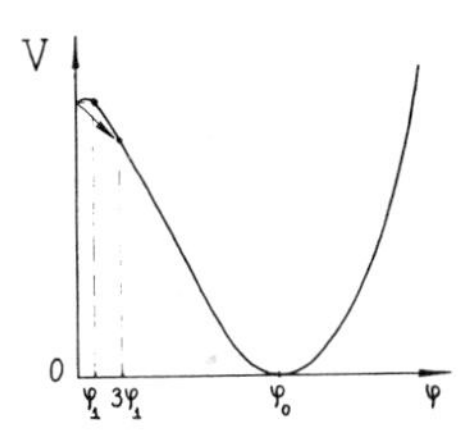

Fig. 1. Effective potential in the Coleman–Weinberg theory for $T \ll \varphi_0$. The arrow indicates the direction of the tunneling with bubble formation.

Volume 108B, number 6 PHYSICS LETTERS 4 February 1982

field φ inside the bubble is

$$-m^2 = -\tfrac{2}{15} d^2 V/d\varphi^2 \lesssim 75 g^2 T_c^2 \sim 25 T_c^2 . \qquad (6)$$

After the bubble formation the field φ inside the bubble gradually grows up to its equilibrium value $\varphi(T_1) \sim \varphi_0$. At the first stages of this process the field φ grows approximately as e^{mt}. Therefore it approaches its equilibrium value $\varphi(T_1)$ only after some period of time $\tau \gtrsim m^{-1} \sim 0.2 T_c^{-1}$. A more complete investigation shows that τ is several times greater than m^{-1}; here for simplicity we shall take as an estimate

$$\tau \sim T_c^{-1} . \qquad (7)$$

It can also be easily shown that during most of this period the field φ inside the bubble remains much less than φ_0. Therefore during some time of the order of $\tau \sim T_c^{-1}$ the vacuum energy density $V(\varphi)$ remains almost equal to $V(0)$, and the part of the universe inside the bubble expands exponentially just as it expanded before the bubble creation. This simple observation has very important consequences for the theory of the phase transitions in the Coleman–Weinberg model.

Let us suppose that the phase transition in the Coleman–Weinberg SU(5) theory occurs at $T_c \sim 2 \times 10^6$ GeV, as it was claimed in ref. [15]. From eq. (3) it follows that with the parameters of the theory used in ref. [15] ($M_X \sim 6 \times 10^{14}$ GeV) the Hubble constant

$$H = [(8\pi/3M_P^2) V(0)]^{1/2}$$

is equal to 1.5×10^{10} GeV. Therefore during the exponential expansion period $\tau \sim T_c^{-1}$ the universe should grow $e^{H\tau}$ times, where

$$e^{H\tau} \sim e^{H/T_c} \sim e^{7500} \sim 10^{3260} . \qquad (8)$$

A typical size of the bubble at the moment of its creation is $O(T_c^{-1}) \sim 10^{-20}$ cm [25]. After the period of the exponential expansion this bubble will have a size of

$$10^{-20} \cdot e^{H\tau} \text{ cm} \sim 10^{3240} \text{ cm} ,$$

which is much greater than the size of the observable part of the universe $l \sim 10^{28}$ cm. Therefore the whole observable part of the universe is contained *inside one bubble*, so we see no inhomogeneities caused by the wall collisions. After some time of the order of τ after the bubble creation all the vacuum energy density $V(0)$ transforms into thermal energy $\sim T_1^4$, where in our model $T_1 \approx 0.15 M_X \sim 10^{14}$ GeV. However the thermalization occurs now not due to the wall collisions, but due to the interactions of particles created by the classical homogeneous field φ, convergently oscillating near its equilibrium value $\varphi(T_1) \approx \varphi_0$ with a frequency of about 10^{14} GeV.

One can easily verify that the size of the particle horizon at the time of the phase transition was much greater than the size of the bubble $\sim T_c^{-1}$, i.e. all points inside the bubble were causally connected. After the exponential expansion period this causally connected domain covers the whole observable part of the universe, which solves the horizon problem [12].

Now let us remember that particle creation in the very early universe in general cannot make it completely isotropic, but makes it quasi-isotropic [10], i.e. locally isotropic in small domains of space of the size of the same order as or greater than the Planck length $l_P \sim 10^{-33}$ cm $\sim M_P^{-1}$ at the Planck time $t_P \sim 10^{-43}$ s, when the temperature was $T_P \sim M_P \sim 10^{19}$ GeV. Since before the phase transition the quantity aT was constant inside each isotropic domain of the universe, at the moment of the phase transition the typical size of the isotropic domain exceeds the bubble size $\sim T_c^{-1}$. Therefore the space–time inside the bubble was isotropic and the exponential expansion extends this isotropy to the whole observable part of the universe. (Moreover, the remaining small anisotropy inside the bubble decreases rapidly during the exponential expansion period [18].) This may solve the long-standing problem of the space–time isotropy in our universe.

Density fluctuations inside the bubble immediately after its formation are negligibly small as compared with $V(0)$, i.e. the space inside the bubble is almost homogeneous. Then the exponential expansion extends this homogeneity to the whole observable part of the universe, which explains the large-scale homogeneity of the universe.

One may argue that it is not very good to obtain an absolutely homogeneous universe, since in that case it would be difficult to understand the origin of galaxies. However, as will be explained in a separate publication (see also refs. [7,26]), the necessary inhomogeneities may be generated after a subsequent phase transition with a smaller degree of supercooling. Moreover, as is shown in ref. [27], the perturbations necessary for galaxy formation arise due to quantum gravity effects just after phase transitions of the type considered above in GUTs with the unification scale $\Lambda \sim 10^{17}-10^{18}$ GeV, which is not unrealistic [28].

From our results it follows that the size of the universe l_1 after the phase transition should exceed 10^{3240} cm, and the temperature T_1 is of the order of 10^{14} GeV. Therefore the total entropy of the universe should exceed $(l_1 T_1)^3 \sim 10^{10000}$, which explains why the total entropy of the universe exceeds 10^{85} and simultaneously solves the flatness problem [12,21].

It is known that the primordial monopoles in GUTs are created only in the points, in which bubbles with different types of Higgs field φ collide [13]. Therefore in our scenario no monopoles are created in the observable part of the universe, which solves the primordial monopole problem in GUTs [13]. For the same reason there will be no domain walls in the observable part of the universe in the theories with broken discrete symmetries [29], and in particular in the theories with spontaneously broken *CP* invariance. This helps to solve the problem of the baryon asymmetry of the universe [30]. Moreover, in the scenario under consideration there appears an additional source of baryon asymmetry. The standard mechanism is connected with the decay of the X, Y bosons and Higgs mesons, which appear in the course of the reheating of the universe during the phase transition. It is clear, however, that the baryon asymmetry generated by the decay of the Higgs mesons may be generated by the decay of the classical Higgs field vacuum $\varphi = 0$ as well. Note also that, whereas in a standard scenario the particles created by the decay of the X, Y and Higgs mesons are only a few percent of the total amount of particles [30], in our case all particles which appear after the phase transition are created by the oscillating classical Higgs field φ during the phase transition.

The new inflationary universe scenario discussed above is, of course, oversimplified. To get a complete scenario, one should analyse the phase transitions in the Coleman–Weinberg model more accurately taking into account the renormalization group equation [21] and the nonperturbative effects [15]. This analysis should be performed simultaneously with the investigation of the effects connected with the nonvanishing curvature and rapid expansion of the universe, which become important for $H > T_c$. One may ask e.g. whether it is possible for the universe to be in a state with temperature T_c smaller than the Hawking temperature $T_H = H/2\pi$, to which consequences the terms $\sim R\varphi^2$ in the effective potential may lead etc. [23]. An investigation of these problems is rather involved and will be contained in a separate publication. Our preliminary result is that there exists an *improved Coleman–Weinberg theory*, in which $d^2V/d\varphi^2 = 0$ at $\varphi = 0$ not in Minkowski space, but rather in de Sitter space with the curvature R determined by the vacuum energy density $V(\varphi)$ at the symmetric point $\varphi = 0$. In some versions of this theory the phase transition with symmetry breaking occurs due to nonperturbative effects [15]. The kinetics of this phase transition is somewhat more complicated than that described above, but the main feature of the new inflationary universe scenario remains intact: The field φ approaches its equilibrium value $\varphi(T_1) \sim \varphi_0$ during the period $\tau \gg H^{-1}$, which just leads to the desirable inflation of the universe discussed in the present paper.

I would like to express my deep gratitude to G.V. Chibisov, P.C.W. Davies, V.P. Frolov, L.P. Grishchuk, S.W. Hawking, R.E. Kallosh, D.A. Kirzhnits, V.F. Mukhanov, V.A. Rubakov, A.A. Starobinsky, A.V. Veryaskin and Ya.B. Zeldovich for many enlightening discussions.

References

[1] D.A. Kirzhnits, JETP Lett. 15 (1972) 529;
D.A. Kirzhnits and A.D. Linde, Phys. Lett. [illegible]B (1972) 471;
S. Weinberg, Phys. Rev. D9 (1974) 3357;
L. Dolan and R. Jackiw, Phys. Rev. D9 (1974) 3320;
D.A. Kirzhnits and A.D. Linde, Zh. Eksp. Teor. Fiz. 67 (1974) 1263 [Sov. Phys. JETP 40 (1975) 628].

[2] D.A. Kirzhnits and A.D. Linde, Ann. Phys. (NY) 101 (1976) 195.

[3] A.D. Linde, Rep. Prog. Phys. 42 (1979) 389.

[4] A.D. Linde, in: Statistical mechanics of quarks and hadrons, ed. H. Satz (North-Holland, Amsterdam, 1981) p. 385; Phys. Lett. 99B (1981) 391.

[5] M. Daniel, Phys. Lett. 98B (1981) 371.

[6] A.D. Linde, Phys. Lett. 70B (1977) 306; 100B (1981) 37.

[7] K. Sato, Mon. Not. R. Astron. Soc. 195 (1981) 467.
[8] A.H. Guth and E. Weinberg, Phys. Rev. D23 (1981) 876.
[9] R. Tolman, Relativity, thermodynamics and cosmology (Clarendon, Oxford, 1969).
[10] Ya.B. Zeldovich and I.D. Novikov, Structure and evolution of the universe (Nauka, Moscow, 1975).
[11] A.A. Starobinsky, Phys. Lett. 91B (1980) 100;
Ya.B. Zeldovich, Pis'ma Astron. Zh. 7 (1981) 579.
[12] A.H. Guth, Phys. Rev. D23 (1981) 347.
[13] Ya.B. Zeldovich and M.Yu. Khlopov, Phys. Lett. 79B (1978) 239;
J.P. Preskill, Phys. Rev. Lett. 43 (1979) 1365.
[14] G.P. Cook and K.T. Mahanthappa, Phys. Rev. D23 (1981) 1321.
[15] A. Billoire and K. Tamvakis, CERN preprint TH. 3019 (1981);
K. Tamvakis and C.E. Vayonakis, CERN preprints TH. 3108, TH. 3128 (1981).
[16] Q. Shafi, CERN preprint TH. 3143 (1981).
[17] S.W. Hawking, I.G. Moss and J.M. Stewart, DAMTP preprint (1981).
[18] J.D. Barrow and M.S. Turner, Nature 292 (1981) 35.
[19] H. Georgi and S.L. Glashow, Phys. Rev. Lett. 32 (1974) 389.
[20] S. Coleman and E. Weinberg, Phys. Rev. D7 (1973) 1888.
[21] V.G. Lapchinsky, V.A. Rubakov and A.V. Veryaskin, IYaI preprint (1981).
[22] M. Sher, Univ. of California preprint NSF-ITP-81-15 (1981).
[23] L.F. Abbott, Nucl. Phys. B185 (1981) 233;
P. Hut and F.R. Klinkhamer, Phys. Lett. 104B (1981) 439.
[24] E. Brezin and G. Parisi, J. Stat. Phys. 19 (1978) 269.
[25] A.D. Linde, Decay of the false vacuum at finite temperature, Lebedev Phys. Inst. preprint (1981).
[26] Ya.B. Zeldovich, Mon. Not. R. Astron. Soc. 192 (1980) 663;
A. Vilenkin, Phys. Rev. Lett. 46 (1981) 1169; Tufts Univ. preprint (1981).
[27] G.V. Chibisov and V.P. Mukhanov, Lebedev Phys. Inst. preprint No. 198 (1981); Mon. Not. R. Astron. Soc., to be published;
D.A. Kompaneets, V.N. Lukash and I.D. Novikov, Space Research Inst. preprint No. 652 (1981).
[28] S. Dimopoulos, S. Raby and F. Wilczek, Stanford Univ. preprint NSF-ITP-81-31 (1981);
S. Dimopoulos and H. Georgi, Harvard Univ. preprint HUTP-81/A022 (1981).
[29] Ya.B. Zeldovich, I.Yu. Kobzarev and L.B. Okun, Sov. Phys. JETP 40 (1975) 1.
[30] A.D. Sakharov, Pis'ma Zh. Eksp. Teor. Fiz. 5 (1967) 32;
M. Yoshimura, Phys. Rev. Lett. 41 (1978) 281;
S. Dimopoulos and L. Susskind, Phys. Rev. D18 (1978) 4500;
J. Ellis, M.K. Gaillard and D.V. Nanopoulos, Phys. Lett. 80B (1979) 360;
S. Weinberg, Phys. Rev. Lett. 42 (1979) 859.

VOLUME 48, NUMBER 17 PHYSICAL REVIEW LETTERS 26 APRIL 1982

Cosmology for Grand Unified Theories with Radiatively Induced Symmetry Breaking

Andreas Albrecht and Paul J. Steinhardt
Department of Physics, University of Pennsylvania, Philadelphia, Pennsylvania 19104
(Received 25 January 1982)

The treatment of first-order phase transitions for standard grand unified theories is shown to break down for models with radiatively induced spontaneous symmetry breaking. It is argued that proper analysis of these transitions which would take place in the early history of the universe can lead to an explanation of the cosmological homogeneity, flatness, and monopole puzzles.

PACS numbers: 98.80.Bp, 11.15.Ex, 12.10.En

Hot big-bang cosmology depends upon special conditions for the early universe to explain the high degree of homogeneity (the "homogeneity puzzle")[1] and the nearly critical mass density (the "flatness puzzle")[2] found in the universe today. In addition, it has been shown that in typical grand unified theories (GUT's) phase transitions should occur in the early history of the universe which lead to many more magnetic monopoles being produced and surviving to the present epoch than are consistent with experiment (the "monopole puzzle").[3]

In this paper we will argue that first-order phase transitions in a special class of GUT's —models in which the GUT symmetry is broken by radiatively induced corrections to the tree approximation to the effective potential—can lead to a solution to these and other cosmological puzzles. (Models with radiatively induced symmetry breaking, a mechanism discovered by Coleman and Weinberg,[4] will be referred to as CW models.) In particular, we will present results for the standard GUT with a finite-temperature effective (scalar) potential:

$$V_T(\varphi) = (2A - B)\sigma^2\varphi^2 - A\varphi^4 + B\varphi^4 \ln(\varphi^2/\sigma^2) + 18(T^4/\pi^2)\int_0^\infty dx\, x^2 \ln\{1 - \exp[-(x^2 + 25g^2\varphi^2/8T^2)^{1/2}]\}, \tag{1}$$

where the adjoint Higgs field, Φ, has been reexpressed as $\varphi(1,1,1,-\frac{3}{2},-\frac{3}{2})$ (the fundamental Higgs field will be irrelevant for this discussion); g is the gauge coupling constant; σ is chosen to be 4.5×10^{14} GeV; $B = 5625g^4/1024\pi^2$; and A is a free parameter. Equation (1) includes the one-loop quantum and thermal corrections to the effective potential. For a CW model, the coefficient of the quadratic term, $2A - B$, is set equal to zero and the Higgs mass is $m_{CW} = 2.7\times 10^{14}$ GeV. We will also present results for non-CW models in which $2A - B$ is small and therefore the Higgs mass, m_H, is such that $\Delta_H \equiv (m_H^2 - m_{CW}^2)/m_{CW}^2$ is small.

As for more general GUT models, the process of the first-order phase transition from the SU(5) *symmetric* phase to the SU(3)⊗SU(2)⊗U(1) *symmetry-breaking* phase for the CW model can be understood by studying the shape of the effective potential as a function of the scalar field for various values of the temperature, as shown in Fig. 1. For temperatures above the critical temperature (T_{GUT}) for the transition, the symmetric phase ($\varphi = 0$) is the global stable minimum of the effective potential. At $T = T_{GUT}$, the symmetric phase and the symmetry-breaking phase have equal energy densities. As the temperature drops below T_{GUT}, the symmetric phase becomes *metastable*—it has a higher-energy density than *stable* symmetry-breaking phase but a potential barrier prevents it from becoming unstable.

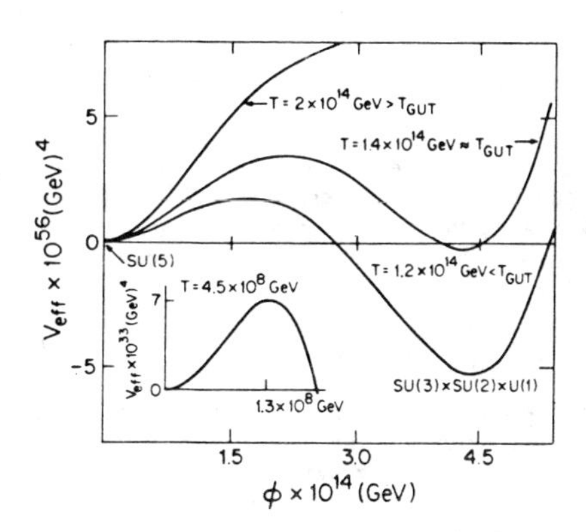

FIG. 1. Effective potential vs φ for various values of T.

The decay of a metastable phase to a stable phase has been compared to a classical nucleation process. At $T < T_{GUT}$ there is a rate per unit time per unit volume, $\Gamma(t)$, for producing finite-sized fluctuations containing stable phase—bubbles—within the metastable system. Once produced, the bubbles grow, coaelesce and convert the system to the stable phase. For cases where the barrier is sufficiently large, Coleman[5] and Linde[6] have found methods for computing $\Gamma(t)$ using a steepest descent (SD) approximation and found it to be of the form $S\exp[-F_f(T)/kT]$. S has the dimensions of $(\text{length})^{-4}$ and k is Boltzmann's constant. $F_f(T)$ is the free energy associated with the bubble computed by SD approximation to be the dominant path across the potential barrier. For this SD bubble fluctuation, the value of $\langle\varphi\rangle$ varies from φ_f (on the stable-phase side of the barrier) in the center of the bubble to $\varphi = 0$ far from the bubble center. As T decreases, $\Gamma(t)$ increases and φ_f/σ decreases, where $\varphi = \sigma$ corresponds to the symmetry-breaking minimum.

The maintenance of a system in a metastable phase during a first-order phase transition as $T \lesssim T_{GUT}$ continues to decrease is known as supercooling. For phase transitions in the early universe, Guth and Tye,[7] taking into account the expansion of the universe, found the expression for the fractional volume $[p(T)]$ of the universe which at temperature T has decayed to the stable phase during supercooling. When $p(T)$, which depends upon $\Gamma(t)$, is of order unity, the decay is said to be *terminated*.

Guth[8] recently suggested that supercooling of first-order phase transitions of typical GUT models can lead to a solution of the cosmological puzzles. His idea was that the energy density of the universe, ρ, during supercooling is dominated (once $T \lesssim T_{GUT}/10$) by the $\rho_0 \approx T_{GUT}^4$, the difference in energy density between the metastable and stable phases. Then, ρ_0 can act as a cosmological constant in Einstein's equation for standard cosmology described by a Robertson-Walker metric:

$$\left(\frac{\dot{R}}{R}\right)^2 = \frac{8\pi}{3M_P^2}\rho \approx \frac{8\pi}{3M_P^2}T_{GUT}^4 \equiv t_{exp}^{-2}, \tag{2}$$

where M_P is the Planck mass. The result is exponential growth of the scale factor, $R(t) \sim R_0 \times \exp(t/t_{exp})$. If the growth is continued long enough, each nearly homogeneous region of the universe experiences an expansion which, Guth showed, could explain the cosmological homogeneity, flatness, and monopole puzzles. The problem with Guth's scenario is that when this high degree of expansion can be arranged in typical GUT's (by adjusting free parameters) the rate of expansion of the universe dominates the rate of production and growth of bubbles; the bubbles never coalesce to complete the transition. Since our own universe exhibits the symmetry breaking of the stable phase, it would have to lie within a single, rare bubble, in which case it is difficult to understand how the high entropy found in our universe could have been generated.[9]

We claim that CW models and near-CW models in which $\Delta_H < 7\times 10^{-6}$ possess special properties that result in completion of the transition to the symmetry-breaking phase along with tremendous expansion. Initially, the analysis of the supercooling for $T < T_{GUT}$ proceeds as in more general GUT models. However, as has been pointed out previously,[9] two important features must be taken into account. Firstly, the GUT fine-structure constant, α $(=\frac{1}{45}$ at $T = T_{GUT})$, increases as a function of temperature[10] until at $T_\alpha \approx 10^6$ GeV, it is of order unity and the one-loop approximation to $V_T(\varphi)$ is no longer valid. Secondly, the prefactor in the expression for $\Gamma(t)$, S, is given by T^4 for CW models since, near $\varphi = 0$, the only parameter with the dimensions of length that affects the barrier and, thus, the decay, is T^{-1}.[9]

When these features are taken into account in the standard SD analysis, they combine to yield the peculiar behavior for $F_f(T)/kT$ and $p(T)$ that is illustrated in Fig. 2. The curves for $F_f(T)/kT$ have been terminated at temperatures, T_{term},

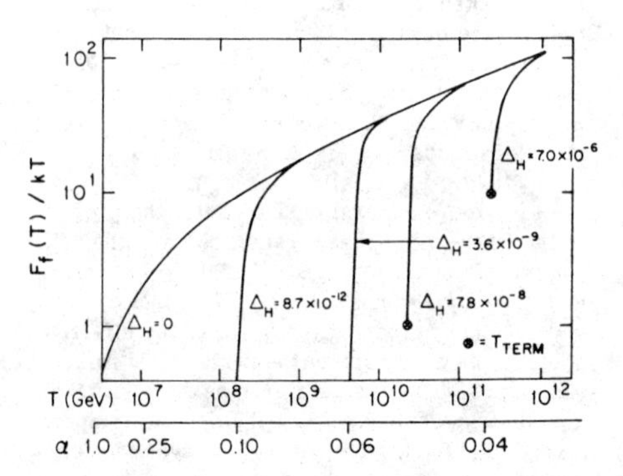

FIG. 2. $F_f(T)/kT$ vs T. Curves terminate for $p(T) \approx 1$. $\Delta_H = (M_H^2 - M_{CW}^2)/M_{CW}^2$.

for which the fractional volume of stable phase, $p(T)$, is of order unity. For $\Delta_H > 7.0\times 10^{-6}$, $F_f(T_{term})/kT_{term}$ is large, but there is insufficient expansion to solve the cosmological puzzles.

For smaller values of Δ_H, including CW models where $\Delta_H = 0$, $F_f(T_{term})/kT_{term}$ is of order unity or less, values which are too small to trust the SD approximation. The value of the temperature for which the SD approximation fails, T_{SD}, is a function of Δ_H and is roughly 10^8 GeV for $\Delta_H = 0$. T_{SD} is larger than T_α in all cases though, so that one can discuss the breakdown of the SD approximation while still considering only the one-loop approximation to the effective potential, Eq. (1).

For the CW and near-CW models, the fact that T reaches T_{SD} before the completion of the transition means that many other types of fluctuations (i.e., paths across the barrier) besides the SD bubble become important. The effect of the barrier becomes negligible. The system (the universe) can be thought of as balancing at a point of unstable (not metastable) equilibrium near $\varphi = 0$. Thermal fluctuations drive different regions of the universe away from the SU(5) symmetric phase but towards different symmetry-breaking minima. Since T is the only dimensional parameter relevant for the fluctuations, the average size of a fluctuation region should be of order T^{-1} and the (roughly constant) value of $\langle\varphi\rangle$ within a fluctuation region is of order T.

Even though $\langle\varphi\rangle \sim T$ corresponds to a point of the effective potential on the stable-phase side of the barrier, slightly to the right of the barrier in Fig. 1, a crucial feature of CW models and near-CW models is that the effective potential is extremely flat from $\varphi = 0$ up to value of $\varphi \sim T_{GUT}$. Thus, even though each fluctuation region has a value of φ that corresponds to a point of classical instability for the effective potential, the motion of φ towards the stable-phase minimum is characterized by a time constant τ that can be very large. Since φ within each fluctuation is of order $T_{SD} \ll T_{GUT}$, the energy density within each fluctuation region is still roughly constant $\sim T_{GUT}^4$. As a result, as concluded independently by Linde,[11] exponential expansion in which φ has a value much less than T_{GUT} continues for a time τ.

We have determined an estimate of τ by considering the evolution according to the classical field equations of a state with $\langle\varphi\rangle = T$ throughout space (presumably a similar method to what was used in Ref. 11) for a range of temperatures for fluctuation production. We found τ for CW models to be large compared to t_{exp}. This means that each fluctuation region undergoes many e foldings in spatial expansion before $\langle\varphi\rangle$ changes appreciably. Multiplying the scale factor after time τ by the average size of the initial fluctuation region (the size of an SD bubble was used for $T > T_{SD}$) we obtain the size of the fluctuation region, R_U, at time τ, after which $\langle\varphi\rangle \sim \sigma$ and the exponential expansion ceases. In Table I are shown the results for this computation for a range of temperatures. Column 2 shows R_U(flat) computed with use of the ordinary Klein-Gordon equation. Column 3 shows R_U(exp) derived by using the same equation with an extra drag term $(3\dot{R}\dot{\varphi}/R)$ included to account for the time dependence of the scale factor.[12] Column 4 shows R_U for $\Delta_H = 3.6\times 10^{-11}$, where the time dependence of the scale factor has been included. For this value of Δ_H, the barrier disappears at $T = 3.7\times 10^8$ GeV and the maximum value that R_U can achieve is $10^{-0.2}$ cm. Since the observed universe, according to the standard model, had a radius of ~1 cm for $T = 10^{14}$ GeV, the choice of Δ_H must be tuned to a value less than 3.6×10^{-11} in order for the observed universe to fit inside a single fluctuation region. Similarly, we have shown that if our scalar field couples to the curvature through a term $bC\varphi^2$ (C = curvature), $|b|$ must be $\lesssim 10^{-2}$. Please note, we have treated this calculation as if it were in flat space; curvature effects will be discussed in future publications.[12]

The result is that the size of a fluctuation region once $T < T_{SD}$ and the SD approximation breaks down is much greater than the size of our present observed universe (10^{28} cm). If one considers the "observed universe" as lying within such a fluctuation region of the "total universe" the special conditions of hot big-bang cosmology can be satisfied. Because the observed universe would be only a small portion of the total universe result-

TABLE I. Fluctuation radius vs T.

T (GeV)	R_U (flat) (cm)	R_U (exp) (cm)	R_U ($\Delta_H = 3.6\times 10^{-11}$) (cm)
4.5×10^6	10^{81}	$\gg 10^{470}$	...
4.5×10^7	$10^{-4.6}$	10^{470}	...
1×10^8	10^{-13}	10^{98}	...
3×10^8	10^{-19}	$10^{-4.4}$	...
4.5×10^8	10^{-20}	10^{-13}	$10^{-7.2}$
4.5×10^9	10^{-23}	10^{-23}	10^{-23}

ing from the extreme expansion of a small homogeneous region, the homogeneity puzzle is solved. The exponentially large value of scale factor accounts for the flatness puzzle.[8] Because $\langle\varphi\rangle \approx$ const over a fluctuation region, there is no reason to expect to find any monopoles (beyond a small thermally produced number) in the observed universe. Even though there is a discrete symmetry ($\varphi \to -\varphi$) in the theory, the distance between domain walls (separating regions with $\langle\varphi\rangle$ of opposite sign) produced in the transition should be greater than 10^{28} cm, and hence, unobservable. The potential energy stored in the scalar field is eventually converted to thermal energy,[13] thus producing a sizable entropy density inside each region. Thus, it appears that all the fundamental cosmological puzzles are solved.

A more complete discussion of these results[12] and an analysis of how a fluctuation evolves to thermal equilibrium and produces baryon asymmetry[13] will appear in forthcoming publications.

The authors thank A. Guth and E. Witten for useful discussions and G. Segre for his advice and support during the preparation of the manuscript This work has been supported in part by the U.S. Department of Energy under Contract No. EY-76-C-02-3071.

[1]S. Weinberg, *Gravitation and Cosmology* (Wiley, New York, 1972).

[2]R. H. Dicke and P. J. E. Peebles, in *General Relativity: An Einstein Centenary Survey,* edited by S. W. Hawking and W. Israel (Cambridge Univ. Press, Cambridge, 1979).

[3]J. Preskill, Phys. Rev. Lett. 43, 1365 (1979).

[4]S. Coleman and E. Weinberg, Phys. Rev. D 7, 788 (1973).

[5]S. Coleman, Phys. Rev. D 15, 2929 (1977).

[6]A. Linde, Phys. Lett. 70B, 306 (1977).

[7]A. Guth and S. H. H. Tye, Phys. Rev. Lett. 44, 631 963(E) (1980).

[8]A. Guth, Phys. Rev. D 23, 347 (1981).

[9]P. Steinhardt, in Proceedings of the 1981 Banff Summer Institute (to be published).

[10]M. Sher, Phys. Rev. D 24, 1699 (1981).

[11]A. Linde, Lebedev Institute Report No. 229, 1981 (to be published). Differences between our results and those of Linde will be discussed in a later paper. (A. Albrecht and P. Steinhardt, to be published).

[12]Albrecht and Steinhardt, Ref. 11.

[13]A. Albrecht, P. Steinhardt, M. Turner, and F. Wilczek, to be published.

XV. *Hierarchy, Technicolor, Supersymmetry, and Variations*

1) Dynamics of spontaneous symmetry breaking in the Weinberg-Salam theory by L. Susskind, *Phys. Rev.* D**20,** 2619 (1979) 920
2) Implications of dynamical symmetry breaking: An addendum by S. Weinberg, *Phys. Rev.* D**19,** 1277 (1979) 927
3) Composite/fundamental Higgs mesons I: Dynamical speculations by H. Georgi and I. N. McArthur, *Nucl. Phys.* B**202,** 382 (1982) 931
4) Dynamical breaking of supersymmetry by E. Witten, *Nucl. Phys.* B**188,** 513 (1981) 946

PHYSICAL REVIEW D VOLUME 20, NUMBER 10 15 NOVEMBER 1979

Dynamics of spontaneous symmetry breaking in the Weinberg-Salam theory

Leonard Susskind*
Stanford Linear Accelerator Center, Stanford University, Stanford, California 94305
(Received 5 July 1978)

We argue that the existence of fundamental scalar fields constitutes a serious flaw of the Weinberg-Salam theory. A possible scheme without such fields is described. The symmetry breaking is induced by a new strongly interacting sector whose natural scale is of the order of a few TeV.

I. WHY NOT FUNDAMENTAL SCALARS?

The need for fundamental scalar fields in the theory of weak and electromagnetic forces[1] is a serious flaw. Aside from the subjective esthetic argument, there exists a real difficulty connected with the quadratic mass divergences which always accompany scalar fields.[2] These divergences violate a concept of naturalness which requires the observable properties of a theory to be stable against minute variations of the fundamental parameters.

The basic underlying framework of discussion of naturalness assumes the existence of a fundamental length scale κ^{-1} which serves as a real cutoff. Many authors[3] have speculated that κ should be of order 10^{19} GeV corresponding to the Planck gravitational length. The basic parameters of such a theory are some set of dimensionless bare couplings g_0 and masses. The dimensionless bare masses are defined as the ratio of bare mass to cutoff:

$$\mu_0 = m_0/\kappa . \tag{1}$$

The principle of naturalness requires the physical properties of the output at low energy to be stable against very small variations of g_0 and μ_0. One such striking property is the existence of a "light" mass spectrum of order 1 GeV. From a dimensionless viewpoint the light spectrum has mass 10^{19} times smaller than the fundamental scale. It is in order to ask what kind of special adjustments of parameters must be made in order to ensure such a gigantic ratio of mass scales.

To illustrate a case of an unnatural adjustment, consider a particle which receives a self-energy which is quadratic in κ. To make the discussion simple, suppose the form of the mass correction is

$$\begin{aligned} m^2 &= m_0{}^2 + \Delta m^2 \\ &= m_0{}^2 + \kappa^2 g_0{}^2 . \end{aligned} \tag{2}$$

Solving for $\mu_0{}^2$ gives

$$\mu_0{}^2 = \frac{m_0{}^2}{\kappa^2} = \frac{m^2}{\kappa^2} - g_0{}^2 . \tag{3}$$

Now if m is a physical mass of order 1 GeV and $\kappa \sim 10^{19}$ GeV, then

$$\mu_0{}^2 = -g_0{}^2 (1 - 10^{-38}) . \tag{4}$$

Equation (4) means that $\mu_0{}^2$ must be adjusted to the 38th decimal place. What happens if it is not? Then the mass will come out to be of order 10^{19} GeV.

Such adjustments are unnatural and will be assumed absent in the correct theory. Unfortunately, all present theories contain such unnatural adjustments because of the quadratic divergences in the scalar-particle masses.

Not all theories in which the physical and cutoff scales are vastly different are unnatural. Fortunately there exists a class of non-Abelian gauge theories where enormous scale ratios may occur naturally. These are the asymptotically free theories.[4]

In asymptotically free theories the scale-dependent "running" coupling constant satisfies

$$q\frac{\partial g}{\partial q} = -Cg^3 + O(g^5) , \tag{5}$$

where q is the momentum scale at which the coupling is measured. The constant C is positive so that the coupling increases toward the infrared. It is believed that such theories spontaneously generate masses corresponding to values of q for which g becomes large. Integrating Eq. (5) gives

$$\frac{1}{g^2(q)} = 2C \ln(q/\kappa) + \frac{1}{g_0{}^2} , \tag{6}$$

where $g^2(K)$ is identified with the bare coupling g_0. Of course, Eq. (6) is inaccurate when g becomes large, but we may use it as a guide to where the coupling becomes large. It is evident that the value of m which makes $g(m)$ become large satisfies

$$\frac{m}{\kappa}=\exp\left(-\frac{1}{2Cg_0^2}\right). \tag{7}$$

Evidently, to make $m/\kappa \sim 10^{-19}$ requires the bare coupling g_0 to be

$$g_0^2=\frac{1}{38C\ln 10}=\frac{0.012}{C}. \tag{8}$$

As an example, consider pure non-Abelian SU_3 Yang-Mills theory. The constant C is given by[4]

$$C=\frac{11}{16\pi^2} \tag{9}$$

so that

$$g_0^2 \sim 0.2\,.$$

This is hardly an unnatural value for g_0^2. Furthermore, the value of m/κ is not violently sensitive to small variations of g_0.

II. A NATURAL SCENARIO

Let us assume that at the smallest distances (Planck length) nature is described by a very symmetric "grand unified" theory. The grand unifying group is called G. Let us suppose that G is spontaneously broken. This might occur for a variety of reasons, including the existence of scalar fields or gravitational attraction. Since we shall forbid unnatural adjustments of constants we must assume that any masses which are generated in the first round of symmetry breaking are of order 10^{19} GeV.

At a somewhat larger distance scale, say 10^{17} GeV, a phenomenological description should exist. It will contain those survivors of the first symmetry breakdown which gained no mass. Furthermore, it will have a symmetry group

$$G_1 \otimes G_2 \otimes G_3 \otimes \cdots$$

consisting of factors which are not broken by the first breakdown.

The survivor fields will include

(1) the gauge bosons for the group $G_1 \otimes G_2 \otimes \cdots$,

(2) some subset of fermions which were protected by unbroken γ_5 symmetries,

(3) some Goldstone bosons. In what follows we assume no such Goldstone bosons are present.

If not for the fermions, the different G_i would define uncoupled gauge sectors. These sectors are, in general, coupled by fermions having nontrivial transformation properties under more than one G_i. For example, quarks form the bridge which couples quantum-chromodynamic (QCD) gluons with the photon, intermediate-vector-boson sector in the standard theory.

Thus, to specify a theory we must give a set of G_i and a set of fermion fields along with their transformation properties under all G_i. Furthermore, we will also need a set of coupling constants g_i. These may be taken to be the running couplings at a low enough energy so that the effects of the very heavy masses have disappeared. Henceforth we assume this to be 10^{+17} GeV. Henceforth we define $K=10^{17}$.

Consider next the evolution of the running couplings. Some of them will increase and some will decrease as the energy scale is lowered. From studying examples it is clear that different g_i may blow up and produce masses at rather different scales. To see why this is so consider the case of two uncoupled gauge theories G_1 and G_2. Each will have its bare coupling g_1, g_2 and will evolve to give a mass scale

$$\frac{m_1}{K}\approx\exp\left(-\frac{1}{2C_1g_1^2}\right),\qquad \frac{m_2}{K}\approx\exp\left(-\frac{1}{2C_2g_2^2}\right), \tag{10}$$

and

$$\frac{m_1}{m_2}=\exp\left[-\frac{1}{2}\left(\frac{1}{C_2g_2^2}-\frac{1}{C_1g_1^2}\right)\right]. \tag{11}$$

If we now assume $1/C_ig_i^2$ is large enough to make m_i/K very small then a few present differences between C_1 and C_2 or g_1 and g_2 can easily make $m_1/m_2 \sim 10^{-3}$.

Thus our expectation is for a $G_1 \otimes G_2 \otimes G_3 \cdots$ gauge theory with a set of Fermi fields connecting the G_i and a set of dynamically produced mass scales fairly well separated. The question to which this paper is addressed is: Can this type of theory produce the required kinds of spontaneous symmetry breakdown needed to understand weak, electromagnetic, and strong interactions?

III. A WARMUP EXAMPLE

The set of subgroups G_i must include SU_3 (color) and the electromagnetic-weak group $SU_2 \otimes U_1$ which we will call flavor. The fermion content must include quarks and leptons. As our simplest example we consider a theory with the massless flavor doublet (u,d) of color-triplet quarks. The quarks interact with an octet of color gluons and the four flavor gauge fields W^α and B. The coupling constants are chosen as they would be in realistic models so that the QCD coupling becomes ~1 at 1 GeV and the electromagnetic charge is $\sim\frac{1}{3}$. We call the SU_3, SU_2, and U_1 coupling constants g_3, g_2, and g_1. This theory is the standard theory of a single quark doublet with the exception that no fundamental scalar Higgs field are included.

The Lagrangian for our warmup model is

$$\mathcal{L} = -\tfrac{1}{4}F_{\mu\nu}F_{\mu\nu} - \tfrac{1}{4}W_{\mu\gamma}W_{\mu\gamma} - \tfrac{1}{4}B_{\mu\nu}B_{\mu\nu} + i\bar{\psi}\gamma_\mu(\partial_\mu + ig_3\hat{f} + ig_2\hat{W}_\mu + ig_1\hat{B}_\mu)\psi . \tag{12}$$

The objects $\hat{F}_\mu$, $\hat{W}_\mu$, and $\hat{B}_\mu$ are constructed from the SU_3, SU_2, U_1 vector potentials, Dirac matrices, 3×3 color matrices, and 2×2 flavor matrices. For example,

$$\hat{W}_\mu = W_\mu^\alpha \frac{\tau^\alpha}{2}\left(\frac{1-\gamma_5}{2}\right), \tag{13}$$

where τ^α are flavor Pauli matrices. Similarly,

$$\hat{B}_\mu = \left[\frac{1-\gamma_5}{4} - \frac{1}{3} + \left(\frac{1+\tau_3}{2}\right)\left(\frac{1+\gamma_5}{2}\right)\right]B_\mu . \tag{14}$$

In analyzing the above model we will make use of a number of standard assumptions. We now list them:

(1) The weak-electromagnetic sector can be treated as a small perturbation. The remaining assumptions apply to the pure SU_3 (color) sector when g_1 and g_2 are switched off.

(2) The strong interactions are invariant under chiral $SU_2\otimes SU_2$ in the limit of vanishing bare quark mass.[6]

(3) Chiral $SU_2\times SU_2$ is spontaneously broken and realized in the Nambu-Goldstone mode. The pion is the Goldstone boson. The "order parameter" signaling the spontaneous breakdown is $\langle 0|\bar{\psi}\psi|0\rangle = \langle 0|\bar{u}u + \bar{d}d|0\rangle$, which is nonzero.

(4) The chiral limit ($m_\pi^2 \to 0$) is a smooth one in which all strong-interaction quantities (other than m_π) change by only a few percent. In particular, this includes f_π—the pion decay constant.

Our problem is to determine the behavior of the W, Z, and photon masses in this theory. In particular, we would like to know if the strong interactions can somehow replace the Higgs scalars and provide masses for the intermediate vector bosons. To this end we must examine the effects of quarks and SU_3 gluons on the W and B propagators. The relevant processes are shown in Figs. 1(a)–1(c).

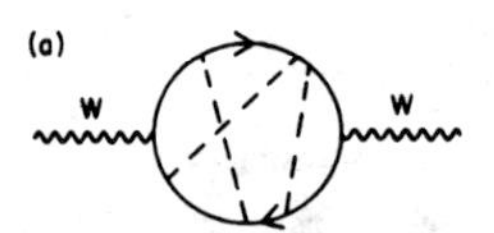

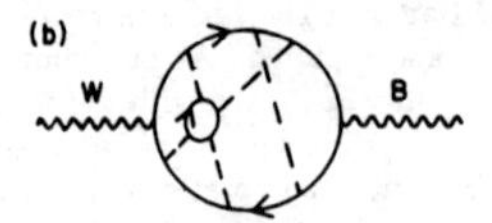

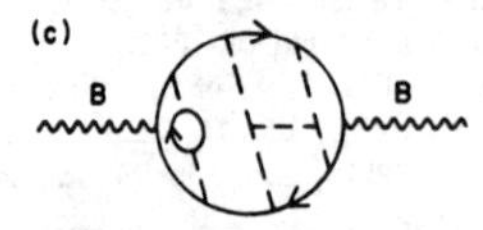

FIG. 1. Contributions to $\pi_{\mu\nu}$ from quark-gluon states. Solid lines indicate quarks. Broken lines are gluons.

Let us first ignore the B field and concentrate on the class of processes illustrated in Fig. 1(a). Invoking the familiar arguments of gauge invariance, we write the one-particle-irreducible vacuum polarization as

$$\pi_{\mu\nu}^{\alpha\beta} = \delta^{\alpha\beta}\left(\frac{g_2}{4}\right)^2 (k^2 g_{\mu\nu} - k_\mu k_\nu)\pi(k^2), \tag{15}$$

where α,β indicate SU_2 indices. Evidently the W propagator is modified from

$$\delta^{\alpha\beta}\left(\frac{k^2 g_{\mu\nu} - k_\mu k_\nu}{k^4}\right) \tag{16}$$

to

$$\frac{\delta^{\alpha\beta}(g_{\mu\nu} - k_\mu k_\nu/k^2)}{k^2[1 + g_2^2\pi(k^2)/4]} . \tag{17}$$

Unless $\pi(k^2)$ is singular at $k^2=0$ the W propagator will have a pole at $k^2=0$ indicating a massless vector boson. From this point of view, the role of the fundamental scalar Goldstone bosons is to provide a pole in π at $k^2=0$.

Now consider the contribution of the pion to $\pi(k^2)$. Since no explicit scalars are included, the quarks must be massless. (Recall that the only source of quark mass in the Weinberg-Salam theory is the Yukawa couplings.) It then follows that the pion is massless, at least insofar as it is regarded as an unperturbed state of the pure strong interaction. Thus we can immediately write the pion contribution to π as a massless pole in the vicinity of $k^2=0$:

$$\pi(k^2) \approx f_\pi^2/k^2 . \tag{18}$$

Accordingly, the pion replaces the usual scalar fields and shifts the mass of the W to

$$M_W^2 = \left(\frac{g_2}{2}\right)^2 f_\pi^2 \approx (30\ \text{MeV})^2 . \tag{19}$$

Next consider the contribution of the pion to the processes in Figs. 1(b) and 1(c). For this we need to know the coupling of the pion to the Abelian U_1 current. From Eq. (14) we see that this current is

$$\bar{\psi}\gamma_\mu\left[\frac{1}{2}\left(\frac{1-\gamma_5}{2}\right) - \frac{1}{3} + \left(\frac{1+\tau_3}{2}\right)\left(\frac{1+\tau_5}{2}\right)\right]\psi . \tag{20}$$

The term which couples to the neutral pion is

$$-\tfrac{1}{4}\bar{\psi}\tau_3\gamma_5\gamma_\mu\psi . \tag{21}$$

Thus Figs. 1(b) and 1(c) receive pion-pole contributions

$$\pi_{WB}=\frac{g_1 g_2}{4k^2}f_\pi^{\,2}, \tag{22}$$

$$\pi_{BB}=\left(\frac{g_1}{4k^2}\right)^2 f_\pi^{\,2}. \tag{23}$$

All of this is summarized by a mass matrix

$$M^2=\begin{bmatrix} g_2^{\,2} & 0 & 0 & 0 \\ 0 & g_2^{\,2} & 0 & 0 \\ 0 & 0 & g_2^{\,2} & g_1 g_2 \\ 0 & 0 & g_1 g_2 & g_1^{\,2} \end{bmatrix} \tfrac{1}{4} f_\pi^{\,2}, \tag{24}$$

where the labeling of the rows and columns is (W^+, W^-, W^0, B).

The mass matrix in Eq. (24) is identical to that in the Weinberg-Salam (WS) theory with the exception that f_π would be replaced by the vacuum expectation value of the scalar field ϕ. Thus the masses of Z and $W^\pm$ are in the same ratio as in the WS theory but are scaled down by the factor

$$\frac{f_\pi}{\langle\phi\rangle}\approx\frac{1}{3000}. \tag{25}$$

Naturally, in the model we are considering the pion is absent from the real spectrum, being replaced by the longitudinal $W^\pm$ and Z.

The correspondence between the pion and the usual scalar doublet ϕ may be made manifest. Define

$$\pi^\alpha=\bar\psi\gamma_5\tau^\alpha\psi, \qquad \sigma=\bar\psi\psi. \tag{26}$$

The two component field ϕ of WS may be replaced by

$$\begin{pmatrix}\phi_1\\ \phi_2\end{pmatrix}\longrightarrow\begin{pmatrix}\pi_1+i\pi_2\\ \pi_3+i\sigma\end{pmatrix}. \tag{27}$$

It is easily seen that such a two-component object transforms as a spinor under left-handed SU_2 and has the same Abelian charge as ϕ. Lastly, the spontaneous breaking of the symmetry is accomplished by the usual strong interactions which (we believe) give rise to $\langle\sigma\rangle\neq 0$.

An interesting point we wish to emphasize before attempting a realistic example involves the $SU_2\times SU_2$ symmetry of the hadron sector (before g_2 and g_1 are switched on). In general, to consistently couple a sector to the weak-electromagnetic interaction that sector need only have $SU_2\times U_1$ symmetry. The extra symmetry under $SU_2\times SU_2$ is also present in the WS model. To see it we write

$$\begin{pmatrix}\phi_1\\ \phi_2\end{pmatrix}=\begin{pmatrix}\alpha_1+i\alpha_2\\ \alpha_3+i\alpha_4\end{pmatrix} \tag{28}$$

and then note that the scalar field Lagrangian in WS has symmetry under the four-dimensional rotations ($=SU_2\times SU_2$) in the ($\alpha_1, \alpha_2, \alpha_3, \alpha_4$,) space.

In the WS model the additional symmetry is accidental and may be eliminated if nonrenormalizable interaction or additional scalar multiplets are introduced. In our case it is entirely natural, following from the symmetries of a multiplet of Dirac fermions.

It is interesting to ask what evidence exists for the $SU_2\times SU_2$ symmetry. In our example the extra symmetry implies ordinary isospin [SU_2 (left) + SU_2 (right)] symmetry and guarantees that f_π is the same for neutral and charged pions. If the symmetry were reduced to $SU_2\times U_1$ then in general $f_{\pi^\pm}\neq f_{\pi^0}$. The result would be a modification of the structure of the mass matrix in Eq. (24). The success of the WS model in neutral-current phenomenology is rather sensitive to this structure. Therefore, a large deviation from $SU_2\times SU_2$ symmetry in the scalar field sector will be inconsistent with observed neutral currents.

A final point involves the existence of more than one quark multiplet. If the number of quark doublets is increased from one to N, the hadronic chiral symmetry becomes $SU_{2N}\times SU_{2N}$. Since mass terms are forbidden when the fundamental scalars are absent this will necessarily be a symmetry of the hadronic sector. The number of Goldstone bosons will be N^2-1. The longitudinal Z and $W^\pm$ will again absorb three of these, leaving N^2-4 spin-zero objects. These objects will gain mass because the weak interactions explicitly violate the symmetries which correspond to them. In other words, they are what Weinberg calls pseudo-Goldstone bosons. Their mass will in general be of the same order of magnitude as that of Z and W.

IV. A MORE REALISTIC EXAMPLE

Let us now consider the possible existence of a new undiscovered strongly interacting sector, similar to ordinary strong interactions except with a mass scale of order 10^3 GeV. To be specific we introduce a new family of fermions called "heavy-color" quarks and an associated field χ. The heavy-color quarks form a flavor $SU_2\times U_1$ doublet and an n-tuple in a new SU_n symmetry space called heavycolor. Heavycolor is a gauge symmetry requiring a multiplet of gauge bosons G_μ. The symmetry of the theory is then

$$SU_n(\mathrm{HC})\otimes SU_3(C)\otimes SU_2\otimes U_1$$

with couplings g_n, g_3, g_2, g_1. The fermion content includes

(1) *Leptons.* These are flavor doublets and color–heavy-color singlets.

(2) *Quarks.* These are flavor doublets, color triplets, heavy-color singlets.

(3) *Heavy-color quarks.* These are flavor doublets, color singlets, and heavy-color n-tuples.

The coupling g_n is chosen so that a mass scale of order 1 TeV—the heavy-color (HC) interaction—becomes strong. To make this precise we first consider the pure HC theory ignoring quarks, leptons, color, and flavor. The bare g_n is then adjusted so that the lightest nonzero mass of a heavy-color hadron is ~1 TeV.

The Lagrangian of our model is

$$\mathcal{L} = -\tfrac{1}{4}G_{\mu\nu}G_{\mu\nu} - \tfrac{1}{4}F_{\mu\nu}F_{\mu\nu} - \tfrac{1}{4}W_{\mu\nu}W_{\mu\nu} - \tfrac{1}{4}B_{\mu\nu}B_{\mu\nu} + \bar{\chi}\gamma_\mu(\partial_\mu + ig_n\hat{G}_\mu + ig_2\hat{W}_\mu + ig_1\hat{B}_\mu)\chi + \bar{\psi}\gamma_\mu(\partial_\mu + ig_3\hat{F}_\mu + ig_2\hat{W}_\mu + ig_1\hat{B}_\mu)\psi. \tag{29}$$

In speculating about the solution of this model we will make use of the following observations and assumptions.

(1) The evolutions of the heavy-color and color couplings with scale are only slightly different from what they would be if each sector were completely isolated. The justification for this is that heavy color and color are only coupled by their *weak* interactions with B and W. If g_1 and g_2 were zero, g_n and g_3 would evolve completely separately. In fact, the quark-gluon and heavy-color-quark–heavy-color-gluon worlds would be completely noninteracting.

(2) The isolated heavy-color sector is essentially similar to the color sector except scaled up in energy by ~3000. This means that $\langle 0|\bar{\chi}\chi|0\rangle \neq 0$. This implies the existence of a family of massless heavy-color pions with decay constants $F_\pi \sim f_\pi \times 3000$. It also means that there exists a rich spectrum of heavy-color hadrons.

As in our warmup example the Z and $W^\pm$ gain a mass. This time the mass is mainly due to the mixing of the HC-pion with W and B. The ordinary pion becomes very slightly mixed with the HC-pion but remains exactly massless. This is so because the ordinary and HC axial-vector currents are separately conserved. The longitudinal components of Z and W can only absorb one linear combination of the Goldstone bosons associated with these currents.

The model described here is certainly incomplete. As it stands it cannot account for the masses of leptons and quarks. We shall discuss this further in Sec. V.

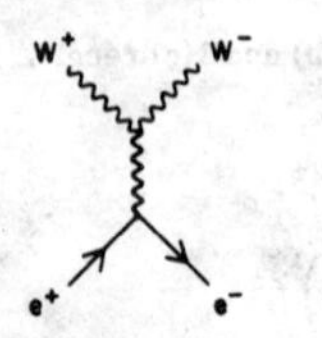

FIG. 2. The process $e^+e^- \to W^+W^-$.

V. IMPLICATIONS OF HEAVY COLOR

The behavior of processes at and above the TeV range is very different in the usual and present theories. By the usual theory I will always mean two things. First, symmetry breaking is caused by fundamental scalar fields. Second, all coupling constants including the scalar self-coupling are small so that perturbation theory is applicable.

Our first problem is to determine the mass scale of the heavy-color hadrons. To this end we observe that Eq. (19) will be replaced by

$$M_W^2 = \frac{g_2^2}{4}F_\pi^2, \tag{30}$$

where F_π is the HC-pion decay constant. Since we know that $M_W \sim 90$ GeV, we find

$$F_\pi \sim 250 \text{ GeV}. \tag{31}$$

Since we have assumed that the heavy-color sector is simply a scaled-up version of the usual strong interactions, it follows that heavy-color-hadron masses are F_π/f_π times their hadronic counterparts. Since $F_\pi/f_\pi \sim 3\times 10^3$, the mass of a low-lying HC-hadron will be ~3 TeV.

The main differences in behavior of this and the usual model involve processes in which longitudinally polarized Z's and W's are produced at energies above a TeV. For example, consider e^+e^- annihilation. As in the usual theory the e^+e^- pair can form a virtual photon or Z boson which can then materialize as a pair of transverse $W^\pm$ bosons. This process is illustrated in Fig. 2. The transverse $W^\pm$ are weakly coupled and therefore contribute a smooth nonresonant contribution to R that can be computed in perturbation theory. (See Fig. 3.)

In the usual weakly coupled scalar version of the Weinberg-Salam theory the virtual γ or Z can also materialize as a pair of charged scalars disguised as longitudinal $W^\pm$. Since the scalars are also weakly coupled the contribution to R is smooth, nonresonant, and similar to Fig. 3.

In the present theory the scalar sector is replaced by the heavy-color sector and the virtual γ-Z may decay into a pair of heavy-color quarks. The resulting behavior of R will exhibit all the

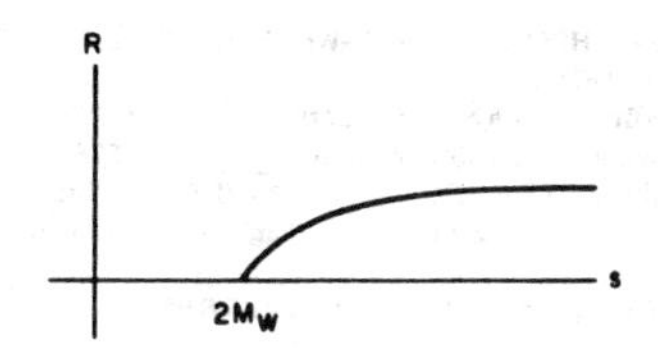

FIG. 3. R as a function of energy for transverse gluon production.

characteristics of resonances and final-state interaction which characterized R at ordinary energies ~0–3 GeV. It should exhibit the bumps of the HC-ρ, HC-Ω, and so on. (See Fig. 4.) The only difference is that the entire scale of masses, widths, and level separations will be of order 1 TeV instead of 1 GeV.

The final states of such processes will involve increasing multiplicities of HC-hadrons. If our experience in hadron physics is a good guide then most of the final HC-hadrons will be HC-pions. Of course real HC-pions do not exist, having been replaced by the longitudinal $W^{\pm}$, Z states. Indeed, the following theorem is easy to prove: To lowest order in α the amplitude for producing a given state including some set of $Z_{\rm long}$ and $W_{\rm long}$ is equal to the amplitude for a state in which the Z_L, W_L are replaced by HC-pions.

The longitudinal bosons decay in a conventional way to leptons and hadrons. However, the distribution of the longitudinal bosons will not resemble the usual theory. In the usual theory each boson, longitudinal or transverse, costs a factor of α since all couplings are small. In the present theory longitudinal bosons proliferate like pions once the energy exceeds a few TeV.

Perhaps the most interesting consequence of the new theory is the existence of a new conservation law—heavy-color baryon number $=\int \chi^*\chi\, d^3x$. The lightest HC-hadron carrying HC-baryon number will be stable and have a mass ~1–2 TeV. If the heavy-color group SU_n has odd (even) n this particle will be a fermion (boson). Its only interaction with ordinary matter will be weak-electromagnetic. It may be charged like the proton or neutral like the neutron. If, like the proton, it is charged and found in any abundance in the universe, it may be detectable as a component of cosmic rays.

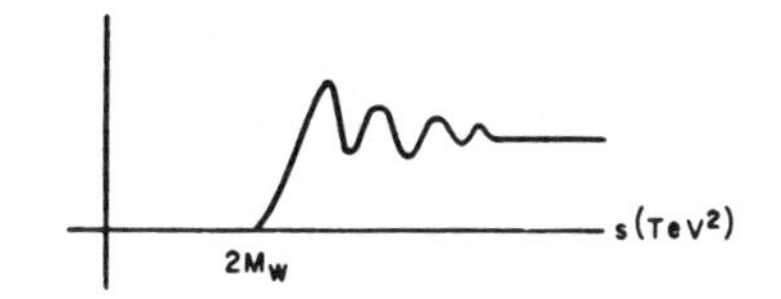

FIG. 4. Conbributions to R from the heavy-color sector.

VI. CONCLUSIONS

One aspect of the scalar-boson problem has not been mentioned in this paper. The usual scalar-boson mechanism provides masses not only for the vectors but also the leptons and quarks. In the present example the only way to mimic the fermion mass mechanism would involve four-Fermi nonrenormalizable interactions. Indeed, if the scalar field ϕ is replaced by HC-quark bilinears in the Yukawa couplings, a quartic coupling of the form $\chi^\dagger\chi\,\psi^\dagger\psi$ is generated. This coupling produces the conventional fermion mass matrix when $\langle\bar{\chi}\chi\rangle$ gets a nonvanishing value.

The inability to generate mass without four-Fermi couplings is due to the chiral γ_5 symmetry of vector couplings. In the present theory this symmetry is a continuous symmetry if we ignore weak instantons. Therefore, any dynamical fermion mass generation would require massless Goldstone bosons.

In general, by adding more sectors, including a gauge group which mixes e and μ as well as strange and nonstrange quarks, we can reduce the γ_5 symmetry to a discrete symmetry. In order to do this we must make use of the instantons of this new sector which means the coupling must be significantly greater than α. If this theory can be made to work then no Goldstone bosons would be required by dynamical mass generation.

Notes added.

(1) After submitting this paper for publication I became aware of the work of S. Weinberg in which many of the motivations for heavy-color are described.[5]

(2) Very recently Weinberg[6] and Georgi, Lane, and Eichten have been led to consider a very similar proposal for replacing Higgs bosons by dynamically bound pionlike objects originating in a strong interaction whose scale is ~1 TeV. I would like to thank S. Weinberg for a copy of his report and H. Georgi for communicating his interest in the problem to me.

ACKNOWLEDGMENT

I would like to thank K. Wilson for explaining the reasons why scalar fields require unnatural adjustments of bare constants.

*Permanent address: Dept. of Physics, Stanford University.

[1]For a review of the conventional theory see S. Weinberg, in *Proceedings of the International Symposium on Lepton and Photon Interactions at High Energies, Hamburg, 1977*, edited by F. Gutbrod (DESY, Hamburg, 1977).

[2]The particular concept of naturalness and the objections to scalar fields described in this paper are due to K. Wilson, private communication.

[3]H. Georgi, H. Quinn, and S. Weinberg, Phys. Rev. Lett. 33, 451 (1974).

[4]H. D. Politzer, Phys. Rev. Lett. 30, 1346 (1973); D. Gross and F. Wilczek, *ibid*. 30, 1343 (1973).

[5]S. Weinberg, Phys. Rev. D 13, 974 (1975). This paper contains many of the conceptual motivations described here.

[6]S. Weinberg, Phys. Rev. D 19, 1277 (1979).

PHYSICAL REVIEW D VOLUME 19, NUMBER 4 15 FEBRUARY 1979

Implications of dynamical symmetry breaking: An addendum

S. Weinberg
Lyman Laboratory of Physics, Harvard University, Cambridge, Massachusetts 02138
(Received 2 June 1978)

It is shown that the dynamical symmetry breakdown of a gauge symmetry can in some cases lead to simple relations among the masses of intermediate vector bosons.

This note is an addendum to a general survey[1] of the physical implications of dynamical symmetry breaking.[2] By a "dynamical" symmetry breaking is meant any spontaneous breakdown of a global or gauge symmetry for which the associated Goldstone bosons are composite rather than elementary particles. Formation of such bound states requires strong forces among the constituent particles, and for this reason it has generally been supposed that a dynamical breakdown of gauge symmetries would not naturally provide any simple relations among the intermediate-vector-boson masses generated by the symmetry breakdown. The possibility of such mass relations was not considered in Ref. 1.

In this note I wish to show that a dynamical breakdown of a gauge symmetry can indeed lead to simple relations among intermediate-vector-boson masses. In particular, a dynamical breakdown of SU(2) × U(1) in the gauge theory[3] of weak and electromagnetic interactions can lead to the same successful relation $M_Z/M_W = \sec\theta$ that is found when the symmetry breakdown is due to vacuum expectation values of scalar-field doublets. For a dynamical symmetry breakdown this is not as automatic as for symmetry breaking by scalar-doublet vacuum expectation values, but depends on the assumption of a specific pattern of spontaneous symmetry breaking. The source of this mass relation can be traced, both for symmetry breaking by scalar doublets and in the dynamical case to be considered here, to the same simple property of the Goldstone bosons which mix with the SU(2) × U(1) gauge fields. However, there remain severe difficulties in developing realistic detailed models of elementary particles in which the symmetry breaking is purely dynamical.

First, let us consider a gauge model that is illustrative, though its quark content is quite unrealistic. The gauge group of the weak, electromagnetic, and strong interactions is as usual taken as SU(2) × U(1) × G_S, but G_S is arbitrary. The coupling constants g, g', and g_s associated with the subgroups SU(2), U(1), and G_S are assumed to have the orders of magnitude $g \approx g' \approx e$ and $g_s \approx 1$. [Strictly speaking, g_s becomes of order unity at a renormalization scale Λ, which, as we shall see in this example, would have to be of the order of 200 GeV. Hence, G_S could not consist solely of the usual color group SU(3).] The model contains just two G_S multiplets of quarks, U and D, which form a left-handed SU(2) × U(1) doublet $(1+\gamma_5)(U,D)$ and right-handed singlets $(1-\gamma_5)U$ and $(1-\gamma_5)D$, but no scalar fields. (The "color" indices associated with G_S are dropped everywhere.) In the limit $e \to 0$, the strong interactions will automatically be invariant not only under the gauge group G_S, but also under an "accidental" global SU(2) × SU(2) symmetry,[4] consisting of independent unitary unimodular transformations on the doublets $(1+\gamma_5)(U,D)$ and $(1-\gamma_5)(U,D)$. We assume that the strong forces associated with G_S produce a dynamical breakdown of SU(2) × SU(2), which for $e=0$ leaves the global "isospin" subgroup SU(2) unbroken. [At the same time, G_S itself may also break down to some gauge subgroup, perhaps SU(3).] It is the residual global invariance of the strong interactions for $e=0$ that leads in this model to a simple relation between M_W and M_Z.

To see this, we use the general lowest-order formula[5] for the intermediate-vector-boson matrix

$$\mu^2_{\alpha\beta} = \tfrac{1}{64}\sum_a F_a^{\,2}\,\mathrm{Tr}(t_\alpha x_a)\,\mathrm{Tr}(t_\beta x_a)\,. \qquad (1)$$

Here x_a are the generators of all spontaneously broken global symmetries, in a suitably orthonormalized basis[6]; F_a are the couplings of the corresponding Goldstone bosons to the associated currents; and t_α are the generators of the weak and electromagnetic gauge groups, including all coupling-constant factors. In the present case, we have

$$\begin{aligned} &x_a = \gamma_5\tau_a\,, \quad a=1,2,3 \\ &t_i = \tfrac{1}{4}g(1+\gamma_5)\tau_i\,, \quad i=1,2,3 \\ &t_0 = -g'\left[\tfrac{1}{4}(1-\gamma_5)\tau_3+\tfrac{1}{6}\right], \end{aligned} \qquad (2)$$

with τ_a the Pauli isospin matrices. The residual global SU(2) symmetry prevents the appearance of any positive-parity Goldstone bosons, and also imposes a relation among the F_a,

$$F_1 = F_2 = F_3 \equiv F\,. \qquad (3)$$

This is the same relation as is satisfied by the F's in the nondynamical case, where the symmetry breakdown is due to vacuum expectation values of scalar doublets, and it leads to the same relation among intermediate-vector-boson masses. From Eq. (1), we then find the nonvanishing elements of the intermediate-vector-boson mass matrix,

$$\mu^2_{11}=\mu^2_{22}=\mu^2_{33}=\tfrac{1}{16}F^2g^2\,,$$
$$\mu^2_{30}=\mu^2_{03}=\tfrac{1}{16}F^2gg'\,, \qquad (4)$$
$$\mu^2_{00}=\tfrac{1}{16}F^2g'^2\,.$$

(All Goldstone bosons are eliminated by the Higgs mechanism here.) It is easy to see that the nonvanishing mass eigenvalues are

$$M_W{}^2=\tfrac{1}{16}F^2g^2\,,\quad M_Z{}^2=\tfrac{1}{16}F^2(g^2+g'^2)\,. \qquad (5)$$

With $g'/g\equiv\tan\theta$, these have the usual ratio $M_Z/M_W=\sec\theta$.

The U and D of this model cannot, of course, be identified with any known particles. Their mass must be of order F, because for $e=0$ there are no free parameters in the theory except for the G_S renormalization scale Λ, so that M_U, M_D, and F must all be of order Λ. But if we identify $\sqrt{2}\,g^2/8M_W{}^2$ with the usual Fermi coupling constant G_F, then F takes the value $2^{-3/4}G_F{}^{-1/2}$, or 175 GeV. Also, U and D are nearly degenerate, since their masses are split by the weak and electromagnetic interactions only in order α.

In order to construct a slightly more realistic model, we consider the same gauge group $SU(2)\times U(1)\times G_S$, but now we suppose that there are *four* quark flavors U_1,D_1,U_2,D_2, which form two left-handed $SU(2)\times U(1)$ doublets $(1+\gamma_5)(U_1,D_1)$ and $(1+\gamma_5)(U_2,D_2)$, plus four right-handed singlets. In this model, the strong interactions for $e\to 0$ will automatically have an accidental global $SU(4)\times SU(4)$ symmetry.[4] We assume that this symmetry suffers a spontaneous dynamical breakdown in such a way that two quarks, U and D, receive equal masses, while the other two, u and d, remain massless. The subgroup of $SU(4)\times SU(4)$ which remains unbroken is assumed to be the largest subgroup consistent with such masses, with the generators (in a U,D,u,d basis) given by

$$\begin{pmatrix}\vec{\tau} & 0\\ 0 & 0\end{pmatrix},\quad \begin{pmatrix}0 & 0\\ 0 & \vec{\tau}\end{pmatrix},\quad \gamma_5\begin{pmatrix}0 & 0\\ 0 & \vec{\tau}\end{pmatrix},\quad \begin{pmatrix}1 & 0\\ 0 & -1\end{pmatrix}. \qquad (6)$$

(Also, parity is assumed to be not spontaneously broken.) The orthonormalized generators of the broken part of $SU(4)\times SU(4)$ can then be taken as

$$\vec{x}_A=\gamma_5\begin{pmatrix}\vec{\tau} & 0\\ 0 & 0\end{pmatrix},\quad x_B=\frac{\gamma_5}{\sqrt{2}}\begin{pmatrix}1 & 0\\ 0 & -1\end{pmatrix},$$

$$\vec{x}_C=\frac{1}{\sqrt{2}}\begin{pmatrix}0 & -i\vec{\tau}\\ i\vec{\tau} & 0\end{pmatrix},\quad \vec{x}'_C=\gamma_5\vec{x}_C\,,$$
$$\vec{x}_D=\frac{1}{\sqrt{2}}\begin{pmatrix}0 & \vec{\tau}\\ \vec{\tau} & 0\end{pmatrix},\quad \vec{x}'_D=\gamma_5\vec{x}_D\,,$$
$$x_E=\frac{1}{\sqrt{2}}\begin{pmatrix}0 & 1\\ 1 & 0\end{pmatrix},\quad x'_E=\gamma_5x_E\,, \qquad (7)$$
$$x_F=\frac{1}{\sqrt{2}}\begin{pmatrix}0 & -i\\ i & 0\end{pmatrix},\quad x'_F=\gamma_5x_F\,.$$

Using the unbroken part of $SU(4)\times SU(4)$, we easily see that there are only three independent F_a parameters; they are $F_{Ai}\equiv F_A, F_B$ and $F_{Ci}=F'_{Ci}=F_{Di}=F'_{Di}=F_E=F'_E=F_F=F'_F\equiv F_C$. (Here $i=1,2,3$.) The quark fields U,D,u,d of definite mass are not necessarily the same as those in the original weak doublets $(1+\gamma_5)(U_1,D_1)$ and $(1+\gamma_5)(U_2,D_2)$. However, we can always put the weak doublet into the form

$$(1+\gamma_5)\begin{pmatrix}u\\ d\cos\phi+D\sin\phi\end{pmatrix},$$
$$(1+\gamma_5)\begin{pmatrix}U\\ -d\sin\phi+D\cos\phi\end{pmatrix}. \qquad (8)$$

The angle ϕ must be determined by minimizing a "potential" $V(\phi)$, given to lowest order in e by the sum of graphs in which W or Z is emitted and absorbed by a strong-interaction vacuum fluctuation.[7] By using the unbroken $SU(2)\times SU(2)\times SU(2)\times U(1)$ subgroup of $SU(4)\times SU(4)$, we easily see that the Z contribution is ϕ independent, while the W contribution is a sum of terms proportional to $\cos^2\phi$ or $\sin^2\phi$. The whole potential therefore has the ϕ dependence $V(\phi)=A+B\cos^2\phi$. Thus, depending on the sign of B, the angle ϕ at which $V(\phi)$ is a minimum must take the values $\phi=\pi/2$ or $\phi=0$. Let us consider these two cases in turn:

(a) $\phi=\pi/2$: The gauge generators (in a U,D,u,d, basis) here take the form

$$t_1=\tfrac{1}{4}g(1+\gamma_5)\begin{pmatrix}0 & -i\tau_2\\ i\tau_2 & 0\end{pmatrix},$$
$$t_2=\tfrac{1}{4}g(1+\gamma_5)\begin{pmatrix}0 & i\tau_1\\ -i\tau_1 & 0\end{pmatrix},$$
$$t_3=\tfrac{1}{4}g(1+\gamma_5)\begin{pmatrix}\tau_3 & 0\\ 0 & \tau_3\end{pmatrix}, \qquad (9)$$
$$t_0=-g'\left\{\tfrac{1}{4}(1-\gamma_5)\begin{pmatrix}\tau_3 & 0\\ 0 & \tau_3\end{pmatrix}+\tfrac{1}{6}\right\}.$$

From Eq. (1), we find that the nonvanishing elements of the intermediate-vector-boson mass matrix are

$$\begin{aligned}&\mu^2_{11}=\mu^2_{22}=\tfrac{1}{4}F_C{}^2g^2\,,\\&\mu^2_{33}=\tfrac{1}{16}F_A{}^2g^2\,,\\&\mu^2_{30}=\mu^2_{03}=\tfrac{1}{16}F_A{}^2gg'\,,\\&\mu^2_{00}=\tfrac{1}{16}F_A{}^2g'^2\,.\end{aligned}\tag{10}$$

The nonvanishing intermediate-vector-boson masses are then

$$M_W{}^2=\tfrac{1}{4}F_C{}^2g^2,\quad M_Z{}^2=\tfrac{1}{16}F_A{}^2(g^2+g'^2)\,.\tag{11}$$

Hence no simple mass relation arises in this case.

(b) $\phi=0$: The gauge generators (in a $U,D,\,u,d$ basis) here take the form

$$\begin{aligned}&t_i=\tfrac{1}{4}g(1+\gamma_5)\begin{pmatrix}\tau_i&0\\0&\tau_i\end{pmatrix},\\&t_0=-g'\left\{\tfrac{1}{4}(1-\gamma_5)\begin{pmatrix}\tau_3&0\\0&\tau_3\end{pmatrix}+\tfrac{1}{6}\right\}.\end{aligned}\tag{12}$$

From Eq. (1), we find the nonvanishing elements of the intermediate-vector-boson mass matrix are

$$\begin{aligned}&\mu^2_{11}=\mu^2_{22}=\mu^2_{33}=\tfrac{1}{16}F_A{}^2g^2\,,\\&\mu^2_{30}=\mu^2_{03}=\tfrac{1}{16}F_A{}^2gg',\quad\mu^2_{00}=\tfrac{1}{16}F_A{}^2g'^2\,,\end{aligned}\tag{13}$$

so the masses are

$$M_W{}^2=\tfrac{1}{16}F_A{}^2g^2,\quad M_Z{}^2=\tfrac{1}{16}F_A{}^2(g^2+g'^2)\tag{14}$$

and have the same ratio $M_Z/M_W=\sec\theta$ as in the simpler two-flavor model. The reason for this is just that the only Goldstone bosons which mix with W and Z are those associated with $\vec{x}_A$, and the unbroken subgroup of SU(4) × SU(4) requires these to have equal F_a values.

This model is still far from realistic, whether $\phi=\pi/2$ or $\phi=0$. In both cases, the u and d quarks remain massless to all orders in e. In addition for $\phi=\pi/2$, the two light quarks u,d are in SU(2) × U(1) doublets with D and U, not with each other. Furthermore, in both cases the model contains 17 physical Goldstone bosons, some of them "true" Goldstone bosons of zero mass. It is not clear how the light quarks could get reasonable masses or how the "true" Goldstone bosons could be eliminated with a dynamical symmetry breaking.

Note added in proof. After this paper was submitted for publication, I received a report by L. Susskind, which deals with similar questions. [The motivation in his paper for dynamical symmetry breaking, in terms of grand unified gauge theories, is the same as that described by H. Georgi, H. Quinn, and S. Weinberg, Phys. Rev. Lett. 33, 451 (1974).] The undiscovered new strong interaction of Susskind is a special case of what was called an "extra strong" interaction in Ref. 1, and a "superstrong" interaction by S. Weinberg, Phys. Today 30 (No. 4), 42 (1977). Susskind independently observes that the relation $M_Z/M_W=\sec\theta$ follows if dynamical symmetry breaking leaves an "isospin" subgroup unbroken. In addition, he points out that the origin for this relation is essentially the same as that for symmetry breaking by vacuum expectation values of scalar doublets. In the latter case, the part of the Lagrangian which is relevant to the calculation of gauge boson masses is the "kinematic" Lagrangian $\mathcal{L}_\phi=-\frac{1}{2}\sum_n(D_\mu\phi_n)^\dagger(D_\mu\phi_n)$, the sum running over N scalar doublets (ϕ_n^0,ϕ_n^-). By setting $\phi_n^0=\phi_{n1}+i\phi_{n2}$, $\phi_n^-=\phi_{n3}+i\phi_{n4}$, one finds that in the limit $e=0$, $\mathcal{L}_\phi$ has an $O(4)^N=[SU(2)\times SU(2)]^N$ symmetry, with an O(3) = SU(2) subgroup which is automatically left unbroken by the vacuum expectation values of the ϕ_{n1} fields, and which transforms the weak SU(2) generators as a three-vector. As shown both here and in Susskind's paper, this is the same feature that allows one to derive the Z-W mass ratio in the case of dynamical symmetry breaking.

I also wish to comment here on the problems of developing a grand unified theory of strong as well as weak and electromagnetic interactions in which the spontaneous symmetry breaking at all levels is due to vacuum expectation values of elementary scalar fields. As is well known, such theories require constraints on the parameters in the Lagrangian. However, it is *not* true that these constraints necessarily incorporate extremely small parameters, such as 10^{-19}, or that they need to involve quadratic divergences at all. It is only necessary to suppose that at a stationary point of the potential where the SU(3) × SU(2) × U(1) subgroup is unbroken, some of the non-Goldstone eigenvalues of the scalar mass matrix vanish. Nonperturbative effects will then produce a minimum of the potential very near this stationary point, at which SU(2) × U(1) is spontaneously broken, with W and Z masses which are automatically less than the superheavy gauge boson masses by a factor $\exp(-C/e^2)$, where C is a numerical constant of order unity. (Also, any quadratic divergences are always an artifact of the cutoff procedure; they do not appear if we use dimensional regularization.) These matters are discussed by E. Gildener and S. Weinberg, Phys.

Rev. D 13, 3333 (1976), Sec. VI, and in a paper now in preparation.

I am very grateful to Kenneth Lane for remarks which led me to take up this problem, and for valuable conversations on various aspects of this subject. This research was supported in part by the National Science Foundation under Grant No. PHY77-22864.

---

[1]S. Weinberg, Phys. Rev. D 13, 974 (1976).

[2]Dynamical mechanisms for spontaneous symmetry breaking were first discussed by Y. Nambu and G. Jona-Lasinio, Phys. Rev. 122, 345 (1961); J. Schwinger, *ibid.* 125, 397 (1962); 128, 2425 (1962). The application to modern gauge theories is due to R. Jackiw and K. Johnson, Phys. Rev. D 8, 2386 (1973); J. M. Cornwall and R. E. Norton, *ibid.* 8, 3338 (1973).

[4]It is of course understood that the whole theory is also invariant under the global U(1) symmetry associated with fermion number conservation. However, the full global symmetry group for N quark flavors is $SU(N) \times SU(N) \times U(1)$, and not $U(N) \times U(N)$ (as supposed in Ref. 1), because invariance under the *chiral* U(1) symmetry is broken by triangle anomalies in the presence of instantons; see G. 't Hooft, Phys. Rev. Lett. 37, 8 (1976), and references cited therein.

[5]See Ref. 1, Eq. (7.11). The factor $\frac{1}{64}$ appears here because all traces include sums over Dirac indices.

[6]See Ref. 1, Sec. IV.

[7]See Ref. 1, Sec. VI.

Nuclear Physics B202 (1982) 382–396

COMPOSITE/FUNDAMENTAL HIGGS MESONS
(I). Dynamical speculations*

Howard GEORGI and Ian N. McARTHUR[1]

Lyman Laboratory of Physics, Harvard University, Cambridge, MA 02138, USA

Received 29 December 1981
(Corrected version received 21 January 1982)

We argue that a suitable technicolor model with a non-trivial ultraviolet fixed point may be equivalent to an asymptotically free effective technicolor (eTC) model with the same global symmetries. The eTC model may contain scalar mesons with the quantum numbers of the technifermion mass operators which can be interpreted as the Higgs mesons.

1. Introduction

What is the Higgs meson? What breaks the SU(2)×U(1) symmetry of the weak interactions? The original suggestion that it is a *fundamental* scalar field seems unsatisfying because it yields no insight into two deep puzzles of contemporary particle physics. The two puzzles are the flavor puzzle (Why does Nature repeat Herself?) and the gauge hierarchy puzzle (Why is the SU(2)×U(1) breaking scale so much smaller than the unification scale?). Both these mysterious features of the world can be *described* in the context of models with a fundamental Higgs. But the description seems forced and unnatural. It is only a *description*, not an explanation.

Attempts to demystify the second puzzle by incorporating supersymmetry have not been very successful. More ambitious attempts to address both puzzles go under the name of "extended technicolor" or ETC. Technicolor (TC) is a generic name for the strong gauge interactions which are (allegedly) responsible for the dynamical breakdown of the SU(2)×U(1) symmetry [1, 2]. The mechanism is the same as the mechanism of chiral symmetry breakdown in the color SU(3) interactions of QCD. Technifermions, which feel the TC force, transform according to a chiral representation of SU(2)×U(1), the left-handed (LH) and right-handed (RH) components transforming differently. When these technifermions are bound by the TC interactions, they develop dynamical masses which break SU(2)×U(1). The Higgs mesons are, in a sense, composites.

Technicolor is a very attractive idea. It directly addresses the gauge hierarchy puzzle. If the TC gauge coupling is small at the unification scale, the TC confinement

* This research is supported in part by the National Science Foundation under grant no. PHY77-22864.
[1] Supported in part by a Hackett Studentship from the University of Western Australia.

scale Λ_{TC} (which determines the SU(2)×U(1) breaking) is exponentially smaller than the unification scale, because the coupling constant changes slowly [3]. The trouble with technicolor is that it is hard, without introducing light fundamental scalars, to get the SU(2)×U(1) breaking into the normal quark and lepton mass matrix. It is hard to break all the global chiral symmetries which keep the quarks and leptons massless.

In ETC models, the TC gauge group is extended so that quarks and leptons are in ETC multiplets with technifermions [4, 5]. If the ETC symmetry breaks down to TC at an ETC scale larger than Λ_{TC}, the ETC gauge interactions can break all the global chiral symmetries and generate quark and lepton masses without any fundamental scalar mesons.

The ETC idea is extremely ambitious. A successful ETC model would describe physics in terms of a small number of gauge couplings and some strong interaction parameters which are calculable in principle, if not with present technology. Unfortunately, there are difficulties. It is not clear what breaks ETC. Does this require yet another TC mechanism? In many toy models, there are unacceptably light Goldstone or pseudo-Goldstone bosons associated with TC chiral symmetry breaking [4, 5]. Most toy models produce too large a θ parameter in QCD or else yield an axion [6]. Most toy models produce unacceptably large flavor changing neutral current effects [5, 7]. Finally, the most serious difficulty is that there are no models which are not toys. No one has been able to build a model with anywhere near enough structure to describe the complicated quark and lepton mass matrix, at least not without putting the complication in by hand in the form of unreasonable dynamical assumptions [8].

Holdom had an idea which solves some of these problems by raising the ETC scale [9], M_{ETC}. The standard estimate of M_{ETC} is based on the fact that the quark and lepton masses are proportional to the vacuum expectation value (VEV) of the technifermion mass operator, $\bar{\Psi}\Psi$, which, if TC is asymptotically free, scales with dimension three. Thus on dimensional grounds, we expect the VEV to be

$$\langle \bar{\Psi}\Psi \rangle \simeq \Lambda_{TC}^3 . \tag{1.1}$$

Then quark and lepton masses induced by the ETC interactions are of order

$$m \simeq \Lambda_{TC}^3 / M_{ETC}^2 . \tag{1.2}$$

Holdom noted that if TC is not asymptotically free, but instead has a non-trivial ultraviolet (UV) fixed point, then $\bar{\Psi}\Psi$ scales with some anomalous dimension D at short distances and (1.2) is modified to

$$m \simeq \Lambda_{TC}^D / M_{ETC}^{D-1} . \tag{1.3}$$

If D is less than three, M_{ETC} can be large, which makes the ETC interactions less phenomenologically troublesome. But the difficulties of producing a realistic mass matrix remain.

Glashow and one of us suggested a radical extension of Holdom's idea [10]. If D is precisely equal to one, M_{ETC} can be arbitrarily large. In particular, it can be identified with the grand unification scale M_{G}. This identification requires a complete reassessment of the ETC idea. It is quite natural to have fundamental scalar mesons and additional fermions with a mass of order M_{G}. The global chiral symmetries of the light quark and lepton fields can all be broken through their Yukawa couplings to the fields associated with heavy (mass $\sim M_{\mathrm{G}}$) particles. Thus there is no need for ETC as an additional gauge interaction at all. The role of ETC can be played by the interactions of light quarks and leptons with heavy particles at M_{G}.

This is the first of a pair of papers in which we explore this idea in more detail. In this paper, we speculate about the dynamics of gauge theories with non-trivial UV fixed points. We suggest an explanation for the magic value $D=1$ for the anomalous dimension, and propose a calculational scheme based on an effective technicolor interaction (eTC) which allows us to organize our ignorance of the strongly coupled TC theory. We will find that the Higgs meson is both composite and fundamental, depending upon how you look at it.

2. Effective technicolor

Suppose that the TC gauge coupling constant is G and that the β function for the TC theory has the form shown in fig. 1 [9]. The point $G=G^*$ is a non-trivial UV fixed point. We are eventually going to be interested in the strong coupling phase $G>G^*$, but before discussing this phase, we consider what happens if $G=G^*$. Since $\beta(G^*)=0$, if this is true at one scale, it is true at all scales and the theory is scale invariant. In general, we might expect the $G=G^*$ theory to be a complicated strongly interacting theory which does not admit a simple particle interpretation because all the fields scale with anomalous dimensions. There is some evidence that such behavior is possible in theories with non-trivial infrared fixed points [11].

There is, however, a different possibility which in some ways is much simpler. The $G=G^*$ theory may be a free theory (certainly a massless free theory is scale

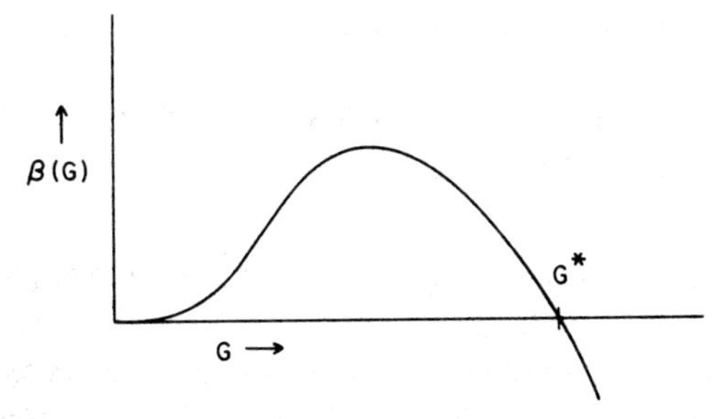

Fig. 1. Naive β function for the TC gauge group.

invariant). But it need not be the same free theory as the $G = 0$ TC theory. Indeed, we would not expect it to be. Instead, the theory may be described by some effective fields.

If this is the nature of the TC theory at $G = G^*$, we might expect that for $G > G^*$, we can rewrite the TC theory in terms of the effective fields which become free for $G = G^*$. Then, in terms of these effective fields, the theory should be described by some effective coupling constant $g(G)$, such that $g(G^*) = 0$. At short distances, this effective theory should be much easier to deal with than the original theory because the coupling is small. The effective theory is asymptotically free.

What we would like to do, then, is to find the rules for going from the original theory with a non-trivial UV fixed point at $G = G^*$ to the effective asymptotically free theory with a UV fixed point at $g = 0$. Before discussing this in detail, we will make one general comment. In the effective theory, since the coupling is small near the fixed point, we can calculate the β function in perturbation theory, and we will find, in any theory, that it vanishes like some power of g. $\beta(g)$ has a multiple zero at $g = 0$. Then unless the function $g(G)$ has very special properties, $\beta(G)$ must also have a multiple zero. Thus fig. 1 is not a good representation of $\beta(G)$. Instead it has the shape illustrated in fig. 2. As discussed in ref. [10], this behavior is important to our understanding of the gauge hierarchy puzzle. Above (and below), we are tacitly assuming that the eTC theory is interacting. Since we are dealing with a quantum field theory, this statement is not entirely trivial. See the postscript for a more careful analysis.

What is the effective theory? We cannot answer the question with certainty, or even prove that an effective theory exists. But if we assume the existence of an asymptotically free effective theory, we can formulate a set of consistency conditions which the effective theory must satisfy. For definiteness, let us assume that the technifermions of the TC theory are N Dirac fermions transforming according to a complex representation of the TC gauge group. The generalization to Majorana fermions transforming according to an arbitrary anomaly free representation of TC should be obvious.

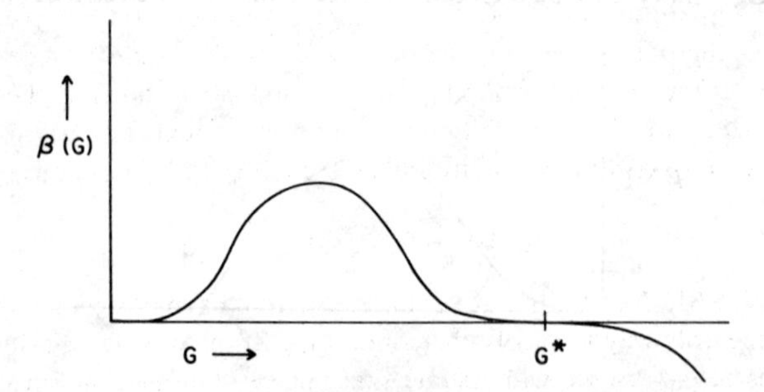

Fig. 2. More sensible β function for the TC gauge group.

The theory with N massless Dirac fermions has an $SU(N)_L \times SU(N)_R \times U(1)$ symmetry. If we write all the fermions as LH fields, we can characterize their transformation properties under $SU(N)_L \times SU(N)_R \times TC$ as follows: the LH technifermions transform like an N of $SU(N)_L$ and like some representation R of TC; the antiparticles of the RH technifermions transform like an $\bar{N}$ of $SU(N)_R$ and like $\bar{R}$ of TC. In an obvious notation, the LH fermions are

$$(N, 1, R) + (1, \bar{N}, \bar{R}) . \tag{2.1}$$

The effective theory should satisfy the following consistency conditions:

(1) It is renormalizable, since we have not introduced any scale.

(2) It is a non-abelian gauge theory, since it is asymptotically free [12]. (See also the postscript.) We will call the non-abelian gauge symmetry of the effective theory "effective technicolor" or eTC.

(3) It should have the same $SU(N) \times SU(N) \times U(1)$ global symmetry as the TC theory.

(4) It must contain fermions with the same anomalies with respect to the $SU(N) \times SU(N) \times U(1)$ generators as the anomalies of (2.1).

The fourth condition may require some explanation. We can imagine, following 't Hooft [13], gauging the global symmetries of the original TC theory and cancelling the anomalies of the fermions of (2.1) with spectator fermions which transform trivially under TC. The effective theory must also be anomaly free, thus the eTC fermions must have the same anomalies as the TC fermions. Note that we do not have to worry about the fact that the global symmetries may (and indeed will) be spontaneously broken, because the effective theory should be equivalent to the original at all momenta, including momenta large compared to the confinement scale, Λ_{TC}, where dynamical symmetry breaking takes place.

Unfortunately, conditions 1–4 are not sufficient to completely determine the effective theory. But in a class of TC theories, there is a particularly simple way of satisfying these conditions. Suppose that the representation R of the TC group has dimension r, but the TC group is some subgroup of $SU(r)$ such that the TC theory is not asymptotically free. For example, we might have $r = 3$ with TC the $U(1) \times U(1)$ subgroup of $SU(3)$ generated by T_3 and T_8, with discrete symmetries which enforces the equality of the two $U(1)$ coupling constants. If the eTC gauge group is swollen to $SU(r)$, and if this swelling is enough to restore asymptotic freedom, then the conditions 1–4 are satisfied if the LH eTC fermions transform as

$$(N, 1, r) + (1, \bar{N}, \bar{r})$$

under $SU(N) \times SU(N) \times SU(r)$. It is not clear, by any means, that this kind of swelling is the only way to satisfy the conditions. But it is by far the simplest, and in this series of papers, we will assume that this swelling mechanism takes place. We note in passing, that the swelling solution satisfies elementary decoupling requirements obtained by allowing some of the technifermions to have masses which can be varied from zero to values large compared to $\Lambda_{\rm TC}$ [13, 14].

One* might object to the SU(3) swelling solution for the eTC gauge group of a U(1)×U(1) TC theory because the 3-dimensional representation of the U(1)×U(1) gauge group is reducible. The TC theory with N massless Dirac 3's thus has a larger symmetry than SU(N)×SU(N)×U(1). There is a separate SU(N)×$SU(N)$ for each component of the 3. What happens to the extra symmetries?

The SU(N)×SU(N) symmetry which we have discussed and imposed on the eTC theory is the one which acts equally on all the components of the 3 of TC. In other words, it commutes with the discrete symmetries which mix up the various components of the three. The generators of the extra symmetries transform non-trivially under the discrete symmetries. This is a clue to the answer.

In the eTC theory, the discrete symmetries are promoted to become part of the eTC gauge symmetry. And the eTC gauge theory is confining. Thus the physical Hilbert space contains only states which are singlets under the eTC SU(3) and thus under any discrete subgroups. Thus the extra symmetry generators cannot possibly be relevant. They take physical states to unphysical states.

This is a particularly clear case of a very general feature of swelling solutions. Whenever the TC gauge group can be regarded as a subgroup of the eTC group, there are TC singlet states which are not eTC singlets. We must assume that the dynamics pushes these states out of the physical Hilbert space (like colored states in confining QCD). In the U(1)×U(1) example, these dynamics spontaneously break the extra SU(N) symmetries (which are not present in the eTC theory). The Goldstone bosons decouple from the physical states for the same reasons.

The extra SU(N) symmetries are analogous to the chiral U(1) symmetry in QCD. The extra SU(N) currents are not invariant under the discrete symmetries of the TC theory just as the conserved chiral U(1) is not color gauge invariant. The Goldstone bosons associated with the spontaneous breaking of the extra SU(N) symmetries decouple from the physical states just as the Goldstone boson of the axial U(1) decouples.

3. Effective Higgs scalars

So far we have discussed the gauge symmetry and the fermions of the effective theory. But there may also be scalar mesons. These may be of three kinds. There may be scalars which transform non-trivially under the eTC gauge group but trivially under the global symmetries. There may be scalars which transform non-trivially under both the eTC gauge group and the global symmetries. Finally, there may be scalars which transform trivially under the eTC gauge group but non-trivially under the global symmetry group. It is in this last kind of effective scalar multiplet that we will find the physical Higgs scalar.

* "One", in this instance, was John Preskill.

The representations of effective scalars which can appear are constrained by conditions 1–4. The strongest constraint is that the scalar self couplings not spoil asymptotic freedom. This constraint is particularly strong if there are scalars of the last and most interesting type, which transform trivially under the eTC gauge group. We must assume that this last type of scalar exists, because scalars transforming non-trivially under eTC are of no use as Higgs mesons. They cannot couple to quarks and leptons which are eTC singlets. We will call the eTC singlet scalars, the "effective Higgs mesons".

The self-couplings of the scalar mesons could not be asymptotically free were it not for their eTC gauge couplings and their Yukawa couplings to the effective eTC fermions. For the effective Higgs mesons, which have no eTC gauge couplings, the Yukawa couplings are crucial. The effective Higgs mesons must, therefore, have Yukawa couplings to eTC singlet mass operators of the fermions. In the swelling scheme, this means that the effective Higgs scalars, if there are any, must transform like $(N, \bar{N}, 1)$ under $SU(N)\times SU(N)\times eTC$. This finally, is the connection to the speculation of ref. [10]. The technifermion mass operators of the TC theory scale with dimension $D=1$ at short distances because in the effective eTC theory, they are associated with canonically scaling effective Higgs scalars.

Even when the effective Higgs scalars have the appropriate Yukawa couplings, the constraint of asymptotic freedom is not trivial. Indeed, the effective theory is *not* asymptotically free in the strong sense that the origin of the multidimensional coupling constant space is a stable fixed point, because for generic values of the eTC gauge coupling g, the Yukawa coupling f and the Higgs scalar couplings λ, the Yukawa coupling goes to zero at short distances too fast, faster than the gauge coupling, leaving the scalar couplings to fend for themselves [15]. The couplings can go to zero at short distances only if the ratios f/g and λ/g^2 are fixed at special values. This is called an eigenvalue condition [16]. It should, perhaps, not surprise us that these couplings are related by an eigenvalue condition. After all, in the original TC theory, there is only one coupling, G. All the couplings in the effective theory must be functions of G and therefore be related to one another. An eigenvalue condition is just a particularly simple example of such a relation.

In the appendix, we will give an example, based on the swelling scheme, of an asymptotically free eTC theory with effective Higgs scalars.

4. Effective Higgs couplings

In the applications of effective technicolor to grand unification, the eTC theory (and the TC theory to which it is equivalent) are low-energy approximations, valid for scales small compared to the unification scale M_G. The full complexity of the theory shows up only at the scale M_G. But of course, even in the low-energy theory, there are fields which do not take part directly in the eTC interactions, the quark and lepton fields and the $SU(3)\times SU(2)\times U(1)$ gauge fields. If we are to justify our

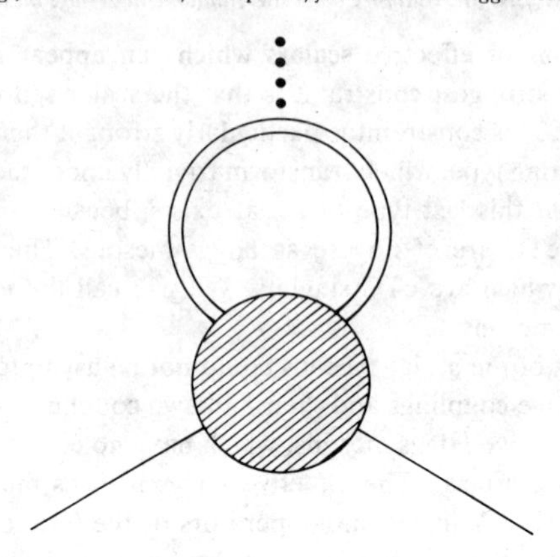

Fig. 3. Diagramatic interpretation of the effective Higgs (dotted line) coupling to quarks or leptons (solid line). The double solid line is a technifermion. The cross-hatched circle represents the interactions of particles with mass of order M_G.

name for the effective Higgs scalars, we must argue that they can couple approximately to the low-energy fields. To the extent that these couplings are weak, we can ignore the more complicated question of how they modify the eTC theory itself. Note in particular that we must assume that these interactions do not modify the form (fig. 2) of $\beta(G)$.

The gauge couplings are simple. They are determined by the gauge symmetry, which just incorporates some subgroup of the $SU(N)\times SU(N)\times U(1)$ global symmetry of the eTC theory.

More interesting are the renormalizable couplings of the effective Higgs to quarks and leptons and the $SU(N)\times SU(N)\times U(1)$ breaking self couplings of the effective Higgs. These are related to $SU(N)\times SU(N)\times U(1)$ breaking terms induced in the TC theory through the couplings of the light technifermions and the light quarks and leptons to heavy fundamental bosons and fermions with mass of order M_G. These couplings also produce a variety of non-renormalizable interactions, but these we ignore because they are unimportant at low energy.

The effective Higgs Yukawa couplings to quarks and leptons can be thought of as arising from the diagram shown in fig. 3. This type of diagram is a handy way of keeping track of the $SU(N)\times SU(N)\times U(1)$ symmetry properties of the effective Higgs couplings, which is really all we can determine without more detailed information about the TC dynamics.

The $SU(N)\times SU(N)\times U(1)$ breaking effective scalar self-couplings in the eTC scalar self-couplings in the eTC theory come from the interactions of one, two, three or four technifermions with heavy particles. Some sample diagrams are shown in fig. 4.

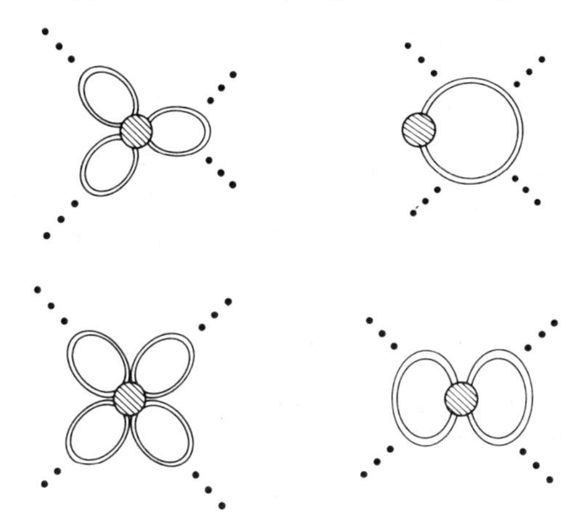

Fig. 4. Diagrams contributing to $SU(N)\times SU(N)\times U(1)$ breaking self-couplings of the effective Higgs mesons.

5. Effective Higgs VEV's

In the TC theory, there are no explicit mass terms. The technicolor scale Λ_{TC} arises dynamically through dimensional transmutation [17]. We expect a similar behavior in the eTC theory. But despite the absence of explicit mass terms for the effective Higgs scalars, they develop a VEV of order Λ_{TC}. In the language of the effective theory, this happens because of their Yukawa couplings to the effective technifermion mass operators, which develop VEVs of order Λ_{TC}^3, like $\bar{\Psi}\Psi$ in the color SU(3) theory. The Yukawa couplings then induce a VEV for the effective Higgs scalars*.

We expect this mechanism to break the $SU(N)\times SU(N)\times U(1)$ global symmetry down to $SU(N)\times U(1)$. Thus there are N^2-1 Goldstone or pseudo-Goldstone bosons. In the simplest possible case in which $N=2$ and the LH (or RH) technifermions transform like a doublet under the weak SU(2), there are no pseudo-Goldstone bosons at all. All the surviving Higgs scalars (the true Higgs, two more neutral scalars and a charged scalar) have mass of order Λ_{TC}.

6. Conclusions

We have argued that a suitable technicolor model with a non-trivial UV fixed point can be reinterpreted in terms of an asymptotically free effective technicolor

* This is true even if, as discussed in the postscript, there is an effective Higgs scalar mass term of order Λ_{TC}, so long as the Yukawa coupling is non-zero. Of course, if there is a negative mass-squared term, the VEV develops classically. But the Yukawa couplings induce a non-zero VEV even if the mass-squared is positive.

model with light effective Higgs scalars whose couplings are related to the effective technicolor gauge coupling by an eigenvalue condition. Yukawa couplings of these effective Higgs mesons to quarks and leptons can be induced by interactions at the unification scale, M_G.

Thus the effective Higgs meson is both composite and fundamental. In the TC theory, it must be interpreted as a bound state of technifermion and antitechnifermion. But in the equivalent eTC theory, the effective Higgs scalar is a "fundamental" field, scaling canonically at short distances.

Clearly, much work remains to be done to validate and quantify the dynamical speculations made in this paper and to pin down the connection between the TC theory and the eTC theory. But the eTC language allows us to organize our ignorance into a few strong interaction parameters. This is enough to allow us to use these ideas in model building. In the next paper in this series, we apply these ideas to construct explicit, realistic, grand unified theories.

One of us (HG) acknowledges the support of the theory group at the University of Washington for a visit during which much of this work was done (despite their warm hospitality). We are grateful to S. Barr, S. Ellis and A. Zee at Washington, S. Dimopoulos, P. Ginsparg, S. Glashow and J. Preskill at Harvard, and to M. Peskin and H.D. Politzer for useful conversations.

Appendix

Here we give an example of an asymptotically free eTC theory with effective Higgs scalars. The eTC gauge group is $SU(r)$ and the eTC fermions are Dirac fermions with the LH parts and the charge conjugate of the RH parts transforming under $SU(N)_L \times SU(N)_R \times SU(r)$ as $(N, 1, r)$ and $(1, \bar{N}, \bar{r})$ respectively. This set of eTC fermions satisfies conditions 1–4 of sect. 2 if the TC theory has LH fermions transforming under $SU(N)_L \times SU(N)_R \times TC$ as

$$(N, 1, \mathrm{R}) + (1, \bar{N}, \bar{\mathrm{R}}),$$

with dim $\mathrm{R} = r$.

The eTC theory will also contain scalar meson eTC singlets which can couple to quarks and leptons. These are the effective Higgs mesons. In order for the self-couplings of the scalar mesons to be asymptotically free, they must have Yukawa couplings to the eTC fermions and hence must transform like $(N, \bar{N}, 1)$ under $SU(N)_L \times SU(N)_R \times SU(r)$. We will assume that the eTC theory contains no other scalars which are eTC singlets. However, it will be seen shortly that scalars which have eTC quantum numbers are required in certain circumstances. We assume that there is no coupling between eTC singlet and eTC non-singlet scalars.

The Yukawa coupling of the effective Higgs meson to the eTC fermion can be written as

$$f\bar{\Psi}_L \phi \Psi_R + \text{h.c.}\,,$$

where f is the Yukawa coupling constant, ϕ is an $N\times N$ matrix of complex scalars which carry $SU(N)_L$ and $SU(N)_R$ indices, and Ψ_L (Ψ_R) is a column vector with $SU(N)_L$ ($SU(N)_R$) and eTC indices. The effective Higgs mesons admit two quartic self-couplings, which we write as

$$-\tfrac{1}{2}\lambda_1 \operatorname{Tr}[(\phi^\dagger\phi)^2] - \tfrac{1}{2}\lambda_2[\operatorname{Tr}(\phi^\dagger\phi)]^2\,.$$

If $N=2$ or 4, there are additional determinental interactions which are possible, but we will set them to zero for simplicity.

Then, in the absence of scalars which are non-singlets under eTC, the one-loop β functions for the coupling constants are:

$$\beta_g = -\frac{1}{16\pi^2}[\tfrac{11}{3}r - \tfrac{2}{3}N]g^3 \equiv -\frac{1}{16\pi^2} b_0 g^3\,,$$

$$\beta_f = \frac{1}{16\pi^2}\left[(N+1)f^3 - \frac{6(r^2-1)}{2r} f g^2\right],$$

$$\beta_{\lambda_1} = \frac{1}{16\pi^2}[4N\lambda_1^2 + 12\lambda_1\lambda_2 + 4\lambda_1 f^2 - 4f^4]\,,$$

$$\beta_{\lambda_2} = \frac{1}{16\pi^2}[(2N^2+8)\lambda_2^2 + 8N\lambda_1\lambda_2 + 6\lambda_1^2 + 4\lambda_2 f^2]\,.$$

We require that the eTC coupling constants be asymptotically free. The gauge coupling is asymptotically free if $b_0 > 0$. As explained in sect. 3, the other couplings only approach zero at short distance if the ratios f/g, λ_1/g^2 and λ_2/g^2 have fixed values (i.e. f, λ_1 and λ_2 are slaved to g which is asymptotically free). Then

$$\begin{aligned} 0 &= \mu\frac{\partial}{\partial_\mu}\left(\frac{f}{g}\right) \\ &= \frac{g^2}{16\pi^2}\left(\frac{f}{g}\right)\left[(N+1)\left(\frac{f}{g}\right)^2 - \frac{3(r^2-1)}{r} + b_0\right], \end{aligned}$$

$$\begin{aligned} 0 &= \mu\frac{\partial}{\partial_\mu}\left(\frac{\lambda_1}{g^2}\right) \\ &= \frac{g^2}{16\pi^2}\left[4N\left(\frac{\lambda_1}{g^2}\right)^2 + 12\left(\frac{\lambda_1}{g^2}\right)\left(\frac{\lambda_2}{g^2}\right) + 4\left(\frac{f}{g}\right)^2\left(\frac{\lambda_1}{g^2}\right) - 4\left(\frac{f}{g}\right)^4 + 2b_0\left(\frac{\lambda_1}{g^2}\right)\right], \end{aligned}$$

$$\begin{aligned} 0 &= \mu\frac{\partial}{\partial_\mu}\left(\frac{\lambda_2}{g^2}\right) \\ &= \frac{g^2}{16\pi^2}\left[(2N^2+8)\left(\frac{\lambda_2}{g^2}\right)^2 + 8N\left(\frac{\lambda_1}{g^2}\right)\left(\frac{\lambda_2}{g^2}\right) + 6\left(\frac{\lambda_1}{g^2}\right)^2 + 4\left(\frac{f}{g}\right)^2\left(\frac{\lambda_2}{g^2}\right) + 2b_0\left(\frac{\lambda_2}{g^2}\right)\right]. \end{aligned}$$

The first equation shows that there exists a non-trivial fixed point for f/g only if $N>r+9/2r$. In this case, the fixed point is at

$$\left(\frac{f}{g}\right)^2=\frac{[-2/3r-3/r+2/3N]}{(N+1)},$$

to one-loop order. If $N\leqslant r+9/2r$ (this regime includes some possible phenomenologically interesting cases), the only fixed point is at $f/g=0$. Then if the Yukawa coupling is non-zero, it becomes large relative to g at short distances and causes β_f to become positive. This situation can be remedied by introducing scalars which are singlets under $SU(N)_L\times SU(N)_R$ (which do not couple to ϕ or ψ_L and ψ_R) but which transform non-trivially under the eTC gauge group. These change b_0 to

$$b_0=\left(\tfrac{11}{3}r-\tfrac{2}{3}N-\sum_{R}\tfrac{1}{6}T(R)\right),$$

where the sum is over all representations R of the eTC non-singlet scalars and $\delta_{ab}T(R)=\mathrm{Tr}\,(t_a(R)t_b(R))$ for the matrices $t_a(R)$ of the representation R. This leads to a modification of the β functions for g, f/g, λ_1/g^2 and λ^2/g^2. By suitable choice of the representation R, it is possible to produce a non-trivial fixed point for f/g for $N\leqslant r+9/2r$ and maintain the asymptotic freedom of g.

Now consider the eigenvalue conditions for λ_1 and λ_2. $\mu\,\partial/\partial\mu\,(\lambda_2/g^2)$ can vanish only if λ_1 and λ_2 have opposite signs, so we must check that the positivity constraints $\lambda_1+\lambda_2\geqslant 0$ and $\lambda_1+N\lambda_2\geqslant 0$ can be satisfied at non-trivial fixed points. We can rewrite the eigenvalue condition for λ_1/g^2 and λ_2/g^2 in terms of λ_1/f^2 and λ_2/f^2:

$$0=4N\left(\frac{\lambda_1}{f^2}\right)^2+12\left(\frac{\lambda_1}{f^2}\right)\left(\frac{\lambda_2}{f^2}\right)-4+K\left(\frac{\lambda_1}{f^2}\right),$$

$$0=(2N^2+8)\left(\frac{\lambda_2}{f^2}\right)^2+8N\left(\frac{\lambda_1}{f^2}\right)\left(\frac{\lambda_2}{f^2}\right)+6\left(\frac{\lambda_1}{f^2}\right)^2+K\left(\frac{\lambda_2}{f^2}\right)^2,$$

where $K=4+2b_0(g/f)^2$. We have verified numerically that for given N, there exist solutions in the positivity domain if K is greater than a number of order N (for example, if $N=2$, $K>7$; if $N=3$, $K>8$; if $N=10$, $K>14$). This is in accord with the observation that if $K\sim N$ and $K\gg 1$, the equations have the approximate solution $\lambda_1/f^2\approx 4/K$ and $\lambda_2/f^2\approx -96/K^3$, which is in the positivity domain if $K^2>24N$.

It remains to check that for given N and r, it is possible to choose K large enough to allow a solution in the positivity domain. If $N\leqslant r+9/2r$, this follows because g/f can be made arbitrarily large by a suitable choice of the representations R of the non-singlet eTC scalars. If $N > r + 9/2r$, then

$$b_0\left(\frac{g}{f}\right)^2 < (\tfrac{11}{3}r - \tfrac{2}{3}N)\frac{(N+1)}{(\tfrac{2}{3}N - \tfrac{2}{3}r - 3/r)},$$

which is sufficient to allow values of K in the domain of our solutions.

Postscript

In sect. 2, we argued that the asymptotic freedom of the eTC theory makes it plausible that the TC theory has a multiple zero at its UV fixed point. This in turn is crucial to the solution of the gauge hierarchy problem. If $\beta(G)$ had a simple zero at G^*, there would be a power law relation between $G(M) - G^*$ and Λ_{TC}/M (where M is the renormalization scale and $G(M)$ is the running coupling constant). Then the very small value of Λ_{TC}/M_G would imply a very small value of $G^* - G(M_G)$ and thus a fine tuning of the coupling constant. But if, as we believe, $\beta(G)$ has a multiple zero at G^*, then Λ_{TC}/M depends exponentially on $[G(M) - G^*]^{-1}$ so that a large hierarchy can occur without fine tuning.

In this postscript, we give a (probably biased) summary of a counter argument, which is our interpretation of a series of discussions with M. Peskin.

In the TC theory, the theory at the fixed point is scale invariant, but presumably the theory for $G > G^*$ is not. Instead, dimensional transmutation [17] takes place and the theory is characterized by the single dimensional parameter Λ_{TC}. Because scale invariance is broken in the TC theory, the mapping from the TC theory to the eTC theory may involve superrenormalizable couplings. In particular, one might expect that the effective Higgs mesons in the eTC theory may have a non-zero mass, because a mass term

$$\tfrac{1}{2}m^2\phi^\dagger\phi \tag{P.1}$$

is not forbidden by any global symmetry of the eTC theory*. Then the mass m must be of order Λ_{TC}, by dimensional analysis.

Let us then discuss the mapping from the TC theory to the eTC theory in terms of two dimensionless parameters for the eTC theory, a generic coupling g (representing all the couplings, presumably all related by an eigenvalue condition as discussed in the appendix so that all non-zero couplings are proportional to g or g^2) and a parameter

$$\mu = m/M, \tag{P.2}$$

where M is the renormalization point.

* Note, however, that it might be possible for the eTC theory to develop global supersymmetry, in which case a Higgs mass could be forbidden by symmetry. We thank S. Dimopoulos and P. Ginsparg for bringing this bizarre option to our attention.

The fixed point $G = G^*$ maps into $g = \mu = 0$. $G(M)$ for $M \gg \Lambda_{TC}$ must map into some trajectory $g(M)$, $\mu(M)$ near the origin. But for small g, g and μ obey simple renormalization group equations in the eTC theory,

$$M\frac{\partial}{\partial M}g \simeq -Bg^3\,, \qquad M\frac{\partial}{\partial M}\mu \simeq -\mu\,. \tag{P.3}$$

There are three types of non-trivial solutions to (P.3) for large M:

$$g(M) \simeq 1/\sqrt{2B\ln(M/\Lambda_{TC})}\,, \qquad \mu(M) = 0\,, \tag{P.4A}$$

$$g(M) \simeq 1/\sqrt{2B\ln(M/\Lambda_{TC})}\,, \qquad \mu(M) \simeq x\Lambda_{TC}/M\,, \tag{P.4B}$$

$$g(M) = 0\,, \qquad \mu(M) \simeq x\Lambda_{TC}/M\,. \tag{P.4C}$$

In (P.4A, B), we have used $g(M)$ to define Λ_{TC}. In (P.4B) the quantity x is the ratio of the scalar mass term to Λ_{TC} as defined by $g(M)$,

$$x = m/\Lambda_{TC} = O(1)\,. \tag{P.5}$$

In (P.4C), we leave the x in for convenience. It is not really a meaningful parameter because it can be absorbed into the definition of Λ_{TC}.

Type A is just what we discussed in sects. 2–4 and the appendix, when we ignored superrenormalizable interactions. If this mapping is picked out by the TC dynamics, we would expect $G(M) - G^*$ to be proportional to $g(M)$, in which case the zero at $G = G^*$ will be multiple. Type B is physically very similar to type A. The small $(O(\Lambda_{TC}))$ mass of the scalar should be irrelevant for $M \gg \Lambda_{TC}$, so the same argument should apply.

Type C is a genuinely new possibility. This describes a free theory with a non-zero scalar meson mass term. If this mapping is picked out by the TC dynamics, we would expect a simple zero of $\beta(G)$ at $G = G^*$ because we would expect $G(M) - G^*$ to be proportional to $\mu(M)$.

In fact, the type C mapping is unacceptable for other reasons. There is no way to induce a non-zero VEV for the effective Higgs meson in such a theory. The parameter m^2 cannot be negative (x in (P.4C) cannot be imaginary) because the theory would not be bounded below. And there is no eTC coupling to induce a non-zero VEV. Thus if the type C mapping is realized, this is not a theory with effective Higgs mesons, because the scalars have no VEV. We see no reason, however, to believe that C is a more reasonable mapping than A or B.

There is a definition of Λ_{TC} which allows us to treat cases A, B and C all on the same footing. Consider the trajectory

$$g(M) = 1\sqrt{2B\ln(M/\Lambda_{TC}\cos\phi)}\,, \qquad \mu(M) = \Lambda_{TC}\frac{\sin\phi}{M}\,, \tag{P.6}$$

for $0 \leq \phi \leq \frac{1}{2}\pi$. Here Λ_{TC} is defined as the root mean square of the dimensional parameters in $g(M)$ and $\mu(M)$,

$$\Lambda_{TC}^2 = M^2\{\mu(M)^2 + \exp[-1/Bg(M)^2]\}, \tag{P.7}$$

and $\sin\phi$ is the ratio of m to Λ_{TC},

$$\sin\phi = m/\Lambda_{TC}, \tag{P.8}$$

Now A corresponds to $\phi = 0$, C to $\phi = \frac{1}{2}\pi$ and B to everything else. The requirement that x is order 1 in (P.4B) translates into the statement that ϕ is not close to 0 $(x = 0)$ or $\frac{1}{2}\pi$ $(x = \infty)$. This is the physical requirement that the dimensional parameters in the eTC theory,

$$\begin{aligned} \sin\phi \cdot \Lambda_{TC} &= m, \\ \cos\phi \cdot \Lambda_{TC} &= M \exp[-1/2Bg(M)^2], \end{aligned} \tag{P.9}$$

are of the same order of magnitude.

References

[1] L. Susskind, Phys. Rev. D20 (1979) 2619;
E. Farhi and L. Susskind, Phys. Reports 74 (1981) 277
[2] S. Weinberg, Phys. Rev. D13 (1976) 974; D19 (1979) 1277
[3] H. Georgi, H.R. Quinn and S. Weinberg, Phys. Rev. Lett. 33 (1974) 451
[4] S. Dimopoulos and L. Susskind, Nucl. Phys. B155 (1979) 237
[5] E. Eichten and K. Lane, Phys. Lett. 90B (1980) 125
[6] E. Eichten, K. Lane and J. Preskill, Phys. Rev. Lett. 45 (1980) 225
[7] S. Dimopoulos and J. Ellis, Nucl. Phys. B182 (1981) 505
[8] B. Holdom, Phys. Rev. D23 (1981) 1637
[9] B. Holdom, Phys. Rev. D24 (1981) 1441
[10] H. Georgi and S.L. Glashow, Phys. Rev. Lett. 47 (1981) 1511
[11] W. Caswell, Phys. Rev. Lett. 33 (1974) 244
[12] A. Zee, Phys. Rev. D7 (1973) 3630;
S. Coleman and D.J. Gross, Phys. Rev. Lett. 31 (1973) 851
[13] G. 't Hooft, Lecture at Cargèse Summer Inst., 1979;
S. Coleman and B Grossman, in preparation.
[14] J. Preskill and S. Weinberg, Phys. Rev. D24 (1981) 1059
[15] T.P. Cheng, E. Eichten and L.F. Li, Phys. Rev. D9 (1974) 2259
[16] N.P. Chang, Phys. Rev. D10 (1974) 2706
[17] S. Coleman and E. Weinberg, Phys. Rev. D7 (1973) 1888

Nuclear Physics B185 (1981) 513–554

DYNAMICAL BREAKING OF SUPERSYMMETRY*

Edward WITTEN

Joseph Henry Laboratories, Princeton University, Princeton, New Jersey 08544, USA

Received 29 April 1981

General conditions for dynamical supersymmetry breaking are discussed. Very small effects that would usually be ignored, such as instantons of a grand unified theory, might break supersymmetry at a low energy scale. Examples are given (in 0 + 1 and 2 + 1 dimensions) in which dynamical supersymmetry breaking occurs. Difficulties that confront such a program in four dimensions are described.

1. Introduction

Supersymmetry has fascinated particle physicists since it was first discovered [1]. It is an outstanding example of a known mathematical structure which may plausibly be absorbed in the future into our understanding of particle physics.

Of course, if nature really is described by a supersymmetric theory, the symmetry must be spontaneously broken [2]. At what energies does the symmetry breaking occur? It might very well occur at energies of order the Planck mass. In that case supersymmetry would be relevant to particle physics at "ordinary" energies only indirectly, in as much as the broken supersymmetry might make predictions concerning particle quantum numbers and relations among masses and coupling constants.

On the other hand, it is possible that supersymmetry breaking occurs at "ordinary" energies like a few hundred GeV or a few TeV. In this case, ordinary particle physics, at energies much less than the Planck mass, is presumably described by a renormalizable, globally supersymmetric model. There has been some success [3] in constructing realistic models of this sort for ordinary particle interactions.

If supersymmetry breaking does occur at ordinary particle physics energies, we must ask why the energy scale of supersymmetry breaking is so tiny compared to the natural energy scale of gravity and supergravity [4], which is presumably the Planck mass. This is a variant of the "hierarchy problem" [5]: why is the mass scale of ordinary particle physics so much less than the mass scale of grand unification or gravitation?

* Supported in part by NSF Grant PHY80-19754.

Presumably, if supersymmetry is spontaneously broken at the tree level, the breaking will have a strength of the same order as the natural mass scale of the theory. For supersymmetry to be broken only at, say, 10^3 GeV, which is 10^{-16} times the Planck mass of 10^{19} GeV, we require a theory in which supersymmetry is unbroken at the tree level and is broken only by extremely small corrections. These quantum corrections are presumably non-perturbative. We are looking for a theory in which supersymmetry is "dynamically broken" by non-perturbative effects.

If dynamical breaking of supersymmetry can occur, this could not only explain how supersymmetry could survive down to low energies and then be spontaneously broken. It might also resolve the usual hierarchy problem of understanding why the W and Z mesons are so light compared to the mass scale of possible grand unification and to the Planck mass. Once one can understand the existence of a "'low" mass scale of supersymmetry breaking, of order perhaps 10^3 GeV, it is perfectly possible that $SU(2) \times U(1)$ breaking could be part of this low energy symmetry breaking. In fact, one of the rather few phenomenological motivations for supersymmetry is precisely this $SU(2) \times U(1)$ hierarchy problem. For $SU(2) \times U(1)$ breaking to occur at a low energy scale, we need the usual Higgs doublet to be massless on the scale of grand unification or the Planck mass. According to our best understanding, masslessness of elementary charged scalars is not natural, except in supersymmetric theories. In supersymmetric theories (with supersymmetry not spontaneously broken) massless scalars occur naturally because scalars that are in the same supermultiplet with massless fermions or vector mesons must be massless. We may therefore imagine that the $SU(2) \times U(1)$ Higgs doublet is kept massless down to low energies by unbroken supersymmetry. Once supersymmetry is spontaneously broken the Higgs doublet need no longer be massless. The same non-perturbative effects that trigger supersymmetry breaking could therefore give the Higgs doublet a mass squared, which if it is negative will lead to the spontaneous breaking of $SU(2) \times U(1)$.

This scenario is made slightly more plausible by the fact that the Higgs doublet $\begin{pmatrix} \phi^0 \\ \phi^- \end{pmatrix}$ has the same $SU(3) \times SU(2) \times U(1)$ quantum numbers as the lepton doublet $\begin{pmatrix} \nu \\ e^- \end{pmatrix}_L$. This suggests that they may be supersymmetry partners. The fact that the leptons get their masses only from $SU(2) \times U(1)$ breaking is, of course, a consequence of gauge invariance provided that, as suggested by experiment, right-handed leptons transform differently under $SU(2) \times U(1)$ from the way the left-handed leptons transform. Therefore, if the Higgs doublet is related by supersymmetry to the leptons, it must be massless as long as supersymmetry and $SU(2) \times U(1)$ are unbroken, and its expectation value cannot be larger than the mass scale of supersymmetry breaking.

(I have tacitly used the fact that in global supersymmetry, which we are presuming to be what is relevant at low energies, a boson and its fermionic partner always have the same quantum numbers under gauge transformations. This is so because in global supersymmetry the supercharges always commute with gauge symmetries; the commutator, if not zero, would be a supersymmetry transformation depending on the infinite number of parameters of the gauge symmetry, so would have to be a local supersymmetry.)

In this paper I will discuss in a general way what would be involved in dynamical breaking of supersymmetry. I wish to warn the reader in advance that no essential problems are solved in this paper. I do not have a realistic model or even a realistic mechanism by which dynamical supersymmetry breaking can occur in four dimensions, and the discussion will be very qualitative. I hope, however, to raise some relevant issues. The plan of this paper is as follows. In sect. 2 some well-known facts are reviewed. In sect. 3 it is argued that unbroken supersymmetry is potentially unstable. In sect. 4 some grand unified models are described which could be realistic if dynamical supersymmetry breaking occurs. In sect. 5 the "breakdown" of naturalness in supersymmetric theories is discussed. In sect. 6 two models are presented (in less than four dimensions) in which dynamical supersymmetry breaking occurs. In sect. 7 the grand unified models of sect. 4 are re-examined, in the light of lessons from the discussions of naturalness and of soluble models. In sect. 8 it is suggested that gravity could play a role in dynamical supersymmetry breaking. In sect. 9 it is suggested that supersymmetry could cure the problems that plague theories of dynamical SU(2)×U(1) symmetry breaking. Some conclusions are drawn in sect. 10.

2. Supersymmetry and the vacuum energy

In this section some long-established aspects of spontaneously broken supersymmetry will be reviewed.

One of the central features of globally supersymmetric theories is that the hamiltonian H is the sum of the squares of the supersymmetry charges. In the simplest supersymmetry algebra there is a single Majorana spinor of four hermitian supersymmetry charges Q_α, $\alpha = 1\ldots4$. In terms of these we may write

$$H = \sum_{\alpha=1}^{4} Q_\alpha^2, \tag{1}$$

provided that a possible additive constant in H is chosen properly.

Since H is the sum of squares of hermitian operators, the energy of any state is positive or zero [6]. A state can have zero energy only if it is annihilated by each of the Q, since if $H|0\rangle = 0$, then $0 = \langle 0|H|0\rangle = \Sigma_\alpha \langle 0|Q_\alpha^2|0\rangle = \Sigma_\alpha |Q_\alpha|0\rangle|^2$, which is possible only if $Q_\alpha|0\rangle = 0$. If, conversely, a state is annihilated by the Q_α, then its energy is zero because $Q_\alpha|0\rangle = 0$ implies $H|0\rangle = \Sigma_\alpha Q_\alpha^2|0\rangle = 0$.

If there exists a supersymmetrically invariant state – that is, a state annihilated by the Q_α – then it is automatically the true vacuum state, since it has zero energy and any state that is not invariant under supersymmetry has positive energy. Thus, if a supersymmetric state exists, it is the ground state and supersymmetry is not spontaneously broken. Only if there does not exist a state invariant under supersymmetry is supersymmetry spontaneously broken. In this case the ground-state energy is positive.

In this respect supersymmetry is quite different from ordinary symmetries. With an ordinary internal symmetry, a symmetric state may exist without being the ground state. The situation has been illustrated [7] by fig. 1. In fig. 1a is shown a scalar potential which describes a theory with two ground states. In each ground state the scalar field has a vacuum expectation value, possibly breaking some internal symmetry. However, supersymmetry is not spontaneously broken, because the ground-state energy, the minimum value of the potential, is zero for each of the two possible states.

In fig. 1b, on the other hand, there is a unique ground state, the scalar field does not have an expectation value, and no internal symmetry is spontaneously broken. However, supersymmetry is spontaneously broken because the minimum value of the potential is positive. A state with $V=0$ does not exist.

Notice that despite the fact that supersymmetry is a Fermi-Bose symmetry, a non-zero vacuum expectation value of a scalar field does not necessarily mean that supersymmetry is spontaneously broken. Supersymmetry is spontaneously broken if the anticommutator of the supercharge Q_α with some operator X is non-zero:

$$\langle 0|\{Q_\alpha, X\}|0\rangle = \langle 0|(Q_\alpha X + XQ_\alpha)|0\rangle \neq 0, \tag{2}$$

since if $Q_\alpha|0\rangle = 0$ the expression in (2) would evidently have to vanish. However, in supersymmetric theories the elementary scalars ϕ can never be written as $\{Q_\alpha, X\}$, so they can obtain vacuum expectation values without necessarily breaking supersymmetry. (It is possible to write the derivatives $\partial_\mu\phi$ of the elementary scalars in the form $\{Q, \psi\}$, where the ψ are elementary fermions. However, whether supersymmetry is spontaneously broken or not, Lorentz invariance prevents $\partial_\mu\phi$ from having a vacuum expectation value.)

The fact that a positive vacuum energy indicates supersymmetry breaking is a special case of the fact that supersymmetry is spontaneously broken if any $\{Q, X\}$

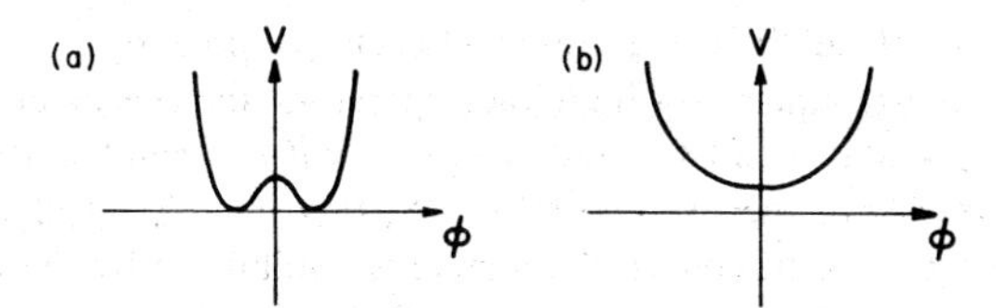

Fig. 1. A classical illustration of the differences between supersymmetry and global symmetries. In (a), the expectation value of the scalar field breaks an internal symmetry, but does not break supersymmetry, because the vacuum energy is zero. In (b), supersymmetry is spontaneously broken.

has a vacuum expectation value. In fact, the vacuum energy density E is defined in terms of the expectation value of the energy-momentum tensor $T_{\mu\nu}$:

$$\langle 0|T_{\mu\nu}|0\rangle = Eg_{\mu\nu}. \tag{3}$$

But in a supersymmetric theory $T_{\mu\nu} = (\gamma_\mu)_{\alpha\beta}\{Q_\alpha, S_{\nu\beta}\}$, where Q_α is the supercharge and $S_{\nu\beta}$ the supersymmetry current, so

$$Eg_{\mu\nu} = (\gamma_\mu)_{\alpha\beta}\langle 0|\{Q_\alpha, S_{\nu\beta}\}|0\rangle, \tag{4}$$

and a non-zero value for this expression means that supersymmetry is spontaneously broken.

The same sort of reasoning leads to a simple proof [8] of the analogue of Goldstone's theorem for supersymmetry. In the standard fashion of current algebra, one writes

$$\langle 0|\{Q_\alpha, S_{\nu\beta}\}|0\rangle = \int d^4x \frac{\partial}{\partial x^\sigma}\langle 0|T(S_{\sigma\alpha}(x)S_{\nu\beta}(0))|0\rangle, \tag{5}$$

where the equality depends upon the fact that $\partial_\sigma S^\sigma = 0$ and upon the definition that $Q_\alpha = \int d^3x\, S_{0\alpha}$. In view of the anticommutation relation (4), the left-hand side of (5) is equal to $(\gamma_\nu)_{\alpha\beta}$ times the vacuum energy E. If (and only if) supersymmetry is spontaneously broken, E is non-zero and the right-hand side of (5) must be non-zero. Being the integral of a total divergence, the right-hand side of (5) can be non-zero only if there is a surface contribution. This means that a massless particle must be present.

A surface contribution can appear in (5) only if the two-point function $\langle 0|T(S(x)S(0))|0\rangle$ vanishes for large $|x|$ only as $1/|x|^3$. The only intermediate state whose contribution would fall off so slowly would be a one-particle state containing a single massless fermion of spin one half. The massless fermion whose existence is so established is known as the Goldstone fermion.

Let us define the coupling f of the supercurrent to the Goldstone fermion by

$$\langle 0|S_{\mu\alpha}|\psi_\beta\rangle = f(\gamma_\mu)_{\alpha\beta}, \tag{6}$$

where $|\psi_\beta\rangle$ is a one fermion state with spin state $|\beta\rangle$. Then from eqs. (4) and (5) and the fact that only the one-fermion state contributes to the right-hand side of (5) follows a simple and fundamental formula,

$$E = f^2, \tag{7}$$

relating the vacuum energy density E to the coupling f.

For the sake of clarity, it should be noted that the supercurrent may create a massless fermion from the vacuum by means of a derivative coupling, $\langle 0|S_{\mu\alpha}|\psi_\beta\rangle =$

$gp_\mu\delta_{\alpha\beta}$, without supersymmetry being broken. Such a derivative coupling gives a matrix element that vanishes too rapidly to give a surface contribution in (5). Supersymmetry breaking is related specifically to the existence of the non-derivative coupling of eq. (6). In this paper, the phrase "the current creates a massless fermion from the vacuum" always refers specifically to a non-derivative coupling, as in eq. (6).

3. Supersymmetry and internal symmetries

The fact reviewed in the last section, that supersymmetry is spontaneously broken if and only if the vacuum energy is non-zero, leads to some basic differences between spontaneous breaking of supersymmetry and spontaneous breaking of ordinary (local or global) internal symmetries.

Let us ask under what conditions quantum corrections can change the pattern of symmetry breaking that one finds at the tree level. We will confine our attention primarily to theories in which the quantum corrections are weak – characterized by small, well-defined coupling constants. Can weak quantum effects change the pattern of symmetry breaking that one observes at the tree level?

A simple argument [9] shows that this ordinarily cannot happen in the case of internal symmetries. If an internal symmetry is unbroken at the tree level, this means (fig. 2a) that the classical potential has its minimum at a symmetrical point. Broken symmetry means (fig. 2b) that the classical potential has its minimum away from the origin. It is not possible by means of arbitrarily small corrections to turn a potential of type 2a into one of type 2b, or vice versa. Therefore, sufficiently weak quantum corrections will not break a symmetry that is unbroken at the tree level, nor will they restore a broken symmetry.

Although there is much truth in the above reasoning, some possible exceptions should be noted. An exception can arise if the tree potential (fig. 3) has a degeneracy between states of broken and unbroken symmetry. In this case the quantum corrections, no matter how small, are crucial in lifting the degeneracy and determining what is the true ground state. However, except by artificially adjusting the parameters, there is no known way to obtain this sort of degeneracy at the tree level. As shown by Coleman and E. Weinberg [10], an exception can also arise if the

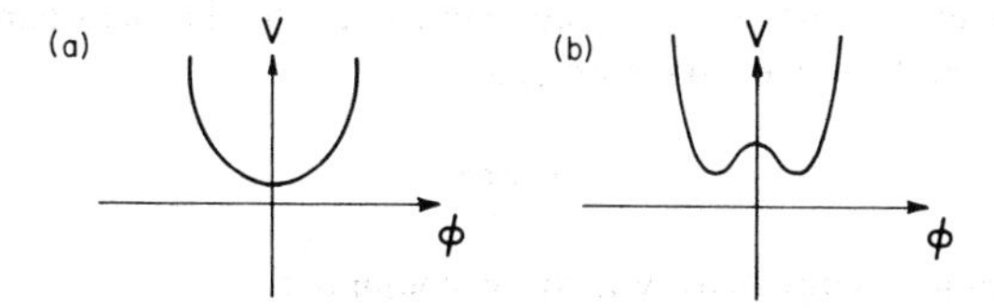

Fig. 2. In the case of internal symmetries, an arbitrarily small change in the parameters cannot produce broken symmetry, as in (b), from unbroken symmetry, as in (a).

curvature of the potential at the origin in field space is constrained to vanish. This vanishing curvature corresponds to an approximate degeneracy, since the energy is almost independent of the field. However, no known mechanism leads to vanishing curvature of the potential in a natural way. An exception might also arise if the quantum corrections to the effective potential were sufficiently singular near the origin in field space. This can occur in two dimensions in theories with massless fermions because of $\phi^2 \ln\phi$ terms in the effective potential, but does not seem to occur in four dimensions. Finally, one might wonder about theories like QCD. In QCD with massless quarks, chiral symmetry is unbroken at the tree level but is believed to be broken by quantum corrections. In this theory, because of the masslessness of the quarks, there are at the tree level states of non-zero chirality with arbitrarily low (although not zero) energy. It is this approximate degeneracy which makes it possible for quantum corrections to spontaneously break the symmetry, no matter how small the coupling is initially.

In the absence of the sort of degeneracy or approximate degeneracy just discussed, small quantum corrections (characterized by sufficiently small coupling constants) cannot change the pattern of symmetry breaking, because they cannot change a potential of type 2a into one of type 2b, or vice versa. Symmetries that are unbroken at the tree level are really unbroken.

In supersymmetry, these issues must be reconsidered, because the criteria for supersymmetry breaking are rather different. Let us refer back to fig. 1. In fig. 1a a scalar field has a vacuum expectation value, possibly breaking some internal symmetry. However, supersymmetry is *not* spontaneously broken, because the vacuum energy, the value of the potential at its minimum, is zero. In fig. 1b, on the other hand, the scalar field has zero vacuum expectation value and internal symmetries are not spontaneously broken. But supersymmetry is spontaneously broken, because the ground-state energy, the minimum of the potential, is positive.

It follows from this that if supersymmetry is broken in the tree approximation, then it really is broken in the exact theory, at least if the coupling is weak enough. Arbitrarily weak corrections cannot shift the minimum of the potential from the non-zero value of fig. 1b to zero. This could occur, if at all, only if the coupling constant exceeds some critical value. In this respect supersymmetry resembles ordinary internal symmetries.

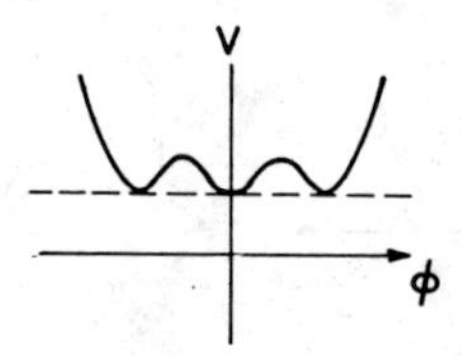

Fig. 3. If the potential possesses an accidental degeneracy at the tree level, quantum corrections will determine what is the ground state.

If we assume instead that supersymmetry is *not* broken at the tree level, the situation is very different, and arbitrarily weak quantum corrections could conceivably induce supersymmetry breaking. If in some approximation the minimum of the potential is zero, an arbitrarily small effect, shifting the potential by a tiny amount, could shift the minimum to a small but non-zero value (fig. 4). Then supersymmetry is spontaneously broken. It is potentially very delicate to claim that, in a given theory, supersymmetry is not spontaneously broken. An approximate calculation including many effects and showing that the vacuum energy is zero in a certain approximation always leaves open the possibility that even smaller effects that have been neglected could raise the ground-state energy slightly above zero. Unbroken symmetry could be unstable.

The above statement actually requires an important qualification. We know that if supersymmetry is spontaneously broken, there must exist a massless fermion, the Goldstone fermion. Weak quantum corrections will not bring into being a massless fermion if one does not already exist. If all fermions have non-zero mass at the tree level, weak corrections will not shift any of the fermion masses to zero. Consequently, in any theory in which supersymmetry is not broken at the tree level and in which all fermions have non-zero masses at the tree level, the supersymmetry must be truly unbroken, at least for weak enough coupling.

However, there are many reasons that a fermion might be massless other than its being a Goldstone fermion. Fermions might be massless because of unbroken chiral symmetries. As long as supersymmetry is unbroken, fermions may be massless because they are related by supersymmetry to massless gauge mesons or to massless Goldstone bosons.

The mere fact that a massless fermion exists does not mean *ipso facto* that supersymmetry is spontaneously broken. A Goldstone fermion is not simply a massless fermion. It is specifically a massless fermion that can be created from the vacuum by the supersymmetry current,

$$\langle 0|S_{\mu\alpha}|\psi_\beta\rangle = f(\gamma_\mu)_{\alpha\beta}, \tag{8}$$

with some non-zero f.

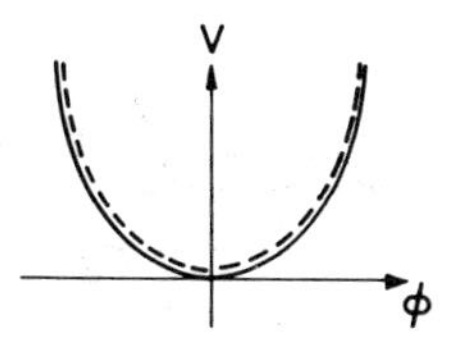

Fig. 4. In the case of supersymmetry, an arbitrarily small distortion of a potential with unbroken symmetry (solid line) can give a positive vacuum energy and therefore broken symmetry (dashed line).

Let us consider theories in which, at the tree level, supersymmetry is unbroken and there exist massless fermions which are not Goldstone fermions; that is, $f=0$ for each massless fermion. Theories of this sort would seem to be potentially unstable against supersymmetry breaking. Could not tiny quantum corrections give a non-zero value of f and thus make the massless fermion that already exists at the tree level into a Goldstone fermion? At the same time, these tiny quantum corrections would shift the minimum of the potential to a small non-zero level, which would be $E=f^2$ in view of the comments at the end of sect. 2.

For instance, even if f vanishes at the tree level, could not one-loop effects give a non-zero f of order α (and hence E of order α^2)? Or if f vanishes at the one-loop order, could not f arise as a two-loop effect, of order α^2?

The question is particularly interesting because any attempt at a realistic supersymmetric theory of nature would have, at the tree level, if supersymmetry is unbroken, massless fermions that could conceivably play the role just described. For instance, the supersymmetric partner of the photon would be a massless neutral fermion which might become a Goldstone fermion. If $SU(2)\times U(1)$ is unbroken at the tree level – as it must be if we are to use supersymmetry to solve the hierarchy problem – the supersymmetric partner of the Z meson is also massless and is another candidate Goldstone fermion. Realistic models might contain still more candidates.

A candidate Goldstone fermion must, of course, have the same quantum numbers as the supersymmetry current under all unbroken symmetries; otherwise the matrix element in eq. (8) would have to vanish. In particular, the would-be Goldstone fermion must be neutral under any gauge symmetries which are to remain unbroken. (Recall that in global supersymmetry, which we assume to be the relevant "low energy" approximation, the supersymmetry current is neutral under all gauge symmetries.) However, the candidate Goldstone fermions suggested above, the supersymmetry partners of the photon and Z meson, satisfy this condition.

The fact that unbroken supersymmetry could be unstable against weak corrections only if a massless neutral fermion exists at the tree level is analogous to the fact that unbroken internal symmetry is unstable only if there are degeneracies or approximate degeneracies at the tree level. But in practice there is a crucial difference. The difference is that massless neutral fermions inevitably exist in supersymmetric models of realistic interest.

Now we return to the previous question. If the coupling f of the supersymmetry current to a massless fermion vanishes at the tree level, can it receive a non-zero contribution from loop corrections?

The answer to this question is a quite remarkable surprise. By a detailed study of Feynman diagrams (done most conveniently in terms of superspace diagrams) it is possible to show that the answer is no. If f vanishes at the tree level, then f remains zero to all finite orders of perturbation theory. In the literature this has been stated in terms of the effective potential; it is stated that if the effective potential vanishes at some point in field space, then it vanishes at that point to all finite orders [11].

Since the effective potential at its minimum is the square of f, the two statements are equivalent.

The fact that f remains zero to all finite orders if it vanishes at the tree level is a special case of a more general phenomenon. In supersymmetric theories, the usual concept of naturalness does not apply, as long as supersymmetry is not spontaneously broken. The known facts in this area, and some corollaries, will be discussed in sect. 5.

No field theoretic reason is known for the fact that f remains zero to all finite orders. The existing proofs are based on details of perturbation theory (and are most tractable by use of the beautiful and efficient method of superspace perturbation theory [12]). Since the known proofs are based on details of perturbation theory, the result is not necessarily valid at the non-perturbative level.

If f became non-zero at, say, the one- or two-loop level, we would obtain a gauge hierarchy of some sort. The ratios of the squared masses of gauge bosons would be proportional to f and would be of order α or α^2. Such a hierarchy would not nearly be great enough to account for the observed ratio of energy scales in physics.

It is therefore very exciting that f is known to vanish to all finite orders, on the basis of arguments that do not necessarily apply non-perturbatively. If non-perturbative effects can give a non-zero f, we may obtain the desired enormous hierarchy. In this paper, an attempt will be made to explore this possibility.

4. "Realistic" grand unified models

Before discussing the status of naturalness in supersymmetric theories, and some possible mechanisms for dynamical breakdown of supersymmetry, let us first discuss some issues that arise in constructing realistic models of supersymmetry and grand unification.

To make a realistic model, we must first choose the symmetry algebra. So far we have discussed the simplest algebra, with a single spinor supercharge Q_α. More generally, there may be several spinor supercharges $Q_{\alpha i}$, $i = 1 \ldots N$. In this case the hamiltonian is obtained by summing the Q^2 over α, for fixed i. Specifically,

$$H = \sum_\alpha Q_{\alpha i}^2 = \sum_\alpha Q_{\alpha j}^2, \tag{9}$$

where one sums over α, but i or j is fixed.

In global supersymmetry (presumably what is relevant at low energies), realistic models are possible only for $N = 1$. This follows from one of the basic observations in particle physics: the massless fermions of helicity $\frac{1}{2}$ do not transform under $SU(3) \times SU(2) \times U(1)$ the same way the helicity $-\frac{1}{2}$ fermions transform. (Equivalently, the massless fermions of given helicity transform in a "complex" representation of the gauge group.) It is easy to see that in global supersymmetry with $N > 1$, the helicity $\frac{1}{2}$ and helicity $-\frac{1}{2}$ fermions necessarily transform identically. For instance, for $N=2$, the supersymmetry representations with massless particles of helicity $\frac{1}{2}$ contain the three helicities $(\frac{1}{2},0,-\frac{1}{2})$ or $(1,\frac{1}{2},0)$. Moreover, all particles in a given multiplet of global supersymmetry transform the same way under $SU(3)\times SU(2)\times U(1)$ (recall that in global supersymmetry, the supersymmetry charges commute with the group generators). The $(\frac{1}{2},0,-\frac{1}{2})$ multiplet relates fermions of helicity $\frac{1}{2}$ and $-\frac{1}{2}$, which would have to have the same quantum numbers. The $(1,\frac{1}{2},0)$ multiplet relates fermions of helicity $\frac{1}{2}$ to massless bosons of helicity one. But massless bosons of helicity one are always gauge bosons, transforming in the adjoint representation, which is real. (And there always are helicity -1 bosons transforming the same way; their partners would have helicity $-\frac{1}{2}$.) So whether we consider the $(1,\frac{1}{2},0)$ or the $(\frac{1}{2},0,-\frac{1}{2})$ multiplet, the fermions in $N=2$ (or $N>2$) global supersymmetry transform in a real representation of the gauge group; helicity $\frac{1}{2}$ and helicity $-\frac{1}{2}$ fermions transform equivalently.

To form a realistic model, we therefore should assume $N=1$.

Is it possible that N is greater than one microscopically, and that a "realistic" $N=1$ theory would be only a low energy approximation? Evidently, in this case, the fermions of the effective $N=1$ theory, transforming in a complex representation of the gauge group, must be absent in the microscopic lagrangian (they do not form representations of the $N>1$ algebra; and they cannot be put into such representations by adding additional particles that could receive mass at energies above $SU(2)\times U(1)$ breaking). If N is really greater than one in nature, one must suppose that the observed fermions have been generated dynamically at energies at which only the $N=1$ algebra (or no supersymmetry at all) is relevant.

Actually, in global supersymmetry, it is impossible for supersymmetry with $N>1$ to be spontaneously broken down to supersymmetry with $N=1$. This follows from eq. (9). If there is an unbroken supersymmetry, say $Q_{\alpha 1}|0\rangle=0$, then $H|0\rangle=\Sigma_\alpha Q_{\alpha 1}^2|0\rangle=0$. From this it follows that all of the supersymmetries are unbroken, because, for any i, $\Sigma_\alpha Q_{\alpha i}^2|0\rangle=H|0\rangle=0$.

In supergravity, it is possible to spontaneously break some supersymmetries without breaking all of them. Examples were first given by Scherk and Schwarz [13]. The possibility of doing this is related to the delicacy in defining global space-time transformations in general relativity. For some additional discussion, including some efforts to generate fermions in a complex representation of $SU(3)\times SU(2)\times U(1)$ after spontaneous breaking of some supersymmetries, see ref. [14].

The net conclusion is that supersymmetry with $N>1$ may be relevant to physics, but only at energies at which gravitation is important. $N=1$ supersymmetry may be relevant at ordinary energies.

Let us now turn to a consideration of some semi-realistic models. We must first recall some facts about construction of models with $N=1$ supersymmetry.

We may in general have an arbitrary gauge group, with gauge mesons A_μ^a and fermionic partners λ^a. In addition, we may introduce left-handed fermions ψ_L^i in an arbitrary multiplet of the gauge group. They form supersymmetry multiplets $\begin{pmatrix} \psi_L^i \\ \phi^i \end{pmatrix}$ with complex scalar bosons ϕ^i. The right-handed fermion fields are the complex conjugates of the left-handed fermion fields, $\psi_{jR} = (\psi_L^j)^*$ and their supersymmetry partners are the complex conjugates of ϕ_j^* of the ϕ^i.

The scalar potential of this theory is a sum of two terms. One term is derived from the gauge couplings. Let us group all of the scalar fields in a vector ϕ. Let T^a be the generators of the gauge group acting on the (possibly reducible) scalar representation. Then the term in the scalar potential coming from the gauge couplings is

$$V_1(\phi^i, \phi_j^*) = \sum_a \left(e_a(\phi^*, T^a\phi)\right)^2, \tag{10}$$

where the sum runs over all generators, and e_a is the coupling constant associated with the generator T^a.

(In one case this formula must be generalized. If the gauge group is not semisimple, but contains a U(1) generator, say Y, with charge e, then it is possible [15] to define a supersymmetric theory in which the contribution $e^2(\phi^*, Y\phi)^2$ in (10) is generalized to be $e^2\,((\phi^*, Y\phi) + \mu^2)^2$, where μ^2 is an arbitrary constant. This "D term" has been used in constructing realistic models [3].)

The second term in the scalar potential is related by supersymmetry not to the gauge couplings but to the Yukawa couplings. One begins by introducing a new function, sometimes called the superspace potential. The superspace potential W is a function of the ϕ^i but not of their complex conjugates the ϕ_j^*. For a renormalizable theory W should be at most cubic in the ϕ^i; otherwise W is restricted only by gauge invariance. The general form of W is

$$W(\phi^i) = a_i\phi^i + a_{ij}\phi^i\phi^j + a_{ijk}\phi^i\phi^j\phi^k,$$

where a_i, a_{ij}, and a_{ijk} are gauge-invariant tensors.

In terms of W, the Yukawa couplings of ϕ and ψ are defined by

$$L^{\text{Yuk}} = (\partial^2 W/\partial\phi^i\partial\phi^j)\psi_L^i\psi_L^j.$$

In addition, there is a new contribution to the scalar potential,

$$V_2 = \sum_i \left|\frac{\partial W}{\partial\phi_i}\right|^2. \tag{11}$$

The total scalar potential is

$$V(\phi, \phi^*) = V_1 + V_2, \tag{12}$$

with V_1 and V_2 given in (10) and (11). Under what conditions is supersymmetry unbroken? It is required that V vanishes when evaluated at the expectation value of the scalar fields. Since V_1 and V_2 are each non-negative, they must both be zero. The condition for supersymmetry to be unbroken at the tree level is that for each a and each i, we require

$$(\phi^*, T^a\phi) = 0, \qquad \partial W/\partial\phi^i = 0. \tag{13}$$

If these equations have a simultaneous solution, supersymmetry is unbroken at the tree level. Otherwise, supersymmetry is spontaneously broken.

Now let us consider some more or less realistic examples of grand unified supersymmetric models. For simplicity we will take the gauge group to be SU(5). We wish a scalar potential which will at the tree level spontaneously break SU(5) down to SU(3) × SU(2) × U(1). If we wish to solve the hierarchy problem via dynamically broken supersymmetry, then supersymmetry should be unbroken at the tree level. However, models in which supersymmetry is broken at the tree level will also be considered below.

It is very easy to make a model in which, at the tree level, SU(5) is spontaneously broken to SU(3) × SU(2) × U(1) but supersymmetry is unbroken. This can be done rather simply if the fields ϕ^i discussed above consist of a single traceless complex matrix A^i_j transforming in the adjoint representation of SU(5). (The hermitian and antihermitian parts of A transform separately under SU(5), but must both be present because of supersymmetry.)

In this case the most general choice for W is

$$W = \tfrac{1}{2}M\,\mathrm{Tr}\,A^2 + \tfrac{1}{3}\lambda\,\mathrm{Tr}\,A^3, \tag{14}$$

where M and λ are constants, which can be taken to be real by redefining the overall phases of A and of W. [The phase of W always cancels out in expressions such as (11) for the physical scalar potential.] The first equation in (13) says that the hermitian and antihermitian parts of A commute and so can be diagonalized simultaneously by an SU(5) transformation.

The second equation in (13) gives

$$\lambda\left(A^2 - \tfrac{1}{5}\mathrm{Tr}\,A^2\right) + MA = 0. \tag{15}$$

There are three solutions, up to a gauge transformation:

(i) $$A = 0,$$

(ii) $$A = \frac{M}{3\lambda}\begin{pmatrix} 1 & & & & \\ & 1 & & & \\ & & 1 & & \\ & & & 1 & \\ & & & & -4 \end{pmatrix}, \tag{16}$$

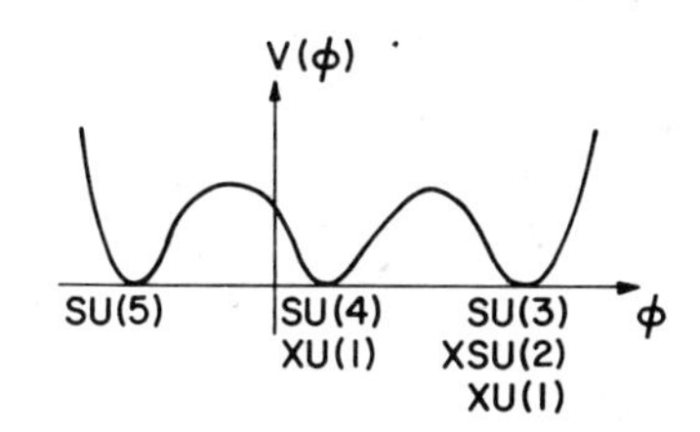

Fig. 5. A schematic illustration of a potential that does not determine the unbroken symmetry group.

(iii)
$$A = \frac{2M}{\lambda}\begin{pmatrix} 1 & & & & \\ & 1 & & & \\ & & 1 & & \\ & & & -\frac{3}{2} & \\ & & & & -\frac{3}{2} \end{pmatrix}.$$

The unbroken gauge symmetry is SU(5), SU(4) × U(1), or SU(3) × SU(2) × U(1), depending on which solution one considers.

Since these three states solve the equations (13) of unbroken supersymmetry, they all have zero energy at the tree level and so are completely degenerate. This is schematically depicted in fig. 5; a Higgs potential is sketched which has three precisely degenerate minima, each at zero energy. To finite orders in perturbation theory, as discussed in the last section, the degeneracy will not be lifted. The three states each remain at zero energy.

It is possible that non-perturbative effects do not change this picture at all. Perhaps supersymmetry remains unbroken – when all non-perturbative effects are included – in each of the three states. In that case this theory has three equally valid vacuum states, with different unbroken gauge symmetries and different spectra of the elementary particles.

It is possible that non-perturbative effects leave supersymmetry unbroken in one or two of the ground states but not in the others. If (fig. 6a) there is precisely one state in which non-perturbative effects leave supersymmetry unbroken, then this is uniquely selected as the true ground state of lowest energy.

On the other hand, it may be (fig. 6b) that non-perturbative effects lift the ground-state energy to tiny but positive values in each of the three vacuum states, so that supersymmetry is spontaneously broken for any choice of the ground state. The true ground state will then be the one in which supersymmetry is broken most weakly and the energy is smallest. If this is the state in which the unbroken gauge symmetry at the tree level was SU(3) × SU(2) × U(1), and if the non-perturbative effects that break supersymmetry also break SU(2) × U(1), one may obtain a good description of nature with a large gauge hierarchy.

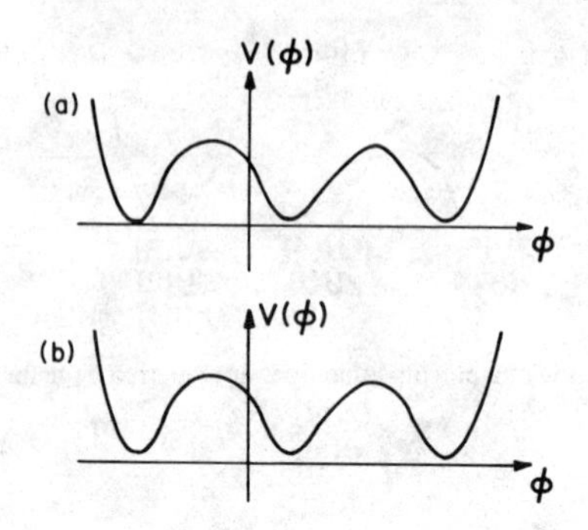

Fig. 6. Small quantum corrections to the potential of the previous drawing have determined the true ground state. If dynamical supersymmetry breaking occurs in each minimum but one, that one is the true vacuum (a). If supersymmetry is broken in each minimum, then the true vacuum (b) is one in which supersymmetry is broken most weakly.

Of course, to make the model realistic we must introduce quarks and leptons as well as the fields considered above. In accordance with the standard SU(5) phenomenology, we should introduce matter multiplets transforming in the 10 and $\overline{5}$ representations of SU(5). So for each quark-lepton generation we have a supersymmetry multiplet $\Phi^{ij} = \begin{pmatrix} \phi^{ij} \\ \psi^{ij} \end{pmatrix}$ in the 10 representation ($\Phi^{ij} = -\Phi^{ji}$), and an additional $\overline{5}$ multiplet $\Phi_i = \begin{pmatrix} \phi_i \\ \psi_i \end{pmatrix}$. In view of some phenomenological problems that will be mentioned shortly, it is probably necessary to include some additional matter multiplets transforming in a real representation of SU(5). One then generalizes the function W of eq. (14) to a gauge-invariant function $W(A^i_j, \phi^{kl}, \phi_m)$ with cubic and lower powers of the fields. This need not disturb the picture described above.

Apart from the question of whether non-perturbative effects really do spontaneously break supersymmetry, there are two rather serious phenomenological problems that arise in a model of this kind.

(i) It may be awkward to obtain a realistic quark-lepton mass spectrum. For instance, if the only matter multiplets other than A are the SU(5) 10 and $\overline{5}$ fields ϕ^{ij}_a and ϕ_{kb} (a and b are flavor indices), then the only term involving these fields that can be added to W is $\Delta W = g_{abc}\phi^{ij}_a\phi_{ib}\phi_{jc}$, where the g_{abc} are arbitrary coupling constants (restricted only by $g_{abc} = -g_{acb}$). If one assumes that along with dynamical supersymmetry breaking one of the SU(2) × U(1) doublet fields in ϕ_{kb} will get an expectation value, then the Yukawa couplings associated with ΔW will give masses to the down quarks and charged leptons. However, the up quarks will be massless at the tree level; the Yukawa couplings required to give them masses are forbidden by supersymmetry. Even if the up quarks receive masses in higher orders, these masses would be unacceptably small, at least in the case of the top quark. Because of the antisymmetry of g_{abc} in the last two indices, in this model the down quark and charged lepton in the same generation as the scalar that has an expectation value would also be massless at the tree level; in view of the lightness of the down quark and electron, this is phenomenologically acceptable.

(ii) It is difficult in this sort of model to obtain a realistically long proton lifetime unless symmetries are assumed that make the proton stable. In ordinary SU(5), the color triplet scalars that are related by SU(5) to the Higgs doublet can mediate baryon decay. However, one can assume an arbitrarily large mass for these scalars. Here, these color triplet scalars cannot be assumed to be arbitrarily heavy, for they are supersymmetry partners of the down, strange, and bottom quarks, and we are assuming that supersymmetry is unbroken down to low energies. The baryon non-conserving couplings of these scalars – which are related by supersymmetry to the ΔW term discussed above – are a serious problem.

To overcome these problems it is necessary to introduce additional matter fields beyond the $\overline{5}$ and 10, and to assume either some global symmetries or a larger gauge group.

We will return in sects. 6, 7 to the question of whether in a model like this one supersymmetry really is dynamically broken. Here let us consider the other logical possibility that arises when one considers supersymmetry together with grand unification. Perhaps supersymmetry is spontaneously broken at the tree level: at the energies of grand unification.

This will occur if and only if eqs. (13) are inconsistent and have no solution. Actually, $\phi = 0$ always satisfies the first equation, so as a necessary condition for supersymmetry to be broken at the tree level, it is necessary that $\partial W/\partial\phi_i = 0$ is not satisfied at $\phi = 0$. This is possible only if W contains a term which is linear in the ϕ fields, which is in turn possible only if at least one of the ϕ fields is a singlet under the gauge group.

For a "minimal" model of this kind, we may introduce one complex singlet field X and two complex fields A^i_j and B^i_j, each in the adjoint representation of SU(5). For W many choices are possible. The following choices, among many others, are interesting. One may take

$$W(X, A, B) = M^2 X - gX \operatorname{Tr} A^2 + \lambda \operatorname{Tr} AB, \tag{17}$$

$$W(X, A, B) = M^2 X - gX \operatorname{Tr} A^2 + \lambda \operatorname{Tr} A^2 B, \tag{18}$$

where M, g, and λ are parameters which by redefining fields can be assumed to be real and positive. These choices of W give models analogous to the O'Raifeartaigh model [2].

The choices of W in (17) and (18) give theories that are technically natural because of global symmetries analogous to those in the O'Raifeartaigh model. One must bear in mind that any change in the fields under which W changes only by an overall phase is a symmetry of the theory, because the phase of W cancels out in the Higgs potential [eq. (11)] and can be removed from the Yukawa couplings by chiral rotations of the Fermi fields. Eq. (17) is technically natural because of symmetries under $X \to e^{i\alpha}X$, $B \to e^{i\alpha}B$ and under $X \to -X$, $A \to -A$. Eq. (18) is technically natural because of symmetry under $X \to e^{i\alpha}X$, $B \to e^{i\alpha}B$ and under $A \to -A$. However, as will be discussed in the next section, the concept of naturalness does not apply in the usual way to supersymmetric theories. One could just as well consider choices of W that are not technically natural.

It is easy to see that either choice of W given above leads to a theory in which supersymmetry is spontaneously broken at the tree level, because eqs. (13) for unbroken supersymmetry are inconsistent and do not have a solution. The equation $\partial W/\partial X = 0$ requires $M^2 = g\,\mathrm{Tr}\,A^2$, but the equation $\partial W/\partial B^i_j = 0$ requires $A^i_j = 0$. These are obviously incompatible, so supersymmetry is spontaneously broken.

The next step is to minimize the potential of the scalar fields, which can be read off from eqs. (10)–(12). For the choice of W in eq. (17) the potential is

$$V = |M^2 - g\,\mathrm{Tr}\,A^2|^2 + \lambda^2 \mathrm{Tr}\,AA^* + \mathrm{Tr}(\lambda B - 2gXA)(\lambda B^* - 2gX^*A^*)$$
$$+ e^2 \mathrm{Tr}(i[A, A^*] + i[B, B^*])^2, \tag{19}$$

where e is the SU(5) coupling constant. Minimization of this potential does not uniquely determine the pattern of symmetry breaking. The general minimum of the potential is as follows: A is any hermitian matrix with $\mathrm{Tr}\,A^2 = (2gM^2 - \lambda^2)/(2g^2)$, X is arbitrary, and $B = 2gXA/\lambda$. There are many massless scalars at the tree level. A one-loop calculation must be performed to lift the degeneracy and determine the expectation values of A and X. While the unbroken symmetry may turn out to be SU(3) × SU(2) × U(1), there are many other possibilities, such as SU(4) × U(1).

If one considers instead the choice of W in eq. (18), the potential is

$$V = |M^2 - g\,\mathrm{Tr}\,A^2|^2 + \lambda^2\left(\mathrm{Tr}\,A^2A^{*2} - \tfrac{1}{5}\mathrm{Tr}\,A^2\,\mathrm{Tr}\,A^{*2}\right)$$
$$+ \mathrm{Tr}|\lambda(AB + BA - \tfrac{2}{5}\mathrm{Tr}\,AB) - 2gXA|^2 + e^2\mathrm{Tr}(i[A, A^*] + i[B, B^*])^2. \tag{20}$$

One now finds that A is uniquely determined to be

$$A = \frac{M}{\sqrt{30g + \lambda^2}} \begin{pmatrix} 2 & & & & \\ & 2 & & & \\ & & 2 & & \\ & & & -3 & \\ & & & & -3 \end{pmatrix} \tag{21}$$

in a suitable basis. The unbroken gauge group is thus uniquely determined to be $SU(3) \times SU(2) \times U(1)$. The vacuum expectation value of X is again not determined at the tree level, and B is proportional to A.

It should be obvious that a theory of this kind, once quarks and leptons are introduced, can give just as good an account of nature as is given by standard grand unified theories. In fact, since supersymmetry is broken at very high energies, the difficulty is really whether it is possible to distinguish these theories in terms of low energy predictions from standard grand unified theories without supersymmetry.

One possibility for distinguishing this sort of theory from conventional grand unified theories arises from the fact that supersymmetry forbids certain Yukawa couplings and therefore one may obtain new relations among quark and lepton masses. (Even though supersymmetry is badly broken, the corrections to supersymmetry constraints on Yukawa couplings, arising from loop diagrams, will be of order α.) However, somewhat as in conventional grand unification, the simplest models give mass relations that are too restrictive. For instance, if one enlarges the models described above by adding left-handed fields in the $\bar{5}$ and 10 representations only (and no other new fields), one finds that (as in the models discussed previously with supersymmetry unbroken at the tree level) all up quarks are massless at the tree level.

The key hurdle that this type of theory must face is, of course, whether the hierarchy problem can be solved. Having obtained strong breaking of supersymmetry and SU(5), can one also obtain the extremely weak breaking of $SU(2) \times U(1)$ that is needed to describe nature?

The models considered above illustrate the fact that in theories with supersymmetry breaking at the tree level, there typically are massless scalars at the tree level. The X particle was massless in both models, and in the first model there also were massless charged scalars. When quarks and leptons are incorporated in the model, the situation becomes far more dramatic. If one includes quarks and leptons in the minimal fashion described above, by introducing supermultiplets $\begin{pmatrix} \psi_L \\ \phi \end{pmatrix}$ in the $\bar{5}$ and 10 representations, it is almost obvious that *all* the scalar partners of the quarks and leptons are automatically massless at the tree level. This is so because it is impossible to write a gauge invariant quadratic or cubic term in W that will couple $\bar{5}$ and 10 fields to the fields A, B, and X considered previously. So at the tree level, even though supersymmetry has been spontaneously broken, the scalar partners of quarks and leptons do not "know" this; they remain degenerate with the fermions, and therefore massless.

Thus, the supersymmetry partners of $\begin{pmatrix} \nu \\ e \end{pmatrix}_L$, which have the quantum numbers of the usual Weinberg-Salam Higgs doublet, are massless at the tree level. Unfortunately, there also are at the tree level many massless colored scalars – partners of the quarks – which are dangerous because their exchange could violate baryon number. One would ordinarily assume that when loop corrections are considered, all of the massless scalars (with or without color) will acquire large mass. If this is so, the danger of scalars mediating baryon decay will be eliminated. But the possibility of low energy SU(2) × U(1) breaking would also be eliminated.

To make a successful model of this sort one must find a reason that the usual Higgs doublet – but not the colored scalars – remains massless in finite orders of perturbation theory. This would violate usual concepts of naturalness. In the next section we will discuss some respects in which usual concepts of naturalness are violated by supersymmetric theories. However, as far as I know, this is restricted to theories in which supersymmetry is not spontaneously broken. I cannot suggest any solution of the hierarchy problem in models where supersymmetry is broken at the tree level.

5. Naturalness in supersymmetric theories

We have mentioned before that, when there are massless fermions at the tree level, unbroken supersymmetry would seem to be potentially unstable against perturbative corrections. Nevertheless, it is known that this does not occur. It is known that if the matrix elements $\langle 0|S_{\mu\alpha}|\psi_\beta\rangle$ vanish at the tree level, they vanish to all finite orders, even if some of the $|\psi_\beta\rangle$ share all quantum numbers with the $S_{\mu\alpha}$.

Stated in this way, the fact that unbroken supersymmetry is stable against perturbative corrections seems to violate our usual concepts of naturalness. Indeed, there are a variety of respects in which supersymmetric theories do not satisfy properties that one would ordinarily regard as consequences of naturalness. The known arguments concerning the breakdown of naturalness in supersymmetric theories all depend on details of perturbation theory. No general argument is known that is based purely on symmetry and invariance principles. It is therefore an open question whether the breakdown of naturalness (in several respects that will be discussed) and the stability of unbroken supersymmetry are valid at the non-perturbative level.

Let us now consider the basic facts in this area. By far the most efficient way to construct supersymmetric invariants in $N=1$ global supersymmetry is the superspace formalism [12]. One has matter superfields

$$\Phi(x)=\phi(x)+\theta^\alpha\psi_\alpha(x)+\theta_\alpha\theta^\alpha F(x),$$

and spinor superfields

$$F_\alpha(x)=\lambda_\alpha(x)+F_{\alpha\beta}(x)\theta^\beta+\theta_\alpha D(x)+\theta^\beta\theta_\beta D_{\alpha\alpha'}\tilde{\lambda}^{\alpha'}$$

for the gauge fields. The most general superfield invariant under arbitrary superspace gauge transformations is a product of superfields of the above-mentioned type and their covariant derivatives.

Given a Lorentz-invariant superfield Q, one can always make a supersymmetric invariant (perhaps zero) by integrating it over all of superspace:

$$I=\int d^4x\, d^2\theta\, d^2\tilde{\theta}\, Q(x,\theta,\tilde{\theta}\,). \tag{22}$$

In this expression θ and $\tilde{\theta}$ are the anticommuting coordinates of negative and positive chirality. In addition, if one has a Lorentz-invariant superfield $R(x,\theta)$ which is a function of x and θ only (but not $\tilde{\theta}$) one can form a supersymmetric invariant by integrating over x and θ:

$$I=\int d^4x\, d^2\theta\, R(x,\theta). \tag{23}$$

Obviously, any invariant which can be written in the form (22) can also be written in the form (23). It is enough to define

$$R=\int d^2\tilde{\theta}\, Q. \tag{24}$$

However, it is important to realize that there exist supersymmetry invariants which can be written in the form (23) but cannot be written in the form (22) with any superfield Q.

For instance, all mass terms and all Yukawa couplings in renormalizable supersymmetric theories come from operators of type (23) which cannot be written in the form (22). (The integrand is the function W, discussed in the previous section. The only exception is the D term, to be discussed.) On the other hand, the kinetic energy terms, both for matter fields and for gauge fields, can be written in the form (22). The status of the gauge field kinetic energy is actually somewhat delicate. It can be written in the form (23) with a gauge-invariant integrand; it can also be written in the form (22), but with an integrand that in this case is not gauge invariant under arbitrary superspace gauge transformations, but only under those that preserve the Wess-Zumino gauge condition (these include ordinary space-dependent gauge transformations). To complete our survey of supersymmetric invariants, it should be noted that there is one operator whose status is somewhat anomalous. This is the "D term" which can appear when the gauge group is not semisimple but has a U(1) factor. It cannot be written in the form (23), and when it is written in the form (22), the integrand is only invariant under Wess-Zumino gauge transformations. We will at first assume that the gauge group is semisimple, and postpone to the end the complications associated with the D term.

Now, it has been proved [16] that to any finite order of perturbation theory, quantum corrections to the effective potential generate only operators of type (22), never operators that can only be written in the form (23). (The technique of superspace perturbation theory [12] is a great aid in simplifying these arguments.) The fact that only operators of type (22) appear in the effective potential has a number of interesting corollaries, many of which have been noted in the literature. (In discussing these corollaries, it must be kept in mind that the gauge kinetic energy, and also the D term, to which we will return later, are to be regarded as operators of type (22); this depends on the details of how the Wess-Zumino gauge enters in the formalism used in proving that only operators of type (22) are generated.)

First of all, because the kinetic energy operators are of type (22), wave-function renormalization occurs for both matter and gauge fields. But because the mass terms and Yukawa couplings come from operators of type (23), there is no renormalization of these parameters independent of the wave-function renormalization. The statement "no renormalization" refers to finite as well as infinite contributions. The bare mass terms and Yukawa couplings come from terms in the bare potential of type (23), and no additional finite or infinite contributions of type (23) are generated in perturbation theory. In supersymmetric theories one can, if one wishes, impose arbitrary relations among mass and Yukawa coupling parameters or set some of them equal to zero.

The fact that only operators of type (22) are generated in perturbation theory has another consequence that is very important for our purposes. This is the fact that supersymmetry remains unbroken to any finite order if it is unbroken at the tree level. This has often been expressed [6] as the statement that the vacuum energy is zero to all orders if it is zero at the tree level.

Let us review a few facts. Given any elementary fermion ψ_α, from its commutator we can form a scalar field $\{Q_\alpha, \psi^\alpha\}$ which is known as the auxiliary field and is denoted as F or D depending on whether ψ is related by supersymmetry to a boson of spin zero or spin one. If the auxiliary field has a vacuum expectation value supersymmetry is spontaneously broken, since obviously supersymmetry is spontaneously broken if $\langle\{Q_\alpha, \psi^\alpha\}\rangle \neq 0$. The converse is also true in the context of perturbation theory. If supersymmetry is spontaneously broken in finite orders of perturbation theory, one of the elementary fermions must be a Goldstone fermion, $\langle 0|S_{\mu\alpha}|\psi_\beta\rangle = \gamma_{\mu\alpha\beta} f$, with $f \neq 0$. (If supersymmetry breaking is a non-perturbative effect, the Goldstone fermion might be a bound state.) But if ψ appears as a one-particle pole in the two-point function $\langle 0|\mathrm{T}(S_{\mu\alpha}\psi_\beta)|0\rangle$, then standard current algebra (by considering the divergence of that matrix element) implies that the auxiliary field has an expectation value (equal, in fact, to f).

To obtain supersymmetry breaking in finite orders of perturbation theory, we must give an expectation value to F or D. This means that we must obtain in the supersymmetric effective potential an operator linear in F or D, times fields with vacuum expectation values. The only fields with vacuum expectation values are elementary scalar fields, since neither gauge fields nor Fermi fields nor derivatives of fields have expectation values. When one constructs superfields whose integrals would be gauge invariant, if one ignores terms involving Fermi fields, gauge fields, and derivatives, each power of F is accompanied by at most two powers of the anticommuting coordinates θ^α. The situation for D is more complicated and will be discussed momentarily.

Operators linear in F, which are at most quadratic in the θ_α, can survive an integral of the type $\int d^2\theta$ that appears in eq. (23). However, as we have noted, no operators of (23) are generated by loop corrections to the effective potential, to any finite order. Non-zero contributions in the integral $\int d^2\theta d^2\tilde{\theta}$ appearing in (22), if they involve F at all, involve F quadratically (or involve F times D) so as to be quartic in θ and $\tilde{\theta}$. So a term linear in F does not appear, and F does not get an expectation value.

Let us now consider D. While there is an auxiliary field for every generator of the gauge group, the D fields that could get vacuum expectations values are those associated with *unbroken* gauge symmetries. Only those D fields are related by supersymmetry to massless gauge mesons and to massless fermions that might become Goldstone fermions. A D field associated with a broken symmetry is not in lowest order related by supersymmetry to any massless fermion and therefore it could not get a vacuum expectation value in the lowest order of perturbation theory in which supersymmetry is spontaneously broken and some fermion becomes a Goldstone fermion.

Let us consider first semisimple groups (an important complication that arises otherwise will be described later). Operators of type (22) can contain terms linear in D (the matter field kinetic energy is an example). However, the linear terms have the following structure: a D field associated with a given generator of the gauge group always multiplies scalar fields that are non-singlets under the action of that generator. A D field associated with an unbroken gauge symmetry is always multiplied by charged scalar fields that have zero expectation value precisely because the symmetry is unbroken. Hence no expectation value of the D term is induced.

The non-generation of operators of type (23) has some other interesting corollaries. In a broad class of theories, one can prove [11] that if supersymmetry is unbroken at the tree level, then loop corrections do not induce shifts in the vacuum expectation values of the scalar fields*. This can be proved, in those cases where it is true, by showing that all operators of type (22) are stationary when the fields are set equal to their tree approximation vacuum expectation values. For instance, in all theories of spin 0 and spin $\frac{1}{2}$ fields only, we have seen that the effective potential is at least quadratic in the F fields. So it is stationary at any point where the F fields vanish. But the vanishing of the F fields is the condition that defines the tree level vacuum expectation values of the fields.

Another interesting corollary of the non-generation of operators of type (23) is that in a broad class of supersymmetric theories, one can show that if supersymme-

* However, B. Ovrut and J. Wess have recently found a class of theories in which this is not so (private communication).

try is unbroken at the tree level, then any particle massless at the tree level remains massless to all finite orders of perturbation theory, even if its masslessness resulted from arbitrary adjustment of parameters. (I do not know if this is true in all supersymmetric theories.) Let us again for simplicity concentrate on the case of theories with spin 0 and spin $\frac{1}{2}$ fields only. The F fields in general contain terms linear and quadratic in the elementary scalar fields. But *massless* scalars do not appear in the linear terms in F; if they did they would not be massless, since at the tree level the potential is $\frac{1}{2}\Sigma|F_i|^2$. Since the effective potential, at every order, is at least quadratic in the F fields, the massless scalars appear only in terms that are at least quartic, and do not get masses.

So far we have been discussing theories with semisimple gauge groups. We must now finally discuss how this picture changes if the gauge group is not semisimple. Assume that the gauge group contains a U(1) factor. Let D be the auxiliary field related by supersymmetry to the U(1) gauge boson. Then D itself is gauge invariant and $\int d^4x D(x)$ is supersymmetric. It can be written in the form (22), although with an integrand that has only restricted (Wess-Zumino) gauge invariance. Being of the form (22) this "D term" can be generated by loops, even if it is absent at the tree level. In fact, having dimension two it can be generated with a coefficient that is quadratically divergent. A simple calculation shows that if the U(1) symmetry is not spontaneously broken at the tree level, then the D term is generated with coefficient

$$\sum_i e_i \int d^4k \frac{1}{k^2}, \tag{25}$$

where the sum runs over all massless left-handed Fermi fields, e_i being the U(1) quantum number of the ith such field. In any theory with a U(1) factor in the gauge group, the D term will be generated in loops unless it is forbidden by a discrete parity symmetry (D is pseudoscalar).

The D term violates many of the statements made above and thereby restores naturalness. The D term can break supersymmetry, can give mass to massless particles, and can shift vacuum expectation values of scalars. Because the D term is subject to renormalization, however, no particular value of its coefficient is natural. If a small change in the coefficient can trigger supersymmetry breaking, the unbroken supersymmetry was not natural. Simply generating a D term by loops is not the way to get a natural gauge hierarchy.

What about theories that we may be most interested in, in which the gauge group is semisimple (perhaps simple) but is spontaneously broken to a non-semisimple group? For instance, suppose SU(5) is spontaneously broken to SU(3) × SU(2) × U(1) at the tree level, but without breaking supersymmetry. One might then expect that the D term of the unbroken U(1) subgroup could be generated in loops, restoring naturalness (at least in some respects) and perhaps spontaneously breaking supersymmetry. This does not occur. To any finite order of perturbation theory we can work with an SU(5) invariant effective potential. Remarkably, there does not exist a supersymmetric and SU(5) invariant operator that after supersymmetry breaking reduces to the ordinary D term. In fact we have already seen that supersymmetric, SU(5) invariant operators that can be generated in loops and are linear in the D field associated with an unbroken symmetry always contain charged fields with zero vacuum expectation value. They cannot reduce after breaking of SU(5) (but not supersymmetry) to the simple D term.

If SU(5) (or some other group) is spontaneously broken down to SU(3) × SU(2) × U(1), then it would seem natural, from the point of view of the low energy physics, for the D term to be generated. The masslessness of fields which would get mass if the D term appeared, and also unbroken supersymmetry, if this would be spoiled by appearance of D, would not seem natural from the point of view of the low energy theory. Nonetheless, the mere fact that at some energy SU(3) × SU(2) × U(1) have been unified in SU(5) is enough to ensure that the D term is not generated in any finite order. It simply does not have a supersymmetric SU(5) invariant generalization. An illustration of this is the one-loop expression of eq. (25) which clearly vanishes if U(1) is part of a semisimple group.

This is possibly a unique case in physics in which an operator allowed by the unbroken symmetry group of a theory is prevented from appearing by a *broken* symmetry that is relevant only at higher energies. It sounds, superficially, like precisely "what the doctor ordered" in order to solve the gauge hierarchy problem. Nevertheless, in trying to apply this idea we will run into difficulties.

6. Some models

The previous sections have aimed at convincing the reader that dynamical breaking of supersymmetry is plausible. In this section, two simple models will be presented in which dynamical supersymmetry breaking really does occur. The models involve systems of less than four dimensions. As will become clear in the following sections, I do not know whether a workable mechanism for dynamical supersymmetry breaking exists in four dimensions.

The first model is not a field theory model at all but a model in potential theory – supersymmetric quantum mechanics. A supersymmetric quantum mechanical system is one in which there are operators Q_i that commute with the hamiltonian,

$$[Q_i, H] = 0, \qquad i = 1 \cdots N, \tag{26}$$

and satisfy the algebra

$$\{Q_i, Q_j\} = \delta_{ij} H. \tag{27}$$

The simplest such system has $N = 2$ and involves a spin one half particle moving on the line. The wave function is therefore a two-component Pauli spinor,

$$\psi(x)=\begin{pmatrix}\phi_1(x)\\ \phi_2(x)\end{pmatrix}.$$

The Q_i are defined as

$$Q_1=\tfrac{1}{2}(\sigma_1 p+\sigma_2 W(x)),$$

$$Q_2=\tfrac{1}{2}(\sigma_2 p-\sigma_1 W(x)), \tag{28}$$

where the σ_i are the usual Pauli spin matrices, where as usual $p=-i\hbar\mathrm{d}/\mathrm{d}x$, and where W is an arbitrary function of x. It is straightforward to check that the algebra (26) and (27) is satisfied with

$$H=\tfrac{1}{2}\left(p^2+W^2(x)+\hbar\sigma_3\frac{\mathrm{d}W}{\mathrm{d}x}\right). \tag{29}$$

We will assume $|W|\to\infty$ as $|x|\to\infty$ so that the spectrum of the hamiltonian is discrete.

Now let us ask under what conditions supersymmetry is spontaneously broken. In the weak coupling (small $\hbar$) limit, one ignores the zero-point motion, and one ignores the last term, which is explicitly proportional to $\hbar$ (and which is a potential theory analogue of what in field theory would be a Yukawa coupling). At the tree level, the ground-state energy is therefore simply the minimum of W^2. The number of supersymmetrically invariant, zero-energy states is therefore, at the tree level, equal to the number of solutions of the equation $W(x)=0$.

Several interesting choices of W can be considered. In fig. 7 several choices of $W(x)$ and of the corresponding potential energy $V(x)=W^2(x)$ have been plotted. In fig. 7a W has a single zero, so the potential has a single supersymmetric minimum of zero energy. In fig. 7b, W has two zeros, so there are at the tree level two supersymmetrically invariant states. In fig. 7c, W has no zeros, the minimum of the potential is at non-zero energy, and supersymmetry is spontaneously broken at the tree level.

Before considering the exact spectrum, it is interesting to discuss the $O(\hbar)$ corrections to the vacuum energy, in the case that this energy vanishes classically. Suppose that W has a simple zero at $x=x_0$, so $W(x)=\lambda(x-x_0)+O((x-x_0)^2)$ for some λ. The hamiltonian is then $H=\frac{1}{2}(p^2+\lambda^2(x-x_0)^2+\hbar\lambda\sigma_3)$ plus higher order terms in $(x-x_0)$ that will not affect the $O(\hbar)$ terms in the ground-state energy. The first two terms are a harmonic oscillator hamiltonian with zero-point energy $\frac{1}{2}\hbar|\lambda|$. The last term, $\frac{1}{2}\hbar\lambda\sigma_3$, has eigenvalues $\pm\frac{1}{2}\hbar|\lambda|$. Choosing the eigenvalue of σ_3 to minimize the energy, the ground-state energy is $\frac{1}{2}\hbar|\lambda|-\frac{1}{2}\hbar|\lambda|=0$.

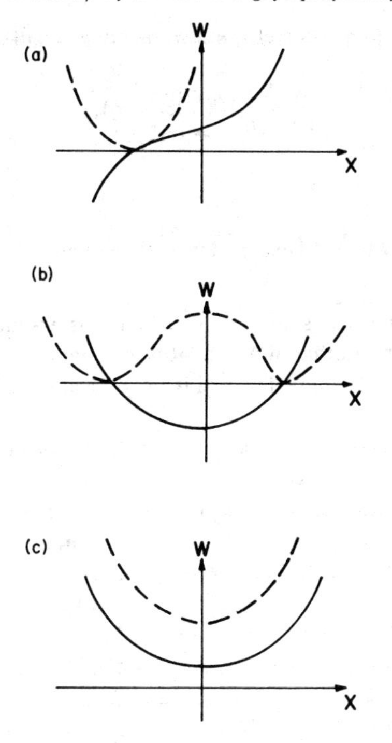

Fig. 7. Some choices of W (solid line) and W^2 (dashed line). In (a) W has a single zero and there is a single supersymmetric state in perturbation theory. In (b) W has two zeros and there are two such states. In (c) W has no zeros and supersymmetry is spontaneously broken at the tree level.

This is an example of the famous cancellation between the Bose and Fermi contributions to the ground-state energy of supersymmetric theories.

Rather than considering higher order terms in perturbation theory, let us now consider the exact spectrum of the theory. We will find that while at the tree level the number of supersymmetric states equals the number of zeros of W, in the exact spectrum the number of supersymmetric states is equal to one if W has an odd number of zeros, and is equal to zero if W has an even number of zeros. Therefore, in the theory of fig. 7a the exact ground state is supersymmetric, and in fig. 7c supersymmetry is really spontaneously broken. But in fig. 7b non-perturbative effects dynamically break the supersymmetry, which is unbroken at the tree level.

To establish these results, we will simply study the equation $Q_i\psi=0$ which must be satisfied by a supersymmetric state. Actually, because of the general relation $Q_1^2=Q_2^2=\frac{1}{2}H$ (or the fact that in this particular model $Q_2=-i\sigma_3 Q_1$, as can easily be checked), it is enough to satisfy $Q_1\psi=0$, or $\sigma_1 p\psi=-\sigma_2 W\psi$. Multiplying by σ_1

and using the facts that $p=-i\hbar \mathrm{d}/\mathrm{d}x$, $\sigma_1\sigma_2=i\sigma_3$, this equation becomes

$$\frac{\mathrm{d}\psi}{\mathrm{d}x}=\frac{1}{\hbar}W(x)\sigma_3\psi(x), \tag{30}$$

and the solution is

$$\psi(x)=\left(\exp\int_0^x \mathrm{d}y\frac{W(y)}{\hbar}\sigma_3\right)\psi(0). \tag{31}$$

This defines a supersymmetric state provided that it is a state – provided that ψ is normalizable. We must recall that in non-relativistic quantum mechanics, the quantization of energy levels and eigenvalues holds only because wave functions must be normalizable.

We have assumed that $|W(x)|\to\infty$ as $|x|\to\infty$. If W has an odd number of zeros, then the sign of W for $x\to+\infty$ is opposite from the sign of W for $x\to-\infty$. In this case there is a unique choice of $\psi(0)$ that makes (31) normalizable. For instance, if $W(x)\sim x^3$ for large x, then $\int_0^x \mathrm{d}y W(y)\sim\frac{1}{4}x^4$, and if $\sigma_3\psi(0)=-\psi(0)$, then (31) is normalizable. Since in fig. 7a W has precisely one zero, in this theory there is a supersymmetric ground state, given by (31) with $\sigma_3\psi(0)=-\psi(0)$.

On the other hand, if W has an even number of zeros, then the sign of W is the same for $x\to+\infty$ as for $x\to-\infty$. In this case no choice of $\psi(0)$ makes (31) normalizable. For instance, if $W(x)\sim x^4$ for large x, then $\int_0^x \mathrm{d}y W(y)\sim\frac{1}{5}x^5$. In this case, if $\sigma_3\psi(0)=-\psi(0)$, then (31) behaves badly for large negative x, and if $\sigma_3\psi(0)=+\psi(0)$, then (31) behaves badly for large positive x. Thus we find that both in figs. 7b and c, where W has an even number of zeros (two or zero, respectively), the exact spectrum does not contain a state invariant under supersymmetry. In the case of fig. 7c, where supersymmetry is spontaneously broken even at the tree level, this is no surprise. But in the case of fig. 7b, we have learned that the exact spectrum does not contain a state invariant under supersymmetry even though two such states exist at the tree level. Thus, quantum corrections dynamically break supersymmetry in this model.

We have not yet established explicitly that it is *non-perturbative* effects (rather than effects of some finite order in perturbation theory) that break supersymmetry in the model of fig. 7b, but this can easily be seen. In finite orders of perturbation theory one expands around a zero of W (if one exists). We have seen that supersymmetry is always unbroken if W possesses precisely one zero. Perturbation theory, in any finite order, is not sensitive to the existence or non-existence of a second zero of W. The vacuum energy, therefore, must vanish to all finite orders if it vanishes at the tree level.

Only a calculation which reveals the existence of more than one zero of W can possibly reveal the dynamical breaking of supersymmetry, in the model of fig. 7b. Such a calculation would be an instanton calculation – a calculation in which one evaluates the tunneling from one zero of W (one minimum of the potential) to the other. An instanton calculation in the model of fig. 7b does, in fact, demonstrate the dynamical symmetry breaking. This calculation will not be considered here in detail, because this would lead to the sort of considerations that we will consider anyway in the model discussed next.

Our next and last model of dynamically broken supersymmetry is a field theory in 2 + 1 dimensions. It will be, in fact, a minimal supersymmetric generalization of one of the original instanton models considered by Polyakov [17].

This will be a model with a SU(2) gauge symmetry spontaneously broken down to U(1) at the tree level. Supersymmetry will be unbroken at the tree level but will be spontaneously broken due to non-perturbative effects, instantons.

The Bose fields of this model will be the gauge field $\boldsymbol{A}_\mu$ and a Higgs triplet $\boldsymbol{\phi}$; both are isovectors. The supersymmetry partners of the bose fields are Majorana Fermi fields ψ_A and χ_A, respectively. A suitable supersymmetric lagrangian involves, in addition to the conventional gauge-invariant kinetic energy, the following Yukawa couplings and scalar self-couplings:

$$L_{\text{int}} = -\tfrac{1}{2}\lambda^2\phi^2(\phi^2 - a^2)^2 + e\varepsilon_{ijk}\bar{\psi}^i\chi^j\phi^k$$
$$-\lambda\bar{\chi}\chi(\phi^2 - a^2) - 2\lambda(\bar{\chi}\cdot\phi)(\phi\cdot\chi), \tag{32}$$

where e is the gauge coupling and λ is a Yukawa coupling parameter. (A simple quartic potential for the scalars is forbidden by supersymmetry unless additional fields are introduced. The sixth-order potential considered above is renormalizable in three dimensions.)

The supersymmetry transformation laws can be written down most easily if one bears in mind that in three dimensions a vector V_i is equivalent to a second-rank symmetric spinor V_{AB} ($= V_{BA}$). The connection between V_{AB} and V_i is as follows: $V_{12} = V_z$, $V_{11} = \sqrt{\tfrac{1}{2}}(V_x + iV_y)$, $V_{22} = \sqrt{\tfrac{1}{2}}(V_x - iV_y)$. Since the gauge field $\boldsymbol{A}_i$, the field strength $\boldsymbol{B}_i$ and the covariant derivative of the Higgs field $\boldsymbol{D}_i\phi$ are all vectors, they can be represented as symmetric spinors $\boldsymbol{A}_{AB}$, $\boldsymbol{B}_{AB}$, and $\boldsymbol{D}_{AB}\phi$. With this understanding, the supersymmetry transformation laws are

$$\delta \boldsymbol{A}_{AB} = \tfrac{1}{2}(\psi_A\varepsilon_B + \psi_B\varepsilon_A),$$
$$\delta\psi_A = \boldsymbol{B}_{AB}\varepsilon^B,$$
$$\delta\phi = \chi_A\varepsilon^A,$$
$$\delta\chi_A = D_{AB}\phi\varepsilon^B + \varepsilon_A\lambda\phi(\tfrac{1}{2}\phi^2 - a^2), \tag{33}$$

where ε_A is a parameter; indices are raised or lowered with the antisymmetric spinor ε^{AB}.

Looking now at the potential $\frac{1}{2}\lambda^2\phi^2(\phi^2-a^2)$ of this theory, we note that there are two possible choices for the ground state: $\phi=0$ and $|\phi|=a$. The first choice leads to a strongly coupled theory with an SU(2) gauge group. Dynamical supersymmetry breaking may or may not occur in this theory; tools for investigating this do not exist. We will instead analyze the vacuum state with $|\phi|=a$ at the tree level. We will see that in this vacuum, dynamical supersymmetry breaking occurs.

The $|\phi|=a$ vacuum has SU(2) broken down to U(1). The only massless boson is the U(1) gauge boson. The only massless fermion is the partner of the U(1) gauge boson. It can be represented in a gauge invariant way as $\boldsymbol{\phi}\cdot\boldsymbol{\psi}_A$. If supersymmetry is to be spontaneously broken, this massless fermion must become a Goldstone fermion; we must find a non-zero coupling of the massless fermion to the supersymmetry current. Actually, it follows from current algebra that $\boldsymbol{\phi}\cdot\boldsymbol{\psi}_A$ creates a Goldstone fermion if and only if the operator $\varepsilon^{AB}\{Q_A,\boldsymbol{\phi}\cdot\boldsymbol{\psi}_B\}=\{Q_A,\boldsymbol{\phi}\cdot\boldsymbol{\psi}^A\}$ has a non-zero vacuum expectation value. (Here ε^{AB} is the invariant, anti-symmetric second-rank spinor.) From (34) it is easy to see that $\{Q_A,\boldsymbol{\phi}\cdot\boldsymbol{\psi}^A\}=\boldsymbol{\chi}_A\cdot\boldsymbol{\psi}^A$ (note that $\{Q_A,\boldsymbol{\psi}^A\}=0$ because B_{AB} is symmetric and $\varepsilon^{AB}\boldsymbol{B}_{AB}=0$). We will demonstrate supersymmetry breaking by showing that the order parameter $\boldsymbol{\chi}_A\cdot\boldsymbol{\psi}^A$ has a vacuum expectation value.

After spontaneous breaking of SU(2) to U(1), this model has instantons, as in the model (without fermions) originally discussed by Polyakov. The instantons are just 't Hooft-Polyakov magnetic monopoles in one dimension less. They have a non-zero value of the topological charge $(e/2\pi a)\int \mathrm{d}^3x\,\partial_i(\boldsymbol{B}_i\cdot\boldsymbol{\phi})$, which in 3 + 1 dimensions would be regarded as magnetic charge but in 2 + 1 dimensions is the instanton number.

In the supersymmetric theory considered here, the instanton contribution to the vacuum to vacuum transition amplitude vanishes. This is because there exist two fermion zero modes*. These modes are unrelated to chiral symmetries but can be obtained by acting with a supersymmetry transformation on the classical solution. The zero modes can be read off from the transformation laws (33), and have the following form:

$$\boldsymbol{\psi}_A=\boldsymbol{B}^{\mathrm{cl}}_{AB}\varepsilon^B,$$

$$\boldsymbol{\chi}_A=D_{AB}\boldsymbol{\phi}^{\mathrm{cl}}\varepsilon^B+\varepsilon_A\lambda\boldsymbol{\phi}^{\mathrm{cl}}\left(\tfrac{1}{2}\boldsymbol{\phi}^{\mathrm{cl}^2}-a^2\right). \tag{34}$$

Here $\boldsymbol{B}^{\mathrm{cl}}_{AB}$ and $\boldsymbol{\phi}^{\mathrm{cl}}$ are the classical instanton solution, and ε_B is a constant.

Although the one-instanton contribution to the ground-state energy vanishes because of the zero modes (34), the instanton gives an expectation value to the supersymmetry-breaking order parameter $\boldsymbol{\chi}_A\cdot\boldsymbol{\psi}^A$, which can be easily calculated. In general, one should expand the χ and ψ fields as a sum of zero and non-zero modes

* For related discussions, see Jackiw and Rebbi [18].

of the Dirac operator. However, a non-zero contribution to the expectation value of $\chi_A \cdot \psi^A$ arises only from the zero-mode terms in χ and ψ; contributions in which χ and ψ do not "absorb" the two zero modes vanish. So we replace $\chi \cdot \psi(x)$ by

$$\sum_{\text{zero modes}} \chi_A(x)^{(\text{zero mode})} \cdot \psi^A(x)^{(\text{zero mode})} = \boldsymbol{B}^{\text{cl}}_{AB}(x) \cdot \boldsymbol{D}^{\text{cl}}_{AB} \phi^{\text{cl}}(x), \tag{35}$$

where eq. (34) has been used to evaluate the right-hand side. After integrating over the position x, which is equivalent to integrating over the location of the instanton, we learn that one can forget about the existence of zero modes if one replaces $\chi \cdot \psi$ by $\int d^3x\, \boldsymbol{B}^{\text{cl}}_{AB}(x) \cdot D^{\text{cl}}_{AB} \phi(x)$. Returning to a vector notation and using the Bianchi identity $D_i B_i = 0$, one finds that $\chi \cdot \psi$ should be replaced by $\int d^3x\, \partial_i(\boldsymbol{B}_i \cdot \boldsymbol{\phi})$, which is precisely $2\pi a/e$ times the topological charge.

Since this quantity is certainly non-zero both for instantons and anti-instantons, dynamical supersymmetry breaking is within reach. However, instantons and anti-instantons have opposite topological charge so a cancellation may occur. The reason for the cancellation is that $\chi \cdot \psi$ is pseudoscalar and cannot get a vacuum expectation value if the discrete space-time symmetries are conserved. (Also, $\chi \cdot \psi$ is odd under the transformation $\boldsymbol{\phi} \to -\boldsymbol{\phi}$, $\chi \to -\chi$, which, in conjunction with a certain discrete gauge transformation, is a symmetry of the theory.) The cancellation between instantons and anti-instantons can be avoided if the lagrangian contains a term that violates the discrete symmetries.

One might expect that one could stop the cancellation between instantons and anti-instantons by including in the action the topological charge density

$$\Delta L = \frac{\theta e}{4\pi^2 a} \partial_i(\boldsymbol{\phi} \cdot \boldsymbol{B}_i), \tag{36}$$

analogous to $\theta F\tilde{F}$ in four dimensions. This respects supersymmetry because being a total divergence it does not change the equations of motion or the conservation of the supercurrent. This addition to the action would appear to weight instantons and anti-instantons by factors $e^{\pm i\theta}$, respectively, and so to break the cancellation between them. There are, however, some strong arguments that physical amplitudes are not really θ dependent in $2+1$ dimensions, because the topological charge density is the divergence of a *gauge-invariant* operator.

If this is so, one may still expect to remove the cancellation between instantons and anti-instantons, and so obtain dynamical supersymmetry breaking, by including in the lagrangian other operators that violate the discrete symmetries. An operator with the right quantum numbers is

$$\Delta L = g\phi^2 \partial_i(\boldsymbol{\phi} \cdot \boldsymbol{B}_i), \tag{37}$$

which is not a total divergence and has a supersymmetric generalization. In the classical instanton field, (37) has an expectation value opposite in sign to its value in the anti-instanton field, so the cancellation is lifted. [Although (37) is unrenormalizable, this is presumably only a technicality. Eq. (37) could be obtained as a low energy effective lagrangian in renormalizable theories with additional massive fields that have been integrated out.] Thus, the cancellation between instantons and anti-instantons is not universal, and dynamical supersymmetry breaking can occur in models of this class.

These models have some interesting generalizations, which will not be explored here in detail. The potential theory model can be generalized to systems with an arbitrary finite number of degrees of freedom. The phenomenon just noted that supersymmetry is spontaneously broken only at non-zero θ occurs also in some potential theory models and in some two dimensional gauge theories; it may be very widespread. The most important generalization of the above models would be, of course, their extension to four dimensions.

7. Return to four dimensions

Let us now return to the more or less "realistic" grand unified theories considered in sect. 4 – the theories in which the gauge group is spontaneously broken to $SU(3) \times SU(2) \times U(1)$ at the tree level, without spontaneous breaking of supersymmetry. From sect. 5 (and the previous literature on supersymmetry on which most of sect. 5 is based) we know that in these theories supersymmetry will remain unbroken to all finite orders of perturbation theory. Can these theories develop a natural gauge hierarchy, via a non-perturbative mechanism that would spontaneously break supersymmetry and $SU(2) \times U(1)$ at very low energies?

A non-perturbative mechanism might involve gauge couplings which become strong at low energies as a result of a negative β function. This leads to the question of supersymmetric technicolor, which will be considered in sect. 9. However, in this section and the next one, I wish to consider the possibility that dynamical supersymmetry breaking can occur as a *small* effect due to *weak* couplings, as occurred in the models of sect. 6. Can dynamical sypersymmetry breaking occur in four dimensions when all couplings are weak?

This seems plausible because, as has been discussed, unbroken supersymmetry is linked to the exact vanishing of the vacuum energy. Perhaps supersymmetry can be spontaneously broken by extremely weak quantum corrections, which would give a small but non-zero value to the vacuum energy. Perhaps this occurs in the "realistic" grand unified models of sect. 4.

Even without considering detailed mechanisms, there are two necessary conditions that must be satisfied, if this is to work. One of these has been discussed in the preceding sections, and one has not been discussed explicitly.

First of all, if supersymmetry is to be spontaneously broken, there will have to be a Goldstone fermion, that is, a massless neutral fermion created from the vacuum by the supersymmetry current,

$$\langle 0|S_{\mu\alpha}|\psi_\beta\rangle = f\gamma_{\mu\alpha\beta}, \tag{38}$$

with $f \neq 0$. Since weak quantum corrections will not manufacture a massless fermion as a bound state, dynamical supersymmetry breaking could be induced by weak corrections only in theories in which massless neutral fermions already exist at the tree level. However, as has been mentioned above, all theories of realistic phenomenological interest have massless neutral fermions at the tree level. For instance, the supersymmetric partners of the photon and Z mesons are massless as long as supersymmetry and $SU(2)\times U(1)$ are unbroken. So this constraint is satisfied automatically.

Second, if we are to solve the gauge hierarchy problem via weak corrections that dynamically break supersymmetry, we need a theory in which the low energy physics, in perturbation theory, is unstable, in a sense that will now be described. At the tree level, some particles get masses, and others remain massless. (The massless particles include the gauge mesons, the quarks and leptons, and their supersymmetric partners.) The effective lagrangian for the massless particles is, at the tree level, something of the general form

$$L^{\text{eff}} = \sum C_i O_i^{(4)}, \tag{39}$$

where the C_i are numerical coefficients and the $O_i^{(4)}$ are supersymmetric operators of dimension four, invariant under all unbroken gauge symmetries.

Notice that L^{eff} contains operators of dimension four only. Operators of dimension less than four would give masses to some particles, but L^{eff} is by definition the low energy effective lagrangian for the massless particles only.

Now, a non-perturbative effect which is supposed to trigger supersymmetry breaking can always be interpreted, at low energies, as a correction to the effective lagrangian (39). Even if supersymmetry is going to be spontaneously broken, we can always work with an effective lagrangian that is invariant under supersymmetry (although its minimum is not invariant). Hence, supersymmetry can be spontaneously broken by quantum corrections only if there are supersymmetric, gauge-invariant operators whose addition to (39) would induce supersymmetry breaking.

Actually, we can be far more specific. Non-perturbative effects will modify (39) by terms with coefficients smaller than any power of α, let us say of order $\exp(-1/\alpha)$ to be definite. Effects of order $\exp(-1/\alpha)$ will be far too small to be of realistic interest if they multiply operators of dimension four or more. Even if they triggered supersymmetry breaking, this would give far too *large* a hierarchy – too *large* a ratio of mass scales. Effects of order $\exp(-1/\alpha)$ are of interest only if they involve operators of dimension less than four and are hence *enhanced* by positive powers of the grand unified mass scale. For instance, if non-perturbative effects cause a dimension-two operator $O^{(2)}$ to appear in the effective lagrangian, this would contribute something like

$$\Delta L^{\text{eff}} = M^2 \exp\left(-\frac{1}{\alpha}\right) O^{(2)} \tag{40}$$

to the effective lagrangian, where M is a mass that would be expected to be of order the unified mass scale. If $O^{(2)}$ triggers $SU(2)\times U(1)$ breaking and supersymmetry breaking, (40) could be just right to solve the hierarchy problem.

The net effect of this is the following. For dynamical supersymmetry breaking to be possible as a solution of the hierarchy problem, there must exist in the theory an operator of dimension less than four which respects all relevant symmetries (supersymmetry and the unbroken gauge symmetries) yet is not present in the low energy effective lagrangian in perturbation theory. (If it appeared in finite orders of perturbation theory, it would generate excessively large masses and too weak a gauge hierarchy.) If the operator is generated non-perturbatively in the effective lagrangian, we may solve the hierarchy problem.

It sounds like a very tall order to ask that in a theory there should exist operators that are invariant under all relevant symmetries, yet are not generated in perturbation theory. But this criterion is automatically satisfied in our "realistic" grand unified theories! We have seen that there is always at least one operator, the D term of the U(1) subgroup of $SU(3)\times SU(2)\times U(1)$, which satisfies all relevant symmetries of the low energy theory but is not generated in perturbation theory. It has dimension two. In addition, depending on details of the model, there may be low dimension operators of type (23) which are absent at the tree level for one reason or another but which satisfy all the low energy symmetries. These operators, too, will not be generated in perturbation theory but might be generated non-perturbatively.

Our more or less "realistic" grand unified theories, with unbroken supersymmetry at the tree level and the gauge group spontaneously broken to $SU(3)\times SU(2)\times U(1)$, therefore satisfy all the preliminary conditions that must be satisfied in order that non-perturbative effects might be able to break supersymmetry. It is therefore appropriate to look for particular mechanisms.

Actually, with the present state of knowledge in theoretical physics, there is in the weak coupling domain only one known non-perturbative mechanism that could be relevant, instantons. We are therefore led to ask whether instantons could induce dynamical supersymmetry breaking, as in the models of the previous section. (See also the discussion in ref. [18].)

Suppose that SU(5) is broken to $SU(3)\times SU(2)\times U(1)$ at the tree level. The embedding of $SU(3)\times SU(2)$ in SU(5) can be visualized by breaking a 5×5 matrix up in blocks:

$$\left(\begin{array}{c|c} 3\times 3 & 3\times 2 \\ \hline 2\times 3 & 2\times 2 \end{array}\right). \tag{41}$$

One may consider instantons lying in one of the unbroken subgroups. Instantons of the Weinberg-Salam SU(2) might dynamically break supersymmetry, but their effects would be unreasonably small, as in 't Hooft's calculation of baryon number violation in the SU(2) × U(1) model [19]. Instantons of color SU(3) might spontaneously break supersymmetry. The mass scale in this case would be a few hundred MeV, which is the scale at which the QCD coupling becomes strong. If instantons of strongly coupled theories induce supersymmetry breaking, this would be of interest in connection with technicolor theories. However, we will postpone the discussion of theories with strong gauge coupling to sect. 9.

Here we wish to consider the possibility that supersymmetry is spontaneously broken by instantons that do *not* lie in the unbroken gauge group. For instance, in the theory of SU(5) broken to SU(3) × SU(2) × U(1), one may consider an instanton in an SU(2) subgroup that lies partly in color SU(3) and partly in weak interaction SU(2). The instanton may lie in the SU(2) subgroup of SU(5) indicated in parenthesis in (42). Instantons of such a subgroup are characterized by a natural

$$\left(\begin{array}{ccc|cc} & & & & \\ & & & & \\ & & X & X & \\ \hline & & X & X & \\ & & & & \end{array}\right) \tag{42}$$

mass scale, the mass M_X of the heavy bosons of the theory. They are also characterized by a small coupling constant, the coupling α_{GUT} at energies of grand unification. Their effects are therefore explicitly calculable, without infrared divergences or ambiguities.

The action of one of these instantons is $2\pi/\alpha_{GUT}$. Consequently, if they do cause one of the appropriate dimension two operators to appear in the effective lagrangian, then it will appear, as in eq. (40), with a coefficient of order

$$m^2 = M_X^2\left(\frac{2\pi}{\alpha_{GUT}}\right)^k \exp(-2\pi/\alpha_{GUT}), \tag{43}$$

where k is the number of collective coordinates. Assuming M_X to be in the range of 10^{15} GeV to 10^{19} GeV, this gives a reasonable low energy mass scale, $m \sim 10^3$ GeV, if α_{GUT} is about $\frac{1}{20}$ or $\frac{1}{25}$. This is larger than the conventional value by about a factor of 2. But because supersymmetric theories will contain extra fields not included in the conventional calculation, we should expect that α_{GUT} would be larger. Therefore, eq. (43) seems quantitatively reasonable.

Unfortunately, instanton-induced supersymmetry breaking either does not work in four dimensions or at least does not work as straightforwardly as it works in three dimensions. A simple way to explain this is the following. To obtain supersymmetry breaking we need a non-zero matrix element of the form

$$\langle 0|S_{\mu\alpha}|\psi_\beta\rangle . \tag{44}$$

[In sect. 6 we calculated, instead, an order parameter that is related to (44) by current algebra.] In an instanton field there will always be fermion zero modes, obtained by acting on the classical solution with a supersymmetry transformation. In three dimensions a Majorana fermion has two components. There are therefore two zero modes. Two zero modes is the proper number to give a non-zero matrix element (44). One zero mode is absorbed by the current $S_{\mu\alpha}$. The second couples to the fermion ψ_β. However, in four dimensions, a Majorana fermion has four components, so instead of two zero modes one has four zero modes. With four zero modes the instanton gives non-zero values to four fermion matrix elements but not to (44). (Although a classical instanton solution does not quite exist in the cases of interest here, one can count zero modes as if a solution did exist. For a discussion of theories with no classical instanton solution, see recent work by Affleck [20].)

One should not necessarily accept the above reasoning at face value. For instance, why couldn't $S_{\mu\alpha}$ couple to a three-fermion state (and absorb three zero modes)? However, detailed analysis appears to show that, at least in reasonably simple models, $S_{\mu\alpha}$ does not couple to the relevant three fermion state, and the simple argument above – too many zero modes for supersymmetry breaking – is correct. This is disappointing, because on the face of things eq. (43) is an appealing answer to the hierarchy problem.

To have precisely two zero modes in an instanton field would be possible if two of the four supersymmetry generators annihilated the instanton, rather than creating zero modes. This does not occur for instantons that lie in a spontaneously broken part of the gauge group. Instantons of an unbroken gauge symmetry are annihilated by two of the four sypersymmetry charges, but we will postpone disçussing dynamical effects of unbroken non-abelian gauge symmetries until sect. 9.

The theoretical situation regarding the "realistic" grand unified models is very unsatisfactory. These theories appear to have the potential for arbitrarily small effects to break supersymmetry. If this does not occur, an argument independent of perturbation theory and independent of detailed instanton analysis should be found. Until it is found, one must suspect that these theories may develop huge gauge hierarchies.

8. Gravitational instantons?

The instanton mechanism considered in the last section would be rather appealing, if it worked, not only because eq. (43) for the mass scale seems reasonable, but also because this mechanism would provide a perhaps unique opportunity to observe non-perturbative effects of grand unification. In eq. (43), the supersymmetry breaking effects are *enhanced*, rather than suppressed, by powers of the enormous mass M_X.

It is intriguing to ask whether, instead, supersymmetry breaking could result from non-perturbative effects of another force usually regarded as weak: gravitation.

Could supersymmetry breaking result from gravitational instantons? Could the mass scale of supersymmetry breaking be of order $M_p^2 \exp(-1/\lambda)$, where M_p is the Planck mass and λ is a dimensionless coupling constant – as yet unknown – for gravitation? If λ is small, this could solve the hierarchy problem. If λ is large, this dynamical supersymmetry breaking would not be relevant to the hierarchy problem, but we wish to understand dynamical supersymmetry breaking if it occurs in nature even if it is not relevant to the hierarchy problem.

Let us first discuss in general terms what would be involved in supersymmetry breaking by gravitational instantons. By a gravitational instanton one means, in this context, an asymptotically euclidean solution of the positive signature Einstein equations, perhaps with suitable matter fields present. Actually, several proofs have been given that there are no asymptotically flat gravitational instantons in the absence of matter fields [21]. However, such solutions do exist if a second-rank antisymmetric tensor field is present [22], and probably in other theories. Also, it may not really be necessary to work with exact classical solutions.

In supergravity, there is a spin $\frac{3}{2}$ particle, the Rarita-Schwinger field $\psi_{\mu\alpha}$. It is related by supersymmetry to the graviton. Local supersymmetry is spontaneously broken if and only if the spin $\frac{3}{2}$ particle has non-zero mass. This can be seen as follows. A non-zero mass for the spin $\frac{3}{2}$ particle means that it is no longer degenerate with the graviton (which is presumably still massless), so supersymmetry is broken. On the other hand, S-matrix theory arguments [23] (which have been worked out only in special cases but are presumably valid in general) show that as long as the spin $\frac{3}{2}$ particle is massless, supersymmetry must be unbroken.

If instantons are to break supersymmetry, they must give a mass to the previously massless Rarita-Schwinger field. Instantons must, therefore, contribute to the two-point function $\langle 0|\psi_{\mu\alpha}\psi_{\nu\beta}|0\rangle$. In an instanton field, there will be fermion zero modes. To give a mass to a previously massless Rarita-Schwinger field, two zero modes is the requisite number. Thus, although the rationale is different, we reach the same conclusion as in the previous section. An instanton in the field of which there are precisely two zero modes could very plausibly induce spontaneous supersymmetry breaking.

For illustrative purposes, let us ask whether in pure supergravity theory (only the graviton and Rarita-Schwinger field) there may exist instantons with precisely two zero modes. This theory will be considered for the sake of discussion only. While dynamical supersymmetry breaking may or may not occur in pure supergravity theory, instantons cannot be the mechanism for it. The Rarita-Schwinger field, to get a mass, must have four helicity states instead of two. Instantons will not create the extra two helicity states "out of thin air"; they can induce dynamical supersymmetry breaking only in theories in which there is, at the tree level, a massless spin $\frac{1}{2}$ fermion which can supply the extra two helicity states to make the Rarita-Schwinger field massive. (This is the "super-Higgs" mechanism [24].) What is more, asymptotically euclidean instantons do not exist in pure supergravity, as has already been noted. Nonetheless, the reader is invited to temporarily suspend disbelief and ask whether a hypothetical instanton in pure supergravity could admit precisely two zero modes.

Zero modes will be obtained by making a supersymmetry transformation. For any spinor field $\varepsilon(x)$, the Rarita-Schwinger equations are satisfied by

$$\psi_\mu = D_\mu \varepsilon, \tag{45}$$

since under supersymmetry $\delta\psi_\mu = D_\mu \varepsilon$. For what choices of ε is (45) a "zero mode"?

If ε vanishes at infinity, then ψ_μ should be gauged away, in the fashion of Faddeev-Popov. If ε grows at infinity, ψ_μ is not normalizable and should not be included. However, if ε is asymptotic to a constant at infinity, then ε is not normalizable so (45) is not gauged away in the Faddeev-Popov construction, but ψ_μ is normalizable and is a valid zero mode. Only the asymptotic behavior of ε for large $|x|$ is relevant, because given two choices of ε with the same asymptotic behavior, their difference vanishes at infinity, so the difference between the two ψ fields can be gauged away.

In general, there will be four zero modes, since four linearly independent choices can be made for the asymptotic behavior of ε in (45). The number of zero modes can be less than four only if some of the ψ_μ in (45) vanish (or equivalently, can be gauged away with a normalizable gauge parameter; in this case $\psi_\mu = 0$ if ε is defined properly). Hence, the number of zero modes is less than four if and only if the equation $D_\mu \varepsilon = 0$ has solutions.

That equation is extremely restrictive because it implies the integrability condition $[D_\mu, D_\nu]\varepsilon = 0$ or

$$R_{\mu\nu\alpha\beta}\sigma^{\alpha\beta}\varepsilon = 0. \tag{46}$$

In general, except in flat space there are no solutions. Non-trivial solutions exist only if the Riemann tensor is self-dual or anti-self-dual,

$$R_{\mu\nu\alpha\beta} = \pm\tfrac{1}{2} R_{\mu\nu\gamma\delta}\varepsilon_{\alpha\beta}{}^{\gamma\delta}, \tag{47}$$

in which case there are two solutions, of definite chirality. [Since $\sigma^{\alpha\beta} = \frac{1}{2}\varepsilon^{\alpha\beta\gamma\delta}\sigma^{\gamma\delta}\gamma_5$, (46) follows from (47) if $\gamma_5\varepsilon = \mp\varepsilon$.]

For a self-dual gravitational instanton there are therefore exactly two non-vanishing zero modes in the form (45), corresponding to the two choices of the asymptotic behavior of ε for which $D_\mu \varepsilon = 0$ *cannot* be satisfied.

These two zero modes have the same chirality [positive or negative depending on the sign in (47)]. Two zero modes of the same chirality are just what we would need to give the Rarita-Schwinger field a mass and thus break supersymmetry.

However, while there are many solutions of (47), none of them is asymptotically euclidean [25]. While this can be proved in a variety of ways, the previous comments may be regarded as a "physics proof." An asymptotically euclidean solution of (47) would give the Rarita-Schwinger field a mass, which is impossible, since two helicity states are missing.

When additional fields are included, the formula $\delta\psi_\mu = D_\mu\varepsilon$ is replaced by $\delta\psi_\mu = \bar{D}_\mu\varepsilon$, where $\bar{D}_\mu$ is an operator that includes couplings to the other fields. The integrability condition (46) is replaced by $[\bar{D}_\mu, \bar{D}_\nu]\varepsilon = 0$. In some theories of gravity plus matter, there may be asymptotically euclidean instantons for which the integrability condition has solutions. In this case, dynamical supersymmetry breaking would be likely to follow. For instance, theories with torsion would seem to be logical candidates, since Regge and Hanson [26] have shown that in the presence of torsion, an asymptotically euclidean space may satisfy (46). Supersymmetry breaking would appear to be one of few areas where non-perturbative effects of gravity might plausibly be observable.

9. Supersymmetric technicolor?

In sects. 7 and 8, we have considered the issues that arise if one tries to attribute dynamical supersymmetry breaking to instantons. There is one other place in the current body of knowledge about high energy physics where one could hope to find a mechanism for dynamical supersymmetry breaking. In gauge theories with negative beta function, the gauge coupling becomes strong at low energies. Various non-perturbative effects may occur, including confinement, binding of color singlet hadrons, and dynamical mass generation. In the case of QCD, one of the bound states that forms is the pion – a Goldstone boson for spontaneously broken chiral symmetry. In a supersymmetric theory with a strong gauge coupling, could not the bound states include a massless fermion – a Goldstone fermion of broken supersymmetry?

In the last few years it has been suggested [27] that in addition to the SU(3) × SU(2) × U(1) gauge forces, there may exist even stronger gauge forces – "hypercolor" or "technicolor". These forces are suggested to become strong at energies of hundreds of GeV, and to spontaneously break the weak SU(2) × U(1).

In the absence of elementary scalars, this sort of theory can lead to a natural resolution of the gauge hierarchy problem. The principal difficulty is that because there are no elementary scalars, there are unwanted chiral symmetries which prevent fermion masses. The inclusion of scalars light enough to matter would ruin the solution of the problem of naturalness. The unwanted chiral symmetries can be explicitly broken by means of additional gauge interactions, known as "extended technicolor" [28], but this leads to further difficulties.

These problems could conceivably be solved by a supersymmetric generalization of the usual hypercolor or technicolor scenario. Suppose that supersymmetry is unbroken down to the technicolor mass scale. Suppose that strong technicolor forces bind a Goldstone fermion as well as binding various Goldstone bosons. Then the strong technicolor forces can dynamically break both supersymmetry and SU(2) × U(1).

On the other hand, because of supersymmetry, the theory will automatically contain elementary scalars with masses no greater than the technicolor mass scale. Consequently, the difficulties of conventional technicolor will not be present in supersymmetric technicolor theories. Yukawa couplings of the elementary scalars can explicitly break the unwanted chiral symmetries and give masses to quarks and leptons.

The most serious obstacle to a successful theory of this sort is probably the question of whether strong gauge forces really do break supersymmetry. On this score there is one weak but encouraging indication. In QCD, the vacuum expectation value of $\bar{\psi}\psi$ serves as an order parameter for spontaneously broken chiral symmetry. It is therefore plausible to suppose that fermion bilinears have vacuum expectation values in supersymmetric gauge theories. However, in supersymmetric gauge theories with chiral matter multiplets, one finds formulas of the sort $\bar{\psi}\psi = \{Q_\alpha, \bar{\psi}^\alpha \phi\}$ so that a non-zero vacuum expectation value of $\bar{\psi}\psi$ would spontaneously break supersymmetry as well as spontaneously breaking chiral symmetry.

Actually, we know that dynamical breaking of various global symmetries occurs in supersymmetric gauge theories, as long as confinement occurs. This follows from recent arguments based on triangle anomalies [29], which are also valid in the supersymmetric context. However, the fact that chiral symmetry is spontaneously broken in these theories does not guarantee that $\bar{\psi}\psi$ has a vacuum expectation value. In fact, an alternative order parameter for chiral symmetry breaking in these theories would be a scalar bilinear of the general type $\bar{\phi}\phi$ ($\bar{\phi}$ and ϕ are two scalar fields that transform differently under global symmetries; they are not complex conjugates of each other). A non-zero $\langle\bar{\phi}\phi\rangle$ does *not* mean that supersymmetry is spontaneously broken. It is perfectly possible that the breaking of global symmetries in these theories leads to non-zero $\langle\bar{\phi}\phi\rangle$, yet does not lead to non-zero $\langle\bar{\psi}\psi\rangle$ – precisely because of unbroken supersymmetry.

A convincing argument pro or con the question of supersymmetry breaking by strong gauge forces would be very welcome.

10. Conclusions

The potential advantages of the scenarios described in this paper should seem obvious. I wish, however, to draw the reader's attention to several difficulties (some of them mentioned previously) that will exist even if one mechanism or another for dynamically broken supersymmetry can be shown to work.

It may be difficult to account for the relative stability of the proton, since the supersymmetric partners of the quarks are color triplet scalars which may mediate proton decay.

One may encounter unwanted relations among quark and lepton masses, similar to those in the simplest SU(5) and O(10) models but perhaps even more severe.

It is hard to unify technicolor, with or without supersymmetry, with SU(3) × SU(2) × U(1).

In global supersymmetry, the vacuum energy is positive definite (and equal to f^2) if supersymmetry is spontaneously broken. When gravity is included, this translates into an unacceptably large, positive cosmological constant. In supergravity, but not in global supersymmetry, the Bose-Fermi symmetry can be spontaneously broken while the vacuum energy remains zero. This suggests that, if supersymmetry is relevant to nature, it must be spontaneously broken under circumstances such that gravity cannot be ignored; only if gravity cannot be ignored can supersymmetry be broken without the generation of a positive vacuum energy. This may in turn suggest that supersymmetry should *not* survive to low energies, for if it is broken only at low energies one might expect gravity to be irrelevant. However, this line of reasoning leaves many alternatives open.

On a more optimistic note, let us recall that, since gravity does exist in nature, the "Goldstone fermion" that we have discussed at length is not a true, physical, massless spin $\frac{1}{2}$ particle in the world we live in. If supersymmetry exists and has been spontaneously broken, the Goldstone fermion has been absorbed to become the helicity $\pm\frac{1}{2}$ components of the massive spin $\frac{3}{2}$ Rarita-Schwinger particle [24]. Just as the mass of the W boson is equal, in theories of dynamical symmetry breaking, to eF_π, the mass of the Rarita-Schwinger particle will be f/M_p, where M_p is the Planck mass (replace F_π by f and e by $1/M_p$). As f ranges from $(10^5\ \text{GeV})^2$ to $(10^6\ \text{GeV})^2$, which is a plausible range at least in the case of supersymmetric technicolor, the spin $\frac{3}{2}$ particle varies in mass from 1 eV to 100 eV. This is the required range to account for the dark matter that is observed to exist [30] in galactic halos and in clusters of galaxies, so it is just barely possible that most of the mass in the universe consists of massive Rarita-Schwinger particles of spin $\frac{3}{2}$!

In conclusion, let us note that while various questions have been addressed in this paper, the most important question has not been answered. Is dynamical supersymmetry breaking in four dimensions a myth or a reality?

I would like to acknowledge comments by K. Bardakci and D.J. Gross, and to thank B. Ovrut and J. Wess for discussions of their work.

Recent discussions of models of the type suggested in sect. 9 have been made by M. Dine, W. Fischler, and M. Srednicki (preprint, Princeton University) and by S. Dimopoulos and S. Raby (preprint, Stanford University).

References

[1] J. Wess and B. Zumino, Nucl. Phys. B70 (1974) 39;
D.V. Volkov and V.P. Akulov, Phys. Lett. 46B (1973) 109;
Y.A. Gol'fand and E.P. Likhtam, JETP Lett. 13 (1971) 323

[2] P. Fayet and J. Iliopoulos, Phys. Lett. 51B (1974) 461;
P. Fayet, Phys. Lett. 58B (1975) 67;
L. O'Raifeartaigh, Nucl. Phys. B96 (1975) 331
[3] P. Fayet, Nucl. Phys. B90 (1975) 104; Phys. Lett. 69B (1977) 489;
84B (1979) 421;
G.R. Farrar and P. Fayet, Phys. Lett. 76B (1978) 575; 79B (1978) 442; 89B (1980) 191
[4] D.Z. Freedman, P. van Nienwenhuizen and S. Ferrara, Phys. Rev. D13 (1976) 3214;
S. Deser and B. Zumino, Phys. Lett. 62B (1976) 335
[5] S. Weinberg, Phys. Lett. 62B (1976) 111;
E. Gildener and S. Weinberg, Phys. Rev. D13 (1976) 333
[6] J. Iliopoulos and B. Zumino, Nucl. Phys. B76 (1974) 310
[7] P. Fayet and S. Ferrara, Phys. Reports 32 (1977) 249
[8] A. Salam and J. Strathdee, Phys. Lett. 49B (1974) 465
[9] H. Georgi and A. Pais, Phys. Rev. D10 (1974) 1246
[10] S. Coleman and E. Weinberg, Phys. Rev. D7 (1973) 1888
[11] B. Zumino, Nucl. Phys. B89 (1975) 535;
P. West, Nucl. Phys. B106 (1976) 219;
D.M. Capper and M. Ramón Medrano, J. Phys. G2 (1976) 269;
W. Lang, Nucl. Phys. B114 (1976) 123;
S. Weinberg, Phys. Lett. 62B (1976) 111
[12] A. Salam and J. Strathdee, Nucl. Phys. B76 (1974) 477; B86 (1975) 142; Phys. Rev. D11 (1975) 1521;
J. Honerkamp, F. Krause, M. Scheunert and M. Schlindwein, Phys. Lett. 53B (1974) 60;
L. Mezencescu and V.I. Ogievetsky, Usp. Fiz. Nauk 117 (1975) 637;
S. Ferrara and O. Piguet, Nucl. Phys. B93 (1975) 261
[13] J. Scherk and J.H. Schwarz, Phys. Lett. 82B (1979) 60; Nucl. Phys. B153 (1979) 61;
E. Cremmer, J. Scherk and J.H. Schwarz, Phys. Lett. 84B (1979) 83
[14] E. Witten, Nucl. Phys. B186 (1981) 412
[15] P. Fayet and J. Iliopoulos, Phys. Lett. 51B (1974) 461
[16] M.T. Grisaru, W. Siegel and M. Rocek, Nucl. Phys. B159 (1979) 429;
R. Jackiw and C. Rebbi, Phys. Rev. D16 (1977) 1053
[17] A. Polyakov, Phys. Lett. 59B (1975) 82
[18] L.F. Abbott, M.T. Grisaru and H.J. Schnitzer, Phys. Rev. D16 (1977) 3002
[19] G. 't Hooft, Phys. Rev. Lett. 37 (1976) 8
[20] I. Affleck, On constrained instantons, Harvard preprint (1980)
[21] P. Schoen and S.T. Yau, Phys. Rev. Lett. 42 (1979) 547;
E. Witten, A simple proof of the positive energy theorem, Princeton preprint, (1980) Comm. Math. Phys., to be published
[22] E. Witten, unpublished
[23] M.T. Grisaru, H.N. Pendleton and P. van Nieuwenhuizen, Phys. Rev. D15 (1977) 996
[24] D.V. Volkov and V.A. Soroka, JETP Lett. 18 (1973) 22;
B. Zumino, Coral Gables Lectures (January, 1977), Lectures at the 1976 Scottish Universities Summer School (Edinburgh Univ. Press, Edinburgh, Scotland, 1976);
E. Cremmer, B. Julia, J. Scherk, P. van Nieuwenhuizen, S. Ferrara and L. Girardello, Phys. Lett. 79B (1978) 231
[25] S. Hawking, Phys. Rev. D14 (1976) 2460;
N. Hitchin, J. Diff. Geom. 9 (1974) 435
[26] T. Regge, private communication
[27] S. Weinberg, Phys. Rev. D13 (1976) 974; D19 (1978) 1277;
L. Susskind, Phys. Rev. D20 (1979) 2619
[28] S. Dimopoulos and L. Susskind, Nucl. Phys. B155 (1979) 237;
E. Eichten and K. Lane, Phys. Lett. 90B (1980) 125
[29] G. 't Hooft, Lecture at Cargèse Summer Inst., 1979, *in* Recent developments in gauge theories, ed. G. 't Hooft et al. (Plenum, 1980);
S. Dimopoulos, S. Raby and L. Susskind, Nucl. Phys. B173 (1980) 208;

S. Coleman and E. Witten, Phys. Rev. Lett. 45 (1980) 100;
Y. Frishman, A. Schwimmer, T. Banks and S. Yankielowicz, Nucl. Phys. B177 (1981) 157;
A. Zee, Pennsylvania preprint (1980);
R. Barbieri, L. Maiani and R. Petronzio, CERN preprint TH. 2900 (1980);
G. Farrar, CERN preprint TH. 2909 (1980);
R. Chanda and P. Roy, CERN preprint TH. 2923 (1980);
J. Preskill and S. Weinberg, Harvard preprint (1981);
Dan-di Wu, Harvard preprint HUTP-81/A001 (1980);
I. Bars and S. Yankielowicz, Yale preprint YTP81-04 (1981)

[30] S.M. Faber and J.S. Gallagher, *in* Annual reviews of astronomy and astrophysics, ed. G. Burbridge, D. Layzer and J.G. Philips (1979)

XVI. *Invisible Axions*

1) Weak-interaction singlet and strong *CP* invariance
by J. E. Kim, *Phys. Rev. Lett.* **43,** 103 (1979) 990

2) A simple solution to the strong *CP* problem with a harmless axion
by M. Dine, W. Fischler and M. Srednicki, *Phys. Lett.* **104**B, 199 (1981) 995

3) *SU*(5) and the invisible axion
by M. B. Wise, H. Georgi and S. L. Glashow, *Phys. Rev. Lett.* **47,** 402 (1981) 999

VOLUME 43, NUMBER 2 PHYSICAL REVIEW LETTERS 9 JULY 1979

Weak-Interaction Singlet and Strong *CP* Invariance

Jihn E. Kim
Department of Physics, University of Pennsylvania, Philadelphia, Pennsylvania 19104
(Received 16 February 1979)

Strong *CP* invariance is *automatically* preserved by a spontaneously broken chiral $U(1)_4$ symmetry. A weak-interaction singlet heavy quark Q, a new scalar meson σ^0, and *a very light axion* are predicted. Phenomenological implications are also included.

Recent attempts[1-4] to incorporate the observed *CP*-invariance violation[5] in unified gauge theories of the weak and electromagnetic interactions can be classified into three broad categories:

(i) *Hard CP-invariance violation.*—The Lagrangian itself violates *CP* invariance, i.e., it contains complex Yukawa or scalar self-coupling constants. Though the exact meaning of *CP*-invariance property can only be given after all the physical particles acquire masses, one may include theories with complex coupling constants in this category. Effective complex gauge-boson couplings to quarks[1] and the complex Higgs scalar couplings[2] of Weinberg belong to this category.

(ii) *Dynamical CP-invariance violation.*—The theory conserves *CP* in the tree approximation, but it is not conserved when radiative corrections are included.[3] This theory is attractive in the sense that it automatically gives a small *CP*-invariance–violating phase, but has a drawback in the uncertainty of estimating radiative corrections. Further, there is a danger of too small[6] a *CP*-invariant–violating phase in the final result.

(iii) *Spontaneous CP-invariance violation.*—Before spontaneous symmetry breaking, the Lagrangian conserves *CP*, i.e., one can always find appropriate *CP* phases for various fields which made the Lagrangian *CP* invariant. In general, we can start with real coupling constants. The spontaneous symmetry breaking introduces complex vacuum expectation values which make the final Lagrangian *CP*-invariance violating.[4]

I wish to point out that a *simple* theory should belong to one of these classifications for *completeness* of the gauge theory of weak and electromagnetic interactions. (Of course, one can treat the *CP*-invariance violation separately from the known weak- and electromagnetic-interaction theory by introducing a superweak interaction.[7])

For some time, the above picture of weak and electromagnetic interactions had not seemed to induce any difficulties when quantum chromodynamics (QCD) was considered as the gauge theory of strong interactions. However, the recently discovered instanton solutions of QCD and associated quantum effects[8] lead to an effective interaction

$$\mathcal{L}_{\rm int} = (\theta/32\pi^2) F_{\mu\nu}{}^a \tilde{F}^{a\mu\nu}, \tag{1}$$

which violates both P and T invariance but preserves C invariance. A limit on the parameter θ is about $<10^{-9}$–10^{-10} from a recently given bound on the electric dipole moment of neutron, $d_n < 1.24 \times 10^{-25}$ $e\cdot$cm. To guarantee strong *CP* invariance *automatically*, Peccei and Quinn observed[9] that the effective Lagrangian (1) can be rotated away if the Lagrangian has a chiral U(1) invariance denoted as $U(1)_A$, becuase the corresponding chiral current satisfies $\partial_\mu J_5{}^\mu = (g^2/16\pi^2) F_{\mu\nu}{}^a \tilde{F}^{a\mu\nu}$. Other alternatives[10] starting from $\theta = 0$ are not in general automatic because one cannot remove (1) by insisting upon *CP* invariance in the Lagrangian. Though the Lagrangian conserves *CP*, Eq. (1) will still come about at higher orders of weak interactions.

In this paper, I wish to discuss an *automatic theory for strong CP conservation* introducing the $U(1)_A$ symmetry in the Lagrangian. If this $U(1)_A$ is not broken, the symmetry is realized in the Wigner-Weyl manner and the only possible way of relating this unbroken $U(1)_A$ symmetry with flavor-conserving gluons is to have at least one massless quark. This can be easily checked by introducing additional spinless mesons that do not generate vacuum expectation values. The massless-quark possibility has been recently discussed.[11] On the other hand, the broken $U(1)_A$ symmetry implies a Goldstone boson (the so-called axion).[12] However, the Peccei-Quinn-Weinberg-Wilczek axion (PQWW axion) seems not to exist.[13]

This leads me to consider a phenomenologically different axion. I will consider the gauge group $SU(2)_L \otimes U(1) \otimes SU(3)_C$ for weak, electromagnetic, and strong interactions. The known six leptons and six quarks are represented in the Kobayashi-Maskawa picture.[1] One Higgs doublet is sufficient to complete this picture. *CP* is not conserved by scheme (i). In addition, I will introduce a weak-interaction–singlet quark Q and a weak-interaction–singlet, complex Higgs Scalar σ with zero weak hypercharge so as to have a symmetry $U(1)_A$. Further, a discrete symmetry R is introduced as

$$R:\ Q_L \to -Q_L,\ Q_R \to +Q_R,\ \sigma \to -\sigma;$$

all the other fields are invariant. (2)

This guarantees the absence of a bare-mass term $m\bar{Q}Q$. The invariant Yukawa coupling of Q and the Higgs potential V are taken to be

$$\mathcal{L}_y = f\bar{Q}_L\sigma Q_R + f^*\bar{Q}_R\sigma^* Q_L, \tag{3}$$

$$V(\varphi,\sigma) = -\mu_\varphi{}^2\varphi^+\varphi - \mu_\sigma{}^2\sigma^*\sigma + \lambda_\varphi(\varphi^+\varphi)^2 + \lambda_\sigma(\sigma^*\sigma)^2 + \lambda_{\varphi\sigma}\varphi^+\varphi\sigma^*\sigma. \tag{4}$$

It is trivial to see that (3) and (4) are invariant under a $U(1)_A$ transformation

$$Q \to e^{i\gamma_5\alpha}Q,\quad \sigma \to e^{-2i\alpha}\sigma, \tag{5}$$

which can be used to rotate away the interaction (1) provided Q belongs to a nontrivial representation of $SU(3)_C$. The remaining U(1) invariance from (3) and (4) gives the Q-type baryon-number conservation. For specific illustrations Q is assumed to be a color triplet. Also for a finite range of parameters we have $\langle\varphi\rangle_0 \neq 0$ and $\langle\sigma\rangle_0 \neq 0$ to generate quark masses. Therefore, the $U(1)_A$ is spontaneously broken and is realized by the existence of the axion a that does not couple to ordinary quarks at the tree level. With a nonvanishing $\lambda_{\varphi\sigma}$, the standard Higgs is mixed with $\mathrm{Re}\sigma$, but not with a. In the following, $\lambda_{\varphi\sigma} = 0$ is assumed.

Though R (or gauge) symmetry forbids the light-quark coupling $\bar{q}\sigma q$, R symmetry is broken spontaneously and such a coupling will be present at higher orders. This induced coupling can be estimated by a diagram given inside the box of Fig. 1,

$$-i\left(\frac{g_s{}^2}{4\pi^2}\right)^2 \frac{m_q}{v'} \ln\left(\frac{m_Q}{m_q}\right)\bar{q}\gamma_5 q a, \tag{6}$$

where g_s is the color coupling constant, $v' = \langle\sigma\rangle_0$. The mass of the axion is estimated[12,14] by the current-algebra approach (cf. Fig. 1 for the role of an instanton),

$$m_a = \frac{\sqrt{Z}}{1+Z}\frac{\alpha_s{}^2}{\pi^2}\frac{f_\pi}{v'} m_\pi \ln\left(\frac{m_Q{}^2}{m_u m_d}\right), \tag{7}$$

with $Z = \langle m_u\bar{u}u\rangle/\langle m_d\bar{d}d\rangle$. If $m_Q = 100$ GeV, $f = 0.001$ (or $v' = 100$ TeV), $\alpha_s = 0.15$, and $2m_u \approx m_d \approx 10$ MeV, the axion mass is about 2.7 eV. For this particular set of values the lifetime of axion is about 1.0×10^{16} (0.95×10^{12}) yr for $e_Q = 0$ $(e_Q = -\frac{1}{3})$.

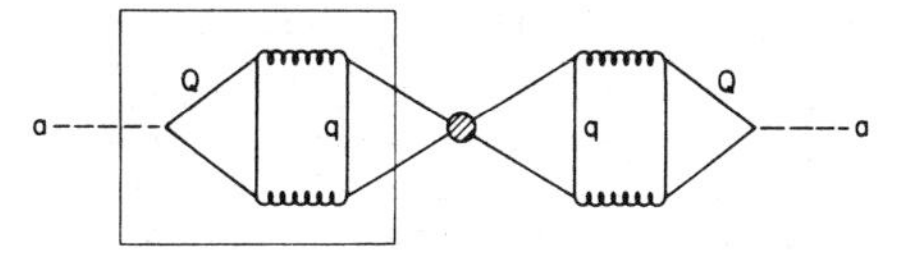

FIG. 1. Diagram for axion coupling to an ordinary quark q through the new quark Q and the gluon (curly lines) loops. The blob in the center is the 't Hooft instanton interaction which could be used for computation of m_a if a reliable method were available for estimating its effects. In the absence of such a method, the diagram is not to be interpreted as an orthodox Feynman diagram.

The main decay mode is $a \to 2\gamma$. For this kind of heavy quark Q, another possible diagram (Fig. 2) does not contribute significantly to the axion mass compared to (7), being presumably down by (a rough estimate) $\sim \pi^2(\mu/m_Q)^{3/2}[\alpha_s^2 \ln(m_Q^2/m_u m_d)]^{-1} \sim 0.023$ for $\mu \approx 1$ GeV. In the following, I will neglect the direct heavy-quark coupling to instantons.

There can be many variations of the present model even for the case of nontrivial weak interactions, but they are more complicated and the essential features are contained in the present example.

With the above estimate of the axion properties, I proceed to discuss the phenomenological implications of the heavy quark Q, a new scalar σ^0, and the axion a.

The new heavy quark Q.—In principle, the color content and the charge of Q can be arbitrary. Color content is assumed to be the same as light quarks (i.e., 3). If its charge is $\frac{2}{3}$ or $-\frac{1}{3}$, the heavy-quark system can be observed in high-energy e^+e^- machines, such as PEP and PETRA, for $10 < m_Q \lesssim 20$ GeV. If its charge is 0, there can be fractionally charged color-singlet hadrons such as $Q\bar{u}$, $Q\bar{d}$, Quu, Qud, etc. Hence, the observation of fractionally charged matter[15] will not disprove the idea of quark confinement. The lightest of these fractionally charged hadrons, $Q\bar{u}$, Quu, and QQu, and a neutral baryon QQQ are absolutely stable. (For this stability consideration, the case of charged Q also applies.) The search for stable heavy particles by Cutts *et al.*[16] at Fermilab sets a limit of 10 GeV and hence $m_Q \gtrsim$ GeV. These stable particles can be observed at a higher-mass search. The implications of a stable charged baryon for hypernuclei and x-ray spectra have been discussed in the literature.[17] If the stable charged particles are produced in e^+e^- annihilation, they can be identified by the presence of a high-momentum track with a very low v/c as measured by time-of-flight information. Heavy mesons formed out of $Q\bar{Q}$ are not stable and they decay to ordinary hadrons through gluon emission. If there is an asymmetry in the new baryon number B_Q in the universe, this will show up as stable particles through cosmic rays. If this asymmetry is hidden somewhere in the universe as neutral stable particles (QQQ), it will be very hard to observe them but they still contribute to the mass of the universe.

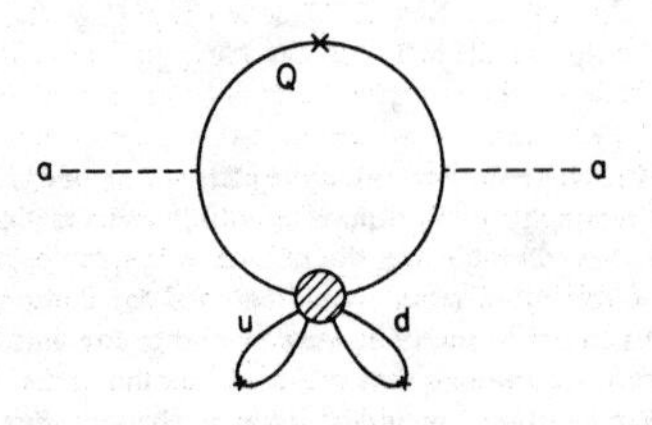

FIG. 2. Possible 't Hooft instanton interaction with a heavy quark.

The new scalar σ^0.—By the spontaneous symmetry breaking of $U(1)_A$, σ will be split into a scalar boson σ^0 of mass $(2\mu_0)^{1/2}$ and an axion a. This σ^0 is *not* a Higgs meson, because it does not break the gauge symmetry, but the phenomenology of it is similar to the Higgs because of its coupling to quark as m_Q/v'. If this scalar mass is $\gtrsim 2m_Q$, we will see spectacular final state of stable particles such as $(Q\bar{u})$ and $(\bar{Q}u)$. If its mass is $\lesssim 2m_Q$, the effective interaction through loops $(c/v')F_{\mu\nu}{}^a F^{a\mu\nu}\sigma^0$, with numerical constant c, will describe the decay $\sigma^0 \to$ ordinary hadrons. The order of magnitude of its lifetime is $\tau(\sigma^0) \approx \tau(\pi^0)(v'/f_\pi)^2(m_{\pi^0}/m_{\sigma^0})^3 \approx 2\times10^{-10}$ sec for $v' \approx 10^5$ GeV and $m_{\sigma^0} \approx 10$ GeV. This kind of particle can be identified as a jet in pp high-energy collisions, since there can be no missing leptons. The estimate of production cross section for this kind of jet is given in the literature.[18] If Q is charged, σ^0 can be seen by producing $\bar{Q}Q$ resonance in e^+e^- annihilation and looking for a hard monochromatic γ from its decay.[19]

The axion a.—Newton's law of gravitation will not be affected if the mass of axion, m_a, is heavy enough so that[20]

$$g_{aN\bar{N}}{}^{2} e^{-m_a r} \lesssim \tfrac{1}{10} G_N m_p{}^2,$$

where $g_{aN\bar{N}}$ is the axion-nucleon coupling and r is the typical distance scale ~ 1 cm for the measurement of Newton's constant $G_N \cong 3.6\times10^{-38}$ GeV^{-2}. With use of Eq. (6) as a rough guess on

$$g_{aN\bar{N}} \sim (\alpha_s{}^2/\pi^2)(f_\pi/v')\ln(m_Q/m_q) g_{\pi N\bar{N}} \sim 1\times10^{-8} g_{\pi N\bar{N}}$$

(for $\alpha_s \approx 0.15$, $v' \approx 10^5$ GeV, $m_Q = 100$ GeV, and $m_q \approx 10$ MeV), a very crude bound on the axion mass is $m_a \gtrsim 10^{-3}$ eV. Reasonable estimates of axion mass satisfy this bound.

Because of a suppression of axion coupling to hadrons by a factor of $\alpha_s{}^2/\pi^2$ (also v' can be adjusted to suppress it) compared to the PQWW axion, it will be relatively harder to observe by hadrons of light quarks. I guess that the axion coupling is suppressed by about a factor of 100.

The decay process $K^+ \to \pi^+ a$ or $\psi \to a\gamma$ is suppressed by a factor of 10^4 relative to the previous[13] estimate of the branching ratios, and at present they give no useful information.

Any processes involving leptons are almost completely suppressed. Such processes can only occur for neutral quark Q by (Q-quark loop)-(gluon loop)-(light-quark loop)-(W or Z boson)-lepton, or by Higgs charged-lepton loops. Therefore, the axion does not contribute to the anomalous magnetic moment of the muon, cannot give spin-spin interaction in atoms and molecules, and cannot mimic $\nu_e e \to \nu_e e$ and $\bar{\nu}_e e \to \bar{\nu}_e e$ elastic scatterings in reactor experiment[21] which was one of the serious evidence against the PQWW axion. There can, however, be a very small contribution to this 1974 experiment, $\bar{\nu}_e D \to np\bar{\nu}_e$ by $aD \to np$, of Reines *et al.*[21] but this is about 10^{-5} times smaller than the measured background.[13]

The axion production and interaction cross section in beam-dump experiments is 10^8 times smaller than the estimate given before,[13] and hence, 10^5 times smaller than the best experimental bound,[22] 10^{-68} cm^4/nucleon.

Cosmological and astrophysical considerations limit[23] the mass of the PQWW axion. In the present case, since the axion is not supposed to have decayed to two photons after the decoupling from $ap \rightleftarrows \gamma p$, it could not have changed the entropy of the universe (number of photons/number of nucleons), or not distort the radiation background spectrum. Further, the axion would not change the standard prediction of deuterium abundance because $aD \to pn$ would have ceased much earlier than $\gamma D \to pn$. Stellar evolution may be affected by the existence of a light axion by the Primakoff process, but it is not known at present what kind of upper limit on the axion mass (a region of a zero axion mass which allows no Primakoff process will be allowed) will upset the present stellar evolution theory.

What are the possible experiments to prove the present scheme? Probably high-precision experiments of the axion search will do. But the easier verification of the weak-interaction singlets Q and σ^0 in pp or $\bar{p}p$ machines (e^+e^- annihilation machine also if Q is charged) will shed light on the whole idea of the spontaneously broken chiral $U(1)_A$ invariance and the multiple vacuum structure of QCD.

I am grateful for advice and encouragement from Henry Primakoff and Anthony Zee. I have also benefitted from discussions with S. M. Barr, L. N. Chang, P. Langacker, G. Segrè, and A. Weldon. This research was supported in part by the U. S. Department of Energy under Contract No. EY-76-C-02-3071.

---

[1] R. N. Mohapatra, Phys. Rev. D 6, 2023 (1972); M. Kobayashi and K. Maskawa, Prog. Theor. Phys. 49, 652 (1973).

[2] S. Weinberg, Phys. Rev. Lett. 37, 657 (1976).

[3] See, for example, H. Georgi and A. Pais, Phys. Rev. D 10, 1246 (1974).

[4] T. D. Lee, Phys. Rev. D 8, 1226 (1973).

[5] J. H. Christenson, J. W. Cronin, V. L. Fitch, and R. Turlay, Phys. Rev. Lett. 13, 138 (1964).

[6] Weinberg's theory (Ref. 2) and Lee's theory (Ref. 4) already give a correct magnitude of *CP*-invariance violation when it is maximally violated. If radiative corrections introduce a small *CP*-invariance–violating phase to an effective potential, the chance is great that the resultant *CP* parameter is too small.

[7] L. Wolfenstein, Phys. Rev. Lett. 13, 592 (1964).

[8] A. Belavin *et al.*, Phys. Lett. 59B, 85 (1975); G.'t Hooft, Phys. Rev. D 14, 3432 (1976); R. Jackiw and C. Rebbi, Phys. Rev. Lett. 37, 172 (1976); C. G. Callan, R. F. Dashen, and D. Gross, Phys. Lett. 63B, 334 (1976).

[9] R. D. Peccei and H. Quinn, Phys. Rev. D 16, 1791 (1977).

[10] R. N. Mohapatra and G. Senjanovic, Phys. Lett. 79B, 283 (1978); H. Georgi, Hadron. J. 1, 155 (1978); M. A. B. Bég and H.-S. Tsao, Phys. Rev. Lett. 41, 278 (1978); G. Segrè and H. A. Weldon, Phys. Rev. Lett. 42, 1191 (1979).

[11] L. N. Chang *et al.*, Pennsylvania University Report No. UPR-0114T, 1979 (to be published); S. Dimopoulos and L. Susskind, to be published.

[12] S. Weinberg, Phys. Rev. Lett. 40, 223 (1978); F. Wilczek, Phys. Rev. Lett. 40, 279 (1978).

[13] See, for example, R. D. Peccei, in *Proceedings of the Nineteenth International Conference on High Energy Physics, Tokyo, Japan, August 1978*, edited by S. Homma, M. Kawaquchi, and H. Miyazawa (Physical Society of Japan, Tokyo, 1979).

[14] Estimates of the axion mass have been given by Weinberg (Ref. 12), Wilczek (Ref. 12), W. A. Bardeen and S.-H. H. Tye [Phys. Lett. 74B, 229 (1978)], and J. Kandaswamy *et al.* [Phys. Rev. D 17, 1430 (1978)]. The effective coupling of the axion to the light quarks enables one to obtain, by the current-algebra approach of Bardeen and Tye, the estimate of the axion mass which is given in Eq. (7). Another possible diagram (Fig. 2) which has a direct heavy-quark coupling to an instanton will not be important since the effective instanton interaction to a sufficiently heavy quark Q will be down by $(m_Q \rho)^2$ compared to the ones without the heavy quark. A rough estimate shows that the contribution of the diagram in Fig. 2 is comparable to that in Fig. 1 for a 5–10-GeV heavy quark.

[15] G. S. LaRue *et al.*, Phys. Rev. Lett. 38, 1011 (1977),

and 42, 142, 1019(E) (1979).
[16]D. Cutts *et al.*, Phys. Rev. Lett. 41, 363 (1978).
[17]R. Cahn, Phys. Rev. Lett. 40, 80 (1978).
[18]H. M. Georgi *et al.*, Phys. Rev. Lett. 40, 692 (1978).
[19]F. Wilczek, Phys. Rev. Lett. 39, 1304 (1977).
[20]L. Resnick *et al.*, Phys. Rev. D 8, 172 (1973).
[21]F. Reines *et al.*, Phys. Rev. Lett. 37, 315 (1976).
[22]Harvard University–University of Pennsylvania–Brookhaven National Laboratory experiment cited by H. H. Williams, Bull. Am. Phys. Soc. 24, 60 (1979).
[23]K. Sato and H. Sato, Prog. Theor. Phys. 54, 912 (1975); D. A. Dicus *et al.*, Phys. Rev. D 18, 1829 (1978).

Volume 104B, number 3 PHYSICS LETTERS 27 August 1981

A SIMPLE SOLUTION TO THE STRONG *CP* PROBLEM WITH A HARMLESS AXION

Michael DINE [1]
The Institute for Advanced Study, Princeton, NJ 08540, USA

Willy FISCHLER [2]
University of Pennsylvania, Philadelphia, PA 19104, USA
and The Institute for Advanced Study, Princeton, NJ 08540, USA

Mark SREDNICKI [3]
Joseph Henry Laboratories, Princeton University, Princeton, NJ 08540, USA

Received 15 May 1981

We describe a simple generalization of the Peccei–Quinn mechanism which eliminates the strong *CP* problem at the cost of a very light, very weakly coupled axion. The mechanism requires no new fermions and is easily implemented in grand unified theories.

Since the discovery of instantons [1], the problem of strong *CP* violation has been a vexing one [2]. In a gauge theory like quantum chromodynamics (QCD), it is possible to add to the lagrangian a term

$$\mathcal{L}_\theta = (\theta g^2/32\pi^2)\, F\tilde{F}\,. \qquad (1)$$

Such a term is a pure gradient, and one might argue that it can have no physical consequences. The presence of instantons, however, shows that such a term can have physical effects [2,3]. In particular, it will induce *CP* violation. QCD estimates [4], and the experimental limits on the neutron electric dipole moment [5] suggest $\theta < 10^{-8}$. It would be very surprising if this small value of θ was simply an accident.

Under certain circumstances, however, θ is an unobservable parameter and strong interactions automatically conserve *CP*. In particular, this is the case if the classical lagrangian possesses a U(1) symmetry which is explicitly broken by color anomalies. Several mechanisms for obtaining such a U(1) symmetry have been proposed. One, setting $M_{\mathrm{u}} = 0$ [2], does not seem consistent with successful current algebra predictions. A second, proposed by Peccei and Quinn [6], requires that at least two scalar doublets be included in the Weinberg–Salam model. The chief difficulty with the Peccei–Quinn scheme is that it predicts a light pseudoscalar particle called the axion [7]. The properties of the axion can be determined with some confidence [7,8], and no such particle has been observed.

In this note we describe a simple generalization of the Peccei–Quinn scheme with a harmless axion. Other generalizations of the Peccei–Quinn scheme have been proposed; most seek to increase the mass of the axion [9] [‡1]. Kim [10], however, has proposed a mechanism for decreasing the mass of the axion while simultaneously decreasing the axion's couplings to ordinary matter. His model requires the existence of both additional quarks and additional scalars, all of which are neutral with respect to electroweak interactions.

In this paper, we discuss a model in which the axion is also very light and very weakly coupled to ordinary

[1] Research supported by the Department of Energy under Grant #DE-AC02-76ER02220.
[2] Research supported by the Department of Energy under Grant #DE-AC02-76ER03071.
[3] Research supported by the National Science Foundation under Grant #PHY80-19754.

[‡1] For models with heavy axions, see for example, ref. [9]. Note that in the model proposed by Dimopoulos, the Fermi constant comes out too small.

matter. Our model is quite economical. It requires the addition of just one scalar field to the Peccei–Quinn model. Moreover, this scheme is easily implemented in grand unified theories. Our model was suggested by a study of supersymmetric versions of technicolor [11], but it may be possible to implement the mechanism in other dynamical symmetry breaking schemes, such as extended technicolor models.

The model is identical to that of Peccei and Quinn, except for the addition of a complex scalar field, ϕ, which is a singlet under SU(2) × U(1). The model has two scalar doublets, ϕ_u and ϕ_d, with hypercharge −1 and +1, respectively. ϕ_u couples only to right-handed charge 2/3 quarks, ϕ_d only to right-handed charge −1/3 quarks and to right-handed charged leptons. Thus the Yukawa couplings have the structure

$$\mathcal{L}_Y = G_u(\bar{u}\bar{d})_L \phi_u u_R + G_d(\bar{u}\bar{d})_L \phi_d d_R + \text{h.c.}\,, \quad (2)$$

and similarly for other quarks and leptons. The restriction of the couplings to this form is required by the symmetry we describe below.

We will demand that the lagrangian possesses a global symmetry at the classical level under which the scalar fields transform as

$$\phi_u \to \exp(i\alpha X_u)\phi_u\,, \quad \phi_d \to \exp(i\alpha X_d)\phi_d\,,$$
$$\phi \to \exp(i\alpha X_\phi)\phi\,, \quad (3)$$

where

$$X_u + X_d = -2X_\phi = 1\,. \quad (4)$$

The most general scalar potential consistent with this symmetry as well as the SU(2) × U(1) symmetry of electroweak interactions is

$$V(\phi, \phi_u, \phi_d) = \lambda_u(|\phi_u|^2 - V_u^2)^2 + \lambda_d(|\phi_d|^2 - V_d^2)^2$$
$$+ \lambda(|\phi|^2 - V^2)^2 + (a|\phi_u|^2 + b|\phi_d|^2)|\phi|^2 \quad (5)$$
$$+ c(\phi_u^i \epsilon_{ij} \phi_d^j \phi^2 + \text{h.c.}) + d|\phi_u^i \epsilon_{ij} \phi_d^j|^2 + e|\phi_u^* \phi_d|^2\,.$$

Here ϵ_{ij} is the completely antisymmetric symbol of SU(2).

For a finite range of values of the parameters in the potential, the desired symmetry breakdown occurs. In particular, we demand

$$\langle\phi_u\rangle = 2^{-1/2}\begin{pmatrix} f_u \\ 0 \end{pmatrix}, \quad \langle\phi_d\rangle = 2^{-1/2}\begin{pmatrix} 0 \\ f_d \end{pmatrix}, \quad (6)$$

giving mass to the W and Z bosons and quarks and leptons. Note that, since only doublets are used, one has, at tree level, the important relation

$$M_W/M_Z = \cos\theta_W\,. \quad (7)$$

Moreover, this structure insures the absence of dangerous strangeness-changing neutral currents.

We also require that the potential give ϕ a large vacuum expectation value,

$$\sqrt{2}\langle\phi\rangle = f_\phi \gg (f_u^2 + f_d^2)^{1/2} \equiv f\,. \quad (8)$$

These expectation values, in addition to providing the breaking of weak isospin, also break the X-symmetry of eq. (1). This symmetry is anomalous, however, and the corresponding Goldstone boson gets a small mass due to instanton effects [more precisely, through those QCD effects which break the U(1) symmetry]. We can study the properties of this particle, which we refer to as the axion, using standard current algebraic techniques. In particular, the methods used by Bardeen and Tye [8] to study the axion of the Peccei–Quinn model allow us to determine the axion's mass, lifetime, and couplings.

Before instanton effects are considered (more generally, QCD effects which violate X-symmetry), the model possesses two conserved U(1) currents, the hypercharge current and the X-current. Writing the fields as

$$\phi_u = 2^{-1/2}\begin{pmatrix} f_u + \eta^u + i\xi_1^u \\ \xi_2^u + i\xi_3^u \end{pmatrix},$$
$$\phi = 2^{-1/2}\begin{pmatrix} \xi_2^d + i\xi_3^d \\ f_d + \eta^d + i\xi_1^d \end{pmatrix}, \quad (9)$$
$$\phi = 2^{-1/2}(f_\phi + \eta^\phi + i\xi^\phi)\,,$$

the field eaten by the Z boson is

$$\phi^Y = (f_u\xi_1^u - f_d\xi_1^d)/f\,. \quad (10)$$

The X-current is given by

$$j_\mu^X = X_\phi \phi^\dagger \overleftrightarrow{\partial}_\mu \phi + X_u \phi_u^\dagger \overleftrightarrow{\partial}_\mu \phi_u + X_d \phi_d^\dagger \overleftrightarrow{\partial}_\mu \phi_d$$
$$+ X_u(\bar{u}\gamma_\mu\gamma_5 u + \bar{c}\gamma_\mu\gamma_5 c + \bar{t}\gamma_\mu\gamma_5 t + \ldots)$$
$$+ X_d(\bar{d}\gamma_\mu\gamma_5 d + \bar{s}\gamma_\mu\gamma_5 s + \bar{b}\gamma_\mu\gamma_5 b + \ldots) \quad (11)$$
$$+ X_d(\bar{e}\gamma_\mu\gamma_5 e + \bar{\mu}\gamma_\mu\gamma_5 \mu + \bar{\tau}\gamma_\mu\gamma_5 \tau + \ldots)\,.$$

Volume 104B, number 3 PHYSICS LETTERS 27 August 1981

In order to isolate the axion and study its properties it is convenient to choose X_u and X_d so that j_μ^X does not couple to ϕ^Y, i.e.

$$\langle 0|j_\mu^X|\phi^Y\rangle = 0\,. \tag{12}$$

This fixes

$$X_u = f_d^2/f^2\,,\quad X_d = f_u^2/f^2\,. \tag{13}$$

At the tree level, this current creates a massless particle, the axion. The axion field is easily determined to be

$$A = [2f_u f_d(f_u\xi^d + f_d\xi^u) - f^2 f_\phi \xi^\phi] \times [f(f^2 f_\phi^2 + 4f_u^2 f_d^2)^{1/2}]^{-1}\,. \tag{14}$$

Note that in the limit $f_\phi \gg f_u, f_d$,

$$A \simeq -\xi^\phi + (2f_u f_d/f_\phi f^2)(f_u\xi^\phi + f_d\xi^u)\,, \tag{15}$$

i.e. the axion is primarily composed of ϕ field. The axion decay constant, f_A, is defined by

$$\langle 0|j_\mu^X|A\rangle = f_A q_\mu\,. \tag{16}$$

Recalling that $X_\phi = -1/2$ [eq. (4)], and using eqs. (11), (13) and (14) one finds

$$f_A = (2f)^{-1}(4f_u^2 f_d^2 + f^2 f_\phi^2)^{1/2}\,. \tag{17}$$

In the limit that f_ϕ is very large

$$f_A = \tfrac{1}{2} f_\mu + O(f^2/f_\phi)\,. \tag{18}$$

The current j_μ^X is anomalous [12],

$$\partial^\mu j_\mu^X = N(g^2/32\pi^2) F_{\mu\nu}{}^a \tilde{F}_{\mu\nu}{}^a\,, \tag{19}$$

where N is the number of quark doublets and $F_{\mu\nu}{}^a$ is the color field strength tensor. This anomaly will lead to a mass for the axion. In order to calculate this mass, we follow Bardeen and Tye and define an anomaly-free, almost-conserved current. We do this by adding to the current j_μ^X a piece involving the light quarks (which we take to be u and d; including the strange quark would only change the results slightly). Specifically, we study the current

$$\tilde{J}_\mu^X = j_\mu^X - N\{(1+Z)^{-1}\bar{u}\gamma_\mu\gamma_5 u + [Z/(1+Z)]\,\bar{d}\gamma_\mu\gamma_5 d\}\,. \tag{20}$$

where Z will be defined below. This current is obviously anomaly-free, and its divergence is proportional to light quark masses. The axion mass may now be calculated in terms of the pion mass using Dashen's theorem [13] and the fact that A_μ^3 (the usual axial τ_3 current) is almost conserved. If we demand

$$\langle 0|[\tilde{Q}_X, [Q_3^5, \mathcal{H}]]|0\rangle = 0\,, \tag{21}$$

where

$$\tilde{Q}_X = \int d^3x\, \tilde{J}_0^X(x)\,, \tag{22}$$

$$Q_3^5 = \int d^3x\, A_0^3(x)\,, \tag{23}$$

and $\mathcal{H}$ is the hamiltonian density, Dashen's formula gives

$$m_A^2 f_A^2 = \langle 0[\tilde{Q}^X, [\tilde{Q}^X, \mathcal{H}]]|0\rangle\,. \tag{24}$$

It is a straightforward matter to evaluate the commutators in these equations. One finds that eq. (19) implies (value of m_u/m_d from ref. [14])

$$Z \simeq m_u/m_d \simeq 0.56\,. \tag{25}$$

The axion mass may finally be determined by computing

$$m_\pi^2 f_\pi^2 = \langle 0|[Q_5^3, [Q_5^3, \mathcal{H}]]|0\rangle\,, \tag{26}$$

and comparing with eq. (24).

One finds

$$m_A^2 = (f_\pi^2/f_A^2) m_\pi^2 N^2 Z(1+Z)^{-2}\,, \tag{27}$$

or

$$m_A \simeq [74\ \text{KeV}\ (250\ \text{GeV}/f_A)] \tag{28}$$

(where we have taken $N = 3$).

The axion lifetime can be computed through the two-photon anomaly, analogous to the calculation for the π^0. One obtains

$$\tau(A \to 2\gamma) = \tau(\pi^0 \to 2\gamma)(m_\pi/m_a)^5 N^4 Z^3/(1+Z)^4 = 41\ \text{s} \times (f_A/250\ \text{GeV})^5\,. \tag{29}$$

A crucial feature of this axion is the strength of its couplings to ordinary matter. These are suppressed by a factor

$$r = f/f_A\,, \tag{30}$$

relative to those of the axion of the Peccei–Quinn model. Thus production of axions in any process (hadronic collisions, nuclear decays, etc.) is reduced by a factor r^2; the probability of subsequent detection of these axions is reduced by the same factor. In fact, if r is greater than about 10, no terrestrial experiment which is currently used to set limits on axions could have observed ours [15].

An axion of this type might have cosmological or astrophysical implications. If the axion decays after recombination time and before the present era, i.e. if

$$10^5\,\mathrm{yr} \lesssim \tau \lesssim 10^{10}\,\mathrm{yr}\,, \tag{31}$$

then axion decays might give an observable distortion of the cosmic microwave radiation background. A more stringent restriction arises from the existence of red giant stars [16]. Unless the axion is *extraordinarily* weakly coupled, it will be copiously produced in these stars (and for that matter, the sun), and rapidly carry off all of their thermal energy. The authors of ref. [16] find a limit

$$m_A \lesssim 0.01\ \mathrm{eV}\,. \tag{32}$$

This corresponds to

$$\langle\phi\rangle > 10^9\ \mathrm{GeV}\,. \tag{33}$$

Such a huge vacuum expectation value might be expected in grand unified models [17]. For example, if one constructs an SU(5) model with two 5's of Higgs and a scalar singlet, one can implement our scheme for strong *CP* conservation explicitly. The natural scale for $\langle\phi\rangle$ is then 10^{15} GeV. In fact, our mechanism suggests that the *CP* problem may not be a problem of weak-interaction physics at all; rather it may be resolved only by physics at extremely short distances. The only requirement is the existence of an anomalous U(1) symmetry broken at a large energy scale.

References

[1] A.A. Belavin, A.M. Polyakov, A.S. Schwartz and Yu. S. Tyupkin, Phys. Lett. 59B (1975) 85.

[2] G. 't Hooft, Phys. Rev. Lett. 37 (1976) 172.

[3] R. Jackiw and C. Rebbi, Phys. Rev. Lett. 37 (1976) 172; C.G. Callan, R.F. Dashen and D.J. Gross, Phys. Lett. 63B (1976) 334.

[4] V. Baluni, Phys. Rev. D19 (1979) 227; R. Crewther, P. DiVecchia, G. Veneziano and E. Witten, Phys. Lett. 88B (1979) 123.

[5] N.F. Ramsey, Phys. Rep. 43C (1978) 409.

[6] R.D. Peccei and H.R. Quinn, Phys. Rev. Lett. 38 (1977) 1440; Phys. Rev. D16 (1977) 1791.

[7] S. Weinberg, Phys. Rev. Lett. 40 (1978) 223; F. Wilczek, Phys. Rev. lett. 40 (1978) 279.

[8] W.A. Bardeen and S.-H.H. Tye, Phys. Lett. 74B (1978) 229.

[9] S. Treiman and F. Wilczek, Phys. Lett. 74B (1978) 381; S.-H.H. Tye, Cornell preprint CLNS-81/479 (1981); S. Dimopoulos, Phys. Lett. 84B (1979) 435.

[10] J.E. Kim, Phys. Rev. Lett. 43 (1979) 103; see also: M.A. Shifman, A.I. Vainshtein and V.I. Zakharov, Nucl. Phys. B166 (1980) 493.

[11] M. Dine, W. Fischler and M. Srednicki, Princeton Univ. preprint.

[12] S. Adler, Phys. Rev. 177 (1969) 2426; J.S. Bell and R. Jackiw, Nuovo Cimento 60A (1960) 49; W.A. Bardeen, Phys. Rev. 184 (1969) 1848.

[13] R. Dashen, Phys. Rev. 183 (1969) 1245.

[14] S. Weinberg, Harvard Univ. preprint HUTP-77, A057.

[15] T.W. Donnelly, S.J. Freedman, R.S. Lytel, R.D. Peccei and M. Schwartz, Phys. Rev. D18 (1978) 1607.

[16] D.A. Dicus, E.W. Kolb, V.L. Teplitz and R.V. Wagoner, Phys. Rev. D18 (1978) 1829; Phys. Rev. D22 (1980) 839.

[17] H. Georgi and S. Glashow, Phys. Rev. Lett 32 (1974) 438.

VOLUME 47, NUMBER 6 PHYSICAL REVIEW LETTERS 10 AUGUST 1981

SU(5) and the Invisible Axion

Mark B. Wise and Howard Georgi
Lyman Laboratory of Physics, Harvard University, Cambridge, Massachusetts 02138

and

Sheldon L. Glashow[a]
Center for Theoretical Physics, Massachusetts Institute of Technology, Cambridge, Massachusetts 02138
(Received 18 May 1981)

Dine, Fischler, and Srednicki have proposed a solution to the strong CP puzzle in which the mass and couplings of the axion are suppressed by an inverse power of a large mass. We construct an explicit SU(5) model in which this mass is the vacuum expectation value which breaks SU(5) down to SU(3)⊗SU(2)⊗U(1).

PACS numbers: 14.80.Kx, 11.30.Er, 12.20.Hx

The standard SU(3)⊗SU(2)⊗U(1) gauge theory appears to be adequate to *describe* all of the phenomenology of the strong, electromagnetic, and weak interactions. Moreover, much of the structure of these interactions is *explained* by the theory in the sense that it follows directly from the form of the gauge interactions. However, there are a number of features which can be *described* in the context of SU(3)⊗SU(2)⊗U(1) but which are in no sense *explained.* Some of these features, such as charge quantization and the observed value of the weak mixing angle, are *explained* by the extension of SU(3)⊗SU(2)⊗U(1) to the grand unifying group SU(5).[1] The rest comprise the fundamental puzzles of contemporary particle physics: Why SU(3)⊗SU(2)⊗U(1) [or SU(5)] and not some other gauge group? How many generations of quarks and leptons exist and why? Why do the quark masses and mixing angles take their observed values? Why is the CP nonconservation in the SU(3) strong interactions so small? Finally, in the context of grand unified theories, there is the hierarchy puzzle. Why are the mass scales associated with the electroweak and strong interactions so small compared to the unification mass scale $M_u \simeq 10^{15}$ GeV?[2] Some or all of these questions may not have answers. The world may just be the way it is.

Our penultimate question, the puzzle of the smallness of strong CP nonconservation, is particularly tantalizing. Several different mechanisms have been proposed to *explain* the smallness. Soft CP nonconservation[3] or a massless up quark[4] might do it at a price in elegance. The Peccei-Quinn[5] symmetry would do it, but the predicted axion[6] is not seen.[7] Some workers[8] have suggested scenarios in which the axion is heavy and hard to see. Dine, Fischler, and Srednicki[9] (DFS) have recently suggested a clever variant of the Peccei-Quinn scheme in which the axion mass and its coupling to normal matter are inversely proportional to a large and arbitrary vacuum expectation value (VEV) of an SU(2) singlet scalar field. If this VEV is large enough, their axion is invisible.

In this paper, we comment on the DFS idea. We first note that the singlet VEV must be greater than 10^9 GeV to satisfy astrophysical constraints.[10] In the SU(3)⊗SU(2)⊗U(1) theory, such a large mass scale is unnatural. Thus, in the context of SU(3)⊗SU(2)⊗U(1), the DFS idea is a trade-off. It explains the smallness of strong CP nonconservation at the cost of introducing a hierarchy puzzle.

In a grand unified theory, it seems reasonable to imagine that the singlet VEV is of order M_u. Our main purpose in this paper is to describe a model in which it is more than reasonable, it is automatic, because the DFS singlet field is precisely the field whose VEV breaks SU(5) down to SU(3)⊗SU(2)⊗U(1). In our model, the hierarchy puzzle is still with us, but the strong CP puzzle is solved at no addtional cost.

The astrophysical constraints on a light axion have been discussed by Dicus, Kolb, Teplitz, and Wagoner.[10] They find that for a light axion with conventional couplings, the power radiated in axions by the helium core of a red supergiant star would exceed the power in photon emission by about 10^{13}. Consistency with the usual stellar models can only be achieved if the axion couplings are reduced by at least $10^{6.5}$. In the DFS model, the axion coupling is reduced by the ratio of the usual Higgs VEV, $u \simeq 250$ GeV, to the singlet VEV. Thus the singlet VEV must be of order 10^9 GeV or larger.

Our main concern is the construction of an explicit SU(5) model which solves the strong CP puzzle. The fermion fields are the usual left-handed $\underline{10}$'s (T_L) and right-handed $\underline{5}$'s (F_R). The spinless fields are two $\underline{5}$'s, represented by column vectors H_1 and H_2, and a *complex* $\underline{24}$, represented by a traceless 5×5 matrix Σ. The Yukawa couplings are (schematically)

$$g_1 \overline{T}_L{}^c T_L H_1 + g_2 \overline{T}_L F_R H_2, \quad (1)$$

where c denotes charge conjugation. These are invariant under the Peccei-Quinn symmetry

$$T_L \to e^{-i\alpha/2} T_L, \quad F_R \to e^{i\alpha/2} F_R,$$
$$H_1 \to e^{i\alpha} H_1, \quad H_2 \to e^{-i\alpha} H_2. \quad (2)$$

We demand that this be a symmetry of the scalar meson self-interactions with the addition of the following transformation law for the Σ field:

$$\Sigma \to e^{-i\alpha}\Sigma. \quad (3)$$

Then the most general potential for the scalars is

$$V(H_1, H_2, \Sigma) = V_1(\Sigma) + V_2(H) + V_3(H, \Sigma), \quad (4)$$

where

$$V_1(\Sigma) = -\tfrac{1}{2}\mu^2 \operatorname{Tr}(\Sigma^\dagger\Sigma) + \tfrac{1}{4}a[\operatorname{Tr}(\Sigma^\dagger\Sigma)]^2 + \tfrac{1}{2}b\operatorname{Tr}(\Sigma^\dagger\Sigma\Sigma^\dagger\Sigma) + \tfrac{1}{4}c[\operatorname{Tr}(\Sigma^2)][\operatorname{Tr}(\Sigma^\dagger)^2] + \tfrac{1}{2}d\operatorname{Tr}(\Sigma\Sigma\Sigma^\dagger\Sigma^\dagger), \quad (5)$$

$$V_2(H) = -\tfrac{1}{2}\mu_1^2(H_1^\dagger H_1) - \tfrac{1}{2}\mu_2^2(H_2^\dagger H_2) + \tfrac{1}{4}\alpha_1(H_1^\dagger H_1)^2 + \tfrac{1}{4}\alpha_2(H_2^\dagger H_2)^2 + \tfrac{1}{4}\alpha_3(H_1^\dagger H_1)(H_2^\dagger H_2) + \tfrac{1}{4}\alpha_4(H_1^\dagger H_2)(H_2^\dagger H_1), \quad (6)$$

$$V_3(H, \Sigma) = \gamma_1(H_1^\dagger H_1)\operatorname{Tr}(\Sigma^\dagger\Sigma) + \gamma_2(H_2^\dagger H_2)\operatorname{Tr}(\Sigma^\dagger\Sigma) + \beta_1 H_1^\dagger\Sigma\Sigma^\dagger H_1 + \beta_2 H_2^\dagger\Sigma\Sigma^\dagger H_2 + \delta_1 H_1^\dagger\Sigma^\dagger\Sigma H_1 + \delta_2 H_2^\dagger\Sigma^\dagger\Sigma H_2$$
$$+ gH_2^\dagger\Sigma^2 H_1 + g^* H_1^\dagger\Sigma^{\dagger 2}H_2 + hH_2^\dagger H_1\operatorname{Tr}(\Sigma^2) + h^* H_1^\dagger H_2\operatorname{Tr}(\Sigma^{\dagger 2}), \quad (7)$$

where all constants except g and h are real.

For a range of parameters, the VEV's will take the form

$$\langle\Sigma\rangle = \begin{bmatrix} 2 & 0 & 0 & 0 & 0 \\ 0 & 2 & 0 & 0 & 0 \\ 0 & 0 & 2 & 0 & 0 \\ 0 & 0 & 0 & -3-\epsilon & 0 \\ 0 & 0 & 0 & 0 & -3+\epsilon \end{bmatrix} \lambda_0/2 \quad (8)$$

$$\langle H_{1,2}\rangle = \begin{bmatrix} 0 \\ 0 \\ 0 \\ 0 \\ \lambda_{1,2}/\sqrt{2} \end{bmatrix}. \quad (9)$$

The SU(3)⊗SU(2)⊗U(1) singlet component of Σ is the DFS singlet field in this model. Its VEV, λ_0, must be of order M_u while

$$|\lambda_1|^2 + |\lambda_2|^2 = u^2. \quad (10)$$

It follows that ϵ is very small:

$$|\epsilon| = O(u^2/M_u^2). \quad (11)$$

The axion is primarily the antihermetian part of the singlet component of Σ. But, it contains a small admixture (of order λ_j/λ_0) of the neutral components of H_j through which it couples to fermions.

One might worry that by enlarging the Higgs structure of our SU(5) theory we may have made the hierarchy puzzle more severe than in the standard SU(5) model. We can quantify this worry by counting the number of unnatural constraints which must be imposed to insure that the VEV's satisfy the desired hierarchy

$$|\lambda_0| \gg |\lambda_{1,2}| \gg |\epsilon\lambda_0|. \quad (12)$$

In the standard SU(5) model there is only one constraint in the sense that only one combination of large numbers must cancel to make the theory work.

The most straightforward way to minimize V is to require that (8) and (9) is an extremum, and that the second derivative matrix is positive semidefinite, so that (8) and (9) is at least a local minimum. Alternatively, for a range of the parameters, we can rewrite V as a sum of positive semidefinite terms, all of which vanish at the VEV, (8) and (9), which is thus an absolute minimum.

With either method, we find that there is a single unnatural condition which must be satisfied. As in the standard SU(5) model, the condition is that the square of mass of the true Higgs doublet (in the sense of Georgi and Nanopoulos[11]) be small.[12] The true Higgs boson is the SU(2) doublet component of

$$\lambda_1^* H_1 + \lambda_2^* H_2. \quad (13)$$

The orthogonal doublet typically has a mass of order M_u. This is very different from the usual Peccei-Quinn scheme in which the extra charged Higgs is light. In our version of the DFS model, the only extra particle with mass small compared to M_u is the invisible axion.

The invisible axion is a curious beast. Although it is very light, it does not really belong to the effective low-energy field theory that describes our world. Because it is a pseudo-Goldstone boson associated with symmetry breaking at M_u, all of its interactions are suppressed by inverse powers of M_u. This solution to the strong CP puzzle simply has no other consequences in low-energy particle physics. However, there may be cosmological implications of this idea.

Guth and Pi[13] point out a cosmological problem of conventional SU(5) with no trilinear coupling of the 24. It is associated with the discrete symmetry $\Sigma \to -\Sigma$ which leads to a twice degenerate vacuum. Our model has no trilinear couplings; however, the discrete symmetry $\Sigma \to -\Sigma$ is embedded within the continuous Peccei-Quinn symmetry of the Higgs potential.

The invisibility of our axion is established by the following order-of-magnitude estimates of its properties: axion mass $\sim f_\pi m_\pi / M_u \sim 10^{-8}$ eV; lifetime for 2γ decay $\sim (M_u/f_\pi)^5 \tau_{\pi^0} \sim 10^{56}$ yr; pseudoscalar couplings $\sim f_\pi/M_u \sim 10^{-16}$; scalar couplings $\sim \bar{\theta} f_\pi / M_u \sim 10^{-31}$.

CP-nonconserving scalar couplings of the axion are induced by the nonperturbative breaking of the Peccei-Quinn symmetry. In principle, $\bar{\theta}$ is calculable in our model, and we estimate it to be about 10^{-15}. The scalar couplings will lead to a "long-range" attraction of baryons by axion exchange.[14] The effect is about 10^{-24} of the universal gravitational attraction. The contribution of $\bar{\theta}$ to the electric dipole moment of the neutron is about 10^{-31} $e\cdot$cm.[15] The most disquieting aspect of this solution to the strong CP problem is the predicted existence of an almost massless particle which is in practice unobservable.[16]

One of us (M.B.W.) is a recipient of a Harvard Society of Fellows, Junior Fellowship. This research was supported in part by the National Science Foundation under Grant No. PHY77-22864 and in part by the U. S. Department of Energy under Contract No. DE-AC02-76ER0-3069.

Note added.—Our argument that only a single unnatural condition is needed to produce the hierarchy is rather general. It applies, for example, to the SU(5) model in Ref. 10, where a real 24, a complex singlet, and two 5's of Higgs are used. The astrophysical constraints[10] on the SU(2)⊗U(1) singlet VEV are also mentioned in Ref. 9.

---

[a] On leave from Lyman Laboratory of Physics, Harvard University, Cambridge, Mass. 02138.

[1] H. Georgi and S. L. Glashow, Phys. Rev. Lett. 32, 438 (1974).

[2] H. Georgi, H. R. Quinn, and S. Weinberg, Phys. Rev. Lett. 33, 451 (1974).

[3] H. Georgi, Hadron J. 1, 156 (1978); R. N. Mohapatra and G. Senjanović, Phys. Lett. 79B, 283 (1978); G. Segrè and H. A. Weldon, Phys. Rev. Lett. 42, 1191 (1979); S. Barr and D. Langacker, Phys. Rev. Lett. 42, 1654 (1979); M. Bég and H. Tsao, Phys. Rev. Lett. 41, 278 (1978); V. Corrin, G. Segrè, and H. A. Weldon, Phys. Rev. D 21, 1410 (1980).

[4] H. Georgi and I. McArthur, Harvard University Report No. HUTP-81/A011, 1981 (unpublished); A. Zepeda, Phys. Rev. Lett. 41, 139 (1978); N. Deshpande and D. Soper, Phys. Rev. Lett. 41, 735 (1978).

[5] R. Peccei and H. Quinn, Phys. Rev. Lett. 38, 1440 (1977).

[6] S. Weinberg, Phys. Rev. Lett. 40, 223 (1978); F. Wilczek, Phys. Rev. Lett. 46, 279 (1978).

[7] J. Donnely, S. Freedman, R. Lytel, R. Peccei, and M. Schwartz, Phys. Rev. D 18, 1607 (1978).

[8] S. H. Tye, Cornell University Report No. CLNS/81-489, 1981 (unpublished); E Cohen, Weizmann Institute Report No. WIS 2/81, 1981 (unpublished); S. Dimopoulos, Phys. Lett. 84B, 435 (1979).

[9] M. Dine, W. Fischler, and M. Srednicki, "A Simple Solution to the Strong CP Problem with a Harmless Axion" (to be published). S. Raby has informed us that similar ideas will appear in a forthcoming paper by him and H. P. Nilles. Related ideas can be gleaned from an earlier paper by J. Kim, Phys. Rev. Lett. 43, 103 (1979).

[10] D. A. Dicus, E. W. Kolb, V. L. Teplitz, and R. V. Wagoner, Phys. Rev. D 18, 1829 (1978).

[11] H. Georgi and D. V. Nanopoulos, Phys. Lett. 82B, 95 (1979).

[12] S. Weinberg, Phys. Lett. 82B, 387 (1979).

[13] A. Guth and S.-Y. Pi, private communication.

[14] J. Preskill and E. Witten, private communication.

[15] V. Baluni, Phys. Rev. D 19, 2227 (1979); R. Crewther, P. Di Vecchia, and G. Veneziano, Phys. Lett. 88B, 123 (1979).

[16] There are other models with essentially unobservable light particles. See, for example, Y. Chikashige, R. Mohapatra, and R. Peccei, Phys. Lett. 98B, 265 (1981).

XVII. *Composite Quarks and Leptons*

1) Naturalness, chiral symmetry, and spontaneous chiral symmetry breaking
by G. 't Hooft, in *Recent Developments in Gauge Theories,* p. 135 (Plenum Press, 1980) 1004

2) Family structure with composite quarks and leptons
by I. Bars, *Phys. Lett.* **106**B, 105 (1981) 1027

NATURALNESS, CHIRAL SYMMETRY, AND SPONTANEOUS CHIRAL SYMMETRY BREAKING

G. 't Hooft

Institute for Theoretical Fysics

Utrecht, The Netherlands

ABSTRACT

A property called "naturalness" is imposed on gauge theories. It is an order-of-magnitude restriction that must hold at all energy scales μ. To construct models with complete naturalness for elementary particles one needs more types of confining gauge theories besides quantum chromodynamics. We propose a search program for models with improved naturalness and concentrate on the possibility that presently elementary fermions can be considered as composite. Chiral symmetry must then be responsible for the masslessness of these fermions. Thus we search for QCD-like models where chiral symmetry is not or only partly broken spontaneously. They are restricted by index relations that often cannot be satisfied by other than unphysical fractional indices. This difficulty made the author's own search unsuccessful so far. As a by-product we find yet another reason why in ordinary QCD chiral symmetry must be broken spontaneously.

III1. INTRODUCTION

The concept of causality requires that macroscopic phenomena follow from microscopic equations. Thus the properties of liquids and solids follow from the microscopic properties of molecules and atoms. One may either consider these microscopic properties to have been chosen at random by Nature, or attempt to deduce these from even more fundamental equations at still smaller length and time scales. In either case, it is unlikely that the microscopic equations contain various free parameters that are carefully adjusted by Nature to give cancelling effects such that the macroscopic systems have some special properties. This is a philosophy which we would like to apply to the unified gauge theories: the effective interactions at a large length scale, corresponding to a low energy scale μ_1, should follow from the properties at a much smaller length scale, or higher energy scale μ_2, without the requirement that various different parameters at the energy scale μ_2 match with an accuracy of the order of μ_1/μ_2. That would be unnatural. On the other hand, if at the energy scale μ_2 some parameters would be very small, say

$$\alpha(\mu_2) = \mathcal{O}(\mu_1/\mu_2) , \qquad \text{(III1)}$$

then this may still be natural, provided that this property would not be spoilt by any higher order effects. We now conjecture that the following dogma should be followed:
- at any energy scale μ, a physical parameter or set of physical parameters $\alpha_i(\mu)$ is allowed to be very small only if the replacement $\alpha_i(\mu) = o$ would increase the symmetry of the system. -
In what follows this is what we mean by naturalness. It is clearly a weaker requirement than that of P. Dirac[1] who insists on having no small numbers at all. It is what one expects if at any mass scale $\mu > \mu_o$ some ununderstood theory with strong interactions determines a spectrum of particles with various good or bad symmetry properties. If at $\mu = \mu_o$ certain parameters come out to be small, say 10^{-5}, then that cannot be an accident; it must be the consequence of a near symmetry.

For instance, at a mass scale

$\mu = 50$ GeV,

the electron mass m_e is 10^{-5}. This is a small parameter. It is acceptable because $m_e = o$ would imply an additional chiral symmetry corresponding to separate conservation of left handed and right handed electron-like leptons. This guarantees that all renormalizations of m_e are proportional to m_e itself. In sects. III2 and III3 we compare naturalness for quantum electrodynamics and ϕ^4 theory.

Gauge coupling constants and other (sets of) interaction constants may be small because putting them equal to zero would turn the gauge bosons or other particles into free particles so that they are separately conserved.

If within a set of small parameters one is several orders of magnitude smaller than another then the smallest must satisfy our "dogma" separately. As we will see, naturalness will put the severest restriction on the occurrence of scalar particles in renormalizable theories. In fact we conjecture that this is the reason why light, weakly interacting scalar particles are not seen.

It is our aim to use naturalness as a new guideline to construct models of elementary particles (sect. III4). In practice naturalness will be lost beyond a certain mass scale μ_o, to be referred to as "Naturalness Breakdown Mass Scale" (NBMS). This simply means that unknown particles with masses beyond that scale are ignored in our model. The NBMS is only defined as an order of magnitude and can be obtained for each renormalizable field theory. For present "unified theories", including the existing grand unified schemes, it is only about 1000 GeV. In sect. 5 we attempt to construct realistic models with an NBMS some orders of magnitude higher.

One parameter in our world is unnatural, according to our definition, already at a very low mass scale ($\mu_o \sim 10^{-2}$ eV). This is the cosmological constant. Putting it equal to zero does not seem to increase the symmetry. Apparently gravitational effects do not obey naturalness in our formulation. We have nothing to say about this fundamental problem, accept to suggest that *only* gravitational effects violate naturalness. Quantum gravity is not understood anyhow so we exclude it from our naturalness requirements.

On the other hand it is quite remarkable that all other elementary particle interactions have a high degree of naturalness. No unnatural parameters occur in that energy range where our popular field theories could be checked experimentally. We consider this as important evidence in favor of the general hypothesis of naturalness. Pursuing naturalness beyond 1000 GeV will require theories that are immensely complex compared with some of the grand unified schemes.

A remarkable attempt towards a natural theory was made by Dimopoulos and Susskind [2)]. These authors employ various kinds of confining gauge forces to obtain scalar bound states which may substitute the Higgs fields in the conventional schemes. In their model the observed fermions are still considered to be elementary.

Most likely a complete model of this kind has to be constructed step by step. One starts with the experimentally accessible aspects of the Glashow-Weinberg-Salam-Ward model. This model is natural if one restricts oneself to mass-energy scales below 1000 GeV. Beyond 1000 GeV one has to assume, as Dimopoulos and Susskind do, that the Higgs field is actually a fermion-antifermion composite field. Coupling this field to quarks and leptons in order to produce their mass, requires new scalar fields that cause naturalness to break down at 30 TeV or so. Dimopoulos and Susskind speculate further on how to remedy this. To supplement such ideas, we toyed with the idea that (some of) the presently "elementary" fermions may turn out to be bound states of an odd number of fermions when considered beyond 30 TeV. The binding mechanism would be similar to the one that keeps quarks inside the proton. However, the proton is not particularly light compared with the characteristic mass scale of quantum chromodynamics (QCD). Clearly our idea is only viable if something prevented our "baryons" from obtaining a mass (eventually a small mass may be due to some secondary perturbation).

The proton ows its mass to spontaneous breakdown of chiral symmetry, or so it seems according to a simple, fairly successful model of the mesonic and baryonic states in QCD: the Gell-Mann-Lévy sigma model[3]. Is it possible then that in some variant of QCD chiral symmetry is not spontaneously broken, or only partly, so that at least some chiral symmetry remains in the spectrum of fermionic bound states? In this article we will see that in general in SU(N) binding theories this is not allowed to happen, i.e. chiral symmetry must be broken spontaneously.

III2. NATURALNESS IN QUANTUM ELECTRODYNAMICS

Quantum Electrodynamics as a renormalizable model of electrons (and muons if desired) and photons is an example of a "natural" field theory. The parameters α, m_e (and m_μ) may be small independently. In particular m_e (and m_μ) are very small at large μ. The relevant symmetry here is chiral symmetry, for the electron and the muon separately. We need not be concerned about the Adler-Bell-Jackiw anomaly here because the photon field being Abelian cannot acquire non-trivial topological winding numbers[4].

There is a value of μ where Quantum Electrodynamics ceases to be useful, even as a model. The model is not asymptotically free, so there is an energy scale where all interactions become strong:

$$\mu_o \simeq m_e \exp(6\pi^2/e^2N_f) , \qquad \text{(III2)}$$

where N_f is the number of light fermions. If some world would be described by such a theory at low energies, then a replacement of the theory would be necessary at or below energies of order μ_o.

III3. ϕ^4-THEORY

A renormalizable scalar field theory is described by the Lagrangian

$$\mathcal{L} = -\tfrac{1}{2}(\partial_\mu\phi)^2 - \tfrac{1}{2}m^2\phi^2 - \frac{1}{4!}\lambda\phi^4 . \qquad \text{(III3)}$$

the interactions become strong at

$$\mu \simeq m \exp(16\pi^2/3\lambda) , \qquad \text{(III4)}$$

but is it still natural there?

There are two parameters, λ and m. Of these, λ may be small because $\lambda = o$ would correspond to a non-interacting theory with total number of ϕ particles conserved. But is small m allowed? If we put m = o in the Lagrangian (III3) then the symmetry is not enhanced*). However we can take both m and λ to be small, because if $\lambda = m = o$ we have invariance under

$$\phi(x) \rightarrow \phi(x) + \Lambda \ . \qquad \text{(III5)}$$

This would be an approximate symmetry of a new underlying theory at energies of order μ_o. Let the symmetry be broken by effects described by a dimensionless parameter ε. Both the mass term and the interaction term in the effective Lagrangian (III3) result from these symmetry breaking effects. Both are expected to be of order ε. Substituting the correct powers of μ_o to account for the dimensions of these parameters we have

$$\begin{aligned} \lambda &= \mathcal{O}(\varepsilon) \ , \\ m^2 &= \mathcal{O}(\varepsilon\mu_o^2) \ . \end{aligned} \qquad \text{(III6)}$$

Therefore,

$$\mu_o = \mathcal{O}(m/\sqrt{\lambda}) \ . \qquad \text{(III7)}$$

This value is much lower than eq. (III4). We now turn the argument around: if any "natural" underlying theory is to describe a scalar particle whose *effective* Lagrangian at low energies will be eq. (III3), then its energy scale cannot be given by (III4) but at best by (III7). We say that naturalness breaks down beyond $m/\sqrt{\lambda}$. It must be stressed that these are orders of magnitude. For instance one might prefer to consider λ/π^2 rather than λ to be the relevant parameter. μ_o then has to be multiplied by π. Furthermore, λ could be much smaller than ε because $\lambda = o$ separately also enhances the symmetry. Therefore, apart from factors π, eq. (III7) indicates a maximum value for μ_o.

Another way of looking at the problem of naturalness is by comparing field theory with statistical physics. The parameter m/μ would correspond to $(T-T_c)/T$ in a statistical ensemble. Why would the temperature T chosen by Nature to describe the elementary particles be so close to a critical temperature T_c? If $T_c \neq o$ then T may not be close to T_c just by accident.

III4. NATURALNESS IN THE WEINBERG-SALAM-GIM MODEL

The difficulties with the unnatural mass parameters only occur in theories with scalar fields. The only fundamental scalar

---

*) Conformal symmetry is violated at the quantum level.

field that occurs in the presently fashionable models is the Higgs field in the extended Weinberg-Salam model. The Higgs mass-squared, m_H^2, is up to a coefficient a fundamental parameter in the Lagrangian. It is small at energy scales $\mu \gg m_H$. Is there an approximate symmetry if $m_H \to 0$? With some stretch of imagination we might consider a Goldstone-type symmetry:

$$\phi(x) \to \phi(x) + \text{const.} \tag{III8}$$

However we also had the local gauge transformations:

$$\phi(x) \to \Omega(x)\,\phi(x)\ . \tag{III9}$$

The transformations (III8) and (III9) only form a closed group if we also have invariance under

$$\phi(x) \to \phi(x) + C(x)\ . \tag{III10}$$

But then it becomes possible to transform ϕ away completely. The Higgs field would then become an unphysical field and that is not what we want. Alternatively, we could have that (III8) is an approximate symmetry only, and it is broken by all interactions that have to do with the symmetry (III9) which are the weak gauge field interactions. Their strength is $g^2/4\pi = \mathcal{O}(1/137)$. So at best we can have that the symmetry is broken by $\mathcal{O}(1/137)$ effects. Therefore

$$m_H^2/\mu^2 \gtrsim \mathcal{O}(1/137)\ .$$

Also the $\lambda\phi^4$ term in the Higgs field interactions breaks this symmetry. Therefore

$$m_H^2/\mu^2 \gtrsim \mathcal{O}(\lambda) \gtrsim \mathcal{O}(1/137)\ . \tag{III11}$$

Now

$$m_H^2 = \mathcal{O}(\lambda F_H^2)\ , \tag{III12}$$

where F_H is the vacuum expectation value of the Higgs field, known to be*)

$$F_H = (2G\sqrt{2})^{-1/2} = 174\ \text{GeV}\ . \tag{III13}$$

We now read off that

$$\mu \lesssim \mathcal{O}(F_H) = \mathcal{O}(174\ \text{GeV})\ . \tag{III14}$$

*) Some numerical values given during the lecture were incorrect. I here give corrected values.

This means that at energy scales much beyond F_H our model becomes more and more unnatural. Actually, factors of π have been omitted. In practice one factor of 5 or 10 is still not totally unacceptable. Notice that the actual value of m_H dropped out, except that

$$m_H = \mathcal{O}\left(\frac{\sqrt{\lambda}}{g} M_W\right) \gtrsim \mathcal{O}(M_W) \ . \qquad \text{(III15)}$$

Values for m_H of just a few GeV are unnatural.

III5. EXTENDING NATURALNESS

Equation (III14) tells us that at energy scales much beyond 174 GeV the standard model becomes unnatural. As long as the Higgs field H remains a fundamental scalar nothing much can be done about that. We therefore conclude, with Dimopoulos and Susskind[2] that the "observed" Higgs field must be composite. A non-trivial strongly interacting field theory must be operative at 1000 GeV or so. An obvious and indeed likely possibility is that the Higgs field H can be written as

$$H = Z\bar{\psi}\psi \ , \qquad \text{(III16)}$$

where Z is a renormalization factor and ψ is a new quark-like object, a fermion with a new color-like interaction [2]. We will refer to the object as meta-quark having meta-color. The theory will have all features of QCD so that we can copy the nomenclature of QCD with the prefix "meta-". The Higgs field is a meta-meson.

It is now tempting to assume that the meta-quarks transform the same way under weak SU(2) x U(1) as ordinary quarks. Take a doublet with left-handed components forming one gauge doublet and right handed components forming two gauge singlets. The meta-quarks are massless. Suppose that the meta-chiral symmetry is broken spontaneously just as in ordinary QCD. What would happen?

What happens is in ordinary QCD well described by the Gell-Mann-Lévy sigma model. The lightest mesons form a quartet of real fields, ϕ_{ij}, transforming as a

$$2^{\text{left}} \otimes 2^{\text{right}}$$

representation of

$$SU(2)^{\text{left}} \otimes SU(2)^{\text{right}}.$$

Since the weak interaction only deals with $SU(2)^{\text{left}}$ this quartet can also be considered as one complex doublet representation of weak SU(2). In ordinary QCD we have

$$\phi_{ij} = \sigma\delta_{ij} + i\tau^a_{ij}\pi^a \ , \qquad \text{(III17)}$$

and

$$<\sigma>_{vacuum} = \frac{1}{\sqrt{2}} f_\pi = 91 \text{ MeV} \ . \qquad \text{(III18)}$$

The complex doublet is then

$$\phi_i = \frac{1}{\sqrt{2}} \begin{pmatrix} \sigma + i\pi^3 \\ \pi^2 + i\pi^1 \end{pmatrix} , \qquad \text{(III19)}$$

and

$$<\phi_i>_{vacuum} = \begin{pmatrix} 1 \\ 0 \end{pmatrix} \times 64 \text{ MeV} \ . \qquad \text{(III20)}$$

We conclude that if we transplant this theory to the TeV range then we get a scalar doublet field with a non-vanishing vacuum expectation value for free. All we have to do now is to match the numbers. If we scale all QCD masses by a scaling factor κ then we match

$$F_H = 174 \text{ GeV} = \kappa \ 64 \text{ MeV} \ ;$$

$$\kappa = 2700 \ . \qquad \text{(III21)}$$

Now the mesonic sector of QCD is usually assumed to be reproduced in the 1/N expansion [5] where N is the number of colors (in QCD we have N = 3). The 4-meson coupling constant goes like 1/N. Then one would expect

$$f_\pi \propto \sqrt{N} \ . \qquad \text{(III22)}$$

Therefore

$$\kappa = 2700 \sqrt{\frac{3}{N}} \ , \qquad \text{(III23)}$$

if the metacolor group is SU(N).

Thus we obtain a model that reproduces the W-mass and predicts the Higgs mass. The Higgs is the meta-sigma particle. The ordinary sigma is a wide resonance at about 700 MeV[3], so that we predict

$$m_H = \kappa m_\sigma = 1900 \sqrt{\frac{3}{N}} \text{ GeV} \ , \qquad \text{(III24)}$$

and it will be extremely difficult to detect among other strongly interacting objects.

III6. WHAT NEXT?

The model of the previous section is to our mind nearly inevitable, but there are problems. These have to do with the observed fermion masses. All leptons and quarks owe their masses to an interaction term of the form

$$g\bar{\psi}H\psi\ , \qquad \text{(III25)}$$

where g is a coupling constant, ψ is the lepton or quark and H is the Higgs field. With (III16) this becomes a four-fermion interaction, a fundamental interaction in the new theory. Because it is non-renormalizable further structure is needed. In ref. 2 the obvious choice is made: a new "meta-weak interaction" gauge theory enters with new super-heavy intermediate vector bosons. But since H is a scalar this boson must be in the crossed channel, a rather awkward situation. (See option a in Figure 1.) A simpler theory is that a new scalar particle is exchanged in the direct channel. (See option b in Figure 1.)

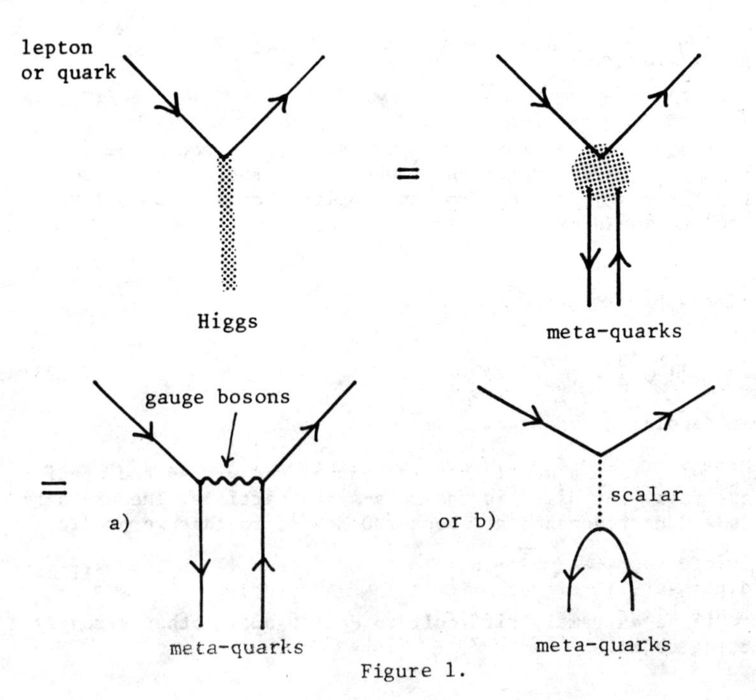

Figure 1.

Notice that in both cases new scalar fields are needed because in case a) something must cause the "spontaneous breakdown" of the new gauge symmetries. Therefore choice b) is simpler.
We removed a Higgs scalar and we get a scalar back. Does naturalness improve? The answer is yes. The coupling constant g in the interaction (III25) satisfies

$$g = g_1 g_2 / M_s^2 Z \ . \tag{III26}$$

Here g_1 and g_2 are the couplings at the new vertices, M_s is the new scalar's mass, and Z is from (III16) and is of order

$$Z \sim \frac{1}{\sqrt{\frac{N}{3}}\,(\kappa\, m_\rho)^2} = \frac{\sqrt{N/3}}{(1800\ \mathrm{GeV})^2} \ . \tag{III27}$$

Suppose that the heaviest lepton or quark is about 10 GeV. For that fermion the coupling constant g is

$$g = \frac{m_f}{F} \simeq 1/20 \ .$$

We get

$$g_1 g_2 \simeq \left(\frac{M_s}{1800\ \mathrm{GeV}}\right)^2 \sqrt{\frac{N}{3}} \cdot \frac{1}{20} \ .$$

Naturalness breaks down at

$$\mu = \mathcal{O}\left(\frac{M_s}{g_{1,2}}\right) = 8000 \sqrt[4]{\frac{3}{N}}\ \mathrm{GeV} \ ,$$

an improvement of about a factor 50 compared with the situation in sect. III4. Presumably we are again allowed to multiply by factors like 5 or 10, before getting into real trouble.

Before speculating on how to go on from here to improve naturalness still further we must assure ourselves that all other alleys are blind ones. An intriguing possibility is that the presently observed fermions are composite. We would get option c), Figure 2.

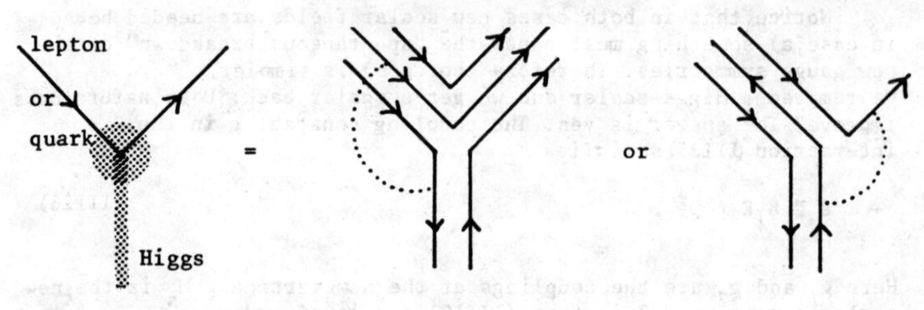

Fig. 2

The dotted line could be an ordinary weak interaction W or photon, that breaks an internal symmetry in the binding force for the new components. The new binding force could either act at the 1 TeV or at the 10-100 TeV range. It could either be an extension of metacolor or be a (color)" or paracolor force. Is such an idea viable?

Clearly, compared with the energy scale on which the binding forces take place, the composite fermions must be nearly massless. Again, this cannot be an accident. The chiral symmetry responsible for this must be present in the underlying theory. Apparently then, the underlying theory will possess a chiral symmetry which is not (or not completely) spontaneously broken, but reflected in the bound state spectrum in the Wigner mode: some massless chiral objects and parity doubled massive fermions. This possibility is most clearly described by the σ-model as a model for the lowest bound states occurring in ordinary quantum chromodynamics.

III7. THE σ MODEL

The fermion system in quantum chromodynamics shows an axial symmetry. To illuminate our problem let us consider the case of two flavors. The local color group is $SU(3)_c$. The subscript c here stands for color. The flavor symmetry group is $SU(2)_L \otimes SU(2)_R \otimes U(1)$ where the subscripts L and R stands for left and right and the group elements must be chosen to be space-time independent. We split the fermion fields ψ into left and right components:

$$\psi = \tfrac{1}{2}(1+\gamma_5)\psi_L + \tfrac{1}{2}(1-\gamma_5)\psi_R \;. \qquad \text{(III28)}$$

ψ_L transforms as a $3_c \otimes 2_L \otimes 1_R \otimes 2_{\mathcal{L}}$ (III29)

and ψ_R transforms as a $3_c \otimes 1_L \otimes 2_R \otimes \bar{2}_{\mathcal{L}}$ (III30)

where the indices refer to the various groups. $\mathcal{L}$ stands for the Lorentzgroup SO(3,1), locally equivalent to SL(2,c) which has two different complex doublet representations $2_{\mathcal{L}}$ and $\bar{2}_{\mathcal{L}}$ (corresponding to the transformation law for the neutrino and antineutrino, respectively). The fields ψ_L and ψ_R have the same charge under U(1), whereas axial U(1) group (under which they would have opposite charges) is absent because of instanton effects[4].

The effect of the color gauge fields is to bind these fermions into mesons and baryons all of which must be color singlets. It would be nice if one could describe these hadronic fields as representations of $SU(2)_L \otimes SU(2)_R \otimes U(1)$ and the Lorentz group, and then cast their mutual interactions in the form of an effective Lagrangian, invariant under the flavor symmetry group. In the case at hand this is possible and the resulting construction is a successful and one-time popular model for pions and nucleons: the σ model[3]. We have a nucleon doublet

$$N = \tfrac{1}{2}(1+\gamma_5)N_L + \tfrac{1}{2}(1-\gamma_5)N_R \, , \tag{III31}$$

where

$$N_L \text{ transforms as a } 1_c \otimes 2_L \otimes 1_R \otimes 2 \quad , \tag{III32a}$$

$$\text{and } N_R \text{ transforms as a } 1_c \otimes 1_L \otimes 2_R \otimes \bar{2}_{\mathcal{L}} \quad . \tag{III32b}$$

Further we have a quartet of real scalar fields $(\sigma, \vec{\pi})$ which transform as a $1_c \otimes 2_L \otimes 2_{\mathcal{L}} \otimes 1_{\mathcal{L}}$. The Lagrangian is

$$\mathcal{L} = -\bar{N}[\gamma\partial + g_o(\sigma + i\vec{\tau}.\vec{\pi}\gamma_5)]N - \tfrac{1}{2}(\partial\pi)^2 - \tfrac{1}{2}(\partial\sigma)^2 - V(\sigma^2 + \vec{\pi}^2) \; . \tag{III33}$$

Here V must be a rotationally invariant function.

Usually V is chosen such that its absolute minimum is away from the origin. Let V be minimal at $\sigma = v$ and $\vec{\pi} = o$. Here v is just a c-number. To obtain the physical particle spectrum we write

$$\sigma = v + s \tag{III34}$$

and we find

$$\mathcal{L} = -\tfrac{1}{2}\bar{N}(\gamma\partial + g_o v)N - \tfrac{1}{2}(\partial\vec{\pi})^2 - \tfrac{1}{2}(\partial s)^2 - 2v^2V''(v^2)s^2 + \text{interaction terms} \; . \tag{III35}$$

Clearly, in this case the nucleons acquire a mass term $m_s = g_o v$ and the s particle has a mass $m_s^2 = 4v^2V''(v^2)$, whereas the pion remains strictly massless. The entire mass of the pion must be due to effects that explicitly break $SU(2)_L \times SU(2)_R$, such as a small mass term $m_q\bar{\psi}\psi$ for the quarks (III28). We say that in this case the flavor group $SU(2)_L \otimes SU(2)_R$ is spontaneously broken into the isospin group SU(2).

Another possibility however, apparently not realised in ordinary quantum chromodynamics, would be that $SU(2)_L \otimes SU(2)_R$ is *not* spontaneously broken. We would read off from the Lagrangian (III33) that the nucleons N would form a massless doublet and that the four fields $(\sigma,\vec{\pi})$ could be heavy. The dynamics of other confining gauge theories could differ sufficiently from ordinary QCD so that, rather than a spontaneous symmetry breakdown, massless "baryons" develop. The principle question we will concentrate on is why do these massless baryons form the representation (III32), and how does this generalize to other systems. We would let future generations worry about the question where exactly the absolute minimum of the effective potential V will appear.

III8. INDICES

We now consider any color group G_c. The fundamental fermions in our system must be non-trivial representation of G_c and we assume "confinement" to occur: all physical particles are bound states that are singlets under G_c. Assume that the fermions are all massless (later mass terms can be considered as a perturbation). We will have automatically some global symmetry which we call the flavor group G_F. (We only consider exact flavor symmetries, not spoilt by instanton effects.) Assume that G_F is not spontaneously broken. Which and how many representations of G_F will occur in the massless fermion spectrum of the baryonic bound states? We must formulate the problem more precisely. The massless nucleons in (III33) being bound states, may have many massive excitations. However, massive Fermion fields cannot transform as a $2_{\mathcal{L}}$ under Lorentz transformations; they must go as a $2_{\mathcal{L}} \oplus \bar{2}_{\mathcal{L}}$. That is because a mass term being a Lorentz invariant product of two fields at one point only links $2_{\mathcal{L}}$ representations with $\bar{2}_{\mathcal{L}}$ representations. Consider a given representation r of G_F. Let p be the number of field multiplets transforming as $r \otimes 2_{\mathcal{L}}$ and q be the number of field multiplets $r \otimes \bar{2}_{\mathcal{L}}$. Mass terms that link the $2_{\mathcal{L}}$ with $\bar{2}_{\mathcal{L}}$ fields are completely invariant and in general to be expected in the effective Lagrangian. But the absolute value of

$$\ell = p - q \tag{III36}$$

is the minimal number of surviving massless chiral field multiplets. We will call ℓ the index corresponding to the representation r of G_F. By definition this index must be a (positive or negative) integer. In the sigma model it is postulated that

$$\text{index } (2_L \otimes 1_R) = 1 \tag{III37}$$

$$\text{index } (1_L \otimes 2_R) = -1$$

index (r) = o for all other representations r.

This tells us that if chiral symmetry is not broken spontaneously one massless nucleon doublet emerges. We wish to find out what massless fermionic bound states will come out in more general theories. Our problem is: how does (III37) generalize?

III9. ABSENCE OF MASSLESS BOUND STATES WITH SPIN 3/2 OR HIGHER

In the foregoing we only considered spin o and spin 1/2 bound states. Is it not possible that fundamentally massless bound states develop with higher spin? I believe to have strong arguments that this is indeed not possible. Let us consider the case of spin 3/2. Massive spin 3/2 fermions are described by a Lagrangian of the form

$$\mathcal{L} = \tfrac{1}{2}\bar{\psi}_\mu [\sigma_{\mu\nu}(\gamma\partial+m) + (\gamma\partial+m)\sigma_{\mu\nu}]\psi_\nu \ . \tag{III38}$$

Just like spin-one particles, this has a gauge-invariance if $m \to o$:

$$\psi_\mu \to \psi_\mu + \partial_\mu \eta(x) \ , \tag{III39}$$

where $\eta(x)$ is arbitrary. Indeed, massless spin 3/2 particles only occur in locally supersymmetric field theories. The field $\eta(x)$ is fundamentally unobservable.

Now in our model ψ_μ would be shorthand for some composite field: $\psi_\mu \to \psi\psi\psi$. However, then all components of this, including η, would be observables. If m = o we would be forced to add a gauge fixing term that would turn η into an unacceptable ghost particle*).

We believe, therefore, that unitarity and locality forbid the occurrence of massless bound states with spin 3/2. The case for higher spin will not be any better. And so we concentrate on a bound state spectrum of spin 1/2 particles only.

---

*) Note added: during the lectures it was suggested by one attendant to consider only gauge-invariant fields as $\Psi_{\mu\nu} = \partial_\mu\psi_\nu - \partial_\nu\psi_\mu$. However, such fields must satisfy constraints: $\partial[\alpha\Psi_{\mu\nu}] = o$. Composite field will never automatically satisfy such constraints.

III10. SPECTATOR GAUGE FIELDS AND -FERMIONS

So far, our model consisted of a strong interaction color gauge theory with gauge group G_c, coupled to chiral fermions in various representations r of G_c but of course in such a way that the anomalies cancel. The fermions are all massless and form multiplets of a global symmetry group, called G_F. For QCD this would be the flavor group. In the metacolor theory G_F would include all other fermion symmetries besides metacolor.

In order to study the mathematical problem raised above we will add another gauge connection field that turns G_F into a local symmetry group. The associated coupling constants may all be arbitrarily small, so that the dynamics of the strong color gauge interactions is not much affected. In particular the massless bound state spectrum should not change. One may either think of this new gauge field as a completely quantized field or simply as an artificial background field with possibly non-trivial topology. We will study the behavior of our system in the presence of this "spectator gauge field". As stated, its gauge group is G_F.

Note however, that some flavor transformations could be associated with anomalies. There are two types of anomalies:

i) those associated with $G_c \times G_F$, only occurring where the color field has a winding number. Only U(1) invariant subgroups of G_F contribute here. They simply correspond to small explicit violations of the G_F symmetry. From now on we will take as G_F only the anomaly-free part. Thus, for QCD with N flavors, G_F is not U(N) × U(N) but

$$G_F = SU(N) \otimes SU(N) \otimes U(1) .$$

ii) those associated with G_F alone. They only occur if the spectator gauge field is quantized. To remedy these we simply add "spectator fermions" coupled to G_F alone. Again, since these interactions are weak they should not influence the bound state spectrum.

Here, the spectator gauge fields and fermions are introduced as mathematical tools only. It just happens to be that they really do occur in Nature, for instance the weak and electromagnetic SU(2) × U(1) gauge fields coupled to quarks in QCD. The leptons then play the role of spectator fermions.

III11. ANOMALY CANCELLATION FOR THE BOUND STATE SPECTRUM

Let us now resume the particle content of our theory. At small distances we have a gauge group $G_c \otimes G_F$ with chiral fermions in several representations of this group. Those fermions which are trivial under G_c are only coupled weakly and are called "spectator fermions". All anomalies cancel, by construction.

At low energies, much lower than the mass scale where color binding occurs, we see only the G_F gauge group with its gauge fields. Coupled to these gauge fields are the massless bound states, forming new representations r of G_F, with either left- or right handed chirality. The numbers of left minus right handed fermion fields in the representations r are given by the as yet unknown indices $\ell(r)$. And finally we have the spectator fermions which are unchanged.

We now expect these very light objects to be described by a new local field theory, that is, a theory local with respect to the large distance scale that we now use. The central theme of our reasoning is now that this new theory must again be anomaly free. We simply cannot allow the contradictions that would arise if this were not so. Nature must arrange its new particle spectrum in such a way that unitarity is obeyed, and because of the large distance scale used the effective interactions are either vanishingly small or renormalizable. The requirement of anomaly cancellation in the new particle spectrum gives us equations for the indices $\ell(r)$, as we will see.

The reason why these equations are sometimes difficult or impossible to solve is that the new representations r must be different from the old ones; if $G_c = SU(N)$ then r must also be faithful representations of $G_F/Z(N)$. For instance in QCD we only allow for octet or decuplet representations of $(SU(3))_{flavor}$, whereas the original quarks were triplets.

However, the anomaly cancellation requirement, restrictive as it may be, does not fix the values of $\ell(r)$ completely. We must look for additional limitations.

III12 APPELQUIST-CARAZZONE DECOUPLING AND N-INDEPENDENCE

A further limitation is found by the following argument. Suppose we add a mass term for one of the colored fermions.

$$\Delta\mathcal{L} = m\, \bar{\psi}_{1L}\, \psi_{1R} + \text{h.c.}$$

Clearly this links one of the left handed fermions with one of the right handed ones and thus reduces the flavor group G_F into $G_F' \subset G_F$. Now let us gradually vary m from o to infinity. A famous theorem [5] tells us that in the limit $m \to \infty$ all effects due to this massive quark disappear. All bound states containing this quark should also disappear which they can only do by becoming very heavy. And they can only become heavy if they form representations r' of G_F' with total index $\ell'(r') = o$. Each representation r of G_F forms an array of representations r' of G_F'. Therefore

$$\ell'(r') = \sum_{r \text{ with } r' \subset r} \ell(r) \ . \tag{III40}$$

Apparently this expression must vanish.

Thus we found another requirement for the indices $\ell(r)$. The indices will be nearly but not quite uniquely determined now. Calculations show that this second requirement makes our indices $\ell(r)$ practically independent of the dimensions n_i of G_F. For instance, if $G_c = SU(3)$ and if we have left- and righthanded quarks forming triplets and sextets then

$$G_F = SU(n_1)_L \otimes SU(n_2)_R \otimes SU(n_3)_L \otimes SU(n_4)_R \otimes U(1)^3 \tag{III41}$$

where $n_{1,2}$ refer to the triplets and $n_{3,4}$ to the sextets. G_c is anomaly-free if

$$n_1 - n_2 + 7(n_3 - n_4) = 0 \ . \tag{III42}$$

Here we have three independent numbers n_i.
If we write the representations r as Young tableaus then $\ell(r)$ could still depend explicitly on n_i.

However, suppose that someone would start as approximation of Bethe-Salpeter type to discover the zero mass bound state spectrum. He would study diagrams such as Fig. 3

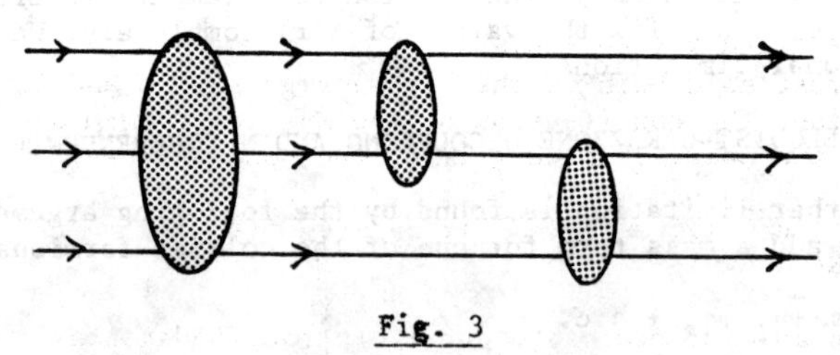

Fig. 3

The resulting indices $\ell(r)$ would follow from topological properties of the interactions represented by the blobs. It is unlikely that this topology would be seriously effected by details such as the contributions of diagrams containing additional closed fermion loops. However, that is the only way in which explicit n-dependence enters. It is therefore natural to assume $\ell(r)$ to be n-independent. This latter assumption fixes $\ell(r)$ completely. What is the result of these calculations?

III13. CALCULATIONS

Let G be any (reducible or irreducible) gauge group. Let chiral fermions in a representation r be coupled to the gauge fields by the covariant derivative

$$D_\mu = \partial_\mu + i\,\lambda^a(r)\,A^a_\mu \,, \tag{III43}$$

where A^a_μ are the gauge fields and $\lambda^a(r)$ a set of matrices depending on the representation r. Let the left-handed fermions be in the representations r_L and the right-handed ones in r_R. Then the anomalies cancel if

$$\sum_L \mathrm{Tr}\{\lambda^a(r_L),\ \lambda^b(r_L)\}\ \lambda^c(r_L) = \sum_R \mathrm{Tr}\{\lambda^a(r_R),\ {}^b(r_R)\}\ \lambda^c(r_R)\ . \tag{III44}$$

The object $d^{abc}(r) = \mathrm{Tr}\{\lambda^a(r),\ \lambda^b(r)\}\ \lambda^c(r)$ can be computed for any r. In table 1 we give some examples. The fundamental representation r_o is represented by a Young tableau: $\square$. Let it have n components. We take the case that $\mathrm{Tr}\ \lambda(r_o) = o$. Write

$$\mathrm{Tr}\ I(r_o) = n\ , \qquad \mathrm{Tr}\ I(r) = N(r)\ ,$$

$$\mathrm{Tr}\ \lambda(r) = o\ ,$$

$$\mathrm{Tr}\ \lambda^a(r)\ \lambda^b(r) = C(r)\ \mathrm{Tr}\ \lambda^a(r_o)\ \lambda^b(r_o)\ ,$$

$$d^{abc}(r) = K(r)\ d^{abc}(r_o)\ . \tag{III45}$$

We read off C and K from table 1.
Now III44 must hold both in the high energy region and in the low energy region. The contribution of the spectator fermions in both regions is the same. Thus we get for the bound states

$$\left(\sum_L - \sum_R\right) d^{abc}(r) = n_c\left(d^{abc}(r_{oL}) - d^{abc}(r_{oR})\right) \tag{III46}$$

where a,b,c are indices of G_F and r_o is the fundamental representation of G_F. We have the factor n_c written explicitly, being the number of color components.

Let us now consider the case $G_c = SU(3)$;
$G_F = SU_L(n) \otimes SU_R(n) \otimes U(1)$. We have n "quarks" in the fundamental representations. The representations r of the bound states must be in $G_F/Z(3)$. They are assumed to be built from three quarks, but we are free to choose their chirality. The expected representations

Table 1

r	N(r)	C(r)	K(r)
$\square$	n	1	1
$\square$	n	1	-1
$\square\square$, $\begin{smallmatrix}\square\\\square\end{smallmatrix}$	$\frac{n(n\pm1)}{2}$	$n\pm2$	$n\pm4$
$\square\square\square$, $\begin{smallmatrix}\square\\\square\\\square\end{smallmatrix}$	$\frac{n(n\pm1)(n\pm2)}{6}$	$\frac{(n\pm2)(n\pm3)}{2}$	$\frac{(n\pm3)(n\pm6)}{2}$
$\begin{smallmatrix}\square\square\\\square\end{smallmatrix}$	$\frac{n(n^2-1)}{3}$	n^2-3	n^2-9
$A \otimes B$	N(A)N(B)	C(A)N(B) + C(B)N(A)	K(A)N(B) + K(B)N(A)

are given in table 2, where also their indices are defined. Because of left-right symmetry these numbers change sign under interchange of left ↔ right.

Table 2

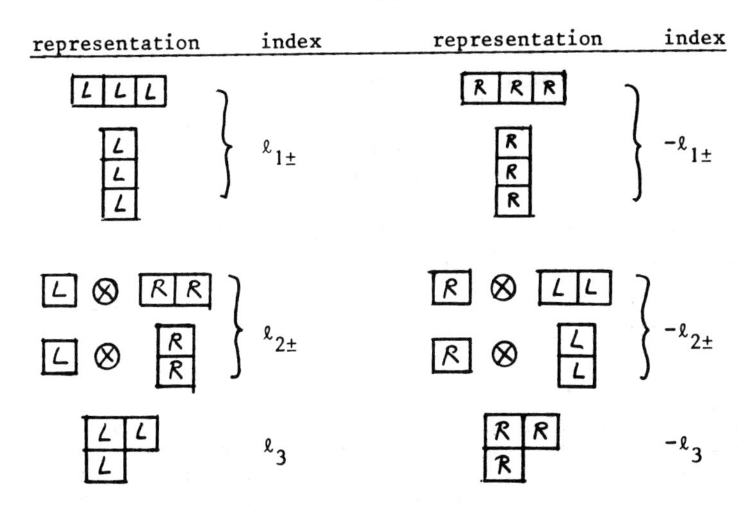

For the time being we assume no other representations. In eq. III46 we may either choose a, b and c all to be $SU(n)_L$ indices, or choose a and b to be $SU(n)_L$ indices and c the U(1) index. We get two independent equations:

$$\sum_{\pm} \tfrac{1}{2}(n\pm3)(n\pm6)\ell_{1\pm} - \sum_{\pm} \tfrac{1}{2}n(n\pm7)\ell_{2\pm} + (n^2-9)\ell_3 = 3, \text{ if } n > 2 ,$$

and

$$\sum_{\pm} \tfrac{1}{2}(n\pm2)(n\pm3)\ell_{1\pm} - \sum_{\pm} \tfrac{1}{2}n(n\pm3)\ell_{2\pm} + (n^2-3)\ell_3 = 1, \text{ if } n > 1 . \tag{III47}$$

The Appelquist-Carazzone decoupling requirement, eq. (III40), gives us in addition two other equations:

$$\ell_{1+} - \ell_{2+} + \ell_3 = 0 ,$$

$$\ell_{1-} - \ell_{2-} + \ell_3 = 0 , \text{ both if } n > 2 . \tag{III48}$$

For $n > 2$ the general solution is

$$\ell_{1+} = \ell_{1-} = \ell ,$$
$$\ell_{2+} = \ell_{2-} = 3\ell - \frac{1}{3} ,$$
$$\ell_{3} = 2\ell - \frac{1}{3} . \qquad \text{(III49)}$$

Here ℓ is still arbitrary. Clearly this result is unacceptable. We cannot allow any of the indices ℓ to be non-integer. Only for the case n = 2 (QCD with just two flavors) there is another solution. In that case ℓ_{2-} and ℓ_3 describe the same representation, and ℓ_{1-} an empty representation. We get

$$\ell_{2-} + \ell_3 = k = 1 - 10\,\ell_{1+} + 5\,\ell_{2+} . \qquad \text{(III50)}$$

According to the σ-model, $\ell_{1+} = \ell_{2+} = 0$; k = 1. The σ-model is therefore a correct solution to our equations.

In the previous section we promised to determine the indices completely. This is done by imposing n-independence for the more general case including also other color representations such as sextets besides triplets. The resulting equations are not very illuminating, with rather ugly coefficients. One finds that in general no solution exists except when one assumes that all mixed representations have vanishing indices. With mixed representations we mean a product of two or more non-trivial representations of two or more non-Abelian invariant subgroups of G_F. If now we assume n-independence this must also hold if the number of sextets is zero. So ℓ_{2+} and ℓ_{2-} must vanish. We get

$$\ell_{1+} = \ell_{1-} = 1/9 ,$$
$$\ell_3 = -1/9 . \qquad \text{(III51)}$$

If all quarks were sextets, not triplets, we would get

$$\ell_{1+} = \ell_{1-} = 2/9 ,$$
$$\ell_3 = -2/9 . \qquad \text{(III52)}$$

In the case G_c = SU(5) the indices were also found. See table 3.

Table 3
indices for G_c = SU(5)

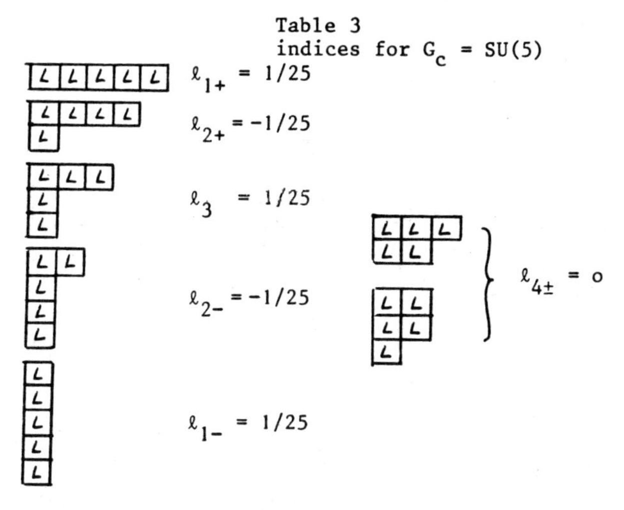

This clearly suggests a general tendency for SU(N) color groups to produce indices $\pm 1/N^2$ or o.

III14. CONCLUSIONS

Our result that the indices we searched for are fractional is clearly absurd. We nevertheless pursued this calculation in order to exhibit the general philosophy of this approach and to find out what a possible cure might be. Our starting point was that chiral symmetry is not broken spontaneously. Most likely this is untenable, as several authors have argued[6]. We find that explicit chiral symmetry in QCD leads to trouble in particular if the number of flavors is more than two. A daring conjecture is then that in QCD the strange quark, being rather light, is responsible for the spontaneous breakdown of chiral symmetry.
An interesting possibility is that in some generalized versions of QCD chiral symmetry is broken only partly, leaving a few massless chiral bound states. Indeed there are examples of models where our philosophy would then give integer indices, but since we must' drop the requirement of n-dependence our result was not unique and it was always ugly. No such model seems to reproduce anything resembling the observed quark-lepton spectrum.

Finally there is the remote possibility that the paradoxes associated with higher spin massless bound states can be resolved. Perhaps the Δ(1236) plays a more subtle role in the σ-model than assumed so far (we took it to be a parity doublet).

We conclude that we are unable to construct a bound state theory for the presently fundamental fermions along the lines suggested above.

We thank R. van Damme for a calculation yielding the indices in the case $G_c = SU(5)$.

REFERENCES

1. P.A.M. Dirac, Nature 139 (1937) 323, Proc. Roy. Soc. A165 (1938) 199, and in: Current Trends in the Theory of Fields, (Tallahassee 1978) AIP Conf. Proc. No 48, Particles and Fields Subseries No 15, ed. by Lannuti and Williams, p. 169.
2. S. Dimopoulos and L. Susskind, Nucl. Phys. B155 (1979) 237.
3. M. Gell-Mann and M. Lévy, Nuovo Cim. 16 (1960) 705.
   B.W. Lee, Chiral Dynamics, Gordon and Breach, New York, London, Paris 1972.
4. G. 't Hooft, Phys. Rev. Lett. 37 (1976) 8; Phys. Rev. D14 (1976) 3432.
   S. Coleman, "The Uses of Instantons", Erice Lectures 1977.
   R. Jackiw and C. Rebbi, Phys. Rev. Lett. 37 (1976) 172.
   C. Callan, R. Dashen and D. Gross, Phys. Lett. 63B (1976) 334.
5. G. 't Hooft, Nucl. Phys. B72 (1974) 461.
6. T. Appelquist and J. Carazzone, Phys. Rev. D11 (1975) 2856.
7. A. Casher, Chiral Symmetry Breaking in Quark Confining Theories, Tel Aviv preprint TAUP 734/79 (1979).

Volume 106B, number 1,2 PHYSICS LETTERS 29 October 1981

FAMILY STRUCTURE WITH COMPOSITE QUARKS AND LEPTONS ☆

Itzhak BARS
J.W. Gibbs Laboratory, Department of Physics, Yale University, New Haven, CT 06520, USA

Received 29 May 1981
Revised manuscript received 26 August 1981

A theory of composite quarks and leptons as composites of three spin-1/2 preons is proposed. It is argued that the persistent mass condition is a natural requirement in such models. A specific realistic model in constructed that predicts up to 8 SO(10) families in the spinor representation, for which masses are generated by explicit small breaking of the underlying chiral symmetries. The small masses of the families are explained as being due only to $SU(3)_C \otimes SU(2)_W \otimes U(1)_W$ symmetry.

The possibility that quarks and leptons are composites of more elementary spin-1/2 preons bound together by hypercolor forces has been investigated recently. An essential ingredient is the existence of an exact chiral symmetry [1–3] to explain massless composite fermions in the absence of certain interactions. The formation of the massless bound states can be studied in the absence of the color and electroweak forces which, thanks to asymptotic freedom, are negligible at the large scale where hypercolor forces become strong ($\Lambda_H \gtrsim$ few TeV). If these and any other small forces (such as Yukawa couplings) are neglected, one is left with a fundamental lagrangian describing a renormalizable and asymptotically free gauge theory based on a hypercolor group G_H containing massless left-handed spin-1/2 preons in representations R_i of G_H. Each distinct representation $R_1, R_2, ..., R_n$ may appear in multiplicities $N_1, N_2, ... N_n$, respectively. The fundamental lagrangian of this theory then automatically possesses a chiral "flavor" global symmetry

$$G_F = SU(N_1) \otimes SU(N_2) \otimes ... \otimes SU(N_n)$$
$$\otimes [U(1)]^{n-1} \otimes Z,$$

where Z is a discrete unbroken subgroup of the U(1) broken by hypercolor instantons [2,3]. All bound states of the theory will fall into representations of G_F.

The massive bound states will have masses of the order Λ_H, since this is the only scale in the theory. Such particles are irrelevant at low energies ($E \ll \Lambda_H$) and can be assumed to decouple from low-energy physics, where we do experiments presently.The G_H singlet and G_F nonsinglet *massless* bound states of such a theory, if there are any, can have only spin-0 or -1/2 [4] and they will fall into definite representations of the chiral group G_F. Note that in such a model G_F nonsinglet spin-1 massless vector mesons are not allowed [4]. This implies that the color–electroweak gauge bosons must be taken as elementary as the preons. (This does not apply to ternon theories or to supergravity.) In order to have massless spin-1/2 composites, some of the chiral symmetries must not break spontaneously because such a breaking would generate dynamical masses of order Λ_H. The currents of unbroken chiral symmetries must satisfy 't Hooft's [1] anomaly constraints as well as decoupling conditions which will be discussed further. In a model that satisfies these requirements the chiral symmetries may survive spontaneous breaking. Then, small masses can be generated for the composite quarks and leptons by turning on the small parameters that were neglected before which break explicitly the chiral symmetries of the preon lagrangian [‡1]. These include the color and electroweak forces [3], Yukawa couplings as well

☆ Research supported in part by the US Department of Energy under contract No. DE-AC-02-76-ER03075.

Volume 106B, number 1,2 PHYSICS LETTERS 29 October 1981

as possible small bare masses for the preons.

To analyze further the properties of the composite quarks and leptons, it is convenient to construct a low-energy effective lagrangian which is consistent with the (slightly broken) chiral symmetries as well as color and electroweak gauge interactions embedded in G_F. This effective lagrangian in general contains negligible non-renormalizable terms (such as anomalous magnetic moment of the electron $\delta a \approx m_e^2/\Lambda_H^2$) with appropriate powers of Λ_H^{-1} consistent with dimensions, and (slightly broken) chiral symmetries. It is then easy to see that the scheme leads to a picture consistent with low-energy SU(3) × SU(2) × U(1) phenomenology, with quarks and leptons behaving as if they were elementary fields, including properties such as negligible anomalous magnetic moments [7] etc., provided $\Lambda_H \gtrsim$ few TeV. If Λ_H is really only a few TeV, the theory must be constructed in such a way that baryon and lepton numbers are conserved at low energies to prevent too large rates for proton decay or other rare processes. The success of this approach hinges on constructing a theory with (i) unbroken chiral symmetry and (ii) realistic degrees of freedom. These will now be discussed in more detail.

In the absence of methods for constructing the massless spin-1/2 bound states directly we must guess their classification under the chiral group

$$G_F = SU(N_1) \otimes ... \otimes SU(N_n) \otimes [U(1)]^{n-1} \otimes Z$$

guided by certain requirements. In this paper three conditions will be imposed. The first two are 't Hooft's [1] anomaly constraints and "decoupling" conditions. The third could be called the principle of "economy", which will require that, for a given hypercolor group G_H and a set of preons P_i, the composite massless fermions should contain the minimum number of "valence" preons in a hypercolor singlet state. This is an intuitive requirement carried over from QCD where one notices that the lowest mass spin-1/2 baryons are constructed from 3 quarks and that exotic color singlet combinations of 5 quarks, 7 quarks, etc., do not appear either because they do not bind or because they are heavy and wide. Note that in QCD this does not prevent a quark–antiquark "sea" in a singlet representation of flavor quantum numbers by gluons emitting and absorbing $q\bar{q}$ pairs. The "restriction" in QCD is really on the "valence" quantum numbers and not on the number of quarks in a baryon. By analogy, in the preon theories, it is reasonable to assume that hypercolor singlet spin-1/2 fermions containing more than the minimum number of "valence" preons do not bind or have masses of order Λ_H. This requirement will play a role in selecting the hypercolor group as will be seen below.

We recall that 't Hooft's "decoupling" condition [1] – which more properly may be called the "persistent mass condition" [8] – requires that when some chiral symmetry is broken in the lagrangian by giving intrinsic mass to some preon, the *remaining* chiral symmetry should not prevent the composite fermion containing any number of massive preons from acquiring a mass. The group theoretical meaning is that, with respect to the *remaining* chiral symmetries, the relevant composite fermions must appear in representations that are parity doubled such as to make chirally symmetric mass terms in the effective low-energy lagrangian. If we deal only with left-handed fermions by writing every ψ_R as $(\psi^c)_L$, then the composites that must acquire mass must appear either in real representations or for every complex representation, its complex conjugate must also be present. This is a very strong requirement that limits greatly the representation content of the composite fermions with respect to G_F. After a few toy models were found [1] a general class of theories satisfying both the anomaly constraints and the persistent mass conditions has been obtained [2]. These are reproduced here in tables 1 and 2.

The persistent mass condition has been criticized recently [2,8,9] on the grounds that there may be theories in which phase transitions may occur as a function of the preon mass. Below a critical preon mass there would be no objection to composite massless fermions that contain massive preons and prevented from acquiring a mass by a residual chiral symmetry [‡2]. However, the model must pass the test of the Appelquist–Carrazone theorem in a "gedanken

[‡1] Estimates of one-loop radiative corrections to symmetry relations can be done in terms of symmetry currents following a general formalism developed some time ago [5] which should be modified to include the breaking due to instantons and associated anomalies. One may also use an effective lagrangian approach [6].

[‡2] Examples of such theories are provided by $k = 4$ in table 1. See ref. [2] for more detailed arguments.

Volume 106B, number 1,2 PHYSICS LETTERS 29 October 1981

Table 1
First model: $k \geqslant 5$ satisfies the persistent mass requirements while $k = 4$ requires a phase transition as a function of the P_1 preon mass. $N \leqslant \frac{11}{2}k - 1$ for asymptotic freedom. All P_i and f_i are left-handed spinors. A dotted box indicates the complex conjugate representation.

Group	$SU(k)_H \otimes SU(N+4-k) \otimes SU(N)$	$U(1)_1$	$U(1)_2$
preons	$P_1 = (\begin{smallmatrix}\boxdot\\\boxdot\end{smallmatrix}, 1, 1)$	$-2N$	$2(N+4-k)$
	$P_2 = (\square, \boxdot, 1)$	$-N$	$N+4-2k$
	$P_3 = (\square, 1, \square)$	$N+k$	$-(N+4-k)$
composite fermions	$f_1 = (1, \begin{smallmatrix}\square\\\square\end{smallmatrix}, 1) = P_1P_2^*P_2^*$	0	$2k$
	$f_2 = (1, \boxdot, \boxdot) = P_1^*P_2P_3^*$	$-k$	$-k$
	$f_3 = (1, 1\square\square) = P_1P_3P_3$	$2k$	0

Table 2
Second model: $k \geqslant 3$ satisfies the persistent mass requirements. $N \leqslant \frac{11}{2}k + 1$ for asymptotic freedom. All P_i and f_i are left-handed spinors. A dotted box indicates the complex conjugate representation.

Group	$SU(k)_H \otimes SU(N-4-k) \otimes SU(N)$	$U(1)_1$	$U(1)_2$
preons	$P_1 = (\boxdot\boxdot, 1, 1)$	$-2N$	$2(N-4-k)$
	$P_2 = (\boxdot, \boxdot, 1)$	$-N$	$N-4-2k$
	$P_3 = (\square, 1, \square)$	$N+k$	$-(N-4-k)$
composite fermions	$f_1 = (1, \square\square, 1) = P_1P_2^*P_2^*$	0	$2k$
	$f_2 = (1, \boxdot, \boxdot) = P_1^*P_2P_3^*$	$-k$	$-k$
	$f_3 = (1, 1\begin{smallmatrix}\square\\\square\end{smallmatrix}) = P_1P_3P_3$	$2k$	0

experiment" when the mass of any preon is sent to infinity [8,9]. In this limit the preon really decouples (at least in perturbation theory, to all orders) and this effect must become apparent in the bound-state spectrum by developing infinite masses for those bound states that contain the massive preons. In a model that violates the persistent mass condition this is possible only by breaking spontaneously the residual chiral symmetries ‡3. However, this requires that the remaining theory (which is missing the massive preons) has its chiral symmetries spontaneously broken. In this way one can complete an argument, as in the case of QCD [9], that in *certain* theories that violate the persistent mass requirement it is natural to expect the chiral symmetries to break spontaneously by a dynamical mechanism. For example, a multicolor gauge theory which is a direct extension of QCD except for the number of colors, would break its chiral symmetries dynamically and would not be acceptable as a preon theory for this reason [9]. Thus, the persistent mass requirement is a reasonable condition that avoids the Appelquist–Carrazone "disaster", in a preon theory that satisfies it. The classes of models in tables 1 and 2 meet this requirement [2] and therefore are "safe" ‡4.

In this paper we concentrate on the models in table 2. For $k = 3$ there are $SU(3)_H$ singlet combina-

‡3 More complicated scenarios in which the chiral symmetries restore themselves by undergoing first order phase transitions or by multiple phase transitions cannot be ruled out. But in certain theories one can argue that such scenarios are unlikely, as in the case of QCD discussed in ref. [9].

‡4 In the special cases mentioned in footnote 2 the Appelquist–Carrazone "disaster" can be avoided. This is because the persistent mass requirement is violated only by one preon (P_1) and in the finite mass limit all composite fermions disappear from the spectrum after the spontaneous breaking of the residual chiral symmetries. Then the type of argument of ref. [9] cannot be given for the complete theory (including P_1). The residual theory (P_2, P_3) is a QCD-like theory with an *even* number of colors which must undergo spontaneous breakdown since it has only mesons and could not satisfy the anomaly constraints. Thus, these special cases provide nice examples that illustrate the self-consistency of the results arrived at independently via the Appelquist–Carrazone theorem or via the anomaly constraints.

tions of 3 preons which compete with those listed in the table as candidates for massless composite fermions ($P_2P_2P_2$, $P_2^*P_2^*P_3$, $P_2P_3^*P_3^*$, $P_3P_3P_3$). It is difficult to see how the dynamics would make these combinations higher in mass than those in table 2. Therefore we avoid $k = 3$. For even values of $k = 4, 6$, ... etc., *the only possible composite fermions are those in table 2*, plus exotic combinations involving more preons. On the basis of the principle of "economy" mentioned above, the combinations in the table are the only ones that could be massless. For higher odd values of $k = 5, 7, 9$, ... etc., there are singlet combinations (such as $P_2P_2P_2P_2P_2$ for $k = 5$ etc.) which may be eliminated again because of "economy". However, it seems that k = even is on safer grounds. Clearly $k = 4$ is the most economical from the point of view of the overall number of elementary preons.

The electroweak and color interactions may be embedded in G_F in many ways. One possibility was discussed in ref. [2]. Here a more elegant and realistic model will be proposed. In table 2, take $N = 16$ and $k = 4$. In SU($N = 16$) embed (but do not necessarily gauge) the usual flavor group SO(10) such that $P_3 = (4, 1, \overline{16})$ transforms like the spinor representation $\overline{16}$ of SO(10). In addition, a family group, e.g. Sp(8) will be embedded in SU($N - 4 - k$) = SU(8) such that $P_2 = (\bar{4}, 8, 1)$. Thus P_3 is trivial under family transformations and P_2 is trivial under flavor transformations, while $P_1 = (\overline{10}, 1, 1)$ is trivial under both. The composite fermions now can be classified under Sp(8) × SO(10) as

$$f_1 = (36, 1), \quad f_2 = (8, 16), \quad f_3 = (1, 120).$$

The color and electroweak interactions are identified with the *gauged* group $SU(3)_C \otimes SU(2)_W \otimes U(1)_W$ embedded in SO(10) in the usual way. Note that if only this subgroup is gauged in SU(16), then baryon (B) and lepton (L) numbers remain as unbroken global symmetries. On the other hand, if SO(10) or SU(5) is gauged, there should occur a breaking (could be independent of Λ_H) at $\approx 10^{16}$ GeV down to the color–electroweak group. However, the simplest possibility is to gauge the color–electroweak group only. Independently, any part of Sp(8) or other subgroups of SU(8) [such as SO(8) or SU(4), etc.] can be gauged in such a way as to be consistent with the discussion that follows.

The composites f_1 and f_3 are real representations and, therefore, they can develop Sp(8) × SO(10) invariant masses, while f_2 must remain massless since the quadratic term $f_2f_2 \sim (8, 16) \times (8, 16)$ does not contain (1, 1). Actually, f_2 remains massless as long as the $SU(3)_C \otimes SU(2)_W \otimes U(1)_W \otimes B \otimes L$ invariance embedded in SO(10) remains exact, while f_1 and f_3 are still massive. Therefore, if the original automatic symmetry SU(8) × SU(16) ⊗ U(1) × U(1) breaks down by *any mechanism* to $SU(3)_C \otimes SU(2)_W \otimes U(1)_W \otimes B \otimes L$, f_2 remains massless and f_1, f_3 acquire mass of order of a fraction of Λ_H. Thus, in our model, the familiar low-energy symmetries are solely responsible for the explanation of low-mass fermions that fit into the usual family patterns that have been observed for quarks and leptons: *At this symmetry level there are no other massless composite fermions.* Note that the preons cannot receive a mass as long as $SU(4)_H \otimes SU(3)_C \otimes SU(2)_W \otimes U(1)_W \otimes B \otimes L$ or even its subgroup $SU(4)_H \otimes SU(3)_C \otimes U(1)_{EM} \otimes B \otimes L$ remains exact. Therefore, any mass for the composite, including the families, must come from interactions and not from bare preon masses. *This, then, is a new mass generating mechanism for the quarks and leptons.* What causes the symmetry breaking from SU(8) ⊗SU(16) ⊗U(1) ⊗U(1) down to $SU(3)_C \otimes SU(2)_W \otimes U(1)_W \otimes B \otimes L$? Part of the breaking is, of course, the presence of the color–electroweak gauge bosons. However, there could be, in addition, dynamical breaking via the strong hypercolor force as well as explicit breaking via Yukawa couplings. For an example of explicit breaking, see footnote 5.

With f_1 and f_3 heavy, this cheme provides $f_2 = (8, 16)$ which contains 8 families of composite quarks and leptons. The families differ from each other because they contain a different preon of P_2-type. The 16-dimensional family structure is due to the classification of P_3. As long as $SU(2)_W \otimes U(1)_W$ is exact all 8 families are massless even if the family group Sp(8) is completely broken (see example in footnote 5). Masses can be generated by spontaneous breakdown of $SU(2)_W \otimes U(1)_W$ with appropriate interactions due to explicit Higgs [‡6], technicolor [‡7] or a new mechanism [3]. Thus a family hierarchy is easy to obtain in this scheme.

It is, of course, possible to construct models with higher values of k, leading to fewer families ($16 - 4 - k$). But note that as k increases, the ratio of composite fermions to preons decreases. Furthermore, instead of $N = 16$, we could take $N = 15$ leading to an SU(5) group embedded in SU(15) as $\bar{5} + 10$, instead of the spinor representation of SO(10) embedded in SU(16). The results are very similar except for the right-handed neutrinos missing completely. With $k = 4$, we found another very different model containing SU(5) but it contains low-mass V + A quarks and leptons appearing in higher dimensional SU(5) representations [‡8]. Thus if we require the lowest value of k, our model is essentially unique in its simplicity and the desired physical features.

In the present analysis it was shown that it is possible to construct a realistic model of composite quarks and leptons consistent with low-energy SU(3) × SU(2) × U(1) phenomenology. Details of the mass matrix, including Cabibbo angles, are yet to be analyzed. In principle, they are completely calculable [‡1] in terms of the interactions that break the original symmetry. Grand unification groups such as SU(5) or SO(10) are again relevant, but they were used to classify preons which in turn led to a classification of quarks and leptons. The model predicts up to 8 families $f_2 = (8,16)$, plus heavier composite fermions $f_1 = (36,1)$ and $f_3 = (1,120)$ with masses below Λ_H. Of course we expect even heavier particles of various spins around Λ_H. Families are distinguished by the different preons they contain and the family structure follows from the structure of preon interactions with the gauge bosons of $SU(3)_C \times SU(2)_W \times U(1)_W$. If Λ_H is indeed only a few TeV, experiments done within a decade may show all the residual strong interactions (analogous to nuclear or particle strong interactions) that follow from hypercolor with preons playing a role analogous to present-day quarks. The presence of such interactions, which are negligible at present energies but become very important near $E \sim \Lambda_H$, with all their details, such as resonances, will distinguish our model from elementary quarks and leptons as well as other possible competing composite models. An important qualitative prediction of the present model is that parity will be strongly violated at energies comparable to Λ_H since our hypercolor forces are not parity conserving.

‡5 One must insure that these interactions break also both U(1) symmetries listed in table 2 and the discrete symmetry Z described in ref. [2], since they prevent f_1 and f_3 from acquiring masses. For example, Sp(8) × SO(10) gauge interactions are not sufficient, because there are left over three discrete symmetries (after Sp(8) × SO(10) instanton effects) which still prevent the desired masses. An example of Yukawa couplings which, just by their presence and without any spontaneous breakdown, break explicitly the $SU(4)_H \otimes SU(8) \otimes SU(16) \otimes U(1)_1 \otimes U(1)_2 \otimes Z$ symmetry all the way down to $SU(4)_H \otimes SU(3)_C \otimes SU(2)_W \otimes U(1)_W \otimes B \otimes L$ is the following

$$\alpha \phi_1 P_1 P_1 + \beta^{ij} \phi_2 P_2^i P_2^j + \gamma \phi_3 \begin{pmatrix} U^c \\ D^c \end{pmatrix} U + \delta \phi_3^c \begin{pmatrix} U^c \\ D^c \end{pmatrix} D + \sigma \phi_3 \begin{pmatrix} E^c \\ N^c \end{pmatrix} E + \epsilon \phi_3^c \begin{pmatrix} E^c \\ N^c \end{pmatrix} N,$$

where the 16 components of $P_3 = (4, 1, \overline{16})$ are denoted in the "up", "down", "electron", "neutrino" notation. The scalars transform with respect to $SU(4)_H \otimes SU(3)_C \otimes SU(2)_W$ as $\phi_1 = (20', 1, 1)$, $\phi_2 = (6, 1, 1)$, $\phi_3 = (6, 1, 2)$. Note that these are real representations (or pseudoreal for ϕ_3) and their couplings break all unwanted U(1)'s on P_1, P_2, P_3 except the usual baryon and lepton numbers which are harmless. β^{ij} is an 8 × 8 antisymmetric matrix of constants which breaks the family symmetry. These scalars do not develop vacuum expectation values since $SU(4)_H$ remains exact. Also, even though they can form composite mesons together with the preons, they do not lead to new composite fermions because of their hypercolor classification.

‡6 A Higgs doublet $\phi_4 = (1, 1, 2)$ can be added to the example of footnote 5 which would couple as $\phi_4\phi_2\phi_3 + \phi_4\phi_1\phi_2\phi_3 + (\phi_4\phi_3)^2$ etc. but cannot couple to the preons. Then $\langle \phi_4 \rangle \neq 0$ will generate the appropriate masses for W, Z, quarks and leptons, while the preons remain massless.

‡7 Some anomaly-free part of the SU(8) family group can be gauged and interpreted as technicolor: e.g. gauge SO(5) and take SO(3) as the group of three families.

‡8 The model is the following: With $N = 13$ and $k = 4$, we embed in SU(5) × SU(13) an SU(5) (×) SU(3) so that the preons transform as $P_1 = (\overline{10}, 1, 1)$, $P_2 = (\bar{4}, \bar{5}, 1)$, $P_3 = (4, 10, 1) + (4, 1, 3)$. Then the composite fermions transform under SU(5) × SU(3) as $f_1 = (15, 1)$; $f_2 = (10, 1) + (\overline{40}, 1) + (\bar{5}, \bar{3})$; $f_3 = (45, 1) + (10, 3) + (1, \bar{3})$. The pieces that are triplets under the family group SU(3) (10, 3) + $(\bar{5}, \bar{3})$ + $(1, \bar{3})$ can be identified with the usual three families. The others couple exactly to each other to become massive when SU(5) is broken down to SU(4) or all the way to SU(3) × U(1).

References

[1] G. 't Hooft, lecture Cargese Summer Institute (1979); see also: S. Dimopoulos, S. Raby and L. Susskind, Stanford preprint ITP-662;

S. Coleman and E. Witten, Phys. Rev. Lett. 45 (1980) 100;
R. Barbieri, L. Maiani and R. Petronzio, Phys. Lett. 96B (1980) 63;
T. Banks, S. Yankielowicz and A. Schwimmer, Phys. Lett. 96B (1980) 67;
Y. Frishman, S. Schwimmer, T. Banks and S. Yankielowicz, Weizmann preprint WIS80-27 (1980);
G. Farrar, Phys. Lett. 96B (1980) 2731;
R. Chanda and P. Roy, CERN preprint TH2923 (1980);
T. Banks and A. Schwimmer, Trieste preprint ICTP/80/81-6.

[2] I. Bars and S. Yankielowicz, Phys. Lett. 101B (1981) 159.
[3] S. Weinberg, Texas preprint (March 1981).
[4] E. Witten and S. Weinberg, Phys. Lett. 96B (1980) 59.
[5] I. Bars, Phys. Lett. 51B (1974) 267.
[6] S. Weinberg, Phys. Rev. D13 (1976) 974.
[7] S. Brodsky and S. Drell, Phys. Rev. D22 (1980) 2236;
G.L. Shaw, D. Silverman and R. Slansky, Phys. Lett. 94B (1980) 57.
[8] J. Preskill and S. Weinberg, Texas preprint (January 1981).
[9] I. Bars, Yale preprint YTP 81-09 (February 1981).

XVIII. *Gravity and Grand Unification*

1) The $N=8$ supergravity theory. I. The Lagrangian
by E. Cremmer and B. Julia, *Phys. Lett.*
80B, 48 (1978) 1034

2) A grand unified theory obtained from
broken supergravity
by J. Ellis, M. K. Gaillard and B. Zumino,
Phys. Lett. **94**B, 343 (1980) 1039

3) Search for a realistic Kaluza-Klein Theory
by E. Witten, *Nucl. Phys.* B**186,** 412 (1981) 1045

4) Gravity as a dynamical consequence of the strong,
weak, and electromagnetic interactions
by A. Zee, in the *Proceedings of the Erice
Conference* (1981) 1062

Volume 80B, number 1, 2 PHYSICS LETTERS 18 December 1978

THE $N = 8$ SUPERGRAVITY THEORY. I. THE LAGRANGIAN

E. CREMMER and B. JULIA
Laboratoire de Physique Théorique de l'Ecole Normale Supérieure [1], *75231 Paris Cedex 05, France*

Received 25 September 1978

The SO(8) supergravity action is constructed in closed form. A local SU(8) group as well as the exceptional group E_7 are invariances of the equations of motion and of a new first order lagrangian.

Introduction. Supersymmetry became popular in elementary particle physics after the construction of the dual spinor model where the fermionic extension of the two dimensional conformal algebra was used to decouple the negative norm states from the physical sector of the Hilbert space. But the main appeal of its generalization to four dimensional theories lies in its predictive power, even if it remains to be seen whether this quality will manifest itself at the phenomenological level, or turn into a drawback and force us to study more complicated schemes. To answer this question, it is necessary to construct supersymmetric Lagrangians explicitly and to study their properties as classical and quantum field theories. Like all other continuous symmetries, supersymmetry has a local version; it is realized in the so-called extended supergravity theories all of which generalize bosonic gravity theories and unify gravitational interactions with other interactions, mediated by lower spin fields, like colour, weak and electromagnetic forces. The largest of these models has an SO(8) internal symmetry if one restricts oneself to one spin two elementary field, and no derivatives higher than two in the lagrangian. It contains all SO(N) modesl $N < 8$ including Einstein's theory ($N = 0$); its classical spectrum is the following: 1 graviton, 8 Rarita–Schwinger spin 3/2 fields, 28 vectors, 56 Majorana spin 1/2 fields, and 70 scalars, all massless. A very strong constraint arises for $N = 8$ because there is no "matter multiplet" to couple to the supergravity Lagrangian, i.e. one is forced to a true unification of all particles. However, this theory has a serious defect from the point of view of conservative quantum field theorists, it contains a coupling constant K with dimension of the inverse of a mass, and hence is not obviously renormalizable (one of the reasons being the non-polynomiality of the action: KA is dimensionless for all scalar fields A and appears to all powers). It is thus rather encouraging that supersymmetry partly eliminates this difficulty. There are some indications that all pure extended supergravity theories are one loop and probably two loop finite; the symmetry not only forbids the appearance of unwanted counterterms but restricts the lagrangians so much that the smallest non-polynomial theory, the SO(4) supergravity, has been constructed in closed form. It also permits in principle a step by step construction of the SO(8) theory, but algebraic complexity prevented the authors of ref. [1] to go beyond the second order in K. One must use more powerful techniques to obtain the full lagrangian in order to investigate classical spontaneous symmetry breaking or the finiteness of higher order quantum corrections.

Here come the idea of dimensional reduction and surprisingly the (two-dimensional) dual spinor model again. This remarkable model is though to be supersymmetric (in the 10 dim. embedding space-time), various truncated versions have been proved to be supersymmetric in the zero-slope limit [2], and the complete theory has been constructed in the same approximation by assuming supersymmetry [3]: this is the 11 dimensional

[1] Laboratoire propre du C.N.R.S. associé à l'Ecole Normale Supérieure et à l'Université de Paris Sud.

Volume 80B, number 1, 2 PHYSICS LETTERS 18 December 1978

supergravity, and the largest containing at most spin 2 particles after reduction from 11 to 4 dimensions [4]. The dimensional reduction implies a second low-energy approximation, beyond the zero-slope limit (which neglects extended modes of the string), it ignores (in its simplest version) all states with non-vanishing momentum in the internal directions x^i ($4 \leqslant i \leqslant 10$, say), that is (high) 4-dimensional mass.

To put our results in perspective let us recall two properties of models suggested by this technique. Both of these have extended $N = 4$ supersymmetry because of usual ($N = 1$) supersymmetry in 10 dimensions. The first one is the Yang–Mills supersymmetry theory for any gauge group G, it has been shown [5] to have vanishing β-function up to 2-loop order, this suggests extra-symmetries. The second model is the SU(4) supergravity theory [6], its equations of motion are the same as those of SO(4) theory, the SO(6) ~ SU(4) algebra of symmetries comes naturally from the 6 dimensional internal structure and is realized in the lagrangian. There is a further invariance of the equations of motion, it is a SU(1, 1) group of transformations realized non-linearly on the scalar fields [7].

Now the theory of ref. [3] has analogous promising features, it is polynomial (up to metric coefficients required for covariance), it admits a large group of invariances: GL(11, **R**) global, included in the local coordinate reparametrization group, SO(1, 10) local thanks to auxiliary non propagating fields $\omega_M{}^{AB}$, 55 local commuting one-parameter groups corresponding to the gauge fields A_{MNP}, and local supersymmetry invariance N=1 in 11 dimensions. We are interested in the four-dimensional reduction for phenomenological purposes and because one expects better ultra-violet behaviour there. It was shown in [3] that the reduced theory has the same field content as both the $N = 7$ and $N = 8$ extended supergravity theories. It must have $N = 8$ local supersymmetry invariance but has only obvious O(7) local and GL(7, **R**) global invariances, hence it is clear that one must restore the O(8) symmetry of the lagrangian after reduction, keeping in mind that SU(N) or U(N) symmetries usually hold for massless theories [8]. The larger the symmetry group, the easier it will be to make a careful phenomenological analysis, for some attractive properties of the physics see [9]. Invariances will also play a crucial role in the quantum theory, and for its eventual renormalizability.

Let us first sketch the reduction procedure. The local O(1, 10) invariance of 11 dimensional supergravity can be used to put the moving frame $e_M{}^A$ in block triangular form; $e_m{}^\alpha = 0$; α, a, A (μ, m, M) denote flat (curved) indices between 0 and 3, 4 and 10, i.e. "internal", or general between 0 and 10. This operation preserves local O(7) invariance. GL(7, **R**) invariance will also hold in the sector with no internal coordinate dependence. As shown in [3] the spectrum coincides with the one of ref. [1], we obtain 35 pseudo scalars A_{mnp} but 28 + 7 scalars coming in 2 disjoint representations of O(7) [10], this means that the SO(7) of $N = 7$ supergravity is distinct from our obvious O(7). After a Weyl transformation which eliminates the factor det $[e_m{}^a]$, duality transformations [2] can be done in four dimensions: the 7 scalars φ^i are classically equivalent to the tensor fields $F_{\mu\nu\rho i}$, the 21 abelian vectors are replaced by their "magnetic" duals $B_\mu{}^{ij}$, and together with $B_\mu{}^i = -\frac{1}{2} e_\mu{}^a e_a{}^i$ they will form the representation 28 of SO(8) and gauge 28 U(1) groups. At this stage the lagrangian has become a monster: by duality non polynomial terms in the pseudoscalar fields have appeared.

Let us recall that Majorana spinors in eleven dimensions have 32 components and split, after reduction, into 8 four-component Majorana spinors. These internal indices between 1 and 8 will be written in capital letters A, B ... also (but they appear only in 4 dimensions), they are SU(8) indices [8]. The eleven Γ matrices can be chosen in a factorized form: $\gamma^\alpha \otimes \mathbf{1}_A{}^B$ and $\gamma^5 \otimes \Gamma_A{}^a{}^B = \Gamma^a_{(11\,\mathrm{dim.})}$, and the 28 (8 by 8 matrices) $\Gamma^a, \frac{1}{2}[\Gamma^a, \Gamma^b]$ form the chiral spinor representation of the algebra SO(8). Appropriate linear combinations of ψ_μ and ψ_m have the canonical kinetic terms and form an octet $\psi_{\mu A}$ and the 56-plet of spin $\frac{1}{2}$ χ_{ABC} (cf. [1]).

The form of the SO(8) algebra suggests a unification of the 7-bein $e_a{}^i$ with the 7φ^i's. It can be done by enlarging GL(7, **R**) to SL(8, **R**) and local SO(8). This operation will be used once more, so let us formulate it carefully. ($e_a{}^i$, φ^i) parametrize a matrix of SL(8, **R**): $v_a{}^i$ (indices now run between 4 and 11), but the physical degrees of freedom are described by the "metric" $g^{ij} = v^t \eta v$, $\eta = \mathrm{diag}(-1, 8 \text{ times})$; superfluous components are eliminated by a *local* invariance SO(8) acting on the left of v. Using 7 of these invariances (preserving SO(7)) one can set 7 components of $v_a{}^i$ to zero and arrive at $e_a{}^i$ and φ^i. Mathematically, the scalars live on the homogeneous space SL(8**R**)/SO(8), the physicists are used to coset spaces in the context of gravitation and non linear chiral lagrangians. It is tantalizing to try a generalization including the pseudoscalars. The simplest way to find it is to guess an invariance SU(8)[‡1] extending SO(8) for the full theory and to look for a group containing SL(8, **R**) and SU(8), such that its coset space $S = G/\mathrm{SU}(8)$ be 70-dimensional; 63 + 70 = 133: here is the exceptional group E_7 of Killing–Cartan [11], coming in the theory quite naturally. Like the φ^i's the A_{ijk} appear only polynomially, the reason being also that they parametrize S in a triangular form. To be pedantic this is an Iwasawa decomposition of a non compact real form of E_7 with respect to its maximal compact subgroup SU(8). Concretely, the 11-dimensional theory led to a simple parametrization of the real representation 56 of E_7:

$$\mathcal{V} = \exp P \begin{bmatrix} v^{[c}{}_i v^{d]}{}_j & 0 \\ 0 & v_{[c}{}^i v_{d]}{}^j \end{bmatrix}, \quad P = \begin{bmatrix} 0 & {}^*A^{abcd} \\ A_{abcd} & 0 \end{bmatrix}, \tag{1}$$

with

$${}^*A^{abcd} = \tfrac{1}{24}\epsilon^{abcdefgh} A_{efgh} = \tfrac{1}{6}\epsilon^{abcdefg}_{(7\,\mathrm{dim.})} e_e{}^m e_f^n e_g^p A^{(7\mathrm{dim.})}_{mnp},$$

$$v_c{}^i = e^{1/4}\begin{bmatrix} e_c{}^i & 0 \\ -\varphi^i e^{-1}, & e^{-1} \end{bmatrix} \text{ the achtbein and } e = \det[e_i^c]$$

The positive definite scalar lagrangian boils down to:

$$L^0 = -\frac{\sqrt{g_{\mu\nu}}}{192K^2}\,\mathrm{Tr}\,(\partial_\mu \mathcal{G}\cdot\partial^\mu \mathcal{G}^{-1}) \text{ with } \mathcal{G} = \mathcal{V}^{\mathrm{t}}\cdot\mathcal{V}. \tag{2}$$

The group E_7 acts on the right of $\mathcal{V}$ and the "local" SU(8) eliminating the superfluous degrees of freedom, on the left (like GL(4, **R**) and SO(1, 3) for the vierbein $e_\alpha{}^\mu$). E_7 also acts on the fundamental representation 56 $\mathcal{F}_{\mu\nu}$ formed with the field strengths: $G_{\mu\nu}{}^{ij} = \partial_\mu B_\nu{}^{ij} - \partial_\nu B_\mu{}^{ij}$ and the 28, $H_{\mu\nu ij} = \tilde{G}_{\mu\nu}{}^{ij}$ + non-linear terms, appearing in the equations of motion, in such a way that the latter are unified with the Bianchi identities in the single equation $\partial_\mu \tilde{\mathcal{F}}^{\mu\nu} = 0$, where $\tilde{\mathcal{F}}^{\mu\nu} = \frac{1}{2}\epsilon^{\mu\nu\rho\sigma}\mathcal{F}_{\rho\sigma}$. Non trivial dynamics is generated by the covariant constraint:

$$\Omega\cdot\mathcal{V}\cdot\tilde{\mathcal{F}} = \mathcal{V}\cdot\mathcal{F}, \tag{3}$$

where Ω is the symplectic form invariant under E_7 [11]. It is interesting to note that all E_7 symmetries (including electric-magnetic dualities) can be realized at the lagrangian level by adding 56 abelian potentials $\mathcal{A}'_\mu$ in $L^1 = \frac{1}{4}\mathcal{A}'_\nu\cdot\Omega\cdot\partial_\mu\tilde{\mathcal{F}}^{\mu\nu}$; elimination of 28 of them leads to the usual dissymmetric lagrangians. But quantization is often simpler in "first-order formalism", hence this new result might be useful beyond the classical approximation in any such "dual" theory.

The analogy with general relativity extends to the fermionic Lagrangian: the fermionic fields $\psi_{\mu A}$ and χ_{ABC} transform linearly under the local SU(8), all their couplings to the bosonic fields involve the moving frame $(e_\alpha{}^\mu, \mathcal{V})$ and constant Clebsch–Gordan coefficients: the γ^α and Γ^a matrices. Let us first consider the couplings to the scalar fields alone. In analogy with the connection fields $\omega_{\mu\alpha}{}^\beta$ we can introduce auxiliary (non-propagating) fields $\Omega_{\mu A}{}^B$, they gauge SU(8) but have no kinetic terms. In the following D_μ will denote the covariant derivative in the appropriate representation of SU(8) (28_{C} for $\mathcal{V}$). In first order formalism L^0 becomes:

$$L'^0 = \frac{\sqrt{g_{\mu\nu}}}{48K^2}\,\mathrm{Tr}\,[(D_\mu\mathcal{V}\cdot\mathcal{V}^{-1})\cdot(D^\mu\mathcal{V}\cdot\mathcal{V}^{-1})], \tag{4}$$

and these fermionic terms L_{F} can be rewritten as:

$$L_{\mathrm{F}} = \frac{\epsilon^{\mu\nu\rho\sigma}}{2}\bar{\psi}_{\mu A}\gamma_\sigma\gamma_5 D_\nu\psi_{\rho A} + \frac{\mathrm{i}\sqrt{g_{\mu\nu}}}{24}\bar{\chi}_{ABC}\bar{\gamma}^\nu D_\nu\chi_{ABC}$$

$$+ \frac{\sqrt{g_{\mu\nu}}}{192}\bar{\psi}_{\mu D}\gamma^\nu\gamma^\mu P_\nu^{ABCD}\chi_{ABC}\,. \tag{5}$$

Here D_μ includes the Lorentz connection $\omega_{\mu\alpha}{}^\beta$ in the spin 1/2 representation, and $P_\nu{}^{ABCD}$ is the component of $\partial_\nu\mathcal{V}\cdot\mathcal{V}^{-1}$ perpendicular to SU(8), i.e. the propagating part of the scalar fields. The other component drops out of L'_0 after elimination of $\Omega_{\mu A}{}^B$, thanks to SU(8) invariance.

The fermionic couplings $K\Theta(G)$ of the fields $G_{\mu\nu}{}^{ij}$ are found by reduction in L_{R}, the second order lagrangian; they respect the E_7 and SU(8) invariances if one uses the first order formalism. Indeed the field strengths and $H_{\mu\nu ij} = -4(\delta\tilde{L}_R/\delta G^{ij}_{\mu\nu})$, i.e. $\mathcal{F}_{\mu\nu}$ still belong to a 56-plet of E_7, but (3) no longer holds. However, one can add uniquely to this multiplet another one, quadratic in the fermion fields, such that the new fields $(\hat{G}_{\mu\nu}{}^{ij}, \hat{H}_{\mu\nu ij}) = \hat{\mathcal{F}}$ satisfy the duality constraint (3). L_1 acquires an additional E_7 invariant term:

‡1 U(8) is suggested by ref. [8], however the self-duality condition of ref. [1]:

$$A_{ijkl} + \mathrm{i}B_{ijkl} = z_{ijkl} = (1/4!)\epsilon_{ijklmnop}\bar{z}^{mnop}$$

reduces it to SU(8).

$$L'1 = \frac{1}{4}\mathcal{A}'_{\nu}\cdot\Omega\cdot\partial_{\mu}\tilde{\mathcal{F}}^{\mu\nu} + \frac{1}{2}K\Theta(\hat{\mathcal{F}}) , \tag{6}$$

where

$$32\Theta/\sqrt{g_{\mu\nu}} = \bar{\psi}_{\mu A}\gamma^{[\nu}\hat{\mathcal{F}}_{AB}\gamma^{\mu]}\psi_{\nu B} - \frac{1}{2}i\bar{\psi}_{\mu C}\hat{\mathcal{F}}_{AB}\gamma^{\mu}\chi_{ABC}$$
$$-(\eta/144)\epsilon^{ABCDEFGH}\bar{\chi}_{ABC}\hat{\mathcal{F}}_{DE}\chi_{FGH} , \tag{7}$$

where $\eta = \pm 1$ [12] depends on the explicit representation of the real Clifford algebra Γ^a $(a = 1, ... 7)$. We have used the SU(8) covariant tensor $\hat{\mathcal{F}}_{AB}$ constructed from $\hat{\mathcal{F}}_{\mu\nu}$ by:

$$\mathcal{F}_{AB} = ((\Gamma_{rs})_{AB} ;\ i\gamma_5(\Gamma^{rs})_{AB})\mathcal{V}\cdot\mathcal{F}_{\mu\nu}\gamma^{\mu\nu} . \tag{8}$$

To summarize, the lagrangian of SO(8) supergravity admits global E_7 and local SU(8) invariances and reads:

$$L = -\frac{\sqrt{g_{\mu\nu}}}{4K^2}R(e_{\mu}{}^{\alpha}, \omega_{\mu}{}^{\alpha}{}_{\beta}) + L'_1 + L'_0 + L_F +$$
$$+ \text{eventual quartic terms} . \tag{9}$$

We have two constraints: $\mathcal{V}$ is a matrix representing E_7 in the fundamental representation and (3) holds for $\hat{\mathcal{F}}$; both are compatible with the symmetries. We believe that as in $N = 4$ supergravity all quartic terms can be absorbed in supercovariant expressions (for example $\hat{\mathcal{F}}$ appearing in (7)), or generated by the elimination of auxiliary fields like $\Omega_{\mu A}{}^{B}$. It is pointless to give them without describing the supersymmetry algebra [13].

In order to reproduce the results of [1], one must choose a specific parametrization of the coset space $E_7/SU(8)$ (i.e. a choice of SU(8) gauge). It coincides only to first order in the scalar fields with the symmetric gauge $\mathcal{V} = \mathcal{V}^t$ where $\mathcal{V}$ is generated by the Lie generators perpendicular to SU(8), we have called it the "analytic gauge". Details of the reduction procedure and important, if sophisticated, algebraic identities will be the object of a longer article [12].

In this letter we have worked our way from O(7) to E_7, without paying much attention to supersymmetry, it is conjectured that the shrinking of the lagrangian accompanying the growth of the group will continue when one uses supercovariance. Several generalizations must be investigated further, first of all the use of other non-compact symmetry groups without coupled ghosts, then the introduction of a non-vanishing gauge coupling and the associated potentials, and also the quantized version of this theory. Indeed the non-trivial topology of our phase space suggests a large number of non perturbative effects, which would be used to cure some of the phenomenological difficulties of the classical model [9]. One recalls that in 2 dimensions the vector auxiliary field of the non-linear σ-model may become dynamical [13].

We are grateful to Joel Scherk for his collaboration to the first computations of this work.

References

[0] An extensive list of references can be found in: S. Ferrara, Erice Lectures (1978), CERN TH-2514.
[1] B. De Wit and D.Z. Freedman, Nucl. Phys. B130 (1977) 105.
[2] F. Gliozzi, J. Scherk and D. Olive, Nucl. Phys. B122 (1977) 253;
L. Brink, J.H. Schwarz and J. Scherk, Nucl. Phys. B121 (1977) 77.
[3] E. Cremmer, B. Julia and J. Scherk, Phys. Lett. 76B (1978) 409.
[4] W. Nahm, Nucl. Phys. B135 (1978) 149.
[5] D.R.T. Jones, Phys. Lett. 72B (1977) 199;
E.C. Poggio and H.N. Pendleton, Phys. Lett. 72B (1977) 200.
[6] E. Cremmer, J. Scherk and S. Ferrara, Phys. Lett. 74B (1978) 61.
[7] E. Cremmer, J. Scherk, Phys. Lett. 74B (1978) 341.
[8] R. Haag, J.T. Lopuszanski, M. Sohnius, Nucl. Phys. B88 (1975) 257.
[9] M. Gell-Mann, Lecture at the Washington meeting of the APS (April 1977).
[10] We thank D. Freedman for bringing this fact to our attention.
[11] E. Cartan, Oeuvres Complètes T1, Part I (Thèse) (Gauthier–Villars, Paris, 1952).
[12] E. Cremmer and B. Julia, to be published.
[13] A. D'Adda, M. Lüscher and P. Di Vecchia, Nordita preprint;
E. Witten, Harvard preprint.

More detailed and pedagogical account of this work can be found in:

E. Cremer and B. Julia, Nucl. Phys. B159, (1979) 141.

E. Cremer, "Dimensional reduction and hidden symmetries in extended supergravity", in Proc. of Spring School in Supergravity, Trieste, 1981, edited by S. Ferrara, J.G. Taylor, and P. van Nieuwenhuizen, Cambridge University Press. In press.

B. Julia, "Applications of supergravity to gravitation theory" in Proc. of International School on Gravitation and Cosmology, Erice, 1982. In preparation.

Volume 94B, number 3 PHYSICS LETTERS 11 August 1980

A GRAND UNIFIED THEORY OBTAINED FROM BROKEN SUPERGRAVITY

John ELLIS
CERN, Geneva, Switzerland

Mary K. GAILLARD
LAPP, Annecy-le-Vieux, France

and

Bruno ZUMINO
CERN, Geneva, Switzerland

Received 12 May 1980

To our friend Jacques Prentki on the occasion of his 60th birthday

We examine the possibility that the "fundamental" particles appearing in grand unified theories are a subset of the SU(8) bound states of preons belonging to the SO(8) extended supergravity, selected by the requirement that they form a renormalizable gauge theory. Analysis of the SU(8) Higgs potential given by supersymmetry suggests that the maximal grand unification symmetry is SU(5). A maximal subset of fermions free of SU(5) anomalies, and hence renormalizable, contains three generations of $\bar{5} + 10$ left-handed helicity states. The unbroken SU(5) theory may also contain 5 and 24 Higgs fields which are massless at the tree level.

At present there are two main approaches to the problem of unifying all particle interactions: grand unified theories (GUTs) [1,2] and extended supergravities [3]. On the one hand, interactions between "fundamental" particles of matter are often thought to be described by a renormalizable grand unified gauge theory. These GUTs have some phenomenological successes such as calculations [4–6] of $\sin^2\theta_W$ and the mass of the bottom quark [5–7], while also being sufficiently flexible to accommodate the full spectrum of fermions and interactions observed at low energies. The bottom quark mass calculation furnishes a phenomenological suggestion [6,7] that there are only three generations (u, d, e, ν_e), (c, s, μ, ν_μ), (t, b, τ, ν_τ) of "light" fermions with masses $\ll$ the grand unification mass of order 10^{15} GeV. A key problem [8] with GUTs is to understand the hierarchy of scales of symmetry breaking and masses, which apparently requires the unnatural fine-tuning of at least one [6,9] parameter in the Higgs potential. In fact these so-called "grand" unified theories have many ugly free parameters, mostly associated with their untidy Higgs systems. They are also grossly incomplete in that, although they become exact symmetries only at energies within about four orders of magnitude of the Planck mass, they do not attempt to include gravity. Extended supergravities [3] attempt to provide a viable theory of the coupling of matter fields to gravity without [10] many (all?) of the loop divergences usually encountered. Unfortunately there is so far no phenomenological evidence for supersymmetry, let alone supergravity. The observed "light" particles have no obvious supersymmetric partners [11] and the largest known extended supergravity has an SO(8) internal symmetry group which is too small [12] ‡1 to contain the small-

‡1 The first attempt to embed low-energy gauge theories in SO(8) supergravity using the basic supermultiplet was reported by Gell-Mann [13].

Volume 94B, number 3 PHYSICS LETTERS 11 August 1980

est GUT based [2] on the group SU(5). Furthermore, it is not known whether the internal symmetry group of SO(8) extended supergravity can be gauged.

A possible solution to these last two problems was discovered by Cremmer and Julia [14] ‡2 who pointed out that each extended supergravity theory contained a concealed local unitary symmetry [SU(8) in the case of SO(8) supergravity, U(4, 5, 6) for SO(4, 5, 6) supergravity] associated with composite fields. They conjectured [14] ‡2 that some unknown dynamical mechanism might cause the composite adjoint vector fields of this symmetry to become physical, propagating gauge bosons ‡3. We have further conjectured with Maiani [17] ‡4 that the supersymmetric partners of these vector fields might also become dynamical, with some of them being the fermions and Higgs bosons "observed" at low energies. Thus the fields appearing explicitly in a supergravity theory would be preons of which all the "fundamental" particles of GUTs, but not the graviton, are composites. Unfortunately this program runs into two essential problems [17] ‡4. The gauge currents have anomalies which would prevent the construction of a renormalizable theory, and conventional methods are unable to give masses to all unwanted fermionic states due to the lack of the necessary helicity partners with identical symmetry properties under the exact gauge subgroup $SU(3)_{colour} \times U(1)_{em}$. These two problems arise from the essential chirality of the composite supermultiplet. We want some chirality in the theory to explain the left-handedness of weak currents and the low (zero?) masses of the neutrinos, but one can have too much of a good thing.

In this paper we propose a solution to the problem of embedding GUTs in supergravity, based on an unpublished argument of Veltman [19] ‡5 which we understand as follows. If composite states are to have masses much smaller than their inverse size and to have effective interactions described by perturbation theory at energy scales much less than their inverse size, then their interactions must be renormalizable. Otherwise singularities would arise in the computation of vertex functions for which the only cut-off is the inverse size, which would result either in masses of order the inverse size for those states which have nonrenormalizable couplings, or a breakdown of perturbation theory for their effective interactions.

Examples of this "theorem" are provided by the pseudo-Goldstone bosons (PGBs) of exact gauge symmetries broken dynamically, such as colour and technicolour [21]. In these theories there are low-mass bound states (pions, technipions) whose masses would be zero in the absence of explicit chiral symmetry breaking in the lagrangian. At energy scales much less than their inverse sizes [$(1\ \mathrm{fm})^{-1}$, 1 TeV], the interactions of these PGBs are described by an effective, renormalizable low-energy gauge theory [$U(1)_{em}$, $SU(3)_{colour} \times SU(2)_L \times U(1)$]. The strong or extra-strong non-renormalizable interactions of these PGBs are suppressed by powers of (E/f) at low energies E, where f is of order $(1\ \mathrm{fm})^{-1}$ for pions and 250 GeV for technipions.

On the basis of Veltman's "theorem" we conjecture that some unknown dynamical mechanism causes the symmetries of extended supergravity to break down at or near the Planck mass, giving rise to an effective lower energy GUT which is renormalizable and realized by some subset of the supermultiplet containing the composite gauge fields found by Cremmer and Julia [14] ‡2. The most general renormalizable theory contains an anomaly-free representation of spin 1/2 fermions and fundamental scalar fields, as well as gauge vector bosons. Therefore the invocation of Veltman's "theorem" circumvents the two problems found [17] ‡4 in our previous analysis with Maiani. This new philosophy gives us constraints on the symmetries, particle content and couplings of GUTs, and we obtain the following results.

The only supergravity theory large enough to contain a realistic GUT is based on SO(8) and its composite symmetry SU(8), while the only GUT that it contains is based on the SU(5) group of Georgi and Glashow [2]. Two arguments then indicate that a direct breakdown of SU(8) to SU(5) is plausible. Preliminary studies of the SU(8) scalar field potential suggest that it may break the symmetry down at or near the Planck mass to a symmetry which cannot be larger than SU(5). Analysis of the anomalies associated with the SU(8) gauge group reveals that it must have

‡2 A previous attempt to relate this work to phenomenology has been made by Curtright and Freund [15].

‡3 This hypothesized phenomenon would be analogous to the behaviour of CP^{N-1} models in two dimensions, see ref. [16].

‡4 Telegraphic accounts of this work had previously been reported [18].

‡5 Arguments similar in spirit have probably occurred to many people such as those quoted in ref. [20]. We understand that A. Kabelschacht is preparing a rigorous proof of this "theorem".

Volume 94B, number 3 PHYSICS LETTERS 11 August 1980

anomalies, and that the only possible anomaly-free sub-theories based on the SU(7) and SU(6) subgroups which are vector-like with respect to $SU(3)_{colour} \times U(1)$ are actually completely vector-like. By contrast, the basic composite supermultiplet contains a maximal subset which is free of SU(5) anomalies and has left-handed spin 1/2 fields in the following reducible representation:

$$(\mathbf{45} + \overline{\mathbf{45}}) + 4(\mathbf{24}) + 9(\mathbf{10} + \overline{\mathbf{10}}) + 3(\mathbf{5} + \overline{\mathbf{5}}) + 3(\overline{\mathbf{5}} + \mathbf{10}) + 9(\mathbf{1}) . \quad (1)$$

Most of the spectrum (1) is vector-like, and these particles can acquire masses of order 10^{15} GeV to 10^{19} GeV. However, the spectrum (1) is slightly chiral, containing three generations of SU(5) left-handed $\overline{\mathbf{5}} + \mathbf{10}$ fermions.

We find that if some components of the traceless scalar fields develop a vacuum expectation value, then the maximal residual symmetry is SU(5). One appealing pattern breaks SU(8) down to SU(5) in such a way that the unbroken SU(5) theory contains at least some $(\mathbf{5} + \overline{\mathbf{5}})$ and adjoint **24** representations of scalar fields, which are massless at the tree level and may be suitable for generating the successive lower energy spontaneous breakdowns of $SU(5) \to SU(3)_{colour} \times SU(2)_L \times U(1) \to SU(3)_{colour} \times U(1)$ through radiative corrections [9, 22]. In a normal pattern of spontaneous symmetry breakdown the massless $(\mathbf{5} + \overline{\mathbf{5}})$ scalars would have been "eaten" by SU(8) vector bosons in a prior breakdown to SU(5). However, in our scenario the SU(8)/SU(5) gauge bosons could already have acquired masses through dynamical effects associated with anomalies, so that the $(\mathbf{5} + \overline{\mathbf{5}})$ scalars may survive to lower energies as pseudo-Goldstone bosons instead of being "eaten" at the Planck mass as Higgs fields.

We now describe the derivation of the results outlined above. The minimal GUT is based [2] on the group SU(5) with rank 4. The candidate effective gauge group G_E from supergravity should be at least as large as this, and should in fact presumably contain a non-trivial generation group G_G at least as large as SU(2):

$$G_E \supseteq SU(5) \times G_G . \quad (2)$$

Only the largest supergravity based on SO(8) has a G_E = SU(8) large enough to satisfy the condition (2). The rank of SU(8) is 7 and one might imagine that it could contain a GUT larger than SU(5), but this is not the case. The only [5,23] realistic GUT group of rank 5 is SO(10), and this is not a subgroup of SU(8). The decomposition SU(8)/SU(2) does not contain a suitable [24] rank 6 GUT such as E_6.

To proceed further we need the supermultiplet of composite fields containing the gauge fields of G_E. Arguments based on N = 1 and 2 supermultiplets and on superconformal theories have led us [17] ‡4 to expect the generic structure indicated in table 1. The representations listed in table 1 are reducible; they contain a traceless part and traces such as $S^A_{[ACD]}$ etc. In the case of N = 8 supergravity the V^A_A field does not correspond to a gauge symmetry of the theory, whereas it does in the cases of N = 4, 5, 6. In the case of N = 6 supergravity the helicity ±1/2 fermions appear in the representations (trace representations are shown in parentheses)

$$-1/2: \quad \mathbf{84}\,(+\mathbf{6}) = \mathbf{45} + \mathbf{24} + \mathbf{10} + \mathbf{5}\,(+\mathbf{5} + \mathbf{1}) ,$$

$$+1/2: \quad \overline{\mathbf{70}}\,(+\mathbf{20}) = \overline{\mathbf{40}} + \overline{\mathbf{15}} + \overline{\mathbf{5}} + \overline{\mathbf{10}}\,(+\,\mathbf{10} + \overline{\mathbf{10}}) . \quad (3)$$

where we have also indicated SU(5) decompositions. Adding the *TCP* conjugates of the −1/2 fermions to the +1/2 fermions we get a total set of left-handed fermions:

$$-1/2: \quad \mathbf{84} + \mathbf{70}\,(+\mathbf{6} + \mathbf{20}) = \mathbf{45} + \mathbf{40} + \mathbf{24} + \mathbf{15} + 2\,\mathbf{10} + 2\,\mathbf{5} \; (+\mathbf{10} + \overline{\mathbf{10}} + \mathbf{5} + \mathbf{1}) , \quad (4)$$

Table 1
A composite supermultiplet from SO(N) supergravity [a].

helicity	3/2	1	1/2	0	–	$(4-N)/2$	$(3-N)/2$	+ *TCP*
tensor representation	J^A	V^A_B	$F^A_{[BC]}$	$S^A_{[BCD]}$	–	$R^A_{[BCD...F]}$	$\tilde{R}^A$	conjugates

[a] The lower indices are totally antisymmetrized 1, 2, 3, ..., (N − 1) times.

which is not sufficiently large to contain the "observed" spectrum of SU(5) fermions whether or not we include the trace representations. The same is true a fortiori for the $N = 5$ case. We see again that we are forced to consider [‡6] $N = 8$, whose spin 1/2 fields have the SU(5) × SU(3) decompositions:

$$-1/2: \quad \mathbf{216}(+8) = (\overline{\mathbf{45}}, \mathbf{1}) + (\mathbf{24}, \mathbf{3}) + (\mathbf{10}, \overline{\mathbf{3}}) + (\overline{\mathbf{5}}, \overline{\mathbf{3}}) + (\mathbf{5}, \mathbf{8}) + (\mathbf{5}, \mathbf{1}) + (\mathbf{1}, \overline{\mathbf{6}}) + (\mathbf{1}, \mathbf{3})\,(+(\mathbf{5}, \mathbf{1}) + (\mathbf{1}, \mathbf{3}))\,,$$

$$+1/2: \quad \mathbf{504}(+56) = (\overline{\mathbf{45}}, \overline{\mathbf{3}}) + (\overline{\mathbf{40}}, \mathbf{3}) + (\mathbf{24}, \mathbf{1}) + (\overline{\mathbf{15}}, \mathbf{1}) + (\overline{\mathbf{10}}, \mathbf{8}) + (\mathbf{10}, \overline{\mathbf{6}}) + (\mathbf{10}, \mathbf{3}) + (\overline{\mathbf{10}}, \mathbf{1}) + (\mathbf{5}, \mathbf{3}) + (\overline{\mathbf{5}}, \overline{\mathbf{3}})\,(+(\mathbf{10}, \overline{\mathbf{3}}) + (\overline{\mathbf{10}}, \mathbf{1}) + (\mathbf{5}, \overline{\mathbf{3}}) + (\mathbf{1}, \mathbf{1}))\,, \tag{5}$$

which are clearly sufficient to accommodate the "observed" quarks and leptons. Since the composite spin 1 trace field V^A_A is not a gauge field of SO(8), it seems possible that none of the trace fields shown in parentheses in (5) actually become dynamical at low energies.

Before looking for anomaly-free subsets of the fermions (5) which Veltman's "theorem" suggests are candidates for low energy effective gauge theories, we first analyze the breakdown of SU(8) by the scalar fields $S^A_{[BCD]}$ of table 1 and their conjugates. They have the SU(7) decomposition

$$0: \quad \mathbf{420}(+28) = \mathbf{224} + \mathbf{140} + \mathbf{35} + \mathbf{21}\,(+21 + 7)\,, \tag{6}$$

which contains no SU(7) singlet. Therefore, if any component of $S^A_{[BCD]}$ can acquire a vacuum expectation value, SU(7) must be broken. Continuing to the SU(6) decomposition

$$0: \quad \mathbf{420}(+28) = \overline{\mathbf{105}} + 2\;\mathbf{84} + \mathbf{35} + 2\;\mathbf{20} + 4\;\mathbf{15} + 2\;\mathbf{6}\,(+15 + 2\;6 + 1)\,, \tag{7}$$

we also find no SU(6) singlet in the untraced representation. From our hypothesis that the SU(8) supermultiplet trace fields are not dynamical at low energies, it follows that SU(6) must also be broken. Finally, we have the SU(5) decomposition

$$0: \quad \mathbf{420}(+28) = 3\;\mathbf{45} + \overline{\mathbf{40}} + 3\;\mathbf{24} + 9\;\mathbf{10} + 3\;\overline{\mathbf{10}} + 9\;\mathbf{5} + \overline{\mathbf{5}} + 3\;\mathbf{1}\,(+10 + 3\;5 + 3\;1)\,, \tag{8}$$

in which a singlet appears in the untraced representation for the first time. This means that group theory allows SU(5) as a maximal unbroken subgroup of SU(8).

We therefore focus on the SU(5) subgroup of SU(8), which is now the largest group about whose anomalies we need worry. There are many subsets of the fermions (5) which are free of SU(5) anomalies. Low energy phenomenology requires an $SU(3)_{colour} \times U(1)_{em}$ subgroup of SU(5) to be vector-like. With this constraint there is a unique subset of (5) which has the maximum number of left-handed fermions: lumping +1/2 fermions and *TCP* conjugates of the −1/2 fermions together it is that displayed in eq. (1). Perturbative symmetry breaking by radiative corrections may favour [9,22] the choice of a representation with the largest possible number of fermions. The sight of three chiral generations of $(\mathbf{10} + \overline{\mathbf{5}})$ fermions in eq. (1) fills us with joy.

If we step back and try to find a subset of the fermions (5) which has no SU(6) anomalies and has a vector-like $SU(3)_{colour} \times U(1)$ subgroup, then the only solutions are completely vector-like. Furthermore, the "maximal" such theory which has the fermion content

$$(\mathbf{84} + \overline{\mathbf{84}}) + 2\;\mathbf{35} + 4\;\mathbf{20} + 2(\mathbf{15} + \overline{\mathbf{15}}) + (\mathbf{6} + \overline{\mathbf{6}})\,, \tag{9}$$

clearly has fewer fermions than the "maximal" SU(5) theory (1). Conversely, if we try to embed the SU(5) representation (1) in an SU(6) representation of fermions it will automatically have SU(6) anomalies. Thus the combined requirements of chirality, a vector-like $SU(3)_{colour} \times U(1)$ and maximality strongly favour the SU(5) anomaly-free representation (1), and strongly suggest that the SU(6) subgroup of SU(8) is broken down by anomalies.

It is in principle possible to go further in the analysis of symmetry breaking by analyzing the scalar field lagrangian using supersymmetry constraints [25] which can determine a subset of couplings, namely

$$\mathcal{L}((\mathbf{420})^4) + \mathcal{L}((\mathbf{216})^2 \times \mathbf{420}) + \mathcal{L}(\mathbf{216} \times \overline{\mathbf{504}} \times \mathbf{420})\,. \tag{10}$$

‡6 The supergravity theories for $N = 7$ and $N = 8$ are almost certainly identical with both having an SU(8) gauge symmetry.

Volume 94B, number 3 PHYSICS LETTERS 11 August 1980

One might hope that despite the dynamical breaking of supersymmetry, some or all of these constraints may remain valid at high momentum scale at or near the Planck mass [‡7]. They would then provide boundary conditions for computing the evolution of the coupling constants down to lower momenta using the renormalization group. It seems likely that the supersymmetric scalar field lagrangian may provide a breakdown of SU(8) to SU(5) at or near the Planck mass. This could either be directly through the presence of a term quadratic in the **420** fields $S^A_{[BCD]}$ or indirectly through the presence of a zero in the quartic term $[(\mathbf{420})^4]$ for some field direction $S^A_{[BCD]}$ which could then be exploited by radiative corrections to give $\langle 0|S^A_{[BCD]}|\ \rangle \neq 0$.

One possibility suggested by preliminary studies of the scalar field potential is that some components $S^i_{[iCD]}$, where the upper index is identical with one of the lower indices, may develop non-zero values. These vev's have components in the following representations of the SU(6) subgroup of SU(8):

$$\{\langle 0|S^i_{[iCD]}|0\rangle\} \in \mathbf{105}, \mathbf{84}, \mathbf{35}, \mathbf{15}, \mathbf{6}\ . \qquad (11)$$

If $\langle 0|^i_{[iCD]}|0\rangle \neq\neq 0$ for just one pair $[C, D]$ then the symmetry breaking would be through an adjoint representation of some SU(6) subgroup of SU(8). The **35** and the **6** in (11) contain singlets of SU(5) which are what is needed if scalar fields are to break SU(8) down to SU(5) at or near the Planck mass.

In achieving this Higgs breaking, the scalar field potential would necessarily generate some zero mass fields which would normally be eaten by the corresponding SU(8)/SU(5) vector fields thereby giving them a mass. However, in our framework these vectors are supposed to have already acquired masses because of the dynamical symmetry breaking due to anomalies. They no longer have any appetite left, and the would-be massless higgses are left as physical states. Presumably they need only be massless at the tree level as there is no reason to expect the loop corrections to the SU(8) scalar field potential to respect all the SU(8) supersymmetry. Since when an SU(8) adjoint representation is decomposed with respect to SU(5):

$$\mathbf{63} = \mathbf{24} + 3(\mathbf{5} + \bar{\mathbf{5}}) + 9\,\mathbf{1}\ , \qquad (12)$$

one encounters three $(\mathbf{5} + \bar{\mathbf{5}})$ respresentations, it seems that the uneaten "pseudo-higgses" may include several fields suitable for subsequent stages of SU(5) symmetry breaking through radiative corrections [22]. In order for a complete scenario [9] of SU(5) symmetry breaking by radiative corrections to be valid, one also needs at least one **24** of scalar fields massless at least at the tree level, to take care of the first stage of SU(5) symmetry breaking. Preliminary analysis of the SU(6) × (N = 2 supersymmetry) constraints on the couplings of SU(6) adjoint scalars on (11) suggests than an SU(5) adjoint **24** subset of scalars may indeed be massless at the tree level, though this requires further analysis. Having **24** and **5** higgses massless at the tree level in the manner discussed here would be sufficient to evade theorems [26] about sequential breaking of symmetries through radiative corrections. It remains to be seen whether this will in fact be possible.

There are clearly very many other questions about the program we have outlined. They are all in principle answerable, since our approach has no free parameters in the absence of topological effects. Answering these questions may in practice prove non-trivial.

We would like to thank M. Gell-Mann, L. Maiani, D.V. Nanopoulos, P. Sikivie and M. Veltman for encouragement and instructive discussions.

[‡7] This would be analogous to using SU(5) symmetric initial conditions to calculate $\sin^2\theta_W$ from the measured values of α and α_s if we knew the renormalization group equations but did not know about the Higgs mechanism for spontaneous symmetry breaking.

References

[1] J.C. Pati and A. Salam, Phys. Rev. Lett. 31 (1973) 661; Phys. Rev. D8 (1973) 1240.
[2] H. Georgi and S.L. Glashow, Phys. Rev. Lett. 32 (1974) 438.
[3] D.Z. Freedman, P. Van Nieuwenhuizen and S. Ferrara, Phys. Rev. D13 (1976) 3214;
S. Deser and B. Zumino, Phys. Lett. 62B (1976) 335;
S. Ferrara and P. Van Nieuwenhuizen, Phys. Rev. Lett. 37 (1976) 1669;
S. Ferrara, J. Scherk and B. Zumino, Phys. Lett. 66B (1977) 35;
D.Z. Freedman, Phys. Rev. Lett. 38 (1977) 105;
A. Das, Phys. Rev. D15 (1977) 2805;
E. Cremmer, J. Scherk and S. Ferrara, Phys. Lett. 68B (1977) 234;
E. Cremmer and J. Scherk, Nucl. Phys. B127 (1977) 259;
B. de Wit and D.Z. Freedman, Nucl. Phys. B130 (1977) 105.

[4] H. Georgi, H.R. Quinn and S. Weinberg, Phys. Rev. Lett. 33 (1974) 451.
[5] M.S. Chanowitz, J. Ellis and M.K. Gaillard, Nucl. Phys. B128 (1977) 506.
[6] A.J. Buras, J. Ellis, M.K. Gaillard and D.V. Nanopoulos, Nucl. Phys. B135 (1978) 66.
[7] D.V. Nanopoulos and D.A. Ross, Nucl. Phys. B157 (1979) 273.
[8] E. Gildener, Phys. Rev. D14 (1976) 1667.
[9] J. Ellis, M.K. Gaillard, A. Peterman and C.T. Sachrajda, Nucl. Phys. B164 (1980) 253.
[10] For a review, see: M. Grisaru and P. Van Nieuwenhuizen, in: Deeper pathways in high-energy physics, Proc. Orbis Scientiae (Coral Gables, 1977) (Plenum Press, New York, 1977).
[11] See, for example, P. Fayet, in: New frontiers in high energy physics, Proc. Orbis Scientiae (Coral Gables, 1978) (Plenum Press, New York, 1978) p. 413; G. Barbiellini et al., DESY preprint 79/67 (1979).
[12] See, e.g., B. Zumino, in: Proc. Einstein Symp. (Berlin, 1979), Lecture Notes in Physics, Vol. 100, eds. H. Nelkowski, A. Herman, H. Poser, R. Schrader and R. Seiler (Springer, Berlin, 1979) p. 114, and references therein.
[13] M. Gell-Mann, Talk at the 1977 Washington Meeting of the Am. Phys. Soc.
[14] E. Cremmer and B. Julia, Phys. Lett. 80B (1978) 48; Nucl. Phys. B159 (1979) 141.
[15] T. Curtright and P.G.O. Freund, Supergravity, Proc. Supergravity Workshop at Stony Brook (Sept. 1979), eds. P. Van Nieuwenhuizen and D.Z. Freedman (North-Holland, Amsterdam, 1979) p. 197.
[16] A. D'Adda, P. Di Vecchia and M. Lüscher, Nucl. Phys. B146 (1978) 63; B152 (1979) 125; E. Witten, Nucl. Phys. B149 (1979) 285.
[17] J. Ellis, M.K. Gaillard, L. Maiani and B. Zumino, LAPP preprint TH-15/CERN preprint TH. 2481 (1980), to be published in Proc. Europhysics Study Conf. on Unification of fundamental interactions (Erice, 1980).
[18] J. Ellis, Proc. EPS Intern. Conf. on High energy physics (Geneva, June 1979) (CERN, 1979) p. 940; M.K. Gaillard and L. Maiani, 1979 Cargèse Summer Institute, LAPP preprint TH-09 (1979).
[19] M. Veltman, private communication.
[20] G.'t Hooft, Cargèse Summer Institute Lecture Notes (1979); G.L. Kane, G. Parisi, S. Raby, L. Susskind and K.G. Wilson, private communications.
[21] L. Susskind, Phys. Rev. D20 (1979) 2619.
[22] S. Coleman and E. Weinberg, Phys. Rev. D7 (1977) 1888; S. Weinberg, Phys. Lett. 82B (1979) 387.
[23] H. Georgi and D.V. Nanopoulos, Nucl. Phys. B155 (1979) 52, and references therein.
[24] R. Barbieri and D.V. Nanopoulos, CERN preprint TH. 2810 (1980), and references therein.
[25] L. Brink, J.H. Schwarz and J. Scherk, Nucl. Phys. B121 (1977) 77; S. Ferrara and B. Zumino, Nucl. Phys. B79 (1974) 413; L. O'Raifeartaigh, Phys. Lett. 56B (1975) 41.
[26] H. Georgi and A. Pais, Phys. Rev. D16 (1977) 3520.

Nuclear Physics B186 (1981) 412–428

SEARCH FOR A REALISTIC KALUZA–KLEIN THEORY*

Edward WITTEN

Joseph Henry Laboratories, Princeton University, Princeton, New Jersey 08544, USA

Received 12 January 1981

An attempt is made to construct a realistic model of particle physics based on eleven-dimensional supergravity with seven dimensions compactified. It is possible to obtain an SU(3)×SU(2)×U(1) gauge group, but the proper fermion quantum numbers are difficult to achieve.

In 1921 Kaluza suggested [1] that gravitation and electromagnetism could be unified in a theory of five-dimensional riemannian geometry. The idea was further developed by Klein [2] and was the subject of considerable interest during the classical period of work on unified field theories [3]. Readable expositions of some of the classical work have been given in text books by Bergmann and by Lichnerowicz [4]; more recent discussions have been given by Rayski and by Thirring [5].

While the Kaluza–Klein approach has always been one of the most intriguing ideas concerning unification of gauge fields with general relativity, it has languished because of the absence of a realistic model with distinctive and testable predictions. Yet the urgency of the unification of gauge fields with general relativity has surely greatly increased with the growing importance of gauge fields in physics. Moreover, the Kaluza–Klein theory has generalizations to non-abelian gauge fields which actually were first proposed [6] well before real applications were known for Yang–Mills fields in physics.

In the last few years this approach has been revived by Scherk and Schwarz and by Cremmer and Scherk, originally in connection with dual models [7]. These authors introduced many new ideas as well as new focus. In contrast to much of the classical literature, they advocated that the extra dimensions should be regarded as true, physical dimensions, on a par with the four observed dimensions. Cremmer and Scherk suggested that the obvious differences between the four observed dimensions and the extra microscopic ones could arise from a spontaneous breakdown of the vacuum symmetry, or, as they called it, from a process of "spontaneous compactification" of the extra dimensions.

These ideas have motivated much recent work. The idea of spontaneous compactification has been developed in more detail by Luciani [8]. An interesting idea by Palla [9] about massless fermions in theories with extra compact dimensions

* Research partially supported by NSF grant PHY78-01221.

will figure in some of the discussion below. Manton [10] has discussed some questions that arise in trying to generate Higgs fields as components of the gauge field in extra dimensions. The idea of extra hidden dimensions has stimulated much work in supersymmetry theory, including the successful construction of the $N=8$ supergravity theory by Cremmer, Julia and Scherk and by Cremmer and Julia [11]. This work has been generalized to give models with broken sypersymmetry [12].

In many respects, of course, the modern approaches to this subject tend to differ from the classical point of view. In view of the proliferation of new particles in the last thirty years, one may be more willing today than in the past to postulate the infinite number of new degrees of freedom that must exist if extra dimensions really exist. Much of the classical literature focussed on the need to eliminate a massless spin-zero particle that naturally exists in the original Kaluza–Klein theory; the question seems less urgent today because the obvious answer is that quantum mechanical mass renormalization could easily account for the failure to observe this particle (a mass of 10^{-4} eV would make it undetectable). Some of the early work was motivated by the hope that the fifth dimension could provide the hidden variables that would eliminate indeterminacy from quantum mechanics. Despite the many generalizations and changes in emphasis that have occurred, I will refer generically to theories in which gauge fields are unified with gravitation by means of extra, compact dimensions as Kaluza–Klein theories.

It has often been suggested that spontaneous compactification and supergravity could be usefully combined together. The $N=8$ supergravity theory was constructed by "dimensional reduction" starting from an eleven-dimensional theory. In this context, "dimensional reduction" just means that the fields are taken to be independent of seven of the original eleven coordinates, to which physical reality need not be attributed. However, Cremmer and Julia [11] suggested that one might wish to consider seriously the eleven dimensions and interpret seven of them as compact dimensions in the spirit of Kaluza and Klein. This idea has been raised, on occasion, by various other theorists. In this paper, I will describe an attempt – not completely successful, but not completely unsuccessful either – to construct a realistic theory of Kaluza–Klein type, based on eleven-dimensional supergravity.

As discussed by some of the authors mentioned above, from a modern point of view the Kaluza–Klein unified theory of gravitation and electromagnetism is probably best understood as a theory of spontaneous symmetry breaking in which the group of general coordinate transformations in five dimensions is spontaneously broken to the product of the four-dimensional general coordinate transformation group and a local U(1) gauge group.

Let us review how this arises. One considers standard general relativity in five dimensions with the standard Einstein–Hilbert action

$$A=\int \mathrm{d}^5x\sqrt{g}R. \tag{1}$$

Instead of assuming that the ground state of this system is five-dimensional Minkowski space, which we will denote as M^5, one takes the ground state to be the product $M^4 \times S^1$ of four-dimensional Minkowski space M^4 with the circle S^1. The space $M^4 \times S^1$ is, like M^5, a solution of the five-dimensional Einstein equations. Classically it is difficult to decide which of the spaces M^5 and $M^4 \times S^1$ is a more appropriate choice as the ground state, since they both have zero energy, insofar as energy can be defined in general relativity*. Conventionally, one might assume that the ground state is M^5. In the Kaluza–Klein approach one assumes, instead, that the ground state is $M^4 \times S^1$, and the physical spectrum is determined by studying small oscillations around this ground state. One assumes that the radius of the circle S^1 is microscopically small, perhaps of order of the Planck length, and this accounts for why the existence of this fifth dimension is not noted in everyday experience.

The symmetries of the Kaluza–Klein ground state $M^4 \times S^1$ are the four-dimensional Poincaré symmetries, acting on M^4, and a U(1) group of rotations of the circle S^1. These symmetries would be observed as local or gauge symmetries in the apparent four-dimensional world because the whole theory started with the Einstein action (1) which is generally covariant. In fact, if one considers small oscillations around the "ground state" $M^4 \times S^1$, one finds an infinite number of massive excitations, the masses being of order the inverse of the circumference of S^1. One finds also a finite number of massless modes, which presumably would constitute the low-energy physics. The massless modes turn out to be a spin-two graviton and a spin-one photon, which are gauge particles of the symmetries of $M^4 \times S^1$, and a Brans–Dicke scalar.

The ansatz which exhibits the massless modes is the following. The metric tensor of this theory is a five by five matrix $g_{AB}(x^\mu, \phi)$ which in general may depend on the four coordinates x^μ, $\mu = 1 \cdots 4$, of M^4, and on the angular coordinate ϕ of S^1. The massless modes are those for which g_{AB} is a function of x^μ only. One can then write g_{AB} in block form

$$g_{AB}(x^\mu, \phi) = \left(\begin{array}{c|c} g_{\mu\nu}(x) & A_\mu(x) \\ \hline A_\mu(x) & \sigma(x) \end{array}\right), \tag{2}$$

where $g_{\mu\nu}$ is a four by four matrix (the first four rows and columns of g_{AB}), $A_\mu = g_{\mu 5}$, and $\sigma = g_{55}$. Then $g_{\mu\nu}$ is the ordinary metric tensor of the apparent four-dimensional world, and describes a massless spin-two particle; A_μ is the gauge field of the U(1) symmetry, and σ is the Brans-Dicke scalar.

In the classical work on the Kaluza–Klein theory, it is shown that the five-dimensional Einstein action (1), when expanded in terms of $g_{\mu\nu}$, A_μ, and σ (and the other modes, which decouple from these at low energies) contains a four-dimensional Einstein action $\sqrt{g}R^{(4)}$ for $g_{\mu\nu}$, a Maxwell action $F^2_{\mu\nu}$ for A_μ, and the usual

* The definition of energy in general relativity depends on the boundary conditions, so while both M^5 and $M^4 \times S^1$ have zero energy, a comparison between them is meaningless, like comparing zero apples to zero oranges.

kinetic energy for σ. Also, one can readily check that A_μ transforms as a gauge field $A_\mu \to A_\mu + \partial_\mu \varepsilon$ under coordinate transformations of the special type $(x^i, \phi) \to (x^i, \phi + \varepsilon(x^i))$ if the metric g_{AB} is transformed by the standard rule

$$g_{AB} \to g_{A'B'} \frac{\partial x'^{A'}}{\partial x^A} \frac{\partial x'^{B'}}{\partial x^B} .$$

The Kaluza–Klein theory thus unifies the metric tensor $g_{\mu\nu}$ and a gauge field A_μ into the unified structure of five-dimensional general relativity. This theory is surely one of the most remarkable ideas ever advanced for unification of electromagnetism and gravitation.

The Kaluza–Klein theory, as noted above, also has a non-abelian generalization, which has been extensively discussed over the years. In this generalization, one starts with general relativity in $4+n$ dimensions, possibly with additional matter fields or with a cosmological constant. Instead of assuming the ground state to be M^{4+n}, Minkowski space of $4+n$ dimensions, one assumes the ground state to be a product space $M^4 \times B$, where B is a compact space of dimension n. $M^4 \times B$ should be a solution of the classical equations of motion, or possibly, as will be discussed later, a minimum of some effective potential.

As in the previous discussion, symmetries of B will be observed as gauge symmetries in the effective four dimensional world. With a suitable choice of B, one may unify an arbitrary gauge group, abelian or non-abelian, with ordinary general relativity, in a $4+n$ dimensional theory.

The ansatz which generalizes (2) is the following. Let ϕ_i, $i = 1 \cdots n$, be coordinates for the internal space B. Let T^a, $a = 1 \cdots N$, be the generators of the symmetry group G of B. Let the action of the symmetry generator T^a on the ϕ_i be $\phi_i \to \phi_i + K_i^a(\phi)$, where $K_i^a(\phi)$ is the "Killing vector" associated with the symmetry T^a. Then the massless excitations of the candidate "ground state" $M^4 \times B$ correspond to an ansatz of the following form:

$$g_{AB}(x^\alpha, \phi^k) = \left(\begin{array}{c|c} g_{\mu\nu}(x^\alpha) & \sum_a A_\mu^a(x^\alpha) K_i^a(\phi^k) \\ \hline \sum_a A_\mu^a(x^\alpha) K_i^a(\phi^k) & \gamma_{ij}(\phi^k) \end{array} \right), \tag{3}$$

where γ_{ij} is the metric tensor of the internal space B. The fields $A_\mu^a(x^\alpha)$ are massless gauge fields of the group G. In this way one may obtain the gauge fields of an arbitrary abelian or non-abelian gauge group as components of the gravitational field in $4+n$ dimensions.

One may verify that the $4+n$ dimensional gravitational action really contains the proper kinetic energy term $\sum_a (F_{\mu\nu}^a)^2$. It is also straightforward to check that under infinitesimal coordinate transformations of the special form $(x^\alpha, \phi_i) \to (x^\alpha, \phi_i + \sum_a \varepsilon^a(x^\alpha) K_i^a(\phi))$, which is an x-dependent symmetry transformation of the internal space B, the field $A_\mu^a(x)$ transforms in the expected fashion, $A_\mu^a(x) \to A_\mu^a(x) + D_\mu \varepsilon^a(x)$. Thus, A_μ^a really has the properties expected of an ordinary four-dimensional gauge field. This gauge field is a remnant of the original coordinate invariance group in $4+n$ dimensions, which has been spontaneously broken down to the symmetries of $M \times B$.

As has been noted before, there is a fairly extensive literature on this construction. The case which has been discussed most widely is the case in which B is itself the manifold of some group H. It should be noted that, if H is a non-abelian group, the symmetry group G of the group manifold is not H but $H \times H$, since the group manifold can be transformed by either left or right multiplication. If one starts with general relativity in $4+n$ dimensions, the ansatz (3) will automatically give massless gauge mesons of the full symmetry group $H \times H$.

What problems arise if we try to construct a realistic theory along these lines? Known particle interactions can be described by the gauge group $SU(3) \times SU(2) \times U(1)$. So the symmetry group G of the compact space B must at least contain this as a subgroup,

$$SU(3) \times SU(2) \times U(1) \subset G\,. \tag{4}$$

So B must at least have $SU(3) \times SU(2) \times U(1)$ as a symmetry group.

To be as economical as possible, we may wish to choose B to be a manifold of minimum dimension with an $SU(3) \times SU(2) \times U(1)$ symmetry. What is the minimum dimension of a manifold which can have $SU(3) \times SU(2) \times U(1)$ symmetry?

$U(1)$ is the symmetry group of the circle S^1, which has dimension one. The lowest dimension space with symmetry $SU(2)$ is the ordinary two-dimensional sphere S^2. The space of lowest dimension with symmetry group $SU(3)$ is the complex projective space CP^2, which has real dimension four. (CP^2 is the space of three complex variables (Z^1, Z^2, Z^3), not all zero, with the identification $(Z^1, Z^2, Z^3) \simeq (\lambda Z^1, \lambda Z^2, \lambda Z^3)$ for any non-zero complex number λ. CP^2 can also be defined as the homogeneous space $SU(3)/U(2)$.) Therefore, the space $CP^2 \times S^2 \times S^1$ has $SU(3) \times SU(2) \times U(1)$ symmetry, and it has $4+2+1=7$ dimensions.

As we will see below, seven dimensions is in fact the minimum dimensionality of a manifold with $SU(3) \times SU(2) \times U(1)$ symmetry, although $CP^2 \times S^2 \times S^1$ is not the only seven-dimensional manifold with this symmetry. If, therefore, we wish to construct a theory in which $SU(3) \times SU(2) \times U(1)$ gauge fields arise as components of the gravitational field in more than four dimensions, we must have at least seven extra dimensions. With also four non-compact "space-time"dimensions, the total dimensionality of our world must be at least $4+7=11$.

This last number is most remarkable, because eleven dimensions is probably the maximum for supergravity. Eleven-dimensional supergravity has been explicitly constructed, and it is strongly believed that supergravity theories do not exist in dimensions greater than eleven. (The reason for this belief is that, on purely algebraic grounds [13], a supergravity theory in $d > 11$ would have to contain massless particles of spin greater than two. But there are excellent reasons, both S-matrix theoretic [14] and field theoretic [15], to believe that consistent field theories with gravity coupled to massless particles of spin greater than two do not exist.) It is consequently just barely possible to obtain $SU(3) \times SU(2) \times U(1)$ gauge fields as part of the gravitational field in a supergravity theory, if we use the unique, maximal, eleven-dimensional supergravity theory.

It is certainly a very intriguing numerical coincidence that eleven dimensions, which is the maximum number for supergravity, is the minimum number in which one can obtain $SU(3)\times SU(2)\times U(1)$ gauge fields by the Kaluza–Klein procedure. This coincidence suggests that the approach is worth serious consideration.

Let us now discuss in more detail the question of why seven dimensions is the minimum number of dimensions for a space with $SU(3)\times SU(2)\times U(1)$ symmetry – and the related matter of determining all seven-dimensional manifolds with this symmetry.

The space of lowest dimension with any symmetry group G is always a homogeneous space G/H, where H is a maximal subgroup of G. (The space G/H is defined as the set of all elements g of G, with two elements g and g' regarded as equivalent, $g\simeq g'$, if they differ by right multiplication by an element of H, that is, if $g=g'$ with $h\in$ H.) The dimension of G/H is always equal to the dimension of G minus the dimension of H.

In the case $G=SU(3)\times SU(2)\times U(1)$, the largest dimension subgroup that is suitable is $SU(2)\times U(1)\times U(1)$. Any larger subgroup of G would contain as a subgroup one of the three factors SU(3), SU(2), or U(1) of G, and this factor would then not have any non-trivial action on G/H – it would not really be a symmetry group of G/H. Since the dimension of $SU(3)\times SU(2)\times U(1)$ is $8+3+1=12$ and the dimension of $SU(2)\times U(1)\times U(1)$ is $3+1+1=5$, the dimension of $(SU(3)\times SU(2)\times U(1))/(SU(2)\times U(1)\times U(1))$ is $12-5=7$. It is for this reason that a space with $SU(3)\times SU(2)\times U(1)$ symmetry must have at least seven dimensions. However, there are many ways to embed $SU(2)\times U(1)\times U(1)$ in $SU(3)\times SU(2)\times U(1)$, and as a result there are many seven-dimensional manifolds with $SU(3)\times SU(2)\times U(1)$ symmetry.

To embed $SU(2)\times U(1)\times U(1)$ in $SU(3)\times SU(2)\times U(1)$ we first embed SU(2). SU(2) can be embedded in $SU(3)\times SU(2)\times U(1)$ in a variety of ways. The only embedding that turns out to be relevant is for SU(2) to be embedded in SU(3) as an "isospin" subgroup, so that the fundamental triplet of SU(3) transforms as $2+1$ under SU(2). [Other embeddings of SU(2) lead to spaces G/H on which some of the $SU(3)\times SU(2)\times U(1)$ symmetries act trivially, as discussed in the previous paragraph.] We still must embed $U(1)\times U(1)$ in $SU(3)\times SU(2)\times U(1)$.

$SU(3)\times SU(2)\times U(1)$ has three commuting U(1) generators which commute with the SU(2) subgroup of SU(3) that we have just chosen. There is a "hypercharge" generator of SU(3), which we may call λ_8, which commutes with the "isospin" subgroup. Also, we have the U(1) factor of $SU(3)\times SU(2)\times U(1)$, which will be called Y, and we may choose an arbitrary U(1) generator of the SU(2) factor, which will be called T_3.

So $SU(3)\times SU(2)\times U(1)$ contains an essentially unique subgroup $SU(2)\times U(1)\times U(1)\times U(1)$, where the three U(1) factors are λ_8, T_3, and Y. We do not want to divide

SU(3)×SU(2)×U(1) by the full SU(2)×U(1)×U(1)×U(1) subgroup because this would yield a space ($CP^2 \times S^2$, to be precise) on which the U(1) of SU(3)×SU(2)×U(1) would act trivially and would not really be a symmetry. So we delete one of the three U(1) factors, and divide only by SU(2)×U(1)×U(1).

The U(1) factor that is deleted may be an arbitrary linear combination $p\lambda_8+qT_3+rY$ of λ_8, T_3, and Y where p, q, and r are any three integers which have no common divisor*. So we define H as SU(2)×U(1)×U(1), where the SU(2) is our "isospin" subgroup of SU(3), and the two U(1)'s are the two linear combinations of λ_8, T_3, and Y which are orthogonal to $p\lambda_8+qT_3+rY$. The space G/H is then a seven-dimensional space with SU(3)×SU(2)×U(1) symmetry, which we may call M^{pqr}.

In a few cases the M^{pqr} are familiar spaces. M^{001} is our previous example $CP^2\times S^2\times S^1$. But in most cases the M^{pqr} are not familiar spaces, and are not products.

In a few cases the M^{pqr} have greater symmetry than SU(3)×SU(2)×U(1). M^{101} is $S^5\times S^2$, which has the symmetry O(6)×SU(2). M^{011} is $CP^2\times S^3$, whose full symmetry is SU(3)×SU(2)×SU(2). Except for these two cases, one cannot obtain from seven extra dimensions a symmetry "larger" than SU(3)×SU(2)×U(1). Therefore, the observed gauge group in nature is practically the "largest" group one could obtain from a Kaluza–Klein theory with seven extra dimensions.

Although the M^{pqr} for general values of p, q, and r are not familiar spaces, it is possible to give a rather explicit description of them. Consider first the eight dimensional space $S^5\times S^3$ [S^n is the n-dimensional sphere, with symmetry group $O(n+1)$]. The symmetry group of $S^5\times S^3$ is O(6)×O(4). Let us introduce a particular generator of O(6),

$$K=\begin{pmatrix} 0 & 1 & 0 & 0 & 0 & 0\\ -1 & 0 & 0 & 0 & 0 & 0\\ 0 & 0 & 0 & 1 & 0 & 0\\ 0 & 0 & -1 & 0 & 0 & 0\\ 0 & 0 & 0 & 0 & 0 & 1\\ 0 & 0 & 0 & 0 & -1 & 0 \end{pmatrix}, \tag{5}$$

and a particular generator of O(4),

$$L=\begin{pmatrix} 0 & 1 & 0 & 0\\ -1 & 0 & 0 & 0\\ 0 & 0 & 0 & 1\\ 0 & 0 & -1 & 0 \end{pmatrix}. \tag{6}$$

Then the subgroup of O(6) that commutes with K is SU(3)×U(1) [the U(1) being

* And r should be non-zero to avoid obtaining a space on which U(1) is realized as the identity.

generated by K itself] and the subgroup of O(4) that commutes with L is SU(2)$\times$ U(1) [the U(1) being generated by L].

For any non-zero p and q, we now define $N=-qK+pL$. Then N generates a U(1) subgroup of O(6)$\times$O(4), consisting of elements of the form $\exp tN$, $0\leq t\leq 2\pi$. We may now form from $S^5\times S^3$ a seven-dimensional space $M^{pq}=(S^5\times S^3)/U(1)$, where two points in $S^5\times S^3$ are considered to be identical if they are mapped into each other by the action of the U(1) subgroup generated by N.

This space M^{pq} is equal to the $r=1$ case of what we have previously called M^{pqr}. The M^{pq} are actually the most general simply connected seven-dimensional manifolds with SU(3)$\times$SU(2)$\times$U(1) symmetry. To obtain M^{pqr} for $r\neq 1$ one must factor out from $S^5\times S^3$ an additional discrete subgroup consisting of elements of the form $\exp(2\pi qK/r)$ $(q=0,1,2,\ldots,r-1)$. We define $M^{pqr}=M^{pq}/Z^r=(S^5\times S^3)/(U(1)\times Z^r)$.

To verify that the construction of the M^{pqr} just presented is equivalent to the previous definition as (SU(3)$\times$SU(2)$\times$U(1))/(SU(2)$\times$U(1)$\times$U(1)), one uses the fact that SU(3)/SU(2) is S^5, while SU(2) is S^3, so (SU(3)$\times$SU(2)$\times$U(1))/(SU(2)$\times$ U(1)$\times$U(1)) is $(S^5\times S^3\times U(1))/(U(1)\times U(1))$. Dividing out the two U(1) factors, one arrives at the above definition of M^{pqr} as $(S^5\times S^3)/(U(1)\times Z^r)$.

The M^{pqr} are not quite the most general seven-dimensional manifold with SU(3)$\times$ SU(2)$\times$U(1) symmetry, because for special values of p, q, and r it is possible to supplement SU(2)$\times$U(1)$\times$U(1) with an additional twofold discrete symmetry. One obtains in this way some non-orientable manifolds with one of the M^{pqr} as a double covering space. These spaces are the following. Dividing M^{001} by a discrete symmetry one can get $CP^2\times P^2\times S^1$ (P^k is real projective space of dimension k), or $CP^2\times(S^2\times S^1)/Z_2$, where Z_2 is a simultaneous inversion of S^2 and S^1. From M^{101} one gets $S^5\times P^2$ and $(S^5\times S^2)/Z_2$, where the Z_2 is a simultaneous inversion of S^5 and S^2. Likewise, by dividing M^{10r} by an additional two-fold symmetry one can make $S^5/Z^r\times P^2$ and $(S^5/Z^r\times S^2)/Z_2$. These spaces are non-orientable. This completes the list of seven-dimensional manifolds with SU(3)$\times$SU(2)$\times$U(1) symmetry.

If one is willing to suppose that the ground state of eleven-dimensional supergravity is a product of four-dimensional Minkowski space with one of the M^{pqr}, one can obtain an SU(3)$\times$SU(2)$\times$U(1) gauge group, the gauge fields being components of the gravitational field, according to the ansatz of eq. (3). Of course, to describe nature, it is not sufficient to have the gauge group. It is also necessary to have quarks and leptons of essentially zero mass [very light compared to the energy scale of gravitation; massless in any approximation in which SU(3)$\times$SU(2)$\times$U(1) is not spontaneously broken] which should be in the appropriate representation of the gauge group. And it is necessary to find Higgs bosons whose vacuum expectation value could ultimately trigger SU(2)$\times$U(1) breaking.

How can one obtain massless quarks and leptons in the Kaluza–Klein framework? To understand the basic idea*, suppose that in a $4+n$ dimensional theory we have a

* See also a discussion by Palla [9].

massless spin one half fermion. It satisfies the $4+n$ dimensional Dirac equation,

$$\not{D}\psi = 0\,, \tag{7}$$

or explicitly

$$\sum_{i=1}^{4+n} \gamma^i D_i \psi = 0\,. \tag{8}$$

This Dirac operator can be written in the form

$$\not{D}^{(4)}\psi + \not{D}^{(\text{int})}\psi = 0, \tag{9}$$

where $\not{D}^{(4)} = \sum_{i=1}^{4} \gamma^i D_i$ is the ordinary four-dimensional Dirac operator, and $\not{D}^{(\text{int})} = \sum_{i=5}^{4+n} \gamma^i D_i$ is the Dirac operator in the internal space of n compact dimensions.

The expression (9) immediately shows that the eigenvalue of $\not{D}^{(\text{int})}$ will be observed in practice as the four-dimensional mass. If $\not{D}^{(\text{int})}\psi = \lambda\psi$, then ψ will be observed by four-dimensional observers who are unaware of the existence of the extra microscopic dimensions as a fermion of mass $|\lambda|$.

The operator $\not{D}^{(\text{int})}$ acts on a compact space, so its spectrum is discrete. Its eigenvalues either are zero or are of order $1/R$, R being the radius of the extra dimensions. Since $1/R$ is, in the Kaluza–Klein approach, presumably of order the Planck mass, the non-zero eigenvalues of $\not{D}^{(\text{int})}$ correspond to extremely massive fermions which would not have been observed. The observed quarks and leptons must correspond to the zero modes of $\not{D}^{(\text{int})}$.

If, in eleven-dimensional supergravity, the ground state is a product of four-dimensional Minkowski space with one of the M^{pqr}, then the zero modes of the Dirac operator in the internal space will, if there are any zero modes at all, automatically form multiplets of $SU(3)\times SU(2)\times U(1)$, since this is the symmetry of the internal space. It therefore is reasonable to wonder whether for an appropriate choice of p, q, and r, zero modes could exist and form the appropriate representation of the symmetry group, so as to reproduce the observed spectrum of quarks and leptons.

Of course, to reproduce what is observed in nature, we would need quite a few zero modes of the internal space Dirac operator. If the top quark exists, there are in nature at least 45 fermion degrees of freedom of given helicity, counting all colors and flavors of quarks and leptons. We would therefore need at least 45 Dirac zero modes. However, when a Dirac operator has zero modes, the number usually depends on topological invariants. Perhaps by choosing suitable values of p, q, and r we could suitably "twist" the topology and obtain the required 45 zero modes lying in the appropriate representation of $SU(3)\times SU(2)\times U(1)$.

Actually, if one has in mind eleven-dimensional supergravity, one must modify this program slightly. In eleven-dimensional supergravity, there is no fundamental spin one half field. The only fundamental Fermi field in that theory is the Rarita–Schwinger field $\psi_{\mu\alpha}$, of spin $\frac{3}{2}$ (μ is a vector index, α a spinor index).

Although this field has spin $\frac{3}{2}$ from the point of view of eleven dimensions, the components of ψ_μ with $5 \leq \mu \leq 11$ are spin one half fields from the point of view of ordinary four-dimensional physics. For $\mu \geq 5$, μ would be observed as an internal symmetry index, not a space-time index; it carries spin zero. Although the components ψ_μ with $\mu = 1 \cdots 4$ are spin-$\frac{3}{2}$ fields in the four-dimensional sense, the components with $\mu = 5 \cdots 11$ are spin one half fields. So zero-mode solutions of the spin-$\frac{3}{2}$ wave equation in the extra dimensions would be observed as massless spin-$\frac{1}{2}$ fermions in four dimensions. These would be the ordinary light fermions of the spontaneously compactified eleven-dimensional theory.

In one sense, it is an advantage to have to consider the Rarita–Schwinger operator rather than the Dirac operator. The Rarita–Schwinger operator can have zero modes more easily and in more abundance than the Dirac operator, because the Dirac operator has positivity properties which tend to suppress the number of zero modes. For instance, with four extra dimensions, it is known [16] that there is only a single non-flat compact solution of Einstein's equations on which the Dirac operator has zero modes. This is the Kahler manifold K3 (which has no Killing vectors). On this space there are two zero modes of the Dirac operator – but 42 zero modes of the Rarita–Schwinger operator. The large discrepancy is caused, in this case, by a much larger coefficient of the axial anomaly for Rarita–Schwinger fields. This example shows, incidentally, that the rather large number of zero modes that would be required to describe what is observed in physics is not necessarily out of reach.

In the approach considered here, the solution of the problem of flavor – the problem of the existence of several "generations" of fermions with the same quantum numbers – would be that the extra dimensions have a sufficiently complex topology that there are several zero modes with the same $SU(3)\times SU(2)\times U(1)$ quantum numbers. When an operator has several zero modes, they are not necessarily related by any symmetry. For instance, the isospinor Dirac operator in a Yang–Mills instanton of topological number K has K modes; these modes are not related by any symmetry. This is fortunate, because the various generations of fermions have very different masses and are not obviously related by any symmetry.

Unfortunately, there is a basic reason that this idea does not work, at least not in the form described above. The reason for this is related to one of the most basic facts about the observed quarks and leptons: the fermions of given helicity transform in a complex representation of the gauge group, or, to put it differently, right-handed fermions do not transform the same way that the left-handed fermions transform. For instance, left-handed color triplets (quarks) are SU(2) doublets, but right-handed color triplets are SU(2) singlets. This is the reason that quarks and leptons do not have bare masses but receive their mass from the Higgs mechanism – from $SU(2)\times U(1)$ symmetry breaking. This is a very important fact theoretically, because it is the basis for our theoretical understanding of why the quarks and leptons are very light compared to the mass scale of grand unification or the Planck mass. If left- and right-handed fermions transformed the same way under the the gauge group, bare masses would have been possible and could have been arbitrarily large.

In the framework that has been described above, right- and left-handed fermions would inevitably transform the same way under SU(3)×SU(2)×U(1). The reason for this is that low mass fermions are supposed to arise as zero modes of the Rarita–Schwinger operator in the extra dimensions. But the Rarita–Schwinger operator in the seven extra dimensions does not "know" whether a spinor field is left- or right-handed with respect to four-dimensional Lorentz transformations. It treats four-dimensional left- and right-handed fermions in the same way. One therefore could not get the observed SU(3)×SU(2)×U(1) representation. One would inevitably get vector-like rather than V–A weak interactions, with bare masses being possible for all fermions. (Indeed, precisely because bare masses would be possible for all fermions, it is not natural to get any massless fermions at all.)

There is an intriguing mechanism by which, at first sight, it seems that the internal space Rarita–Schwinger equation could treat left and right fermions differently. Eleven-dimensional spinors are constructed with eleven gamma matrices γ_i, $i = 1 \cdots 11$. Let us define an operator $\Gamma_{11} = i\gamma_1 \cdots \gamma_{11}$ which is a sort of eleven-dimensional helicity operator. Let us also define an operator $\Gamma_4 = i\gamma_1\gamma_2\gamma_3\gamma_4$ which measures the ordinary four-dimensional helicity, and an operator $\Gamma_7 = \gamma_5 \cdots \gamma_{11}$ which one might think of as "helicity" in the internal eleven-dimensional space. Then $\Gamma_{11}^2 = \Gamma_4^2 = \Gamma_7^2 = 1$ and $\Gamma_{11} = \Gamma_4\Gamma_7$.

The Rarita–Schwinger field ψ of eleven-dimensional supergravity satisfies a Weyl condition $\psi = \Gamma_{11}\psi$. (This condition must be imposed; otherwise there would be more Fermi than Bose degrees of freedom and supersymmetry would not be possible.) This identity may equivalently be written $\Gamma_4\psi = \Gamma_7\psi$.

The latter equation shows that in eleven-dimensional supergravity the four-dimensional helicity of fermions is correlated with the seven-dimensional "helicity". Components with $\Gamma_4 = +1$ (or -1) have $\Gamma_7 = +1$ (or -1). If the quantum numbers of zero modes of the seven-dimensional Rarita–Schwinger equation depended on Γ_7, as one might intuitively expect, they would also depend on Γ_4.

Unfortunately, the spectrum of the seven-dimensional Rarita–Schwinger operator does not depend on Γ_7. The reason for this is very simple (and depends only on the fact that the number of extra dimensions is odd). In defining how spinors transform under coordinate transformations in riemannian geometry one needs the matrices $\sigma_{ij} = [\gamma_i, \gamma_j]$. One does not (on an orientable manifold) need the γ_i themselves. The transformation $\gamma_i \leftrightarrow -\gamma_i$ does not change the σ_{ij}, so it does not affect the definition of spinors. It does, however, change the sign of $\Gamma_7 = \gamma_1\gamma_2 \cdots \gamma_7$. Consequently, spinors with opposite values of Γ_7 transform the same way under coordinate transformations. Since, in the approach discussed here, SU(3)×SU(2)×U(1) transformations are coordinate transformations, spinors with opposite values of Γ_7 have the same SU(3)×SU(2)×U(1) quantum numbers.

One could try to avoid this conclusion by taking the extra seven dimensions to be a non-orientable manifold. In a non-orientable manifold, the definition of spinors is subtle and involves the γ_i as well as σ_{ij}. However, seven-dimensional non-orientable manifolds with $SU(3)\times SU(2)\times U(1)$ symmetry are not abundant (they have all been listed above), and it is not difficult to show that none of them are suitable.

One might also try to avoid the above stated conclusion by going beyond riemannian geometry to include some variant of torsion. What possibilities this would offer is not very clear; the matter will be discussed at the end of this paper.

Obtaining the right quantum numbers for quarks and leptons is, of course, not the only problem that must be faced in order to obtain a realistic theory, although it may be the most difficult problem. We must also worry about spontaneous breaking of supersymmetry, spontaneous breaking of *CP*, spontaneous breaking of $SU(2)\times U(1)$ gauge symmetry, and obtaining the proper values of the low-energy parameters (coupling constants, masses, and mixing angles); and we must worry about what the true ground state of the theory really is. These questions will now be briefly discussed in turn.

For spontaneous breaking of supersymmetry the prospects are very bright; in fact, supersymmetry almost inevitably is spontaneously broken as part of any scheme in which there are compact dimensions with a non-abelian symmetry.

The reason for this is the following. Unbroken supersymmetry means that under a supersymmetry transformation the vacuum expectation values of the fields do not change. The vacuum expectation values of the Bose fields automatically are invariant under supersymmetry, since their supersymmetric variation would be proportional to the (vanishing) vacuum expectation values of the Fermi fields. The delicate question is whether the vacuum expectation values of the fermi fields change under supersymmetry.

To illustrate the point, let us ignore the possible presence in the theory of Bose fields other than the gravitational field. Then the transformation law for the Rarita–Schwinger field is $\delta\psi_\mu = D_\mu\varepsilon$, ε being the gauge parameter. An unbroken supersymmetry – a symmetry of the vacuum – must have $\delta\psi_\mu = 0$, so unbroken supersymmetry transformations correspond to solutions of $D_\mu\varepsilon = 0$.

On a curved manifold, this equation will almost certainly not have solutions, since $D_\mu\varepsilon = 0$ implies the integrability condition $[D_\mu, D_\nu]\varepsilon = 0$ or $R_{\mu\nu\alpha\beta}[\gamma^\alpha, \gamma^\beta]\varepsilon = 0$, which on most curved manifolds is not satisfied by any non-zero ε. For instance, on none of the M^{pqr} does a solution exist. (The properties of seven-dimensional manifolds admitting solutions of $D_\mu\varepsilon = 0$ have been discussed in the mathematical literature [17], but non-trivial examples do not seem to be known.) So in theories with curved extra dimensions, there will generally not be any unbroken supersymmetries.

The picture does not change greatly when one includes Bose fields other than the gravitational field. We now have $\delta\psi_\mu = \bar{D}_\mu\varepsilon$, where $\bar{D}_\mu = D_\mu$ plus non-minimal terms involving the vacuum expectation values of other Bose fields (and possibly involving the expectation values of fermion bilinears, as discussed below). Unbroken supersymmetries are now solutions of $\bar{D}_\mu\varepsilon = 0$, but solutions will still typically not exist because the integrability condition $[\bar{D}_\mu, \bar{D}_\nu]\varepsilon = 0$ will still not have solutions.

Although solutions will generally not exist, the extra dimensions and the vacuum expectation values of the fields may be just such that one or more solutions of $\bar{D}_\mu \varepsilon = 0$ would exist. Each solution of $\bar{D}_\mu \varepsilon = 0$ in the internal space would correspond to an unbroken supersymmetry charge in four dimensions. If there is precisely one such solution, and so only one unbroken supersymmetry generator, this corresponds to a theory in which $N = 8$ supersymmetry has been spontaneously broken down to $N = 1$ supersymmetry. If there are K solutions, there is an unbroken $N = K$ supersymmetry.

A particularly attractive possibility would be a theory in which the equation $\bar{D}_\mu \varepsilon = 0$ has precisely one solution in the extra dimensions, corresponding to unbroken $N = 1$ supersymmetry. With $N = 1$ supersymmetry it is possible to construct more or less realistic models of observed particle physics. With $N \geq 2$ it is not possible to make a realistic model, because the supersymmetry algebra for $N \geq 2$ forces left- and right-handed fermions to transform in the same way under the gauge group, in contrast with what is observed. It is attractive to believe that $N = 1$ supersymmetry might survive after compactification of seven dimensions because this would severely constrain the theory, would make many predictions that might be testable in accelerators, and [19] might shed light on $SU(2) \times U(1)$ breaking and the gauge hierarchy problem. Of course, we would then have to explain how $N = 1$ supersymmetry is eventually spontaneously broken at low energies.

In addition to supersymmetry breaking, we must also explain P and CP breaking in order to construct a realistic theory. The eleven-dimensional supergravity langrangian is invariant under inversions of space (or time) combined with a change of sign of the antisymmetric tensor gauge field that exists in this theory. After compactification of seven dimensions, the eleven-dimensional symmetry could be manifested as both P (inversion of space) and C (inversion of the compact dimensions). These potential invariances must be spontaneously broken.

A natural mechanism for spontaneous breaking of P, C, and CP involves the antisymmetry tensor gauge field of the eleven-dimensional supergravity theory. The curl $F_{\alpha\beta\gamma\delta}$ of this field may have a vacuum expectation value without breaking Lorentz invariance or $SU(3) \times SU(2) \times U(1)$. In fact, as discussed recently by several authors [20], a vacuum expectation value of F_{1234} is Lorentz invariant. It would violate P and CP but conserve C. The components F_{ijkm} for $i \cdots m \geq 5$ may also have expectation values, which would spontaneously break C and CP but conserve P. It is not difficult to see (by considering the little group of a point on M^{pqr}) that on any of the M^{pqr}, the most general $SU(3) \times SU(2) \times U(1)$ invariant vacuum expectation value of F_{ijkl} depends on two real parameters.

Although the eleven-dimensional theory can have spontaneous breaking of C, P, and CP, the strong interaction angle θ will inevitably vanish at the tree level. The reason for this is that in the eleven-dimensional theory, there is no operator which might be added to the lagrangian which reduces in four dimensions to $\theta \int d^4x F_{\mu\nu} \tilde{F}_{\mu\nu}$. There simply does not exist in eleven dimensions any topological invariant that can be written as the integral of a lagrangian density. Of course, the question of how large a vacuum angle might be generated by quantum corrections must wait until we understand how to do calculations in this (presumably) non-renormalizable theory.

It is also necessary, of course, to obtain $SU(2) \times U(1)$ symmetry breaking; this presumably means that we must find, at the tree level, a massless Higgs doublet which could later obtain a very tiny negative mass squared.

There are various ways that, in a Kaluza–Klein theory, one might obtain massless charged scalars. In the original Kaluza–Klein theory, with a single compact dimension (a circle) there is a massless scalar (at least at the tree level) because the classical field equations do not determine the radius of the circle. Space-time dependent fluctuations of this radius would be observed as a massless scalar degree of freedom.

If the equations that determine our hypothetical ground state $M^4 \times M^{pqr}$ admit not a unique solution for the metric of M^{pqr} but a whole family of solutions, then oscillations within this family would be observed as massless scalars. Some of these oscillations might involve departures from $SU(2) \times U(1)$ symmetry and could be the desired Higgs bosons.

One might also obtain massless scalars as components of the antisymmetric tensor gauge field. In fact, massless scalars can be obtained in this way, but tend to be neutral under the gauge group.

Regardless of where the scalars come from, why would they be massless? The most plausible explanation would be an unbroken supersymmetry relating the massless bosons to massless fermions. This could involve the possibility discussed above that the equation $\bar{D}_\mu \varepsilon = 0$ has a unique non-trivial solution, leaving $N = 1$ supersymmetry unbroken. In this case, of course, we must hope to find a non-perturbative mechanism spontaneously breaking the supersymmetry and giving a small vacuum expectation value to the scalar bosons. (Some relevant issues will be discussed in a future paper [21].)

Without understanding the Higgs bosons and the low-energy symmetry breaking, it is of course not possible to predict the quark and lepton masses and mixing angles. If we understood the dynamics that determines the metric of M^{pqr} (assuming that the ground state really is $M^4 \times M^{pqr}$), we could predict the strong, weak, and electromagnetic coupling constants, since the gauge fields all arise, by the ansatz of eq. (3), as part of the metric tensor in eleven dimensions, and the gauge field kinetic energy is part of the Einstein action. [The most general $SU(3) \times SU(2) \times U(1)$ invariant metric on M^{pqr} depends on three arbitrary parameters. If we understood the dynamics and could calculate the three parameters, we could predict the SU(3), SU(2), and U(1) coupling constants.] Even though we do not understand this dynamics (see below), it is possible to make a useful comment.

In a theory of this kind, the gauge coupling constants, which are determined by integrating the action over the compact dimensions, would scale as a rather high power of $1/(M_P R)$, where M_P is the Planck mass and R is the radius of the extra dimensions. The fact that the observed gauge coupling constants in nature differ from one by only one or two orders of magnitude shows that R cannot be too much greater than $1/M_\mathrm{p}$; the extra dimensions really have a radius not too different from 10^{-33} cm.

The eleven-dimensional supergravity theory has no global symmetry that could be interpreted as baryon number, so in this theory nucleons are almost surely unstable. The mass scale in nucleon decay, however, would probably be $1/R$, which is the mass scale of the heavy quanta in this theory. Since, as just noted, $1/R$ cannot be much less than M_p, the nucleon lifetime will probably be very long, perhaps 10^{45} years, which is far too long for nucleon decay to be observable. If the present nucleon decay experiments give a positive result, the approach described in this paper would become significantly less attractive.

It is now time to finally discuss the question of whether one can really sensibly expect $\mathrm{M}^4 \times \mathrm{M}^{pqr}$ to be the ground state of this theory.

The most attractive possibility would be that $\mathrm{M}^4 \times \mathrm{M}^{pqr}$ might be a solution of the classical equations of motion, possibly with a suitable vacuum expectation assumed for $F_{\mu\nu\alpha\beta}$. Unfortunately, a straightforward calculation shows that this is not true (regardless of what vacuum expectation value one assumes). If one arbitrarily adds to the lagrangian a cosmological constant (with a sign corresponding to a positive energy density) then $\mathrm{M}^4 \times \mathrm{M}^{pqr}$ can be a solution. However, local supersymmetry does not permit a cosmological constant in the eleven-dimensional lagrangian.

This problem is not necessarily fatal, since one can always hope that $\mathrm{M}^4 \times \mathrm{M}^{pqr}$, although not a solution of the classical equations of motion, is the minimum of the appropriate effective potential. In eleven-dimensional supergravity, there is no small dimensionless parameter whose smallness could justify the use of the classical field equations as an approximation. So the fact that $\mathrm{M}^4 \times \mathrm{M}^{pqr}$ does not satisfy the classical equations, while not encouraging, is not necessarily critical.

In any case, there is absolutely no obvious reason that $\mathrm{M}^4 \times \mathrm{M}^{pqr}$, rather than the more obvious possibility of eleven dimensional Minkowski space, should be the ground state of this theory.

It will be shown in a separate paper that even when Kaluza–Klein vacuum states are stable classically, they can be destabilized by quantum mechanical tunneling [22]. However, unbroken supersymmetry (plus a technical requirement that the extra dimensions be simply connected; this is not satisfied in the original Kaluza–Klein theory) seems to be a sufficient condition for stability. This is another reason that theories in which $\bar{D}_\mu \varepsilon = 0$ has a solution and there is an unbroken supersymmetry at the energies of compactification would be attractive.

As has been pointed out above, the most serious obstacle to a realistic model of the type considered in this paper is that the fermion quantum numbers do not turn out right. It is conceivable that this problem could be overcome if instead of riemannian geometry one considered geometry with torsion or some generalization of torsion; in such a theory the fermion transformation laws might be different.

How can one obtain torsion in eleven-dimensional supergravity? As has been noted [11], the theory formally contains torsion in the sense that certain fermion bilinears enter, formally, in the way that torsion would appear. Of course, a "torsion" that is bilinear in Fermi fields does not have a classical limit. However, by analogy with QCD, in which q̄q has a vacuum expectation value, one may be willing in supergravity to assume a vacuum expectation value for the "torsion field" $K \sim \bar{\psi}\psi$ (or perhaps for some other bilinears). Perhaps in this way the predictions for fermion quantum numbers can be modified. This possibility is under study.

I wish to acknowledge discussions with V. Bargmann and J. Wolf.

Note added in proof

For a recent discussion of Dirac zero modes in Kaluza–Klein theories, see ref. [23].

References

[1] Th. Kaluza, Sitzungsber. Preuss. Akad. Wiss. Berlin, Math. Phys. K1 (1921) 966
[2] O. Klein, Z. Phys. 37 (1926) 895; Arkiv. Mat. Astron. Fys. B 34A (1946); Contribution to 1955 Berne Conf., Helv. Phys. Acta Suppl. IV (1956) 58
[3] A. Einstein and W. Mayer, Preuss. Akad. (1931) p. 541, (1932) p. 130;
A. Einstein and P. Bergmann, Ann. Math. 39 (1938) 683;
A. Einstein, V. Bargmann, and P.G. Bergmann, Theodore von Kármán Anniversary Volume (Pasadena, 1941) p. 212;
O. Veblen, Projektive Relativitats Theorie (Springer, Berlin, 1933);
W. Pauli, Ann. der Phys. 18 (1933) 305; 337;
P. Jordan, Ann. der Physik (1947) 219;
Y. Thirz, C. R. Acad. Sci. 226 (1948) 216
[4] P.G. Bergmann, Introduction to the theory of relativity (Prentice-Hall, New York, 1942);
A. Lichnerowicz, Theories relativistes de la gravitation et de l'electromagnetisme (Masson, Paris, 1955)
[5] J. Rayski, Acta Phys. Pol. 27 (1965) 89;
W. Thirring, *in* 11th Schladming Conf., ed. P. Urban (Springer-Verlag, Wien, New York, 1972)
[6] B. DeWitt, *in* Lectures at 1963 Les Houches School, Relativity, groups, and topology, ed. B. DeWitt and C. DeWitt (New York, Gordon and Breach, 1964), published separately under the title Dynamical theory of groups and fields (New York, Gordon and Breach, 1965);
R. Kerner, Ann. Inst. H. Poincaré 9 (1968) 143;
A. Trautman, Rep. Math. Phys. 1 (1970) 29;
Y.M. Cho, J. Math. Phys. 16 (1975) 2029;
Y.M. Cho and P.G.O. Freund, Phys. Rev. D12 (1975) 1711;
Y.M. Cho and P.S. Jang, Phys. Rev. D12 (1975) 3789;
L.N. Chang, K.I. Macrae, and F. Mansouri, Phys. Rev. D13 (1976) 235
[7] J. Scherk and J.H. Schwarz, Phys. Lett. 57B (1975) 463;
E. Cremmer and J. Scherk, Nucl. Phys. B103 (1976) 393; B108 (1976) 409
[8] J.-F. Luciani, Nucl. Phys. B135 (1978) 111
[9] L. Palla, Proc. 1978 Tokyo Conf. on High-energy physics, p. 629
[10] N. Manton, Nucl. Phys. B158 (1979) 141
[11] E. Cremmer, B. Julia and J. Scherk, Phys. Lett. 76B (1978) 409;
E. Cremmer and B. Julia, Phys. Lett. 80B (1978) 48; Nucl. Phys. B159 (1979) 141
[12] J. Scherk and J.H. Schwarz, Phys. Lett. 82B (1979) 60; Nucl. Phys. B153 (1979) 61;
E. Cremmer, J. Scherk, and J.H. Schwarz, Phys. Lett. 84B (1979) 83

[13] W. Nahm, Nucl. Phys. B135 (1978) 149
[14] M.T. Grisaru, H.N. Pendleton and P. van Nieuwenhuysen, Phys. Rev. D15 (1977) 996
[15] F.A. Berends, J.W. van Holten, B. deWit and P. van Nieuwenhuizen, Nucl. Phys. B154 (1979) 261; Phys. Lett. 83B (1979) 188; J. Phys. A13 (1980) 1643;
C. Aragone and S. Deser, Phys. Lett. 85B (1979) 161; Phys. Rev. D21 (1980) 352;
K. Johnson and E.C.G. Sudarshan, Ann. of Phys. 13 (1961) 126;
G. Velo and D. Zwanziger, Phys. Rev. 186 (1967) 1337
[16] N. Hitchin, J. Diff. Geom. 9 (1974) 435;
S.T. Yau, Proc. Nat. Acad. Sci. US 74 (1977) 1798;
S. Hawking, Nucl. Phys. B146 (1978) 381
[17] R. Brown and A. Gray, Trans. Amer. Math. Soc. 141 (1969) 465
[18] P. Fayet, Phys. Lett. 69B (1977) 489; 84B (1979) 421;
G.R. Farrar and P. Fayet, Phys. Lett. 76B (1978) 575; 79B (1978) 442; 89B (1980) 191
[19] S. Weinberg, Phys. Lett. 62B (1976) 111
[20] A. Aurilia, H. Nicolai and P.K. Townsend, preprint, CERN Th. 2884 (1980);
P. Freund and M.A. Rubin, Phys. Lett. 97B (1980) 233
[21] E. Witten, preprint, Princeton Univ. (April, 1981)
[22] E. Witten, in preparation
[23] W. Mecklenburg, preprint, ICTP (March, 1981)

GRAVITY AS A DYNAMICAL CONSEQUENCE OF THE STRONG, WEAK, AND ELECTROMAGNETIC INTERACTIONS*

A. Zee

Department of Physics, FM-15

University of Washington, Seattle, WA 98195

ABSTRACT

La lumière fut, donc la pomme a chu.[1]

Is Newton's gravitational constant G a fundamental constant of Nature or it is calculable in terms of the other fundamental constants?

We would like to argue here that G is indeed calculable by physicists. After outlining the general philosophy and motivation behind this statement, we present a specific calculation of G, unfortunately not in the real world but for a class of gauge theories which presumably does not describe the real world.

That G is positive is certainly one of the experimental facts in physics least open to doubt. As physicists we should try to understand why.

In gauge theories, we insist that the coefficient of $F_{\mu\nu}F^{\mu\nu}$ is positive (and write it as $1/g^2$) so that the Euclidean action is positive definite and hence bounded below. This also insures that the gauge bosons propagate normally and not as ghosts. The situation for gravity is somewhat murkier. The scalar curvature R, even in Euclidean space, can be either positive or negative. We thus can no longer rely on the condition that the Euclidean path integral

*To be published in the Proceedings of the Erice Conference October 1981

The importance of scale invariance was realized as long ago as 1938 by Heisenberg[3] who remarked that theories with dimensionless coupling constants are interestingly located on the boundary between theories with "nasty" and theories with "nice" short-distance behavior. Of the four fundamental interactions, Fermi's theory of weak interaction and Einstein's theory of gravity are both characterized by couplings of dimension of inverse mass squared. A great triumph of physics over the last twenty years is the discovery that the weak interaction is actually secretly characterized by a dimensionless coupling and that Fermi's constant is given by the vacuum expectation value of some scalar field: $G_F = \langle\phi\rangle^{-2}$. It is natural to ask if gravity is also secretly dimensionless and if Newton's constant is also determined by the vacuum expectation value of some scalar field. The physics of the weak interaction and of gravity is of course quite different. Gravitation is long-ranged and so the vacuum expectation value cannot be associated with the mass of a mediating particle. Indeed, $1/G$ appears in the Einstein-Hilbert action multiplying what is essentially the kinetic energy term for the graviton. G and the Yang-Mills coupling g^2 measure the "stiffness" of the graviton and the gluon field respectively against excitation.

It is trivially easy to realize the idea that G^{-1} is given by the vacuum expectation of a scalar field. One merely has to replace the Einstein-Hilbert action by the action

$$\int d^4x\sqrt{-g}\,[\tfrac{1}{2}\,\varepsilon\phi^2 R + \tfrac{1}{2}\,\partial_\mu\phi\partial_\nu\phi g^{\mu\nu} - V(\phi)].$$

This was done independently by a large number of people.[4-9] Here we follow the treatment given in Ref. (4). The potential $V(\phi)$ is assumed to attain its minimum value when $\phi = V$. Then

$$G_N = \frac{1}{16\pi}\,(\tfrac{1}{2}\,\varepsilon V^2)^{-1}.$$

The introduction of scalar fields into gravity has a long history. Here, the crucial feature is the incorporation of spontaneous symmetry breaking. As a consequence the scalar field is "anchored" in a deep potential well $V(\phi)$ and thus the physical consequences of the present theory are indistinguishable from Einstein's theory except under extreme conditions of space-time curvature. This is in sharp contrast to earlier work such as that of Brans and Dicke. The coupling ε is dimensionless and has to be taken to be positive. Thus, this theory does not shed any light on the sign of G.

One is tempted to identify ϕ as the Higgs field responsible for the breaking of grand unified theory into strong, weak and electromagnetic interactions. In that case, gravity is weak because the other three coupling constants move so slowly (logarithmically) under the renormalization group. This "modern" view of why gravity is weak was in fact known to Landau. Unfortunately, the relevant symmetry scale of the SU(5) grand unified theory is only about 10^{-4} $M_{P\ell}$ ($M_{P\ell}$ denotes the Planck mass $\sim 10^{19}$ GeV). The SU(5) theory is however incomplete in a number of ways and we may hope that eventually the present theories will be extended to a theory set at the Planck mass.

As has already been remarked, the present theory is indistinguishable from Einstein's theory except under extreme conditions such as may exist in the early Universe. Newton's gravitational "constant" may then vary with temperature. It is not inconceivable that this phenomenon of a varying gravitational "constant" may be relevant for the horizon problem.[10] Any serious discussion is necessarily highly speculative,[11,12] however.

This theory is not only both trivial to construct and dull in its consequences but also rather unattractive. Our motivating philosophy is based on scale invariance. It was suggested in Ref. (10) that the theory of the world including gravitation should contain no dimensional parameter. We may take $V(\phi) = -\lambda\phi^4$ and, following Ref. (2), generate the symmetry breaking by radiative one-loop effects.

The next logical step was taken by Adler[13] who asked what would happen if there are no elementary scalar fields at all? After all, elementary scalar fields are generally regarded with some repugnance by physicists. Remarkably enough, in the absence of elementary scalar fields, gauge invariance and scale invariance combine to forbid terms proportional to the scalar curvature R in the Lagrangian. Thus, the term $\bar{\psi}\psi R$ has dimension five and is forbidden by scale invariance. On the other hand, the term $A_\mu A^\mu R$ does have dimension four but is not gauge invariant. We find it rather satisfying that terms proportional to the scalar curvature are excluded by scale invariance and gauge invariance, two fundamental symmetries with deep geometrical significance.

Since scale invariance is "automatically" broken by the renormalization procedure, or if one prefers, by quantum fluctuations, terms proportional to R will be induced.[14] Thus, we have the possibility that gravity is generated by the dynamical breaking of some grand unified gauge theory near the Planck mass.

It may well turn out that gravity is actually, in some sense, a consequence of the other three interactions. "La lumière fut, donc la pomme a chu". We find this philosophy rather appealing in that it obviates the need for a marriage of gravity with the other three interactions. Gravity, mediated by a spin-two field, does look quite different from the other three interactions, which have now been revealed to be all mediated by spin-one fields. It is true that various ingenious individuals have invented clever symmetries relating particles of different spins. However, the ensuing marriage arranged by local supersymmetry is attended by unwanted, or at least so far unobserved, spin -3/2 particles. This is not to deny the great beauty of supergravity. The philosophy behind supergravity is in some sense the exact opposite of the one advocated here in that it seeks to determine the other three interactions starting with Einstein's theory of gravity. Regrettably, the philosophy of induced gravity also appears to be incompatible with the beautiful idea of Kaluza that gauge invariance is merely the manifestation of general coordinate invariance in higher dimensions, an idea which is in turn not unrelated to supergravity.[15] Perhaps some way could be found to reconcile induced gravity and the Kaluza-Klein theory.[16]

Strictly speaking, the remarks above must be amended somewhat if R^2 and $R^2_{\mu\nu}$ terms are included in the Lagrangian. It is then a matter of semantics whether one regards gravity as just as fundamental as the other three interactions. We still prefer to say, perhaps somewhat too picturesquely, that gravity is generated by the other three interactions in the sense that the known propagation of the graviton at long distance is a consequence of the other three interactions.

We remark parenthetically that in certain supergravity theories the sign of G is correlated[17] to the normal propagation of the gauge bosons. This should count as a plus for supergravity.

Before proceeding further, we would like to remark briefly on the question of quantizing gravity. There are two possible points of view. (I) Perhaps gravity should not be quantized at all. Some people feel that gravity does have some mysterious connection with space-time geometry[18] and the graviton should not be treated as just any particle, as particle physicists are wont to do. We are also so pitifully ignorant of physics at the Planck scale and beyond. (Strictly speaking, this argument actually suggests that if gravity is to be quantized one should not worry too much about renormalizability of quantum gravity.) The price one has to pay for this view is that the action principle which leads to Einstein's equation is then ad hoc. Unfortunately, this view, that $g_{\mu\nu}$ describes a classical arena in which quantized matter fields play, is probably inconsistent.[19] For instance, by measuring the classical gravitational field we could in principle determine our distance from a massive object to arbitrary accuracy, in contradiction of the uncertainty principle. (II) Gravity should be quantized. In path integral lingo, the metric $g_{\mu\nu}(x)$ is to be integrated over like any other fields. In this case, we are obliged to put in the R^2 and $R^2_{\mu\nu}$ terms, while, with view (I), it is arguably optional.

For simplicity and as a first try, we adopt view (I) in what follows, keeping in mind that it is most likely inconsistent. Many of our remarks apply equally well to both views but some of the formulas below have to be modified. To the extent that semi-classical radiation theory has a limited but well-defined domain of validity, we expect view (I) to represent an approximation to the full theory with gravity itself fully quantized.

In some sense, the roots of gravity lie in Lorentz invariance. To write down Lorentz-invariant interactions between fields we have to introduce the Minkowski metric $\eta_{\mu\nu}$. Once we admit the possibility of $\eta_{\mu\nu}$ depending on space-time, we promote the metric to a field $g_{\mu\nu}(x) = \eta_{\mu\nu} + h_{\mu\nu}(x)$. In field-theory language, $h_{\mu\nu}$ is then a field without a proper kinetic energy term. In the view discussed here, the appropriate kinetic energy term arises as a consequence of dynamical scale-symmetry breaking. It is amusing to ask whether there are other hitherto unobserved interactions which are generated in the same way.[20]

Scale invariance also forbids the appearance of a cosmological constant term $\int d^4x\sqrt{g}\,\Lambda$ in the action which would in general appear upon symmetry breaking. At the moment, no one knows how to avoid generating this undesirable term. This is perhaps the weakest point in the program to generate gravity spontaneously, and, indeed, this problem afflicts all current theories in particle physics which utilize the notion of spontaneous symmetry breaking. We imagine that the ultimate correct theory of dynamical symmetry breaking will not produce a cosmological term.

The discussion here represents, in some sense, the modern realization of ideas of Sakharov[21] who identified gravity as the "elasticity of space-time". Very similar ideas have been discussed under the name of "pre-geometry".[22] The main distinction of the discussion here and in Ref. (21,22) is our insistence that the other three interactions be described by a renormalizable and scale invariant theory so that G is finite and calculable and hence is not cut-off dependent.

After all this discussion, we now give an actual formula for G derived independently by Adler[13] and by the present author.[23] We follow here the derivation given in Ref. (23). Starting with the Lagrangian $\mathcal{L} = \int d^4x(-\, 1/2\; T^{\mu\nu})h_{\mu\nu} + \cdots$ (which defines $T_{\mu\nu}$)

we expand the generating functional $\langle 0|T^* e^{+i\int\mathcal{L}d^4x}|0\rangle$ to extract the term quadratic in h. We treat $h_{\mu\nu}$ as a c-number classical field and thus the effective order $-h^2$ Lagrangian is given by

$$i\int d^4x\mathcal{L}_{eff}(x) = \frac{1}{2!}\left(\frac{-i}{2}\right)^2 \int d^4xd^4yh_{\mu\nu}(x)h_{\lambda\rho}(y) \langle 0|T^*T^{\mu\nu}(x)T^{\lambda\rho}(y)|0\rangle . \tag{1}$$

The T* product is understood to be the connected part. We specialize to the form $h_{\mu\nu} = 1/4\ \eta_{\mu\nu}h$ and choose h(x) to be a slowly varying function so that we can expand

$$h(x) = h(y) + z^\mu\partial_\mu h(y) + \frac{1}{2!} z^\mu z^\nu\ \partial_\mu\partial_\nu h(y) +\cdots . \tag{2}$$

$$z = x - y.$$

Defining $T(x) = \eta_{\mu\nu}T^{\mu\nu}(x)$ we find

$$i\mathcal{L}_{eff}(x) = \frac{-1}{2^7} h^2(x) \int d^4z\langle 0|T^*T(z)T(0)|0\rangle + \frac{1}{2^{10}}[\partial h(x)]^2 \int d^4z\ z^2\langle 0|T^*T(z)T(0)|0\rangle + \cdots.$$

A simple computation shows the order $-h^2$ term in $\sqrt{-g}\,R$ to be $- 3/32(\partial h)^2$. Thus we find the following representation for Newton's coupling constant:

$$\frac{1}{16\pi G} = \frac{i}{96} \int d^4x\ x^2\langle 0|T^*T(x)T(0)|0\rangle . \tag{3}$$

It is worth emphasizing that the right-hand side of Eq. (3) is a purely flat-space quantity determined completely by the other three interactions. If we understand the other three interactions thoroughly, we could in principle calculate $\langle 0|T^*T(x)T(0)|0\rangle$ precisely, evaluate the integral in Eq. (3), and thus obtain G.

If the R^2 and $R^2_{\mu\nu}$ terms are included and if gravity is quantized, there will be an extra term on the right-hand side of Eq. (3) due to the contribution of virtual gravitons. This extra term has recently been worked out by Adler.[24]

If the strong, electromagnetic, and weak interactions are described by a grand unified gauge theory with massless fermions, the operator T(x) is determined via the trace anomaly[25] to be

$$T = [2\beta(g)/g]\,\frac{1}{4}\,F^{\alpha}_{\mu\nu}F^{\mu\nu\alpha}.$$

The formula for Newton's constant in Eq. (3) expresses in precise terms our philosophy that gravity is induced as a result of quantum fluctuations. A heuristic understanding of this is not difficult to find and is readily suggested by[26] the Feynman diagram representing $\langle 0|T^*T(x)T(0)|0\rangle$ which is shown here:

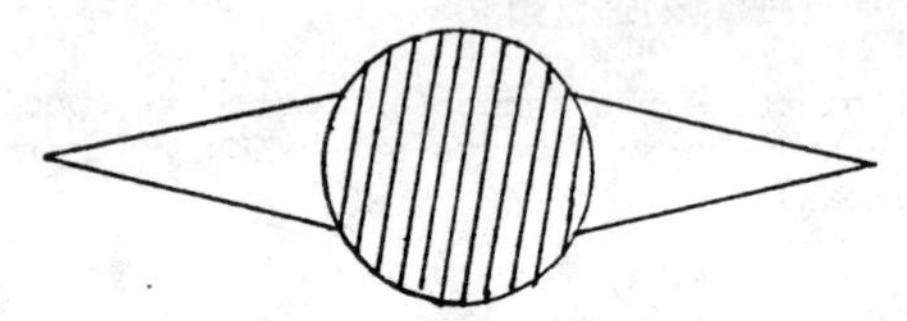

(The trace operator $T(x)$ couples to a pair of particles which interact in the shaded blob.) In the absence of $(\partial h)^2$ term in the action, a spatially varying gravitational field contains no energy. While this may be true classically it cannot be true quantum mechanically. With quantum fluctuation a pair, say e^-e^+ pair (actually a gauge boson pair), could always be created. With $\vec{\nabla}h_{\mu\nu} \neq 0$ this pair will then fall, gaining kinetic energy. By bringing the e^-e^+ pair together and annihilating them we can always extract energy out of the gravitational field. Another version given by Adler[14] involves the legendary Einstein's elevator. The usual statement is that for a small enough elevator one cannot say whether the elevator is uniformly accelerating or whether it is falling in a gravitational field (say that of the earth). But with quantum fluctuations an e^+e^- pair may be created and the e^- could tunnel out of the elevator to make a grand tour sampling the curvature tensor before coming back to annihilate the positron. Clearly, these two arguments represent out intuitive understanding of the Feynman diagram displayed above.

An evaluation of Eq. (3) will give G in terms of the scale mass μ. We envisage that a calculation of the proton mass m_N, say, in terms of μ is done independently so that eliminating μ one then determines the dimensionless ratio $G\,m_N^2$. As was remarked in the introduction, the sign of G is of great interest. If in a realistic calculation we should obtain a negative G this would clearly indicate that either the idea of induced gravity or the theory we are calculating with is wrong.

Define

$$\psi(-x^2) = \langle 0|T^*T(x)T(0)|0\rangle_{\text{connected}}\,.$$

Then the formula for G may be rotated to Euclidean space to read

$$\frac{1}{16\pi G} = -\frac{1}{96}\int_E d^4x\, x^2\, \psi(x^2). \qquad (3')$$

A crude calculation of G_{ind} taking into account the short-distance ultraviolet region was given in Ref. (23) and motivated the derivation of Eq. (3'). Unfortunately, we were led to conclude that the sign of G_{ind} depends on the long-distance infrared region of which we are totally ignorant. There was also a calculation[27] of G using a dilute instanton gas approximation, but again, due to our ignorance of the long-distance physics, the infrared region was excluded by an artificial cut-off on the instanton size. Thus, neither of these calculations is conclusive as regards the sign of G_{ind}. In order to include the infrared region, Adler has outlined[24] a program based on numerical lattice calculations.

In a recent paper,[28] we remarked that there is a class of gauge theories in which the infrared region is completely known -- in fact, the function ψ may be computed explicitly. These are gauge theories with the property that in the expansion of the renormalization group function $\beta(g) = -1/2\, g^3(b_0 + b_1 g^2 + \cdots)$ the coefficient b_0 is positive and small (so that the theory is barely asymptotically free) while the coefficient b_1 is negative. (Such theories are well-known to exist; quantum chromodynamics with sixteen quark triplets provides an example.[29] There is then an infrared stable fixed point given by $g_*^2 = (-b_0/b_1)$. By choosing appropriately the gauge group and the fermion representations, we can make g_*^2 arbitrarily small. The calculation described below is "exact" to the extent that g_*^2 is small.

We refer the interested reader to Ref. (28) for details of the calculation. Suffice it to say here that after some manipulations we obtain

$$\frac{1}{16\pi G\mu^2} = -\frac{\pi^2}{96}\left(\frac{c b_0^2 g_*^2}{16}\right)\int_0^\infty \frac{d\tau}{\tau^2} F(\tau). \qquad (4)$$

Here we have introduced a dimensionless distance variable $\tau = \mu^2 x^2$. c is a known numerical constant. The scale mass μ has been fixed by the condition $g^2(\mu^2) = 1/2\, g_*^2$ and thus has physical meaning.

It turns out the integrand $F(\tau)$ is positive. Naively, this would imply that G will be negative. The fact that $F(\tau)$ is positive simply confirms a general spectral function analysis. One may naively write a Källen-Lehmann representation for the two-point function $\psi(x^2)$ and conclude from the positivity of the spectral function that G will always come out negative. The fallacy in the argument is that the Källen-Lehmann representation actually does not exist[13] due to the fact that $T(x)$ is a dimension four operator and so $\psi(x^2)$ behaves at short distances like x^{-8}. There is thus no general theorem about the sign of G in induced gravity. The short-distance behavior of $\psi(x^2)$ is reflected here by the behavior of $F(\tau)$ as $\tau \to 0$. One verifies easily by a renormalization group analysis that

$$F(\tau) \underset{\tau \to 0}{\to} \left(\frac{1}{\log \tau} + \cdots\right)^2 .$$

Thus, the integral diverges in the ultraviolet region and must be regulated.

In Ref. (28) we dimensionally regulated the integral by continuing to space-time with dimension = 2ω and taking the limit $\omega \to 2$. The continuation produces a function analytic in the complex ω plane with a cut along the positive real axis. We approach the cut symmetrically. A justification for this prescription has been given by Adler.[24]

It is well known that within dimensional regularization, to any finite perturbative order, one only encounters poles in the complex ω plane. Here, however, the trace anomaly incorporates effects to all orders in perturbation theory. Also, we have not continued the β function and the logarithmic behavior of $F(\tau)$. Had we done that, our integral would have a cut on the real axis to the right of $\omega = 2$ and an infinite number of poles to the left of $\omega = 2$. These poles coalesce to form a cut in the suitable limit.

We extract in this way a finite value for G. The sign of G turns out to be given by

$$\text{sign } G = (-1)^{[\gamma]}$$

where $[x]$ = largest integer less than or equal to x. The parameter γ is defined as $2/(b_0 g_*^2)$. As is well known, dimensional regularization often reverses the naive sign of integrals. Precisely this phenomenon occurs here. The ultraviolet region contributes positively to G and may or may not overwhelm the infrared region, depending on the parameter γ. This toy calculation suggests that a general argument on the sign of G may not exist.

We claim that this calculation verifies general arguments given by Adler that G should be finite and calculable. We do not have to subtract off any poles. This is because quadratic divergences do not occur in dimensional regularization and the potential logarithmic divergence is softened logarithmically by asymptotic freedom.

One might feel that the regularization procedure is a bit delicate. Certainly, the calculation should be performed with another regularization scheme, namely the higher derivative regularization,[30] as a check. However, if Adler's rigorous argument[13] that G is finite and calculable in the context of induced gravity is correct, then the value of G obtained should be regularization independent. We see no reason to doubt Adler's argument.

Eventually, one would like to do a calculation for the real world. Realistically ψ is not known. Adler[24] has proposed that the integral K in Eq. (4) be divided into two integrals, one from 0 to τ_0, the other from τ_0 to ∞. Let us refer to these two pieces as K_{UV} and K_{IR} so that $K = K_{UV}(\tau_0) + K_{IR}(\tau_0)$. Adler envisages the calculation of the integrand in K_{IR} by Monte-Carlo methods on the lattice. No regularization is needed to evaluate K_{IR}. In contrast, we must dimensionally regulate the ultraviolet piece K_{UV} so we need to have an analytic form for the integrand. For τ_0 small enough, we can use the asymptotic freedom form for the integrand. We define $K^n_{UV}(\tau_0)$ as the value of the ultraviolet integral in Eq. (4) with $F(\tau)$ replaced by the first n terms of an asymptotic freedom expansion of $F(\tau)$ (with the integral dimensionally regulated). The plan is to match this into a numerical calculation of $K_{IR}(\tau_0)$. This matching needs to be done with care, however. The reason is that in the expansion of $F(\tau)$, as $\tau \to 0$ each succeeding term is only logarithmically less singular. Thus, for any finite n

$$\lim_{\tau_0 \to 0} [K^n_{UV}(\tau_0) + K_{IR}(\tau_0)] = \infty .$$

The correct procedure[31] is to evaluate the limit

$$\lim_{n \to \infty} [K^n_{UV}(\tau_0) + K_{IR}(\tau_0)]$$

for a fixed but small enough τ_0.

Without actually doing a realistic calculation, we could anticipate what the value of G will turn out to be. Consider a scale invariant gauge theory which does not have a small infrared fixed point but which has a growing coupling constant in the infrared limit. Then G should come out to be of order of the square of the mass scale at which the theory becomes strongly coupling. This is essentially the same conclusion as reached in the version of induced gravity involving Higgs fields. Thus, if the standard SU(5) theory minus its Higgs fields actually manages to generate its own breaking at a scale of M_X, then the value of G induced in such a theory will be of order $M_X{}^2$.

In the derivation of the formula for G, we made the simplification of taking $h_{\mu\nu} = 1/4\ \eta_{\mu\nu}h$. (See Eqs. (1,2).) This is of course not a necessary assumption.[32] Consider Eq. (1). Introducing the Fourier transform

$$h_{\mu\nu}(x) = \int \frac{d^4k}{(2\pi)^4} e^{ikx}\ h_{\mu\nu}(k)$$

we find

$$i\int d^4x\ \mathcal{L}_{eff} = \frac{1}{2}\left(\frac{-i}{2}\right)^2 \int \frac{d^4k}{(2\pi)^4} h_{\mu\nu}(k)h_{\lambda\rho}(-k)\int d^4y\ e^{iky} \tag{5}$$
$$\langle 0|T^*T^{\mu\nu}(y)T^{\lambda\rho}(0)|0\rangle\ .$$

To determine the general form of $\langle 0|T^*T_{\mu\nu}(x)T_{\lambda\rho}(0)|0\rangle$ we must derive the appropriate divergence condition it satisfies. To do this, one notes that[33] the stress-energy density tensor $\mathcal{T}_{\mu\nu} = \sqrt{-g}\ T^{\mu\nu}$ in curved space-time is covariantly conserved

$$\partial^\mu\langle\mathcal{T}_{\mu\nu}\rangle + \Gamma^\nu_{\sigma\tau}\langle\mathcal{T}^{\sigma\tau}\rangle = 0. \tag{6}$$

The expectation value is taken in the curved manifold. Varying this equation with respect to $g_{\lambda\rho}$ and taking the flat space limit we obtain

$$i\partial^\mu\langle 0|T^*T_{\mu\nu}(x)T_{\lambda\rho}(0)|0\rangle = \Lambda(\eta_{\nu\rho}\partial_\lambda + \eta_{\nu\lambda}\partial_\rho - \eta_{\lambda\rho}\partial_\nu)\delta^{(4)}(x). \tag{7}$$

The expectation value here is taken in flat space. We can now derive[32] the general form

$$\int d^4x\ e^{ikx}\ i\langle 0|T^*T_{\mu\nu}(x)T_{\lambda\rho}(0)|0\rangle$$
$$= L_{\mu\nu\lambda\rho}\int_0^\infty ds\left[\rho_2(s)\frac{k^2}{k^2-s} + \frac{3}{4}\rho_0(s)\right] \tag{8}$$
$$+ \Pi_{\mu\nu}\Pi_{\lambda\rho}\int_0^\infty ds\frac{\left(\frac{4}{3}\rho_2(s) + \rho_0(s)\right)}{k^2-s} - \Lambda(\eta_{\nu\rho}\eta_{\lambda\mu} + \eta_{\nu\lambda}\eta_{\rho\mu} - \eta_{\lambda\rho}\eta_{\mu\nu}).$$

We define the projection operator

$$\Pi_{\mu\nu} = \eta_{\mu\nu} k^2 - k_\mu k_\nu$$

$$L_{\mu\nu\lambda\rho} = k^{-2}(\Pi_{\mu\lambda}\Pi_{\nu\rho} + \Pi_{\mu\rho}\Pi_{\nu\lambda} - 2\Pi_{\mu\nu}\Pi_{\lambda\rho}) \ .$$

The operator $L_{\mu\nu\lambda\rho}$ contracted with $h_{\mu\nu}h_{\lambda\rho}$ gives an expression proportional to the quadratic part of Einstein's Lagrangian. The presence of the last term in Eq. (8) is required by the divergence condition, Eq. (7). The spectral functions ρ_2 and ρ_0 are the same as those defined in Ref. 33. The representation is formal in the sence that the integration over s may not converge but it indicates the correct tensor structure for each contribution to the spectral function.

Using this general representation one could readily[32] derive the formula for G (Eq. (3)) without the simplifying assumption $h_{\mu\nu} = 1/4\ \eta_{\mu\nu}h$.

Next, we make a number of miscellaneous remarks.

Suppose the theory is exactly scale invariant (that is to say, the theory is at a fixed point $g = g^*$, $\beta(g^*) = 0$, so $T = 0$). In this case, we have $1/G = 0$ or $G = \infty$. This is exactly what one would expect: The Einstein-Hilbert term is not generated if scale invariance holds. N = 4 supersymmetric Yang-Mills theory may have $\beta(g) = 0$.

For the induced gravity idea to work we have to assume that we know the other three interactions over all distance scales. One possible view may be that one should cut off the integral in Eq. (3') at the Planck length. This is unacceptable in that, as is explained above, G will always come out negative.

Creatures living in six-dimensional space-time will find the Yang-Mills action rather peculiar and note that it is forbidden by scale invariance. A philosophy of induced Yang-Mills would be proposed by physicists in this six-dimensional world[23] and a discussion similar to ours could be given.

The induced gravity idea is also reminiscent of the situation in the Gross-Neveu[34] model and the CP^N model. In these two dimensional models, kinetic energy terms for boson fields, forbidden by scale invariance and renormalizability, are dynamically induced. However, it is also instructive to note that the generation of these terms are infrared in the sense that the terms in question have dimensions four, greater than the dimension two of space-time. In contrast, the Einstein-Hilbert term has dimension less than the space-time dimension. Incidentally, formulas analogous to Eq. (3) can be readily written down. For instance, in the Gross-Neveu model

$$\mathcal{L} = \bar{\psi}\, i\not{\partial}\psi - \sigma\bar{\psi}\psi - \frac{1}{2g^2}\sigma^2$$

the missing kinetic energy term for the σ field $1/2\,(\partial_\mu\sigma)^2$ will be induced. The coefficient of this term is given by

$$\frac{1}{M^2} = -\frac{1}{4}\, i \int d^2x\, x^2 \langle 0|T^*\, \bar{\psi}\psi(x)\, \bar{\psi}\psi(0)|0\rangle \; .$$

Since $\bar{\psi}\psi$ has dimension one this integral is not singular in the short-distance limit. It is perfectly finite. The dispersion argument which did not go through for induced gravity goes through here. We find

$$\frac{1}{M^2} = \int_0^\infty d\sigma^2\, \rho(\sigma^2)/\sigma^4 .$$

The spectral function ρ is positive-definite and so $1/M^2$ is positive as it should be. Alas, things are not so simple in the real world.

We remark in passing that even in the presence of scalar fields the term $\phi^2 R$ may be forbidden if the theory has a global supersymmetry.[24,25]

Induced gravity may have amusing implications for cosmology. Presumably, when the temperature is high enough, the induced term will disappear. The effect of temperature on the induced gravitational constant deserves to be investigated.

Finally, we mention that in the induced gravity framework the Weyl-Eddington action

$$\int d^4x \sqrt{g}\, (-)(\rho R^2 + \gamma\, C^2_{\mu\nu\sigma\tau})$$

will also be induced. Here $C_{\mu\nu\sigma\tau}$ denotes the Weyl tensor. The representation in Eq. (8) allows us to derive formulas[32] for ρ and γ.

In general, one would expect that ρ and γ would be logarithmically divergent. A more careful analysis reveals that if the other three interactions are described by an asymptotically free theory the induced ρ will in fact be finite.[36] Also its sign is determined by a general argument and comes out to be just the right sign so a tachyon does not appear in the theory.[36]

In conclusion, we feel that a coherent and reasonable account of gravitational physics may be given along the following line. The three non-gravitational interactions are described by a scale and conformal invariant and asymptotically free Yang-Mills theory with massless fermions. We insist on conformal invariance so that the gravitational sector of theory is given by the Weyl action $\int d^4x\, C^2_{\mu\nu\sigma\tau}$. The theory is renormalizable[37] and, thanks to the Lee-Wick mechanism,[38] has a unitary S-matrix. Possible breakdown of causality is observable only at the Planck length. In this theory, Einstein's theory of gravity is induced as an effective long-distance theory. An R^2 term is also induced with a finite and physically desired sign.

ACKNOWLEDGMENTS

We have benefited over the years from comments by many colleagues, including S. Adler, R. Barbieri, S. Barr, J. Barrows, S. Bludman, D. Boulware, L. Brown, N. Cabibbo, E. Cremmer, R. Dashen, J. Ellis, F. Englert, J. Iliopolous, A. Linde, L. McLerran, L. Maiani, A. Neveu, P. Ramond, D. Reiss, P. Rossi, L. Smollin, H. Terazawa, M. Testa, M. Tonin, H.A. Weldon.

We are especially indebted to S. Adler and L. Brown for many useful discussions.

This work supported in part by the U.S. Department of Energy.

REFERENCES

1. E. Cremmer and H. Lubatti, private communication.
2. S. Coleman and E. Weinberg, Phys. Rev. D7, 1888 (1973).
3. W. Heisenberg, Z. Physik 110, 251 (1938); S. Sakata, H. Umezawa, and S. Kamefuchi, Prog. Theo. Phys. 7, 327 (1952).
4. A. Zee, Phys. Rev. Lett. 42, 417 (1979).
5. L. Smollin, Nucl. Phys. B160, 253 (1979).
6. Y. Fujii, Phys. Rev. D9, 874 (1974).
7. P. Minkowski, Phys. Lett. 71B, 419 (1977).
8. T. Matsuki, Prog. Theor. Phys. 59, 235 (1978).
9. A.D. Linde, Pis'ma Zh. Eksp. Teor. Fiz. 30, 479 (1979)(JETP Lett. 30, 447 (1979)].
10. A. Zee, Phys. Rev. Lett. 44, 703 (1980).
11. A. Linde, Phys. Lett. 93B, 394 (1980); H. Sato, Prof. Theo. Phys. 64, 1498 (1980); M.D. Pollock, Padova preprint IFPD 40/81.

12. H. Fleming, Sao Paulo preprint; B. Meyer, preprint.
13. S. Adler, Phys. Rev. Lett. 44, 1567 (1980); Phys. Lett. 95B, 241 (1980).
14. For further references and for a review of the physics and philosophy behind induced gravity see S.L. Adler, Rev. of Mod. Phys. (to be published) and in The High Energy Limit, ed. by A. Zichichi; A. Zee, "Grand Unification and Gravity", in the Proceedings of the 4th Kyoto Summer School (to be published) and in Proceedings of the Erice Conference (October 1981, to be published).
15. E. Cremmer, Lectures at the Trieste and at the Seattle School, 1981.
16. H.Y. Guo, Beijing preprint.
17. E. Cremmer, K. Stelle, and P. Townsend, private communications.
18. See, for example, E. Wigner's remark cited in the second paper in Ref. 14.
19. We thank T. Banks, F. Englert, and S. Mandelstam for conversations on this point.
20. A. Zee, to be published.
21. A. Sakharov, Dokl. Akad. Nauk. 177, 70 (1967); O. Klein, Phys. Scr. 9, 69 (1974); C. Misner, K. Thorne, and J. Wheeler, Gravity, p. 426.
22. K. Akama, Y. Chikashige, T. Matsuki, and H. Terazawa, Prog. Theo. Phys. 60, 1900 (1980); see also recent work by G. Veneziano and collaborators (these Proceedings).
23. A. Zee, Phys. Rev. D23, 858 (1981).
24. S.L. Adler, Rev. of Mod. Phys. (to be published) and in The High Energy Limit, ed. A. Zichichi.
25. R.J. Crewther, Phys. Rev. Lett. 28, 1421 (1972).
26. Historically, the diagram suggested the formula.
27. B. Hasslacher and E. Mottola, Phys. Lett. 95B, 237 (1980).
28. A. Zee, to be published.
29. W. Caswell, Phys. Rev. Lett. 33, 244 (1974).
30. B. Lee and J. Zinn-Justin, Phys. Rev. D5, 3121 (1972); A. Slavnov, Nucl. Phys. B31, 301 (1971).
31. S. Adler, private communication.
32. L. Brown and A. Zee, to be published.
33. D. Boulware and S. Deser, J. of Math. Phys. 8, 1468 (1967), Eq. (13).
34. D. Gross and A. Neveu, Phys. Rev. D10, 3235 (1974).
35. M. Kaku and P. Townsend, Phys. Lett. 76B, 54 (1978).
36. A. Zee, preprint.
37. K. Stelle, Phys. Rev. D16, 953 (1977) and references therein.
38. T.D. Lee and G. Wick, Phys. Rev. D2, 1033 (1970) and references therein.